Advances in Mathematical Modeling and Analysis

Advances in Mathematical Modeling and Analysis

Editor: Casper Geller

NY RESEARCH PRESS

New York

Published by NY Research Press
118-35 Queens Blvd., Suite 400,
Forest Hills, NY 11375, USA
www.nyresearchpress.com

Advances in Mathematical Modeling and Analysis
Edited by Casper Geller

International Standard Book Number: 978-1-63238-875-9 (Hardback)

Cataloging-in-Publication Data

Advances in mathematical modeling and analysis / edited by Casper Geller.
    p. cm.
Includes bibliographical references and index.
ISBN 978-1-63238-875-9
1. Mathematical models. 2. Mathematical analysis. 3. Mathematics. I. Geller, Casper.
QA401 .A38 2022
511.8--dc23

Contents

Preface

This book has been a concerted effort by a group of academicians, researchers and scientists, who have contributed their research works for the realization of the book. This book has materialized in the wake of emerging advancements and innovations in this field. Therefore, the need of the hour was to compile all the required researches and disseminate the knowledge to a broad spectrum of people comprising of students, researchers and specialists of the field.

The process of developing a mathematical model is known as mathematical modeling. It is characterization of system that uses mathematical concepts and languages. Mathematical models are used in natural science such as physics, chemistry, earth science and biology. They are also used in social sciences such as political science, sociology, economics and psychology. Dynamical systems, statistical models, differential equations or game theoretic models are some forms of mathematical models. There are two primary mathematical modeling problems which are classified as black box models or white box models. A black-box model is a system of model in which there is no priori information available, whereas a white-box model is a system where all necessary information is available. This book contains some path-breaking studies in the field of mathematical modeling. The various advancements in mathematical modeling and analysis are glanced at and their applications as well as ramifications are looked at in detail. For all those who are interested in this discipline, this book can prove to be an essential guide.

At the end of the preface, I would like to thank the authors for their brilliant chapters and the publisher for guiding us all-through the making of the book till its final stage. Also, I would like to thank my family for providing the support and encouragement throughout my academic career and research projects.

Editor

Stationary Distribution and Dynamic Behaviour of a Stochastic SIVR Epidemic Model with Imperfect Vaccine

Driss Kiouach (ID) **and Lahcen Boulaasair**

MSTI (Modelling, Systems and Technologies of Information) Team, High School of Technology, Ibn Zohr University, Agadir, Morocco

Correspondence should be addressed to Driss Kiouach; d.kiouach@uiz.ac.ma

Academic Editor: Oluwole D. Makinde

We consider a stochastic SIVR (susceptible-infected-vaccinated-recovered) epidemic model with imperfect vaccine. First, we obtain critical condition under which the disease is persistent in the mean. Second, we establish sufficient conditions for the existence of an ergodic stationary distribution to the model. Third, we study the extinction of the disease. Finally, numerical simulations are given to support the analytical results.

1. Introduction

Mathematical models have been an unavoidable tool in analyzing the mechanisms of infectious diseases. Most modelers were inspired by the works in [1–3]. For controlling the spread of diseases, vaccination is considered the most effective way to reduce both the morbidity and mortality of individuals (see [4–6]). In this paper, we consider the deterministic SIVR (susceptible-infected-vaccinated-recovered) model:

$$dS = [\mu - (\mu + \phi) S - \beta SI] \, dt,$$

$$dI = [\beta (S + \eta V) I - (\mu + \gamma) I] \, dt,$$

$$dV = [\phi S - \eta \beta VI - \mu V] \, dt, \tag{1}$$

$$dR = [\gamma I - \mu R] \, dt,$$

where S, I, and R denote the densities of susceptible, infected, and recovered individuals, respectively. V denotes the density of individuals who are immune to an infection as result of vaccination.

The parameters involved in the system are described below:

β: the average number of contacts per infected per unit time

μ: the recruitment rate and the death rate

γ: recovery rate of infected individuals

ϕ: the rate at which susceptible individuals are moved into the vaccination process

η: a positive factor satisfying $0 \leq \eta \leq 1$, and $\eta = 0$ means that the vaccine is perfectly effective and $\eta = 1$ means that the vaccine has no effect

In model (1), the fundamental parameter that governs the spread of the disease into a population is the basic reproduction number denoted by R_0. It can be thought of as the number of cases one case generates on average over the course of its infectious period in an otherwise uninfected population (see [7]).

Let us denote by $R_0 = \beta/(\mu + \gamma)$ the basic reproduction number of model (1) and by $R_\phi = R_0((\mu + \eta\phi)/(\mu + \phi))$ the basic reproduction number in a population in which a proportion ϕ had been vaccinated. It is known that, in the absence of the disease, there is a unique disease-free equilibrium $(S_0, I_0, V_0, R_0) = (\mu/(\mu+\phi), 0, \phi/(\mu+\phi), 0)$ which is globally asymptotically stable if $R_\phi < 1$. If $R_\phi > 1$ for some parameters values, the model exhibits a backward bifurcation leading to the existence of multiple endemic equilibria and news subthreshold, which may be important when it comes to designing vaccination strategies (see [8, 9]).

It is well known that epidemic models are inevitably affected by the environmental noise that influences the dynamic behaviour of the epidemic models (see [10–14]). For this, inspired by the works in [15–17], Tornatore et al. [18] have formulated and studied a stochastic version of model (1) by replacing the contact rate β in system (1) by $\beta + \sigma(dW(t)/dt)$, where $W(t)$ denotes the standard Brownian motion and σ^2 denotes the intensity of the white noise. Then, system (1) becomes as follows:

$$dS = [\mu - (\mu + \phi) S - \beta SI] dt - \sigma SIdW(t),$$

$$dI = [\beta (S + \eta V) I - (\mu + \gamma) I] dt$$
$$\quad + \sigma (S + \eta V) IdW(t), \qquad (2)$$

$$dV = [\phi S - \eta \beta VI - \mu V] dt - \sigma \eta VIdW(t),$$

$$dR = [\gamma I - \mu R] dt.$$

Dianli Zhao and Sanling Yuan [19] have obtained conditions ensuring the persistence and extinction of model (2) and have found a threshold $R_\phi^S = (1/(\mu + \gamma))[\beta((\mu + \eta\phi)/(\mu + \phi)) - (\sigma^2/2)((\mu + \eta\phi)/(\mu + \phi))^2]$ whose value below 1 or above 1 can determine the extinction and persistence of the epidemic under mild extra conditions.

In this paper, we assume that the multiplicative noise sources are linear in $S(t), I(t)$ and $V(t)$, according to [20]. Note that recovered population has no effect on the dynamics of $S, I,$ and V. Then, following this approach, we obtain the following reduced stochastic SIVR model:

$$dS = [\mu - (\mu + \phi) S - \beta SI] dt + \sigma_1 SdB_1(t),$$

$$dI = [\beta (S + \eta V) I - (\mu + \gamma) I] dt + \sigma_2 IdB_2(t), \qquad (3)$$

$$dV = [\phi S - \eta \beta VI - \mu V] dt + \sigma_3 VdB_3(t),$$

where $B_1(t), B_2(t),$ and $B_3(t)$ are standard independent Brownian motions and σ_i^2 is a positive constant, for all $i \in \{1, 2, 3\}$.

This paper is organized as follows. In Section 2, we present some lemmas concerning the existence of a global positive solution and ergodic stationary distribution. In Section 3, we prove that the disease is persistent under one condition. In Section 4, we establish sufficient conditions for the existence of a unique ergodic stationary distribution. In Section 5, we determine a condition under which the disease goes to extinction. In the last section, we introduce some examples and numerical simulations to confirm our results.

2. Preliminaries

Throughout this paper, let $(\Omega, \mathscr{F}, (\mathscr{F}_t)_{t \geq 0}, \mathbb{P})$ be a complete probability space with a filtration $(\mathscr{F}_t)_{t \geq 0}$ satisfying the usual conditions (i.e., it is increasing and right continuous, while $\mathscr{F}_0$ contains all $\mathbb{P}$-null sets). Moreover, let $\mathbb{R}_+^d = \{(x_1, \ldots, x_d) \in \mathbb{R}^d : x_i > 0, i = 1, \ldots, d\}$. If f is a bounded function on $\mathbb{R}_+$, define $\check{f} = \sup_{t \in [0, \infty)} f(t)$ and $\hat{f} = \inf_{t \in [0, \infty)} f(t)$. We define also $\langle h \rangle = (1/t) \int_0^t h(s) ds$.

The following theorem concerns the existence and uniqueness of the global positive solution. Since the proof is

almost the same as that in [21, Lemma 2.1 and Lemma 2.2], we omit it here.

Theorem 1. *For any initial value $(S(0), I(0), V(0)) \in \mathbb{R}_+^3$, there is a unique solution $(S(t), I(t), v(t)) \in \mathbb{R}_+^3$ of system (3) on $t \geq 0$ and the solution will remain in $\mathbb{R}_+^3$ with probability one. The solution $(S(t), I(t), v(t))$ has the following properties:*

$$\lim_{t \to \infty} \frac{S(t)}{t} = 0,$$

$$\lim_{t \to \infty} \frac{I(t)}{t} = 0,$$

$$\lim_{t \to \infty} \frac{V(t)}{t} = 0,$$

$$\text{a.s.,} \qquad (4)$$

$$\limsup_{t \to \infty} \frac{\log S(t)}{t} \leq 0,$$

$$\limsup_{t \to \infty} \frac{\log I(t)}{t} \leq 0,$$

$$\limsup_{t \to \infty} \frac{\log V(t)}{t} \leq 0,$$

$$\text{a.s.}$$

Furthermore, if $\mu > \max(\sigma_1^2, \sigma_2^2, \sigma_3^2)/2$, then

$$\lim_{t \to \infty} \frac{\int_0^t S(r) dB_1(r)}{t} = 0,$$

$$\lim_{t \to \infty} \frac{\int_0^t I(r) dB_2(r)}{t} = 0, \qquad (5)$$

$$\lim_{t \to \infty} \frac{\int_0^t V(r) dB_3(r)}{t} = 0,$$

$$\text{a.s.}$$

Now, we consider the d-dimensional stochastic differential equation:

$$dx(t) = f(x(t), t) dt + g(x(t), t) dB(t)$$
$$\text{for all } t \geq t_0, \qquad (6)$$

with initial value $x(0) = x_0 \in \mathbb{R}^d$. $B(t)$ denotes an n-dimensional standard Brownian motion defined on the complete probability space $(\Omega, \mathscr{F}, \{\mathscr{F}_t\}_{t \geq 0}, \mathbb{P})$. Let us denote by $\mathscr{C}^{2,1}(\mathbb{R}^d \times [t_0, \infty]; \mathbb{R}_+)$ the family of all nonnegative functions $V(x, t)$ defined on $\mathbb{R}^d \times [t_0, \infty]\mathbb{R}^d \times [t_0, \infty]$ such that they are continuously twice differentiable in x and once in t. The differential operator L of (6) is defined by [22]

$$
\begin{aligned}
L = \frac{\partial}{\partial t} + \sum_{i=1}^{d} f_i(x,t) \frac{\partial}{\partial x_i} \\
+ \frac{1}{2} \sum_{i,j=1}^{d} \left[g^T(x,t) g(x,t) \right]_{ij} \frac{\partial^2}{\partial x_i \partial x_j}.
\end{aligned}
\tag{7}
$$

If L acts on a function $V \in \mathscr{C}^{2,1}(\mathbb{R}^d \times [t_0, \infty]; \mathbb{R}_+)$, then

$$
\begin{aligned}
LV(x,t) = V_t(x,t) + V_x(x,t) f(x,t) \\
+ \frac{1}{2} trace \left[g^T(x,t) V_{xx}(x,t) g(x,t) \right],
\end{aligned}
\tag{8}
$$

where

$$
\begin{aligned}
V_t &= \frac{\partial V}{\partial t}, \\
V_x &= \left(\frac{\partial V}{\partial x_1}, \ldots, \frac{\partial V}{\partial x_d} \right), \\
V_{xx} &= \left(\frac{\partial^2 V}{\partial x_i \partial x_j} \right)_{d \times d}.
\end{aligned}
\tag{9}
$$

By Itô's formula, we get

$$
\begin{aligned}
dV(x(t),t) = LV(x(t),t) \, dt \\
+ V_x(x(t),t) g(x(t),t) \, dB(t).
\end{aligned}
\tag{10}
$$

Next, we shall present a lemma that gives a criterion for the existence of an ergodic stationary distribution to system (3).

Let $X(t)$ be a homogeneous Markov process in E_d (E_d denotes d-dimensional Euclidian space) and it is described by the following stochastic differential equation:

$$
dX(t) = h(X(t)) \, dt + \sum_{r=1}^{k} g_r(X(t)) \, dB_r(t)
\tag{11}
$$

The diffusion matrix is defined as follows:

$$
\begin{aligned}
A(X) &= \left(a_{ij}(x) \right), \\
a_{ij}(X) &= \sum_{r=1}^{k} g_r^i(X) g_r^j(X).
\end{aligned}
\tag{12}
$$

Lemma 2 (see [23], Chapter 4). *The Markov process $X(t)$ has a unique ergodic stationary distribution $\pi(.)$ if there exists a bounded domain $D \subset E_d$ with regular boundary Γ and*

(H.1): there is a positive number λ such that $\sum_{i,j=1}^{d} a_{ij}(X)\xi_i\xi_j \geq \lambda|\xi|^2, X \in D, \xi \in \mathbb{R}^d$,

(H.2): there exists a nonnegative $\mathscr{C}^2$-function V such that LV is negative for any $E_d \setminus D$. Then

$$
\mathbb{P}_x \left\{ \lim_{t \to \infty} \frac{1}{t} \int_0^t \Phi(X(s)) \, ds = \int_{E_d} \Phi(x) \pi(x) \, dx \right\}
\tag{13}
$$

$$
= 1,
$$

for all $x \in E_d$, where $\Phi(.)$ is a function integrable with respect to the measure π.

Now, we consider the following one-dimensional homogeneous Markov process:

$$
\begin{aligned}
dX(t) = b(X(t)) \, dt + \sigma(X(t)) \, dB(t), \\
X(t_0) = X_0, \quad \text{for all } t \geq t_0,
\end{aligned}
\tag{14}
$$

where $b(X)$ and $\sigma(X)$ are measurable functions from $\mathbb{R}$ to $\mathbb{R}$ and $B(t)$ is the Brownian motion. It is assumed that the functions $b(X)$, $\sigma(X)$, and $1/\sigma(X)$ are locally bounded.

(H.3): the functions $b(X)$ and $\sigma(X)$ are such that

$$
Y(x) = \int_0^x \exp\left[-2 \int_0^s \frac{b(u)}{\sigma^2(u)} du \right] ds \longrightarrow \pm\infty
$$

$$
\text{as } x \longrightarrow \pm\infty, \tag{15}
$$

$$
C = \int_{-\infty}^{+\infty} \frac{1}{\sigma^2(x)} \exp\left[2 \int_0^x \frac{b(u)}{\sigma^2(u)} du \right] dx < \infty.
$$

Lemma 3 (see [24], Theorem 1.13). *If (H.3) is satisfied, then the stochastic process (14) has ergodic properties with the density given by*

$$
\pi(x) = \frac{1}{C\sigma^2(x)} \exp\left[2 \int_0^x \frac{b(u)}{\sigma^2(u)} du \right].
\tag{16}
$$

3. Persistence

Definition 4. System (3) is said to be persistent in the mean, if

$$
\liminf_{t \to \infty} \langle I(t) \rangle > 0 \quad \text{a.s.}
\tag{17}
$$

Theorem 5. *If $R_0^S > 1$, then the disease will be persistent in mean; that is,*

$$
\liminf_{t \to \infty} \beta(a + b + \eta c) \langle I(t) \rangle
$$

$$
\geq \left(\mu + \gamma + \frac{\sigma_2^2}{2} \right) \left(R_0^S - 1 \right), \quad a.s.,
\tag{18}
$$

where

$$
\begin{aligned}
a &= \frac{\beta\mu}{(\mu + \phi + \sigma_1^2/2)^2}, \\
b &= \frac{\eta\beta\mu\phi}{(\mu + \phi + \sigma_1^2/2)^2 (\mu + \sigma_3^2/2)}, \\
c &= \frac{\eta\beta\mu\phi}{(\mu + \phi + \sigma_1^2/2)(\mu + \sigma_3^2/2)^2}, \\
R_0^S &= \frac{\beta\mu}{(\mu + \phi + \sigma_1^2/2)(\mu + \gamma + \sigma_2^2/2)} \\
&\quad + \frac{\eta\beta\mu\phi}{(\mu + \phi + \sigma_1^2/2)(\mu + \gamma + \sigma_2^2/2)(\mu + \sigma_3^2/2)}.
\end{aligned}
\tag{19}
$$

Proof. Set

$$V_1 = -\log I - (a+b)\log S - c\log V, \qquad (20)$$

where positive constants a, b, and c will be determined later. We apply Itô's formula on V_1; then we get

$$
\begin{aligned}
dV_1 = & \left[-\frac{a\mu}{S} - \frac{b\mu}{S} + (a+b)(\mu+\phi) + \beta(a+b)I \right. \\
& - \beta S - \eta\beta V + (\mu+\gamma) - \frac{c\phi S}{V} + cn\beta I + \mu c + \frac{a}{2}\sigma_1^2 \\
& \left. + \frac{b}{2}\sigma_1^2 + \frac{\sigma_2^2}{2} + \frac{c}{2}\sigma_3^2 \right] dt - \left[\sigma_1(a+b)dB_1 \right. \\
& + \sigma_2 dB_2 + c\sigma_3 dB_3 \right] = \left[-\beta S - \frac{a\mu}{S} - \frac{b\mu}{S} - \eta\beta V \right. \\
& - \frac{c\phi S}{V} + \left(\mu+\gamma+\frac{\sigma_2^2}{2}\right) + \beta(a+b+\eta c)I \\
& + a\left(\mu+\phi+\frac{\sigma_1^2}{2}\right) + b\left(\mu+\phi+\frac{\sigma_1^2}{2}\right) \\
& \left. + c\left(\mu+\frac{\sigma_3^2}{2}\right) \right] dt - \left[\sigma_1(a+b)dB_1 + \sigma_2 dB_2 \right. \\
& + c\sigma_3 dB_3 \right] \le \left[-2\sqrt{a\beta\mu} - 3\sqrt[3]{bc\eta\beta\mu\phi} \right. \\
& + \beta(a+b+\eta c)I + \left(\mu+\gamma+\frac{\sigma_2^2}{2}\right) \\
& + a\left(\mu+\phi+\frac{\sigma_1^2}{2}\right) + b\left(\mu+\phi+\frac{\sigma_1^2}{2}\right) \\
& \left. + c\left(\mu+\frac{\sigma_3^2}{2}\right) \right] dt - \left[\sigma_1(a+b)dB_1 + \sigma_2 dB_2 \right. \\
& + c\sigma_3 dB_3 \right]
\end{aligned}
\qquad (21)
$$

Let

$$
\begin{aligned}
a &= \frac{\beta\mu}{\left(\mu+\phi+\sigma_1^2/2\right)^2}, \\
b &= \frac{\eta\beta\mu\phi}{\left(\mu+\phi+\sigma_1^2/2\right)^2 \left(\mu+\sigma_3^2/2\right)}, \\
c &= \frac{\eta\beta\mu\phi}{\left(\mu+\phi+\sigma_1^2/2\right)\left(\mu+\sigma_3^2/2\right)^2}.
\end{aligned}
\qquad (22)
$$

Then

$$
\begin{aligned}
dV_1 \le & \left[-\frac{\beta\mu}{\mu+\phi+\sigma_1^2/2} - \frac{\eta\beta\mu\phi}{\left(\mu+\phi+\sigma_1^2/2\right)\left(\mu+\sigma_3^2/2\right)} \right. \\
& \left. + \left(\mu+\gamma+\frac{\sigma_2^2}{2}\right) + \beta(a+b+\eta c)I \right] dt \\
& - \left[\sigma_1(a+b)dB_1 + \sigma_2 dB_2 + c\sigma_3 dB_3 \right].
\end{aligned}
\qquad (23)
$$

By integrating the last inequality, we obtain

$$
\begin{aligned}
& \frac{V_1\left(S(t), I(t), V(t)\right) - V_1\left(S(0), I(0), V(0)\right)}{t} \\
& \le -\left(\mu+\gamma+\frac{\sigma_2^2}{2}\right)\left(R_0^S - 1\right) + \beta(a+b+\eta c)\langle I\rangle \\
& - \frac{1}{t}\int_0^t \sigma_1(a+b)dB_1(s) + \sigma_2 dB_2(s)_2 \\
& + c\sigma_3 dB_3(s).
\end{aligned}
\qquad (24)
$$

By applying Theorem 1 and the strong law of large numbers for martingales [22], we get the desired result. □

4. Stationary Distribution

In this section, by using the theory of Has'minski [18], we prove the existence of a unique ergodic stationary distribution, which indicates that the disease is persistent.

Theorem 6. *If $R_0^S > 1$, then system (3) has a unique stationary distribution and it has the ergodic property.*

Proof. To prove Theorem 6, it is sufficient to verify assumptions (H.1) and (H.2).

(i) To verify (H.2), we show that there exists a neighborhood $D \subset \mathbb{R}_+^3$ and a nonnegative $\mathscr{C}^2$-function V such that $LV(X)$ is negative for any $X \in \mathbb{R}_+^3 \setminus D$. To this end, we define a $\mathscr{C}^2$-function in the form

$$
\begin{aligned}
W(S, I, V) = & MV_1(S, I, V) + V_2(S, I, V) - \log S \\
& - \log V,
\end{aligned}
\qquad (25)
$$

where V_1 is the same function defined in Section 3, $V_2 =: (1/(1+\theta))(S+I+V)^{1+\theta}$, and M, θ are positive constants that satisfy

$$
\begin{aligned}
& -M\left(\mu+\gamma+\frac{\sigma_2^2}{2}\right)\left(R_0^S - 1\right) + A + 2\mu + \phi + \frac{\sigma_1^2}{2} + \frac{\sigma_3^2}{2} \\
& < 0,
\end{aligned}
\qquad (26)
$$

$$\mu > \frac{\theta}{2}\max\left(\sigma_1^2, \sigma_2^2, \sigma_3^2\right),$$

where

$$A = \sup_{(S,I,V)\in\mathbb{R}_+^3}\left\{ \mu(S+I+V)^\theta - \frac{\rho}{2}(S+I+V)^{\theta+1} \right\}. \qquad (27)$$

It is easy to see that

$$\lim_{(S^2+I^2+V^2)\longrightarrow\infty} W(S,I,V) = \lim_{(S^2+I^2+V^2)\longrightarrow 0} W(S,I,V)$$

$$= +\infty. \tag{28}$$

Moreover, $W(S,I,V)$ is a continuous function. Hence, $W(S,I,V)$ must have a minimum point $(\overline{S},\overline{I},\overline{V})$ in the interior of $\mathbb{R}^3_+$. Then we define a nonnegative $\mathscr{C}^2$-function V as

$$V(S,I,V) = W(S,I,V) - W(\overline{S},\overline{I},\overline{V}). \tag{29}$$

From the proof of Theorem 5, we have

$$LV_1(S,I,V) \le -\left(\mu+\gamma+\frac{\sigma_2^2}{2}\right)\left(R_0^S - 1\right)$$
$$+ \beta(a+b+\eta c)I \tag{30}$$

The operator L defined in Section 2 acts on V_2, $-\log S$, and $-\log V$ as follows:

$$L(-\log S) = -\frac{\mu}{S} + \mu + \phi + \beta I + \frac{\sigma_1^2}{2},$$

$$L(-\log V) = -\frac{\phi S}{V} + \mu + \eta\beta I + \frac{\sigma_3^2}{2},$$

$$LV_2(S,I,V)$$

$$= (S+I+V)^\theta [\mu - \mu(S+I+V) - \gamma I]$$
$$+ \frac{\theta}{2}(S+I+V)^{\theta-1}\left(\sigma_1^2 S^2 + \sigma_2^2 I^2 + \sigma_3^2 V^2\right)$$

$$\le \mu(S+I+V)^\theta - \mu(S+I+V)^{\theta+1} \tag{31}$$
$$+ \frac{\theta}{2}\left(\sigma_1^2 \vee \sigma_2^2 \vee \sigma_3^2\right)(S+I+V)^{\theta+1}$$

$$= \mu(S+I+V)^\theta$$
$$- \left[\mu - \frac{\theta}{2}\left(\sigma_1^2 \vee \sigma_2^2 \vee \sigma_3^2\right)\right](S+I+V)^{\theta+1}$$

$$= \mu(S+I+V)^\theta - \rho(S+I+V)^{\theta+1}$$

$$\le A - \frac{\rho}{2}\left(S^{1+\theta} + I^{1+\theta} + V^{1+\theta}\right),$$

where

$$\rho = \mu - \frac{\theta}{2}\left(\sigma_1^2 \vee \sigma_2^2 \vee \sigma_3^2\right). \tag{32}$$

Therefore

$$LV(S,I,V) \le f(S) + g(I) - \frac{\rho}{2}V^{1+\theta} - \frac{\phi S}{V}$$
$$- M\left(\mu+\gamma+\frac{\sigma_2^2}{2}\right)\left(R_0^S - 1\right) + A + 2\mu \tag{33}$$
$$+ \phi + \frac{\sigma_1^2}{2} + \frac{\sigma_3^2}{2},$$

where

$$f(S) = -\frac{\mu}{S} - \frac{\rho}{2}S^{1+\theta},$$
$$g(I) = \beta[M(a+b+\eta c)+1+\eta]I - \frac{\rho}{2}I^{1+\theta}. \tag{34}$$

Next, we construct the following compact subset:

$$D = \left\{\epsilon \le S \le \frac{1}{\epsilon},\ \epsilon^2 \le V \le \frac{1}{\epsilon^2},\ \epsilon \le I \le \frac{1}{\epsilon}\right\}, \tag{35}$$

where ϵ is a sufficiently small positive number, satisfying the following inequalities:

$$\overset{\vee}{g} - \frac{\mu}{\epsilon} < 0, \tag{36}$$

$$\overset{\vee}{f} + \beta[M(a+b+\eta c)+1+\eta]\epsilon < 0, \tag{37}$$

$$\overset{\vee}{f} + \overset{\vee}{g} - \frac{\phi}{\epsilon} < 0, \tag{38}$$

$$\overset{\vee}{g} - \frac{\rho}{2}\frac{1}{\epsilon^{1+\theta}} < 0, \tag{39}$$

$$\overset{\vee}{f} + B - \frac{\rho}{4}\frac{1}{\epsilon^{1+\theta}} < 0, \tag{40}$$

$$\overset{\vee}{f} + \overset{\vee}{g} - \frac{\rho}{2}\frac{1}{\epsilon^{2(1+\theta)}} < 0, \tag{41}$$

where the constant B will be determined later.
Then

$$\mathbb{R}^3_+ \setminus D = D_1^c \cup D_2^c \cup D_3^c \cup D_4^c \cup D_5^c \cup D_6^c, \tag{42}$$

with

$$D_1^c = \left\{(S,I,V) \in \mathbb{R}^3_0 < S < \epsilon\right\},$$

$$D_2^c = \left\{(S,I,V) \in \mathbb{R}^3_0 < I < \epsilon\right\},$$

$$D_3^c = \left\{(S,I,V) \in \mathbb{R}^3_0 < V < \epsilon^2, S > \epsilon\right\}, \tag{43}$$

$$D_4^c = \left\{(S,I,V) \in \mathbb{R}^3_+ \frac{1}{S} > \frac{1}{\epsilon}\right\},$$

$$D_5^c = \left\{(S,I,V) \in \mathbb{R}^3_+ \frac{1}{I} > \frac{1}{\epsilon}\right\},$$

$$D_6^c = \left\{(S,I,V) \in \mathbb{R}^3_+ \frac{1}{V} > \frac{1}{\epsilon^2}\right\}.$$

Now, we will show that LV is negative for any $(S, I, V) \in \mathbb{R}^3_+ \setminus D$.

Case 1. If $(S, I, V) \in D_1^c$, we obtain from (36) that

$$
LV(S, I, V) \leq -M\left(\mu + \gamma + \frac{\sigma_2^2}{2}\right)\left(R_0^S - 1\right) + A + 2\mu
$$
$$
+ \phi + \frac{\sigma_1^2}{2} + \frac{\sigma_3^2}{2} + \check{g} - \frac{\mu}{\epsilon} < 0. \tag{44}
$$

Case 2. If $(S, I, V) \in D_2^c$, (37) implies that

$$
LV(S, I, V) \leq -M\left(\mu + \gamma + \frac{\sigma_2^2}{2}\right)\left(R_0^S - 1\right) + A + 2\mu
$$
$$
+ \phi + \frac{\sigma_1^2}{2} + \frac{\sigma_3^2}{2} + \check{f} \tag{45}
$$
$$
+ \beta\left[M(a + b + \eta c) + 1 + \eta\right]\epsilon < 0.
$$

Case 3. If $(S, I, V) \in D_3^c$, from (38) it follows that

$$
LV(S, I, V) \leq -M\left(\mu + \gamma + \frac{\sigma_2^2}{2}\right)\left(R_0^S - 1\right) + A + 2\mu
$$
$$
+ \phi + \frac{\sigma_1^2}{2} + \frac{\sigma_3^2}{2} + \check{f} + \check{g} - \frac{\phi}{\epsilon} < 0. \tag{46}
$$

Case 4. If $(S, I, V) \in D_4^c$, (39) implies that

$$
LV(S, I, V) \leq -M\left(\mu + \gamma + \frac{\sigma_2^2}{2}\right)\left(R_0^S - 1\right) + A + 2\mu
$$
$$
+ \phi + \frac{\sigma_1^2}{2} + \frac{\sigma_3^2}{2} + \check{g} - \frac{\rho}{2}\frac{1}{\epsilon^{1+\theta}} < 0. \tag{47}
$$

Case 5. If $(S, I, V) \in D_5^c$, from (40) we get

$$
LV(S, I, V) \leq -M\left(\mu + \gamma + \frac{\sigma_2^2}{2}\right)\left(R_0^S - 1\right) + A + 2\mu
$$
$$
+ \phi + \frac{\sigma_1^2}{2} + \frac{\sigma_3^2}{2} + \check{f} + B - \frac{\rho}{4}\frac{1}{\epsilon^{1+\theta}} < 0, \tag{48}
$$

where

$$
B = \sup_{I \in (0, \infty)}\left\{\beta\left[M(a + b + \eta c) + 1 + \eta\right]I - \frac{\rho}{4}I^{1+\theta}\right\}. \tag{49}
$$

Case 6. If $(S, I, V) \in D_6^c$, from (41) we obtain

$$
LV(S, I, V) \leq -M\left(\mu + \gamma + \frac{\sigma_2^2}{2}\right)\left(R_0^S - 1\right) + A + 2\mu
$$
$$
+ \phi + \frac{\sigma_1^2}{2} + \frac{\sigma_3^2}{2} + \check{f} + \check{g} - \frac{\rho}{2}\frac{1}{\epsilon^{2(1+\theta)}} \tag{50}
$$
$$
< 0.
$$

From the previous discussion, we have

$$
LV(S, I, V) < 0, \quad (S, I, V) \in \mathbb{R}^3_+ \setminus D. \tag{51}
$$

(ii) Now, we verify assumption (*H.1*). The diffusion matrix of system (3) is

$$
\tilde{A} = \begin{pmatrix} \sigma_1^2 S^2 & 0 & 0 \\ 0 & \sigma_2^2 I^2 & 0 \\ 0 & 0 & \sigma_3^2 V^2 \end{pmatrix}. \tag{52}
$$

There is $\lambda = \min(\sigma_1^2 S^2, \sigma_2^2 I^2, \sigma_3^2 V^2) > 0$ such that

$$
\sum_{i,j=1}^{3} \tilde{a}_{ij}(X)\xi_i\xi_j = \sigma_1^2 S^2 \xi_1^2 + \sigma_2^2 SI^2 \xi_2^2 + \sigma_3^2 V^2 \xi_3^2
$$
$$
\geq \lambda |\xi|^2, \tag{53}
$$

for $(S, I, V) \in \overline{D}$ and $\xi \in \mathbb{R}^3_+$. That is to say, assumption (*H.1*) holds.

Consequently, system (3) has an ergodic stationary distribution. □

5. Extinction

Theorem 7. *We assume that* $\mu > \max(\sigma_1^2, \sigma_2^2, \sigma_3^2)/2$. *Let* $(S(t), I(t), V(t)) \in \mathbb{R}^3_+$ *be the solution of system (3). If* $\tilde{R}_0^S < 1$, *then the disease dies out with probability one.*

The threshold $\tilde{R}_0^S$ *is defined as follows:*

$$
\tilde{R}_0^S = \frac{\beta\mu}{(\mu + \phi)(\mu + \gamma + \sigma_2^2/2)}
$$
$$
+ \frac{\eta\beta\phi}{(\mu + \phi)(\mu + \gamma + \sigma_2^2/2)} \tag{54}
$$

Proof. By integrating system (3), we obtain

$$
\frac{S(t) - S(0)}{t} + \frac{I(t) - I(0)}{t} + \frac{V(t) - V(0)}{t} = \mu
$$
$$
- \mu\langle S \rangle - (\mu + \gamma)\langle I \rangle - \mu\langle V \rangle
$$
$$
+ \frac{1}{t}\left[\sigma_1 \int_0^t S(r)\,dB_1(r) + \sigma_2 \int_0^t I(r)\,dB_2(r) \right. \tag{55}
$$
$$
\left. + \sigma_3 \int_0^t v(r)\,dB_3(r)\right].
$$

It follows that

$$\langle V \rangle = 1 - \langle S \rangle - \frac{\mu + \gamma}{\mu} \langle I \rangle + \psi_1(t), \tag{56}$$

where

$$\psi_1(t) = -\left(\frac{S(t) - S(0)}{t} + \frac{I(t) - I(0)}{t} \right.$$

$$+ \frac{V(t) - V(0)}{t} \right) + \frac{1}{\mu t} \left[\sigma_1 \int_0^t S(r) \, dB_1(r) \right. \tag{57}$$

$$\left. + \sigma_2 \int_0^t I(r) \, dB_2(r) + \sigma_3 \int_0^t v(r) \, dB_3(r) \right].$$

From the first equation of system (3), we obtain

$$\langle S \rangle = \frac{\mu}{\mu + \phi} - \frac{\beta}{\mu + \phi} \langle SI \rangle + \psi_2(t), \tag{58}$$

where

$$\psi_2(t) = -\frac{S(t) - S(0)}{(\mu + \phi)t} + \frac{\sigma_1}{(\mu + \phi)t} \int_0^t S(s) \, dB_1(s). \tag{59}$$

Applying Itô's formula to system (3), one obtains

$$\frac{\log I(t) - \log I(0)}{t} = \beta \langle S \rangle + \eta \beta \langle V \rangle - (\mu + \gamma) - \frac{\sigma_2^2}{2}$$

$$+ \frac{\sigma_2 B_2(t)}{t}. \tag{60}$$

By injecting (56) and (58) into (60), we have

$$\frac{\log I(t) - \log I(0)}{t} = \beta \frac{\mu + \eta \phi}{\mu + \phi} - \left(\mu + \gamma + \frac{\sigma_2^2}{2} \right)$$

$$+ \beta(1 - \eta) \psi_1(t) + \eta \beta \psi_2(t)$$

$$- \beta^2 \frac{1 - \eta}{\mu + \phi} \langle SI \rangle$$

$$- \beta \frac{\eta(\mu + \gamma)}{\mu} \langle I \rangle + \frac{\sigma_2 B_2(t)}{t} \tag{61}$$

$$\leq \beta \frac{\mu + \eta \phi}{\mu + \phi} - \left(\mu + \gamma + \frac{\sigma_2^2}{2} \right)$$

$$+ \beta(1 - \eta) \psi_1(t) + \eta \beta \psi_2(t)$$

$$+ \frac{\sigma_2 B_2(t)}{t}.$$

From the strong law of large numbers for martingales [22], we get

$$\lim_{t \to \infty} \frac{B_2(t)}{t} = 0, \quad \text{a.s.} \tag{62}$$

By Theorem 1, we get

$$\lim_{t \to \infty} \psi_1(t) = \lim_{t \to \infty} \psi_2(t) = 0. \tag{63}$$

By taking the superior limit of both sides of (61), we obtain

$$\limsup_{t \to \infty} \frac{\log I(t)}{t} \leq \frac{\beta \mu}{\mu + \phi} + \frac{\eta \beta \phi}{\mu + \phi} - \left(\mu + \gamma + \frac{\sigma_2^2}{2} \right)$$

$$= \left(\mu + \gamma + \frac{\sigma_2^2}{2} \right) \left(\tilde{R}_0^S - 1 \right). \tag{64}$$

This finishes the proof. □

Theorem 8. *We assume that $\sigma_1 = \sigma_3 = \sigma$ and $B_1(t) = B_3(t) = B(t)$. Let $(S(t), I(t), V(t)) \in \mathbb{R}_+^3$ be the solution of system (3). If $\sigma^2 > 2\eta \phi$, then*

$$\limsup_{t \to \infty} \frac{\log I(t)}{t}$$

$$\leq (\mu + \gamma) \left(\frac{\beta}{\mu + \gamma} \int_0^\infty x \pi(x) \, dx - 1 \right) \quad a.s. \tag{65}$$

If $(\beta / (\mu + \gamma)) \int_0^\infty x \pi(x) dx < 1$, then

$$\lim_{t \to \infty} I(t) = 0 \quad a.s., \tag{66}$$

and the distribution of $S(t) + \eta V(t)$ converges weakly to the measure that has the density given by

$$\pi(x) = \frac{Q}{\sigma^2} x^{-2(1 - \eta \phi / \sigma^2)} e^{-2\mu / \sigma^2 x}, \quad x \in (0, \infty), \tag{67}$$

where

$$Q = \sigma^2 \left(\frac{2\mu}{\sigma^2} \right)^{1 - 2(\eta \phi / \sigma^2)} \Gamma^{-1} \left(1 - 2 \frac{\eta \phi}{\sigma^2} \right). \tag{68}$$

Proof. We assume that $\sigma_1 = \sigma_3 = \sigma$ and $B_1(t) = B_3(t) = B(t)$. From system (3), we have

$$d(S(t) + \eta V(t)) \leq [\mu + \eta \phi (S(t) + \eta V(t))] \, dt$$

$$+ \sigma_1 S(t) \, dB_1(t)$$

$$+ \eta V(t) \sigma_3 dB_3(t), \tag{69}$$

$$= dX(t),$$

where

$$dX(t) = [\mu + \eta \phi X(t)] \, dt + \sigma X(t) \, dB(t), \tag{70}$$

with the initial value $X(0) = S(0) + \eta V(0)$.

We compute that

$$\int \frac{\mu + \eta \phi x}{\sigma^2 x^2} dx = \frac{1}{\sigma^2} \int \left(\frac{\mu}{x^2} + \frac{\eta \phi}{x} \right) dx$$

$$= \frac{1}{\sigma^2} \left[-\frac{\mu}{x} + \eta \phi \ln x \right] + B, \tag{71}$$

where B is a constant.

One can see that

$$\int_0^\infty x^{-2(1-\eta\phi/\sigma^2)} e^{-2\mu/\sigma^2 x} dx$$

$$= \left(\frac{2\mu}{\sigma^2}\right)^{2(\eta\phi/\sigma^2)-1} \int_0^\infty y^{(1-2(\eta\phi/\sigma^2))-1} e^{-y} dy \qquad (72)$$

$$= \left(\frac{2\mu}{\sigma^2}\right)^{2(\eta\phi/\sigma^2)-1} \Gamma\left(1 - 2\frac{\eta\phi}{\sigma^2}\right),$$

where Γ is the Gamma function defined by

$$\Gamma(Z) = \int_0^\infty x^{z-1} e^{-x} dx. \qquad (73)$$

If $\sigma^2 > 2\eta\phi$, then

$$\int_0^\infty x^{-2(1-\eta\phi/\sigma^2)} e^{-2\mu/\sigma^2 x} dx < \infty. \qquad (74)$$

Thus, condition $(H.3)$ in Lemma 3 is verified. So, system (70) has the ergodic property and the invariant density given by

$$\pi(x) = \frac{Q}{\sigma^2} x^{-2(1-\eta\phi/\sigma^2)} e^{-2\mu/\sigma^2 x}, \quad x \in (0,\infty). \qquad (75)$$

Let us compute Q.
According to

$$\int_0^\infty \pi(x)\,dx = 1, \qquad (76)$$

one has

$$Q = \sigma^2 \left(\frac{2\mu}{\sigma^2}\right)^{1-2(\eta\phi/\sigma^2)} \Gamma^{-1}\left(1 - 2\frac{\eta\phi}{\sigma^2}\right). \qquad (77)$$

From the ergodic theorem, it follows that

$$\lim_{t\to\infty} \frac{1}{t}\int_0^t X(r)\,dr = \int_0^\infty x\pi(x)\,dx \quad \text{a.s.} \qquad (78)$$

On the other hand, by applying Itô's formula to $\log I(t)$ and then integrating, one has

$$\frac{\log I(t) - \log I(0)}{t}$$

$$= \frac{\beta}{t}\int_0^t (S(r) + \eta V(r))\,dr - (\mu + \gamma) \qquad (79)$$

$$- \frac{\sigma^2}{2}\int_0^t (S(r) + \eta V(r))^2\,dr + \frac{1}{t}M(t),$$

where $M(t) = \sigma\int_0^t (S(r) + \eta V(r))dB(r)$ whose quadratic variation is

$$\langle M, M\rangle(t) = \sigma^2 \int_0^t (S(r) + \eta V(r))^2\,dr. \qquad (80)$$

In view of the exponential martingales inequality [22], for any positive constants T, α, and ν, we have

$$\mathbb{P}\left\{\sup_{0\le t\le T}\left[M(t) - \frac{\alpha}{2}\langle M, M\rangle(t)\right] > \nu\right\} \le e^{-\alpha\nu}. \qquad (81)$$

Choosing $T = k, \alpha = 1$ and $\nu = 2\log k$, one has

$$\mathbb{P}\left\{\sup_{0\le t\le k}\left[M(t) - \frac{1}{2}\langle M, M\rangle(t)\right] > 2\log k\right\} \le \frac{1}{k^2}. \qquad (82)$$

Applying the Borel-Cantelli Lemma [22] leads to the fact that, for almost all $\omega \in \Omega$, there exists a random integer $k_0 = k_0(\omega)$ such that, for any $k \ge k_0$, we obtain

$$\sup_{0\le t\le k}\left[M(t) - \frac{1}{2}\langle M, M\rangle(t)\right] \le 2\log k. \qquad (83)$$

That is,

$$M(t) \le \frac{1}{2}\langle M, M\rangle(t) + 2\log k \quad \text{for all } t \le k. \qquad (84)$$

Substituting this inequality into (79) yields

$$\frac{\log I(t) - \log I(0)}{t}$$

$$\le \frac{\beta}{t}\int_0^t (S(r) + \eta V(r))\,dr - (\mu + \gamma) + 2\frac{\log k}{t} \qquad (85)$$

$$\text{for all } t \le k.$$

Applying the comparison theorem of stochastic differential equations on inequality (69) gives

$$S(t) + \eta V(t) \le X(t) \quad \text{a.s.} \qquad (86)$$

It follows that

$$\frac{\log I(t) - \log I(0)}{t} \le \frac{\beta}{t}\int_0^t X(r)\,dr - (\mu + \gamma)$$

$$+ 2\frac{\log k}{t}, \quad \text{for all } t \le k. \qquad (87)$$

Taking the superior limit on both sides of the previous inequality leads to

$$\limsup_{t\to\infty} \frac{\log I(t) - \log I(0)}{t}$$

$$\le \beta\int_0^\infty x\pi(x)\,dx - (\mu + \gamma) \quad \text{a.s.} \qquad (88)$$

If $(\beta/(\mu+\gamma))\int_0^\infty x\pi(x)dx < 1$, we conclude that

$$\lim_{t\to\infty} I(t) = 0 \quad \text{a.s.} \qquad (89)$$

As a result, for any small $\epsilon > 0$, there exist t_0 and a set $\Omega_\epsilon \subset \Omega$ such that $\mathbb{P}(\Omega_\epsilon) > 1 - \epsilon$ and $\beta(S + \eta^2 V)I \le \epsilon(S + \eta^2 V) \le \epsilon(S + \eta V)$ for all $t \ge t_0$ and $\omega \in \Omega_\epsilon$. From

$$[\mu - \mu(S(t) + \eta V(t)) - \epsilon(S(t) + \eta V(t))]\,dt$$

$$+ \sigma(S(t) + \eta V(t))dB(t) \le d(S(t) + \eta V(t))$$

$$\le [\mu - \mu(S(t) + \eta V(t))]\,dt \qquad (90)$$

$$+ \sigma(S(t) + \eta V(t))dB(t),$$

it follows that the distribution of the process $S(t) + \eta V(t)$ converges weakly to the measure that has the density π. This completes the proof. $\qquad\square$

6. Numerical Simulations

In this section, we present the numerical simulations to support the above analytical results, illustrating persistence in mean and extinction of the disease.

In the two following examples, we choose the initial value as $(S(0), I(0), V(0)) = (0.2, 0.2, 0.2)$.

Example 9. In model (3), we choose the parameters as follows:

$$
\begin{aligned}
\sigma_1 &= 0.2, \\
\sigma_2 &= 0.2, \\
\sigma_3 &= 0.2, \\
\beta &= 10, \\
\mu &= 1, \\
\phi &= 15, \\
\gamma &= 4, \\
\eta &= 0.6.
\end{aligned}
\tag{91}
$$

We compute that

$$
R_0^S = 1.2215 > 1.
\tag{92}
$$

Then, according to Theorem 5, the disease is persistent (see Figure 1). Furthermore, by Theorem 6, system (3) has a unique stationary distribution (see Figure 2).

Example 10. We choose the parameters as follows:

$$
\begin{aligned}
\sigma_1 &= 0.6, \\
\sigma_2 &= 1, \\
\sigma_3 &= 0.8, \\
\beta &= 8, \\
\mu &= 1, \\
\phi &= 15, \\
\gamma &= 4, \\
\eta &= 0.6.
\end{aligned}
\tag{93}
$$

Then

$$
\tilde{R}_0^S = 0.9091 < 1
\tag{94}
$$

Theorem 7 implies that system (3) has disease extinction (see Figure 3).

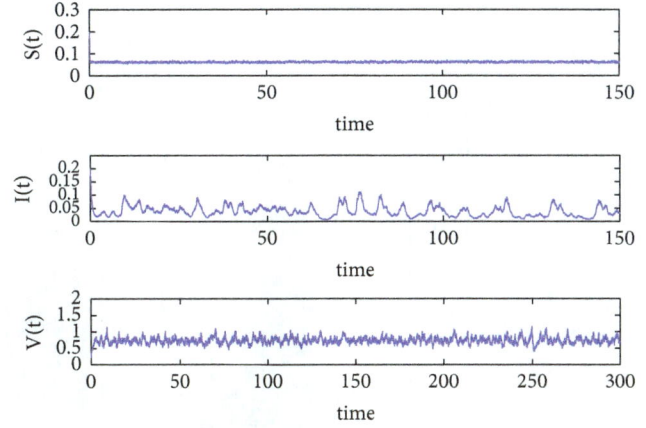

FIGURE 1: Simulation of the paths $S(t)$, $I(t)$, and $V(t)$ for model (3).

Example 11. We choose the parameters as follows:

$$
\begin{aligned}
\sigma_1 &= \sigma_3 = \sigma = 1.15, \\
\beta &= 10, \\
\mu &= 1, \\
\phi &= 1, \\
\gamma &= 4, \\
\eta &= 0.2.
\end{aligned}
\tag{95}
$$

Then

$$
\sigma^2 > 2\eta\phi,
$$

$$
\frac{\beta}{\mu + \gamma} \int_0^\infty x\pi(x)\,dx < 1.
\tag{96}
$$

According to Theorem 8, we get

$$
\lim_{t \to \infty} I(t) = 0 \quad \text{a.s.}
\tag{97}
$$

It follows that the distribution of the process $S(t) + \eta V(t)$ converges weakly to the measure that has the density π (see Figure 4).

7. Conclusion

This paper is concerned with the dynamics of a stochastic SIVR epidemic model with multiplicative noise sources. By constructing a convenient positive function, we establish sufficient conditions for the existence of a unique ergodic stationary distribution to model (3) and we prove that the disease will be permanent. In addition, we also establish sufficient conditions for extinction of the disease.

Many works have been done to study the continuous stochastic models by using the white noise. Sometimes, population systems may suffer sudden environmental perturbations, such as SARS, floods, and toxic pollutants. These phenomena cannot be modeled by stochastic continuous

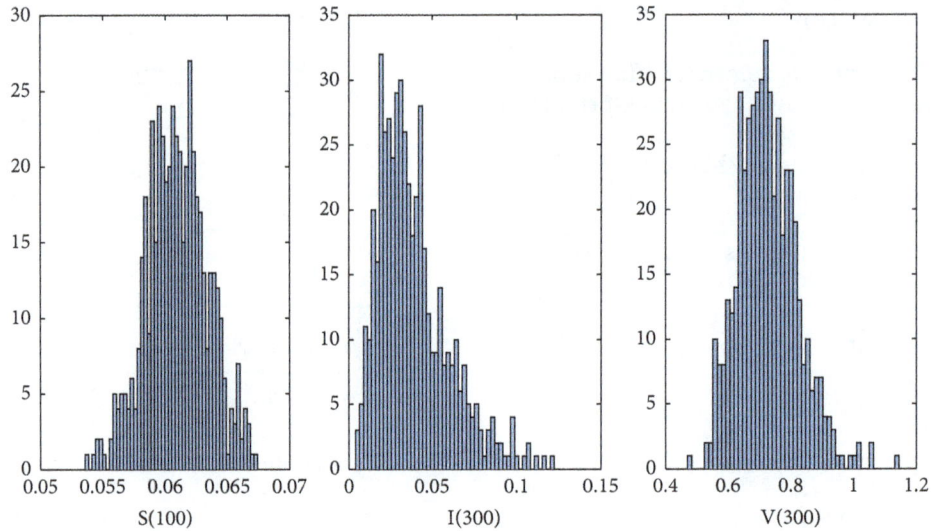

FIGURE 2: Distribution of $S(300)$, $I(300)$, and $V(300)$.

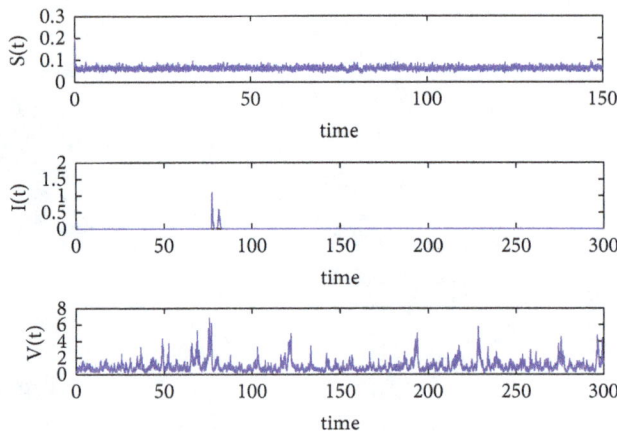

FIGURE 3: Simulation of the paths $S(t)$, $I(t)$, and $V(t)$ for model (3).

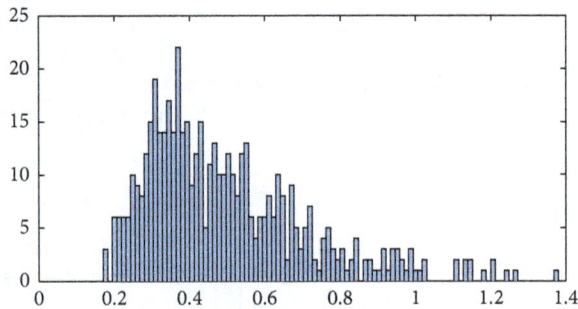

FIGURE 4: Distribution of $S(t) + \eta V(t)$ at time t =10 000.

Conflicts of Interest

The authors declare that there are no conflicts of interest regarding the publication of this paper.

References

[1] R. M. Anderson and R. M. May, "Population biology of infectious disease: Part I," *Nature*, vol. 280, pp. 361–376, 1979.

[2] R. M. May and R. M. Anderson, "Population biology of infectious disease: Part II," *Nature*, vol. 280, no. 5722, pp. 455–461, 1979.

[3] W. O. Kermack and A. G. McKendrick, "A contribution to the mathematical theory of epidemics," *Proceedings of the Royal Society A Mathematical, Physical and Engineering Sciences*, vol. 115, no. 772, pp. 700–721, 1927.

[4] H. W. Hethcote, "The mathematics of infectious diseases," *SIAM Review*, vol. 42, no. 4, pp. 599–653, 2000.

[5] A. Korobeinikov, P. K. Maini, and W. J. Walker, "Estimation of effective vaccination rate: pertussis in New Zealand as a case study," *Journal of Theoretical Biology*, vol. 224, no. 2, pp. 269–275, 2003.

[6] J. Li and Z. Ma, "Qualitative analyses of SIS epidemic model with vaccination and varying total population size," *Mathematical and Computer Modelling*, vol. 35, no. 11-12, pp. 1235–1243, 2002.

[7] C. Fraser, C. A. Donnelly, S. Cauchemez et al., "Pandemic potential of a strain of influenza A (H1N1): early findings," *Science*, vol. 324, no. 5934, pp. 1557–1561, 2009.

[8] J. Arino, K. L. Cooke, P. Van den Driessche, and J. Velasco-Hernández, "An epidemiology model that includes a leaky vaccine with a general waning function," *Discrete and Continuous Dynamical Systems - Series B*, vol. 4, no. 2, pp. 479–495, 2004.

[9] F. Brauer, "Backward bifurcations in simple vaccination models," *Journal of Mathematical Analysis and Applications*, vol. 298, no. 2, pp. 418–431, 2004.

models. During the last years, there was an increasing interest in Lévy noise that has non-Gaussian statistics. A lot of works have been realized and have shown the effectiveness of the Lévy noise in studying such phenomena (see [25–27]). We investigate in our future works the impact of the Lévy noise on the dynamic of complicated population systems.

[10] Y. Cai, Y. Kang, M. Banerjee, and W. Wang, "A stochastic epidemic model incorporating media coverage," *Communications in Mathematical Sciences*, vol. 14, no. 4, pp. 893–910, 2016.

[11] T. C. Gard, "Persistence in stochastic food web models," *Bulletin of Mathematical Biology*, vol. 46, no. 3, pp. 357–370, 1984.

[12] C. Ji, D. Jiang, Q. Yang, and N. Shi, "Dynamics of a multigroup SIR epidemic model with stochastic perturbation," *Automatica*, vol. 48, no. 1, pp. 121–131, 2012.

[13] E. Renshaw, *Modelling Biological Populations in Space and Time*, vol. 11, Cambridge University Press, Cambridge, UK, 1991.

[14] D. Kiouach and Y. Sabbar, "Stability and threshold of a stochastic SIRS epidemic model with vertical transmission and transfer from infectious to susceptible individuals," *Discrete Dynamics in Nature and Society*, Article ID 7570296, 13 pages, 2018.

[15] O. A. Chichigina, A. A. Dubkov, D. Valenti, and B. Spagnolo, "Stability in a system subject to noise with regulated periodicity," *Physical Review E: Statistical, Nonlinear, and Soft Matter Physics*, vol. 84, no. 2, 2011.

[16] O. Chichigina, D. ValentI, and B. Spagnolo, "A simple noise model with memory for biological systems," *Fluctuation and Noise Letters*, vol. 52, pp. 243–250, 2005.

[17] B. Spagnolo, D. Valenti, and A. Fiasconaro, "Noise in ecosystems: a short review," *Mathematical Biosciences and Engineering*, vol. 1, no. 1, pp. 185–211, 2004.

[18] E. Tornatore, P. Vetro, and S. M. Buccellato, "SIVR epidemic model with stochastic perturbation," *Neural Computing and Applications*, vol. 24, no. 2, pp. 309–315, 2014.

[19] D. Zhao and S. Yuan, "Persistence and stability of the disease-free equilibrium in a stochastic epidemic model with imperfect vaccine," *Advances in Difference Equations*, vol. 2016: Article ID 280, 14 pages, 2016.

[20] R. Rudnicki, "Long-time behaviour of a stochastic prey-predator model," *Stochastic Processes and Their Applications*, vol. 108, no. 1, pp. 93–107, 2003.

[21] Y. Zhao and D. Jiang, "The threshold of a stochastic SIS epidemic model with vaccination," *Applied Mathematics and Computation*, vol. 243, pp. 718–727, 2014.

[22] X. Mao, *Stochastic Differential Equations and Applications*, Horwood, Chichester, UK, 2nd edition, 1997.

[23] R. Khasminskii, *Stochastic Stability of Differential Equations*, vol. 66, Springer, Berlin, Germany, 2nd edition, 2012.

[24] Y. A. Kutoyants, *Statistical Inference for Ergodic Diffusion Processes*, Springer, London, UK, 2004.

[25] A. La Cognata, D. Valenti, A. A. Dubkov, and B. Spagnolo, "Dynamics of two competing species in the presence of Lévy noise sources," *Physical Review E: Statistical, Nonlinear, and Soft Matter Physics*, vol. 82, no. 1, Article ID 011121, 2010.

[26] X. Zhang, D. Jiang, T. Hayat, and B. Ahmad, "Dynamics of a stochastic SIS model with double epidemic diseases driven by Lévy jumps," *Physica A: Statistical Mechanics and its Applications*, vol. 471, pp. 767–777, 2017.

[27] J. Yu and M. Liu, "Stationary distribution and ergodicity of a stochastic food-chain model with Lévy jumps," *Physica A: Statistical Mechanics and its Applications*, vol. 482, pp. 14–28, 2017.

2

A Mathematical Model of Malaria Transmission with Structured Vector Population and Seasonality

Bakary Traoré, Boureima Sangaré, and Sado Traoré

Department of Mathematics, Polytechnic University of Bobo Dioulasso, 01 BP 1091, Bobo-Dioulasso 01, Burkina Faso

Correspondence should be addressed to Boureima Sangaré; mazoul979@yahoo.fr

Academic Editor: Sabri Arik

In this paper, we formulate a mathematical model of nonautonomous ordinary differential equations describing the dynamics of malaria transmission with age structure for the vector population. The biting rate of mosquitoes is considered as a positive periodic function which depends on climatic factors. The basic reproduction ratio of the model is obtained and we show that it is the threshold parameter between the extinction and the persistence of the disease. Thus, by applying the theorem of comparison and the theory of uniform persistence, we prove that if the basic reproduction ratio is less than 1, then the disease-free equilibrium is globally asymptotically stable and if it is greater than 1, then there exists at least one positive periodic solution. Finally, numerical simulations are carried out to illustrate our analytical results.

1. Introduction

Malaria is an infectious disease caused by plasmodium parasite which is transmitted to humans through the bites of infectious female mosquitoes. According to the estimations of World Health Organization (WHO) in 2015, 3.2 billion persons were at risk of infection and 2.4 million new cases were detected with 438,000 cases of deaths. However sub-Saharan Africa remains the most vulnerable region with high rate of deaths due to malaria.

To reduce the impact of malaria in the world, many scientific efforts were done including mathematical models construction. The first model of malaria transmission was developed by Ross [1]. According to Ross, if the mosquito population can be reduced to below a certain threshold, then malaria can be eradicated. Later, Macdonald did some modifications to the model and included superinfection. He showed that reducing the number of mosquitoes has little effect on the epidemiology of malaria in areas of intense transmission [2]. Nowadays, several mathematical models have been developed in order to reduce the malaria death rate in the world [3, 4]. In spite of the efforts made, it is still difficult to predict future malaria intensity, particularly in view of climate change.

It must be noticed that transmission and distribution of vector-borne diseases are greatly influenced by environmental and climatic factors. Seasonality and circadian rhythm of mosquito population, as well as other ecological and behavioural features, are strongly influenced by climatic factors such as temperature, rainfall, humidity, wind, and duration of daylight [5]. Moreover, in most mathematical models, the mosquito life cycle is generally ignored because eggs, larvae, and pupae are not involved in the transmission cycle. That is a useful simplification of the system but unfortunately the results of these models do not predict malaria intensity in most endemic regions. Thus, it is necessary to consider the life cycle of mosquitoes and the seasonality effect, which are very important aspects of the dynamics of malaria transmission.

Recently, Moulay et al. [6] have formulated a mathematical model describing the mosquito population dynamics which takes into account autoregulation phenomena of eggs and larvae stages. They have defined a threshold and proved that the growth of the mosquito population is governed by that threshold. Considering the climatic factors and the mosquitoes life cycle, we formulate a mathematical model describing the dynamics of malaria transmission. We analyze the impact of the model describing the mosquito population dynamics on the model of malaria transmission. Besides, by using the comparison theorem and the theory of uniform persistence, we, respectively, study the global stability of the nontrivial disease-free equilibrium [7–10] and the existence of positive periodic solutions.

This paper is organized as follows. In Section 2, we formulate the mathematical model of our problem. Section 3 provides the mathematical analysis of the model. Computational simulations are performed in Section 4 in order to illustrate our mathematical results. In the last section, Section 5, we conclude and give some remarks and future works.

2. Model Formulation

Motivated by the compartmental models in [6, 11], we derive an age-structured malaria model with seasonality to account for the cross infection between mosquitoes and humans. The human population is divided into four epidemiological categories representing the state variables: *the susceptible* class S_h, *exposed* class E_h, *infectious* class I_h, and *recovered* class R_h (immune and asymptomatic, but slightly infectious humans). In the life cycle of anopheles, there are mainly two major stages: mature stage and aquatic stage. Therefore, we divide the mosquitoes population into these stages: immature and mature. The immature stage is divided in two compartments: *eggs* class E, larvae and pupae class L. In the mature stage, we have three compartments: *the susceptible* class S_m, *exposed* class E_m, and *infectious* class I_m. At any time, the total number of humans and mature mosquitoes is given, respectively, by

$$N_h(t) = S_h(t) + E_h(t) + I_h(t) + R_h(t),\qquad(1)$$

$$A(t) = S_m(t) + E_m(t) + I_m(t).\qquad(2)$$

It is assumed throughout this paper that

(**H1**) all vector population measures refer to densities of female mosquitoes,

(**H2**) the mosquitoes bite only humans,

(**H3**) there is no vertical transmission of malaria,

(**H4**) all the new recruits are susceptible.

2.1. Interactions between Humans and Mosquitoes. When an infectious mosquito bites a susceptible human, the parasite enters the body of the human with a probability c_{mh} and the human moves into the exposed class E_h. Some time after, he leaves from class E_h to class I_h with rate α. Infectious humans migrate into the class R_h after acquisition of their immunity with rate r_h. The immunized lose their immunity with rate γ if they do not have continuous exposure to infection. Humans leave the total population through natural death rate d_h and malaria death rate d_p.

Similarly, when a susceptible mosquito bites an infectious human, it enters the class E_m with a probability c_{hm}. Some time after, it leaves from class E_m to infective class I_m with rate ν_m where it remains for life. Mature mosquitoes leave the population through natural mortality d_m.

Using the standard incidence as in the model of Ngwa and Shu [4], we define, respectively, the infection incidence from mosquitoes to humans, $k_h(t)$, and from humans to mosquitoes, $k_m(t)$:

$$k_h(t) = c_{mh}\beta(t)\frac{I_m(t)}{N_h(t)},\qquad(3)$$

$$k_m(t) = c_{hm}\beta(t)\frac{I_h(t)}{N_h(t)} + \bar{c}_{hm}\beta(t)\frac{R_h(t)}{N_h(t)}.\qquad(4)$$

Furthermore, using the above assumptions, we obtain the transfer diagram (Figure 1) of the model.

2.2. The Mathematical Model. Using the above assumptions and by making a balance of the movements in each class, we obtain the following system:

$$\frac{dE}{dt}(t) = b\left(1 - \frac{E(t)}{K_E}\right)A(t) - (s+d)E(t),$$

$$\frac{dL}{dt}(t) = s\left(1 - \frac{L(t)}{K_L}\right)E(t) - (s_L + d_L)L(t),$$

$$\frac{dS_h}{dt}(t) = \Lambda + \gamma R_h(t) - (d_h + k_h(t))S_h(t),$$

$$\frac{dE_h}{dt}(t) = k_h(t)S_h(t) - (d_h + \alpha)E_h(t),$$

$$\frac{dI_h}{dt}(t) = \alpha E_h(t) - (d_h + d_p + r_h)I_h(t),\qquad(5)$$

$$\frac{dR_h}{dt}(t) = r_h I_h(t) - (d_h + \gamma)R_h(t),$$

$$\frac{dS_m}{dt}(t) = s_L L(t) - (d_m + k_m(t))S_m(t),$$

$$\frac{dE_m}{dt}(t) = k_m(t)S_m(t) - (\nu_m + d_m)E_m(t),$$

$$\frac{dI_m}{dt}(t) = \nu_m E_m(t) - d_m I_m(t).$$

The growth of the whole human population and mature vector is, respectively, described by the following equations:

$$\frac{dN_h}{dt}(t) = \Lambda - d_h N_h(t) - d_p I_h(t),$$

$$\frac{dA}{dt}(t) = s_L L(t) - d_m A(t).\qquad(6)$$

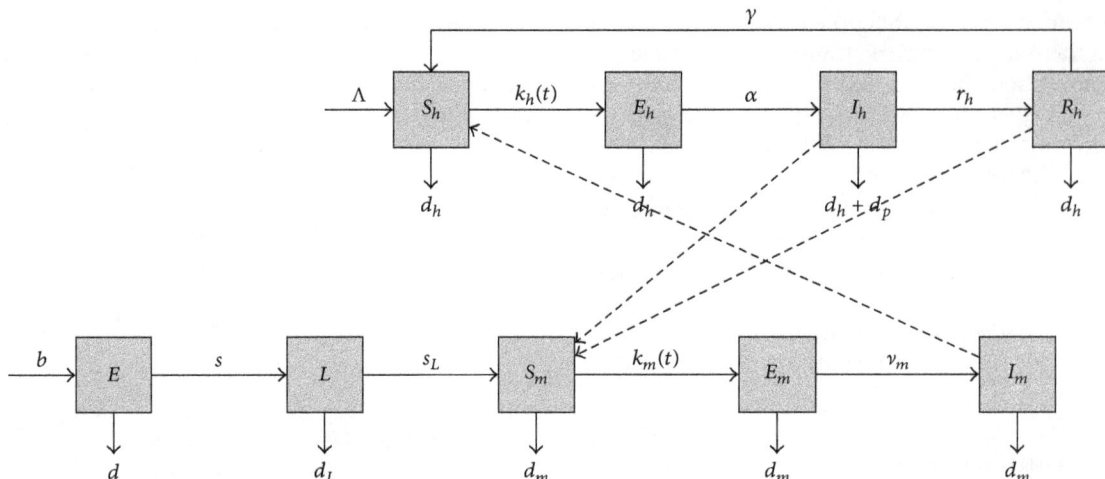

FIGURE 1: The dashed arrows indicate the direction of the infection and the solid arrows represent the transition from one class to another.

Using (2), we get $S_m(t) = A(t) - E_m(t) - I_m(t)$ and then the model can be rewritten as follows:

$$\frac{dE}{dt}(t) = b\left(1 - \frac{E(t)}{K_E}\right)A(t) - (s + d)E(t),$$

$$\frac{dL}{dt}(t) = s\left(1 - \frac{L(t)}{K_L}\right)E(t) - (s_L + d_L)L(t),$$

$$\frac{dA}{dt}(t) = s_L L(t) - d_m A(t),$$

$$\frac{dS_h}{dt}(t) = \Lambda + \gamma R_h(t) - (d_h + k_h(t))S_h(t),$$

$$\frac{dE_h}{dt}(t) = k_h(t)S_h(t) - (d_h + \alpha)E_h(t),$$

(7)

$$\frac{dI_h}{dt}(t) = \alpha E_h(t) - (d_h + d_p + r_h)I_h(t),$$

$$\frac{dR_h}{dt}(t) = r_h I_h(t) - (d_h + \gamma)R_h(t),$$

$$\frac{dE_m}{dt}(t) = k_m(t)A(t) - k_m(t)I_m(t)$$

$$- (\nu_m + d_m + k_m(t))E_m(t),$$

$$\frac{dI_m}{dt}(t) = \nu_m E_m(t) - d_m I_m(t).$$

Mathematically model (7) can be written as follows:

$$\dot{X}(t) = F(t, X(t)),$$

(8)

where $X(t) = (E(t), L(t), A(t), S_h(t), E_h(t), I_h(t), R_h(t), E_m(t), I_m(t))^T$. The function $F : \mathbb{R}_+ \times \mathbb{R}^9 \to \mathbb{R}^9$ is $C^\infty(\mathbb{R}^9)$ and defined by

$F(t, X(t))$

$$= \begin{pmatrix} b\left(1 - \dfrac{E(t)}{K_E}\right)A(t) - (s + d)E(t) \\ s\left(1 - \dfrac{L(t)}{K_L}\right)E(t) - (s_L + d_L)L(t) \\ s_L L(t) - d_m A(t) \\ \Lambda + \gamma R_h(t) - (d_h + k_h(t))S_h(t) \\ k_h(t)S_h(t) - (d_h + \alpha)E_h(t) \\ \alpha E_h(t) - (d_h + d_p + r_h)I_h(t) \\ r_h I_h(t) - (d_h + \gamma)R_h(t) \\ k_m(t)A(t) - k_m(t)I_m(t) - (\nu_m + d_m + k_m(t))E_m(t) \\ \nu_m E_m(t) - d_m I_m(t) \end{pmatrix}$$

(9)

Let us consider $F = (F_1, F_2)^T$ and $X(t) = (X_1(t), X_2(t))^T$ with $X_1(t) = (E(t), L(t), A(t))^T$ and $X_2(t) = (S_h(t), E_h(t), I_h(t), R_h(t), E_m(t), I_m(t))^T$. Then system (8) can be rewritten as follows:

$$\dot{X}_1(t) = F_1(X_1(t), X_2(t)),$$

(10)

$$\dot{X}_2(t) = F_2(t, X_1(t), X_2(t)),$$

(11)

with the functions F_1 and F_2 defined as follows:

$F_2(t, X_1(t), X_2(t))$

$$= \begin{pmatrix} \Lambda + \gamma R_h(t) - (d_h + k_h(t))S_h(t) \\ k_h(t)S_h(t) - (d_h + \alpha)E_h(t) \\ \alpha E_h(t) - (d_h + d_p + r_h)I_h(t) \\ r_h I_h(t) - (d_h + \gamma)R_h(t) \\ k_m(t)A(t) - k_m(t)I_m(t) - (\nu_m + d_m + k_m(t))E_m(t) \\ \nu_m E_m(t) - d_m I_m(t) \end{pmatrix},$$

$$F_1\left(X_1\left(t\right), X_2\left(t\right)\right) = \begin{pmatrix} b\left(1 - \dfrac{E\left(t\right)}{K_E}\right) A\left(t\right) - \left(s + d\right) E\left(t\right) \\ s\left(1 - \dfrac{L\left(t\right)}{K_L}\right) E\left(t\right) - \left(s_L + d_L\right) L\left(t\right) \\ s_L L\left(t\right) - d_m A\left(t\right) \end{pmatrix}.$$

$$(12)$$

System (10) describes the maturation cycle of mosquitoes and system (11) describes the dynamics of malaria transmission. System (10) is biologically well defined in

$$\Delta := \left\{ (E, L, A) \in \mathbb{R}^3_+ \; E \le K_E, \; L \le K_L, \; A \le \frac{s_L}{d_m} K_L \right\} \quad (13)$$

and system (11) is biologically well defined in

$$\Omega = \left\{ \left(S_h, E_h, I_h, R_h, E_m, I_m\right) \in \mathbb{R}^6_+ \mid \right.$$

$$\left. S_h + E_h + I_h + R_h \le \frac{\Lambda}{d_h}, \; E_m + I_m \le \frac{s_L K_L}{d_m} \right\};$$

$$(14)$$

then model (7) is biologically well defined in $\Gamma := \Delta \times \Omega$.

3. Mathematical Analysis

3.1. Positivity and Boundedness of Solutions

Lemma 1 (see [6]). *The set Δ is a positive invariant region under the flow induced by (10).*

We assume that

(H5) $\beta(t)$ is a ω-periodic positive function with $\omega = 12$ months,

(H6) all the parameters of the model are positive except the disease-induced death rate, d_p, which is assumed to be nonnegative.

Theorem 2. *For any initial condition $\phi \in \mathbb{R}^9_+$, system (8) has a unique solution. Further, the compact Γ is a positively invariant set, which attracts all positive orbits in $\mathbb{R}^9_+$.*

Proof. For all $\phi \in \mathbb{R}^9_+$, the function F is locally Lipschitzian in $X(t)$. It then follows through Cauchy-Lipschitz theorem that system (8) has a unique local solution.

Furthermore, according to (6), we have

$$\frac{dN_h}{dt}(t) = \Lambda - d_h N_h(t) - d_p I_h(t) \le \Lambda - d_h N_h(t),$$

$$\frac{dA}{dt}(t) = s_L L(t) - d_m A(t) \le s_L K_L - d_m A(t).$$

$$(15)$$

It then follows that if $N_h(t) > \Lambda/d_h$ and $A(t) > s_L K_L/d_m$, then $dN_h/dt(t) < 0$ and $dA/dt(t) < 0$.

Let us consider the following differential equations:

$$\frac{dN_h}{dt}(t) = \Lambda - d_h N_h(t),$$

$$\frac{dA}{dt}(t) = s_L K_L - d_m A(t)$$

$$(16)$$

with general solutions:

$$N_h(t) = \frac{\Lambda}{d_h} + \left(N_h(0) - \frac{\Lambda}{d_h}\right) e^{-d_h t},$$

$$A(t) = \frac{s_L K_L}{d_m} + \left(A(0) - \frac{s_L K_L}{d_m}\right) e^{-d_m t}.$$

$$(17)$$

By applying the standard comparison theorem, we obtain, for all $t \ge 0$, $N_h(t) \le \Lambda/d_h$ and $A(t) \le s_L K_L/d_m$ if $N_h(0) \le \Lambda/d_h$ and $A(0) \le s_L K_L/d_m$. Thus, the set Ω is positively invariant with respect to system (11). Therefore, from Lemma 1, the set Δ is positively invariant with respect to system (10). Then, we conclude that the compact set $\Gamma = \Delta \times \Omega$ is positively invariant. Thus, all the solutions of system (8) are nonnegative and bounded. $\qquad\square$

3.2. Disease-Free Equilibriums.
Let us consider the following threshold parameter: $r = (b/(s+d))(s/(s_L+d_L))(s_L/d_m)$. Then we have the following result.

Proposition 3 (see [6]). *System (10) always has the mosquito-free equilibrium $P_0 = (0, 0, 0)$.*

(i) *If $r \le 1$, then system (10) has no other equilibrium.*

(ii) *If $r > 1$, there is a unique endemic equilibrium*

$$P_1 = \left(E^*, L^*, A^*\right) = \left(1 - \frac{1}{r}\right)\left(\frac{K_E}{\gamma_E}, \frac{K_L}{\gamma_L}, \frac{s_L}{d_m}\frac{K_L}{\gamma_L}\right), \quad (18)$$

where

$$\gamma_E = 1 + \frac{(s+d) d_m K_E}{b s_L K_L},$$

$$\gamma_L = 1 + \frac{(s_L + d_L) K_L}{s K_E}.$$

$$(19)$$

Lemma 4. *Model (7) has*

(i) *trivial disease-free equilibrium $E_0 = (0, 0, 0, S_h^*, 0, 0, 0, 0, 0)$ if $r \le 1$,*

(ii) *nontrivial disease-free equilibrium $E_1 = (E^*, L^*, A^*, S_h^*, 0, 0, 0, 0, 0)$ if $r > 1$, where $S_h^* = \Lambda/d_h$, $A^* = S_m^* = s_L L^*/d_m$, and $E^*, L^*,$ and A^* are given above.*

Proof. By solving the system $F_2(t, X_1(t), X_2(t)) = 0$ at the disease-free equilibrium, $E_h(t) = I_h(t) = R_h(t) = E_m(t) = I_m(t) = 0$, $\forall t \ge 0$, we get the equilibrium point $E_1^+ = (S_h^*, 0, 0, 0, 0, 0)$ for system (11), with $S_h^* = \Lambda/d_h$. Moreover, thanks to Proposition 3, system (10) has a unique mosquito-free equilibrium $(0, 0, 0)$ if $r \le 1$ and a unique endemic equilibrium (E^*, L^*, A^*) if $r > 1$. Thus, we conclude that system (7) has a trivial disease-free equilibrium $E_0 = (0, 0, 0, S_h^*, 0, 0, 0, 0, 0)$ if $r \le 1$ and a nontrivial disease-free equilibrium $E_1 = (E^*, L^*, A^*, S_h^*, 0, 0, 0, 0, 0)$ if $r > 1$. $\qquad\square$

Remark 5. We will only consider the equilibrium state E_1 because it is more biologically realistic. So, in the rest of the paper, we assume that $r > 1$.

3.3. Threshold Dynamics. Linearizing system (8) at the equilibrium state E_1, we obtain the following system (here we write down only the equations for the "diseased" classes):

$$\frac{dE_h}{dt}(t) = c_{mh}\beta(t) I_m(t) - (d_h + \alpha) E_h(t),$$

$$\frac{dI_h}{dt}(t) = \alpha E_h(t) - (d_h + d_p + r_h) I_h(t),$$

$$\frac{dR_h}{dt}(t) = r_h I_h(t) - (d_h + \gamma) R_h(t),$$

$$\frac{dE_m}{dt}(t) = c_{hm}\beta(t) \frac{A^*}{S_h^*} I_h(t) + \bar{c}_{hm}\beta(t) \frac{A^*}{S_h^*} R_h(t)$$
$$\qquad - (\nu_m + d_m) E_m(t),$$

$$\frac{dI_m}{dt}(t) = \nu_m E_m(t) - d_m I_m(t). \tag{20}$$

This system can be rewritten as

$$\frac{dZ(t)}{dt} = (F(t) - V(t)) Z(t), \tag{21}$$

where $Z(t) = (E_h(t), I_h(t), R_h(t), E_m(t), I_m(t))^T$ and $F(t)$ and $V(t)$ are 5×5 matrix defined as follows:

$F(t)$

$$= \begin{pmatrix} 0 & 0 & 0 & 0 & c_{mh}\beta(t) \\ 0 & 0 & 0 & 0 & 0 \\ 0 & 0 & 0 & 0 & 0 \\ 0 & c_{hm}\beta(t)\dfrac{A^*}{S_h^*} & \bar{c}_{hm}\beta(t)\dfrac{A^*}{S_h^*} & 0 & 0 \\ 0 & 0 & 0 & 0 & 0 \end{pmatrix}, \tag{22}$$

$V(t)$

$$= \begin{pmatrix} d_h + \alpha & 0 & 0 & 0 & 0 \\ -\alpha & d_h + d_p + r_h & 0 & 0 & 0 \\ 0 & -r_h & d_h + \gamma & 0 & 0 \\ 0 & 0 & 0 & \nu_m + d_m & 0 \\ 0 & 0 & 0 & -\nu_m & d_m \end{pmatrix}.$$

Let us assume that $Y(t,s), t \geq s$, is the matrix solution of the linear ω-periodic system

$$\frac{dy}{dt} = -V(t) y. \tag{23}$$

That is, for each $s \in \mathbb{R}$, the 5×5 matrix $Y(t,s)$ satisfies the equation

$$\frac{d}{dt} Y(t,s) = -V(t) Y(t,s), \quad \forall t \geq s, \ Y(s,s) = I, \tag{24}$$

where I is the 5×5 identity matrix. Thus, the monodromy matrix $\Phi_{-V}(t)$ of (23) is equal to $Y(t,0), \forall t \geq 0$.

Let C_ω be the ordered Banach space of all ω-periodic functions from $\mathbb{R}$ to $\mathbb{R}^5$ which is equipped with the maximum norm $\|\cdot\|$ and the positive cone $C_\omega^+ := \{\phi \in C_\omega : \phi(t) \geq 0, \ \forall t \in \mathbb{R}\}$. Then, we can define a linear operator $\mathscr{L} : C_\omega \to C_\omega$ by

$$(\mathscr{L}\phi)(t) = \int_0^\infty Y(t, t-a) F(t-a) \phi(t-a) \, da, \tag{25}$$

$$\forall t \in \mathbb{R}, \ \phi \in C_\omega.$$

It then follows from [12] that $\mathscr{L}$ is the next infection operator, and the basic reproduction ratio is $\mathscr{R}_0 = \rho(\mathscr{L})$, the spectral radius of $\mathscr{L}$.

In order to calculate $\mathscr{R}_0$, we consider the following linear ω-periodic system:

$$\frac{dw(t)}{dt} = \left[\frac{1}{\lambda} F(t) - V(t)\right] w(t), \tag{26}$$

$$\forall t \in \mathbb{R}_+, \ \lambda \in (0, \infty).$$

Let $W(t, s, \lambda), t \geq s, s \in \mathbb{R}$, be the evolution operator of system (26) on $\mathbb{R}^5$. Clearly $W(t, 0, 1) = \Phi_{F-V}(t), \forall t \geq 0$. The following result will be used in our numerical calculation of the basic reproduction ratio.

Lemma 6 (see [12]). *(i) If $\rho(W(\omega, 0, \lambda)) = 1$ has a positive solution λ_0, then λ_0 is an eigenvalue of $\mathscr{L}$, and hence $\mathscr{R}_0 > 0$.*
(ii) If $\mathscr{R}_0 > 0$, then $\lambda = \mathscr{R}_0$ is the unique solution of $\rho(W(\omega, 0, \lambda)) = 1$.
(iii) $\mathscr{R}_0 = 0$ if and only if $\rho(W(\omega, 0, \lambda)) < 1$, for all $\lambda > 0$.

3.4. Stability of Equilibrium State E_1. In this section, we will study the asymptotic behaviour of the nontrivial equilibrium E_1; thus we have the following result, which will be used in the proofs of our main results.

Lemma 7 (see [12]). *The following statements are valid:*

(i) $\mathscr{R}_0 = 1$ *if and only if* $\rho(\Phi_{F-V}(\omega)) = 1$.

(ii) $\mathscr{R}_0 < 1$ *if and only if* $\rho(\Phi_{F-V}(\omega)) < 1$.

(iii) $\mathscr{R}_0 > 1$ *if and only if* $\rho(\Phi_{F-V}(\omega)) > 1$.

Lemma 8 (see [6]). *If $r > 1$, then P_1 is globally asymptotically stable in $\mathrm{int}(\Delta)$, with respect to system (10).*

Theorem 9. *The nontrivial equilibrium E_1 is locally asymptotically stable if $\mathscr{R}_0 < 1$ and unstable if $\mathscr{R}_0 > 1$.*

Proof. Let $\mathscr{A}(t)$ be the Jacobian matrix of (8) evaluated at E_1. Then we have

$$\mathscr{A}(t) = \begin{pmatrix} \mathscr{A}_{11} & \mathscr{A}_{12} \\ \mathscr{A}_{21} & \mathscr{A}_{22}(t) \end{pmatrix}, \tag{27}$$

where

$$\mathscr{A}_{12} = \begin{pmatrix} 0 & 0 & 0 & 0 & b\left(1 - \dfrac{E^*}{K_E}\right) & b\left(1 - \dfrac{E^*}{K_E}\right) \\ 0 & 0 & 0 & 0 & 0 & 0 \\ 0 & 0 & 0 & 0 & -d_m & -d_m \end{pmatrix},$$

$$\mathscr{A}_{21} = \begin{pmatrix} 0 & 0 & 0 \\ 0 & 0 & 0 \\ 0 & 0 & 0 \\ 0 & 0 & 0 \\ 0 & 0 & 0 \\ 0 & 0 & 0 \end{pmatrix},$$

(28)

$$\mathscr{A}_{11}$$
$$= \begin{pmatrix} -(s+d) - \dfrac{bA^*}{K_E} & 0 & (s+d)\dfrac{E^*}{A^*} \\ (s_L + d_L)\dfrac{L^*}{E^*} & -(s_L + d_L) - \dfrac{sE^*}{K_L} & 0 \\ 0 & s_L & -d_m \end{pmatrix},$$

$$\mathscr{A}_{22}(t) = \begin{pmatrix} -d_h & C(t) \\ \hat{0} & F(t) - V(t) \end{pmatrix}$$

with

$$C(t) = \left(0 \ \ 0 \ \ \gamma \ \ 0 \ \ -c_{mh}\beta(t)\right),$$

$$\hat{0} = (0,0,0,0,0)^T.$$

(29)

E_1 is locally asymptotically stable if $\rho(\Phi_{\mathscr{A}}(\omega)) < 1$. The matrix $\mathscr{A}_{11}$ is a constant matrix and its characteristic equation is given by $\pi(z) = z^3 + a_1 z^2 + a_2 z + a_3$, where

$$a_1 = \left(1 - \dfrac{1}{r}\right)\left(\dfrac{sK_E}{\gamma_E K_L} + \dfrac{bs_L K_L}{d_m \gamma_L K_E}\right) + s_L + d_L + s + d$$
$$\quad + d_m,$$

$$a_2 = \left[\dfrac{bs_L K_L}{d_m \gamma_L K_E}\left(1 - \dfrac{1}{r}\right)\right]\left[\dfrac{sK_E}{\gamma_E K_L}\left(1 - \dfrac{1}{r}\right) + s_L + d_L\right.$$
$$\quad \left. + d_m\right] + \dfrac{d_m s K_E}{\gamma_E K_L}\left(1 - \dfrac{1}{r}\right) + d_m(s_L + d_L),$$

(30)

$$a_3 = d_m\left(1 - \dfrac{1}{r}\right)\left[\dfrac{bss_L}{d_m \gamma_L \gamma_E}\left(1 - \dfrac{1}{r}\right) + (s+d)\dfrac{sK_E}{\gamma_E K_L}\right.$$
$$\quad \left. + (s_L + d_L)\dfrac{bs_L K_L}{d_m \gamma_L K_E}\right].$$

If $r > 1$, then a_1, a_2, a_3 and $a_1 a_2 - a_3$ are clearly positive. So, thanks to Routh-Hurwitz criterion, all eigenvalues of $\mathscr{A}_{11}$ have negative real part. It then follows that $\rho(\Phi_{\mathscr{A}_{11}}(\omega)) < 1$. Thus, the stability of E_1 depends on $\Phi_{\mathscr{A}_{22}}(\omega)$.

Thus, if $\rho(\Phi_{F-V}(\omega)) < 1$, then $\rho(\Phi_{\mathscr{A}_{22}}(\omega)) < 1$ and then E_1 is stable. If $\rho(\Phi_{F-V}(\omega)) > 1$ then E_1 is unstable. So, thanks to Lemma 7, E_1 is locally asymptotically stable if $\mathscr{R}_0 < 1$ and unstable if $\mathscr{R}_0 > 1$. □

Lemma 10 (see [13]). *Let $\theta = (1/\omega)\ln \rho(\Phi_{A(\cdot)}(\omega))$; then there exists a positive ω-periodic function $v(t)$ such that $e^{\theta t}v(t)$ is a solution of $\dot{x}(t) = A(t)x(t)$.*

Theorem 11. *If $\mathscr{R}_0 < 1$ and $d_p = 0$, then E_1 is globally asymptotically stable.*

Proof. If $d_p = 0$, we can rewrite (6) as follows:

$$\dfrac{dN_h}{dt}(t) = \Lambda - d_h N_h(t),$$
$$\dfrac{dA}{dt}(t) = s_L L(t) - d_m A(t).$$

(31)

Thus, there exists a period ω' such that $\forall t \geq \omega'$, $N_h(t) \geq N_h^* - \epsilon$ and $A(t) \leq A^* + \epsilon$, $\forall \epsilon > 0$.

At disease-free equilibrium, we have $N_h^* = S_h^*$ and $S_m^* = A^*$. So, $A(t)/N_h(t) \leq (A^* + \epsilon)/(S_h^* - \epsilon)$. It then follows from system (11) that

$$\dfrac{dE_h}{dt}(t) \leq c_{mh}\beta(t) I_m(t) - (d_h + \alpha) E_h(t),$$

(32a)

$$\dfrac{dI_h}{dt}(t) = \alpha E_h(t) - (d_h + r_h) I_h(t),$$

(32b)

$$\dfrac{dR_h}{dt}(t) = r_h I_h(t) - (d_h + \gamma) R_h(t),$$

(32c)

$$\dfrac{dE_m}{dt}(t) \leq c_{hm}\beta(t)\dfrac{A^* + \epsilon}{S_h^* - \epsilon} I_h(t)$$
$$\quad + \bar{c}_{hm}\beta(t)\dfrac{A^* + \epsilon}{S_h^* - \epsilon} R_h(t)$$
$$\quad - (\nu_m + d_m) E_m(t),$$

(32d)

$$\dfrac{dI_m}{dt}(t) = \nu_m E_m(t) - d_m I_m(t).$$

(32e)

Let us consider the following auxiliary system:

$$\dfrac{d\bar{E}_h}{dt}(t) = c_{mh}\beta(t)\bar{I}_m(t) - (d_h + \alpha)\bar{E}_h(t),$$

$$\dfrac{d\bar{I}_h}{dt}(t) = \alpha\bar{E}_h(t) - (d_h + r_h)\bar{I}_h(t),$$

$$\dfrac{d\bar{R}_h}{dt}(t) = r_h\bar{I}_h(t) - (d_h + \gamma)\bar{R}_h(t),$$

$$\dfrac{d\bar{E}_m}{dt}(t) = c_{hm}\beta(t)\dfrac{A^* + \epsilon}{S_h^* - \epsilon}\bar{I}_h(t)$$

(33)

$$\quad + \bar{c}_{hm}\beta(t)\dfrac{A^* + \epsilon}{S_h^* - \epsilon}\bar{R}_h(t)$$
$$\quad - (\nu_m + d_m)\bar{E}_m(t),$$

$$\dfrac{d\bar{I}_m}{dt}(t) = \nu_m\bar{E}_m(t) - d_m\bar{I}_m(t),$$

which can be rewritten as follows:

$$\dfrac{d\bar{h}}{dt}(t) = M_\epsilon(t)\bar{h}(t);$$

(34)

$$\bar{h}(t) = \left(\bar{E}_h(t), \bar{I}_h(t), \bar{R}_h(t), \bar{E}_m(t), \bar{I}_m(t)\right)^T$$

with

$$M_\epsilon(t) = \begin{pmatrix} -(d_h + \alpha) & 0 & 0 & 0 & c_{mh}\beta(t) \\ \alpha & -(d_h + r_h) & 0 & 0 & 0 \\ 0 & r_h & -(d_h + \gamma) & 0 & 0 \\ 0 & c_{hm}\beta(t)\dfrac{A^* + \epsilon}{S_h^* - \epsilon} & \bar{c}_{hm}\beta(t)\dfrac{A^* + \epsilon}{S_h^* - \epsilon} & -(\nu_m + d_m) & 0 \\ 0 & 0 & 0 & \nu_m & -d_m \end{pmatrix}. \tag{35}$$

From Lemma 7, if $\mathscr{R}_0 < 1$, then $\rho(\Phi_{F-V}(\omega)) < 1$. Clearly, $\lim_{\epsilon \to 0^+} \Phi_{M_\epsilon}(\omega) = \Phi_{F-V}(\omega)$ and, by continuity of the spectral radius, we have $\lim_{\epsilon \to 0^+} \rho(\Phi_{M_\epsilon}(\omega)) = \rho(\Phi_{F-V}(\omega)) < 1$. Thus, there exists $\epsilon_1 > 0$ such that $\rho(\Phi_{M_\epsilon}(\omega)) < 1, \forall \epsilon \in [0, \epsilon_1[$.

From Lemma 10, there exists a positive ω-periodic function $v(t)$ such that $\bar{h}(t) = e^{\theta t} v(t)$ is a solution of (34). Since $\rho(\Phi_{M_\epsilon}(\omega)) < 1$, $\theta < 0$. The ω-periodic function $v(t)$ is bounded and it then follows that $\lim_{t\to\infty} \bar{h}(t) = 0$. Applying comparison theorem on system (32a)–(32e), we get $\lim_{t\to\infty}(E_h(t), I_h(t), R_h(t), E_m(t), I_m(t)) = (0,0,0,0,0)$. Using the theory of asymptotically periodic semiflow [[14], Theorem 3.2.1], we have $\lim_{t\to\infty} S_h(t) = S_h^*$, $\lim_{t\to\infty} A(t) = A^* = S_m^*$. From Lemma 8, if $r > 1$ then P_1 is globally asymptotically stable, so $\lim_{t\to\infty} E(t) = E^*$ and $\lim_{t\to\infty} L(t) = L^*$. Hence, the equilibrium E_1 is globally attractive. $\square$

3.5. Existence of Positive Periodic Solutions. System (8) is constructed by coupling two subsystems. The term coupling these two systems is given by the function $s_L L(t)$. The coupling takes place only in one direction because the dynamics of system (11) depend on the dynamics of system (10). The asymptotic behaviour of system (10) is given by Lemma 8. Now we are going to study the existence of positive periodic solutions of system (11):

$$\frac{dS_h}{dt}(t) = \Lambda + \gamma R_h(t) - (d_h + k_h(t)) S_h(t),$$

$$\frac{dE_h}{dt}(t) = k_h(t) S_h(t) - (d_h + \alpha) E_h(t),$$

$$\frac{dI_h}{dt}(t) = \alpha E_h(t) - (d_h + d_p + r_h) I_h(t),$$

$$\frac{dR_h}{dt}(t) = r_h I_h(t) - (d_h + \gamma) R_h(t), \tag{36}$$

$$\frac{dE_m}{dt}(t) = k_m(t) A(t) - k_m(t) I_m(t)$$
$$\qquad - (\nu_m + d_m + k_m(t)) E_m(t),$$

$$\frac{dI_m}{dt}(t) = \nu_m E_m(t) - d_m I_m(t).$$

Model (11) is well defined in Ω and if $r > 1$ it has a disease-free equilibrium $E_1^+ = (S_h^*, 0, 0, 0, 0, 0)$ with $S_h^* = \Lambda/d_h$.

Let us consider the following sets:

$$X := \mathbb{R}_+^6,$$

$$X_0 := \{(S_h, E_h, I_h, R_h, E_m, I_m) \in X \mid E_h > 0, I_h > 0, R_h > 0, E_m > 0, I_m > 0\}, \tag{37}$$

$$\partial X_0 := X \setminus X_0.$$

Let $u(t, \psi)$ be the unique solution of (11) with initial conditions ψ, $\Phi(t)$ the periodic semiflow generated by periodic system (11), and $P : X \to X$ the Poincaré map associated with system (11); namely,

$$P(\psi) = \Phi(\omega)\psi = u(\omega, \psi), \quad \forall \psi \in X,$$
$$P^m(\psi) = \Phi(m\omega)\psi = u(m\omega, \psi), \quad \forall m \geq 0. \tag{38}$$

Proposition 12. *The sets X_0 and ∂X_0 are positively invariant under the flow induced by (11).*

Proof. Note that if X_0 is positively invariant, then ∂X_0 is positively invariant. Thus we only need to prove that X_0 is positively invariant.

For any initial condition $\psi \in X_0$, solving the equations of system (11) we derive that

$$S_h(t) = \exp\left(-\int_0^t (k_h(s) + d_h)\, ds\right)\left[S_h(0)\right.$$
$$+ \int_0^t (\Lambda + I_h(s) + \gamma R_h(s))$$
$$\cdot \exp\left(\int_0^s (k_h(c) + d_h)\, dc\right) ds\Big]$$
$$\geq \exp\left(-\int_0^t (k_h(s) + d_h)\, ds\right)$$
$$\cdot \left[\int_0^t (\Lambda + I_h(s) + \gamma R_h(s))\right.$$
$$\cdot \exp\left(\int_0^s (k_h(c) + d_h)\, dc\right) ds\Big] > 0, \quad \forall t > 0,$$

$$E_h(t) = e^{-(d_h+\alpha)t}\left(E_h(0) + \int_0^t k_h(s)S_h(s)e^{(d_h+\alpha)s}ds\right)$$

$$\geq e^{-(d_h+\alpha)t}\left(\int_0^t k_h(s)S_h(s)e^{(d_h+\alpha)s}ds\right) > 0,$$

$$\forall t > 0,$$

$$I_h(t) = e^{-(d_h+d_p+r_h)t}\left(I_h(0) + \int_0^t \alpha E_h(s)e^{(d_h+d_p+r_h)s}ds\right)$$

$$\geq e^{-(d_h+d_p+r_h)t}\left(\int_0^t \alpha E_h(s)e^{(d_h+d_p+r_h)s}ds\right) > 0,$$

$$\forall t > 0,$$

$$R_h(t) = e^{-(d_h+\gamma)t}\left(R_h(0) + \int_0^t r_h I_h(s)e^{(d_h+\gamma)s}ds\right)$$

$$\geq e^{-(d_h+\gamma)t}\left(\int_0^t r_h I_h(s)e^{(d_h+\gamma)s}ds\right) > 0, \quad \forall t > 0,$$

$$E_m(t) = e^{\int_0^t -(k_m(s)+d_m+v_m)ds}\left[E_m(0) + \int_0^t k_m(s)\right.$$

$$\left. \cdot (A(s) - I_m(s))e^{\int_0^s (k_m(c)+d_m+v_m)dc}ds\right]$$

$$\geq e^{\int_0^t -(k_m(s)+d_m+v_m)ds}\left[\int_0^t k_m(s)(A(s) - I_m(s))\right.$$

$$\left. \cdot e^{\int_0^s (k_m(c)+d_m+v_m)dc}ds\right] > 0, \quad \forall t > 0,$$

$$I_m(t) = e^{-d_m t}\left(I_m(0) + \int_0^t v_m E_m(s)e^{d_m s}\right)$$

$$\geq e^{-d_m t}\left(\int_0^t v_m E_m(s)e^{d_m s}\right) > 0, \quad \forall t > 0.$$

$$(39)$$

Thus, X_0 is positively invariant. So, ∂X_0 is also positively invariant. $\square$

Note that, from Theorem 2, Ω is a compact set which attracts all positive orbits in X, which implies that the discrete-time system $P : X \to X$ is point dissipative. Moreover, $\forall n_0 \geq 1$, P^{n_0} is compact; it then follows from Theorem 2.9 in [15] that P admits a global attractor in X.

Lemma 13. *If $\mathcal{R}_0 > 1$, there exists $\eta > 0$ such that when $\|\psi - E_1^+\| \leq \eta$, $\forall \psi \in X_0$, we have $\limsup_{m\to\infty}\|P^m(\psi) - E_1^+\| \geq \eta$.*

Proof. Suppose by contradiction that $\limsup_{m\to\infty}\|P^m(\psi) - E_1^+\| < \eta$ for some $\psi \in X_0$. Then, there exists an integer $n \geq 1$

such that, for all $m \geq n$, $\|P^m(\psi) - M\| < \eta$. By the continuity of the solution $u(t,\psi)$, we have $\|u(t, P^m(\psi)) - u(t, E_1^+)\| \leq \sigma$ for all $t \geq 0$ and $\sigma > 0$. For all $t \geq 0$, let $t = m\omega + t_1$, where $t_1 \in [0,\omega]$ and $m = [t/\omega]$. $[t/\omega]$ is the greatest integer less than or equal to t/ω. If $\|\psi - E_1^+\| \leq \eta$, then by the continuity of the solution $u(t,\psi)$ we have

$$\|u(t,\psi) - u(t, E_1^+)\|$$

$$= \|u(t_1 + m\omega, \psi) - u(t_1 + m\omega, M)\|$$

$$= \|\Phi(t_1 + m\omega)\psi - \Phi(t_1 + m\omega)E_1^+\|$$

$$= \|\Phi(t_1)\Phi(m\omega)\psi - \Phi(t_1)\Phi(m\omega)E_1^+\| \quad (40)$$

$$= \|\Phi(t_1)P^m(\psi) - \Phi(t_1)P^m(E_1^+)\|$$

$$= \|\Phi(t_1)P^m(\psi) - \Phi(t_1)E_1^+\| \leq \sigma.$$

It then follows that $S_h^* - \sigma \leq S_h(t) \leq S_h^* + \sigma$ and $A^* - \sigma \leq A(t) \leq A^* + \sigma$. So, there exists $\sigma^* > 0$ such that $S_h(t)/N_h(t) \geq 1 - \sigma^*$ and $A(t)/N_h(t) \geq A^*/N_h^* - \sigma^*$.

From (11) we have

$$\frac{dE_h}{dt}(t) \geq c_{mh}\beta(t)(1-\sigma^*)I_m(t) - (d_h + \alpha)E_h(t),$$

$$\frac{dI_h}{dt}(t) = \alpha E_h(t) - (d_p + d_h + r_h)I_h(t),$$

$$\frac{dR_h}{dt}(t) = r_h I_h(t) - (d_h + \gamma)R_h(t),$$

$$\frac{dE_m}{dt}(t) \geq \beta(t)\left(\frac{A^*}{N_h^*} - \sigma^*\right)[c_{hm}I_h(t) + \bar{c}_{hm}R_h(t)]$$

$$\qquad - (v_m + d_m)E_m(t),$$

$$\frac{dI_m}{dt}(t) = v_m E_m(t) - d_m I_m(t).$$

$$(41)$$

Let us consider the following auxiliary linear system:

$$\frac{d\hat{h}}{dt}(t) = M_{\sigma^*}(t)\hat{h}(t);$$

$$\hat{h}(t) = \left(\hat{E}_h(t), \hat{I}_h(t), \hat{R}_h(t), \hat{E}_m(t), \hat{I}_m(t)\right)^T$$

$$(42)$$

with

$$M_{\sigma^*}(t) = \begin{pmatrix} -(d_h+\alpha) & 0 & 0 & 0 & (1-\sigma^*)c_{mh}\beta(t) \\ \alpha & -(d_h+r_h) & 0 & 0 & 0 \\ 0 & r_h & -(d_h+\gamma) & 0 & 0 \\ 0 & c_{hm}\beta(t)\left(\frac{A^*}{S_h^*}-\sigma^*\right) & \bar{c}_{hm}\beta(t)\left(\frac{A^*}{S_h^*}-\sigma^*\right) & -(v_m+d_m) & 0 \\ 0 & 0 & 0 & v_m & -d_m \end{pmatrix}. \quad (43)$$

By applying the same method as above, if $\mathcal{R}_0 > 1$ then $\rho(\Phi_{M_{\sigma^*}}(\omega)) > 1$. In this case θ is positive, and then $\hat{h}(t) \to \infty$ as $t \to \infty$. Moreover, since X_0 is positively invariant, then there exists an integer $q \geq n$ and a real number $\kappa > 0$ such that

$$\begin{aligned} &(E_h(q\omega), I_h(q\omega), R_h(q\omega), E_m(q\omega), I_m(q\omega)) \\ &\geq \kappa \hat{h}(0). \end{aligned} \tag{44}$$

Applying the theorem of comparison principle, we get

$$\begin{aligned} &(E_h(q\omega + t), I_h(q\omega + t), R_h(q\omega + t), E_m(q\omega + t), \\ &I_m(q\omega + t)) \geq \kappa \hat{h}(t), \quad \forall t \geq 0. \end{aligned} \tag{45}$$

It then follows that $\lim_{t\to\infty} |E_h(t), I_h(t), R_h(t), E_m(t), I_m(t)| = \infty$, which contradicts the fact that solutions are bounded. $\square$

Theorem 14. *If $\mathcal{R}_0 > 1$, then system (7) has at least one positive periodic solution.*

Proof. We first prove that P is uniformly persistent with respect to $(X_0, \partial X_0)$.

We define the following sets:

$$\begin{aligned} M_\partial &= \{\psi \in \partial X_0 \mid P^m(\psi) \in \partial X_0, \text{ for any } m \geq 0\}, \\ \mathcal{D} &= \{(S_h, 0, 0, 0, 0, 0) \in X \mid S_h \geq 0\}. \end{aligned} \tag{46}$$

Let us prove that $M_\partial = \mathcal{D}$.

It is easy to remark that $\mathcal{D} \subset M_\partial$. We only need to prove that $M_\partial \subset \mathcal{D}$.

Let $\psi \in \partial X_0 \setminus \mathcal{D}$. If

(i) $I_h(0) > 0$, $I_m(0) > 0$, and $E_h(0) = E_m(0) = R_h(0) = 0$, then we have $S_h(t) > 0$, $I_h(t) > 0$, $I_m(t) > 0$, $E_m(t) > 0$, $E_h(t) > 0$, $R_h(t) > 0$, $\forall t > 0$,

(ii) $I_h(0) = I_m(0) = 0$ and $E_h(0) > 0$, $E_m(0) > 0$, $R_h(0) > 0$, then we have $S_h(t) > 0$, $I_h(t) > 0$, $I_m(t) > 0$, $E_m(t) > 0$, $E_h(t) > 0$, $R_h(t) > 0$, $\forall t > 0$.

For any cases, it follows that $(S_h(t), E_h(t), I_h(t), R_h(t), E_m(t), I_m(t)) \notin \partial X_0$ for $t > 0$ sufficiently small, which contradicts the fact that ∂X_0 is positively invariant. Hence, $M_\partial \subset \mathcal{D}$. Thus, it then follows that $M_\partial = \mathcal{D}$.

The equality $M_\partial = \mathcal{D}$ implies that E_1^+ is a fixed point of P and acyclic in M_∂; every solution in M_∂ approaches to E_1^+. Moreover, Lemma 13 implies that E_1^+ is an isolated invariant set in X and $W^s(E_1^+) \cap X_0 = \emptyset$. By the acyclicity theorem on uniform persistence for maps, Theorem 1.3.1 and Remark 1.3.1 in [14], it follows that P is uniformly persistent with respect to X_0. Thus, Theorem 3.1.1 in [14] implies that the periodic semiflow $\Phi(t) : X \to X$ is also uniformly persistent with respect to X_0. Thanks to Theorem 1.3.6 in [14], model (11) has at least one ω-periodic solution $\tilde{u}(t, \psi^*)$ with $\psi^* \in X_0$ and $t \geq 0$. Now, we show that $\tilde{u}(t, \psi^*)$ is positive.

Suppose that $\psi^* = 0$; then, for all $t > 0$, we obtain $\tilde{u}_i(t, \psi^*) > 0$, for $i = 1, 2, 3, 4, 5, 6$. By using the periodicity of

the solution, we have $S_h^*(0) = S_h^*(n\omega) = 0$, $E_h^*(0) = E_h^*(n\omega) = 0$, $I_h^*(0) = I_h^*(n\omega) = 0$, $R_h^*(0) = R_h^*(n\omega) = 0$, $E_m^*(0) = E_m^*(n\omega) = 0$, $I_m^*(0) = I_m^*(n\omega) = 0$, $\forall n \geq 1$, which contradicts the fact that $\tilde{u}_i(t, \psi^*) > 0$ for $i = 1, 2, 3, 4, 5, 6$. So, the periodic solution is positive. $\square$

4. Numerical Simulation

In this section, we will present a series of numerical simulations of model (11) in order to support our theoretical results, to predict the trend of the disease, and to explore some control measures.

4.1. Initial Conditions and Estimation of $\beta(t)$. To validate our results, we choose the following initial conditions: $E(0) = 2400$, $L(0) = 1200$, $S_h(0) = 1500$, $E_h(0) = 50$, $I_h(0) = 200$, $R_h(0) = 50$, $S_m(0) = 3000$, $E_m(0) = 100$, $I_m(0) = 500$, and $A(0) = 3600$. Our numerical simulation will be performed using the MATLAB technical computing software with the fourth-order Runge–Kutta method [16].

Using the method developed in [11], we express the biting rate as follows:

$$\begin{aligned} \beta(t) = \ &\alpha_0 - 1.83692\cos(0.523599t) \\ &- 0.175817\cos(1.0472t) \\ &- 0.166233\cos(1.5708t) \\ &- 0.16485\cos(2.0944t) \\ &- 0.17681\cos(2.61799t) \\ &- 1.37079\sin(0.523599t) \\ &+ 0.296267\sin(1.0472t) \\ &+ 0.2134\sin(1.5708t) \\ &- 0.295228\sin(2.0944t) \\ &- 0.201712\sin(2.61799t), \end{aligned} \tag{47}$$

with $\alpha_0 \geq 3$.

4.2. The Model Parameters and Their Dimensions. Numerical values of parameters are given in Table 1.

4.3. Numerical Results. Using the above initial conditions, we now simulate model (11) in order to illustrate our mathematical results.

By taking $\alpha_0 = 7$, $d_p = 0.0028$, $c_{mh} = 0.022$, $c_{hm} = 0.48$, $\bar{c}_{hm} = 0.048$, $b = 180$, $s = 15$, $d = 6$, $d_L = 7.5$, $s_L = 15$, $d_m = 3.4038$ and considering the above initial conditions, we get $r = 25.1819$, $\mathcal{R}_0 = 1.3310 > 1$ and Figures 2, 3, and 4.

Figure 2 describes the evolution of infected (exposed and infectious) humans. Figure 3 describes the evolution of infected (exposed and infectious) mosquitoes and Figure 4 describe the evolution of susceptible humans and mosquitoes. Figures 2 and 3 show that malaria remains

TABLE 1: Values for constant parameters for the malaria model.

Parameter	Description	Value	Reference	Dimension
Λ	Constant recruitment rate for humans	400	Estimated	Humans/month
d_h	Human death rate	0.019	Estimated	/month
α	Transmission rate of humans from E_h to I_h	3.04	[17]	/month
d_p	Disease-induced death rate for humans	0.0028	[11]	/month
r_h	Recovery rate of humans	0.0159	[11]	/month
γ	Per capita rate of loss of immunity for humans	0.0167	[11]	/month
s_L	Transfer rate from L to adult	15	[6]	/month
d_m	Death rate for adult vectors	3.4038	[11]	/month
ν_m	Transmission rate of mosquitoes from E_m to I_m	2.523	[11]	/month
c_{mh}	Probability of transmission of infection from I_m to S_h	0.022	[17]	Dimensionless
c_{hm}	Probability of transmission of infection from I_h to S_m	0.48	[17]	Dimensionless
$\bar{c}_{hm}$	Probability of transmission of infection from R_h to S_m	0.048	[17]	Dimensionless
K_E	Available breeder sites occupied by eggs	30000	Estimated	Space
K_L	Available breeder sites occupied by larvae	18000	Estimated	Space
s	Transfer rate from E to L	15	[6]	/month
b	Eggs laying rate	180	[6]	/month
d	Death rate of eggs	6	[6]	/month
d_L	Larvae death rate	6	[6]	/month

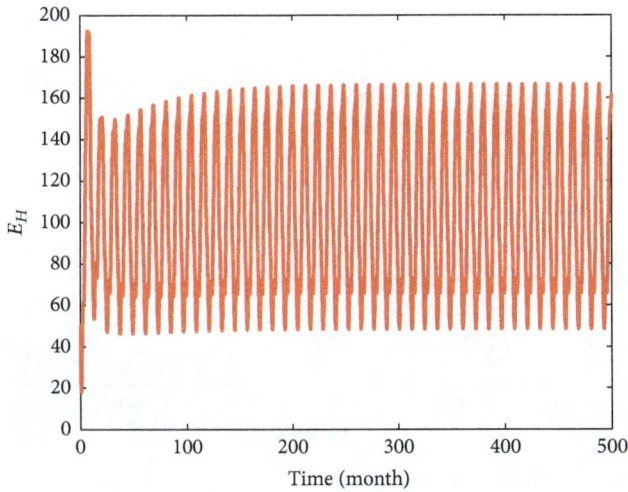

(a)

· Exposed humans

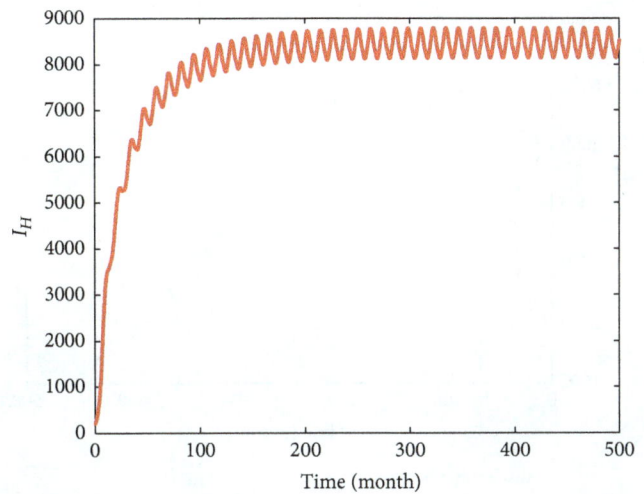

(b)

· Infectious humans

FIGURE 2: Distribution of infected humans.

persistent in the two populations. Besides, we observe that system (11) has one positive periodic solution. So, these numerical results illustrate the result of our Theorem 14.

In order to understand the model behaviour around the disease-free equilibrium, we consider the same above initial conditions and the following values: $\alpha_0 = 4$, $d_p = 0$, $c_{mh} = 0.022$, $c_{hm} = 0.24$, $\bar{c}_{hm} = 0.024$, $b = 180$, $s = 15$, $d = 6$,

$d_L = 7.5$, $s_L = 15$, $d_m = 6$. Then we get $r = 14.2857$ and $\mathscr{R}_0 = 0.2602 < 1$. Figures 5 and 6 illustrate that the disease dies out in both populations. Thus, the numerical results are the same as what we got in Theorem 11.

4.4. Parameters of Control of Malaria. Now, we assume that people became more conscious about the malaria disease and

· Exposed mosquitoes

(a)

· Infectious mosquitoes

(b)

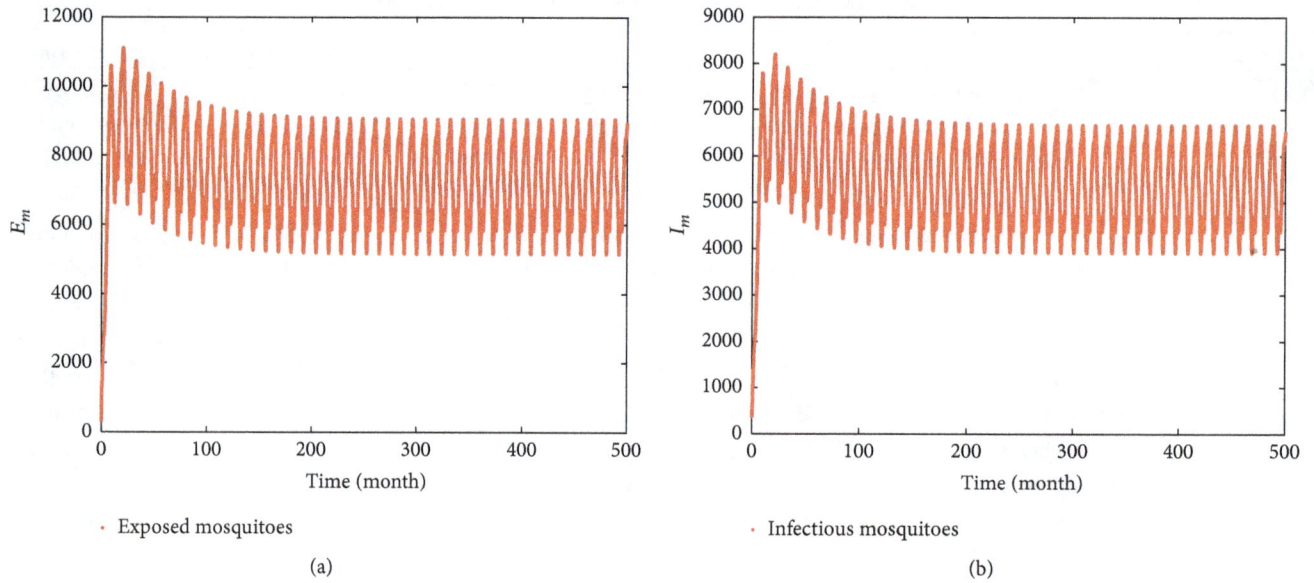

FIGURE 3: Distribution of infected mosquitoes.

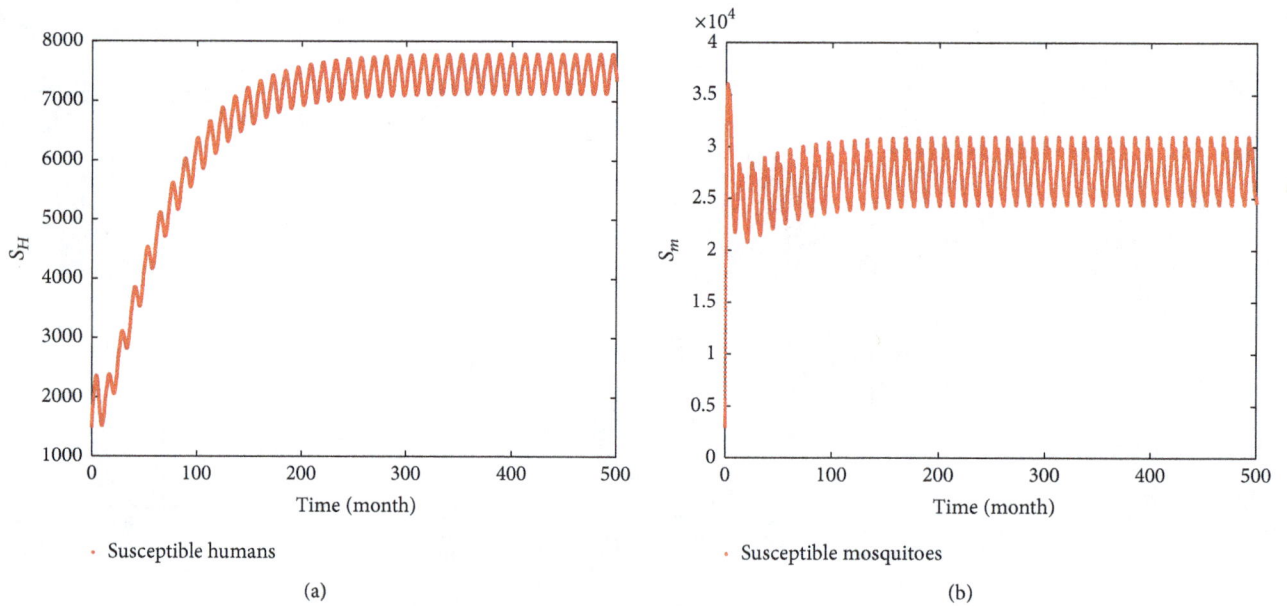

· Susceptible humans

(a)

· Susceptible mosquitoes

(b)

FIGURE 4: Distribution of susceptible humans and mosquitoes.

they use some efficient methods to reduce the proliferation of mosquitoes. That reduction can perhaps consist in fighting against the development of eggs, larvae, and pupa, firstly, by using chemical application methods (larvicide) or by introducing larvivore fish, and secondly, by using ecological methods (cleaning up the environment) to reduce the breeding sites of eggs and larvae. Let $\mu_1, \mu_2 \in [0, 1[$, respectively, be the efficiency of both intervention measures. So, we will use $\tilde{r} = (1 - \mu_1)r$, $\widetilde{K}_E = (1 - \mu_2)K_E$, and $\widetilde{K}_L = (1 - \mu_2)K_L$

in order to evaluate their impact on the dynamics of malaria transmission.

Thus, by considering the above initial conditions and by taking $\alpha_0 = 7$, $d_p = 0.0028$, $c_{mh} = 0.022$, $c_{hm} = 0.48$, $\bar{c}_{hm} = 0.048$, $d_m = 3.4038$, we obtain the following results.

(i) *Numerical Results for* $\mu_1 \simeq 89\%$. For this value, we get $\tilde{r} = 2.8204$ and $\mathcal{R}_0 = 0.6414$. Moreover, according to Figure 7, we notice that the distribution of infected humans

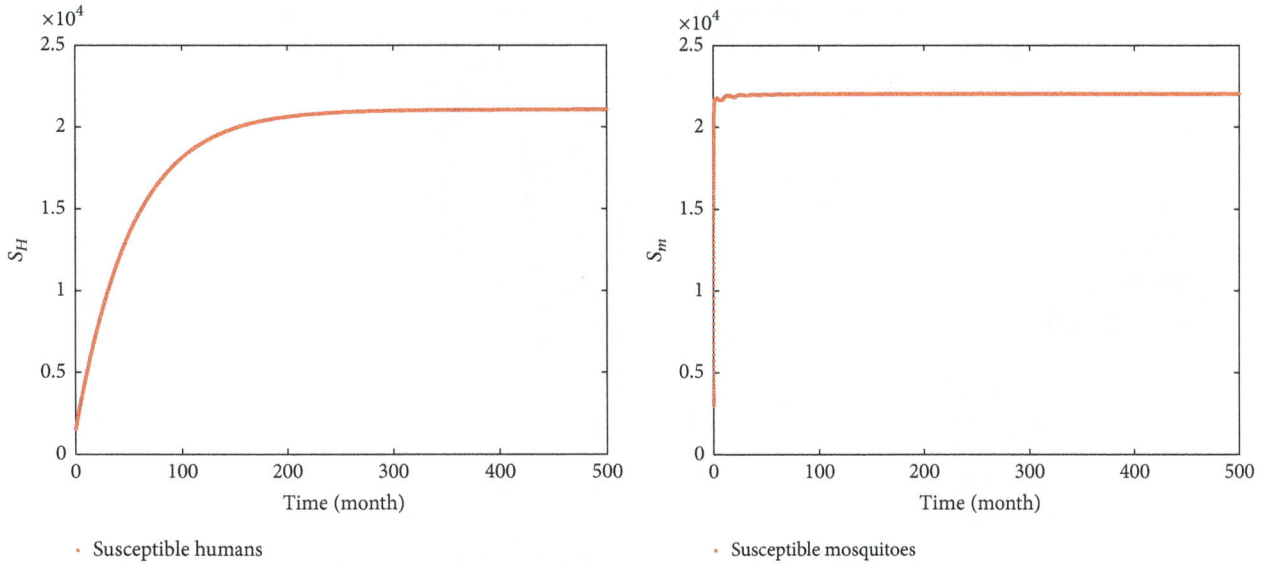

FIGURE 5: Distribution of susceptible humans and mosquitoes.

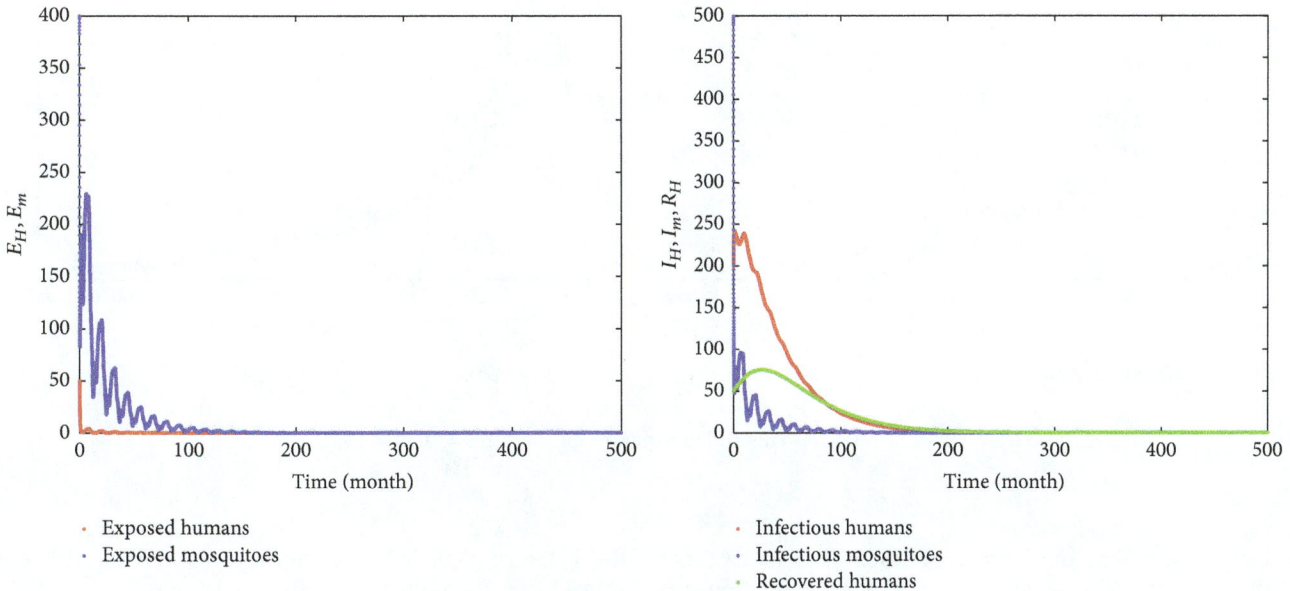

FIGURE 6: Distribution of infected humans and mosquitoes.

and mosquitoes has highly reduced and the malaria is progressively dying out in the populations.

(ii) Numerical Results for $\mu_2 = 80\%$. Using $\mu_2 = 0.8$, we get $\widetilde{K}_E = 6000$, $\widetilde{K}_L = 3600$, and $\mathscr{R}_0 = 0.5953$. Further, Figure 8 clearly shows that the disease is quickly disappearing from the populations.

Remark 15. We must notice that the two parameters are important in the malaria transmission because a little perturbation of those parameters influences the dynamics of malaria transmission. So they can be used to fight against the persistence of the disease. The control μ_1 is efficient but its action is very slow in finite time, but the control μ_2 is the best because it is more optimal and its action is very quick. Thus cleaning up the environment can be a very good mean of controlling malaria in the populations.

5. Conclusion

In this paper, we have presented a seasonal determinist model of malaria transmission. From the theoretical point of view,

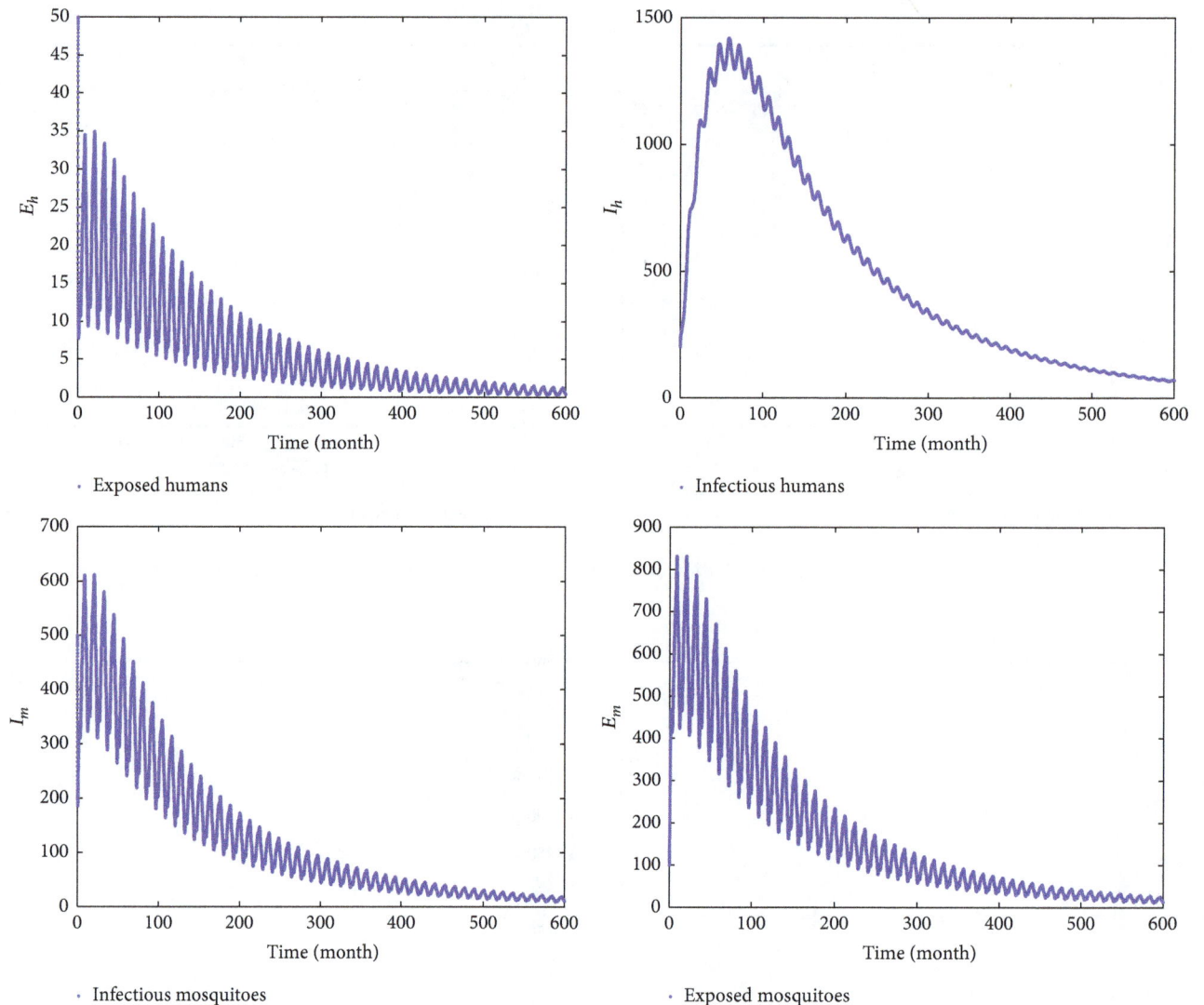

FIGURE 7: Distribution of infected humans and mosquitoes for $b = 80$, $s = 10$, $d = 15$, $s_L = 6$, and $d_L = 14$.

we have shown that the basic reproduction ratio, $\mathscr{R}_0$, is the distinguishing threshold parameter of the extinction or the persistence of the disease: if $\mathscr{R}_0$ is less than 1 malaria disappears in the human and mosquito populations and if it is greater than 1 malaria persists.

It also emerges from our study that the transmission of malaria is highly influenced by the dynamics of immature mosquitoes and depends on the regulatory threshold parameter of the mosquito population, r. Thus, the severity of malaria increases with this parameter. So, the life cycle of the anopheles is a very important aspect that must be taken into account in malaria modeling.

Moreover, we have shown that malaria transmission can be controlled by fighting against the proliferation of the mosquitoes, namely, by reducing the value of r or by reducing

the value of available breeder sites, K_E and K_L. We have proved that the reduction of the available breeder sites is a very efficient and more ecological method in fighting against malaria transmission. It then follows that environmental sanitation can be a very good means to control malaria in the endemic regions.

However, it must be noticed that our model is limited due to the following reasons: (i) we have not considered the effect of climate change on the life cycle of mosquitoes. (ii) The larva and pupa class were not distinguished.

In the future, one can develop a more realistic model by incorporating the above important factors and by considering the general force of infection. In addition, we can also take into account the degree of vulnerability of human populations in the model.

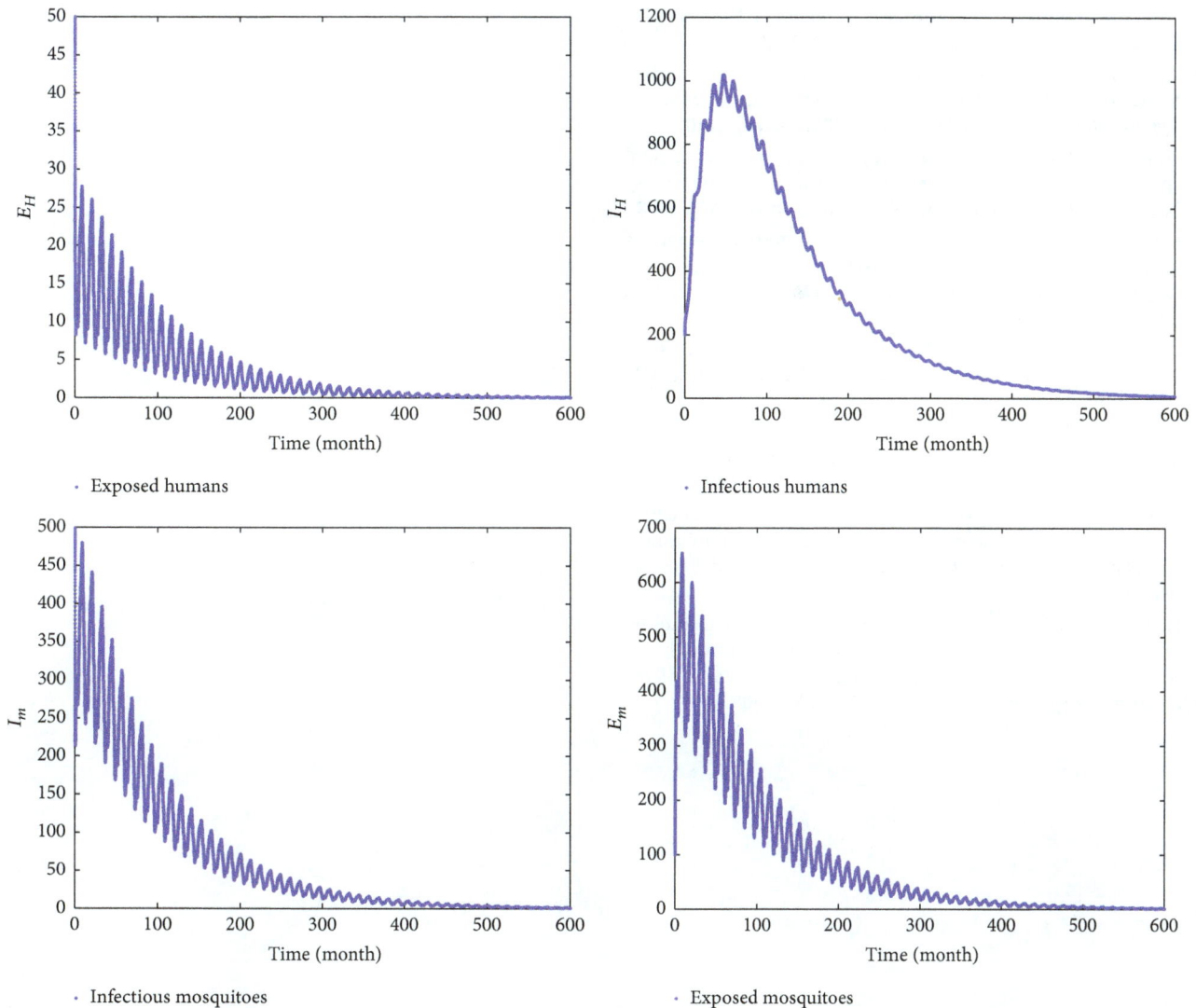

· Exposed humans

· Infectious humans

· Infectious mosquitoes

· Exposed mosquitoes

FIGURE 8: Distribution of infected humans and mosquitoes for $b = 180$, $s = 15$, $d = 6$, $s_L = 15$, and $d_L = 7.5$.

Conflicts of Interest

The authors declare that there are no conflicts of interest regarding the publication of this paper.

References

[1] R. Ross, *The prevention of malaria*, London, John Murray, 1911.

[2] G. Macdonald, *The epidemiology and control of malaria*, Oxford University press, London, 1957.

[3] C. Chiyaka, J. M. Tchuenche, W. Garira, and S. Dube, "A mathematical analysis of the effects of control strategies on the transmission dynamics of malaria," *Applied Mathematics and Computation*, vol. 195, no. 2, pp. 641–662, 2008.

[4] G. A. Ngwa and W. S. Shu, "A mathematical model for endemic malaria with variable human and mosquito populations," *Mathematical and Computer Modelling*, vol. 32, no. 7-8, pp. 747–763, 2000.

[5] L. M. Beck-Johnson, W. A. Nelson, K. P. Paaijmans, A. F. Read, M. B. Thomas, and O. N. Bjørnstad, "The effect of temperature on *Anopheles* mosquito population dynamics and the potential for malaria transmission," *PLoS ONE*, vol. 8, no. 11, Article ID e79276, 2013.

[6] D. Moulay, M. A. Aziz-Alaoui, and M. Cadivel, "The chikungunya disease: modeling, vector and transmission global dynamics," *Mathematical Biosciences*, vol. 229, no. 1, pp. 50–63, 2011.

[7] E. N. Chukwu, "On the boundedness and stability properties of solutions of some differential equations of the fifth order," *Annali di Matematica Pura ed Applicata. Serie Quarta*, vol. 106, pp. 245–258, 1975.

[8] A. S. Sinha, "Stability result of a sixth order non-linear system," vol. 7, pp. 641–643, 1971.

[9] C. Tunç, "On the stability and boundedness of solutions in a class of nonlinear differential equations of fourth order with constant delay," *Vietnam Journal of Mathematics*, vol. 38, no. 4, pp. 453–466, 2010.

[10] C. Tunç, "New results on the stability and boundedness of nonlinear differential equations of fifth order with multiple

deviating arguments," *Bulletin of the Malaysian Mathematical Sciences Society*, vol. 36, no. 3, pp. 671–682, 2013.

[11] Y. Lou and X.-Q. Zhao, "A climate-based malaria transmission model with structured vector population," *SIAM Journal on Applied Mathematics*, vol. 70, no. 6, pp. 2023–2044, 2010.

[12] W. Wang and X. Zhao, "Threshold dynamics for compartmental epidemic models in periodic environments," *Journal of Dynamics and Differential Equations*, vol. 20, no. 3, pp. 699–717, 2008.

[13] J. Wang, S. Gao, Y. Luo, and D. Xie, "Threshold dynamics of a huanglongbing model with logistic growth in periodic environments," *Abstract and Applied Analysis*, vol. 2014, Article ID 841367, 10 pages, 2014.

[14] Z. Xiao-Qiang, CMS Books in mathematics/Ouvrages de mathématiques de la SMC, Springer Verlag, New York, NY, USA, 16th edition, 2003.

[15] P. Magal and X.-Q. Zhao, "Global attractors and steady states for uniformly persistent dynamical systems," *SIAM Journal on Mathematical Analysis*, vol. 37, no. 1, pp. 251–275, 2005.

[16] W. Ouedraogo, B. Sangaré, and S. Traoré, "Some mathematical problems arising in biological models: a predator-prey model fish-plankton," *Journal of Applied Mathematics and Bioinformatics*, vol. 5, no. 4, pp. 1–27, 2015.

[17] N. Chitnis, J. M. Hyman, and J. M. Cushing, "Determining important parameters in the spread of malaria through the sensitivity analysis of a mathematical model," *Bulletin of Mathematical Biology*, vol. 70, no. 5, pp. 1272–1296, 2008.

3

Basic Properties and Qualitative Dynamics of a Vector-Borne Disease Model with Vector Stages and Vertical Transmission

Sansao A. Pedro (iD)

Departamento de Matemática e Informática, Universidade Eduardo Mondlane, 257, Maputo, Mozambique

Correspondence should be addressed to Sansao A. Pedro; sansaopedro@gmail.com

Academic Editor: Zhidong Teng

This work systematically discusses basic properties and qualitative dynamics of vector-borne disease models, particularly those with vertical transmission in the vector population. Examples of disease include Dengue and Rift Valley fever which are endemic in Sub-Saharan Africa, and understanding of the dynamics underlying their transmission is central for providing critical informative indicators useful for guiding control strategies. Of particular interest is the applicability and derivation of relevant population and epidemic thresholds and their relationships with vertical infection. This study demonstrates how the failure of R_0 derived using the next-generation method compounds itself when varying vertical transmission efficiency, and it shows that the host type reproductive number gives the correct R_0. Further, novel relationships between the host type reproductive number, vertical infection, and ratio of female mosquitoes to host are established and discussed. Analytical results of the model with vector stages show that the quantities Q_0, Q_0^v, and R_0^c, which represent the vector colonization threshold, the average number of female mosquitoes produced by a single infected mosquito, and effective reproductive number, respectively, provide threshold conditions that determine the establishment of the vector population and invasion of the disease. Numerical simulations are also conducted to confirm and extend the analytical results. The findings imply that while vertical infection increases the size of an epidemic, it reduces its duration, and control efforts aimed at reducing the critical thresholds Q_0, Q_0^v, and R_0^c to below unity are viable control strategies.

1. Introduction

Vector-borne diseases have been the scourge of man and animals since the beginning of time [1]. Today, vector-borne diseases account for over 17% of all infectious diseases causing more than 1 million deaths annually, and their distribution is mainly determined by a complex dynamic of environmental and social factors [2]. In spite of all these inherent complexities, mathematical models have been used to translate assumptions concerning biological, environmental, and social aspects into mathematical structures, linking biological processes of transmission and dynamics of infection at population level. Such dynamic models have impacted both our understanding of epidemic spread and public health planning (for more details see [3–5] and references therein).

In this study our particular interest is in investigating qualitative properties of epidemic models of mosquito-borne diseases in which the vector is of genera *Aedes*.

These mosquito species are known to transmit many vector-borne diseases of vast epidemiological importance including Dengue fever and Rift Valley fever (RVF), just to mention a few. These diseases are endemic in Sub-Saharan Africa with pronounced health and economic impacts on domestic animals and humans. In fact, losses due to RVF can reach millions of dollars during periods of disease outbreaks [6, 7]. An interesting phenomenon underlying many vector-borne diseases is their ability to persist year round, fluctuating seasonally but only falling to zero during some months. Hence, an important question is, how does the virus maintain itself in nature? For RVF it has been hypothesized that RVF virus (RVFV) is maintained through transovarial transmission in *Aedes* mosquito eggs [8]. *Aedes* eggs need be dry for several days before they can mature. After maturing, they hatch during the next flooding event large enough to cover them with water [9–11]. The eggs have high desiccation resistance and can survive dry conditions in a dormant form for months to years [12–15]. Thus, the emergence of adult *Aedes* mosquitoes from infected eggs can reintroduce RVF in livestock at the beginning of the rainy season, before other mosquitoes species amplify it further [16]. For instance, in eastern and southern Africa there is more and more evidence of disease activities between outbreaks [7, 17–20] highlighting the role of vertical transmission for initial disease spread and endemicity.

In epidemiology disease spread and persistence are measured through quantities known as epidemic thresholds. Their derivation and characterization are one of the most important results of mathematical epidemic models. The basic reproductive number, R_0, is the most critical epidemic threshold given its applicability and suitability for deciding whether an outbreak will occur or fade out, making it essential for guiding disease control efforts. However, the derivation of this epidemic threshold in vector-borne disease models in particular suffers from a lack of uniqueness and it fails to give the correct average number of expected secondary infections produced by one infected individual [21]. This failure is more likely to compound itself when vertical transmission mode is included in the transmission model, since the resulting R_0 comes as the sum of the vertical and horizontal transmission components, if the next-generation method is used. Previous mathematical models have made a significant attempt in including vertical infection in modelling vector-borne diseases [22–28], but none of them discuss how the failure of R_0 compounds itself in the presence of vertical transmission. Therefore, the present work aims to discuss some relevant basic properties of vector-borne disease models when vertical infection is taken into account and their implications for disease control efforts. Further, our goal is to derive new epidemic thresholds useful for guiding control efforts in the settings of vector-borne disease models with vector stages that include vertical transmission mode.

We formulate two models, one simple but realistic and the other more complex with vector stages. The first is an extension of the one proposed by Ross [29] and popularized by Macdonald [30] and Anderson [31]. The model is used to discuss system properties such as the asymmetric relationship between the host-to-vector and vector-to-host reproductive numbers. In addition, we highlight how to derive epidemic thresholds useful for guiding disease control efforts and discuss their relationships with vertical transmission efficiency. It is shown that the model has two model equilibria, namely, the disease-free and the endemic, and Lyapunov function theory is used to establish their global qualitative dynamics. The second is an extension of the basic model, where the dynamics of both aquatic and adult mosquitoes are modelled explicitly. In this model we let the populations of aquatic and adults vary with time but be limited by their respective carrying capacity. The inclusion of the explicit vector submodel allows for derivation of critical thresholds such as the reproductive number for both the vector population and the disease system. Then, these thresholds are used to determine global qualitative dynamics of both the disease-free and endemic equilibria.

The paper is set out as follows. In Section 2 we formulate and discuss the two model systems. In Section 3 we provide the epidemic threshold theorems regarding both the vector population and disease equilibria. We also discuss important model properties and how to derive and identify model epidemic thresholds useful for guiding disease control efforts. Furthermore, numerical simulations are carried out to investigate the influence of the key parameters on the spread of the disease (taking RVF as disease example), to support analytical analyses and conclusions and illustrate possible behavioural scenarios of the model with vector stages. Finally, in Section 4 we present a short discussion of the results and their biological implications.

2. Materials and Methods

For human and animal diseases, horizontal transmission typically occurs through direct or indirect physical contact with infectious hosts, or through disease vectors such as mosquitoes, ticks, or other biting insects. Among mosquito vectors vertical transmission is often through eggs. Of particular interest are female mosquitoes of genera *Aedes* which transmit the virus to their eggs. These eggs have some adaptive behaviour which allows them to stay dormant in nature for relatively long periods. Although vertical transmission also occurs among vectors involved in the transmission of Dengue disease, RVF is the disease for which the model is a good approximation. In particular, parameter values related to RVF are used to illustrate the dynamics of the disease numerically.

2.1. Host-Vector Basic Model with Vertical Transmission. Let $N_h = S_h + I_h + R_h$ and $N_v = S_v + I_v$ denote the total host and vector populations sizes, respectively. We assume that individuals at each compartment mix homogeneously and each mosquito bites each individual host at a constant rate α/N_h, where α is the biting rate per unit time. Let p_{hv} be a probability of successful infection transmission from an infected mosquito to a susceptible host and p_{vh} be a probability of successful infection transmission from an infected host to a susceptible mosquito per bite. Thus, the forces of infection are as follows, $\lambda_{hv} = p_{hv}(\alpha/N_h)I_v$ and $\lambda_{vh} = p_{hv}(\alpha/N_h)I_h$. Hosts are recruited into the population at per capita rate μ_h which is proportional to the total population and leave each compartment through death. Noninfected mosquitoes join the susceptible compartment at rate $\mu_v(N_v - q_v I_v)$ while vertically infected mosquitoes join the infected class at rate $\mu_v q_v I_v$. Assuming constant population sizes, that is, births equal to deaths, R_h can be obtained when both S_h and I_h are known. In the same way, S_v can be obtained when I_v is known. Thus, the expressions for both R_h and S_v can be omitted and the system can be written in terms of proportions:

$$\frac{dS_h}{dt} = \mu_h\left(1 - S_h\right) - p_{hv}\alpha m I_v S_h,$$

$$\frac{dI_h}{dt} = p_{hv}\alpha m I_v S_h - \left(\gamma_h + \mu_h\right)I_h, \qquad (1)$$

$$\frac{dI_v}{dt} = p_{vh}\alpha I_h\left(1 - I_v\right) - \mu_v\left(1 - q_v\right)I_v,$$

where $q_v \in (0, 1)$ is the probability of vertical infection, γ_h the rate at which infected hosts recover from infection, and m denotes the ratio of female mosquitoes to hosts.

2.2. Model with Vector Stages and Vertical Transmission.

Here we extend the basic model to include vector stages. Partial results of the resulting model without vertical infection have been obtained in [32]. Our aim is to extend their model and analysis by investigating global dynamics of all model equilibria and examine the extent to which vertical infection alters the dynamics of the system. The mosquito population is divided into aquatic (eggs, larvae, and pupae) and terrestrial (adults) subpopulations with K_1 and K_2 being their carrying capacity, respectively. The parameter K_1 represents the larval maximal capacity limited by the availability of breeding sites while K_2 is the maximal capacity of adult mosquitoes limited by factors conditioning their survival such as high altitudes and high temperatures. Further, the aquatic subpopulation is divided into epidemiological classes, susceptible (L_s), and infected (L_i) while adults are divided also into susceptible (S_v) and infected (I_v). The per capita oviposition rate is $r(1-L/K_1)$, where r is the intrinsic oviposition rate and $L = L_s + L_i$. Aquatic mosquitoes emerge as adults at a per capita rate θ where the proportion $1 - q_v$ emerge noninfected while the remainder are infected. Disease transmission dynamics between vector and host populations remain the same as in the basic model. As a result the following nondimensional system of ordinary differential equations represents the model that governs the temporal evolution of the disease:

$$\frac{dL_s}{dt} = \delta r (1 - L)(N_v - q_v I_v) - (\mu_L + \theta) L_s,$$

$$\frac{dL_i}{dt} = \delta r (1 - L) q_v I_v - (\mu_L + \theta) L_i,$$

$$\frac{dS_v}{dt} = \frac{\theta}{\delta} L_s (1 - N_v) - \lambda_{vh} I_h S_v - \mu_v S_v,$$

$$\frac{dI_v}{dt} = \frac{\theta}{\delta} L_i (1 - N_v) + \lambda_{vh} I_h S_v - \mu_v I_v, \tag{2}$$

$$\frac{dS_h}{dt} = \mu_h (1 - S_h) - \lambda_{hv} I_v S_h,$$

$$\frac{dI_h}{dt} = \lambda_{hv} I_v S_h - (\gamma_h + \mu_h) I_h,$$

where $L = L_s + L_i$, $\delta = K_2/K_1$, $\lambda_{vh} = p_{vh}\alpha$, $\lambda_{hv} = p_{hv}\alpha m$, $m = K_2/N_h$, $S_v = S_v/K_2$, and $I_v = I_v/K_2$.

Let $\{\{L_s(0), L_i(0), S_v(0), I_v(0), S_h(0), I_h(0) \geqslant 0\} \in \mathbb{R}_+^6\}$ be the initial conditions of system (2). It is easy to check that the feasible region for (2) is the positive orthant of $\mathbb{R}^6$ and that the closed set

$$\Phi = \{(L_s, L_i, S_v, I_v, S_h, I_h) \in \mathbb{R}_+^6 : L_s + L_i \leqslant 1, S_v + I_v \tag{3}$$

$$\leqslant 1, S_h + I_h \leqslant 1\}$$

is positively invariant for system (2).

2.2.1. Positivity of Solutions

Lemma 1 (see [33]). *Let us denote* $u = (u_i)_{i=\overline{1,n}}$ *and consider the function* $f : \mathbb{R}_+ \times \mathbb{R}^n \longrightarrow \mathbb{R}^n$ *continuous with respect to* t, u *and Lipschitz with respect to* u. *If* $f(t, u) \geq 0$ *for* $(t, u) \in \mathbb{R}_+ \times \mathbb{R}^n$, *with* $u_i = 0$, *then, for every* $u_0 \in \mathbb{R}_+^n$, *there exists* $T > 0$ *such that the solution to*

$$\frac{du}{dt} = f(t, u),$$

$$u(t_0) = u_0 \tag{4}$$

exists and is unique and positive with value in $\mathbb{R}_+^n$ *and defined on some interval* $[0, T]$. *If* $T < \infty$, *then*

$$\limsup_{t \to T} \sum_{i=1}^n u_i = +\infty. \tag{5}$$

Theorem 2. *The solution set* $\{L_s(t), L_i(t), S_v(t), I_v(t), S_h(t), I_h(t)\}$ *of the model (2) exists and is unique and positive for* $t > 0$.

Proof. Let $X = (L_s(t), L_i(t), S_v(t), I_v(t), S_h(t), I_h(t))$ and denote the function $f = (f_i)_{i=\overline{1,6}}$ such that

$$f_1(L_s(t), L_i(t), S_v(t), I_v(t), S_h(t), I_h(t))$$
$$= \delta r (1 - L)(N_v - q_v I_v) - (\mu_L + \theta) L_s,$$

$$f_2(L_s(t), L_i(t), S_v(t), I_v(t), S_h(t), I_h(t))$$
$$= \delta r (1 - L) q_v I_v - (\mu_L + \theta) L_i,$$

$$f_3(L_s(t), L_i(t), S_v(t), I_v(t), S_h(t), I_h(t))$$
$$= \frac{\theta}{\delta} L_s (1 - N_v) - \lambda_{vh} I_h S_v - \mu_v S_v,$$

$$f_4(L_s(t), L_i(t), S_v(t), I_v(t), S_h(t), I_h(t)) \tag{6}$$
$$= \frac{\theta}{\delta} L_i (1 - N_v) + \lambda_{vh} I_h S_v - \mu_v I_v,$$

$$f_5(L_s(t), L_i(t), S_v(t), I_v(t), S_h(t), I_h(t))$$
$$= \mu_h (1 - S_h) - \lambda_{hv} I_v S_h,$$

$$f_6(L_s(t), L_i(t), S_v(t), I_v(t), S_h(t), I_h(t))$$
$$= \lambda_{hv} I_v S_h - (\gamma_h + \mu_h) I_h.$$

Since the function f is continuous and Lipschitz continuous with respect to X, according to Picard's theorem, there exists $T_0 > 0$ such that the solution to (2) exists locally at least on an interval of this form $[0, T_0]$. Further, considering the initial condition $X_1 = X(T_0)$ at $t_0 = T_0$ and using Picard's theorem, it follows that there exists $T_0 \leq T_1 \in R_+$ such that the solution to (2) exists and is unique on $[T_0, T_1]$. Since f is continuous and differentiable, the solution of (2) with a given initial condition is unique. Therefore, the solutions of (2) obtained on $[0, T_0]$ and on $[T_0, T_1]$ form the unique solution of (2) on $\mathbb{I}_{T_1} = [0, T_1]$ with the initial condition X_0 at

$t_0 = 0$. Repeating this process again and again, we end up with the maximal forward interval of existence for the solutions of (2), say $\mathbb{I}_{\widehat{T}} = [0, \widehat{T})$ with $\widehat{T} > 0$. Furthermore, for $X \in \mathbb{R}_+^6$,

$$f_i \left(L_s(t), L_i(t), S_v(t), I_v(t), S_h(t), 0 \right) \geq 0. \tag{7}$$

Therefore, the solutions of (2) on $\mathbb{I}_{\widehat{T}}$ are positive, according to Lemma 1.

Finally, according to Theorem 2, the solutions to (2) are bounded on $[0, \widehat{T})$. In other words, they do not blow up on any finite interval of $\mathbb{R}_+$. It follows that, according to Lemma 1, the solution of (2) exists for all time. Hence for any initial condition in $\mathbb{R}_+^6$, system (2) possesses a unique and positive solution in $\mathbb{R}_+^6$. $\qquad\square$

3. Results

3.1. Analysis of the Basic Host-Vector Model

3.1.1. Model Equilibria and Stability Analysis. The basic host-vector model with vertical transmission exhibits two equilibria, namely, the disease-free E_1^0 and the endemic E_1^*, respectively. At the disease-free equilibrium,

$$E_1^0 = \left(S_h^0, I_h^0, I_v^0 \right) = (1, 0, 0), \tag{8}$$

both vector and host populations persist but with no disease. The prevalence of the disease is denoted by

$$
\begin{aligned}
E_1^* &= \left(S_h^*, I_h^*, I_v^* \right) \\
&= \left(\frac{a + \mu_h R_0^e}{(a + \mu_h) R_0^e}, \frac{\mu_h (R_0^e - 1)}{c R_0^e + b \mu_h / V}, \frac{\mu_h (R_0^e - 1)}{a + \mu_h R_0^e} \right),
\end{aligned} \tag{9}
$$

where $a = p_{hv} \alpha m$, $b = p_{vh} \alpha$, $c = \gamma_h + \mu_h$, $V = \mu_v (1 - q_v)$, and

$$R_0^e = \frac{ab}{Vc} = \frac{p_{hv} p_{vh} \alpha^2 m}{\mu_v (1 - q_v)(\gamma_h + \mu_h)}. \tag{10}$$

The Jacobian matrix of system (1) at E_1^0 is given by

$$J_1' = \begin{pmatrix} -\mu_h & 0 & -a \\ 0 & -c & a \\ 0 & b & -V \end{pmatrix}, \tag{11}$$

such that the characteristic polynomial of matrix (11) is then given as

$$P_1(\lambda) = \lambda^3 + a_2 \lambda^2 + a_1 \lambda + a_0, \tag{12}$$

with $a_2 = V + c + \mu_h$, $a_1 = Vc(1 - R_0^e) + (V + c)\mu_h$, and $a_0 = \mu_h Vc(1 - R_0^e)$. The coefficient $a_2 > 0$ and both a_1, a_0 are nonnegative if and only if $R_0^e < 1$. Hence, all Routh stability criteria are satisfied; that is, the three eigenvalues of matrix (11) are negative or have negative real parts. Furthermore, for $R_0^e = 1$, (9) becomes the disease-free equilibrium. Therefore, the following result holds.

Theorem 3. *The disease-free equilibrium $E_1^0 = (1, 0, 0)$ exists and it is globally asymptotically stable if $R_0^e \leq 1$.*

Alternatively, the global stability of the disease-free equilibrium E_1^0 can be established using the following Lyapunov function:

$$V(S_h, I_h, I_v) = u_1 \left(S_h - S_h^0 - S_h^0 \ln \frac{S_h}{S_h^0} \right) + u_2 I_h + u_3 I_v \tag{13}$$

where u_1, u_2, u_3 are some positive constants. Calculating the derivative of V along the solutions of system (1), we obtain

$$
\begin{aligned}
V' &= u_1 \frac{S_h - S_h^0}{S_h} \left[\mu_h \left(S_h^0 - S_h \right) - a S_h I_v \right] \\
&\quad + u_2 \left[a S_h I_v - (\gamma_h + \mu_h) I_h \right] \\
&\quad + u_3 \left[b I_h (1 - I_v) - \mu_v (1 - q_v) I_v \right],
\end{aligned} \tag{14}
$$

$$
= -u_1 \mu_h \frac{\left(S_h^0 - S_h \right)^2}{S_h}
$$

$$
- u_3 \mu_v (1 - q_v) \left[1 - \frac{u_1 a}{u_3 \mu_v (1 - q_v)} \right] I_v \tag{15}
$$

$$
- u_2 (\gamma_h + \mu_h) \left[1 - \frac{u_3 b}{u_2 (\gamma_h + \mu_h)} \right] I_h
$$

$$
- u_3 b I_h I_v + (u_2 - u_1) a S_h I_v. \tag{16}
$$

Choosing $u_1 = u_2 = b/(\gamma_h + \mu_h)$ and $u_3 = 1$, (16) becomes

$$
V' = -u_1 \mu_h \frac{\left(S_h^0 - S_h \right)^2}{S_h} - \mu_v (1 - q_v) \left[1 - R_0^e \right] I_v
$$

$$
- b I_h I_v. \tag{17}
$$

Thus, V' is negative for $R_0^e \leq 1$. Note also that $V' = 0$ if and only if $S_h = S_h^0$ and $I_h = I_v = 0$. Therefore, the largest invariant set for (1) is the singleton $\{E_1^0\}$. Hence, by LaSalle's invariance principle [34], E_1^0 is globally asymptotically stable when $R_0^e \leq 1$ and Theorem 3 is valid.

Remark 4. Clearly, the endemic equilibrium E_1^* exists and is unique for $R_0^e > 1$. This excludes the possibility of occurrence of backward bifurcation. This result is of great epidemiological significance in guiding efforts for disease control as it indicates that $R_0^e = 1$ is the critical epidemic threshold.

To establish the local stability of the endemic equilibrium we evaluate the Jacobian of the system at E_1^*, which gives

$$J_1^* = \begin{pmatrix} -(\mu_h + a I_v^*) & 0 & -a S_h^* \\ a I_v^* & -c & a S_h^* \\ 0 & b(1 - I_v^*) & -b I_h^* - V \end{pmatrix}. \tag{18}$$

The characteristic polynomial of matrix (18) is then given by

$$P_2(\lambda) = \lambda^3 + b_2 \lambda^2 + b_1 \lambda + b_0, \tag{19}$$

where

$$b_2 = V^2 c R_0^e (\mu_h + a) + V c^2 R_0^e \left(R_0^e \mu_h^2 + a \right)$$
$$+ V a R_0^e \mu_h \left(v R_0^e + b \right) + V b \left(R_0^e \right)^2 \mu_h$$
$$+ b \mu_h^2 R_0^e (a + c + \mu_h) + abc\mu_h,$$

$$b_1 = V^2 c \left(R_0^e \right)^3 \mu_h (a + c + \mu_h) + V b \left(R_0^e \right)^3 \mu_h^2 (a + \mu_h)$$
$$+ V c^2 \left(R_0^e \right)^3 \mu_h (a + \mu_h)$$
$$+ V b c \left(R_0^e \right)^2 \mu_h^2 \left(R_0^e - 1 \right) \qquad (20)$$
$$+ abc R_0^e \mu_h^2 \left(R_0^e - \frac{V}{\mu_h} \right) + bc \left(R_0^e \right)^2 \mu_h^3,$$

$$b_0 = V^2 c^2 \left(R_0^e \right)^3 \mu_h + V b c \left(R_0^e \right)^2 \mu_h^2 \left(R_0^e - 1 \right)$$
$$+ V^2 a c^2 \left(R_0^e \right)^2 \left(R_0^e - 1 \right)$$
$$+ V a b c R_0^e \mu_h \left[R_0^e \left(R_0^e - 1 \right) - 1 \right].$$

The coefficient $b_2 > 0$ and both b_1, b_0 are nonnegative if and only if $R_0^e > 1$. Hence, all Routh criteria are satisfied; that is, the three eigenvalues of matrix (11) are negative or have negative real parts. Therefore, the following results holds.

Theorem 5. *The endemic equilibrium $E_1^* = (S_h^*, I_h^*, I_v^*)$ exists and it is locally asymptotically stable if $R_0^e > 1$.*

A global stability result for the endemic equilibrium E_1^* of system (1) is given below.

Theorem 6. *If $R_0^e > 1$, the endemic equilibrium $E_1^* = (S_h^*, I_h^*, I_v^*)$ is globally asymptotically stable.*

Proof. Let $D_1 = v_1 (S_h - S_h^* - S_h^* \ln(S_h/S_h^*))$, $D_2 = v_2 (I_h - I_h^* - I_h^* \ln(I_h/I_h^*))$, and $D_3 = v_3 (I_v - I_v^* - I_v^* \ln(I_v/I_v^*))$ be the components of the Lyapunov function

$$U(S_h, I_h, I_v) = D_1 + D_2 + D_3, \qquad (21)$$

where v_1, v_2, v_3 are some positive parameters to be chosen later. Differentiating U along the solutions of system (1), we obtain

$$D_1' = v_1 \frac{S_h - S_h^*}{S_h} \left[\mu_h - \mu_h S_h - a S_h I_v \right]$$

$$= -v_1 \mu_h \frac{(S_h - S_h^*)^2}{S_h} + v_1 \frac{S_h - S_h^*}{S_h} (a S_h^* I_v^* - a S_h I_v) \qquad (22)$$

$$= -v_1 \mu_h \frac{(S_h - S_h^*)^2}{S_h} + I_v^* - \frac{S_h}{S_h^*} I_v - \frac{S_h^*}{S_h} I_v^* - I_v,$$

for $v_1 = 1/a S_h^*$ and $\mu_h = \mu_h S_h^* - a S_h^* I_v^*$ at equilibrium.

$$D_2' = v_2 \frac{I_h - I_h^*}{I_h} \left[a S_h I_v - (\gamma_h + \mu_h) I_h \right]$$

$$= v_2 \frac{I_h - I_h^*}{I_h} \left[a S_h I_v - \frac{a S_h^* I_v^*}{I_h^*} I_h \right]$$

$$= \frac{S_h}{S_h^*} I_v - \frac{I_h}{I_h^*} I_v^* - \frac{I_h^* S_h}{I_h S_h^*} I_v + I_v^*, \qquad (23)$$

for $v_2 = v_1$ and $\gamma_h + \mu_h = a S_h^* I_v^* / I_h^*$ at equilibrium.

$$D_3' = v_3 \frac{I_v - I_v^*}{I_v} \left[b I_h (1 - I_v) - \mu_v (1 - q_v) I_v \right]$$

$$= b v_3 \frac{I_v - I_v^*}{I_v} \left[b I_h (1 - I_v) - \frac{I_h^* (1 - I_v^*)}{I_v^*} I_v \right]$$

$$= b v_3 \frac{I_v - I_v^*}{I_v} \left[I_h - I_h I_v - I_v I_h^* \left(\frac{1 - I_v^*}{I_v^*} \right) \right], \qquad (24)$$

$$\leqslant \frac{I_v - I_v^*}{I_v} \left[I_h - I_h I_v - I_v I_h^* \right],$$

for $v_3 = 1/b$ and $\mu_v (1 - q_v) = I_h^* (1 - I_v^*)/I_v^*$ at equilibrium. Now all together,

$$U' \leqslant -v_1 \mu_h \frac{(S_h - S_h^*)^2}{S_h} - I_v^* \left(\frac{I_v}{I_v^*} + \frac{I_h}{I_h^*} - 2 \right) - \frac{S_h^*}{S_h} I_v^*$$

$$- \frac{I_h^* S_h}{I_h S_h^*} I_v - I_v^* \left(\frac{I_v}{I_v^*} + \frac{I_h}{I_h^*} - 2 \right) - 2 I_v^* + 2 I_v + I_h$$

$$- I_h I_v - I_v I_h^* - \frac{I_v^*}{I_v} I_h + I_v^* + I_v^* I_h^*,$$

$$\leqslant -v_1 \mu_h \frac{(S_h - S_h^*)^2}{S_h} - 2 I_v^* \left(\frac{I_v}{I_v^*} + \frac{I_h}{I_h^*} - 2 \right) \qquad (25)$$

$$+ 2 I_v \left(1 - \frac{I_v^*}{I_v} \right) + I_h \left(1 - \frac{I_v^*}{I_v} \right)$$

$$+ I_v^* I_h^* \left(1 - \frac{I_v}{I_v^*} \right)$$

$$+ I_v^* \left[1 - \frac{S_h^*}{S_h} \left(1 + \frac{I_h^* S_h^2 I_v}{I_h (S_h^*)^2 I_v^*} \right) \right].$$

Using the inequality $1 - x + \ln x \leqslant 0$ for $x > 0$ with equality holding if and only if $x = 1$ and the fact that the arithmetic mean is greater than or equal to the geometric mean, we obtain $U'(S_h, I_h, I_v) \leqslant 0$ for all $S_h, I_h, I_v > 0$. Furthermore, we obtain that $U'(S_h, I_h, I_v) = 0$ holds only when $S_h = S_h^*, I_h = I_h^*, I_v = I_v^*$ and that E_1^* is the only equilibrium state of these systems on this plane (line). Therefore, by LaSalle's invariance principle [34], the positive equilibrium E_1^* is globally asymptotically stable. $\square$

3.1.2. Epidemic Thresholds, Vertical Infection, and Basic Properties. One of the most important critical thresholds in epidemic models is the basic reproductive number, R_0, which is usually found using the next-generation method, as the dominant eigenvalue of the *next-generation matrix* [35, 36]. Following the method in [36] we write system (1) consisting only of infectious compartments as the difference between new infection and transfer rates and the resulting Jacobian matrices evaluated at the disease-free equilibrium $E^0 = (1, 0, 0)$. For system (1) we have two infected classes, namely, I_v and I_h. It follows that the transmission and the transfer matrices F and F, respectively, are defined as

$$
F = \begin{pmatrix} q_v \mu_v & p_{vh}\alpha \\ p_{hv}\alpha m & 0 \end{pmatrix},
$$

$$
V = \begin{pmatrix} \mu_v & 0 \\ 0 & \gamma_h + \mu_h \end{pmatrix}.
$$

(26)

Unlike in host-vector models without vertical transmission, the diagonal elements of the transmission matrix F are nonzero. This stems from the fact that in the presence of vertical transmission there is vector to vector transmission, which completely changes the nature of the basic reproductive number. Thus, the next-generation matrix, K, is then given by

$$
K = FV^{-1} = \begin{pmatrix} q_v & \dfrac{p_{vh}\alpha}{\gamma_h + \mu_h} \\ \dfrac{p_{hv}\alpha m}{\mu_v} & 0 \end{pmatrix},
$$

(27)

and the resulting dominant eigenvalue of the spectral radius FV^{-1}, which is the basic reproductive number, R_0, is given by

$$
R_0 = \frac{1}{2}q_v + \frac{1}{2}\sqrt{q_v^2 + 4R_0^H},
$$

(28)

$$
\text{with } R_0^H = \frac{p_{vh}p_{hv}\alpha^2 m}{\mu_v(\gamma_h + \mu_h)}.
$$

When there is no vertical transmission, $q_v = 0$, as in the case of malaria, $R_0 = R_0^H$ is simply the geometric mean of the product of the number of new infections in hosts from one infected vector and the number of new infections in vectors from one infected host, in the limiting case that both populations are fully susceptible. The interpretation and epidemiological significance of R_0^H are well established. It is easy to see that the transmission of infection is increased with efficiency of vector biting α and probabilities of successful infection transmission, but it is hindered by high mosquito death rates and faster host recovery. The biting rate α appears as α^2 because it enters twice in the transmission cycle [31]. An important parameter is the ratio of female mosquitoes to hosts $m = N_v/N_h$, which is central for disease spread according to model settings. R_0 increases with the number (or density) of mosquitoes but decreases with the number (or density) of host population. This results from the asymmetry in the dependence of the vector's biting rate on the sizes of the host and vector populations such that when there are many more hosts compared to mosquitoes, sustained transmission may be impossible. Therefore, in the absence of

vertical transmission, for the infection to successfully spread and invade, the ratio of mosquitoes to hosts needs to be sufficiently large so that double bites are common [37]:

$$
\frac{N_v}{N_h} > \frac{\mu_v(\gamma_h + \mu_h)}{p_{vh}p_{hv}\alpha^2},
$$

(29)

where the critical ratio is given by $m_c = \mu_v(\gamma_h + \mu_h)/p_{vh}p_{hv}\alpha^2$. Note that each mosquito could infect less than one host on average, and yet R_0 could still be more than unity. To elucidate this fact we write R_0^H as a product of each host type single-step reproductive number; that is,

$$
R_0^H = R_{hv} \times R_{vh} = \frac{p_{hv}\alpha m}{\mu_v} \times \frac{p_{vh}\alpha}{\gamma_h + \mu_h},
$$

(30)

where R_{vh} represents the number of new infections in mosquitoes from a single infected host while R_{hv} represents the number of new infections in hosts from a single infected mosquito. Clearly, R_0^H can be greater than unity even when one of these reproduction numbers is less than unity, and it can also be less than unity even if one of its components is greater than unity. In Figure 1(a) we depict contours plots corresponding to the overall R_0 in (31) along the plane (R_{hv}, R_{vh}) in the presence of vertical transmission. Its effects in this asymmetric relationship between R_0 and its components is not very pronounced and the ratio of mosquitoes to hosts remains one of the leading factors when there is a large disparity between the sizes of the host and vector populations.

Clearly, the geometric mean is less than the average expected number of new infections per generation. This is the case where the next-generation method fails to produce the correct R_0 if transmission between hosts is intermediated by another host (for more discussion about the failure of this method see [21]). Instead, it gives the weighted average lying between the number of new infections each individual produces in the next infection event. Note that if the number of hosts is increased, this deficit is compounded. This can have serious implications for guiding disease control efforts as it fails to provide the actual severity of the infection. Therefore, in this study for application purpose we propose the use of another epidemic threshold. This is R_0 from (10) derived in Section 3.1.1, which is hereby referred to as the 'effective' reproductive number:

$$
R_0^e = \frac{1}{1 - q_v} \frac{p_{vh}p_{hv}\alpha^2 m}{\mu_v(\gamma_h + \mu_h)},
$$

(31)

as (1) it satisfies the property that the endemic equilibrium E_1^* only persists if R_0^e is greater than unity and (2) the endemic equilibrium exists without occurrence of backward bifurcation, meaning that $R_0^e = 1$ is the correct critical value. At this point, it is of particular interest to establish the critical ratio of mosquitoes to hosts in the settings of vertical transmission. From (31) we obtain

$$
\frac{N_v}{N_h} > (1 - q_v)\frac{\mu_v(\gamma_h + \mu_h)}{p_{vh}p_{hv}\alpha^2},
$$

(32)

that is, a new critical ratio $m_c^* = (1 - q_v)m_c < m_c$ for $0 < q_v < 1$. This result indicates that, in the presence of

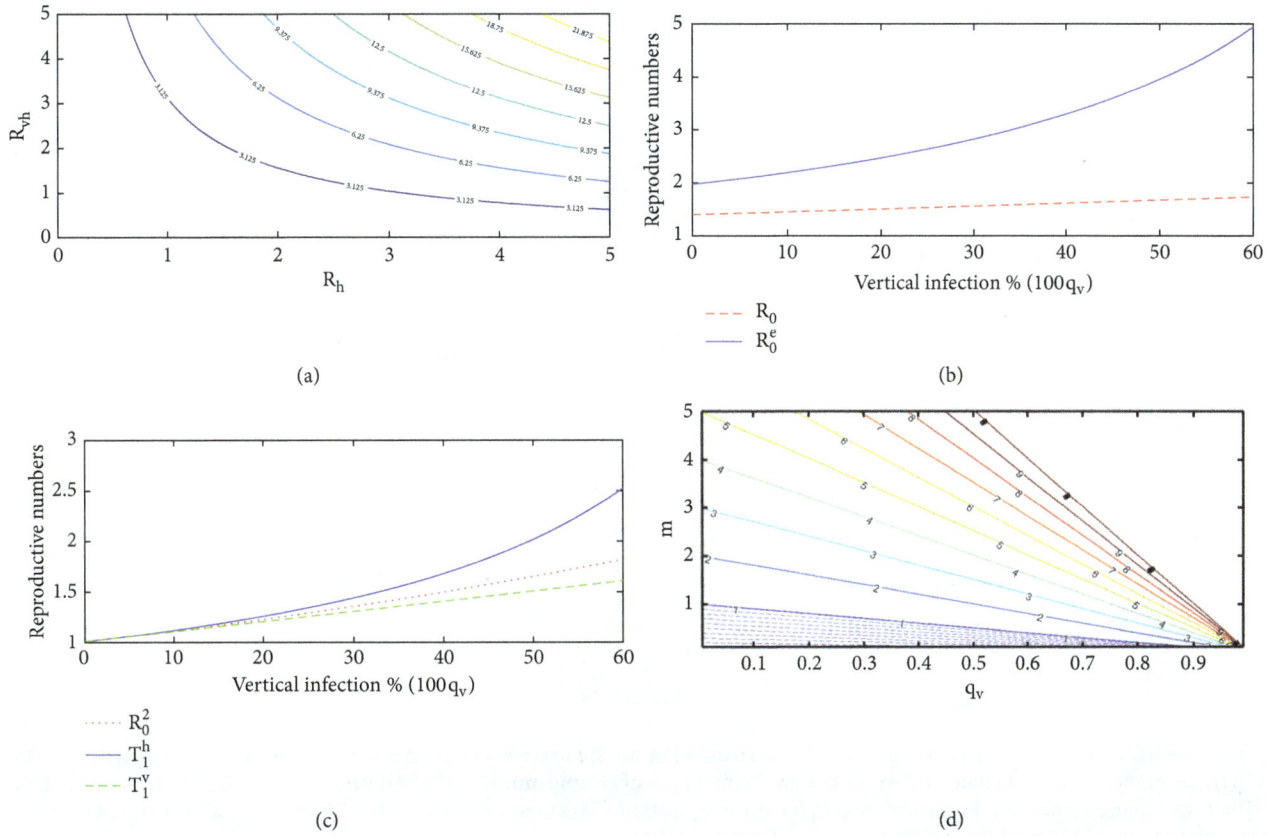

(a)

(b)

R_0^2

T_1^h

T_1^v

(c)

(d)

FIGURE 1: Relationships among epidemic thresholds with respect to vertical transmission and ratio of mosquitoes to hosts. (a) The relationship between number of new infections in hosts and number of new infections in vectors. (b) Comparison between basic reproductive numbers R_0 and R_0^e derived from two different methods with respect to vertical infection. (c) Comparison of behaviours among the complete transmission cycle R_0^2 and host type and vector type reproductive numbers with respect to vertical transmission efficiency away from the common epidemic criterion. (d) How the ratio of mosquitoes to hosts m and vertical infection q_v contribute to patterns of R_0^e. Dashed lines indicated contours lines for $R_0^e < 1$ while continuous lines are for $R_0^e \geqslant 1$.

vertical transmission, the critical ratio of female mosquitoes to hosts reduces at a rate proportional to vertical transmission efficiency. This indicates an earlier occurrence of the point where an epidemic is just possible as above this level the equilibrium prevalence is expected to rapidly increase to its asymptotic value.

Others are epidemic thresholds that provide a direct measure of the control effort required for disease eradication. These measures are known as "type" reproductive numbers and are determined based on the next-generation matrix [38, 39] such that for our case we have host and vector type reproductive numbers T_1^h and T_1^v, respectively. Note that in the absence of vertical transmission the host and vector type reproductive numbers are both equal to the square of the basic reproductive number, R_0 in (41), which turns out to be R_0^e in (31) for $q_v = 0$. In the presence of vertical infection, R_0^2 gives the expected number of secondary infections after one average, complete (host-vector-host or vector-host-vector) transmission cycle but does not correspond to a specific population type [23]. For the case of host population, one infected host leads to some secondary host infections in the next host-vector-host transmission cycle. This results exactly from the horizontal transmission mechanism and it is given

by R_0^H. However, further secondary host infections may also occur after any number of vector-vector transmission cycles as a result of transovarial transmission. From the expression of R_0 in (31) it can be seen that new infections resulting from vector-vector transmission cycle are given by $1/(1 - q_v)$; hence the host type reproductive number is then given by

$$T_1^h = \frac{1}{1 - q_v} R_0^H = R_0^e. \tag{33}$$

From (33), $R_0^H = R_0^e$ if $1/(1 - q_v) = 1$; that is, $q_v^* = 0$ which is the boundary condition. An important question would be, what is the proportion of vertical infection that doubles R_0^H? For $1/(1 - q_v) = 2$ we obtain $q_v^* = 0.5$, which means that if about 50% of infected mosquitoes produce infected offspring, we expect T_1^h to double. Note that all secondary infections resulting from a single infected vector must occur in the next vector-host-vector transmission cycle; hence, the vector type reproductive number takes the form

$$T_1^v = R_0^H + q_v. \tag{34}$$

For details on the derivation of both types of reproductive numbers, see Supplementary Materials, Section A-1. A

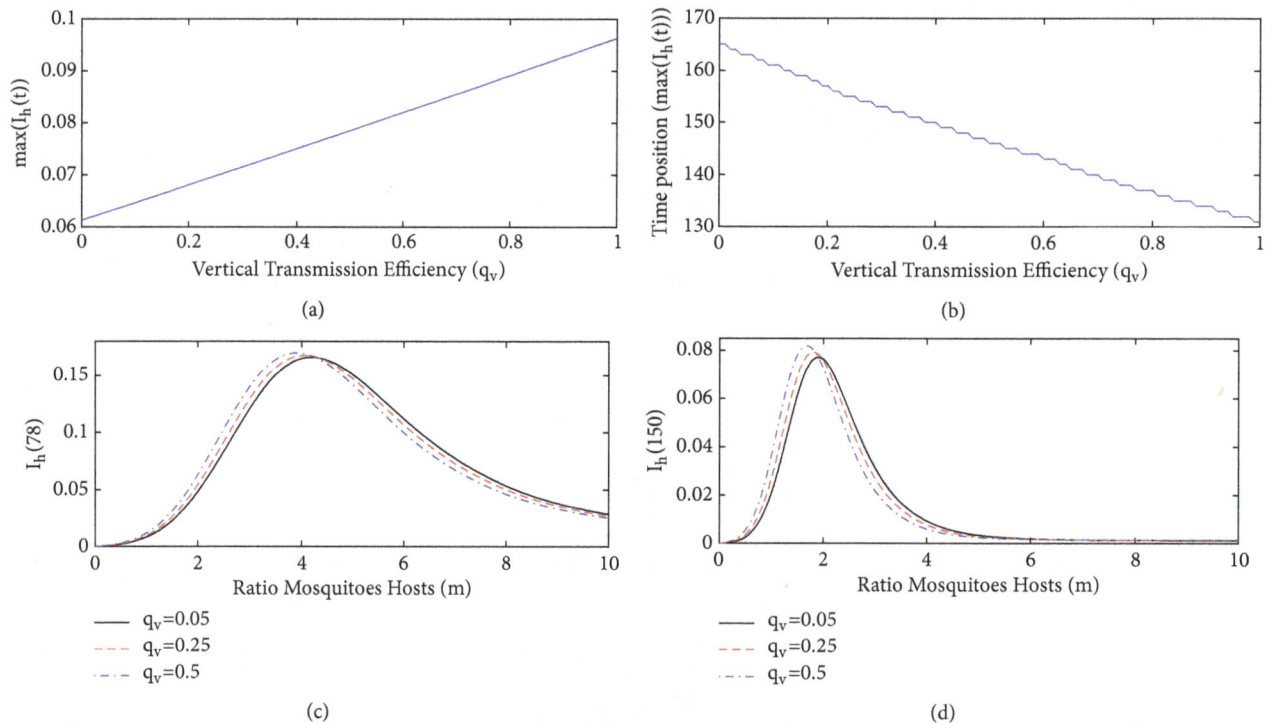

FIGURE 2: The left figure (a) explores effects of vertical transmission on the prevalence of the disease on hosts, where $(I_h(q_v))$ is plotted at $\max(I_h(t))$ for every q_v. The right figure (b) explores how the first peak of an epidemic $(\max(I_h(t)))$ varies with vertical transmission efficiency (q_v). The lower figures represent the dynamics of $I_h(t)$ at time position (78) corresponding to day 27 (figure c) and at time position (150) corresponding to day 50 (figure d) for different values of vertical infection.

number of methods for deriving R_0 exist and in [21] it is demonstrated that the resulting R_0 is not unique but they may have a common threshold at $R_0 = 1$. Hence, their behaviour below and above unity may show significant differences. In Figure 1© we examine how vertical transmission efficiency contributes to the qualitative behaviour of R_0^2, T_1^h, and T_1^v above the unity threshold criterion. An important characteristic pattern of their relationship is that below a certain percentage of vertical transmission efficiency the three reproductive numbers are undistinguishable but beyond the relationship between vertical infection and host type reproductive number becomes exponential. This behaviour results from the fact that in each generation, the number of host infections is proportional to the number of infected mosquitoes, which decreases proportionally to the vertical infection efficiency. Hence, for low vertical infection efficiency, possibly the first generations make a significant contribution to subsequent transmission cycles (for more discussion see [23, 27]).

In Figure 1(b) we compare the basic reproductive number R_0 derived through the next-generation method with the one derived from the endemic equilibrium R_0^e with respect to vertical transmission efficiency. Clearly, the two give different relationships and it is easy to see that the effect of vertical infection is completely diluted on R_0. Hence, this measure if used for disease control effort is more likely to provide misleading guidance. The relationship of the basic reproductive number R_0^e with mosquito to host ratio m and vertical

infection q_v shows some interesting patterns. For $m > 1$ the relationship between R_0^e and q_v is exponential as described in Figure 1(c); however for $m < 1$ it is linear rendering q_v having little effect if not negligible for $q_v \in (0, 0.8)$ (see Figure 1(d) dashed lines). This suggests that when there are more hosts compared to mosquitoes, vertical infection efficiency becomes completely negligible as in the model settings, and if $m < 1$, sustained host-vector transmissions may be impossible. On the other hand, we observe that for $q_v \ggg 0.8$ and even for $m < 1$, R_0^e becomes much larger highlighting the importance of the interplay between vertical infection and the ratio of mosquitoes to hosts.

Remark 7. In the settings of vector-borne diseases both host type reproductive number T_1^h or (R_0^e) and vector type reproductive number T_1^v are critical for providing useful informative indicator for guiding disease control efforts.

3.2. Effects of Ratio of Mosquitoes to Hosts and Vertical Transmission on Disease Prevalence. To fully capture effects of vertical transmission efficiency on vector-borne transmission models, we explore how this parameter influences the disease spread patterns.

Results are shown in Figure 2. The left figure explores how the first peak of an epidemic denoted by $\max(I_h(t))$ varies with vertical transmission efficiency, q_v. The results show that $\max(I_h(t))$ increases linearly with q_v. The right-hand figure plots the changes on time of occurrence of each epidemic peak at each value of vertical transmission efficiency. The results suggest that as vertical transmission efficiency increases the time position of each epidemic peak decreases linearly. This suggests that if vertical infection efficiency is increased, the first peak of an epidemic is likely to be attained earlier than the expected time. This suggests that while vertical transmission efficiency increases the epidemic size of an outbreak, it linearly decreases the time of the peak of an outbreak. The lower figures, (c) and (d), describes how the prevalence of infected hosts, I_h, varies with the ratio of female mosquitoes to hosts for different values of vertical transmission efficiency. Figure 2(c) indicates that the prevalence of the disease saturates at larger values of the ratio at earlier stage of the initial epidemic while Figure 2(d) indicates that the prevalence of the disease saturates at lower values of the ratio at later stage of the initial epidemic. This result stems from the fact that at earlier stage of the initial epidemic the size of the vector population is still very high compared to later stages of evolution of the epidemic.

3.3. Analysis of the Model with Vector Stages and Vertical Transmission

3.3.1. Model Equilibria. Model equilibria are obtained by setting the right-hand side of system (2) equal to zero. The model has three equilibria, two disease-free and one endemic equilibrium. Details about the derivation of the components of the endemic equilibrium are given in Section A-2 of Supplementary Materials. The first trivial equilibrium is both mosquito-free and disease-free:

$$X^0 = \left(L_s^0, L_i^0, S_v^0, I_v^0, S_h^0, I_h^0\right) = (0, 0, 0, 0, 1, 0), \quad (35)$$

which corresponds only to the presence of hosts (this can be livestock for the case of RVF). However, this equilibrium is not biologically feasible or relevant in particular in tropical regions where vector-borne diseases such as Dengue and RVF are endemic. The second equilibrium corresponds to the coexistence of both vectors and hosts but without infection. This is the disease-free equilibrium:

$$X^1 = \left(L_s^1, L_i^1, S_v^1, I_v^1, S_h^1, I_h^1\right) = (l^*, 0, P^*, 0, 1, 0), \quad (36)$$

which can be used for computation of the basic reproduction number when using the next-generation method [36]. The nonzero components are

$$l^* = \frac{\delta r\left(1 - Q_0^{-1}\right)}{\delta r + \theta + \mu_L},$$

$$P^* = \frac{\theta\left(1 - Q_0^{-1}\right)}{\theta + \delta\mu_v}, \quad (37)$$

$$\text{with } Q_0 = \frac{r\theta}{\mu_v\left(\theta + \mu_L\right)}.$$

Hence, X^1 is biologically significant if $Q_0 > 1$; that is, the mosquito population exists and can establish itself if on average each adult female mosquito produces at least one female mosquito at oviposition rate r during her life time period $1/\mu_v$ after successfully surviving the aquatic stage at probability $\theta/(\theta + \mu_L)$.

The endemic equilibrium which represents the prevalence of the disease in both host and vector populations is given by

$$X^* = \left(L_s^*, L_i^*, S_v^*, I_v^*, S_h^*, I_h^*\right), \quad (38)$$

where

$$L_s^* = l^* - L_i^*,$$

$$S_v^* = P^* - I_v^*,$$

$$I_v^* = \frac{\mu_h\left(1 - S_h^*\right)}{\lambda_{hv}S_h^*}, \quad (39)$$

$$S_h^* = 1 - \frac{\gamma_h + \mu_h}{\mu_h}I_h^*,$$

$$I_h^* = \frac{(1 - q_v)\,\mu_v\left(\gamma_h + \mu_h\right)\left(R_0^c - 1\right)}{\lambda_{vh}\lambda_{hv}P^*\left(\gamma_h + \mu_h\right)/\mu_h + \lambda_{vh}/\left(1 - q_v\right)\mu_v}, \quad (40)$$

where

$$R_0^c = \frac{\lambda_{vh}\lambda_{hv}P^*}{\left(1 - q_v\right)\mu_v\left(\gamma_h + \mu_h\right)}. \quad (41)$$

Similar to the discussion in Section 3.1.2 the above epidemic threshold in (41) is the "effective" or "control" reproductive number since it satisfies the property that the endemic equilibrium E^* only persists if R_0^c is greater than unity and there is no backward bifurcation.

Remark 8. The relationship between the endemic equilibrium and R_0 which depends on the vector population threshold indicates that the existence of this equilibrium is governed by the norms of the vector population threshold. Therefore, this vector population threshold is an important parameter in vector control efforts as it provides a means for identifying key factors for reducing the vector population.

In the following section we derive R_0 using the next-generation method.

3.3.2. The Basic Reproductive Number and Other Control Thresholds. The basic reproductive number, R_0, is a concept of great epidemiological significance. Following the method of [36], we write system (2) considering only disease compartments, L_i, I_v and I_h as $\dot{x} = \mathcal{F} - \mathcal{V}$ with $x^T = (L_i, I_v, I_h)$, where

$$\mathcal{F} = \begin{pmatrix} \delta r\left(1 - L\right)q_v I_v \\ \frac{\theta}{\delta}\left(1 - N_v\right)L_i + \lambda_{vh}I_h S_v \\ \lambda_{hv}I_v S_h \end{pmatrix},$$

$$\mathcal{V} = \begin{pmatrix} \left(\theta + \mu_L\right)L_i \\ \mu_v I_v \\ \left(\gamma_h + \mu_h\right)I_h \end{pmatrix}. \quad (42)$$

Evaluating the Jacobian of above matrices at the disease-free equilibrium $X^1 = (l^*, 0, P^*, 0, 1, 0)$, the basic reproduction number is then the spectral radius of the next-generation matrix; that is, $R_0 = \rho(\mathcal{F}(X^1)\mathcal{V}(X^1)^{-1})$,

$$R_0 = \frac{1}{2}q_v + \frac{1}{2}\sqrt{q_v^2 + 4R_0^H}, \quad \text{with } R_0^H = \frac{\lambda_{vh}\lambda_{hv}P^*}{\mu_v(\gamma_h + \mu_h)}. \quad (43)$$

Following the discussion in Section 3.1.2 this epidemic threshold (43) can lead to an underestimate of the control effort required if used to guide disease control strategies. To overcome this inconsistency we use type reproductive numbers to estimate efforts required to provide informative indicators in control of vector-borne diseases. To avoid repetition we omit the steps required in the derivation process, such that the host type reproductive number is

$$T_1^h = \frac{R_0^H}{1 - q_v} = R_0^c, \quad (44)$$

and the vector type reproductive number is then given by

$$T_1^v = R_0^H + q_v. \quad (45)$$

In the majority of mosquito-borne diseases, treatment or adequate vaccine is not likely to be feasible, either due to financial constraints or nonavailability, in particular in African communities with low resilience to economic challenges. In such cases the main preventive measures are individual protection against mosquito bites and the control of the proliferation of both the larval and adult mosquitoes. This can be achieved by lowering appropriate epidemic thresholds below unity, that is, forcing the reproductive number of the pathogen below 1. In the context of our model this implies reducing either of or both type reproductive numbers (T_1^h and/or T_1^v) below unity. Hence, we define the eradication effort to be the percentage reduction in the mosquito population size required to prevent endemic transmission of the pathogen [23], such that if using T_1^h the eradication effort is $100(1 - 1/T_1^h) = 100(R_0^H + q_v - 1)/R_0^H$ and for T_1^v the effort required to eradicate the disease is $100(1 - 1/T_1^v) = 100(R_0^H + q_v - 1)/(R_0^H + q_v)$. Therefore, vertical infection at $q_v\%$ can be responsible for at most $q_v\%$ of the required eradication effort, provided $R_0^H > 1$.

3.3.3. Stability of the Host-Only System.
This fixed point $X^0 = (L_s^0, L_i^0, S_v^0, I_v^0, S_h^0, I_h^0) = (0, 0, 0, 0, 1, 0)$ is an equilibrium point of the host-only system given by

$$\frac{dS_h}{dt} = \mu_h(1 - S_h), \quad (46)$$

which is local and globally asymptotically stable for all $t > 0$ as the derivative $(\mu_h(1 - S_h))'$ is negative.

3.3.4. Local Stability of the Disease-Free Equilibrium.
The local stability of the disease-free equilibrium point $X^1 = (l^*, 0, P^*, 0, 1, 0)$ is established by analysing the eigenvalues of the Jacobian matrix at X^1:

$$J(E^1) = \begin{pmatrix} -\delta r P^* - (\theta + \mu_L) & \delta r P^* & \delta r(1 - l^*) & \delta r(1 - l^*)(1 - q_v) & 0 & 0 \\ 0 & -(\theta + \mu_L) & 0 & \delta r(1 - l^*)q_v & 0 & 0 \\ \frac{\theta}{\delta}(1 - P^*) & 0 & -\frac{\theta}{\delta}l^* - \mu_v & -\frac{\theta}{\delta}l^* & 0 & -\lambda_{vh}P^* \\ 0 & \frac{\theta}{\delta}(1 - P^*) & 0 & -\mu_v & 0 & \lambda_{vh}P^* \\ 0 & 0 & 0 & -\lambda_{hv} & -\mu_h & 0 \\ 0 & 0 & 0 & \lambda_{hv} & 0 & -(\gamma_h + \mu_h) \end{pmatrix}. \quad (47)$$

Direct computation shows that $\lambda_1 = -\mu_h$ is an eigenvalue of matrix (47) and the remaining are solutions of equation

$$\lambda^5 + c_4\lambda^4 + c_3\lambda^3 + c_2\lambda^2 + c_1\lambda + c_0 = 0 \quad (48)$$

where

$$c_4 = \delta r P^* + 2(\theta + \mu_L) + (\gamma_h + \mu_h) + \frac{\theta}{\delta}l^* + 2\mu_v > 0,$$

$$c_3 = (1 - q_v)(\gamma_h + \mu_h)\mu_v(1 - R_0^c) + \delta r P^*\left(\gamma_h + \mu_h\right.$$

$$+ \left(\frac{\theta l^*}{\delta}\right) + 2\mu_v\right) + (\theta + \mu_L + \mu_v)^2 + \frac{\theta}{\delta}$$

$$\cdot l^*(2(\theta + \mu_L) + \gamma_h + \mu_h + \mu_v) + 2(\gamma_h + \mu_h)(\theta$$

$$+ \mu_L + \mu_v) + (1 - q_v)\mu_v(\theta + \mu_L - (\gamma_h + \mu_h)) > 0,$$

$$c_2 = (1 - q_v)(\gamma_h + \mu_h)\mu_v(1 - R_0^c)\left(\delta r P^* + 2(\theta + \mu_L)\right.$$

$$+ \left(\frac{\theta l^*}{\delta}\right) + \mu_v\right) + (1 - q_v)\mu_v(\theta + \mu_L - (\gamma_h + \mu_h))$$

$$\cdot \left(\delta r P^* + \theta + \mu_L + \left(\frac{\theta l^*}{\delta}\right) + \mu_v\right) + (\gamma_h + \mu_h)(\theta$$

$$+ \mu_L + \mu_v)^2$$

$$+ \delta r P^*\left[(\theta + \mu_L)\left(\gamma_h + \mu_h + \left(\frac{\theta l^*}{\delta}\right) + \mu_v\right)\right.$$

$$+ (\gamma_h + \mu_h)\left(\left(\frac{\theta l^*}{\delta}\right) + 2\mu_v\right) + \mu_v\left(\left(\frac{\theta l^*}{\delta}\right) + \mu_v\right)\right]$$

$$+ \frac{\theta}{\delta} l^* \left(\theta + \mu_L\right) \left(\theta + \mu_L + 2 \left(\gamma_h + \mu_h\right) + \mu_v\right) + \frac{\theta}{\delta}$$

$$\cdot l^* \left(\gamma_h + \mu_h\right) \mu_v > 0,$$

$$c_1 = \left(1 - q_v\right) \left(\gamma_h + \mu_h\right) \mu_v \left(1 - R_0^c\right)$$

$$\cdot \left[\delta r P^* \left(\theta + \mu_L + \left(\frac{\theta l^*}{\delta}\right) + \mu_v\right) + \left(\theta + \mu_L\right)^2\right.$$

$$\left. + \left(\theta + \mu_L\right) \mu_v + 2 \left(\theta + \mu_L\right) \left(\frac{\theta l^*}{\delta}\right)\right]$$

$$+ \left[\delta r P^* \left(\left(\frac{\theta l^*}{\delta}\right) + \mu_v\right) + \left(\theta + \mu_L\right) \left(\frac{\theta l^*}{\delta}\right)\right)$$

$$\cdot \left[\left(1 - q_v\right) \mu_v \left(\theta + \mu_L - \left(\gamma_h + \mu_h\right)\right) + \left(\gamma_h + \mu_h\right) \mu_v\right]$$

$$+ \left(\theta + \mu_L\right) \left(\gamma_h + \mu_h\right) \left(\frac{\theta l^*}{\delta}\right) \left(\delta r P^* + \left(\theta + \mu_L\right)\right) > 0,$$

$$c_0 = \left(1 - q_v\right) \left(\theta + \mu_L\right) \left(\gamma_h + \mu_h\right) \mu_v \left(1 - R_0^c\right)$$

$$\cdot \left[\delta r P^* \left(\left(\frac{\theta l^*}{\delta}\right) + \mu_v\right) + \left(\frac{\theta l^*}{\delta}\right) \mu_v\right].$$

$$(49)$$

By Descartes' rule of signs, since $c_4 > 0$, $c_3 > 0$, $c_2 > 0$, and $c_1 > 0$, then all roots of (48) are negative or have a negative real part for $c_0 > 0$. The coefficient c_0 is nonnegative if and only if $R_0^c < 1$; hence the following result holds.

Theorem 9. *The disease-free equilibrium point $X^1 = (l^*, 0, P^*, 0, 1, 0)$ is locally asymptotically stable whenever $R_0^c < 1$.*

3.3.5. Global Stability of the Disease-Free Equilibrium. To establish the global asymptotic stability (GAS) of X^1 we use results obtained by Kamgang and Sallet [40], which are the extension of some results in [36]. Around the disease-free equilibrium system (2) can be written as

$$\dot{x}_S = A_1 \left(x\right) \left(x_S - x_{E^1, S}\right) + A_{12} \left(x\right) x_I$$

$$\dot{x}_I = A_2 \left(x\right) x_I \tag{50}$$

where x_S is the vector representing disease-free compartments (L_s, S_v, S_h) and the vector x_I represents the state of infected compartments (L_i, I_v, I_h). This requires rewriting system equation around X^1, as follows:

$$\frac{dS_h}{dt} = \mu_h - \lambda_{hv} I_v S_h - \mu_h S_h$$

$$= -\mu_h \left(S_h - S_h^1\right) - \lambda_{hv} I_v S_h,$$

$$\frac{dL_s}{dt} = \delta r \left(1 - L_s\right) \left(S_v + \left(1 - q_v\right) I_v\right) \tag{51}$$

$$- \delta r L_i \left(S_v + \left(1 - q_v\right) I_v\right) - \left(\theta + \mu_L\right) L_s,$$

$$= - \left(\delta r S_v + \theta + \mu_L\right) \left(L_s - L_s^1\right)$$

$$+ \delta r \left(1 - L_s^1\right) \left(S_v - S_v^1\right) - \delta r S_v L_i$$

$$+ \delta r \left(1 - q_v\right) \left(1 - L\right) I_v.$$

Note that the latter arises from $(\theta + \mu_L) L_s^1 = \delta r (1 - L_s^1) S_v^1$. Then, the following matrices are obtained,

$$A_1 \left(x\right) = \begin{pmatrix} -\left(\delta r S_v + \theta + \mu_L\right) & \delta r \left(1 - L_s^1\right) & 0 \\ \frac{\theta}{\delta} \left(1 - N_v\right) & -\mu_v & 0 \\ 0 & 0 & -\mu_h \end{pmatrix}, \tag{52}$$

$$A_{12} \left(x\right) = \begin{pmatrix} -\delta r S_v & \delta r \left(1 - q_v\right) \left(1 - L\right) & 0 \\ 0 & 0 & -\lambda_{vh} S_v \\ 0 & -\lambda_{vh} S_v & 0 \end{pmatrix},$$

$$A_2 \left(x\right)$$

$$= \begin{pmatrix} -\left(\theta + \mu_L\right) & \delta r \left(1 - L\right) q_v & 0 \\ \frac{\theta}{\delta} \left(1 - N_v\right) & -\mu_v & \lambda_{vh} S_v \\ 0 & \lambda_{hv} & -\left(\gamma_h + \mu_h\right) \end{pmatrix}. \tag{53}$$

From basic matrix properties and direct algebraic computation it follows that all eigenvalues of matrix A_1 are real and negative and both A_1 and A_2 are Metzler matrices. Hence, system $\dot{x}_S = A_1(x)(x_S - x_{X^1, S})$ is GAS at the disease-free equilibrium $x_{X^1, S}$. To establish the global stability of overall system (50) at X^1 conditions of the following theorem in [40] must be satisfied.

Theorem 10. *Let $\Phi \subset \mathcal{U} = \mathbb{R}_+^3 \times \mathbb{R}_+^3$. System (50) is of class C^1, defined on $\mathcal{U}$ if*

(1) *$\mathcal{U}$ is positively invariant relative to (50).*

(2) *The system $\dot{x}_S = A_1(x)(x_S - x_{DFE, S})$ is GAS at $x_{DFE, S}$.*

(3) *For any $x \in \Phi$, matrix $A_2(x)$ is Metzler irreducible.*

(4) *There exists a matrix $\overline{A}_2$, which is an upper bound of the set $\mathcal{M} = \{A_2(x) \in \mathcal{M}_3(\mathbb{R}) \mid x \in \overline{\Phi}\}$, with the property that if $\overline{A}_2 \in \mathcal{M}$, for any $\overline{x} \in \overline{\Phi}$, such that $A_2(\overline{x}) = \overline{A}_2$, then $\overline{x} \in \mathbb{R}^3 \times \{0\}$.*

(5) *The stability modulus of $\overline{A}_2$, $\alpha(\overline{A}_2) = \max_{\lambda \in S_p(\overline{A}_2)} \text{Re}(\lambda)$, satisfies $\alpha(\overline{A}_2) \le 0$.*

Then, X^1 is GAS in $\overline{\Phi}$.

Proof. Clearly, conditions (1)-(3) of the theorem have been satisfied. For all $x \in \Phi$, $A_2(x)$ is irreducible because $(I + |A_2(x)|)^2 > 0$. An upper bound of the set of matrices $\mathcal{M}$, which is the matrix $\overline{A}_2$, is given by matrix $A_2(\overline{x})$, where $\overline{x} = (\overline{L}_s^1, 0, \overline{S}_v^1, 0, 1, 0) \in \mathbb{R}^3 \times \{0\}$, with $\overline{L}_s^1 = K_1$ and $\overline{S}_v^1 = K_2$. Similarly matrix $\overline{A}_2$ is irreducible. Recall that from the Perron-Frobenius theorem for an irreducible matrix you get that one of the matrix eigenvalues is positive and greater than or equal to all others, that is, the dominant eigenvalue. Matrix A_2 is exactly the matrix used to compute the basic reproductive number, i.e., the dominant eigenvalue. For more details or proof in general settings see [40]. □

Now conditions (1)-(4) have been verified. To check the last condition, we make use of the following Lemma [40].

Lemma 11. *Let H be a square Metzler matrix written in block form $H = \begin{pmatrix} A & B \\ C & D \end{pmatrix}$, with A and D square matrices. H is Metzler stable if and only if matrices A and $D - CA^{-1}B$ are Metzler stable.*

Matrix $A_2(x)$ in block matrices takes the following components:

$$A = (-(\theta + \mu_L)),$$

$$B = (\delta r (1 - L) q_v \quad 0),$$

$$C = \left(\frac{\theta}{\delta} (1 - N_v) \quad 0 \right)^T, \tag{54}$$

$$D = \begin{pmatrix} -\mu_v & \lambda_{vh} P^* \\ \lambda_{hv} & -(\gamma_h + \mu_h) \end{pmatrix}.$$

Clearly, A is a Metzler stable matrix and

$$D - CA^{-1}B = \begin{pmatrix} -\mu_v (1 - q_v) & \lambda_{vh} P^* \\ \lambda_{hv} & -(\gamma_h + \mu_h) \end{pmatrix} \tag{55}$$

is also Metzler stable if $\mu_v(1 - q_v)(\gamma_h + \mu_h) - \lambda_{hv}\lambda_{vh}P^* \geq 0$, that is, $1 - R_0^c \geq 0$. Therefore, from Theorem 10 and Lemma 11 the following result holds.

Theorem 12. *For $Q_0 > 1$ there exists a unique disease-free equilibrium $X^1 = (l^*, 0, P^*, 0, 1, 0)$, which is globally asymptotically stable whenever $R_0^c \leq 1$.*

Remark 13. The above result is of great epidemiological importance. It highlights two fundamental indicators: (1)

Although vector control is central for control of vector-borne diseases, it does not mean eliminating all the vectors. Note that the effect of this include both infected and noninfected mosquitoes. (2) Disease eradication efforts are independent of the initial sizes of both vector and host populations; however, the ratio between the two populations is a key factor.

3.3.6. Local Stability of the Endemic Equilibrium. Results of the local stability of the disease-free equilibrium X^1 suggest that for $R_0^c = 1$ the Jacobian matrix (47) has zero eigenvalue while the remaining eigenvalues are negative or have a negative real part. Since the algebraic computation involved when establishing the stability of the endemic equilibrium through linearization is quite extensive, we employ the centre manifold theory [41]. This theory is used to examine existence of backward or forward bifurcation. The bifurcation occurs at $R_0^c = 1$, and choosing $q_v = q_v^*$ as a bifurcation parameter, then

$$q_v^* = 1 - \frac{\lambda_{hv}\lambda_{vh}P^*}{\mu_v(\gamma_h + \mu_h)}, \Longrightarrow R_0^H \leq 1. \tag{56}$$

The Jacobian matrix $J(X^1, q_v^*)$ is the same as matrix (47); hence, $J(X^1, q_v^*)$ has a simple zero eigenvalue $\lambda = 0$ when $R_0^c = 1$ and others given as $\lambda = -\mu_h$ and roots of the polynomial $\lambda^4 + c_4\lambda^3 + c_3\lambda^2 + c_2\lambda + c_1 = 0$, where c_1, c_2, c_3, c_4 are as defined in Section 3.3.4. Let $\omega = (\omega_1, \omega_2, \omega_3, \omega_4, \omega_5, \omega_6)$ be the right eigenvector associated with zero eigenvalue of the Jacobian matrix $J(X^1, q_v^*)$. Its components are derived by solving $J(X^1, q_v^*) \times \omega = 0$, which gives

$$\omega = \left(F^*\omega_4, H^*\omega_4, G^*\omega_4, \omega_4, -\frac{\lambda_{hv}}{\mu_h}\omega_4, \frac{\lambda_{hv}}{\gamma_h + \mu_h} \right) \tag{57}$$

where

$$F^* = \frac{\left[r\theta\mu_h \left(Q_0^{-1} - q_v^* \right) + (1 - q_v^*) \mu_h (\gamma_h + \mu_h)(\theta + \mu_L) \right] \delta r (\theta + \delta\mu_v)}{r\theta^2 \mu_h (1 - Q_0^{-1})(\delta r + \theta + \mu_L)},$$

$$G^* = \frac{(1 - q_v^*) \mu_v \theta (\theta + \mu_L) \left[(\gamma_h + \mu_h)(\delta r + \theta + \mu_L) - \mu_h(\theta + \mu_L)(Q_0^v - 1) \right] + \mu_h\mu_v(\theta + \mu_L)^2 \left[\theta + \delta\mu_v (1 - Q_0 q_v^*) \right]}{r\theta(\theta + \delta\mu_v)(1 - Q_0^{-1})},$$

$$\tag{58}$$

$$H^* = \frac{\delta r q_v^* (\theta + \delta\mu_v)}{\theta (\delta r + \theta + \mu_L)},$$

$$Q_0^v = \frac{r\theta}{(1 - q_v^*) \mu_v (\theta + \mu_L)}.$$

Clearly, F^* and G^* are nonnegative if and only if $q_v^* \leq 1/Q_0$, $Q_0^v > 1$, and $Q_0 > 1$. Note that Q_0^v gives the average number of emerged infected female mosquitoes produced by one infected female mosquito. Hence, for $Q_0^v > 1$ infected mosquitoes will persist in the population, leading to disease persistence. Let us denote $\upsilon = (\upsilon_1, \upsilon_2, \upsilon_3, \upsilon_4, \upsilon_5, \upsilon_6)$ as the left eigenvector associated with zero eigenvalue of the Jacobian

matrix $J(X^1, q_v^*)$. Its components are obtained when solving $[J(X^1, q_v^*)]^T \times \omega = 0$, ($T$ represents transpose), which gives

$$\upsilon = \left(0, M^*\upsilon_4, 0, \upsilon_4, 0, \frac{\lambda_{vh}P^*}{\gamma_h + \mu_h}\upsilon_4 \right) \tag{59}$$

where $M^* = \theta\mu_v(\delta r + \theta + \mu_L)/\delta r(\theta + \mu_L)(\theta + \delta\mu_v)$.

TABLE 1: Existence and stability of model equilibria.

		X^0	X^1 if $(0 < q_v < 1)$	X^* if $(0 < q_v < 1)$
I.	$Q_0 < 1$ & $R_0^c < 1$	GS	DNE	DNE
II.	$Q_0 > 1$ & $R_0^c < 1$	DNE	GS	DNE
III.	$Q_0 > 1$ & $R_0^c > 1$	DNE	US	LS
IV.	$Q_0 < 1$ & $R_0^c > 1$	US	DNE	DNE

DNE: does not exist, US: unstable, LS: locally stable, GS: globally stable.

Computation of Bifurcation Parameters a^* and b^*. Because $v_1 = v_3 = v_5 = 0$ the associated nonzero partial derivatives of f_1, f_3, f_5 are not considered and the remaining nonzero partial derivatives of system $f = (f_1, f_2, f_3, f_4, f_5, f_6)$ at (X^1, q_v^*) are given by $\partial^2 f_2/\partial x_1 \partial x_4 = \partial^2 f_2/\partial x_2 \partial x_4 = -\delta r q_v^*$, $\partial^2 f_4/\partial x_2 \partial x_4 = \partial^2 f_4/\partial x_3 \partial x_2 = -\theta/\delta$, $\partial^2 f_4/\partial x_3 \partial x_6 = \lambda_{vh}$, and $\partial^2 f_6/\partial x_4 \partial x_5 = \lambda_{hv}$.

Thus, $a^* = 2v_2\omega_1\omega_4(\partial^2 f_2/\partial x_1 \partial x_4) + 2v_2\omega_2\omega_4(\partial^2 f_2/\partial x_2 \partial x_4) + 2v_4\omega_2\omega_4(\partial^2 f_4/\partial x_2 \partial x_4) + 2v_4\omega_3\omega_2(\partial^2 f_4/\partial x_3 \partial x_2) + 2v_4\omega_3\omega_6(\partial^2 f_4/\partial x_3 \partial x_6) + 2v_6\omega_4\omega_5(\partial^2 f_6/\partial x_4 \partial x_5)$, which after some algebraic simplification gives

$$a^* = -2v_4\omega_4^2 \left(\delta r q_v^* F^* M^* + \delta r q_v^* H^* M^* + \frac{\theta}{\delta} H^* \right.$$

$$\left. + G^* H^* + (1 - q_v^*) \mu_v \left(\frac{\lambda_{vh}}{\mu_h} - \frac{G^*}{P^*} \right) \right). \tag{60}$$

Given that the vital dynamics of the host are much more slower $\lambda_{vh}/\mu_h \gg G^*/P^*$ and using the property that $v\omega = 1$, then $a^* < 0$ for $\omega_4 > 0$. For b^* the nonzero partial derivative of $f = (f_1, f_2, f_3, f_4, f_5, f_6)$ at (X^1, q_v^*) is $\partial^2 f_2/\partial x_4 \partial q_v^* = \delta r(1 - l^*)$, such that $b^* = v_2\omega_4(\partial^2 f_2/\partial x_4 \partial q_v^*) = \mu_v v_2\omega_4$. Using the property that $v\omega = 1$ we obtain that $b^* > 0$. Since $a^* < 0$ for $\omega_4 > 0$ and $b^* > 0$, the model (2) exhibits a forward bifurcation at $R_0^c = 1$. Therefore, the following result holds.

Theorem 14. *The endemic equilibrium point* $X^* = (L_s^*, L_i^*, S_v^*, I_v^*, S_h^*, I_h^*)$ *exists and is locally asymptotically stable for* $R_0^c > 1$.

Remark 15. The condition $Q_0^v > 1$ is an indication that vertical transmission is vital for disease long-term persistence. A particular example is the case of Dengue fever and RVF. Upon failure of this condition, it is much more likely that results of Theorem 14 do not hold. Another important feature of Q_0^v is that vertical infection is proportional to Q_0, and hence vector control is still a viable control strategy even for vector-borne disease with transovarial transmission.

The above stability results and those in Theorems 9, 10, 12, and 14 can be summarized in Table 1 and in the bifurcation diagram in Figure 3.

3.4. Numerical Results of the Model with Vector Stages and Vertical Transmission. Numerical analysis using reasonable parameter values for RVF is carried out. In Figure 4 we plot time series of infected larval, adult mosquitoes and hosts, and their respective phase portraits. The latter are plotted

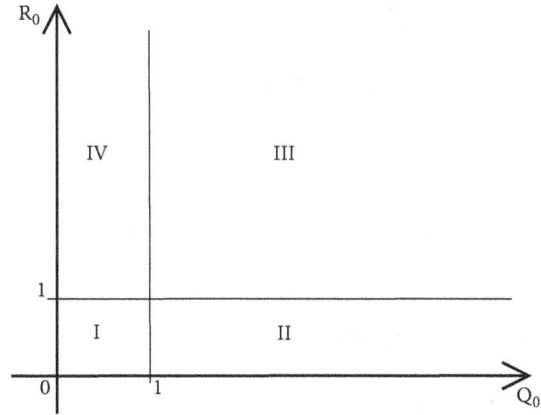

FIGURE 3: The bifurcation diagram in (Q_0, R_0^c) plane. The stability of equilibria in each region is given in Table 1.

for $t \in [0 \ 50]$ years for observing long-term dynamics. The parameter values used give $R_0^c > 1$ and existence of a stable endemic equilibrium. These results not only confirm findings by analytical analysis (Theorems 9 and 12) but suggest that this equilibrium is also global asymptotic stable for the chosen parameter values. This also implies that the endemic equilibrium is stable not only close to the bifurcation point but also for R_0^c beyond unity (see inner figures, Figures 4(a)–4(c)).

4. Discussion and Conclusion

One of the main contributions of mathematical epidemic models is that they enhance our understanding of disease transmission and public health planning. This is possible because with models we are able to derive critical epidemic thresholds in terms of model parameters which carry important disease features and key players. One of these measures is the basic reproduction number R_0, which if greater than unity implies that the disease will persist in the population and fade out otherwise. However, a major challenge is to derive the correct R_0 for disease models in which infection transmission is intermediated by another host. This is the case of vector-borne diseases, where practical methods such as the next-generation matrix for deriving R_0 fails to give the correct value [21]. Rather, it gives the geometric mean of secondary infection per generation. Using the basic model we discussed alternative measures, known as host and vector type reproductive numbers (T_1^h and T_1^v, respectively) [38, 39], to be used for guiding disease control efforts. Further, for

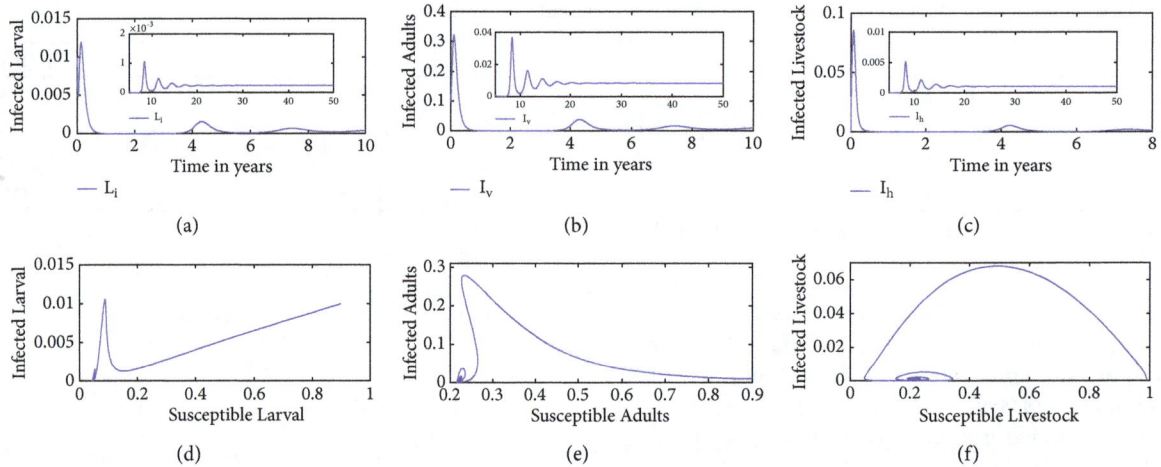

FIGURE 4: Time series and phase diagrams plot. The inner figures describe the temporal evolution of L_i, I_v, I_h after the transient period for $R_0^c = 2.1002$, while for the outer figures $R_0^c = 1.0012$. Parameter values used for $R_0^c = 1.0012$ are $\mu_v = 1/25, \mu_h = 1/(8*360), p_{hv} = 0.14, p_{vh} = 0.35, \alpha = 256/360, \gamma_h = 1/4, m = 1.5, q_v = 0.2, \theta = 1/10, \mu_L = 1/18, r = 1/12, \delta = 0.4$ and parameter values used for $R_0^c = 2.1002$ are given in Table 1 of Section A-3 of Supplementary Materials.

both the basic and the complex model we showed that the basic reproductive number derived from the existence of endemic equilibrium is the correct epidemic measure and it is equivalent to the host type reproductive number T_1^h. In fact, this is an alternative method for finding R_0 when it is possible to derive the explicit components of the endemic equilibrium point [21, 42]. Then we established important relationships between the reproductive numbers R_0^2, T_1^h, T_1^v and vertical infection efficiency. It follows that there is a critical threshold of vertical infection from which the contribution of this transmission route becomes significant. Otherwise, if vertical infection efficiency is very low, its contribution to disease persistence may be insignificant. The work in [23, 27] discusses this feature in more detail and their findings suggest that these results stem from the fact that for low vertical infection efficiency the virus is rapidly lost unless there is a regular disease amplification in the host population. Recent entomological studies have confirmed this situation in particular for RVF where there has been more and more evidence of disease interepidemic activities [7, 17–20], underlined by sporadic outbreaks on livestock at low levels. During this period vertical infection plays a key role for providing protection against chance extinction, but only if it is efficient enough to span several mosquito generations [23], since at the end of unfavourable season the horizontally transmitted epidemic is declining.

Also, our results demonstrated that while the relationship between R_0 and vertical infection efficiency is linear, the relationship between the correct threshold R_0^e and vertical infection efficiency is exponential. This shows that the failure of R_0 as a critical epidemic threshold compounds itself when varying vertical transmission. Therefore, more caution is needed when deriving the basic reproductive number in vector-borne disease systems that include vertical transmission as one of the transmission modes. Another important property to highlight in our models is the female mosquitoes

to host ratio which is more likely to be responsible for the asymmetric relationship between the host-to-vector and vector-to-host reproductive numbers as discussed in other studies [28, 43]. We found that in the presence of vertical infection the critical ratio of female mosquitoes to hosts decreases linearly when increasing vertical transmission efficiency. This result indicates an earlier occurrence of the point where an epidemic is just possible as above this level the equilibrium prevalence is expected to rapidly increase to its asymptotic value [37]. Further we analysed how vertical infection affects the prevalence of the disease at the first peak of an epidemic. The results showed that the prevalence at the first peak is positively correlated to vertical transmission efficiency while the time position of occurrence of each peak is negatively correlated to vertical transmission efficiency. These findings suggest that while vertical transmission efficiency increases the epidemic size of an outbreak, it reduces the duration of an outbreak. Additionally, the relationship between the prevalence of the disease and ratio of female mosquitoes to hosts was accessed for different values of vertical infection. Analysis showed that there is an optimal value of the ratio above which the prevalence of the disease decreases. However, such optimal value changes according to the stage of epidemic development, such that at early stage of the epidemic the optimal value is attained much later than that at later stage of the epidemic. Another interesting result is that after the optimal value of the ratio the effect of vertical transmission changes where higher values have less effect compared to lower values of vertical transmission efficiency.

Extending the basic model to include larval compartments with maximal capacities for both larval and adult populations we derived important critical thresholds for both vector population and disease system. We found that the model exhibits three boundary equilibria, namely, a mosquito-free system, a mosquito-host system but disease-free, and one where the disease persists in both populations.

A detailed stability analysis is presented and numerical simulations are conducted using parameter values relevant to Rift Valley fever. The mathematical results are used to interpret the biological implications of the relationships between R_0, vertical infection, and the ratio of female mosquitoes to hosts for assessing viable control measures. Our analysis shows that the qualitative behaviours of the system are completely determined by three key quantities: Q_0, the vector population threshold; Q_0^v, the average number of female mosquitoes produced by a single infected mosquito; and R_0^c, the effective reproductive number (see the bifurcation diagram in Figure 3 and Table 1). These results provide important qualitative understanding of the interaction between vertical infection and ratio of female mosquitoes to hosts on the prevalence of the disease.

The mosquito population persists if each female adult mosquito produces at least one larval mosquito after surviving the larval stage. Then, if an infected individual is introduced in the community without any form of protection, the disease will invade the system leading to appearance of an endemic equilibrium if the epidemic criterion is satisfied ($R_0^e > 1$). The biological implication of this result is that the disease can only invade an area if already colonized by the mosquito population as expected. Hence, measures targeting critical features of the life cycle of the vector are viable disease control strategies. Furthermore, we found that the persistence of the disease is subjected to vertical transmission efficiency governed by the epidemic criterion $Q_0^v > 1$. This implies that for the disease to persist for each larval mosquito produced, $1/(1 - q_v)$ should be infected. This explicit formulation of the threshold conditions in terms of parameters governing the infection transmission process and vector life history is of great epidemiological significance as it allows disease control efforts to be targeted at specific disease and vector stages. We have also shown that the disease-free equilibrium is locally asymptotically stable for $R_0^c < 1$ and globally asymptotically stable if $R_0^c \leqslant 1$. The latter result is of outmost importance, because it shows that if at any time, through appropriate interventions (e.g., destruction of breeding sites, use of insecticides and repellents on the host, or even vaccination), we are able to lower R_0^c below 1, then the disease will disappear. When the vertical transmission route is possible, this requires also lowering the contribution from this transmission mode. That is, lowering Q_0^v which implies keeping q_v very close to zero. Therefore, the design of control programs should take into account the implications of this mode of transmission in the case of vector-borne diseases that include vertical transmission in the vector population.

The goal of this study was to use mathematical models for discussing important properties of vector-borne disease systems and their implications in attempts for control strategies. Though we have used simple but realistic models, we believe that they remain valuable tools useful in the context of research and for providing qualitative understanding of complex processes underlying vector-borne disease transmission systems in particular in the context of transovarial transmission in the vector population. In conclusion, our results show that (1) including larval stages in the vector population when modelling vector-borne disease systems has benefits, as it allows determining conditions for colonization in terms of vector life cycle features; (2) the failure of R_0 derived from the next-generation method compounds itself in the presence of vertical transmission; (3) host type reproductive number gives the correct R_0 useful for guiding disease control strategies; (4) while vertical transmission efficiency is likely to increase the size of an epidemic, it decreases its duration; and (5) in the presence of vertical infection the critical ratio of female mosquitoes to hosts reduces linearly when increasing vertical transmission efficiency.

Conflicts of Interest

The author declares that he has no conflicts of interest.

References

[1] D. J. Gubler, "Vector-borne diseases," *Revue Scientifique et Technique de l'OIE*, vol. 28, no. 2, pp. 583–588, 2009.

[2] World Health Organization, *WHO—Vector-Borne Diseases*, 2016.

[3] A. Huppert and G. Katriel, "Mathematical modelling and prediction in infectious disease epidemiology," *Clinical Microbiology and Infection*, vol. 19, no. 11, pp. 999–1005, 2013.

[4] F. Brauer, "Mathematical epidemiology: Past, present, and future," *Infectious Disease Modelling*, vol. 2, no. 2, pp. 113–127, 2017.

[5] C. E. Walters, M. M. Meslé, and I. M. Hall, "Modelling the global spread of diseases: A review of current practice and capability," *Epidemics*, 2018.

[6] R. M. Murithi, P. Munyua, P. M. Ithondeka et al., "Rift Valley fever in Kenya: History of epizootics and identification of vulnerable districts," *Epidemiology and Infection*, vol. 139, no. 3, pp. 372–380, 2011.

[7] C. Sindato, E. D. Karimuribo, D. U. Pfeiffer et al., "Spatial and temporal pattern of rift valley fever outbreaks in Tanzania; 1930 to 2007," *PLoS ONE*, vol. 9, no. 2, 2014.

[8] K. J. Linthicum, F. G. Davies, A. Kairo, and C. L. Bailey, "Rift Valley fever virus (family Bunyaviridae, genus Phlebovirus). Isolations from diptera collected during an inter-epizootic period in Kenya," *Epidemiology & Infection*, vol. 95, no. 1, pp. 197–209, 1985.

[9] W. Horsfall, H. Fowler, M. Lj, and L. Jr, *Bionomics and Embryology of the Inland Flood Water Mosquito Aedes Vexans*, University of Illinois Press, 1974.

[10] F. F. Ludueña Almeida and D. E. Gorla, "The biology of Aedes (Ochlerotatus) albifasciatus Macquart, 1838 (Diptera: Culicidae) in central Argentina," *Memórias do Instituto Oswaldo Cruz*, vol. 90, no. 4, pp. 463–468, 1995.

[11] N. Becker, D. Petric, M. Zgomba et al., *Mosquitoes and Their Control*, Springer-Verlag Berlin Heidelberg 2003, 2010.

[12] K. J. Linthicum, T. M. Logan, C. L. Bailey, D. J. Dohm, and J. R. Moulton, "Transstadial and horizontal transmission of Rift Valley fever virus in Hyalomma truncatum," *The American Journal of Tropical Medicine and Hygiene*, vol. 41, no. 4, pp. 491–496, 1989.

[13] X. Guo, T. Zhao, Y. Dong, and B. Lu, "Survival and replication of dengue-2 virus in diapausing eggs of Aedes albopictus (diptera: Culicidae)," *Journal of Medical Entomology*, vol. 44, no. 3, pp. 492–497, 2007.

[14] B. Angel and V. Joshi, "Distribution and seasonality of vertically transmitted dengue viruses in Aedes mosquitoes in arid and semi-arid areas of Rajasthan, India," *Journal of Vector Borne Diseases*, vol. 45, no. 1, pp. 56–59, 2008.

[15] M. Pepin, M. Bouloy, B. H. Bird, A. Kemp, and J. Paweska, "Rift Valley fever virus (Bunyaviridae: Phlebovirus): An update on pathogenesis, molecular epidemiology, vectors, diagnostics and prevention," *Veterinary Research*, vol. 41, no. 6, 2010.

[16] Y. Ba, D. Diallo, C. M. F. Kebe, I. Dia, and M. Diallo, "Aspects of bioecology of two rift valley fever virus vectors in Senegal (West Africa): Aedes vexans and Culex poicilipes (Diptera: Culicidae)," *Journal of Medical Entomology*, vol. 42, no. 5, pp. 739–750, 2005.

[17] A. D. Labeaud, E. M. Muchiri, M. Ndzovu et al., "Interepidemic Rift Valley fever virus seropositivity, northeastern Kenya," *Emerging Infectious Diseases*, vol. 14, no. 8, pp. 1240–1246, 2008.

[18] R. D. Sumaye, E. Geubbels, E. Mbeyela, and D. Berkvens, "Inter-epidemic transmission of rift valley fever in livestock in the Kilombero river valley, Tanzania: a cross-sectional survey," *PLOS Neglected Tropical Diseases*, vol. 7, no. 8, Article ID e2356, 2013.

[19] N. O. Owange, W. O. Ogara, H. Affognon et al., "Occurrence of rift valley fever in cattle in Ijara district, Kenya," *Preventive Veterinary Medicine*, vol. 117, no. 1, pp. 121–128, 2014.

[20] J. K. Lichoti, A. Kihara, A. A. Oriko et al., "Detection of rift valley fever virus interepidemic activity in some hotspot areas of kenya by sentinel animal surveillance, 2009-2012," *Veterinary Medicine international*, vol. 2014, Article ID 379010, 9 pages, 2014.

[21] J. Li, D. Blakeley, and R. J. Smith, "The failure of R_0," *Computational and Mathematical Methods in Medicine*, vol. 2011, Article ID 527610, 17 pages, 2011.

[22] H. D. Gaff, D. M. Hartley, and N. P. Leahy, "An epidemiological model of rift valley fever," *Electronic Journal of Differential Equations*, vol. 115, pp. 1–12, 2007.

[23] B. Adams and M. Boots, "How important is vertical transmission in mosquitoes for the persistence of dengue? Insights from a mathematical model," *Epidemics*, vol. 2, no. 1, pp. 1–10, 2010.

[24] N. Chitnis, J. M. Hyman, and C. A. Manore, "Modelling vertical transmission in vector-borne diseases with applications to Rift Valley fever," *Journal of Biological Dynamics*, vol. 7, no. 1, pp. 11–40, 2013.

[25] S. A. Pedro, S. Abelman, F. T. Ndjomatchoua, R. Sang, and H. E. Z. Tonnang, "Stability, bifurcation and chaos analysis of vector-borne disease model with application to rift valley fever," *PLoS ONE*, vol. 9, no. 10, Article ID e108172, 2014.

[26] F. Chamchod, R. S. Cantrell, C. Cosner, A. N. Hassan, J. C. Beier, and S. Ruan, "A modeling approach to investigate epizootic outbreaks and enzootic maintenance of Rift Valley fever virus," *Bulletin of Mathematical Biology*, vol. 76, no. 8, pp. 2052–2072, 2014.

[27] S. A. Pedro, H. E. Tonnang, and S. Abelman, "Uncertainty and sensitivity analysis of a Rift Valley fever model," *Applied Mathematics and Computation*, vol. 279, pp. 170–186, 2016.

[28] S. A. Pedro, S. Abelman, and H. E. Z. Tonnang, "Predicting Rift Valley Fever Inter-epidemic Activities and Outbreak Patterns: Insights from a Stochastic Host-Vector Model," *PLOS Neglected Tropical Diseases*, vol. 10, no. 12, 2016.

[29] R. Ross, *Murray London*, 2nd edition, 1911.

[30] G. Macdonald, "The analysis of equilibrium in malaria," *Tropical Diseases Bulletin*, vol. 49, no. 9, pp. 813–829, 1952.

[31] R. M. Anderson, *Population Dynamics of Infectious Diseases: Theory and Applications*, vol. 368, 1982.

[32] N. A. Maidana and H. M. Yang, "Describing the geographic spread of dengue disease by traveling waves," *Mathematical Biosciences*, vol. 215, no. 1, pp. 64–77, 2008.

[33] H. R. Thieme, *Mathematics in Population Biology*, vol. 47, 2003.

[34] J. LaSalle, "The stability of dynamical systems," in *Proceedings of the CBMS-NSF Regional Conference Series in Applied Mathematics*, SIAM, Thailand, Philadelphia, USA, 1976.

[35] O. Diekmann and J. A. P. Heesterbeek, *Mathematical Epidemiology of Infectious Diseases: Model Building, Analysis and Interpretation*, John Wiley & Sons, 2000.

[36] P. van den Driessche and J. Watmough, "Reproduction numbers and sub-threshold endemic equilibria for compartmental models of disease transmission," *Mathematical Biosciences*, vol. 180, pp. 29–48, 2002.

[37] M. J. Keeling and P. Rohani, *Modeling Infectious Diseases in Humans and Animals*, vol. 47, 2007.

[38] M. G. Roberts and J. A. P. Heesterbeek, "A new method for estimating the effort required to control an infectious disease," *Proceedings of the Royal Society B Biological Science*, vol. 270, no. 1522, pp. 1359–1364, 2003.

[39] J. A. Heesterbeek and M. G. Roberts, "The type-reproduction number t in models for infectious disease control," *Mathematical Biosciences*, vol. 206, no. 1, pp. 3–10, 2007.

[40] J. C. Kamgang and G. Sallet, "Global asymptotic stability for the disease free equilibrium for epidemiological models," *Comptes Rendus Mathematique*, vol. 341, no. 7, pp. 433–438, 2005.

[41] C. Castillo-Chavez and B. Song, "Dynamical models of tuberculosis and their applications," *Mathematical Biosciences and Engineering*, vol. 1, no. 2, pp. 361–404, 2004.

[42] J. M. Heffernan, R. J. Smith, and L. M. Wahl, "Perspectives on the basic reproductive ratio," *Journal of the Royal Society Interface*, vol. 2, no. 4, pp. 281–293, 2005.

[43] A. L. Lloyd, J. Zhang, and A. M. Root, "Stochasticity and heterogeneity in host-vector models," *Journal of the Royal Society Interface*, vol. 4, no. 16, pp. 851–863, 2007.

4

A Stochastic TB Model for a Crowded Environment

Sibaliwe Maku Vyambwera ⓘ **and Peter Witbooi** ⓘ

Department of Mathematics and Applied Mathematics, University of the Western Cape, Private Bag X17, Bellville 7535, South Africa

Correspondence should be addressed to Peter Witbooi; pwitbooi@uwc.ac.za

Academic Editor: Zhidong Teng

We propose a stochastic compartmental model for the population dynamics of tuberculosis. The model is applicable to crowded environments such as for people in high density camps or in prisons. We start off with a known ordinary differential equation model, and we impose stochastic perturbation. We prove the existence and uniqueness of positive solutions of a stochastic model. We introduce an invariant generalizing the basic reproduction number and prove the stability of the disease-free equilibrium when it is below unity or slightly higher than unity and the perturbation is small. Our main theorem implies that the stochastic perturbation enhances stability of the disease-free equilibrium of the underlying deterministic model. Finally, we perform some simulations to illustrate the analytical findings and the utility of the model.

1. Introduction

Tuberculosis (TB) continues to be a major global health problem that is responsible for 1.5 million deaths worldwide each year [1]. TB is most prevalent in communities with socioeconomical problems but is not confined to such. The authors in [2, 3] associate TB infection with poverty and underdevelopment of some countries. It has been observed globally that one of the major factors driving TB infection is overcrowding. TB mostly occurs in poorest countries that are not developed and particularly where a population is overcrowded and in countries that are influenced by war. Conflict is the most common cause of large population displacement, which often results in relocation to temporary settlements such as camps. Factors including malnutrition and overcrowding in camp settings further increase the exposure to TB infection in these populations. Following up on a paper of Ssematimba et al. [3] regarding internally displaced people's camps in Uganda, Buonomo and Lacitignola [2] proposed a model that considers the dynamics of TB in concentration camps with a case study in Uganda. Another type of crowded environment which provides favourable conditions for TB to flourish is prisons and more so if the prison is full beyond its capacity. There are more than 10 million inmates in prisons all over the world. The United States of America is in the top rank with about 2.2 million

inmates while South Africa is in rank 11 [4]. South African prison has approximately 160000 inmates in custody, of which 120000 are sentenced individuals while the rest are awaiting trial. This means that a large number of inmates are kept in remand population and some of them might not be found guilty at the end of the process, after having been exposed to high risk of TB infection.

Mathematical models have been used to model TB by considering the size of the area and how size and density affect the extent to which TB can invade a certain population [2, 3, 5–7]. Quite obviously, considering the manner in which TB is aerially transmitted from one person to another, the prison situation provides favourable conditions for TB to flourish. TB is an infectious disease caused by bacillus Mycobacterium tuberculosis that most often affects the lungs (pulmonary TB) and can affect other parts as well such as brain, kidneys, and spine (extrapulmonary TB) [8, 9]. The TB infection can take place when an infected individual releases some droplet nuclei which can remain airborne in any indoor area for up to four hours. The tubercle bacillus can persist in a dark area for several hours but it is exceptionally sensitive to sunshine. The risk of infection increases as the length of prison stay increases and the sentenced offenders are more likely to get TB infection as compared to the awaiting trial inmates.

Against this background the paper [10] offers a model for the population dynamics of TB in a prison or prison system.

In particular, it computes the parameters relevant to South Africa for the given model, using publicly available data. The current paper considers a stochastic form of the model in [10]. It is well understood that stochastic differential equation (sde) attempts to reflect the effect of random disturbances in or on a system. A second reason for studying sde models is that it is good to know that a given model carries some resilience against small disturbances. In this case we consider the transmission parameters to be stochastically perturbed, similarly to [11]. Stochastic pertubation has been studied by Yang and Mao [12]; they considered a multigroup SEIR epidemic model. In most cases, it has been observed in [12, 13] that introducing a stochastic perturbation into an unstable disease-free equilibrium model system of ordinary differential equation may lead to a system being stable in sde. Stochastic differential equation models for various diseases have been studied and similar work has been done in [11, 12, 14–16].

Our paper focuses on the analysis of TB in prisons as prisons have been recognized as institutions with very high TB burden as compared to a general population [17]. For a deterministic model of similar type, in [10] we computed parameter values pertaining to South Africa. For the stochastic model in this paper the focus is on mathematical analysis. In Section 2, the model is introduced, based on the paper of Buonomo and Lacitignola [2]. The existence and uniqueness of the solution to the stochastic models is investigated by using the Lyapunov method in Section 3. Stability of the disease-free equilibrium for stochastic models is shown in Section 4. We show our results by means of numerical simulations and conclude in Section 5.

2. The Model

We introduce a stochastic compartmental model which is based on the deterministic model in the paper of Buonomo and Lacitignola [2]. We divide the population, which is of size $N(t)$ at time t, into four compartments, namely, the class $S(t)$ of susceptible individuals $S(t)$, the class $E(t)$ of individuals infected with TB who are not infectious, the class $I(t)$ of individuals infected with active TB who are infectious, and the class $T(t)$ of individuals under treatment. It is important to note that in general populations removal of individuals out of the system is only by death. In this model, as in [10], the removal is by death or by discharge from prison, and the discharge is the dominant factor. This rate of removal is denoted by μ. The disease induced mortality rate is denoted by δ. Individuals are recruited into the susceptible class $S(t)$ at a constant rate μA. Susceptible individuals get infected with active TB at a rate $c_1 SI$, where c_1 is the effective contact rate between the infectious and susceptible individuals. Individuals leave the exposed class $E(t)$ for infectious class $I(t)$ at rate kE. Exposed individuals who are infectious move to the infectious class $I(t)$ at a rate $c_3 EI$, where c_3 is the effective contact rate between the exposed and infectious individuals. Successfully treated individuals who were infectious move to exposed class at a rate $c_2 TI$, where c_2 is the effective contact rate between the treated and infectious individuals. Exposed

and infectious individuals move into class $T(t)$ at the rates r_1 and r_2, respectively.

Let us assume $(\Omega, \mathscr{F}, \{\mathscr{F}_t\}_{t \geq t_0}, \mathbb{P})$ to be a complete probability space with a filtration $\{\mathscr{F}_t\}_{t \geq t_0}$ which is right continuous. Let $W_i(t)$ $(i = 1, 2)$ be two mutually independent Brownian motions. Let us fix a nonnegative number σ, which shall serve as the intensity of the perturbation. We also fix two other positive numbers p and q with $p+q = 1$ that will balance the perturbation. The stochastic perturbations are similar to those in the model of [11].

Model System (1)

$$dS = \left[f_S \mu A - c_1 SI - \mu S \right] dt$$
$$- \sigma \left(pESdW_1(t) + qISdW_2(t) \right),$$
$$dE = \left[f_E \mu A + c_1 SI + c_2 TI - c_3 EI - (\mu + r_1 + k) E \right] dt$$
$$+ \sigma pESdW_1(t), \tag{1}$$
$$dI = \left[f_I \mu A + kE - (\mu + r_2 + \delta) I + c_3 EI \right] dt$$
$$+ \sigma qISdW_2(t),$$
$$dT = \left[r_1 E + r_2 I - c_2 TI - \mu T \right] dt.$$

It is noticed that if $f_E + f_I > 0$ then system (1) does not have a disease-free equilibrium. We will first investigate the model without the inflow of infected cases, i.e., when $f_E = f_I = 0$. In this case the disease-free state

$$E_0 = (S_0, E_0, I_0, T_0) = (A, 0, 0, 0) \tag{2}$$

is an equilibrium point. The underlying deterministic model of (1) is the model given by the same system of equations in the special case $\sigma = 0$, i.e., without stochastic perturbation as in [10]. The underlying deterministic model coincides with the model of Buonomo and Lacitignola [2]. The basic reproduction number of the underlying deterministic model has already been computed in paper [2] and is given by the following formula:

$$R_0 = \frac{kc_1 A}{\mu_1 \mu_2}, \tag{3}$$

where $\mu_1 = \mu + r_1 + k$ and $\mu_2 = \mu + r_2 + \delta$.

We now present the following set:

$$\Delta_A = \left\{ x \in \mathbb{R}^4 : x_1, x_2, x_3, x_4 > 0 \ \ x_1 + x_2 + x_3 + x_4 \right. \tag{4}$$
$$\left. \leq A \right\}.$$

Remark 1. For the rest of the paper we will assume that the sample paths are restricted to Ω_0, which is defined as follows:

$$\Omega_0 = \left\{ w \in \Omega \mid (S(t, w), E(t, w), I(t, w), T(t, w) \right.$$
$$\left. \in \Delta_A \quad \text{for all } t \geq 0 \right\}. \tag{5}$$

Lemma 2 (see [13]). *For $k \in \mathbb{N}$, let $X(t) = (X_1(t), X_2(t), \ldots, X_k(t))$ be a bounded $\mathbb{R}^k$-valued function and let $(t_{0,n})$ be* *any increasing unbounded sequence of positive real numbers. Then there is family of sequences $(t_{l,n})$ such that for each $l \in 1, 2, \dots, k$, $(t_{l,n})$ is a subsequence of $(t_{l-1,n})$ and the sequence $X_l(t_{l,n})$ converges to a chosen limit point of the sequence $X_l(t_{l-1,n})$.*

3. Existence and Uniqueness of Positive Global Solutions

Proposition 3. *Suppose that we have a solution*

$$X(t) = (S(t), E(t), I(t), T(t)) \qquad (6)$$

of system (1) over an interval $t \in [0, \tau)$ with $S(0) + E(0) + I(0) + T(0) < A$ and with $X(t) \in \mathbb{R}_{++}^4$ for all $0 \le t \le \tau$, a.s., then $S(t) + E(t) + I(t) + T(t) \le A$.

Proof. Given any solution in $X(t)$ satisfying the conditions of Proposition 3, then we have the total population in system (1) obeying the following ordinary differential equation:

$$\frac{d(N-A)}{dt} = -\mu(N-A) - \delta I \le -\mu(N-A) \quad \text{a.s.} \qquad (7)$$

Therefore, similarly to [11], for instance, $N(0) < A$ implies that $N(t) < A$ for all $t \in [0, \tau)$. □

In this section, we investigate the existence and uniqueness of global positive solutions of stochastic models by using the Lyapunov method. This method is popularly applied for such problems; see [23, 24], for instance.

Theorem 4. *There is a unique solution $(S(t), E(t), I(t), T(t)) \in \mathbb{R}_+^4$ to system (1) on $t \ge 0$ for any given initial value $(S(0), E(0), I(0), T(0)) \in \mathbb{R}_+^4$, and the solution will remain in $\mathbb{R}_+^4$ with probability one; namely, $(S(t), E(t), I(t), T(t)) \in \mathbb{R}_+^4$ for all $t \ge 0$ almost surely.*

Sketch of the proof. Since the coefficients in (1) satisfy the Lipschitz condition locally, for any given initial value $(S(0), E(0), I(0), T(0))$, there is a unique local solution $(S(t), E(t), I(t), T(t))$ on $t \in [0, \tau_{en})$, where τ_{en} is the explosion time. Our a im is to show that this solution is global and positive almost surely; i.e., $\tau_{en} = \infty$ a.s.

Let $r_0 > 0$ such that $S(0), E(0), I(0), T(0) > r_0$. For each integer $r \le r_0$, we define the stopping times

$$\tau_r = \inf \{ t \in [0, \tau_{en}] : S(t) \le r \text{ or } E(t) \le r \text{ or } I(t) \\ \le r \text{ or } T(t) \le r \}. \qquad (8)$$

Let

$$\tau = \lim_{r \to 0} \tau_r = \inf \{ t \in [0, \tau_{en}) : S(t) \le 0 \text{ or } E(t) \\ \le 0 \text{ or } I(t) \le 0 \text{ or } T(t) \le 0 \}. \qquad (9)$$

For this purpose we introduce a function V as follows:

$$V = \ln \frac{A}{S} + \ln \frac{A}{E} + \ln \frac{A}{I} + \ln \frac{A}{T}. \qquad (10)$$

We note that, by Proposition 3, each of the terms

$$\ln \frac{A}{S},$$
$$\ln \frac{A}{E},$$
$$\ln \frac{A}{I}, \qquad (11)$$
$$\ln \frac{A}{T}$$

is positive, and

$$\lim_{u \to 0^+} \frac{A}{u} = +\infty. \qquad (12)$$

By Itô's formula, for all $t \ge 0$, $s \in [0, t \wedge \tau_r]$, we have

$$\begin{aligned} dV(X(s)) = &-\frac{1}{S(s)} \Big(f_S \mu A - c_1 S(s) I(s) - \mu S(s) \\ &+ \frac{(\sigma p E(s))^2}{2} + \frac{(\sigma q I(s))^2}{2} \Big) ds - \frac{1}{E(s)} \Big(f_E \mu A \\ &+ c_1 S(s) I(s) + c_2 T(s) I(s) - c_3 E(s) I(s) \\ &- (\mu + r_1 + k) E(s) + \frac{(\sigma p S(s))^2}{2} \Big) ds \\ &- \frac{1}{I(s)} \Big(f_I \mu A + k E(s) - (\mu + r_2 + \delta) I(s) \\ &+ c_3 E(s) I(s) + \frac{(\sigma q S(s))^2}{2} \Big) ds - \frac{1}{T(s)} \big(r_1 E(s) \\ &+ r_2 I(s) - c_2 T(s) I(s) - \mu T(s) \big) ds + \sigma p \big(E(s) \\ &- S(s) \big) dW_1(s) + \sigma q \big(I(s) \\ &- S(s) \big) dW_2(s). \end{aligned} \qquad (13)$$

After eliminating some negative terms we have the following inequality:

$$dV(X(s)) \le M_1 ds + dM_2(s), \qquad (14)$$

where

$$\begin{aligned} M_1 = &4\mu + r_1 + r_2 + k + d + I(c_1 + c_2) + c_3(E + I) \\ &+ \frac{\sigma^2}{2}(p^2 E^2 + q^2 I^2) + \frac{1}{2}(\sigma(p+q)S)^2, \end{aligned} \qquad (15)$$

and

$$dM_2(s) = \sigma p(E - S) dW_1(s) + \sigma q(I - S) dW_2(s). \qquad (16)$$

Taking the integral in (14) from 0 to $t \wedge \tau_{r_0}$, we have

$$\int_0^{t \wedge \tau_r} dV(X(s)) \le \int_0^{t \wedge \tau_r} M_1 ds + \int_0^{t \wedge \tau_r} dM_2(s). \qquad (17)$$

By taking expectations, the latter inequality yields

$$\mathbb{E}\left[V\left(S\left(t \wedge \tau_r\right), E\left(t \wedge \tau_r\right), I\left(t \wedge \tau_r\right), T\left(t \wedge \tau_r\right)\right)\right]$$
$$\leq V\left(X\left(0\right)\right) + M_1 t. \tag{18}$$

Now we note that

$$\mathbb{E}V\left[S\left(t \wedge \tau_r\right), E\left(t \wedge \tau_r\right), I\left(t \wedge \tau_r\right), T\left(t \wedge \tau_r\right)\right]$$
$$= \mathbb{E}\left[\Psi_{(\tau_r \leq t)}V\left(S\left(t \wedge \tau_r\right), E\left(t \wedge \tau_r\right), I\left(t \wedge \tau_r\right),\right.\right.$$
$$\left.\left. T\left(t \wedge \tau_r\right)\right)\right] + \mathbb{E}\left[\Psi_{(\tau_r > t)}V\left(S\left(t \wedge \tau_r\right), E\left(t \wedge \tau_r\right),\right.\right. \tag{19}$$
$$\left.\left. I\left(t \wedge \tau_r\right), T\left(t \wedge \tau_r\right)\right)\right] \geq \mathbb{E}\left[\Psi_{(\tau_r \leq t)}V\left(S\left(\tau_r\right), E\left(\tau_r\right),\right.\right.$$
$$\left.\left. I\left(\tau_r\right), T\left(\tau_r\right)\right)\right],$$

where $\Psi_{(.)}$ is the indicator function. If $\tau_r < \infty$, then there are some components of $S(\tau_r), E(\tau_r), I(\tau_r), T(\tau_r)$ equal to r, and therefore $(S(\tau_r), E(\tau_r), I(\tau_r), T(\tau_r)) \geq \ln(A/r)$.
Thus we have

$$\mathbb{E}\left[V\left(S\left(t \wedge \tau_r\right), E\left(t \wedge \tau_r\right), I\left(t \wedge \tau_r\right), T\left(t \wedge \tau_r\right)\right)\right]$$
$$\geq \ln\left(\frac{A}{r}\right)\mathbb{P}\left(\tau_r \leq t\right). \tag{20}$$

Combining (14) and (18) gives, for all $t \geq 0$,

$$\mathbb{P}\left(\tau \leq t\right) \leq \frac{V\left(X\left(0\right)\right) + M_1 t}{\ln\left(A/r\right)} \tag{21}$$

Letting $r \to 0$, we obtain, for all $t \geq 0$, $\mathbb{P}(\tau \leq t) = 0$. Hence $\mathbb{P}(\tau = \infty) = 1$. As $\tau_{en} = \tau = \infty$ a.s. Therefore, the solution of model (1) will not explode at a finite time with probability one. This completes the proof. $\square$

4. Stability of Disease-Free Equilibrium

Let us choose a positive number a_3 and two nonnegative numbers a_1 and a_2. Specific values will be assigned to these numbers in different analyses.
Let us assume that

$$a_3 \geq \frac{k}{\mu_1}. \tag{22}$$

Now we define a stochastic process $Z(X(t))$

$$Z\left(X\left(t\right)\right) = a_1\left(A - S\left(t\right)\right) + a_2 T\left(t\right) + a_3 E\left(t\right) + I\left(t\right) \tag{23}$$

and a process

$$V\left(X\left(t\right)\right) = \ln Z\left(X\left(t\right)\right). \tag{24}$$

For $w \in \Omega_0$, we note that $Z(X(t)) > 0$ and therefore $V(X(t))$ are defined for all $w \in \Omega_0$. For convenience, we introduce the variables:

$$Q_Z = \frac{A - S}{Z},$$
$$T_Z = \frac{T}{Z},$$
$$E_Z = \frac{E}{Z}, \tag{25}$$
$$I_Z = \frac{I}{Z}$$

and for a stochastic process $x(t)$ we shall write

$$\langle x \rangle_s = \frac{1}{s}\int_0^s x\left(u\right) du. \tag{26}$$

4.1. On the Lyapunov Exponent of Z. The Lyapunov exponent of a quantity $q(t), t \geq 0$ is defined as

$$\limsup_{t \to \infty} \frac{1}{t}\ln q\left(t\right). \tag{27}$$

The infinitesimal generator $\mathscr{L}$ of system (1) (see Øksendal [25]) will play an important role in the sequel. Now we can calculate $\mathscr{L}V$ and express it as a function of $X(t)$. From Lemma 2 it follows that for each $w \in \Omega_0$ there is an increasing sequence (t_n^w) with the following properties (but we shall suppress w and write (t_n)):
For every $w \in \Omega$,

$$\lim_{n \to \infty} \langle \mathscr{L}V\left(X\right)\rangle_{t_n} = \limsup_{t \to \infty} \langle \mathscr{L}V\left(X\right)\rangle_t \tag{28}$$

and the limits below, which shall be denoted by q, τ, j, i, do exist:

$$q = \lim_{n \to \infty} \langle Q_Z \rangle_{t_n},$$
$$\tau = \lim_{n \to \infty} \langle T_Z \rangle_{t_n},$$
$$j = \lim_{n \to \infty} \langle E_Z \rangle_{t_n}, \tag{29}$$
$$i = \lim_{n \to \infty} \langle I_Z \rangle_{t_n}.$$

We write

$$\Lambda = \limsup_{t \to \infty} \langle \mathscr{L}V\left(X\right)\rangle_t. \tag{30}$$

Let

$$c_* = \max\left\{c_1, c_2, c_3\left(\frac{\mu}{k} - 1\right)\right\}. \tag{31}$$

We can write

$$\int_0^t dV = \int_0^t \mathscr{L}Vdt + M\left(t\right), \tag{32}$$

where

$$M(t) = \int_0^t \frac{1}{Z} \sigma p (E - S) \, dW_1 + \int_0^t \frac{1}{Z} \sigma q (I - S) \, dW_2, \quad (33)$$

and we note that by the strong law of large numbers [16],

$$\lim_{n \to \infty} \frac{1}{t} M(t) = 0 \quad \text{a.s.} \quad (34)$$

Therefore

$$\limsup_{t \to \infty} \frac{1}{t} V(X(t))$$

$$= \limsup_{t \to \infty} \frac{1}{t} \int_0^t \mathscr{L}V(X(s)) \, ds \,(\text{a.s.}) \quad (35)$$

$$= \lim_{n \to \infty} \frac{1}{t_n} \int_0^{t_n} \mathscr{L}V(X(s)) \, ds \,(\text{a.s.}).$$

Now we expand $\mathscr{L}V$:

$$\mathscr{L}V = \frac{-a_1}{Z} [\mu A - c_1 SI - \mu S]$$

$$- \frac{a_1^2 \sigma^2}{2Z^2} \left(p^2 E^2 S^2 + q^2 I^2 S^2 \right)$$

$$+ \frac{a_2}{Z} [r_1 E + r_2 I - c_2 TI - \mu T]$$

$$+ \frac{a_3}{Z} [c_1 SI + c_2 TI - c_3 EI - (\mu + r_1 + k) E]$$

$$\quad (36)$$

$$- \frac{a_3^2}{2Z^2} \sigma^2 \left(p^2 E^2 S^2 \right)$$

$$+ \frac{1}{Z} [kE - (\mu + r_2 + \delta) I + c_3 EI]$$

$$- \frac{1}{2Z^2} \left(\sigma^2 q^2 I^2 S^2 \right) - a_1 a_3 (\sigma p ES)^2$$

$$- a_1 (\sigma q IS)^2.$$

With regard to the calculation of $\mathscr{L}V$ we note the following:

$$a_3 I_Z \{c_1 S + c_2 T - c_3 E\} + c_3 I_Z E$$

$$= a_3 I_Z \left\{ c_1 S + c_2 T + c_3 \left(\frac{1}{a_3} - 1 \right) E \right\} \quad (37)$$

$$\leq a_3 I_Z c_* (S + T + E) \leq a_3 I_Z c_* A.$$

Therefore,

$$\mathscr{L}V \leq a_3 I_Z c_* A - I_Z (\mu_2 - a_2 r_2)$$

$$+ E_Z (a_2 r_1 - a_3 \mu_1 + k) - a_2 \mu T_Z \quad (38)$$

$$+ I_Z (a_1 c_1 S - a_2 c_2 T) - a_1 \mu Q + B,$$

where

$$B = -\frac{(a_1 \sigma)^2}{2} \left[(pE_Z S)^2 + (qI_Z S)^2 \right] - \frac{a_3^2}{2} \left[(\sigma p E_Z S)^2 \right]$$

$$- \frac{1}{2} \left[(\sigma q I_Z S)^2 \right] - a_1 a_3 (\sigma p E_Z S)^2 - a_1 (\sigma q I_Z S)^2. \quad (39)$$

This yields the inequality:

$$\mathscr{L}V \leq I_Z ((a_1 c_1 + a_3 c_*) A - \mu_2 + a_2 r_2)$$

$$+ E_Z (k - a_3 \mu_1 + a_2 r_1) - a_2 \mu_2 T_Z - a_1 \mu Q_Z \quad (40)$$

$$+ B.$$

In the expression for B, if we ignore the multiples of a_1 (they are negative), then we obtain an inequality:

$$B \leq -\frac{(\sigma S)^2}{2} \left\{ (pa_3 E_Z)^2 + (qI_Z)^2 \right\}. \quad (41)$$

4.2. *Stability Theorems.* We now introduce another invariant R_σ, which enables us to formulate stability theorems for the stochastic model (1). As a corollary of the main theorem we can deduce a global stability theorem for disease-free equilibrium. Let

$$R_\sigma = \frac{kc_* A}{\mu_1 \mu_2}. \quad (42)$$

In the model of Buonomo and Lacitignola [2], we have backward bifurcation at $R_0 = 1$. Therefore, the condition $R_0 < 1$ does not imply global stability of the underlying deterministic model. As a corollary to the main theorem, Theorem 6, will follow the fact that for the model in [2] the disease-free equilibrium is globally asymptotically stable when $R_\sigma < 1$. In preparation for our main theorem we introduce a function $h(x)$ as follows:

$$h(x) = \frac{p^2 (1 - x)^2 + q^2 x^2}{x}; \quad x > 0. \quad (43)$$

Then

$$\lim_{x \to \infty} h(x) = \infty \text{ and if } q \neq 0, \text{ then } \lim_{x \to 0^+} h(x) = \infty. \quad (44)$$

Also we note that

$$h'(x) = \frac{1}{x^2} \left[-p^2 + x^2 \right]. \quad (45)$$

Therefore $h'(x) = 0 \Leftrightarrow x = p$ and we know that $p \leq 1$. Since h has only one critical value on the interval $(0, \infty)$, in view of (44), it follows that the critical point is an absolute minimum of h on the interval $(0, \infty)$.

Therefore the minimum value $h_{\min}$ of h over $[0, 1]$ is

$$h_{\min} = \frac{p^2 (1 - p) + q^2 p}{p} = p(1 - p) + \left(1 - p^2 \right) \quad (46)$$

$$= (1 - p)(p + 1 + p) = (1 - p)(1 + 2p).$$

Proposition 5. *If*

$$R_\sigma - \frac{(\sigma A)^2 h_{\min}}{2\mu_2} < 1, \quad (47)$$

then (I, E) *converges exponentially to zero almost surely.*

Proof. We introduce the function V of (24), with $a_1 = a_2 = 0$. Now note that (47) is equivalent to

$$\frac{kc_* A}{\mu_1} - \frac{(\sigma A)^2 h_{\min}}{2} - \mu_2 < 0. \tag{48}$$

We choose a number $\epsilon > 0$ sufficiently small such that

$$\frac{k + \epsilon}{\mu_1} c_* A - \mu_2 - \frac{(\sigma A)^2}{2} h_{\min} < 0. \tag{49}$$

Now we choose

$$a_3 = \frac{k + \epsilon}{\mu_1}. \tag{50}$$

From inequality (40) it follows that

$$\mathscr{L}V \le [a_3 c_* A - \mu_2] I_Z + [k - a_3 \mu_1] E_Z - B_1, \tag{51}$$

where

$$B_1 = \frac{(\sigma A)^2}{2} \left\{ p^2 (a_3 E_Z)^2 + (qI_Z)^2 \right\}. \tag{52}$$

Now note that we can express B_1 as follows:

$$\begin{aligned} B_1 &= \frac{(\sigma A)^2}{2} \left\{ p^2 (1 - I_Z)^2 + (qI_Z)^2 \right\} \\ &= \frac{(\sigma A)^2}{2} I_Z h(I_Z). \end{aligned} \tag{53}$$

Therefore, we have

$$B_1 \ge \frac{(\sigma A)^2}{2} I_Z h_{\min}, \tag{54}$$

and, consequently,

$$\begin{aligned} \mathscr{L}V \le & \left[a_3 c_* A - \mu_2 - \frac{(\sigma A)^2}{2} h_{\min} \right] I_Z \\ & + [k - a_3 \mu_1] E_Z. \end{aligned} \tag{55}$$

Therefore

$$\Lambda \le \left[a_3 c_* A - \mu_2 - \frac{(\sigma A)^2}{2} h_{\min} \right] i + \epsilon j \tag{56}$$

and since i and j cannot both be zero, it follows that $\Lambda < 0$. This completes the proof. □

Theorem 6. *(a) If $(E(t), I(t))$ almost surely converges exponentially to 0, then*

$$\lim_{t \to \infty} S(t) = A \quad (a.s.) \text{ and } \lim_{t \to \infty} T(t) = 0 \quad (a.s.). \tag{57}$$

(b) If

$$R_\sigma - \frac{(\sigma A)^2 h_{\min}}{2\mu_2} < 1, \tag{58}$$

then disease-free equilibrium is almost surely exponentially stable.

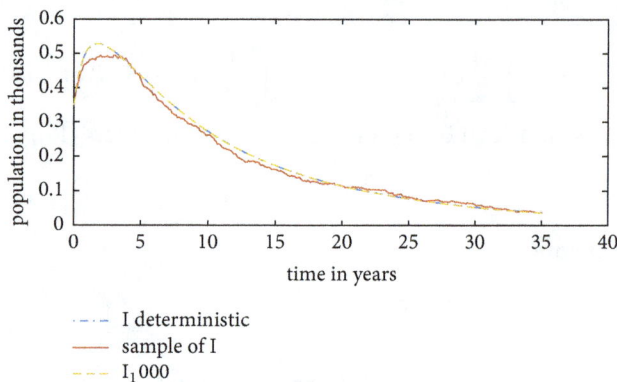

FIGURE 1: $R_0 = 1.3917$, $R_\sigma = 1.1653$, $c_1 = 0.000065$, and $\sigma = 0.04$.

Proof. (a) Suppose to the contrary that we have

$$\lim_{t \to \infty} (A - S(t)) + T(t) > 0 \quad (a.s.). \tag{59}$$

Let Z be the same as that in (23), with $a_1 = a_2 = a_3 = 1$. Since $(E(t), I(t))$ almost surely converges exponentially to 0 while

$$\lim_{t \to \infty} (A - S(t)) + T(t) > 0 \quad (a.s.), \tag{60}$$

it follows that $j = 0$ and $i = 0$ (a.s.). Thus from inequality (40) it follows that

$$\Lambda \le -\mu_2 T_Z - \mu Q_Z \quad (a.s.). \tag{61}$$

Therefore $\Lambda < 0$. This implies that Z converges to 0, and thus

$$\lim_{t \to \infty} (A - S(t)) + T(t) = 0 \quad (a.s.), \tag{62}$$

which is a contradiction. This completes the proof of (a).

(b) This follows from Proposition 5 and Theorem 6(a). □

5. Numerical Simulation

The simulations presented here illustrate the analytical results of our model in (1). The parameter values have already been calculated in the paper [10], by using real data, mostly from [18, 20, 21]. We will now use those parameter values, listed in Table 1, and vary the value of c_1 and σ in order for us to be able to find different values of R_0 and R_σ. We first consider a model without the inflow of infective cases and then with the inflow of infective cases.

We give some numerical simulations to show different dynamic outcomes of the deterministic model and its stochastic version. We illustrate by means of simulations the possible disease eradication in the absence of the inflow of infective cases. This will be shown in Figures 1, 2, and 3. Over these three cases we vary the value of c_1 and σ so as to obtain different values of R_0 and R_σ.

In Figure 1, we present a case in which we take $c_1 = 0.000065$, $\sigma = 0.04$ and then we obtain $R_0 = 1.3917$ and $R_\sigma = 1.1653$. This situation does not satisfy the conditions of Theorem 6, and indeed the I-class does not appear to

TABLE 1: Model parameters and initial conditions.

Parameter	Estimated value	Source
μ	0.18192	[10], data from [18, 19]
d	0.01876	[10], data from [18, 20]
c_1	0.00007893	[10], see also [21]
c_2	$20/A$	[10, 22]
c_3	$k(2A)$	Estimated from [10]
r_1	0.30	[2]
r_2	0.50	[2]
k	0.05	[2, 21]
A	160000	[18]
f_S, f_E, f_I	0.2, 0.74, 0.06	[18]
$S_{t_{15}}$	32000	[10], data from [18]
$E_{t_{15}}$	107000	[10], data from [18]
$I_{t_{15}}$	3500	[10], data from [18]
$T_{t_{15}}$	17100	[10], data from [18]

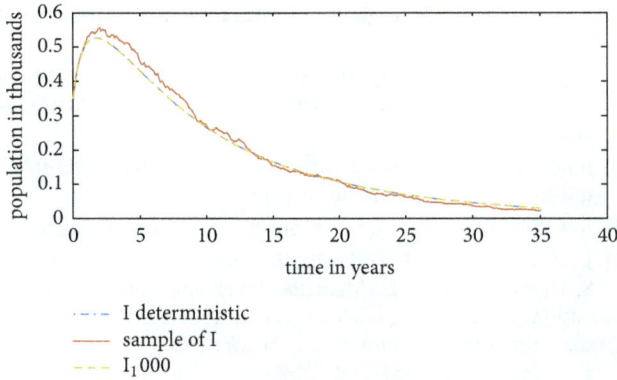

FIGURE 2: $R_0 = 1.3275$, $R_\sigma = 0.9737$, and $c_1 = 0.000062$, $\sigma = 0.05$.

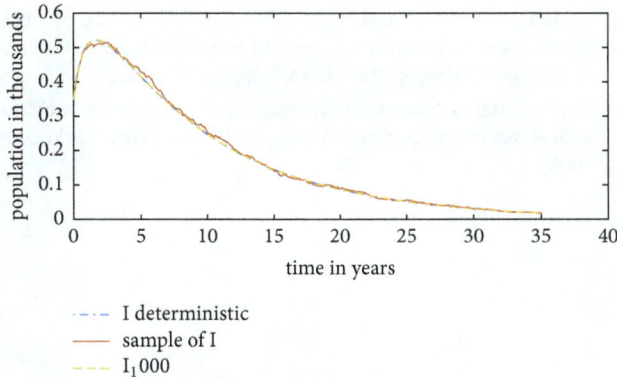

FIGURE 3: $R_0 = 1.1562$, $R_\sigma = 0.0.9298$, and $c_1 = 0.000054$, $\sigma = 0.04$.

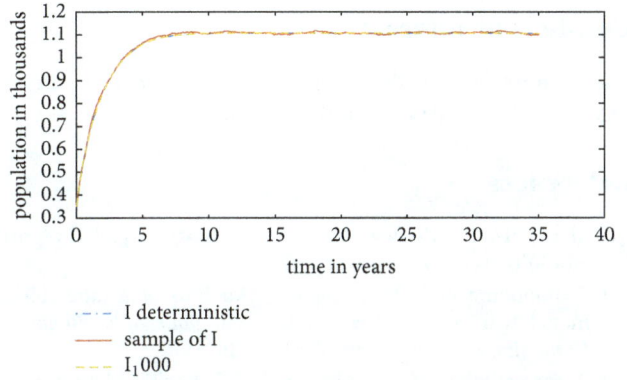

FIGURE 4: $R_0 = 1.6900$, $R_\sigma = 1.4635$, and $c_1 = 0.00007893$, $\sigma = 0.04$ with $f_S = 0.2$, $f_E = 0.74$, and $f_I = 0.06$.

In Figure 3, a choice of $c_1 = 0.000054$ and $\sigma = 0.04$ yields $R_0 = 1.1562$ and $R_\sigma = 0.9298$. This choice of parameters satisfies the conditions in Theorem 6, and surely the infectious class seems to converge to zero.

We now study model (1) with the inflow of infectives and present a sample computation. We choose $c_1 = 0.00007893$ as in Table 1 and $\sigma = 0.04$. Then the values of R_0 and R_σ can be calculated as $R_0 = 1.6900$, $R_\sigma = 1.4635$. In Figure 4, it is observed that when the basic reproduction number for the underlying deterministic model is above unity, then the disease will persist into our prison system. It is also seen that the inflow of infective cases play a part in influencing the number of TB infected cases in the prison system.

6. Conclusion

A stochastic SEIT model was presented and analysed to assess the impact of active TB on a crowded environment, specifically in prisons. We started off by verifying that there is a unique global positive solution for the system of stochastic differential equation in (1). It was noted that whenever the basic reproduction number is significantly greater than

converge to zero. This means that the disease will persist in our prison population.

In Figure 2, we notice that when the perturbation is sufficiently big, then the disease will possibly be eliminated for a stochastic model even if for the deterministic model it does not seem be the case. We have chosen $c_1 = 0.000062$, $\sigma = 0.05$ and then we calculate $R_0 = 1.3275$ and $R_\sigma = 0.9737$.

unity then the disease will persist in the prison population through our simulations in Figures 1 and 2. It has also been observed for a stochastic model that when the perturbation is sufficiently big then the disease tends to vanish and this can be seen in Figure 2. It is more important to study smaller perturbation. It has been observed that whenever $R_\sigma < 1$, then I and E almost surely converge exponentially to zero in step with Theorem 6, in the absence of the inflow of infective. These results can also be seen in Figure 3. By introducing the inflow of infective cases into the prison system, TB remains endemic, as can be seen in Figure 4. By screening the inflow on admission and providing for them a separate accommodation, TB infection in a prison system can be greatly reduced.

Disclosure

This research did not require or request any funding, and no funding was received.

Conflicts of Interest

The authors declare that there are no conflicts of interest regarding the publication of this paper.

References

[1] May 2017, http://www.who.int/tb/publications/global-report/gtbr2016-executive-summary.pdf.

[2] B. Buonomo and D. Lacitignola, "Analysis of a tuberculosis model with a case study in Uganda," *Journal of Biological Dynamics*, vol. 4, no. 6, pp. 571–593, 2010.

[3] A. Ssematimba, J. Y. Mugisha, and L. S. Luboobi, "Mathematical models for the dynamics of tuberculosis in density-dependent populations: the case of internally displaced peoples' camps (IDPCs) in Uganda," *Journal of Mathematics and Statistics*, vol. 1, no. 3, pp. 217–224, 2005.

[4] May 2017, http://www.prisonstudies.org/highest-to-lowest/prison-population-total?field-region-taxonomy-tid=All.

[5] C. Castillo-Chavez and Z. Feng, "To treat or not to treat: the case of tuberculosis," *Journal of Mathematical Biology*, vol. 35, no. 6, pp. 629–656, 1997.

[6] J. P. Aparicio, A. F. Capurro, and C. Castillo-Chavez, "Transmission and dynamics of tuberculosis on generalized households," *Journal of Theoretical Biology*, vol. 206, no. 3, pp. 327–341, 2000.

[7] N. Bacaër, R. Ouifki, C. Pretorius, R. Wood, and B. Williams, "Modeling the joint epidemics of TB and HIV in a South African township," *Journal of Mathematical Biology*, vol. 57, no. 4, pp. 557–593, 2008.

[8] April 2016, http://www.who.int/mediacentre/factsheets/fs104/en/.

[9] Z. Feng, C. Castillo-Chavez, and A. Capurro, "A model for tuberculosis with exogeneous reinfection," *Theoretical Population Biology*, vol. 57, pp. 235–247, 2000.

[10] P. Witbooi and S. M. Vyambwera, "A model of population dynamics of TB in a prison system and application to South Africa," *BMC Research Notes*, vol. 10, no. 1, article no. 643, 2017.

[11] P. J. Witbooi, "An SEIRS epidemic model with stochastic transmission," *Advances in Difference Equations*, Paper No. 109, 16 pages, 2017.

[12] Q. Yang and X. Mao, "Extinction and recurrence of multi-group SEIR epidemic models with stochastic perturbations," *Nonlinear Analysis: Real World Applications*, vol. 14, no. 3, pp. 1434–1456, 2013.

[13] P. J. Witbooi, G. E. Muller, and G. J. Van Schalkwyk, "Vaccination control in a stochastic SVIR epidemic model," *Computational and Mathematical Methods in Medicine*, Article ID 271654, Art. ID 271654, 9 pages, 2015.

[14] P. J. Witbooi, "Stability of an SEIR epidemic model with independent stochastic perturbations," *Physica A: Statistical Mechanics and its Applications*, vol. 392, no. 20, pp. 4928–4936, 2013.

[15] Z. Liu, "Dynamics of positive solutions to SIR and SEIR epidemic models with saturated incidence rates," *Nonlinear Analysis: Real World Applications*, vol. 14, no. 3, pp. 1286–1299, 2013.

[16] A. Lahrouz, L. Omari, and D. Kiouach, "Global analysis of a deterministic and stochastic nonlinear SIRS epidemic model," *Nonlinear Analysis: Modelling and Control*, vol. 16, no. 1, pp. 59–76, 2011.

[17] April 2016, http://www.prisonstudies.org/country/southafrica.

[18] June 2016, http://judicialinsp.dcs.gov.za/Annualreports/annual-report.asp.

[19] Regan Jules-Macquet, *The State of South African Prisons*, NICRO Public Education Series, Pretoria, South Africa, 1st edition edition, 2014.

[20] June 2016, http://www.dailymaverick.co.za/article/2013-07-01-the-losing-battle-against-tb-in-prisons/.

[21] July 2016, https://www.tbfacts.org/tb-statistics-south-africa/.

[22] J. V. Ershova, L. J. Podewils, E. Bronner, H. G. Stockwell, S. Dlamini, and L. D. Mametja, "Evaluation of adherence to national treatment guidelines among tuberculosis patients in three provinces of South Africa," *South African Medical Journal*, vol. 104, no. 5, pp. 362–368, 2014.

[23] L. Wang, H. Huang, A. Xu, and W. Wang, "Stochastic extinction in an SIRS epidemic model incorporating media coverage," *Abstract and Applied Analysis*, vol. 2013, Article ID 891765, 2013.

[24] D. Jiang, J. Yu, C. Ji, and N. Shi, "Asymptotic behavior of global positive solution to a stochastic SIR model," *Mathematical and Computer Modelling*, vol. 54, no. 1-2, pp. 221–232, 2011.

[25] B. Øksendal, *Stochastic Differential Equations, An Introduction with Applications*, Springer-Verlag, Heidelberg New York, USA, 2000.

5

Modeling the Effects of Spatial Heterogeneity and Seasonality on Guinea Worm Disease Transmission

Anthony A. E. Losio[1,2] **and Steady Mushayabasa** ⓘ [1]

[1]*University of Zimbabwe, Department of Mathematics, P.O. Box MP 167, Harare, Zimbabwe*
[2]*University of Juba, Department of Mathematics, P.O. Box 82, Juba, Central Equatoria, South Sudan*

Correspondence should be addressed to Steady Mushayabasa; steadymushaya@gmail.com

Academic Editor: Keshlan S. Govinder

Guinea worm disease is one of the neglected tropical diseases that is on the verge of elimination. Currently the disease is endemic in four countries, namely, Ethiopia, Mali, Chad, and South Sudan. Prior studies have demonstrated that climate factors and limited access to safe drinking water have a significant impact on transmission and control of Guinea worm disease. In this paper, we present a new mathematical model to understand the transmission dynamics of Guinea worm disease in South Sudan. The model incorporates seasonal variations, educational campaigns, and spatial heterogeneity. Both qualitative and quantitative analysis of the model have been carried out. Utilizing Guinea worm disease surveillance data of South Sudan (2007-2013) we estimate the model parameters. Meanwhile, we perform an optimal control study to evaluate the implications of vector control on long-term Guinea worm infection dynamics. Our results demonstrate that vector control could play a significant role on Guinea worm disease eradication.

1. Introduction

Guinea worm disease, also known as Dracunculiasis, is one of the neglected tropical diseases that is on the verge of elimination. The disease is transmitted to human via drinking contaminated water. Since the inception of the Guinea worm eradication program in the 1980s, the number of reported cases reduced from 3.5 million in 20 countries in 1986 to only 22 cases in 2015 from only 4 countries, namely, South Sudan, Chad, Mali, and Ethiopia [1]. Precisely, 99% of these cases were from South Sudan [2]. The disease has neither a vaccine to prevent nor medication to treat it. Despite being rarely fatal, individuals infected with the disease become nonfunctional for about 8.5 weeks [3].

Effective control of the disease can be achieved through provision of safe drinking water, behavioral changes in patients, and communities and vector control. Although access to safe drinking water alone is known to be extremely integral on Guinea worm disease eradication [1], its provision remains a significant challenge in Guinea worm disease endemic countries. Recently the world health organization reported that only 60% of the residence of one village in

Ethiopia had access to safe drinking water [4]. Limited access to safe drinking water was also noted to be a strong factor on the spread of Guinea worm disease in South Sudan [1].

Such heterogeneity in individuals' degree of susceptibility to infection has a strong impact on the transmission and control of Guinea worm disease. So far this aspect of heterogeneity has not been taken into account, leading to inadequate understanding of the influence of the spatial factors in the transmission and spread of Guinea worm disease. Another limitation in Guinea worm disease modeling is that the effects of seasonal variation has been inadequately addressed. In the Sahelian zone, transmission of Guinea worm disease has been observed to occur in the rainy season (May to August) while in the humid and forest zone, the peak occurs in the dry season (September to January) [5]. Such seasonal variations need to be incorporated in models that aim to inform policy makers effective and efficient ways to attain Guinea worm disease eradication.

Mathematical models have become essential tools in mapping the spread and control of infectious diseases. By setting up a suitable epidemic model, we can improve our qualitative and quantitative understanding on the effects of preventative and control measures to aid the elimination of the disease. So far, few mathematical models have been proposed to understand the spread and control of Guinea worm disease (see, for example, [6–9]). Smith et al. [6] proposed a deterministic model for Guinea worm disease transmission and explored the role of disease preventative measures, namely, education, filtration, and chlorination. Utilizing the Latin Hypercube Sampling technique Smith and coworkers established that education is more effective toward the eradication of the disease. This work and the other aforementioned studies have undeniably produced many significant insights and improved our existing knowledge on Guinea worm disease dynamics.

In this article, we propose a new nonautonomous model to explore the transmission and control of Guinea worm disease in a heterogeneous population of South Sudan. While our model takes a clue from the one by Smith et al. [6], in this article we have gone a step further by incorporating spatial heterogeneity and seasonal variations into a single framework for a comprehensive modeling of Guinea worm disease dynamics. To that end, the proposed model subdivides the susceptible human population into two categories, namely, the high risk and low risk. In addition, all epidemiological stages of the disease that are strongly influenced by seasonal variations have been modeled as periodic functions.

We organize the remainder of the paper as follows. In the next section we present the methods and results of the study. In particular, we will formulate the model, compute the basic reproduction number, analyze the stability of the steady states, fit the model with Guinea worm disease data of South Sudan (2007-2013), and perform an optimal control study to explore the effects of vector control on Guinea worm disease eradication. Finally, we conclude the paper with some discussion in Section 3.

2. Methods and Results

2.1. Model Framework. The model proposed in [6] subdivides the host population into categories of susceptible $S(t)$, exposed $E(t)$, infected individuals displaying clinical signs of the disease $I(t)$, and the concentration of the bacteria in the environment $W(t)$. We now extend this model by incorporating seasonal variations and variation in susceptibility to infection:

(i) **Seasonal variations**: to account for seasonal variations on Guinea worm disease dynamics we modeled the contact rate, incubation rate, pathogen shedding rate, and pathogen decay rate by the following periodic functions, respectively:

$$\beta(t) = \beta_0 \left[1 + \sin\left(2\pi t \omega^{-1}\right)\right],$$
$$\gamma(t) = \gamma_0 \left[1 + \sin\left(2\pi t \omega^{-1}\right)\right],$$
$$\alpha(t) = \alpha_0 \left[1 + \sin\left(2\pi t \omega^{-1}\right)\right], \quad (1)$$
$$\delta(t) = \delta_0 \left[1 + \sin\left(2\pi t \omega^{-1}\right)\right],$$

where $\beta_0, \gamma_0, \alpha_0,$ and δ_0 denote the contact rate, incubation rate, pathogen shedding rate, and pathogen

decay rate, respectively, without seasonal forcing and $\omega > 0$ is the period.

(ii) **Variation in susceptibility**: in order to account for heterogeneity in our study, we subdivided the susceptible population into high and low risk individuals. High risk refers to susceptible individuals who have limited access to safe drinking water. As in [13, 14] we assume that low risk susceptible individuals have low chances of acquiring the infection.

Keeping the above facts in mind, the dynamics of Guinea worm disease in this study are governed by the following nonautonomous system:

$$\dot{S}_1(t) = A\pi_1 - \beta(t) W(t) S_1(t) - (\mu + a_1) S_1(t)$$
$$\qquad + a_2 S_2(t) + p\kappa\sigma I(t),$$
$$\dot{S}_2(t) = A\pi_2 - \beta(t) W(t) (1 - \theta) S_2(t)$$
$$\qquad - (\mu + a_2) S_2(t) + a_1 S_1(t)$$
$$\qquad + p\kappa(1 - \sigma) I(t),$$
$$\dot{E}(t) = \beta(t) W(t) [S_1(t) + (1 - \theta) S_2(t)] \qquad (2)$$
$$\qquad - (\gamma(t) + \mu) E(t),$$
$$\dot{I}(t) = \gamma(t) E(t) - (\mu + \kappa) I(t),$$
$$\dot{R}(t) = (1 - p)\kappa I(t) - \mu R(t),$$
$$\dot{W}(t) = \alpha(t) I(t) - \delta(t) W(t).$$

All model variables and parameters are considered positive. Further the variables $S_1(t), S_2(t), E(t), I(t),$ and $R(t)$ represent the susceptible high risk, susceptible low risk, exposed, infectious, and recovered individuals at time t, respectively, such that the total human population is given by $N(t) = S_1(t) + S_2(t) + E(t) + I(t) + R(t)$. Meanwhile $W(t)$ represents the population of the pathogen in the environment. Model parameters defined in (1) retain the same definitions.

Model parameter A is the recruitment rate, π_1 and π_2 represent the fraction of new recruits into classes S_1 and S_2, respectively, μ is the natural mortality rate, a_1 is the transfer from susceptible high risk to susceptible low risk, a_2 is the transfer from susceptible low risk to susceptible high risk, θ is a modification factor that accounts for the impact of access of safe drinking water on disease transmission, and κ^{-1} is the average infectious period.

In addition, p is the proportion of recovered individuals who become susceptible to infection again based on their behavior and the complementary part $(1 - p)$ represent recovered individuals who have extremely minimal chances of reinfection. People with limited access safe drinking water can minimize chances of infection by using a cloth filter or a pipe filter, to remove the copepods. Thus we assume that the pain and experience of recovered individuals have an impact on one's behavior which in turn lead to one being reinfected or not. We further assume that a fraction σ of recovered individuals who move become susceptible to infection join $S_1(t)$ and the complementary $(1 - \sigma)$ join $S_2(t)$.

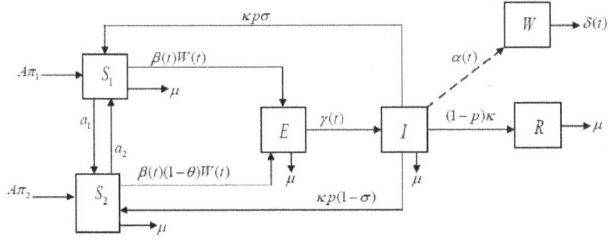

FIGURE 1: Diagram of the model structure.

Figure 1 shows a flowchart depicting the dynamics of Guinea worm disease in this study.

2.2. The Reproduction Number. Now we explore the dynamical behavior of Guinea worm disease. Since the variable $R(t)$ described by differential equation $\dot{R}(t) = (1 - p)\kappa I(t) - \mu R(t)$ does not appear in all the other five equations of system (2) it is sufficient to study the dynamics of Guinea worm disease from system:

$$\dot{S}_1(t) = A\pi_1 - \beta(t) W(t) S_1(t) - (\mu + a_1) S_1(t)$$
$$+ a_2 S_2(t) + p\kappa\sigma I(t),$$

$$\dot{S}_2(t) = A\pi_2 - \beta(t) W(t) (1 - \theta) S_2(t)$$
$$- (\mu + a_2) S_2(t) + a_1 S_1(t)$$
$$+ p\kappa(1 - \sigma) I(t), \tag{3}$$

$$\dot{E}(t) = \beta(t) W(t) [S_1(t) + (1 - \theta) S_2(t)]$$
$$- (\gamma(t) + \mu) E(t),$$

$$\dot{I}(t) = \gamma(t) E(t) - (\mu + \kappa) I(t),$$

$$\dot{W}_1(t) = \alpha(t) I(t) - \delta(t) W(t).$$

System (3) has a disease-free equilibrium given by $\mathscr{E}^0 = (S_1^0, S_2^0, 0, 0, 0)$, with

$$S_1^0 = \frac{A(a_2 + \mu\pi_1)}{\mu(\mu + a_1 + a_2)},$$
$$S_2^0 = \frac{A(a_1 + \mu\pi_2)}{\mu(\mu + a_1 + a_2)}. \tag{4}$$

The severity of a disease can be measured by the basic reproduction number, $\mathscr{R}_0$, which is defined as the average number of secondary infections caused by a single infected individual in a completely susceptible population. Utilizing the next generation matrix approach [15] and adopting the notations therein, the matrices for new infections (denoted by $F(t)$) and the transfer terms (denoted by $V(t)$) at disease-free equilibrium are given by

$$F(t)$$
$$= \begin{pmatrix} 0 & 0 & \dfrac{\beta(t) A [(a_2 + \mu\pi_1) + (1 - \theta)(a_1 + \mu\pi_2)]}{\mu(\mu + a_1 + a_2)} \\ 0 & 0 & 0 \\ 0 & \alpha(t) & 0 \end{pmatrix},$$
$$\tag{5}$$

$$V(t) = \begin{pmatrix} \mu + \gamma(t) & 0 & 0 \\ -\gamma(t) & \mu + \kappa & 0 \\ 0 & 0 & \delta(t) \end{pmatrix}.$$

Thus, the basic reproduction number of the time-averaged autonomous systems is

$$[\mathscr{R}_0] = \rho\left(F(0) V^{-1}(0)\right)$$
$$= \left(\frac{\beta_0\gamma_0\alpha_0 A [(a_2 + \mu\pi_1) + (1 - \theta)(a_1 + \mu\pi_2)]}{\delta_0(\mu + \gamma_0)(\kappa + \mu)\mu(\mu + a_1 + a_2)}\right)^{0.5}. \tag{6}$$

In order to analyze the threshold dynamics of epidemiological models in periodic environments, Wang and Zhao [16] extended the framework in [15] by introducing the next infection operator

$$(L\phi)(t) = \int_0^\infty Y(t, t - s) F(t - s) \phi(t - s) ds \tag{7}$$

where $Y(t, s)$, $t \geq s$, is the evolution operator of the linear ω-periodic system $dy/dt = -V(t)y$ and $\phi(t)$, the initial distribution of infectious humans, is ω-periodic and always positive. The effective reproduction number for a periodic model is then determined by calculating the spectral radius of the next infection operator,

$$\mathscr{R}_0 = \rho(L). \tag{8}$$

Through direct calculation, the evolution operator $Y(t, s)$, for model (3), is given by

$$Y(t, s) = \begin{pmatrix} \widetilde{Y}_1(t, s) & 0 & 0 \\ \widetilde{Y}_2(t, s) & e^{-(\mu + \kappa)(t - s)} & 0 \\ 0 & 0 & \widetilde{Y}_3(t, s) \end{pmatrix}, \tag{9}$$

where

$$\widetilde{Y}_1(t, s) = e^{-(\mu + M_1)(t - s)},$$

$$\widetilde{Y}_2(t, s) = \left[\gamma_0\left(\frac{1}{(M_1 - \kappa)} - \frac{1}{[1 + (M_1 - \kappa)^2]}\right)\right]\left[(M_1\right.$$
$$\left. - \kappa)\sin\left(\frac{2\pi t}{\omega}\right) + \cos\left(\frac{2\pi t}{\omega}\right)\right] \times \left[e^{-(\mu + \kappa)(t - s)}\right.$$
$$\left. - e^{-(\mu + M_1)(t - s)}\right], \tag{10}$$

$$\widetilde{Y}_3(t, s) = \exp\left\{-\delta_0\left((t - s)\right.\right.$$
$$\left.\left. - \left(\frac{\omega}{2\pi}\right)\left[\cos\left(\frac{2\pi t}{\omega}\right) - \cos\left(\frac{2\pi s}{\omega}\right)\right]\right)\right\}$$

with

$$M_1 = \exp - \left\{ \gamma_0 \left(1 - \left(\frac{\omega}{2\pi} \right) \left[\cos \left(\frac{2\pi}{\omega} \right) \right] \right) (t - s) \right\}. \quad (11)$$

We can numerically evaluate the next infection operator by

$$(L\phi)(t) = \int_0^\infty Y(t, t - s) F(t - s) \phi(t - s) \, ds$$
$$= \int_0^\omega G(t - s) \phi(t - s) \, ds \quad (12)$$

where

$$G(t, s) \approx \sum_{k=0}^{M} Y(t, t - s - k\omega) F(t - s - k\omega) \quad (13)$$

$$G(t, s) \approx \sum_{k=0}^{M} \begin{pmatrix} 0 & 0 & \beta(t - s - k\omega) \dfrac{A\left[(a_2 + \mu\pi_1) + (1 - \theta)(a_1 + \mu\pi_2)\right]}{\mu(\mu + a_1 + a_2)} \widetilde{Y}_1(t, s) \\ 0 & 0 & \beta(t - s - k\omega) \dfrac{A\left[(a_2 + \mu\pi_1) + (1 - \theta)(a_1 + \mu\pi_2)\right]}{\mu(\mu + a_1 + a_2)} \widetilde{Y}_2(t, s) \\ 0 & \alpha(t - s - k\omega) \widetilde{Y}_3(t, s) & 0 \end{pmatrix}. \quad (14)$$

2.3. Threshold Dynamics. In this section we will use the basic reproduction number $\mathscr{R}_0$ defined in (8) to establish the threshold results, presented in Theorem 1, for model (3). To that end, we first note that $\mathbb{R}_+^2$ is positively invariant for the following cooperative system:

$$\dot{S}_1(t) = A\pi_1 - (\mu + a_1) S_1(t) + a_2 S_2(t),$$
$$\dot{S}_2(t) = A\pi_2 - (\mu + a_2) S_2(t) + a_1 S_1(t), \quad (15)$$

and that (S_1^0, S_2^0) is the unique equilibrium solution which is globally attractive in $\mathbb{R}_+^2$.

Theorem 1.

(i) *If $R_0 < 1$, then the disease-free equilibrium $\mathscr{E}_0$ of system (3) is globally asymptotically stable.*

(ii) *If $R_0 > 1$, then system (3) admits at least one positive ω-periodic solution, and solutions of system (3) are uniformly persistent.*

Proof. If $(S_1(t), S_2(t), E(t), I(t), W(t))$ is a nonnegative solution of (3), then we have

$$\dot{S}_1(t) \le A\pi_1 - (\mu + a_1) S_1(t) + a_2 S_2(t),$$
$$\dot{S}_2(t) \le A\pi_2 - (\mu + a_2) S_2(t) + a_1 S_1(t). \quad (16)$$

Note that any nonnegative solution $(S_1(t), S_2(t))$ of system (3) approaches (S_1^0, S_2^0) as $t \longrightarrow \infty$. It then follows from the standard comparison theorem (see, e.g., [17, Theorem A.4]) that, for any $\epsilon > 0$, there is a $T > 0$ such that

$$S_i(t) < S_i^0 + \epsilon, \quad i = 1, 2, \text{ for } t > T. \quad (17)$$

Thus, for $t > T$, we have

$$\dot{E}(t) \le \beta(t) W(t) \left[\left(S_1^0 + \epsilon \right) + (1 - \theta) \left(S_2^0 + \epsilon \right) \right]$$
$$\qquad - (\gamma(t) + \mu) E(t),$$
$$\dot{I}(t) \le \gamma(t) E(t) - (\mu + \kappa) I(t), \quad (18)$$
$$\dot{W}(t) \le \alpha(t) I(t) - \delta(t) W(t).$$

Define

$$F_\epsilon(t)$$
$$= \begin{bmatrix} 0 & 0 & \beta(t) \left[\left(S_1^0 + \epsilon \right) + (1 - \theta) \left(S_2^0 + \epsilon \right) \right] \\ 0 & 0 & 0 \\ 0 & \alpha(t) & 0 \end{bmatrix}. \quad (19)$$

By [16, Thorem 2.2], we have $R_0 < 1 \iff \rho(\phi_{F-V}(\omega)) < 1$, where $\rho(\phi_{F-V}(\omega))$ is the spectral radius of $\phi_{F-V}(\omega)$ and $\phi_{F-V}(\omega)$ is the monodromy matrix of the linear ω-periodic system $dy/dt = (F - V)y$. Then we can set ϵ sufficiently small such that $\rho(\phi_{F_\epsilon-V}(\omega)) < 1$. As a consequence, the trivial solution $(0, 0, 0)$ of the following linear ω-periodic system

$$\dot{E}(t) = \beta(t) W(t) \left[\left(S_1^0 + \epsilon \right) + (1 - \theta) \left(S_2^0 + \epsilon \right) \right]$$
$$\qquad - (\gamma(t) + \mu) E(t),$$
$$\dot{I}(t) = \gamma(t) E(t) - (\mu + \kappa) I(t), \quad (20)$$
$$\dot{W}(t) = \alpha(t) I(t) - \delta(t) W(t)$$

is globally asymptotically stable. Again by the comparison theorem, we know that $E(t) \longrightarrow 0$, $I(t) \longrightarrow 0$ and $W(t) \longrightarrow 0$ as $t \longrightarrow \infty$. Finally, from the first and second equations of system (3) it follows that $S_i(t) \longrightarrow S_i^0$ as $t \longrightarrow \infty$ for $i = 1, 2$. This proves the result in part (i) of Theorem 1.

Next we shall focus on the case $R_0 > 1$. We define $X = \mathbb{R}_+^5$, $X_0 = \mathbb{R}_+^2 \times \text{Int}(\mathbb{R}_+^3)$, $\partial X_0 = X \setminus X_0$. It can easily be verified that both X and X_0 are positively invariant. Let $P : \mathbb{R}_+^5 \longrightarrow \mathbb{R}_+^5$ be the Poincaré map associated with system (3); that is, $P(x_0) = u(\omega, x_0)$ for all $x_0 \in \mathbb{R}_+^5$, where $u(t, x_0)$ is the unique solution of (3) with $u(0, x_0) = x_0$. Set

$$M_\partial = \{(S_1(0), S_2(0), E(0), I(0), W(0))$$
$$\in \partial X_0 : P^m(S_1(0), S_2(0), E(0), I(0), W(0))$$
$$\in \partial X_0, \; \forall m \geq 0\}, \tag{21}$$
$$M = \{(S_1, S_2, 0, 0, 0) : S_1 \geq 0, S_2 \geq 0\}.$$

We first demonstrate that

$$M_\partial = M. \tag{22}$$

It is evident that $M \subseteq M_\partial$. For any $(S_1(0), S_2(0), E(0), I(0), W(0)) \in \partial X_0 \setminus M$, if $E(0) = I(0) = 0$ and $W(0) = 0$, then $\dot{E}(0) = \beta(t)W(t)[(S_1^0 + \epsilon) + (1-\theta)(S_2^0 + \epsilon)] = \beta(0)W(0)[S_1^0(0) + (1-\theta)S_2(0)] > 0$. If $E(0) = W(0) = 0$ and $I(0) > 0$, then $\dot{E}(0) > 0$ and $\dot{W}(0) = \alpha(0)I(0) > 0$. If $I(0) = W(0) = 0$ and $E(0) > 0$ then $\dot{E}(0) > 0$ and $\dot{I}(0) = \gamma(0)E(0) > 0$. It follows that $(S_1(t), S_2(t), E(t), I(t), W(t)) \notin \partial X_0$ for $0 < t \ll 1$. By the positive invariance of X_0, we know that $P^m(S_1(0), S_2(0), E(0), I(0), W(0)) \notin \partial X_0$ for $m \geq 1$; hence $(S_1(0), S_2(0), E(0), I(0), W(0)) \notin M_\partial$, and thus (22) holds.

Now consider the fixed point $M_0 = (S_1^0, S_2^0, 0, 0, 0)$ of the Poincaré map P. Define $\mathscr{W}^S(M_0) = \{x_0 : P^m(x_0) \longrightarrow M_0, m \longrightarrow \infty\}$. We show that

$$\mathscr{W}^S(M_0) \cap X_0 = \emptyset. \tag{23}$$

Based on the continuity of solutions with respect to the initial conditions, for any $\epsilon > 0$, there exists $\vartheta > 0$ small enough such that for all $(S_1(0), S_2(0), E(0), I(0), W(0)) \in X_0$ with

$$\|(S_1(0), S_2(0), E(0), I(0), W(0)) - M_0\| \leq \vartheta, \tag{24}$$

we have

$$\|u(t, (S_1(0), S_2(0), E(0), I(0), W(0))) - u(t, M_0)\| \tag{25}$$
$$< \epsilon, \quad \forall t \in [0, \omega].$$

To obtain (23), we claim that

$$\limsup_{m \longrightarrow \infty} \|P^m(S_1(0), S_2(0), E(0), I(0), W(0))$$
$$- M_0\| \geq \vartheta, \tag{26}$$
$$\forall(S_1(0), S_2(0), E(0), I(0), W(0)) \in X_0.$$

We prove this claim by contradiction; that is, we suppose $\limsup_{m \longrightarrow \infty} \|P^m(S_1(0), S_2(0), E(0), I(0), W(0)) - M_0\| < \vartheta$ for some $(S_1(0), S_2(0), E(0), I(0), W(0)) \in X_0$. Without loss of generality, we assume that $\|P^m(S_1(0), S_2(0), E(0), I(0), W(0)) - M_0\| < \vartheta, \; \forall m \geq 0$. Thus,

$$\|u(t, P^m(S_1(0), S_2(0), E(0), I(0), W(0))$$
$$- u(t, M_0)\| < \epsilon, \quad \forall t \in [0, \omega], \; m \geq 0. \tag{27}$$

Moreover, for any $t \geq 0$, we write $t = t_0 + n\omega$ with $t_0 \in [0, \omega)$ and $n = [t/\omega]$, the greatest integer less than or equal to t/ω. Then we obtain

$$\|u(t, (S_1(0), S_2(0), E(0), I(0), W(0)) - u(t, M_0)\|$$
$$= \|u(t_0, P^m(S_1(0), S_2(0), E(0), I(0), W(0)) \tag{28}$$
$$- u(t_0, M_0)\| < \epsilon$$

for any $t \geq 0$. Let $(S_1(t), S_2(t), E(t), I(t), W(t)) = u(t, (S_1(0), S_2(0), E(0), I(0), W(0)))$. It follows that

$$-\epsilon < S_1(t) - S_1^0 < \epsilon,$$
$$-\epsilon < S_2(t) - S_2^0 < \epsilon,$$
$$0 < E(t) < \epsilon, \tag{29}$$
$$0 < I(t) < \epsilon,$$
$$0 < W(t) < \epsilon.$$

$$\dot{E}(t) = \beta(t)W(t)\left[S_1(t) + (1-\theta)S_2(t)\right]$$
$$- (\gamma(t) + \mu)E(t)$$
$$\geq \beta(t)W(t)\left[S_1^0 - \epsilon + (1-\theta)\left(S_2^0 - \epsilon\right)\right] \tag{30}$$
$$- (\gamma(t) + \mu)E(t)$$
$$= -(\gamma(t) + \mu)E(t) - \beta(t)W(t)\epsilon(2-\theta)$$
$$+ \beta(t)W(t)\left(S_1^0 + (1-\theta)S_2^0\right).$$

Hence we obtain

$$\frac{d}{dt}\begin{bmatrix} E \\ I \\ W \end{bmatrix} \geq [F - V - \epsilon K]\begin{bmatrix} E \\ I \\ W \end{bmatrix}, \tag{31}$$

where $F - V$ is given by (5) and

$$\epsilon \cdot K = \epsilon \cdot \begin{pmatrix} 0 & 0 & (2-\theta)\beta(t) \\ 0 & 0 & 0 \\ 0 & 0 & 0 \end{pmatrix}. \tag{32}$$

Again based on [16, Theorem 2.2], $R_0 > 1$ if and only if $\rho(\Phi_{F-V}(\omega)) > 1$. Thus, for ϵ small enough, we have $\rho(\Phi_{F-V-\epsilon \cdot K}(\omega)) > 1$ which immediately yields the contradiction as

$$\lim_{t \longrightarrow \infty} E(t) = \infty,$$
$$\lim_{t \longrightarrow \infty} I(t) = \infty, \tag{33}$$
$$\lim_{t \longrightarrow \infty} W(t) = \infty.$$

Let $P_1 : \mathbb{R}_+^2 \longrightarrow \mathbb{R}_+^2$ be the Poincaré map associated with (15). Then (S_1^0, S_2^0) is globally attractive in $\mathbb{R}_+^2 \setminus \{0\}$ for

P_1. It follows that M_0 is isolated invariant set in X, and notice that $\mathscr{W}^S(M_0) \cap X_0 = \emptyset$. Hence, every orbit in M_∂ converges to M_0 and M_0 is acyclic in M_∂. By [18, Theorem 1.3.1], for a stronger repelling property of ∂X_0, we conclude that P is uniformly persistent with respect to $(X_0, \partial X_0)$, which implies the uniform persistence of the solutions of system (3) with respect to $(X_0, \partial X_0)$ [18, Theorem 3.1.1]. Consequently, based on [18, Theorem 1.3.6], the Poincaré map P has a fixed point $(\tilde{S}_1(0), \tilde{S}_2(0), \tilde{E}(0), \tilde{I}(0), \overline{W}(0)) \in X_0$, with $\tilde{S}_1(0) \neq 0, \tilde{S}_2(0) \neq 0$. Thus, $(\tilde{S}_1(0), \tilde{S}_2(0), \tilde{E}(0), \tilde{I}(0), \overline{W}(0)) \in \text{Int}(\mathbb{R}^5_+)$ and $(\tilde{S}_1(t), \tilde{S}_2(t), \tilde{E}(t), \tilde{I}(t), \overline{W}(t)) = u(t, (\tilde{S}_1(0), \tilde{S}_2(0), \tilde{E}(0), \tilde{I}(0), \overline{W}(0)))$ is a positive ω-periodic solution of the system. □

2.4. Estimation of Parameter Values. In this section, we aim to use the Guinea worm disease surveillance data of South Sudan (2007-2013) to estimate periodic functions defined in model (3), while the baseline values for constant model parameters will be drawn from literature:

(i) The fraction of individuals recruited into susceptible risk population π_1: according to the world health organization report of 2014, more than 90% of the people in South Sudan lives on less than US\$1 a day [10]. Based on this assertion we assume that $\pi_1 = 0.9$ and it follows that $\pi_2 = 0.1$.

(ii) The natural mortality rate $\mu = 1/$life expectancy: the united nations children's fund report of 2013 [11] highlighted that life expectancy in South Sudan is 55 years; thus $\mu = 0.0015$ month^{-1}.

(iii) The mean infectious period κ^{-1} : as discussed in the introduction section, Guinea worm disease infectious individuals shed larvae into the environment during a short time and more often this occurs when they physically submerge their wound in open water sources. Based on findings in prior studies [6], we set $\kappa = 8.3 \times 10^{-2}$ month^{-1}.

(iv) The rate of transfer from high risk susceptible population to low risk susceptible population and vice versa: since more than half of the population live below the international poverty line [11], we assume an extremely low transfer rate of individuals from class S_1 to S_2; thus we set $a_1 = 9.0 \times 10^{-6}$ month^{-1}. In contrast, due to civil unrest which characterize South Sudan population we assume that the rate of transfer from low risk is higher ($a_1 < a_2$); thus we set $a_2 = 0.009$ month^{-1}.

Table 1 presents the description of model parameters and their baseline values.

Now, we need to estimate the periodic functions $\beta(t)$, $\alpha(t)$, $\gamma(t)$, and $\delta(t)$. This will be done via the Fourier series analysis method as in [19, 20]. Observe that the monthly numbers of new Guinea worm cases in system (3) correspond to the term

$$g(t) = \gamma(t) E(t). \tag{34}$$

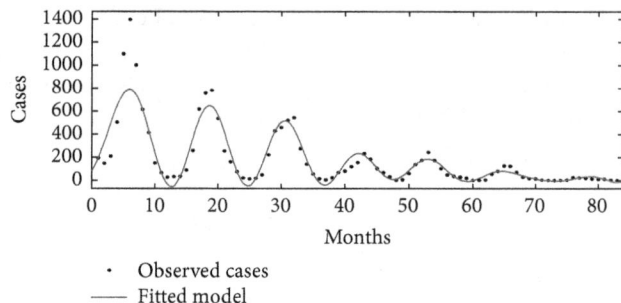

FIGURE 2: The monthly numbers of new Guinea worm cases and its fitted curve.

· Observed cases
— Fitted model

Further, since variables and the periodic parameters in model system (3) are continuous functions of t, we will make use of trigonometric functions to fit $g(t)$. Define

$$g(t) = c_0 + \sum_{j=1}^{8} \left[c_j \cos jLt + d_j \sin jLt \right], \tag{35}$$

where c_j and d_j ($j = 1, 2, \ldots, 8$) represent coefficients to be determined and L is the fundamental frequency. We used MATLAB software to determine the coefficients c_j and d_j ($j = 1, 2, \ldots, 8$) and we obtained

$$g(t) \approx 171.20 - 44.82 \cos(1.78t) + 8.38 \cos(3.56t)$$
$$- 29.07 \cos(5.34t) - 3.08 \cos(7.12t)$$
$$- 16.63 \cos(8.90t) - 51.20 \cos(10.68t)$$
$$+ 172.20 \cos(12.46) + 21.91 \cos(14.24t)$$
$$- 153.3 \sin(1.78t) + 73.84 \sin(3.56t)$$
$$- 59.83 \sin(5.34t) + 68.95 \sin(7.12t)$$
$$- 62.73 \sin(8.90t) + 105.60 \sin(10.68t)$$
$$- 2.17 \sin(12.46t) - 78.16 \sin(14.24t), \tag{36}$$

with $L = 4\pi/7$ and to be precise and exact $\omega = 7/2$. The comparison of the data with curve of $g(t)$ is shown in Figure 2.

Next, we proceed to use (36) to estimate the periodic functions $\beta(t)$ (disease transmission rate), $\gamma(t)$ (incubation rate), $\alpha(t)$ (pathogen shedding rate), and $\delta(t)$ (pathogen decay rate). First, we assume the initial population levels as follows: $S_1(0) = S_2(0) = 10,000$, $W(0) = 5000$. Further, we set $E(0) = I(0) = 192$ based on the number of cases reported in January 2007. Let

$$\beta(t) = \widetilde{\beta_0}\beta_1(t),$$
$$\gamma(t) = \widetilde{\gamma_0}\gamma_1(t),$$
$$\alpha(t) = \widetilde{\alpha_0}\alpha_1(t),$$
$$\delta(t) = \widetilde{\delta_0}\delta_1(t), \tag{37}$$

where $\widetilde{\beta_0}$, $\widetilde{\gamma_0}$, $\widetilde{\alpha_0}$, and $\widetilde{\delta_0}$ are related to the magnitude of seasonal fluctuations and $\beta_1(t)$, $\gamma_1(t)$, $\alpha_1(t)$, and $\delta_1(t)$ are

TABLE 1: Description of model parameters for system (3).

Symbol	Definition	Value	Unit	Source
p	Proportion of recovered individuals who become susceptible to the disease	0.3	-	Assumed
σ	Proportion of recovered individuals who are at high risk to reinfection	0.9	-	[10]
a_1	Rate of transfer from high risk to low risk susceptible population	9.0×10^{-6}	month^{-1}	Assumed
a_2	Rate of transfer from low risk to high risk susceptible population	0.009	month^{-1}	Assumed
π_1	Proportion of new births in high risk susceptible population	0.9	-	[10]
π_2	Proportion of new births in low risk susceptible population	0.1	-	[10]
μ	Natural mortality rate for humans	0.0015	month^{-1}	[11]
κ^{-1}	Average infectious period	8.3×10^{-2}	month	Assumed
θ	Modification factor	0.98	-	Assumed
A	Human birth rate	100	Individuals month^{-1}	Assumed
β_0	Averaged environment–to–host transmission	1.2×10^{-5}	Larvae month^{-1}	Fitting
γ_0	Averaged incubation rate	1.0×10^{-1}	month	[12]
α_0	Averaged parasite shedding rate	2×10^{-1}	Larvae month^{-1} individual^{-1}	Fitting
δ_0	Parasite death rate	2.16	month^{-1}	[6]

periodic functions to be determined. After simulations and comparisons we obtained

$$\beta_1(t) = 10^{-3}[205 - 53.78\cos(1.78t)$$
$$+ 10.06\cos(3.56t) - 34.88\cos(5.34t)$$
$$- 3.70\cos(7.12t) - 19.20\cos(8.90t)$$
$$- 61.44\cos(10.68t) + 206.64\cos(12.46)$$
$$+ 26.29\cos(14.24t) - 183.96\sin(1.78t)$$
$$+ 88.608\sin(3.56t) - 71.80\sin(5.34t)$$
$$+ 82.74\sin(7.12t) - 75.28\sin(8.90t)$$
$$+ 126.72\sin(10.68t) - 2.60\sin(12.46t)$$
$$- 93.79\sin(14.24t)],$$

$$\alpha_1(t) = 10^{-3}[342.4 - 89.64\cos(1.78t)$$
$$+ 16.76\cos(3.56t) - 58.14\cos(5.34t)$$
$$- 6.16\cos(7.12t) - 33.26\cos(8.90t)$$
$$- 102.4\cos(10.68t) + 344.4\cos(12.46)$$
$$+ 43.82\cos(14.24t) - 306.6\sin(1.78t)$$

$$+ 147.68\sin(3.56t) - 119.66\sin(5.34t)$$
$$+ 137.9\sin(7.12t) - 125.46\sin(8.90t)$$
$$+ 211.2\sin(10.68t) - 4.34\sin(12.46t)$$
$$- 156.32\sin(14.24t)],$$

$$\gamma_1(t) = 10^{-2}[256.8 - 67.23\cos(1.78t)$$
$$+ 12.57\cos(3.56t) - 43.61\cos(5.34t)$$
$$- 4.62\cos(7.12t) - 24.95\cos(8.90t)$$
$$- 76.8\cos(10.68t) + 258.3\cos(12.46)$$
$$+ 32.87\cos(14.24t) - 229.95\sin(1.78t)$$
$$+ 110.76\sin(3.56t) - 89.75\sin(5.34t)$$
$$+ 103.43\sin(7.12t) - 94.10\sin(8.90t)$$
$$+ 158.4\sin(10.68t) - 3.26\sin(12.46t)$$
$$- 117.24\sin(14.24t)],$$

$$\delta_1(t) = 10^{-4}[369.79 - 96.81\cos(1.78t) + 18.10$$
$$\cdot \cos(3.56t) - 62.79\cos(5.34t) - 6.65\cos(7.12t)$$
$$- 35.92\cos(8.90t) - 110.59\cos(10.68t) + 371.95$$

$\cos{(12.46)} + 47.33 \cos{(14.24t)} - 331.13$

$\cdot \sin{(1.78t)} + 159.49 \sin{(3.56t)} - 129.23$

$\cdot \sin{(5.34t)} + 148.93 \sin{(7.12t)} - 135.50$

$\cdot \sin{(8.90t)} + 228.10 \sin{(10.68t)} - 4.69$

$\cdot \sin{(12.46t)} - 168,83 \sin{(14.24t)}.$

$$(38)$$

with $\widetilde{\beta_0} = 0.001$, $\widetilde{\gamma_0} = 0.15$, $\widetilde{\alpha_0} = 0.006$, and $\widetilde{\delta_0} = 0.00012$. Using direct calculation and (37) one can easily derive the expressions for $\beta(t)$, $\gamma(t)$, $\alpha(t)$, and $\delta(t)$.

2.5. Optimal Control. Guinea worm disease has no treatment; however, the disease can be prevented by applying a chemical called temephos (which is an organophosphate), to unsafe drinking water sources in order to kill the copepods in water [7]. This approach reduces the parasite population and in turn it minimizes the likelihood of disease transmission. Now, we extend our initial model to incorporate a control function $u(t)$ that represent the effect of vector control through the application of chemicals to water. The extended model takes the form

$$\dot{S}_1(t) = A\pi_1 - \beta(t) W(t) S_1(t) - (\mu + a_1) S_1(t)$$
$$+ a_2 S_2(t) + p\kappa\sigma I(t),$$

$$\dot{S}_2(t) = A\pi_2 - \beta(t) W(t)(1 - \theta) S_2(t)$$
$$- (\mu + a_2) S_2(t) + a_1 S_1(t)$$
$$+ p\kappa(1 - \sigma) I(t), \qquad (39)$$

$$\dot{E}(t) = \beta(t) W(t)[S_1(t) + (1 - \theta) S_2(t)]$$
$$- (\gamma(t) + \mu) E(t),$$

$$\dot{I}(t) = \gamma(t) E(t) - (\mu + \kappa) I(t),$$

$$\dot{W}(t) = \alpha(t) I(t) - [u(t) + \delta(t)] W(t).$$

A successful control strategy is one that reduces the numbers of exposed and infected individuals over a finite time horizon $[0, T]$ at minimal cost. Our objective functional is therefore formulated as

$$J[u(t)] = \int_0^T [E(t) + I(t) + Bu(t)] \, dt \qquad (40)$$

subject to the constraints of the ordinary differential equations in system (39) and the coefficient B (positive) represents a weight in the cost of the control. This objective functional and the differential equations are linear in the control with bounded states, and one can demonstrate by standard results that an optimal control and corresponding optimal states exist [21]. Although majority of applications of optimal control theory to infectious disease control consider controls in a quadratic form due to a number of mathematical advantages (e.g., if the control set is a compact and convex polyhedron it

follows that the Hamiltonian attains its minimum over the control set at a unique point), here we considered a linear control since it is regarded as a more realistic function to use in a biological framework compared to the quadratic [22–24].

By using Pontryagin's maximum principle [21, 25] we derive necessary conditions for our optimal control and corresponding states. The Hamiltonian is

$$H = E(t) + I(t) + Bu(t) + \lambda_1 [A\pi_1$$
$$- \beta(t) W(t) S_1(t) - (\mu + a_1) S_1(t) + a_2 S_2(t)$$
$$+ p\kappa\sigma I(t)] + \lambda_2 [A\pi_2 - \beta(t) W(t)(1 - \theta) S_2(t)$$
$$- (\mu + a_2) S_2(t) + a_1 S_1(t) + p\kappa(1 - \sigma) I(t)] \qquad (41)$$
$$+ \lambda_3 [\beta(t) W(t)[S_1(t) + (1 - \theta) S_2(t)]$$
$$- (\mu + \gamma) E(t)] + \lambda_4 [\gamma E(t) - (\mu + \kappa) I(t)]$$
$$+ \lambda_5 [\alpha I(t) - (u(t) + \delta) W(t)].$$

Given an optimal control $u(t)$, there exist adjoint functions, $\lambda_1, \lambda_2, \lambda_3, \lambda_4, \lambda_5$ corresponding to the states S_1, S_2, E, I, and W such that

$$\dot{\lambda}_1 = \lambda_1 [\beta(t) W + \mu + a_1] - \lambda_2 a_1 - \lambda_3 \beta(t) W,$$

$$\dot{\lambda}_2 = \lambda_2 [\beta(t) W (1 - \theta) + \mu + a_2] - \lambda_1 a_2$$
$$- \lambda_3 \beta(t) W (1 - \theta),$$

$$\dot{\lambda}_3 = -1 + \lambda_3 [\mu + \gamma] - \lambda_4 \gamma(t), \qquad (42)$$

$$\dot{\lambda}_4 = -1 - \lambda_1 p\kappa\sigma - \lambda_2 p\kappa(1 - \sigma) + \lambda_4 (\mu + \kappa) - \lambda_5 \alpha,$$

$$\dot{\lambda}_5 = \lambda_1 \beta(t) S_1 + \lambda_2 \beta(t) S_2 (1 - \theta)$$
$$- \lambda_3 \beta(t) [S_1 + (1 - \theta) S_2] + \lambda_5 (u(t) + \delta),$$

where $\lambda_1(T) = 0$, $\lambda_2(T) = 0$, $\lambda_3(T) = 0$, $\lambda_4(T) = 0$, and $\lambda_5(T) = 0$ are the transversality conditions.

The Hamiltonian is minimized with respect to the control variable u^*. Since the Hamiltonian is linear in the control, we must consider if the optimal control is bang-bang (at it is lower or upper bound), singular, or a combination. The singular case could occur if the slope or the switching function

$$\frac{\partial H}{\partial u} = B - \lambda_5 W \qquad (43)$$

is zero on nontrivial interval of time. Note that the optimal control would be at its upper bound or its lower bound according to

$$\frac{\partial H}{\partial u} < 0 \text{ or } > 0. \qquad (44)$$

To investigate the singular case, let us suppose $\partial H / \partial u = 0$ on some nontrivial interval. In this case we calculate

$$\frac{d}{dt} \left(\frac{\partial H}{\partial u} \right) = 0 \qquad (45)$$

and then we will show that control is not present in that equation. To solve for the value of the singular control, we will further calculate

$$\frac{d^2}{dt^2}\left(\frac{\partial H}{\partial u}\right) = 0. \tag{46}$$

We simplify the time derivative of $\partial H/\partial u$,

$$0 = \frac{d}{dt}\left(\frac{\partial H}{\partial u}\right) = \frac{d}{dt}\left(B - \lambda_5 W\right) = -\lambda_5 W' - \lambda_5' W. \tag{47}$$

We calculate both sums separately and add them together. The first sum can be written as

$$-\lambda_5 W' = [-\lambda_5]\left[\alpha I - (u(t) + \delta) W\right]$$
$$= -\alpha\lambda_5 I + (u(t) + \delta)\lambda_5 W. \tag{48}$$

The second sum can be written as

$$\left[-\lambda_5'\right] W = \left[\lambda_1\left(-\beta(t) S_1\right) + \lambda_2\left[-\beta(t) S_2(1-\theta)\right]\right.$$
$$\left. + \lambda_3\beta(t)\left[S_1 + (1-\theta) S_2\right] + \lambda_5\left(-u(t) - \delta\right)\right] W$$
$$= -\beta(t)\lambda_1 S_1 W - \beta(t)(1-\theta)\lambda_2 S_2 W + \beta(t)$$
$$\cdot \lambda_3 S_1 W + \beta(t)(1-\theta)\lambda_3 S_2 W - (u(t) + \delta)\lambda_5 W. \tag{49}$$

Thus combining, we have

$$0 = \frac{d}{dt}\left(\frac{\partial H}{\partial u}\right)$$
$$= -\alpha\lambda_5 I - \beta\lambda_1 S_1 W - \beta(1-\theta)\lambda_2 S_2 W + \beta\lambda_3 S_1 W$$
$$+ \beta(1-\theta)\lambda_3 S_2 W \tag{50}$$
$$= -\alpha\lambda_5 I + \left[\beta(\lambda_3 - \lambda_1)\right] S_1 W$$
$$+ \left[\beta(1-\theta)(\lambda_3 - \lambda_2)\right] S_2 W.$$

We see that the control does not explicitly show in this expression, so next we calculate the second derivative with respect to time.

$$0 = \frac{d^2}{dt^2}\left(\frac{\partial H}{\partial u}\right)$$
$$= \left[(-\alpha)\lambda_5'\right] I + \left[(-\alpha)\lambda_5\right] I' + \left[\beta(\lambda_3' - \lambda_1')\right] S_1 W$$
$$+ \left[\beta(\lambda_3 - \lambda_1)\right]\left(S_1 W' + S_1' W\right) \tag{51}$$
$$+ \left[\beta(1-\theta)(\lambda_3' - \lambda_2')\right] S_2 W$$
$$+ \left[\beta(1-\theta)(\lambda_3 - \lambda_2)\right]\left(S_2 W' + S_2' W\right).$$

Using systems (3) and (41), we simplify (47) as follows:

$$0 = \frac{d^2}{dt^2}\left(\frac{\partial H}{\partial u}\right) = [\alpha(2\beta(t)$$
$$\cdot [(\lambda_3 - \lambda_1) S_1 + (\lambda_3 - \lambda_2)(1-\theta) S_2] + (\mu + \kappa)\lambda_5$$
$$- \delta\lambda_5)] I - [\alpha\gamma\lambda_5] E + [\beta(t)$$
$$\cdot (A [(\lambda_3 - \lambda_1)\pi_1 + (\lambda_3 - \lambda_2)(1-\theta)\pi_2]$$
$$+ [p\kappa((\lambda_3 - \lambda_1)\sigma + (\lambda_3 - \lambda_2)(1-\theta)(1-\sigma))]$$
$$\cdot I)] W - [[\alpha\lambda_5] I + [\beta(\lambda_3 - \lambda_1)] S_1 W \tag{52}$$
$$+ [\beta(\lambda_3 - \lambda_2)(1-\theta)] S_2 W] u(t) + [\beta(t)(-1$$
$$+ \gamma(\lambda_3 - \lambda_4) - \delta(\lambda_3 - \lambda_1)$$
$$+ a_1 [(\lambda_2 - \lambda_3) + (\lambda_3 - \lambda_2)(1-\theta)])] S_1 W$$
$$+ [\beta(t)((1-\theta)[-1 + \gamma(\lambda_3 - \lambda_4) - \delta(\lambda_3 - \lambda_2)]$$
$$+ a_2 [(\lambda_1 - \lambda_3)(1-\theta) + (\lambda_3 - \lambda_1)])] S_2 W.$$

The above equation can be written in the form

$$\frac{d^2}{dt^2}\left(\frac{\partial H}{\partial u}\right) = \Phi_1(t) u(t) + \Phi_2(t) = 0. \tag{53}$$

and we can solve for the singular control as

$$u_{\text{singular}}(t) = -\frac{\Phi_2(t)}{\Phi_1(t)} \tag{54}$$

if

$$\Phi_1(t) \neq 0,$$
$$a \leq -\frac{\Phi_2(t)}{\Phi_1(t)} \leq b \tag{55}$$

with

$$\Phi_1(t) = -[[\alpha\lambda_5] I + [\beta(t)(\lambda_3 - \lambda_1)] S_1 W$$
$$+ [\beta(t)(\lambda_3 - \lambda_2)(1-\theta)] S_2 W] = -\alpha\frac{B}{W} I - \beta(t) \tag{56}$$
$$\cdot (\lambda_3 - \lambda_1) S_1 W - \beta(t)(\lambda_3 - \lambda_2)(1-\theta) S_2 W$$

and

$$\Phi_2(t)$$
$$= \alpha(2\beta(t)[(\lambda_3 - \lambda_1) S_1 + (\lambda_3 - \lambda_2)(1-\theta) S_2]$$
$$+ (\mu + \kappa)\lambda_5 - \delta\lambda_5) I - \alpha\gamma\frac{B}{W} E + \beta(t)$$
$$\cdot A(\pi_1(\lambda_3 - \lambda_1) + (\lambda_3 - \lambda_2)(1-\theta)\pi_2) W + \beta(t)$$
$$\cdot p\kappa(\sigma(\lambda_3 - \lambda_1) + (\lambda_3 - \lambda_2)(1-\theta)(1-\sigma)) IW$$
$$+ \beta(t)(-1 + \gamma(\lambda_3 - \lambda_4) - \delta(\lambda_3 - \lambda_1)$$

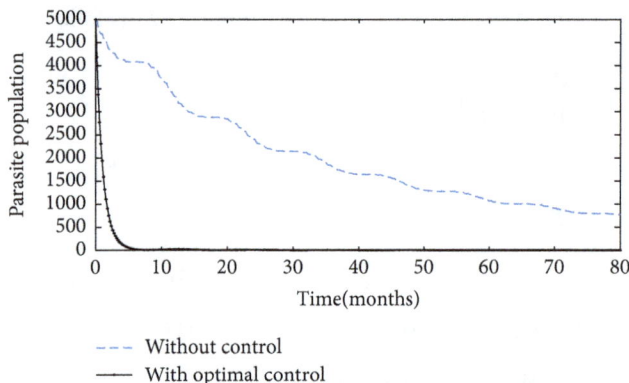

- - - Without control
—— With optimal control

FIGURE 3: The concentration of the pathogen in the environment with and without the optimal control.

$$+ a_1 (\lambda_2 - \lambda_3) + a_1 (\lambda_3 - \lambda_2) (1 - \theta)) S_1 W + \beta (t)$$

$$\cdot ((1 - \theta) [-1 + \gamma (\lambda_3 - \lambda_4) - \delta (\lambda_3 - \lambda_2)]$$

$$+ a_2 [(\lambda_1 - \lambda_3) (1 - \theta) + (\lambda_3 - \lambda_1)]) S_2 W. \tag{57}$$

To check the generalized Legendre-Clebsch condition for the singular control to be optimal, we require $(d/du)(d^2/dt^2)(\partial H/\partial u) = \Phi_1(t)$ to be negative [26, 27]. To summarize, our control characterization is as follows: on a nontrivial interval,

$$\text{if } \frac{\partial H}{\partial u} < 0 \text{ at } t \text{ then } u^*(t) = b$$

$$\text{if } \frac{\partial H}{\partial u} > 0 \text{ at } t \text{ then } u^*(t) = a \tag{58}$$

$$\text{if } \frac{\partial H}{\partial u} < 0 \text{ at } t \text{ then } u_{\text{singular}}(t) = -\frac{\Phi_2}{\Phi_1}$$

Hence, our control is optimal at t provided $\Phi_1(t) < 0$ and $a \leq -\Phi_2(t)/\Phi_1(t) \leq b$.

Using parameter values from Table 1 we investigate the effects of vector control on Guinea worm eradication in South Sudan. We solved the optimality system numerically using forward-backward sweep method [28]. Starting with an initial guess for the control, the state system is solved forward in time. Using those new state values, the adjoint system is solved backward in time. The control is updated using a convex combination of the old control values and the new control values from the characterization. The iterative method is repeated until convergence.

Figure 3 shows the concentration of the pathogen in the environment as a function of time, in both the presence and absence of the control. It is clear that the presence of optimal control can lead to eradication of the pathogen from the environment. More precisely, we note that, in the presence of vector control, the concentration of the pathogen within the environment will be reduced to a level close to 0 when $t \geq 10$ months. We also note that in the absence of the control

the bacteria population has an oscillatory pattern and the amplitude of the oscillations appears to be constant.

Figure 4 illustrates the numbers of exposed and infectious individuals over time with and without optimal control. The results clearly depict that the optimal control strategy significantly reduces the exposed and infectious populations (compared to the case without control), to a level close to 0 when $t > 30$ months. More precisely, the number of exposed and infectious individuals will be reduced to a level close to 0 when $t > 30$ and $t > 60$, respectively. Meanwhile, we observe that the numbers of the pathogen in the environment and the infected population will not converge to zero at the same period. This is due to the long incubation period associated with Guinea worm disease.

Figure 5 shows the optimal control profile as a function of time, with $B = 2 \times 10^{-5}$. As we can observe, the control profile starts from the maximum ($u = 0.8$) initially and stays there for a period of 60 months and then it switches to its minimum ($u = 0.2$) where it remains till the final time. Figure 6 shows the optimal control profile as a function of time, with high costs ($B = 2 \times 10^5$). It is evident that when the cost of the control is high, the control will have to be implemented at its maximum for a short period of time compared to a scenario when the costs are low.

3. Discussion and Conclusions

Guinea worm disease is parasitic waterborne disease that is on the verge of eradication. The numbers of individuals afflicted by this disease has declined from 3.5 million cases in 1986 to only 25 cases in 2016, thanks to the Global Guinea worm eradication program [29].

In this paper, we have proposed, analyzed, and simulated a new mathematical model for Guinea worm disease that incorporates indirect disease transmission, low and high risk susceptible population, and seasonality. Our model is motivated by the fact that observed incidence of Guinea worm disease exhibits strong seasonal fluctuations in the Sahelian zone and the humid and forest zone [5], with morbidity burden of the disease concentrated in a few months each year. In addition, we are also motivated by the fact that in countries where the disease is still a menace (South Sudan, Chad, Mali, and Ethiopia) access to safe water remains a formidable challenge [1].

To account for seasonal fluctuations we modeled the transmission rate, incubation rate, pathogen shedding rate, and pathogen decay rate as periodic functions. To include in heterogeneity on disease transmission we subdivided the susceptible population into two compartments, the low and high risk. We computed the reproduction number $\mathcal{R}_0$ and demonstrated that when $\mathcal{R}_0 < 1$ the model is globally asymptotically stable. We also demonstrated that for $\mathcal{R}_0 > 1$ the disease persists and there exists at least one positive periodic solution. This implies that the disease can be eradicated from the community whenever the basic reproduction number is less than unity; otherwise the disease persists.

Using the Fourier series analysis method we fitted our proposed model to the reported data on symptomatic cases of Guinea worm disease in South Sudan in order to estimate

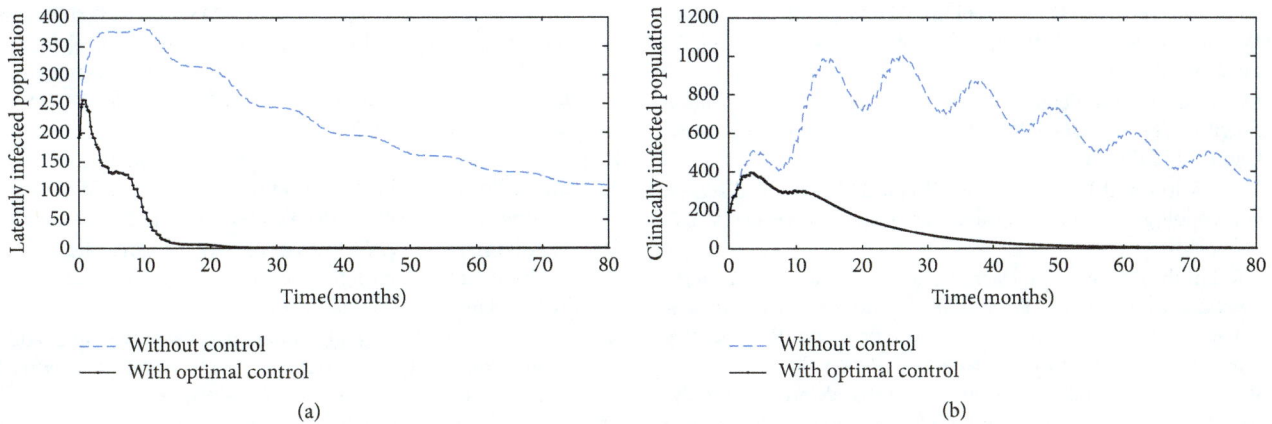

(a)   (b)

FIGURE 4: The numbers of exposed and infectious human population with and without the optimal control.

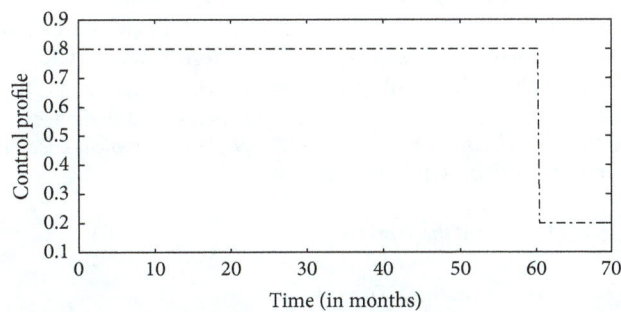

FIGURE 5: The control profile.

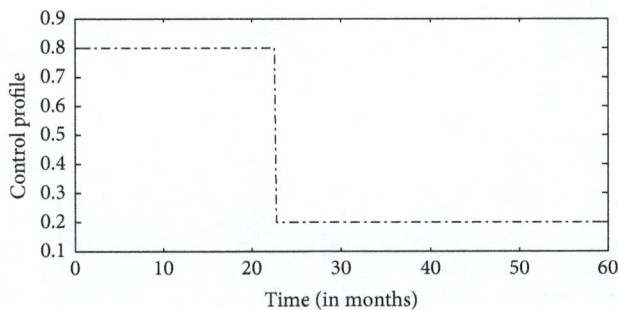

FIGURE 6: The control profile with high costs.

the periodic function for the disease transmission rate, incubation rate, pathogen shedding rate, and pathogen decay rate. Meanwhile, we performed an optimal control study to assess the implications of vector control on Guinea worm disease eradication. Our optimal control aims to minimize the numbers of latently infected and infections individuals at minimal costs. Our results demonstrate that optimal control has the potential to ensure successful and timely elimination of the disease in South Sudan. Further, we also observed that with low costs the control will be carried out at maximum for a longer period of time compared to when the costs are high.

This study has some limitations. In our mathematical model we did not incorporate the life cycle of the parasite, a factor which can significantly enhance our understanding of

Guinea worm disease dynamics. In future we plan to extend this work to incorporate this aspect.

Conflicts of Interest

The authors have no conflicts of interest.

Acknowledgments

The authors are grateful to the handling editor Dr. Keshlan S. Govinder for valuable comments and suggestions that led to an improvement of this paper.

References

[1] B. H. Beyene, A. Bekele, A. Shifara, and Y. Ebsite, "Elimination of Guinea worm in Ethopia; current status of the diseases eradictaion strategies and challenges to the end game. Ethiop Md J," in *current status of the disease's eradictaion strategies and challenges to the end game. Ethiop Md J*, vol. 55, supplement 1, pp. 15–31, 55, 15-31, 2017.

[2] World Health Organization, *World Health Assembly, Resolution WHA 57.9. Elimination of Dracunculiasis: resolution of the 57th World Health Assembly*, Geneva, Switzerland, 2016.

[3] E. Ruiz-Tiben and D. R. Hopkins, "Dracunculiasis (Guinea Worm Disease) Eradication," *Advances in Parasitology*, vol. 61, pp. 275–309, 2006.

[4] O. O. Alew and N. Peters, *Ethiopian Dracunculiasis Eradication Program (EDEP) GOG WOREDA Meeting, National Review*, Gambella, Ethopia, 2015.

[5] K. A. Mitra and R. A. Mawson, "Neglected tropical diseases: epidemiology and global burden," *Trop. Med. Infect. Dis*, vol. 2, p. 36, 2017.

[6] J. R. Smith, P. Cloutier, J. Harrison, and A. Desforges, *A Mathematical Model for the Eradication of Guinea Worm Disease. Understanding the Dynamics of Emerging and Re-Emerging Infectious Diseases Using Mathematical Models*, 2012.

[7] R. Netshikweta and W. Garira, "A multiscale model for the world's first parasitic disease targeted for eradication: guinea worm disease," *Comput Math Methods Med*, vol. 2017, 2017.

[8] I. A. Adetunde, "The Epidemiology of guinea worm infection in Tamale District, in the Northern Region of Ghana," *Journalof Modern Mathematics and Statistics*, pp. 50–54, 2008.

[9] L. Kathryh, "Guinea Worm Disease (Dracunculiasis): Opening a mathematical can of worms," Tech. Rep., Bryn Mawr College, 2012.

[10] World Health Organization, *Public Health Risk Assessment and Interventions - Conflict and Humanitarian Crisis*, Republic of South Sudan, 2014.

[11] United Nations Children's Fund (UNICEF), *State of the World'S Children*, 2013.

[12] G. Biswas, D. P. Sankara, J. Agua-Agum, and A. Maiga, "Dracunculiasis (guinea worm disease): eradication without a drug or a vaccine.," *Philosophical Transactions of the Royal Society B: Biological Sciences*, vol. 368, no. 1623, p. 20120146, 2013.

[13] O. C. Collins and K. S. Govinder, "Incorporating heterogeneity into the transmission dynamics of a waterborne disease model," *Journal of Theoretical Biology*, vol. 356, pp. 133–143, 2014.

[14] S. L. Robertson, M. C. Eisenberg, and J. H. Tien, "Heterogeneity in multiple transmission pathways: Modelling the spread of cholera and other waterborne disease in networks with a common water source," *Journal of Biological Dynamics*, vol. 7, no. 1, pp. 254–275, 2013.

[15] P. van den Driessche and J. Watmough, "Reproduction numbers and sub-threshold endemic equilibria for compartmental models of disease transmission," *Mathematical Biosciences*, vol. 180, pp. 29–48, 2002.

[16] W. Wang and X. Zhao, "Threshold dynamics for compartmental epidemic models in periodic environments," *Journal of Dynamics and Differential Equations*, vol. 20, no. 3, pp. 699–717, 2008.

[17] H. L. Smith and P. Waltman, *The Theory of the Chemostat*, Cambridge University Press, 1995.

[18] X. Q. Zhao, *Dynamical Systems in Population Biology*, Springer, New York, NY, USA, 2003.

[19] L. Liu, X.-Q. Zhao, and Y. Zhou, "A tuberculosis model with seasonality," *Bulletin of Mathematical Biology*, vol. 72, no. 4, pp. 931–952, 2010.

[20] S. Bowong and J. Kurths, "Modeling and analysis of the transmission dynamics of tuberculosis without and with seasonality," *Nonlinear Dynamics*, vol. 67, no. 3, pp. 2027–2051, 2012.

[21] W. Fleming and R. Rishel, *Deterministic and Stochastic Optimal Control*, Springer New York, New York, NY, 1975.

[22] C. J. Silva and D. F. Torres, "A TB-HIV/AIDS coinfection model and optimal control treatment," *Discrete and Continuous Dynamical Systems - Series A*, vol. 35, no. 9, pp. 4639–4663, 2015.

[23] H. Schättler, U. Ledzewicz, and H. Maurer, "Sufficient conditions for strong local optimality in optimal control problems with l2 -type objectives and control constraints, discrete contin," *Discrete and Continuous Dynamical Systems - Series B*, vol. 19, no. 8, pp. 2657–2679, 2014.

[24] C. Silva, H. Maurer, and D. Torres, "Optimal control of a tuberculosis model with state and control delays," *Mathematical Biosciences and Engineering*, vol. 14, no. 1, pp. 321–337, 2017.

[25] L. S. Pontryagin, V. G. Boltyanskii, R. V. Gamkrelidze, and E. F. Mishchenko, *The Mathematical Theory of Optimal Processes*, Wiley, New York, NY, USA, 1962.

[26] A. J. Krener, "The high order maximal principle and its application to singular extremals," *SIAM Journal on Control and Optimization*, vol. 15, no. 2, pp. 256–293, 1977.

[27] H. R. Joshi, S. Lenhart, S. Hota, and F. Agusto, "Optimal control of an SIR model with changing behavior through an education campaign," *Electronic Journal of Differential Equations*, No. 50, 14 pages, 2015.

[28] S. M. Lenhart and J. T. Workman, *Optimal Control Applied to Biological Models*, CRC Press, 2007.

[29] D. Molyneux and D. P. Sankara, "Guinea worm eradication: Progress and challenges— should we beware of the dog?" *PLOS Neglected Tropical Diseases*, vol. 11, no. 4, Article ID e0005495, 2017.

6

A Stochastic Model for Malaria Transmission Dynamics

A Stochastic Model for Malaria Transmission Dynamics

Rachel Waema Mbogo [ORCID], **Livingstone S. Luboobi, and John W. Odhiambo**

Institute of Mathematical Sciences (IMS), Strathmore University, Box 59857 00200, Nairobi, Kenya

Correspondence should be addressed to Rachel Waema Mbogo; rmbogo@strathmore.edu

Academic Editor: Zhidong Teng

Malaria is one of the three most dangerous infectious diseases worldwide (along with HIV/AIDS and tuberculosis). In this paper we compare the disease dynamics of the deterministic and stochastic models in order to determine the effect of randomness in malaria transmission dynamics. Relationships between the basic reproduction number for malaria transmission dynamics between humans and mosquitoes and the extinction thresholds of corresponding continuous-time Markov chain models are derived under certain assumptions. The stochastic model is formulated using the continuous-time discrete state Galton-Watson branching process (CTDSGWbp). The reproduction number of deterministic models is an essential quantity to predict whether an epidemic will spread or die out. Thresholds for disease extinction from stochastic models contribute crucial knowledge on disease control and elimination and mitigation of infectious diseases. Analytical and numerical results show some significant differences in model predictions between the stochastic and deterministic models. In particular, we find that malaria outbreak is more likely if the disease is introduced by infected mosquitoes as opposed to infected humans. These insights demonstrate the importance of a policy or intervention focusing on controlling the infected mosquito population if the control of malaria is to be realized.

1. Introduction

Malaria is an infectious disease caused by the Plasmodium parasite and transmitted between humans through bites of female *Anopheles* mosquitoes. Approximately half of the world's population is at risk of malaria. It remains one of the most prevalent and lethal human infections throughout the world. An estimated 40% of the world's population lives in malaria endemic areas. Most cases and deaths occur in sub-Saharan Africa. It causes an estimated 300 to 500 million cases and 1.5 to 2.7 million deaths each year worldwide. Africa shares 80% of the cases and 90% of deaths [1]. According to the website of the World Health Organization [2] there were approximately 214 million new cases of malaria and 438,000 deaths worldwide in 2015. Most cases were reported in the African region.

Recently, the incidence of malaria has been rising due to drug resistance. Various control strategies have been taken to reduce malaria transmissions. Since the first mathematical model of malaria transmission was introduced by Ross [3], quite a number of mathematical models have been formulated to investigate the transmission dynamics of malaria.

Xiao and Zou [4] used mathematical models to explore a natural concern of possible epidemics caused by multiple species of malaria parasites in one region. They found that epidemics involving both species in a single region are possible.

Li and others [5] considered fast and slow dynamics of malaria model with relapse and analyzed the global dynamics by using the geometric singular perturbation theory. They suggested that a treatment should be given to symptomatic patients completely and adequately rather than to asymptomatic patients. On the other hand, for the asymptomatic patients, their results strongly suggested that, to control and eradicate the malaria, it is very necessary for governments to control the relapse rate strictly. Relapse is when symptoms reappear after the parasites had been eliminated from blood but persist as dormant hypnozoites in liver cells [6]. This commonly occurs between 8 and 24 weeks and is commonly seen with *P. vivax* and *P. ovale* infections. Other papers also consider the influence of relapse in giving up smoking or quitting drinking; please see [7].

Chitnis et al. [8] and Li et al. [5] assumed that the recovered humans have some immunity to the disease and do not get clinically ill but they still harbour low levels of parasite in their blood streams and can pass the infection to mosquitoes. After some period of time, they lose their immunity and return to the susceptible class. Unfortunately, Li and others did not consider that the recovered humans will return to their infectious state because of incomplete treatment.

Stochasticity is fundamental to biological systems. In some situations the system can be treated as a large number of similar agents interacting in a homogeneously mixing environment, and so the dynamics are well-captured by deterministic ordinary differential equations. However, in many situations, the system can be driven by a small number of agents or strongly influenced by an environment fluctuating in space and time [9–12].

Stochastic models incorporate discrete movements of individuals between epidemiological classes and not average rates at which individuals move between classes [13–15]. In stochastic epidemic models, numbers in each class are integers and not continuously varying quantities [13]. A significant possibility is that the last infected individual can recover before the disease is transmitted and the infection can only reoccur if it is reintroduced from outside the population [16]. In contrast, most deterministic models have the flaw that infections can fall to very low levels—well below the point at which there is only one infected individual only to rise up later [17]. In addition, the variability introduced in stochastic models may result in dynamics that differ from the predictions made by deterministic models [16].

For a large population size and a large number of infectious individuals, the deterministic threshold $R_0 > 1$ provides a good prediction of a disease outbreak. However, this prediction breaks down when the outbreak is initiated by a small number of infectious individuals. In this setting, Markov chain (MC) models with a discrete number of individuals are more realistic than deterministic models where the number of individuals is assumed to be continuous-valued [18].

Motivated by these works, in this paper, we propose a model which is an extension of the model formulated by Huo and Qui (2014), who assumed that the pseudorecovered humans can recover and return to the susceptible class or relapse and become infectious again. Using the extended model, we will formulate the basic reproductive number R_0 and use it to compare the disease dynamics of the deterministic and stochastic models in order to determine the effect of randomness in malaria transmission dynamics.

This paper is organized as follows; in Section 2, we present a malaria transmission deterministic model with relapse, which is an extension of the model in [6]. We compute the basic reproduction number, R_0, of the malaria transmission deterministic model using the next-generation matrix approach. The stochastic version of the deterministic model and its underlying assumptions necessary for model formulation are presented and discussed in Section 3. In this section, we also compute the stochastic threshold for disease extinction or invasion by applying the multitype Galton-Watson branching process. In Section 4, we show the relationship between reproductive number of the deterministic model and the thresholds for disease extinction of the stochastic version; we also illustrate our results using numerical simulations. We conclude with a discussion of the results in Section 5.

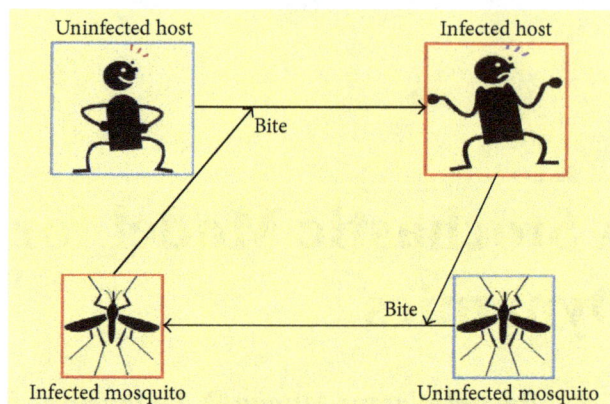

FIGURE 1: The mosquito-human transmission cycle. An infectious mosquito bites a susceptible, uninfected human and transmits the virus via saliva. Once the human becomes infectious (usually accompanied by symptoms), the human can transmit the pathogen to an uninfected mosquito via the blood the mosquito ingests. *Source.* Figure 1 is reproduced from Manore and Hyman (2016) [20].

2. The Malaria Deterministic Model

The interaction of the Mosquito-host is shown in Figure 1.

2.1. Model Formulation. In this section, we introduce a deterministic model of malaria with relapse, an extended form of the model in [6]. The model proposes a more realistic mathematical model of malaria, in which it is assumed that the pseudorecovered humans interact with infected mosquitoes and may acquire more parasites causing reinfection or, due to incomplete treatment, the infection my reoccur (relapse) and return to infectious class. Also the pseudorecovered human may lose their immunity and return to the susceptible class. Given that humans might get repeatedly infected due to not acquiring complete immunity, then the population dynamics are assumed to be described by the SIRS model; hence we consider a deterministic compartmental model which divides the total human population size at time t, denoted by $N(t)$, into susceptible individuals $S_h(t)$ (those who are not currently harbouring the parasite but are liable to be infected), infectious individuals $I_h(t)$ (those already infected and are able to transmit the disease to mosquitoes), and pseudorecovered individuals $R_h(t)$ (those who are treated from the disease but with partial recovery and hence can transmit the disease to mosquitoes). Mosquitoes are assumed not to recover from the parasites so the mosquito population can be described by the SI model.

The structure of model is shown in Figure 2.

2.2. Variable and Parameter Description for the Model. The variables for the model are summarized in description of variables for the malaria transmission model in the Notations.

The parameters for the model are described as in description of parameters for the malaria transmission model in the Notations.

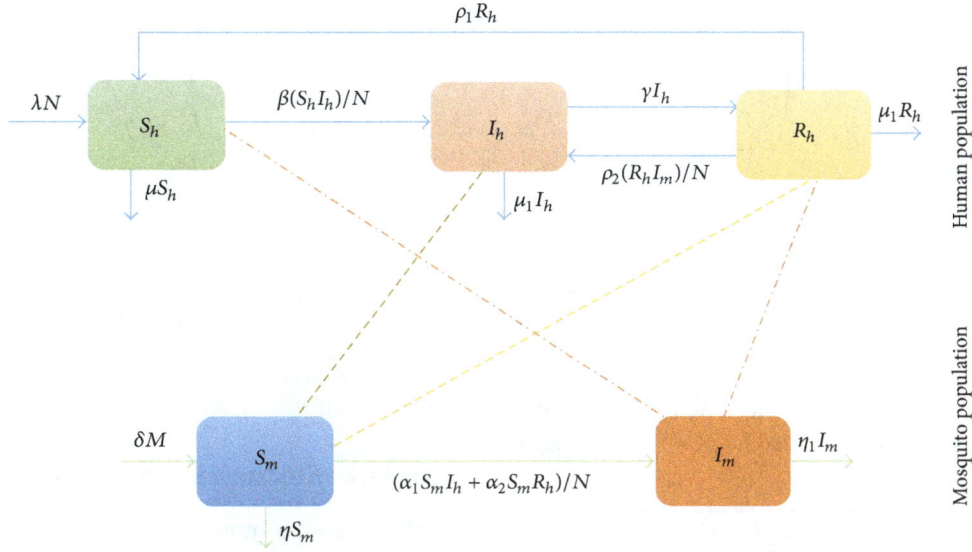

FIGURE 2: Schematic representation of the mosquito-human malaria transmission dynamics.

2.3. Equations of the Model.

Assuming that the disease transmits in a closed system which translates into the simplifying assumption of a constant population size with the birth rate equal to death rate, that is, $\lambda = \mu$ (see [21]) so that $N(t) = N$, hence, the above assumptions lead to the following system of differential equations which describe the interaction between mosquitoes and humans:

$$\frac{dS_h(t)}{dt} = \lambda N - \frac{\beta S_h I_m}{N} + \rho_1 R_h - \mu S_h,$$

$$\frac{dI_h(t)}{dt} = \frac{\beta S_h I_m}{N} + \frac{\rho_2 R_h I_m}{N} - (\gamma + \mu_1) I_h,$$

$$\frac{dR_h(t)}{dt} = \gamma I_h - \left(\rho_1 + \frac{\rho_2 I_m}{N} + \mu_1\right) R_h, \tag{1}$$

$$\frac{dS_m(t)}{dt} = \delta M - \frac{\alpha_1 S_m I_h}{N} - \frac{\alpha_2 S_m R_h}{N} - \eta S_m,$$

$$\frac{dI_m(t)}{dt} = \frac{\alpha_1 S_m I_h}{N} + \frac{\alpha_2 S_m R_h}{N} - \eta_1 I_m,$$

where S_h, I_h, R_h, S_m, and I_m represent the number of susceptible humans, infectious humans, recovered humans, susceptible mosquitoes, and infectious mosquitoes, respectively.

2.4. Computation of R_0 Using the Next-Generation Matrix Approach.

The basic reproduction number, R_0, is defined as the secondary infections produced by one infective agent that is introduced into an entirely susceptible population at the disease-free equilibrium. The next-generation matrix approach is frequently used to compute R_0. The original nonlinear system of ODEs including these compartments can be written as $\partial X_i/\partial t = \mathscr{F} - \mathscr{V}$, where $\mathscr{F} = (\mathscr{F}_i)$ and $\mathscr{V} = (\mathscr{V}_i)$ represent new infections and transfer between compartments, respectively [22–24].

The Jacobian matrices of $\mathscr{F}(X)$ and $\mathscr{V}(X)$ at the disease-free equilibrium E_0 are, respectively,

$$F = \mathscr{F}(E_0) = \begin{pmatrix} 0 & 0 & \beta \\ 0 & 0 & 0 \\ \alpha_1 & \alpha_2 & 0 \end{pmatrix}$$

$$V = \mathscr{V}(E_0) = \begin{pmatrix} \gamma + \mu_1 & -\rho_2 & 0 \\ -\gamma & \mu_1 + \rho_1 + \rho_2 & 0 \\ 0 & 0 & \eta_1 \end{pmatrix}. \tag{2}$$

The matrix $\mathscr{J} = F - V$ is the Jacobian matrix evaluated at the DFE.

$$\mathscr{J} = F - V = \begin{pmatrix} -\gamma - \mu_1 & \rho_2 & \beta \\ \gamma & -\mu_1 - \rho_1 - \rho_2 & 0 \\ \alpha_1 & \alpha_2 & -\eta_1 \end{pmatrix}. \tag{3}$$

The inverse matrix of V is

$$V^{-1} = \begin{pmatrix} \dfrac{\mu_1 + \rho_1 + \rho_2}{(\gamma + \mu_1)(\mu_1 + \rho_1) + \mu_1\rho_2} & \dfrac{\rho_2}{(\gamma + \mu_1)(\mu_1 + \rho_1) + \mu_1\rho_2} & 0 \\ \dfrac{\gamma}{(\gamma + \mu_1)(\mu_1 + \rho_1) + \mu_1\rho_2} & \dfrac{\gamma + \mu_1}{(\gamma + \mu_1)(\mu_1 + \rho_1) + \mu_1\rho_2} & 0 \\ 0 & 0 & \dfrac{1}{\eta_1} \end{pmatrix}.$$

$$FV^{-1} = \begin{pmatrix} 0 & 0 & \dfrac{\beta}{\eta} \\ 0 & 0 & 0 \\ \dfrac{\gamma\alpha_2 + \alpha_1(\mu + \rho_1 + \rho_2)}{(\gamma + \mu)(\mu + \rho_1) + \mu\rho_2} & \dfrac{(\gamma + \mu)\alpha_2 + \alpha_1\rho_2}{(\gamma + \mu)(\mu + \rho_1) + \mu\rho_2} & 0 \end{pmatrix}.$$

$$(4)$$

The matrix $\mathcal{F}\mathcal{V}^{-1}$ is called the next-generation matrix. The (i, k) entry of FV^{-1} indicates the expected number of new infections in compartment i produced by the infected individual originally introduced into compartment k. The model reproduction number, R_0, which is defined as the spectral radius of $\mathcal{F}\mathcal{V}^{-1}$ and denoted by $\rho(\mathcal{F}\mathcal{V}^{-1})$, is given by

$$R_0 = \rho\left(\mathcal{F}\mathcal{V}^{-1}\right) = \sqrt{\frac{\beta\left[\gamma\alpha_2 + \alpha_1(\rho_1 + \rho_2 + \mu_1)\right]}{\eta_1\left[(\gamma + \mu_1)(\rho_1 + \mu_1) + \mu_1\rho_2\right]}}. \quad (5)$$

Here R_0 is associated with disease transmission by infected humans as well as the infection of susceptible humans by infected mosquitoes.

Simplifying (5) we find an equivalent form, which is defined as the product of the transmission from mosquito to human and from human to mosquito as follows:

$$R_0 = \sqrt{\left(\frac{\beta}{\eta_1}\right)\left(\frac{\alpha_1}{(\gamma + \mu_1)} + \frac{\alpha_2}{(\rho_1 + \rho_2 + \mu_1)}\right)}. \quad (6)$$

In (6), R_0 is associated with disease transmission by infected mosquitoes as well as infection of susceptible mosquitoes by infected humans. The term β/η_1 represents the number of infected humans generated by infectious mosquito in its life span. The term $\alpha_1/(\gamma + \mu_1)$ represents the number of infected mosquitoes generated by infectious human during the infectious period of the individual while the term $\alpha_2/(\rho_1 + \rho_2 + \mu_1)$ represents the number of infected mosquitoes generated by pseudorecovered human during his/her infectious period. Susceptible humans acquire infection following effective contacts with infected mosquitoes. Susceptible mosquitoes acquire malaria infection from infected humans in two ways, namely, by infectious human or pseudorecovered humans.

Also, from (6), we can rewrite R_0 as infection between mosquitoes and infectious humans or infection between mosquitoes and pseudorecovered humans.

$$R_0 = \sqrt{R_{01} + R_{02}}, \quad (7)$$

where

$$R_{01} = \left(\frac{\beta}{\eta_1}\right)\left(\frac{\alpha_1}{(\gamma + \mu_1)}\right), \quad (8)$$

$$R_{02} = \left(\frac{\beta}{\eta_1}\right)\left(\frac{\alpha_2}{(\rho_1 + \rho_2 + \mu_1)}\right). \quad (9)$$

R_{01} in (8) represents the product of individuals generated by infectious human and infected mosquito, respectively, during their infectious time.

R_{02} in (9) represents the product of individuals generated by pseudorecovered human and infected mosquito, respectively, during their infectious time.

For model (1), the disease dies out if $R_0 < 1$ and the disease persists if $R_0 > 1$. Hence $R_0 < 1$, iff $R_{01} < 1$ and $R_{02} < 1$.

3. Malaria Stochastic Epidemic Model

For the mosquito-human malaria dynamics, since time is continuous and the disease states are discrete, we derive the stochastic version of the deterministic model (1) using a continuous-time discrete state Galton-Watson branching process (CTDSGWbp), which is a type of stochastic process. The malaria CTDSGWbp model is a time-homogeneous process with the Markov property. The model takes into account random effects of individual birth and death processes, that is, demographic variability. A stochastic process is defined by the probabilities with which different events happen in a small time interval Δt. In our model there are two possible events (production and death/removal) for each population. The corresponding rates in the deterministic model are replaced in the stochastic version by the probabilities that any of these events occur in a small time interval of length Δt [25, 26].

3.1. Model Formulation. Let time be continuous, $t \in [0; \infty)$, and let $S_h(t)$, $I_h(t)$, $R_h(t)$, $S_m(t)$, and $I_m(t)$ represent discrete random variables for the number of susceptible humans, infectious humans, recovered humans, susceptible mosquitoes and infectious mosquitoes, respectively, with finite space.

$$S_h(t), I_h(t), R_h(t), S_m(t), I_m(t)$$
$$\in \{0, 1, 2, 3, \dots, H\}, \quad (10)$$

where H is positive and represents the maximum size of the populations.

If a disease emerges from one infectious group with $R_0 > 1$ and if i infective agents are introduced into a wholly susceptible population, then the probability of a major disease outbreak is approximated by $1 - (1/R_0)^i$ while the probability of disease extinction is approximately $(1/R_0)^i$ [13]. However, this result does not hold if the infection emanates from multiple infectious groups [27]. For multiple infectious groups, the stochastic thresholds depend on two factors, namely, the number of individuals in each group and the probability of disease extinction for each group. Further, the persistence of an infection into a wholly susceptible population is not guaranteed by having R_0 greater than one.

For CTDSGWbp models, the transition from one state to a new state may occur at any time t. If the process begins in

TABLE 1: State transitions and rates for the CTDSGWbp malaria model.

Event	Population components at t	Population components at $t + \Delta t$	Transition probabilities
Birth of humans	$(S_h, I_h, R_h, S_m, I_m)$	$(S_h + 1, I_h, R_h, S_m, I_m)$	$\lambda \Delta t$
Death of susceptible humans	$(S_h, I_h, R_h, S_m, I_m)$	$(S_h - 1, I_h, R_h, S_m, I_m)$	$\mu S_h \Delta t$
Infection of humans	$(S_h, I_h, R_h, S_m, I_m)$	$(S_h - 1, I_h + 1, R_h, S_m, I_m)$	$\beta (S_h/N) I_m \Delta t$
Recovery rate of humans	$(S_h, I_h, R_h, S_m, I_m)$	$(S_h + 1, I_h, R_h - 1, S_m, I_m)$	$\rho_1 R_h \Delta t$
Relapse rate of humans	$(S_h, I_h, R_h, S_m, I_m)$	$(S_h, I_h + 1, R_h - 1, S_m, I_m)$	$\rho_2 R_h \Delta t$
Treatment rate	$(S_h, I_h, R_h, S_m, I_m)$	$(S_h, I_h - 1, R_h + 1, S_m, I_m)$	$\gamma I_h \Delta t$
Death of infected humans	$(S_h, I_h, R_h, S_m, I_m)$	$(S_h, I_h - 1, R_h, S_m, I_m)$	$\mu_1 I_h \Delta t$
Death of recovered humans	$(S_h, I_h, R_h, S_m, I_m)$	$(S_h, I_h, R_h - 1, S_m, I_m)$	$\mu_1 R_h \Delta t$
Birth of mosquitoes	$(S_h, I_h, R_h, S_m, I_m)$	$(S_h, I_h, R_h, S_m + 1, I_m)$	$\delta \Delta t$
Death of susceptible mosquitoes	$(S_h, I_h, R_h, S_m, I_m)$	$(S_h, I_h, R_h, S_m - 1, I_m)$	$\eta S_m \Delta t$
Infection of mosquitoes from humans	$(S_h, I_h, R_h, S_m, I_m)$	$(S_h, I_h, R_h, S_m - 1, I_m + 1)$	$(\alpha_1 I_h + \alpha_2 R_h)(S_m/N) \Delta t$
Death of infected mosquitoes	$(S_h, I_h, R_h, S_m, I_m)$	$(S_h, I_h, R_h, S_m, I_m - 1)$	$\eta_1 I_m \Delta t$

a state $G(0)$, after a random time period τ, it transits to a new state $G(\tau)$. The process remains in state $G(\tau)$ for a random time t after which it moves through to the new state $G(p)$, with $p = \tau + t$ [27]. This process continues throughout the model. The state transitions and rates for the stochastic model are presented in Table 1.

3.2. The Branching Process Approximation. We use branching process to analyze the malaria dynamics near the disease-free equilibrium. Since infectious human, pseudorecovered humans, and infectious mosquitoes are the only sources of infection, we apply the multitype branching process in the three variables $I_h(t)$, $R_h(t)$, and $I_M(t)$. The susceptible humans and mosquitoes are assumed to be at the disease-free state [28]. We use the multitype Galton-Watson branching process (GWbp) to determine disease invasion and extinction probabilities. More review on the GWbp branching theory process can be accessed through [18, 27, 28]. We now define the offspring pgfs for the three variables, where each offspring pgf has a general form

$$G_i (s_1, s_2, \ldots, s_n)$$
$$= \sum_{k_n=0}^{\infty} \sum_{k_{n-1}=0}^{\infty} \cdots \sum_{k_1=0}^{\infty} P_i (k_1, k_2, \ldots, k_n) s_1^{k_1} s_2^{k_2} \cdots s_n^{k_n}, \quad (11)$$

where $P_i(k_1, k_2, \ldots, k_n) = \text{prob}(X_{i1} = k_1, X_{i2} = k_2, \ldots, X_{in} = k_n)$ is the probability that one infected individual of type i gives birth to k_j individuals of type j; see [29].

For one malaria infectious human, there are three possible events: infection of a mosquito, recovery of the infectious human, or death of the infectious human. The offspring pgf for infectious humans define the probabilities associated with the "birth" of secondary infectious mosquito or the "death" of the initial infectious human, given that the process started

with only one infectious human; that is $I_h(0) = 1$, $R_h(0) = 0$, and $I_M(0) = 0$.

The offspring pgf for I_h is given by

$$G_1 (s_1, s_2, s_3) = \sum_{k_1=0}^{\infty} \sum_{k_2=0}^{\infty} \sum_{k_3=0}^{\infty} P_1 (k_1, k_2, k_3) s_1^{k_1} s_2^{k_2} s_3^{k_3}, \quad (12)$$

where $P_1(k_1, k_2, k_3) = \text{prob}(X_{11} = k_1, X_{12} = k_2, X_{13} = k_3)$ is the probability that one infectious human through infection produces another infectious human k_1 or a pseudorecovered human k_2 or an infectious mosquito k_3.

Similarly, the offspring pgf for R_h is given by

$$G_2 (s_1, s_2, s_3) = \sum_{k_1=0}^{\infty} \sum_{k_2=0}^{\infty} \sum_{k_3=0}^{\infty} P_2 (k_1, k_2, k_3) s_1^{k_1} s_2^{k_2} s_3^{k_3}, \quad (13)$$

where $P_2(k_1, k_2, k_3) = \text{prob}(X_{21} = k_1, X_{22} = k_2, X_{23} = k_3)$ is the probability that one pseudorecovered human through infection produces an infectious human k_1 or another pseudorecovered human k_2 or an infectious mosquito k_3.

Lastly, the offspring pgf for I_m is given by

$$G_3 (s_1, s_2, s_3) = \sum_{k_1=0}^{\infty} \sum_{k_2=0}^{\infty} \sum_{k_3=0}^{\infty} P_3 (k_1, k_2, k_3) s_1^{k_1} s_2^{k_2} s_3^{k_3}, \quad (14)$$

where $P_3(k_1, k_2, k_3) = \text{prob}(X_{31} = k_1, X_{32} = k_2, X_{33} = k_3)$ is the probability that one infectious mosquito through infection produces an infections human k_1 or a pseudorecovered human k_2 or another infectious mosquito k_3.

The power to which s_j is raised is the number of infectious individuals generated from one infectious individual. If an individual recovers or dies, then no new infections are generated, hence (s_j^0).

The offspring pgfs for I_h, R_h, and I_m are used to calculate the expected number of offsprings produced by a single infectious human or by pseudorecovered human or infectious mosquito. They are also used to calculate the probability of disease extinction.

The specific offspring pgfs for I_h, R_h, and I_m are defined using the rates in description of parameters for the malaria transmission model in the Notations, when the initial susceptible populations are near disease-free equilibrium, $S_h(0) \approx N$ and $S_m(0) \approx M$.

From (12), the offspring pgf for infectious human, given $I_h(0) = 1$, $R_h(0) = 0$, and $I_M(0) = 0$, is given by

$$G_1(s_1, s_2, s_3) = \frac{\alpha_1 s_1 s_3 + \gamma + \mu_1}{\alpha_1 + \gamma + \mu_1}. \tag{15}$$

For G_1, one infectious human dies or is treated with probability $(\mu_1 + \gamma)/(\alpha_1 + \gamma + \mu_1)$; this means an infectious human dies before infecting a susceptible mosquito. The term $\gamma/(\alpha_1 + \gamma + \mu_1)$ represents the probability that the infectious human is treated and moves to the pseudorecovered class, this results in "$X_{11} = 0$, $X_{12} = 01$, and $X_{13} = 0$, though there is movement of an infectious human to pseudorecovered state due to partial treatment (this is not new offspring)." The infectious human infects a mosquito with probability $\alpha_1/(\alpha_1 + \gamma + \mu_1)$. This means an infectious human infects a susceptible mosquito and remains infectious, which results in "$X_{11} = 1$, $X_{12} = 0$, and $X_{13} = 1$." Note the term $s_1 s_3$ in (15) means one infectious human generates one infectious mosquito (s_3 raised to power one) and remains infectious (s_1 raised to power one).

For one pseudorecovered human, there are four events: infection of a mosquito, relapse to infected class, successful treatment of the pseudorecovered human, or death of the recovered host. Similarly, from (13), the offspring pgf for recovered humans given $I_h(0) = 0$, $R_h(0) = 1$, and $I_M(0) = 0$ is given by

$$G_2(s_1, s_2, s_3) = \frac{\alpha_2 s_2 s_3 + \rho_1 + \rho_2 + \mu_1}{\alpha_2 + \rho_1 + \rho_2 + \mu_1}. \tag{16}$$

For G_2, one pseudorecovered host dies or relapses or is fully treated with probability $(\rho_1 + \rho_2 + \mu_1)/(\alpha_2 + \rho_1 + \rho_2 + \mu_1)$ or infects a mosquito with probability $\alpha_2/(\alpha_2 + \rho_1 + \rho_2 + \mu_1)$. This means a pseudorecovered human infects a susceptible mosquito and remains infectious, which results in "$X_{21} = 0$, $X_{22} = 1$, and $X_{23} = 1$." Note the term $s_2 s_3$ in (16) means one recovered human generates one infectious mosquito (s_3 raised to power one) and remains infectious (s_2 raised to power one).

For one infectious mosquito, there are only two events: infection of a susceptible human or death of the mosquito.

From (14), the offspring pgf for infected mosquito given $I_h(0) = 0$, $R_h(0) = 0$, and $I_M(0) = 1$ is given by

$$G_3(s_1, s_2, s_3) = \frac{\beta s_1 s_3 + \eta_1}{\beta + \eta_1}. \tag{17}$$

For G_3, one infectious mosquito dies with probability $\eta_1/(\beta + \eta_1)$ or infects a human with probability $\beta/(\beta + \eta_1)$. This means an infectious mosquito infects a susceptible human and remains infectious, which results in "$X_{31} = 1$, $X_{32} = 0$, and $X_{33} = 1$." Note the term $s_1 s_3$ in (17) means one infectious mosquito generates one infectious human (s_1 raised to power one) and remains infectious (s_3 raised to power one).

3.3. The Relationship between R_0 and the Stochastic Threshold S_0. The offspring pgfs, evaluated at $(1, 1, 1)$, gives the expectation matrix with elements m_{ji}. Below are the offspring pgfs evaluated at $(1, 1, 1)$.

$$\frac{\partial G_1}{\partial s_1} = \frac{\alpha_1}{\alpha + \gamma + \mu_1},$$

$$\frac{\partial G_1}{\partial s_2} = \gamma,$$

$$\frac{\partial G_1}{\partial s_3} = \frac{\alpha_1}{\alpha + \gamma + \mu_1},$$

$$\frac{\partial G_2}{\partial s_1} = \rho_2,$$

$$\frac{\partial G_2}{\partial s_2} = \frac{\alpha_2}{\alpha_2 + \rho_1 + \rho_2 + \mu_1}, \tag{18}$$

$$\frac{\partial G_2}{\partial s_3} = \frac{\alpha_2}{\alpha_2 + \rho_1 + \rho_2 + +\mu_1},$$

$$\frac{\partial G_3}{\partial s_1} = \frac{\beta}{\beta + \eta_1},$$

$$\frac{\partial G_3}{\partial s_2} = 0,$$

$$\frac{\partial G_3}{\partial s_3} = \frac{\beta}{\beta + \eta_1}.$$

The expectation matrix of the offspring pgfs, evaluated at $(1, 1, 1)$, is given by

$$\mathcal{M} = \begin{pmatrix} \dfrac{\partial G_1}{\partial s_1} & \dfrac{\partial G_2}{\partial s_1} & \dfrac{\partial G_3}{\partial s_1} \\[2mm] \dfrac{\partial G_1}{\partial s_2} & \dfrac{\partial G_2}{\partial s_2} & \dfrac{\partial G_3}{\partial s_2} \\[2mm] \dfrac{\partial G_1}{\partial s_3} & \dfrac{\partial G_2}{\partial s_3} & \dfrac{\partial G_3}{\partial s_3} \end{pmatrix} = \begin{pmatrix} \dfrac{\alpha_1}{\gamma + \mu_1 + \alpha_1} & \dfrac{\rho_2}{\gamma + \mu_1 + \alpha_1} & \dfrac{\alpha_1}{\alpha + \gamma + \mu_1} \\[2mm] \dfrac{\gamma}{\mu_1 + \alpha_2 + \rho_1 + \rho_2} & \dfrac{\alpha_2}{\mu_1 + \alpha_2 + \rho_1 + \rho_2} & \dfrac{\alpha_2}{\mu_1 + \alpha_2 + \rho_1 + \rho_2} \\[2mm] \dfrac{\beta}{\beta + \eta_1} & 0 & \dfrac{\beta}{\beta + \eta_1} \end{pmatrix}. \tag{19}$$

The entries m_{11}, m_{21}, and m_{31} represent the expected number of infectious humans, pseudorecovered humans, and infectious mosquitoes, respectively, produced by one infectious human. Similarly, the entries m_{12}, m_{22}, and m_{32} represent the expected number of infectious humans, pseudorecovered humans, and infectious mosquitoes, respectively, produced by one pseudorecovered human. Lastly, the entries m_{13}, m_{23}, and m_{33} represent the expected number of infectious humans, pseudorecovered humans, and infectious mosquitoes, respectively, produced by one infectious mosquito.

The matrix $\mathscr{C} = \mathscr{W}(\mathscr{M} - I)$ is the Jacobian matrix evaluated for the stochastic version.

$$\mathscr{C} = \mathscr{W}(\mathscr{M} - \mathscr{I})$$

$$= \begin{pmatrix} -\gamma - \mu_1 & \gamma & \alpha_1 \\ \rho_2 & -\mu_1 - \rho_1 - \rho_2 & \alpha_2 \\ \beta & 0 & -\eta_1 \end{pmatrix}, \quad (20)$$

where $\mathscr{W} = \text{diag}(\gamma + \mu_1 + \alpha_1, \mu_1 + \alpha_2 + \rho_1 + \rho_2, \beta + \eta_1)$ is a diagonal matrix and $\mathscr{I}$ is the identity matrix.

From (3) and (20), we show that

$$\mathscr{F} - \mathscr{V} = \mathscr{W}(\mathscr{M} - \mathscr{I})^T$$

$$= \begin{pmatrix} -\gamma - \mu_1 & \rho_2 & \beta \\ \gamma & -\mu_1 - \rho_1 - \rho_2 & 0 \\ \alpha_1 & \alpha_2 & -\eta_1 \end{pmatrix}. \quad (21)$$

The spectral radius of matrix $\mathscr{M}$ obtained by finding the eigenvalues of matrix $\mathscr{M}$ is given by

$$S_0 = \rho(\mathscr{M}) = \max \left\{ \frac{\beta(\gamma + \mu_1) + (2\beta + \eta_1)\alpha_1}{(\beta + \eta_1)(\gamma + \mu_1 + \alpha_1)}, \right.$$

$$\left. \frac{\alpha_2}{\alpha_2 + \rho_1 + \rho_2 + \mu_1} \right\}. \quad (22)$$

Taking the first expression of $S_0 = \rho(\mathscr{M})$ in (24), we have

$$S_0 = \rho(\mathscr{M}) = \frac{\beta(\gamma + \mu_1 + \alpha_1) + \alpha_1(\beta + \eta_1)}{(\beta + \eta_1)(\gamma + \mu_1 + \alpha_1)}$$

$$= \left(\frac{\beta}{\beta + \eta_1} \right) + \left(\frac{\alpha_1}{\gamma + \mu_1 + \alpha_1} \right). \quad (23)$$

This gives the probability of malaria transmission by either infectious human or by infectious mosquito. From (23),

$$S_0 = \rho(\mathscr{M}) = \frac{\beta(\gamma + \mu_1 + \alpha_1) + \alpha_1(\beta + \eta_1)}{(\beta + \eta_1)(\gamma + \mu_1 + \alpha_1)}. \quad (24)$$

The probability of disease extinction is one if $\rho(\mathscr{M}) < 1$. Hence from (24) we have

$$\beta(\gamma + \mu_1 + \alpha_1) + \alpha_1(\beta + \eta_1)$$

$$< (\beta + \eta_1)(\gamma + \mu_1 + \alpha_1). \quad (25)$$

Expanding and simplifying the inequality, we have

$$\beta\alpha_1 < \eta_1(\gamma + \mu_1) \quad (26)$$

which reduces to

$$\left(\frac{\beta}{\eta_1} \right) \left(\frac{\alpha_1}{(\gamma + \mu_1)} \right) < 1 \implies$$

$$R_{01} < 1. \quad (27)$$

When the infection is between infectious humans and infectious mosquitoes, (27) is true. The result in (27) agrees with the deterministic reproduction number for disease elimination. Hence we conclude that the probability of disease elimination in the CTDSGWbp model is one iff

$$\rho(\mathscr{M}) < 1 \implies$$

$$R_0 < 1. \quad (28)$$

3.4. Deriving Probability of Disease Extinction P_0 Using Branching Process Approximation. To find the probability of extinction (no outbreak), we compute the fixed points of the system $(q_1, q_2, q_3) \in (0, 1)$ of the offspring pgfs for the three infectious stages; that is, we solve $G_{I_h} = q_1$, $G_{R_h} = q_2$, and $G_{I_m} = q_3$. The solutions of these systems are $(1, 1, 1)$ and (q_1, q_2, q_3); see [19]. Equating $G_1(s_1, s_2, s_3)$ in (15) to q_1, then letting $s_1 = q_1$ and $s_3 = q_3$, and solving for q_1, we have

$$q_1 = \frac{(\beta + \eta_1)(\gamma + \mu_1)}{\beta(\gamma + \mu_1 + \alpha_1)}$$

$$= \frac{\gamma + \mu_1}{\alpha_1 + \gamma + \mu_1} + \frac{\alpha_1}{\alpha_1 + \gamma + \mu_1} \left(\frac{1}{R_{01}} \right). \quad (29)$$

Equating $G_2(s_1, s_2, s_3)$ in (16) to q_2, letting $s_2 = q_2$ and $s_3 = q_3$, and solving for q_2, we have

$$q_2$$

$$= \frac{(\beta + \eta_1)\alpha_1(\mu_1 + \rho_1 + \rho_2)}{-\eta_1(\gamma + \mu_1)\alpha_2 + \alpha_1(\beta\alpha_2 + (\beta + \eta_1)(\mu_1 + \rho_1 + \rho_2))}. \quad (30)$$

Equating $G_3(s_1, s_2, s_3)$ in (17) to q_3, then letting $s_1 = q_1$ and $s_3 = q_3$, and solving for q_3, we have

$$q_3 = \frac{\eta_1(\gamma + \mu_1 + \alpha_1)}{(\beta + \eta_1)\alpha_1} = \frac{\eta_1}{\beta + \eta_1} + \frac{\beta}{\beta + \eta_1} \left(\frac{1}{R_{01}} \right). \quad (31)$$

The expression for q_1 in (29) has a biological interpretation. Beginning from one infectious human, there is no outbreak if the infectious human recovers or dies with probability $(\gamma + \mu_1)/(\alpha_1 + \gamma + \mu_1)$ or if there is no successful transmission to a susceptible mosquito with probability $(\alpha_1/(\alpha_1 + \gamma + \mu_1))(1/R_{01})$. This implies that if there is successful contact, then the probability of successful transmission from infectious human to susceptible mosquito is $1 - 1/R_{01}$.

The expression for q_3 in (31) has a biological interpretation. Beginning from one infectious mosquito, there is no outbreak if the infectious mosquito dies with probability

TABLE 2: Model parameter values.

Parameter	Description	Units	Parameter values	Source
λ	Birth rate of humans	Per day	0.000039	[5]
μ	Death rate of susceptible humans	Per day	0.000039	[5]
β	Infection rate of humans	Per day	0.02	[19]
μ_1	Death rate of infected humans	Per day	0.00039	[5]
ρ_1	Recovery rate of humans	Per day	0.01	[19]
ρ_2	Relapse rate of humans	Per day	0.002	[5]
γ	Treatment rate	Per day	0.037	[5]
μ_1	Death rate of recovered humans	Per day	0.00034	[5]
δ	Birth rate of mosquitoes	Per day	0.143	[5]
η	Death rate of susceptible mosquitoes	Per day	0.143	[5]
α_1	Infection rate from infectious human	Per day	0.072	[5]
α_2	Infection rate from recovered human	Per day	0.0072	[5]
η_1	Death rate of infected mosquitoes	Per day	0.143	[19]

$\eta_1/(\beta + \eta_1)$ or if there is no successful transmission to a susceptible human with probability $(\beta/(\beta + \eta_1))(1/R_{01})$.

There are some other important relationships; if disease transmission is by infectious mosquitoes as well as infection of susceptible mosquitoes by infectious humans, then

$$q_1 * q_3 = \frac{1}{R_{01}}. \tag{32}$$

From (32), we see that, in both mosquito and human populations, the probability of no successful transmission from infectious human to susceptible mosquito and from infected mosquito to susceptible human is $1/R_{01}$.

If disease transmission is by infectious mosquitoes as well as infection of susceptible mosquitoes by pseudorecovered humans, then

$$q_2 * q_3 = \frac{1}{R_{02}}. \tag{33}$$

From (33), we see that, in both mosquito and human populations, the probability of no successful transmission from pseudorecovered humans to susceptible mosquito and from infected mosquito to pseudorecovered humans is $1/R_{02}$.

To compute the probability of disease extinction and of an outbreak for our malaria model, we recall that, for multiple infectious groups, the stochastic thresholds depend on two factors, namely, the number of initial individuals in each group and the probability of disease extinction for each group. Using q_1, q_2, and q_3 in (29)–(31) and assuming initial individuals for infectious humans, pseudorecovered humans and infected mosquitoes are $I_h(0) = h_0$, $R_h(0) = r_0$, and $I_m(0) = m_0$, respectively. Then the probability of malaria clearance is given by

$$P_0 = q_1^{h_0} * q_2^{r_0} * q_3^{m_0} = \left(\frac{(\beta + \eta_1)(\gamma + \mu_1)}{\beta(\gamma + \mu_1 + \alpha_1)} \right)^{h_0}$$

$$* \left(\frac{\alpha_1(\beta + \eta_1)(\mu_1 + \rho_1 + \rho_2)}{-\eta_1 \alpha_2(\gamma + \mu_1 + \alpha_1) + \alpha_1(\beta + \eta_1)(\alpha_2 + \mu_1 + \rho_1 + \rho_2)} \right)^{r_0} \tag{34}$$

$$* \left(\frac{\eta_1(\gamma + \mu_1 + \alpha_1)}{(\beta + \eta_1)\alpha_1} \right)^{m_0}.$$

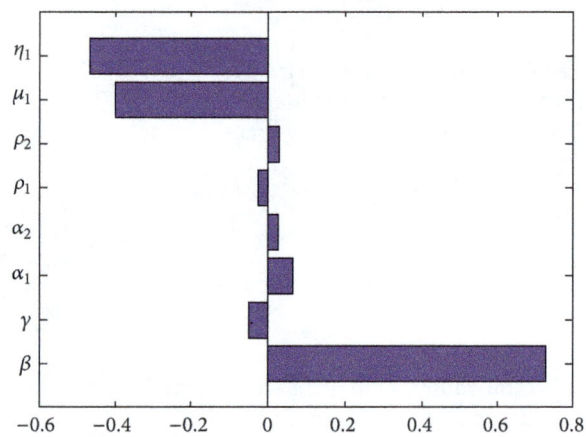

FIGURE 3: Tornado plot for parameters that influence R_0.

4. Numerical Simulations

In this section, we illustrate numerically the disease dynamics of model (1) using parameter values in Table 2. The numerical simulations are done using Maple codes.

4.1. Effects of Model Parameters on R_0. Using parameter values in Table 2, we identified how different input parameters affect the reproduction number R_0 of our model as shown in Figure 3.

From Figure 3, we see that increase in γ, ρ_1, μ_1, and η_1 will decrease R_0. Also decrease in β, α_1, α_2, and ρ_2 will decrease R_0.

From the Tornado plot, the infection of susceptible humans by infected mosquitoes (denoted by β) is a major factor in the malaria transmission dynamics. Reducing β would reduce R_0 significantly hence reducing the possibility of disease outbreak. Vector control is the main way to prevent and reduce malaria transmission. If coverage of vector control interventions within a specific area is high enough, then a measure of protection will be conferred across the community. WHO recommends protection for all people at risk of

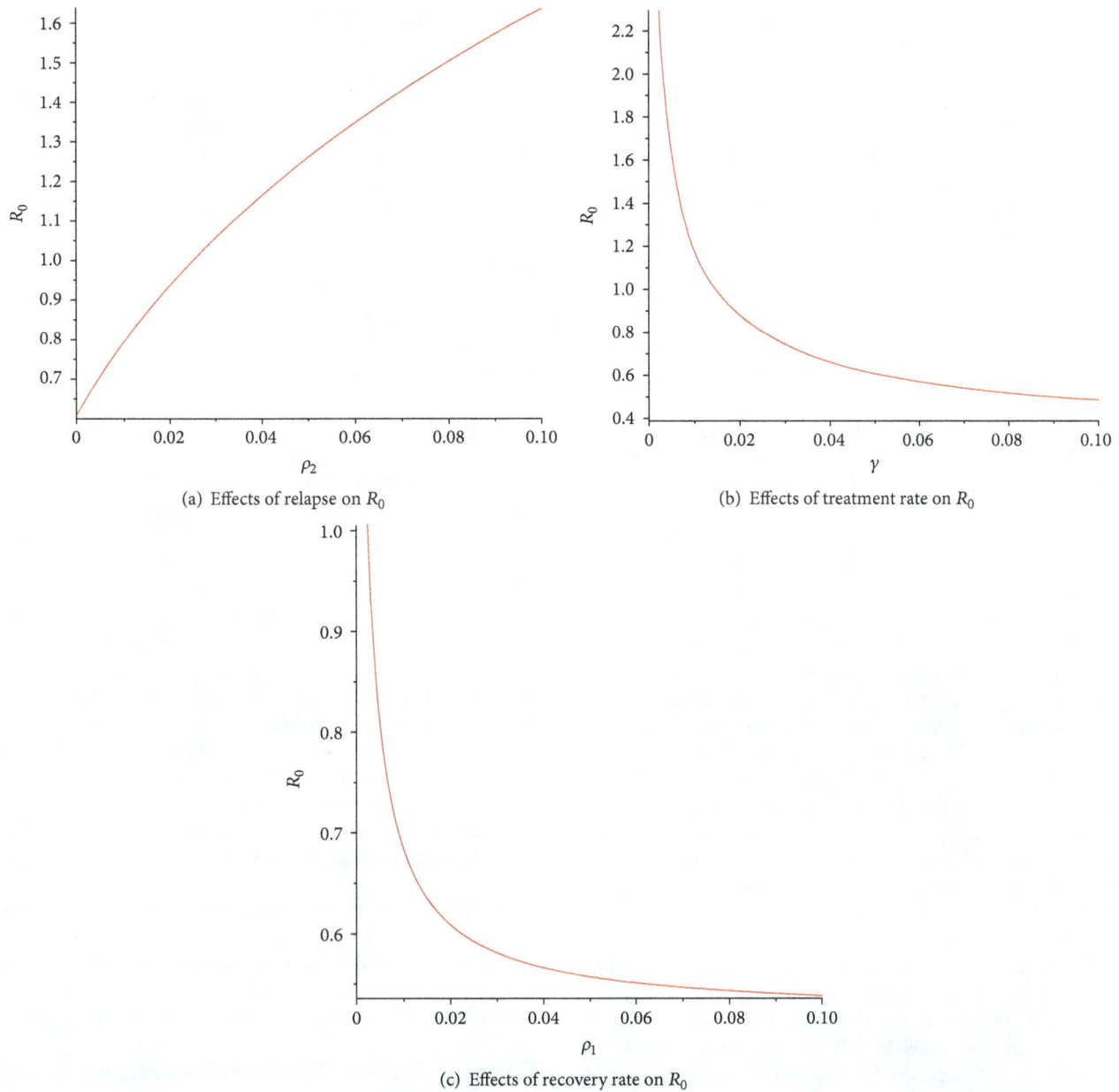

(a) Effects of relapse on R_0

(b) Effects of treatment rate on R_0

(c) Effects of recovery rate on R_0

FIGURE 4: Effects of relapse and recovery rate on R_0.

malaria with effective malaria vector control which agrees with the information from Figure 3. Two forms of vector control insecticide-treated mosquito nets and indoor residual spraying are effective in a wide range of circumstances. Another factor which affects the malaria transmission significantly is the death rate of infected mosquitoes (denoted by η_1). Increasing η_1 would decrease R_0 hence reducing the chances of disease outbreak.

4.1.1. Graphical Representation of Parameters Effects on R_0. Figure 4 shows the effects of relapse and recover rates on the basic reproduction number. From the simulations, we find that R_0 is increasing with increase in relapse rate, while it is decreasing with increase in recovery rate. To control and

eradicate the malaria epidemic, it is important and necessary for governments of endemic countries to decrease the relapse rate and increase the treatment and recovery rate.

4.2. Probability of Disease Extinction. Using parameter values in Table 2, we compute numerically the probability of disease extinction P_0 and of an outbreak $1 - P_0$ for our malaria model using different initial values for the infectious classes.

The probability of disease extinction is high if the disease emerges from infected humans. It is very low if the infection emerges from infected mosquitoes. However, as the initial number of infected humans grows largely, there is a high probability of disease outbreak as illustrated in Table 3. The probability of disease extinction is significantly low if the

TABLE 3: Probability of disease extinction P_0 and of an outbreak $1 - P_0$ for the malaria model.

h_0	r_0	m_0	P_0	$1 - P_0$	R_{01}	R_{02}	R_0
1	0	0	0.6838	0.3162	$R_{01} > 1$	$R_{02} < 1$	$R_0 > 1$
0	1	0	0.9834	0.0166	$R_{01} > 1$	$R_{02} < 1$	$R_0 > 1$
0	0	1	0.1365	0.8635	$R_{01} > 1$	$R_{02} < 1$	$R_0 > 1$
1	1	0	0.6724	0.3276	$R_{01} > 1$	$R_{02} < 1$	$R_0 > 1$
1	1	1	0.0918	0.9082	$R_{01} > 1$	$R_{02} < 1$	$R_0 > 1$
10	0	0	0.0223	0.9777	$R_{01} > 1$	$R_{02} < 1$	$R_0 > 1$
0	10	0	0.8459	0.1541	$R_{01} > 1$	$R_{02} < 1$	$R_0 > 1$
0	0	10	0	1	$R_{01} > 1$	$R_{02} < 1$	$R_0 > 1$
10	10	0	0.0188	0.9812	$R_{01} > 1$	$R_{02} < 1$	$R_0 > 1$
10	10	10	0	1	$R_{01} > 1$	$R_{02} < 1$	$R_0 > 1$

disease emerges from infected mosquitoes. Therefore, the disease dynamics for model system (1) at the beginning of the epidemics are being driven by initial number of infected mosquitoes.

A female mosquito will continue to bite and draw blood until her abdomen is full. If she is interrupted before she is full, she will fly to the next person. After feeding, the mosquito rests for two or three days before laying her eggs, and then it is ready to bite again. Since mosquitoes do not recover from the infection, then a single infected mosquito is capable of biting and infecting so many susceptible humans in their lifespan which may in turn infect so many uninfected mosquitoes there by reducing the probability of disease clearance and increasing the probability of a major disease outbreak. Moreover, mosquitoes are the reservoir host for the parasites that cause malaria and it takes a long time for them to be malaria-free, hence the high probability of malaria outbreak if the parasite is introduced by infected mosquito.

Table 3 depicts that, at the beginning of malaria outbreak, any policy or intervention to control the spread of malaria should focus on controlling the infected mosquito population as well as the infected humans. If more effort to control the disease is only focused on the infected humans, then it is very unlikely that malaria will be eliminated. This is an interesting insight from the stochastic threshold that could not be provided by the deterministic threshold.

4.3. Numerical Simulation of Malaria Model. Using parameter values in Table 2, we numerically simulate the behavior of model (1). Initial conditions are $S_h(0) = 99$, $I_h(0) = 1$, $R_h(0) = 0$, $S_m(0) = 999$, and $I_m(0) = 1$.

4.3.1. Numerical Simulation When $R_0 < 1$. From Figure 5, the analysis shows that when $R_0 < 1$ and $S_0 < 1$, then the probability of disease extinction is $P_0 = 0.9476 \simeq 1$, although this agrees with (30), which points out that the probability of disease elimination in the CTDSGWbp model is one iff

$$S_0 < 1 \implies$$
$$R_0 < 1. \tag{35}$$

There is still a small probability $1 - P_0 = 0.0524$ of disease outbreak. R_0 has been widely used as a measure of disease

dynamics to estimate the effectiveness of control measures and to inform on disease management policy. However, from the analysis in Figure 5, it is evident that R_0 can be flawed and disease can persist with $R_0 < 1$ depending on the kind of disease being modeled.

4.3.2. Numerical Simulation When $R_0 > 1$. Using parameter values in Table 2, we numerically simulate the behavior of model (1) when $R_0 > 1$. Initial conditions are $S_h(0) = 99$, $I_h(0) = 1$, $R_h(0) = 0$, $S_m(0) = 999$, and $I_m(0) = 1$.

The analysis from Figure 6 suggests that when $R_0 > 1$, there is still some probability of disease extinction ($P_0 = 0.3713$). Although the probability is low, there is still a chance to clear the disease. Therefore, from the analysis in Figure 6, disease can be eliminated with $R_0 > 1$ and hence the R_0 threshold should not be the only parameter to consider in quantifying the spread of a disease.

4.3.3. Effects of Relapse on Human Population. Figure 7 shows changing effects of relapse on the human populations.

To control and eradicate the malaria epidemic, it is important and necessary for governments of endemic countries to decrease the relapse rate as can be seen in Figure 7.

5. Discussions and Recommendations

In this study, we investigated the transmission dynamics of malaria using CTDSGWbp model. The disease dynamic extinction thresholds from the stochastic model were compared with the corresponding deterministic threshold. We derived the stochastic threshold for disease extinction S_0 and showed the relationship that exists between R_0 and S_0 in terms of disease extinction and outbreak in both deterministic and stochastic models.

Our analytical and numerical results showed that both deterministic and stochastic models predict disease extinction when $R_0 < 1$ and $S_0 < 1$. However, the predictions by these models are different when $R_0 > 1$. In this case, deterministic model predicts with certainty disease outbreak while the stochastic model has a probability of disease extinction at the beginning of an infection. Hence, with stochastic models, it is possible to attain a disease-free equilibrium even when $R_0 > 1$. Also we noticed that initial conditions do not affect the deterministic threshold while the stochastic thresholds

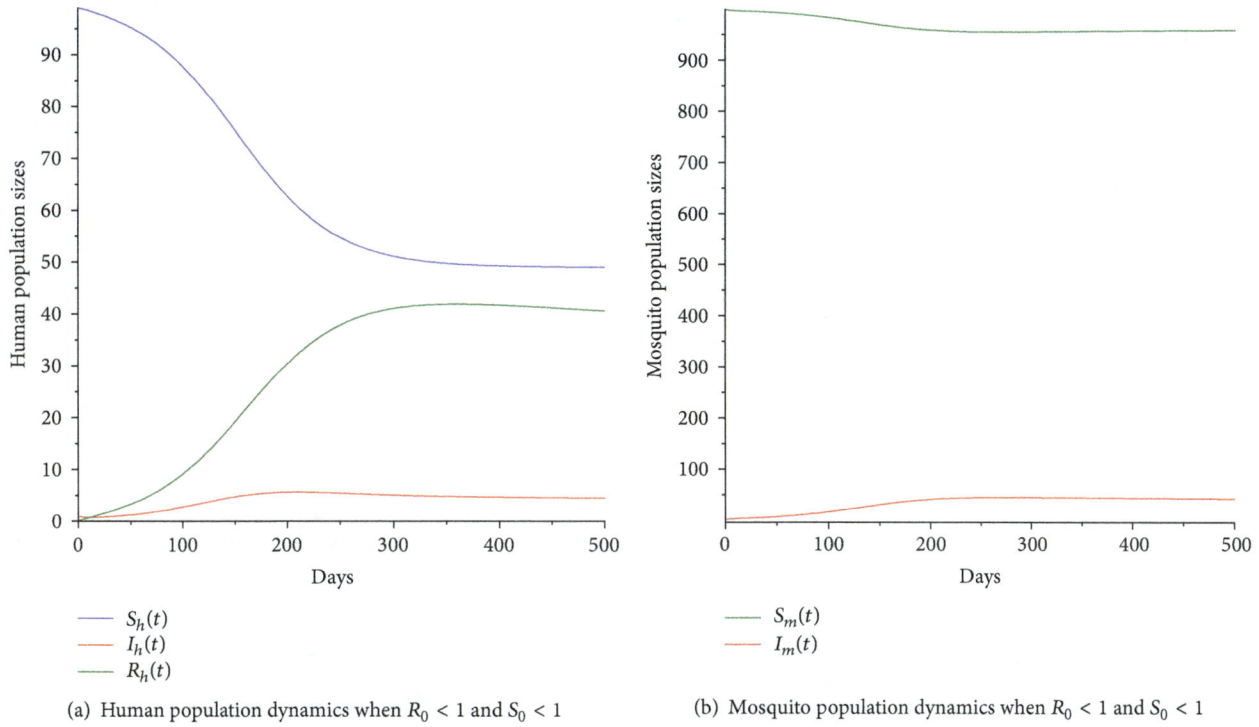

(a) Human population dynamics when $R_0 < 1$ and $S_0 < 1$

(b) Mosquito population dynamics when $R_0 < 1$ and $S_0 < 1$

FIGURE 5: Malaria dynamics when $R_0 = 0.4564$ and $S_0 = 0.5661$ and $P_0 = 0.9476$.

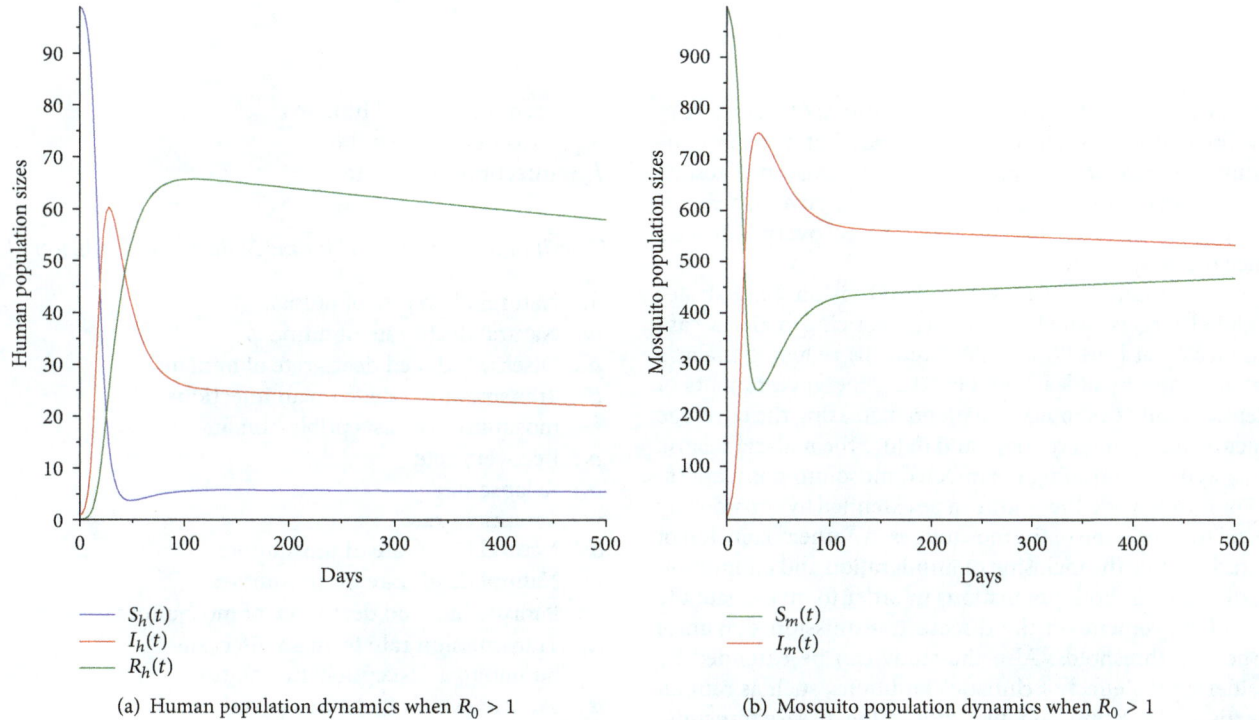

(a) Human population dynamics when $R_0 > 1$

(b) Mosquito population dynamics when $R_0 > 1$

FIGURE 6: Malaria dynamics when $R_0 = 2.4629$ and $P_0 = 0.3713$.

are affected. Thus, the dynamics of the stochastic model are highly dependent on the initial conditions and should not be ignored.

The probabilities of disease extinction for different initial sizes of infected humans and infected mosquitoes were approximated numerically. The results indicate that the probability of eliminating malaria is high if the disease emerges from infected human as opposed to when it emerges from infected mosquito at the beginning of the disease. The analysis has shown that any policy or intervention to

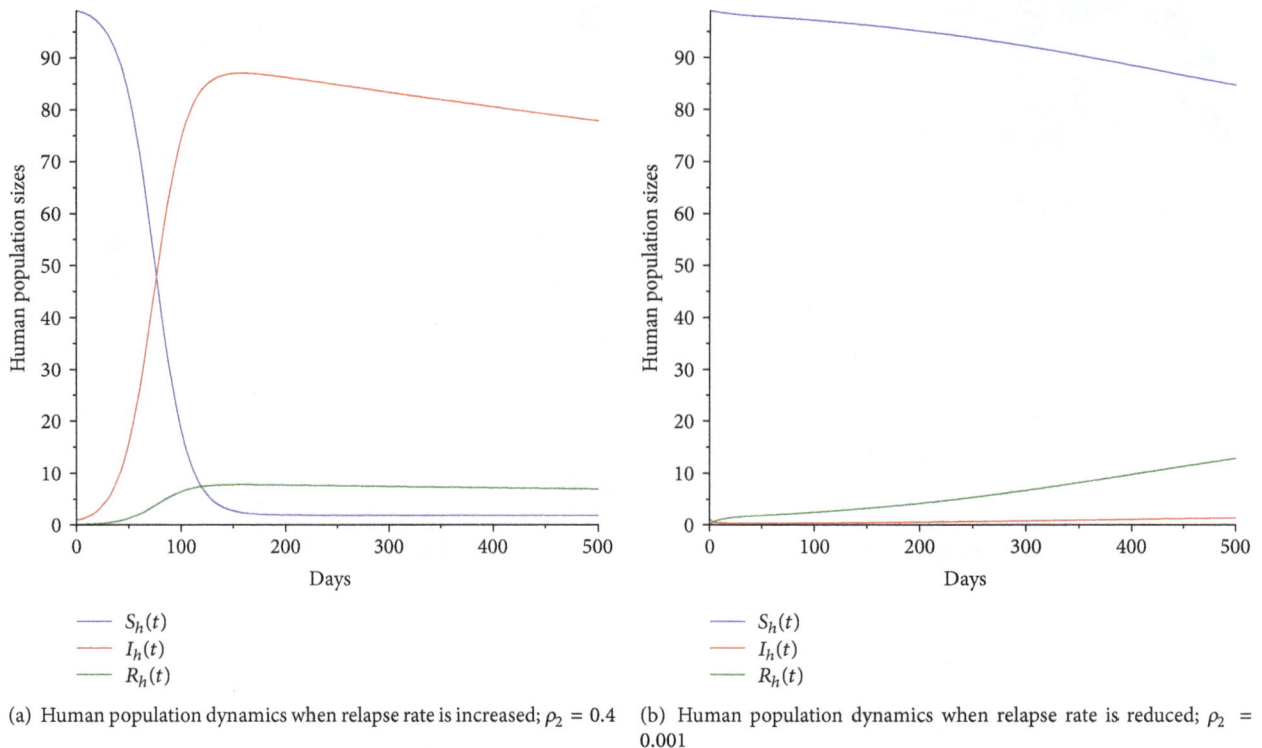

(a) Human population dynamics when relapse rate is increased; $\rho_2 = 0.4$ (b) Human population dynamics when relapse rate is reduced; $\rho_2 = 0.001$

FIGURE 7: Changing effects of relapse on the human populations.

control the spread of malaria at the beginning of an outbreak must focus not only on infected humans but also on controlling the infected mosquito population. Also our results strongly suggest that, to eradicate malaria, governments of endemic countries should increase the recovery rate and reduce the relapse rate.

In conclusion, to achieve the WHO vision 2030 strategic global targets, which include (i) reducing malaria case incidence by at least 90% by 2030 and (ii) reducing malaria mortality rates by at least 90% by 2030, the governments of endemic countries should embark on increasing the coverage of vector control interventions and reduce the malaria relapse rate as well as controlling the infected mosquito population.

For future work, the study can be extended by considering vertical transmission of the mosquitoes. Another extension of the study can be the inclusion of immigration and emigration of individuals in both populations in order to investigate the effect of movement on the disease transmission dynamics on the two thresholds. Also the study can be extended by considering the effect of climatic conditions, such as rainfall patterns, temperature, and humidity on the disease transmission dynamics.

Notations

Description of Variables for the Malaria Transmission Model

S_h: Susceptible humans
I_h: Infectious humans

R_h: Pseudorecovered humans
S_m: Susceptible mosquito
I_m: Infectious mosquito.

Description of Parameters for the Malaria Transmission Model

λ: Natural birth rate of humans
μ: Natural death rate of humans
μ_1: Disease induced death rate of humans
β: Transmission rate from an infectious mosquito to a susceptible human
ρ_1: Recovery rate
ρ_2: Relapse rate
γ: Treatment rate
δ: Natural birth rate of mosquitoes
η: Natural death rate of mosquitoes
η_1: Parasite induced death rate of mosquitoes
α_1: Transmission rate from an infectious human to a susceptible mosquito
α_2: Transmission rate from a pseudorecovered human to a susceptible mosquito
N: The total size of human population
M: The total size of mosquito population.

Conflicts of Interest

All authors declared no conflicts of interest regarding the publication of this paper.

Authors' Contributions

Rachel Waema Mbogo produced the first draft of the manuscript. All authors critically reviewed the paper and approved the final version.

References

[1] WHO, *World Malaria*, World Health Organization, Geneva, Switzerland, 2012.

[2] WHO, *World Malaria*, World Health Organization, Geneva, Switzerland, 2015.

[3] R. Ross, "An application of the theory of probabilities to the study of a priori pathometry," in *Proceedings of the Royal Society A*, vol. 92, pp. 204–230, 1916.

[4] Y. Xiao and X. Zou, "Can multiple malaria species co-persist?" *SIAM Journal on Applied Mathematics*, vol. 73, no. 1, pp. 351–373, 2013.

[5] J. Li, Y. Zhao, and S. Li, "Fast and slow dynamics of malaria model with relapse," *Mathematical Biosciences*, vol. 246, no. 1, pp. 94–104, 2013.

[6] H.-F. Huo and G.-M. Qiu, "Stability of a mathematical model of malaria transmission with relapse," *Abstract and Applied Analysis*, vol. 2014, Article ID 289349, 9 pages, 2014.

[7] H.-F. Huo and C.-C. Zhu, "Influence of relapse in a giving up smoking model," *Abstract and Applied Analysis*, vol. 2013, Article ID 525461, 12 pages, 2013.

[8] N. Chitnis, J. M. Cushing, and J. M. Hyman, "Bifurcation analysis of a mathematical model for malaria transmission," *SIAM Journal on Applied Mathematics*, vol. 67, no. 1, pp. 24–45, 2006.

[9] L. J. S. Allen, "Stochastic population and epidemic models:persistence and extinction," in *Mathematical Biosciences Institute lecture series, Stochastic in Biological systems*, Springer International Publishing, Switzerland, 2015.

[10] T. Britton, "Stochastic epidemic models: A survey," *Mathematical Biosciences*, vol. 225, no. 1, pp. 24–35, 2010.

[11] Y. Cai, Y. Kang, M. Banerjee, and W. Wang, "A stochastic epidemic model incorporating media coverage," *Communications in Mathematical Sciences*, vol. 14, no. 4, pp. 893–910, 2016.

[12] Y. Cai, Y. Kang, and W. Wang, "A stochastic SIRS epidemic model with nonlinear incidence rate," *Applied Mathematics and Computation*, vol. 305, pp. 221–240, 2017.

[13] M. Maliyoni, F. Chirove, H. . Gaff, and K. S. Govinder, "A stochastic tick-borne disease model: exploring the probability of pathogen persistence," *Bulletin of Mathematical Biology*, vol. 79, no. 9, pp. 1999–2021, 2017.

[14] M. S. Bartlett, "The relevance of stochastic models for large-scale epidemiological phenomena," *Journal of Applied Statistics*, vol. 13, no. 13, pp. 2–8, 1964.

[15] M. S. Bartlett, *Stochastic population models*, Methuen, London, UK, 1960.

[16] L. J. S. Allen, "An Introduction to stochastic Epidemic models," in *Mathematical Epidemiology*, Springer, Berlin, Germany, 2008.

[17] L. J. S. Allen and P. Van den Driessche, "The basic reproduction number in some discrete-time epidemic models," *Journal of Difference Equations and Applications*, vol. 14, pp. 11–27, 2008.

[18] L. J. Allen and P. van den Driessche, "Relations between deterministic and stochastic thresholds for disease extinction in continuous- and discrete-time infectious disease models," *Mathematical Biosciences*, vol. 243, no. 1, pp. 99–108, 2013.

[19] L. J. S. Allen, "A primer on stochastic epidemic models: Formulation, numerical simulation, and analysis," *Infectious Disease Modelling*, vol. 243, pp. 1–15, 2017.

[20] C. Manore and M. Hyman, "Mathematical models for fighting zika virus," *SIAM News*, vol. 2016, pp. 1–5, 2016.

[21] S. Olaniyi, M. A. Lawal, and O. S. Obabiyi, "Stability and sensitivity analysis of a deterministic epidemiological model with pseudo-recovery," *IAENG International Journal of Applied Mathematics*, vol. 46, no. 2, pp. 1–8, 2016.

[22] L. Xue and C. Scoglio, "Network-level reproduction number and extinction threshold for vector-borne diseases," *Mathematical Biosciences and Engineering*, vol. 12, no. 3, pp. 565–584, 2015.

[23] P. van den Driessche and J. Watmough, "Reproduction numbers and sub-threshold endemic equilibria for compartmental models of disease transmission," *Mathematical Biosciences*, vol. 180, pp. 29–48, 2002.

[24] O. Diekmann, J. A. Heesterbeek, and J. A. Metz, "On the definition and the computation of the basic reproduction ratio R_0 in models for infectious diseases in heterogeneous populations," *Journal of Mathematical Biology*, vol. 28, no. 4, pp. 365–382, 1990.

[25] W. R. Mbogo, L. S. Luboobi, and J. W. Odhiambo, "Stochastic model for in-host HIV dynamics with therapeutic intervention," *ISRN Biomathematics*, vol. 2013, Article ID 103708, 11 pages, 2013.

[26] W. R. Mbogo, L. S. Luboobi, and J. W. Odhiambo, "Mathematical model for HIV and CD4+ cells dynamics in vivo," *International Journal of Pure and Applied Mathematics*, vol. 6, no. 2, pp. 83–103, 2013.

[27] L. J. S. Allen and G. E. Lahodny Jr., "Extinction thresholds in deterministic and stochastic epidemic models," *Journal of Biological Dynamics*, vol. 6, no. 2, pp. 590–611, 2012.

[28] J. Lahodny, R. Gautam, and R. Ivanek, "Estimating the probability of an extinction or major outbreak for an environmentally transmitted infectious disease," *Journal of Biological Dynamics*, vol. 9, no. 1, pp. 128–155, 2015.

[29] L. J. S. Allen, "Branching processes," in *Encyclopaedia of Theoretical Ecology*, University of California Press, Calif, USA, 2012.

7

Analytical Approach for Solving the Internal Waves Problems Involving the Tidal Force

Jaharuddin [ID]¹ **and Hadi Hermansyah**²

¹*Department of Mathematics, Faculty of Mathematics and Natural Sciences, Bogor Agricultural University, Bogor, West Java 16680, Indonesia*
²*Department of Mechanical Engineering, Balikpapan State Polytechnic, Balikpapan, East Borneo 76126, Indonesia*

Correspondence should be addressed to Jaharuddin; jaharmath@gmail.com

Academic Editor: Jian G. Zhou

The mathematical model for describing internal waves of the ocean is derived from the assumption of ideal fluid; i.e., the fluid is incompressible and inviscid. These internal waves are generated through the interaction between the tidal currents and the basic topography of the fluid. Basically the mathematical model of the internal wave problem of the ocean is a system of nonlinear partial differential equations (PDEs). In this paper, the analytical approach used to solve nonlinear PDE is the Homotopy Analysis Method (HAM). HAM can be applied to determine the resolution of almost any internal wave problem involving tidal forces. The use of HAM in the solution to basic fluid equations is efficient and simple, since it involves only modest calculations using the common integral.

1. Introduction

Internal waves are gravitational waves that exist on two layers of fluid having different densities. Internal waves are formed due to a meeting among layers of seawater that have different densities of generating forces coming from wind, tide, or even movement of ships. The density difference causes the seawater to become layered where water with a larger density will be below that with a smaller density. This condition stimulates the formation of boundary of the two layers fluid (interfaces) where in case of external disturbance (by the existing generating force), an interlayer wave occurs without affecting the waves on the surface. Generating internal waves requires a large force, for instance, generated by the interaction of strong tidal currents, fluid coating, and lower topography. Research on internal waves at sea has previously been applied to various applications and ranges, for example, to detect the strength of offshore oil platform pylons [1] and to measure how the impact of internal waves can affect *Chlorophila* [2]. In addition, this wave can also affect the marine habitat that is the spatial distribution of *Planktothrix rubescens* [3].

Internal waves of the ocean can be modeled in terms of mathematical equations using the ideal fluids assumptions (incompressible and inviscid) of mass conservation laws and the law of momentary vapor. Internal waves are generated through the interaction between the tidal flow and the topography in a nonuniform fluid layer by solving the Navier-Stokes equation in Boussinesq approximation. Basically the mathematical representation of the internal waves of the ocean is a system of nonlinear partial differential equations (PDEs) [4]. In many cases, nonlinear PDE systems are very difficult to be resolved analytically. Thus an analytic approach can provide a solution which is almost needed.

The analytical approach for solving the nonlinear PDEs was first introduced by Liao in 1992, i.e., the homotopy analysis method (HAM). HAM excellence lies in the selection mechanism of initial values and auxiliary parameters so as to extend the convergence region [5]. Earlier version of HAM methods has been applied for various nonlinear problem solving such as the Klein-Gordon equation [6], El Nino Southern Oscillation [7], Huxley [8], Zakharov-Kuznetsov equation [9], and one species growth model in the polluted environment [10]. In this article, we review the internal wave issues in the sea that involve tidal forces using the HAM method. The completion of almost this method will be compared to the numerical settlement of error calculations and graphical visualization of the settlement.

The equation used in this study is the Navier-Stokes equation with Boussinesq approximation, in which it is assumed that the internal waves are generated by the interaction between pairs of currents with two-dimensional topography in a nonuniform fluid layer. In this model, ρ is the density, ρ_0 a reference density, p pressure, and u and w velocity, respectively, in the horizontal and vertical directions

$$\frac{\partial u}{\partial t} + u\frac{\partial u}{\partial x} + g\frac{\partial \eta}{\partial x} - \nu\frac{\partial^2 u}{\partial x^2} - \frac{F_{tide}}{\rho_0} = 0,$$

$$\frac{\partial w}{\partial t} + u\frac{\partial w}{\partial x} + \frac{g\rho}{\rho_0} + gH - \frac{F_{tide}}{\rho_0} = 0, \quad (1)$$

$$\frac{\partial \eta}{\partial t} + u\frac{\partial \eta}{\partial x} + wH + \eta\frac{\partial u}{\partial x} = 0,$$

where t is time, η is the fluid depth, g is the constant of gravity, H represents the mean of fluid depth, ν is the kinematic viscosity, and F_{tide} is the tidal force given by

$$F_{tide} = \rho_0 A\omega^2 \sin \omega t. \quad (2)$$

In (2), A is tidal excursion and it was found that the value of A was less than 10% variation in the measured quantities of this range; the data presented are for $A = 20$ m; ω is the tidal frequency. Another parameter is the height caused by the change in horizontal directional pressure formulated by $H = -(f/g)\overline{U}$, where $\overline{U} = 2.5$ m/s and the Coriolis parameter $f = 2\Omega \sin(\alpha)$ depending on the angular velocity earth rotation Ω.

2. Analysis Method

In this part we illustrate the concept of homotopy method. Suppose that a nonlinear equation is given in the form as below:

$$\mathcal{N}[u(x,t)] = 0, \quad (3)$$

where $\mathcal{N}$ is a nonlinear derivative operator, $u(x,t)$ is an unknown function, x and t are independent variables, and $\mathcal{L}$ is defined as linear operator which satisfies

$$\mathcal{L}[f(x,t)] = 0, \quad \text{when } f(x,t) = 0. \quad (4)$$

Let $u_0(x,t)$ be the initial approach of solving (3); $q \in [0,1]$ is an embedding parameter, $\hbar$ is auxiliary parameter, and $A(x)$ is an additional function. In the frame of the homotopy method, we first construct such a continuous variation (or deformation) $\phi(x,t;q)$ that as q increases from 0 to 1, $\phi(x,t;q)$ varies from the initial approach $u_0(x,t)$ to the solution $u(x,t)$ of (3). Such kind of continuous variation (or mapping) is governed by the so-called zero-order deformation equation

$$(1-q)\mathcal{L}[\phi(x,t;q) - u_0(x,t)]$$
$$= q\hbar A(x)\mathcal{N}[\phi(x,t;q)]. \quad (5)$$

At $q = 0$, the zero-order deformation equation (5) becomes

$$\mathcal{L}[\phi(x,t;0) - u_0(x,t)] = 0, \quad (6)$$

such that

$$\phi(x,t;0) = u_0(x,t). \quad (7)$$

When $q = 1$ and $\hbar \neq 0$, then the zero-order deformation equation (5) becomes

$$\mathcal{N}[\phi(x,t;1)] = 0, \quad (8)$$

which is exactly the same as the original equation (3), provided

$$\phi(x,t;1) = u(x,t). \quad (9)$$

Thus, as q increases from 0 to 1, the solution $\phi(x,t;q)$ varies continuously from the initial approach $u_0(x,t)$ to the exact solution $u(x,t)$. So, (5) defines a homotopy of function $(x,t;q) : u_0(x,t) \sim u(x,t)$. Such kind of continuous variation is called deformation in topology, and this is the reason why we call (5) the zero-order deformation equation. By using the Taylor expansion from $\phi(x,t;q)$ to q, the following is obtained

$$\phi(x,t;q) = u_0(x,t) + \sum_{m=1}^{\infty} u_m(x,t)q^m, \quad (10)$$

where

$$u_m(x,t) = \frac{1}{m!}\frac{\partial^m \phi(x,t;q)}{\partial q^m}\bigg|_{q=0}. \quad (11)$$

Suppose that given the initial value of $u_0(x,t)$, the linear operator $\mathcal{L}$ and the auxiliary parameters $\hbar$ are not equal to zero and the auxiliary function $A(x)$ is chosen so that (10) is from $\phi(x,t;q)$ convergent at $q = 1$. Hence, we may assume the following series solution:

$$u(x,t) = \phi(x,t;1) = u_0(x,t) + \sum_{m=1}^{+\infty} u_m(x,t). \quad (12)$$

According to (10), (5) can be rewritten as follows:

$$(1-q)\mathcal{L}\left[\sum_{m=1}^{\infty} u_m(x,t)q^m\right]$$
$$= q\hbar A(x)\mathcal{N}[\phi(x,t;q)], \quad (13)$$

such that

$$\mathcal{L}\left[\sum_{m=1}^{\infty} u_m(x,t)q^m\right] - q\mathcal{L}\left[\sum_{m=1}^{\infty} u_m(x,t)q^m\right]$$
$$= q\hbar A(x)\mathcal{N}[\phi(x,t;q)]. \quad (14)$$

By deriving (14) as much as m times with respect to q, then the following is obtained:

$$m!\mathcal{L}[u_m(x,t) - u_{m-1}(x,t)]$$
$$= \hbar A(x)m\frac{\partial^{m-1}\mathcal{N}[\phi(x,t;q)]}{\partial q^{m-1}}\bigg|_{q=0}, \quad (15)$$

such that

$$\mathscr{L}\left[u_m(x,t) - \chi_m u_{m-1}(x,t)\right]$$
$$= \hbar A(x)\,\mathfrak{R}_m\left(u_{m-1}(x,t)\right), \tag{16}$$

where

$$\mathfrak{R}_m\left(u_{m-1}(x,t)\right) = \frac{1}{(m-1)!}\frac{\partial^{m-1}\mathscr{N}\left[\phi(x,t;q)\right]}{\partial q^{m-1}}\bigg|_{q=0} \tag{17}$$

and

$$\chi_m = \begin{cases} 0 & m \le 1, \\ 1 & m > 1. \end{cases} \tag{18}$$

3. Application of HAM

In this section we discuss the use of homotopy method to explain the internal wave motion with finite depth. The linear operation in the homotopy method is defined as follows:

$$\mathscr{L}_i\left[\phi_i(x,t;q)\right] = \frac{\partial \phi_i(x,t;q)}{\partial t}, \quad i = 1,2,3. \tag{19}$$

Based on the system in (1), we may have the following linear operators:

$$\mathscr{N}_1[\phi_1,\phi_2,\phi_3] = \frac{\partial \phi_1}{\partial t} + \phi_1\frac{\partial \phi_1}{\partial x} + g\frac{\partial \phi_3}{\partial x} - \nu\frac{\partial^2 \phi_1}{\partial x^2}$$
$$- \frac{F_{tide}}{\rho_0}$$
$$\mathscr{N}_2[\phi_1,\phi_2,\phi_3] = \frac{\partial \phi_2}{\partial t} + \phi_1\frac{\partial \phi_2}{\partial x} + \frac{g\rho}{\rho_0} + gH - \frac{F_{tide}}{\rho_0}, \tag{20}$$
$$\mathscr{N}_3[\phi_1,\phi_2,\phi_3] = \frac{\partial \phi_3}{\partial t} + \phi_1\frac{\partial \phi_3}{\partial x} + \phi_2 H + \phi_3\frac{\partial \phi_1}{\partial x}.$$

Now, the zero-order deformation equation is as follows:

$$(1-q)\mathscr{L}_1\left[\phi_1(x,t;q) - u_0(x,t)\right]$$
$$= q\hbar_1\mathscr{N}_1[\phi_1,\phi_2,\phi_3]$$
$$(1-q)\mathscr{L}_2\left[\phi_2(x,t;q) - w_0(x,t)\right]$$
$$= q\hbar_2\mathscr{N}_2[\phi_1,\phi_2,\phi_3] \tag{21}$$
$$(1-q)\mathscr{L}_3\left[\phi_3(x,t;q) - \eta_0(x,t)\right]$$
$$= q\hbar_3\mathscr{N}_3[\phi_1,\phi_2,\phi_3].$$

According to (21), when $q = 0$ we can write

$$\phi_1(x,t;0) = u_0(x,t) = u(x,0),$$
$$\phi_2(x,t;0) = w_0(x,t) = w(x,0), \tag{22}$$
$$\phi_3(x,t;0) = \eta_0(x,t) = \eta(x,0),$$

and when $q = 1$, we have

$$\phi_1(x,t;1) = u(x,t),$$
$$\phi_2(x,t;1) = w(x,t), \tag{23}$$
$$\phi_3(x,t;1) = \eta(x,t).$$

Thus, we obtain the mth-order deformation equation:

$$\mathscr{L}_1\left[u_m(x,t) - \chi_m u_{m-1}(x,t)\right]$$
$$= \hbar_1 R_{1,m}\left[\vec{u}_{m-1}, \vec{w}_{m-1}, \vec{\eta}_{m-1}\right],$$
$$\mathscr{L}_2\left[w_m(x,t) - \chi_m w_{m-1}(x,t)\right]$$
$$= \hbar_2 R_{2,m}\left[\vec{u}_{m-1}, \vec{w}_{m-1}, \vec{\eta}_{m-1}\right], \tag{24}$$
$$\mathscr{L}_3\left[\eta_m(x,t) - \chi_m \eta_{m-1}(x,t)\right]$$
$$= \hbar_3 R_{3,m}\left[\vec{u}_{m-1}, \vec{w}_{m-1}, \vec{\eta}_{m-1}\right],$$

where

$$\vec{u}_m = (u_0(x,t), u_1(x,t), u_2(x,t), \ldots, u_m(x,t)),$$
$$\vec{w}_m = (w_0(x,t), w_1(x,t), w_2(x,t), \ldots, w_m(x,t)), \tag{25}$$
$$\vec{\eta}_m = (\eta_0(x,t), \eta_1(x,t), \eta_2(x,t), \ldots, \eta_m(x,t)).$$

Now, the solution of the mth-order deformation equation (24) for $m \ge 1$ becomes

$$u_m(x,t) = \chi_m u_{m-1}(x,t)$$
$$+ \hbar_1\int_0^t R_{1,m}\left(\vec{u}_{m-1}, \vec{w}_{m-1}, \vec{\eta}_{m-1}\right)ds$$
$$w_m(x,t) = \chi_m w_{m-1}(x,t)$$
$$+ \hbar_2\int_0^t R_{2,m}\left(\vec{u}_{m-1}, \vec{w}_{m-1}, \vec{\eta}_{m-1}\right)ds \tag{26}$$
$$\eta_m(x,t) = \chi_m \eta_{m-1}(x,t)$$
$$+ \hbar_3\int_0^t R_{3,m}\left(\vec{u}_{m-1}, \vec{w}_{m-1}, \vec{\eta}_{m-1}\right)ds,$$

where

$$R_{1,m}\left(\vec{u}_{m-1}, \vec{w}_{m-1}, \vec{\eta}_{m-1}\right)$$
$$= \frac{\partial u_{m-1}}{\partial t} + \sum_{n=0}^{m-1} u_n\frac{\partial u_{m-1-n}}{\partial x} + g\frac{\partial \eta_{m-1}}{\partial x} - \nu\frac{\partial^2 u_{m-1}}{\partial x^2}$$
$$- \frac{F_{tide}}{\rho_0}$$

$$R_{2,m}\left(\vec{u}_{m-1}, \vec{w}_{m-1}, \vec{\eta}_{m-1}\right)$$

$$= \frac{\partial w_{m-1}}{\partial t} + \sum_{n=0}^{m-1} u_n \frac{\partial w_{m-1-n}}{\partial x} + \frac{g\rho}{\rho_0} + gH - \frac{F_{tide}}{\rho_0}$$

$$R_{3,m}\left(\vec{u}_{m-1}, \vec{w}_{m-1}, \vec{\eta}_{m-1}\right)$$

$$= \frac{\partial \eta_{m-1}}{\partial t}$$

$$+ \sum_{n=0}^{m-1} u_n \frac{\partial \eta_{m-1-n}}{\partial x} + \eta_n \frac{\partial u_{m-1-n}}{\partial x} + w_{m-1}H.$$

$$(27)$$

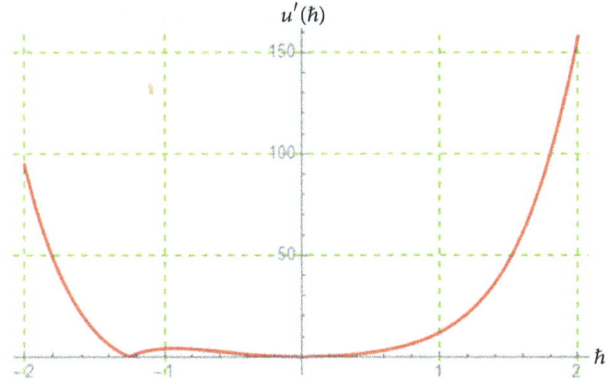

FIGURE 1: The $\hbar$-curves.

According to (10) and (18) we have

$$u(x,t) = u_0(x,t) + \sum_{m=1}^{+\infty} u_m(x,t),$$

$$w(x,t) = w_0(x,t) + \sum_{m=1}^{+\infty} w_m(x,t), \qquad (28)$$

$$\eta(x,t) = \eta_0(x,t) + \sum_{m=1}^{+\infty} \eta_m(x,t).$$

Furthermore, the initial settlement approach is chosen based on the completion of the current wave from the Navier-Stokes equation obtained by the following equation:

$$u_0(x,t) = \omega A \cosh(kx - \omega t),$$

$$w_0(x,t) = \omega A \sinh(kx - \omega t), \qquad (29)$$

$$\eta_0(x,t) = A \cosh(kx - \omega t).$$

For simplification, then select $\hbar_1 = \hbar_2 = \hbar_3$. Further, the boundary conditions used in the solution of (1) are a polynomial determined by the settlement of HAM. The solution of (1) is numerically determined with the aid of a symbolic computing program. The resulting numerical settlement will be compared to the almost-resultant settlement with the HAM. The parameters used for the evaluation need the inclusion of the tidal force parameter $F_{tide} = \rho_0 A \omega^2 \sin \omega t$, where A is the tidal excursion; in this case A should be less than the channel width ($A = 20$ m) $\rho_0 = 1000 \, \text{kg/m}^3$; $\nu = 0.01 \, \text{m}^2/\text{s}$ is kinematic viscosity, tidal frequency ($\omega = \omega_{M2} = 1.4052 \times 10^{-4}$ rad/s), and earth's rotational angle velocity $\Omega = 7.29 \times 10^{-5}$ and $\alpha = \pi/3$ as the constant geostrophic current velocity. Furthermore there is also a constant of gravity $g = 9.8 \, \text{m/s}^2$.

In the HAM application, the completion of high-order deformation is determined by (26). The completion of the high-order deformation obtained is the basis of determining the completion of the series. The result of series completion is a function that depends on the values of x and t. In this section, the completion of the obtained series is evaluated at a certain x and t value to determine the completion of the HAM. Nearly obtained solutions compared to their numerical settlement of the calculation of absolute error and visualization of the completion graph.

Based on the initial approach to (29) and given parameters, the following high-order deformation solutions are obtained:

$$u_0(x,t) = \omega A \cosh(kx - \omega t),$$

$$u_1(x,t) = \hbar \left(-0.00002t \cosh(0.005t - 0.2x)\right.$$

$$- 0.00025t \sinh(0.005t) - 19.59975t$$

$$\cdot \sinh(0.005t - 0.2x) - 0.0005t$$

$$\left. \cdot \cosh(0.005t - 0.2x) \sinh(0.005t - 0.2x)\right),$$

$$w_0(x,t) = \omega A \sinh(kx - \omega t),$$

$$w_1(x,t) = \hbar \left(0.0001975t - 0.00025t\right.$$

$$\cdot \cosh(0.005t - 0.2x)$$

$$+ 0.0005t \cosh(0.005t - 0.2x)^2 - 0.00025t$$

$$\left. \cdot \sinh(0.005t) + \frac{10133.2t}{1034}\right), \qquad (30)$$

$$\eta_0(x,t) = A \cosh(kx - \omega t),$$

$$\eta_1(x,t) = \hbar \left(0.05t \sinh(0.005t - 0.2x)\right.$$

$$\left. - 0.2t \cosh(0.005t - 0.2x) \sinh(0.005t - 0.2x)\right),$$

and so on. The rest of the components of the iteration formulas by HAM can easily be obtained by symbolic computation software. Thus, we obtain the following approximate solution in term of a series up to third order:

$$u(x,t) = u_0(x,t) + u_1(x,t) + \cdots + u_3(x,t)$$

$$w(x,t) = w_0(x,t) + w_1(x,t) + \cdots + w_3(x,t) \qquad (31)$$

$$\eta(x,t) = \eta_0(x,t) + \eta_1(x,t) + \cdots + \eta_3(x,t)$$

Note that (31) contains the auxiliary parameter $\hbar$. To obtain an appropriate range for $\hbar$, we consider the $\hbar$-curves. Based on Figure 1 we get the value of $\hbar = -1$.

TABLE 1: An absolute error between the numerical and HAM solution at $x = 0.1$ and $0 \leq t \leq 1$.

t	$|u_{\text{HAM}} - u_{\text{NUM}}|$	$|w_{\text{HAM}} - w_{\text{NUM}}|$	$|\eta_{\text{HAM}} - \eta_{\text{NUM}}|$
0	0	0.000027	0
0.1	2.039×10^{-6}	2.349×10^{-5}	6.708×10^{-6}
0.2	4.279×10^{-6}	2.001×10^{-5}	1.111×10^{-5}
0.3	6.644×10^{-6}	1.675×10^{-5}	1.503×10^{-5}
0.4	9.168×10^{-6}	1.286×10^{-5}	2.679×10^{-5}
0.5	1.186×10^{-5}	4.272×10^{-6}	7.368×10^{-5}
0.6	1.473×10^{-5}	1.963×10^{-5}	2.242×10^{-5}
0.7	1.778×10^{-5}	8.011×10^{-5}	6.218×10^{-5}
0.8	2.101×10^{-5}	2.143×10^{-4}	1.532×10^{-3}
0.9	2.443×10^{-5}	4.810×10^{-4}	3.406×10^{-3}
1	2.806×10^{-5}	9.682×10^{-4}	6.953×10^{-3}

The comparison of the results of the homotopy analysis method and numerical solution is given in Table 1. The explicit Runge-Kutta method in symbolic computation package has been used to find numerical solution of u, w, and η. Table 1 shows the absolute error between the HAM and the numerical solution for $\hbar = -1$, $x = 0.1$, and $0 \leq t \leq 1$. Based on Table 1, it is found that, in the HAM, the approximate solution has a small absolute error against the numerical settlement evaluated at a certain independent variable value. It can be seen in the table that there exists a very good agreement between HAM result and numerical solutions.

The generation of internal waves by tidal force occurs in the stratified fluid when propagation of barotropic tidal currents interacts with rough surface topography, resulting in vertical movement and of local internal pressure. These local perturbations propagate as waves which are far from the center of generation. Internal wave plays an important role for transferring energy to the deep sea turbulence. When barotropic tidal currents flow on a rough topography sill, part of barotropic energy will vanish directly through dissipation and local mixing, and other part of barotropic energy converts into the generating process of internal tides (baroclinic). The result of baroclinic energy will be dissipated locally or radiated to the open ocean.

4. Conclusions

The problem of internal waves in the ocean can be illustrated by the basic fluid equation. The mathematical representation of the basic fluid equation is a system of nonlinear partial differential equations that are difficult to solve analytically. Homotopy method has been succesfully applied in finding the approximate solution of the internal wave model. Solutions by this methods are then compared with one of numerical method. The use of HAM in the solution to basic fluid equations is efficient and simple, since it involves only modest calculations using the common integral.

Conflicts of Interest

The authors declare that there are no conflicts of interest regarding the publication of this paper.

References

[1] A. R. Osborne and T. L. Burch, "Internal solitons in the Andaman Sea," *Science*, vol. 208, no. 4443, pp. 451–460, 1980.

[2] D. Yang, H. Ye, and G. Wang, "Impacts of internal waves on chlorophyll a distribution in the northern portion of the South China Sea," *Chinese Journal of Oceanology and Limnology*, vol. 28, no. 5, pp. 1095–1101, 2010.

[3] Y. Cuypers, B. Vinçon-Leite, A. Groleau, B. Tassin, and J.-F. Humbert, "Impact of internal waves on the spatial distribution of Planktothrix rubescens (cyanobacteria) in an alpine lake," *The ISME Journal*, vol. 5, no. 4, pp. 580–589, 2011.

[4] T. T. Warner, *Numerical Weather and Climate Prediction*, Cambridge University Press, Cambridge, UK, 2011.

[5] S. J. Liao, "Beyond Perturbation," in *Introduction to Homotopy Analysis Method*, pp. 296–309, A CRC Press Company, New york, NY, USA, 2004.

[6] A. K. Alomari, M. S. Noorani, and M. S. Roslinda, "Approximate analytical solutions of the Klein-Gordon equation by means of the homotopy analysis method," *Journal of Quality Measurement and Analysis*, vol. 4, pp. 45–57, 2008.

[7] J. Mo, W. Lin, and H. Wang, "A class of homotopic solving method for ENSO model," *Acta Mathematica Scientia B*, vol. 29, no. 1, pp. 101–110, 2009.

[8] K. M. Hemida and M. S. Mohamed, "Numerical simulation of the generalized Huxley equation by homotopy analysis method," *Journal of Applied Functional Analysis*, vol. 5, no. 4, pp. 344–350, 2010.

[9] M. Usman, T. Zubair, I. Rashid, and et. al, "Modified homotopy analysis method for zakharov-kuznetsov," *Walailak*, vol. 10, pp. 467–478, 2013.

[10] Jaharuddin, "A single species population model in polluted environment solved by homotopy analysis method," *Applied Mathematical Sciences*, vol. 8, no. 17-20, pp. 951–961, 2014.

8

The Dynamics of Epidemic Model with Two Types of Infectious Diseases and Vertical Transmission

Raid Kamel Naji and Reem Mudar Hussien

Department of Mathematics, College of Science, University of Baghdad, Baghdad, Iraq

Correspondence should be addressed to Reem Mudar Hussien; reem.m.hussien@gmail.com

Academic Editor: Zhen Jin

An epidemic model that describes the dynamics of the spread of infectious diseases is proposed. Two different types of infectious diseases that spread through both horizontal and vertical transmission in the host population are considered. The basic reproduction number R_0 is determined. The local and the global stability of all possible equilibrium points are achieved. The local bifurcation analysis and Hopf bifurcation analysis for the four-dimensional epidemic model are studied. Numerical simulations are used to confirm our obtained analytical results.

1. Introduction

Mathematical models can be defined as a method of emulating real life situations with mathematical equations to expect their future behavior. In epidemiology, mathematical models play role as a tool in analyzing the spread and control of infectious diseases. Although one of the most famous principles of ecology is the competitive exclusion principle that stipulates "two species competing for the same resources cannot coexist indefinitely with the same ecological niche" [1, 2], Volterra was the first scientist who used the mathematical modeling and showed that the indefinite coexistence of two or more species limited by the same resource is impossible [3]. Moreover, Ackleh and Allen [4] were the first who used the competitive exclusion principle of the infectious disease with different levels in single host population.

It is well known that one of the most useful parameters concerning infectious diseases is called basic reproduction number. It can be specific to each strain of an epidemic model. In fact the basic reproduction number of the model is defined as the maximum reproduction numbers of other strains [5–7]. Diekmann et al. [8] had studied epidemic models with one strain, while Martcheva in [9] studied the *SIS*-type of disease with multistrain. However, Ackleh and Allen [10] studied *SIR*-type of disease with **n** strain and vertical transmission.

Keeping the above in view, in our proposed model two strains with two different types of infectious diseases are considered. Accordingly two different reproduction numbers are obtained and then competitive exclusion principle is presented. It is assumed that two different types of diseases transmission, say horizontal and vertical transmission, are used too. The horizontal transmission occurs by direct contact between infected and susceptible individuals, while vertical transmission occurs when the parasite is transmitted from parent to offspring [11–13]. The incidence of an epidemiological model is defined as the rate at which susceptible becomes infectious. Different types of incidence rates are introduced into literatures [14–17]. Finally two types of incidence rates, say bilinear mass action and nonlinear type, are used with the horizontal and vertical transmission, respectively. The local and global stability for all possible equilibria are carried out with the help of Lyapunov function and LaSalle's invariant principle [18]. An application of Sotomayor theorem [19, 20] for local bifurcations is used to study the occurrence of local bifurcations near the equilibria. The Hopf bifurcation [21, 22] conditions are derived. Finally, numerical simulations are used to confirm our obtained analytical results and specify the control set of parameters.

2. Model Formulation

Consider a real world system consisting of a host population $N(t)$ that is divided into four compartments: $S(t)$ which represents the number of susceptible individuals at time t; $I_1(t)$ and $I_2(t)$ that represent the number of infected individuals at time t for SIRS-type of disease and SIS-type of disease, respectively; finally $R(t)$ that represents the number of recovered individuals at time t, thus $N(t) = S(t) + I_1(t) + I_2(t) + R(t)$. Now in order to formulate the dynamics of the above system mathematically, the following assumptions have been adopted:

(1) There is a constant number of the host populations entering to the system with recruitment rate $\Lambda > 0$.

(2) There is a vertical transmission of both of the diseases; that is, the infectious host gives birth to a new infected host of rates $0 \leq p_1 \leq 1$ and $0 \leq p_2 \leq 1$ for the diseases I_1 and I_2, respectively. Consequently $p_1 I_1$ and $p_2 I_2$ individuals enter into infected compartments I_1 and I_2, respectively, and the same quantities are disappearing from recruitment in the susceptible compartment.

(3) The diseases are transmitted by contact, according to the mass action law, between the individuals in the S-compartment and those in I_i ($i = 1, 2$) compartments with nonlinear incidence rate for I_1 that is given by $\beta_1 S I_1 / (1 + I_1)$, in which $\beta_1 > 0$ represents the infection force rate while $1/(1 + I_1)$ represents the inhibition effect of the crowding effect of the infected individuals, and linear incidence rate for I_2 that is given by $\beta_2 S I_2$, where $\beta_2 > 0$ represents the infection rate.

(4) The individuals in the I_1 compartment are facing death due to the disease with infection death rate $\alpha_1 \geq 0$. They recover from disease and get immunity with a recovery rate $\delta > 0$.

(5) The individuals in the I_2 compartment are facing death due to the disease with infection death rate $\alpha_2 \geq 0$. They also recover from the disease but return back to be susceptible with recovery rate $\gamma > 0$.

(6) The individuals in the R compartment are losing the immunity from the I_1 disease and return back to be susceptible again with losing immunity rate $0 \leq \eta < 1$.

(7) There is a natural death rate $\mu > 0$ for the individuals in the host population. Finally, it is assumed that both the diseases cannot be transmitted to the same individual simultaneously.

According to these assumptions the dynamics of the above real world system can be represented mathematically by the following set of differential equations:

$$\frac{dS}{dt} = \Lambda - \left(\frac{\beta_1 I_1}{1 + I_1} + \beta_2 I_2 \right) S + (\gamma - p_2) I_2 - \mu S - p_1 I_1$$
$$+ \eta R,$$
$$\frac{dI_1}{dt} = \frac{\beta_1 S I_1}{1 + I_1} - (\mu + \alpha_1 + \delta - p_1) I_1,$$

$$\frac{dI_2}{dt} = \beta_2 S I_2 - (\mu + \alpha_2 + \gamma - p_2) I_2,$$
$$\frac{dR}{dt} = \delta I_1 - (\eta + \mu) R$$

$$(1)$$

with the initial condition $S(0) > 0$, $I_1(0) > 0$, $I_2(0) > 0$, and $R(0) > 0$. Moreover to insure that the recruitment Λ in the susceptible compartment is always positive the following hypotheses are assumed to be holding always:

$$\delta \geq p_1,$$
$$\gamma \geq p_2.$$
$$(2)$$

Theorem 1. *The closed set $\Omega = \{(S, I_1, I_2, R) \in \mathbb{R}_+^4 : N \leq \Lambda/\mu\}$ is positively invariant and attracting with respect to model (1).*

Proof. Let $(S(t), I_1(t), I_2(t), R(t))$ be any solution of system (1) with any given initial condition. Then by adding all the equations in system (1) we obtain that

$$\frac{dN}{dt} = \Lambda - \mu S - (\mu + \alpha_1) I_1 - (\mu + \alpha_2) I_2 - \mu R$$
$$\leq \Lambda - \mu N.$$
$$(3)$$

Thus, from standard comparison theorem [20], we obtain

$$N(t) \leq N(0) e^{-\mu t} + \frac{\Lambda}{\mu} \left(1 - e^{-\mu t} \right).$$
$$(4)$$

Consequently it is easy to verify that

$$N(t) \leq \frac{\Lambda}{\mu}, \quad \text{when} \quad (0) \leq \frac{\Lambda}{\mu}.$$
$$(5)$$

Thus, Ω is positively invariant. Further, when $N(0) > \Lambda/\mu$, then either the solution enters Ω in finite time, or $N(t)$ approaches Λ/μ as $t \to \infty$. Hence, Ω is attracting (i.e., all solutions in $\mathbb{R}_+^4$ eventually approach, enter, or stay in Ω). $\square$

Therefore, the system of equations given in model (1) is mathematically well-posed and epidemiologically reasonable, since all the variables remain nonnegative $\forall t \geq 0$. Further since the equations of model (1) are continuous and have continuously partial derivatives then they are Lipschitzian. In addition to that from Theorem 1, model (1) is uniformly bounded. Therefore the solution of it exists and is unique. Hence, from now onward it is sufficient to consider the dynamics of model (1) in Ω.

3. Equilibrium Points and Basic Reproduction Number

Model (1) has four equilibrium points that are obtained by setting the right hand sides of this model equal to zero. The first equilibrium point is the disease-free equilibrium (DFE) point that is denoted by $E_0 = (S_0, 0, 0, 0)$ with $S_0 = \Lambda/\mu$. Moreover the basic reproduction number of model (1), which is denoted by R_0, is the maximum eigenvalue of the next generation matrix (i.e., the maximum of the reproduction numbers, those computed of each disease). That is,

$$R_0 = \max\{R_1, R_2\}. \tag{6}$$

Here $R_1 = (\beta_1 S_0 + p_1)/(\mu + \alpha_1 + \delta)$ and $R_2 = (\beta_2 S_0 + p_2)/(\mu + \alpha_2 + \gamma)$.

The other three equilibrium points can be described as follows.

The first disease-free equilibrium point, which is located in the boundary SI_2-plane, is denoted by $E_1 = (\bar{S}, 0, \bar{I}_2, 0)$ where

$$\bar{S} = \frac{S_0}{T_2},$$
$$\bar{I}_2 = \frac{\Lambda}{(\mu + \alpha_2)}\left(1 - \frac{1}{T_2}\right), \tag{7}$$

and here $T_2 = \beta_2 S_0/(\mu + \alpha_2 + \gamma - p_2)$. Clearly E_1 exists uniquely in the interior of SI_2-plane provided that

$$T_2 > 1. \tag{8}$$

The second disease-free equilibrium point that is located in the boundary $SI_1 R$-space is given by $E_2 = (\tilde{S}, \tilde{I}_1, 0, \tilde{R})$ where

$$\tilde{S} = \frac{S_0(1 + \tilde{I}_1)}{T_1};$$
$$\tilde{I}_1 = \frac{\Lambda(1 - 1/T_1)}{(\mu + \alpha_1 + \delta) - \eta\delta/(\eta + \mu) + \Lambda/T_1}, \tag{9}$$
$$\tilde{R} = \frac{\tilde{I}_1}{\eta + \mu},$$

and here $T_1 = \beta_1 S_0/(\mu + \alpha_1 + \delta - p_1)$. Obviously E_2 exists uniquely in the interior of positive octant of $SI_1 R$-space provided that

$$T_1 > 1. \tag{10}$$

Finally, the endemic equilibrium point, which is denoted by $E_3 = (S^*, I_1^*, I_2^*, R^*)$, where

$$S^* = \frac{S_0}{T_2};$$
$$I_1^* = \left(\frac{T_1}{T_2} - 1\right);$$

$$R^* = \frac{\delta((T_1/T_2) - 1)}{\eta + \mu},$$

$$I_2^* = \frac{\Lambda}{(\mu + \alpha_2)}\left(1 - \frac{1}{T_2}\right)$$
$$- \frac{((T_1/T_2) - 1)}{(\mu + \alpha_2)}\left[(\mu + \alpha_1 + \delta) - \frac{\eta\delta}{\eta + \mu}\right], \tag{11}$$

exists uniquely in the interior of Ω provided that the following conditions hold:

$$T_1 > T_2 > 1,$$
$$\Lambda > \frac{\Lambda}{T_2} + I_1^*\left((\mu + \alpha_1 + \delta) - \frac{\eta\delta}{\eta + \mu}\right). \tag{12}$$

Keeping the above in view, it is easy to verify with the help of condition (2) that

$$T_i > 1 \,(T_i < 1) \iff R_i > 1 \,(R_i < 1), \quad i = 1, 2. \tag{13}$$

Then directly we obtain $T_i > 1 \,(T_i < 1) \Leftrightarrow R_0 > 1 \,(R_0 < 1)$. Consequently, T_i represent the threshold parameters for the existence of the last three equilibrium points of model (1). Moreover, it is well known that the basic reproduction number (R_0) is representing the average number of secondary infections that occur from one infected individual in contact with susceptible individuals. Therefore if $R_0 < 1$, then each infected individual in the entire period of infectivity will produce less than one infected individual on average, which shows the disease will be wiped out of the population. However, if $R_0 > 1$, then each infected individual in the entire infection period having contact with susceptible individuals will produce more than one infected individual; this leads to the disease invading the susceptible population.

4. Local Stability Analysis

In this section, the local stability analyses of all possible equilibrium points of model (1) are discussed by determining the Jacobian matrix with their eigenvalues. Now the general Jacobian matrix of model (1) can be written:

$$J = \begin{bmatrix} -\dfrac{\beta_1 I_1}{1 + I_1} - \beta_2 I_2 - \mu & -\dfrac{\beta_1 S}{(1 + I_1)^2} - p_1 & -\beta_2 S + (\gamma - p_2) & \eta \\[2ex] \dfrac{\beta_1 I_1}{1 + I_1} & -\dfrac{\beta_1 S I_1}{(1 + I_1)^2} + f(S, I_1) & 0 & 0 \\[2ex] \beta_2 I_2 & 0 & \beta_2 S - (\mu + \alpha_2 + \gamma - p_2) & 0 \\[2ex] 0 & \delta & 0 & -(\eta + \mu) \end{bmatrix}, \tag{14}$$

where $f(S, I_1) = \beta_1 S/(1 + I_1) - (\mu + \alpha_1 + \delta - p_1)$. Therefore the local stability results near the above equilibrium points can be presented in the following theorems.

Theorem 2. *The disease-free equilibrium $E_0 = (S_0, 0, 0, 0)$ is locally asymptotically stable when $R_0 < 1$ and unstable for $R_0 > 1$.*

Proof. The characteristic equation of the Jacobian matrix of model (1) at the disease-free equilibrium can be written as

$$(\lambda + \mu)(\lambda - \beta_1 S_0 + (\mu + \alpha_1 + \delta - p_1))$$
$$\cdot (\lambda - \beta_2 S_0 + (\mu + \alpha_2 + \gamma - p_2))(\lambda + (\eta + \mu)) = 0. \tag{15}$$

So, if $R_0 < 1$, then according to (6), (15) has four negative real roots (eigenvalues). Hence, the DFE is locally asymptotically stable. Further, for $R_0 > 1$ (15) has at least one positive eigenvalue and then the DFE is a saddle point. □

Theorem 3. *The first disease-free equilibrium point $E_1 = (\bar{S}, 0, \bar{I}_2, 0)$ of model (1) is locally asymptotically stable provided that*

$$T_2 > 1 > T_1. \tag{16}$$

Proof. The characteristic equation of the Jacobian matrix of model (1) at E_1 can be written as

$$\left[\bar{\lambda} - \beta_1 \bar{S} + (\mu + \alpha_1 + \delta - p_1)\right]\left[\bar{\lambda} + (\eta + \mu)\right]$$
$$\cdot \left[\bar{\lambda}^2 - Tr_1\bar{\lambda} + D_1\right] = 0; \tag{17}$$

here $Tr_1 = -(\beta_2 \bar{I}_2 + \mu) < 0$ and $D_1 = \beta_2 \bar{I}_2(\beta_2 \bar{S} - (\gamma - p_1)) > 0$ due to condition (16). Hence both the eigenvalues $\bar{\lambda}_S$ and $\bar{\lambda}_{I_2}$, which describe the dynamics in the S-direction and I_2-direction, respectively, have negative real parts. Moreover, from (17), the eigenvalue in the I_1-direction can be written as

$$\bar{\lambda}_{I_1} = \beta_1 \bar{S} - (\mu + \alpha_1 + \delta - p_1) = \left(1 - \frac{T_2}{T_1}\right)\beta_1 \bar{S}. \tag{18}$$

Thus under the given condition (16), we have $\bar{\lambda}_{I_1} < 0$, while $\bar{\lambda}_R = -(\eta + \mu)$ is always negative. Hence, E_1 is locally asymptotically stable. □

Theorem 4. *The second disease-free equilibrium point $E_2 = (\tilde{S}, \tilde{I}_1, 0, \tilde{R})$ of model (1) is locally asymptotically stable provided that*

$$T_1 > 1 > T_2\left(1 + \tilde{I}_1\right), \tag{19a}$$

$$\tilde{P}_2 > \delta, \tag{19b}$$

where $\tilde{P}_2$ is given in the proof.

Proof. The characteristic equation of the Jacobian matrix of model (1) at E_2 can be written as

$$\left[\tilde{\lambda} - \beta_2 \tilde{S} + (\mu + \alpha_2 + \gamma - p_2)\right]\left[\tilde{\lambda}^3 + A\tilde{\lambda}^2 + B\tilde{\lambda} + C\right]$$
$$= 0; \tag{20}$$

here

$$A = \left(\tilde{P}_1 + \tilde{P}_2\right) + (\eta + \mu),$$
$$B = \left(\tilde{P}_1 + \tilde{P}_2\right)(\eta + \mu) + \left(\tilde{P}_1\tilde{P}_2 + \tilde{P}_3\right), \tag{21}$$
$$C = (\eta + \mu)\left(\tilde{P}_1\tilde{P}_2 + \tilde{P}_3\right) - \tilde{P}_4$$

with

$$\tilde{P}_1 = \left(\frac{\beta_1\tilde{I}_1}{1 + \tilde{I}_1} + \mu\right) > 0,$$

$$\tilde{P}_2 = \frac{\beta_1\tilde{S}\tilde{I}_1}{\left(1 + \tilde{I}_1\right)^2} > 0,$$

$$\tilde{P}_3 = \left(\frac{\beta_1\tilde{I}_1}{1 + \tilde{I}_1}\left(\frac{\beta_1\tilde{I}_1}{\left(1 + \tilde{I}_1\right)^2} + p_1\right)\right) > 0, \tag{22}$$

$$\tilde{P}_4 = \frac{\eta\delta\beta_1\tilde{I}_1}{1 + \tilde{I}_1} > 0.$$

Clearly, the eigenvalue $\tilde{\lambda}_{I_2}$ in the I_2-direction can be written as

$$\tilde{\lambda}_{I_2} = \beta_2\tilde{S} - (\mu + \alpha_2 + \gamma - p_2)$$
$$= \left(1 - \frac{T_1}{T_2\left(1 + \tilde{I}_1\right)}\right)\beta_2\tilde{S}, \tag{23}$$

and thus $\tilde{\lambda}_{I_2} < 0$ under the condition (19a). In addition from (20) we have $A > 0$ always, while C can be written as

$$C = \tilde{P}_5\frac{\tilde{P}_2}{\delta} + \mu\tilde{P}_5\frac{\tilde{P}_2}{\delta} + \tilde{P}_2\left(\beta_2 I_2^*(\eta + \mu)\right)$$
$$+ \tilde{P}_2\mu(\eta + \mu) + \tilde{P}_3(\eta + \mu) + \tilde{P}_4\left(\tilde{P}_2 + (\eta + \mu)\right) \tag{24}$$
$$- \tilde{P}_5.$$

Hence, $C > 0$ provided that the sufficient condition (19b) holds. Further it is easy to verify that

$$AB - C = \left(\tilde{P}_1 + \tilde{P}_2\right)\left(B + (\eta + \mu)^2\right) + \tilde{P}_4 > 0. \tag{25}$$

Hence, due to the Routh-Hurwitz criterion the third-degree polynomial term in (20) has roots (eigenvalues) with negative real parts. Hence E_2 is locally asymptotically stable. □

Theorem 5. *The endemic equilibrium point $E_3 = (S^*, I_1^*, I_2^*, R^*)$ of model (1) is locally asymptotically stable provided that*

$$T_1 > T_2 > 1, \tag{26a}$$

$$\overline{\overline{Q}} > q_5 > \delta, \tag{26b}$$

where $\overline{\overline{Q}}$ and q_5 are given in the proof.

Proof. The characteristic equation of the Jacobian matrix of model (1) at E_3 can be written as

$$\lambda^{*4} + A_1\lambda^{*3} + B_1\lambda^{*2} + C_1\lambda^* + D_1 = 0. \qquad (27)$$

Here

$$A_1 = (q_1 + q_2) + (\eta + \mu) > 0,$$

$$B_1 = (\eta + \mu)(q_1 + q_2) + q_1 q_2 + q_3 + q_4 > 0,$$

$$C_1 = (q_1 q_2 + q_3)(\eta + \mu) + q_4(q_2 + (\eta + \mu)) - q_5, \qquad (28)$$

$$D_1 = q_4 q_2(\eta + \mu) > 0$$

with

$$q_1 = \frac{\beta_1 I_1^*}{1 + I_1^*} + \beta_2 I_2^* + \mu;$$

$$q_2 = \frac{\beta_1 S^* I_1^*}{\left(1 + I_1^*\right)^2};$$

$$q_3 = \frac{\beta_1 I_1^*}{1 + I_1^*}\left(\frac{\beta_1 S^*}{\left(1 + I_1^*\right)^2} + p_1\right), \qquad (29)$$

$$q_4 = \beta_2 I_2^*[\beta_2 S^* - (\gamma - p_2)];$$

$$q_5 = \frac{\eta \delta \beta_1 I_1^*}{1 + I_1^*}.$$

Obviously, $q_i > 0$, $i = 1, 2, 3, 5$, while q_4 is positive under condition (26a). Now, by using the values of q_i and the sufficient condition (26b), then straightforward computation gives

$$C_1 = \overline{\overline{Q}} + q_4(q_2 + (\eta + \mu)) - q_5 > 0, \qquad (30)$$

and here $\overline{\overline{Q}} = (q_1 q_2 + q_3)(\eta + \mu)$. Moreover we have

$$A_1 B_1 C_1 = \left[Q + (q_1 + q_2)^2(\eta + \mu) + q_4 q_1\right.$$

$$\left. + \left(q_1 q_2 + q_3 + (\eta + \mu)^2\right)(q_1 + q_2)\right][Q - q_5], \qquad (31)$$

$$C_1^2 + A_1^2 D_1 = (Q - q_5)^2 + \left[(q_1 + q_2) + (\eta + \mu)\right]^2$$

$$\cdot (q_4 q_2(\eta + \mu)),$$

where $Q = \overline{\overline{Q}} + q_4(q_2 + (\eta + \mu))$. Therefore we obtain that

$$A_1 B_1 C_1 - C_1^2 - A_1^2 D_1$$

$$= F\overline{\overline{Q}} + q_4 \overline{\overline{Q}}(q_1 + q_2) + q_4 q_1 A_1(\eta + \mu)^2$$

$$+ q_4^2 q_1(q_2 + (\eta + \mu)) \qquad (32)$$

$$+ q_4 q_2(q_1 + q_2)(q_1 q_2 + q_3) - Fq_5 + C_1 q_5.$$

Here $F = (q_1 + q_2)^2(\eta + \mu) + (q_1 q_2 + q_3 + (\eta + \mu)^2)(q_1 + q_2) + q_4 q_1$.

Hence, according to condition (26b) it is easy to verify that $A_1 B_1 C_1 - C_1^2 - A_1^2 D_1 > 0$. Therefore, all the coefficients

of (27) are positive and $A_1 B_1 C_1 - C_1^2 - A_1^2 D_1 > 0$. Hence, due to the Routh-Hurwitz criterion all the eigenvalues $(\lambda_S^*, \lambda_{I_1}^*, \lambda_{I_2}^*$ and $\lambda_R^*)$ of the Jacobian matrix near the endemic equilibrium point E_3 have negative real parts. Thus, the proof is complete. $\qquad \square$

5. Global Stability Analysis

This section deals with the global stability of the equilibrium points of model (1) using Lyapunov methods with LaSalle's invariant principle. The obtained results are presented in the following theorems.

Theorem 6. *Assume that DFE $E_0 = (S_0, 0, 0, 0)$ of model (1) is locally asymptotically stable; then it is global asymptotically stable in Ω.*

Proof. Consider $\widehat{V} : \Omega \to \mathbb{R}$ that is defined by

$$\widehat{V}(S, I_1, I_2, R) = \sum_{i=1}^{2} I_i. \qquad (33)$$

Computing the derivative of this positive semidefinite function with respect to time along the solution of model (1) and then simplifying the resulting terms give

$$\frac{d\widehat{V}}{dt} = \left(\frac{\beta_1 S}{1 + I_1} - (\mu + \alpha_1 + \delta - p_1)\right)I_1$$

$$+ (\beta_2 S - (\mu + \alpha_2 + \gamma - p_2))I_2. \qquad (34)$$

Since the solution of model (1) is bounded by $S_0 = \Lambda/\mu$ as $t \to \infty$,

$$\frac{d\widehat{V}}{dt} \le (\beta_1 S_0 - (\mu + \alpha_1 + \delta - p_1))I_1$$

$$+ (\beta_2 S_0 - (\mu + \alpha_2 + \gamma - p_2))I_2 \qquad (35)$$

$$= (\mu + \alpha_1 + \delta - p_1)(T_1 - 1)I_1$$

$$+ (\mu + \alpha_2 + \gamma - p_2)(T_2 - 1)I_2.$$

Since $T_i < 1$, $i = 1, 2$, due to the local stability condition of E_0 then $d\widehat{V}/dt < 0$. Also we have that $d\widehat{V}/dt = 0$ on the set $\{(S, I_1, I_2, R) \in \Omega : I_1 = I_2 = 0\}$, so $d\widehat{V}/dt$ is negative semidefinite and hence according to Lyapunov first theorem E_0 is globally stable point. Now, since on this set we have

$$\frac{dS}{dt} = \Lambda - \mu S + \eta R = 0, \qquad (36)$$

if and only if $S = S_0$, $R = 0$, thus the largest invariant set contained in this set is reduced to the disease-free equilibrium point E_0. Hence according to LaSalle's invariant principle [18], E_0 is attractive point and hence it is globally asymptotically stable in Ω. $\qquad \square$

Theorem 7. *Assume that the first disease-free equilibrium point $E_1 = (\overline{S}, 0, \overline{I}_2, 0)$ is locally asymptotically stable; then it is global asymptotically stable in Ω provided that*

$$I_2 > \overline{I}_2. \qquad (37)$$

Proof. Consider that $L : \Omega \rightarrow \mathbb{R}$ that is defined by

$$
L(S, I_1, I_2, R) = \frac{1}{2}\left[(S - \bar{S}) + I_1 + (I_2 - \bar{I}_2) + R\right]^2
$$
$$
+ \frac{(2\mu + \alpha_1)}{\beta_1} I_1
$$
$$
+ \frac{(2\mu + \alpha_2)}{\beta_2}\left(I_2 - \bar{I}_2 - \bar{I}_2 \ln \frac{I_2}{\bar{I}_2}\right) \qquad (38)
$$
$$
+ \frac{(2\mu + \alpha_1)}{2\delta} R^2.
$$

Clearly L is continuous and positive definite function. Now by taking the derivative of L with respect to time along the solution of model (1), we get after simplifying the resulting terms that

$$
\frac{dL}{dt} = -\mu(S - \bar{S})^2 - \mu R^2 - 2\mu R(S - \bar{S})
$$
$$
- (2\mu + \alpha_1)(S - \bar{S}) I_1
$$
$$
- (2\mu + \alpha_2)(S - \bar{S})(I_2 - \bar{I}_2)
$$
$$
- (2\mu + \alpha_1 + \alpha_2)(I_2 - \bar{I}_2) I_1
$$
$$
- (2\mu + \alpha_1) I_1 R - (\mu + \alpha_2)(I_2 - \bar{I}_2)^2
$$
$$
- (2\mu + \alpha_2)(I_2 - \bar{I}_2) R - (\mu + \alpha_1) I_1^2, \qquad (39)
$$
$$
\frac{dL}{dt} \le -\mu\left[(S - \bar{S}) + R\right]^2 - \mu\left[I_1 + (I_2 - \bar{I}_2)\right]^2
$$
$$
- 2\mu(I_2 - \bar{I}_2) R
$$
$$
+ \frac{2\mu(\mu + \alpha_1 + \delta - p_1)}{\beta_1}\left[\frac{T_1}{T_2} - 1\right] I_1
$$
$$
- \frac{\mu(\eta + \mu)}{\delta} R^2.
$$

Hence according to local stability condition (16) along with the sufficient condition (37) it obtains that dL/dt is negative definite function. Thus due to Lyapunov second theorem E_1 is global asymptotically stable in Ω. $\qquad\square$

Theorem 8. *Assume that the second disease-free equilibrium point $E_2 = (\bar{S}, \bar{I}_1, 0, \bar{R})$ of model (1) is locally asymptotically stable; then it is global asymptotically stable in Ω if*

$$
q_{12}^2 < q_{11} q_{22}, \qquad (40a)
$$
$$
q_{14}^2 < q_{11} q_{44}, \qquad (40b)
$$
$$
q_{23}^2 < q_{22} q_{33}, \qquad (40c)
$$
$$
q_{34}^2 < q_{33} q_{44}, \qquad (40d)
$$

where

$$
q_{11} = \mu;
$$
$$
q_{22} = (\mu + \alpha_1) + \frac{\bar{S}}{(1 + \bar{I}_1)(1 + I_1)};
$$
$$
q_{33} = (\mu + \alpha_2);
$$
$$
q_{44} = \frac{(2\mu + \alpha_1)(\eta + \mu)}{2\delta} + \mu; \qquad (41)
$$
$$
q_{12} = \frac{1}{1 + I_1} - (2\mu + \alpha_1);
$$
$$
q_{14} = 2\mu;
$$
$$
q_{23} = (2\mu + \alpha_1 + \alpha_2);
$$
$$
q_{34} = (2\mu + \alpha_2).
$$

Proof. Consider the function $W : \Omega \rightarrow \mathbb{R}$ that is defined by

$$
W(S, I_1, I_2, R)
$$
$$
= \frac{1}{2}\left[(S - \bar{S}) + (I_1 - \bar{I}_1) + I_2 + (R - \bar{R})\right]^2
$$
$$
+ \frac{1}{\beta_1}\left(I_1 - \bar{I}_1 - \bar{I}_1 \ln \frac{I_1}{\bar{I}_1}\right) + \frac{(2\mu + \alpha_2)}{\beta_2} I_2 \qquad (42)
$$
$$
+ \frac{(2\mu + \alpha_1)}{2\delta}(R - \bar{R})^2.
$$

Clearly W is continuous and positive definite function. Now by taking the derivative of W with respect to time along the solution of model (1), we get after simplifying the resulting terms that

$$
\frac{dW}{dt} = -q_{11}(S - \bar{S})^2 + q_{12}(S - \bar{S})(I_1 - \bar{I}_1)
$$
$$
- q_{14}(S - \bar{S})(R - \bar{R}) - q_{22}(I_1 - \bar{I}_1)^2
$$
$$
- q_{23}(I_1 - \bar{I}_1) I_2 - q_{33} I_2^2 - q_{34}(R - \bar{R}) I_2 \qquad (43)
$$
$$
- q_{44}(R - \bar{R})^2
$$
$$
+ \frac{2\mu(\mu + \alpha_2 + \gamma - p_2)}{\beta_2}\left[\frac{T_2(1 + \bar{I}_1)}{T_1} - 1\right] I_2.
$$

Now by using the given conditions (40a)–(40d) we get that

$$
\frac{dW}{dt} \le -\left[\sqrt{\frac{q_{11}}{2}}(S - \bar{S}) - \sqrt{\frac{q_{22}}{2}}(I_1 - \bar{I}_1)\right]^2
$$
$$
- \left[\sqrt{\frac{q_{22}}{2}}(I_1 - \bar{I}_1) + \sqrt{\frac{q_{33}}{2}} I_2\right]^2
$$

$$-\left[\sqrt{\frac{q_{11}}{2}}\left(S-\tilde{S}\right)+\sqrt{\frac{q_{44}}{2}}\left(R-\tilde{R}\right)\right]^2$$

$$-\left[\sqrt{\frac{q_{44}}{2}}\left(R-\tilde{R}\right)+\sqrt{\frac{q_{33}}{2}}I_2\right]^2$$

$$+\frac{2\mu\left(\mu+\alpha_2+\gamma-p_2\right)}{\beta_2}\left[\frac{T_2\left(1+\tilde{I}_1\right)}{T_1}-1\right]I_2. \tag{44}$$

Hence according to local stability condition (19a) it obtains that dW/dt is negative definite function. Thus due to Lyapunov second theorem E_1 is global asymptotically stable in Ω. $\square$

Theorem 9. *Assume that the endemic equilibrium point $E_3 = (S^*, I_1^*, I_2^*, R^*)$ of model (1) is locally asymptotically stable; then it is global asymptotically stable in Ω if*

$$\alpha_{14}^2 < \alpha_{11}\alpha_{44}, \tag{45a}$$

$$\alpha_{12}^2 < \alpha_{11}\alpha_{22}, \tag{45b}$$

$$\alpha_{24}^2 < \alpha_{22}\alpha_{44}, \tag{45c}$$

where

$$\alpha_{11} = \left(\frac{\beta_1 I_1}{1+I_1}+\beta_2 I_2+\mu\right),$$

$$\alpha_{22} = \beta_1 S^*,$$

$$\alpha_{44} = \left(\eta+\mu\right),$$

$$\alpha_{12} = \frac{\beta_1}{1+I_1}-\left(\frac{\beta_1 S^*}{\left(1+I_1\right)\left(1+I_1^*\right)}+p_1\right),$$

$$\alpha_{24} = \delta,$$

$$\alpha_{14} = \eta. \tag{46}$$

Proof. Consider the function $V: \Omega \to \mathbb{R}$ that is defined by

$$V\left(S, I_1, I_2, R\right) = \frac{\left(S-S^*\right)^2}{2}+\left(I_1-I_1^*-I_1^*\ln\frac{I_1}{I_1^*}\right)$$

$$+\frac{\left(2\mu+\alpha_2\right)}{\beta_2}\left(I_2-I_2^*-I_2^*\ln\frac{I_2}{I_2^*}\right) \tag{47}$$

$$+\frac{\left(R-R^*\right)^2}{2}.$$

Clearly the function V is continuous and positive definite function. By taking the derivative of V with respect to time

along the solution of model (1), we get after simplifying the resulting terms that

$$\frac{dV}{dt} = -\alpha_{11}\left(S-S^*\right)^2-\alpha_{22}\left(I_1-I_1^*\right)^2$$

$$-\alpha_{44}\left(R-R^*\right)^2+\alpha_{12}\left(S-S^*\right)\left(I_1-I_1^*\right)$$

$$+\alpha_{14}\left(R-R^*\right)\left(S-S^*\right) \tag{48}$$

$$+\alpha_{24}\left(I_1-I_1^*\right)\left(R-R^*\right).$$

Then by using the given conditions (45a)–(45c) we obtain that

$$\frac{dV}{dt} < -\left[\sqrt{\alpha_{11}}\left(S-S^*\right)+\sqrt{\alpha_{22}}\left(I_1-I_1^*\right)\right]^2$$

$$-\left[\sqrt{\alpha_{11}}\left(S-S^*\right)+\sqrt{\alpha_{44}}\left(R-R^*\right)\right]^2 \tag{49}$$

$$-\left[\sqrt{\alpha_{44}}\left(R-R^*\right)+\sqrt{\alpha_{22}}\left(I_1-I_1^*\right)\right]^2.$$

Hence, dV/dt is negative semidefinite, and $dV/dt = 0$ on the set $\{(S, I_1, I_2, R) \in \Omega : S = S^*, I_1 = I_1^*, I_2 > 0, R = R^*\}$, so according to Lyapunov first theorem E_3 is globally stable point. Further, since on this set we have

$$\frac{dS}{dt} = \Lambda-\left(\frac{\beta_1 I_1^*}{1+I_1^*}+\beta_2 I_2\right)S^*+\left(\gamma-p_2\right)I_2-\mu S^*$$

$$-p_1 I_1^*+\eta R^* = 0, \tag{50}$$

if and only if $I_2 = I_2^*$, then the largest compact invariant set contained in this set is reduced to the endemic equilibrium point E_3. Hence according to LaSalle's invariant principle [18], E_3 is attractive point and hence it is globally asymptotically stable in Ω. $\square$

6. Bifurcation Analysis

In this section the local bifurcations near the equilibrium points of model (1) are investigated as shown in the following theorems with the help of Sotomayor theorem [20]. Note that model (1) can be rewritten in a vector form $dX/dt = f(X)$, where $X = (S, I_1, I_2, R)^T$ and $f = (f_1, f_2, f_3, f_4)^T$ with f_i, $i = 1, 2, 3, 4$, are given in the right hand side of model (1). Moreover, straightforward computation gives that the general second derivative of the Jacobian matrix (14) can be written:

$$D^2 f\left(X, \beta\right)\left(U, U\right)$$

$$=\begin{bmatrix} \dfrac{-2\beta_1\zeta_1\zeta_2}{\left(1+I_1\right)^2}-2\beta_2\zeta_1\zeta_3+\dfrac{2\beta_1 S\left(\zeta_2\right)^2}{\left(1+I_1\right)^3} \\[2ex] \dfrac{2\beta_1\zeta_1\zeta_2}{\left(1+I_1\right)^2}-\dfrac{2\beta_1 S\left(\zeta_2\right)^2}{\left(1+I_1\right)^3} \\[2ex] 2\beta_2\zeta_1\zeta_3 \\[2ex] 0 \end{bmatrix}, \tag{51}$$

where β is any bifurcation parameter and $U = (\zeta_1, \zeta_2, \zeta_3, \zeta_4)^T$ is any eigenvector.

Theorem 10. *Assume that $T_2 < 1$; then as T_1 passes through the value $T_1 = 1$, model (1) near the disease-free equilibrium E_0 has*

(1) *no saddle-node bifurcation;*

(2) *a transcritical bifurcation;*

(3) *no pitchfork bifurcation.*

Proof. Since $T_1 = \beta_1 S_0/(\mu + \alpha_1 + \delta - p_1) = 1$; then $\beta_1 = (\mu + \alpha_1 + \delta - p_1)/S_0 = \beta_1^*$. Now straightforward computation shows that the Jacobian matrix of model (1) at E_0 with $\beta_1 = \beta_1^*$ has zero eigenvalue ($\lambda_{I_1} = 0$) and can be written as follows:

$$J_0$$

$$= \begin{bmatrix} -\mu & -\beta_1^* S_0 - p_1 & -\beta_2 S_0 + (\gamma - p_2) & \eta \\ 0 & 0 & 0 & 0 \\ 0 & 0 & \beta_2 S_0 - (\mu + \alpha_2 + \gamma - p_2) & 0 \\ 0 & \delta & 0 & -(\eta + \mu) \end{bmatrix}. \quad (52)$$

Let $V = (v_1, v_2, v_3, v_4)^T$ be the eigenvector corresponding to $\lambda_{I_1} = 0$. Thus $J_0 V = 0$ gives

$$V = \begin{bmatrix} av_2 \\ v_2 \\ 0 \\ bv_2 \end{bmatrix}, \quad (53)$$

where v_2 is any nonzero real number, $a = (\eta\delta/(\eta + \mu) - (\mu + \alpha_1 + \delta))/\mu$, and $b = \delta/(\eta + \mu)$.

Similarly, $W = (w_1, w_2, w_3, w_4)^T$ represents the eigenvector corresponding to eigenvalue $\lambda_{I_1} = 0$ of J_0^T. Hence $J_0^T W = 0$ gives that

$$W = \begin{bmatrix} 0 \\ w_2 \\ 0 \\ 0 \end{bmatrix}, \quad (54)$$

and here w_2 is any nonzero real number. Now, since

$$\frac{df}{d\beta_1} = f_{\beta_1}(X, \beta_1) = \left(\frac{-I_1 S}{1 + I_1}, \frac{I_1 S}{1 + I_1}, 0, 0\right)^T, \quad (55)$$

thus $f_{\beta_1}(E_0, \beta_1^*) = (0, 0, 0, 0)^T$, which gives $W^T f_{\beta_1}(E_0, \beta_1^*) = 0$.

Thus, according to Sotomayor's theorem for local bifurcation, model (1) has no saddle-node bifurcation near DFE at $\beta_1 = \beta_1^*$.

Now since

$$Df_{\beta_1}(E_0, \beta_1^*) = \begin{bmatrix} 0 & -S_0 & 0 & 0 \\ 0 & S_0 & 0 & 0 \\ 0 & 0 & 0 & 0 \\ 0 & 0 & 0 & 0 \end{bmatrix}, \quad (56)$$

then, $W^T(Df_{\beta_1}(E_0, \beta_1^*)V) = S_0 v_2 w_2 \neq 0$. Now, by substituting E_0 and β_1^* in (51) we get

$$D^2 f(E_0, \beta_1^*) \cdot (V, V)$$

$$= \begin{bmatrix} -2\beta_1^* v_1 v_2 - 2\beta_2 v_1 v_3 + 2\beta_1^* S_0 (v_2)^2 \\ 2\beta_1^* v_1 v_2 - 2\beta_1^* S_0 (v_2)^2 \\ 2\beta_2 v_1 v_3 \\ 0 \end{bmatrix}. \quad (57)$$

Therefore,

$$W^T \left(D^2 f(E_0, \beta_1^*) \cdot (V, V)\right)$$

$$= \left(2\beta_1^* v_1 v_2 - 2\beta_1^* S_0 (v_2)^2\right) w_2 \quad (58)$$

$$= \left[2\beta_1^* (v_2)^2 (a - S_0)\right] w_2 \neq 0.$$

So, according to Sotomayor's theorem model (1) has a transcritical bifurcation at E_0 with parameter $\beta_1 = \beta_1^* \approx (T_1 = 1)$, while the pitchfork bifurcation cannot occur. $\square$

Note that similar results as those of Theorem 10 are obtained at $T_2 = 1$ or $\beta_2 = \beta_2^* = (\mu + \alpha_2 + \gamma - p_2)/S_0$.

Theorem 11. *Assume that $T_1 = T_2$; then model (1) near the first disease-free equilibrium point $E_1 = (\bar{S}, 0, \bar{I}_2, 0)$ has*

(1) *no saddle-node bifurcation;*

(2) *a transcritical bifurcation;*

(3) *no pitchfork bifurcation.*

Proof. There are two cases; in the first case, it is assumed that $T_1 = T_2 = 1$; then straightforward computation shows that $J(E_1) = J(E_0)$; that is, $\bar{I}_2 = 0$. So (by Theorem 10) model (1) has no bifurcation and then the proof is complete.

Now in the second case it is assumed that $T_1 = T_2 \neq 1$ or equivalently $\beta_1 = \beta_1^{**} = \beta_2(\mu + \alpha_1 + \delta - p_1)/(\mu + \alpha_2 + \gamma - p_2)$. So straightforward computation shows that the Jacobian matrix of model (1) at E_1 with $\beta_1 = \beta_1^{**}$ has zero eigenvalue ($\lambda_{I_1} = 0$) and can be written as follows:

$$J_1 = \begin{bmatrix} -\beta_2 \bar{I}_2 - \mu & -\beta_1^{**}\bar{S} - p_1 & -\beta_2\bar{S} + (\gamma - p_2) & \eta \\ 0 & 0 & 0 & 0 \\ \beta_2\bar{I}_2 & 0 & \beta_2\bar{S} - (\mu + \alpha_2 + \gamma - p_2) & 0 \\ 0 & \delta & 0 & -(\eta + \mu) \end{bmatrix}. \quad (59)$$

Let $X = (x_1, x_2, x_3, x_4)^T$ be the eigenvector corresponding to $\lambda_{I_1} = 0$, which satisfies $J_1 X = 0$, so we get

$$X = \begin{bmatrix} 0 \\ x_2 \\ a_1 x_2 \\ b x_2 \end{bmatrix}, \qquad (60)$$

where x_2 is any nonzero real number, $a_1 = -[(\beta_1 \bar{S} + p_1) - \eta\delta/(\eta + \mu)]/\beta_1 \bar{S} - (\gamma - p_2)$, and $b = \delta/(\eta + \mu)$.

Similarly the eigenvector $Y = (y_1, y_2, y_3, y_4)^T$ that is corresponding to the eigenvalue $\lambda_{I_1} = 0$ of J_1^T satisfies $J_1^T Y = 0$, so we get

$$Y = \begin{bmatrix} 0 \\ y_2 \\ 0 \\ 0 \end{bmatrix}, \qquad (61)$$

and here y_2 is any nonzero real number. Now, since

$$\frac{df}{d\beta_1} = f_{\beta_1}(X, \beta_1) = \left(\frac{-I_1 S}{1 + I_1}, \frac{I_1 S}{1 + I_1}, 0, 0 \right)^T \qquad (62)$$

then by substituting the values of E_1 and β_1^{**} we obtain that $f_{\beta_1}(E_1, \beta_1^{**}) = (0, 0, 0, 0)^T$ and hence we get that $Y^T f_{\beta_1}(E_1, \beta_1^{**}) = 0$.

Thus according to Sotomayor's theorem for local bifurcation, model (1) has no saddle-node bifurcation near E_1 at $\beta_1 = \beta_1^{**}$. Now since

$$Df_{\beta_1}(E_1, \beta_1) = \begin{bmatrix} 0 & -\bar{S} & 0 & 0 \\ 0 & \bar{S} & 0 & 0 \\ 0 & 0 & 0 & 0 \\ 0 & 0 & 0 & 0 \end{bmatrix}, \qquad (63)$$

then $Y^T(Df_{\beta_1}(E_1, \beta_1^{**})X) = \bar{S}x_2 y_2 \neq 0$. Thus by substituting E_1 and β_1^{**} in (51) we get

$$D^2 f(E_1, \beta_1^{**}) \cdot (X, X)$$
$$= \begin{bmatrix} -2\beta_1^{**} x_1 x_2 - 2\beta_2 x_1 x_3 + 2\beta_1^{**} \bar{S}(x_2)^2 \\ 2\beta_1^{**} x_1 x_2 - 2\beta_1^{**} \bar{S}(x_2)^2 \\ 2\beta_2 x_1 x_3 \\ 0 \end{bmatrix}. \qquad (64)$$

Therefore, $[Y^T(D^2 f(E_1, \beta_1^{**}) \cdot (X, X))] = -2\beta_1^{**} \bar{S}(x_2)^2 y_2 \neq 0$.

So, model (1) has a transcritical bifurcation at E_1 with parameter $\beta_1 = \beta_1^{**} \approx (T_1 = T_2 = 1)$, while the pitchfork bifurcation cannot occur and hence the proof is complete. $\square$

Theorem 12. *Assume that condition (19b) holds and let* $T_1 = T_2(1 + \tilde{I}_1)$; *then model (1) near the second disease-free equilibrium point* $E_2 = (\tilde{S}, \tilde{I}_1, 0, \tilde{R})$ *undergoes*

(1) *no saddle-node bifurcation;*

(2) *a transcritical bifurcation;*

(3) *no pitchfork bifurcation.*

Proof. From $T_1 = T_2(1 + \tilde{I}_1)$ it is obtained that

$$\beta_2 = \beta_2^{**} = \frac{\beta_1(\mu + \alpha_2 + \gamma - p_2)}{(\mu + \alpha_1 + \delta - p_1)(1 + \tilde{I}_1)}, \qquad (65)$$

and then straightforward computation shows that the Jacobian matrix of model (1) at E_2 with $\beta_2 = \beta_2^{**}$ has zero eigenvalue ($\lambda_{I_2} = 0$) and can be written as follows:

$$J_2 = \begin{bmatrix} \dfrac{-\beta_1 \tilde{I}_1}{1 + \tilde{I}_1} - \mu & \dfrac{-\beta_1 \tilde{S}}{(1 + \tilde{I}_1)^2} - p_1 & -\beta_2^{**}\tilde{S} + (\gamma - p_2) & \eta \\ \dfrac{\beta_1 \tilde{I}_1}{1 + \tilde{I}_1} & \dfrac{-\beta_1 \tilde{S}\tilde{I}_1}{(1 + \tilde{I}_1)^2} & 0 & 0 \\ 0 & 0 & 0 & 0 \\ 0 & \delta & 0 & -(\eta + \mu) \end{bmatrix}. \qquad (66)$$

Further the eigenvector $L = (l_1, l_2, l_3, l_4)^T$ that is corresponding to $\lambda_{I_2} = 0$ satisfies $J_2 L = 0$, so we get

$$L = \begin{bmatrix} \dfrac{\tilde{S}}{1 + \tilde{I}_1} l_2 \\ l_2 \\ \dfrac{-E}{\beta_2^{**}\tilde{S} - (\gamma - p_2)} l_2 \\ b l_2 \end{bmatrix}, \qquad (67)$$

where l_2 is any nonzero real number and $E = (\beta_1\tilde{S}/(1 + \tilde{I}_1)^2)(\tilde{I}_1 + 1) + \mu\tilde{S}/1 + \tilde{I}_1 + p_1 - \eta b$.

Similarly the eigenvector $K = (k_1, k_2, k_3, k_4)^T$ that is corresponding to eigenvalue $\lambda_{I_2} = 0$ of J_2^T satisfies $J_2^T K = 0$, so we get

$$K = \begin{bmatrix} 0 \\ 0 \\ k_3 \\ 0 \end{bmatrix} \qquad (68)$$

and here k_3 is any nonzero real number. Now, since $df/d\beta_2 = f_{\beta_2}(X, \beta_2) = (-I_2 S, I_2 S, 0, 0)^T$, therefore $f_{\beta_2}(E_2, \beta_2^{**}) = (0, 0, 0, 0)^T$, which yields $K^T f_{\beta_2}(E_2, \beta_2^{**}) = 0$. Consequently according to Sotomayor's theorem for local bifurcation, model (1) has no saddle-node bifurcation near E_2 at $\beta_2 = \beta_2^{**}$.

Now since

$$Df_{\beta_2}(E_2, \beta_2) = \begin{bmatrix} 0 & 0 & -\tilde{S} & 0 \\ 0 & 0 & 0 & 0 \\ 0 & 0 & \tilde{S} & 0 \\ 0 & 0 & 0 & 0 \end{bmatrix} \qquad (69)$$

then $K^T(Df_{\beta_2}(E_2, \beta_2^{**})L) = \tilde{S}l_3 k_3 \neq 0$. Now, by substituting E_2 and β_2^{**} in (51) we get

$$D^2 f(E_2, \beta_2^{**}) \cdot (L, L)$$

$$= \begin{bmatrix} -\dfrac{\beta_1}{\left(1+\tilde{I}_1\right)^2} l_1 l_2 - 2\beta_2^{**} l_1 l_3 - \dfrac{\beta_1 l_1 l_2}{\left(1+\tilde{I}_1\right)^2} + \dfrac{2\beta_1 S(l_2)^2}{\left(1+\tilde{I}_1\right)^3} \\[4mm] \dfrac{\beta_1 l_2}{\left(1+\tilde{I}_1\right)^2} l_1 l_2 + \dfrac{\beta_1}{\left(1+\tilde{I}_1\right)^2} l_1 l_2 - \dfrac{2\beta_1 S(l_2)^2}{\left(1+\tilde{I}_1\right)^3} \\[4mm] 2\beta_2^{**} l_1 l_3 \\[4mm] 0 \end{bmatrix}. \tag{70}$$

Therefore, $[K^T(D^2 f(E_2, \beta_2^{**}) \cdot (L, L))] = 2\beta_2^{**} l_1 l_3 k_3 \neq 0$. Thus model (1) undergoes a transcritical bifurcation at E_2 with parameter $\beta_2 = \beta_2^{**} \approx (T_1 = T_2(1 + \tilde{I}_1))$, while the pitchfork bifurcation cannot occur. □

Moreover, the following results are obtained too:

(1) Although $T_1 = T_2 \neq 1$ gives $\beta_2 = \grave{\beta}_2^{**}$, model (1) does not undergo any of the above types of bifurcation near the equilibrium point E_1 with parameter $\beta_2 = \grave{\beta}_2^{**}$.

(2) Although $T_1 = T_2(1 + \tilde{I}_1)$ gives $\beta_1 = \grave{\beta}_1^{**}$, model (1) does not undergo any of the above types of bifurcation near the equilibrium point E_2 with parameter $\beta_1 = \grave{\beta}_1^{**}$.

(3) The determinant of the Jacobian matrix at E_3, say $J(E_3)$, cannot be zero and hence it has no real zero eigenvalue. So there is no bifurcation near E_3.

Keeping the above in view, in the following theorem we detect of the possibility of having Hopf bifurcation.

Theorem 13. *Assume that condition (26a) holds and let the following conditions be satisfied. Then model (1) undergoes Hopf bifurcation around the endemic equilibrium point when the parameter β_2 crosses a critical positive value $\grave{\beta}_2^*$,*

$$N_2 > 0, \tag{71a}$$
$$N_4 < 0,$$
$$\sigma_1 > 2\sigma_2, \tag{71b}$$
$$\delta < q_5 < \{q_4 q_2, q_4 q_1\}, \tag{71c}$$
$$q_2 > 1, \tag{71d}$$

and here $\sigma_1 = q_1^2 + q_2^2 + (\eta + \mu)^2$ and $\sigma_2 = (q_1 + q_2)(\eta + \mu) + 2q_1 q_2 + 2q_3 + 2q_4$, while N_2 and N_4 are given in the proof.

Proof. It is well known that, in order for Hopf bifurcation in four-dimensional systems to occur, the following conditions should be satisfied [21, 22]:

(1) The characteristic equation given in (28) has two real and negative eigenvalues and two complex eigenvalues, say, $\lambda^*(\beta_2) = \tau_1(\beta_2) \pm \tau_2(\beta_2)$.

(2) $\tau_1(\grave{\beta}_2^*) = 0$.

(3) $(d/d\beta_2)\tau_1(\beta_2)|_{\beta_2 = \grave{\beta}_2^*} \neq 0$ (The transversality condition).

Accordingly the first two points are satisfied if and only if

$$\Delta_2\left(\grave{\beta}_2^*\right) = A_1 B_1 C_1 - C_1^2 - A_1^2 D_1 = 0, \tag{72}$$
$$A_1^3 - 4\Delta_1 > 0,$$

while the third condition holds provided that

$$\left. \frac{d}{d\beta_2} \tau_1(\beta_2) \right|_{\beta_2 = \grave{\beta}_2^*} \\ = \frac{-(\Psi(\beta_2)\Theta(\beta_2) + \Phi(\beta_2)\Gamma(\beta_2))}{\Psi(\beta_2)^2 + \Phi(\beta_2)^2} \neq 0. \tag{73}$$

That means $\Psi(\beta_2)\Theta(\beta_2) + \Phi(\beta_2)\Gamma(\beta_2) \neq 0$.
Here

$$\Psi(\beta_2) = \left(4\tau_1^3 - 12\tau_1\tau_2^2\right) + 3A_1\left(\tau_1^2 - \tau_2^2\right) + 2B_1\tau_1 \\ + C_1,$$

$$\Phi(\beta_2) = \left(12\tau_1^2\tau_2 - 4\tau_2^3\right) + 6A_1\tau_1\tau_2 + 2B_1\tau_2,$$

$$\Theta(\beta_2) = \dot{A}_1\left(\tau_1^3 - 3\tau_1\tau_2^2\right) + \dot{B}_1\left(\tau_1^2 - \tau_2^2\right) + \dot{C}_1\tau_1 \\ + \dot{D}_1, \tag{74}$$

$$\Gamma(\beta_2) = \dot{A}_1\left(3\tau_1^2\tau_2 - \tau_2^3\right) + 2\dot{B}_1\tau_1\tau_2 + \dot{C}_1\tau_2.$$

Now, straightforward computation shows the condition

$$\Delta_2\left(\grave{\beta}_2^*\right) = 0 \tag{75}$$

gives that $N_1\beta_2^3 + N_2\beta_2^2 + N_3\beta_2 + N_4 = 0$,

where

$$N_1 = [q_2(\eta + \mu)((\eta + \mu) + q_2) + (\eta + \mu)(\mu + \alpha_2)((\eta + \mu) + 2q_2 + (\mu + \alpha_2))] I_2^{*3} > 0,$$

$$N_2 = \left[(\eta + \mu)\left[q_2(\eta + \mu)(3q_6 + 2 + (\eta + \mu)) + q_3((\eta + \mu) + 2(\mu + \alpha_2) + 2q_2) + (\mu + \alpha_2) \right. \right. \\ \left. \cdot [q_6(2(\eta + \mu) + (\mu + \alpha_2)) + q_2((\eta + \mu) + 4q_6 + q_2)] + q_2^2(3q_6 + q_2) + (\mu + \alpha_2)(\eta + \mu)^2 - q_5\right] + (\mu + \alpha_2)\left[q_2^2(1 + q_2) + q_2(q_3 + q_6(\mu + \alpha_2))\right] - q_2 q_5\right] I_2^{*2},$$

$$N_3 = \left[(\eta + \mu)^2 \left[q_2^3 + q_3 \left(2q_6 + 2q_2 + (\eta + \mu) \right) \right. \right.$$

$$+ 3q_6 q_2 - q_5 + (q_6 + q_2) q_6 (\mu + \alpha_2) \right] + (\eta + \mu)$$

$$\cdot \left[q_6 q_2^2 (3q_6 + 2q_2) + (2q_6 + q_2) \left[2q_2 q_3 \right. \right.$$

$$+ q_2 (\eta + \mu)^2 \right] + (\mu + \alpha_2) \left[1 + q_6 (\eta + \mu)^2 \right.$$

$$+ (2q_6 + q_2)(q_6 q_2 + q_3 - 2q_6 q_5) - q_2 q_5 \right] + q_2 (\mu$$

$$+ \alpha_2) \left[q_6 q_2 (1 + q_2) + (q_6 + q_2) q_3 + 1 \right] - (q_2 (2q_6$$

$$\left. \left. + q_2) + q_3 \right) q_5 \right] I_2^*,$$

$$N_4 = \left[(\eta + \mu)^2 \left[q_6 q_2 \left(q_6^2 + q_6 q_2 + q_2^2 \right) + q_3 q_6 (q_6 \right. \right.$$

$$+ 2q_2) + (q_6 + q_2)(\eta + \mu)(q_6 q_2 + q_3) + q_3 q_2^2$$

$$- q_2 \right] + (q_6 + q_2)(\eta + \mu) \left[q_6^2 q_2^2 + 2q_6 q_2 q_3 + q_3^2 \right]$$

$$+ \left((\eta + \mu)(q_3 - q_6 q_2) - q_6 (\eta + \mu)(q_6 + (\eta + \mu)) \right.$$

$$\left. \left. - q_2^2 (\eta + \mu) - (q_6 + q_2)(q_6 q_2 + q_3) \right) q_5 - q_5^2 \right]$$

$$\tag{76}$$

with q_1, q_2, q_3, q_4, and q_5 given in (28) and $q_6 = q_1 - \beta_2 I_2^*$. Clearly, from condition (71a) there is unique positive root, say, $\beta_2 = \grave{\beta}_2^*$. Consequently by using $\Delta_2 = 0$ in the characteristic equation and then doing some algebraic computation we get four roots,

$$\lambda_{1,2} \left(\grave{\beta}_2^* \right) = \pm i \sqrt{\frac{C_1}{A_1}} = \tau_2 \left(\grave{\beta}_2^* \right),$$

$$\lambda_{3,4} = \frac{1}{2} \left(-A_1 \pm \sqrt{A_1^2 - 4\frac{\Delta_1}{A_1}} \right).$$

$$\tag{77}$$

Now, it is easy to verify that λ_3 and λ_4 are real and negative provided that (71b).

Further for $\beta_2 \in (\grave{\beta}_2^* - \epsilon, \grave{\beta}_2^* + \epsilon)$ the general form of complex eigenvalues can be written as

$$\lambda_1 = \tau_1 (\beta_2) + i\tau_2 (\beta_2),$$

$$\lambda_2 = \tau_1 (\beta_2) - i\tau_2 (\beta_2).$$

$$\tag{78}$$

Substituting $\lambda(\beta_2) = \tau_1(\beta_2) + i\tau_2(\beta_2)$ into characteristic equation and after that calculating the derivative with respect to β_2 and then comparing the real and imaginary parts give that

$$\Psi (\beta_2) \, \dot{\tau}_1 (\beta_2) - \Phi (\beta_2) \, \dot{\tau}_2 (\beta_2) = -\Theta (\beta_2),$$

$$\Phi (\beta_2) \, \dot{\tau}_1 (\beta_2) + \Psi (\beta_2) \, \dot{\tau}_2 (\beta_2) = -\Gamma (\beta_2).$$

$$\tag{79}$$

Moreover, by solving the above linear system for $\dot{\tau}_1(\beta_2)$ and $\dot{\tau}_2(\beta_2)$ then we get that

$$\Psi (\beta_2) \Theta (\beta_2) + \Phi (\beta_2) \Gamma (\beta_2) = q_1 q_2 \left[q_3 (\mu + \alpha_2) \right.$$

$$+ 2 (\eta + \mu)^3$$

$$+ (\mu + \alpha_2)(\eta + \mu) \left((\eta + \mu) + (3q_2 + 3q_1 - 1) \right) \right]$$

$$+ q_2 q_3 \left[q_2 (\mu + \alpha_2) \right.$$

$$+ (\eta + \mu) \left((\mu + \alpha_2) + (2q_2 + 2q_1 - 1) \right) \right] + q_1 (\eta$$

$$+ \mu)(\mu + \alpha_2) \left[(\eta + \mu) + 2q_3 + (\eta + \mu)^2 + q_4 \right] + (\eta$$

$$+ \mu)^3 \left[q_2 (\mu + \alpha_2) + q_2^2 + q_3 + q_4 \right] + q_5 (\mu + \alpha_2)$$

$$\cdot \left[2q_2 + (\eta + \mu) \right] + q_1 q_3 (\eta + \mu)^2 + q_2 (\eta + \mu)(\mu$$

$$+ \alpha_2) \left[q_2 (q_2 - 1) + (\eta + \mu)(2q_2 + 2q_1 - 1) \right]$$

$$+ q_1 (\mu + \alpha_2) \left[q_1 q_2^2 + q_2^3 + q_2 q_4 - q_5 \right] + q_1 (\eta + \mu)$$

$$\cdot \left[2q_1 q_2 (1 + q_2) + 2q_2^3 + q_2 q_4 - q_5 \right] + (q_2 q_4 - q_5)$$

$$\cdot \left(q_1 q_2 + q_2^2 \right) + (\eta + \mu)^2 \left[3q_1 q_2^2 + q_2^3 + q_1 q_4 - q_5 \right].$$

$$\tag{80}$$

Thus it is easy to verify that $\Psi(\beta_2)\Theta(\beta_2) + \Phi(\beta_2)\Gamma(\beta_2) \neq 0$ provided that (71c) and (71d). Thus, the proof is complete. $\square$

7. Numerical Simulations

In this section, the global dynamics of model (1) is investigated numerically for different sets of initial values and different sets of parameters values. The objectives of such investigation are to determine the effect of varying the parameters values and confirm our obtained results. It is observed that, for the following biologically feasible set of hypothetical parameters values

$$\Lambda = 20,$$

$$\beta_1 = 0.75,$$

$$\beta_2 = 0.1,$$

$$\gamma = 0.75,$$

$$p_1 = 0.01,$$

$$p_2 = 0.01$$

$$\mu = 0.3,$$

$$\eta = 0.5,$$

$$\alpha_1 = 0.1,$$

$$\alpha_2 = 0.6,$$

$$\delta = 0.7,$$

$$\tag{81}$$

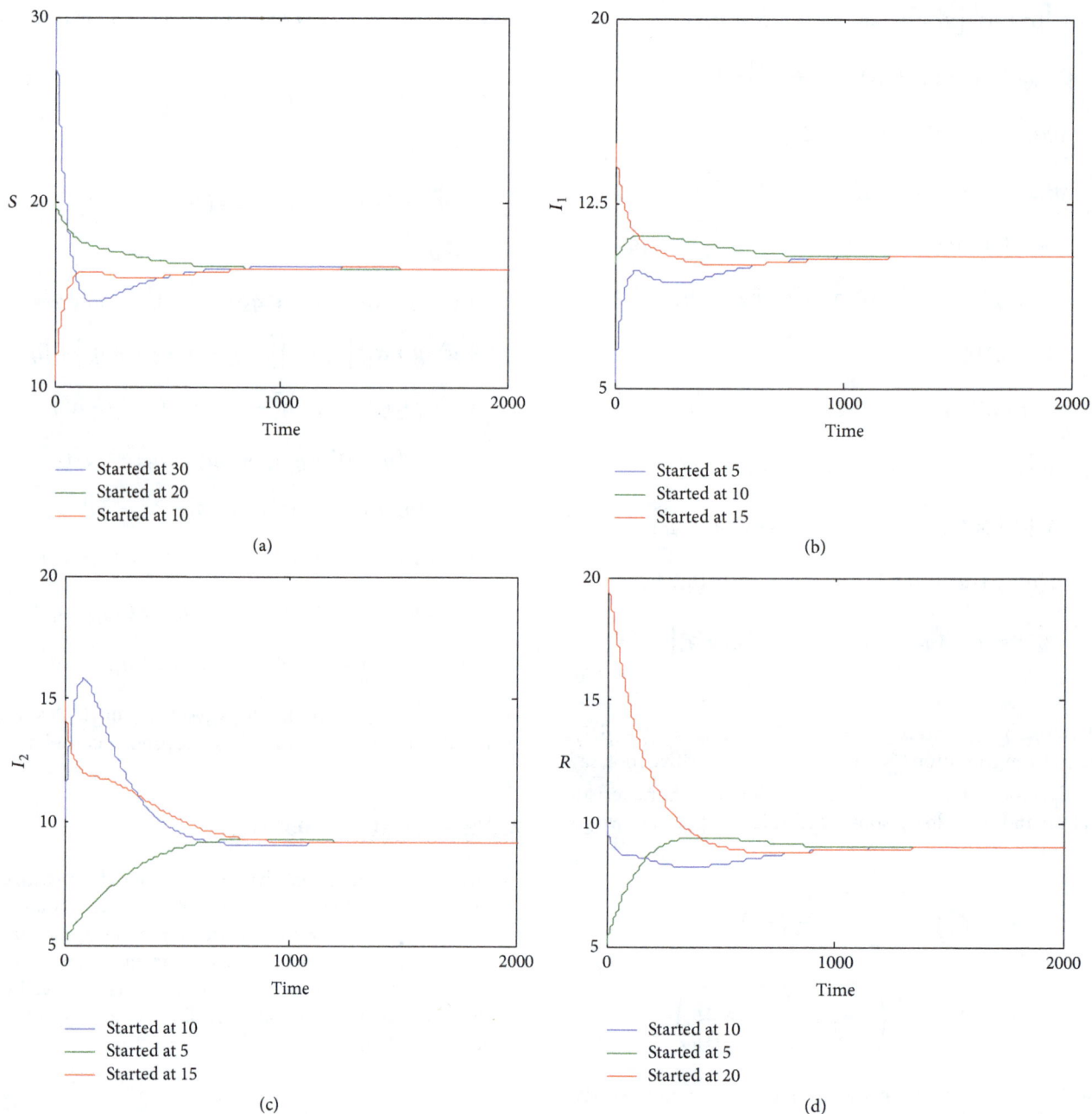

FIGURE 1: Globally asymptotically stable positive equilibrium point of model (1) for the parameters set (81), started from different sets of initial point.

the solution of model (1) approaches asymptotically to the endemic equilibrium point $E_3 = (16.4, 10.28, 9.18, 8.99)$ as shown in Figure 1, started from different sets of initial points.

Clearly Figure 1 confirms our obtained analytical results regarding existence of a globally asymptotically stable positive equilibrium point when the parameters values are satisfying $R_1 > R_2 > 1$. On the other hand, model (1) for the following set of hypothetical data approaches asymptotically to the DFE as shown in Figure 2:

$$\Lambda = 20,$$
$$\beta_1 = 0.05,$$

$$\beta_2 = 0.05,$$
$$\gamma = 0.6,$$
$$p_1 = 0.01,$$
$$p_2 = 0.03,$$
$$\eta = 0.5,$$
$$\mu = 0.9,$$
$$\alpha_1 = 0.1,$$
$$\alpha_2 = 0.1,$$
$$\delta = 0.3.$$

$$(82)$$

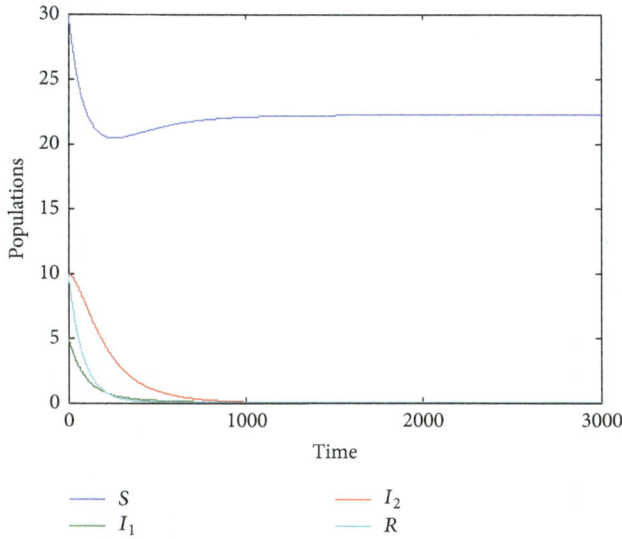

FIGURE 2: Time series of the solution of model (1) that approaches asymptotically to DFE for the data (82).

It is easy to verify that for the data (82) we have $R_0 = 0.86 < 1$, and the solution approaches to $E_0 = (22.22, 0, 0, 0)$.

Now in order to investigate the effect of varying one parameter value at a time on the dynamical behavior of model (1), the following results are observed.

(i) Varying of the parameters values $(\Lambda, \gamma, p_2, \eta, \alpha_1, \alpha_2, \delta)$ does not affect the dynamical behavior of model (1); that is, the system still approaches to coexistence equilibrium point.

(ii) For the data (81) with $\beta_1 \leq 0.015$, the solution of model (1) approaches asymptotically to $E_1 = (S, 0, I_2, 0)$ in the interior of positive quadrant of SI_2-plane with $R_1 < 1 < R_2$. However for $0.015 < \beta_1 < 0.066$ the solution of the model still approaches asymptotically to E_1 even when $1 < R_1 < R_2$. Finally when $\beta_1 \geq 0.066$ the solution of model (1) approaches to coexistence equilibrium point with $1 < R_2 < R_1$ as shown in Figure 3.

(iii) Similar results are obtained in case of varying the parameter β_2 keeping the rest of parameters values in (81) fixed. In fact for $\beta_2 \leq 0.024$, we have $R_2 < 1 < R_1$, and the solution of model (1) approaches asymptotically to $E_2 = (S, I_1, 0, R)$. However for $0.024 < \beta_2 < 0.06$, we have $1 < R_2 < R_1$, and it is observed that the solution of model (1) approaches asymptotically to $E_2 = (S, I_1, 0, R)$ too, while when $\beta_2 \geq 0.06$ the solution of model (1) approaches asymptotically to $E_3 = (S, I_1, I_2, R)$ as shown in Figure 4.

(iv) Now decreasing the parameter μ, keeping the rest of the parameters values in (81) fixed, gives similar dynamical behavior as that of varying β_2. Further it is observed that when $\mu > 0.9$, for which $R_2 < 1 < R_1$, the solution of model (1) approaches asymptotically to $E_2 = (S, I_1, 0, R)$; however when $0.57 < \mu < 0.9$,

we have $1 < R_2 < R_1$, and the solution of model (1) still approaches to $E_2 = (S, I_1, 0, R)$. Finally, for $\mu \leq 0.57$ that satisfies $1 < R_2 < R_1$, the solution of model (1) approaches asymptotically to $E_3 = (S, I_1, I_2, R)$. Clearly this confirmed our obtained existence conditions (10) and (12) as well as stability conditions of these points.

(v) Finally, varying the parameter p_1, keeping the rest of the parameters values in (81) fixed, showed that, for $p_1 \geq 0.6$, that satisfies $1 < R_2 < R_1$, the solution of model (1) approaches asymptotically to $E_2 = (S, I_1, 0, R)$; however when $p_1 < 0.6$, which satisfies $1 < R_2 < R_1$ too, the solution of model (1) approaches asymptotically to $E_3 = (S, I_1, I_2, R)$ as shown in Figure 5.

8. Conclusion

In this paper, we proposed and analyzed an epidemic model involving vertical and horizontal transmission of infection with nonlinear incidence rate. It is assumed that the rates of infections p_1, p_2 are less than the recovery rates δ and γ, respectively. According to the diseases in model (1) the population is divided into four subclasses: susceptible individuals that are represented by $S(t)$, infected individuals for $SIRS$-type of disease that are represented by $I_1(t)$, infected individuals for SIS-type of disease that are represented by $I_2(t)$, and recovery individuals that are denoted by $R(t)$. The boundedness and invariant of the model are discussed. The basic reproduction number of the model and the associated threshold parameter values, namely, T_i, $i = 1, 2$, are determined. It is observed that if the basic reproduction number is less than unity then the diseases are eradicated from the model. The competitive exclusion principle occurred in model (1) such that only the second disease-free equilibrium point appeared in case of $R_2 < 1 < R_1$. However only the first disease-free equilibrium point appeared in case of $R_1 < 1 < R_2$. Finally the coexistence of both the diseases occurred in case of $1 < R_2 < R_1$ and the sufficient condition (12) holds. The dynamical behavior of model (1) has been investigated locally as well as globally using Routh-Hurwitz criterion and Lyapunov function, respectively. The local bifurcations of model (1) and the Hopf bifurcation around the endemic equilibrium point are studied. Finally to understand the effect of varying each parameter on the global dynamics of system (1) and to confirm our obtained analytical results, model (1) has been solved numerically and the following results are obtained for the set of hypothetical parameters values given by (81).

(1) Model (1) approaches asymptotically to a globally asymptotically stable point $E_3 = (16.4, 10.28, 9.18, 8.99)$.

(2) Varying one of the parameters values $(\Lambda, \gamma, p_2, \eta, \alpha_1, \alpha_2, \delta)$ at a time keeping other parameters fixed has no effect on the dynamical behavior of the model.

(3) As the infection rate of the first disease (β_1) decreases keeping other parameters fixed as in (81) the solution

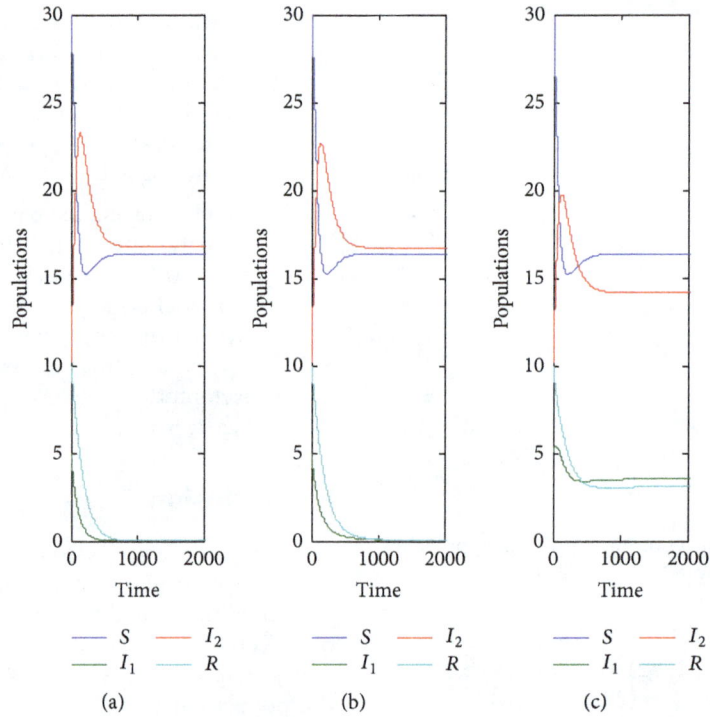

FIGURE 3: Time series of the solution of model (1) for the data given by (81). (a) For $\beta_1 = 0.015$ the model approaches to $E_1 = (16.4, 0, 16.75, 0)$. (b) For $\beta_1 = 0.06$ the model approaches to the same point. (c) For $\beta_1 = 0.3$ the model approaches to $E_3 = (16.4, 3.51, 14.1, 3.07)$.

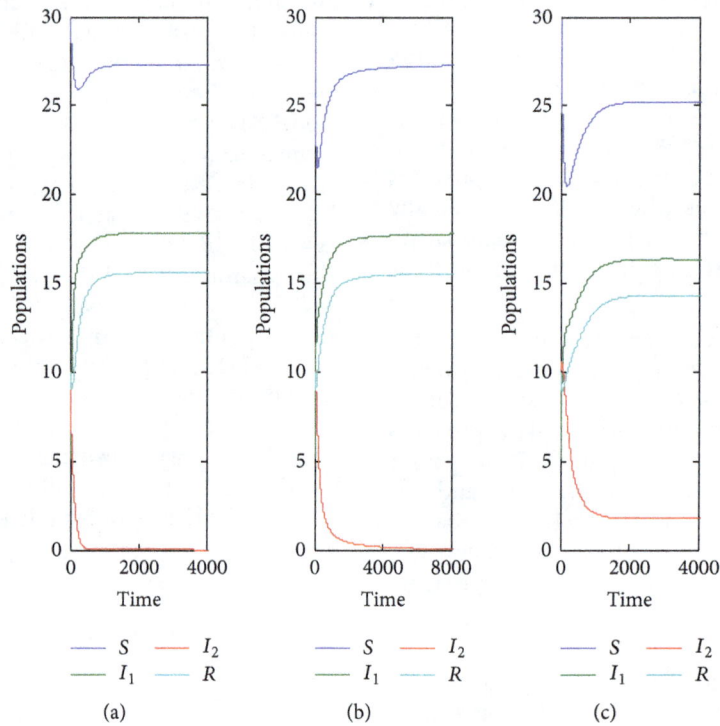

FIGURE 4: Time series of the solution of model (1) for the data given by (81). (a) For $\beta_2 = 0.024$ the model approaches to $E_2 = (27.33, 17.80, 0, 15.5)$. (b) For $\beta_2 = 0.059$ the model approaches to the same point. (c) For $\beta_2 = 0.065$ the model approaches to $E_3 = (25.2, 16.3, 1.76, 14.3)$.

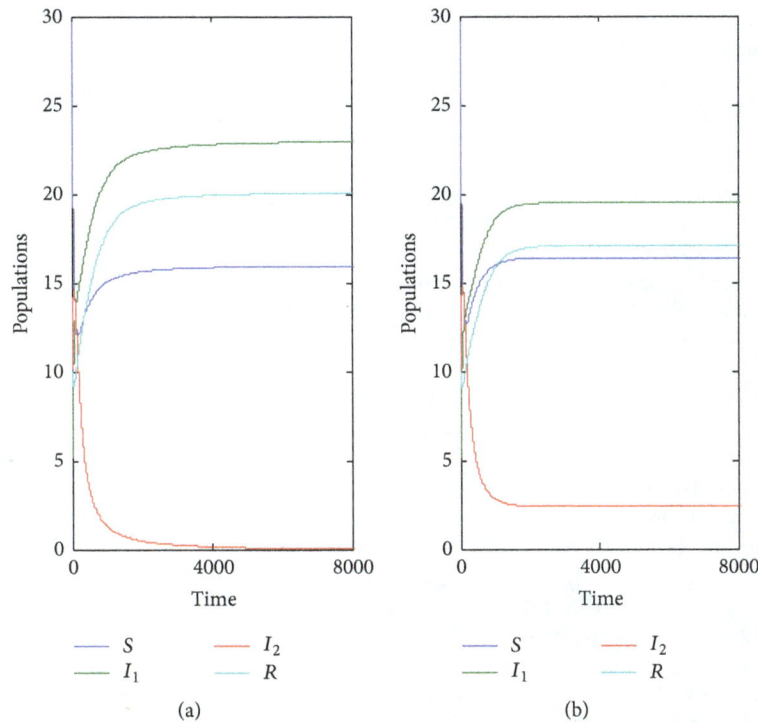

FIGURE 5: Time series of the solution of model (1) for the data given by (81). (a) For $p_1 = 0.6$ the model approaches to $E_2 = (15.9, 22.9, 0, 20.08)$. (b) For $p_1 = 0.5$ the model approaches to $E_3 = (16.4, 19.5, 2.40, 17.06)$.

of model (1) approaches asymptotically to the equilibrium point $E_1 = (16.4, 0, 16.75, 0)$. However in case of increasing this parameter the model is still globally asymptotically stable in the interior of $\mathbb{R}_+^4$.

(4) As the infection rate of the second disease (β_2) decreases keeping other parameters fixed as in (81) the solution of model (1) approaches asymptotically to the equilibrium point $E_2 = (27.33, 17.80, 0, 15.5)$. However in case of increasing this parameter the model is still globally asymptotically stable in the interior of $\mathbb{R}_+^4$.

(5) As the mortality rate (μ) increases keeping other parameters fixed as in (81) the solution of model (1) approaches asymptotically to the equilibrium point $E_2 = (13.8, 5.16, 0, 2.5)$ and when μ decreases the model is still globally asymptotically stable in the interior of $\mathbb{R}_+^4$. Further, it is observed that p_1 has the same effect as μ on the dynamical behavior of model (1).

(6) For the parameter set given in (82) the solution of model (1) approaches asymptotically to DFE.

(7) Finally, although for our selected parameters values model (1) does not undergo periodic dynamics, the model still has possibility to have periodic dynamics for other sets of parameters, especially Hopf bifurcation existing analytically.

Conflict of Interests

The authors declare that there is no conflict of interests regarding the publication of this paper.

References

[1] G. Hardin, "The competitive exclusion principle," *Science*, vol. 131, no. 3409, pp. 1292–1297, 1960.

[2] A. Iggidr, J.-C. Kamgang, G. Sallet, and J.-J. Tewa, "Global analysis of new malaria intrahost models with a competitive exclusion principle," *SIAM Journal on Applied Mathematics*, vol. 67, no. 1, pp. 260–278, 2006.

[3] V. Volterra, "Variations and fluctuations of the number of individuals in animal species living together," *Journal du Conseil International pour l'Exploration de la Mer*, vol. 3, no. 1, pp. 3–51, 1928.

[4] A. S. Ackleh and L. J. Allen, "Competitive exclusion in SIS and SIR epidemic models with total cross immunity and density-dependent host mortality," *Discrete and Continuous Dynamical Systems—Series B*, vol. 5, no. 2, pp. 175–188, 2005.

[5] F. Brauer, J. Wu, and P. van den Driessche, *Mathematical Epidemiology*, Springer, Berlin, Germany, 2008.

[6] P. van den Driessche and J. Watmough, "Reproduction numbers and sub-threshold endemic equilibria for compartmental models of disease transmission," *Mathematical Biosciences*, vol. 180, no. 1-2, pp. 29–48, 2002.

[7] J. M. Heffernan, R. J. Smith, and L. M. Wahl, "Perspectives on the basic reproductive ratio," *Journal of the Royal Society Interface*, vol. 2, no. 4, pp. 281–293, 2005.

[8] O. Diekmann, J. A. P. Heesterbeek, and J. A. J. Metz, "On the definition and the computation of the basic reproduction ratio R_o in models for infectious diseases in heterogeneous populations in models for infectious diseases in heterogeneous populations," *Journal of Mathematical Biology*, vol. 28, no. 4, pp. 365–382, 1990.

[9] M. Martcheva, "A non-autonomous multi-strain SIS epidemic model," *Journal of Biological Dynamics*, vol. 3, no. 2-3, pp. 235–251, 2009.

[10] A. S. Ackleh and L. J. S. Allen, "Competitive exclusion and coexistence for pathogens in an epidemic model with variable population size," *Journal of Mathematical Biology*, vol. 47, no. 2, pp. 153–168, 2003.

[11] M. Lipsitch, S. Siller, and M. A. Nowak, "The evolution of virulence in pathogens with vertical and horizontal transmission," *Evolution*, vol. 50, no. 5, pp. 1729–1741, 1996.

[12] D. Bichara, A. Iggidr, and G. Sallet, "Global analysis of multi-strains SIS, SIR and MSIR epidemic models," *Journal of Applied Mathematics and Computing*, vol. 44, no. 1-2, pp. 273–292, 2014.

[13] M. Y. Li, H. L. Smith, and L. Wang, "Global dynamics an SEIR epidemic model with vertical transmission," *SIAM Journal on Applied Mathematics*, vol. 62, no. 1, pp. 58–69, 2001.

[14] W. M. Liu, H. W. Hethcote, and S. A. Levin, "Dynamical behavior of epidemiological models with nonlinear incidence rates," *Journal of Mathematical Biology*, vol. 25, no. 4, pp. 359–380, 1987.

[15] O. Adebimpe, A. A. Waheed, and B. Gbadamosi, "Modeling and analysis of an SEIRS epidemic model with saturated incidence," *Journal of Engineering Research and Application*, vol. 3, no. 5, pp. 1111–1116, 2013.

[16] R. K. Naji and A. N. Mustafa, "The dynamics of an eco-epidemiological model with nonlinear incidence rate," *Journal of Applied Mathematics*, vol. 2012, Article ID 852631, 24 pages, 2012.

[17] C. V. De-León, "On the global stability of SIS, SIR and SIRS epidemic models with standard incidence," *Chaos, Solitons & Fractals*, vol. 44, no. 12, pp. 1106–1110, 2011.

[18] J. P. LaSalle, "Stability theory for ordinary differential equations," *Journal of Differential Equations*, vol. 4, no. 1, pp. 57–65, 1968.

[19] E. Shim, *An epidemic model with immigration of infectives and vaccination [M.S. thesis]*, Department of Mathematics, Institute of Applied Mathematics, University of British Columbia, Vancouver, Canada, 2004.

[20] L. Perko, *Differential Equation and Dynamical Systems*, Springer, New York, NY, USA, 3rd edition, 2001.

[21] M. M. A. El-Sheikh and S. A. A. El-Marouf, "On stability and bifurcation of solutions of an SEIR epidemic model with vertical transmission," *International Journal of Mathematics and Mathematical Sciences*, vol. 2004, no. 56, pp. 2971–2987, 2004.

[22] X. Zhou and J. Cui, "Analysis of stability and bifurcation for an SEIV epidemic model with vaccination and nonlinear incidence rate," *Nonlinear Dynamics*, vol. 63, no. 4, pp. 639–653, 2011.

9

Solvability of the Brinkman-Forchheimer-Darcy Equation

Piotr Skrzypacz and Dongming Wei

School of Science and Technology, Nazarbayev University, 53 Kabanbay Batyr Ave., Astana 010000, Kazakhstan

Correspondence should be addressed to Piotr Skrzypacz; piotr.skrzypacz@nu.edu.kz

Academic Editor: Malgorzata Peszynska

The nonlinear Brinkman-Forchheimer-Darcy equation is used to model some porous medium flow in chemical reactors of packed bed type. The results concerning the existence and uniqueness of a weak solution are presented for nonlinear convective flows in medium with variable porosity and for small data. Furthermore, the finite element approximations to the flow profiles in the fixed bed reactor are presented for several Reynolds numbers at the non-Darcy's range.

1. Introduction

In this section we introduce the mathematical model describing incompressible isothermal flow in porous medium without reaction. The considered equations for the velocity and pressure fields are for flows in fluid saturated porous media. These problems are of importance for example in oil reservoir optimization; see [1]. Most research results for flows in porous media are based on the Darcy equation which is considered to be a suitable model at a small range of Reynolds numbers. However, there are restrictions of Darcy equation for modelling some porous medium flows; that is, in closely packed media, saturated fluid flows at slow velocity but with relatively large Reynolds numbers. The flows in such closely packed medium behave nonlinearly and cannot be modelled accurately by the Darcy equation which is linear. The deficiency can be circumvented with the Brinkman–Forchheimer-Darcy law for flows in closely packed media, which leads to the following model: let $\Omega \subset \mathbb{R}^n$, $n = 2, 3$, represent the reactor channel. We denote its boundary by $\Gamma = \partial\Omega$. The conservation of volume-averaged values of momentum and mass in the packed reactor reads as follows

$$-\operatorname{div}\left(\varepsilon\nu\nabla\mathbf{u} - \varepsilon\mathbf{u} \otimes \mathbf{u}\right) + \frac{\varepsilon}{\varrho}\nabla p + \sigma\left(\mathbf{u}\right) = \mathbf{f} \quad \text{in } \Omega,$$

$$\operatorname{div}\left(\varepsilon\mathbf{u}\right) = 0 \quad \text{in } \Omega,$$
(1)

where $\mathbf{u} : \Omega \to \mathbb{R}^n$ and $p : \Omega \to \mathbb{R}$ denote the unknown velocity and pressure, respectively. The positive quantity $\varepsilon = \varepsilon(\mathbf{x})$ stands for porosity which describes the proportion of the nonsolid volume to the total volume of material and varies spatially in general. The expression $\sigma(\mathbf{u})$ represents the friction forces caused by the packing of spherical particles with one constant diameter. The right hand side $\mathbf{f}$ represents an outer force (e.g., gravitation), ϱ the constant fluid density, and ν the constant kinematic viscosity of the fluid, respectively. The expression $\mathbf{u} \otimes \mathbf{u}$ symbolizes the dyadic product of $\mathbf{u}$ with itself.

The formula given by Ergun [2] will be used to model the influence of the packing on the flow inertia effects

$$\sigma\left(\mathbf{u}\right) = 150\nu\frac{(1 - \varepsilon)^2}{\varepsilon^2 d_p^2}\mathbf{u} + 1.75\frac{1 - \varepsilon}{\varepsilon d_p}\mathbf{u}\left|\mathbf{u}\right|.$$
(2)

Thereby d_p stands for the diameter of pellets and $|\cdot|$ denotes the Euclidean vector norm. The linear term in (2) accounts for the head loss according to Darcy and the quadratic term is the Forchheimer law. The model (1)-(2) with the pressure term of the form $\nabla(\varepsilon p)$ has been proposed in [3, Section 2.2]. For the derivation of the equations, their limitations, modelling, and homogenization questions in porous media we refer to [4–9] and [10, Chapter 1].

To close the system (1) we prescribe Dirichlet boundary condition

$$\mathbf{u}|_\Gamma = \mathbf{g},$$
(3)

where

$$\int_{\Gamma_i} \varepsilon \mathbf{g} \cdot \mathbf{n} \, ds = 0 \tag{4}$$

has to be fulfilled on each connected component Γ_i of the boundary Γ. The distribution of porosity ε is assumed to satisfy the bounds

$$0 < \varepsilon_0 \le \varepsilon(\mathbf{x}) \le \varepsilon_1 \le 1 \quad \forall \mathbf{x} \in \Omega, \tag{A1}$$

with some constants $0 < \varepsilon_0, \varepsilon_1 \le 1$.

A comprehensive account of fluid flows through porous media beyond Darcy law's valid regimes and classified by the Reynolds number can be found in, for example, [11]. Also, see [12] for simulating pumped water levels in abstraction boreholes using such nonlinear Darcy-Forchheimer law and [13–15] for recent references on this model.

In the next section we use the porosity distribution which is estimated for packed beds consisting of spherical particles and takes the near wall channeling effect into account. This kind of porosity distribution obeys assumption (A1).

Let us introduce dimensionless quantities

$$\mathbf{u}^* = \frac{\mathbf{u}}{U_0},$$

$$p^* = \frac{p}{\varrho U_0^2},$$

$$\mathbf{x}^* = \frac{\mathbf{x}}{d_p}, \tag{5}$$

$$\mathbf{g}^* = \frac{\mathbf{g}}{U_0},$$

where U_0 denotes the magnitude of some reference velocity. For simplicity of notation we omit the asterisks. Then, the reactor flow problem reads in dimensionless form as follows:

$$-\mathrm{div}\left(\frac{\varepsilon}{\mathrm{Re}} \nabla \mathbf{u} - \varepsilon \mathbf{u} \otimes \mathbf{u}\right) + \varepsilon \nabla p + \frac{\alpha}{\mathrm{Re}} \mathbf{u} + \beta \mathbf{u} |\mathbf{u}| = \mathbf{f}$$

$$\text{in } \Omega,$$

$$\mathrm{div}(\varepsilon \mathbf{u}) = 0$$

$$\text{in } \Omega, \tag{6}$$

$$\mathbf{u} = \mathbf{g}$$

$$\text{on } \Gamma,$$

where

$$\alpha(\mathbf{x}) = 150 \kappa^2(\mathbf{x}),$$

$$\beta(\mathbf{x}) = 1.75 \kappa(\mathbf{x}), \tag{7}$$

with

$$\kappa(\mathbf{x}) = \frac{1 - \varepsilon(\mathbf{x})}{\varepsilon(\mathbf{x})}, \tag{8}$$

and the Reynolds number is defined by

$$\mathrm{Re} = \frac{U_0 d_p}{\nu}. \tag{9}$$

The existence and uniqueness of the solution of a system for flow, temperature, and solute transport containing the nonlinear flow model (6) but with constant porosity and without the convective term have been established in [16]. We will extend this result for the flow component to the case (6) when the variable porosity depends on the location and the convective term is included. The recent existence results for the linear Brinkman problem with spatially varying α can be found in [17].

Remark 1. Equation (6) becomes a Navier-Stokes problem if $\varepsilon \equiv 1$.

Notation. Throughout the work we use the following notations for function spaces. For $m \in \mathbb{N}_0$, $p \ge 1$ and bounded subdomain $G \subset \Omega$; let $W^{m,p}(G)$ be the usual Sobolev space equipped with norm $\| \cdot \|_{m,p,G}$. If $p = 2$, we denote the Sobolev space by $H^m(G)$ and use the standard abbreviations $\| \cdot \|_{m,G}$ and $| \cdot |_{m,G}$ for the norm and seminorm, respectively. We denote by $D(G)$ the space of $C^\infty(G)$ functions with compact support contained in G. Furthermore, $H_0^m(G)$ stands for the closure of $D(G)$ with respect to the norm $\| \cdot \|_{m,G}$. The counterparts spaces consisting of vector valued functions will be denoted by bold faced symbols like $\mathbf{H}^m(G) := [H^m(G)]^n$ or $\mathbf{D}(G) := [D(G)]^n$. The L^2 inner product over $G \subset \Omega$ and $\partial G \subset \partial \Omega$ will be denoted by $(\cdot, \cdot)_G$ and $\langle \cdot, \cdot \rangle_{\partial G}$, respectively. In the case $G = \Omega$ the domain index will be omitted. In the following we denote by C the generic constant which is usually independent of the model parameters; otherwise dependence will be indicated.

2. Existence and Uniqueness Results

In the following the porosity ε is assumed to belong to $W^{1,3}(\Omega) \cap L^\infty(\Omega)$. We start with the weak formulation of problem (6) and look for its solution in suitable Sobolev spaces that are adjusted to the modified momentum and mass balances in (6) and to the smoothness of the weighting function ε.

2.1. Variational Formulation. Let

$$L_0^2(\Omega) := \left\{ v \in L^2(\Omega) : (v, 1) = 0 \right\} \tag{10}$$

be the space consisting of L^2 functions with zero mean value. We define the spaces

$$\mathbf{X} := \mathbf{H}^1(\Omega),$$

$$\mathbf{X}_0 := \mathbf{H}_0^1(\Omega),$$

$$Q := L^2(\Omega), \tag{11}$$

$$M := L_0^2(\Omega),$$

$$\mathbf{V} := \mathbf{X}_0 \times M.$$

Let us introduce the following bilinear forms:

$$a: \mathbf{X} \times \mathbf{X} \longrightarrow \mathbb{R}, \quad a(\mathbf{u}, \mathbf{v}) = \frac{1}{Re}(\varepsilon \nabla \mathbf{u}, \nabla \mathbf{v}),$$

$$b: \mathbf{X} \times Q \longrightarrow \mathbb{R}, \quad b(\mathbf{u}, q) = (\operatorname{div}(\varepsilon \mathbf{u}), q), \quad (12)$$

$$c: \mathbf{X} \times \mathbf{X} \longrightarrow \mathbb{R}, \quad c(\mathbf{u}, \mathbf{v}) = \frac{1}{Re}(\alpha \mathbf{u}, \mathbf{v}).$$

Furthermore, we define the semilinear form

$$d: \mathbf{X} \times \mathbf{X} \times \mathbf{X} \longrightarrow \mathbb{R}, \quad d(\mathbf{w}; \mathbf{u}, \mathbf{v}) = (\beta |\mathbf{w}| \mathbf{u}, \mathbf{v}), \quad (13)$$

and trilinear form

$$n: \mathbf{X} \times \mathbf{X} \times \mathbf{X} \longrightarrow \mathbb{R}, \quad n(\mathbf{w}, \mathbf{u}, \mathbf{v}) = ((\varepsilon \mathbf{w} \cdot \nabla) \mathbf{u}, \mathbf{v}). \quad (14)$$

We set

$$A(\mathbf{w}; \mathbf{u}, \mathbf{v}) := a(\mathbf{u}, \mathbf{v}) + c(\mathbf{u}, \mathbf{v}) + n(\mathbf{w}, \mathbf{u}, \mathbf{v}) \\ + d(\mathbf{w}; \mathbf{u}, \mathbf{v}). \quad (15)$$

Multiplying momentum and mass balances in (6) by test functions $\mathbf{v} \in \mathbf{X}_0$ and $q \in M$, respectively, and integrating by parts implies the weak formulation

$$\begin{aligned} &\text{Find} \quad (\mathbf{u}, p) \in \mathbf{X} \times M \quad \text{with} \quad \mathbf{u}|_\Gamma = \mathbf{g} \\ &\text{such that} \quad A(\mathbf{u}; \mathbf{u}, \mathbf{v}) - b(\mathbf{v}, p) = (\mathbf{f}, \mathbf{v}), \\ &\qquad\qquad b(\mathbf{u}, q) = 0 \\ &\qquad\qquad \forall \mathbf{v} \in \mathbf{X}_0, \ \forall q \in M. \end{aligned} \quad (16)$$

First, we recall the following result from [18].

Theorem 2. *The mapping $u \mapsto \varepsilon u$ is an isomorphism from $H^1(\Omega)$ onto itself and from $H_0^1(\Omega)$ onto itself. It holds for all $u \in H^1(\Omega)$ that*

$$\|\varepsilon u\|_1 \le C \{\varepsilon_1 + |\varepsilon|_{1,3}\} \|u\|_1,$$
$$\left\|\frac{u}{\varepsilon}\right\|_1 \le C \{\varepsilon_0^{-1} + \varepsilon_0^{-2} |\varepsilon|_{1,3}\} \|u\|_1. \quad (17)$$

In the following the closed subspace of $\mathbf{H}_0^1(\Omega)$ defined by

$$\mathbf{W} = \left\{\mathbf{w} \in \mathbf{H}_0^1(\Omega): b(\mathbf{w}, q) = 0 \ \forall q \in L_0^2(\Omega)\right\} \quad (18)$$

will be employed. Next, we establish and prove some properties of trilinear form $n(\cdot, \cdot, \cdot)$ and nonlinear form $d(\cdot; \cdot, \cdot)$.

Lemma 3. *Let $\mathbf{u}, \mathbf{v} \in \mathbf{H}^1(\Omega)$ and $\mathbf{w} \in \mathbf{H}^1(\Omega)$ with $\operatorname{div}(\varepsilon \mathbf{w}) = 0$ and $\mathbf{w} \cdot \mathbf{n}|_\Gamma = 0$. Then we have*

$$n(\mathbf{w}, \mathbf{u}, \mathbf{v}) = -n(\mathbf{w}, \mathbf{v}, \mathbf{u}). \quad (19)$$

Furthermore, the trilinear form $n(\cdot, \cdot, \cdot)$ and the nonlinear form $d(\cdot; \cdot, \cdot)$ are continuous; that is,

$$|n(\mathbf{u}, \mathbf{v}, \mathbf{w})| \le C_\varepsilon \|\mathbf{u}\|_1 \|\mathbf{v}\|_1 \|\mathbf{w}\|_1 \quad \forall \mathbf{u}, \mathbf{v}, \mathbf{w} \in \mathbf{H}^1(\Omega), \quad (20)$$

$$|d(\mathbf{u}, \mathbf{v}, \mathbf{w})| \le C_\varepsilon \|\mathbf{u}\|_1 \|\mathbf{v}\|_1 \|\mathbf{w}\|_1 \quad \forall \mathbf{u}, \mathbf{v}, \mathbf{w} \in \mathbf{H}^1(\Omega). \quad (21)$$

and for $\mathbf{u} \in \mathbf{W}$ and for a sequence $\mathbf{u}^k \in \mathbf{W}$ with $\lim_{k\to\infty} \|\mathbf{u}^k - \mathbf{u}\|_0 = 0$, we have also

$$\lim_{k\to\infty} n(\mathbf{u}^k, \mathbf{u}^k, \mathbf{v}) = n(\mathbf{u}, \mathbf{u}, \mathbf{v}) \quad \forall \mathbf{v} \in \mathbf{W}. \quad (22)$$

Proof. We follow the proof of [19, Lemma 2.1, §2, Chapter IV] and adapt it to the trilinear form

$$n(\mathbf{w}, \mathbf{u}, \mathbf{v}) = ((\varepsilon \mathbf{w} \cdot \nabla) \mathbf{u}, \mathbf{v}) = \sum_{i,j=1}^n (\varepsilon w_j \partial_j u_i, v_i), \quad (23)$$

which has the weighting factor ε. Hereby, symbols with subscripts denote components of bold faced vectors, for example, $\mathbf{u} = (u_i)_{i=1,...,n}$. Let $\mathbf{u} \in \mathbf{H}^1$, $\mathbf{v} \in \mathbf{D}(\Omega)$, and $\mathbf{w} \in \mathbf{W}$. Integrating by parts and employing density argument, we obtain immediately (19).

$$\begin{aligned} \sum_{i,j=1}^n (\varepsilon w_j \partial_j u_i, v_i) &= -\sum_{i,j=1}^n (\partial_j(\varepsilon w_j v_i), u_i) \\ &\quad + \sum_{i,j=1}^n \langle \varepsilon w_j n_j u_i, v_i \rangle \\ &= -\sum_{i,j=1}^n (\varepsilon w_j \partial_j v_i, u_i) \\ &\quad - (\operatorname{div}(\varepsilon \mathbf{w}) \mathbf{u}, \mathbf{v}) + \langle (\varepsilon \mathbf{w} \cdot \mathbf{n}) \mathbf{u}, \mathbf{v} \rangle \\ &= -n(\mathbf{w}, \mathbf{v}, \mathbf{u}). \end{aligned} \quad (24)$$

From Sobolev embedding $H^1(\Omega) \hookrightarrow L^4(\Omega)$ (see [20]) and Hölder inequality follows

$$\begin{aligned} |(\varepsilon w_j \partial_j u_i, v_i)| &\le |\varepsilon|_{0,\infty} \|w_j\|_{0,4} \|\partial_j u_i\|_0 \|v_i\|_{0,4} \\ &\le C |\varepsilon|_{0,\infty} \|w_j\|_1 |u_i|_1 \|v_i\|_1, \end{aligned} \quad (25)$$

and consequently the proof of (20) is completed. Since $\lim_{k\to\infty} \|u_i^k u_j^k - u_i u_j\|_{0,1} = 0$ and $\varepsilon \partial_j v_i \in L^\infty(\Omega)$, the continuity estimate (20) implies

$$\begin{aligned} \lim_{k\to\infty} n(\mathbf{u}^k, \mathbf{u}^k, \mathbf{v}) &= -\lim_{k\to\infty} n(\mathbf{u}^k, \mathbf{v}, \mathbf{u}^k) \\ &= -\lim_{k\to\infty} \sum_{i,j=1}^n (\varepsilon u_j^k \partial_j v_i^k, u_i^k) \\ &= -\sum_{i,j=1}^n (\varepsilon u_j \partial_j v_i, u_i) = -n(\mathbf{u}, \mathbf{v}, \mathbf{u}) \\ &= n(\mathbf{u}, \mathbf{u}, \mathbf{v}). \end{aligned} \quad (26)$$

The continuity of $d(\cdot; \cdot, \cdot)$ follows from Hölder inequality and Sobolev embedding $H^1(\Omega) \hookrightarrow L^4(\Omega)$ (see [20]).

$$\begin{aligned} |d(\mathbf{u}; \mathbf{v}, \mathbf{w})| &\le |\beta|_\infty \|\mathbf{u}\|_{0,4} \|\mathbf{v}\|_{0,4} \|\mathbf{w}\|_0 \\ &\le C_\varepsilon \|\mathbf{u}\|_1 \|\mathbf{v}\|_1 \|\mathbf{w}\|_1. \end{aligned} \quad (27)$$

$\square$

In the next stage we consider the difficulties caused by prescribing the inhomogeneous Dirichlet boundary condition. Analogous difficulties are already encountered in the analysis of Navier-Stokes problem. We will carry out the study of three-dimensional case. The extension in two dimensions can be constructed analogously. Since $\mathbf{g} \in \mathbf{H}^{1/2}(\Gamma)$, we can extend $\mathbf{g}$ inside of Ω in the form of

$$\mathbf{g} = \varepsilon^{-1} \operatorname{curl} \mathbf{h}, \qquad (28)$$

with some $\mathbf{h} \in \mathbf{H}^2(\Omega)$. The operator curl is defined then as

$$\operatorname{curl} \mathbf{h} = (\partial_2 h_3 - \partial_3 h_2, \partial_3 h_1 - \partial_1 h_3, \partial_1 h_2 - \partial_2 h_1). \qquad (29)$$

We note that in the two-dimensional case the vector potential $\mathbf{h} \in \mathbf{H}^2(\Omega)$ can be replaced by a scalar function $h \in H^2(\Omega)$ and the operator curl is then redefined as $\operatorname{curl} h = (\partial_2 h, -\partial_1 h)$. Our aim is to adapt the extension of Hopf (see [21]) to our model. We recall that for any parameter $\mu > 0$ there exists a scalar function $\varphi_\mu \in C^2(\overline{\Omega})$ such that

(i) $\varphi_\mu = 1$ in some neighborhood of Γ (depending on μ),

(ii) $\varphi_\mu(\mathbf{x}) = 0$ if $d_\Gamma(\mathbf{x}) \geq 2 \exp\left(-\dfrac{1}{\mu}\right)$, where $d_\Gamma(\mathbf{x}) := \inf_{\mathbf{y} \in \Gamma} |\mathbf{x} - \mathbf{y}|$ denotes the distance of $\mathbf{x}$ to Γ, $\qquad$ (Ex)

(iii) $\left| \partial_j \varphi_\mu(\mathbf{x}) \right| \leq \dfrac{\mu}{d_\Gamma(\mathbf{x})}$ if $d_\Gamma(\mathbf{x}) < 2 \exp\left(-\dfrac{1}{\mu}\right)$, $j = 1, \ldots, n$.

For the construction of φ_μ see also [19, Lemma 2.4, §2, Chapter IV].

Let us define

$$\mathbf{g}_\mu := \varepsilon^{-1} \operatorname{curl}(\varphi_\mu \mathbf{h}). \qquad (30)$$

In the following lemma we establish bounds which are crucial for proving existence of velocity.

Lemma 4. *The function* $\mathbf{g}_\mu$ *satisfies the following conditions:*

$$\operatorname{div}(\varepsilon \mathbf{g}_\mu) = 0,$$

$$\mathbf{g}_\mu \big|_\Gamma = \mathbf{g}, \qquad (31)$$

$$\forall \mu > 0,$$

and for any $\delta > 0$ *there exists sufficiently small* $\mu > 0$ *such that*

$$\left| d\left(\mathbf{u} + \mathbf{g}_\mu; \mathbf{g}_\mu, \mathbf{u}\right) \right| \leq \delta \|\beta\|_{0,\infty} |\mathbf{u}|_1 \left(|\mathbf{u}|_1 + \|\mathbf{g}_\mu\|_0 \right) \qquad (32)$$

$$\forall \mathbf{u} \in \mathbf{X}_0,$$

$$\left| n\left(\mathbf{u}, \mathbf{g}_\mu, \mathbf{u}\right) \right| \leq \delta |\mathbf{u}|_1^2 \quad \forall \mathbf{u} \in \mathbf{W}. \qquad (33)$$

Proof. The relations in (31) are obvious. We follow [16] in order to show (32). Since $\mathbf{h} \in \mathbf{H}^2(\Omega)$ Sobolev's embedding theorem implies $\mathbf{h} \in \mathbf{L}^\infty(\Omega)$, so we get according to the properties of φ_μ in (Ex) the following bound:

$$\left| \mathbf{g}_\mu \right| \leq C \varepsilon_0^{-1} \left\{ |\nabla \mathbf{h}| + \dfrac{\mu}{d_\Gamma(\mathbf{x})} |\mathbf{h}| \right\}$$

$$\leq C \left\{ \dfrac{\mu}{d_\Gamma(\mathbf{x})} + |\nabla \mathbf{h}| \right\}. \qquad (34)$$

Defining

$$\Omega_\mu := \left\{ \mathbf{x} \in \Omega : d_\Gamma(\mathbf{x}) < 2 \exp\left(-\dfrac{1}{\mu}\right) \right\}, \qquad (35)$$

we obtain from Cauchy-Schwarz and triangle inequalities

$$\left| \left(\beta \left| \mathbf{u} + \mathbf{g}_\mu \right|, \mathbf{g}_\mu \cdot \mathbf{u} \right) \right| \leq \|\beta\|_{0,\infty} \|\mathbf{u}\|_0 \left\| \mathbf{u} \cdot \mathbf{g}_\mu \right\|_{0,\Omega_\mu}$$
$$+ \|\beta\|_{0,\infty} \|\mathbf{g}_\mu\|_0 \left\| \mathbf{u} \cdot \mathbf{g}_\mu \right\|_{0,\Omega_\mu}, \qquad (36)$$

$$\left\| \mathbf{u} \cdot \mathbf{g}_\mu \right\|_{0,\Omega_\mu}^2 \leq \int_{\Omega_\mu} |\mathbf{u}|^2 |\mathbf{g}_\mu|^2 \, d\mathbf{x} \leq C \int_{\Omega_\mu} |\mathbf{u}|^2$$

$$\cdot \left\{ \left(\dfrac{\mu}{d_\Gamma(\mathbf{x})} \right)^2 + \dfrac{2\mu}{d_\Gamma(\mathbf{x}) |\nabla \mathbf{h}|} + |\nabla \mathbf{h}|^2 \right\} d\mathbf{x}$$

$$\leq C \left\{ \mu^2 \left\| \dfrac{\mathbf{u}}{d_\Gamma} \right\|_{0,\Omega_\mu}^2 + 2\mu \left\| \dfrac{\mathbf{u}}{d_\Gamma} \right\|_{0,\Omega_\mu} \|\mathbf{u}\|_{0,4,\Omega_\mu} \qquad (37) \right.$$

$$\cdot \|\nabla \mathbf{h}\|_{0,4,\Omega_\mu} + \|\mathbf{u}\|_{0,4,\Omega_\mu}^2 \|\nabla \mathbf{h}\|_{0,4,\Omega_\mu}^2 \Big\}$$

$$\leq C \left\{ \mu \left\| \dfrac{\mathbf{u}}{d_\Gamma} \right\|_{0,\Omega_\mu} + \|\mathbf{u}\|_{0,4} \|\nabla \mathbf{h}\|_{0,4,\Omega_\mu} \right\}^2,$$

and consequently

$$\left\| \mathbf{u} \cdot \mathbf{g}_\mu \right\|_{0,\Omega_\mu} \leq C \left\{ \mu \left\| \dfrac{\mathbf{u}}{d_\Gamma} \right\|_{0,\Omega_\mu} + \|\mathbf{u}\|_{0,4} \|\nabla \mathbf{h}\|_{0,4,\Omega_\mu} \right\}. \qquad (38)$$

Applying Hardy inequality (see [20])

$$\left\| \dfrac{v}{d_\Gamma} \right\|_0 \leq C |v|_1 \quad \forall v \in H_0^1(\Omega), \qquad (39)$$

and using Sobolev embedding $H^1(\Omega) \hookrightarrow L^4(\Omega)$, estimate (38) becomes

$$\left\| \mathbf{u} \cdot \mathbf{g}_\mu \right\|_{0,\Omega_\mu} \leq C\lambda(\mu) \|\mathbf{u}\|_1, \qquad (40)$$

where

$$\lambda(\mu) := \max\left\{\mu, \|\nabla \mathbf{h}\|_{0,4,\Omega_\mu}\right\}. \qquad (41)$$

From (36) and (40), Poincaré inequality, and the fact that $\lim_{\mu\to 0}\lambda(\mu) = 0$ we conclude that for any $\delta > 0$ we can choose sufficiently small $\mu > 0$ such that

$$\left|\left(\beta\,|\mathbf{u} + \mathbf{g}_\mu|\,\mathbf{g}_\mu, \mathbf{u}\right)\right| \le \delta\,\|\beta\|_{0,\infty}\,|\mathbf{u}|_1\left(|\mathbf{u}|_1 + \|\mathbf{g}_\mu\|_0\right), \qquad (42)$$

holds. Therefore the proof of estimate (32) is completed. Now, we take a look at the trilinear convective term

$$n\left(\mathbf{u}, \mathbf{g}_\mu, \mathbf{u}\right) = \left((\varepsilon\mathbf{u}\cdot\nabla)\mathbf{g}_\mu, \mathbf{u}\right)_{\Omega_\mu}$$
$$= \left((\varepsilon\mathbf{u}\cdot\nabla)\left\{\varepsilon^{-1}\mathrm{curl}\left(\varphi_\mu\mathbf{h}\right)\right\}, \mathbf{u}\right)_{\Omega_\mu}$$
$$= \left((\mathbf{u}\cdot\nabla)\left\{\mathrm{curl}\left(\varphi_\mu\mathbf{h}\right)\right\}, \mathbf{u}\right)_{\Omega_\mu}$$
$$\quad - \left((\mathbf{u}\cdot\nabla\varepsilon)\mathbf{g}_\mu, \mathbf{u}\right)_{\Omega_\mu}. \qquad (43)$$

The first term of above difference becomes small due to [19, Lemma 2.3, §2, Chapter IV], and it satisfies

$$\left|\left((\mathbf{u}\cdot\nabla)\left\{\mathrm{curl}\left(\varphi_\mu\mathbf{h}\right)\right\}, \mathbf{u}\right)_{\Omega_\mu}\right| = \left|\left((\mathbf{u}\cdot\nabla)\left(\varepsilon\mathbf{g}_\mu\right), \mathbf{u}\right)_{\Omega_\mu}\right|$$
$$\le \delta\,|\mathbf{u}|_1^2, \qquad (44)$$

as long as $\mu > 0$ is chosen sufficiently small. Using Hölder inequality, Sobolev embedding $H^1(\Omega) \hookrightarrow L^6(\Omega)$ yields

$$\left|\left((\mathbf{u}\cdot\nabla\varepsilon)\mathbf{g}_\mu, \mathbf{u}\right)_{\Omega_\mu}\right| \le C\,\|\varepsilon\|_{1,3}\,\|\mathbf{g}_\mu\cdot\mathbf{u}\|_0\,\|\mathbf{u}\|_1, \qquad (45)$$

which together with (40) implies for sufficiently small $\mu > 0$ the bound

$$\left|\left((\mathbf{u}\cdot\nabla\varepsilon)\mathbf{g}_\mu, \mathbf{u}\right)_{\Omega_\mu}\right| \le \delta\,|\mathbf{u}|_1^2. \qquad (46)$$

From (44) and (46) follows the desired estimate (33). □

While the general framework for linear and nonsymmetric saddle point problems can be found in [18], our problem requires more attention due to its nonlinear character. Setting $\mathbf{w} := \mathbf{u} - \mathbf{g}_\mu$, the weak formulation (16) is equivalent to the following problem:

Find $(\mathbf{w}, p) \in \mathbf{V}$

such that $A\left(\mathbf{w} + \mathbf{g}_\mu; \mathbf{w} + \mathbf{g}_\mu, \mathbf{v}\right) - b(\mathbf{v}, p) \qquad (47)$
$$+ b\left(\mathbf{w} + \mathbf{g}_\mu, q\right) = (\mathbf{f}, \mathbf{v}) \quad \forall (\mathbf{v}, q) \in \mathbf{V}.$$

Let us define the nonlinear mapping $G: \mathbf{W} \to \mathbf{W}$ with

$$[G(\mathbf{w}), \mathbf{v}] := a\left(\mathbf{w} + \mathbf{g}_\mu, \mathbf{v}\right) + c\left(\mathbf{w} + \mathbf{g}_\mu, \mathbf{v}\right) - (\mathbf{f}, \mathbf{v})$$
$$+ n\left(\mathbf{w} + \mathbf{g}_\mu, \mathbf{w} + \mathbf{g}_\mu, \mathbf{v}\right)$$
$$+ d\left(\mathbf{w} + \mathbf{g}_\mu; \mathbf{w} + \mathbf{g}_\mu, \mathbf{v}\right), \qquad (48)$$

where $[\cdot, \cdot]$ defines the inner product in $\mathbf{W}$ via $[u, v] := (\nabla u, \nabla v)$. Then, the variational problem (47) reads in the space $\mathbf{W}$ as follows.

Find $\mathbf{w} \in \mathbf{W}$ such that

$$[G(\mathbf{w}), \mathbf{v}] = 0 \quad \forall \mathbf{v} \in \mathbf{W}. \qquad (49)$$

2.2. Solvability of Nonlinear Saddle Point Problem. We start our study of the nonlinear operator problem (49) with the following lemma.

Lemma 5. *The mapping G defined in (48) is continuous and there exists $r > 0$ such that*

$$[G(\mathbf{u}), \mathbf{u}] > 0 \quad \forall \mathbf{u} \in \mathbf{W} \text{ with } |\mathbf{u}|_1 = r. \qquad (50)$$

Proof. Let $(\mathbf{u}^k)_{k\in\mathbb{N}}$ be a sequence in $\mathbf{W}$ with $\lim_{k\to\infty}\|\mathbf{u}^k - \mathbf{u}\|_1 = 0$. Then, applying Cauchy-Schwarz inequality and (33), we obtain for any $\mathbf{v} \in \mathbf{W}$

$$\left|\left[G\left(\mathbf{u}^k\right) - G(\mathbf{u}), \mathbf{v}\right]\right|$$
$$\le \frac{1}{\mathrm{Re}}\left|\left(\varepsilon\nabla\left(\mathbf{u}^k - \mathbf{u}\right), \nabla\mathbf{v}\right)\right| + \frac{1}{\mathrm{Re}}\left|\left(\alpha\left(\mathbf{u}^k - \mathbf{u}\right), \mathbf{v}\right)\right|$$
$$+ \left|\left(\beta\,|\mathbf{u}^k + \mathbf{g}_\mu|\left(\mathbf{u}^k - \mathbf{u}\right), \mathbf{v}\right)\right|$$
$$+ \left|\left(\beta\left(|\mathbf{u}^k + \mathbf{g}_\mu| - |\mathbf{u} + \mathbf{g}_\mu|\right)\left(\mathbf{u} + \mathbf{g}_\mu\right), \mathbf{v}\right)\right|$$
$$+ \left|n\left(\mathbf{u}^k, \mathbf{u}^k, \mathbf{v}\right) - n(\mathbf{u}, \mathbf{u}, \mathbf{v})\right| + \left|n\left(\mathbf{u}^k - \mathbf{u}, \mathbf{g}_\mu, \mathbf{v}\right)\right|$$
$$+ \left|n\left(\mathbf{g}_\mu, \mathbf{u}^k - \mathbf{u}, \mathbf{v}\right)\right| \qquad (51)$$
$$\le \frac{\varepsilon_1}{\mathrm{Re}}\,|\mathbf{u}^k - \mathbf{u}|_1\,|\mathbf{v}|_1 + \frac{1}{\mathrm{Re}}\,\|\alpha\|_{0,\infty}\,\|\mathbf{u}^k - \mathbf{u}\|_0\,\|\mathbf{v}\|_0$$
$$+ \|\beta\|_{0,\infty}\,\|\mathbf{u}^k + \mathbf{g}_\mu\|_{0,4}\,\|\mathbf{u}^k - \mathbf{u}\|_0\,\|\mathbf{v}\|_{0,4}$$
$$+ \|\beta\|_{0,\infty}\,\|\mathbf{u} + \mathbf{g}_\mu\|_{0,4}\,\|\mathbf{u}^k - \mathbf{u}\|_0\,\|\mathbf{v}\|_{0,4}$$
$$+ \left|n\left(\mathbf{u}^k, \mathbf{u}^k, \mathbf{v}\right) - n(\mathbf{u}, \mathbf{u}, \mathbf{v})\right|$$
$$+ C\,\|\mathbf{u}^k - \mathbf{u}\|_1\,\|\mathbf{g}_\mu\|_1\,\|\mathbf{v}\|_1.$$

The boundedness of $\mathbf{u}^k$ in $\mathbf{W}$, (22), the Poincaré inequality, and the above inequality imply that

$$\left|\left[G\left(\mathbf{u}^k\right) - G(\mathbf{u}), \mathbf{v}\right]\right| \longrightarrow 0 \quad \text{as } k \longrightarrow \infty\ \forall \mathbf{v} \in \mathbf{W}. \qquad (52)$$

Thus, employing

$$\left|G\left(\mathbf{u}^k\right) - G(\mathbf{u})\right|_1 = \sup_{\substack{\mathbf{v}\in\mathbf{W}\\\mathbf{v}\neq 0}} \frac{\left[G\left(\mathbf{u}^k\right) - G(\mathbf{u}), \mathbf{v}\right]}{|\mathbf{v}|_1}, \qquad (53)$$

we state that G is continuous. Now, we note that for any $\mathbf{u} \in \mathbf{W}$ we have

$$
\begin{aligned}
[G(\mathbf{u}), \mathbf{u}] &= \frac{1}{\mathrm{Re}} \left(\varepsilon \nabla \left(\mathbf{u} + \mathbf{g}_\mu \right), \nabla \mathbf{u} \right) \\
&\quad + \frac{1}{\mathrm{Re}} \left(\alpha \left(\mathbf{u} + \mathbf{g}_\mu \right), \mathbf{u} \right) \\
&\quad + \left(\beta \left| \mathbf{u} + \mathbf{g}_\mu \right| \left(\mathbf{u} + \mathbf{g}_\mu \right), \mathbf{u} \right) \\
&\quad + n \left(\mathbf{u} + \mathbf{g}_\mu, \mathbf{u} + \mathbf{g}_\mu, \mathbf{u} \right) - (\mathbf{f}, \mathbf{u}) \\
&\geq \frac{\varepsilon_0}{\mathrm{Re}} |\mathbf{u}|_1^2 - \frac{\varepsilon_1}{\mathrm{Re}} \left| \left(\nabla \mathbf{g}_\mu, \nabla \mathbf{u} \right) \right| + \frac{1}{\mathrm{Re}} \left(\alpha \mathbf{u}, \mathbf{u} \right) \\
&\quad - \frac{1}{\mathrm{Re}} \left| \left(\alpha \mathbf{g}_\mu, \mathbf{u} \right) \right| + \left(\beta \left| \mathbf{u} + \mathbf{g}_\mu \right|, |\mathbf{u}|^2 \right) \\
&\quad - \left| \left(\beta \left| \mathbf{u} + \mathbf{g}_\mu \right| \mathbf{g}_\mu, \mathbf{u} \right) \right| + n \left(\mathbf{u}, \mathbf{g}_\mu, \mathbf{u} \right) \\
&\quad + n \left(\mathbf{g}_\mu, \mathbf{g}_\mu, \mathbf{u} \right) - \|\mathbf{f}\|_0 \|\mathbf{u}\|_0 \\
&\geq \frac{\varepsilon_0}{\mathrm{Re}} |\mathbf{u}|_1^2 - \frac{\varepsilon_1}{\mathrm{Re}} |\mathbf{g}_\mu|_1 |\mathbf{u}|_1 \\
&\quad - \frac{1}{\mathrm{Re}} \|\alpha\|_{0,\infty} \|\mathbf{g}_\mu\|_0 \|\mathbf{u}\|_0 \\
&\quad - \left| \left(\beta \left| \mathbf{u} + \mathbf{g}_\mu \right| \mathbf{g}_\mu, \mathbf{u} \right) \right| - \left| n \left(\mathbf{u}, \mathbf{g}_\mu, \mathbf{u} \right) \right| \\
&\quad - C \|\mathbf{g}_\mu\|_1^2 \|\mathbf{u}\|_1 - \|\mathbf{f}\|_0 \|\mathbf{u}\|_0 .
\end{aligned}
\tag{54}
$$

leads to the desired assertion (50).   □

The following lemma plays a key role in the existence proof.

Lemma 6. *Let Y be finite dimensional Hilbert space with inner product $[\cdot, \cdot]$ inducing a norm $\| \cdot \|$ and $T : Y \rightarrow Y$ be a continuous mapping such that*

$$
[T(x), x] > 0 \quad \text{for } \|x\| = r_0 > 0.
\tag{59}
$$

Then there exists $x \in Y$, with $\|x\| \leq r_0$, such that

$$
T(x) = 0.
\tag{60}
$$

Proof. See [22].   □

Now we are able to prove the main result concerning existence of velocity.

Theorem 7. *The problem (49) has at least one solution $\mathbf{u} \in \mathbf{W}$.*

Proof. We construct the approximate sequence of Galerkin solutions. Since the space $\mathbf{W}$ is separable, there exists a

From the Poincaré inequality, we infer the estimate

$$
\|v\|_1 \leq C |v|_1 \quad \forall v \in H_0^1(\Omega),
\tag{55}
$$

which together with (32), (33), and (54) results in

$$
\begin{aligned}
[G(\mathbf{u}), \mathbf{u}] &\geq \left\{ \frac{\varepsilon_0}{\mathrm{Re}} - \delta \left(1 + \|\beta\|_{0,\infty} \right) \right\} |\mathbf{u}|_1^2 - \left\{ \frac{\varepsilon_1}{\mathrm{Re}} |\mathbf{g}_\mu|_1 \right. \\
&\quad + C_1 \frac{1}{\mathrm{Re}} \|\alpha\|_{0,\infty} \|\mathbf{g}_\mu\|_0 + \delta \|\beta\|_{0,\infty} \|\mathbf{g}_\mu\|_0 + C_2 \|\mathbf{g}_\mu\|_1^2 \\
&\quad \left. + C_3 \|\mathbf{f}\|_0 \right\} |\mathbf{u}|_1 .
\end{aligned}
\tag{56}
$$

Choosing δ such that

$$
0 < \delta < \delta_0 := \frac{\varepsilon_0}{\mathrm{Re}} \left(1 + \|\beta\|_{0,\infty} \right)^{-1},
\tag{57}
$$

and $r > r_0$ with

$$
r_0 := \frac{(\varepsilon_1/\mathrm{Re}) |\mathbf{g}_\mu|_1 + (1/\mathrm{Re}) C_1 \|\alpha\|_{0,\infty} \|\mathbf{g}_\mu\|_0 + \delta \|\beta\|_{0,\infty} \|\mathbf{g}_\mu\|_0 + C_2 \|\mathbf{g}_\mu\|_1^2 + C_3 \|\mathbf{f}\|_0}{\varepsilon_0/\mathrm{Re} - \delta \left(1 + \|\beta\|_{0,\infty} \right)},
\tag{58}
$$

sequence of linearly independent elements $(\mathbf{w}^i)_{i \in \mathbb{N}} \subset \mathbf{W}$. Let $\mathbf{X}_m$ be the finite dimensional subspace of $\mathbf{W}$ with

$$
\mathbf{X}_m := \mathrm{span} \left\{ \mathbf{w}^i, \ i = 1, \ldots, m \right\},
\tag{61}
$$

and endowed with the scalar product of $\mathbf{W}$. Let $\mathbf{u}^m = \sum_{j=1}^m a_j \mathbf{w}^j$, $a_j \in \mathbb{R}$, be a Galerkin solution of (49) defined by

$$
\left[G(\mathbf{u}^m), \mathbf{w}^j \right] = 0, \quad \forall j = 1, \ldots, m.
\tag{62}
$$

From Lemmas 5 and 6 we conclude that

$$
[G(\mathbf{u}^m), \mathbf{w}] = 0 \quad \forall \mathbf{w} \in \mathbf{X}_m
\tag{63}
$$

has a solution $\mathbf{u}^m \in \mathbf{X}_m$. The unknown coefficients a_j can be obtained from the algebraic system (62). On the other hand, multiplying (62) by a_j and adding the equations for $j = 1, \ldots, m$ we have

$$0 = [G(\mathbf{u}^m), \mathbf{u}^m] \geq \left\{ \frac{1}{\mathrm{Re}} - \delta \left(1 + \|\beta\|_{0,\infty} \right) \right\} |\mathbf{u}^m|_1^2$$

$$- \left\{ \frac{1}{\mathrm{Re}} |\mathbf{g}_\mu|_1 + C_1 \frac{1}{\mathrm{Re}} \|\alpha\|_{0,\infty} \|\mathbf{g}_\mu\|_0 \right. \tag{64}$$

$$\left. + \delta \|\beta\|_{0,\infty} \|\mathbf{g}_\mu\|_0 + C_2 \|\mathbf{g}_\mu\|_1^2 + C_3 \|\mathbf{f}\|_0 \right\} |\mathbf{u}^m|_1.$$

This gives together with (58) the uniform boundedness in $\mathbf{W}$

$$|\mathbf{u}^m|_1 \leq r_0; \tag{65}$$

therefore there exists $\mathbf{u} \in \mathbf{W}$ and a subsequence $m_k \to \infty$ (we write for the convenience m instead of m_k) such that

$$\mathbf{u}^m \rightharpoonup \mathbf{u} \quad \text{in } \mathbf{W}. \tag{66}$$

Furthermore, the compactness of embedding $H^1(\Omega) \hookrightarrow L^4(\Omega)$ implies

$$\mathbf{u}^m \longrightarrow \mathbf{u} \quad \text{in } \mathbf{L}^4(\Omega). \tag{67}$$

Taking the limit in (63) with $m \to \infty$ we get

$$[G(\mathbf{u}), \mathbf{w}] = 0 \quad \forall \mathbf{w} \in \mathbf{X}_m. \tag{68}$$

Finally, we apply the continuity argument and state that (68) is preserved for any $\mathbf{w} \in \mathbf{W}$; therefore $\mathbf{u}$ is the solution of (49). $\square$

For the reconstruction of the pressure we need inf-sup-theorem.

Theorem 8. *Assume that the bilinear form $b(\cdot, \cdot)$ satisfies the inf-sup condition*

$$\inf_{q \in M} \sup_{\mathbf{v} \in \mathbf{X}_0} \frac{b(\mathbf{v}, q)}{|\mathbf{v}|_1 \|q\|_0} \geq \gamma > 0. \tag{69}$$

Then, for each solution $\mathbf{u}$ of the nonlinear problem (49) there exists a unique pressure $p \in M$ such that the pair $(\mathbf{u}, p) \in \mathbf{V}$ is a solution of the homogeneous problem (47).

Proof. See [19, Theorem 1.4, §1, Chapter IV]. $\square$

We end up this subsection by proving the existence of the pressure.

Theorem 9. *Let $\mathbf{w}$ be solution of problem (49). Then, there exists unique pressure $p \in M$.*

Proof. We verify the inf-sup condition (69) of Theorem 8 by employing the isomorphism of Theorem 2. From [19, Corollary 2.4, Section 2, Chapter I] follows that for any q in $L_0^2(\Omega)$ there exists $\mathbf{v}$ in $\mathbf{H}_0^1(\Omega)$ such that

$$(\operatorname{div} \mathbf{v}, q) \geq \gamma^* \|\mathbf{v}\|_1 \|q\|_0, \tag{70}$$

with a positive constant γ^*. Setting $\mathbf{u} = \mathbf{v}/\varepsilon$ and applying the isomorphism in Theorem 2, we obtain the estimate

$$b(\mathbf{u}, q) = (\operatorname{div} \mathbf{v}, q) \geq \gamma^* \|\mathbf{v}\|_1 \|q\|_0 \geq \gamma_\varepsilon \|\mathbf{u}\|_1 \|q\|_0, \tag{71}$$

where $\gamma_\varepsilon = \gamma^*/C\{\varepsilon_0^{-1} + \varepsilon_0^{-2} |\varepsilon|_{1,3}\}$. From the above estimate we conclude the inf-sup condition (69). $\square$

2.3. Uniqueness of Weak Solution.
We exploit a priori estimates in order to prove uniqueness of weak velocity and pressure.

Theorem 10. *If $\|\mathbf{g}_\mu\|_1$, $\|\mathbf{f}\|_{-1} := \sup_{0 \neq \mathbf{v} \in \mathbf{H}^1(\Omega)} ((\mathbf{f}, \mathbf{v})/\|\mathbf{v}\|_1)$ are sufficiently small; then the solution of (49) is unique.*

Proof. Assume that $(\mathbf{u}_1, p_1)$ and $(\mathbf{u}_2, p_2)$ are two different solutions of (47). From (19) in Lemma 3 we obtain $n(\mathbf{w}, \mathbf{u}, \mathbf{u}) = 0 \; \forall \mathbf{w}, \mathbf{u} \in \mathbf{W}$. Then, we obtain

$$0 = [G(\mathbf{u}_1) - G(\mathbf{u}_2), \mathbf{u}_1 - \mathbf{u}_2] = a(\mathbf{u}_1 - \mathbf{u}_2, \mathbf{u}_1 - \mathbf{u}_2)$$
$$+ c(\mathbf{u}_1 - \mathbf{u}_2, \mathbf{u}_1 - \mathbf{u}_2) - (\mathbf{f}, \mathbf{u}_1 - \mathbf{u}_2)$$
$$+ n(\mathbf{u}_1 + \mathbf{g}_\mu, \mathbf{u}_1 + \mathbf{g}_\mu, \mathbf{u}_1 - \mathbf{u}_2)$$
$$- n(\mathbf{u}_2 + \mathbf{g}_\mu, \mathbf{u}_2 + \mathbf{g}_\mu, \mathbf{u}_1 - \mathbf{u}_2)$$
$$+ (\beta |\mathbf{u}_1 + \mathbf{g}_\mu| (\mathbf{u}_1 + \mathbf{g}_\mu), \mathbf{u}_1 - \mathbf{u}_2)$$
$$- (\beta |\mathbf{u}_2 + \mathbf{g}_\mu| (\mathbf{u}_2 + \mathbf{g}_\mu), \mathbf{u}_1 - \mathbf{u}_2)$$
$$\geq \frac{\varepsilon_0}{\mathrm{Re}} |\mathbf{u}_1 - \mathbf{u}_2|_1^2 - \|\mathbf{f}\|_{-1} \|\mathbf{u}_1 - \mathbf{u}_2\|_1$$
$$+ n(\mathbf{u}_1 - \mathbf{u}_2, \mathbf{u}_2 + \mathbf{g}_\mu, \mathbf{u}_1 - \mathbf{u}_2) \tag{72}$$
$$+ (\beta |\mathbf{u}_1 + \mathbf{g}_\mu| (\mathbf{u}_1 - \mathbf{u}_2), \mathbf{u}_1 - \mathbf{u}_2)$$
$$+ (\beta (|\mathbf{u}_1 + \mathbf{g}_\mu| - |\mathbf{u}_2 + \mathbf{g}_\mu|) (\mathbf{u}_2 + \mathbf{g}_\mu), \mathbf{u}_1 - \mathbf{u}_2)$$
$$\geq \frac{\varepsilon_0}{\mathrm{Re}} |\mathbf{u}_1 - \mathbf{u}_2|_1^2 - \|\mathbf{f}\|_{-1} \|\mathbf{u}_1 - \mathbf{u}_2\|_1$$
$$- |n(\mathbf{u}_1 - \mathbf{u}_2, \mathbf{u}_2, \mathbf{u}_1 - \mathbf{u}_2)|$$
$$- |n(\mathbf{u}_1 - \mathbf{u}_2, \mathbf{g}_\mu, \mathbf{u}_1 - \mathbf{u}_2)| - \|\beta\|_{0,\infty}$$
$$\cdot |(|\mathbf{u}_1 + \mathbf{g}_\mu| \cdot |\mathbf{u}_1 - \mathbf{u}_2|, |\mathbf{u}_1 - \mathbf{u}_2|)| - \|\beta\|_{0,\infty}$$
$$\cdot |(|\mathbf{u}_1 + \mathbf{g}_\mu| - |\mathbf{u}_2 + \mathbf{g}_\mu|| \cdot |\mathbf{u}_2 + \mathbf{g}_\mu|, |\mathbf{u}_1 - \mathbf{u}_2|)|.$$

From Cauchy-Schwarz inequality and Sobolev embedding $H^1(\Omega) \hookrightarrow L^4(\Omega)$ we deduce

$$|(|\mathbf{u}_1 + \mathbf{g}_\mu| \cdot |\mathbf{u}_1 - \mathbf{u}_2|, |\mathbf{u}_1 - \mathbf{u}_2|)|$$
$$\leq C \{\|\mathbf{u}_1\|_0 + \|\mathbf{g}_\mu\|_0\} \|\mathbf{u}_1 - \mathbf{u}_2\|_1^2,$$
$$|(|\mathbf{u}_1 + \mathbf{g}_\mu| - |\mathbf{u}_2 + \mathbf{g}_\mu|| \cdot |\mathbf{u}_2 + \mathbf{g}_\mu|, |\mathbf{u}_1 - \mathbf{u}_2|)| \tag{73}$$
$$\leq C \{\|\mathbf{u}_2\|_0 + \|\mathbf{g}_\mu\|_0\} \|\mathbf{u}_1 - \mathbf{u}_2\|_1^2,$$

and according to (20) we have

$$|n(\mathbf{u}_1 - \mathbf{u}_2, \mathbf{u}_2, \mathbf{u}_1 - \mathbf{u}_2)| \leq C \|\mathbf{u}_2\|_1 \|\mathbf{u}_1 - \mathbf{u}_2\|_1^2, \tag{74}$$

and by (31) we can find μ such that

$$|n(\mathbf{u}_1 - \mathbf{u}_2, \mathbf{g}_\mu, \mathbf{u}_1 - \mathbf{u}_2)| \leq \frac{\varepsilon_0}{4\mathrm{Re}} \|\mathbf{u}_1 - \mathbf{u}_2\|_1^2. \tag{75}$$

Now, we find upper bounds for $\mathbf{u}_1$ and $\mathbf{u}_2$. Testing (47) with $\mathbf{u}$ results in

$$\frac{\varepsilon_0}{\mathrm{Re}} \|\mathbf{u}\|_1^2 \leq \|\mathbf{f}\|_{-1} \|\mathbf{u}\|_1 + \frac{\varepsilon_0}{\mathrm{Re}} \|\mathbf{g}_\mu\|_1 \|\mathbf{u}\|_1 + C \|\mathbf{g}_\mu\|_0 \|\mathbf{u}\|_0$$

$$+ C \|\mathbf{g}_\mu\|_1^2 \|\mathbf{u}\|_1 + C \|\beta\|_{0,\infty} \|\mathbf{g}_\mu\|_0 \|\mathbf{u}\|_1^2 \qquad (76)$$

$$+ C \|\beta\|_{0,\infty} \|\mathbf{g}_\mu\|_{0,4}^2 \|\mathbf{u}\|_1 .$$

From Sobolev embedding $H^1(\Omega) \hookrightarrow L^4(\Omega)$ we deduce for sufficiently small $\| \mathbf{g}_\mu \|_1$

$$\|\mathbf{u}\|_1 \leq \frac{\|\mathbf{f}\|_{-1} + C_1 \|\mathbf{g}_\mu\|_1 + C_2 \|\mathbf{g}_\mu\|_1^2}{\varepsilon_0/\mathrm{Re} - C_3 \|\beta\|_{0,\infty} \|\mathbf{g}_\mu\|_1} \qquad (77)$$

$$=: C\left(\|\mathbf{g}_\mu\|_1, \|\mathbf{f}\|_{-1} \right).$$

Putting (73), (74), (75), and (77) into (72) and using the inequality

$$\|\mathbf{f}\|_{-1} \|\mathbf{u}_1 - \mathbf{u}_2\|_1 \leq \frac{\varepsilon_0}{4\mathrm{Re}} \|\mathbf{u}_1 - \mathbf{u}_2\|_1^2 + \frac{2\mathrm{Re}}{\varepsilon_0} \|\mathbf{f}\|_{-1}^2, \qquad (78)$$

we obtain

$$0 \geq \frac{\varepsilon_0}{2\mathrm{Re}} \|\mathbf{u}_1 - \mathbf{u}_2\|_1^2 - \frac{2\mathrm{Re}}{\varepsilon_0} \|\mathbf{f}\|_{-1}^2$$

$$- C\left(\|\mathbf{g}_\mu\|_1, \|\mathbf{f}\|_{-1} \right) \|\beta\|_{0,\infty} \|\mathbf{u}_1 - \mathbf{u}_2\|_1^2 \qquad (79)$$

$$- \frac{\varepsilon_0}{4\mathrm{Re}} \|\mathbf{u}_1 - \mathbf{u}_2\|_1^2 - C\left(\|\mathbf{g}_\mu\|_1, \|\mathbf{f}\|_{-1} \right) \|\mathbf{u}_1 - \mathbf{u}_2\|_1^2 .$$

For sufficiently small $\|\mathbf{g}_\mu\|_1, \|\mathbf{f}\|_{-1}$ the constant $C(\|\mathbf{g}_\mu\|_1, \|\mathbf{f}\|_{-1})$ in (77) gets small and consequently the right hand side of (79) is nonnegative. This implies $\mathbf{u}_1 = \mathbf{u}_2$ and according to Theorem 9 $p_1 - p_2 = 0$. □

3. A Channel Flow Problem in Packed Bed Reactors

In this section, we provide an example of the flow problem in packed bed reactors with numerical solutions at small and relatively large Reynolds numbers to show the nonlinear behavior of the velocity solutions. Our numerical tests were conducted using the noncommercial object-oriented finite element package MooNMD [23] that was originally developed by the research group in Magdeburg and used for several benchmarks. The numerical results generated by MooNMD have been also verified by commercial software packages FLUENT® and FEMLAB®; see [24] and [3, Chapter 2.6].

Let the reactor channel with Newtonian-fluid be represented by the plain domain $\Omega = (0, L) \times (-R, R)$ where $R = 5$ and $L = 60$. In all computations we use the porosity distribution from [3, Section 2.2] which is determined experimentally and takes into account the effect of wall channeling in packed bed reactors

$$\varepsilon(x, y) = \varepsilon(y) = \varepsilon_\infty \left\{ 1 + \frac{1 - \varepsilon_\infty}{\varepsilon_\infty} e^{-6(R-|y|)} \right\}, \qquad (80)$$

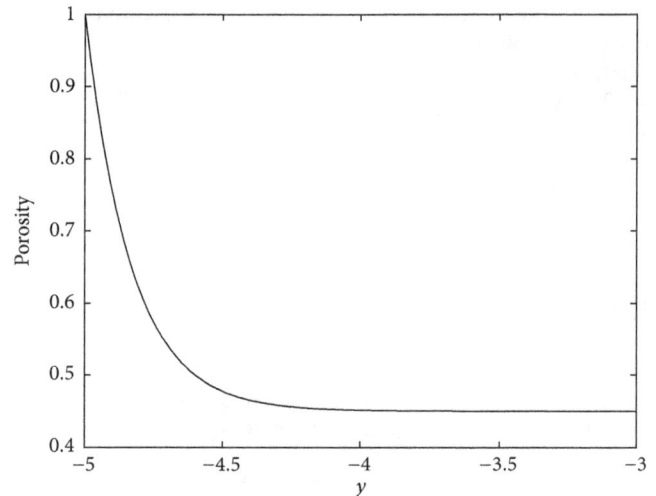

FIGURE 1: Varying porosity.

where $\varepsilon_\infty = 0.45$. The distribution of the porosity is presented in Figure 1. We distinguish between the inlet, outlet, and membrane parts of domain boundary Γ and denote them by Γ_{in}, Γ_{out}, and Γ_w, respectively. Let

$$\Gamma_{\mathrm{in}} = \{(x, y) \in \Gamma : x = 0\},$$

$$\Gamma_{\mathrm{out}} = \{(x, y) \in \Gamma : x = L\}, \qquad (81)$$

$$\Gamma_w = \{(x, y) \in \Gamma : y = -R, \ y = R\}.$$

At the inlet Γ_{in} and at the membrane wall Γ_w we prescribe Dirichlet boundary conditions, namely, the plug flow conditions

$$\mathbf{u}|_{\Gamma_{\mathrm{in}}} = \mathbf{u}_{\mathrm{in}} = (u_{\mathrm{in}}, 0)^T,$$

$$\mathbf{u}|_{\Gamma_w} = \mathbf{u}_w = \begin{cases} (0, u_w)^T & \text{for } y = -R, \\ (0, -u_w)^T & \text{for } y = R, \end{cases} \qquad (82)$$

where $u_{\mathrm{in}} > 0$ and $u_w \geq 0$. We consider the case of fixed bed reactor, that is, $u_w = 0$. At the outlet Γ_{out} we set the following outflow boundary condition:

$$-\frac{1}{\mathrm{Re}} \frac{\partial \mathbf{u}}{\partial \mathbf{n}} + p\mathbf{n} = \mathbf{0}, \qquad (83)$$

where $\mathbf{n}$ denotes the outer normal. This boundary condition results from the integration by parts when deriving the weak formulation, and it is called the *do-nothing* boundary condition. In order to avoid discontinuity between the inflow and wall conditions we replace constant profile by trapezoidal one with zero value at the corners. Our computations are carried out on the Cartesian mesh which consists of 100 stretched rectangular cells on the coarse level (see Figure 2) and will be three times uniformly refined. In order to approximate the weak solution of the dimensionless system from (6) subject to the Dirichlet and *do-nothing* boundary conditions, we apply biquadratic conforming and discontinuous piecewise linear finite elements for the velocity and pressure, respectively.

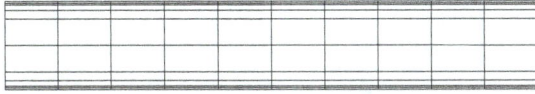

FIGURE 2: Initial mesh for reactor flow problem.

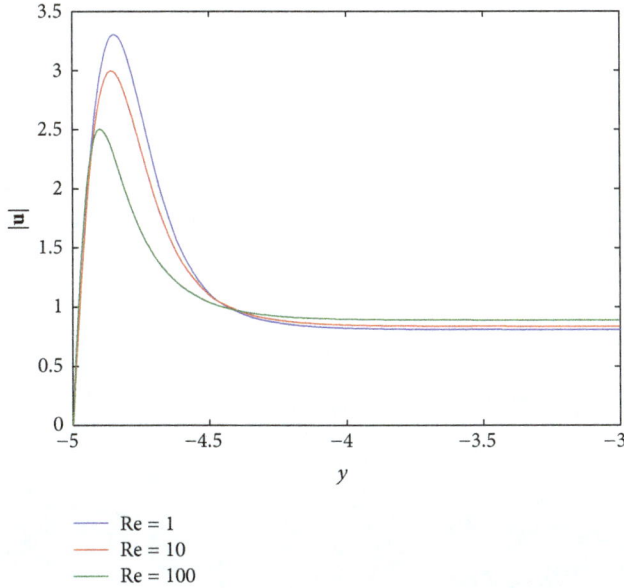

FIGURE 3: Flow profiles in fixed bed reactor at $x = 50$.

The detailed finite element analysis of the Brinkman-Forchheimer-Darcy equation will be presented in the forthcoming work. The plots of velocity magnitude in fixed bed reactor ($u_w = 0$) are presented along the vertical axis $x = 50$. In the investigated reactor the inlet velocity is assumed to be normalized ($u_{in} = 1$). Due to the variation of porosity we might expect higher velocity at the reactor walls Γ_w. This tunneling effect can be well observed in Figure 3 which shows the velocity profiles for different Reynolds numbers. This effect is not present when applying standard Brinkman equation with constant porosity. We remark that the maximum of velocity magnitude decreases with increasing Reynolds numbers in the considered cases Re $= 1, 10, 100$. Our preliminary results are comparable with those ones obtained in [3, Chapter 4.3]. The weighted areas under each velocity profile $u = u(y)$ are the same due to conservation of mass. In fact, we have according to the divergence theorem

$$\int_{(0,50)\times(-R,R)} \operatorname{div}(\varepsilon \mathbf{u})\, d\mathbf{x} = 0 \implies$$

$$\int_{-R}^{R} \varepsilon(y)\, u(y)\, dy = \int_{-R}^{R} \varepsilon(y)\, v_{in} dy. \tag{84}$$

However, we expect increasing maximum of the velocity in the case of high Reynolds numbers. This case requires stabilized finite element methods and will be considered in the forthcoming works.

4. Conclusion

In this work, we have extended the existence and uniqueness of solution result in literature for the porous medium flow problem based on the nonlinear Brinkman-Forchheimer-extended Darcy law. The existing result is valid only for constant porosity and without the considered convection effects, and our result holds for variable porosity and it includes convective effects. We also provided a numerical solution to demonstrate the nonlinear velocity solutions at moderately large Reynolds numbers for which case the Brinkman-Forchheimer-Darcy law applies.

Conflicts of Interest

The authors declare that there are no conflicts of interest regarding the publication of this paper.

References

[1] M. g. Peszynska, A. Trykozko, and W. Sobieski, "Forchheimer law in computational and experimental studies of flow through porous media at porescale and mesoscale," in *Current advances in nonlinear analysis and related topics*, vol. 32 of *GAKUTO Internat. Ser. Math. Sci. Appl.*, pp. 463–482, Gakkotosho, Tokyo, 2010.

[2] S. Ergun, "Fluid flow through packed columns," *Chemical Engineering Progress*, vol. 48, no. 2, pp. 89–94, 1952.

[3] A. Seidel-Morgenstern, Ed., *Membrane Reactors: Distributing Reactants to Improve Selectivity and Yield*, Wiley-VCH, 2010.

[4] O. Bey, *Strömungsverteilung und Wärmetransport in Schüttungen Nummer 570 in Fortschritt-Berichte*, vol. 3, VDI Verlag, Düsseldorf, Germany, 1998.

[5] K. Vafai and S. J. Kim, "On the limitations of the Brinkman-Forchheimer-extended Darcy equation," *International Journal of Heat and Fluid Flow*, vol. 16, no. 1, pp. 11–15, 1995.

[6] U. Hornung, *Homogenization and Porous Media*, Springer, New York, NY, USA, 1997.

[7] C. R. Garibotti and M. Peszyńska, "Upscaling non-Darcy flow," *Transport in Porous Media*, vol. 80, no. 3, pp. 401–430, 2009.

[8] V. Matossian, V. Bhat, M. Parashar et al., "Autonomie oil reservoir optimization on the grid," *Concurrency Computation Practice and Experience*, vol. 17, no. 1, pp. 1–26, 2005.

[9] M. Winterberg and E. Tsotsas, "Modelling of heat transport in beds packed with spherical particles for various bed geometries and/or thermal boundary conditions," *International Journal of Thermal Sciences*, vol. 39, no. 5, pp. 556–570, 2000.

[10] D. A. Nield and A. Bejan, *Convection in Porous Media*, Springer, New York, NY, USA, 2nd edition, 1999.

[11] T. Zhao, *Investigation of Landslide-Induced Debris Flows by the DEM and CFD [Ph.D. thesis]*, University of Oxford, 2014.

[12] K. Upton, *Multi-scale modelling of borehole yields in chalk aquifers [Ph.D. thesis]*, Imperial College London, 2015.

[13] A. Grillo, M. Carfagnay, and S. Federicoz, "The Darcy-Forchheimer law for modelling fluid flow in biological tissues," *Theoretical and Applied Mechanics*, vol. 41, no. 4, pp. 283–322, 2014.

[14] W. Sobieski and A. Trykozko, "Darcy's and Forchheimer's law in practice. Part 1. the experiment," *Technical Sciences*, vol. 17, no. 4, pp. 321–335, 2014.

[15] A. S. Lal and A. C. Menon, "Design of a new porous medium heat exchanger for an aircraft refrigeration system," *IJTEL*, vol. 3, pp. 545–548, 2014.

[16] P. N. Kaloni and J. Guo, "Steady nonlinear double-diffusive convection in a porous medium based upon the Brinkman-Forchheimer model," *Journal of Mathematical Analysis and Applications*, vol. 204, no. 1, pp. 138–155, 1996.

[17] P. S. Vassilevski and U. Villa, "A mixed formulation for the Brinkman problem," *SIAM Journal on Numerical Analysis*, vol. 52, no. 1, pp. 258–281, 2014.

[18] C. Bernardi, F. Laval, B. Metivet, and B. Pernaud-Thomas, "Finite element approximation of viscous flows with varying density," *SIAM Journal on Numerical Analysis*, vol. 29, no. 5, pp. 1203–1243, 1992.

[19] V. Girault and P. Raviart, *Finite Element Methods for Navier-Stokes Equations, Theory and Algorithms*, vol. 5 of *Springer Series in Computational Mathematics*, Springer, Berlin, Germany, 1986.

[20] R. A. Adams, *Sobolev Spaces*, Pure and Applied Mathematics, Academic Press, New York, NY, USA, 1995.

[21] E. Hopf, "Ein allgemeiner Endlichkeitssatz der Hydrodynamik," *Mathematische Annalen*, vol. 117, pp. 764–775, 1941.

[22] J. L. Lions, *Quelques Méthodes de Résolution des Problèmes aux Limites Non Linéaires*, Dunod, Paris, Farnce, 1969.

[23] V. John and G. Matthies, "MooNMD—a program package based on mapped finite element methods," *Computing and Visualization in Science*, vol. 6, no. 2-3, pp. 163–169, 2004.

[24] A. Tota, K. Georgieva, M. Mangold, and P. Skrzypacz, "Comparative numerical simulations of a packed bed membrane reactor," in *Proceedings of the 6th International Conference on Catalysis in Membrane Reactor (ICCMR-6 '04)*, Lahnstein, Germany, 2004.

10

A Study of a Diseased Prey-Predator Model with Refuge in Prey and Harvesting from Predator

Ahmed Sami Abdulghafour ⓘ[1] **and Raid Kamel Naji** ⓘ[2]

[1]*Scientific Affairs Department, Aliraqia University, Baghdad, Iraq*
[2]*Department of Mathematics, College of Science, University of Baghdad, Baghdad, Iraq*

Correspondence should be addressed to Ahmed Sami Abdulghafour; alania961@gmail.com

Academic Editor: Oluwole D. Makinde

In this paper, a mathematical model of a prey-predator system with infectious disease in the prey population is proposed and studied. It is assumed that there is a constant refuge in prey as a defensive property against predation and harvesting from the predator. The proposed mathematical model is consisting of three first-order nonlinear ordinary differential equations, which describe the interaction among the healthy prey, infected prey, and predator. The existence, uniqueness, and boundedness of the system' solution are investigated. The system's equilibrium points are calculated with studying their local and global stability. The persistence conditions of the proposed system are established. Finally the obtained analytical results are justified by a numerical simulation.

1. Introduction

The interaction between the prey and predator species has a long history since Lotka-Volterra model; see [1]. Similarly, the interaction of the susceptible–infected–recovered population is an interesting subject of research work since the pioneering work of Kermack and McKendrick [2]. The dynamics of disease within the ecological systems now becomes an important subject of research. In fact Anderson and May [3] were the first who combined these two systems, while Chattopadhyay and Arino [4] were the first who used the term "eco-epidemiology" for such type of models. On one hand several studies of prey-predator dynamics have been done within the last decades taking into account the effects of variety of biological factors; see, for example, [5–8] and the references therein. On the other hand variety of mathematical models have been proposed and studied in the field of epidemiology taking into account different types of incidence rates and disease; see, for example, [9–12] and the references therein.

The existence of disease in the prey population, predator population, or both is real situation in the ecological species. It has been observed that this type of incidence occurs through infection by some viral disease, bacterial disease, or parasite disease. Many of these studies were focused on the study of disease in prey population only [13–15]. Other researchers were interested in the study of disease within the predator population only [16–19]. There are also some studies where both the prey and predator populations are infected with the disease [20–23].

It is well known that the harvesting of the species is necessary for the coexistence of the species and hence it took a lot of interest from the researchers in their proposed ecological models. Different types of harvesting have been proposed and studied including constant harvesting, density dependent proportional harvesting, and nonlinear harvesting [8, 24–26]. The refuge of prey species is also a biological factor necessary for the coexistence of the species and hence it is another factor of great interest as defensive properties of the prey against the predation; see [8, 27, 28].

Keeping the above in view, in this paper we propose and study an eco-epidemiological prey-predator model involving a prey refuge and harvesting from the predator. It is assumed that the disease exists only in prey and it will not transfer to predator through the feeding process. The paper is organized as follows. In Section 2, the mathematical model is formulated and its dimensionless variables and parameters are determined; moreover the existence, uniqueness, and uniform boundedness of its solution are discussed. In Section 3, the local stability of all possible equilibrium points and persistence of the system are studied. Section 4 deals with the global stability analysis of the system using suitable Lyapunov functions. However Section 5 provides a numerical simulation of the proposed system for suitable chose of parameters values. Finally Section 6 gives some conclusions and discussed the obtained results.

2. Mathematical Model

In this section, a prey-predator system involving infected disease in prey is proposed for study. It is assumed that there is a harvesting effort applied on the predator individuals only. Accordingly, the following hypotheses are adopted to formulate the mathematical model.

(1) The prey population is divided into two compartments, susceptible with density at time T given by $S(T)$ and infected with density at time T represented by $I(T)$, while the predator consists of only one compartment with density at time T denoted by $Y(T)$.

(2) The prey population grows logistically with intrinsic growth rate $r > 0$ and carrying capacity $K > 0$; it is assumed that the infected cannot reproduce; rather it competes with the susceptible individuals for food and space.

(3) There is a type of protection of prey population from the predation by predator, represented by a constant prey's refuge rate $m \in (0, 1)$ that leaves $(1 - m)S$ of prey available to be hunted by the predator.

(4) The susceptible prey population becomes infected by contact with the infected prey according to the simple mass action kinetics with $\beta > 0$ as the rate of infection.

(5) The predator population consumes both the prey populations according to modified Holling type II functional response for the predation [29] with half saturation constant $b > 0$ and maximum attack rates $a_1 > 0$ and $a_2 > 0$ for susceptible prey and infected prey, respectively. Since there is a vulnerability of infected prey relative to susceptible prey the vulnerability constant rate $\theta > 0$ is used in the functional response. Moreover the constants $e_1 \in (0, 1)$ and $e_2 \in (0, 1)$ are the conversion rates from susceptible and infected preys to predator, respectively.

(6) The disease causes a death in the infected population represented by diseased death rate $d_1 > 0$, while in the absence of prey, the predator decays exponentially with natural death rate $d_2 > 0$.

(7) Only the predator population is assumed to be harvested with the Michael-Mentence harvesting function [30], where $E > 0$ represents hunting effort, $c > 0$ is the catchability coefficient of the predator, and l_i, $i = 1, 2$, are suitable positive constants.

According to the above set of hypotheses the dynamics of an eco-epidemic prey-predator model with refuge in prey and harvesting in predator can be described in the following set of first-order nonlinear differential equations.

$$\frac{dS}{dT} = rS\left(1 - \frac{S+I}{K}\right) - \frac{a_1(1-m)SY}{b+(1-m)S+\theta(1-m)I}$$
$$- \beta SI$$

$$\frac{dI}{dT} = \beta SI - \frac{a_2(1-m)IY}{b+(1-m)S+\theta(1-m)I} - d_1 I \quad (1)$$

$$\frac{dY}{dT} = \left(\frac{e_1 a_1(1-m)S + e_2 a_2(1-m)I}{b+(1-m)S+\theta(1-m)I}\right)Y$$
$$- \frac{cEY}{l_1 E + l_2 Y} - d_2 Y$$

where $S(0) \geq 0$, $I(0) \geq 0$, and $Y(0) \geq 0$.

Now in order to reduce the number of parameters and specify the control set of parameters the following dimensionless variables and parameters are used:

$$t = rT,$$
$$x = \frac{X}{K},$$
$$i = \frac{I}{K},$$
$$y = \frac{a_1 Y}{rK},$$
$$w_1 = \frac{b}{K},$$
$$w_2 = \frac{\beta K}{r},$$
$$w_3 = \frac{a_2}{a_1},$$
$$w_4 = \frac{d_1}{r}, \quad (2)$$
$$w_5 = \frac{e_1 a_1}{r},$$
$$w_6 = \frac{e_2 a_2}{r},$$
$$w_7 = \frac{a_1 l_1 E}{r l_2 K},$$
$$w_8 = \frac{c a_1 E}{r^2 l_2 K},$$
$$w_9 = \frac{d_2}{r}$$

System (1) reduces to the following dimensionless system:

$$\frac{ds}{dt}$$

$$= s\left[1 - (s+i) - \frac{(1-m)\,y}{w_1 + (1-m)\,s + \theta\,(1-m)\,i} - w_2 i\right]$$

$$= sf_1(s,i,y)$$

$$\frac{di}{dt} = i\left[w_2 s - \frac{w_3\,(1-m)\,y}{w_1 + (1-m)\,s + \theta\,(1-m)\,i} - w_4\right] \qquad (3)$$

$$= if_2(s,i,y)$$

$$\frac{dy}{dt} = y\left[\frac{w_5\,(1-m)\,s + w_6\,(1-m)\,i}{w_1 + (1-m)\,s + \theta\,(1-m)\,i} - \frac{w_7}{w_8 + y} - w_9\right]$$

$$= yf_3(s,i,y)$$

According to the above dimensionless form, it is easy to verify that the differential equations are continuous and have continuous partial derivatives on the domain $\mathbb{R}^3_+ = \{(s,i,y) \in \mathbb{R}^3 : s(0) \geq 0,\ i(0) \geq 0;\ y(0) \geq 0\}$ and hence they are Lipschitzian functions. Therefore system (3) has a unique solution. Furthermore the solution of system (3) that initials in $\mathbb{R}^3_+$ is uniformly bounded as shown in the following theorem.

Theorem 1. *All solutions of system (3) are uniformly bounded.*

Proof. According to the prey population which is given in the first equation of system (3), it is observed that

$$\frac{ds}{dt} \leq s\,[1-s] \qquad (4)$$

Then by solving this differential inequality, we obtain that $s(t) \leq s(0)/(e^{-t}[1 - s(0)] + s(0))$, and then for $t \longrightarrow \infty$, we have $s(t) \leq 1$. Now assume that $\Omega = s + i + y$; then

$$\frac{d\Omega}{dt} = \frac{ds}{dt} + \frac{di}{dt} + \frac{dy}{dt} \xrightarrow{yields} \frac{d\Omega}{dt} \leq 2 - \mu\Omega \qquad (5)$$

where $\mu = \min\{1, w_4, w_9\}$. Therefore by solving the last differential inequality we obtain that

$$\Omega(t) \leq \Omega(0)\,e^{-\mu t} - \frac{2}{\mu}\left(e^{-\mu t} - 1\right) \qquad (6)$$

Hence for $t \longrightarrow \infty$, we get that $\Omega(t) \leq 2/\mu$. Thus all the solutions of system (3) are uniformly bounded. $\qquad \square$

3. The Local Stability and Persistence

In the following the existence and local stability of all possible equilibrium points are investigated and then the persistence conditions of the system are established. Obviously system (3) has at most five nonnegative equilibrium points. The vanishing equilibrium point $E_0 = (0,0,0)$ and the axial equilibrium point $E_1 = (1,0,0)$ always exist. The first planar $E_2 = (\bar{s}, 0, \bar{y})$, where $\bar{s}$ represents a unique positive root of the following third-order polynomial equation:

$$A_1 s^3 + A_2 s^2 + A_3 s + A_4 = 0 \qquad (7a)$$

where $A_1 = (w_9 - w_5)R^2$, $A_2 = [(w_9 - w_5)(w_1 - R) + w_9 w_1]R$, $A_3 = (w_5 w_8 - w_7)R^2 + [w_1 w_5 - 2w_1 w_9 - w_8 w_9]R + w_1^2 w_9$, $A_4 = -w_1[w_1 w_9 + (w_7 + w_8 w_9)R] < 0$, while

$$\bar{y} = \frac{(1-\bar{s})\,(w_1 + R\bar{s})}{R} \qquad (7b)$$

from now onward $R = (1-m)$. Straightforward computation shows that E_2 exists uniquely in the interior of positive quadrant of sy −plane provided that the following sufficient conditions hold:

$$\bar{s} < 1 \qquad (8a)$$

$$w_9 > w_5;$$

$$A_2 > 0 \qquad (8b)$$

$$or\ w_9 > w_5;$$

$$A_3 < 0$$

The second planar equilibrium point $E_3 = (\hat{s}, \hat{i}, 0)$, where

$$\hat{s} = \frac{w_4}{w_2};$$

$$\hat{i} = \frac{w_2 - w_4}{w_2\,(1 + w_2)} \qquad (9)$$

exists uniquely in the interior of positive quadrant of si −plane provided that

$$w_2 > w_4 \qquad (10)$$

Finally the coexistence equilibrium point is denoted by $E_4 = (s^*, i^*, y^*)$ where

$$i^* = \frac{w_3 + w_4 - (w_2 + w_3)\,s^*}{w_3\,(1 + w_2)} \qquad (11a)$$

$$y^* = \frac{(w_2 s^* - w_4)\,[(w_1 w_3\,(1 + w_2) + (w_3 + w_4)\,\theta R) + (w_3\,(1 + w_2)\,R - (w_2 + w_3)\,\theta R)\,s]}{w_3^2\,(1 + w_2)\,R} \qquad (11b)$$

while s^* is a unique positive root of the following third-order polynomial equation:

$$D_1 s^3 + D_2 s^2 + D_3 s + D_4 = 0 \qquad (11c)$$

Here

$$D_1 = (\sigma_3 - (w_2 + w_3)\sigma_6) w_2 \sigma_5$$

$$D_2 = (w_2 \sigma_4 - w_4 \sigma_5)\sigma_3 - w_2(w_2 + w_3)\sigma_4 \sigma_6$$
$$+ [w_2(w_3 + w_4) + w_4(w_2 + w_3)]\sigma_5 \sigma_6$$
$$- w_2 \sigma_5 \sigma_8$$

$$D_3 = [\sigma_1 - \sigma_2(w_2 + w_3)] - w_4 \sigma_3 \sigma_4 \qquad (12)$$
$$+ [w_2(w_3 + w_4) + w_4(w_2 + w_3)]\sigma_4 \sigma_6$$
$$- w_4(w_3 + w_4)\sigma_5 \sigma_6 - w_2 \sigma_4 \sigma_8 + w_4 \sigma_5 \sigma_8$$

$$D_4 = \sigma_2(w_3 + w_4) - \sigma_7 - w_4(w_3 + w_4)\sigma_4 \sigma_6$$
$$+ w_4 \sigma_4 \sigma_8$$

with

$$\sigma_1 = [w_5 w_8 - (w_7 + w_8 w_9)] w_3{}^3 (1 + w_2)^2 R^2$$

$$\sigma_2 = [w_6 w_8 - \theta(w_7 + w_8 w_9)] w_3{}^2 (1 + w_2) R^2$$

$$\sigma_3 = (w_5 - w_9) w_3 (1 + w_2) R^2$$

$$\sigma_4 = [w_1 w_3 (1 + w_2) + (w_3 + w_4)\theta R]$$

$$\sigma_5 = [w_3 (1 + w_2) R - (w_2 + w_3)\theta R] \qquad (13)$$

$$\sigma_6 = (w_6 - w_9 \theta) R$$

$$\sigma_7 = w_1(w_7 + w_8 w_9) w_3{}^3 (1 + w_2)^2 R$$

$$\sigma_8 = w_1 w_9 w_3 (1 + w_2) R$$

Clearly (11c) has a unique positive root provided that one set of the following sets of conditions holds.

$$D_1 > 0;$$

$$D_2 > 0; \qquad (14a)$$

$$D_4 < 0$$

$$D_1 > 0;$$

$$D_3 < 0; \qquad (14b)$$

$$D_4 < 0$$

$$D_1 < 0;$$

$$D_2 < 0; \qquad (14c)$$

$$D_4 > 0$$

$$D_1 < 0;$$

$$D_3 > 0; \qquad (14d)$$

$$D_4 > 0$$

Consequently, the coexistence equilibrium point $E_4 = (s^*, i^*, y^*)$ exists uniquely in the interior of $\mathbb{R}^3_+$ provided that in addition to any one of conditions (14a), (14b), (14c), and (14d) the following sufficient conditions hold:

$$s^* < \frac{w_3 + w_4}{(w_2 + w_3)} \qquad (15a)$$

$$(w_2 s^* - w_4)[\sigma_4 + \sigma_5 s] > 0 \qquad (15b)$$

Now in order to study the local stability of these equilibrium points, the Jacobian matrix of system (3) at the point $E = (s, i, y)$ is computed as follows:

$$J(E) = \left(C_{ij} \right)_{3\times3} \qquad (16)$$

where

$$C_{11} = f_1(s, i, y) + s\left(\frac{R^2 y}{N_1{}^2} - 1 \right);$$

$$C_{12} = s\left(\frac{\theta R^2 y}{N_1{}^2} - (1 + w_2) \right);$$

$$C_{13} = s\left(-\frac{R}{N_1} \right);$$

$$C_{21} = i\left(w_2 + \frac{w_3 R^2 y}{N_1{}^2} \right);$$

$$C_{22} = f_2(s, i, y) + i\left(\frac{\theta w_3 R^2 y}{N_1{}^2} \right); \qquad (17)$$

$$C_{23} = i\left(-\frac{w_3 R}{N_1} \right);$$

$$C_{31} = y\left(\frac{w_1 w_5 R + (w_5 \theta - w_6) R^2 i}{N_1{}^2} \right);$$

$$C_{32} = y\left(\frac{w_1 w_6 R + (w_6 - \theta w_5) R^2 s}{N_1{}^2} \right);$$

$$C_{33} = y\left(\frac{w_7}{N_2{}^2} \right) + f_3(s, i, y).$$

Here $N_1 = w_1 + Rs + \theta Ri$ and $N_2 = w_8 + y$. Consequently the Jacobian matrix at vanishing equilibrium point $E_0 = (0, 0, 0)$ is written as

$$J(E_0) = \begin{bmatrix} 1 & 0 & 0 \\ 0 & -w_4 & 0 \\ 0 & 0 & -\left(\dfrac{w_7}{w_8} + w_9 \right) \end{bmatrix} \qquad (18)$$

Obviously the eigenvalues of $J(E_0)$ are given by $\lambda_{01} = 1 > 0$, $\lambda_{02} = -w_4 < 0$, $\lambda_{03} = -(w_7/w_8 + w_9) < 0$. Therefore E_0 is a saddle point.

The Jacobian matrix at axial equilibrium point $E_1 = (1, 0, 0)$ is written as

$$J(E_1) = \begin{bmatrix} -1 & -(1 + w_2) & -\dfrac{R}{w_1 + R} \\ 0 & w_2 - w_4 & 0 \\ 0 & 0 & \dfrac{w_5 R}{w_1 + R} - \dfrac{w_7}{w_8} - w_9 \end{bmatrix} \quad (19)$$

Then the eigenvalues of $J(E_1)$ are given by $\lambda_{11} = -1 < 0$, $\lambda_{12} = w_2 - w_4$ and $\lambda_{13} = w_5 R/(w_1 + R) - w_7/w_8 - w_9$. Accordingly the axial point E_1 is locally asymptotically stable if and only if

$$w_2 < w_4 \quad (20a)$$

$$\frac{w_5 R}{w_1 + R} < \frac{w_7}{w_8} + w_9 \quad (20b)$$

Now the Jacobian matrix at first planar equilibrium point $E_2 = (\bar{s}, 0, \bar{y})$ can be written

$J(E_2)$

$$= \begin{bmatrix} \dfrac{R^2 \bar{s}\,\bar{y}}{(w_1 + R\bar{s})^2} - \bar{s} & \dfrac{\theta R^2 \bar{s}\,\bar{y}}{(w_1 + R\bar{s})^2} - (1 + w_2)\bar{s} & -\dfrac{R\bar{s}}{w_1 + R\bar{s}} \\ 0 & w_2 \bar{s} - \dfrac{w_3 R\bar{y}}{w_1 + R\bar{s}} - w_4 & 0 \\ \dfrac{w_1 w_5 R\bar{y}}{(w_1 + R\bar{s})^2} & \dfrac{w_1 w_6 R\bar{y} + (w_6 - \theta w_5) R^2 \bar{s}\,\bar{y}}{(w_1 + R\bar{s})^2} & \dfrac{w_7 \bar{y}}{(w_8 + \bar{y})^2} \end{bmatrix} \quad (21)$$

Clearly The eigenvalues of $J(E_2)$ can be determined as follows:

$$\left(w_2 \bar{s} - \frac{w_3 R\bar{y}}{w_1 + R\bar{s}} - w_4 - \lambda \right) \left(\lambda^2 - T_2 \lambda + D_2 \lambda \right) = 0 \quad (22)$$

where

$$T_2 = \frac{R^2 \bar{s}\,\bar{y}}{(w_1 + R\bar{s})^2} - \bar{s} + \frac{w_7 \bar{y}}{(w_8 + \bar{y})^2}$$

$$D_2 = \left(\frac{R^2 \bar{s}\,\bar{y}}{(w_1 + R\bar{s})^2} - \bar{s} \right) \left(\frac{w_7 \bar{y}}{(w_8 + \bar{y})^2} \right) \quad (23)$$

$$+ \left(\frac{R\bar{s}}{w_1 + R\bar{s}} \right) \left(\frac{w_1 w_5 R\bar{y}}{(w_1 + R\bar{s})^2} \right)$$

Therefore the eigenvalues of $J(E_2)$ are easily determined by

$$\lambda_{21} = \frac{T_2}{2} - \frac{1}{2}\sqrt{T_2{}^2 - 4D_2}$$

$$\lambda_{22} = w_2 \bar{s} - \frac{w_3 R\bar{y}}{w_1 + R\bar{s}} - w_4 \quad (24)$$

$$\lambda_{23} = \frac{T_2}{2} + \frac{1}{2}\sqrt{T_2{}^2 - 4D_2}$$

Straightforward computation shows that these eigenvalues have negative real parts and hence the first planar equilibrium point E_2 is locally asymptotically stable if and only if the following sufficient conditions hold:

$$w_2 \bar{s} < \frac{w_3 R\bar{y}}{w_1 + R\bar{s}} + w_4 \quad (25a)$$

$$\frac{R^2 \bar{s}\,\bar{y}}{(w_1 + R\bar{s})^2} + \frac{w_7 \bar{y}}{(w_8 + \bar{y})^2} < \bar{s} \quad (25b)$$

$$w_7 R^2 \bar{y}(w_1 + R\bar{s}) + w_1 w_5 R^2 (w_8 + \bar{y})^2$$
$$> w_7 (w_1 + R\bar{s})^3 \quad (25c)$$

Similarly the Jacobian matrix at second planar equilibrium point $E_3 = (\hat{s}, \hat{i}, 0)$ can be written

$J(E_3)$

$$= \begin{bmatrix} -\hat{s} & -\hat{s}(1 + w_2) & -\dfrac{R\hat{s}}{w_1 + R\hat{s} + \theta R\hat{i}} \\ w_2 \hat{i} & 0 & -\dfrac{w_3 R\hat{i}}{w_1 + R\hat{s} + \theta R\hat{i}} \\ 0 & 0 & \dfrac{w_5 R\hat{s} + w_6 R\hat{i}}{w_1 + R\hat{s} + \theta R\hat{i}} - \dfrac{w_7}{w_8} - w_9 \end{bmatrix} \quad (26)$$

Then the eigenvalues of $J(E_3)$ are given by

$$\left(\frac{w_5 R\hat{s} + w_6 R\hat{i}}{w_1 + R\hat{s} + \theta R\hat{i}} - \frac{w_7}{w_8} - w_9 - \lambda \right) \left(\lambda^2 - T_3 \lambda + D_3 \right)$$
$$= 0 \quad (27)$$

where

$$T_3 = -\hat{s} < 0$$
$$D_3 = w_2(1 + w_2)\hat{s}\hat{i} > 0 \quad (28)$$

Clearly the first two eigenvalues resulting from second term in (27) have negative real parts, while the third eigenvalues in the y −direction which are written as

$$\lambda_{33} = \frac{w_5 R\hat{s} + w_6 R\hat{i}}{w_1 + R\hat{s} + \theta R\hat{i}} - \frac{w_7}{w_8} - w_9 \quad (29)$$

have negative real part and then the second planar equilibrium point is locally asymptotically stable if and only if

$$\frac{w_5 R\hat{s} + w_6 R\hat{i}}{w_1 + R\hat{s} + \theta R\hat{i}} < \frac{w_7}{w_8} + w_9 \quad (30)$$

Finally the Jacobian matrix at coexistence equilibrium point $E_4 = (s^*, i^*, y^*)$ can be written in the form

$$J(E_4) = [a_{ij}] \quad i, j = 1, 2, 3 \quad (31)$$

where

$$a_{11} = \frac{R^2 s^* y^* - N_1^{*2} s^*}{N_1^{*2}};$$

$$a_{12} = \frac{\theta R^2 s^* y^* - (1 + w_2) N_1^{*2} s^*}{N_1^{*2}};$$

$$a_{13} = -\frac{R s^*}{N_1^*} < 0;$$

$$a_{21} = w_2 i^* + \frac{w_3 R^2 i^* y^*}{N_1^{*2}} > 0;$$

$$a_{22} = \frac{\theta w_3 R^2 i^* y^*}{N_1^{*2}} > 0;$$

$$a_{23} = -\frac{w_3 R i^*}{N_1^*} < 0;$$

$$a_{31} = \frac{w_1 w_5 R y^* + (w_5 \theta - w_6) R^2 i^* y^*}{N_1^{*2}};$$

$$a_{32} = \frac{w_1 w_6 R y^* + (w_6 - \theta w_5) R^2 s^* y^*}{N_1^{*2}};$$

$$a_{33} = \frac{w_7 y^*}{N_2^{*2}} > 0. \tag{32}$$

Here $N_1^* = w_1 + R s^* + \theta R i^*$ and $N_2^* = w_8 + y^*$. The characteristic equation of $J(E_4)$ can be written in the form

$$\lambda^3 + A_1 \lambda^2 + A_2 \lambda + A_3 = 0 \tag{33}$$

where

$$A_1 = -(a_{11} + a_{22} + a_{33})$$

$$A_2 = a_{11}a_{22} - a_{12}a_{21} + a_{11}a_{33} - a_{13}a_{31} + a_{22}a_{33}$$

$$- a_{23}a_{32} \tag{34}$$

$$A_3 = a_{33}(a_{12}a_{21} - a_{11}a_{22}) - a_{23}(a_{12}a_{31} - a_{11}a_{32})$$

$$- a_{13}(a_{21}a_{32} - a_{22}a_{31})$$

while

$$\Delta = A_1 A_2 - A_3$$

$$= -(a_{11} + a_{22})[a_{11}a_{22} - a_{12}a_{21}] - 2a_{11}a_{22}a_{33}$$

$$- (a_{11} + a_{33})[a_{11}a_{33} - a_{13}a_{31}] \tag{35}$$

$$- (a_{22} + a_{33})[a_{23}a_{32} - a_{22}a_{33}]$$

Therefore straightforward computation shows that

$$A_1 = -\frac{1}{N_1^{*2} N_2^{*2}} \left[N_2^{*2} s^* \left(R^2 y^* - N_1^{*2} \right) \right.$$

$$\left. + \theta w_3 R^2 N_2^{*2} i^* y^* + w_7 N_1^{*2} y^* \right]$$

$$A_3 = \frac{w_7 s^* i^* y^*}{N_1^{*2} N_2^{*2}} \left[(w_2 + w_3) \theta R^2 y^* - (1 + w_2) \right.$$

$$\left. \cdot (w_2 N_1^{*2} + w_3 R^2 y^*) \right] + \frac{w_3 R s^* i^* y^*}{N_1^{*5}} \left[(w_5 \theta - w_6) \right.$$

$$\cdot R^2 N_1^* \left[R y^* - N_1^* (s^* + (1 + w_2) i^*) \right] \tag{36}$$

$$\left. + w_1 (w_6 - w_5(1 + w_2)) R N_1^{*2} \right]$$

$$+ \frac{R^2 s^{*2} y^*}{N_1^{*4}} \left[w_1 w_2 w_6 N_1^* \right.$$

$$+ w_6 R (w_2 N_1^* s^* + w_3 R y^*)$$

$$\left. - w_5 \theta R (w_2 N_1^* s^* + w_3 R y^*) \right]$$

while

$$\Delta = -M_1^* + M_2^* + M_3^* - M_4^* \tag{37}$$

with

$$M_1^* = \frac{1}{N_1^{*6}} \left[N_1^{*2} s^* i^* y^* \left(R^2 s^* y^* - N_1^{*2} s^* \right. \right.$$

$$\left. + w_3 \theta R^2 i^* y^* \right) \left((1 + w_2) \left[w_2 N_1^{*2} + w_3 R^2 y^* \right] \right.$$

$$\left. \left. - (w_2 + w_3) \theta R^2 y^* \right) \right]$$

$$M_2^* = \frac{w_3 w_7 \theta R^2 s^* i^* y^{*2}}{N_1^{*4} N_2^{*2}} \left[N_1^{*2} - R^2 y^* \right]$$

$$M_3^* = \frac{s^* y^*}{N_1^{*6} N_2^{*4}} \left[\left(-R^2 N_2^{*2} s^* y^* + N_1^{*2} N_2^{*2} s^* \right. \right. \tag{38}$$

$$\left. - w_7 N_1^{*2} y^* \right) \left(w_7 R^2 N_1^{*2} y^* - w_7 N_1^{*4} \right.$$

$$\left. \left. + w_1 w_5 R^2 N_2^{*2} + (w_5 \theta - w_6) R^3 N_2^{*2} i^* \right) \right]$$

$$M_4^* = \frac{w_3 R^2 i^* y^{*2}}{N_1^{*6} N_2^{*2}} \left[\left(w_3 \theta R^2 N_2^{*2} i^* + w_7 N_2^{*2} \right) \right.$$

$$\left. \cdot \left(w_1 w_6 N_1^* + (w_6 - \theta w_5) R N_1^* s^* + \theta w_7 y^* \right) \right]$$

Keeping the above in view, it is well known that from Routh-Hurwitz criterion (33) has three roots with negative real parts provided that $A_1 > 0$, $A_3 > 0$ and $\Delta > 0$. Consequently the following theorem can be proved easily.

Theorem 2. *The coexistence equilibrium point E_4 of system (3) is locally asymptotically stable in the interior of positive cone provided that the following sufficient conditions hold:*

$$\max\left\{(1+w_2)\,w_5, \frac{\theta w_5 Rs^*}{w_1 + Rs^*}\right\} < w_6 < w_5\theta \tag{39a}$$

$$\max\left\{\frac{(1+w_2)\left(w_2 N_1^{*2} + w_3 R^2 y^*\right)}{\theta\,(w_2 + w_3)}, \right.$$

$$\left. RN_1^*\left(s^* + (1+w_2)\,i^*\right)\right\} < R^2 y^* < \frac{(1+w_2)}{\theta} \tag{39b}$$

$$\cdot N_1^{*2}$$

$$R^2 N_2^{*2} s^* y^* + \theta w_3 R^2 N_2^{*2} i^* y^* + w_7 N_1^{*2} y^*$$

$$< N_1^{*2} N_2^{*2} s^* \tag{39c}$$

$$w_1 w_2 w_6 N_1^* + w_6 R\left(w_2 N_1^* s^* + w_3 Ry^*\right)$$

$$> w_5 \theta R\left(w_2 N_1^* s^* + w_3 Ry^*\right) \tag{39d}$$

$$w_7 R^2 N_1^{*2} y^* + w_1 w_5 R^2 N_2^{*2} + (w_5\theta - w_6)\,R^3 N_2^{*2} i^*$$

$$> w_7 N_1^{*4} \tag{39e}$$

$$M_2^* + M_3^* > (M_1^* + M_4^*) \tag{39f}$$

Now before we start studying the global stability of the system, the persistence of system (3), which represents biologically the coexistence of all the species for all the time while mathematically it indicates that the solution of system (3) for all $t > 0$ does not have omega limit set on the boundary planes, is investigated in the following theorem when there are no periodic dynamics on the boundary planes.

Clearly system (3) has two subsystems. The first subsystem exists in case of absence of predator species, while the second subsystem exists in case of absence of disease. These subsystems can be written, respectively, as follows:

$$\frac{ds}{dt} = s\left[1 - (s+i) - w_2 i\right] = f_{11}\,(s,i)$$
$$\tag{40}$$
$$\frac{di}{dt} = i\left[w_2 s - w_4\right] = f_{12}\,(s,i)$$

$$\frac{ds}{dt} = s\left[1 - s - \frac{Ry}{w_1 + Rs}\right] = f_{21}\,(s,y)$$
$$\tag{41}$$
$$\frac{dy}{dt} = y\left[\frac{w_5 Rs}{w_1 + Rs} - \frac{w_7}{w_8 + y} - w_9\right] = f_{23}\,(s,y)$$

Note that it is easy to verify that these two subsystems have unique positive equilibrium points in the interior of their positive domains $\Sigma_1 = \{(s,i) \in \mathbb{R}^2 : s(0) \geq 0, i(0) \geq 0\}$ and $\Sigma_2 = \{(s,y) \in \mathbb{R}^2 : s(0) \geq 0, y(0) \geq 0\}$, respectively. These two positive equilibrium points are given by $e_1 = (\widehat{s}, \widehat{i})$ and $e_2 = (\overline{s}, \overline{y})$ for subsystems (40) and (41), respectively. In

fact they coincide with the predator free equilibrium point E_3 and disease free equilibrium point E_2 of system (3), respectively, and have the same existence conditions.

Therefore according to the Bendixson–Dulac theorem on dynamical system (40) "if there exists a C^1 function $\varphi(s,i)$, called the Dulac function, such that the expression $(\partial/\partial s)(\varphi f_{11}) + (\partial/\partial i)(\varphi f_{12})$ has the same sign and is not identically zero ($\neq 0$) almost everywhere in the simply connected region of the plane, then system (40) has no nonconstant periodic dynamic lying entirely in the Σ_1. Thus by choosing $\varphi(s,i) = 1/si$ we obtain that $(\partial/\partial s)(\varphi f_{11}) + (\partial/\partial i)(\varphi f_{12}) = -1/i < 0$. Therefore subsystem (40) has no periodic dynamic lying entirely in interior Σ_1 and hence $e_1 \equiv E_3$ is a globally asymptotically stable in the interior of Σ_1. Similarly by choosing $\varphi(s,i) = 1/sy$ we can show that the positive equilibrium point of the second subsystem (41) given by $e_2 \equiv E_2$ is globally asymptotically stable in the interior of the Σ_2 provided that one of the following two conditions holds.

$$\frac{R^2}{(w_1 + Rs)^2} + \frac{w_7}{s\,(w_8 + y)^2} > \frac{1}{y} \tag{42}$$

$$\frac{R^2}{(w_1 + Rs)^2} + \frac{w_7}{s\,(w_8 + y)^2} < \frac{1}{y} \tag{43}$$

Theorem 3. *Assume that condition (42) or (43) holds. Then system (3) is persistent provided that*

$$w_2 > w_4 \tag{44a}$$

$$\frac{w_5 R}{w_1 + R} > \frac{w_7}{w_8} + w_9 \tag{44b}$$

$$w_2 \overline{s} > \frac{w_3 R\overline{y}}{w_1 + R\overline{s}} + w_4 \tag{44c}$$

$$\frac{w_5 R\widehat{s} + w_6 R\widehat{i}}{w_1 + R\widehat{s} + \theta R\widehat{i}} > \frac{w_7}{w_8} + w_9 \tag{44d}$$

Proof. Consider the following function $\varphi(s,i,y) = s^\alpha i^\beta y^\gamma$, where α, β, and γ are positive constants; clearly $\varphi(s,i,y) > 0$ for all $(s,i,y) \in \text{Int}\,\mathbb{R}_+^3$ and $\varphi(s,i,y) \longrightarrow 0$ when s, i or $y \longrightarrow 0$. Furthermore, it is clear that

$$\frac{\varphi'}{\varphi} = \alpha\left[1 - (s+i) - \frac{Ry}{w_1 + Rs + \theta Ri} - w_2 i\right]$$

$$+ \beta\left[w_2 s - \frac{w_3 Ry}{w_1 + Rs + \theta Ri} - w_4\right] \tag{45}$$

$$+ \gamma\left[\frac{w_5 Rs + w_6 Ri}{w_1 + Rs + \theta Ri} - \frac{w_7}{w_8 + y} - w_9\right]$$

Now the proof follows if $\varphi'/\varphi > 0$ for all the boundary equilibrium points, for suitable choice of constants $\alpha > 0$, $\beta > 0$, and $\gamma > 0$.

$$\frac{\varphi'}{\varphi}(E_0) = \alpha - w_4\beta - \left[\frac{w_7}{w_8} + w_9\right]\gamma$$

$$\frac{\varphi'}{\varphi}(E_1) = [w_2 - w_4]\beta + \left[\frac{w_5 R}{w_1 + R} - \frac{w_7}{w_8} - w_9\right]\gamma$$

$$\frac{\varphi'}{\varphi}(E_2) = \left[w_2\bar{s} - \frac{w_3 R\bar{y}}{w_1 + R\bar{s}} - w_4\right]\beta$$
(46)

$$\frac{\varphi'}{\varphi}(E_3) = \left[\frac{w_5 R\hat{s} + w_6 R\hat{i}}{w_1 + R\hat{s} + \theta R\hat{i}} - \frac{w_7}{w_8} - w_9\right]\gamma$$

Clearly $(\varphi'/\varphi)(E_0) > 0$ for suitable choice of positive constant α sufficiently large with respect to the constants $\beta > 0$ and $\gamma > 0$, while $(\varphi'/\varphi)(E_1)$ is positive under conditions (44a) and (44b). Moreover $(\varphi'/\varphi)(E_2) > 0$ under condition (44c), while $(\varphi'/\varphi)(E_3)$ is positive too under condition (44d). Hence the proof is complete. □

4. Global Stability

In this section the global stability of each equilibrium point of system (3) is studied using suitable Lyapunov function as given in the following theorems.

Theorem 4. *Assume that the axial equilibrium point $E_1 = (1,0,0) = (s_0, 0, 0)$ is locally asymptotically stable in $\mathbb{R}_+^3$. Then, it is globally asymptotically stable, provided that the following conditions are outstanding:*

$$\frac{w_6}{w_3}w_2 < w_5(1 + w_2) < \frac{w_6}{w_3}w_4$$
(47a)

$$w_9 > w_5\frac{R}{w_1}$$
(47b)

Proof. Recognize the following function:

$$V_1 = c_1\int_{s_0}^{s}\frac{\tau - s_0}{\tau}d\tau + c_2 i + c_3 y$$
(48)

where c_i, $i = 1, 2, 3$, are positive constants to be determined later on. Clearly $V_1(s, i, y) > 0$ is a continuously differentiable real valued function for all $(s, i, y) \in \mathbb{R}_+^3$ with $(s, i, y) \neq (s_0, 0, 0)$ and $V_1(s_0, 0, 0) = 0$. Moreover we have that

$$\frac{dV_1}{dt} = c_1\left(\frac{s - s_0}{s}\right)\frac{ds}{dt} + c_2\frac{di}{dt} + c_3\frac{dy}{dt}$$

$$= -c_1(s - s_0)^2 - c_1\frac{Rsy}{N_1} + c_1\frac{Rs_0 y}{N_1}$$

$$- c_1(1 + w_2)si + c_1(1 + w_2)s_0 i + c_2 w_2 si$$
(49)

$$- c_2\frac{w_3 Riy}{N_1} - c_2 w_4 i + c_3\frac{w_5 Rs + w_6 Ri}{N_1}y$$

$$- \frac{c_3 w_7 y}{w_8 + y} - c_3 w_9 y$$

Here $R = (1 - m)$, $N_1 = w_1 + Rs + \theta Ri$. Further simplification gives that

$$\frac{dV_1}{dt} \le -c_1(s - s_0)^2 - [c_1 - c_3 w_5]\frac{Rsy}{N_1}$$

$$- [c_1(1 + w_2) - c_2 w_2]si - [c_2 w_3 - c_3 w_6]\frac{Riy}{N_1}$$
(50)

$$- [c_2 w_4 - c_1(1 + w_2)s_0]i$$

$$- \left[c_3 w_9 - c_1\frac{Rs_0}{w_1}\right]y$$

So by choosing the positive constants c_i, $i = 1, 2, 3$, as given below, we get

$$c_1 = w_5;$$

$$c_2 = \frac{w_6}{w_3};$$

$$c_3 = 1$$

$$\frac{dV_1}{dt} \le -w_5(s - 1)^2 - \left[w_5(1 + w_2) - \frac{w_6}{w_3}w_2\right]si$$
(51)

$$- \left[\frac{w_6}{w_3}w_4 - w_5(1 + w_2)\right]i$$

$$- \left[w_9 - w_5\frac{R}{w_1}\right]y$$

Clearly conditions (47a)-(47b) guarantee that $dV_1/dt < 0$; hence dV_1/dt is negative definite, and hence the axial equilibrium point E_1 is globally asymptotically stable and the proof is complete. □

Theorem 5. *Assume that the disease free equilibrium point $E_2 = (\bar{s}, 0, \bar{y})$ is locally asymptotically stable in $\mathbb{R}_+^3$. Then, it is a globally asymptotically stable, provided that the following conditions hold:*

$$w_4 w_1(1 + w_2)\overline{N}_1 > w_4 R^2\theta\bar{y} + w_1 w_2(1 + w_2)\overline{N}_1\bar{s}$$
(52a)

$$w_5\theta\left(\frac{R\bar{s}}{\overline{N}_1}\right)^2 < w_6$$
(52b)

$$< \frac{w_1 w_3 w_5(1 + w_2)\bar{s} + w_4 w_5\theta R\bar{s}}{w_4\overline{N}_1}$$

$$w_1\overline{N}_1 > R^2\bar{y}$$
(52c)

$$\frac{w_7\overline{N}_1}{w_1 w_5 w_8\overline{N}_2}(y - \bar{y})^2 < \left[1 - \frac{R^2\bar{y}}{w_1\overline{N}_1}\right](s - \bar{s})^2$$
(52d)

where $\overline{N}_1 = w_1 + R\bar{s}$.

Proof. Consider the function

$$V_2 = \bar{c}_1\int_{\bar{s}}^{s}\frac{\tau - \bar{s}}{\tau}d\tau + \bar{c}_2 i + \bar{c}_3\int_{\bar{y}}^{y}\frac{\tau - \bar{y}}{\tau}d\tau$$
(53)

where $\bar{c}_i$, $i = 1, 2, 3$, are positive constants to be determined later on. Clearly $V_2(s, i, y) > 0$ is a continuously differentiable real valued function for all $(s, i, y) \in \mathbb{R}_+^3$ with $(s, i, y) \neq (\bar{s}, 0, \bar{y})$ and $V_2(\bar{s}, 0, \bar{y}) = 0$. Moreover we have that

$$\frac{dV_2}{dt} = \bar{c}_1 (s - \bar{s}) \left[1 - (s + i) - \frac{Ry}{N_1} - w_2 i \right]$$

$$+ \bar{c}_2 i \left[w_2 s - \frac{w_3 Ry}{N_1} - w_4 \right] \tag{54}$$

$$+ \bar{c}_3 (y - \bar{y}) \left[\frac{w_5 Rs + w_6 Ri}{N_1} - \frac{w_7}{w_8 + y} - w_9 \right]$$

Further simplification gives

$$\frac{dV_2}{dt} = \bar{c}_1 (s - \bar{s}) \left[-(s - \bar{s}) - (1 + w_2) i - \frac{R}{N_1} (y - \bar{y}) \right.$$

$$+ \frac{R^2 \bar{y}}{N_1 \overline{N}_1} (s - \bar{s}) + \frac{R^2 \theta \bar{y}}{N_1 \overline{N}_1} i \right] + w_2 \bar{c}_2 i s - \frac{w_3 R \bar{c}_2}{N_1} i y$$

$$- w_4 \bar{c}_2 i + \bar{c}_3 (y - \bar{y}) \left[\frac{w_5 Rs}{N_1} - \frac{w_5 R\bar{s}}{\overline{N}_1} + \frac{w_6 Ri}{N_1} \right. \tag{55}$$

$$\left. + \frac{w_7}{N_2 \overline{N}_2} (y - \bar{y}) \right]$$

Here $N_2 = w_8 + y$ and $\overline{N}_2 = w_8 + \bar{y}$.

Now rearranging the terms of the last equation further gives

$$\frac{dV_2}{dt} \leq -\bar{c}_1 \left[1 - \frac{R^2 \bar{y}}{w_1 \overline{N}_1} \right] (s - \bar{s})^2$$

$$- \frac{R}{N_1} \left(\bar{c}_1 - \bar{c}_3 \frac{w_1 w_5}{\overline{N}_1} \right) (s - \bar{s})(y - \bar{y})$$

$$+ \frac{\bar{c}_3 w_7}{w_8 \overline{N}_2} (y - \bar{y})^2$$

$$- \left[\bar{c}_1 \left((1 + w_2) - \frac{R^2 \theta \bar{y}}{w_1 \overline{N}_1} \right) - \bar{c}_2 w_2 \right] si \tag{56}$$

$$- \frac{R}{N_1} \left[\bar{c}_2 w_3 + \bar{c}_3 \left(\frac{w_5 \theta R \bar{s}}{\overline{N}_1} - w_6 \right) \right] iy$$

$$- \frac{R}{N_1} \left[\bar{c}_1 \frac{R \theta \bar{s} y}{\overline{N}_1} - \bar{c}_3 \left(\frac{w_5 R \theta \bar{s} y}{\overline{N}_1} - w_6 \bar{y} \right) \right] i$$

$$- (\bar{c}_2 w_4 - \bar{c}_1 (1 + w_2) \bar{s}) i$$

So by choosing the positive constants as

$$\bar{c}_1 = 1,$$

$$\bar{c}_2 = \frac{(1 + w_2) \bar{s}}{w_4}, \tag{57}$$

$$\bar{c}_3 = \frac{\overline{N}_1}{w_1 w_5}$$

we get that

$$\frac{dV_2}{dt} \leq - \left[1 - \frac{R^2 \bar{y}}{w_1 \overline{N}_1} \right] (s - \bar{s})^2 + \frac{w_7 \overline{N}_1}{w_1 w_5 w_8 \overline{N}_2} (y - \bar{y})^2$$

$$- \frac{R \bar{y}}{w_1 N_1} \left[\frac{w_6 \overline{N}_1^2 - w_5 \theta (R \bar{s})^2}{w_5 \overline{N}_1} \right] i$$

$$- \left[\frac{w_4 \left[w_1 (1 + w_2) \overline{N}_1 - R^2 \theta \bar{y} \right] - w_1 w_2 (1 + w_2) \overline{N}_1 \bar{s}}{w_1 w_4 \overline{N}_1} \right] \tag{58}$$

$$\cdot si - \frac{R}{N_1} \left[\frac{w_1 w_3 w_5 (1 + w_2) \bar{s} + w_4 w_5 \theta R \bar{s} - w_4 w_6 \overline{N}_1}{w_1 w_4 w_5} \right]$$

$$\cdot iy$$

Accordingly, using the given conditions (52a)–(52d) we obtain

$$\frac{dV_2}{dt} \leq - \left[1 - \frac{R^2 \bar{y}}{w_1 \overline{N}_1} \right] (s - \bar{s})^2$$

$$+ \frac{w_7 \overline{N}_1}{w_1 w_5 w_8 \overline{N}_2} (y - \bar{y})^2 \tag{59}$$

$$- \frac{R \bar{y}}{w_1 N_1} \left[\frac{w_6 \overline{N}_1^2 - w_5 \theta (R \bar{s})^2}{w_5 \overline{N}_1} \right] i$$

Clearly $dV_2/dt < 0$ is negative definite, and hence the disease free equilibrium point E_2 is globally asymptotically stable under the given conditions and hence the proof is complete. $\square$

Theorem 6. *Assume that the free predator equilibrium point $E_3 = (\hat{s}, \hat{i}, 0)$ is locally asymptotically stable in $\mathbb{R}_+^3$. Then it is globally asymptotically stable, provided that the following condition holds:*

$$\frac{w_9}{w_5} > \frac{R \left[w_2 \hat{s} + w_3 (1 + w_2) \hat{i} \right]}{w_1 w_2} \tag{60a}$$

$$w_3 w_5 (1 + w_2) > w_2 w_6 \tag{60b}$$

Proof. Consider the next function

$$V_3 = \hat{c}_1 \int_{\hat{s}}^s \frac{\tau - \hat{s}}{\tau} d\tau + \hat{c}_2 \int_{\hat{i}}^i \frac{\tau - \hat{i}}{\tau} d\tau + \hat{c}_3 y \tag{61}$$

where $\hat{c}_i$, $i = 1, 2, 3$, are positive constants to be determined later on. Clearly $V_3(s, i, y) > 0$ is a continuously differentiable real valued function for all $(s, i, y) \in \mathbb{R}_+^3$ with $(s, i, y) \neq (\hat{s}, \hat{i}, 0)$ and $V_3(\hat{s}, \hat{i}, 0) = 0$. Moreover we have that

$$\frac{dV_3}{dt} = \hat{c}_1 (s - \hat{s}) \left[1 - (s + i) - \frac{Ry}{N_1} - w_2 i \right]$$

$$+ \hat{c}_2 (i - \hat{i}) \left[w_2 s - \frac{w_3 Ry}{N_1} - w_4 \right] \tag{62}$$

$$+ \hat{c}_3 y \left[\frac{w_5 Rs + w_6 Ri}{N_1} - \frac{w_7}{N_2} - w_9 \right]$$

Here R, N_1, and N_2 are given above. Now straightforward computations give

$$\frac{dV_3}{dt} \leq -\hat{c}_1 (s - \hat{s})^2$$

$$- [\hat{c}_1 (1 + w_2) - \hat{c}_2 w_2] (i - \hat{i})(s - \hat{s}) - \hat{c}_3 \frac{w_7}{N_2} y$$

$$- \left[\hat{c}_3 w_9 - \hat{c}_2 \frac{w_3 R}{w_1} \hat{i} - \hat{c}_1 \frac{R\hat{s}}{w_1} \right] y$$

$$- \frac{R}{N_1} [\hat{c}_1 - \hat{c}_3 w_5] sy - \frac{R}{N_1} [\hat{c}_2 w_3 - \hat{c}_3 w_6] iy$$

(63)

By choosing the positive constants as

$$\hat{c}_1 = 1,$$

$$\hat{c}_2 = \frac{(1 + w_2)}{w_2},$$

(64)

$$\hat{c}_3 = \frac{1}{w_5}$$

then we obtain that

$$\frac{dV_3}{dt} \leq - (s - \hat{s})^2$$

$$- \left[\frac{w_9}{w_5} - \frac{R \left(w_2 \hat{s} + w_3 (1 + w_2) \hat{i} \right)}{w_1 w_2} \right] y$$

(65)

$$- \frac{R}{N_1} \left[\frac{w_3 w_5 (1 + w_2) - w_2 w_6}{w_2 w_5} \right] iy$$

Accordingly, using the given conditions (60a) and (60b) we obtain

$$\frac{dV_3}{dt} \leq - (s - \hat{s})^2$$

$$- \left[\frac{w_9}{w_5} - \frac{R \left(w_2 \hat{s} + w_3 (1 + w_2) \hat{i} \right)}{w_1 w_2} \right] y$$

(66)

Clearly $dV_3/dt \leq 0$, which means it is negative semi-definite, and hence the predator free equilibrium point E_3 is globally stable (but not asymptotically stable) under the given conditions. Moreover, since system (3) has the maximum invariant set for $dV_3/dt = 0$ if and only if conditions (60a)-(60b) hold and $(s, i, y) = (\hat{s}, \hat{i}, 0)$, by Lyapunov-Lasalle's theorem, all the solutions starting in $\mathbb{R}^3_+$ approach the singleton set $\{E_3\}$, which is the positively invariant subset of the set where $dV_3/dt = 0$. Hence E_3 becomes attracting too; hence it is globally asymptotically stable and that completes the proof. □

Theorem 7. *Assume that the coexistence equilibrium point $E_4 = (s^*, i^*, y^*)$ of system (3) is locally asymptotically stable*

in $\mathbb{R}^3_+$. Then it is globally asymptotically stable, provided that the following conditions hold:

$$R^2 y^* < \min \left\{ w_1 N_1^*, \frac{(1 + w_2) w_1 N_1^*}{\theta} \right\}$$

(67a)

$$\frac{\theta w_5 Rs^*}{(w_1 + Rs^*)} < w_6 < \frac{w_5 (w_1 + \theta Ri^*)}{Ri^*}$$

(67b)

$$\Lambda_1$$
$$= c_1^* \left[(1 + w_2) N_1 N_1^* - \theta R^2 y^* \right]$$

(67c)
$$- c_2^* \left[w_2 N_1 N_1^* + w_3 R^2 y^* \right] > 0$$

$$2\Lambda_2 - \Lambda_1 > 0$$

(67d)

$$\left[\frac{2\Lambda_3 + \Lambda_1}{N_1 N_1^*} \right] (i - i^*)^2 + \frac{w_7}{N_2 N_2^*} (y - y^*)^2$$
$$< \left[\frac{2\Lambda_2 - \Lambda_1}{N_1 N_1^*} \right] (s - s^*)^2$$

(67e)

Here Λ_i, $i = 1, 2, 3$, c_1^*, and c_2^* are given in the proof.

Proof. Consider the real valued function

$$V_4 = c_1^* \int_{s^*}^{s} \frac{\tau - s^*}{\tau} d\tau + c_2^* \int_{i^*}^{i} \frac{\tau - i^*}{\tau} d\tau$$
$$+ c_3^* \int_{y^*}^{y} \frac{\tau - y^*}{\tau} d\tau.$$

(68)

Here c_i^*, $i = 1, 2, 3$, are positive constants to be determined. Clearly $V_4(s, i, y) > 0$ is a continuously differentiable real valued function for all $(s, i, y) \in \mathbb{R}^3_+$ with $(s, i, y) \neq (s^*, i^*, y^*)$ and $V_3(s^*, i^*, y^*) = 0$. Moreover we have that

$$\frac{dV_4}{dt} = c_1^* (s - s^*)$$

$$\cdot \left[-(s - s^*) - (1 + w_2)(i - i^*) - \frac{Ry}{N_1} + \frac{Ry^*}{N_1^*} \right]$$

(69)
$$+ c_2^* (i - i^*) \left[w_2 (s - s^*) - \frac{w_3 Ry}{N_1} + \frac{w_3 Ry^*}{N_1^*} \right]$$

$$+ c_3^* (y - y^*) \left[\frac{N_3}{N_1} - \frac{N_3^*}{N_1^*} - \frac{w_7}{N_2} + \frac{w_7}{N_2^*} \right]$$

Here R, N_1, and N_2 are given above, while $N_1^* = w_1 + Rs^* + \theta Ri^*$, $N_2^* = w_8 + y^*$, $N_3 = w_5 Rs + w_6 Ri$, and $N_3^* = w_5 Rs^* + w_6 Ri^*$. Further simplification for the above equation leads to

$$\frac{dV_4}{dt} \leq -c_1^* \left(\frac{w_1 N_1^* - R^2 y^*}{w_1 N_1^*} \right) (s - s^*)^2 + c_2^*$$

$$\cdot \frac{\theta w_3 R^2 y^*}{w_1 N_1^*} (i - i^*)^2 + c_3^* \frac{w_7}{w_8 N_2^*} (y - y^*)^2 - (s$$

$$- s^*)(i - i^*) \left[c_1^* \left(\frac{(w_2 + 1) N_1 N_1^* - \theta R^2 y^*}{N_1 N_1^*} \right) \right.$$

$$\left. - c_2^* \left(\frac{w_2 N_1 N_1^* + w_3 R^2 y^*}{N_1 N_1^*} \right) \right] - (s - s^*)(y - y^*)$$

$$\cdot \frac{R}{N_1} \left[c_1^* - \frac{c_3^*}{N_1^*} (w_1 w_5 + \theta w_5 Ri^* - w_6 Ri^*) \right] - (i$$

$$- i^*)(y - y^*) \frac{R}{N_1} \left[c_2^* w_3 \right.$$

$$\left. - \frac{c_3^*}{N_1^*} (w_1 w_6 - \theta w_5 Rs^* + w_6 Rs^*) \right]$$

$$(70)$$

Consequently, by choosing the positive constants c_i^*, $i = 1, 2, 3$, as

$$c_1^* = \frac{w_1 w_5 + \theta w_5 Ri^* - w_6 Ri^*}{N_1^*};$$

$$c_2^* = \frac{w_1 w_6 - \theta w_5 Rs^* + w_6 Rs^*}{w_3 N_1^*}$$

$$(71)$$

and $c_3^* = 1$

the following is obtained:

$$\frac{dV_4}{dt} = -\frac{\Lambda_2}{N_1 N_1^*} (s - s^*)^2 + \frac{\Lambda_3}{N_1 N_1^*} (i - i^*)^2$$

$$+ \frac{w_7}{N_2 N_2^*} (y - y^*)^2 - \frac{\Lambda_1}{N_1 N_1^*} (s - s^*)(i - i^*)$$

$$(72)$$

where $\Lambda_2 = c_1^* (N_1 N_1^* - R^2 y^*) > 0$ and $\Lambda_3 = c_2^* \theta w_3 R^2 y^* > 0$. Further simplification for the last equation gives

$$\frac{dV_4}{dt} = - \left[\frac{2\Lambda_2 - \Lambda_1}{N_1 N_1^*} \right] (s - s^*)^2$$

$$+ \left[\frac{2\Lambda_3 + \Lambda_1}{N_1 N_1^*} \right] (i - i^*)^2 + \frac{w_7}{N_2 N_2^*} (y - y^*)^2$$

$$- \left[\sqrt{\frac{\Lambda_1}{2N_1 N_1^*}} (s - s^*) + \sqrt{\frac{\Lambda_1}{2N_1 N_1^*}} (i - i^*) \right]^2$$

$$(73)$$

Clearly, $dV_4/dt < 0$ is negative definite function under the given conditions and hence the coexistence equilibrium point $E_4 = (s^*, i^*, y^*)$ is globally asymptotically stable and this completes the proof. □

5. Numerical Simulation

In this section the global dynamics of system (3) is studied numerically to verify the obtained analytical results in addition to specifying the control set of parameters. For the following hypothetical set of parameters, system (3) is solved

numerically and the obtained trajectories are drawn in the form of phase portrait and time series.

$$w_1 = 0.4,$$

$$w_2 = 0.5,$$

$$w_3 = 1,$$

$$w_4 = 0.1,$$

$$w_5 = 0.4$$

$$w_6 = 0.7, \qquad (74)$$

$$w_7 = 0.2,$$

$$w_8 = 1,$$

$$w_9 = 0.1,$$

$$m = 0.4,$$

$$\theta = 1$$

It is observed that, for the set of data (74), system (3) has a globally asymptotically stable positive equilibrium point $E_4 = (0.45, 0.28, 0.17)$ as shown in Figure 1.

Now in order to discover the impact of varying the parameters values on the dynamics of system (3), the system is solved numerically with varying one parameter each time and then the attractors of the obtained trajectories are present in the form of figures as shown in Figures 2–7.

It is observed that for the set of data (74) with $w_1 \geq 0.47$ the trajectory of system (3) approaches asymptotically predator free equilibrium point, while it approaches disease free point when $0.15 < w_1 \leq 0.25$; see Figure 2. Further the trajectory of system (3) approaches periodic dynamics in the sy −plane for data (74) with $w_1 \leq 0.15$ as shown in Figure 3; however it approaches asymptotically the positive equilibrium point otherwise.

On the other hand, for the data (74) with $w_2 \geq 0.88$ and $0.1 < w_2 \leq 0.17$, the solution of system (3) approaches asymptotically $E_3 = (\hat{s}, \hat{i}, 0)$ as shown in Figures 4(a)-4(b), while it approaches asymptotically $E_1 = (1, 0, 0)$ when $w_2 \leq 0.1$ as shown in Figure 4(c). Otherwise the system still has a globally asymptotically stable positive equilibrium point. Note that although it looks confusing as the trajectory of system (3) approaches the predator free equilibrium point E_3 too with decreasing the infection rate w_2, that depends on our hypothetical set of data in which we assumed that the conversion rate of predator from susceptible (w_5) is less than that from infected species (w_6) and once (w_5) enter the range $w_5 \geq 0.47$ the system approaches disease free point $E_2 = (\bar{s}, 0, \bar{y})$ as shown in the typical figure given by Figure 4(d).

Moreover it is observed that varying the parameter w_3 with the rest of parameters as in (74) has a quantitative effect on the dynamics of system (3) and the solution still approaches a positive equilibrium point that depends on the value of w_3. Now for the parameters values given by (74) with $0.2 \leq w_4 < 0.5$ and $0.5 \leq w_4$ the trajectory of system (3) approaches asymptotically $E_3 = (\hat{s}, \hat{i}, 0)$ and $E_1 = (1, 0, 0)$,

(a)

(b)

(c)

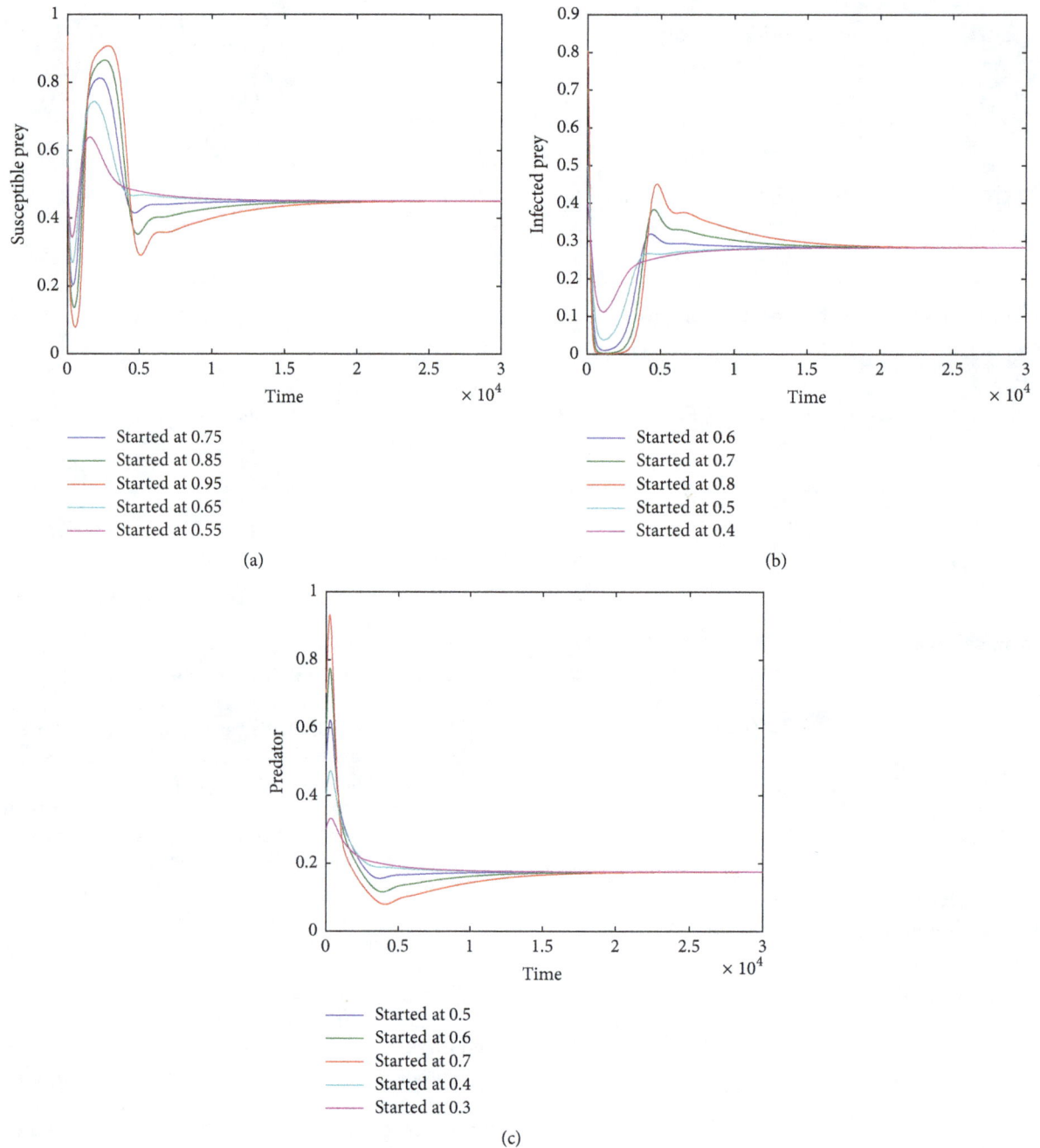

FIGURE 1: Time series for the trajectories of system (3) using data (74) with different sets of initial points. (a) Trajectories of Susceptible prey. (b) Trajectories of infected prey. (c) Trajectories of predator.

respectively, as given in Figure 5. Otherwise the system still is stable at positive equilibrium point. There is similar behavior of the parameter θ in the range $\theta \geq 1.3$ to that of system (3) using data (74) with $0.2 \leq w_4 < 0.5$, while the system still is stable at positive equilibrium point otherwise.

For the parameters (74) with $w_5 \geq 0.48$ (similarly when $w_8 \geq 1.5$) the trajectory of system (3) approaches asymptotically $E_2 = (\bar{s}, 0, \bar{y})$, while it approaches $E_3 = (\hat{s}, \hat{i}, 0)$ for data (74) with $w_5 \leq 0.25$ (similarly when $w_6 \leq 0.65$ or $w_8 \leq 0.8$) as shown in Figure 6. Otherwise the system has a globally asymptotically stable positive point.

Finally for the parameters (74) with $w_7 \geq 0.23$ (similarly when $w_9 \geq 0.13$ or $m \geq 0.49$) the trajectory of system (3) approaches asymptotically $E_3 = (\hat{s}, \hat{i}, 0)$, while it approaches $E_2 = (\bar{s}, 0, \bar{y})$ for data (74) with $w_7 \leq 0.15$ (similarly when $w_9 \leq 0.06$ or $m \leq 0.06$) as shown in Figure 7. Otherwise the system has a globally asymptotically stable positive point.

6. Discussion and Conclusions

The dynamics of a refuged prey-predator system, involving infectious disease in prey species and harvesting from a

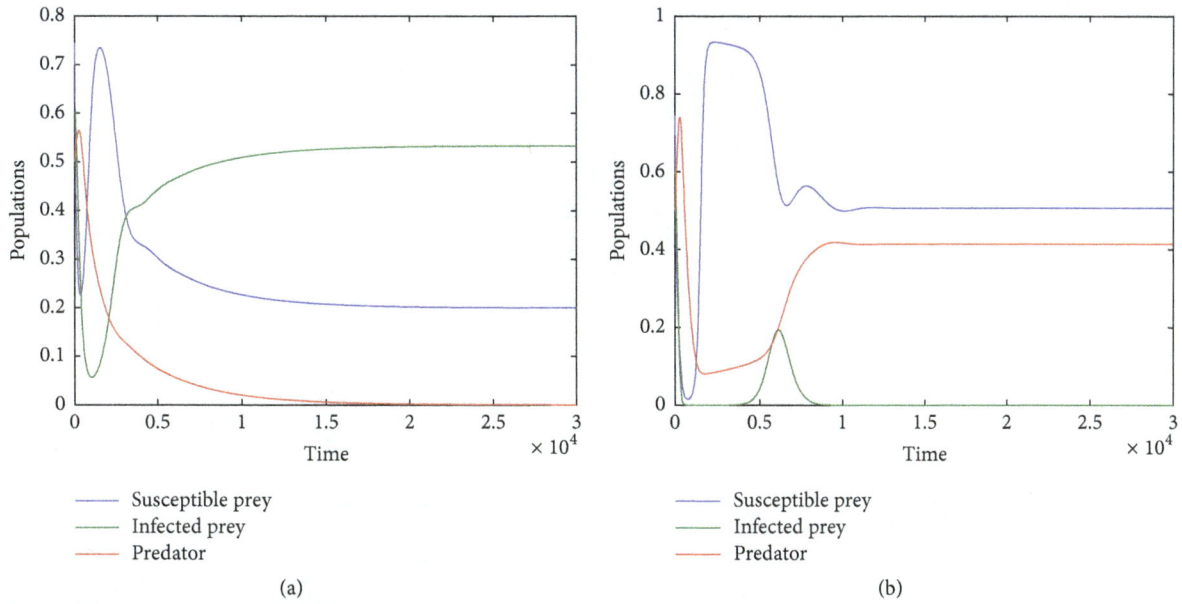

FIGURE 2: Time series for the trajectory of the system (3) using data (74) with typical values of w_1. (a) Trajectories of three species approach asymptotically to $E_3 = (0.2, 0.53, 0)$ when $w_1 = 0.55$. (b) Trajectories of three species approach asymptotically to $E_2 = (0.5, 0, 0.41)$ when $w_1 = 0.2$.

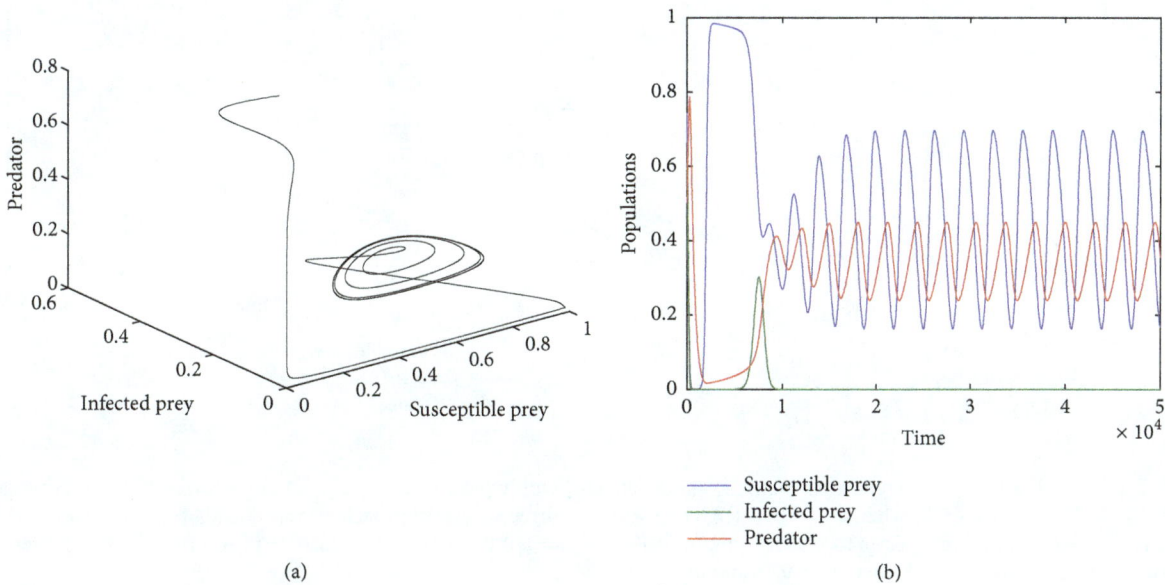

FIGURE 3: System (3) approaches asymptotically to periodic dynamics using data (74) with $w_1 = 0.1$. (a) Periodic attractor in the sy −plane. (b) Trajectories of the three species approach to periodic in the sy −plane.

predator species, is mathematically simulated through a mathematical model consisting of three nonlinear ordinary differential equations of the first order. The existence, uniqueness, and boundedness of the solution of the proposed model are discussed analytically. All feasible equilibrium points are determined and then the local stability analysis for them is carried out. The persistence conditions of the system are established. Suitable Lyapunov functions are used to show the global stability of the system's equilibrium points. Finally the proposed dynamical system is solved numerically in order to

confirm the obtained analytical results and specify the control set of parameters too. It is observed that for the hypothetical set of parameters given by (74) the following results are obtained; different sets of parameters values may be used too.

(1) For the data (74), system (3) has a globally asymptotically stable positive equilibrium point in the interior of $\mathbb{R}^3_+$.

(2) The system has no periodic dynamics lying in the interior of $\mathbb{R}^3_+$; rather it either persists at the positive

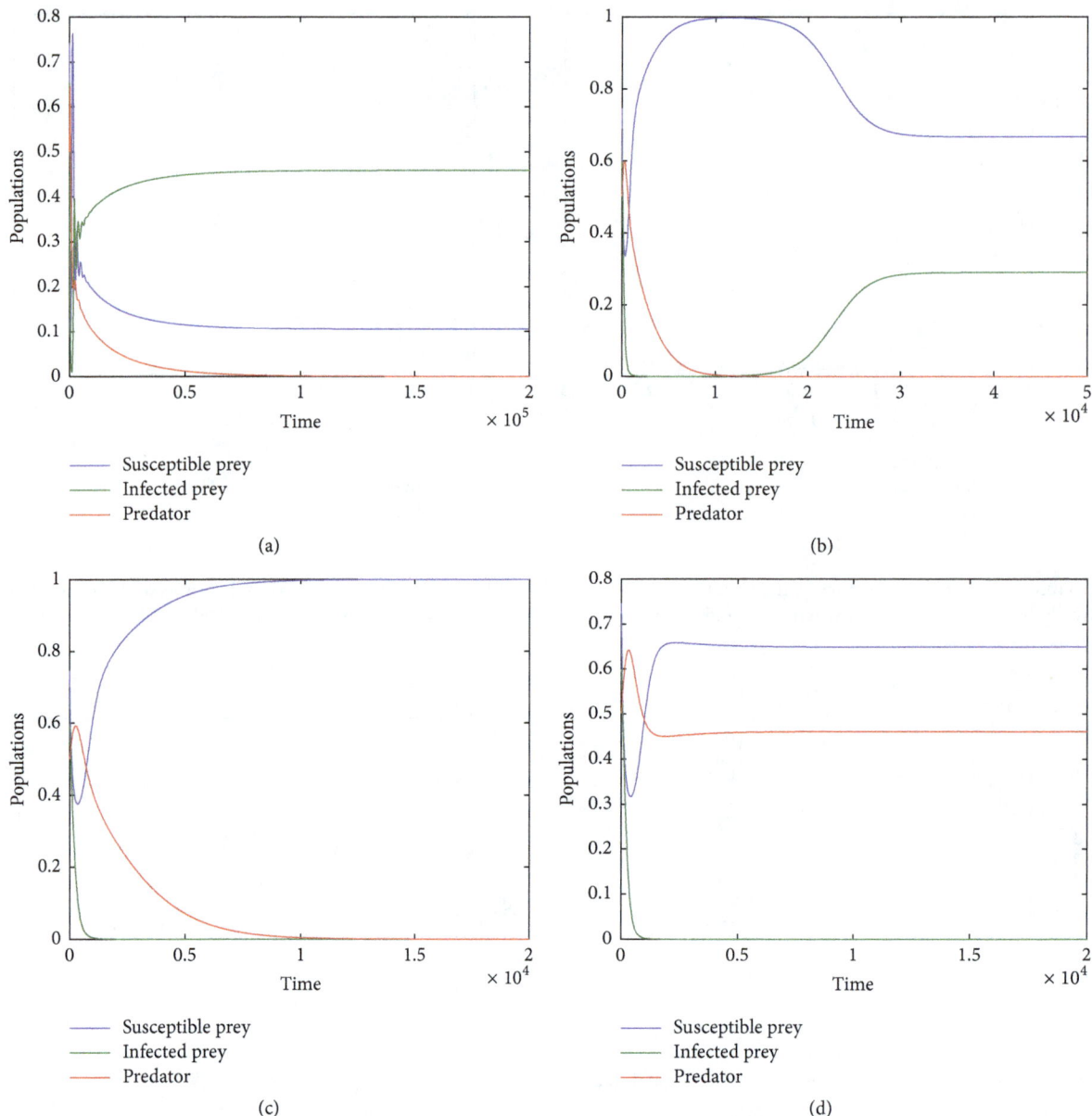

FIGURE 4: Time series for the trajectory of the system (3) using data (74) with typical values of w_2. (a) Trajectories of three species approach asymptotically to $E_3 = (0.1, 0.45, 0)$ when $w_2 = 0.95$. (b) Trajectories of three species approach asymptotically to $E_3 = (0.66, 0.29, 0)$ when $w_2 = 0.15$. (c) Trajectories of three species approach asymptotically to $E_1 = (1, 0, 0)$ when $w_2 = 0.05$. (d) Trajectories of three species approach asymptotically to $E_2 = (0.65, 0, 0.46)$ when $w_2 = 0.15$ and $w_5 = 0.48$.

equilibrium point or else loses its persistence and the system approaches asymptotically a specific attractor in the boundary planes.

(3) Increasing the half saturation constant represented by w_1 above 0.47 leads to losing the persistence of system (3) and the system approaches asymptotically the predator free equilibrium point, while decreasing half saturation constant in the range $0.15 < w_1 \leq 0.25$ makes the solution approaches asymptotically the disease free equilibrium point. Finally further decreasing of w_1 below the value 0.15 leads to losing the stability

of the disease free point and the solution approaches periodic dynamics in the interior of sy −plane.

(4) Increasing the infection rate parameter (w_2) above the value 0.88 leads to losing the persistence of system (3) too and the solution of system (3) approaches asymptotically the predator free equilibrium point. However decreasing the value of this parameter to the range $0,1 < w_2 \leq 0.17$ makes the system approaches asymptotically either disease free equilibrium point or predator free equilibrium point depending on the values of conversion rates from susceptible prey and

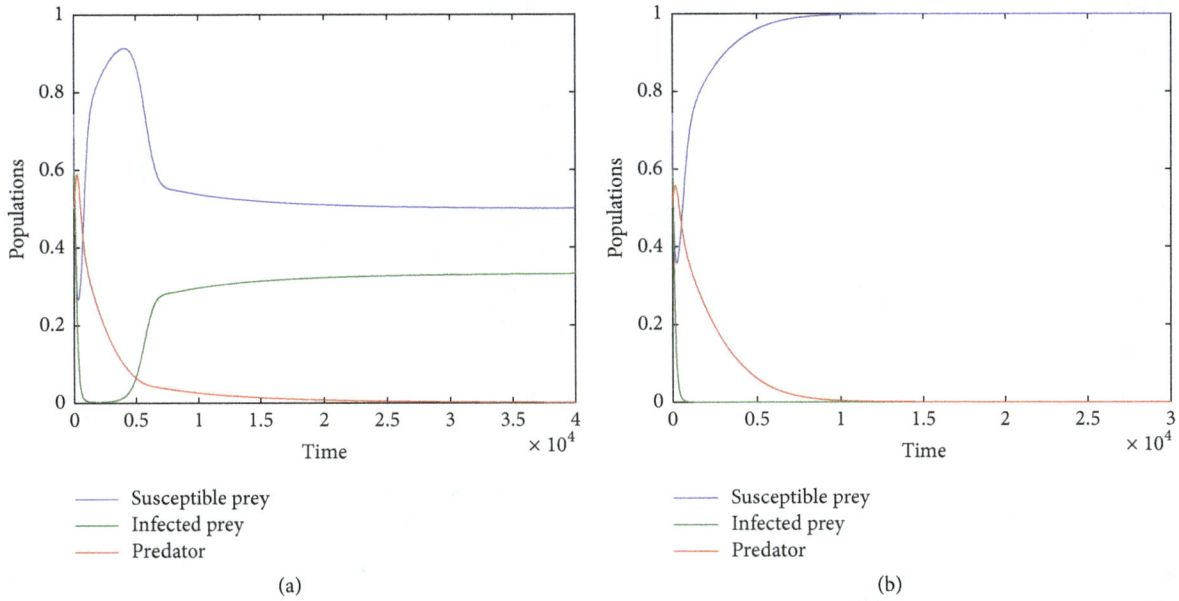

FIGURE 5: Time series for the trajectory of the system (3) using data (74) with typical values of w_4. (a) Trajectories of three species approach asymptotically to $E_3 = (0.5, 0.33, 0)$ when $w_4 = 0.25$. (b) Trajectories of three species approach asymptotically to $E_2 = (1, 0, 0)$ when $w_4 = 0.55$.

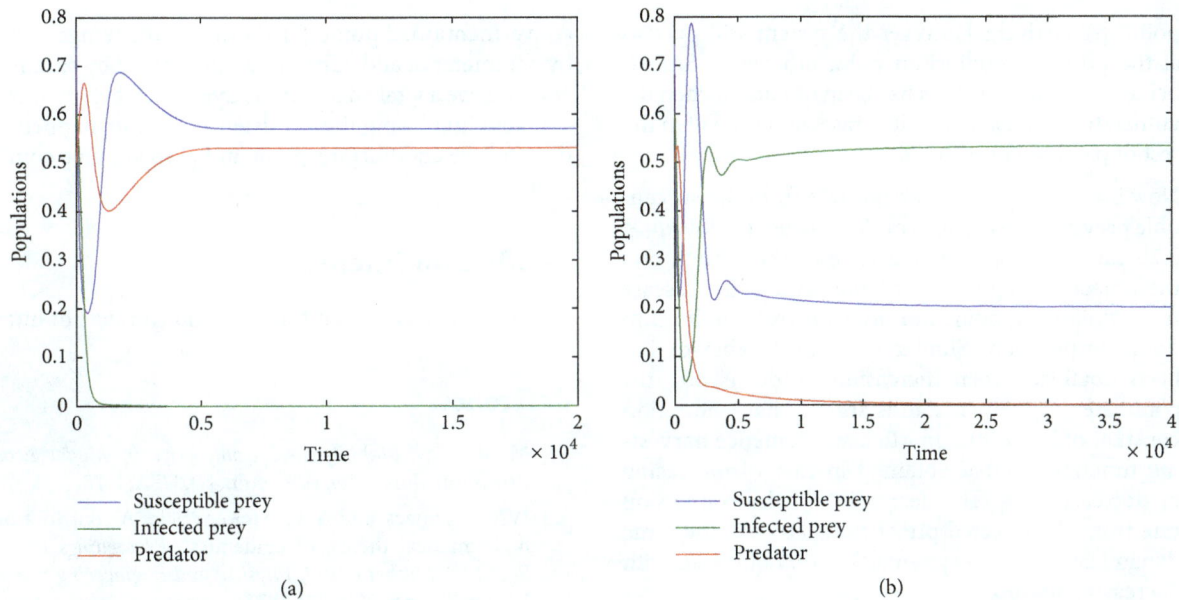

FIGURE 6: Time series for the trajectory of the system (3) using data (74) with typical values of w_5. (a) Trajectories of three species approach asymptotically to $E_2 = (0.57, 0, 0.53)$ when $w_5 = 0.5$. (b) Trajectories of three species approach asymptotically to $E_3 = (0.2, 0.53, 0)$ when $w_5 = 0.15$.

infected prey represented by w_5 and w_6, respectively. Finally decreasing the infection rate further ($w_2 \leq 0.1$) leads to approaches to axial equilibrium point E_1.

(5) It is observed that varying the parameter w_3, which represents the ratio of the predator's attack rate of infected prey to predator's attack rate of susceptible prey, has a quantitative effect on the dynamics of

system (3) and the system still persists at the positive equilibrium point that depends on the value of w_3.

(6) Increasing the death rate of the infected prey (w_4) so that $0.2 \leq w_4 < 0.5$ and $0.5 \leq w_4$ leads to losing the persistence of system (3) and the trajectory of system approaches asymptotically the predator free equilibrium point and axial equilibrium

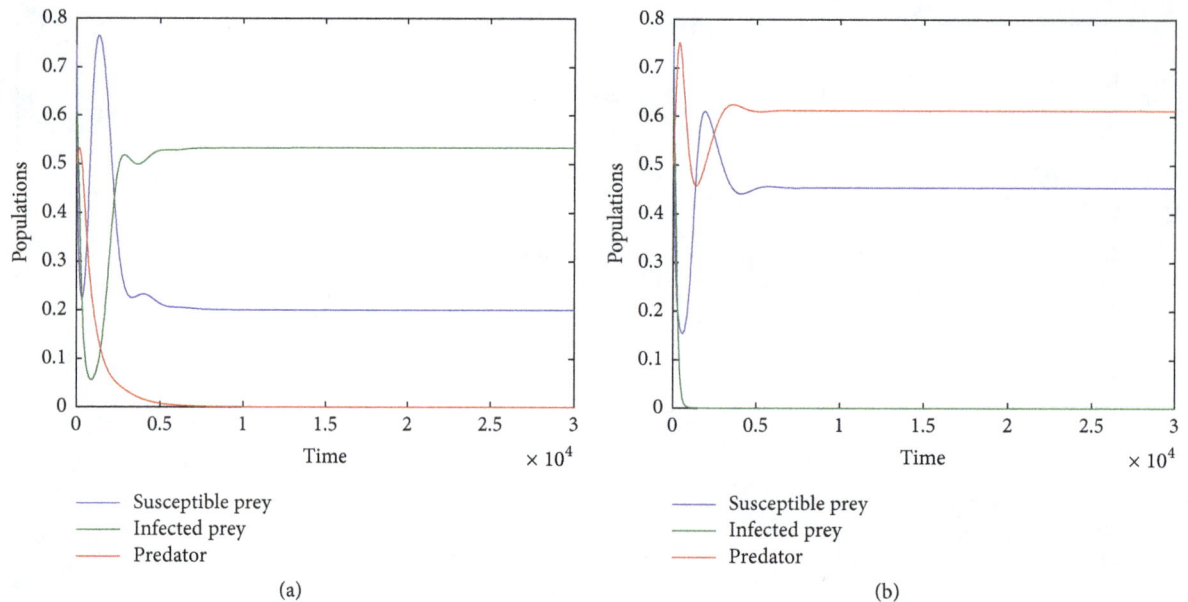

(a)

(b)

FIGURE 7: Time series for the trajectory of the system (3) using data (74) with typical values of w_7. (a) Trajectories of three species approach asymptotically to $E_3 = (0.2, 0.53, 0)$ when $w_7 = 0.3$. (b) Trajectories of three species approach asymptotically to $E_2 = (0.45, 0, 0.61)$ when $w_7 = 0.1$.

point, respectively. However the system still persists at the positive equilibrium point otherwise. Similar behavior of the dynamics has been obtained when the vulnerability constant rate increases above 1.3 with the rest of parameters as in (74).

(7) Now increasing the conversion rate from the susceptible prey above $w_5 \geq 0.48$ or decreasing it below $w_5 \leq 0.25$ causes losing of persistence of the system and the trajectory approaches asymptotically the disease free equilibrium point and predator free equilibrium point, respectively. Similar dynamical behavior has been obtained when increasing or decreasing the parameter w_8, which stands for the half saturation constant of harvesting in Michael-Mentence harvesting function, as that obtained in case of increasing or decreasing w_5. The decreasing of the conversion rate from the infected prey below 0.65 has the same dynamical effects on system (3) as that obtained with decreasing w_5 too.

(8) Finally increasing the parameter w_7, which stands for the maximum harvesting rate in Michael-Mentence harvesting function, above 0.23 leads to losing the persistence of system (3) and the solution approaches asymptotically the predator free equilibrium point, while it approaches asymptotically the disease free equilibrium point with decreasing the parameter w_7 below 0.15. Similar effect on the dynamical behavior of system (3) is obtained in case of increasing or decreasing the refuge rate m and predator death rate w_9 as that effect obtained with varying w_7.

Keeping the above in view, it is easy to verify that all the analytical stability conditions are satisfied for each case in the

above-mentioned point. Furthermore, the refuge represented by parameter m and harvesting represented by parameters w_7 and w_8 have a vital effect on the dynamical behavior of system (3) including losing the persistence and moving between the disease free equilibrium point and predator free equilibrium point.

Conflicts of Interest

The authors declare that they have no conflicts of interest.

References

[1] M. R. May, *Stability and Complexity in Model Ecosystems*, Princeton University, Princeton, NJ, USA, 1973.

[2] W. O. Kermack and A. G. McKendrick, "A contribution to the mathematical theory of epidemics," *Proceedings of the Royal Society A Mathematical, Physical and Engineering Sciences*, vol. 115, no. 772, pp. 700–721, 1927.

[3] R. M. Anderson and R. M. May, "The invasion, persistence and spread of infectious diseases within animal and plant communities," *Philosophical Transactions of the Royal Society B: Biological Sciences*, vol. 314, no. 1167, pp. 533–570, 1986.

[4] J. Chattopadhyay and O. Arino, "A predator-prey model with disease in the prey," *Nonlinear Analysis: Theory, Methods & Applications*, vol. 36, pp. 747–766, 1999.

[5] Z. Z. Ma, F. D. Chen, C. Q. Wu, and W. L. Chen, "Dynamic behaviors of a Lotka-Volterra predator-prey model incorporating a prey refuge and predator mutual interference," *Applied Mathematics and Computation*, vol. 219, no. 15, pp. 7945–7953, 2013.

[6] P. Jyoti Pal and P. K. Mandal, "Bifurcation analysis of a modified Leslie-Gower predator-prey model with Beddington-DeAngelis

functional response and strong Allee effect," *Mathematics and Computers in Simulation*, vol. 97, pp. 123–146, 2014.

[7] R. K. Naji and S. J. Majeed, "The dynamical analysis of a prey-predator model with a refuge-stage structure prey population," *International Journal of Differential Equations*, vol. 2016, Article ID 2010464, 10 pages, 2016.

[8] A. S. Abdulghafour and R. K. Naji, "The impact of refuge and harvesting on the dynamics of prey-predator system," *Science International (Lahore)*, vol. 30, no. 2, pp. 315–323, 2018.

[9] D. Greenhalgh, Q. J. Khan, and F. I. Lewis, "Hopf bifurcation in two SIRS density dependent epidemic models," *Mathematical and Computer Modelling*, vol. 39, no. 11-12, pp. 1261–1283, 2004.

[10] T. Zhang, J. Liu, and Z. Teng, "Stability of Hopf bifurcation of a delayed SIRS epidemic model with stage structure," *Nonlinear Analysis: Theory, Methods & Applications*, vol. 11, no. 1, pp. 293–306, 2010.

[11] P. Yongzhen, S. Li, C. Li, and S. Chen, "The effect of constant and pulse vaccination on an SIR epidemic model with infectious period," *Applied Mathematical Modelling*, vol. 35, no. 8, pp. 3866–3878, 2011.

[12] R. K. Naji and B. H. Abdulateef, "The dynamics of SICIR model with nonlinear incidence rate and saturated treatment function," *Science International (Lahore)*, vol. 29, no. 6, pp. 1223–1236, 2017.

[13] M. S. Rahman and S. Chakravarty, "A predator-prey model with disease in prey," *Nonlinear Analysis: Modelling and Control*, vol. 18, no. 2, pp. 191–209, 2013.

[14] S. Jana and T. K. Kar, "Modeling and analysis of a prey-predator system with disease in the prey," *Chaos, Solitons & Fractals*, vol. 47, pp. 42–53, 2013.

[15] S. Kant and V. Kumar, "Dynamics of a prey-predator system with infection in prey," *Electronic Journal of Differential Equations*, vol. 2017, no. 209, pp. 1–27, 2017.

[16] M. Haque and E. Venturino, "Increase of the prey may decrease the healthy predator population in presence of a disease in predator," *HERMIS*, vol. 7, pp. 38–59, 2006.

[17] M. Haque, "A predator-prey model with disease in the predator species only," *Nonlinear Analysis: Real World Applications*, vol. 11, no. 4, pp. 2224–2236, 2010.

[18] K. P. Das, "A mathematical study of a predator-prey dynamics with disease in predator," *ISRN Applied Mathematics*, vol. 2011, Article ID 807486, 16 pages, 2011.

[19] M. V. R. Murthy and D. K. Bahlool, "Modeling and Analysis of a Prey-Predator System with Disease in Predator," *IOSR Journal of Mathematics*, vol. 12, no. 1, pp. 21–40, 2016.

[20] Y. Hsieh and C. Hsiao, "Predator-prey model with disease infection in both populations," *Mathematical Medicine and Biology*, vol. 25, no. 3, pp. 247–266, 2008.

[21] K. P. Das, K. Kundu, and J. Chattopadhyay, "A predator-prey mathematical model with both the populations affected by diseases," *Ecological Complexity*, vol. 8, no. 1, pp. 68–80, 2011.

[22] K. P. Das, S. K. Sasmal, and J. Chattopadhyay, "Disease control through harvesting - conclusion drawn from a mathematical study of a predator-prey model with disease in both the population," *International Journal of Biomathematics and Systems Biology*, vol. 1, no. 1, pp. 1–29, 2014.

[23] S. Kant and V. Kumar, "Stability analysis of predator-prey system with migrating prey and disease infection in both species," *Applied Mathematical Modelling: Simulation and Computation for Engineering and Environmental Systems*, vol. 42, pp. 509–539, 2017.

[24] N. Bairagi, S. Chaudhuri, and J. Chattopadhyay, "Harvesting as a disease control measure in an eco-epidemiological system—a theoretical study," *Mathematical Biosciences*, vol. 217, no. 2, pp. 134–144, 2009.

[25] R. Bhattacharyya and B. Mukhopadhyay, "On an eco-epidemiological model with prey harvesting and predator switching: local and global perspectives," *Nonlinear Analysis: Real World Applications*, vol. 11, no. 5, pp. 3824–3833, 2010.

[26] S. Gakkhar and K. B. Agnihotri, "The dynamics of disease transmission in a prey predator system with harvesting of prey," *International Journal of Advanced Research in Computer Engineering and Technology*, vol. 1, pp. 229–239, 2012.

[27] Q. Yue, "Dynamics of a modified Leslie–Gower predator–prey model with Holling-type II schemes and a prey refuge," *SpringerPlus*, vol. 5, no. 1, 2016.

[28] J. Ghosh, B. Sahoo, and S. Poria, "Prey-predator dynamics with prey refuge providing additional food to predator," *Chaos, Solitons & Fractals*, vol. 96, pp. 110–119, 2017.

[29] S. Gakkhar and R. K. Naji, "Existence of chaos in two-prey, one-predator system," *Chaos, Solitons & Fractals*, vol. 17, no. 4, pp. 639–649, 2003.

[30] C. W. Clark, "Aggregation and fishery dynamics: a theoretical study of schooling and the Purse Seine tuna fisheries," *NOAA Fisheries Bulletin*, vol. 77, no. 2, pp. 317–337, 1979.

11

Analysis of a Heroin Epidemic Model with Saturated Treatment Function

Isaac Mwangi Wangari[1] and Lewi Stone[1,2]

[1]*Royal Melbourne Institute of Technology School of Mathematics and Geospatial Sciences, Melbourne, VIC, Australia*
[2]*Biomathematics Unit, Department of Zoology, Faculty of Life Sciences, Tel Aviv University, Tel Aviv, Israel*

Correspondence should be addressed to Isaac Mwangi Wangari; mwangiisaac@aims.ac.za

Academic Editor: Mehmet Sezer

A mathematical model is developed that examines how heroin addiction spreads in society. The model is formulated to take into account the treatment of heroin users by incorporating a realistic functional form that "saturates" representing the limited availability of treatment. Bifurcation analysis reveals that the model has an intrinsic backward bifurcation whenever the saturation parameter is larger than a fixed threshold. We are particularly interested in studying the model's global stability. In the absence of backward bifurcations, Lyapunov functions can often be found and used to prove global stability. However, in the presence of backward bifurcations, such Lyapunov functions may not exist or may be difficult to construct. We make use of the geometric approach to global stability to derive a condition that ensures that the system is globally asymptotically stable. Numerical simulations are also presented to give a more complete representation of the model dynamics. Sensitivity analysis performed by Latin hypercube sampling (LHS) suggests that the effective contact rate in the population, the relapse rate of heroin users undergoing treatment, and the extent of saturation of heroin users are mechanisms fuelling heroin epidemic proliferation.

1. Introduction

In 1897, Germany's Bayer pharmaceutical company synthesised heroin and soon after marketed the product as a nonaddictive miracle drug, for use as a cough syrup and pain reliever [1]. Cough medicine was in fact in high demand, since tuberculosis and pneumonia were fast-spreading diseases of the time. As such, the miracle drug heroin was rapidly disseminated across the globe. Fast forward to today, and we know that addiction to heroin is an extremely common phenomenon among heroin users; some 23% of individuals who consume the drug become dependent on it. Worldwide, many countries are affected by the heroin drug-trafficking industry and its growing number of users. America is currently in the midst of another heroin epidemic [2] with approximately 700,000 Americans using heroin in the past year [2]. The number of people using heroin for the first time is increasing at an alarming rate, with >150,000 Americans engaging in heroin use in 2012, which is almost double that recorded in 2006 [2]. Heroin also leads to other diseases

and is considered a major pathway responsible for fuelling proliferation of human immunodeficiency virus (HIV) and Hepatitis B and Hepatitis C virus (HBV, HCV) [3, 4].

The development of heroin habituation and addiction has similar characteristics to an epidemic, in terms of its disturbingly contagious spread through a susceptible population. In the last decades, a whole range of mathematical models have been developed to forecast how diseases spread in time and space and how they can be controlled. Recently, the same mathematical modelling techniques have been extended for the purpose of understanding and combating drug addiction problems. The aim of the present study is to propose a novel heroin epidemic model and make use of it to study issues arising with treatment and establish conditions that may signal heroin persistence within the community.

The ultimate goal of mathematical epidemiology is to understand how to control and eliminate infectious diseases and these ideas have a place for also dealing with social problems. In epidemic theory the basic reproduction number, usually denoted by R_0, is one of the most important concepts,

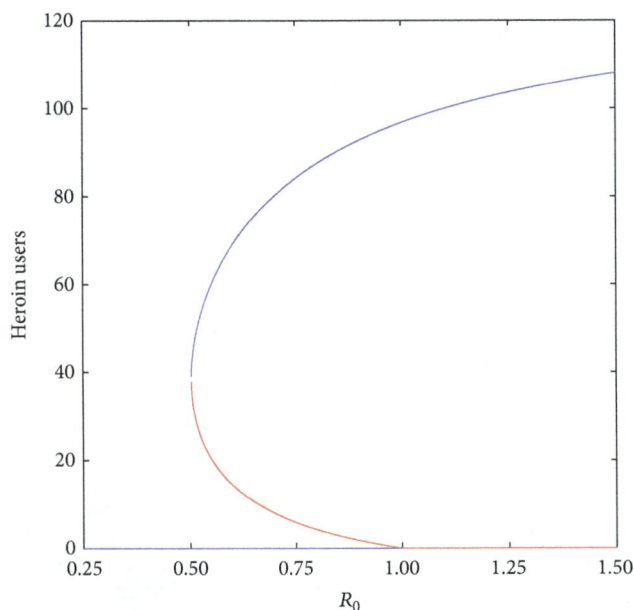

FIGURE 1: Illustration of the qualitative features of backward bifurcation. The red line represents the unstable equilibria (i.e., unstable endemic equilibria and unstable heroin-free equilibrium) while the blue line represents stable equilibria (i.e., stable endemic equilibria and stable heroin-free equilibrium).

given its ability to predict the course of an epidemic. It will also prove invaluable in our study of heroin dynamics in society. R_0 is defined as the number of secondary infections that are likely to occur when a single infectious individual is introduced into an entirely susceptible population [5]. Until recently, it has been widely accepted that the condition $R_0 < 1$ is an essential requirement for the eradication of a disease. However, this viewpoint has been recently challenged with a number of theoretical studies demonstrating that this criterion may not always be sufficient. Instead, the phenomenon of backward bifurcation offers a different interpretation since it shows that although the basic reproduction number is below unity and the infection free equilibrium is stable, there might still be another stable endemic equilibrium and unstable endemic equilibrium coexisting simultaneously. Thus even though $R_0 < 1$, a population may still reside at an endemic equilibrium in which the disease persists indefinitely. In a scenario where multiple equilibria concurrently exist, the extinction or persistence of an epidemic is dependent on initially infected size of subpopulations. The qualitative features of backward bifurcation are illustrated in Figure 1.

A variety of both behavioural and pharmacological medications can be administered to effectively treat heroin addiction. The side effects associated with quitting using heroin (such as pain, diarrhoea, nausea, and vomiting) are very severe and very often compel heroin addicts to relapse. To prevent such cases there are available medications that can be administered during the detoxification stage to relieve craving and physical symptoms. A number of studies have established that pharmacological therapy has positive impact in facilitating drug addicts to remain in treatment programs.

Furthermore, it has been noted that during addiction treatment there is a decline in drug consumption, infectious disease transmission, and crime rates [2]. In our present study we shall incorporate a saturated treatment function and derive threshold conditions that indicate when heroin is able to persist within a community. Besides incorporation of a saturated treatment function, our model also included an extra class of individuals, namely, those who have been successfully treated from heroin using. This class has been neglected in previous heroin epidemic models [6–9]. Much of our work will be focused on exploring the conditions for global stability of the heroin model with treatment. Our work deals with global stability of a heroin model with bilinear incidence rate, self-cure, relapse, and saturated treatment function using the Bendixson criterion.

With this in mind we will extend the SIR (Susceptible-Infected-Recovered) model by [8] to represent a heroin epidemic model and investigate global stability properties. To be precise we study the conditions of global stability for the nontrivial equilibrium states by using two distinct approaches: the Lyapunov direct method and the Li and Muldowney's geometric approach to global stability. It is with no doubt that the famous Lyapunov direct method is a powerful tool for nonlinear stability analysis [10]. One of the main advantages of Lyapunov direct technique is that it is directly applicable to nonlinear systems [11]. However, the major challenge with using Lyapunov direct method is that it requires an auxiliary function which is often hard to construct. And this difficulty is exacerbated especially if the model exhibits backward bifurcation phenomena because Lyapunov functions for such models may not exist. To address these difficulties another powerful tool, the geometric technique due to Li and Muldowney, was developed in the middle of nineties [11–13]. Their method involves generalization of Bendixson's criterion to systems of any finite dimensions and applies compound matrices. Presently this method has gained popularity due to its vast range of applications, in particular to mathematical models that are of biological interest. Although this method is mainly applied in epidemic models (e.g., see, [14–19]) its use can be found in other population dynamics contexts (see [20]). It has been shown in [16] that the geometric technique is more appropriate in mathematical models of SEIR-like structure since their analysis can be easily reduced to a three-dimensional system. Nevertheless, the method has been extended to four-dimensional systems that may be difficult to reduce. In the sequel applications to four-dimensional systems are rare because the procedure becomes mathematically involved when $n \geq 4$. Examples of four-dimensional systems can be traced in the work of Ballyk and coworkers who applied compound matrices to a four-dimensional population model [21] and also by Gumel and coworkers [22] who studied a SVEIR (Susceptible-Vaccinated-Exposed-Infected-Recovered) model of severe acute respiratory syndrome (SARS) epidemic spread.

The four-dimensional model studied here can be reduced to a three-dimensional system. Both Lyapunov direct method and geometric approach are applied to investigate global properties of a four-dimensional heroin epidemic model. Lyapunov direct method will be applied in a special case in particular where the parameter that triggers bistability phenomena is switched off. On the other hand geometric approach will be applied in the general model where all parameters are present including the one that causes bistability. Here we follow the procedure in [11, 16] to obtain sufficient condition for global stability.

2. Model Formulation

In the spirit of the SIR (Susceptible-Infected-Recovered) model in the literature (i.e., [23]), we formulate a heroin epidemic model based on the assumption that heroin use follows a process that can be modelled similar to infectious diseases [24, 25]. The general population is stratified into four mutually exclusive classes, namely, susceptibles (S), individuals successfully treated from heroin use (U_3), heroin users undergoing treatment (U_2), and heroin users not in treatment (U_1). The proposed heroin epidemic model is based on key assumptions which include the following:

(i) Uniform mixing: individuals in the above-mentioned classes freely interact with each other.

(ii) Individuals undergoing treatment are still often using drugs [26].

(iii) Heroin users in treatment relapse to heroin users not in treatment as a result of the self-decision to terminate treatment [27].

(iv) Heroin users in treatment do not infect susceptibles.

Given these assumptions the heroin model may be described by the processes illustrated in Figure 2, which can be written in terms of the following set of equations:

$$\frac{dS}{dt} = \Lambda - \beta U_1 S - \mu S,$$

$$\frac{dU_1}{dt} = \beta U_1 S + p U_2 - (\mu + \delta_1 + \xi) U_1 - T(U_1),$$

$$\frac{dU_2}{dt} = T(U_1) - (p + \sigma + \delta_2 + \mu) U_2, \qquad (1)$$

$$\frac{dU_3}{dt} = \sigma U_2 + \xi U_1 - \mu U_3.$$

In brief, the susceptible subpopulation $S(t)$ is generated at a constant rate through immigration and birth at rate Λ. Some susceptible individuals who come into contact with heroin users $U_1(t)$ may begin to use heroin. Hence, the susceptible population is diminished due to contact with heroin users at rate $\beta U_1 S$, while heroin users increase at the same rate. Heroin users also increase when those undergoing treatment relapse at rate $p U_2$ and return to their heroin using lifestyle. Heroin users reduce in number as a result of treatment which is represented by the treatment function $T(U_1)$. Moreover, the user subpopulation is reduced by heroin-induced death at rate $\delta_1 U_1$ as well as a result of the self-decision to cease using heroin (also referred to as "self-cure") at rate ξU_1. Individuals undergoing treatment are diminished through the following processes: relapse to heroin using at rate $p U_2$,

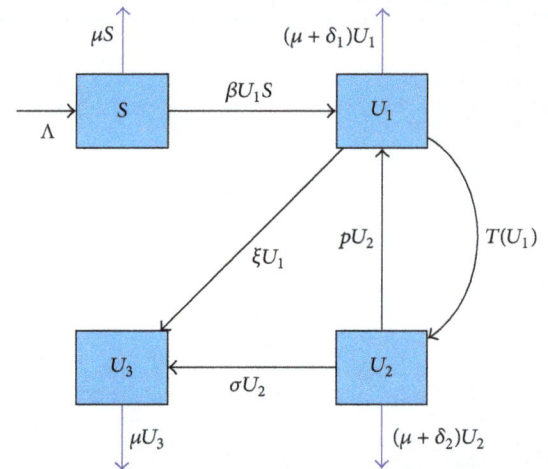

FIGURE 2: A heroin epidemic model with bilinear incidence rate and saturated treatment function. The blue solid arrows represent deaths due to either heroin or natural causes while the black solid arrows represent change of status from one compartment to another.

heroin-induced death at rate $\delta_2 U_2$, and successful treatment at rate σU_2. Finally, the recovered/successfully treated subpopulation $U_3(t)$ is generated when heroin users undergoing treatment are successfully cured and also through "self-cure." All subpopulations are decreased by natural death via the background mortality parameter μ.

Heroin epidemic models studied to date [6–9] assume the classical view that the treatment rate of the infective population should be proportional to the number of infective individuals [28]. This viewpoint was criticised during the SARS (Severe Acute Respiratory Syndrome) outbreaks in 2003. The dramatic increase of SARS cases in Beijing challenged the normal public-health system because it was only possible to treat a limited number of SARS patients at a given time. The experience with SARS epidemic sparked a renewed interest among modellers to investigate the implication of the capacity of the healthcare system. Authors in [29] considered an SIR epidemic model and assumed a Heaviside treatment function while Wang [30] restudied the same SIR model but assumed a piecewise linear treatment function. Here we will assume that the heroin users $U_1(t)$ receive treatment based on the following more general saturated treatment function:

$$T(U_1) \triangleq \frac{\alpha U_1}{1 + \omega U_1}, \qquad (2)$$

where α is positive and ω is nonnegative. In our present model the parameter ω accounts for the extent of saturation of heroin users. Note that for small U_1 the treatment function reduces to $T(U_1) \approx \alpha U_1$ while for large U_1 it reduces to $T(U_1) \approx \alpha/\omega$ which actually characterizes the saturated phenomena of the treatment. Further, if $\omega = 0$, the treatment function becomes $T(U_1) = \alpha U_1$ which is the usual linear treatment rate. $1/(1 + \omega U_1)$ is a measure of inhibition due to a saturation of heroin users who are usually too many to be dealt with given the limited available treatment.

TABLE 1: Description of variables and parameters of model (1).

Variable	Description
S	Number of susceptible individuals at time t
U_3	Number of heroin users who have been successfully treated from heroin use, as well as individuals who have voluntarily stopped using the heroin (and have withdrawal symptoms) "self-cure" at time t
U_2	Number of heroin users undergoing treatment at time t
U_1	Number of drug users not undergoing treatment at time t, that is, the initial and relapsed heroin users
N	Total population at time t ($N = S(t) + U_3(t) + U_2(t) + U_1(t)$)

Parameter	Description
Λ	Recruitment rate of individuals in the general population entering the susceptible population
β	Effective contact rate
μ	Death rate due to natural causes
p	Rate at which individuals undergoing treatment relapse to heroin use
ξ	"Self-cure rate": rate at which heroin users stop using heroin and join the successfully cured class of individuals not taking heroin
δ_1	Heroin-related death rate of heroin users not in treatment
δ_2	Heroin-related death rate of individuals undergoing treatment
σ	Rate at which heroin users in treatment are successfully cured (i.e., completely detoxicated) from heroin use
α	Rate at which heroin users are treated
ω	The extent of saturation of heroin users within the community

A summary of the model variables and parameters is given in Table 1.

3. Basic Properties and Basic Reproduction Number

Since we are studying a human population, the model must be able to ensure that all the associated parameters and the state variables S, U_3, U_2, U_1 are nonnegative for all time $t > 0$. Hence, the following result.

Theorem 1. *Let the initial conditions supplied to model (1) be such that $S(0) > 0, U_3(0) > 0, U_2(0) > 0$, and $U_1(0) > 0$. Then the trajectories $(S(t), U_3(t), U_2(t), U_1(t))$ of the model (1), with positive initial conditions, will remain positive for all time t.*

Proof. Let $t_1 = \sup\{t > 0 : S(t) > 0, U_3(t) > 0, U_2(t) > 0, U_1(t) > 0\} > 0$. Now from the first equation of model (1), it follows that

$$\frac{dS}{dt} = \Lambda - \Phi S - \mu S \geq \Lambda - \Phi S - \mu S \tag{3}$$

$$(\text{where } \Phi = \beta U_1),$$

which can be written as

$$\frac{d}{dt}\left\{ S(t) \exp\left[\mu t + \int_0^t \Phi(\tau)\, d\tau\right]\right\}$$
$$\geq \Lambda \left\{ \exp\left[\mu t + \int_0^t \Phi(\tau)\, d\tau\right]\right\}. \tag{4}$$

Hence,

$$S(t_1) \exp\left[\mu t_1 + \int_0^{t_1} \Phi(\tau)\, d\tau\right] - S(0)$$
$$\geq \int_0^{t_1} \Lambda \left\{ \exp\left[\mu x + \int_0^x \Phi(\tau)\, d\tau\right]\right\} dx, \tag{5}$$

so that

$$S(t_1) \geq S(0) \exp\left[-\mu t_1 - \int_0^{t_1} \Phi(\tau)\, d\tau\right]$$
$$+ \left\{ \exp\left[-\mu t_1 - \int_0^{t_1} \Phi(\tau)\, d\tau\right]\right\} \tag{6}$$
$$\times \int_0^{t_1} \Lambda \left\{ \exp\left[\mu x + \int_0^x \Phi(\tau)\, d\tau\right]\right\} dx > 0.$$

Following a similar procedure we can show that $U_3(t) > 0$, $U_2(t) > 0$, and $U_1(t) > 0$ for all time $t > 0$. Thus, all trajectories of model (1) remain positive for all nonnegative initial conditions, as required. □

Now in what follows we establish the region where model (1) is considered to be biologically feasible. Summing all the equations of the basic model (1) yields

$$\frac{dN(t)}{dt} = \Lambda - \mu N(t) - \delta_1 U_1(t) - \delta_2 U_2(t). \tag{7}$$

Considering that $0 < U_1(t) < N(t)$, $0 < U_2(t) < N(t)$ and letting $\bar{\delta} = \max\{\delta_1, \delta_2\}$, it follows from (7) that

$$\Lambda - \left(\mu + 2\bar{\delta}\right) N(t) \leq \frac{dN(t)}{dt} < \Lambda - \mu N(t). \qquad (8)$$

Therefore

$$\frac{\Lambda}{\mu + 2\bar{\delta}} \leq \liminf_{t\to\infty} N(t) \leq \limsup_{t\to\infty} N(t) \leq \frac{\Lambda}{\mu}$$

$$(9)$$

such that $\limsup\limits_{t\to\infty} N(t) \leq \dfrac{\Lambda}{\mu}$.

Theorem 2. *The closed set*

$$\mho = \Big\{ \left(S, U_3, U_2, U_1\right) : 0 \leq S, U_3, U_2, U_1; S + U_3 + U_2$$

$$(10)$$

$$+ U_1 \leq \frac{\Lambda}{\mu} \Big\}$$

is positively invariant and absorbing with respect to the set of nonlinear differential equation (1).

Proof. Here we show that the feasible solutions of model (1) are uniformly bounded in the region $\mho$. Suppose S, U_3, U_2, and U_1 are any solution of system (1) supplied with nonnegative initial conditions. Then it is straightforward to note that the total population N satisfies the inequality

$$\frac{dN}{dt} = \Lambda - \mu N - \delta_1 U_1 - \delta_2 U_2 \leq \Lambda - \mu N. \qquad (11)$$

From (11) it follows that $dN/dt \leq \Lambda - \mu N$ which implies $dN/dt \leq 0$ if $N \geq \Lambda/\mu$. The standard comparison theorem [31] can be used to deduce that $N(t) \leq N(0)e^{-\mu t} + (\Lambda/\mu)(1 - e^{-\mu t})$. In particular $N(t) \leq \Lambda/\mu$ if $N(0) \leq \Lambda/\mu$ for all $t > 0$. Thus, under the flow induced by system (1), the region $\mho$ is positively invariant. Furthermore, for $N(0) > \Lambda/\mu$ the trajectory solutions $N(t)$ either enter in the region $\mho$ in finite time or asymptotically approach Λ/μ. Thus, in the region $\mho$, model (1) is said to be mathematically and epidemiologically well posed [32] and the solution of all the trajectories generated by model (1) is considered in a biologically feasible region $\mho$. $\qquad \square$

Clearly system (1) has an intrinsic Heroin-free equilibrium (HFE) given by $D^0 = (S_0, 0, 0, 0)$, a scenario representing a heroin-free state in the community. $S_0 = \Lambda/\mu$ represents the number of susceptibles when no one is using heroin. The basic reproduction number denoted by R_0 is defined as the number of secondary infections that are likely to be triggered by a single infectious individual when introduced into a wholly susceptible population [32]. Here R_0 is interpreted as the mean number of secondary cases of heroin users generated by a typical heroin user not in treatment during his/her duration of heroin use in a population of potential drug users.

To obtain the basic reproduction number we observe that the average time an individual spends as a heroin user without

treatment is $T_0 = 1/(\mu + \delta_1 + \xi + \alpha)$ and the probability of surviving this compartment and moving to the treatment compartment is $T_1 = \alpha/(\mu + \delta_1 + \xi + \alpha)$. Now the probability of surviving heroin users in treatment class and then returning to the heroin users class not in treatment is $T_2 = p/(p + \sigma + \delta_2 + \mu)$. Thus, the total average time spent by the heroin users not in the treatment compartment on multiple passes can be obtained as

$$T = T_0 \left[1 + T_1 \cdot T_2 + \left(T_1 \cdot T_2\right)^2 + \cdots \right]. \qquad (12)$$

Clearly, the terms inside the square brackets in (12) constitute a geometric sequence (see Appendix A for detailed derivation) and therefore expression (12) can be written as

$$T = \frac{(p + \sigma + \delta_2 + \mu)}{(\mu + \delta_1 + \xi)(p + \delta_2 + \sigma + \mu) + \alpha(\sigma + \delta_2 + \mu)}. \qquad (13)$$

Multiplying (13) with the effective contact rate β and the average recruitment rate Λ/μ we obtain heroin basic reproduction number as

$$R_0$$

$$= \frac{\beta\Lambda(p + \delta_2 + \mu + \sigma)}{\mu\alpha(\mu + \delta_2 + \sigma) + \mu(\mu + \delta_1 + \xi)(p + \delta_2 + \mu + \sigma)}. \qquad (14)$$

It is easy to observe that R_0 is inversely proportional to treatment α, which implies that if treatment rate is maintained sufficiently high it can control a heroin epidemic (by reducing R_0 to less than one). However, as we will see later when parameter ω (representing the extent of saturation of heroin users) is accounted for, this control is no longer guaranteed.

Theorem 3. *The HFE is locally asymptotically stable provided $R_0 < 1$; otherwise it is unstable.*

This general result has been reviewed in [33] and thus not proved again here. The theorem implies that heroin users will disappear from the community when $R_0 < 1$ if the initial sizes of the subpopulations of system (1) are in the basin of attraction of the heroin-free equilibrium.

Remark 4. It is instructive to note that the basic reproduction number does not include the parameter ω that accounts for the extent of saturation of heroin users. In the next subsection we investigate endemic equilibria of the model where the parameter ω plays a key role in emergence of bistability.

3.1. Endemic Equilibria. Within the context of our heroin model the endemic equilibrium refers to a state when heroin addiction is maintained over long time scales in the population. Therefore $S^*, U_3^*, U_2^*, U_1^* > 0$ hold. To obtain the endemic equilibria $E^* = (S^*, U_3^*, U_2^*, U_1^*)$ we set (1) to zero and solve for the equilibrium quantities S^*, U_3^*, and U_2^* in terms of U_1^*. That is,

$$S^* = \frac{\Lambda}{\beta U_1^* + \mu},$$

$$U_3^* = \frac{\sigma \alpha U_1^* + \xi (p + \sigma + \delta_2 + \mu)(1 + \omega U_1^*) U_1^*}{\mu (p + \sigma + \delta_2 + \mu)(1 + \omega U_1^*)}, \quad (15)$$

$$U_2^* = \frac{\alpha U_1^*}{(p + \sigma + \delta_2 + \mu)(1 + \omega U_1^*)}.$$

Substituting (15) into the second equation of (1) yields

$$f(U_1^*) = A U_1^{*2} + B U_1^* + C = 0, \quad (16)$$

where

$$A = (\mu + \delta_1 + \xi)(p + \sigma + \delta_2 + \mu)\beta\omega,$$

$$\begin{aligned} B &= (\mu + \delta_1 + \xi)(p + \sigma + \delta_2 + \mu)\beta + \mu\omega(\mu + \delta_1 + \xi) \\ &\cdot (p + \sigma + \delta_2 + \mu) + \alpha(\sigma + \delta_2 + \mu)\beta - \beta\Lambda\omega(p \\ &+ \sigma + \delta_2 + \mu), \end{aligned} \quad (17)$$

$$\begin{aligned} C &= [(\mu + \delta_1 + \xi)(p + \sigma + \delta_2 + \mu)\mu \\ &+ \alpha\mu(\sigma + \delta_2 + \mu)](1 - R_0). \end{aligned}$$

The quadratic equation (16) can be analysed to investigate the existence of multiple equilibria when the basic reproduction number is below unity.

If the parameter that accounts for the extent of saturation of heroin users in model (1) is excluded, that is, $\omega = 0$, (16) reduces to a linear equation

$$U_1^* \widetilde{B} + C = 0, \quad (18)$$

where

$$\begin{aligned} \widetilde{B} &= (\mu + \delta_1 + \xi)(p + \sigma + \delta_2 + \mu)\beta \\ &+ \alpha(\sigma + \delta_2 + \mu)\beta \end{aligned} \quad (19)$$

so that model (1) has the unique solution

$$U_1^* = \frac{-C}{\widetilde{B}} \quad (20)$$

which is nonnegative if and only if $R_0 > 1$. Hence, if $\omega = 0$, model (1) has a unique endemic equilibrium whenever $R_0 > 1$ and this equilibrium approaches zero as R_0 tends to one $(R_0 \to 1+)$ because $C \to 0$. But there are no positive endemic equilibria if $R_0 < 1$. These results are summarized in the following lemma.

Lemma 5. *The epidemic model (1) when $\omega = 0$ has a unique positive endemic equilibrium $E_1^* = (S^*, U_3^*, U_2^*, U_1^*)$ whenever $R_0 > 1$ and no positive endemic equilibrium otherwise.*

In what follows, we investigate the global stability for both the HFE and the unique endemic equilibrium E_1^* for the case $\omega = 0$.

3.2. Global Stability for Heroin-Free Equilibrium When $\omega = 0$. To investigate global stability we apply the method presented by Castillo-Chavez et al. [34]. First let $\mathcal{X} = (S, U_3)$ and $\mathcal{Y} = (U_2, U_1)$ with $\mathcal{X} \in \mathbb{R}^2$ representing the number of individuals not using heroin and $\mathcal{Y} \in \mathbb{R}^2$ representing the number of individuals using heroin (i.e., heroin users in treatment and heroin users not in treatment). Now suppose

$$\mathcal{X}' = F(\mathcal{X}, \mathcal{Y}),$$

$$\mathcal{Y}' = G(\mathcal{X}, \mathcal{Y}), \quad (21)$$

$$G(\mathcal{X}, 0) = 0,$$

where $\mathcal{X}'$ and $\mathcal{Y}'$ denote differentiation with respect to time. The HFE is now denoted by $H_0 = (\mathcal{X}^0, 0)$, where $\mathcal{X}^0 = (S^0, 0)$. The following conditions (H1) and (H2) have to be met to guarantee a local asymptotic stability:

(H1) For $\mathcal{X}' = F(\mathcal{X}, 0)$, $\mathcal{X}^0$ is globally asymptotically stable (g.a.s).

(H2) $G(\mathcal{X}, \mathcal{Y}) = \mathcal{B}\mathcal{Y} - \widehat{G}(\mathcal{X}, \mathcal{Y})$, where $\widehat{G}(\mathcal{X}, \mathcal{Y}) \geq 0$, for $(\mathcal{X}, \mathcal{Y}) \in \mho$.

$\mathcal{B} = D_{\mathcal{Y}} G(\mathcal{X}, 0)$ and $\mho$ is the region where model (1) is biologically realistic. Then, Castillo-Chavez and Song [35] have shown that the following lemma is satisfied.

Lemma 6. *The fixed point $H_0 = (\mathcal{X}^0, 0, 0)$ is g.a.s equilibrium of model (1) provided that $R_0 < 1$ (locally asymptotically stable) and that assumptions (H1) and (H2) hold.*

Now consider the following theorem.

Theorem 7. *Suppose $R_0 < 1$. Then the HFE D^0 is g.a.s.*

Proof. Let $\mathcal{X} = (S, U_3)$ and $\mathcal{Y} = (U_2, U_1)$, and $H_0 = (\mathcal{X}^0, 0, 0)$, where $\mathcal{X}^0 = (\Lambda/\mu, 0)$.

Then we have

$$\mathcal{X}' = F(\mathcal{X}, \mathcal{Y}) = \begin{pmatrix} \Lambda - \beta U_1 S - \mu S \\ \sigma U_2 + \xi U_1 - \mu U_3 \end{pmatrix}. \quad (22)$$

It is straightforward to see that, at the heroin-free equilibrium (HFE) $S = S_0 = \Lambda/\mu$, $F(\mathcal{X}, 0) = \binom{0}{0}$. Thus,

$$\mathcal{X}' = F(\mathcal{X}, 0) = \begin{pmatrix} \Lambda - \mu S \\ -\mu U_3 \end{pmatrix}. \quad (23)$$

Now, as $t \to \infty$, $\mathcal{X} \to \mathcal{X}^0$. Hence, $\mathcal{X}^0$ is globally asymptotically stable (i.e., condition (H1) is satisfied).

Now consider

$$G(\mathcal{X}, \mathcal{Y})$$

$$= \begin{bmatrix} -(p + \sigma + \delta_2 + \mu) & \alpha \\ p & \beta S_0 - (\mu + \delta_1 + \xi + \alpha) \end{bmatrix} \begin{bmatrix} U_2 \\ U_1 \end{bmatrix} \quad (24)$$

$$- \begin{bmatrix} 0 \\ \beta U_1 (S_0 - S) \end{bmatrix}$$

so that

$$\mathscr{B} = \begin{bmatrix} -(p + \sigma + \delta_2 + \mu) & \alpha \\ p & \beta S_0 - (\mu + \delta_1 + \xi) \end{bmatrix}, \tag{25}$$

$$\widehat{G}(\mathcal{X}, \mathcal{Y}) = \begin{bmatrix} 0 \\ \beta U_1 (S_0 - S) \end{bmatrix}.$$

Since the total population is bounded by $N = S + U_3 + U_2 + U_1 \le \Lambda/\mu$, we have $S \le N \le \Lambda/\mu$. Thus, $\widehat{G}(\mathcal{X}, \mathcal{Y}) \ge 0$, which now implies that conditions $(H1)$ and $(H2)$ are satisfied. Consequently by Lemma 6 the fixed point H_0 is globally asymptotically stable when $R_0 < 1$, which indicates nonexistence of multiple nontrivial equilibria when $\omega = 0$. The epidemiological implication of HFE being g.a.s is that heroin epidemic will be eliminated from the community if the threshold quantity R_0 is decreased to (or maintained at) a value below unity.   □

Now for $\omega > 0$ we establish the following theorem.

Theorem 8. *For $\omega > 0$ model (1) has*

(i) *a unique positive endemic equilibrium if $B < 0$ and either $C = 0$ or $B^2 - 4AC = 0$,*

(ii) *a unique positive endemic equilibrium if $C < 0$ (i.e., $R_0 > 1$) and $B < 0$,*

(iii) *two positive endemic equilibria if $C > 0, B < 0$, and $B^2 - 4AC > 0$,*

(iv) *no positive endemic equilibrium if $B > 0$ and either $C > 0$ or $B^2 < 4AC$.*

The theorem may be proved as follows. It is obvious to note that in quadratic equation (16) A is always positive and C is either positive or negative depending on whether the basic reproduction number is less than or greater than one, respectively.

For Case (i) where $B < 0$ and $C = 0$ (i.e., $R_0 = 1$) (16) becomes linear $AU_1^* + B = 0$ and has a unique nonzero solution $U_1^* = -B/A$ which is positive if $B < 0$ and negative if $B > 0$. Referring to (15) we see that if U_1^* is unique then so are S^*, U_2^*, and U_3^*.

For Case (ii) where $C < 0$ (i.e., $R_0 > 1$) and $B < 0$, (16) is quadratic and according to Descartes Rule of Signs (see [36]), (16) has one change of signs indicating (16) has a unique positive root and therefore there is a unique endemic equilibrium.

In Case (iii) where $B < 0$ there is a nonnegative endemic equilibrium at $R_0 = 1$. However, because (16) is quadratic and since the equilibrium is continuously determined by R_0 then there must be an interval to the left of $R_0 = 1$ on which two nonnegative equilibria coexist. That is,

$$U_{1,1}^* = \frac{-B - \sqrt{B^2 - 4AC}}{2A},$$

$$U_{1,2}^* = \frac{-B + \sqrt{B^2 - 4AC}}{2A}. \tag{26}$$

For Case (iv) where $B > 0$ and $C > 0$ or $B^2 < 4AC$, (16) has no positive real root as can be seen in (26), implying nonexistence of a positive endemic equilibrium.

Case (iii) suggests that model (1) exhibits the phenomenon of backward bifurcation since the classical requirement for the occurrence of the phenomenon of backward bifurcation is satisfied, that is, the existence of multiple equilibria when the basic reproduction number is less than one. Thus, we have the following lemma.

Theorem 9. *Model (1) has backward bifurcation at $R_0 = 1$ if and only if $B < 0$ (i.e., $\omega > \omega_c$).*

Proof. Consider (16), $f(U_1^*) = AU_1^{*2} + BU_1^* + C = 0$. Note that, at $R_0 = 1$, $C = 0$ implies that the graph $f(U_1^*)$ passes through the origin. If $B < 0$ it follows that $f(U_1^*) = 0$ has a nonnegative root. Since $f(U_1^*)$ is a continuous function of C, if we increase C such that $C > 0$, there is some open interval of C say $(0, \psi)$ on which $f(U_1^*) = 0$ has two nonnegative roots. That is, there exist two nonnegative endemic equilibria when $R_0 < 1$. This is indeed true since Case (iv) of Theorem 8 has already shown that for $B \ge 0$ model (1) does not have positive real roots when $R_0 < 1$. Note that, at $R_0 = 1$, $C = 0$ the following equality holds:

$$(\mu + \delta_1 + \xi)(p + \sigma + \delta_2 + \mu)\mu + \alpha\mu(\mu + \delta_2 + \sigma)$$
$$= (p + \sigma + \delta_2 + \mu)\beta\Lambda. \tag{27}$$

This together with condition $B < 0$ implies that

$$\omega$$
$$> \frac{(\mu + \delta_1 + \xi)(p + \sigma + \delta_2 + \mu)\beta + \alpha(\sigma + \delta_2 + \mu)\beta}{\mu\alpha(\sigma + \delta_2 + \mu)} \tag{28}$$
$$\triangleq \omega_c.$$

□

Thus, the phenomenon of backward bifurcation (referring to Case (iii), a situation where there are two endemic equilibria) occurs at the left of $R_0 = 1$ if and only if condition (28) is satisfied. This suggests that backward bifurcation will only occur if the parameter ω that accounts for the extent of saturation of heroin users exceeds a certain threshold (i.e., $\omega > \omega_c$). However, if $\omega < \omega_c$ backward bifurcation cannot occur. Thus, parameter ω plays a critical role in the formation of backward bifurcation for model (1). It is instructive to note that similar results as the one shown in inequality (28) can be obtained by center manifold theory (see Appendix B), where it is emphasized that if $\omega > \omega_c$ the bifurcation coefficient a is positive indicating that the model system (1) undergoes the phenomenon of backward bifurcation. The epidemiological implication of backward bifurcation is that although it is necessary to reduce the basic reproduction number below one it is not sufficient to eradicate a heroin epidemic; rather R_0 should be reduced further below a certain threshold which we shall denote by R_0^C (see Figure 3(b)).

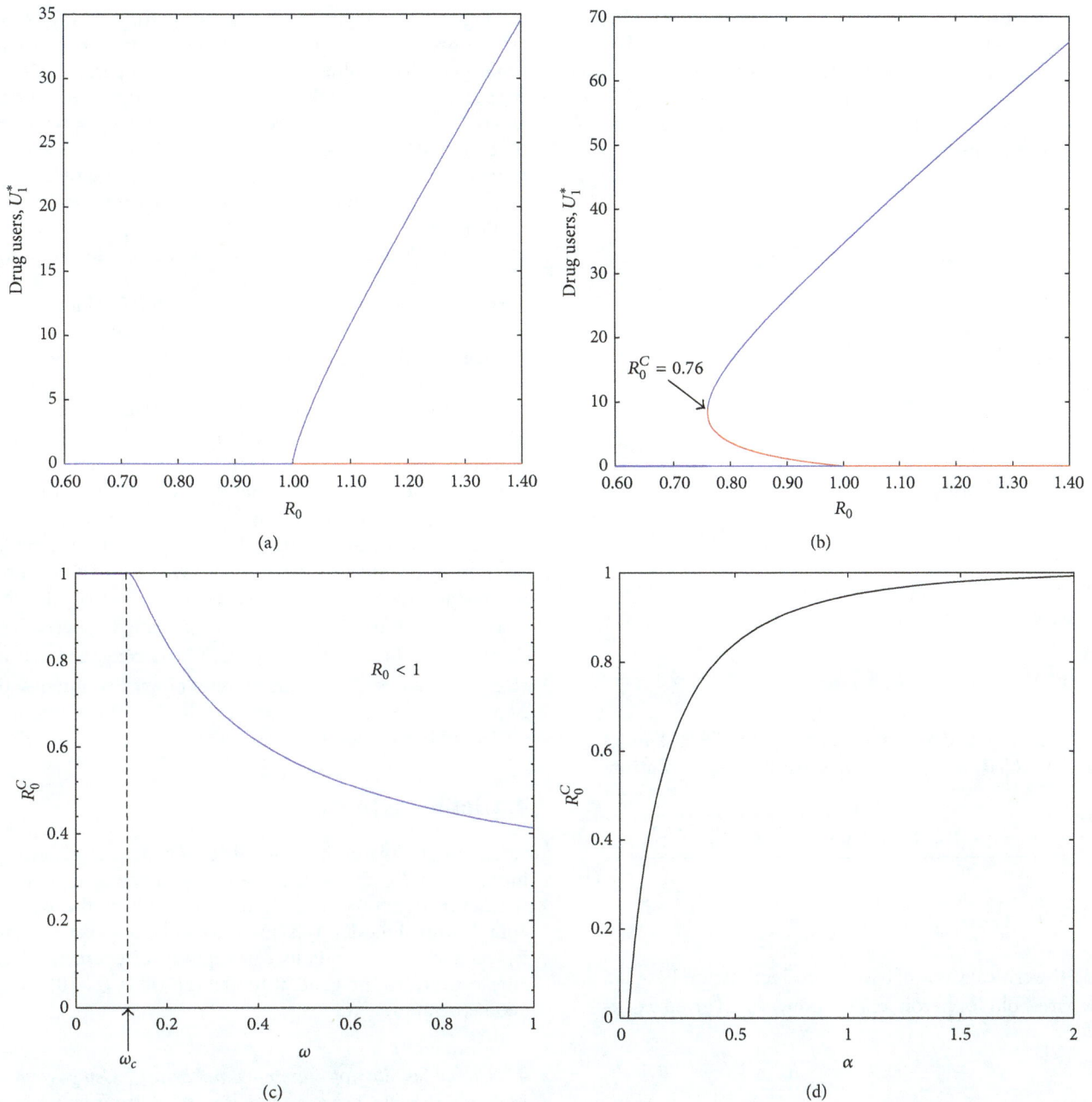

FIGURE 3: (a) and (b) represent bifurcations where drug users, U_1^*, not in treatment equilibrium are plotted as a function of R_0. The blue solid line represents the stable equilibrium while the red solid line represents the unstable equilibrium. (a) represents forward bifurcation and the parameters used include $\omega = 0.11 < \omega_c = 0.1156$, $\mu = 0.01$, $\beta = 0.001$, $\delta_1 = 0.002$, $\delta_2 = 0.001$, $\xi = 0.015$, $\alpha = 0.9$, $p = 0.467$, $\sigma = 0.1$, and $\Lambda \in \{1, 3\}$ while (b) represents backward bifurcation and parameters used are the same as in (a) except $\omega = 0.25 > \omega_c = 0.1156$. (c) depicts the critical value R_0^C as a function of saturation parameter ω. The black dotted line represents the threshold ω_c which if exceeded gives rise to backward bifurcation. (d) depicts the critical value R_0^C as a function of treatment rate α.

3.3. Computation of New Threshold for Heroin Eradication R_0^C.

Here we compute the value of the basic reproduction number where the two nontrivial endemic equilibria (both stable and unstable) collide and annihilate each other leaving only the heroin-free equilibrium point as the stationary solution. This is R_0^C in Figure 3(b). We choose Λ as the parameter of backward bifurcation. As we have noted in Case (iii) of

Theorem 8, (16) has nonnegative roots corresponding to two endemic equilibria if and only if $C > 0$ (i.e., $R_0 < 1$) and $B < 0$, $B^2 > 4AC$. It follows that if $B = -2\sqrt{AC}$, (16) has one nonnegative root $-B/2A$. Supposing there is backward bifurcation at $R_0 = 1$, then there are two endemic equilibria for an interval of values of basic reproduction number starting from a threshold R_0^C defined by $B = -2\sqrt{AC}$

to a point where $R_0 = 1$. To obtain this threshold R_0^C which is often referred to as the critical value of the basic reproduction number we replace the values of A, B, and C into the equality $B^2 = 4AC$ to obtain a quadratic equation in terms of Λ. For mathematical tractability we redefine coefficients B and C as $B = B_1 - B_2\Lambda$ and $C = C_1 - C_2\Lambda$, where

$$B_1 = (\mu + \delta_1 + \xi)(p + \sigma + \delta_2 + \mu)\beta$$
$$+ \mu\omega(\mu + \delta_1 + \xi)(p + \sigma + \delta_2 + \mu)$$
$$+ \alpha(\sigma + \delta_2 + \mu)\beta$$
$$B_2 = \beta\omega(p + \sigma + \delta_2 + \mu),$$
$$C_1 = \mu(\mu + \delta_1 + \xi)(p + \sigma + \delta_2 + \mu)$$
$$+ \alpha\mu(\sigma + \delta_2 + \mu),$$
$$C_2 = \beta(p + \sigma + \delta_2 + \mu),$$
$$B_1, B_2, C_1, C_2 > 0. \tag{29}$$

Note that $A > 0$ remains as previously defined in (16). Now the quadratic equation in terms of Λ can be obtained as

$$B_2^2\Lambda^2 + (4AC_2 - 2B_1B_2)\Lambda + (B_1^2 - 4AC_1) = 0 \tag{30}$$

Since we are considering the scenario where $B < 0$ and $R_0 < 1$, we have $C_1B_2 > C_2B_1$ and thus just the single solution

$$\Lambda^c = \frac{B_1B_2 - 2AC_2 + 2\sqrt{A^2C_2^2 + AB_2(C_1B_2 - C_2B_1)}}{B_2^2}$$
$$> 0. \tag{31}$$

Thus, the critical value of basic reproduction number (i.e., the new threshold for heroin eradication), R_0^C, is given as

$$R_0^C$$
$$= \frac{\beta\Lambda^c(p + \delta_2 + \mu + \sigma)}{\mu\alpha(\mu + \delta_2 + \sigma) + \mu(\mu + \delta_1 + \xi)(p + \delta_2 + \mu + \sigma)}. \tag{32}$$

Consequently, from the above analysis of computation of threshold for heroin eradication we can deduce the following lemma.

Lemma 10. *(a) If $R_0 > 1$, then model (1) has a unique endemic equilibrium point E^*. In this case heroin epidemic will persist in the community.*

(b) If $R_0^C < R_0 < 1$, then model (1) has two endemic equilibria $\overline{E}_1$ and $\overline{E}_2$ and signals that model (1) has backward bifurcation.

(c) If $R_0 < R_0^C < 1$, then model (1) has only the heroin-free equilibrium point D^0 and in this case heroin users will disappear.

Figure 3 exhibits typical bifurcation diagrams for model (1). To obtain the graphs we vary recruitment rate Λ while other parameter values are held fixed. The parameters used for the numerical simulation that leads to Figure 3(a) include $\omega = 0.11$, $\mu = 0.01$, $\beta = 0.001$, $\delta_1 = 0.002$, $\delta_2 = 0.001$, $\xi = 0.015$, $\alpha = 0.9$, $p = 0.467$, $\sigma = 0.1$, and $1 \leq \Lambda \leq 3$. Figure 3(a) represents the forward bifurcation scenario where if $R_0 < 1$ the heroin-free equilibrium is globally asymptotically stable while when $R_0 > 1$ the heroin epidemic can persist. However, as we note from Figure 3(b), increasing parameter ω from $\omega = 0.11$ to $\omega = 0.25$ such that $\omega > \omega_c$, a heroin epidemic can persist once established for a range of R_0 values that are below unity which indicates the occurrence of backward bifurcation. This implies that reducing R_0 below one will not necessarily be sufficient for eradication of heroin usage from the community. If R_0 is sufficiently decreased such that $R_0 < R_0^C$ the positive equilibrium no longer exists and heroin usage will cease to thrive and will eventually fall from its relatively high endemic level to the heroin-free equilibrium. From Figure 3(b) we note that when $R_0^C \leq R_0 \leq 1$ there are a stable endemic equilibrium, an unstable endemic equilibrium, and a stable heroin-free equilibrium. When $R_0 > 1$ there is only one stable endemic equilibrium. Figure 3(c) shows the effect of the saturation parameter ω on R_0^C; namely, increasing ω decreases R_0^C. Figure 3(d) shows that increasing the treatment rate α increases R_0^C which epidemiologically implies that high cure rates of heroin users can lead to shrinking of the backward bifurcation regime.

4. Global Stability

According to Theorem 8, model (1) may have multiple equilibria when $R_0 < 1$ and a unique endemic equilibrium whenever $R_0 > 1$. First, global stability of the endemic equilibrium of model (1) is investigated for a special case, that is, when $\omega = \xi = \sigma = 0$, using Lyapunov direct method, and later proven for the general model (i.e., $\omega, \sigma, \xi > 0$) using a geometric approach.

4.1. Global Stability of Endemic Equilibria E_1^* Using Lyapunov Method (Special Case $\omega = \xi = \sigma = 0$). Lyapunov functions have previously been used in proving global stability of epidemic models; for instance, see [37–40] and the references therein. Now consider the following theorem.

Theorem 11. *If $\omega = \xi = \sigma = 0$ the unique endemic equilibrium E_1^* of model (1) is globally asymptotically stable in the interior of Ω if $R_0 > 1$.*

Proof. Defining the following Lyapunov candidate function:

$$W(S, U_1, U_2) = \frac{(S - S^*)^2}{2S^*} + \left(U_1 - U_1^* - U_1^* \ln \frac{U_1}{U_1^*}\right)$$
$$+ \frac{pU_2^*}{\alpha U_1^*}\left(U_2 - U_2^* - U_2^* \ln \frac{U_2}{U_2^*}\right). \tag{33}$$

Now computing the time derivative of $W(S, U_1, U_2)$ along the solutions of system (1) results in

$$W'(S, U_1, U_2)$$

$$= \left(\frac{S - S^*}{S^*}\right)\frac{dS}{dt} + \left(\frac{U_1 - U_1^*}{U_1}\right)\frac{dU_1}{dt}$$

$$+ \frac{pU_2^*}{\alpha U_1^*}\left(1 - \frac{U_2^*}{U_2}\right)\frac{dU_2}{dt}$$

$$= \left(\frac{S - S^*}{S^*}\right)[\Lambda - \beta U_1 S - \mu S] \tag{34}$$

$$+ \left(\frac{U_1 - U_1^*}{U_1}\right)[\beta U_1 S + pU_2 - (\mu + \delta_1 + \alpha)U_1]$$

$$+ \frac{pU_2^*}{\alpha U_1^*}\left(1 - \frac{U_2^*}{U_2}\right)[\alpha U_1 - (p + \delta_2 + \mu)U_2].$$

Because (S^*, U_2^*, U_1^*) is an endemic steady point of model (1) when $\omega = \xi = \sigma = 0$, then it follows that

$$\Lambda = \beta U_1^* S^* + \mu S^*,$$

$$(p + \delta_2 + \mu) = \frac{\alpha U_1^*}{U_2^*}, \tag{35}$$

$$(\mu + \delta_1 + \alpha) = \beta S^* + \frac{pU_2^*}{U_1^*}.$$

Using (35) in (34) yields

$$W'(S, U_1, U_2)$$

$$= -\left(\frac{S - S^*}{S^*}\right)(\mu(S - S^*) + \beta(SU_1 - S^* U_1^*))$$

$$+ (U_1 - U_1^*)\left(\beta(S - S^*) + p\left(\frac{U_2}{U_1} - \frac{U_2^*}{U_1^*}\right)\right) \tag{36}$$

$$+ \frac{pU_2^*}{\alpha U_1^*}\left(1 - \frac{U_2^*}{U_2}\right)\left(\alpha U_1 - \frac{\alpha U_2 U_1^*}{U_2^*}\right).$$

Note that

$$SU_1 - S^* U_1^* = S^*(U_1 - U_1^*) + U_1(S - S^*). \tag{37}$$

Replacing the above equality in (36) results in

$$W'(S, U_1, U_2) = -\left(\frac{(S - S^*)^2}{S^*}\right)(\mu + \beta U_1)$$

$$+ pU_2^*\left(2 - \frac{U_1^* U_2}{U_1 U_2^*} - \frac{U_2^* U_1}{U_2 U_1^*}\right) \tag{38}$$

$$= -\left(\frac{(S - S^*)^2}{S^*}\right)(\mu + \beta U_1)$$

$$- pU_2^*\left[\sqrt{\frac{U_1^* U_2}{U_1 U_2^*}} - \sqrt{\frac{U_2^* U_1}{U_2 U_1^*}}\right]^2.$$

Hence, $W'(S, U_1, U_2) \leq 0$ for all $S, U_1, U_2 > 0$. Hence, the heroin endemic equilibrium E_1^* is stable and $W'(S, U_1, U_2) = 0$ if and only if $S = S^*, U_2 = U_2^*, U_1 = U_1^*$. The largest compact invariant set when $\omega = \xi = \sigma = 0$ in $\{(S^*, U_1^*, U_2^*) \in \mathbb{R}_+^3\}$ is the singleton $\{E_1^*\}$. Therefore, by LaSalle's invariance principle, the endemic steady state E_1^* is globally asymptotically stable in the interior of $\mathbb{R}_+^3$. □

The previous global stability analysis was only relevant for very specific case. In the subsequent subsection we use the geometric approach by Li and Muldowney [11, 13, 41] to obtain sufficient condition that ensures that the unique endemic equilibrium is globally asymptotically stable for a wide range of parameter values.

4.2. A Geometric Approach to Global Stability. For the general model, global stability is investigated using the Li and Muldowney [11, 13, 41] generalizations of the Poincaré-Bendixson approach for systems of $n > 2$ ordinary differential equations. This criterion is sometimes referred to as a *geometric approach to global stability* [14, 42].

To apply the geometric approach on model (1), consider the autonomous dynamical system $dy/dt = f(y)$, where $f = (f_1, f_2, f_3, f_4)^T$ and f_1, f_2, f_3, f_4 represent the right-hand side of system (1), respectively. We first briefly outline the general mathematical framework of the procedure developed in Li and Muldowney [13, 16].

Suppose that the map $y \mapsto f(y)$ is a C^1 function for y in an open subset $D \subset \mathbb{R}^n$ and consider the following autonomous dynamical system:

$$y' = f(y). \tag{39}$$

Let $y(t, y_0)$ be the solution to (39) satisfying $y(0, y_0) = y_0$. Now we make the following basic assumptions:

(H3) D is simply connected.

(H4) There exists a compact absorbing set $K \subset D$.

(H5) Equation (39) has a unique equilibrium y^* in D.

Now under the stated assumptions (H3)–(H5), y^* is said to be globally stable in D if it is locally stable and all trajectories in D converge to the same equilibrium y^*. That is, system (39) has no nonconstant periodic solutions. It is important to mention that global stability can be tested by *Bendixson criteria*. For $n \geq 2$ a *Bendixson criterion* refers to a condition satisfied by field f which precludes the existence of nonconstant periodic solutions of (39). When $n = 2$, (i.e., the planar case) the classical results (Poincaré-Bendixson theorem and Dulac criteria; see [43]) adequately provide such global conditions. For $n \geq 3$ a remarkable approach for proving global stability is given by the work of Li and Muldowney [11, 13, 16]. They showed that if conditions (H3)–(H5) hold and differential equation (39) fulfils a Bendixson criterion that is robust under C^1 local ε-perturbations (a function $g \in C^1(D \to \mathbb{R}^n)$ is called a C^1 local ε-perturbation of f at $y_0 \in D$ if there exists an open neighbourhood U of y_0 in D such that the support $\text{supp}(f - g) \subset U$ and $|f - g|_{C^1} < \varepsilon$, where $|f - g|_{C^1} = \sup\{|f(y) - g(y)| + |f_y(y) - g_y(y)|$:

$y \in D\}$) of f at all nonequilibrium nonwandering (a point $y_0 \in D$ is said to be nonwandering for system (39) if for any neighbourhood U of y_0 in D there exists arbitrary large t such that $U \cap y(t, U) \neq \emptyset$. As an example, any equilibrium, alpha limit point, or omega limit point is nonwandering) points for system (39), then y^* is globally stable in D provided that it is stable. We now state the new Bendixson criterion based on the use of the Lozinskiĭ measure as developed in [13]. Consider the differential equation (39) under the stated assumptions $(H3)$–$(H5)$. Let $y \mapsto P(y) \binom{n}{2} \times \binom{n}{2}$ be a matrix-valued function which is C^1 for $y \in D$ and consider

$$A = P_f P^{-1} + P J^{[2]} P^{-1}, \qquad (40)$$

where P_f is the directional derivative of P in the direction of the vector field f in system (39) and it is defined as

$$\left(p_{i,j}(y) \right)_f = \left(\frac{\partial p_{i,j}(y)}{\partial x} \right)^T \cdot f(y) = \nabla p_{i,j} \cdot f(y), \qquad (41)$$

and $J^{[2]}$ represents the second additive compound matrix J (where $J(y) = Df(y)$). In [44] the relation of compound matrices to differential equations is established. It is shown that, for an arbitrary $n \times n$ matrix $J = (J_{i,j})$, $J^{[2]}$ is an $\binom{n}{2} \times \binom{n}{2}$ matrix. Now define the following quantity:

$$\bar{q}_2 = \limsup_{t \to \infty} \; \sup_{y_0 \in \mho} \frac{1}{t} \int_0^t \rho\left(A(s, y_0) \right) ds, \qquad (42)$$

where $\rho(A)$ is the Lozinskiĭ measure of A with respect to vector norm $|\cdot|$ in $\mathbb{R}^N$, $N = \binom{n}{2}$, and $\rho(A)$ is defined as

$$\rho(A) = \lim_{h \to 0^+} \frac{|1 + hA| - 1}{h} \qquad (43)$$

(see [45, 46]). In paper [13] it is proved that if conditions $(H3)$ and $(H4)$ are satisfied then $\bar{q}_2 < 0$, indicating that there are no orbits giving rise to simple closed rectifiable curve in D that is invariant for system (39) (i.e., periodic orbits, homoclinic orbits, and heteroclinic cycles). Furthermore, it has been demonstrated in [13] that, under the stated assumptions

$(H3)$–$(H5)$, quantity $\bar{q}_2 < 0$ implies the local stability of equilibrium point y^*. As a result the following theorem is true.

Theorem 12 (see [13]). *Assuming that conditions $(H3)$–$(H5)$ hold, then, the equilibrium point y^* is globally asymptotically stable in D if a function $P(y)$ and a Lozinskiĭ measure ρ exist such that quantity $\bar{q}_2 < 0$.*

Observe that whenever $R_0 > 1$, there exists a unique and positive endemic equilibrium E^* (see Lemma 10) for model system (1). The method outlined above requires that (i) the endemic equilibrium E^* is unique in the interior of $\mho$ (i.e., condition $(H5)$ holds) and (ii) in the interior of $\mho$ there exists an absorbing compact set (condition $(H4)$ holds). The heroin model studied here with the assumption that $R_0 > 1$ fulfils conditions $(H4)$–$(H5)$. It is easy to prove that when $R_0 > 1$, the heroin-free equilibrium D^0 is unstable (see Theorem 3). The instability of the heroin-free equilibrium D^0 combined with $D^0 \in \delta\mho$ signals uniform persistence [47]. That is, there exists a positive constant $c_0 > 0$ such that for every solution $(S(t), U_1(t), U_2(t), U_3(t))$ of system (1) with $(S(0), U_1(0), U_2(0), U_3(0))$ in the interior of biologically feasible region $\mho$ satisfies

$$\liminf_{t \to \infty} |S(t), U_1(t), U_2(t), U_3(t)| \geq c_0. \qquad (44)$$

Because of boundedness of the region $\mho$, uniform persistence is equivalent to the existence of a compact set in the interior of $\mho$ which is absorbing for (1) (see [48]). Hence, condition $(H4)$ is satisfied. Also it is shown that whenever $R_0 > 1$ the model system (1) has only one equilibrium E^* in the interior of $\mho$, so that condition $(H5)$ is verified. Now for the heroin model system (1) the task involves verifying the Bendixson criterion (65). Note that the variable U_3 does not affect first, second, and third equation of system (1). Thus, the fourth equation can be dropped from the analysis, and we only need to consider the following subsystem:

$$\frac{dS}{dt} = \Lambda - \beta U_1 S - \mu S,$$

$$\frac{dU_1}{dt} = \beta U_1 S + p U_2 - (\mu + \delta_1 + \xi) U_1 - \frac{\alpha U_1}{1 + \omega U_1}, \qquad (45)$$

$$\frac{dU_2}{dt} = \frac{\alpha U_1}{1 + \omega U_1} - (p + \sigma + \delta_2 + \mu) U_2.$$

The Jacobian matrix of subsystem (45) is found to be

$$J = \begin{pmatrix} -(\beta U_1 + \mu) & -\beta S & 0 \\ \beta U_1 & \beta S - (\mu + \delta_1 + \xi) - \dfrac{\alpha}{\left(1 + \omega U_1\right)^2} & p \\ 0 & \dfrac{\alpha}{\left(1 + \omega U_1\right)^2} & -(p + \sigma + \delta_2 + \mu) \end{pmatrix}. \qquad (46)$$

In working with Theorem 12 one needs to make use of additive compound matrices. For an arbitrary 3×3 matrix B, the second additive compound matrix $B^{[2]}$ is defined as

$$B = \begin{pmatrix} b_{11} & b_{12} & b_{13} \\ b_{21} & b_{22} & b_{23} \\ b_{31} & b_{32} & b_{33} \end{pmatrix}, \tag{47}$$

$$B^{[2]} = \begin{pmatrix} b_{11} + b_{22} & b_{23} & -b_{13} \\ b_{32} & b_{11} + b_{33} & b_{12} \\ -b_{31} & b_{21} & b_{22} + b_{33} \end{pmatrix}.$$

Thus, the second additive compound matrix of Jacobian matrix J of system (45) is given as

$$J^{[2]} = \begin{pmatrix} J_{11} & p & 0 \\ \dfrac{\alpha}{(1 + \omega U_1)^2} & J_{22} & -\beta S \\ 0 & \beta U_1 & J_{33} \end{pmatrix}, \tag{48}$$

where

$$J_{11} = -\beta U_1 - (2\mu + \delta_1 + \xi) - \dfrac{\alpha}{(1 + \omega U_1)^2} + \beta S,$$

$$J_{22} = -\beta U_1 - (2\mu + p + \sigma + \delta_2), \tag{49}$$

$$J_{33} = -\dfrac{\alpha}{(1 + \omega U_1)^2} - (2\mu + p + \sigma + \delta_1 + \delta_2 + \xi)$$
$$+ \beta S.$$

For the model system (45) a suitable vector norm $| \cdot |$ in $\mathbb{R}^3$ and a 3×3 matrix-valued function $P(y)$ are given by

$$P(S, U_1, U_2) = \begin{pmatrix} 1 & 0 & 0 \\ 0 & \dfrac{U_1}{U_2} & 0 \\ 0 & 0 & \dfrac{U_1}{U_2} \end{pmatrix},$$

Thus $P_f P^{-1} = \begin{pmatrix} 0 & 0 & 0 \\ 0 & \dfrac{U_1'}{U_1} - \dfrac{U_2'}{U_2} & 0 \\ 0 & 0 & \dfrac{U_1'}{U_1} - \dfrac{U_2'}{U_2} \end{pmatrix}, \tag{50}$

$$P J^{[2]} P^{-1}$$

$$= \begin{pmatrix} J_{11} & \dfrac{pU_2}{U_1} & 0 \\ \dfrac{\alpha U_1}{(1 + \omega U_1)^2 U_2} & J_{22} & -\beta S \\ 0 & \beta U_1 & J_{33} \end{pmatrix}.$$

Note that upper prime ($'$) denotes differentiation with respect to time, $d./dt$. Thus, $A = P_f P^{-1} + P J^{[2]} P^{-1}$ can be obtained as

$$A$$
$$= \begin{pmatrix} J_{11} & \dfrac{pU_2}{U_1} & 0 \\ \dfrac{\alpha U_1}{(1 + \omega U_1)^2 U_2} & J_{22} + \dfrac{U_1'}{U_1} - \dfrac{U_2'}{U_2} & -\beta S \\ 0 & \beta U_1 & J_{33} + \dfrac{U_1'}{U_1} - \dfrac{U_2'}{U_2} \end{pmatrix}. \tag{51}$$

It is helpful to write matrix A in block form as

$$A = \begin{bmatrix} A_{11} & A_{12} \\ A_{21} & A_{22} \end{bmatrix}, \tag{52}$$

where

$$A_{11} = -\beta U_1 - (2\mu + \delta_1 + \xi) - \dfrac{\alpha}{(1 + \omega U_1)^2} + \beta S,$$

$$A_{12} = \begin{bmatrix} \dfrac{pU_2}{U_1} & 0 \end{bmatrix},$$

$$A_{21} = \begin{bmatrix} \dfrac{\alpha U_1}{(1 + \omega U_1)^2 U_2} & 0 \end{bmatrix}^T, \tag{53}$$

$$A_{22} = \begin{bmatrix} J_{22} + \dfrac{U_1'}{U_1} - \dfrac{U_2'}{U_2} & -\beta S \\ \beta U_1 & J_{33} + \dfrac{U_1'}{U_1} - \dfrac{U_2'}{U_2} \end{bmatrix}.$$

Following [13], let (u, v, w) represent the vectors in $\mathbb{R}^3 \cong \mathbb{R}^{\binom{3}{2}}$. Now for the norm $| \cdot |$ in $\mathbb{R}^3$ select

$$|(u, v, w)| = \max\{|u|, |v| + |w|\} \tag{54}$$

and let ρ represent the Lozinskiĭ measure with respect to this norm. Applying the method of approximating $\rho(A)$ as given in [46] leads to

$$\rho(A) \leq \sup\{g_1, g_2\}, \tag{55}$$

where

$$g_1 = \rho_1(A_{11}) + |A_{12}|$$
$$g_2 = |A_{21}| + \rho_1(A_{22}). \tag{56}$$

Here $|A_{12}|$ and $|A_{21}|$ are operator norms of A_{12} and A_{21} with respect to the l_1 vector norm, where they are both regarded as mapping from $\mathbb{R}^2$ to $\mathbb{R}$. $\rho_1(A_{22})$ represents the Lozinskiĭ measure of the 2×2 matrix A_{22} with respect to the l_1 norm in $\mathbb{R}^2$. To obtain $\rho_1(A_{22})$ we sum the absolute value of the off-diagonal elements to the diagonal one in each column of

A_{22} and then take the maximum of two sums. Assuming that $(1/2)(\delta_1 + \xi + \alpha/(1 + \omega U_1)^2) > \beta S$, it follows that

$$\rho_1\left(A_{11}\right) = -\beta U_1 - (2\mu + \delta_1 + \xi) - \frac{\alpha}{\left(1 + \omega U_1\right)^2} + \beta S,$$

$$\left|A_{12}\right| = \max\left\{\frac{pU_2}{U_1}, 0\right\} = \frac{pU_2}{U_1},$$

$$\left|A_{21}\right| = \max\left\{\frac{\alpha U_1}{\left(1 + \omega U_1\right)^2 U_2}, 0\right\}^T = \frac{\alpha U_1}{\left(1 + \omega U_1\right)^2 U_2},$$

$$\rho_1\left(A_{22}\right) = \max\left\{\frac{U_1'}{U_1} - \frac{U_2'}{U_2} - (2\mu + p + \sigma + \delta_2), \frac{U_1'}{U_1}\right. \tag{57}$$

$$- \frac{U_2'}{U_2} - (2\mu + p + \sigma + \delta_2) - \delta_1 - \xi - \frac{\alpha}{\left(1 + \omega U_1\right)^2}$$

$$\left. + 2\beta S \right\} = \frac{U_1'}{U_1} - \frac{U_2'}{U_2} - (2\mu + p + \sigma + \delta_2).$$

Thus, g_1 and g_2 are, respectively, given as

$$g_1 = \rho_1\left(A_{11}\right) + \left|A_{12}\right|$$

$$= \beta S + \frac{pU_2}{U_1} - \beta U_1 - \frac{\alpha}{\left(1 + \omega U_1\right)^2} - (2\mu + \delta_1 + \xi) \tag{58}$$

$$g_2 = \left|A_{21}\right| + \rho_1\left(A_{22}\right)$$

$$= \frac{\alpha U_1}{\left(1 + \omega U_1\right)^2 U_2} + \frac{U_1'}{U_1} - \frac{U_2'}{U_2} - (2\mu + p + \sigma + \delta_2). \tag{59}$$

Now from second and third equation of (45) it is easy to obtain the following:

$$\frac{U_1'}{U_1} = \beta S + \frac{pU_2}{U_1} - \frac{\alpha}{\left(1 + \omega U_1\right)} - (\mu + \delta_1 + \xi) \tag{60}$$

$$\frac{U_2'}{U_2} = \frac{\alpha U_1}{\left(1 + \omega U_1\right) U_2} - (p + \sigma + \delta_2 + \mu). \tag{61}$$

Substituting (60) into (58) and (61) into (59), respectively, leads to

$$g_1 = \frac{U_1'}{U_1} - \mu - \beta U_1 + \frac{\alpha \omega U_1}{\left(1 + \omega U_1\right)^2}$$

$$\leq \frac{U_1'}{U_1} - \mu + \frac{\alpha \omega U_1}{\left(1 + \omega U_1\right)^2},$$

$$g_2 = \frac{U_1'}{U_1} - \mu - \frac{\alpha \omega U_1^2}{\left(1 + \omega U_1\right)^2 U_2} \tag{62}$$

$$\leq \frac{U_1'}{U_1} - \mu.$$

Now based on the definition of the method of approximating Lozinskiĭ measure of A, $\rho(A)$ as given in [46], we now approximate the supremum of both g_1 and g_2. Hence,

$$\rho\left(A\right) \leq \sup\left(g_1, g_2\right)$$

$$= \sup\left\{\frac{U_1'}{U_1} - \mu + \frac{\alpha \omega U_1}{\left(1 + \omega U_1\right)^2}, \frac{U_1'}{U_1} - \mu\right\}$$

$$= \left(\frac{U_1'}{U_1} - \mu\right) + \sup\left\{\frac{\alpha \omega U_1}{\left(1 + \omega U_1\right)^2}, 0\right\} \tag{63}$$

$$\leq \frac{U_1'}{U_1} - \mu + \frac{\alpha \omega U_1}{\left(1 + \omega U_1\right)^2}.$$

Thus we get the following inequality:

$$\rho\left(A\right) \leq \frac{U_1'}{U_1} - \mu + \frac{\alpha \omega U_1}{\left(1 + \omega U_1\right)^2}. \tag{64}$$

Now the next step involves substituting $\rho(A)$ in

$$\bar{q}_2 = \limsup_{t \to \infty} \sup_{y_0 \in \mho} \frac{1}{t} \int_0^t \rho\left(A\left(s, y_0\right)\right) ds \tag{65}$$

and deducing whether $\bar{q}_2 < 0$. And if the inequality $\bar{q}_2 < 0$ does not hold we will need to establish a condition that leads to $\bar{q}_2 < 0$ being fulfilled.

Considering uniform persistence, there exist $c_0 > 0$ and $T > 0$ such that, for $t > T$, the following is implied:

$$S(t) \geq c_0,$$

$$U_1(t) \geq c_0,$$

$$U_2(t) \geq c_0, \tag{66}$$

$$U_3(t) \geq c_0.$$

Now by letting $\Gamma_1 = \alpha \omega c_0/(1 + \omega c_0)^2$ and $\Gamma_2 = \mu$ the following claim is made: if

$$\Gamma_1 < \Gamma_2, \tag{67}$$

then it follows that

$$\rho\left(A\right) \leq \frac{U_1'}{U_1} - \widetilde{V}, \tag{68}$$

where

$$\widetilde{V} = \mu - \frac{\alpha \omega c_0}{\left(1 + \omega c_0\right)^2} > 0. \tag{69}$$

Now, for $t > T$, it can be deduced that

$$\bar{q}_2 = \frac{1}{t} \int_0^t \rho\left(A\right) ds = \frac{1}{t} \int_0^T \rho\left(A\right) ds + \frac{1}{t} \int_T^t \rho\left(A\right) ds$$

$$\leq \frac{1}{t} \log\frac{U_1(t)}{U_1(T)} + \frac{1}{t} \int_0^T \rho\left(A\right) ds - \widetilde{V}\frac{t - T}{t} < 0 \tag{70}$$

when $\widetilde{V} > 0$,

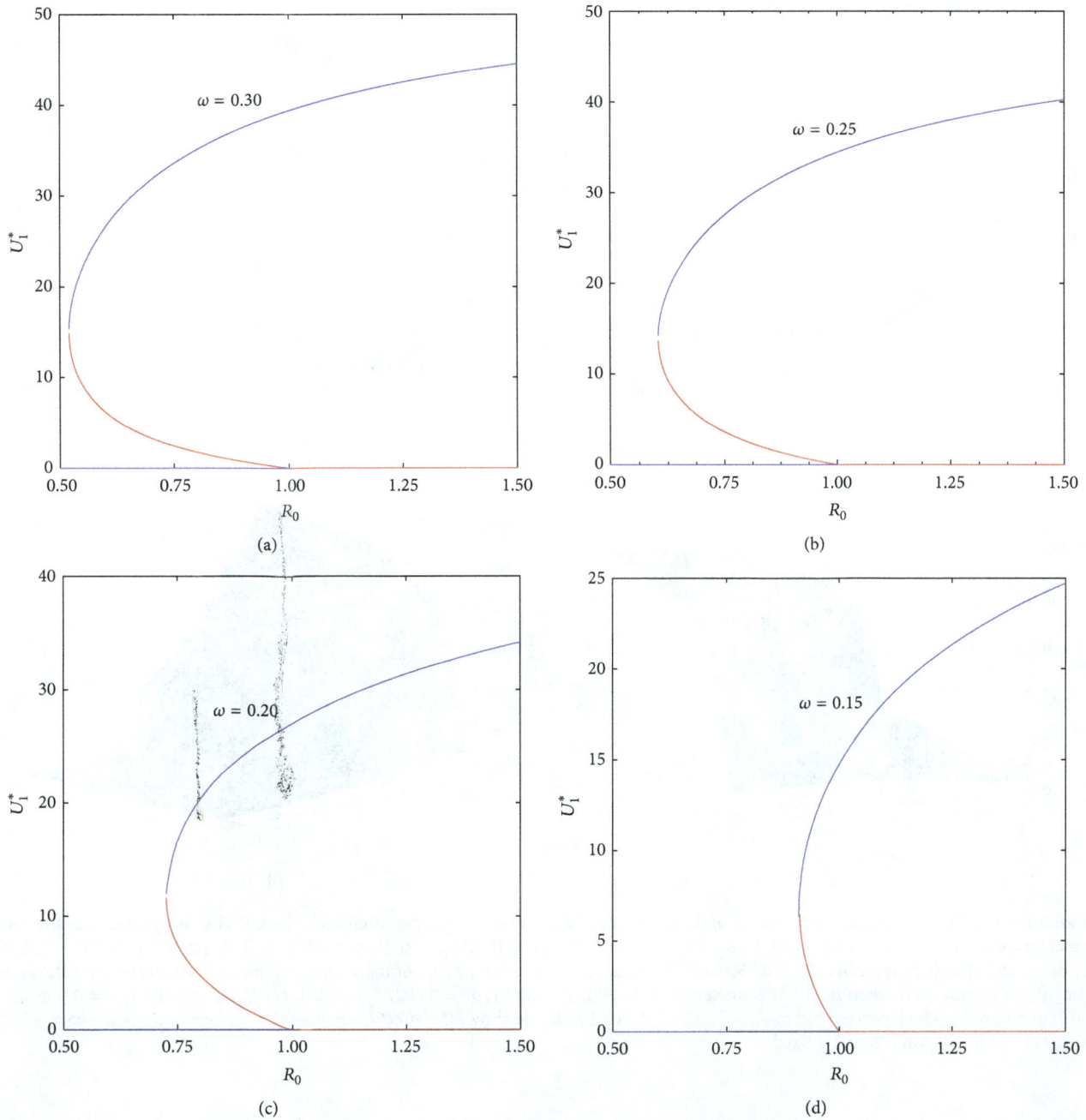

FIGURE 4: Illustration of the effect of increasing parameter ω that accounts for the saturation of heroin users. Here the parameters remain as in caption of Figure 3 except $\Lambda = 2$ while ω is shown in the figure. The heroin eradication thresholds (i.e., R_0^C) corresponding to (a)–(d) are now 0.5204, 0.6038, 0.7314, and 0.9131, respectively.

and, thus, the Bendixson criterion given by (42) is verified. However, it is important to observe that $\bar{q}_2 < 0$ if and only if condition (67) holds true. Thus, the following theorem is established.

Theorem 13. *Provided that $R_0 > 1$, if $\Gamma_1 < \Gamma_2$, then system (1) has a unique endemic equilibrium E^* which is globally asymptotically stable with respect to solutions of (1) originating in the interior of $\mho$.*

The validity of Theorem 13 will be shortly verified numerically.

5. Numerical Examples

In this section numerical simulations of the heroin epidemic model are presented to support theoretical findings. Figure 4 which shows backward bifurcation is obtained by plotting heroin users equilibrium as a function of R_0. The figures

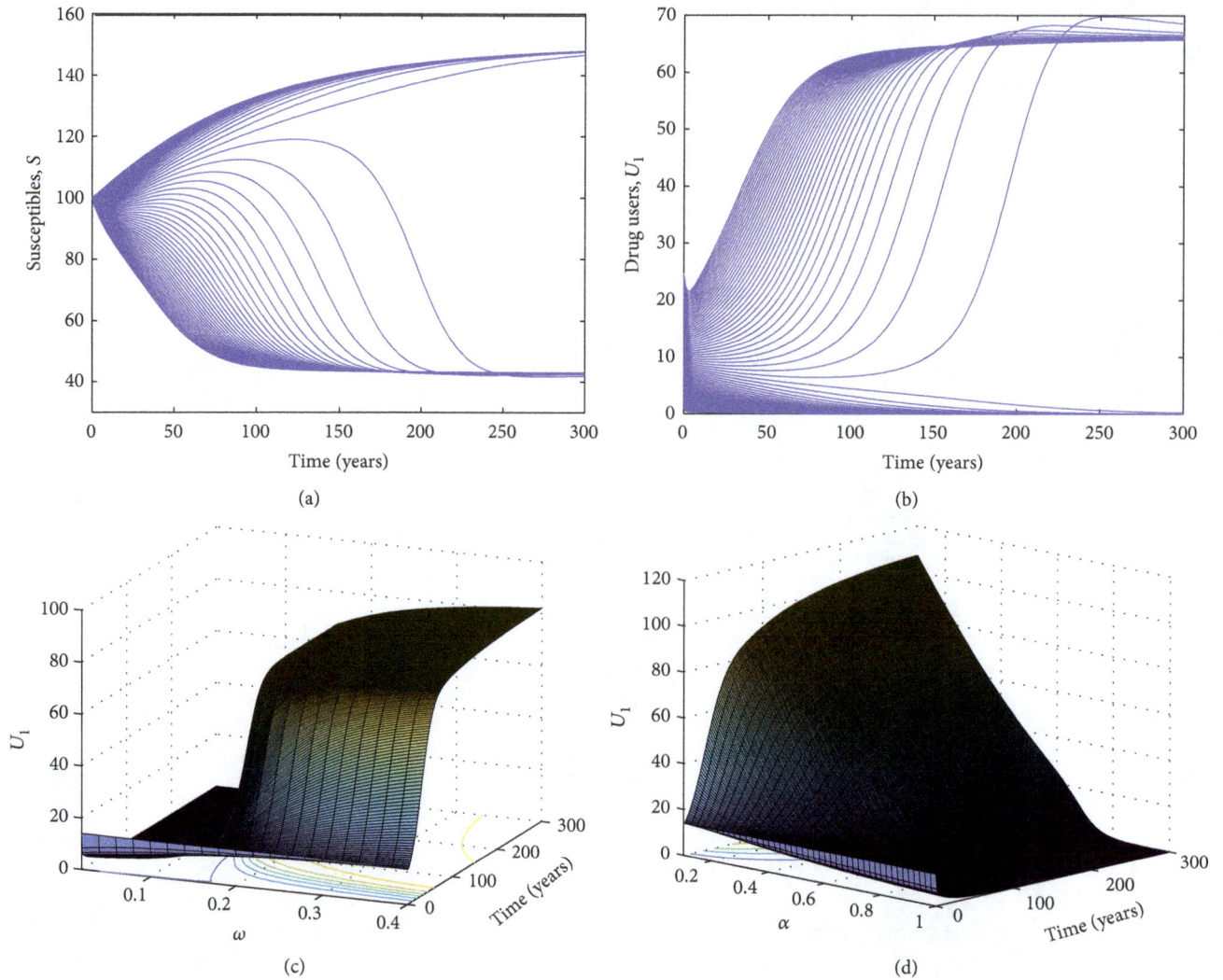

FIGURE 5: (a) and (b) illustrate the dependence of the heroin epidemic extinction or persistence on the initial states provided to the model. Parameters used include $\omega = 0.15$, $\beta = 0.001$, $\delta_1 = 0.002$, $\delta_2 = 0.001$, $\xi = 0.015$, $\mu = 0.01$, $\alpha = 0.9$, $p = 0.467$, $\sigma = 0.1$, and $\Lambda = 1.5$. With these parameters the corresponding $R_0 = 0.7506$. (c) The figure illustrates the impact of increasing parameter ω (the extent of saturation of treatment) on heroin users when $R_0 < 1$. Parameters used include $\beta = 0.001$, $\delta_1 = 0.002$, $\delta_2 = 0.001$, $\xi = 0.015$, $\mu = 0.01$, $\alpha = 0.9$, $p = 0.46$, $\sigma = 0.1$, and $\Lambda = 1.5$ which correspond to $R_0 = 0.7427 < 1$. (d) The figure shows the effect of treatment α on heroin users U_1 when all other parameters and initial conditions are fixed.

present a scenario where R_0 is varied via parameter β (i.e., $0.0005 \leq \beta \leq 0.0015$) and other parameters are fixed. Figures 4(a)–4(d) show that increasing the parameter ω leads to the expansion of the region of bistability while decreasing ω results in contraction of the bistability region. The heroin eradication threshold (also referred to as critical reproduction number) R_0^C shifts from right to left when ω increases and vice versa when ω decreases. High ω value implies not enough treatment for a large population of heroin users, thus favouring a situation where there will always be heroin users within the community even though $R_0 < 1$.

Figures 5(a) and 5(b) exhibit the time course of the heroin endemic in a parameter regime where there is a backward bifurcation. In both Figures 5(a) and 5(b) $R_0 = 0.7506 < 1$. The figures show the dependence of heroin usage on the size of the initial conditions supplied to the system,

which is a common characteristic of models that have a bistability region. If the model is supplied with initial conditions that are below the unstable curve (see the red solid line in Figure 3(b)) the solution trajectories are attracted to the heroin-free equilibrium while if initial conditions are chosen such that they are above the unstable curve, then the solution trajectories are attracted to a stable nontrivial equilibrium. Thus, in the case there is backward bifurcation, the initial number of people engaging in heroin use governs the course of the heroin epidemic.

Figure 5(c) shows the time course of the heroin users when the parameter ω that accounts for the extent of saturation of heroin users is varied while the initial states and all other parameter values are fixed to constant values. It can be seen that not all values of ω will trigger rapid growth towards an endemic equilibrium when $R_0 < 1$. Indeed, parameter ω

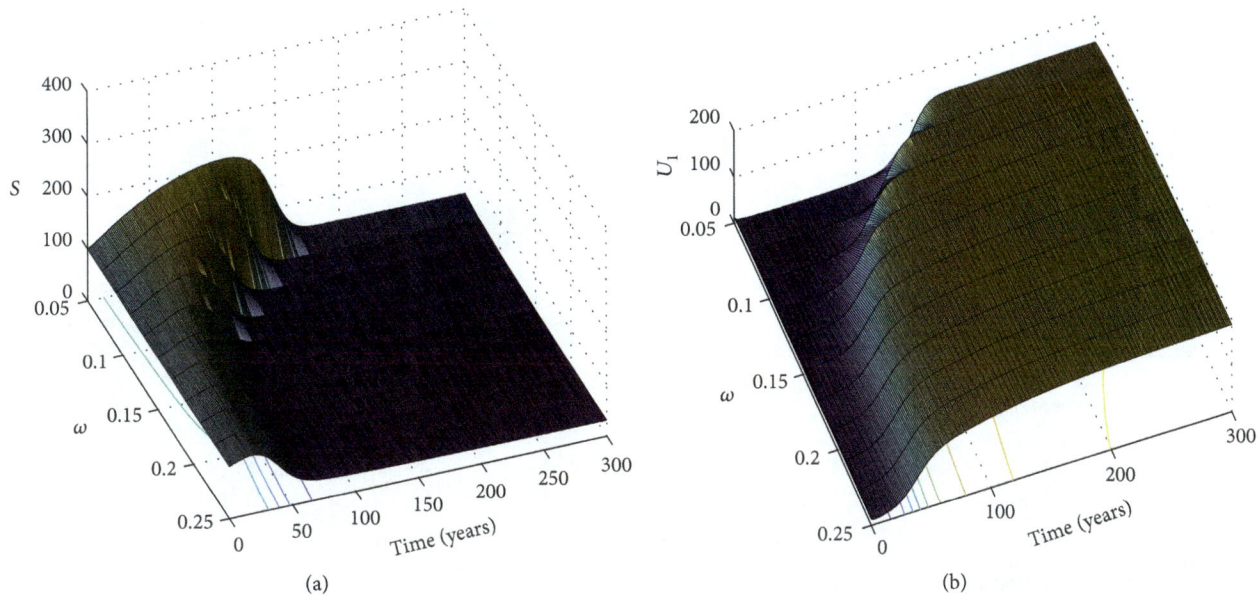

FIGURE 6: (a) and (b) show the effect of increasing the parameter ω that accounts for the extent of saturation of treatment when $R_0 > 1$. Parameters used are the same as in Figure 5(a) except $p = 0.46$ and $\Lambda = 3$ corresponding to $R_0 = 1.4855 > 1$.

has to exceed a certain fixed threshold ω_c, hence supporting our theoretical findings that a nonzero equilibrium when $R_0 < 1$ can only be maintained when ω is greater than ω_c; see (28). Figure 5(d) shows the effect of treatment α on heroin users. High treatment leads to a steady decline of heroin users.

Figure 6 presents a scenario where $R_0 > 1$. In this scenario we expect that when a heroin user enters a heroin-free community there will be rapid growth of heroin users until a globally stable equilibrium point is reached. Recalling that parameter ω does not appear in R_0, nevertheless it does affect the model dynamics. The impact of ω when $R_0 > 1$ is different from the case where $R_0 < 1$. For $R_0 < 1$ it plays a key role in inducing bistability while for $R_0 > 1$ the parameter ω impacts the heroin dynamics by determining the time taken for an epidemic to occur. For relatively high ω values there is a sudden decrease of susceptible subpopulation while for relatively low values of ω there is a gradual decrease of susceptible subpopulation. Moreover, Figure 6(b) depicts that for any given value of ω the heroin users gradually approach a stable endemic equilibrium point. The only striking difference is the time taken to reach the heroin endemic equilibrium. At high values of parameter ω the heroin endemic will rapidly approach an equilibrium.

We now verify the global stability condition obtained using the geometric approach based on the following parameter values:

(i) $\beta = 0.001, \delta_1 = 0.002, \delta_2 = 0.001, \xi = 0.015, \mu = 0.01, \alpha = 0.9, p = 0.467, \sigma = 0.1, \omega = 2, \Lambda = 3, c_0 = 50$. With these parameters the corresponding $R_0 = 1.5012 > 1$ and $\Gamma_1 = 0.0088 < \Gamma_2 = 0.0100$. In this scenario condition (67) is satisfied and the model system (1) should be globally asymptotically stable. Figures 7(a), 7(b), 7(c), and 7(d) show existence of an apparently stable equilibrium.

(ii) We now use the same set of parameter values as Case (i) except $\omega = 0.05$. This leads to $\Gamma_1 = 0.1837 > \Gamma_2 = 0.0100$. In this case the asymptotic stability condition is not satisfied and unsurprisingly model system (1) has periodic solutions as shown in Figures 7(e), 7(f), 7(g), and 7(h). The epidemiological interpretation of this is that the heroin epidemic will fluctuate between low and high endemic levels. The cycles are induced by time delays of the transmission processes.

5.1. Uncertainty and Sensitivity Analysis. Here we conduct sensitivity analysis so as to identify critical inputs of our heroin epidemic model and gain insights on how input uncertainty influences model outcome [49]. To achieve this we make use of the Latin hypercube sampling (LHS) technique which provides a comprehensive method of assessing model sensitivity to parameters over multidimensional parameter space. One of the advantages of using the LHS technique is that it requires fewer samples of parameters than simple random sampling to achieve the same accuracy (see [49] and the references therein for in-depth discussion on LHS). In our heroin epidemic model the LHS technique is important due to the relatively large uncertainty of the model parameter estimates we have used. The technique works in combination with the partial rank correlation coefficient (PRCC) which estimates the sign and strength of the relationship that exists between each model parameter and any specified output variable [50, 51]. The PRCC values are bounded between 1 and -1, with a PRCC value close to 1 (-1) indicating very strong positive (or negative) correlation. The relative importance of the model parameters can be directly evaluated by comparing the values of the PRCC [51]. The uncertainty and sensitivity analysis using the LHS technique involves first selecting a baseline value and a range for each parameter of the heroin

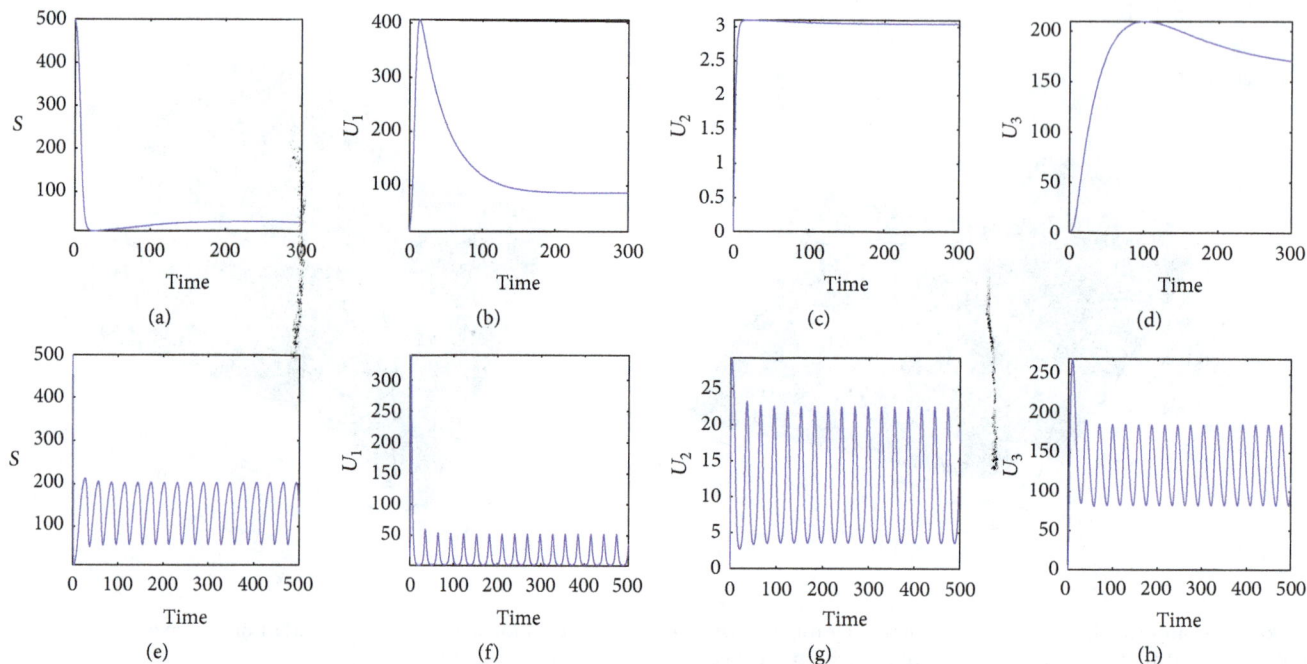

FIGURE 7: Illustration of validity of global stability condition. (a), (b), (c), and (d) represent a scenario where condition (67) holds true and global stability is predicted. (e), (f), (g), and (h) represent a scenario where condition (67) does not hold and oscillations are expected.

TABLE 2: Parameter baseline values and ranges used in sensitivity analysis.

Parameter	Baseline value	Range	Source
Λ	2	1–5	Assumed
β	0.001	0.0005–0.015	[7]
μ	$\frac{1}{80}$	[0.01125, 0.01375]	[52]
p	0.467	0.1–0.8	[7]
α	0.5	0.2–0.95	[23]
δ_1	0.002	0.0008–0.0025	Assumed
δ_2	0.001	0.00095–0.002	Assumed
ξ	0.5	0.05–0.5	Assumed
σ	0.1	0.1–0.7	Assumed
ω	0.01	0.008–0.25	[23]

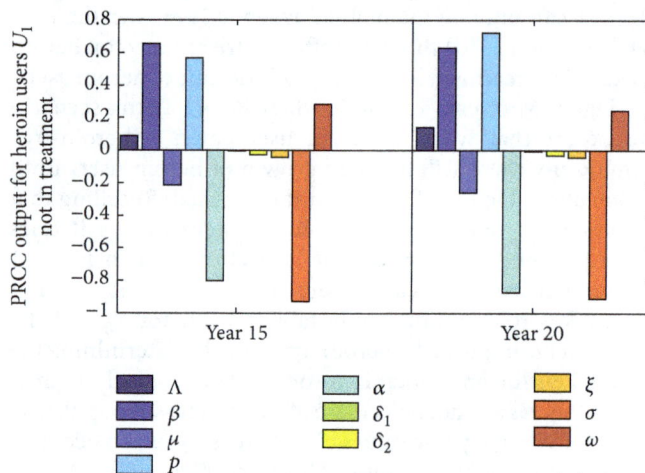

FIGURE 8: The PRCC output for heroin users $U_1(t)$ not in treatment.

epidemic model (1) (see Table 2) and then performing multiple runs for a given outcome variable or response function. To enhance accuracy, 1500 random samples of parameter values were used for the sensitivity analysis and significant levels are set for $p_{\text{value}} < 0.05$.

Figure 8 displays the sensitivity analysis results for the heroin users not in treatment $U_1(t)$. It is straightforward to see that recruitment rate Λ, effective contact rate β, relapsing rate of heroin users in treatment to heroin users not in treatment p, and saturation parameter ω are positively correlated while natural death μ, treatment rate α, heroin-induced death rates (δ_1, δ_2), "self-cure" rate ξ, and successful recovery rate of heroin users in treatment σ are negatively correlated. Among the positively correlated PRCC values the parameters β, p, and ω are strongly positively correlated to heroin users not in

treatment as evidenced by their high PRCC values. However, at time point year 15 the effective contact rate β has a slightly higher PRCC value than relapsing rate p suggesting that at initial stage of a heroin epidemic effective contact between heroin users $U_1(t)$ and susceptibles significantly contributes to the emergence of a heroin epidemic. On the other hand, at time point year 30 the situation observed at time point year 15 is reversed. That is, relapsing of heroin users in treatment has slightly higher PRCC values than the effective contact rate β. Hence, long-term relapsing of heroin users in treatment back to the heroin users not in treatment also plays a role in ensuring that there will always be heroin users within the community. Thus, attempting to control heroin usage within the community measures that ensure that heroin users undergoing treatment do not relapse should be of great importance. The extent of saturation of heroin users as a result of failure to treat heroin users promptly which is accounted by parameter ω also contributes to sustaining heroin epidemic. As suggested by the strongly negatively correlated PRCC value of parameter σ, ensuring that heroin users in treatment are successfully treated (i.e., they do not relapse) can substantially reduce the subpopulation of heroin users. In general the sensitivity analysis results suggest that, to combat heroin epidemic, policy makers and clinicians should target controlling effective contact rate β, relapsing rate p, and the extent of saturation of heroin users rate ω parameters.

6. Concluding Remarks

In this study we formulated a heroin epidemic model with bilinear and saturated treatment function. The threshold parameter R_0 usually referred to as the basic reproduction number plays a key role in the prediction of disease persistence or extinction. Epidemiologically, when R_0 exceeds one, an epidemic persists and if it is below unity the disease will die out. This classical viewpoint has recently been challenged by many researchers since it is not always true that a disease will disappear if R_0 is decreased below one. In the present heroin epidemic model the analytical results indicate that R_0 is indeed the threshold when the parameter $\omega = 0$. However, when a saturated treatment function (i.e., $\omega > 0$) rather than a linear treatment rate is used, the heroin model exhibits the phenomenon of backward bifurcation where a heroin-free equilibrium and two nontrivial equilibria coexist even though the basic reproduction number is below unity (see Theorem 8—Case (iii)). The appearance of backward bifurcation indicates that it is not sufficient to decrease the basic reproduction number below unity for the eradication of heroin users within the community. Thus, to effectively control the spread of heroin users one has to reduce R_0 below another threshold referred to as the critical value of the basic reproduction number R_0^C. That is, heroin users can be eradicated if $R_0 < R_0^C < 1$. It is important to note that although the parameter ω might be present in the model, not every value of ω will lead to bistability. Instead ω has to be greater than a certain threshold ω_c which is an aggregate of model parameters (see (28)). In general both analytical (see Appendix A for center manifold) and numerical results suggest that the saturation parameter ω is the one responsible for backward bifurcation. Failure to intervene before heroin users have accumulated in the community will lead to a situation where a heroin epidemic exists even though basic reproduction number is below one. Improvement of existing medical technology as well as channelling sufficient resources in medicines can significantly facilitate early intervention by ensuring that heroin users receive treatment promptly.

In addition global stability properties using both the Lyapunov direct method and geometric approach by Li and Muldowney have been investigated. We note that, even for a four-dimensional model, the use of the two nonlinear stability techniques becomes nontrivial. In fact when all the parameters of the model are accounted for, it is difficult if not impossible to design a Lyapunov function. Using the geometric approach we establish a global condition that accounts for all parameters that if satisfied signals that heroin persistence within the community is globally stable. However, if the global condition is not satisfied heroin users can oscillate periodically in number (see Figures 7(e), 7(f), 7(g), and 7(h)). Moreover, sensitivity and uncertainty analysis using LHS results indicate that effective contact between susceptibles and heroin users β, the relapsing of heroin users in treatment p, and the extent of saturation of heroin users parameter ω are the ones which contribute to persistence of heroin epidemic within the community.

Appendix

A. Derivation of R_0

Generally the geometric sequence is given as $\{a, ar, ar^2, ar^3, \ldots\}$ and the sum of a certain number of terms of the geometric sequence is given as $S_n = a(1 - r^n)/(1 - r)$, where S_n is the sum of n terms (nth partial sum), a is the first term, and r is the common ratio. Now considering the geometric sequence from expression (12), we note that $a = 1$ and $r = p\alpha/(\mu + \delta_1 + \xi + \alpha)(p + \sigma + \delta_2 + \mu) < 1$.

$$S_n = \frac{a(1 - r^n)}{1 - r} = \frac{a}{1 - r} \quad \text{since } r < 1. \tag{A.1}$$

Substituting a and r in (A.1) yields

$$S_n = \frac{(\mu + \delta_1 + \xi + \alpha)(p + \sigma + \delta_2 + \mu)}{(\mu + \delta_1 + \xi)(p + \sigma + \delta_2 + \mu) + \alpha(\sigma + \delta_2 + \mu)}. \tag{A.2}$$

Multiplying (A.2) by $1/(\mu + \delta_1 + \xi + \alpha)$ gives the required expression

$$\frac{p + \sigma + \delta_2 + \mu}{(\mu + \delta_1 + \xi)(p + \sigma + \delta_2 + \mu) + \alpha(\sigma + \delta_2 + \mu)} \tag{A.3}$$

which if multiplied by the effective contact rate β and the average recruitment rate Λ/μ yields the basic reproduction number R_0.

B. Proof of Existence of Backward Bifurcation

Proof. To prove existence of backward bifurcation in model equations (1), the center manifold approach as outlined by Castillo-Chavez and Song in [35] is used. First for clarity and understanding of the center manifold theory the model equations (1) variables are transformed as follows: $y_1 = S$, $y_2 = U_1$, $y_3 = U_2$, $y_4 = U_3$, and the total population $N = \sum_{j=1}^4 y_j$. Define $Y = (y_1, y_2, y_3, y_4)^T$ (T denotes transpose), such that the model equations (1) can be rewritten as $dY/dt = F(y)$, where $F = (f_1, f_2, f_3, f_4)$. Hence, it follows that

$$\frac{dy_1}{dt} = f_1 = \Lambda - \beta y_1 y_2 - \mu y_1$$

$$\frac{dy_2}{dt} = f_2$$

$$= \beta y_1 y_2 + p y_3 - (\mu + \delta_1 + \xi) y_2 - \frac{\alpha y_2}{1 + \omega y_2} \quad \text{(B.1)}$$

$$\frac{dy_3}{dt} = f_3 = \frac{\alpha y_2}{1 + \omega y_2} - (p + \sigma + \delta_2 + \mu) y_3$$

$$\frac{dy_4}{dt} = f_4 = \sigma y_3 + \xi y_2 - \mu y_4.$$

Now let $\beta = \beta^*$ be the bifurcation parameter. Observe that, at $R_0 = 1$,

$$\beta = \beta^*$$

$$= \frac{\alpha(\mu + \delta_2 + \sigma) + (\mu + \delta_1 + \xi)(p + \delta_2 + \mu + \sigma)}{y_1^*(p + \delta_2 + \mu + \sigma)}, \quad \text{(B.2)}$$

where $y_1^* = \Lambda/\mu = S_0$. With $\beta = \beta^*$ the transformed model equations (B.1) have a simple eigenvalue with zero real part and all other eigenvalues are negative (i.e., they have a hyperbolic equilibrium point). Thus, we can apply center manifold theory to investigate the local dynamics of the transformed system (B.1) near $\beta = \beta^*$. Now we proceed to obtain the Jacobian matrix of the transformed system evaluated at heroin-free equilibrium HFE as

$$J_{\text{HFE}}$$

$$= \begin{pmatrix} -\mu & -\beta S_0 & 0 & 0 \\ 0 & \beta S_0 - (\mu + \delta_1 + \xi + \alpha) & p & 0 \\ 0 & \alpha & -(p + \sigma + \delta_2 + \mu) & 0 \\ 0 & \xi & \sigma & -\mu \end{pmatrix}. \quad \text{(B.3)}$$

It is easy to obtain the right eigenvectors of this Jacobian matrix as $\overline{V} = (\tilde{v}_1, \tilde{v}_2, \tilde{v}_3, \tilde{v}_4)^T$, where

$$\begin{pmatrix} \tilde{v}_1 \\ \tilde{v}_2 \\ \tilde{v}_3 \\ \tilde{v}_4 \end{pmatrix} = \begin{pmatrix} \dfrac{-\beta S_0}{\mu} \\ 1 \\ \dfrac{\alpha}{p + \sigma + \delta_2 + \mu} \\ \dfrac{\xi(p + \sigma + \delta_2 + \mu) + \sigma\alpha}{\mu(p + \sigma + \delta_2 + \mu)} \end{pmatrix} \tilde{v}_2, \quad \text{(B.4)}$$

where $\tilde{v}_2 > 0$. Similarly, it is possible to obtain the left eigenvectors which we denote by $\overline{W} = (\tilde{w}_1, \tilde{w}_2, \tilde{w}_3, \tilde{w}_4)$ as

$$\tilde{w}_1 = 0,$$

$$\tilde{w}_2 = \tilde{w}_2 > 0,$$

$$\tilde{w}_3 = \frac{p\tilde{w}_2}{p + \sigma + \delta_2 + \mu}, \quad \text{(B.5)}$$

$$\tilde{w}_4 = 0.$$

Now we proceed to obtain the bifurcation coefficients a and b as defined in Theorem 4.1 of [35].   □

Calculation of Coefficient a. First the nonvanishing partial derivatives of the transformed model (B.1) evaluated at heroin-free equilibrium are obtained as

$$\frac{\partial^2 f_1(0,0)}{\partial y_1 \partial y_2} = -\beta^*,$$

$$\frac{\partial^2 f_2(0,0)}{\partial y_1 \partial y_2} = \beta^*,$$

$$\frac{\partial^2 f_2(0,0)}{\partial^2 y_2^2} = 2\omega\alpha, \quad \text{(B.6)}$$

$$\frac{\partial^2 f_3(0,0)}{\partial^2 y_2^2} = -2\omega\alpha$$

so that

$$a = \sum_{k,i,j=1}^{4} \tilde{w}_k \tilde{v}_i \tilde{v}_j \frac{\partial^2 f_k(0,0)}{\partial y_i \partial y_j} = \tilde{w}_2 \tilde{v}_1 \tilde{v}_2 \frac{\partial^2 f_2(0,0)}{\partial y_1 \partial y_2} + \tilde{w}_2 \tilde{v}_2 \tilde{v}_2 \frac{\partial^2 f_2(0,0)}{\partial^2 y_2^2} + \tilde{w}_3 \tilde{v}_2 \tilde{v}_2 \frac{\partial^2 f_3(0,0)}{\partial^2 y_2^2}$$

$$= 2\tilde{w}_2 \tilde{v}_2^2 \left[\frac{\mu\omega\alpha(\sigma + \delta_2 + \mu) - \beta^*(\alpha(\mu + \delta_2 + \sigma) + (\mu + \delta_1 + \xi)(p + \delta_2 + \mu + \sigma))}{\mu(p + \sigma + \delta_2 + \mu)} \right] \quad \text{(B.7)}$$

$$= \frac{2\tilde{w}_2 \tilde{v}_2^2 \alpha(\sigma + \delta_2 + \mu)}{p + \sigma + \delta_2 + \mu}[\omega - \omega_c] \quad \text{where } \omega_c \text{ remains as previously defined in (28)}.$$

Calculation of Coefficient b. The bifurcation coefficient b is obtained as

$$b = \sum_{k,i=1}^{4} \tilde{w}_k \tilde{v}_i \frac{\partial^2 f_k(0,0)}{\partial y_i \partial \beta^*} = \frac{\tilde{w}_2 \tilde{v}_2 \Lambda}{\mu} > 0. \quad \text{(B.8)}$$

According to Theorem 4.1 of [35] if both bifurcation coefficients a and b are positive then model (1) will exhibit backward bifurcation. Observe that b is always positive while $a > 0$ if and only if $\omega > \omega_c$. Thus, if $\omega > \omega_c$ then model (1) will exhibit the phenomenon of backward bifurcation.

Conflicts of Interest

The authors declare that there are no conflicts of interest regarding the publication of this paper.

Acknowledgments

The authors are grateful to Royal Melbourne Institute of Technology (RMIT) University for supporting this research.

References

[1] C. Munsey, Heroin® and aspirin® the connection! & the collection!-part i. 2005.

[2] Research Report Series: Heroine, National Institute on Drug Abuse and United States of America,.

[3] X. Li, Y. Zhou, and B. Stanton, "Illicit drug initiation among institutionalized drug users in China," *Addiction*, vol. 97, no. 5, pp. 575–582, 2002.

[4] R. J. Garten, S. Lai, J. Zhang et al., "Rapid transmission of hepatitis C virus among young injecting heroin users in Southern China," *International Journal of Epidemiology*, vol. 33, no. 1, pp. 182–188, 2004.

[5] R. M. Anderson and R. M. May, "Population biology of infectious diseases: part I," *Nature*, vol. 280, no. 5721, pp. 361–367, 1979.

[6] E. White and C. Comiskey, "Heroine epidemics, treatment and ode modelling," *Mathematical Biosciences*, vol. 208, no. 1, pp. 312–324, 2007.

[7] G. Mulone and B. Straughan, "A note on heroin epidemics," *Mathematical Biosciences*, vol. 218, no. 2, pp. 138–141, 2009.

[8] X. Wang, J. Yang, and X. Li, "Dynamics of a Heroin epidemic model with very population," *Applied Mathematics*, vol. 2, no. 6, pp. 732–738, 2011.

[9] G. P. Samanta, "Dynamic behaviour for a nonautonomous heroin epidemic model with time delay," *Journal of Applied Mathematics and Computing*, vol. 35, no. 1-2, pp. 161–178, 2011.

[10] A. M. Lyapunov, "The general problem of the stability of motion," *International Journal of Control*, vol. 55, no. 3, pp. 521–790, 1992.

[11] Y. Li and J. S. Muldowney, "On Bendixson's criterion," *Journal of Differential Equations*, vol. 106, no. 1, pp. 27–39, 1993.

[12] M. Y. Li and J. S. Muldowney, "On R. A. Smith's autonomous convergence theorem," *The Rocky Mountain Journal of Mathematics*, vol. 25, no. 1, pp. 365–379, 1995.

[13] M. Y. Li and J. S. Muldowney, "A geometric approach to global-stability problems," *SIAM Journal on Mathematical Analysis*, vol. 27, no. 4, pp. 1070–1083, 1996.

[14] J. Arino, C. C. McCluskey, and P. van den Driessche, "Global results for an epidemic model with vaccination that exhibits backward bifurcation," *SIAM Journal on Applied Mathematics*, vol. 64, no. 1, pp. 260–276, 2003.

[15] M. Y. Li, J. R. Graef, L. Wang, and J. Karsai, "Global dynamics of a seir model with varying total population size," *Mathematical Biosciences*, vol. 160, no. 2, pp. 191–213, 1999.

[16] M. Y. Li and J. S. Muldowney, "Global stability for the seir model in epidemiology," *Mathematical Biosciences*, vol. 125, no. 2, pp. 155–164, 1995.

[17] M. Y. Li, H. L. Smith, and L. Wang, "Global dynamics of an seir epidemic model with vertical transmission," *SIAM Journal on Applied Mathematics*, vol. 62, no. 1, pp. 58–69, 2001.

[18] J. Tumwiine, J. Y. Mugisha, and L. S. Luboobi, "A host-vector model for malaria with infective immigrants," *Journal of Mathematical Analysis and Applications*, vol. 361, no. 1, pp. 139–149, 2010.

[19] L. Wang and M. Y. Li, "Mathematical analysis of the global dynamics of a model for hiv infection of cd4+ t cells," *athematical Biosciences*, vol. 200, no. 1, pp. 44–57, 2006.

[20] E. Beretta, F. Solimano, and Y. Takeuchi, "Negative criteria for the existence of periodic solutions in a class of delay-differential equations," *Nonlinear Analysis: Theory, Methods & Applications*, vol. 50, no. 7, pp. 941–966, 2002.

[21] M. M. Ballyk, C. C. McCluskey, and G. . Wolkowicz, "Global analysis of competition for perfectly substitutable resources with linear response," *Journal of Mathematical Biology*, vol. 51, no. 4, pp. 458–490, 2005.

[22] A. B. Gumel, C. C. McCluskey, and J. Watmough, An SVEIR model for assessing potential impact of an imperfect ant-SARS vaccine. 2006.

[23] X. Zhang and X. Liu, "Backward bifurcation of an epidemic model with saturated treatment function," *Journal of Mathematical Analysis and Applications*, vol. 348, no. 1, pp. 433–443, 2008.

[24] N. T. J. Bailey, *The Mathematical Theory of Infectious Diseases and its Application*, Griffin, London, UK, 1975.

[25] Z. Ma, Y. Zhou, W. Wang, and Z. JIn, *Mathematical models and dynamics of infectious diseases. China sci*, press, Beijing, 2004.

[26] *Research outcome study in ireland (ROSIE): Evaluating drug treatment effectiveness, baseline findings*, 26 Research outcome study in ireland (ROSIE, Evaluating drug treatment effectiveness, March 2005.

[27] B. Fang, X.-Z. Li, M. Martcheva, and L.-M. Cai, "Global asymptotic properties of a heroin epidemic model with treat-age," *Applied Mathematics and Computation*, vol. 263, pp. 315–331, 2015.

[28] R. M. Anderson, R. M. May, and B. Anderson, *Infectious diseases of humans: dynamics and control, volume 28*, Wiley Online Library, volume 28, 2013.

[29] W. Wang and S. Ruan, "Bifurcation in an epidemic model with constant removal rate of the infectives," *Journal of Mathematical Analysis and Applications*, vol. 291, no. 2, pp. 775–793, 2004.

[30] W. Wang, "Backward bifurcation of an epidemic model with treatment," *Mathematical Biosciences*, vol. 201, no. 1-2, pp. 58–71, 2006.

[31] H. L. Smith and P. Waltman, *The Theory of the Chemostat*, Cambridge University Press, 1995.

[32] H. W. Hethcote, "The mathematics of infectious diseases," *SIAM Review*, vol. 42, no. 4, pp. 599–653, 2000.

[33] P. van den Driessche and J. Watmough, "Reproduction numbers and sub-threshold endemic equilibria for compartmental models of disease transmission," *Mathematical Biosciences*, vol. 180, pp. 29–48, 2002.

[34] C. Castillo-Chavez, S. Blower, P. Driessche, D. Kirschner, A. Yakubu, and C. Castillo-Chávez, *Mathematical approaches for*

emerging and reemerging infectious diseases: models, methods, and theory, volume 126, Springer, volume 126, 2002.

[35] C. Castillo-Chavez and B. Song, "Dynamical models of tuberculosis and their applications," *Mathematical Biosciences and Engineering (MBE)*, vol. 1, no. 2, pp. 361–404, 2004.

[36] S. A. Levin, Descartes rule of signs-how hard can it be? http://sepwww.stanford.edu/oldsep/stew/descartes.pdf, November 2002.

[37] B. Fang, X. Li, M. Martcheva, and L. Cai, "Global stability for a heroin model with age-dependent susceptibility," *Journal of Systems Science & Complexity*, vol. 28, no. 6, pp. 1243–1257, 2015.

[38] C. C. McCluskey, "Lyapunov functions for tuberculosis models with fast and slow progression," *Mathematical Biosciences and Engineering. MBE*, vol. 3, no. 4, pp. 603–614, 2006.

[39] N. Bame, S. Bowong, J. Mbang, G. Sallet, and J.-J. Tewa, "Global stability analysis for seis models with *n* latent classes," *Mathematical Biosciences and Engineering*, vol. 5, no. 1, pp. 20–33, 2008.

[40] A. Korobeinikov and G. C. Wake, "Lyapunov functions and global stability for sir, sirs, and sis epidemiological models," *Applied Mathematics Letters*, vol. 15, no. 8, pp. 955–960, 2002.

[41] M. Y. Li and J. S. Muldowney, "Dynamics of Differential Equations on Invariant Manifolds," *Journal of Differential Equations*, vol. 168, no. 2, pp. 295–320, 2000.

[42] E. Beretta, R. Kon, and Y. Takeuchi, "Nonexistence of periodic solutions in delayed Lotka-Volterra systems," *Nonlinear Analysis. Real World Applications. An International Multidisciplinary Journal*, vol. 3, no. 1, pp. 107–129, 2002.

[43] J. Guckenheimer and P. Holmes, *Nonlinear Oscillations, Dynamical Systems, and Bifurcation of Vector Fields*, Springer, New York, NY, USA, 1983.

[44] J. S. Muldowney, "Compound matrices and ordinary differential equations," *The Rocky Mountain Journal of Mathematics*, vol. 20, no. 4, pp. 857–872, 1990.

[45] W. A. Coppel, *Stability and Asymptotic Behavior of Differential Equations*, D. C. Heath, Boston, Mass, USA, 1965.

[46] J. Martin, "Logarithmic norms and projections applied to linear differential systems," *Journal of Mathematical Analysis and Applications*, vol. 45, pp. 432–454, 1974.

[47] H. I. Freedman, S. G. Ruan, and M. X. Tang, "Uniform persistence and flows near a closed positively invariant set," *Journal of Dynamics and Differential Equations*, vol. 6, no. 4, pp. 583–600, 1994.

[48] V. Hutson and K. Schmitt, "Permanence and the dynamics of biological systems," *Mathematical Biosciences*, vol. 111, no. 1, pp. 1–71, 1992.

[49] S. Marino, I. B. Hogue, C. J. Ray, and D. E. Kirschner, "A methodology for performing global uncertainty and sensitivity analysis in systems biology," *Journal of Theoretical Biology*, vol. 254, no. 1, pp. 178–196, 2008.

[50] F. Chirove Sutimin, E. Soewono, and N. Nuraini, "Analysis of combined langerhans and cd 4+ t cells hiv infection," *SIAM Journal on Applied Mathematics*, vol. 74, no. 4, pp. 1174–1193, 2014.

[51] S. M. Blower and H. Dowlatabadi, "Sensitivity and uncertainty analysis of complex models of disease transmission: an HIV model, as an example," *International Statistical Review*, vol. 62, no. 2, pp. 229–243, 1994.

[52] L. Simpson and A. B. Gumel, "Mathematical assessment of the role of pre-exposure prophylaxis on hiv transmission dynamics," *Applied Mathematics and Computation*, vol. 293, pp. 168–193, 2017.

12

Modelling In Vivo HIV Dynamics under Combined Antiretroviral Treatment

B. Mobisa ⓘ,[1] G. O. Lawi,[2] and J. K. Nthiiri[2]

[1]*Department of Mathematics, Laikipia University, P.O. Box 1100-20300, Nyahururu, Kenya*
[2]*Department of Mathematics, Masinde Muliro University of Science and Technology, P.O. Box 190-50100, Kakamega, Kenya*

Correspondence should be addressed to B. Mobisa; bmobisa@laikipia.ac.ke

Academic Editor: Qingdu Li

In this paper a within host mathematical model for Human Immunodeficiency Virus (HIV) transmission incorporating treatment is formulated. The model takes into account the efficacy of combined antiretroviral treatment on viral growth and T cell population in the human blood. The existence of an infection free and positive endemic equilibrium is established. The basic reproduction number R_0 is derived using the method of next generation matrix. We perform local and global stability analysis of the equilibria points and show that if $R_0 < 1$, then the infection free equilibrium is globally asymptotically stable and theoretically the virus is cleared and the disease dies out and if $R_0 > 1$, then the endemic equilibrium is globally asymptotically stable implying that the virus persists within the host. Numerical simulations are carried out to investigate the effect of treatment on the within host infection dynamics.

1. Introduction

Human Immunodeficiency Virus (HIV) remains a major threat to human life for the last three and half decades. HIV infection in humans causes Acquired Immunodeficiency Virus (AIDS), a disease that has ravaged human population all over the world. Since its discovery in the early 1980s, there has been tremendous research work on how to contain or eradicate the disease. Mathematical modelling of viral infections has led to greater understanding of virus dynamics and helped in predicting and controlling the spread of viral diseases such as HIV, Hepatitis B Virus (HBV), Hepatitis C Virus (HCV), and Dengue Fever. One of the early models of HIV infection known as the basic model was used by Nowak and May [1] and by Perelson and Nelson (1999) and was successful in numerically reproducing the dynamics of the early stages of HIV and its target CD4+ cells following an infection event. Recent studies have focused on HIV viral and cellular infections incorporating dynamics such as intracellular delays, latent infection and viral mutation, and spatial heterogeneity [2–5]. For instance, [6] investigated the global stability of within host virus models with cell-to-cell viral transmission and obtained a complete analytic description of equilibria. A four-dimensional system of delayed differential equations, where the production and removal rates of the virus and cells are given by general nonlinear functions, was proposed by [4]. Their model investigated the dynamical behaviour of virus target and cell target incidences incorporating humoral immune response. They established three key equilibrium results, an infection free equilibrium, a chronic free equilibrium with inactive humoral immune response, and chronic infection equilibrium with active humoral immune response. With dynamics governed by two bifurcation parameters basic reproduction numbers $\widetilde{R}_0^c$ and the humoral immunity numbers $\widetilde{R}_1^G$ and using Lyapunov functionals and Lasalle's invariance principle, the authors proved the global stability of the equilibria.

The inclusion of treatment, at within and between host levels in mathematical modelling, has gained considerable attention in recent years. For instance, at between host levels, epidemiological models with saturated treatment function have been proposed by [7, 8]. Research on within host models that incorporate treatment has been carried out over the years, with early models highlighting the effects of AZT on viral replication [9]. Among the key findings, viral decline is drug dependent. A study of combined drug therapy of HIV infection was conducted by [10]; the mathematical model developed was used to simulate chemotherapy treatment of HIV infection. The simulations were based on clinical data of treatment with combinations of antiviral drugs involving reverse transcriptase inhibitors (RTI) and protease inhibitors (PI) and focused on the timing of treatment. The findings revealed that the success of treatment is based on longer survival times equated to the CD4+ T cells. Global dynamics of delay distributed HIV infection models with differential drug efficacy in cocirculating target cells was investigated [3]. Recent work by [11] sought to mathematically analyze the potential of Prophylaxis treatment in preventing and slowing the spread of HIV/AIDS in the population. In this study early use of Prophylaxis drug was shown to slow the rate of HIV transmission.

Whereas extensive research on HIV transmission dynamics has been carried out, mathematical modelling of HIV with combined treatment still remains an area of active research among mathematicians and biologists.

In this paper, we propose a within host HIV infection model with a logistic incidence rate that explicitly incorporates the two levels of antiretroviral treatment, namely, the reverse transcriptase inhibitors (RTIs) which prevent the reverse transcription of viral RNA into DNA. In this way the RTIs serve to reduce the rate of infection of activated $CD4^+T$ cells. The other category is protease inhibitors (PIs) which prevent HIV-1 protease from clearing the HIV polyprotein into functional units, thereby causing infected cells to produce immature virus particles that are not capable of infecting additional cells; hence PIs decrease the number of newly infectious virus (virions) that are produced [12]. A mathematical analysis of the effects of treatment on the within host infection dynamics is carried out.

2. Model Description and Formulation

A mathematical model of within host HIV infection dynamics is considered. The model is composed of three interacting variables, namely, uninfected CD4+ T cells $T(t)$, actively infected cells $T^*(t)$, and free virus particles $V(t)$. The uninfected CD4+ T cells are produced at rate r and die naturally at the rate μ. The total number of T cells in the body remains bounded; thus the growth of T cells is governed by the logistic proliferation term $(1 - T/T_{max})$ which limits T cell growth as the cell population approaches the limit T_{max}. The uninfected CD4+ T cells become infected by free virus and actively infected cells according to the simple mass infection terms βVT and αTT^*, respectively. This generates actively infected cells, T^*, which die naturally at the rate κ. The infected cells produce free viruses V at the rate $\omega\kappa T^*$ and are cleared from circulation at rate c per virus. This viral decline is a function of the efficiency of the combined treatment of Reverse Transcriptase Inhibitor (RTI) and Protease Inhibitor (PI), which are represented by the parameters ρ and ϑ, respectively. From the description and definitions made, the infection dynamics are summarized by the following system of ODEs:

$$\frac{dT}{dt} = rT\left(1 - \frac{T}{T_{max}}\right) - (1-\rho)\beta VT - (1-\vartheta)\alpha TT^* - \mu T$$
$$\frac{dT^*}{dt} = (1-\rho)\beta VT + (1-\vartheta)\alpha TT^* - \kappa T^* \quad (1)$$
$$\frac{dV}{dt} = \omega\kappa T^* - cV$$

3. Analysis of the Model

Since the model describes cell and virus populations dynamics, all the model variables are nonnegative for $t \geq 0$. In the absence of the virus, the T cell population has a steady state value T_0; hence the initial conditions for the model (1) are $T(0) = T_0 > 0$, $T^*(0) = T_0^* \geq 0$, and $V(0) = V_0 > 0$. It can be shown that with positive initial data the solutions of model (1) will remain positive and bounded in the feasible region $\Omega = \{(T, T^*, V) \in \mathbb{R}_+^3 : T + T^* + V \leq T_{max}(1 - \mu/r)\}$, $\forall t$.

3.1. Basic Reproduction Number. The basic reproduction number R_0 is defined as the average number of secondary infections produced by one infectious virion and one infected cell over the course of their infectious period in uninfected CD4+ T cell population. We compute R_0 for model (1) using the next generation matrix method as used in [13, 14]. Model (1) has two infected compartments T^* and V. Let f_i be the rate of appearance of new infections in compartment i and v_i as the transfer of individuals out of compartment i for the two compartments, respectively, and are given in partitioned form as follows:

$$f = \begin{pmatrix} (1-\rho)\beta VT + \alpha(1-\vartheta)TT^* \\ 0 \end{pmatrix} \quad (2)$$

and

$$\nu = \begin{pmatrix} \kappa T^* \\ -\omega\kappa T^* + cV \end{pmatrix} \quad (3)$$

The Jacobian of f and ν evaluated at the Infection Free Equilibrium $E_0 = (T_{max}, 0, 0)$ yields

$$F = \begin{pmatrix} \alpha(1-\vartheta)T_0 & \beta(1-\rho)T_0 \\ 0 & 0 \end{pmatrix} \quad (4)$$

$$V = \begin{pmatrix} \kappa & 0 \\ -\omega\kappa & c \end{pmatrix} \quad (5)$$

where F is nonnegative and V is nonsingular. The basic reproduction number is thus given by $R_0 = \rho(FV^{-1})$, where $\rho(FV^{-1})$ is the spectral radius of the matrix FV^{-1}. Hence

$$R_0 = T_{max}\left(1 - \frac{\mu}{r}\right)(1-\vartheta)\frac{\alpha}{\kappa}$$
$$+ T_{max}\left(1 - \frac{\mu}{r}\right)\frac{\omega\beta}{c}(1-\rho) \quad (6)$$

3.2. Local Stability Analysis of the Infection Free Equilibrium. We investigate the local stability properties of the infection free equilibrium by approximating the nonlinear system of the differential equations (1) with the linear system at the infection free equilibrium $E_0 = (T_{max}(1 - \mu/r), 0, 0)$.

Theorem 1. *The infection free equilibrium E_0 is locally asymptotically stable if and only if $R_0 < 1$.*

Proof. Evaluating the Jacobian of model (1) at E_0, we obtain

$$J(E_0) = \begin{pmatrix} \mu - r & -\alpha T_{max}\left(1-\frac{\mu}{r}\right)(1-\vartheta) & -\beta T_{max}\left(1-\frac{\mu}{r}\right)(1-\rho) \\ 0 & \alpha T_{max}\left(1-\frac{\mu}{r}\right)(1-\vartheta) - \kappa & \beta T_{max}\left(1-\frac{\mu}{r}\right)(1-\rho) \\ 0 & \omega\kappa & -c \end{pmatrix} \quad (7)$$

and clearly

$$\lambda_1 = \mu - r \quad (8)$$

is one of the eigenvalues of the matrix in (7), which is negative because for a population that is growing in numbers; the rate of production (birth rate) is greater than the death rate, that is $r > \mu$. The nature of the remaining roots of (7) can be determined from the reduced matrix:

A

$$= \begin{pmatrix} \alpha T_{max}\left(1-\frac{\mu}{r}\right)(1-\vartheta) - \kappa & \beta T_{max}\left(1-\frac{\mu}{r}\right)(1-\rho) \\ \omega\kappa & -c \end{pmatrix} \quad (9)$$

Using Routh-Hurwitz stability criteria, matrix A in (9) will have negative real roots if and only if the tr $A < 0$ and det $A > 0$; thus

$$\text{tr } A = R_0 - 1 - T_{max}\left(1-\frac{\mu}{r}\right)(1-\rho)\frac{\beta\omega}{c} - \frac{c}{k} \quad (10)$$

and

$$\det A = -\left[\frac{\alpha}{\kappa}T_{max}\left(1-\frac{\mu}{r}\right)(1-\vartheta) \right.$$
$$\left. + \frac{\beta\omega}{c}T_{max}\left(1-\frac{\mu}{r}\right)(1-\rho)\right] + 1 \quad (11)$$

and using (6), (11) reduces to

$$\det A = 1 - R_0 \quad (12)$$

From (10) and (12), tr $A < 0$ and det $A > 0$ if and only if $R_0 < 1$. Thus E_0 is locally asymptotically stable whenever $R_0 < 1$ and unstable otherwise. □

This means that if a small number of free virus particles enter the blood stream, each virus will infect on average less than one uninfected cell in its entire period of infectivity whenever $R_0 < 1$. Theoretically this shows that the virus is cleared from the body if $R_0 < 1$.

3.3. Global Stability Analysis of the Infection Free Equilibrium. In this section we study the global stability of the infection free equilibrium of model (1) using the theorem by Castillo-Chavez et al. [13]. We rewrite model (1) in the form

$$\frac{dX}{dt} = H(X,Z)$$

$$\frac{dZ}{dt} = G(X,Z), \quad (13)$$

$$G(X,0) = 0$$

where $X \in \mathbb{R}$ denotes the number of uninfected cells and $Z \in \mathbb{R}^2$ denotes the number of actively infected cells and free virus particles, respectively. The infection free equilibrium (IFE) is now denoted by

$$U_0 = (X^0,0), \quad X^0 = T_{max}\left(1-\frac{\mu}{r}\right) \quad (14)$$

The conditions H_1 and H_2 below must be met in order to guarantee global asymptotic stability:

(i) (H_1) For $dX/dt = H(X,0)$, X^0 is Globally Asymptotically Stable (GAS)

(ii) (H_2) $G(X,Z) = PZ - \widehat{G}(X,Z), \widehat{G}(X,Z) \geq 0$, for $(X,Z) \in \Omega$

where $P = D_Z G(X^0,0)$ is an M-matrix (the off-diagonal elements of P are nonnegative) and Ω is the region where the model makes biological sense. If system (13) satisfies conditions H_1 and H_2, then the following theorem holds.

Theorem 2. *The fixed point* $U_0 = (X^0,0)$ *is Globally Asymptotically Stable equilibrium of (13) provided that $R_0 < 1$ and that assumptions (H_1) and (H_2) are satisfied.*

Proof. Let $X(t) = T(t)$, $Z = (T^*(t),V(t))$, $H(X,0) = \begin{pmatrix} rT(1-T/T_{max})-\mu T \\ 0 \end{pmatrix}$, and $G(X,Z) = PZ - \widehat{G}(X,Z)$ where

$$P = \begin{pmatrix} \alpha T_{max}(1-\vartheta) - \kappa & \beta T_{max}(1-\rho) \\ \omega\kappa & -c \end{pmatrix} \quad (15)$$

$$\widehat{G}(X,Z) = \begin{pmatrix} \widehat{G_1}(X,Z) \\ \widehat{G_2}(X,Z) \end{pmatrix} = \begin{pmatrix} 0 \\ 0 \end{pmatrix} \quad (16)$$

From (16) $\widehat{G_1}(X,Z) = \widehat{G_2}(X,Z) = 0$ this implies that $\widehat{G}(X,Z) \geq 0$. Therefore, E_0 is globally asymptotically stable when $R_0 < 1$. □

This means that, at any perturbation of the equilibrium point by the introduction of free virus particles, the model solutions will always converge to the IFE, whenever $R_0 < 1$.

3.4. Existence of the Endemic Equilibrium (EE)

Theorem 3. *A positive endemic equilibrium EE exists provided $R_0 > 1$.*

Proof. The endemic equilibrium $EE = (T_e, T_e^\star, V_e)$ satisfies

$$rT_e\left(1 - \frac{T_e}{T_{max}}\right) - (1-\rho)\,\beta V_e T_e - \alpha\,(1-\vartheta)\,T_e T_e^\star$$
$$- \mu T_e = 0 \tag{17}$$

$$(1-\rho)\,\beta V_e T_e + \alpha\,(1-\vartheta)\,T_e T_e^\star - \kappa T_e^\star = 0 \tag{18}$$

$$\omega\kappa T_e^\star - c V_e = 0 \tag{19}$$

From (19) we have

$$V_e = \frac{\omega\kappa T_e^\star}{c} \tag{20}$$

Substituting (20) in (18) we get

$$T_e = \frac{1}{(1-\rho)\,(\beta\omega/c) + (1-\vartheta)\,(\alpha/\kappa)}$$
$$= \frac{T_{max}\,(1-\mu/r)}{R_0} \tag{21}$$

Substituting V_e and T_e in (17) we obtain

$$T_e^\star = \frac{T_{max}\,(1-\mu/r)\,(R_0 - 1)\,(r - \mu)}{\kappa R_0^2} \tag{22}$$

The endemic equilibrium (EE) is given as

$$EE = \left(\frac{T_{max}\,(1-\mu/r)}{R_0},\ \frac{T_e\,(R_0 - 1)\,(r - \mu)}{\kappa R_0},\ \frac{\omega\kappa T_e^\star}{c}\right) \tag{23}$$

Clearly $T_e^\star > 0$ if and only if $R_0 > 1$. $\qquad\square$

3.5. Local Stability Analysis of the Endemic Equilibrium

Theorem 4. *The endemic equilibrium $EE = (T_e, T_e^\star, V_e)$ is locally asymptotically stable whenever $R_0 > 1$.*

Proof. The Jacobian matrix of model (1) at EE is as follows:

$$J(EE) = \begin{pmatrix} \dfrac{(r-2)\,(r-\mu)}{rR_0} & \dfrac{-\alpha\,(1-\vartheta)\,T_{max}\,(1-\mu/r)}{R_0} & \dfrac{-\beta\,(1-\rho)\,T_{max}\,(1-\mu/r)}{R_0} \\[2ex] \dfrac{(R_0 - 1)\,(r-\mu)}{R_0} & \dfrac{\alpha\,(1-\vartheta)\,T_{max}\,(1-\mu/r)}{R_0} - \kappa & \dfrac{(1-\rho)\,\beta T_{max}\,(1-\mu/r)}{R_0} \\[2ex] 0 & \omega\kappa & -c \end{pmatrix} \tag{24}$$

The characteristic equation of (24) is in the form

$$\lambda^3 + a_0\lambda^2 + a_1\lambda + a_2 = 0 \tag{25}$$

where

$$a_0 = \kappa + c - \frac{\alpha\,(1-\vartheta)\,T_{max}\,(1-\mu/r)}{R_0}$$
$$- \frac{(r-2)\,(r-\mu)}{rR_0}$$

$$a_1 = \left(\frac{(1-\vartheta)\,T_{max}\,(1-\mu/r)}{R_0}\right)$$
$$\cdot \left(\frac{(r-2)\,(r-\mu)}{rR_0} + \frac{(R_0 - 1)\,(r-\mu)}{R_0}\right) \tag{26}$$
$$+ \left(\frac{(r-2)\,(r-\mu)}{rR_0}\right)(c - \kappa) + c\kappa - 1$$

$$a_2 = \frac{(R_0 - 1)\,(r-\mu)}{R_0} + \frac{(r-2)\,(r-\mu)}{rR_0}\,(1 - c\kappa)$$

The number of possible negative real roots of (25) depends on the signs of a_0, a_1, and a_2. This can be established by applying Descartes Rule of Signs as used in [11].

$$P(\lambda) = a_1\lambda^2 + a_2\lambda + a_3 \tag{27}$$

According to this rule the number of negative real zeros of $P(\lambda)$ is either equal to the number of sign changes of $P(-\lambda)$ or less by an even number, as shown in Table 1.

From Table 1 the maximum number of sign changes in $P(-\lambda)$ is 2; hence the characteristic polynomial (27) has two negative roots. Thus

$$P(-\lambda) = -\lambda^3 + a_0\lambda^2 - a_1\lambda + a_2 = 0 \tag{28}$$

has negative real roots. Hence for $r > \mu$ and if cases 1 to 8 are satisfied then the endemic equilibrium EE is locally asymptotically stable. $\qquad\square$

Therefore if $R_0 > 1$ and given a small number of free virus particles, each virus, in the entire period of its infectivity, will produce on average more than one infected cell, implying viral persistence.

3.6. Global Stability Analysis of the Endemic Equilibrium.

In this section we investigate the global stability of the endemic equilibrium using geometric approach, as developed by Li and Muldowney in [15]. For a brief description of this approach, see [6, 8]. Consider the autonomous dynamical system

$$y' = f(x) \tag{29}$$

TABLE 1: Roots of characteristic equation (27).

cases	a_0	a_1	a_2	$R_0 > 1$	no. of sign changes	no. of real -ve roots
1	-	+	-	$R_0 > 1$	2	2,0
2	-	-	-	$R_0 > 1$	0	0
3	+	+	-	$R_0 > 1$	1	0
4	+	-	-	$R_0 > 1$	1	0
5	-	+	+	$R_0 > 1$	1	0
6	-	-	+	$R_0 > 1$	1	0
7	+	+	+	$R_0 > 1$	0	0
8	+	-	+	$R_0 > 1$	2	2,0

where $f : \Omega \longrightarrow \mathbb{R}^n$, $\Omega \subset \mathbb{R}^n$ is an open set and is simply connected and $y \in \Omega$, $y \longmapsto f(x) \in \mathbb{R}^n$, $f(x) \in C'(\Omega)$. Let y^* be an equilibrium point, then y^* is said to be globally stable in Ω if it is locally stable in Ω and all trajectories in Ω converge to y^*. In this method the equilibrium y^* is locally asymptotically stable provided the following conditions hold:

(i) (H1) Ω is simply connected

(ii) (H2) There exists a compact absorbing set $K \subset \Omega$

(iii) (H3) Equation (29) has a unique equilibrium y^* in Ω

Let $P(y)$ be a $\binom{n}{2} \times \binom{n}{2}$ matrix-valued function that is C' on Ω and consider $B = P_f P^{-1} + P(\partial f^2/\partial y)P^{-1}$ where the matrix P_f is $(\partial P_{ij}^*/\partial y)f = dP_{ij}/dt$ and let the matrix $J^{(2)}$ be the second additive compound matrix of the Jacobian matrix J; that is, $J_{(y)} = (J_{ij})$, $J^{(2)}$ is a $\binom{n}{2} \times \binom{n}{2}$ matrix, and in our case n=3; hence

$$J^{(2)} = \begin{pmatrix} J_{11} + J_{22} & J_{23} & -J_{13} \\ J_{32} & J_{11} + J_{33} & J_{12} \\ -J_{31} & J_{21} & J_{22} + J_{33} \end{pmatrix} \quad (30)$$

Consider the Lonziskii measure μ of B with respect to a vector norm $|\cdot|$ in $\mathbb{R}^N$ where $N = \binom{n}{2}$

$$\mu(B) = \lim_{t \to 0^+} \frac{\|I + hB\| - 1}{h} \quad (31)$$

It is proved in [15] that if (H1), (H2), and (H3) hold and condition

$$\bar{q} = \lim_{t \to \infty} \sup \sup_{y_0 \in K} \frac{1}{t} \int_0^t \mu(B(x(s, x_0))) \, ds < 0 \quad (32)$$

is satisfied, then the unique equilibrium y^* is globally asymptotically stable

Lemma 5. Assume that conditions (H1), (H2), and (H3) hold, then y^* is globally asymptotically stable in Ω provided that a function $P(x)$ and a Lonziskii measure [16] μ exist such that condition (32) is satisfied.

Theorem 6. The endemic equilibrium EE is globally asymptotically stable in Ω if $R_0 > 1$.

Proof. Consider the Jacobian of model (1)

$$J = \begin{pmatrix} r - \dfrac{2rT}{T_{max}} - (1-\rho)\beta V - (1-\vartheta)\alpha T^* - \mu & -(1-\vartheta)\alpha T & -(1-\rho)\beta T \\ (1-\rho)\beta V + (1-\vartheta)\alpha T^* & (1-\vartheta)\alpha T - \kappa & (1-\rho)\beta T \\ 0 & \omega\kappa & -c \end{pmatrix} \quad (33)$$

and the second compound additive matrix of (33) is given as

$$J^{(2)} = \begin{pmatrix} a + b - \kappa & d & d \\ \omega\kappa & a - c & -b \\ 0 & e & b - \kappa - c \end{pmatrix} \quad (34)$$

where

$$a = r - \frac{2rT}{T_{max}} - (1-\rho)\beta V - \alpha(1-\vartheta)T^* - \mu$$

$$b = \alpha T(1-\vartheta) \quad (35)$$

$$d = (1-\rho)\beta T$$

$$e = (1-\rho)\beta V + \alpha T^*(1-\vartheta)$$

We define an auxiliary matrix function Q on Ω as

$$Q = \text{diag}\left(\frac{1}{T^*}, \frac{1}{V}, \frac{1}{V}\right) \quad (36)$$

setting $T^*, V > 0$ everywhere in Ω, and Q is smooth and nonsingular. Q_f and $Q_f Q^{-1}$ are given as

$$Q_f = \text{diag}\left(-\frac{\dot{T}^*}{(T^*)^2}, -\frac{\dot{V}}{V^2}, -\frac{\dot{V}}{V^2}\right) \quad (37)$$

$$Q_f Q^{-1} = \text{diag}\left(-\frac{\dot{T}^*}{T^*}, -\frac{\dot{V}}{V}, -\frac{\dot{V}}{V}\right) \quad (38)$$

where $\dot{T^\star} = dT^\star/dt$ and $\dot{V} = dV/dt$. Matrix $QJ^{(2)}Q^{-1}$ is given as

$$QJ^{(2)}Q^{-1} = \begin{pmatrix} a+b-\kappa & \dfrac{dV}{T^\star} & \dfrac{dV}{T^\star} \\ \dfrac{\omega\kappa T^\star}{V} & a-c & -b \\ 0 & e & b-\kappa-c \end{pmatrix} \qquad (39)$$

Thus the matrix $M = Q_f Q^{-1} + QJ^{(2)}Q^{-1}$ as defined in (4.4) of [15] can be written in block form as

$$M = \begin{pmatrix} m_{11} & m_{12} \\ m_{21} & m_{22} \end{pmatrix} \qquad (40)$$

where $m_{11} = a + b - \kappa - \dot{T}\star/T^\star$, $m_{12} = dV/T^\star, dV/T^\star$, $m_{21} = \left(\begin{smallmatrix} \omega\kappa T^\star/V \\ 0 \end{smallmatrix} \right)$, and $m_{22} = \left(\begin{smallmatrix} a-c-\dot{V}/V & -b \\ e & b-c-\kappa-\dot{V}/V \end{smallmatrix} \right)$. Let the vector norm $|\cdot|$ in $\mathbb{R}^3 \cong \mathbb{R}^{\binom{3}{2}}$ be chosen as

$$|(u,v,w)| = \sup\{|u|, |v| + |w|\} \qquad (41)$$

The Lozinskii measure $\mu(M)$ with respect to $|\cdot|$ can be estimated as follows:

$$\mu(M) \le \sup\{g_1, g_2\} \qquad (42)$$

where

$$g_1 = \mu_1(m_{11}) + \|m_{12}\| \qquad (43)$$

$$g_2 = \mu_1(m_{22}) + \|m_{21}\| \qquad (44)$$

$\|m_{12}\|$ and $\|m_{21}\|$ are operator norms associated with the linear mappings $m_{12} : \mathbb{R}^2 \longrightarrow \mathbb{R}$ and $m_{12} : \mathbb{R} \longrightarrow \mathbb{R}^2$, respectively, where $\mathbb{R}$ is endowed with the ℓ_1 vector norm in both cases. Specifically $\mu_1(m_{11}) = a + b - \kappa - \dot{T}\star/T^\star$, $\|m_{12}\| = dV/T^\star$, $\|m_{21}\| = \sup\{\omega\kappa T^\star/V, 0\}$, and $\mu_1(m_{22}) = -c - \dot{V}/V + \sup\{a+e, -\kappa\}$. From model (1) we find

$$\frac{\dot{T^\star}}{T^\star} = (1-\rho)\beta VT + (1-\vartheta)\alpha T - \kappa \qquad (45)$$

$$\frac{\dot{V}}{V} = \frac{\omega\kappa T^\star}{V} - c \qquad (46)$$

and recalling expressions for $a, b, d,$ and e we obtain

$$g_1 = r - \frac{2rT}{T_{max}} - (1-\rho)\beta V - (1-\vartheta)\alpha T^\star - \mu \qquad (47)$$

$$g_2 = \sup\left\{ r - \frac{2rT}{T_{max}} - \mu, -\kappa \right\} \qquad (48)$$

Since $r - 2rT/T_{max} - (1-\rho)\beta V - (1-\vartheta)\alpha T^\star - \mu \le \sup\{r - 2rT/T_{max} - \mu, -\kappa\}$ we find that $g_1 \le g_2$, then (42) implies that $\mu(M) \le g_2$; thus

$$\mu(M) \le r - \frac{2rT}{T_{max}} - \mu$$

$$+ \sup\{-(1-\rho)\beta V - (1-\vartheta)\alpha T^\star, 0\} \qquad (49)$$

$$\le r - \frac{2rT}{T_{max}} - \mu$$

Since T_{max} is the limiting value of $T(t)$, then this implies that $\limsup_{t\to\infty} T(t) \le T_{max}$; therefore expression (49) reduces to

$$\mu(M) \le r - 2r - \mu \qquad (50)$$

Integrating (50) we obtain

$$\frac{1}{t} \int_0^t \mu(M)\, ds \le \frac{1}{t} \int_0^t (r - 2r - \mu)\, ds < 0 \qquad (51)$$

Hence the endemic equilibrium (EE) is globally asymptotically stable whenever $R_0 > 1$. $\qquad \square$

This implies that, regardless of any starting solution, the solution of the model will converge to EE whenever $R_0 > 1$. Immunologically, it means that, at any perturbation of the equilibrium point as a result of the introduction of the free virus particles, the model solutions will converge to the endemic state.

4. Numerical Simulation of the Model

In this section we perform a numerical simulation of model (1) using MATLAB with the parameter values given in Table 1. The primary purpose of the numerical simulation is to analyze the change in state of virus progression with time and also to outline the impact of the variation of treatment efficacy on the transmission dynamics of HIV. This is achieved by varying the parameter values ρ and ϑ while keeping the other parameters constant.

4.1. Effect of Variations in RTI and PI Treatment Efficacy on the Asymptotic Behaviour of the Equilibrium Points. The effect of variations in combined treatment efficacy on the stability of equilibria is investigated using the parameter values given in Table 2; this is achieved by choosing four different initial conditions

(i) IC1: $T_1(t) = 1000, T_1^\star(t) = 0, V_1(t) = 10^{-3}$

(ii) IC2: $T_2(t) = 800, T_2^\star(t) = 10, V_2(t) = 0.001$

(iii) IC3: $T_3(t) = 900, T_3^\star(t) = 5, V_3(t) = 0.01$

(iv) IC4: $T_4(t) = 1000, T_4^\star(t) = 1, V_4(t) = 0.1$

and varying parameters ρ and ϑ while keeping other parameters constant. From Theorems 2 and 6, the corresponding stabilities of model (1) are as follows:

(1) If $R_0 < 1$, the IFE is globally asymptotically stable.

(2) If $R_0 > 1$, the EE is globally asymptotically stable.

(i) $\rho = 0.1$ and $\vartheta = 0.1$. For this set of parameters we obtain $R_0 = 4.545 > 1$. By Theorem 6 the endemic equilibrium is globally asymptotically stable and the states of the system converge to $EE = (110.011, 3.575, 357.53)$, for the four initial conditions IC1-IC4, implying that the virus persists in the host. This shows that the low treatment efficacy of 0.1 cannot effectively combat the virus. This is illustrated by Figures 1, 2, 3, and 4.

TABLE 2: Parameters values for model (1).

Parameters	units	Description	source
T_{max}	1500 cells mm^{-3}	Maximum $CD4+$ cell population level	[9]
r	0.03 cells day^{-1}	Production rate of uninfected T cells	[9]
μ	0.02 cells day^{-1}	Natural death rate of uninfected T cells	[9]
κ	0.24 cells day^{-1}	Death rate of actively infected T cells	[9]
c	2.4 day^{-1}	Shedding rate of virions	[9]
β	$2.4 \times 10^{-5} mm^{-3}$	Viral infection rate by free virions	[9]
α	$2.4 \times 10^{-5} mm^{-3}$	cellular infection rate	[9]
ω	varies: $\omega \geq 0$	Burst rate of actively infected T cells	-
ρ	varies: $0 < \rho < 1$	Efficacy of RT Inhibitor	-
ϑ	varies: $0 < \vartheta < 1$	efficacy of Protease Inhibitor	-

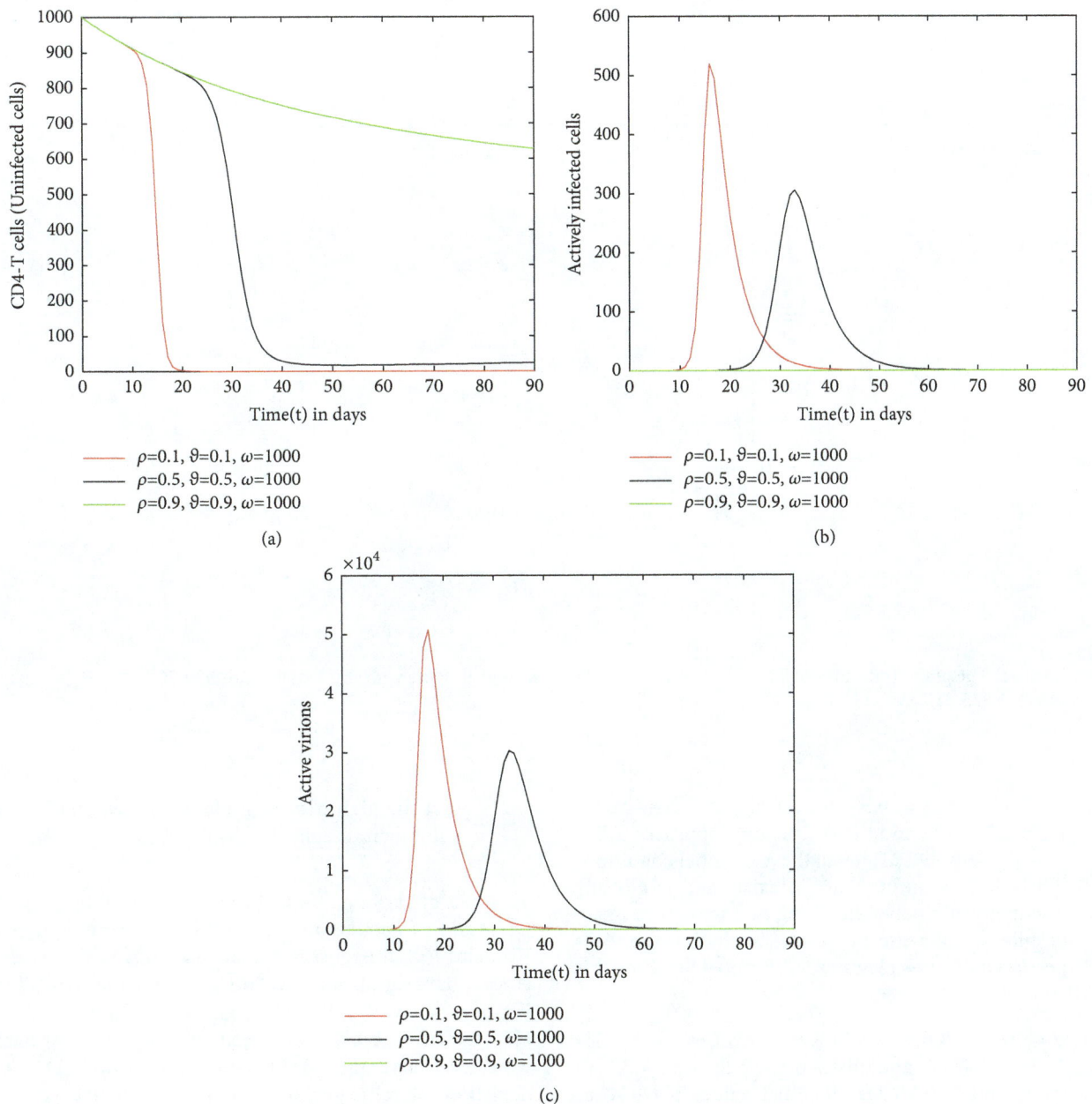

(a)

$\rho=0.1, \vartheta=0.1, \omega=1000$
$\rho=0.5, \vartheta=0.5, \omega=1000$
$\rho=0.9, \vartheta=0.9, \omega=1000$

(b)

$\rho=0.1, \vartheta=0.1, \omega=1000$
$\rho=0.5, \vartheta=0.5, \omega=1000$
$\rho=0.9, \vartheta=0.9, \omega=1000$

(c)

$\rho=0.1, \vartheta=0.1, \omega=1000$
$\rho=0.5, \vartheta=0.5, \omega=1000$
$\rho=0.9, \vartheta=0.9, \omega=1000$

FIGURE 1: Effect of treatment efficacy on $T(t)$, $T^\star(t)$, and $V(t)$.

(a)

(b)

(c)

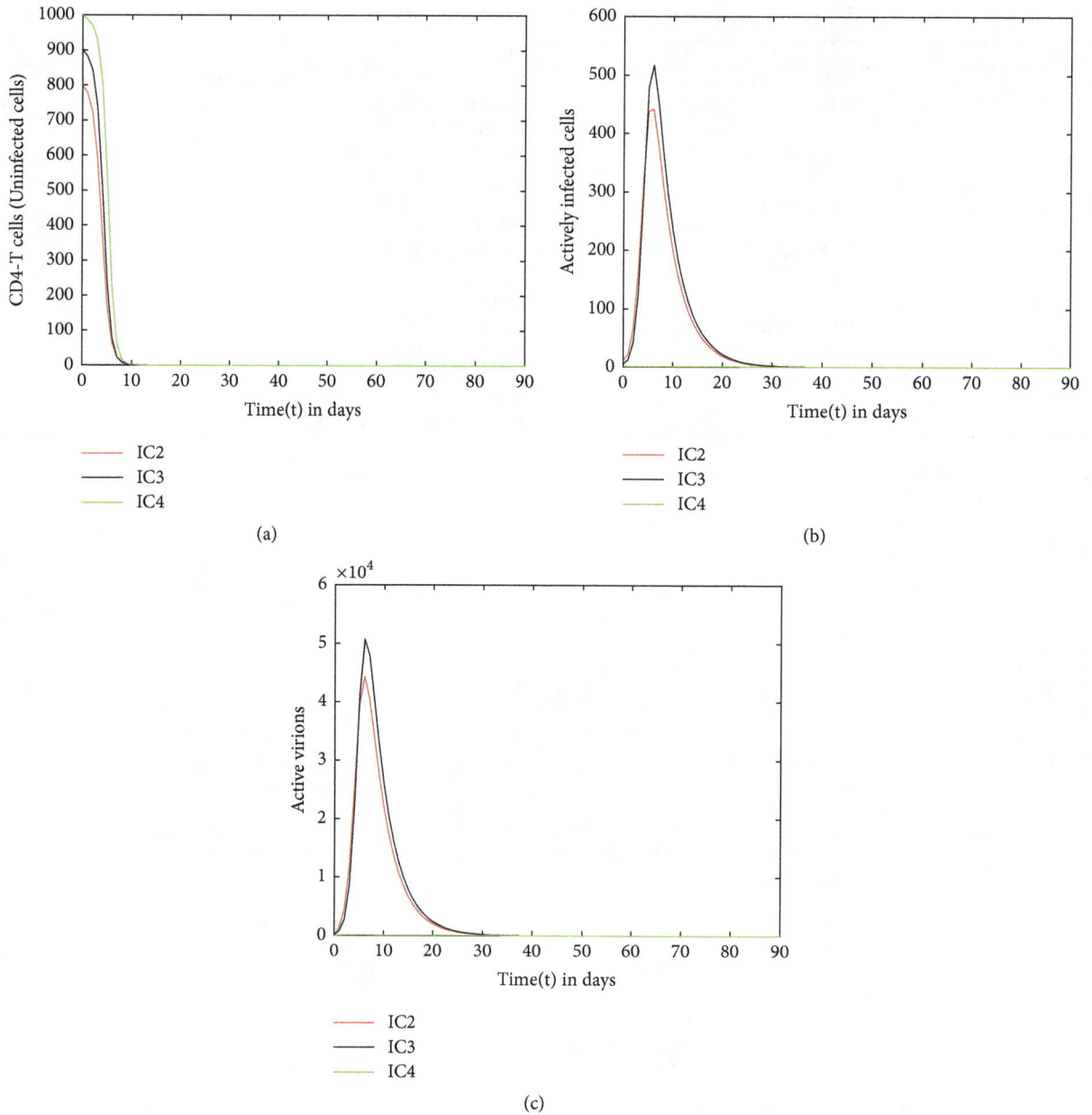

FIGURE 2: (a)-(c) The graph trajectory of $T(t), T^*(t)$, and $V(t)$ for $\rho = 0.1$ and $\vartheta = 0.1$ with the initial conditions IC2-IC4. $R_0 = 4.545$ and $EE = (110.011, 3.575, 357.53)$ is GAS.

(ii) $\rho = 0.5$ and $\vartheta = 0.5$. This set of parameters yields $R_0 = 2.525 > 1$ and the system state approach $EE = (198.02, 4.98, 498.31)$ for all the four initial conditions IC1-IC4. This means that EE exists and is globally asymptotically stable according to Theorem 6. Again despite the increase in treatment efficacy the virus persists in the host. Figures 1, 2, 3, and 4 illustrate this scenario.

(iii) $\rho = 0.82$ and $\vartheta = 0.9$. This treatment efficacy yields $R_0 = 0.101 < 1$ and by Theorem 2 $E_0 = (1500, 0, 0)$ exists and is GAS for all initial values IC1-IC4, as can be observed from Figures 1, 2, 3, and 4 implying

that the high treatment efficacy of over 0.82 helps in stifling viral replication within host.

4.2. Effect of Variations in RTI and PI Treatment Efficacy on the CD4+ T Cell Population and Viral Load. From Figures 1(a), 1(b), and 1(c), it is observed that variations in the treatment efficacy have significant effect on the number of CD4+ T cells. For instance, with a treatment efficacy of 0.1 at the onset of HIV infection, the CD4+ T cells undergo a sharp decline, while treatment efficacy of 0.5 only slows down CD4+ T cell depletion and takes longer (approximately 40 days) to reduce to zero. However with a drug efficacy of 0.9 the number of

(a)

(b)

(c)

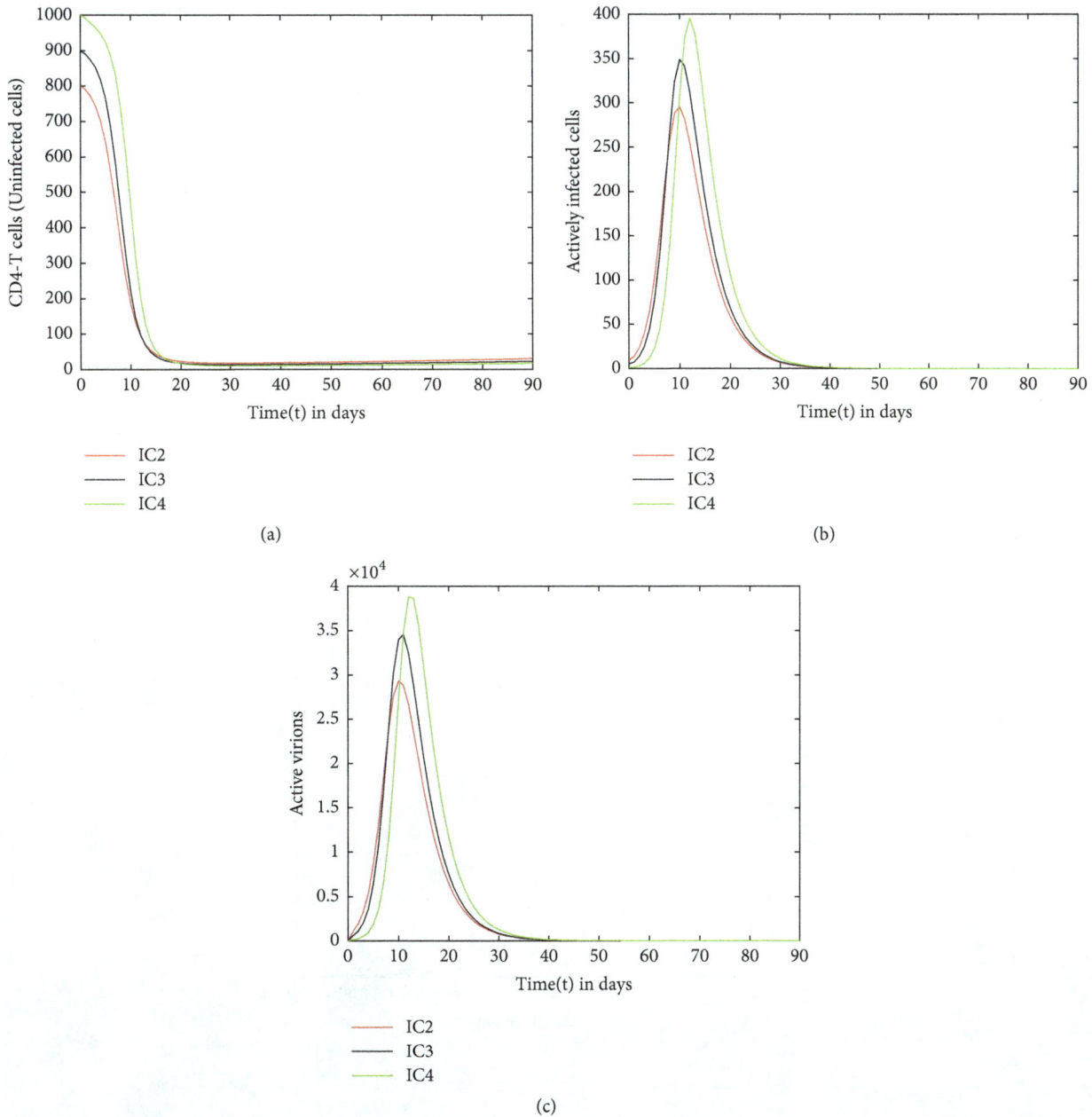

FIGURE 3: (a)-(c) The graph trajectory of $T(t), T^*(t)$, and $V(t)$ for $\rho = 0.5$ and $\vartheta = 0.5$ with the initial conditions IC2-IC4. $R_0 = 2.525 > 1$ and $EE = (198.02, 4.98, 498.31)$ is GAS.

CD4+ T cells is kept above 700 mm^{-3}. This demonstrates that treatment efficacy is directly proportional to the number of CD4+ T cells.

Figures 1(b) and 1(c) illustrate the effect of treatment efficacy on the number of actively infected cells and free virions, respectively. It is clear that from the time of infection if treatment efficacy is kept at a minimum of 0.1, then both actively infected cells and free virions replicate rapidly attaining a primary peak of 519 infected cells within the first 17 days, while the virions reach a peak of 5.074×10^4 within the same period of time. With an increase in treatment efficacy

both infected cells and viral replication rates are significantly reduced; for instance, at 0.9 efficacy level the number of virus and infected cells approach zero within the first 90 days.

5. Conclusion

In this paper, a mathematical model for within host HIV infection with virus-to-cell and cell-to-cell treatment has been formulated and analyzed. Target cell production has been modelled by a logistic incidence rate. The global dynamics have been shown to be dependent on R_0, in which for

(a)

(b)

(c)

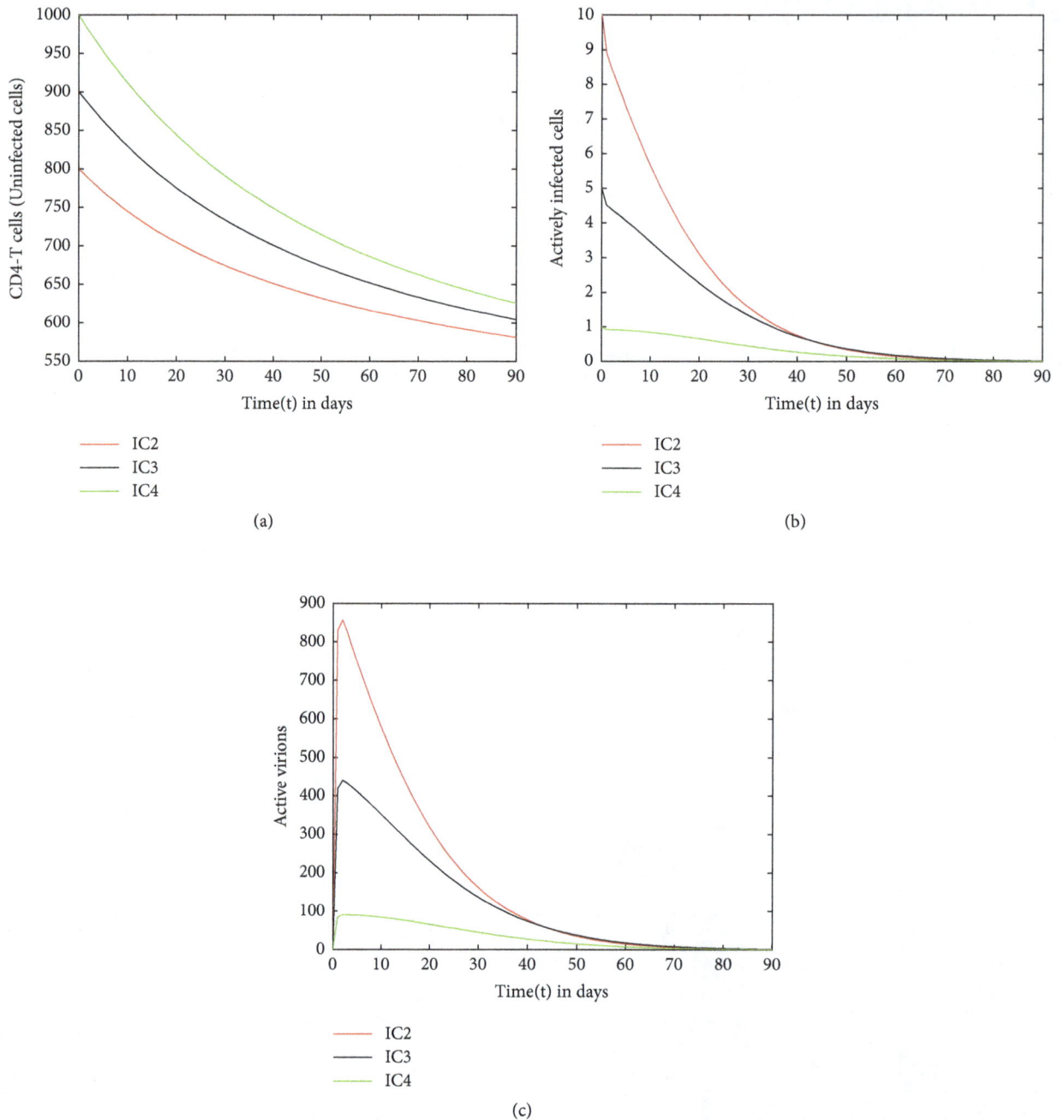

FIGURE 4: (a)-(c) The graph trajectory of $T(t), T^*(t)$, and $V(t)$ for $\rho = 0.9$ and $\vartheta = 0.9$ with the initial conditions IC2-IC4. $R_0 = 0.101 < 1$ and $E_0 = (1500, 0, 0)$ is globally asymptotically stable.

values of $R_0 < 1$ the infection free equilibrium is globally asymptotically stable, and for $R_0 > 1$ the endemic equilibrium is globally asymptotically stable. The global asymptotic stability of IFE and EE has been proved using the method by Castillo-Chavez and the geometric approach, respectively. Numerical simulations have shown that combined treatment has significant effect on the dynamics of HIV infection, in which high levels of treatment efficacy for both RTI and PI treatment was observed to stifle viral replication. This observation is in agreement with the findings of [9] which showed that if $R_0 > 1$ before treatment then the virus will increase as would be the number of infected cells, but after treatment if $R_0 < 1$, then T^* and V would both decline.

Conflicts of Interest

The authors declare that there are no conflicts of interest regarding the publication of this paper.

References

[1] M. A. Nowak and R. M. May, *Virus Dynamics: Mathematics Principles of Immunology and Virology*, Oxford University Press, London, UK, 2000.

[2] S.-S. Chen, C.-Y. Cheng, and Y. Takeuchi, "Stability analysis in delayed within-host viral dynamics with both viral and cellular infections," *Journal of Mathematical Analysis and Applications*, vol. 442, no. 2, pp. 642–672, 2016.

[3] A. M. Elaiw and N. A. Almuallem, "Global dynamics of delay-distributed HIV infection models with differential drug efficacy in cocirculating target cells," *Mathematical Methods in the Applied Sciences*, vol. 39, no. 1, pp. 4–31, 2016.

[4] A. M. Elaiw and A. A. Raezah, "Stability of general virus dynamics models with both cellular and viral infections and delays," *Mathematical Methods in the Applied Sciences*, vol. 40, no. 16, pp. 5863–5880, 2017.

[5] Y. Yang, L. Zou, and S. Ruan, "Global dynamics of a delayed within-host viral infection model with both virus-to-cell and cell-to-cell transmissions," *Mathematical Biosciences*, vol. 270, no. part B, pp. 183–191, 2015.

[6] H. Pourbashash, S. S. Pilyugin, P. De Leenheer, and C. McCluskey, "Global analysis of within host virus models with cell-to-cell viral transmission," *Discrete and Continuous Dynamical Systems - Series B*, vol. 19, no. 10, pp. 3341–3357, 2014.

[7] X. Zhang and X.-N. Liu, "Backward bifurcation of an epidemic model with saturated treatment function," *Journal of Mathematical Analysis and Applications*, vol. 348, no. 1, pp. 433–443, 2008.

[8] J. Zhang, J. Jia, and X. Song, "Analysis of an SEIR epidemic model with saturated incidence and saturated treatment function," *The Scientific World Journal*, vol. 2014, 2014.

[9] A. S. Perelson, D. E. Kirschner, and R. D. Boer, "Dynamics of HIV infection of CD4$^+$ T cells," *Mathematical Biosciences*, vol. 114, no. 1, pp. 81–125, 1993.

[10] D. E. Kirschner and G. F. Webb, "A Mathematical Model of Combined Drug Therapy of HIV Infection," *Journal of Theoretical Medicine*, vol. 1, no. 1, pp. 25–34, 1997.

[11] F. K. Tireito, G. O. Lawi, and C. O. Okaka, "Mathematical analysis of HIV/AIDS prophylaxis treatment model," *Applied Mathematical Sciences*, vol. 12, no. 18, pp. 893–902, 2018.

[12] S. Pankavich, "The effects of latent infection on the dynamics of HIV," *Differential Equations and Dynamical Systems*, vol. 24, no. 3, pp. 281–303, 2016.

[13] C. Castillo-Chavez, Z. Feng, and W. Huang, "On the computation of R0 and its role on global stability," in *Mathematical Approaches for Emerging and Reemerging Infectious Diseases: An Introduction*, C. Castillo-Chavez, S. Blower, P. van den Driessche, D. Kirschner, and A.-A. Yakubu, Eds., vol. 125, pp. 229–250, Springer, New York, NY, USA, 2002.

[14] Van Den Driessche and J. Wu, *Mathematical Epidemiology*, Springer-Varleg, Berlin Heidelberg, Germany, 2008.

[15] M. Y. Li and J. S. Muldowney, "A geometric approach to global-stability problems," *SIAM Journal on Mathematical Analysis*, vol. 27, no. 4, pp. 1070–1083, 1996.

[16] W. A. Coppel, *Stability and Asymptotic Behavior of Differential Equations*, D. C. Heath, Boston, Mass, USA, 1965.

13

A Dynamic Model of PI3K/AKT Pathways in Acute Myeloid Leukemia

Yudi Ari Adi ⓘ,[1,2] Fajar Adi-Kusumo ⓘ,[2] Lina Aryati,[2] and Mardiah S. Hardianti[3]

[1]*Department of Mathematics, Faculty of Mathematics and Natural Sciences, Ahmad Dahlan University, Yogyakarta 55166, Indonesia*
[2]*Department of Mathematics, Faculty of Mathematics and Natural Sciences, Universitas Gadjah Mada, Yogyakarta 55281, Indonesia*
[3]*Department of Internal Medicine, Faculty of Medicine, Universitas Gadjah Mada, Yogyakarta 55281, Indonesia*

Correspondence should be addressed to Yudi Ari Adi; yudi.adi@math.uad.ac.id

Academic Editor: Oluwole D. Makinde

Acute myeloid leukemia (AML) is a malignant hematopoietic disorder characterized by uncontrolled proliferation of immature myeloid cells. In the AML cases, the phosphoinositide 3-kinases (PI3K)/AKT signaling pathways are frequently activated and strongly contribute to proliferation and survival of these cells. In this paper, a mathematical model of the PI3K/AKT signaling pathways in AML is constructed to study the dynamics of the proteins in these pathways. The model is a 5-dimensional system of the first-order ODE which describes the interaction of the proteins in AML. The interactions between those components are assumed to follow biochemical reactions, which are modelled by Hill's equation. From the numerical simulations, there are three potential components targets in PI3K/AKT pathways to therapy in the treatment of AML patient.

1. Introduction

Acute myeloid leukemia (AML) is a hematological malignancy originating in the bone marrow. It is characterized by the infiltration of the bone marrow, blood, and other tissues by proliferative, abnormally differentiated, and sporadic poorly differentiated cells of the hematopoietic system [1, 2]. The AML is the most common malignancy of hematological system, illustrated by the accumulation of acquired somatic genetic alterations in hematopoietic progenitor cells. This alteration modifies the normal mechanisms of cells proliferation, self-renewal, and differentiation [3, 4]. Based on the validated cytogenetics and molecular abnormalities, the National Comprehensive Cancer Network (NCCN) classifies patients into three risk categories, which are better risk, intermediate risk, and poor risk. Patients with the NPM1 mutation in the absence of FLT3-ITD and CEPBA mutations are classified as favorable risk and patients with FLT3-ITD mutated CN-AML and with P53 mutations are classified as poor risk [5]. Untreated AML patient results in fatal infection, bleeding, or organ infiltration within 1 year of diagnosis but often within weeks to months [6]. The standard therapeutic strategies

in a patient with AML are chemotherapy, irradiation, and hematopoietic stem cell transplantation (HSCT) [1, 5, 7–10]. The main objective of those treatments is inducing remission and preventing the relapse [11]. In recent years, despite the potential gain of HSCT, the posttransplantation outcome remains dismal, especially those with high-risk category [10]. Currently, the development of the new therapies has been challenging to further improve the clinical outcome of AML, such as cytotoxic agent, small molecule inhibitor, and targeted therapies [12].

During the last decade, the PI3K/AKT signaling pathway has been studied extensively in human diseases. This pathway plays a significant role in a number of cellular functions, including differentiation, apoptosis, and cell cycle progression [8]. Aberrant PI3K/AKT activation is reported in 50-80% of AML cases [6]. The PI3K/AKT/mTOR network is activated in AML cells through a variety of mechanisms including upstream oncogenes such as FLT3-ITD, KIT, NRAS, and KRAS or autocrine/paracrine growth factors such as VEGF and IGF-1. It can also be activated by altered expression of p110d or phosphorylation of PTEN of pathway components and microenvironmental signals

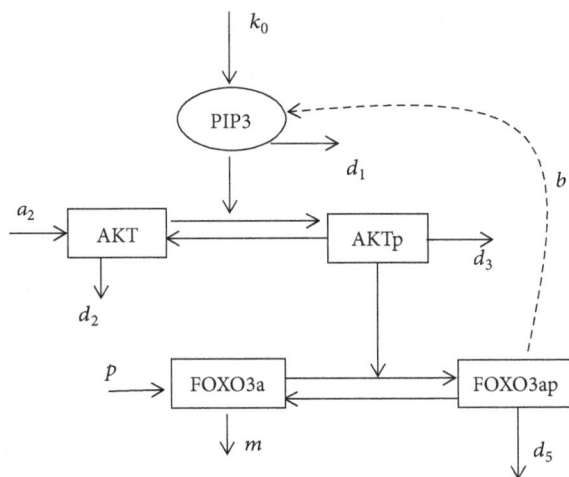

FIGURE 1: Simplified diagram of PI3K/AKT/FOXO3a pathway.

including chemokines and adhesion molecules [13, 14]. The AKT activation is associated with significantly elevated levels of phosphorylation FOXO3a in AML blast cells, supressing its normal function in induction apoptosis and cell cycle regulation [4, 15, 16]. Normally, FOXO3a transcriptionally activates several genes as the target. The FOXO3a binds to the promoter of apoptosis-inducing genes, such as Bim, FasL, and TRAIL, and to the promoter of cell cycle inhibitors, such as p27 and p21. The FOXO3a also activates the autophagy genes Gabarapl1, ATG12, and so forth [7, 9]. Researchers show that phosphorylation of FOXO3a is an adverse prognostic factor in AML associated with increased proliferation and overall survival [9, 16].

In a study of mathematical modelling of cancer, the dynamics of autologous immune system in chronic myeiloid leukemia have been constructed by Clapp et al. [17]. For mathematical models in leukemia and lymphoma, we refer the reader to Clapp and Levy [15]. Hovewer, these models are cellular level modelling, which study the interactions between cells and there is no known published model that studies the dynamics of acute myeloid leukemia (AML) in molecular level or biopathway. For modelling in biopathway, the model of PI3K/AKT pathway has been conducted in [18, 19]. The modelling of the interaction protein-protein in cell repair regulation also has been constructed in [20]. However, such models have not specifically studied a specific disease. In a previous work, Adi et al. in [21] studied the mathematical model of PI3K/AKT signaling pathway in AKT phosphorylation. The model did not include the FOXO3a protein that has been known as prognostic factor in AML. This model also does not follow Hill's equation that describes the substrate-enzyme interaction that has multiple ligand-binding sites. In this paper, we construct a mathematical model of PI3K/AKT pathway in AML which considers FOXO3a, the most important downstream pathway of AKT. With deep understanding of PI3K/AKT signaling pathways, the important parameters that play a role in the development of AML will be identified. Furthermore, a strategy can be determined in the treatment of AML disease through targeted therapy. For the model, the Michaelis Menten kinetics and Hill's equation in some components of biochemical reactions are used.

2. Model Development

The mathematical model is constructed by extending the AKT phosphorylation model in [21]. The extended model is defined by adding FOXO3a, which is a potential downstream pathway of AKT in the leukemic progression. Figure 1 shows the simplification of the complex network diagram of PI3K/AKT signaling pathways which drives the AML cells. Our model is focused on discussing the activities of five proteins in PI3K/AKT pathways that have been observed to be significant players in AML cells. We assume that the interactions of the protein follow the Michaelis Menten and Hill's equation. The mutation of the growth regulatory genes such as FLT3-ITD is common in AML cases and results in the activation of PI3K [13]. The activation of PI3K then catalyzes the phosphorylation of phosphatidylinositol bisphosphate (PIP2) which can be phosphorylated at the D3 position of the inositol ring on extracellular stimulation, resulting in the formation of phosphatidylinositol trisphosphate (PIP3). In this model, the formation is grouped as a single process, called the PI3K level, and denoted by k_0. This formation can be reversed by tumor suppressor PTEN that catalyzes PIP3 dephosphorylation into PIP2. The PIP3 dephosphorylation that is assumed follows Hill's equation with coefficient 4. It is according to the fact that PIP3 has four binding sites of PH domain, that is, with PTEN, SHIP1, InsP4, and AKT [1].

Inactive AKT binds PIP3 which enables 3-phosphoinositide-dependent kinase-1 (PDK1) to phosphorylate AKT at Thr308. For full activation, AKT is also phosphorylated at Ser473 by mTORC2. PDK1 and mTORC2 are grouped as a single enzyme catalyzing the phosphorylation (activation) of AKT. Activated AKT is regulated by protein phosphatase 2A (PP2A) and pleckstrin homology domain leucine-rich repeat

TABLE 1: Initial concentration of the molecular component.

Protein	Concentration (μM)	References
PI3K	0.01 – 0.1	[21]
PIP3	0.7 – 0.8	[21]
AKT	0.01 – 1.0	[20, 21]
PP2A	0.004 – 0.15	[21]
FOXO3a	0.01 – 1.4	Assumed

protein phosphatase (PHLPP). The phosphate PP2A preferentially dephosphorylates AKT on the Thr308 site, while PHLPP specifically dephosphorylates AKT on Ser473 site. In this model, the two phosphatases are grouped into a single enzyme catalyzing the dephosphorylation of AKT. According to the fact that AKT has two binding sites, this protein activity is assumed to follow Hill's equation with coefficient 2. In the next downstream pathways, the activated AKT phosphorylates and inhibits the forkhead transcription factor, FOXO3a. FOXO3a phosphorylation promotes its translocation from the nucleus to the cytoplasm. Phosphorylation by AKTp on Thr 24, Ser 256, and Ser 318 inhibits FOXO3a activities by increasing nuclear export and this in turn increases proliferation. The FOXO3a in the cytoplasm, denoted by FOXO3ap, is the interaction with the 14-3-3 nuclear export protein. This interaction preventing nuclear reimport by concealing nuclear localization signals and promotes the FOXO3ap degradation by the proteasome [7, 16]. In this model, the growth of FOXO3a is assumed to follow the logistic model as well as the translocation and relocation of the cytoplasm and nucleus by phosphorylation and dephosphorylation. According to the fact that FOXO3a has three binding sites, Hill's equation with coefficient 3 is used. The FOXO3ap enhances the expression and phosphorylation of RTKs which could in turn activate and sustain this pathways [3]. Thus, FOXO3a indirectly interacts and enhances PIP3 activity, resulting in a positive feedback loop. The reactivation or dephosphorylation FOXO3a, which is mediated by PP2A, promotes the relocation to the nucleus. Based on the diagram in Figure 1, a mathematical model is defined as follows:

$$\frac{dx_1}{dt} = k_0 + bx_5 - \frac{k_1 x_1^4}{K_1^4 + x_1^4} - d_1 x_1 \tag{1}$$

$$\frac{dx_2}{dt} = a_2 - \frac{k_2 x_1 x_2^2}{K_2^2 + x_2^2} + \frac{k_3 x_3^2}{K_3^2 + x_3^2} - d_2 x_2 \tag{2}$$

$$\frac{dx_3}{dt} = \frac{k_2 x_1 x_2^2}{K_2^2 + x_2^2} - \frac{k_3 x_3^2}{K_3^2 + x_3^2} - d_3 x_3 \tag{3}$$

$$\frac{dx_4}{dt} = x_4 \left(p - mx_4 \right) - \frac{k_4 x_3 x_4^3}{K_4^3 + x_4^3} + \frac{k_5 x_5^3}{K_5^3 + x_5^3} \tag{4}$$

$$\frac{dx_5}{dt} = \frac{k_4 x_3 x_4^3}{K_4^3 + x_4^3} - \frac{k_5 x_5^3}{K_5^3 + x_5^3} - d_5 x_5 \tag{5}$$

The variables x_1, x_2, x_3, x_4, and x_5 represent the concentration of PIP3, AKT, AKT phosphorylation (AKTp), FOXO3a, and FOXO3a phosphorylation (FOXO3ap), respectively.

In the next section, we do some numerical simulation to understand the dynamics of protein in PI3K/AKT pathways. The numerical simulation will be run in two different situations based on the existence of the FOXO3a translocation from the nucleus to the cytoplasm as a normal cell or AML cell to understand the dynamics of AKT/FOXO pathways. In the normal cell, the activities of AKTp do not induce the translocation of FOXO3a from the nucleus to the cytoplasm. In the AML cell, aberrant PI3K/AKT signaling pathway results in phosphorylation of FOXO3a leading to cytoplasmic mislocalization and consequent degradation of these proteins [9].

The situation of normal and AML cells is distinguished based on differences in some parameter values. First, the constant rate of PIP3 dephosphorylation in AML cells is lower than normal cells as a result of various abnormal mechanisms in the PI3K signal upstream pathway, for example, PTEN deletion [8]. Second, the value of dephosphorylation rate of AKT in AML cells is smaller than the one in the normal cells. This is due to the fact that, in AML, there is a decrease of PIP3 level, a protein phosphatase that plays a role in the dephosphorylation of AKTp [8]. Furthermore, in normal cells, it is assumed that AKTp does not induce the translocation of FOXO3a from the nucleus to the cytoplasm so that the parameter value of phosphorylation of FOXO3a is tending to zero. The last difference is that the rate of dephosphorylation of FOXO3ap in AML is smaller than that in normal cells. This is due to the degradation of FOXO3ap in AML cells in the cytoplasm [7, 9].

3. Results and Discussion

The model equations (1)–(5) are not sufficiently accessible to allow us to conduct the mathematical analysis. Therefore, in this paper, we only provide numerical simulations. In this section, the numerical results of system (1)–(5) are simulated by employed the Runge-Kutta method of order 4 to provide the integration in some cases depending on the parameter values. The parameter values used in the system are based on the clinical data that can be obtained in some medical literature as in Tables 1 and 2.

Tables 1 and 2 show the kinetic rates and the initial concentration levels of the various proteins in PI3K and AKT pathways. For AML cells the parameter values are $k_0 = 0.01$; $b = 0.0083$; $k_1 = 0.005$; $K_1 = 0.2$; $a_2 = 0.09$; $k_2 = 1$; $d_1 = 0.0083$; $K_2 = 0.1$; $k_3 = 0.36$; $K_3 = 0.2$; $d_2 = 0.08$; $d_3 = 0.1$; $p = 0.3$; $m = 0.25$; $k_4 = 0.3$; $K_4 = 0.1$; $k_5 = 0.1$; $K_5 = 0.1$; and $d_5 = 0.1$. For normal cells, similar parameter values are used, except $k_1 = 0.017$; $k_3 = 0.67$; $k_4 = 0.001$; and $k_5 = 0.033$.

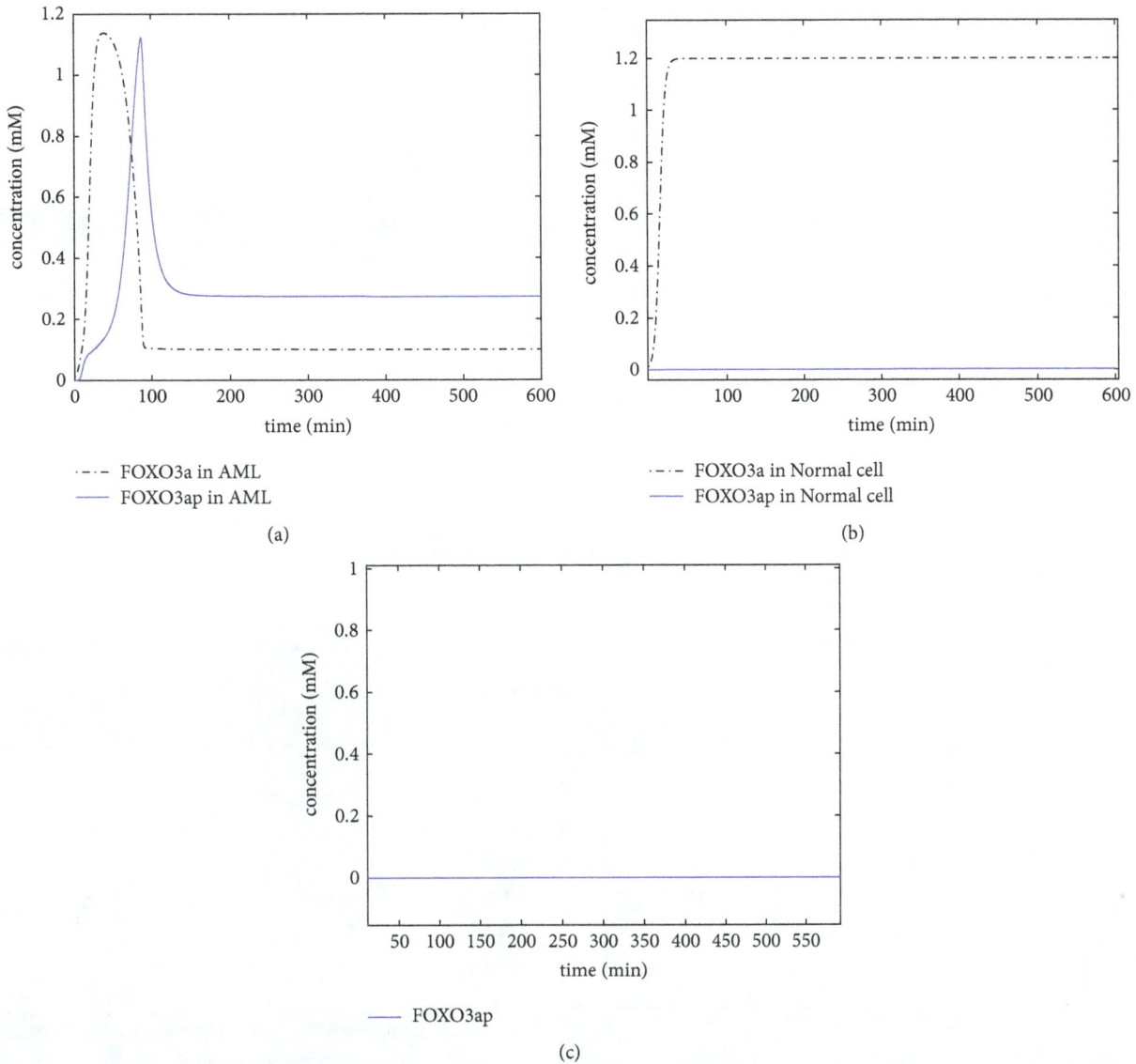

FIGURE 2: Dynamic of FOXO3a and FOXO3ap concentration in AML cell (a), dynamic of FOXO3a and FOXO3ap concentration in the normal cell (b), and FOXO3ap concentration in the absence of FOXO3a phosphorylation (c).

The dynamics concentrations of FOXO3a and FOXO3ap in the AML and the normal cell are given in Figure 2. Figure 2(a) shows that existence of the FOXO3a phosphorylation plays a role in the increasing of FOXO3ap concentration. The presence of inactive FOXO3a in the cytoplasm, FOXO3ap in our model, means that the apoptosis mechanism is not working properly so that there is no cell death. The FOXO3ap also enhances proliferation cell, leading to accumulation of abnormal cells because they do not stop growing when they should. The lifespan of the white blood cell in a myeloid lineage is about 3-12 days [23]. Therefore, there should be apoptosis between 3-12 days characterized by low-level FOXO3ap after that time, which does not occur in the AML cell. It can be seen that FOXO3ap reaches a peak in 100 minutes and then decreases and oscillates to a certain level (see Figure 2(a)).

The increasing concentration of the FOXO3ap would affect the decreasing concentration of FOXO3a. It is illustrated in Figure 2(a) that the FOXO3a concentration initially increases and peaks within 38 minutes. Moreover, the FOXO3a concentration immediately decreases and oscillates in low concentration with small amplitude. Under the normal condition, the concentration of FOXO3a transcription factor in the nucleus is much higher than those in the cytoplasm, FOXO3ap; see Figure 2(b). The concentration of FOXO3a in the normal cell reaches the maximal level in short time. The increasing of FOXO3a is followed by the slightly increasing FOXO3ap in much lower concentration. It indicates that FOXO3a is not translocated to the cytoplasm. It shows that FOXO3a promotes apoptosis and cell cycle regulation as well. Thus the balancing of cell cycle regulation can be well

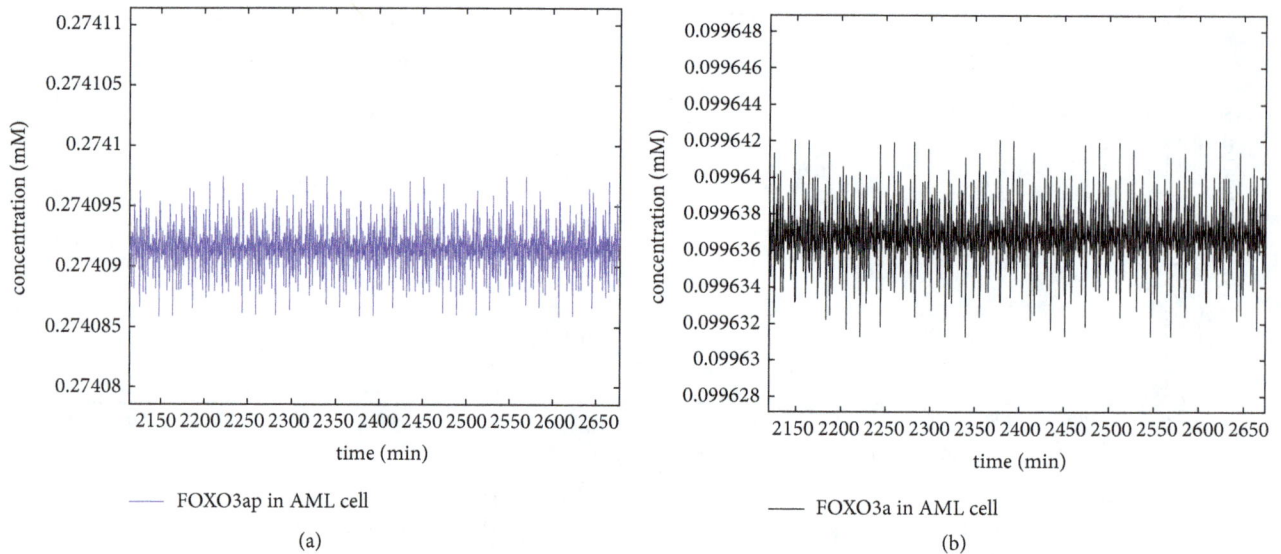

FIGURE 3: The oscillation of the FOXO3ap concentration (a) and FOXO3a concentration (b) in AML cell.

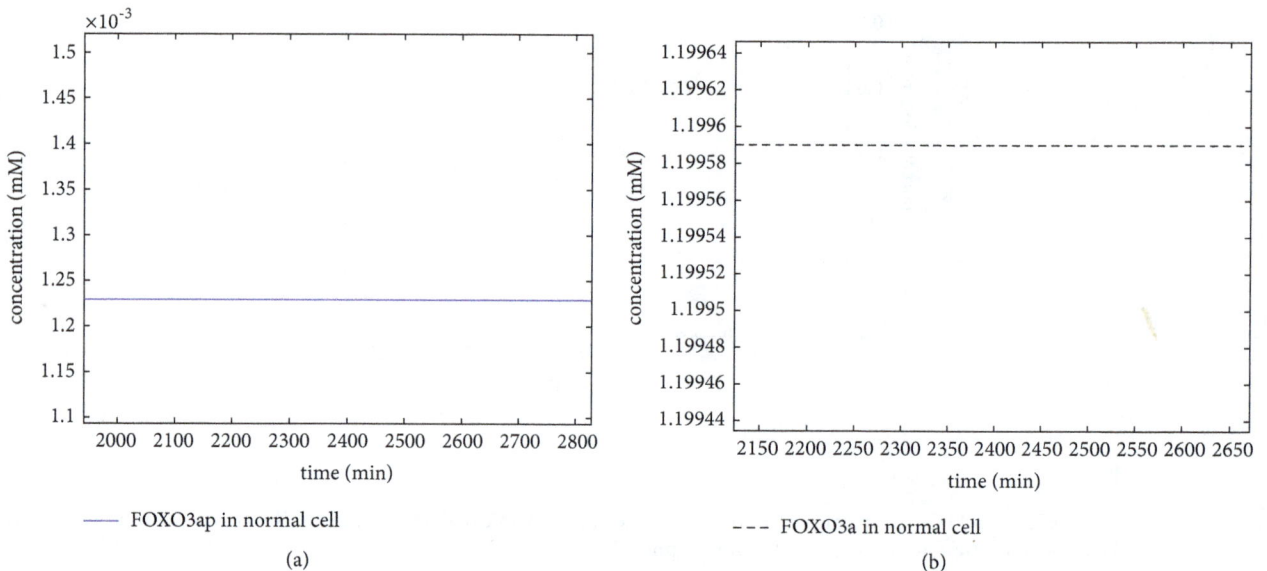

FIGURE 4: The FOXO3ap (a) and FOXO3a (b) concentration in the normal cell.

preserved. We note that if the constant rate of FOXO3a phosphorylation is set to be zero, the concentration of FOXO3ap will be zero (see Figure 2(c)).

Figure 3 shows the oscilation of FOXO3ap and FOXO3a concentration in AML cell. The oscilation of FOXO3ap concentration is given in Figure 3(a) and the oscilation of FOXO3a concentration is given in Figure 3(b). The oscillation indicates that cell cycle and proliferation continue to occur and there is no maturation of these cells. The behavior of FOXO3a can be used to identify the existence of AML disease. In addition, from the simulation, it is known that high level of FOXO3a and low level of FOXO3ap in the normal cell do not oscillate as in AML cell; see Figure 4. It indicates that FOXO3a works properly in cell cycle regulation and apoptosis.

Next, we will see the effect of phosphorylation of FOXO3a on the other proteins, such as PIP3, AKT, and AKTp. Figures 5 and 6 show the comparison of PIP3, AKT, and AKTp concentration in the normal cell and AML cell. It can be seen that, in the longtime behavior, the concentrations of PIP3 and AKTp in the AML cells are higher than those in the normal cells, while AKT in the AML cells is lower than that in the normal cells. In the AML cells, the concentration of PIP3 is gradually increasing and oscillates at a certain level as a response to FOXO3a phosphorylation. The PIP3 concentration reaches the maximum level and remains at that level for a long time and oscillates with small amplitude; see Figure 5(b). Figure 5(c) illustrates the concentration of PIP3 in the normal cells is at a low level without oscillation.

(a)

(b)

(c)

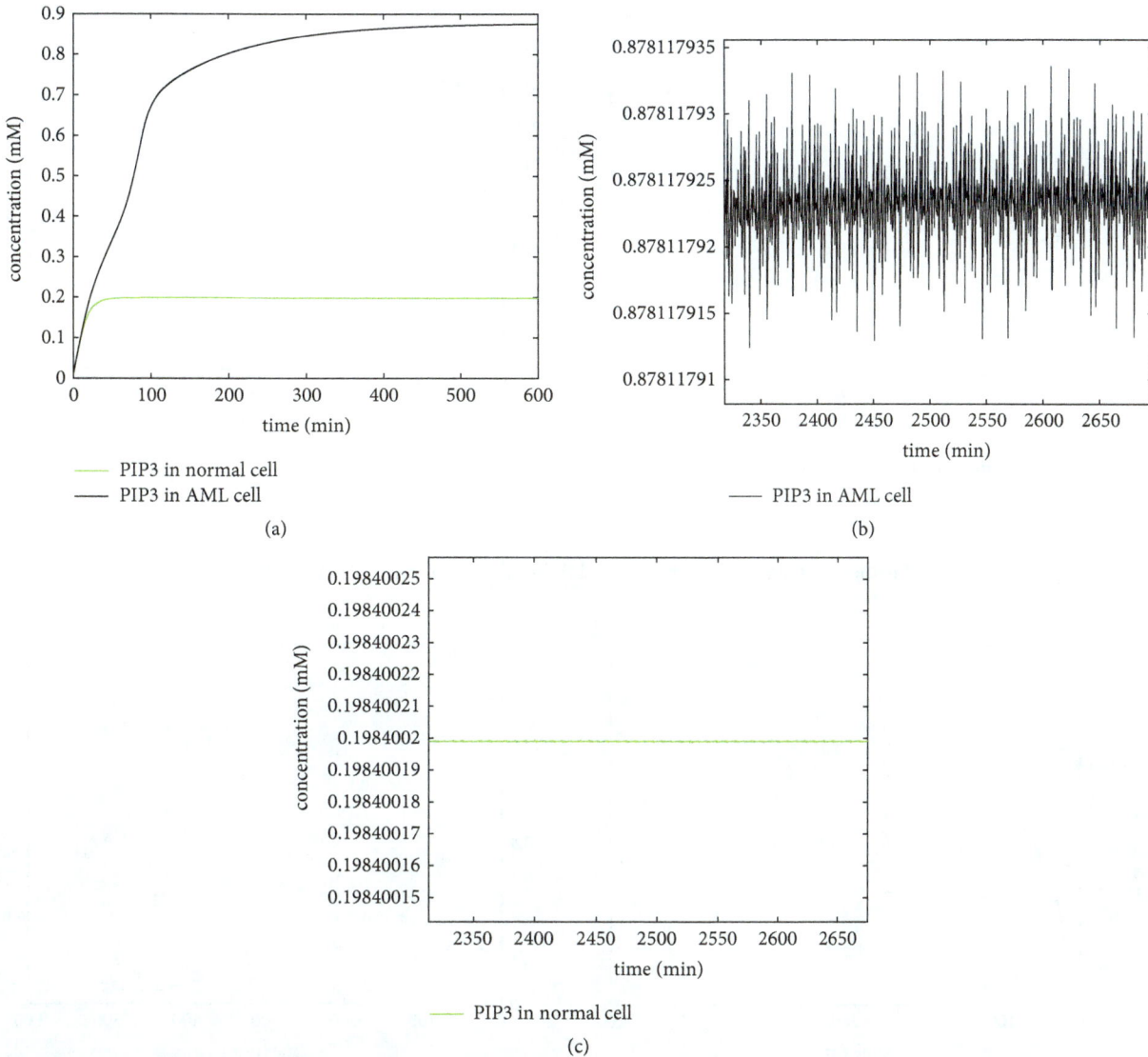

FIGURE 5: Dynamics of PIP3 in the normal cell and AML cell (a). Concentration of PIP3 in AML cell oscillates at a certain level (b), while PIP3 in normal cell does not oscillate (c).

Figure 6 shows the differences between AKT and AKTp behavior in the normal cell and AKT and AKTp in the AML cell, respectively. Under normal conditions, when the levels of PIP3 decrease, the AKT activity is attenuated by dephosphorylation by phosphatase. Figure 6(a) shows the AKT in AML cell subsequently sustained at lower concentrations than in the normal cell. In the AML cell, as the result of FOXO3a translocation from the nucleus to the cytoplasm, the concentration level of AKT decreases quickly, while AKTp immediately increases and remains at a certain level. Figure 6(b) shows that the AKTp in AML cell is subsequently sustained at a higher concentration than in the normal cell. Figure 7 tested the model by increasing the constant rate of FOXO3a phosphorylation from 1 (Figure 7(a)) to 2 (Figure 7(b)), while keeping all other parameter values the same as in Figure 2. The effect of this increase is that the greater the rate of FOXO3a phosphorylation, the lower

concentrations of FOXO3a and FOXO3ap. This condition is due to the fact that the greater value of FOXO3a phosphorylation rate will accelerate the translocation of FOXO3a from the nucleus to the cytoplasm and lead to proteasome degradation. The increase of FOXO3a phosphorylation rate does not extremely affect the dynamics of AKT and AKTp, while the PIP3 concentration becomes slightly lower.

4. Conclusions

As shown in the numerical simulation, the key components in driven AML cell are high levels of PIP3, AKTp, and FOXO3ap, that is, inactive FOXO3a in the cytoplasm. These results suggest that these three components are potential targets for AML therapy, of course with due regard to the other proteins that mediated protein interactions. For example, the parameter values associated with FOXO3a are taken from

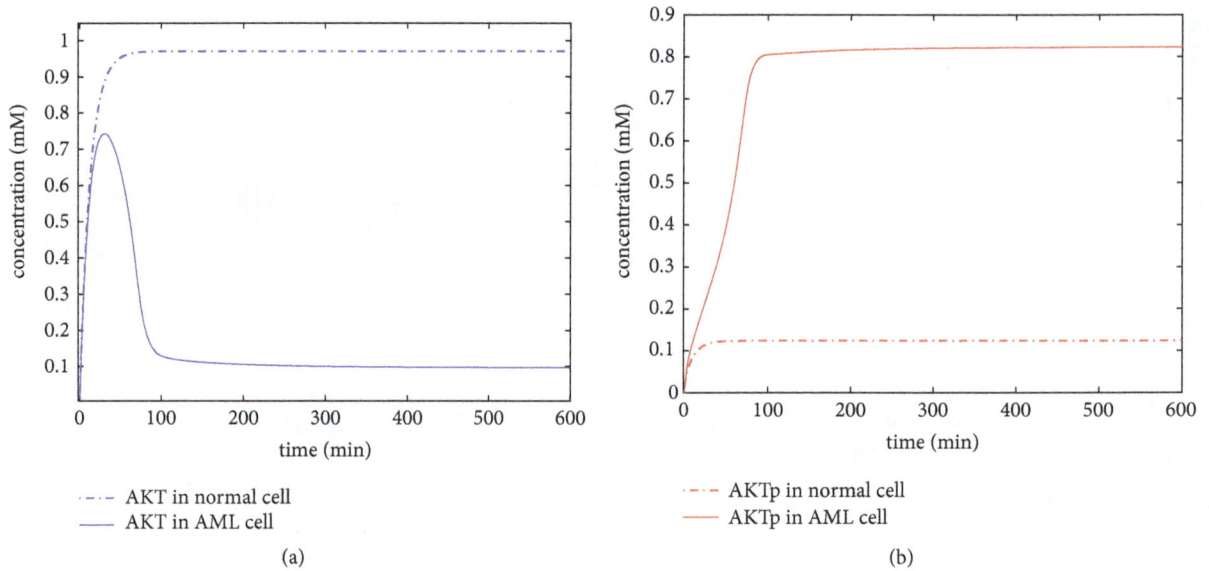

FIGURE 6: Dynamics of AKT (a) and AKTp (b) in normal and AML cell.

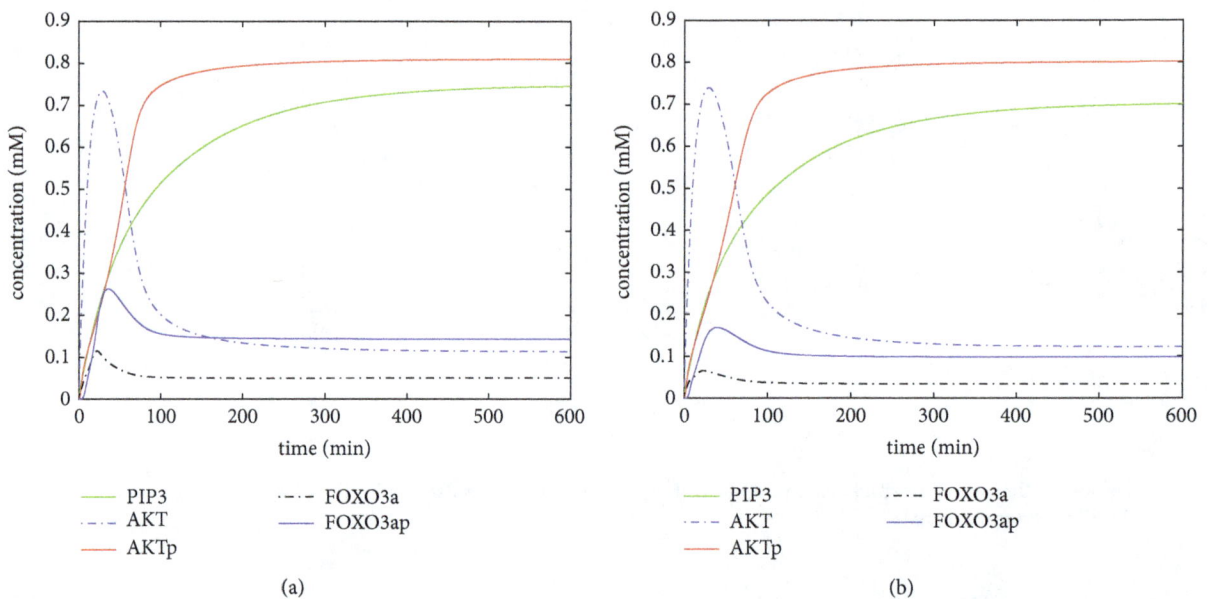

FIGURE 7: Dynamics of PIP3, AKT, AKTp, FOXO3a, and FOXO3ap in the AML cell with increasing rate of FOXO3a phosphorylation from 1.0 (a) to 2.0 (b).

proteins that have a similar function to FOXO3a, such as MDM2. It could be very useful for determining reasonable ranges for the rate of various biochemical reactions involved. As more medical facts are known about the PI3K/AKT signaling pathways in AML, the model may be needed to be modified. It is possible that other equations representing rates of change of other proteins or other signaling pathways that integrated with PI3K/AKT pathways may have to be added to the system. Mathematical analysis of the model may also be useful in understanding protein interactions in this pathway. In the future studies, we will analyze mathematically the dynamics of the system and study the bifurcation related to the variation of its parameter values.

Conflicts of Interest

The authors declare that there are no conflicts of interest regarding the publication of this paper.

Acknowledgments

This work was supported by the Ministry of Research and Higher Education (Kemenristek DIKTI) of Indonesia (Grant no. 109/SP2H/LT/DPRM/2018). Special thanks are due to the Cancer Modelling Team UGM for the discussions during the research.

TABLE 2: Parameter values and kinetic rates being used.

Parameter	Description	Unit	Value	References
k_0	PI3K level	$\mu M min^{-1}$	0.01 – 0.1	[21]
b	Increase activation PIP3 by FOXO3ap	min^{-1}	0.0083	*
k_1	Constant rate of PIP3 dephosphorylation by PTEN	$\mu M min^{-1}$	0.0006 – 0.21	[21]
d_1	PIP3 degradation	min^{-1}	0.001 – 0.01	[22]
a_2	AKT production rate	$\mu M min^{-1}$	0.036 – 0.108	[21, 22]
k_2	Constant rate of AKT phosphorylation	min^{-1}	1 – 20	[21]
k_3	Constant rates of AKTp dephosphorylation by PP2A	$\mu M min^{-1}$	0.36 – 13.5	[21, 22]
d_2	AKT degradation rate	min^{-1}	0.063 – 0.08	[21]
d_3	AKTp degradation rate	min^{-1}	0.0008 – 0.1	[21, 22]
p	FOXO3a production rate	min^{-1}	0.002 – 0.5	*
m	FOXO3a degradation rate by 14-3-3 protein	$\mu M^{-1} min^{-1}$	0.004 – 0.28	*
k_4	Constant rate of FOXO3a phosphorylation	min^{-1}	0 – 0.33	*
k_5	Constant rate of FOXO3a dephosphorylation by PP2A	$\mu M min^{-1}$	0.000297 – 2.92	*
d_5	FOXO3ap degradation rate	min^{-1}	0.033 – 0.125	*
K_1	Michaelis constant of PIP3 dephosphorylation	μM	0.01 – 1	[18, 20]
K_2	Michaelis constant of AKT phosphorylation	μM	0.1	[20]
K_3	Michaelis constant of AKTp dephosphorylation	μM	0.08 – 0.4	[20, 22]
K_4	Michaelis constant of FOXO3a phosphorylation	μM	0.1	*
K_5	Michaelis constant of FOXO3ap dephosphorylation	μM	0.1	*

*Parameters values are assumed to be the same as other transcription factors such as MDM2 in [18, 22].

References

[1] P. Faduola, A. Hakim, J. Mansnérus, A. Imai, and R. O'Neill, "Acute myeloid leukaemia - therapy - past, present and future," *Translational Biomedicine*, vol. 4, no. 2, 2013.

[2] G. Mengistu, F. Balcha, and S. Britton, "Clinical presentation of onchocerciasis among indigenous and migrant farmers in Southwest Ethiopia," *East African Medical Journal*, vol. 76, no. 11, pp. 635–638, 1999.

[3] Y. Shen, J. Bai, and A. He, "Role of mTOR signaling pathway in acute myeloid leukemia," *International Journal of Clinical and Experimental Medicine*, vol. 9, no. 2, pp. 637–647, 2016.

[4] S. Takahashi, "Downstream molecular pathways of FLT3 in the pathogenesis of acute myeloid leukemia: biology and therapeutic implications," *Journal of Hematology & Oncology*, vol. 4, article 13, 2011.

[5] NCCN, *NCCN Clinical Practice Guidelines in Oncology*, 2016.

[6] G. A. Horne, R. Kinstrie, and M. Copland, "Novel drug therapies in myeloid leukemia," *Pharmaceutical patent analyst*, vol. 4, no. 3, pp. 187–205, 2015.

[7] R. S. Nho and P. Hergert, "FoxO3a and disease progression," *World Journal of Biological Chemistry*, vol. 5, no. 3, pp. 346–354, 2014.

[8] R. Polak and M. Buitenhuis, "The PI3K/PKB signaling module as key regulator of hematopoiesis: Implications for therapeutic strategies in leukemia," *Blood*, vol. 119, no. 4, pp. 911–923, 2012.

[9] C. M. Santamaría, M. C. Chillón, R. García-Sanz et al., "High FOXO3a expression is associated with a poorer prognosis in AML with normal cytogenetics," *Leukemia Research*, vol. 33, no. 12, pp. 1706–1709, 2009.

[10] G. J. Schiller, P. Tuttle, and P. Desai, "Allogeneic Hematopoietic Stem Cell Transplantation in FLT3-ITD-Positive Acute Myelogenous Leukemia: The Role for FLT3 Tyrosine Kinase Inhibitors Post-Transplantation," *Biology of Blood and Marrow Transplantation*, vol. 22, no. 6, pp. 982–990, 2016.

[11] T. Grafone, M. Palmisano, C. Nicci, and S. Storti, "An overview on the role of FLT3-tyrosine kinase receptor in acute myeloid leukemia: Biology and treatment," *Oncology Reviews*, vol. 6, no. 1, pp. 64–74, 2012.

[12] T. M. Kadia, F. Ravandi, J. Cortes, and H. Kantarjian, "New drugs in acute myeloid leukemia," *Annals of Oncology*, vol. 27, no. 5, pp. 770–778, 2016.

[13] M. Andreeff, *Targeted Therapy of Acute Myeloid Leukemia*, Springer New York, New York, NY, USA, 2015.

[14] O. Lindblad, E. Cordero, A. Puissant et al., "Aberrant activation of the PI3K/mTOR pathway promotes resistance to sorafenib in AML," *Oncogene*, vol. 35, no. 39, pp. 5119–5131, 2016.

[15] G. Clapp and D. Levy, "A review of mathematical models for leukemia and lymphoma," *Drug Discovery Today: Disease Models*, vol. 16, pp. 1–6, 2015.

[16] S. M. Kornblau, N. Singh, Y. Qiu, W. Chen, N. Zhang, and K. R. Coombes, "Highly phosphorylated FOXO3A is an adverse prognostic factor in acute myeloid leukemia," *Clinical Cancer Research*, vol. 16, no. 6, pp. 1865–1874, 2010.

[17] G. D. Clapp, T. Lepoutre, R. El Cheikh et al., "Implication of the autologous immune system in BCR-ABL transcript variations in chronic myelogenous leukemia patients treated with imatinib," *Cancer Research*, vol. 75, no. 19, pp. 4053–4062, 2015.

[18] B. W. Keng and B. D. Aguda, "Akt versus p53 in a network of oncogenes and tumor suppressor genes regulating cell survival and death," *Biophysical Journal*, vol. 91, no. 3, pp. 857–865, 2006.

[19] G. Wang, *Analysis of Complex Diseases*, CRC Press, Taylor & Francis Group, 2014.

[20] F. Adi-Kusumo and A. Wiraya, "Mathematical modeling of the cells repair regulations in nasopharyngeal carcinoma,"

Mathematical Biosciences, vol. 277, pp. 108–116, 2016.

[21] Y. A. Adi, F. Adi-Kusumo, L. Aryati, and M. S. Hardianti, "Modelling inhibition of AKT phosphorylation in acute myeloid leukemia," in *Proceedings of the 2016 Conference on Fundamental and Applied Science for Advanced Technology, ConFAST 2016*, Indonesia, January 2016.

[22] R. Karabekmez, "Modeling of Cancer Signaling Pathways," UWSpace, 2013.

[23] B. Young, G. O'Dowd, and P. Woodford, *Wheater's Functional Histology: A Text and Colour Atlas*, vol. 64, Churchill Livingstone, Elsevier, 2014.

14

Analysis and Optimal Control Intervention Strategies of a Waterborne Disease Model

Obiora Cornelius Collins ⓘ[1,2] **and Kevin Jan Duffy** ⓘ[1]

[1]*Institute of Systems Science, Durban University of Technology, Durban 4000, South Africa*
[2]*School of Mathematics, Statistics and Computer Science, University of KwaZulu-Natal, Durban 4000, South Africa*

Correspondence should be addressed to Obiora Cornelius Collins; obiora.c.collins@gmail.com

Academic Editor: Wan-Tong Li

A mathematical model is formulated that captures the essential dynamics of waterborne disease transmission under the assumption of a homogeneously mixed population. The important mathematical features of the model are determined and analysed. The model is extended by introducing control intervention strategies such as vaccination, treatment, and water purification. Mathematical analyses of the control model are used to determine the possible benefits of these control intervention strategies. Optimal control theory is utilized to determine how to reduce the spread of a disease with minimum cost. The model is validated using a cholera outbreak in Haiti.

1. Introduction

Waterborne diseases which include Cholera, Hepatitis A and Hepatitis E, Giardia, Cryptosporidium, and Rotavirus are among the serious health problems of people globally. This is especially so in developing countries where there is limited access to clean water. Unsafe water supply, poor sanitation and poor hygiene are major causes of waterborne diseases [1]. According to WHO [2], approximately 1.1 billion people globally do not have access to sources of reliable water. About 700,000 children die every year from diarrhoea caused by unsafe water and poor sanitation [3]. The prevalence of waterborne diseases could be controlled especially in developing countries through access to safe water, provision of adequate sanitation facilities, and better hygiene practices [1]. Control measures such as water purification, vaccination, and treatment of infected individuals are among the most effective ways of reducing the spread of these diseases [4–6]. In this study, we investigate the impact of these types of control measures in reducing the spread of waterborne diseases.

Even with the availability of control measures, affordability is often the greatest obstacle for many communities where diseases are endemic. The spread of waterborne diseases are often associated with poverty, limited resources, and low socioeconomic status [2, 7]. Optimal control theory can point to efficient approaches to reduce the spread of a disease with minimum costs [4, 8]. In this study, we consider optimal control theory to investigate how to reduce the spread of waterborne diseases with minimum costs.

Some of the essential factors that influence the dynamics of waterborne diseases include sanitation [9], different transmission pathways [10, 11], water treatment efforts [12,13], pathogen ecology outside of human hosts [14], climatological factors or seasonal fluctuations [15–18], and heterogeneity in disease transmission [19, 20]. Understanding how these factors interact to influence the dynamics of waterborne diseases are challenging, making the dynamics of waterborne diseases complex. Several theoretical studies have taken some of these factors into account to improve the understanding of waterborne disease dynamics and subsequently investigate the possible means of reducing the diseases [7, 11, 19, 21–26]. Even though these studies have contributed immensely in improving the understanding of waterborne disease dynamics, theoretical studies for waterborne disease dynamics and control are not complete. In this study, we consider mathematical models to investigate the dynamics and control of waterborne diseases. The findings complement existing results in the literature on the dynamics and control of waterborne diseases.

The remaining parts of this paper are organized as follows. In Section 2, a waterborne disease model, which underpins the essential dynamics, is presented and analysed. To determine the possible benefits of control measures a multiple control model (with all controls imposed simultaneously) is presented and analysed in Section 3. In Section 4, optimal control analyses are used to investigate how to reduce the spread of a disease with minimum costs. The model is validated by using it to study a cholera outbreak in Haiti in Section 5. We conclude the paper by discussing our results in Section 6.

2. Waterborne Disease Model and Analysis

In this section we present a waterborne disease model that underlies the dynamics for a homogeneous population without any control intervention measures. Analyses of this model are necessary as a comparison to understand the effects of the control intervention strategies included in subsequent sections.

2.1. Formulation of the Control-Free Model. We consider an extension of the standard SIR model under the assumption of constant human population size $N(t)$ by adding a compartment $W(t)$ that measures pathogen concentration in a water reservoir [11, 27]. As usually done, the total human population $N(t)$ is partitioned into susceptible $S(t)$, infected $I(t)$, and recovered individuals $R(t)$ such that $N(t) = S(t) + I(t) + R(t)$. Individuals enter the susceptible class $S(t)$ through birth at a rate μ. Susceptible individuals $S(t)$ become infected with waterborne disease through contact with contaminated water at a rate b. Direct person-to-person transmissions are not considered because water-to-person transmissions have been shown to be the major route of waterborne disease transmissions [6, 23, 28]. Infected individuals $I(t)$ shed pathogens into water at a rate ν and recover naturally at a rate γ. Pathogens are generated naturally in the water at a rate α and decay at a rate ξ. Natural human deaths occur at a rate μ. With these assumptions we obtain the model

$$\dot{S}(t) = \mu N(t) - bS(t)W(t) - \mu S(t),$$
$$\dot{I}(t) = bS(t)W(t) - (\mu + \gamma)I(t),$$
$$\dot{W}(t) = \nu I(t) - \sigma W(t),$$
$$\dot{R}(t) = \gamma I(t) - \mu R(t),$$

(1)

where $\sigma = \xi - \alpha > 0$ is the natural decay rate of pathogens in the water reservoir. Note that our model (1) is in the form considered by Tien and Earn [11] to study the multiple transmission pathways for waterborne diseases. A difference is that they considered infections to be generated through both direct person-to-person and indirect water-to-person contacts. Our approach that considers infections through

indirect water-to-person contacts only is particularly relevant with waterborne diseases such as cholera which are primarily transmitted through contaminated water.

For the qualitative analyses of model (1), we consider a dimensionless version given by

$$\dot{s}(t) = \mu - \beta s(t)w(t) - \mu s(t),$$
$$\dot{i}(t) = \beta s(t)w(t) - (\mu + \gamma)i(t),$$
$$\dot{w}(t) = \sigma(i(t) - w(t)),$$
$$\dot{r}(t) = \gamma i(t) - \mu r(t),$$

(2)

where $s = S/N$, $i = I/N$, $r = R/N$, $w = \sigma W/(\nu N)$, and $\beta = b\nu N/\sigma$.

All parameters are realistically assumed positive and the initial conditions are as follows:

$$r(0) > 0,$$
$$i(0) \geq 0,$$
$$w(0) \geq 0,$$
$$r(0) \geq 0.$$

(3)

All the solutions of model (2) considered are in the feasible region

$$\Phi = \{s, i, w, r > 0 : s + i + r = 1\}.$$ (4)

Φ is positively invariant and the existence and uniqueness of solutions of model (2) hold in this region. Thus, model (2) is well posed mathematically and epidemiologically in Φ.

2.2. Basic Reproduction Number. The control-free model (2) has a unique disease-free equilibrium (DFE) given by

$$(s^0, i^0, w^0) = (1, 0, 0).$$ (5)

The basic reproduction number $\mathscr{R}_0$ of the control-free model (2) is determined using the next generation matrix method [29]:

$$\mathscr{R}_0 = \frac{\beta}{(\gamma + \mu)}.$$ (6)

2.3. Stability Analysis of the Disease-Free Equilibrium. For a dynamical infectious disease model, stability about its disease-free equilibrium (DFE) describes the short-term dynamics of the disease [30]. Therefore to determine the short-term dynamics of the waterborne disease considered here, it is necessary to investigate the stability of the control-free model (2) about its DFE. From Theorem 2 in van den Driessche and Watmough [29], the following result holds.

Theorem 1. *The DFE of the control-free model (2) is locally asymptotically stable if $\mathscr{R}_0 < 1$ and unstable if $\mathscr{R}_0 > 1$.*

Theorem 1 implies that waterborne disease can be eliminated from the entire population (when $\mathscr{R}_0 < 1$) if the initial size of the infected population is in the basin of attraction of the DFE (5). On the other hand, the disease will establish in the population if $\mathcal{R}_0 > 1$.

Theorem 2. *The DFE of the control-free model (2) is globally asymptotically stable provided that $\mathcal{R}_0 < 1$.*

This theorem can be established using a global stability result by Castillo-Chavez et al. [31]. This global stability ensures that disease elimination is independent of the initial size of the population of infected individuals if $\mathcal{R}_0 < 1$. The epidemiological implication is that in this case waterborne disease can be eradicated from the entire community irrespective of the initial number of infected people in the community.

2.4. Outbreak Growth Rate. If $\mathcal{R}_0 > 1$, then the DFE (5) becomes unstable and a disease outbreak occurs in the population. The positive (dominant) eigenvalue of the Jacobian at the DFE is typically referred to as the initial outbreak growth rate [11]. The eigenvalues of the Jacobian matrix of model (2) evaluated at the DFE (5) are

$$\lambda_1 = -\mu,$$

$$\lambda_2 = \frac{1}{2}\left[-(\mu + \gamma + \sigma)\right.$$
$$\left. - \sqrt{(\mu + \gamma + \sigma)^2 + 4\sigma(\mu + \gamma)(\mathcal{R}_0 - 1)}\right], \quad (7)$$

$$\lambda_3 = \frac{1}{2}\left[-(\mu + \gamma + \sigma)\right.$$
$$\left. + \sqrt{(\mu + \gamma + \sigma)^2 + 4\sigma(\mu + \gamma)(\mathcal{R}_0 - 1)}\right].$$

Clearly, $\lambda_1, \lambda_2 < 0$. Thus, the positive (dominant) eigenvalue is given by

$$\lambda^+ = \lambda_3. \quad (8)$$

From the above results, when $\mathcal{R}_0 = 1$ the outbreak growth rate λ^+ vanishes. Also, if $\mathcal{R}_0 < 1$ all three eigenvalues become negative confirming Theorem 1. So the outbreak growth rate (λ^+) exists only when $\mathcal{R}_0 > 1$. Epidemiologically, this result demonstrates how for $\mathcal{R}_0 > 1$ an outbreak will occur in the community and the growth rate of that outbreak is determined by λ^+. In particular, this might occur when there are no control measures. Next it is necessary to obtain the likely magnitude of an outbreak, often called the expected final size of the outbreak [32].

2.5. Final Outbreak Size. Our analyses have shown that when $\mathcal{R}_0 > 1$ a waterborne disease outbreak occurs and grows at the rate λ^+. The final outbreak size of SIR epidemiological models and other similar models are given by the relation

$$Z = 1 - \exp(-\mathcal{R}_0 Z), \quad (9)$$

where Z denotes the proportion of the population who become infected at some point during the outbreak. This

relation applies to our control-free model (2) [11] and so if there is no control intervention and an outbreak occurs, with $\mathcal{R}_0 > 1$ then the final outbreak size of the epidemic can be determined by (9).

2.6. Stability Analysis of the Endemic Equilibrium. The long-term dynamics of a dynamical system is characterized by the stability about its endemic equilibrium [30]. To determine the long-term dynamics of the control-free model (2) we investigate its stability about this endemic equilibrium (EE). Algebraically, it can be demonstrated that when $\mathcal{R}_0 > 1$, a unique EE occurs in model (2) given by

$$(s^e, i^e, w^e) = \left(\frac{1}{\mathcal{R}_0}, \frac{\mu(\mathcal{R}_0 - 1)}{\mathcal{R}_0(\gamma + \mu)}, i^e\right). \quad (10)$$

Obviously, i^e will vanish if $\mathcal{R}_0 \leq 1$. This confirms that the disease cannot be endemic when $\mathcal{R}_0 \leq 1$. The stability analyses of the EE (10) are summarized as follows using [11, 33–36].

Theorem 3. *The unique endemic equilibrium (10) is locally and globally asymptotically stable whenever $\mathcal{R}_0 > 1$.*

The proof of Theorem 3 can be established using the approach in [11] and implies that whenever $\mathcal{R}_0 > 1$ any outbreak in the population will persist in the population (remain endemic). So, to minimize the chances of a disease outbreak the intervention of control measures can be used such that the basic reproduction number is kept below unity (i.e., $\mathcal{R}_0 < 1$).

3. Multiple Control Model and Analyses

In this section we present a control model to investigate the impact of introducing control measures in the spread of diseases. Three different types of control measure are considered: vaccination, water purification, and treatment. The impact of these control measures is investigated by extending the original control-free model (2) to include these control measures.

3.1. Formulation of the Multiple Control Model. A multiple control model is formulated as follows.

Vaccination is one control strategy for reducing the spread of waterborne diseases such as cholera. For example, a cholera vaccine can offer about 60-90% protection against the disease. Thus in the control model we assume that susceptible individuals are vaccinated at rate ϕ with a vaccine whose efficacy is ε. Effectively the model now has new class of individual that are vaccinated.

Effective treatment of waterborne disease is also very important in reducing the spread of the disease. Some waterborne diseases like cholera can kill within hours of contracting the disease if there is no proper treatment. If people infected with cholera are treated quickly and properly, the mortality rate is less than 1% but if they are left untreated, the mortality rate rises to 50-60% [37, 38]. We introduce treatment in the control-free model (2) by assuming that infected individuals are treated at rate τ and treated individuals $T(t)$ recover due to treatment at rate γ_τ.

According to the World Health Organization [1], unsafe water, poor sanitation, and poor hygiene are the major causes of waterborne diseases. A significant number of cases of a disease could be reduced through access to clean water supplies, provision of adequate sanitation facilities, and better hygiene practices. Here we extend the model (2) by assuming that provision of clean water reduces pathogen concentrations at a rate d.

Based on these assumptions, and by introducing these control intervention strategies simultaneously, we obtain the multiple control model

$$\dot{s}(t) = \mu - \beta s(t) w(t) - (\mu + \phi) s(t),$$

$$\dot{v}(t) = \phi s(t) - (1 - \varepsilon) \beta v(t) w(t) - \mu v(t),$$

$$\dot{i}(t) = \beta s(t) w(t) + (1 - \varepsilon) \beta v(t) w(t)$$
$$- (\mu + \gamma + \tau) i(t), \quad (11)$$

$$\dot{\zeta}(t) = \tau i(t) - (\mu + \gamma_\tau) \zeta(t),$$

$$\dot{w}(t) = \sigma i(t) - (d + \sigma) w(t),$$

$$\dot{r}(t) = \gamma i(t) + \gamma_\tau \zeta(t) - \mu r(t),$$

where V are vaccinated individuals, T are treated individuals, $v = V/N$ are the proportion of vaccinated individuals, and $\zeta = T/N$ are the proportion of treated individuals.

All the solutions of model (11) enter the feasible region

$$\Phi_c = \{s, v, i, \zeta, w, r > 0 : s + v + i + \zeta + r = 1\}. \quad (12)$$

The region Φ_c is positively invariant; thus it is sufficient to consider the solutions of model (11) within it.

3.2. Basic Reproduction Number for the Multiple Control Model.
The multiple control intervention strategy model (11) has a DFE given by

$$\left(s_c^0, v_c^0, i_c^0, \zeta_c^0, w_c^0\right) = \left(\frac{\mu}{\mu + \phi}, \frac{\phi}{\mu + \phi}, 0, 0, 0\right), \quad (13)$$

and a basic reproduction number given by

$$\mathcal{R}_0^c = \mathcal{R}_0 E_c, \quad (14)$$

where

$$E_c = \frac{\sigma(\gamma + \mu)(\mu + (1 - \varepsilon)\phi)}{(d + \sigma)(\gamma + \mu + \tau)(\mu + \phi)}. \quad (15)$$

The basic reproduction number can be defined as the expected number of secondary infections that result from introducing a single infected individual into an otherwise susceptible population [29]. Thus, the threshold quantity $\mathcal{R}_0^c$ is a measure of the number of secondary infections of the population in the presence of vaccination, treatment, and water purification. From (15) and (14), we have that

$$E_c < 1 \iff$$
$$\mathcal{R}_0^c < \mathcal{R}_0. \quad (16)$$

This implies that the multiple control measures in the model do have an impact in reducing the number of secondary infections.

3.3. Stability Analysis of the DFE for Multiple Control Measures.
Again, to determine the short-term dynamics in the presence of the multiple control measures we investigate the stability of the multiple control model at the DFE. The results are summarized in the theorem below.

Theorem 4. If $\mathcal{R}_0^c < 1$, the DFE (13) of model (11) is globally asymptotically stable and unstable if $\mathcal{R}_0^c > 1$.

The epidemiological implication of this result is that waterborne diseases can be eradicated from the entire community using multiple control measures irrespective of the initial size of the infected people provided that $\mathcal{R}_0^c < 1$.

3.4. Outbreak Growth Rate for Multiple Control Measures.
Suppose that the multiple control measures are not effective, then $\mathcal{R}_0^c > 1$ and the DFE (13) becomes unstable and a disease outbreak occurs. The outbreak growth rate of the multiple control model is given by

$$\lambda_c^+ = \frac{1}{2}\left[-(\mu + \gamma + \tau + \sigma + d)\right.$$
$$\left. + \sqrt{(\mu + \gamma + \tau + \sigma + d)^2 + 4(\sigma + d)(\mu + \gamma + \tau)(\mathcal{R}_0^c - 1)}\right]. \quad (17)$$

Comparing the outbreak growth rate of the multiple control model with the no control model is summarized in the theorem below.

Theorem 5. If $d \geq 0$, $\phi \geq 0$, $\tau \geq 0$, and $\varepsilon \geq 0$, then $\lambda_c^+ \leq \lambda^+$. Furthermore, $\lambda_c^+ = \lambda^+$ if and only if $d = \phi = \tau = \varepsilon = 0$.

The proof of this theorem can be established by simple algebraic manipulation. This show that introducing multiple controls can reduce the outbreak growth rate.

3.5. Single and Double Control Measures.
Waterborne disease outbreaks are often associated with poverty and limited resources to control the disease [1]. Often such communities cannot afford to introduce more than one control measure such as the three considered here. Thus, it is important to investigate the impact of introducing a single control or double controls. By comparing the impact of a single control, double control, and multiple controls, we determine the control (or combination of controls) that can yield the best results. These comparisons will be done at both the epidemic stage as well as the endemic stage of the outbreak. Theoretically, for the epidemic stage of the outbreak the basic reproduction number is used, while at the endemic stage the outbreak growth rate is used. These results can help in advising communities with limited resources.

3.5.1. *Basic Reproduction Number for a Single Control Measure.* Suppose the community can afford vaccinations only, then the basic reproduction number is

$$\mathcal{R}_0^v = E_v \mathcal{R}_0, \quad E_v = \frac{\mu + (1 - \varepsilon)\,\phi}{\mu + \phi}. \tag{18}$$

This threshold quantity $\mathcal{R}_0^v$ can be understood as a measure of the number of secondary infections in the presence of vaccination [11, 39]. By elementary algebraic manipulations, the equations

$$E_v < 1 \Longleftrightarrow$$
$$\mathcal{R}_0^v < \mathcal{R}_0, \tag{19}$$
$$\forall 0 < \varepsilon, \phi \le 1,$$

$$E_v = 1 \Longleftrightarrow$$
$$\mathcal{R}_0^v = \mathcal{R}_0, \tag{20}$$
$$\varepsilon = \phi = 0$$

hold. From (19) the number of secondary infections is less when vaccination is introduced provided $0 < \varepsilon, \phi \le 1$, and this implies that introducing vaccination decreases the spread of the infection.

For communities that can afford only treatment the basic reproduction number is

$$\mathcal{R}_0^\tau = \mathcal{R}_0 E_\tau, \quad E_\tau = \frac{\mu + \gamma}{\mu + \gamma + \tau}. \tag{21}$$

Clearly, the equations

$$E_\tau < 1 \Longleftrightarrow$$
$$\mathcal{R}_0^\tau < \mathcal{R}_0 \tag{22}$$
$$\tau \neq 0,$$

$$E_i^\tau = 1 \Longleftrightarrow$$
$$\mathcal{R}_0^\tau = \mathcal{R}_0, \tag{23}$$
$$\tau = 0$$

hold. Epidemiologically, the treatment of infected individuals reduces the number of secondary infections in the population provided that $0 < \tau \le 1$.

Finally, we investigate the impact of introducing water purification as the only control measure. The water purification induced basic reproduction number is

$$\mathcal{R}_0^w = \mathcal{R}_0 E_w, \tag{24}$$

where

$$E_w = \frac{\sigma}{\sigma + d}. \tag{25}$$

Again the following equations hold:

$$E_w < 1 \Longleftrightarrow$$
$$\mathcal{R}_0^w < \mathcal{R}_0 \tag{26}$$
$$\forall d \neq 0,$$

$$E_w = 1 \Longleftrightarrow$$
$$\mathcal{R}_0^w = \mathcal{R}_0, \tag{27}$$
$$d = 0,$$

and introducing water purification reduces the number of secondary infections in the community provided that $0 < d \le 1$.

Having shown the impact of each of the single controls using the basic reproduction number, it is important to compare each of these single controls with the multiple control. Since, E_c is the product of E_v, E_τ, and E_w and each of these is less than 1, then using the calculations of each reproduction number (19), (22), (26), and (16) can be written in compact form as

$$\mathcal{R}_0^c < \mathcal{R}_0^v, \mathcal{R}_0^w, \mathcal{R}_0^\tau < \mathcal{R}_0. \tag{28}$$

These results show that even though each of the single controls has some influence in reducing the number of secondary infections, the multiple control always has at least the greatest influence. The above results agree with intuitive expectation and justify why multiple controls are encouraged whenever an outbreak occurs in any community.

3.5.2. *Outbreak Growth Rate for a Single Control Measure.* For communities that consider only treatment as a control measure, our analyses show that, if infected individuals are not properly treated such that $\mathcal{R}_0^\tau > 1$, then an outbreak occurs in the community. The treatment-induced outbreak growth rate is given by

$$\lambda_\tau^+ = \frac{1}{2}\Big[-(\mu + \gamma + \tau + \sigma)$$
$$+ \sqrt{(\mu + \gamma + \tau + \sigma)^2 + 4\sigma(\mu + \gamma + \tau)(\mathcal{R}_0^\tau - 1)}\,\Big]. \tag{29}$$

To determine the strength of this outbreak, we compare it with the outbreak growth rate in the absence of control intervention. The result of the comparison is summarized in the theorem below.

Theorem 6. *If $\tau \ge 0$, then $\lambda_\tau^+ \le \lambda^+$. Furthermore, $\lambda_\tau^+ = \lambda^+$ if and only if $\tau = 0$.*

Theorem 6 can be established by algebraic manipulations. Thus, the outbreak growth rate in the presence of treatment is always lower than that with no control.

Similarly, for communities that can afford only vaccination or water purification the outbreak growth rates are given by

$$\lambda_v^+ = \frac{1}{2}\Big[-(\mu+\gamma+\sigma) + \sqrt{(\mu+\gamma+\sigma)^2 + 4\sigma(\mu+\gamma)(\mathscr{R}_0^v - 1)}\,\Big]. \tag{30}$$

and

$$\lambda_w^+ = \frac{1}{2}\Big[-(\mu+\gamma+d+\sigma) + \sqrt{(\mu+\gamma+d+\sigma)^2 + 4(\sigma+d)(\mu+\gamma)(\mathscr{R}_0^w - 1)}\,\Big], \tag{31}$$

respectively.

Since $\mathscr{R}_0^v \le \mathscr{R}_0$ and $d \ge 0$ then

$$\lambda_v^+ \le \lambda^+. \tag{32}$$

and

$$\lambda_w^+ \le \lambda^+. \tag{33}$$

This shows that vaccination or introducing water purification reduces the outbreak growth rate more than when no control is introduced.

Epidemiologically, these results demonstrate that even when a control does not prevent a disease from invading the population, the outbreak will be less compared to when no control is considered.

We have shown that each of the single control intervention strategies and the multiple control intervention strategy reduce the outbreak growth rate. Next, the multiple control intervention strategy is shown to reduce the outbreak growth rate more than each of the single control intervention strategy. The details are given in Theorem 7,

Theorem 7. *Suppose that $d \ge 0$, $\phi \ge 0$, $\tau \ge 0$, and $\varepsilon \ge 0$, then*

$$\lambda_c^+ \le \lambda_v^+,$$
$$\lambda_c^+ \le \lambda_\tau^+, \tag{34}$$
$$\lambda_c^+ \le \lambda_w^+.$$

Furthermore,

$$\lambda_c^+ = \lambda_v^+ = \lambda_\tau^+ = \lambda_w^+ \iff$$
$$d = \phi = \tau = \varepsilon = 0. \tag{35}$$

which is easily demonstrated.

3.6. Two Control Measures. Suppose a community can only afford two control intervention strategies which, for example, could be (i) vaccination + treatment, (ii) vaccination + treatment, and (iii) treatment + water purification. Qualitative analyses of these cases can be important for communities that can afford up to two control measures. Similar to the

single control, the basic reproduction number and outbreak growth rate are used to investigate the impact of introducing double control measures. Using the same approach, the basic reproduction numbers induced by vaccination + treatment $\mathscr{R}_0^{v\tau}$, vaccination + water purification $\mathscr{R}_0^{vw}$, and treatment + water purification $\mathscr{R}_0^{\tau w}$ are, respectively, given by

$$\mathscr{R}_0^{v\tau} = E_v E_\tau \mathscr{R}_0,$$
$$\mathscr{R}_0^{vw} = E_v E_w \mathscr{R}_0, \tag{36}$$
$$\mathscr{R}_0^{\tau w} = E_\tau E_w \mathscr{R}_0.$$

By a similar reasoning, the outbreak growth rates associated with vaccination + treatment $\lambda_{v\tau}^+$, vaccination + water purification λ_{vw}^+, and treatment + water purification $\lambda_{\tau w}^+$ are, respectively, given by

$$\lambda_{v\tau}^+ = \frac{1}{2}\Big[-(\mu+\gamma+\tau+\sigma) + \sqrt{(\mu+\gamma+\tau+\sigma)^2 + 4\sigma(\mu+\gamma+\tau)(\mathscr{R}_0^{v\tau}-1)}\,\Big],$$

$$\lambda_{vw}^+ = \frac{1}{2}\Big[-(\mu+\gamma+\sigma+d) + \sqrt{(\mu+\gamma+\sigma+d)^2 + 4(\sigma+d)(\mu+\gamma)(\mathscr{R}_0^{vw}-1)}\,\Big], \tag{37}$$

$$\lambda_{\tau w}^+ = \frac{1}{2}\Big[-(\mu+\gamma+\tau+\sigma+d) + \sqrt{(\mu+\gamma+\tau+\sigma+d)^2 + 4(\sigma+d)(\mu+\gamma+\tau)(\mathscr{R}_0^{\tau w}-1)}\,\Big].$$

Again, we compare these results with the cases for no control, a single control, or multiple control measures. In each comparison, the case with less controls must be a subset of the case with more controls. Comparing these basic reproduction numbers and outbreak growth rates show that considering two control measures is always better than no control or a single control in reducing the disease. On the other hand, the multiple control is better than the two control measures. Thus, if an outbreak occurs in any community, multiple controls are highly recommended. However, if a community has limited resources to control an outbreak, then the double or single control methods can be recommended depending on availability of resources.

To compare other combinations of controls (one case not necessarily a subset of the other) we consider numerical simulations. The parameter values for the numerical simulations are given in Table 1. The results are presented in Figures 1(a)–1(d). Again, as expected, in each comparison the multiple control has the greatest impact in reducing infections while no control has the least impact. The multiple control can prevent close to 25% more people from the disease as compared with no control.

In each case introducing a single control is always better than introducing no control (control Figures 1(a)–1(d)). However, depending on the actual parametrisation, one of these single measures will have the best impact (for example water purification in Figure 1(d)). Thus, for a community that can afford only a single control these types of results

TABLE 1: Parameter values used for numerical simulations with reference.

Parameter	Symbol	Value	Reference
Contact rate	β	0.1072 day^{-1}	[11]
Birth/death rate	μ	0.02 day^{-1}	[20]
Recovery rate	γ	11.3 year^{-1}	[4, 24]
net decay rate of pathogen	σ	0.0333	[11]
Efficacy of vaccine	ε	0.78	[40]
Vaccination rate	ϕ	0.07 day^{-1}	[5]
Treatment rate	τ	0.005 day^{-1}	[5]
Recovery are due treatment	γ_τ	0.003 day^{-1}	[5]
vaccination rate	ϕ	0.07 day^{-1}	[5]
Reduction in w due to water purification	d	2σ day^{-1}	Estimate

can inform their choices. These arguments also apply for communities that can afford a double control. For our results vaccination + water purification has the greatest impact in reducing infections (Figures 1(a) and 1(c)) and this would be recommended. Thus, these examples illustrate how the model could guide communities in choosing appropriate control measures. These findings are consistent with results using different models ([4, 5, 39]).

4. Optimal Control Problem

Qualitative analyses of our model have revealed that multiple control intervention strategies are the best under uniform circumstances. Unfortunately, affordability of any multiple control intervention strategy is a major concern as many communities have limited resources. Thus, further analyses are used investigate possible multiple control intervention strategies with minimum costs. The results of these analyses can be helpful in advising communities with limited resources. Optimal control theory has been successfully used in analysing such problems [4, 8, 39, 41–45] and is used here.

To minimize the cost of implementing multiple controls we make the following assumptions. First, we assume that there are control parameters ϕ, τ, and d that are measurable functions of time and formulate an appropriate optimal control functional that minimizes the cost of implementing model (11). For simplicity, we assume that $\phi = u_1(t)$, $\tau = u_2(t)$, and $d = u_3(t)$.

The appropriate optimal control cost functional is

$$
J(u_1, u_2, u_3) = \int_0^{t_f} \left[A_1 i(t) + B_1 w(t) + C_1 u_1^2(t) \right.
$$
$$
\left. + C_2 u_2^2(t) + C_3 u_3^2(t) \right] dt \tag{38}
$$

where the coefficients, A_1, B_1, C_1, C_2, and C_3, are balancing cost factors that transform the integral into money expended over a finite time t_f. The aim is to minimize the number of infected individuals and pathogens in water source as well as the costs for applying the multiple controls. To account for the anticipated nonlinear costs that could arise from the multiple controls, we consider quadratic functions for measuring the control costs [4, 26, 39, 41–47].

The existence of optimal controls (u_1^*, u_2^*, u_3^*) that minimize the cost functional $J(u_1, u_2, u_3)$ follows from [48, 49]. Pontryagin's Maximum Principle [50] introduces adjoint functions and enables us to minimize a Hamiltonian H, with respect to the controls $(u_1(t), u_2(t), u_3(t))$ instead of minimizing the original objective functional. The Hamiltonian associated with the objective functional is given by

$$
\begin{aligned}
H = {} & A_1 i(t) + B_1 w(t) + C_1 u_1^2(t) + C_2 u_2^2(t) \\
& + C_3 u_3^2(t) + \lambda_s \left(\mu - \beta s(t) w(t) - (\mu + u_1) s(t) \right) \\
& + \lambda_v \left(u_1 s(t) - (1 - \varepsilon) \beta v(t) w(t) - \mu v(t) \right) \\
& + \lambda_i \left(\beta s(t) w(t) + (1 - \varepsilon) \beta v(t) w(t) \right. \\
& \left. - (\mu + \gamma + u_2) i(t) \right) + \lambda_\zeta \left(u_2 i(t) - (\mu + \gamma_\tau) \zeta(t) \right) \\
& + \lambda_w \left(\sigma i(t) - (u_3 + \sigma) w(t) \right) + \lambda_r \left(\gamma i(t) + \gamma_\tau \zeta(t) \right. \\
& \left. - \mu r(t) \right),
\end{aligned} \tag{39}
$$

where $\lambda_s, \lambda_v, \lambda_i, \lambda_\zeta, \lambda_w$, and λ_r are the associated adjoints for the states s, v, i, ζ, w, and r, respectively.

Given optimal controls (u_1^*, u_2^*, u_3^*) together with the corresponding states $(s^*, v^*, i^*, \zeta^*, w^*, r^*)$ that minimize $J(u_1, u_2, u_3)$, there exist adjoint variables $\lambda_s, \lambda_v, \lambda_i, \lambda_\zeta, \lambda_w$, and λ_r satisfying

$$
\begin{aligned}
\frac{d\lambda_s}{dt} = {} & \lambda_s \left(\beta w(t) + \mu + u_1(t) \right) - \lambda_v u_1(t) \\
& - \lambda_i \beta w(t),
\end{aligned}
$$

$$
\frac{d\lambda_v}{dt} = \lambda_v \left((1 - \varepsilon) \beta w(t) + \mu \right) - \lambda_i (1 - \varepsilon) \beta w(t),
$$

$$
\begin{aligned}
\frac{d\lambda_i}{dt} = {} & -A_1 + \lambda_i \left(\mu + \gamma + u_2(t) \right) - \lambda_\zeta u_2(t) - \lambda_w \sigma \\
& - \lambda_r \gamma,
\end{aligned}
$$

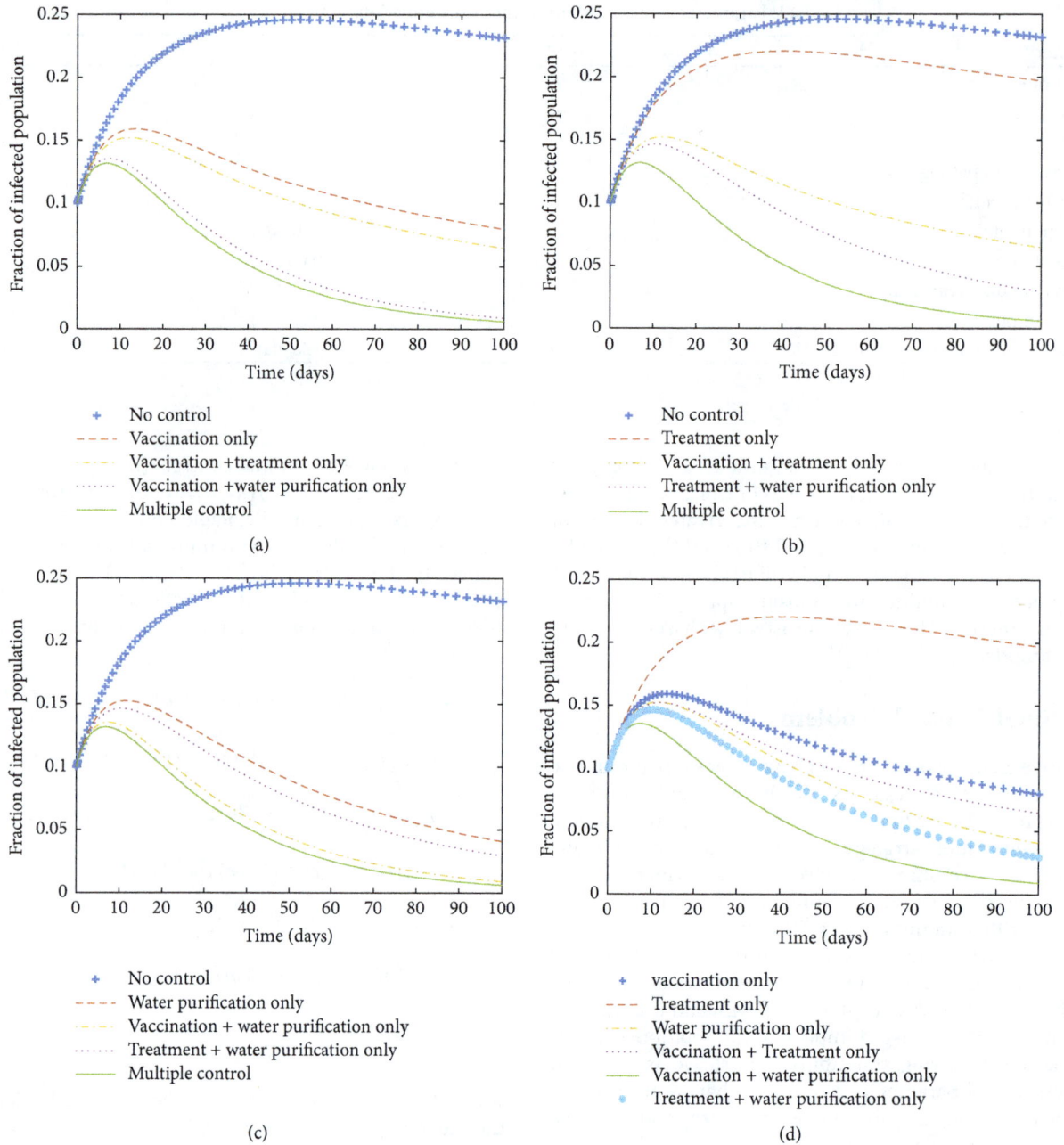

(a)

(b)

(c)

(d)

FIGURE 1: Graphical illustration of the impact of considering a single control, double control, or multiple (three) control measures in reducing the spread of waterborne diseases for different combinations of control (a–d).

$$\frac{d\lambda_\zeta}{dt} = \lambda_\zeta \left(\mu + \gamma_\tau \right) - \lambda_r \gamma_\tau,$$

$$\frac{d\lambda_w}{dt} = -B_1 + \lambda_s \beta s(t) + \lambda_v (1 - \varepsilon) \beta v(t)$$

$$- \lambda_i \left(\beta s(t) + (1 - \varepsilon) \beta v(t) \right)$$

$$+ \lambda_w \left(u_3(t) + \sigma \right),$$

$$\frac{d\lambda_r}{dt} = \lambda_r \mu,$$

(40)

and transversality conditions:

$$\lambda_k \left(t_f \right) = 0 \qquad (41)$$

where $k = s, v, i, \zeta, w, r$.

The differential equations (40) were obtained by differentiating the Hamiltonian function (39) with respect to the corresponding states as follows:

$$\frac{d\lambda_k}{dt} = -\frac{dH}{dk}. \qquad (42)$$

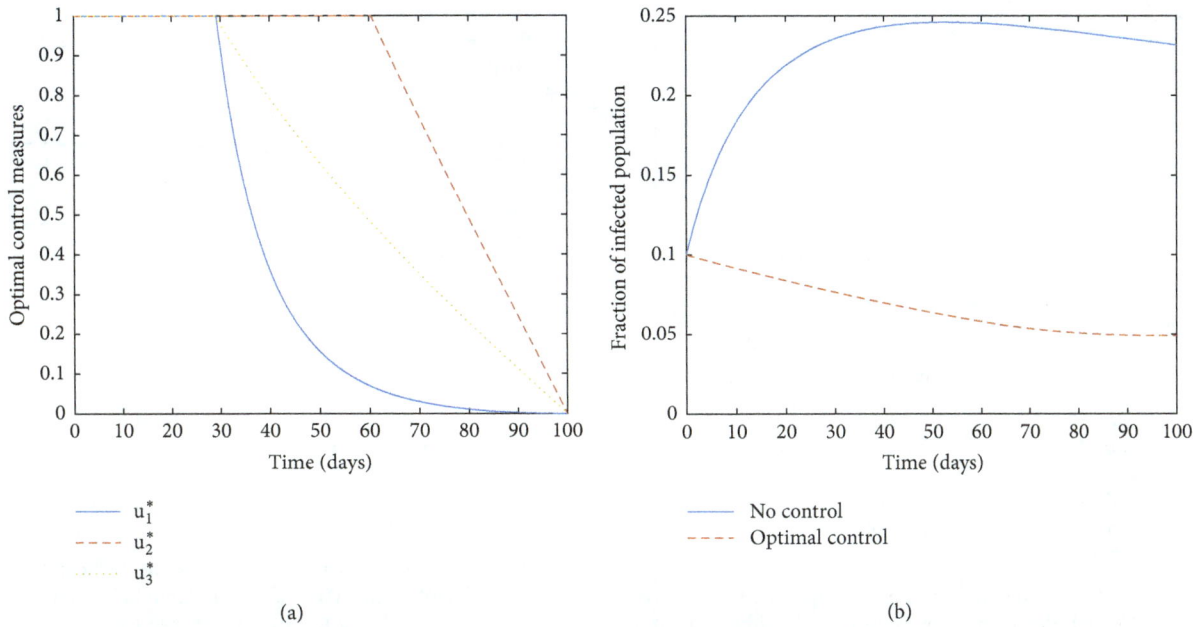

FIGURE 2: Graphical illustration of (a) the optimal control functions u_1^*, u_2^*, u_3^* and (b) impact of considering optimal control over no control in reducing the spread of waterborne disease.

The optimal conditions are

$$0 = \frac{dH}{du_1},$$

$$0 = \frac{dH}{du_2}, \tag{43}$$

$$0 = \frac{dH}{du_3}.$$

Solving for $u_1(t)$ under these optimal conditions (43) and subsequently taking bounds into consideration, we obtain

$$u_1^* = \max\left\{0, \min\left\{1, \frac{s(t)(\lambda_s - \lambda_i)}{2c_1}\right\}\right\}. \tag{44}$$

Using a similar approach, we obtain the remaining optimal controls:

$$u_2^* = \max\left\{0, \min\left\{1, \frac{i(t)(\lambda_i - \lambda_\zeta)}{2c_2}\right\}\right\}, \tag{45}$$

$$u_3^* = \max\left\{0, \min\left\{1, \frac{w(t)\lambda_w}{2c_3}\right\}\right\}. \tag{46}$$

The above results demonstrate the existence of optimal controls (u_1^*, u_2^*, u_3^*) that can reduce the spread of waterborne diseases using multiple control measures with minimum cost. The possible magnitudes and trajectories of each of these optimal controls as well as the optimal solutions are investigated further using numerical simulations.

For the numerical simulations, we consider the parameter values presented in Table 1 together with the following values

for the cost factors: $A_1 = 600.00$, $B_1 = 200.00$, $C_1 = 2$, $C_2 = 6$, and $C_3 = 6$ [4]. Numerical solutions of the optimal system are carried out using the forward-backward algorithm as described in [4, 8]. The numerical results of the optimal control functions u_1^*, u_2^*, and u_3^* that minimize the cost functional subject to the state equations are presented in Figure 2(a). From the results, effective multiple control measures can be achieved with minimum cost by applying controls at the onset of the outbreak. The impact of considering optimal control over no control is investigated by comparing the two cases (Figure 2(b)). The results demonstrate that considering optimal control can prevent about 15% of the total population from getting the disease at minimum cost.

5. A Case Study: The Haiti Cholera Outbreak

A realistic case study, a cholera outbreak in Haiti, is considered to validate the control model (11). In particular, this example is used to demonstrate how this model can be used to study, as well as make future predictions of, cholera outbreaks in cholera endemic communities. According to the Haitian Ministry of Public Health and Population (MSPP), a cholera outbreak was confirmed in Haiti on October 21, 2010 [51, 52]. As of August 4, 2013, about 669,396 cases and 8,217 deaths had been reported since the outbreak started [53]. For this study, the number of reported hospitalized cholera cases in Haiti from October 30, 2010 to December 24, 2012, is considered. Haiti is divided into 10 departments (governing regions) and the capital Port-au-Prince. Using the MSPP data, the cumulative number of reported cholera hospitalizations for each department, from October 30, 2010, to December 24, 2012, is given in Figure 3 [52] (note that these are ordered by the size of the regional populations reported in 2009 before

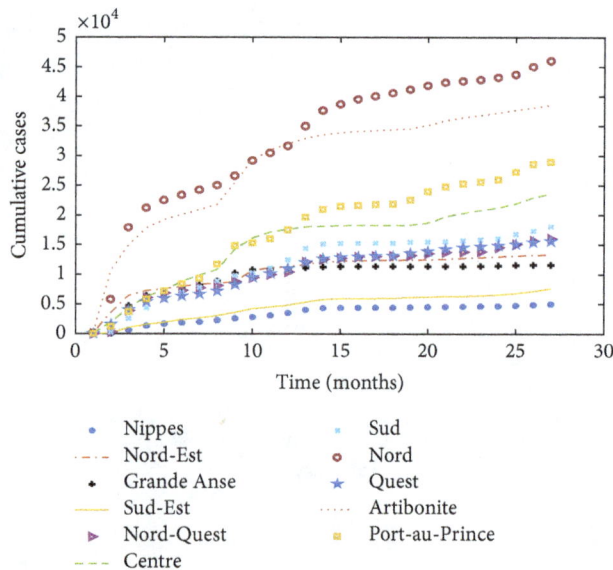

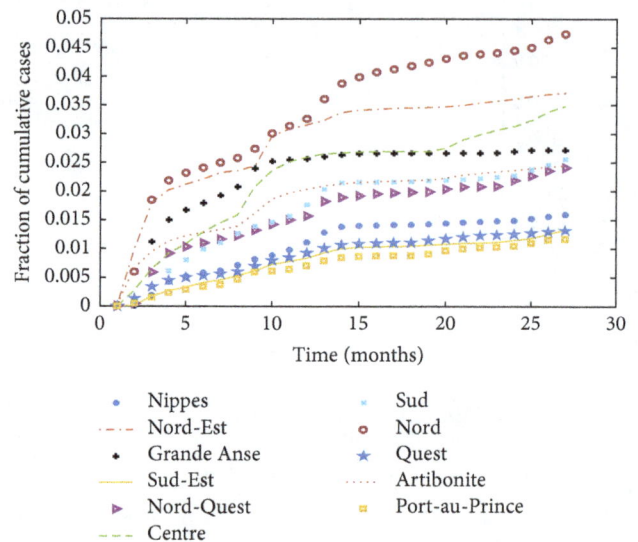

FIGURE 3: A plot showing the cumulative reported hospitalized cholera cases in each department in Haiti from October 30, 2010, to December 24, 2012. The legend is in order of population size (Nippes = 311497, Nord-Est = 358277, Grande Anse = 425878, Sud-Est = 575293, Nord-Quest = 662777, Centre=678626, Sud = 704760, Nord=970495, Quest = 1187833, Artibonite = 1571020, and Port-au-Prince = 2476787).

FIGURE 4: A plot showing the cumulative per capita reported hospitalized cholera cases in each department in Haiti from October 30, 2010, to December 24, 2012. Again, the legend is in order of population size.

the outbreak started [54]). From Figure 3 the cases of cholera largely track population size with the Nord having the greatest number of reported cases followed by Artibonite then Port-au-Prince while the department of Nippes has the smallest number of reported hospitalized cases.

The apparent loose correlation between population size and numbers of cases is expected as larger populations would have more cases overall. However, because the model considered here (model (11)) is scaled to the total population N the data in Figure 3 was also scaled by population, given in Figure 4. This scaling is informative as now some departments with larger populations have less per capita cases of cholera as compared to the others, in particular the capital Port-au-Prince (Figure 4). Alternatively, some departments with smaller populations have higher per capita numbers of cases, in particular Nord-Est (Figure 4). Research globally has shown that cholera is associated with poverty and access to clean water and thus presumably these per capita differences are due to different conditions of access to clean water and treatment in the different locations. For example, these conditions in general would be expected to be comparatively better in the capital of the country, Port-au-Prince, because of better infrastructure and economic conditions. Thus, one might expect the lowest per capita cholera infections as found for Port-au-Prince in Figure 4.

In fitting the model to the data, the parameters $\gamma, \mu, \sigma, \varepsilon$ (values in Table 1) were fixed because they represent growth and death processes expected to be uniform throughout the country. The other parameters that relate to the various control measures were used as fitting parameters. These model fittings were carried out using the built-in MATLAB

(Mathworks, Version, R2012b [55]) least-squares fitting routine fmincon in the optimization tool box. The fits using this strategy were good for half of the departments. For the other five, the model overestimated the levels of initial infections. These departments had the four lowest per capita infection rates in Figure 4 and Sud which was also relatively low. For this reason, σ was included as a fitting parameter in that it represents the rate of disease transmission from infected individuals to the water. Presumably this rate is reduced in well-managed areas by providing treated water and so is variable by region. This new fitting resulted in a better fit for all departments (Figure 5). The one poor fit is the capital Port-au-Prince in which the model again overestimates the initial infections, presumably for similar reasons as those given above. However, here the fit in the final months is still close.

These results demonstrate that our model could be used to study and predict cholera outbreaks in Haiti and other communities where cholera is endemic, consistent with other studies of the Haitian cholera outbreak [51, 56–59].

6. Discussion

Dynamics and control intervention strategies for waterborne diseases in a homogeneous mixed population have been explored. Our analyses have shown that useful information concerning the dynamics and control of waterborne diseases can be obtained by analysing the appropriate epidemiological models.

In the absence of any control intervention strategy, global stability of the disease-free equilibrium shows it is possible for the waterborne disease to be eradicated from the community irrespective of the initial size of infected population provided the basic reproduction number $\mathscr{R}_0 < 1$. This can happen if

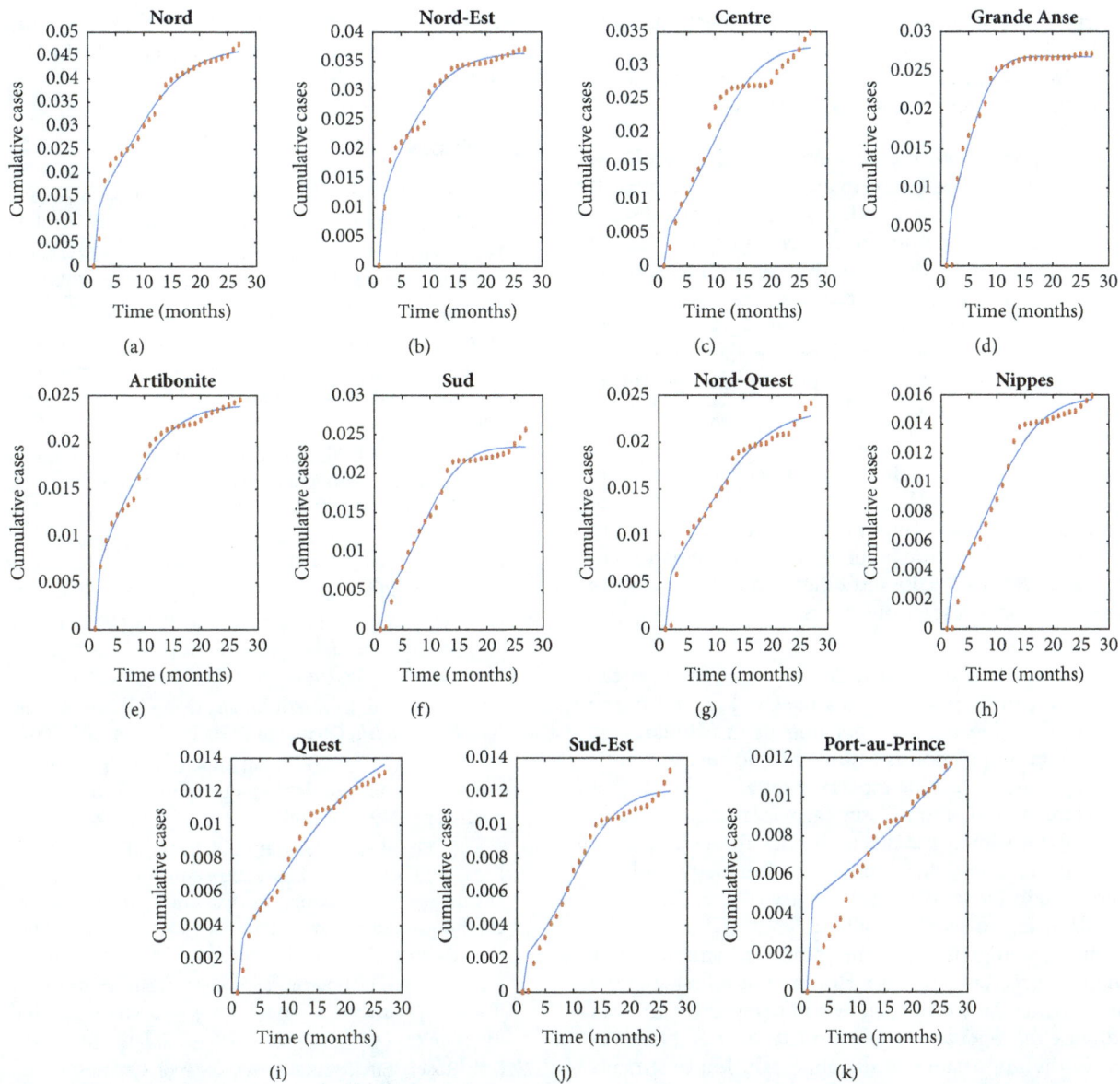

FIGURE 5: Model fitting of the cumulative per capita reported cholera cases for each department in Haiti from October 30, 2010 to December 24, 2012. The bold lines represent the model fit and the stars represent the reported cases in Haiti. These graphs are now ordered according to the per capita overall increases in infections as seen in Figure 4.

the individuals in the community begin to practice healthy living like staying away from contaminated water, boiling water before drinking, proper sewage disposal, etc. On the contrary, if the basic reproduction number is greater than unity the outbreak is likely to persist in the entire population. To determine the likely magnitude of such an outbreak the outbreak growth rate and final outbreak size were computed analytically. In such cases, any outbreak will be difficult to manage without effective control measures.

Next, the model is extended by introducing multiple control intervention strategies: vaccination, treatment, and water purification. Analyses of the multiple control model revealed that introducing multiple control measures can have a significant impact in reducing the spread of disease both at

epidemic and at the endemic stage of an outbreak. Analyses of single and double control strategies showed that considering any control is better than none. However, multiple control measures together have the greatest impact on reducing infections.

Given that waterborne diseases are associated with poverty, we use our model to discuss two methods of reducing the spread of waterborne disease for communities that might not afford multiple control measures. The first method involved finding the most effective control when one or two control measures are introduced. By comparing these different levels of controls, a useful guide can be developed to assist a community choose an affordable alternative. For instance, in the example given above, a single control of water purification can reduce the number of infections more than a combination of vaccination and treatment. Also, the double control measures of vaccination + water purification has almost the same impact as the using the multiple controls.

Another approach to help communities decide on the most effective control method is to use optimal control analyses which investigate how to reduce the spread of a disease with minimum cost. Results from these analyses showed that it is optimal to introduce effective control measures at the onset of an outbreak. This result agrees with realistic expectations because introducing effective control measures at the onset of the outbreak will reduce the number of initially infected people and thus reduce overall expenditure on controlling the outbreak.

Finally, our model was validated using a cholera outbreak in Haiti. Our model fitted the reported cases of cholera in all the departments in Haiti. Thus, our model could be used to study, as well as make future predictions on, the dynamics of waterborne diseases in Haiti and other communities where there are waterborne disease outbreaks.

Even though this study provides new insights into the dynamics and control intervention strategies for waterborne diseases in a homogeneous population, it has limitations. Firstly, the total population is assumed to be constant. This is not always true in real life especially for an outbreak that lasts for a long period of time. Homogeneity in disease transmission is also assumed, but this is also not always true since heterogeneity is an essential part of epidemiology and has been shown to influence disease dynamics [19, 20]. In reality, individuals in any society belong to different socioeconomic classes and can migrate from one locality to another, thus affecting the spread of a disease. However, as a first step the models presented here enable an overall perspective and can guide disease management. These models also point the way forward to the importance of the approach. Improvements through more elaborate models that remove the limiting assumptions can follow from future work by ourselves and other scientists.

Conflicts of Interest

The authors declare that there are no conflicts of interest regarding the publication of this paper.

Acknowledgments

This work is based on research supported in part by the National Research Foundation of South Africa (Grants nos. 98892 and 85494). This work builds in part on previous work from the unpublished Ph.D. thesis [60] of one of the authors. Professor Keshlan S. Govinder is thanked for his input as the Ph.D. supervisor of Obiora Collins and the work in his thesis.

The authors are grateful to Dr. J. Tien, Ohio State University, and his research group, for providing them with the Haiti data to validate their model.

References

[1] World Health Organization, http://www.who.int/water_sanitation_health/diseases/burden/en/index.html, (November, 2013).

[2] World Health Organization (WHO), http://www.who.int/water_sanitation_health/hygiene/en/, (November, 2013).

[3] UNICEF Child Mortality Report, 2012.

[4] R. L. Miller Neilan, E. Schaefer, H. Gaff, K. R. Fister, and S. Lenhart, "Modeling optimal intervention strategies for cholera," *Bulletin of Mathematical Biology*, vol. 72, no. 8, pp. 2004–2018, 2010.

[5] A. Mwasa and J. M. Tchuenche, "Mathematical analysis of a cholera model with public health interventions," *BioSystems*, vol. 105, no. 3, pp. 190–200, 2011.

[6] R. P. Sanches, C. P. Ferreira, and R. A. Kraenkel, "The role of immunity and seasonality in cholera epidemics," *Bulletin of Mathematical Biology*, vol. 73, no. 12, pp. 2916–2931, 2011.

[7] O. C. Collins, S. L. Robertson, and K. S. Govinder, "Analysis of a waterborne disease model with socioeconomic classes," *Mathematical Biosciences*, vol. 269, pp. 86–93, 2015.

[8] S. Lenhart and J. T. Workman, *Optimal Control Applied to Biological Models*, Chapman & Hall, London, UK, 2007.

[9] N. J. Ashbolt, "Microbial contamination of drinking water and disease outcomes in developing regions," *Toxicology*, vol. 198, no. 1-3, pp. 229–238, 2004.

[10] J. N. S. Eisenberg, M. A. Brookhart, G. Rice, M. Brown, and J. M. Colford Jr., "Disease transmission models for public health decision making: Analysis of epidemic and endemic conditions caused by waterborne pathogens," *Environmental Health Perspectives*, vol. 110, no. 8, pp. 783–790, 2002.

[11] J. H. Tien and D. J. Earn, "Multiple transmission pathways and disease dynamics in a waterborne pathogen model," *Bulletin of Mathematical Biology*, vol. 72, no. 6, pp. 1506–1533, 2010.

[12] J. B. Rose, "Environmental ecology of Cryptosporidium and public health implications," *Annual Review of Public Health*, vol. 18, pp. 135–161, 1997.

[13] P. Hunter, M. Waite, and E. Ronchi, *Drinking Water and Infectious Disease: Establishing the Links*, CRC Press, Boca Rato, FL, USA, 2003.

[14] S. M. Faruque, M. J. Islam, Q. S. Ahmad et al., "Self-limiting nature of seasonal cholera epidemics: Role of host-mediated amplification of phage," *Proceedings of the National Acadamy of Sciences of the United States of America*, vol. 102, no. 17, pp. 6119–6124, 2005.

[15] E. Pourabbas, A. d'Onofrio, and M. Rafanelli, "A method to estimate the incidence of communicable diseases under seasonal fluctuations with application to cholera," *Applied Mathematics and Computation*, vol. 118, no. 2-3, pp. 161–174, 2001.

[16] M. Pascual, X. Rodo, S. P. Ellner, R. Colwell, and M. J. Bouma, "Cholera dynamics and El Nino-Southern Oscillation," *Science*, vol. 289, no. 5485, pp. 1766–1769, 2000.

[17] F. C. Curriero, J. A. Patz, J. B. Rose, and S. Lele, "The association between extreme precipitation and waterborne disease outbreaks in the United States, 1948-1994," *American Journal of Public Health*, vol. 91, no. 8, pp. 1194–1199, 2001.

[18] H. Auld, D. MacIver, and J. Klaassen, "Heavy rainfall and waterborne disease outbreaks: The Walkerton example," *Journal of Toxicology and Environmental Health, Part A. Current Issues*, vol. 67, no. 20-22, pp. 1879–1887, 2004.

[19] O. C. Collins and K. S. Govinder, "Incorporating heterogeneity into the transmission dynamics of a waterborne disease model," *Journal of Theoretical Biology*, vol. 356, pp. 133–143, 2014.

[20] S. L. Robertson, M. C. Eisenberg, and J. H. Tien, "Heterogeneity in multiple transmission pathways: Modelling the spread of cholera and other waterborne disease in networks with a common water source," *Journal of Biological Dynamics*, vol. 7, no. 1, pp. 254–275, 2013.

[21] V. Capasso and S. L. Paveri-Fontana, "A mathematical model for the 1973 cholera epidemic in the European Mediterranean region," *Revue d'épidémiologie et de Santé Publiqué*, vol. 27, no. 2, pp. 121–132, 1979.

[22] C. T. Codeço, "Endemic and epidemic dynamics of cholera: the role of the aquatic reservoir," *BMC Infectious Diseases*, vol. 1, article 1, 2001.

[23] D. M. Hartley, J. G. Morris Jr., and D. L. Smith, "Hyperinfectivity: a critical element in the ability of V. cholerae to cause epidemics?" *PLoS Medicine*, vol. 3, no. 1, pp. 63–69, 2006.

[24] A. A. King, E. L. Ionides, M. Pascual, and M. J. Bouma, "Inapparent infections and cholera dynamics," *Nature*, vol. 454, no. 7206, pp. 877–880, 2008.

[25] Z. Mukandavire, S. Liao, J. Wang, H. Gaff, D. L. Smith, and J. G. Morris Jr., "Estimating the reproductive numbers for the 2008-2009 cholera outbreaks in Zimbabwe," *Proceedings of the National Acadamy of Sciences of the United States of America*, vol. 108, no. 21, pp. 8767–8772, 2011.

[26] O. C. Collins and K. S. Govinder, "Stability analysis and optimal vaccination of a waterborne disease model with multiple water sources," *Natural Resource Modeling*, vol. 29, no. 3, pp. 426–447, 2016.

[27] R. M. Anderson and R. M. May, *Infectious Diseases of Humans: Dynamics and Control*, Oxford University Press, 1991.

[28] R. R. Colwell and A. Huq, "Environmental reservoir of vibrio cholerae the causative agent of cholera," *Annals of the New York Academy of Sciences*, vol. 740, no. 1, pp. 44–54, 1994.

[29] P. van den Driessche and J. Watmough, "Reproduction numbers and sub-threshold endemic equilibria for compartmental models of disease transmission," *Mathematical Biosciences*, vol. 180, pp. 29–48, 2002.

[30] S. Liao and J. Wang, "Stability analysis and application of a mathematical cholera model," *Mathematical Biosciences and Engineering*, vol. 8, no. 3, pp. 733–752, 2011.

[31] C. Castillo-Chavez, Z. Feng, and W. Huang, "On the computation of R0 and its role on global stability," in *Mathematical Approaches for Emerging and Reemerging Infectious Diseases: An Introduction*, C. Castillo-Chavez, S. Blower, P. van den Driessche, D. Kirschner, and A.-A. Yakubu, Eds., vol. 125, pp. 229–250, Springer, New York, NY, USA, 2002.

[32] J. L. Ma and D. J. E. Earn, "Generality of the final size formula for an epidemic of a newly invading infectious disease," *Bulletin of Mathematical Biology*, vol. 68, no. 3, pp. 679–702, 2006.

[33] M. Y. Li and J. S. Muldowney, "Global stability for the SEIR model in epidemiology," *Mathematical Biosciences*, vol. 125, no. 2, pp. 155–164, 1995.

[34] A. Korobeinikov and G. C. Wake, "Lyapunov functions and global stability for SIR, SIRS, and SIS epidemiological models," *Applied Mathematics Letters*, vol. 15, no. 8, pp. 955–960, 2002.

[35] A. Korobeinikov, "Lyapunov functions and global properties for SEIR and SEIS epidemic models," *Mathematical Medicine and Biology*, vol. 21, no. 2, pp. 75–83, 2004.

[36] J. P. Tian and J. Wang, "Global stability for cholera epidemic models," *Mathematical Biosciences*, vol. 232, no. 1, pp. 31–41, 2011.

[37] D. A. Sack, R. B. Sack, G. B. Nair, and A. Siddique, "Cholera," *The Lancet*, vol. 363, no. 9404, pp. 223–233, 2004.

[38] K. Todar, "Todar's online textbooks of bacteriology," http://textbookofbacteriology.net/cholera.html, (November, 2013).

[39] J. M. Tchuenche, S. A. Khamis, F. B. Agusto, and S. C. Mpeshe, "Optimal control and sensitivity analysis of an influenza model with treatment and vaccination," *Acta Biotheoretica*, vol. 59, no. 1, pp. 1–28, 2011.

[40] M. E. S. Lucas, J. L. Deen, L. Von Seidlein et al., "Effectiveness of mass oral cholera vaccination in Beira, Mozambique," *The New England Journal of Medicine*, vol. 352, no. 8, pp. 757–767, 2005.

[41] G. Zaman, Y. Han Kang, and I. H. Jung, "Stability analysis and optimal vaccination of an SIR epidemic model," *BioSystems*, vol. 93, no. 3, pp. 240–249, 2008.

[42] T. K. Kar and A. Batabyal, "Stability analysis and optimal control of an SIR epidemic model with vaccination," *BioSystems*, vol. 104, no. 2-3, pp. 127–135, 2011.

[43] K. W. Blayneh, A. B. Gumel, S. Lenhart, and T. Clayton, "Backward bifurcation and optimal control in transmission dynamics of West Nile virus," *Bulletin of Mathematical Biology*, vol. 72, no. 4, pp. 1006–1028, 2010.

[44] H. Laarabi, M. Rachik, O. El Kahlaoui, and E. H. Labriji, "Optimal vaccination strategies of an sir epidemic model with a saturated treatment," *Universal Journal of Applied Mathematics*, vol. 1, no. 3, pp. 185–191, 2013.

[45] F. B. Agusto, "Optimal chemoprophylaxis and treatment control strategies of a tuberculosis transmission model," *World Journal of Modelling and Simulation*, vol. 5, no. 3, pp. 163–173, 2009.

[46] H. R. Joshi, "Optimal control of an HIV immunology model," *Optimal Control Applications & Methods*, vol. 23, no. 4, pp. 199–213, 2002.

[47] O. C. Collins and K. J. Duffy, "Optimal control of maize foliar diseases using the plants population dynamics," *Acta Agriculturae Scandinavica, Section B — Soil & Plant Science*, vol. 66, no. 1, pp. 20–26, 2016.

[48] W. H. Fleming and R. W. Rishel, *Deterministic and Stochastic Optimal Control*, vol. 1, Springer, New York, NY, USA, 1975.

[49] D. L. Lukes, *Differential Equations: Classical to Controlled*, Mathematics in Science and Engineering, Academic Press, New York, NY, USA, 1982.

[50] L. S. Pontryagin, V. G. Boltyanskii, R. V. Gamkrelidze, and E. F. Mishchenko, *The Mathematical Theory of Optimal Control Process*, vol. 4, Gordon and Breach Science Publishers, New York, NY, USA, 1986.

[51] Z. Mukandavire, D. L. Smith, and J. G. Morris Jr., "Cholera in Haiti: Reproductive numbers and vaccination coverage estimates," *Scientific Reports*, vol. 3, 2013.

[52] Ministry of Public Health and Population (MSPP), Haiti, http://mspp.gouv.ht/newsite/, 2013.

[53] Centers for Disease Control and Prevention, http://wwwnc.cdc.gov/travel/notices/watch/haiti-cholera, 2013.

[54] Centers for Disease Control and Prevention (CDC), http://emergency.cdc.gov/situationawareness/haiticholera/data.asp, 2013.

[55] "The MathWorks, Inc., Version 08:37:39, R2012b".

[56] E. Bertuzzo, L. Mari, L. Righetto et al., "Prediction of the spatial evolution and effects of control measures for the unfolding Haiti cholera outbreak," *Geophysical Research Letters*, vol. 38, no. 6, 2011.

[57] D. L. Chao, M. E. Halloran, and I. M. Longini, "Vaccination strategies for epidemic cholera in Haiti with implications for the developing world," *Proceedings of the National Acadamy of Sciences of the United States of America*, vol. 108, no. 17, pp. 7081–7085, 2011.

[58] R. Chunara, J. R. Andrews, and J. S. Brownstein, "Social and news media enable estimation of epidemiological patterns early in the 2010 Haitian cholera outbreak," *The American Journal of Tropical Medicine and Hygiene*, vol. 86, no. 1, pp. 39–45, 2012.

[59] A. R. Tuite, J. Tien, M. Eisenberg, D. J. D. Earn, J. Ma, and D. N. Fisman, "Cholera epidemic in Haiti, 2010: using a transmission model to explain spatial spread of disease and identify optimal control interventions," *Annals of Internal Medicine*, vol. 154, no. 9, pp. 593–601, 2011.

[60] O. C. Collins, "Modelling waterborne diseases," https://research-space.ukzn.ac.za/xmlui/bitstream/handle/10413/11967/Collins_Obiora_C._2013.pdf?sequence=1, (May, 2018).

15

The Holling Type II Population Model Subjected to Rapid Random Attacks of Predator

Jevgeņijs Carkovs, Jolanta Goldšteine, and Kārlis Šadurskis ⓘ

Department of Probability Theory and Mathematical Statistics, Riga Technical University, Kaļķu iela 1, Riga LV-1658, Latvia

Correspondence should be addressed to Kārlis Šadurskis; karlis.sadurskis@gmail.com

Academic Editor: Said R. Grace

We present the analysis of a mathematical model of the dynamics of interacting predator and prey populations with the Holling type random trophic function under the assumption of random time interval passage between predator attacks on prey. We propose a stochastic approximation algorithm for quantitative analysis of the above model based on the probabilistic limit theorem. If the predators' gains and the time intervals between predator attacks are sufficiently small, our proposed method allows us to derive an approximative average dynamical system for mathematical expectations of population dynamics and the stochastic Ito differential equation for the random deviations from the average motion. Assuming that the averaged dynamical system is the classic Holling type II population model with asymptotically stable limit cycle, we prove that the dynamics of stochastic model may be approximated with a two-dimensional Gaussian Markov process with unboundedly increasing variances.

1. Introduction

One of the most popular models of the dynamics of interacting predator and prey populations, such as those found within invertebrate and similar domains of life, in mathematical biology is the system of ordinary differential equations proposed in [1]:

$$\frac{dx}{dt} = rx\left(1 - K^{-1}x\right) - \frac{mx}{A+x}y,$$

$$\frac{dy}{dt} = \left(-\delta + \gamma\frac{mx}{A+x}\right)y, \qquad (1)$$

where phase variables x and y denote the density of prey and predator populations, respectively. In this model it is assumed that in the absence of a predator the prey population has a potential carrying capacity K and develops according to the logistic law with an intrinsic growth rate r, and in the absence of a prey the predator population exponentially decreases to zero with the intrinsic growth rate δ. The mutual influence of changes in the densities of the prey and predator in model (1) is considered by the trophic function $mxy(A+x)^{-1}$, where the positive parameter m is the prey consumption rate by

the predator or, in other words, corresponds to the number of prey individuals that can be "eaten" per unit of time. The positive parameter A reflects the saturation of the amount of prey consumed and, in addition, depends on the rate of reaction of the predator, i.e., the time between attacks on the prey. The parameter γ in formula (1) is a conversion factor that determines the effect of "eaten prey" on the growth rate of the population of the predator. The popularity of the model (1) is explained by the fact that under certain assumptions about positiveness of parameters r, K, m, A, δ, and γ it is structurally stable [2]; that is, there exists a unique asymptotically stable periodic trajectory. This model describes stable fluctuations of the size of the predator and prey populations that are sometimes observed in biological ecosystems. In accordance with the continuous type dynamic system (1), both populations are in permanent contact, and the benefits gained and losses suffered by predator and prey correspondingly in an arbitrary small time interval $[t, t + \Delta)$ are proportional to the length of the interval Δ. In fact, it is clear that changes in the size of both populations are accidental and can only be modelled on average by formula (1). Therefore, subsequently papers were published that took into account the random nature of the model under study by adding stochastic terms of the white noise type in the system of (1) (see the review in [3]). It is important to note that the choice of this type of random perturbation allows preserving the principle of predictability of the behaviour of populations, since stochastic differential equations define random Markov type processes. In the case of modelling by the means of stochastic differential equations, the predator's gain and the prey's loss during time Δ contain not only terms proportional to Δ but also terms that are proportional to the increment of the Brownian motion process $\Delta w := w(t + \Delta) - w(t)$. In most of these papers, the authors manage to prove the possibility of the existence of a stable stationary close to periodic ergodic process that describes the behaviour of the stochastic model under sufficiently small random perturbations.

In the modern literature on mathematical biology many authors use stochastic differential equations as a mathematical model for predator-prey ecological systems (we refer again to the review in [3]). The most typical papers using stochastic models of predator-prey populations are [4–16]. The authors of these papers study the effect of stochastic perturbations of various parameters of classical models, adding either white noise to a chosen parameter [4–10] or the integral over a centered Poisson measure [11–16]. Using the apparatus of modern stochastic analysis, authors study the possibility of the existence of positive solutions, the stability of possible stationary solutions, and the existence of moments of solutions, as well as estimating the asymptotics of the moments of solutions and other properties of solutions. It should be noted that most of the aforementioned papers deal with models with stochastic additives containing no higher than second-order phase variables. If, however, the diffusion coefficients or the integrand of the integral over the Poisson measure has a more complicated form, then the apparatus proposed in the aforementioned papers can hardly be used. At the same time, when analysing the dynamics of some predator-prey biological communities, it may be necessary to investigate the consequences of nonpermanent random contacts that occur at random time moments. In this case, it is natural to assume that at the time of the contact extraction of prey by predator will be a random process that is proportional to the functional response and depends on the phase coordinates in a more complex form.

However, in this kind of model, the predator's gain and the prey's loss during time Δ are proportional to the normally distributed random variable $\Delta w(t)$ with parameter Δ and therefore can be either positive or negative, which is poorly consistent with the definition of these terms in a formula of type (1). In this paper we propose a model that also makes it possible to take into account the stochastic character of the trophic function of the dynamics of populations as considered in the Holling type II model, but the predator's gain and the prey's loss are limited positive random variables. Besides, our model takes into account the possibility that the predator may take some time to attack the prey, and therefore there are intervals when both populations develop independently. Moreover, our model also fulfils the Markov property.

We propose a method of approximate numerical analysis of the probabilistic characteristics of population dynamics, based on application of the asymptotic diffusion approximation algorithm [17]. Application of the diffusion approximation method for the asymptotic estimation of the probability characteristics of the Markovian dynamical system, analogously to the algorithm of the classical central limit theorem, consists of two steps. Initially, using the small parameter *epsilon* and the limit theorem for sequences of Markov processes, we find a deterministic dynamical system for the averaged phase variables. This is followed by finding a stochastic dynamical system for normalized deviations from solutions of the averaged dynamical system. The resulting stochastic differential equation of the Ornstein-Uhlenbeck type is well studied and may be relatively simply analysed [17].

2. The Model

Let us propose a model where the contacts between predator and prey occur very often at random time moments $\{0 = \tau_0 < \tau_1 < \tau_2 < \cdots < \tau_k < \cdots\}$, which are the moments of discontinuity of the trajectories of a stationary piecewise constant Poisson process $\{\xi(t),\ t \geq 0\}$ [18] given on a certain probability space $(\Omega, \mathsf{F}, \mathbf{P})$ with an exponentially distributed length of the constancy intervals $P(\Delta_{k-1} > t) = e^{-\varepsilon^{-1}t}$, where $\Delta_{k-1} = \tau_k - \tau_{k-1}$, and ε is a small positive parameter. Let us denote $\{\mathsf{F}^t,\ t \geq 0\}$ as the minimal filtration [19] to which the process $\{\xi(t),\ t \geq 0\}$ is adapted. Let us also assume that at time moments $\{\tau_k,\ k \in \mathbf{N}\}$ the random variables $\{\xi(\tau_k),\ k \in \mathbf{N}\}$ have a continuous distribution $P(\xi(\tau_k) \leq \xi) = F(\xi)$ and, therefore, without any loss of generality, we may consider that this distribution is uniform on the interval $[0, 1]$. It is known [18] that the probability distributions of a homogeneous Markov process $\{\xi(t),\ t \geq 0\}$ are uniquely determined by its weak infinitesimal operator given by

$$(Q_\varepsilon v)(\xi) := \lim_{t \downarrow 0} \frac{1}{t} E\left\{ v(\xi(t)) - v(\xi) \mid \xi(0) = \xi \right\}. \quad (2)$$

In our case the Markov process is given by an infinitesimal operator

$$(Q_\varepsilon v)(\xi) = \varepsilon^{-1} \int_0^1 (v(z) - v(\xi))\, dz, \quad (3)$$

where ε is a positive parameter.

Let us proceed to the formal definition of the Markov dynamical system dealt with in this paper. Let us suppose that the time of observation starts at $t_0 \geq 0$, and densities of prey and predator populations at this moment are $x(t_0)$, $y(t_0)$, respectively. Suppose that the size of the prey's population $\{x_\varepsilon(t),\ t \in [t_0, t_0 + \tau_1)\}$ changes in accordance with the logistic equation

$$\frac{dx_\varepsilon(t)}{dt} = r x_\varepsilon(t)\left(1 - K^{-1} x_\varepsilon(t)\right). \quad (4)$$

And the size of the predator's population $\{y_\varepsilon(t),\ t \geq 0\}$ changes by the rule

$$\frac{dy_\varepsilon(t)}{dt} = \left[-\delta + \gamma \frac{h(\xi(t)) x_\varepsilon(t)}{A + x_\varepsilon(t)} \right] y_\varepsilon(t), \quad (5)$$

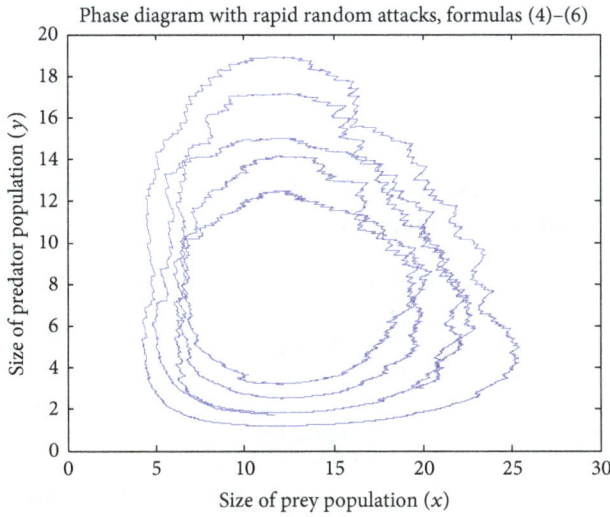

FIGURE 1: Phase trajectory of Markov process (4)-(5)-(6).

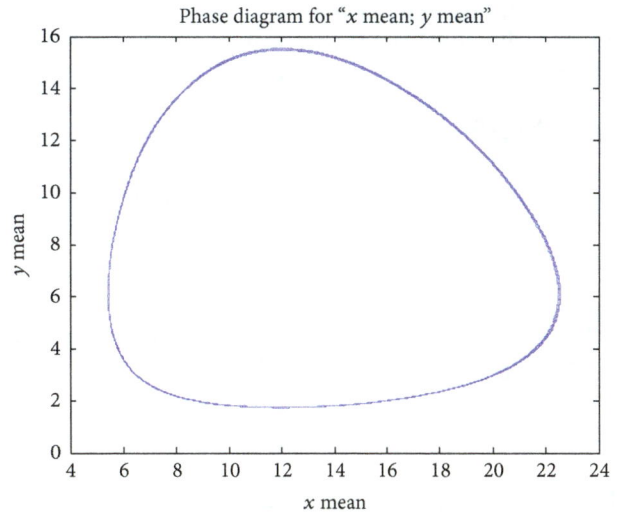

FIGURE 2: Limit cycle of dynamical system (1).

where $\xi(t) \sim R(0,1)$, $h(\xi(t)) > 0$, $\mathbf{E}\{h(\xi(t))\} \equiv m$, and $\mathbf{E}\{h^2(\xi(t))\} \equiv \sigma^2$. At the time moment $t_1 = t_0 + \tau_1$ the collision between predator and prey individuals occurs, and predator's gain is given by the expression

$$x_\varepsilon(t_1) - x_\varepsilon(t_1-) = -\varepsilon y_\varepsilon(t_1) h(\xi(t_1-)) \frac{x_\varepsilon(t_1-)}{A + x_\varepsilon(t_1-)}. \quad (6)$$

Here and further $f(t-) := \lim_{s\uparrow 0} f(t+s)$. Up to the time moment $t_2 = t_1 + \tau_2$, the dynamics of the size of the prey and predator populations are also given by (5)-(6) with initial conditions $x(t_1)$ and $y(t_1)$. At the moment $t_2 = t_1 + \tau_2$ the predator finds the prey again and its gain is given by expression (6) where the argument t_1 is replaced by t_2. Then up to the moment $t_3 = t_2 + \tau_3$ the sizes of predator and prey populations change again according to law (4)-(5), etc.

It is useful to note that, using terminology and results of monograph [20], Markov dynamical system (3)-(4)-(5)-(6) may be defined by the system of stochastic differential equations

$$dx_\varepsilon(t)$$
$$= rx_\varepsilon(t)\left(1 - K^{-1}x_\varepsilon(t)\right) dt$$
$$- \varepsilon \int_0^1 y_\varepsilon(t) h(u) x_\varepsilon(t) \left(A + x_\varepsilon(t)\right)^{-1} \nu_\varepsilon(dt, du), \quad (7)$$
$$dy_\varepsilon(t)$$
$$= \left[-\delta + \gamma x_\varepsilon(t) \left(A + x_\varepsilon(t)\right)^{-1} \int_0^1 h(u) du\right] y_\varepsilon(t) dt,$$

where $\nu_\varepsilon(dt, du)$ is a Poisson random measure with the parameter $\varepsilon^{-1} du\, dt$.

Figure 1 shows the trajectory of the solution of the system of the impulse-differential equations (4)-(5)-(6) with the initial conditions $x_\varepsilon(0) = 20$ and $y_\varepsilon(0) = 3$ and the parameter values $\varepsilon = 0.01$, $\mathbf{E}\{h(\xi(\tau_k))\} = m = 5$, $A = 18$, $\gamma =$

2, $\delta = 4$, $r = 1$, and $K = 1$. Figure 2 shows the trajectory of the solutions of the system of (1) with initial conditions $x(0) = 20$ and $y(0) = 3$ and with the same the parameter values $m = 5$, $A = 18$, $\gamma = 2$, and $\delta = 4$.

As shown in the web-publication [21] and as it can be seen in Figure 2, the system of (1) with the aforementioned parameter values has a unique asymptotically stable periodic trajectory. In Figure 1 the trajectory of the system of (4)-(5)-(6) corresponding to the same initial conditions over time is also grouped within some neighbourhood of the limit cycle described above. Now let us show that the trajectories of the system of (4)-(5)-(6) for positive sufficiently small values of the parameter ε are located within a close neighbourhood of the trajectories of system (1) and coincide in average with them. Let us estimate the deviations from the averaged trajectories. The two-dimensional stochastic process $\{x_\varepsilon(t), y_\varepsilon(t), t \geq 0\}$ is filtration $\{\mathbf{F}^t, t \geq 0\}$ adapted, and for any $t > 0$, $s > 0$, $a \in R$ and $b \in R$ we have the identity:

$$\mathbf{P}\left(x_\varepsilon(t+s) > a, \; y_\varepsilon(t+s)\right.$$
$$\left. > b \mid \{x_\varepsilon(t-u), y_\varepsilon(t-u), u \geq 0\}\right) \equiv \mathbf{P}\left(x_\varepsilon(t+s) \quad (8)\right.$$
$$\left. > a, \; y_\varepsilon(t+s) > b \mid \{x_\varepsilon(t), y_\varepsilon(t)\}\right);$$

i.e., $\{x_\varepsilon(t), y_\varepsilon(t), t \geq 0\}$ has the Markov property [18]. In addition, the dynamic characteristics (4)-(5)-(6) of the process $\{x_\varepsilon(t), y_\varepsilon(t), t \geq 0\}$ are homogeneous in time. Consequently, the dynamical system defined by the Poisson process $\{\xi(t)\}$ and (4)-(5)-(6) defines a homogeneous two-dimensional Markov process with the infinitesimal operator [18] given by

$$(Lv)(x, y) := \lim_{t \downarrow 0} \frac{1}{t} \left[\mathbf{E}\{v(x_\varepsilon(t), y_\varepsilon(t)) \mid x_\varepsilon(0)\right. \quad (9)$$
$$\left. = x, \; y_\varepsilon(0) = y\} - v(x, y)\right].$$

In a sufficiently small time t interval the stationary Poisson process $\xi(t)$ can perform jumps from point $\xi(0) = \eta$ to point

ξ_1, and both random variables are independent and equally uniformly distributed on the segment $[0, 1]$. Therefore, when $t \downarrow 0$ the following asymptotic equalities hold:

$$\mathbf{P}\left(\xi(t) \leq x \mid \xi(0) = \eta, t < \Delta_1\right)$$

$$= \mathbf{P}(\eta \leq x)\,\mathbf{P}(t < \Delta_1) = \mathbf{P}(\eta \leq x)\,e^{-\varepsilon^{-1}t}$$

$$= \mathbf{P}(\eta \leq x) + o(t),$$

$$\mathbf{P}\left(\xi(t) \leq x \mid \xi(0) = \eta, t \geq \Delta_1\right) \qquad (10)$$

$$= \mathbf{P}(\xi_1 \leq x)\left(1 - e^{-\varepsilon^{-1}t}\right) + o(t)$$

$$= \varepsilon^{-1}t\,\mathbf{P}(\xi_1 \leq x) + o(t),$$

where $o(t)$ is an infinitely small quantity with order higher t when $t \to 0$. Consequently, if $v(x, y)$ is a finite sufficiently smooth function, then

$$\mathbf{E}\left\{v\left(x_\varepsilon(t), y_\varepsilon(t)\right) \mid x_\varepsilon(0) = x, y_\varepsilon(0) = y\right\} - v(x, y)$$

$$= \mathbf{E}\left\{v\left(x_\varepsilon(t), y_\varepsilon(t)\right) \mid x_\varepsilon(0) = x, y_\varepsilon(0) = y, \Delta_1\right.$$

$$\leq t\right\}\mathbf{P}(\Delta_1 \leq t) - v(x, y) + \mathbf{E}\left\{v\left(x_\varepsilon(t), y_\varepsilon(t)\right) \mid\right.$$

$$x_\varepsilon(0) = x, y_\varepsilon(0) = y, \Delta_1 > t\right\}\mathbf{P}(\Delta_1 > t) - v(x, y)$$

$$= \mathbf{E}\left\{v\left(x + t\left[rx\left(1 - K^{-1}x\right)\right] - \varepsilon y h(\xi_1)\frac{x}{A + x}\right.\right.$$

$$\left. + o(t), y + t\left[-\delta + \gamma h(\xi_1)\frac{x}{A + x}\right]y + o(t)\right) \qquad (11)$$

$$\left. - v(x, y)\right\}\left(1 - e^{-t\varepsilon^{-1}} + o(t)\right) + \mathbf{E}\left\{v\left(x\right.\right.$$

$$+ trx\left(1 - K^{-1}x\right) + o(t), y$$

$$\left. + t\left[-\delta + \gamma h(\xi_1)\frac{x}{A + x}\right]y + o(t)\right) - v(x, y)\right\}$$

$$\cdot\left(e^{-t\varepsilon^{-1}} + o(t)\right);$$

substituting these asymptotic equations into formula (9), one can obtain an expression for a weak infinitesimal operator for a fixed $\varepsilon > 0$:

$$(L(\varepsilon)v)(x, y) := \lim_{t \downarrow 0}\frac{1}{t}$$

$$\cdot\left[\mathbf{E}\left\{v\left(x_\varepsilon(t), y_\varepsilon(t)\right) \mid x_\varepsilon(0) = x, y_\varepsilon(0) = y\right\}\right.$$

$$\left. - v(x, y)\right] = \varepsilon^{-1}\mathbf{E}\left\{v\left(x - \varepsilon y h(\xi_1)x(A + x)^{-1}, y\right)\right.$$

$$\left. - v(x, y)\right\} + v'_x(x, y)rx\left(1 - K^{-1}x\right) + v'_y(x, y)$$

$$\cdot\left[-\delta + \gamma x\mathbf{E}\{h(\xi_1)\}(A + x)^{-1}\right]y$$

$$= \varepsilon^{-1}\int_0^1 v\left(x - \varepsilon y h(\xi)x(A + x)^{-1}, y\right)d\xi$$

$$- v(x, y) + v'_x(x, y)rx\left(1 - K^{-1}x\right) + v'_y(x, y)$$

$$\cdot\left[-\delta + \gamma mx(A + x)^{-1}\right]y. \qquad (12)$$

3. Asymptotic Analysis of Population Dynamics

As previously mentioned, we assume that the values of the time intervals $\{\Delta_{k-1}, k \in \mathbf{N}\}$, which the predator spends on search for prey, have an exponential distribution with a parameter ε^{-1}; that is, $\mathbf{E}\{\Delta_{k-1}\} = \varepsilon$. If we assume that these intervals are relatively small, that is, the parameter ε is sufficiently small, then it is possible to apply the asymptotic methods proposed in [22] for analysing the Markov process $\{x_\varepsilon(t), y_\varepsilon(t), t \geq 0\}$ given by the infinitesimal operator (12). At first it is necessary to check whether such a limit operator L_0 exists, that, for any continuously differentiable function $v(x, y)$ in some vicinity U_0 of each point from the positive quadrant $\{x \geq 0, y \geq 0\}$, the following equality is true:

$$\lim_{\varepsilon \to 0}\sup_{\{x, y\} \in U_0}\left|(L(\varepsilon)v)(x, y) - (L_0 v)(x, y)\right| = 0. \qquad (13)$$

In our case, this limit exists and it is equal to

$$(L_0 v)(x, y) = \lim_{\varepsilon \to 0}(L(\varepsilon)v)(x, y)$$

$$= v'_x(x, y)rx\left(1 - K^{-1}x\right)$$

$$+ v'_y(x, y)y\left(-\delta + \gamma\frac{mx}{A + x}\right) \qquad (14)$$

$$+ v'_x(x, y)y\frac{mx}{A + x}.$$

Operator (14) corresponds to a deterministic dynamical system of differential equations

$$\frac{d\overline{x}}{dt} = r\overline{x}\left(1 - K^{-1}\overline{x}\right) - \frac{m\overline{x}}{A + \overline{x}}\overline{y},$$

$$\frac{d\overline{y}}{dt} = \left[-\delta + \gamma\frac{m\overline{x}}{A + \overline{x}}\right]\overline{y} \qquad (15)$$

coinciding with the system of (1). Consequently [22], it can be asserted that for any positive number T there exist such positive numbers ε_0 and M that for all $\varepsilon < \varepsilon_0$ the following inequality exists:

$$\sup_{t < T\varepsilon^{-1}}\mathbf{E}\left\{|x_\varepsilon(t) - \overline{x}(t)| + |y_\varepsilon(t) - \overline{y}(t)|\right\} \leq M\varepsilon. \qquad (16)$$

This means that for averaged analysis of the behaviour of the studied population over a sufficiently large time interval one can use the system of (1), i.e., the results given in [2]. As shown in this paper, provided that $A(m\gamma + \delta) < K(m\gamma - \delta)$, the phase portrait of the system of (15) is from the stable limit cycle, the orbit of which all other trajectories approach asymptotically. If the initial data $\{x_\varepsilon(0), y_\varepsilon(0)\}$ for the system of (4)-(5)-(6) are chosen sufficiently close to the coordinates $\{\overline{x}, \overline{y}\}$ of the aforementioned cycle, then the line $\{x_\varepsilon(t), y_\varepsilon(t), t \le T\varepsilon^{-1}\}$, which corresponds to the solutions of these equations, does not leave a sufficiently small vicinity of this cycle. It was shown in [22] that the deviation of the real values $\{x_\varepsilon(t), y_\varepsilon(t)\}$ from the mean values $\{\overline{x}(t), \overline{y}(t)\}$ for each $\{t \in (0, T\varepsilon^{-1})\}$ has the order of smallness $\sqrt{\varepsilon}$ and can be approximated by the solution of the corresponding system of linear stochastic differential equations. To construct these equations, new phase variables are introduced:

$$X_\varepsilon(t) = \frac{x_\varepsilon(t) - \overline{x}(t)}{\sqrt{\varepsilon}},$$

$$Y_\varepsilon(t) = \frac{y_\varepsilon(t) - \overline{y}(t)}{\sqrt{\varepsilon}}. \tag{17}$$

By definition (17) a two-dimensional random process $\{X_\varepsilon(t), Y_\varepsilon(t)\}$ on each interval $t \in (\tau_{k-1}, \tau_k)$, $k \in \mathbf{N}$, is given by the equations

$$\frac{dX_\varepsilon}{dt} = \frac{1}{\sqrt{\varepsilon}} \left[rx_\varepsilon \left(1 - K^{-1}x_\varepsilon\right) - r\overline{x}\left(1 - K^{-1}\overline{x}\right) \right.$$

$$\left. + \frac{m\overline{x}}{A + \overline{x}}\overline{y} \right],$$

$$\frac{dY_\varepsilon}{dt} = \frac{1}{\sqrt{\varepsilon}} \left[-\delta y_\varepsilon + \gamma \frac{h\left(\xi\left(\tau_{k-1}\right)\right) x_\varepsilon}{A + x_\varepsilon} y_\varepsilon + \delta\overline{y} \right.$$

$$\left. - \gamma \frac{m\overline{x}}{A + \overline{x}}\overline{y} \right]$$

but at the moments of contacts τ_k, $k \in \mathbf{N}$, between a predator and a prey it is given by the condition of a jump

$$X_\varepsilon(\tau_k) - X_\varepsilon(\tau_k-) = \frac{1}{\sqrt{\varepsilon}} \left[x_\varepsilon(\tau_k) - x_\varepsilon(\tau_k-) \right]$$

$$= -\frac{1}{\sqrt{\varepsilon}} y_\varepsilon(\tau_k) h\left(\xi\left(\tau_k\right)\right) \frac{x_\varepsilon(\tau_k-)}{A + x_\varepsilon(\tau_k-)}. \tag{19}$$

Substituting in formulas (18)-(19) the expression $\{x_\varepsilon(t), y_\varepsilon(t)\}$ with the expression

$$x_\varepsilon(t) = X_\varepsilon(t)\sqrt{\varepsilon} + \overline{x}(t),$$

$$y_\varepsilon(t) = Y_\varepsilon(t)\sqrt{\varepsilon} + \overline{y}(t) \tag{20}$$

we can obtain the following system of equations:

$$\frac{d}{dt}X_\varepsilon(t) = \frac{1}{\sqrt{\varepsilon}} \left[rX_\varepsilon(t)\sqrt{\varepsilon} - 2rK^{-1}\overline{x}(t) X_\varepsilon(t)\sqrt{\varepsilon} \right.$$

$$\left. - \varepsilon rK^{-1}X_\varepsilon^2(t) + \frac{m\overline{x}(t)}{A + \overline{x}(t)}\overline{y}(t) \right]$$

$$\frac{dY_\varepsilon(t)}{dt} = \frac{1}{\sqrt{\varepsilon}} \left[-\delta\left(Y_\varepsilon(t)\sqrt{\varepsilon} + \overline{y}(t)\right) + \gamma \right.$$

$$\cdot \frac{h\left(\xi\left(\tau_{k-1}\right)\right)\left(X_\varepsilon(t)\sqrt{\varepsilon} + \overline{x}\right)}{A + \left(X_\varepsilon(t)\sqrt{\varepsilon} + \overline{x}(t)\right)}\left(Y_\varepsilon(t)\sqrt{\varepsilon} + \overline{y}(t)\right) \tag{21}$$

$$+ \delta\overline{y}(t) - \gamma\frac{m\overline{x}(t)}{A + \overline{x}(t)}\overline{y}(t) \right]$$

$$X_\varepsilon(\tau_k) - X_\varepsilon(\tau_k-) = -\sqrt{\varepsilon} \left[\left(Y_\varepsilon(\tau_k)\sqrt{\varepsilon} + \overline{y}(\tau_k)\right) \right.$$

$$\left. \cdot h\left(\xi\left(\tau_k\right)\right) \frac{\sqrt{\varepsilon}X_\varepsilon(\tau_k-) + \overline{x}(\tau_k)}{A + \sqrt{\varepsilon}X_\varepsilon(\tau_k-) + \overline{x}(\tau_k)} \right].$$

The process $\{\overline{x}(t), \overline{y}(t), X_\varepsilon(t), Y_\varepsilon(t)\}$ depends only on its state at the time moment t; that is, by definition (17) the two-dimensional random process $\{X_\varepsilon(t), Y_\varepsilon(t)\}$ is filtration $\{F^t, t \ge 0\}$ adapted and, for a given solution of the system of (15), has the Markov property. Henceforward, it is convenient to study the behaviour of solutions of the system of (21) together with the system of (15), investigating the dynamic properties of the multidimensional Markov process $\{\overline{x}(t), \overline{y}(t), X_\varepsilon(t), Y_\varepsilon(t)\}$. A weak infinitesimal operator $L(\varepsilon)$ [18] of this process can be found by calculating the limit expression

$$(L(\varepsilon)V)(x, y, X, Y) := \lim_{t \downarrow 0} \frac{1}{t} \left[\mathbf{E}\left\{ V\left(\overline{x}(t), \overline{y}(t), X_\varepsilon(t), Y_\varepsilon(t)\right)\big|_{\overline{x}(0)=x, \overline{y}(0)=y, X_\varepsilon(0)=X, Y_\varepsilon(0)=Y} \right\} - v(x, y, X, Y) \right] \tag{22}$$

for an arbitrary finite sufficiently smooth function $V(x, y, X, Y)$. Considering the possibility of contact between populations in a sufficiently short time interval $(0, t)$, the following expressions can be used to calculate the limit (22):

$$\mathbf{E}\left\{ V\left(\overline{x}(t), \overline{y}(t), X_\varepsilon(t), \right.\right.$$

$$\left.\left. Y_\varepsilon(t)\right)\big|_{\overline{x}(0)=x, \overline{y}(0)=y, X_\varepsilon(0)=X, Y_\varepsilon(0)=Y} \right\} - V(x, y, X, Y)$$

$$= \mathbf{E}\left\{ V\left(\overline{x}(t), \overline{y}(t), X_\varepsilon(t), \right.\right.$$

$$\left.\left. Y_\varepsilon(t)\right)\big|_{\overline{x}(0)=x, \overline{y}(0)=y, X_\varepsilon(0)=X, Y_\varepsilon(0)=Y} \,|\, \Delta_0 \le t \right\} - V(x, y,$$

$$X, Y) + \mathbf{E}\left\{ V\left(\overline{x}(t), \overline{y}(t), X_\varepsilon(t), \right.\right.$$

$$\left.\left. Y_\varepsilon(t)\right)\big|_{\overline{x}(0)=x, \overline{y}(0)=y, X_\varepsilon(0)=X, Y_\varepsilon(0)=Y} \,|\, \Delta_0 > t \right\} - V(x, y,$$

$$X, Y) = \mathbf{E}\left\{ V\left(x + t\left[rx\left(1 - K^{-1}x\right) - \frac{mx}{A + x}y\right] \right.\right.$$

$$\left.\left. + o(t)\right), y + t\left[-\delta + \gamma m\frac{x}{A + x}\right]y + o(t), X + t \right.$$

$$\cdot \frac{1}{\sqrt{\varepsilon}} \left[rX\sqrt{\varepsilon} - 2rK^{-1}xX\sqrt{\varepsilon} - \varepsilon rK^{-1}X^2 + m \right.$$

$$\left.\left. \cdot \frac{x}{A + x}y\right] - \sqrt{\varepsilon}\left[(Y\sqrt{\varepsilon} + y)h(\xi_1)\frac{\sqrt{\varepsilon}X + x}{A + \sqrt{\varepsilon}X + x}\right] \right.$$

$$+ o(t), Y + t\frac{1}{\sqrt{\varepsilon}}\left[-\delta(Y\sqrt{\varepsilon} + y) + \gamma\right.$$

$$\cdot \frac{h(\xi_1)(X\sqrt{\varepsilon} + x)}{A + (X\sqrt{\varepsilon} + x)}(Y\sqrt{\varepsilon} + y) + \delta y - \gamma\frac{mx}{A+x}y\right]$$

$$+ o(t) - V(x, y, X, Y)\bigg\}\left(1 - e^{-t\varepsilon^{-1}} + o(t)\right)$$

$$+ \mathbf{E}\left\{V\left(x + t\left[rx(1 - K^{-1}x) - \frac{mx}{A+x}y\right]\right.\right.$$

$$+ o(t)\bigg), y + t\left[-\delta + \gamma m\frac{x}{A+x}\right]y + o(t), X + t$$

$$\cdot \frac{1}{\sqrt{\varepsilon}}\left[rX\sqrt{\varepsilon} - 2rK^{-1}xX\sqrt{\varepsilon} - \varepsilon rK^{-1}X^2 + m\right.$$

$$\cdot \frac{x}{A+x}y\right] + o(t), Y + t\frac{1}{\sqrt{\varepsilon}}\left[-\delta(Y\sqrt{\varepsilon} + y) + \gamma\right.$$

$$\cdot \frac{h(\xi_1)(X\sqrt{\varepsilon} + x)}{A + (X\sqrt{\varepsilon} + x)}(Y\sqrt{\varepsilon} + y) + \delta y - \gamma\frac{mx}{A+x}y\right]$$

$$+ o(t) - V(x, y, X, Y)\bigg\}\left(e^{-t\varepsilon^{-1}} + o(t)\right) + \varepsilon(QV)$$

$$\cdot (x, y, X, Y).$$

$$(23)$$

Besides,

$$\mathbf{E}\left\{(\sqrt{\varepsilon}X + x)(a(\Delta_0) + \sqrt{\varepsilon}X + x)^{-1}\right\}$$

$$= \varepsilon^{-1}\int_0^\infty (\sqrt{\varepsilon}X + x)(a(t) + \sqrt{\varepsilon}X + x)^{-1}e^{-t\varepsilon^{-1}}dt$$

$$(24)$$

$$= \int_0^\infty (\sqrt{\varepsilon}X + x)(a(\varepsilon s) + \sqrt{\varepsilon}X + x)^{-1}e^{-s}ds$$

$$= x(A + x)^{-1} + O(\sqrt{\varepsilon}).$$

Substituting the asymptotic by t expansions (23)-(24) into expression (22) after passing to the limit, one can obtain the formula for the weak infinitesimal operator of the homogeneous Markov process $\{\overline{x}(t), \overline{y}(t), X_\varepsilon(t), Y_\varepsilon(t)\}$:

$$(L(\varepsilon)V)(x, y, X, Y, \xi) = \left[rx(1 - K^{-1}x) - \frac{mx}{A+x}\right.$$

$$\cdot y\bigg]\frac{\partial}{\partial x}V(x, y, X, Y) + \bigg[-\delta$$

$$+ \gamma m\frac{x}{A+x}\bigg]y\frac{\partial}{\partial y}V(x, y, X, Y) + \frac{1}{\varepsilon}\mathbf{E}\left\{V\left(x, y, X\right.\right.$$

$$- \sqrt{\varepsilon}\frac{h(\xi_1)(Y\sqrt{\varepsilon} + y)(\sqrt{\varepsilon}X + x)}{A + \sqrt{\varepsilon}X + x}, Y\bigg)$$

$$- V(x, y, X, Y)\bigg\} + \frac{1}{\sqrt{\varepsilon}}\left[rX\sqrt{\varepsilon} - 2rK^{-1}xX\sqrt{\varepsilon}\right.$$

$$- \varepsilon rK^{-1}X^2 + m\frac{x}{A+x}y\right]\frac{\partial}{\partial X}V(x, y, X, Y)$$

$$+ \frac{1}{\sqrt{\varepsilon}}\left[-\delta Y\sqrt{\varepsilon} - \gamma\frac{mx}{A+x}y + \gamma\right.$$

$$\cdot \frac{m(Y\sqrt{\varepsilon} + y)(\sqrt{\varepsilon}X + x)}{A + \sqrt{\varepsilon}X + x}\right]\frac{\partial}{\partial Y}V(x, y, X, Y).$$

$$(25)$$

Since we are discussing the behaviour of populations for sufficiently small ε, then it is necessary to use a passage to the limit in the previous formula for tending the parameter ε to zero on the right. Therefore, one can use the smoothness of the function $V(x, y, X, Y)$ and further use the asymptotic equalities:

$$\mathbf{E}\left\{V\left(x, y, X - \sqrt{\varepsilon}\left[(Y\sqrt{\varepsilon} + y)h(\xi_1)(\sqrt{\varepsilon}X + x)\right.\right.\right.$$

$$\cdot (A + \sqrt{\varepsilon}X + x)^{-1}\bigg], Y\bigg)\bigg\} = V(x, y, X, Y)$$

$$- \sqrt{\varepsilon}yx(A + x)^{-1}m\frac{\partial}{\partial X}V(x, y, X, Y) + \varepsilon\left[-Yx(A\right.$$

$$+ x)^{-1}m - yX(A + x)^{-1}m + yxmX(A$$

$$+ x)^{-2}\bigg]\frac{\partial}{\partial X}V(x, y, X, Y) + \varepsilon\frac{1}{2}$$

$$\cdot yx(A + x)^{-1}\mathbf{E}\left\{h^2(\xi_1)\right\}\frac{\partial^2}{\partial X^2}V(x, y, X, Y)$$

$$(26)$$

$$= V(x, y, X, Y) - \sqrt{\varepsilon}yx(A + x)^{-1}m\frac{\partial}{\partial X}V(x, y, X,$$

$$Y) - \varepsilon\left(Yx + AyX(A + x)^{-1}\right)m(A$$

$$+ x)^{-1}\frac{\partial}{\partial X}V(x, y, X, Y) + \varepsilon\frac{1}{2}$$

$$\cdot yx(A + x)^{-1}\sigma^2\frac{\partial^2}{\partial X^2}V(x, y, X, Y) + o(\varepsilon).$$

Using these expansions, formula (25) can be rewritten in the form convenient for passage to the limit:

$$(L(\varepsilon)V)(x, y, X, Y, \xi) = \left[rx(1 - K^{-1}x) - \frac{mx}{A+x}\right.$$

$$\cdot y\bigg]\frac{\partial}{\partial x}V(x, y, X, Y) + \bigg[-\delta$$

$$+ \gamma m\frac{x}{A+x}\bigg]y\frac{\partial}{\partial y}V(x, y, X, Y)$$

$$+ \left[(rX - 2rK^{-1}xX) - \left(Yx + AyX(A + x)^{-1}\right)\right.$$

$$\cdot m(A + x)^{-1}\bigg]\frac{\partial}{\partial X}V(x, y, X, Y)$$

$$+ \frac{1}{2}yx(A + x)^{-1}\sigma^2\frac{\partial^2}{\partial X^2}V(x, y, X, Y) + \bigg[-\delta Y$$

$$+ \gamma Y \frac{mx}{A + x} + X \gamma y \frac{mA}{(A + x)^2} \Bigg] \frac{\partial}{\partial Y} V(x, y, X, Y)$$

$$+ O\left(\sqrt{\varepsilon}\right).$$

$$(27)$$

It follows that the limit operator $\mathsf{L} = \lim_{\varepsilon \to 0} \mathsf{L}(\varepsilon)$ corresponding to the Markov process $\{\overline{x}(t), \overline{y}(t), X(t), Y(t), t \geq 0\}$, given by a system of ordinary differential equations (15), is a stochastic linear inhomogeneous stochastic differential equation

$$dX(t) = g_{11}(t) X(t) \, dt + g_{12}(t) Y(t) \, dt$$
$$+ f(t) \, dw(t)$$

$$(28)$$

and an ordinary homogeneous differential equation

$$\frac{dY(t)}{dt} = g_{21}(t) X(t) + g_{22}(t) Y(t),$$

$$(29)$$

where

$$g_{21}(t) = \gamma \overline{y}(t) \frac{mA}{(A + \overline{x}(t))^2},$$

$$g_{22}(t) = -\delta + \gamma \frac{m\overline{x}(t)}{A + \overline{x}(t)},$$

$$g_{11}(t) = \left(1 - \frac{2}{K} \overline{x}(t)\right) r - \frac{A\overline{y}(t) m}{(A + \overline{x}(t))^2},$$

$$(30)$$

$$g_{12}(t) = -\frac{\overline{x}(t) m}{A + \overline{x}(t)},$$

$$f(t) = \sigma \sqrt{\frac{\overline{y}(t) \overline{x}(t)}{2(A + \overline{x}(t))}},$$

and $w(t)$ is a standard Brownian process. Since the limit cycle is an asymptotically stable state of the system of equations (15), it is natural to assume that the points $\{\overline{x}(t), \overline{y}(t), t \geq 0\}$ in the system of equations (28)-(29) are the coordinates of the limit cycle mentioned previously. Using the results of [22], one can assert that on an interval with order $1/\sqrt{\varepsilon}$ the probabilistic characteristics of the solutions $\{x_\varepsilon(t), y_\varepsilon(t)\}$ of the impulse-differential equations system (4)-(5)-(6) with accuracy of the order of smallness greater than $\sqrt{\varepsilon}$ can be approximated by random processes $\{\overline{x}(t) + \sqrt{\varepsilon} X(t), \overline{y}(t) + \sqrt{\varepsilon} Y(t)\}$. In Figure 3 a sample trajectory of the process $\{\overline{x}(t) + \sqrt{\varepsilon} X(t), \overline{y}(t) + \sqrt{\varepsilon} Y(t), t \in [0, 20]\}$ is simulated in MATLAB environment with the initial conditions $\overline{x}(0) = 20$, $\overline{y}(0) = 3$, $X(0) = 0$, $Y(0) = 0$, var$\{X(0)\} = 0$, var$\{Y(0)\} = 0$, covar$\{X(0), Y(0)\} = 0$, and $\varepsilon = 0.01$ on the phase plane $\{x, y\}$.

Comparison between Figures 1 and 3 indicates that the proposed approximation sufficiently reflects the dynamics of real processes. However, over a large time interval this approximation can differ significantly from the real process. This fact will be explained using the moments of the solutions of the system of (28)-(29). The probabilistic characteristics of the original process and the proposed approximation should be close [22] over a time interval with order $1/\sqrt{\varepsilon}$. This means

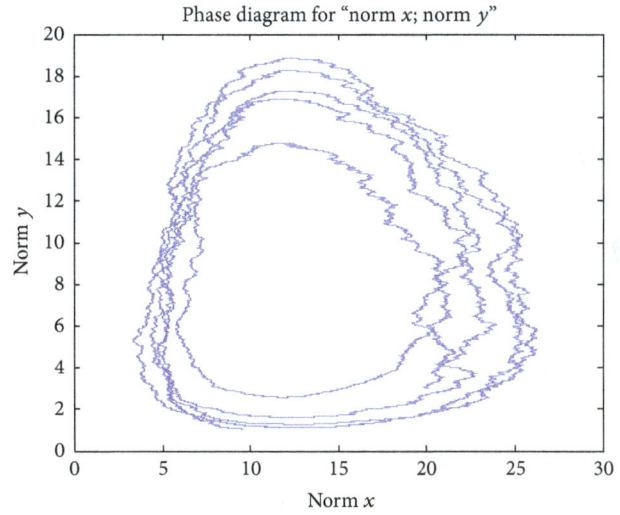

FIGURE 3: Phase trajectory of Markov process $\{\overline{x}(t) + \sqrt{\varepsilon} X(t), \overline{y}(t) + \sqrt{\varepsilon} Y(t), t \in [0, 20]\}$.

that at each time moment $t \in [0, T/\sqrt{\varepsilon}]$ the centered second moments of the process $\{x_\varepsilon(t), y_\varepsilon(t)\}$ can be approximated using the second moments of the process $\{X(t), Y(t)\}$ in the form of asymptotic equalities:

$$\text{var}\{x_\varepsilon(t)\} = \varepsilon q_{11}(t) + o\left(\varepsilon^{3/2}\right),$$

$$\text{var}\{y_\varepsilon(t)\} = \varepsilon q_{22}(t) + o\left(\varepsilon^{3/2}\right),$$

$$(31)$$

$$\text{covar}\{x_\varepsilon(t), y_\varepsilon(t)\} = \varepsilon q_{12}(t) + o\left(\varepsilon^{3/2}\right) q_{11}(t),$$

where $q_{11}(t) = \text{var}\{X(t)\}$, $q_{22}(t) = \text{var}\{Y(t)\}$, and $q_{12}(t) = \text{covar}\{X(t), Y(t)\}$. For further analysis, using the notations

$$G(t) = \begin{pmatrix} g_{11}(t) & g_{12}(t) \\ g_{21}(t) & g_{22}(t) \end{pmatrix},$$

$$\overrightarrow{Z}(t) = \begin{pmatrix} X(t) \\ Y(t) \end{pmatrix},$$

$$(32)$$

$$\overrightarrow{\varphi}(t) = \begin{pmatrix} f(t) \\ 0 \end{pmatrix},$$

let us rewrite the system of (28)-(29) in a vector-matrix form:

$$d\overrightarrow{Z}(t) = G(t) \overrightarrow{Z}(t) \, dt + \overrightarrow{\varphi}(t) \, dw(t).$$

$$(33)$$

Further attention will be focused on the solution of the system of (28)-(29) with zero initial data. It is known [17] that this solution can be represented in the form of a stochastic Ito integral

$$\overrightarrow{Z}(t) = \int_0^t U(t) U^{-1}(s) \overrightarrow{\varphi}(s) \, dw(s),$$

$$(34)$$

where $U(t)$ is the matrix solution of the ordinary differential equation

$$\frac{d}{dt} U(t) = G(t) U(t)$$

$$(35)$$

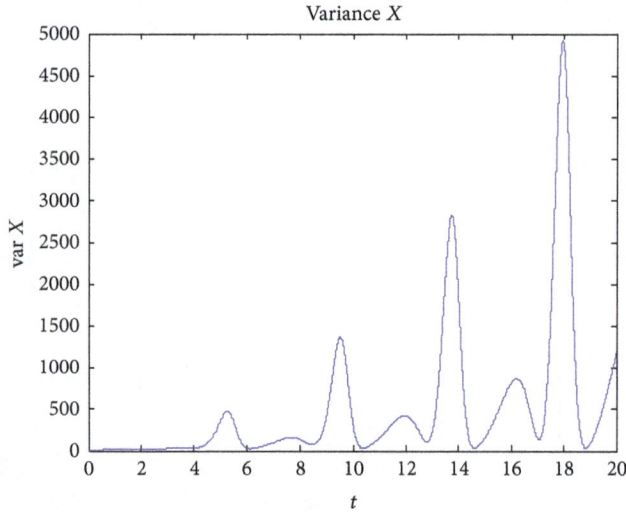

FIGURE 4: Trajectory of the variance $q_{11}(t) = \mathbf{E}\{X^2(t)\}$.

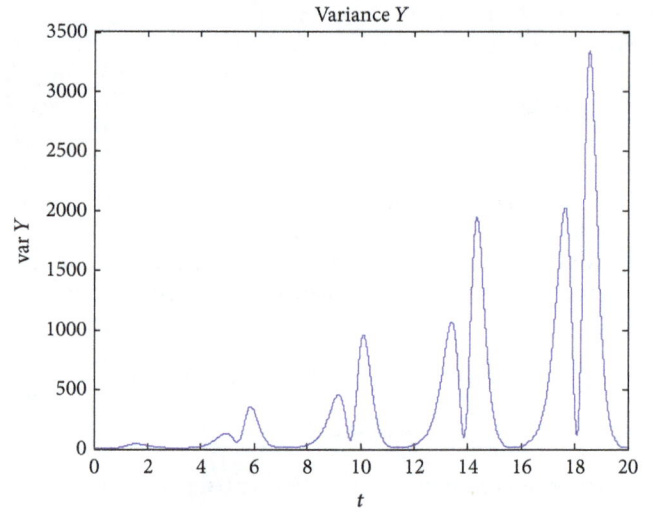

FIGURE 5: Trajectory of the variance $q_{22}(t) = \mathbf{E}\{Y^2(t)\}$.

with respect to the initial value $U(0) = \left(\begin{smallmatrix} 1 & 0 \\ 0 & 1 \end{smallmatrix}\right)$. By the definition of a stochastic integral [18], formula (34) defines a Gaussian process with zero mathematical expectation and a covariance matrix

$$\overrightarrow{Z}(t) = \int_0^t U(t) U^{-1}(s) \overrightarrow{\varphi}(s) \, dw(s). \tag{36}$$

The covariance matrix $Q(t) = \mathbf{E}\{\overrightarrow{Z}(t)\overrightarrow{Z}^T(t)\} := \left(\begin{smallmatrix} q_{11}(t) & q_{12}(t) \\ q_{12}(t) & q_{22}(t) \end{smallmatrix}\right)$ that corresponds to the integral (36) as a function from parameter t can be represented in the form of integral

$$Q(t)$$
$$= \int_0^t U(t) U^{-1}(s) \overrightarrow{\varphi}(s) \overrightarrow{\varphi}^T(s) \left(U^{-1}(s)\right)^T U^T(t) \, ds \tag{37}$$

or in the form of a matrix solution of a system of ordinary differential equations

$$\frac{d}{dt} Q(t) = G(t) Q(t) + Q(t) G^T(t) + \overrightarrow{\varphi}(t) \overrightarrow{\varphi}^t(t) \tag{38}$$

with the initial condition $Q(0) = 0$. Figures 4 and 5 show the time evolution of the variances $q_{11}(t) = \mathbf{E}\{X^2(t)\}$ and $q_{22}(t) = \mathbf{E}\{Y^2(t)\}$.

Figure 6 shows the time evolution of the covariance $q_{12}(t) = \mathbf{E}\{X(t)Y(t)\}$.

Figures 4–6 show that the variances of the normalized deviations increase very rapidly and even at $t \geq 20$ it cannot be recommended to use the most popular applied statistics "3σ rule"; that is, the inequality $|x_\varepsilon(t) - \overline{x}(t)| \leq 3\sqrt{\varepsilon}\sqrt{q_{11}(t)}$, $|y_\varepsilon(t) - \overline{y}(t)| \leq 3\sqrt{\varepsilon}\sqrt{q_{22}(t)}$ because the quantities either $\overline{x}(t) - 3\sqrt{\varepsilon}\sqrt{q_{11}(t)}$ or $\overline{y}(t) - 3\sqrt{\varepsilon}\sqrt{q_{22}(t)}$ can become negative, which contradicts the definition of random processes $x_\varepsilon(t)$ and $y_\varepsilon(t)$. For clarity, Figure 7 shows the process $\{q_{11}(t), q_{22}(t)\}$ trajectory on the time interval $[0, 30]$ on the phase plane $\{x, y\}$.

Here the variable $3\sqrt{\varepsilon}\sqrt{q_{11}(t)} = 0.3\sqrt{q_{11}(t)}$ at some time point already exceeds the value $0.3\sqrt{10000} = 30$, although

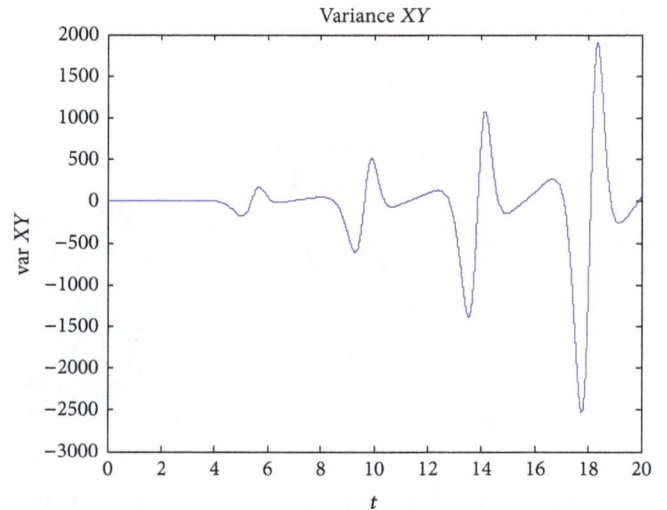

FIGURE 6: Trajectory of the covariance $q_{12}(t) = \mathbf{E}\{X(t)Y(t)\}$.

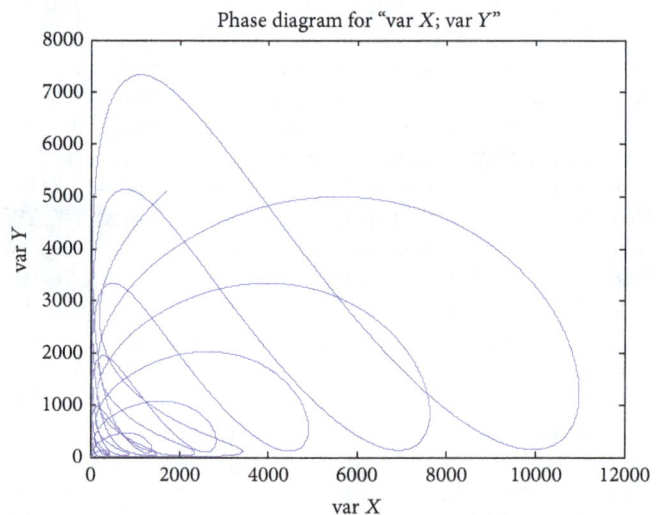

FIGURE 7: Phase-plane trajectory for variances $\{q_{11}(t), q_{22}(t)\}$.

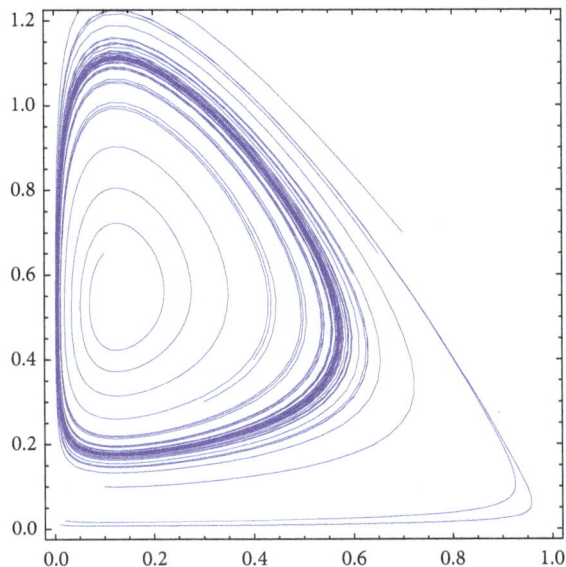

FIGURE 8: Phase portrait for (1), $m = 1$, $r = 1$, $A = 0.5$, $K = 1$, $\gamma = 1$, and $\delta = 0.5$.

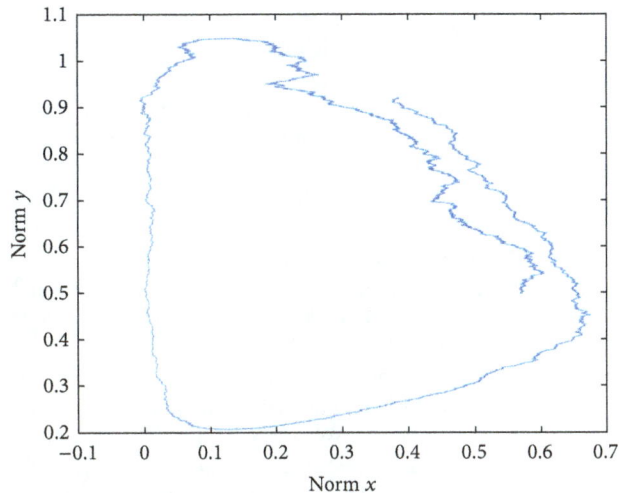

FIGURE 9: Phase trajectory $\{\overline{x}(t) + \sqrt{\varepsilon}X(t), \overline{y}(t) + \sqrt{\varepsilon}Y(t), \ t \in [0, 20]\}$, $m = 1$, $r = 1$, $A = 0.5$, $K = 1$, $\gamma = 1$, and $\delta = 0.2$. $\overline{x}(0) = 0.57$, and $\overline{y}(0) = 0.5$.

the value $\overline{x}(t)$ (the abscissa of the limit cycle in Figure 2) does not exceed 24.

Unbounded increase of variances $q_{11}(t)$, $q_{22}(t)$ is a consequence of a resonance in the equation (38). Let ω be a period of the limit cycle $\{\overline{x}(t), \overline{y}(t)\}$. Then $U(t)U^T(t)$ and $\overrightarrow{\varphi}(t)\overrightarrow{\varphi}^T(t)$ are 2ω-periodic matrix functions and their elements may be decomposed into a Fourier series by $\cos(n\pi/\omega)t$ and $\sin(n\pi/\omega)t$, $n \in N$. Therefore, the integrand in (37) contains a constant component and $\lim_{t\to\infty}\|Q(t)\| = \infty$. At a finite time interval that depends on ε we can use the diffusion approximation

$$\{\overline{x}(t) + \sqrt{\varepsilon}X(t), \overline{y}(t) + \sqrt{\varepsilon}Y(t), t \in [0, T(\varepsilon)]\} \quad (39)$$

for estimation of finite dimensional distributions of the process $\{x_\varepsilon(t), y_\varepsilon(t)\}$.

Even if a deterministic limit cycle closely approaches one of the axes (Figure 8) we can use diffusion approximation $\{\overline{x}(t) + \sqrt{\varepsilon}X(t), \overline{y}(t) + \sqrt{\varepsilon}Y(t)\}$ (Figure 9) for prognoses of the population growth up to time $T = 20$, $\varepsilon = 0.001$.

4. Conclusions

During the analysis of population dynamics models, one must find a compromise between accuracy and complexity of the model. Numerical methods are helpful but their use actualises the stability issues of dynamical systems.

In the case of interacting invertebrate type populations of predator and prey, the fact that the time between predator attacks on prey is random leads to a stochastic model. Even though the deterministic model has an asymptotically stable limit cycle, introducing random parameters into the model changes its trajectories significantly. The algorithm proposed and demonstrated in this paper allows one to analyse the average motion of a system as well as the random deviations

from it and to find all characteristics of the resulting two-dimensional Gaussian Markov process. This process has an unboundedly increasing variance, leading to possible significant deviations even on relatively short time intervals.

Conflicts of Interest

The authors declare that there are no conflicts of interest regarding the publication of this paper.

References

[1] C. S. Holling, "The components of predation as revealed by a study of small-mammal predation of the european pine sawfly," *The Canadian Entomologist*, vol. 91, no. 5, pp. 293–320, 1959.

[2] R. M. May, "Stability and Complexity in Model Ecosystems," p. 235, Princeton University, Princeton, NJ, USA, 1973.

[3] L. J. Allen, *An Introduction to Stochastic Processes with Applications to Biology*, CRC Press, Taylor & Francis Group, 2010.

[4] Q. Liu, L. Zu, and D. Jiang, "Dynamics of stochastic predator-prey models with Holling II functional response," *Communications in Nonlinear Science and Numerical Simulation*, vol. 37, pp. 62–76, 2016.

[5] R. Rudnicki, "Long-time behaviour of a stochastic prey-predator model," *Stochastic Processes and Their Applications*, vol. 108, no. 1, pp. 93–107, 2003.

[6] T. V. Ton and A. Yagi, "Dynamics of a stochastic predator-prey model with the Beddington-DeAngelis functional response," *Communications on Stochastic Analysis*, vol. 5, no. 2, pp. 371–386, 2011.

[7] F. Rao, "Dynamical analysis of a stochastic predator-prey model with an Allee effect," *Abstract and Applied Analysis*, vol. 2013, Article ID 340980, 2013.

[8] C. Ji, D. Jiang, and X. Li, "Qualitative analysis of a stochastic ratio-dependent predator-prey system," *Journal of Computational and Applied Mathematics*, vol. 235, no. 5, pp. 1326–1341, 2011.

[9]   X. Mao, G. Marion, and E. Renshaw, "Environmental Brownian noise suppresses explosions in population dynamics," *Stochastic Processes and Their Applications*, vol. 97, no. 1, pp. 95–110, 2002.

[10] X. Mao, S. Sabanis, and E. Renshaw, "Asymptotic behaviour of the stochastic Lotka-Volterra model," *Journal of Mathematical Analysis and Applications*, vol. 287, no. 1, pp. 141–156, 2003.

[11] J. Lv and K. Wang, "On a stochastic predator-prey system with modified functional response," *Mathematical Methods in the Applied Sciences*, vol. 35, no. 2, pp. 144–150, 2012.

[12] J. Bao and C. Yuan, "Stochastic population dynamics driven by Lévy noise," *Journal of Mathematical Analysis and Applications*, vol. 391, no. 2, pp. 363–375, 2012.

[13] A. Gray, D. Greenhalgh, X. Mao, and J. Pan, "The SIS epidemic model with Markovian switching," *Journal of Mathematical Analysis and Applications*, vol. 394, no. 2, pp. 496–516, 2012.

[14] Q. Luo and X. Mao, "Stochastic population dynamics under regime switching," *Journal of Mathematical Analysis and Applications*, vol. 334, no. 1, pp. 69–84, 2007.

[15] X. Zou and K. Wang, "Optimal harvesting for a stochastic regime-switching logistic diffusion system with jumps," *Nonlinear Analysis: Hybrid Systems*, vol. 13, pp. 32–44, 2014.

[16] X. Li, A. Gray, D. Jiang, and X. Mao, "Sufficient and necessary conditions of stochastic permanence and extinction for stochastic logistic populations under regime switching," *Journal of Mathematical Analysis and Applications*, vol. 376, no. 1, pp. 11–28, 2011.

[17] Y. F. Tsarkov and J. Carkovs, *Random perturbations of Stochastic Differential Equations*, Zinatne, Riga, Latvia, 1989.

[18] A. V. Skorokhod, *Asymptotic Methods in the Theory of Stochastic Differential Equations*, American Mathematical Society, Providence, RI, USA, 1989.

[19] Y. B. Dynkin, *Theory of Markov Processes*, Prentice-Hall, 1961.

[20] I. I. Gihman and A. V. Skorohod, *Stochastic Differential Equations*, Springer, Berlin, Germany, 1972.

[21] M. R. Cullen, Loyola Murymount University, http://www.cengage.com/math/book_content/0495108243_zill/projects_archive/debv4e/AModelingApplication6.pdf.

[22] Y. Tsarkov, "Asymptotic methods for stability analysis of Markov impulse dynamical systems," *Nonlinear Dynamics and Systems Theory. An International Journal of Research and Surveys*, vol. 2, no. 1, pp. 103–115, 2002.

16

A Mathematical Model of Treatment and Vaccination Interventions of Pneumococcal Pneumonia Infection Dynamics

Mohammed Kizito and Julius Tumwiine (iD)

Department of Mathematics, Mbarara University of Science and Technology, P.O. Box 1410, Mbarara, Uganda

Correspondence should be addressed to Julius Tumwiine; jtumwiine@must.ac.ug

Academic Editor: Oluwole D. Makinde

Streptococcus pneumoniae is one of the leading causes of serious morbidity and mortality worldwide, especially in young children and the elderly. In this study, a model of the spread and control of bacterial pneumonia under public health interventions that involve treatment and vaccination is formulated. It is found out that the model exhibits the disease-free and endemic equilibria. The disease-free equilibrium is stable if and only if the basic reproduction number $\mathscr{R}_0 < 1$ and the disease will be wiped out of the population. For $\mathscr{R}_0 \geq 1$, the endemic equilibrium is globally stable and the disease persists. We infer the effect of these interventions on the dynamics of the pneumonia through sensitivity analysis on the effective reproduction number $\mathscr{R}_e$, from which it is revealed that treatment and vaccination interventions combined can eradicate pneumonia infection. Numerical simulation to illustrate the analytical results and establish the long term behavior of the disease is done. The impact of pneumonia infection control strategies is investigated. It is revealed that, with treatment and vaccination interventions combined, pneumonia can be wiped out. However, with treatment intervention alone, pneumonia persists in the population.

1. Introduction

Pneumonia is the major cause of respiratory morbidity of more than 2 million children under 5 years of age mostly in low-income countries [1–6]. It is an infection of the lungs that is caused by bacteria, viruses, fungi, or other pathogens. Most common cause of bacterial pneumonia is the *Streptococcus pneumoniae*, also known as pneumococcus [6–8]. It is characterized primarily by inflammation in the air sacs (alveoli) in the lungs that are filled with fluid or pus making it difficult to breathe.

It is reported that about 30%–70% of young children carry *S. pneumoniae* in their nasopharynx, and up to 40% of the carriers are colonized with penicillin-nonsusceptible *S. pneumoniae*. Pneumococcus spreads through microaspiration of oropharyngeal organisms and inhalation of aerosols containing bacteria or viruses especially in children that carry the bacteria in their throats without being sick. It may also spread via airborne droplets from cough or sneeze of an infected person. Children become severely ill with high fever and rapid breathing. Infants usually suffer convulsions, unconsciousness, hypothermia, lethargy, and feeding problems [9]. The data on carriage among adults [10] is limited and most studies suggest that children are the source of transmission to adults in the family [11].

The risk factors associated with the spread of pneumonia include smoking history and passive smoking, malnutrition, crowded living conditions, lack of exclusive breastfeeding, indoor air pollution, heart disease, alcoholism and drug abuse, acidosis, diabetes, and antecedent viral infection [12, 13]. The overdiagnosis of pneumonia and underdiagnosis of asthma have led significantly to untreated respiratory morbidity and mortality among children less than five years in low-income countries. This has been due to some similarities of symptoms of pneumonia and asthma that often make it difficult to separate the two diseases without proper diagnostic tools [14].

Pneumonia classification depends on its origin and mode of transmission. Some of the pneumonia classifications include community-acquired, health care-associated, hospital-acquired, ventilator-associated, and walking pneumonia [15].

Pneumonia is preventable through vaccination, proper diagnosis, screening, environmental control measures, and appropriate treatment of other diseases [4, 5]. Vaccination is the most effective way to prevent certain bacterial and viral pneumonia in both children and adults. The two types of vaccines available against S. pneumoniae are the pneumococcal polysaccharide vaccine (PPV), based on purified capsular (PS) and pneumococcal conjugate vaccine (PCV), obtained by chemical conjugation of the capsular (PS) to a protein carrier [16]. PCVs were developed for use in children only and PPV for vaccination of the at-risk adults and the elderly [17, 18].

Newborn babies can be protected from pneumonia infection through early recognition and treatment at the level of the community or the primary-care health facility, testing pregnant mothers for Group B streptococcus and chlamydia trachomatis, and giving antibiotic treatment and vaccination with PPV that has a proven record of safety in pregnant and breastfeeding mothers for pneumococcal pneumonia prevention in infants. Suctioning the mouth and throat of babies with meconium-stained amniotic fluid decreases the rate of aspiration pneumonia [19]. Environmental measures for pneumonia prevention include reduction of indoor air pollution by encouraging good hygiene in crowded homes and smoking cessation that reduces risks of pneumonia infections among children and adults. Since the bacteria and viruses can also be spread to your hands and then to your mouth, it is important to wash hands with soap when around a person with pneumonia infection.

Appropriate antibiotics are used for treatment of bacterial pneumonia. Pneumonia treatment depends on the underlying cause of the pneumonia infection. Appropriate antibiotics are used for treatment of bacterial pneumonia. Effective and timely treatment together with better diagnostic tools and education prevents antibiotic resistance [20]. According to Wardlaw et al. [5], treatment alone could save at least 600,000 children's lives annually at a cost of US $600 million if antibiotic treatment is universally delivered to children with pneumonia. Amoxicillin is recommended as a suitable alternative because of its proven efficacy against S. pneumoniae and severe pneumonia cases should be hospitalized.

Vaccines are effective in reduction of the number of new cases and severity of the disease [21–23]. Childhood pneumonia is preventable through immunization with the effective two vaccines: Hib conjugate vaccine (HibCV) against the Haemophilus influenzae type b (Hib) and pneumococcal conjugate vaccine (PCV) against pneumococcus [24]. PCVs have additional protective qualities that enhance their use as they may reduce nasopharyngeal acquisition of vaccine-specific serotypes of S. pneumoniae, and this in turn reduces the incidence of pneumococcal pneumonia among nonvaccinated individuals [25, 26]. This is referred to as indirect or herd immunity. In this study, we focus on treatment and vaccination of S. pneumoniae among children with PCVs.

Mathematical models of infectious diseases have been recognized as powerful tools that can provide important insights into our understanding of epidemiological processes, the course of infection within a host, the transmission dynamics in a host population, and formulation or implementation of disease control programs [27, 28]. Compartmental mathematical models involving vaccination strategy for infectious disease control have been considered in [29, 30]. Greenhalgh et al. [31] and Lamb et al. [32] modeled the transmission of pneumonia among young children to explore the relationship between sequence types and serotypes. Other epidemic models to study pneumonia have been considered (see, e.g., [2, 33–35] and the references therein).

In this study, a deterministic compartmental model to investigate the effect of treatment and vaccination against S. pneumoniae transmission dynamics among children less than five years is formulated. The population studied is divided into a set of distinct compartments according to the disease status. The vaccination strategy consists of vaccination of a proportion of the newborn babies.

This paper is structured as follows. In Section 2, we formulate the model based on the assumptions and definitions of variables and parameters. In Section 3, the pneumonia model with treatment intervention is studied for its boundedness and positivity of solutions and equilibrium points and their stability. In Section 4, the model is extended to investigate the effect of treatment and vaccination interventions combined on the spread of pneumonia. In Section 5, sensitivity analysis of the effective reproduction number $\mathcal{R}_e$ is done. Numerical simulation of the model is carried out in Section 6. Finally, in Section 7, we discuss the results and make a conclusion.

2. Formulation of the Model

The model consists of four compartments categorizing individuals based on their status with respect to the disease. The assumptions and definitions of variables and parameters are given in Sections 2.1 and 2.2, respectively.

2.1. Assumptions

(1) The model assumes a homogeneous mixing of individuals in the population where all individuals have equal likelihood of catching the infection if they are exposed to the disease.

(2) All recovered individuals clear the bacteria from the body and thus do not participate in transmitting the disease.

(3) Newborns are given additional dose of vaccine to elicit booster optimal levels of response.

(4) All treated individuals get vaccinated after completing the dose.

(5) Vaccinated children do not evolve to the susceptible population because of booster vaccine doses.

2.2. Variables and Parameters.
The model variables and parameter definitions represented are given as follows:

$S(t)$: susceptible individuals who are at risk of acquiring pneumonia infection at time t.

$C(t)$: carrier individuals who carry the pneumonia bacteria and can transmit the infection at time t.

$I(t)$: infective individuals capable of transmitting the infection to individuals at risk at time t.

$R(t)$: recovered individuals who have been treated of pneumonia at time t.

$V(t)$: vaccinated individuals at time t.

μ: per capita natural mortality rate of individuals.

Λ: per capita recruitment rate into susceptible population.

θ: proportion of susceptible individuals that joins the carriers.

σ: per capita disease induced mortality rate.

β: per capita recovery rate of carriers.

α: force of infection of susceptible individuals.

τ: per capita recovery rate of infected individuals.

π: rate at which carriers develop symptoms.

η: rate at which treated individuals become susceptible.

γ: rate at which susceptible individuals get vaccinated.

ϕ: rate at which treated individuals are vaccinated.

ω: transmission coefficient for the carrier subgroup.

δ: rate of transmission.

p: probability that a contact is efficient enough to cause infection.

κ: rate of contact.

Based on assumptions and definitions of variables and parameters mentioned above, the following system of ordinary equations is obtained.

3. Pneumonia Model under Treatment

The population dynamics of the pneumonia model with treatment intervention is given by the following system of four ordinary nonlinear differential equations:

$$\frac{dS}{dt} = \Lambda - (\alpha + \mu) S + \eta R,$$

$$\frac{dC}{dt} = \alpha\theta S - (\mu + \beta + \pi) C,$$

$$\frac{dI}{dt} = \alpha (1 - \theta) S + \pi C - (\tau + \mu + \sigma) I, \quad (1)$$

$$\frac{dR}{dt} = \beta C + \tau I - (\mu + \eta) R,$$

together with

$$\frac{dN}{dt} = \Lambda - \mu N - \sigma I. \quad (2)$$

The initial conditions are $S(0) = S_0$, $C(0) = C_0$, $I(0) = I_0$, $R(0) = R_0$, $N(0) = N_0$, and the force of infection is

$$\alpha = \delta \left(\frac{I + \omega C}{N} \right), \quad \text{where } \delta = \kappa p, \quad (3)$$

where δ is the transmission rate, κ is the contact rate, p is the probability that a contact is efficient enough to cause infection, and ω is the transmission coefficient for the carrier subgroup.

3.1. Positivity and Boundedness of the Solutions. The positivity of solutions describes the nonnegativity of solutions of system (1).

Lemma 1. *Let the initial population be*

$$\{S_0, C_0, I_0, R_0 \geq 0\} \in \Omega. \quad (4)$$

Then, the solution set $\{S, C, I, R\}$ of system (1) is positive for all $t > 0$.

Proof. From the first equation of system (1),

$$\frac{dS}{dt} = \Lambda - (\alpha + \mu) S + \eta R \geq - (\alpha + \mu) S. \quad (5)$$

This implies

$$\frac{dS}{dt} \geq - (\alpha + \mu) S. \quad (6)$$

By separation of variables, (6) is integrated to obtain

$$\ln S \geq - (\alpha + \mu) t + K, \quad (7)$$

where K is a constant of integration. Applying the initial conditions $S(0) = S_0$ to (7) gives

$$K = \ln S_0. \quad (8)$$

Hence,

$$S \geq S_0 e^{-(\alpha+\mu)t} \geq 0. \quad (9)$$

$\square$

Similarly, it can be shown that the other equations of system (1) are also positive for all $t > 0$. Thus, the solutions of the model are positive for all values of $t > 0$.

It is important to establish whether system (1) is well-posed and biologically meaningful. Now, we study the invariant region which describes the region in which the solution to system (1) makes biological sense. It is assumed that all the state variables and parameters of the model are nonnegative for all $t \geq 0$.

In the absence of pneumonia,

$$N(t) \leq \frac{\Lambda}{\mu}. \quad (10)$$

Inequality (10) is referred to as the threshold population level. Therefore, the feasible solution set of system (1) enters and remains in the region;

$$\Omega = (S, C, I, R) \in \mathbb{R}_+^4: \quad 0 \le S + C + I + R = N \le \frac{\Lambda}{\mu}, \quad (11)$$

where $\mathbb{R}_+^4$ denotes the nonnegative cone of $\mathbb{R}^4$ including its lower dimensional faces. In this case, whenever $N > \Lambda/\mu$, then $dN/dt \le 0$, implying that the host population reduces asymptotically to the carrying capacity. However, whenever $N \le \Lambda/\mu$, every solution with initial conditions in $\mathbb{R}_+^4$ remains in that region for $t > 0$. Thus, the region Ω is positively invariant, that is, for all values of t, the solution remains positive and thus the model is well-posed and biologically meaningful.

3.2. Equilibria of the Model. We analyze the model for pneumonia transmission to determine the basic reproduction number $\mathscr{R}_0$ and other threshold parameters for pneumonia dynamics. The equilibria of system (1) are obtained by setting the right-hand side of system (1) equal to zero. The disease-free equilibrium is given by

$$E_0 = \left(\frac{\Lambda}{\mu}, 0, 0, 0 \right). \quad (12)$$

Theorem 2. *There is a unique disease-free equilibrium E_0 for the model represented by system (1).*

Proof. This theorem is proved by substituting E_0 into system (1). The results show that all the derivatives are equal to zero, hence the disease-free equilibrium. $\qquad\square$

3.3. Basic Reproduction Number. To establish the linear stability of E_0, we use the next generation operator approach on system (1) to compute the basic reproduction number $\mathscr{R}_0$. This is determined using the approach by van den Driessche and Watmough [38]. For the notation of the matrices F and V, we have

$$F = \begin{bmatrix} \delta\omega\theta & \delta\theta \\ (1-\theta)\omega\delta & (1-\theta)\delta \end{bmatrix},$$

$$V = \begin{bmatrix} \mu + \beta + \pi & 0 \\ -\pi & \tau + \mu + \sigma \end{bmatrix}, \quad (13)$$

where $k_1 = \mu + \beta + \pi$ and $k_2 = \tau + \mu + \sigma$. This gives

$$V = \begin{bmatrix} k_1 & 0 \\ -\pi & k_2 \end{bmatrix}, \quad (14)$$

and, thus,

$$V^{-1} = \frac{1}{k_1 k_2} \begin{bmatrix} k_2 & 0 \\ \pi & k_1 \end{bmatrix},$$

$$FV^{-1} = \begin{bmatrix} \dfrac{\delta\theta(\omega k_2 + \pi)}{k_1 k_2} & \dfrac{\delta k_1 \theta}{k_1 k_2} \\ \dfrac{\delta(1-\theta)(\omega k_2 + \pi)}{k_1 k_2} & \dfrac{\delta k_1(1-\theta)}{k_1 k_2} \end{bmatrix}. \quad (15)$$

Thus, the eigenvalues for the matrix FV^{-1} are

$$\xi = 0,$$

$$\xi = \frac{\delta}{k_1 k_2}(1-\theta)k_1 + \theta(\omega k_2 + \pi). \quad (16)$$

The spectral radius is given by $\xi(FV^{-1}) = (\delta/k_1 k_2)((1-\theta)k_1 + \theta(\omega k_2 + \pi))$, which gives the basic reproduction number

$$\mathscr{R}_0 = \frac{\delta}{k_1 k_2}(k_1(1-\theta) + \theta(\omega k_2 + \pi)). \quad (17)$$

The endemic equilibrium E_e is defined as a steady state solution for system (1). This occurs when there is a persistence of the disease. Hence, the endemic equilibrium $E_e = (S, C, I, R)$ is determined by setting the right-hand side of system (1) equal to zero as follows:

$$\Lambda - (\alpha + \mu)S + \eta R = 0,$$

$$\alpha\theta S - k_1 C = 0,$$

$$\alpha(1-\theta)S + \pi C - k_2 I = 0, \quad (18)$$

$$\beta C + \tau I - (\mu + \eta)R = 0,$$

together with

$$\Lambda - \mu N - \sigma I = 0, \quad (19)$$

$$\alpha = \delta\left(\frac{I + \omega C}{N}\right). \quad (20)$$

From the second equation of system (18)

$$\alpha = \frac{k_1 C}{\theta S}. \quad (21)$$

Plugging α in (21) into the first and third equations of system (18) gives

$$\Lambda - \frac{k_1 C}{\theta} - \mu S + \eta R = 0, \quad (22)$$

$$\frac{k_1(1-\theta)}{\theta}C + \pi C - k_2 I = 0. \quad (23)$$

Substituting for α in (20) into the second equation of system (18) gives

$$\delta\theta IS + \delta\theta\omega SC - k_1 CN = 0. \quad (24)$$

From the fourth equation of system (18), we have

$$R = \frac{\beta C}{\mu + \eta} + \frac{\tau I}{\mu + \eta}. \quad (25)$$

Substituting for R in (22) gives

$$\Lambda + \left(\frac{\eta\beta}{\mu + \eta} - \frac{k_1}{\theta}\right)C + \frac{\eta\tau I}{\mu + \eta} - \mu S = 0. \quad (26)$$

From (23), we obtain

$$C = \frac{\theta k_2 I}{k_1 (1 - \theta) + \pi \theta}. \tag{27}$$

Substituting for C in (24) gives

$$\delta \theta I S + \delta \theta \omega S \frac{\theta k_2 I}{k_1 (1 - \theta) + \pi \theta} - k_1 \frac{\theta k_2 I}{k_1 (1 - \theta) + \pi \theta} N \tag{28}$$
$$= 0.$$

Solving (28) yields

$$I = 0,$$

$$\Lambda + \left(\frac{\eta \theta \beta - k_1 (\mu + \eta)}{\theta (\mu + \eta)} \right) \left(\frac{\theta k_2 I}{k_1 (1 - \theta) + \pi \theta} \right) + \frac{\eta \tau I}{\mu + \eta} - \frac{(\Lambda - \sigma I)}{\mathcal{R}_0} = 0 \implies$$

$$\tag{32}$$

$$I = \frac{\Lambda (\mu + \eta) (k_1 (1 - \theta) + \pi \theta) (\mathcal{R}_0 - 1)}{\mathcal{R}_0 [k_1 k_2 (\mu + \eta) - \eta \theta \beta - \eta \tau (k_1 (1 - \theta) + \pi \theta)] - \sigma (\mu + \eta) (k_1 (1 - \theta) + \pi \theta)}.$$

Finally, we have

$$C = \frac{\theta k_2 \Lambda (\mu + \eta) (\mathcal{R}_0 - 1)}{\mathcal{R}_0 [k_1 k_2 (\mu + \eta) - \eta \theta \beta - \eta \tau (k_1 (1 - \theta) + \pi \theta)] - \sigma (\mu + \eta) (k_1 (1 - \theta) + \pi \theta)},$$

$$\tag{33}$$

$$R = \frac{(\beta \theta k_2 + \tau (k_1 (1 - \theta) + \pi \theta)) \Lambda (\mathcal{R}_0 - 1)}{\mathcal{R}_0 [k_1 k_2 (\mu + \eta) - \eta \theta \beta - \eta \tau (k_1 (1 - \theta) + \pi \theta)] - \sigma (\mu + \eta) (k_1 (1 - \theta) + \pi \theta)}.$$

Therefore, we have the endemic equilibrium $E_e = (S, I, C, R)$, where

$$S = \frac{N}{\mathcal{R}_0},$$

$$C = \frac{\theta k_2 \Lambda (\mu + \eta) (\mathcal{R}_0 - 1)}{\mathcal{R}_0 [k_1 k_2 (\mu + \eta) - \eta \theta \beta - \eta \tau (k_1 (1 - \theta) + \pi \theta)] - \sigma (\mu + \eta) (k_1 (1 - \theta) + \pi \theta)},$$

$$\tag{34}$$

$$I = \frac{\Lambda (\mu + \eta) (k_1 (1 - \theta) + \pi \theta) (\mathcal{R}_0 - 1)}{\mathcal{R}_0 [k_1 k_2 (\mu + \eta) - \eta \theta \beta - \eta \tau (k_1 (1 - \theta) + \pi \theta)] - \sigma (\mu + \eta) (k_1 (1 - \theta) + \pi \theta)},$$

$$R = \frac{(\beta \theta k_2 + \tau (k_1 (1 - \theta) + \pi \theta)) \Lambda (\mathcal{R}_0 - 1)}{\mathcal{R}_0 [k_1 k_2 (\mu + \eta) - \eta \theta \beta - \eta \tau (k_1 (1 - \theta) + \pi \theta)] - \sigma (\mu + \eta) (k_1 (1 - \theta) + \pi \theta)}.$$

Lemma 3. For $\mathcal{R}_0 > 1$, a unique endemic equilibrium E_e exists and there is no endemic equilibrium otherwise.

For the disease to be endemic, $dI/dt > 0$ and $dC/dt > 0$; that is,

$$\text{or } S = \frac{k_1 k_2 N}{\delta (k_1 (1 - \theta) + \theta (\omega k_2 + \pi))}. \tag{29}$$

But

$$\mathcal{R}_0 = \frac{\delta}{k_1 k_2} (k_1 (1 - \theta) + \theta (\omega k_2 + \pi)). \tag{30}$$

Therefore,

$$S = \frac{N}{\mathcal{R}_0}. \tag{31}$$

Substituting for C and S in (26) gives

$$\theta \delta (I + \omega C) \frac{S}{N} - k_1 C > 0, \tag{35}$$

$$(1 - \theta) \delta (I + \omega C) \frac{S}{N} - k_2 I + \pi C > 0. \tag{36}$$

From inequality (35), we have

$$k_1 C < \theta\delta\left(I + \omega C\right)\frac{S}{N}. \tag{37}$$

Using the fact $S/N < 1$, we obtain

$$C < \frac{\theta\delta I}{k_1 - \delta\theta\omega}. \tag{38}$$

From inequality (36), we have

$$I < \frac{(1-\theta)\,\delta I + (1-\theta)\,\delta\omega C + \pi C}{k_2}, \tag{39}$$

and, substituting for C in inequality (38), inequality (39) yields

$$I < \frac{(1-\theta)\,\delta I + (1-\theta)\,\delta\omega\left(\delta\theta I/\left(k_1 - \delta\theta\omega\right)\right) + \pi\left(\delta\theta I/\left(k_1 - \delta\theta\omega\right)\right)}{k_2},$$

$$I < \frac{(1-\theta)\,\delta\left(k_1 - \delta\theta\omega\right)I + (1-\theta)\,\delta^2\theta\omega I + \pi\delta\theta I}{k_2\left(k_1 - \delta\theta\omega\right)}, \tag{40}$$

$$k_1 k_2 - k_2\delta\theta\omega < \delta k_1\left(1 - \theta\right) + \pi\delta\theta,$$

$$1 < \frac{\delta}{k_1 k_2}\left(\left(1 - \theta\right)k_1 + \theta\left(\omega k_2 + \pi\right)\right) = \mathcal{R}_0.$$

Thus, a unique endemic equilibrium exists when $\mathcal{R}_0 > 1$.

3.4. Local and Global Stability of the Disease-Free Equilibrium

Lemma 4. *The disease-free equilibrium E_0 of system (1) is locally asymptotically stable whenever $\mathcal{R}_0 < 1$ and unstable whenever $\mathcal{R}_0 > 1$.*

The threshold quantity $\mathcal{R}_0$ is a measure of the number of secondary infections caused by a single individual in his/her entire lifetime as an infective [39]. It is an important parameter that plays a big role in the control of the disease. The reduction of the disease from the population targets the parameters that will bring its value to less than unity.

When the reproduction number is less than unity, then the disease-free equilibrium is locally asymptotically stable, and thus there is a possibility that the disease will be wiped out of the population.

The Jacobian matrix for system (1) is given by

$$J$$

$$= \begin{bmatrix} -(\alpha + \mu) & 0 & 0 & \eta \\ \alpha\theta & -(\mu + \beta + \pi) & 0 & 0 \\ \alpha(1-\theta) & \pi & -(\tau + \mu + \sigma) & 0 \\ 0 & \beta & \tau & -(\mu + \eta) \end{bmatrix}. \tag{41}$$

Evaluating the Jacobian matrix (41) at the disease-free equilibrium E_0 gives

$$J\left(E_0\right) = \begin{bmatrix} -\mu & 0 & 0 & \eta \\ 0 & -k_1 & 0 & 0 \\ 0 & \pi & -k_2 & 0 \\ 0 & \beta & \tau & -(\mu + \eta) \end{bmatrix}. \tag{42}$$

The disease-free equilibrium E_0 is asymptotically stable if and only if the trace$(J_{E_0}) < 0$ and the det$(J_{E_0}) > 0$.

Thus, from the Jacobian matrix (42),

$$\text{trace}\left(J_{E_0}\right) = -\mu - h_1 - h_2 - \mu - \eta$$

$$= -\left(2\mu + h_1 + h_2 + \eta\right) < 0,$$

$$\det\left(J_{E_0}\right) = -\mu\left(-k_1 k_2\left(\mu + \eta\right)\right) + \eta \times 0 \tag{43}$$

$$= \mu k_1 k_2\left(\mu + \eta\right) > 0.$$

Since the parameters μ, η, k_1, and k_2 are all positive, then $-(2\mu + k_1 + k_2 + \eta) < 0$. Therefore trace$(J_{E_0}) < 0$.

On the other hand, $\mathcal{R}_0$ can never be negative and the numerator $((1-\theta)k_1 + \theta(\omega k_1 + \pi))$ is positive; that is, $k_1 k_2 > 0$. This implies that det$(J_{E_0}) > 0$ since $\mu(\mu + \eta) > 0$ and $k_1 k_2 > 0$. Thus,

$$\mathcal{R}_0 = \frac{\delta}{k_1 k_2}\left(\left(1-\theta\right)k_1 + \theta\left(\omega k_2 + \pi\right)\right) < 1. \tag{44}$$

The conditions trace$(J_{E_0}) < 0$ and det$(J_{E_0}) > 0$ above imply that E_0 is locally asymptotically stable whenever $\mathcal{R}_0 < 1$

Theorem 5. *The disease-free equilibrium is globally asymptotically stable in Ω if $\mathcal{R}_0 \leq 1$ and unstable if $\mathcal{R}_0 > 1$.*

Proof. Consider the Lyapunov function defined by $L = (1/k_1)C + (1/k_2)I$. Its derivative along the solutions to system (1) is

$$\frac{dL}{dt} = \frac{1}{k_1}\frac{dC}{dt} + \frac{1}{k_2}\frac{dI}{dt} = \frac{1}{k_1}\left(\alpha\theta S - k_1 C\right)$$

$$+ \frac{1}{k_2}\left(\alpha\left(1-\theta\right)S + \pi C - k_2 I\right) = \frac{1}{k_1 k_2}\left(k_2\alpha\theta S\right.$$

$$- k_1 k_2 C + \frac{1}{k_1 k_2} (k_1 \alpha (1 - \theta) S + k_1 \pi C - k_1 k_2 I)$$

$$= \frac{1}{k_1 k_2} (k_2 \alpha \theta S - k_1 k_2 C + k_1 \alpha (1 - \theta) S + k_2 \pi C$$

$$- k_1 k_2 I), \tag{45}$$

but

$$\alpha = \delta \left(\frac{I + \omega C}{N} \right) \Longrightarrow$$

$$\frac{dL}{dt} \le \frac{\delta}{k_1 k_2} ((1 - \theta) k_1 + \theta (\omega k_2 + \pi) - 1) \frac{S}{N} (C + I) \tag{46}$$

$$\le (\mathcal{R}_0 - 1)(C + I) \frac{S}{N} \le 0 \quad \text{if } \mathcal{R}_0 \le 1.$$

$\square$

Thus, $\mathcal{R}_0 < 1$ is necessary and sufficient for disease elimination. All the model parameters are positive, so that $dL/dt \le 0$ if $\mathcal{R}_0 \le 1$ with $dL/dt = 0$ if and only if $I = C = 0$. Hence, L is a Lyapunov function on Ω and the largest compact invariant set in $\{(S, C, I, R) \in \Omega : dL/dt \le 0\}$ is the singleton $\{E_0\}$. Therefore, by LaSalle's invariance principle [40], every solution to system (1), with initial conditions in Ω, approaches E_0 as $t \to \infty$ if $\mathcal{R}_0 < 1$. Hence the disease-free equilibrium E_0 of the pneumonia model with treatment intervention is globally asymptotically stable.

3.5. Local and Global Stability Analysis of the Endemic Equilibrium.
We study the local stability of the endemic equilibrium by applying the Routh-Hurwitz criterion.

Theorem 6. If $\mathcal{R}_0 > 1$, the endemic equilibrium E_e of system (1) is locally asymptotically stable in Ω.

Proof. We evaluate the Jacobian matrix (41) at the endemic equilibrium to obtain

$$J(E_e) = \begin{bmatrix} -\overline{\alpha} - \mu & 0 & 0 & \eta \\ \overline{\alpha}\theta & -k_1 & 0 & 0 \\ \overline{\alpha}(1 - \theta) & \pi & -k_2 & 0 \\ 0 & \beta & \tau & -\mu - \eta \end{bmatrix}, \tag{47}$$

where $\overline{\alpha}$ is defined as the force infection at the endemic equilibrium. We obtain a characteristic equation $P(\xi) = |(\xi I - J(E_e)|$, where I is a 4×4 unit matrix

$$P(\xi)$$

$$= \det \begin{bmatrix} \xi + \overline{\alpha} + \mu & 0 & 0 & \eta \\ -\overline{\alpha}\theta & \xi + k_1 & 0 & 0 \\ -\overline{\alpha}(1 - \theta) & -\pi & \xi + k_2 & 0 \\ 0 & -\beta & -\tau & \xi + (\mu + \eta) \end{bmatrix}. \tag{48}$$

Thus, the characteristic equation becomes

$$P(\xi) = \xi^4 + a_1 \xi^3 + a_2 \xi^2 + a_3 \xi + a_4, \tag{49}$$

where

$$a_1 = 2\mu + \eta + k_1 + k_2 + \overline{\alpha},$$

$$a_2 = (\eta + \mu)(k_1 + k_2 + \overline{\alpha} + \mu) + k_1 k_2$$

$$+ (\overline{\alpha} + \mu)(k_1 + k_2),$$

$$a_3 = k_1 k_2 (\eta + \mu) + (k_1 + k_2)(\overline{\alpha} + \mu)(\eta + \mu) \tag{50}$$

$$+ k_1 k_2 (\overline{\alpha} + \mu) + \eta \tau \overline{\alpha} (1 - \theta) + \eta \beta \overline{\alpha} \theta,$$

$$a_4 = k_1 k_2 (\overline{\alpha} + \mu)(\eta + \mu) + \eta \overline{\alpha} \theta \pi \tau + \eta k_1 \overline{\alpha} \tau (1 - \theta)$$

$$+ \eta \beta \overline{\alpha} \theta k_1.$$

Thus, from Routh-Hurwitz criterion [41] we have the matrix

$$\begin{bmatrix} 1 & a_2 & a_4 & \xi^4 \\ a_1 & a_3 & 0 & \xi^3 \\ a_2 - \dfrac{a_3}{a_1} & a_4 & 0 & \xi^2 \\ a_3 - \dfrac{a_1 a_4}{a_2 - a_3/a_1} & 0 & 0 & \xi \\ a_1 & 0 & 0 & 1 \end{bmatrix}. \tag{51}$$

According to the Routh-Hurwitz criterion, for $\mathcal{R}_0 > 1$, the endemic equilibrium E_e is locally asymptotically stable if

$$a_1 > 0,$$

$$\left(a_2 - \frac{a_3}{a_1} \right) > 0,$$

$$\left(a_3 - \frac{a_1 a_4}{a_2 - a_3/a_1} \right) > 0, \tag{52}$$

$$a_4 > 0.$$

$\square$

The global stability of the endemic equilibrium E_e is analyzed using the following constructed Lyapunov function.

Theorem 7. If $\mathcal{R}_0 \ge 1$, the endemic equilibrium E_e of system (1) is globally asymptotically stable.

Proof. Let the Lyapunov function be

$$L(S_e, C_e, I_e, R_e) = \left(S - S_e - S_e \log \left(\frac{S_e}{S} \right) \right)$$

$$+ (C - C_e - C_e \log (C_e C))$$

$$+ \left(I - I_e - I_e \log \left(\frac{I_e}{I} \right) \right) \tag{53}$$

$$+ \left(R - R_e - R_e \log \left(\frac{R_e}{R} \right) \right),$$

$$\frac{dL}{dt} = \left(\frac{S - S_e}{S}\right)\frac{dS}{dt} + \left(\frac{C - C_e}{C}\right)\frac{dC}{dt} + \left(\frac{I - I_e}{I}\right)\frac{dI}{dt}$$

$$+ \left(\frac{R - R_e}{R}\right)\frac{dR}{dt} = \left(\frac{S - S_e}{S}\right)(\Lambda - \alpha S - \mu S + \eta R)$$

$$+ \left(\frac{C - C_e}{C}\right)(\alpha\theta S - k_1 C) + \left(\frac{I - I_e}{I}\right)$$

$$\cdot (\alpha(1 - \theta)S + \pi C - k_2 I) + \left(\frac{R - R_e}{R}\right)$$

$$\cdot (\beta C + \tau I - (\mu + \eta)R) = \left(\frac{S - S_e}{S}\right)$$

$$\cdot (\Lambda - \alpha(S - S_e) - \mu(S - S_e) + \eta(R - R_e))$$

$$+ \left(\frac{C - C_e}{C}\right)(\alpha\theta(S - S_e) - k_1(C - C_e))$$

$$+ \left(\frac{I - I_e}{I}\right)$$

$$\cdot (\alpha(1 - \theta)(S - S_e) + \pi(C - C_e) - k_2(I - I_e))$$

$$+ \left(\frac{R - R_e}{R}\right) \tag{54}$$

$$\cdot (\beta(C - C_e) + \tau(I - I_e) - (\mu + \eta)(R - R_e))$$

$$= \frac{(S - S_e)^2}{S}(-\alpha - \mu) - k_1\frac{(C - C_e)^2}{C} - k_2$$

$$\cdot \frac{(I - I_e)^2}{I} - (\mu + \eta)\frac{(R - R_e)^2}{R} + \Lambda - \frac{\Lambda S_e}{S} + \eta R$$

$$- \frac{\eta R S_e}{S} - \eta R_e + \frac{R_e S_e}{S} + \alpha\theta S - \frac{\alpha\theta C_e S}{C} - \alpha\theta S_e$$

$$+ \frac{\alpha\theta C_e S_e}{C} + \alpha(1 - \theta)S - \alpha(1 - \theta)S_e$$

$$- \frac{\alpha(1 - \theta)I_e S}{I} + \frac{\alpha(1 - \theta)I_e S_e}{I} + \pi C - \pi C_e$$

$$- \frac{\pi I_e C}{I} + \frac{\pi I_e C_e}{I} + \beta C - \beta C_e - \frac{\beta R_e C}{R} + \frac{\beta R_e C_e}{R}$$

$$+ \tau I - \tau I_e - \frac{\tau R_e I}{R} + \frac{\tau R_e I_e}{R};$$

thus collecting positive terms together and negative terms together from the above

$$\frac{dL}{dt} = P - Q, \tag{55}$$

where

$$P = \Lambda + \eta R + \frac{\eta R_e S_e}{S} + \alpha\theta S + \frac{\alpha\theta C_e S_e}{C} + \alpha(1 - \theta)S$$

$$+ \frac{\alpha(1 - \theta)I_e S_e}{I}\pi C + \frac{\pi I - eC_e}{I} + \beta C$$

$$+ \frac{\beta R_e C_e}{R} + \tau I + \frac{\tau R_e I_e}{R},$$

$$Q = (\alpha + \mu)\frac{(S - S_e)^2}{S} - k_1\frac{(C - C_e)^2}{C} - k_2\frac{(I - I_e)^2}{I}$$

$$- (\mu + \eta)\frac{(R - R_e)^2}{R} - \eta R_e - \frac{\Lambda S_e}{S} - \frac{\eta R S_e}{S}$$

$$- \frac{\alpha\theta C_e S}{C} - \alpha\theta S_e - \alpha(1 - \theta)S_e$$

$$- \frac{\alpha(1 - \theta)I_e S}{I} - \pi C_e - \frac{\pi I_e C}{I} - \beta C_e$$

$$- \frac{\beta R_e C}{R} - \tau I_e - \frac{\tau R_e I}{R}. \tag{56}$$

Thus if $P < Q$, then we obtain that $dL/dt \leq 0$, noting that $dL/dt = 0$ if and only if $S = S_e$, $C = C_e$, $I = I_e$, $R = R_e$. Therefore, the largest compact invariant set in $\{(S_e, C_e, I_e, R_e) \in \Omega : dL/dt = 0\}$ is the singleton $\{E_e\}$, where E_e is the endemic equilibrium of system (1).

Thus, by LaSalle's invariance principle [40], it implies that E_e is globally asymptotically stable in Ω if $P < Q$. □

4. Pneumonia Model under Treatment and Vaccination Interventions

In the this section, the model formulated in Section 3 is extended to investigate the impact of treatment and vaccination interventions on the transmission dynamics of pneumonia. The dynamics of the modified pneumonia model is described by the following system of five ordinary nonlinear differential equations:

$$\frac{dS}{dt} = \Lambda - (\alpha + \mu + \gamma)S,$$

$$\frac{dC}{dt} = \alpha\theta S - (\mu + \beta + \pi)C,$$

$$\frac{dI}{dt} = \alpha(1 - \theta)S + \pi C - (\tau + \mu + \sigma)I, \tag{57}$$

$$\frac{dR}{dt} = \beta C + \tau I - (\mu + \phi)R,$$

$$\frac{dV}{dt} = \gamma S - \mu V + \phi R,$$

with initial conditions $S(0) = S_0$, $C(0) = C_0$, $I(0) = I_0$, $R(0) = R_0$, $V(0) = V_0$, $N(0) = N_0$. The total population size is given by $N(t) = S(t) + C(t) + C(t) + R(t) + V(t)$ and is changing at the rate

$$\frac{dN}{dt} = \Lambda - \mu N - \sigma I. \tag{58}$$

The susceptible individuals become infected at rate α which is the force of infection; that is, the number of infected individuals produced by adequate contact and is given by

$$\alpha = \delta\left(\frac{I + \omega C}{N}\right), \tag{59}$$

where $\delta = \kappa p$, δ is the transmission rate, k is the rate of contact, and p is the probability that a contact is efficient enough to cause infection.

4.1. Analysis of the Model.
The equilibria of system (57) is obtained by setting the right-hand side of the equations to be equal to zero. The disease-free equilibrium E_0 is given by

$$\left(\frac{\Lambda}{(\mu+\gamma)}, 0, 0, 0, \frac{\Lambda\gamma}{\mu(\mu+\gamma)}\right). \tag{60}$$

4.2. Effective Reproduction Number.
To establish the stability of E_0, we use the next-generation operator approach on system (57) to compute the effective reproduction number $\mathcal{R}_e$. Using the notation of the matrices F and V, we have

$$F = \begin{bmatrix} \delta\omega S & \delta\theta S \\ (1-\theta)\omega\delta S & (1-\theta)\delta S \end{bmatrix},$$

$$V = \begin{bmatrix} \mu+\beta+\pi & 0 \\ -\pi & \tau+\mu+\sigma \end{bmatrix}. \tag{61}$$

Evaluating F at the disease-free equilibrium we to obtain

$$F = \begin{bmatrix} \delta\omega\theta\dfrac{\Lambda}{(\mu+\gamma)} & \delta\theta\dfrac{\Lambda}{(\mu+\gamma)} \\ (1-\theta)\omega\delta\dfrac{\Lambda}{(\mu+\gamma)} & (1-\theta)\delta\dfrac{\Lambda}{(\mu+\gamma)} \end{bmatrix}. \tag{62}$$

Let $k_1 = \mu+\beta+\pi$ and $k_2 = \tau+\mu+\sigma$; then

$$V = \begin{bmatrix} k_1 & 0 \\ -\pi & k_2 \end{bmatrix}, \tag{63}$$

and, thus,

$$V^{-1} = \frac{1}{k_1 k_2}\begin{bmatrix} k_2 & 0 \\ \pi & k_1 \end{bmatrix}. \tag{64}$$

Now we have

$$FV^{-1} = \frac{1}{k_1 k_2}$$

$$\cdot \begin{bmatrix} \delta\omega\theta\dfrac{\Lambda}{(\mu+\gamma)} & \delta\theta\dfrac{\Lambda}{(\mu+\gamma)} \\ (1-\theta)\omega\delta\dfrac{\Lambda}{(\mu+\gamma)} & (1-\theta)\delta\dfrac{\Lambda}{(\mu+\gamma)} \end{bmatrix}\begin{bmatrix} k_2 & 0 \\ \pi & k_1 \end{bmatrix} \tag{65}$$

$$= \begin{bmatrix} \dfrac{\Lambda\delta\theta(\omega k_2+\pi)}{k_1 k_2(\mu+\gamma)} & \dfrac{\lambda\psi k_1\theta}{k_1 k_2(\mu+\gamma)} \\ \dfrac{\Lambda\psi(1-\theta)(\omega k_2+\pi)}{k_1 k_2(\mu+\gamma)} & \dfrac{\Lambda\delta k_1(1-\theta)}{k_1 k_2(\mu+\gamma)} \end{bmatrix}.$$

The eigenvalues for the matrix FV^{-1} are given by

$$\xi = 0,$$
$$\xi = \frac{\Lambda\delta}{k_1 k_2(\mu+\gamma)}((1-\theta)k_1 + \theta(\omega k_2+\pi)). \tag{66}$$

The spectral radius is given by $\xi(FV^{-1}) = (\Lambda\delta/k_1 k_2(\mu+\gamma))((1-\theta)k_1 + \theta(\omega k_2+\pi))$, which gives the effective reproduction number as

$$\mathcal{R}_e = \frac{\Lambda\delta}{k_1 k_2(\mu+\gamma)}((1-\theta)k_1 + \theta(\omega k_2+\pi)), \tag{67}$$

but

$$\mathcal{R}_0 = \frac{\delta}{k_1 k_2}((1-\theta)k_1 + \theta(\omega k_2+\pi)). \tag{68}$$

Therefore,

$$\mathcal{R}_e = \frac{\Lambda\mathcal{R}_0}{(\mu+\gamma)}. \tag{69}$$

$\mathcal{R}_e$ is referred to as the effective reproduction number rather than the basic reproduction number because vaccination and treatment have been included in the model [32]. It is defined as the expected number of secondary cases caused by a typical infected individual entering an entirely susceptible population at equilibrium.

4.3. Local Stability of the Disease-Free Equilibrium.
The Jacobian matrix for the system is given by

$$J = \begin{bmatrix} -(\alpha+\mu+\gamma) & 0 & 0 & 0 & 0 \\ \alpha\theta & -(\mu+\beta+\pi) & 0 & 0 & 0 \\ \alpha(1-\theta) & \pi & -(\tau+\mu+\sigma) & 0 & 0 \\ 0 & \beta & \tau & -(\mu+\phi) & 0 \\ \gamma & 0 & 0 & \phi & -\mu \end{bmatrix}. \tag{70}$$

The disease-free equilibrium point E_0 is discussed by examining the Jacobian matrix (70) at the steady point E_0. Now, at the disease-free equilibrium

$$\left(\frac{\Lambda}{(\mu + \gamma)}, 0, 0, 0, \frac{\Lambda\gamma}{\mu(\mu + \gamma)} \right), \tag{71}$$

the Jacobian matrix is given by

$$J = \begin{bmatrix} -(\mu + \gamma) & 0 & 0 & 0 & 0 \\ 0 & -k_1 & 0 & 0 & 0 \\ 0 & \pi & -k_2 & 0 & 0 \\ 0 & \beta & \tau & -(\mu + \phi) & 0 \\ \gamma & 0 & 0 & \phi & -\mu \end{bmatrix}. \tag{72}$$

For stability of the disease-free equilibrium, it is required that the $\text{trace}(J_{E_0}) < 0$ and the $\det(J_{E_0}) > 0$. Thus, from the Jacobian matrix (72), it is clearly seen that

$$\text{trace}\left(J_{E_0}\right) = -\left[(\mu + \gamma) + k_1 + k_2 + (\mu + \phi) + \mu\right]$$
$$= -(\gamma + \beta + \pi + \tau + \delta + \phi + 5\mu) < 0. \tag{73}$$

The determinant of the Jacobian matrix is also given by

$$\det\left(J_{E_0}\right)$$

$$= \det \begin{bmatrix} -(\mu + \gamma) & 0 & 0 & 0 & 0 \\ 0 & -k_1 & 0 & 0 & 0 \\ 0 & \pi & -k_2 & 0 & 0 \\ 0 & \beta & \tau & -(\mu + \phi) & 0 \\ \gamma & 0 & 0 & \phi & -\mu \end{bmatrix} \tag{74}$$

$$= -(\mu + \gamma) \det \begin{bmatrix} -k_1 & 0 & 0 & 0 \\ \pi & -k_2 & 0 & 0 \\ \beta & \tau & (\mu + \phi) & 0 \\ 0 & 0 & \phi & -\mu \end{bmatrix}$$

$$= k_1 k_2 \mu (\mu + \phi)(\mu + \gamma) > 0.$$

Therefore, the disease-free equilibrium of the pneumonia model under treatment and vaccination interventions is

locally asymptotically stable. This is established by the fact that the $\text{trace}(J_{E_0}) < 0$ and the $\det(J_{E_0}) > 0$.

Proposition 8. $\mathscr{R}_e < \mathscr{R}_0$ for any given parameters.

Proof.

$$\mathscr{R}_e = \frac{\Lambda\mathscr{R}_0}{(\mu + \gamma)} \tag{75}$$

implies that

$$\mathscr{R}_e = \frac{\Lambda\mathscr{R}_0}{(\mu + \gamma)} < \mathscr{R}_0. \tag{76}$$

Thus $\mathscr{R}_e < \mathscr{R}_0$. $\square$

The above result leads us to the following theorem.

Theorem 9. *The disease-free equilibrium E_0 of the pneumonia model under treatment and vaccination interventions is locally asymptotically stable if $\mathscr{R}_e < 1$ and unstable if $\mathscr{R}_e \geq 1$.*

The proof of the theorem follows from the Jacobian matrix (72).

The endemic equilibrium E_e is defined as a steady state solution for system (57). This occurs when there is a persistence of the disease. Hence, $E_e = (S, C, I, R, V)$ can be determined as below. Consider system (57) with right-hand side equal to zero to obtain

$$\Lambda - (\alpha + \mu + \gamma) S = 0,$$
$$\alpha\theta S - (\mu + \beta + \pi) C = 0,$$
$$\alpha(1 - \theta) S + \pi C - (\tau + \mu + \sigma) I = 0, \tag{77}$$
$$\beta C + \tau I - (\mu + \phi) R = 0,$$
$$\gamma S - \mu V + \phi R = 0,$$

together with

$$\Lambda - \mu N - \sigma I = 0,$$
$$\alpha = \delta\left(\frac{I + \omega C}{N}\right). \tag{78}$$

Solving system (77) together with (78) gives the endemic equilibrium $E_e = (S, I, C, R)$, where

$$S = \frac{\Lambda\sigma k_1 N}{[(k_1 - \sigma\theta\omega)(\Lambda - (\mu + \sigma) N) + \omega\theta(\Lambda - (\mu + \sigma) N)] + (\mu + \sigma) N\sigma k_1},$$

$$C = \frac{\theta(\Lambda - (\mu + \gamma) N)}{k_1},$$

$$I = \frac{(k_1 - \sigma\theta\omega)(\Lambda - (\mu + \gamma) N)}{\sigma k_1},$$

$$R = \frac{\beta\theta\sigma(\Lambda - (\mu + \gamma) N) + \tau(k_1 - \sigma\theta\omega)(\Lambda - (\mu + \gamma) N)}{\sigma k_1(\mu + \phi)},$$

$$V = \frac{\gamma\Lambda\sigma k_1 N}{p} + \frac{\phi\beta\theta\sigma\left(\Lambda - (\mu + \gamma)N\right) + \tau\left(k_1 - \sigma\theta\omega\right)\left(\Lambda - (\mu + \gamma)N\right)}{(\mu + \phi)\sigma k_1},$$

(79)

where

$$k_1 = \mu + \beta + \pi,$$

$$k_2 = \tau + \mu + \sigma,$$

(80)

$$p = [(k_1 - \sigma\rho\omega)\left(\Lambda - (\mu + \sigma)N\right)$$

$$+ \omega\rho\left(\Lambda - (\mu + \sigma)N\right)] + (\mu + \sigma)N\sigma k_1.$$

J

$$= \begin{bmatrix} -(\alpha + \mu + \gamma) & 0 & 0 & 0 & 0 \\ \alpha\theta & -(\mu + \beta + \pi) & 0 & 0 & 0 \\ \alpha(1-\theta) & \pi & -(\tau + \mu + \sigma) & 0 & 0 \\ 0 & \beta & \tau & -(\mu + \phi) & 0 \\ \gamma & 0 & 0 & \phi & -\mu \end{bmatrix}.$$

(81)

4.4. Local Stability of the Endemic Equilibrium.

The Jacobian matrix for system (57) is given by

Evaluating the Jacobian matrix (81) at the endemic equilibrium gives

$$J_{E_e} = \begin{bmatrix} -(\alpha^* + \mu + \gamma) & 0 & 0 & 0 & 0 \\ \alpha^*\theta & -(\mu + \beta + \pi) & 0 & 0 & 0 \\ \alpha^*(1-\theta) & \pi & -(\tau + \mu + \sigma) & 0 & 0 \\ 0 & \beta & \tau & -(\mu + \phi) & 0 \\ \gamma & 0 & 0 & \phi & -\mu \end{bmatrix},$$

(82)

where

$$\alpha^* = \delta\left(\frac{I + \omega C}{N}\right).$$

(83)

We now obtain the characteristic equation $P = |(\xi)I - J(E_e))|$, where I is a 5×5 unit matrix.

$P(\xi)$

$$= \det \begin{bmatrix} \xi + k_1 & 0 & 0 & 0 & 0 \\ -\alpha^*\theta & \xi + k_2 & 0 & 0 & 0 \\ -\alpha^*(1-\theta) & -\pi & \xi + k_3 & 0 & 0 \\ 0 & -\beta & -\tau & \xi + k_4 & 0 \\ -\gamma & 0 & 0 & -\phi & \xi + \mu \end{bmatrix},$$

(84)

where

$$k_1 = \alpha^* + \mu + \gamma,$$

$$k_2 = \mu + \beta + \pi,$$

(85)

$$k_3 = \tau + \mu + \sigma,$$

$$k_4 = \mu + \theta.$$

Thus the characteristic equation becomes

$$P(\xi) = \xi^5 + a_1\xi^4 + a_2\xi^3 + a_3\xi^2 + a_4\xi + a_5,$$

(86)

where

$$a_1 = k_1 + k_2 + k_3 + k_4 + \mu,$$

$$a_2 = k_1 k_2 + k_2 k_3 + k_1 k_4 + k_1\mu + k_2 k_3 + k_2\mu + k_2 k_4$$
$$\quad + k_3\mu + k_3 k_4 + \theta,$$

$$a_3 = k_1 k_2 k_3 + k_1 k_2\mu + k_1 k_2 k_4 + k_1 k_3\mu + k_1 k_3 k_4$$

(87)

$$\quad + k_2 k_3\mu + k_4 k_3\mu + k_1\theta + \theta\mu + k_3\phi,$$

$$a_4 = k_1 k_2 k_3 k_4\mu + k_1 k_3 k_4\mu + k_1\theta\mu + k_1 k_3\phi + k_2 k_3\phi,$$

$$a_5 = k_1 k_2 k_3\phi.$$

The necessary and sufficient conditions for the local asymptotic stability of endemic equilibrium are that the Hurwitz determinants H_i are all positive for the Routh-Hurwitz criteria. For a fifth-degree polynomial [42], these criteria are given by

$$H_1 = a_1 > 0,$$

$$H_2 = a_1 a_2 - a_3 > 0,$$

$$H_3 = a_1 a_2 a_3 + a_1 a_3 + a_1 a_5 - a_1 a_4 - a_3 > 0,$$

(88)

$$H_4 = (a_3 a_4 - a_2 a_5)(a_1 a_2 - a_3) - (a_1 a_4 - a_5) > 0,$$

$$H_5 = a_5 H_4 > 0,$$

from which we can conclude that the endemic equilibrium is locally asymptotically stable.

5. Sensitivity Analysis

Intervention strategies to reduce the mortality and morbidity due to pneumonia should target the parameters that have a high impact on the effective reproduction number, $\mathcal{R}_e$. Sensitivity analysis is used to obtain the sensitivity index that is a measure of the relative change in a state variable when a parameter changes. We compute the sensitivity indices of $\mathcal{R}_e$ to the model parameters with the approach used by Chitnis et al. [43]. These indices show the importance of each individual parameter in the disease transmission dynamics and prevalence.

Definition 10. The normalized forward sensitivity index of a variable, v, that depends differentiability on index on a parameter, p, is defined as

$$\gamma_p^v = \frac{\partial v}{\partial p} * \frac{p}{v}. \tag{89}$$

We use the formula for $\mathcal{R}_e$ to derive an expression for the sensitivity of $\mathcal{R}_e$ given by

$$\gamma_p^{\mathcal{R}_e} = \frac{\partial \mathcal{R}_e}{\partial p} * \frac{p}{\mathcal{R}_e}, \tag{90}$$

to each of the ten parameters given in Table 1. In the following example, we obtain the sensitivity index of $\mathcal{R}_e$ with respect to δ:

$$\gamma_\delta^{\mathcal{R}_e} = \frac{\partial \mathcal{R}_e}{\partial \delta} * \frac{\delta}{\mathcal{R}_e} = 1. \tag{91}$$

The same method is used to obtain the indices of $\gamma_\Lambda^{\mathcal{R}_e}$, $\gamma_\mu^{\mathcal{R}_e}$, $\gamma_\gamma^{\mathcal{R}_e}$, $\gamma_\theta^{\mathcal{R}_e}$, $\gamma_\beta^{\mathcal{R}_e}$, $\gamma_\omega^{\mathcal{R}_e}$, $\gamma_\tau^{\mathcal{R}_e}$, $\gamma_\pi^{\mathcal{R}_e}$, and $\gamma_\sigma^{\mathcal{R}_e}$.

The parameters given in Table 1 are ordered from most sensitive to the least sensitive. The parameter values $\delta = 7.6$, $\mu = 0.0002$, $\Lambda = 10.09$, $\sigma = 0.33$, $\omega = 0.001124$, $\tau = 0.0714$, $\pi = 0.01096$, $\beta = 0.0115$, $\theta = 0.336$, and $\eta = 0.0241$ are used to determine the sensitivity indices.

5.1. Interpretation of Sensitivity Indices. It is noted from the sensitivity indices given in Table 1 that the value of $\mathcal{R}_e$ increases when the parameter values δ, θ, ω, and Λ increase while other parameter values are kept fixed. This implies an increase in the endemicity of the disease since the indices have positive signs. On the other hand, when the parameter values μ, σ, β, τ, π, and η are decreased while the rest of the parameter values are kept fixed, the value of $\mathcal{R}_e$ decreases. This shows a decrease in the disease endemicity because the indices have negative signs. The transmission rate δ and recruitment rate Λ are the most sensitive parameters. The transmission coefficient of the carrier subgroup ω and proportion of susceptible population that become carriers θ are the other key parameters that are sensitive.

6. Numerical Simulation

We illustrate the analytical results of the model by carrying out numerical simulation of the models using a set of

TABLE 1: Numerical values of sensitivity indices of $\mathcal{R}_e$.

Parameter symbols	Sensitivity Index
Λ	+1
δ	+1
ω	+0.874
θ	+0.643
τ	−0.743
η	−0.432
β	−0.0574
μ	−0.0086
π	−0.0045
σ	−0.0014

estimated parameter values obtained from literature. The system is simulated using ODE solvers coded in MATLAB programming language. Simulation of the pneumonia model under treatment intervention alone and the model with treatment and vaccination interventions combined is carried out to investigate the impact of the key parameters on the spread of pneumonia and how their influence can be controlled. The parameter values are presented in Table 2.

7. Discussion and Conclusion

In the study, a deterministic model is formulated and analyzed to investigate the role of treatment and vaccination in the transmission dynamics of pneumonia. The model is well-posed and exists in a feasible region where disease-free and endemic equilibrium are obtained and their stability is investigated.

When the equilibrium is locally stable, all the points near it tend to move towards it over time and when the equilibrium point is globally stable, all the initial starting conditions lead to it over time.

The basic model of pneumonia under treatment intervention alone has a locally and globally asymptotically stable disease-free equilibrium if its associated reproduction number $\mathcal{R}_0 < 1$ and has a unique and globally asymptotically stable endemic equilibrium when the reproduction number exceeds unity.

The disease-free equilibrium is locally stable implying that if initial conditions were to start near it, they would move towards it over time but the initial conditions do not always start at neighborhood of disease-free equilibrium. When the disease-free equilibrium of model is globally stable, it means that all initial starting conditions would lead to it over time; hence treatment would decrease the disease prevalence. The endemic equilibrium of model is globally stable if and only if $\mathcal{R}_0 > 1$, implying that all the points near it tend to move towards it over time.

In order to make the endemic equilibrium unstable so that it switches to disease-free equilibrium, intervention measures like treatment with high efficacy drugs and vaccination programs are necessary.

TABLE 2: Parameter estimates for pneumonia model under interventions.

Symbol	Description	Value	Source
μ	Per capita natural mortality rate	0.0002/day	[2]
Λ	Per capita recruitment rate	10.09/day	Estimated
θ	Fraction of susceptible individuals that join the carriers	0.338/day	Estimated
σ	Per capita disease induced mortality rate	0.33/day	[2]
β	Per capita recovery rate of carriers	0.0115/day	Estimated
α	Force of infection of susceptible individuals	0.0287/day	Estimated
τ	Per capita recovery rate of infective individuals	0.0714/day	[36]
π	Rate at which carriers develop symptoms	0.01096/day	[7]
η	Rate at which treated individuals become susceptible	0.0241/day	Estimated
γ	Rate at which susceptible individuals get vaccinated	0.0621/day	Estimated
ϕ	Rate at which treated individuals are vaccinated	9.4/day	Estimated
ω	Transmission coefficient for the carrier subgroup	0.001124	[37]
δ	Transmission rate	7.6/day	Estimated
p	Probability for a contact to cause infection	0.89–0.99	[37]
κ	Contact rate	1–10/day	[37]

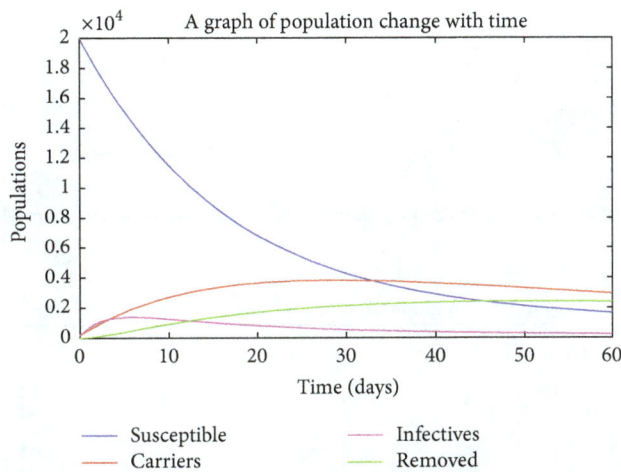

FIGURE 1: Variation of the population under treatment intervention alone.

The pneumonia model under treatment and vaccination interventions have a disease-free equilibrium which is locally asymptotically stable whenever its associated effective reproduction number $\mathcal{R}_e < 1$. This implies that the initial conditions would tend to the disease-free equilibrium point and hence pneumonia will be wiped out of the population.

Sensitivity analysis identifies the transmission rate δ as the key factor in fueling the spread of pneumonia, whereas vaccination rate γ and recovery rate τ are the parameters that inhibit the spread of the disease. From the results obtained, we conclude that a combination of vaccination and treatment interventions programs targeting children can effectively eliminate pneumonia infection from the population.

Numerical simulation of the pneumonia model under treatment strategy τ for the set of parameter estimates presented in Table 2 and initial values of population sizes

are carried out. The results show that when there is a pneumonia outbreak, the population sizes of the infected and carriers increase with time while the susceptible population size decreases with time until an endemic equilibrium is attained as shown in Figure 1. In Figure 2, it is shown that with treatment intervention in place for the different subgroups, the infected population decreases until it equals the treated population. When both treatment and vaccination strategies are applied, numerical simulation reveals a sharp decline in the susceptible population and a rise in both the infected and carrier populations during the initial stages of the epidemic until a disease-free equilibrium is attained as shown in Figure 3. The effect of treatment and vaccination interventions on the population leads to a decrease in the susceptible, infected, carriers, and treated populations and an increase in the vaccinated populations as presented in Figure 4. This confirms that a combination of treatment and

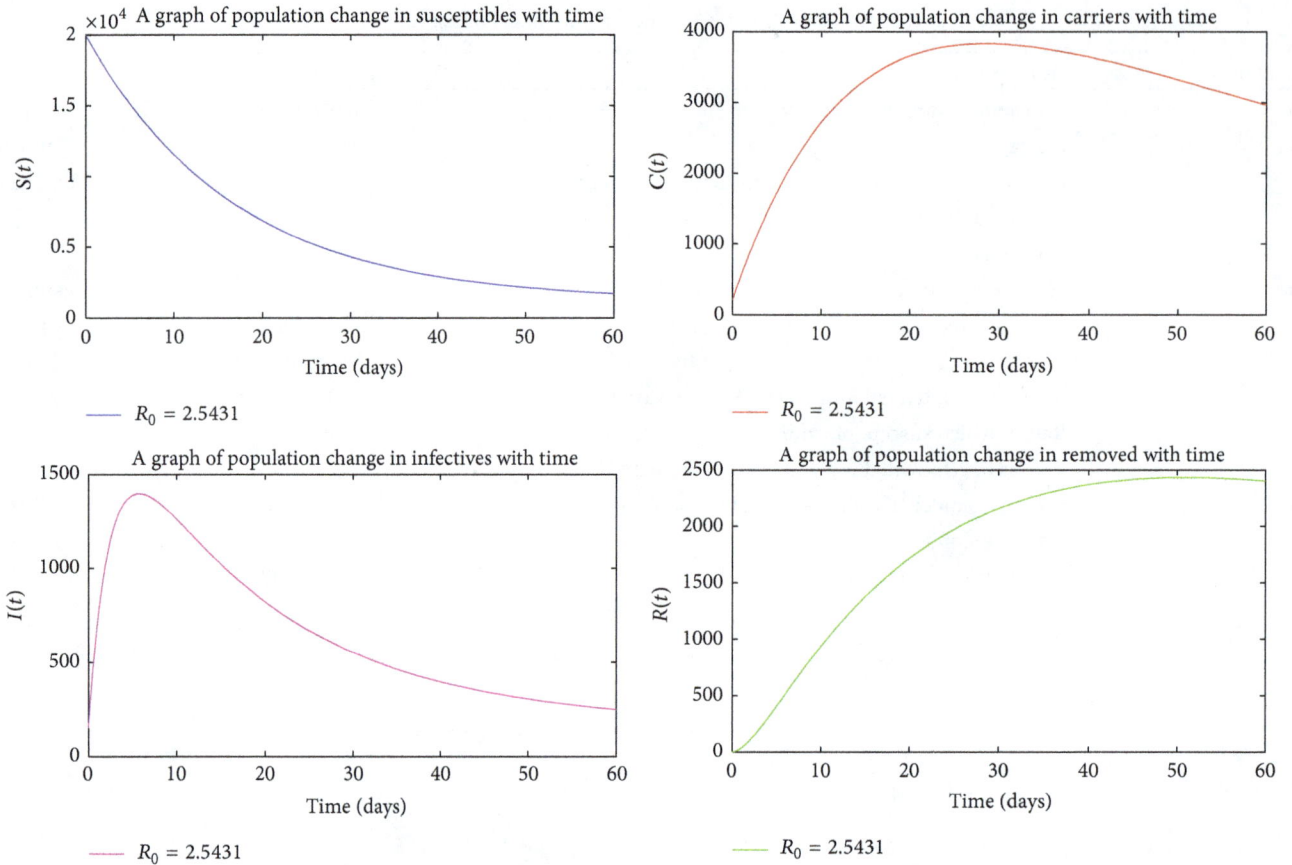

FIGURE 2: Variation of the population under treatment intervention alone.

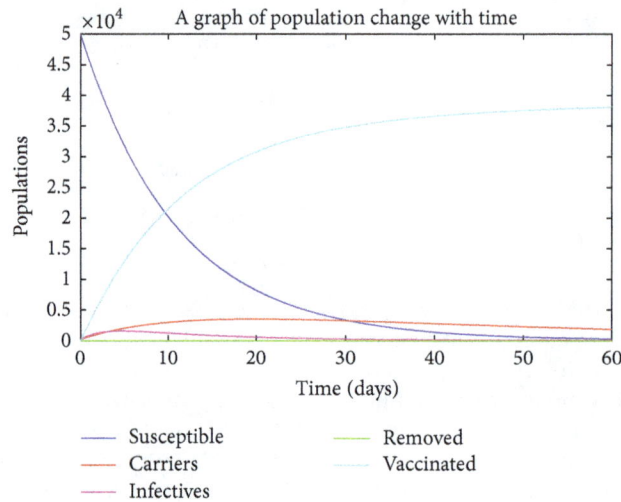

FIGURE 3: Variation of the population under treatment and vaccination interventions.

vaccination interventions can eradicate the disease from the community.

The results have important public health implications since they determine the severity and outcome of the epidemic (i.e., clearance or persistence of infection) and provide a framework for the design of control strategies. The study further shows that a combination of treatment and vaccination has much more impact than treatment alone. Furthermore, analysis of the effective reproduction number $\mathscr{R}_e$ demonstrates that vaccination and treatment reduce the average number of secondary infections when implemented. Thus, in order to control the pneumonia spread, infected

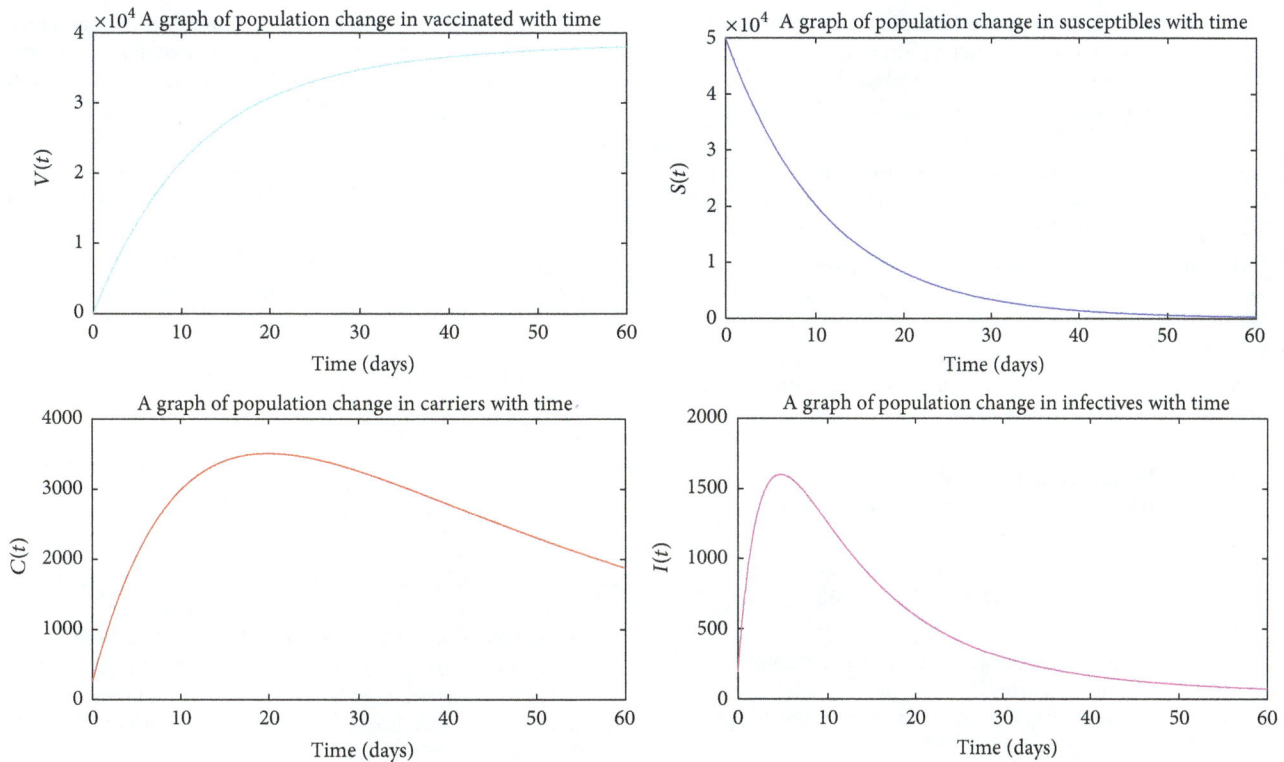

FIGURE 4: Variation of the population under treatment and vaccination interventions.

individuals should be treated immediately; all individuals with compromised immunity including newborn babies and the elderly should be vaccinated.

Conflicts of Interest

The authors declare that no conflicts of interest took place during the preparation of the manuscript.

Acknowledgments

Mohammed Kizito is grateful for the financial support from the Belgium Technical Cooperation (BTC) in the production of this manuscript.

References

[1] S. S. Huang, J. A. Finkelstein, and M. Lipsitch, "Modeling community- and individual-level effects of child-care center attendance on pneumococcal carriage," *Clinical Infectious Diseases*, vol. 40, no. 9, pp. 1215–1222, 2005.

[2] C. Ngari, G. Pokhariyal, and J. Koske, "Analytical Model for Childhood Pneumonia, a Case Study of Kenya," *British Journal of Mathematics & Computer Science*, vol. 12, no. 2, pp. 1–28, 2016.

[3] S. Sazawal and R. E. Black, "Meta-analysis of intervention trials on case-management of pneumonia in community settings," *The Lancet*, vol. 340, no. 8818, pp. 528–533, 1992.

[4] V. Singh and S. Aneja, "Pneumonia - Management in the Developing World," *Paediatric Respiratory Reviews*, vol. 12, no. 1, pp. 52–59, 2011.

[5] T. Wardlaw, P. Salama, E. W. Johansson, and E. Mason, "Pneumonia: the leading killer of children," *The Lancet*, vol. 368, no. 9541, pp. 1048–1050, 2006.

[6] World Health Organization, Pneumonia fact sheet. Media centre, 2013.

[7] C. G. Ngari, D. M. Malonza, and G. G. Muthuri, "A model for childhood pneumonia dynamics," *Journal of Life Sciences Research*, vol. 1, no. 2, pp. 31–40, 2014.

[8] K. A. Poehling, T. R. Talbot, M. R. Griffin et al., "Invasive pneumococcal disease among infants before and after introduction of pneumococcal conjugate vaccine," *The Journal of the American Medical Association*, vol. 295, no. 14, pp. 1668–1674, 2006.

[9] World Health Organization and UNICEF, *Integrated management of childhood illness handbook*, World Health Organization, Geneva, Switzerland, 2006.

[10] G. Regev-Yochay, M. Raz, R. Dagan et al., "Nasopharyngeal carriage of *Streptococcus pneumoniae* by adults and children in community and family settings," *Clinical Infectious Diseases*, vol. 38, no. 5, pp. 632–639, 2004.

[11] T. Leino, K. Auranen, J. Jokinen, M. Leinonen, P. Tervonen, and A. K. Takala, "Pneumococcal carriage in children during their first two years: Important role of family exposure," *The Pediatric Infectious Disease Journal*, vol. 20, no. 11, pp. 1022–1027, 2001.

[12] B. M. Farr, C. L. R. Bartlett, J. Wadsworth, and D. L. Miller, "Risk factors for community-acquire pneumonia diagnosed upon hospital admission," *Respiratory Medicine*, vol. 94, no. 10, pp. 954–963, 2000.

[13] D. Pessoa, *Modelling the Dynamics of Streptococcus Pneumoniae Transmission in Children, Masters thesis [Masters, thesis]*, University of Dehisboa, 2010.

[14] R. Nantanda, J. K. Tumwine, G. Ndeezi, and M. S. Ostergaard, "Asthma and pneumonia among children less than five years with acute respiratory symptoms in Mulago Hospital, Uganda: Evidence of under-diagnosis of asthma," *PLoS ONE*, vol. 8, no. 11, Article ID e81562, 2013.

[15] S. C. Ranganathan and S. Sonnappa, "Pneumonia and Other Respiratory Infections," *Pediatric Clinics of North America*, vol. 56, no. 1, pp. 135–156, 2009.

[16] L. Dunn, "Pneumonia: classification, diagnosis and nursing management.," *Nursing standard (Royal College of Nursing (Great Britain) : 1987)*, vol. 19, no. 42, pp. 50–54, 2005.

[17] S. Moberley, J. Holden, D. P. Tatham, and R. M. Andrews, "Vaccines for preventing pneumococcal infection in adults," *Cochrane Database of Systematic Reviews*, vol. 1, Article ID CD000422, 2013.

[18] F. M. Russell and E. K. Mulholland, "Recent advances in pneumococcal vaccination of children," *Annals of Tropical Paediatrics*, vol. 24, no. 4, pp. 283–294, 2004.

[19] K. Mulholland, "Childhood pneumonia mortality-a permanent global emergency," *The Lancet*, vol. 370, no. 9583, pp. 285–289, 2007.

[20] C. M. C. Rodrigues, "Challenges of Empirical Antibiotic Therapy for Community-Acquired Pneumonia in Children," *Current Therapeutic Research - Clinical and Experimental*, vol. 84, pp. e7–e11, 2017.

[21] M. E. Alexander, S. M. Moghadas, P. Rohani, and A. R. Summers, "Modelling the effect of a booster vaccination on disease epidemiology," *Journal of Mathematical Biology*, vol. 52, no. 3, pp. 290–306, 2006.

[22] O. P. Misra and D. K. Mishra, "Modelling the effect of booster vaccination on the transmission dynamics of diseases that spread by droplet infection," *Nonlinear Analysis: Hybrid Systems*, vol. 3, no. 4, pp. 657–665, 2009.

[23] S. M. Raimundo, H. M. Yang, and A. B. Engel, "Modelling the effects of temporary immune protection and vaccination against infectious diseases," *Applied Mathematics and Computation*, vol. 189, no. 2, pp. 1723–1736, 2007.

[24] A. G. Falade and A. I. Ayede, "Epidemiology, aetiology and management of childhood acute community-acquired pneumonia in developing countries-a review.," *African Journal of Medicine and Medical Sciences*, vol. 40, no. 4, pp. 293–308, 2011.

[25] K. L. O'Brien and R. Dagan, "The potential indirect effect of conjugate pneumococcal vaccines," *Vaccine*, vol. 21, no. 17-18, pp. 1815–1825, 2003.

[26] L. Jódar, J. Butler, G. Carlone et al., "Serological criteria for evaluation and licensure of new pneumococcal conjugate vaccine formulations for use in infants," *Vaccine*, vol. 21, no. 23, pp. 3265–3272, 2003.

[27] H. W. Hethcote, "The mathematics of infectious diseases," *SIAM Review*, vol. 42, no. 4, pp. 599–653, 2000.

[28] S. M. Moghadas, "Gaining insights into human viral diseases through mathematics," *European Journal of Epidemiology*, vol. 21, no. 5, pp. 337–342, 2006.

[29] L. Wang and R. Xu, "Global stability of an SEIR epidemic model with vaccination," *International Journal of Biomathematics*, vol. 9, no. 6, Article ID 1650082, 2016.

[30] Z. Yu, J. Liu, X. Wang, X. Zhu, D. Wang, and G. Han, "Efficient vaccine distribution based on a hybrid compartmental model," *PLoS ONE*, vol. 11, no. 5, Article ID e0155416, 2016.

[31] D. Greenhalgh, K. E. Lamb, and C. Robertson, "A mathematical model for the spread of *Strepotococcus pneumoniae* with transmission dependent on serotype," *Journal of Biological Dynamics*, vol. 6, no. suppl. 1, pp. 72–87, 2012.

[32] K. E. Lamb, D. Greenhalgh, and C. Robertson, "A simple mathematical model for genetic effects in pneumococcal carriage and transmission," *Journal of Computational and Applied Mathematics*, vol. 235, no. 7, pp. 1812–1818, 2011.

[33] X. Liu, Y. Takeuchi, and S. Iwami, "SVIR epidemic models with vaccination strategies," *Journal of Theoretical Biology*, vol. 253, no. 1, pp. 1–11, 2008.

[34] E. J. Ndelwa, M. Kgosimore, E. S. Massawe, and L. Namkinga, "Mathematical modeling and analysis of treatment and screening of pneumonia," *Mathematical Theory and Modeling*, vol. 5, no. 10, pp. 21–39, 2015.

[35] G. T. Tilahun, O. D. Makinde, and D. Malonza, "Modelling and optimal control of pneumonia disease with cost-effective strategies," *Journal of Biological Dynamics*, vol. 11, no. suppl. 2, pp. 400–426, 2017.

[36] A. Melegaro, N. J. Gay, and G. F. Medley, "Estimating the transmission parameters of pneumococcal carriage in households," *Epidemiology and Infection*, vol. 132, no. 3, pp. 433–441, 2004.

[37] K. Doura, J. D. Melendez-Morales, G. G. Meyer, and L. E. Perez, "Biometrics Unit Technical Reports: Number BU-1524-M: An SIS Model of Streptococcal Disease with a Class of Beta-Hemolytic Carriers," Tech. Rep., 2000.

[38] P. van den Driessche and J. Watmough, "Reproduction numbers and sub-threshold endemic equilibria for compartmental models of disease transmission," *Mathematical Biosciences*, vol. 180, pp. 29–48, 2002.

[39] R. M. Anderson and R. M. May, *Infectious diseases of humans: dynamics and control*, Oxford university press, Oxford, UK, 1991.

[40] J. K. Hale, *Ordinary Differential Equations*, John Wiley & Sons, New York, NY, USA, 1969.

[41] F. R. Gantmacher, *The Theory of Matrices*, vol. 1, Chelsea Publishing, New York, NY, USA, 1960.

[42] P. Lancaster, *Theory of Matrices*, Academic Press, New York, NY, USA, 1969.

[43] N. Chitnis, J. M. Hyman, and J. M. Cushing, "Determining important parameters in the spread of malaria through the sensitivity analysis of a mathematical model," *Bulletin of Mathematical Biology*, vol. 70, no. 5, pp. 1272–1296, 2008.

17

Understanding Dengue Control for Short- and Long-Term Intervention with a Mathematical Model Approach

A. Bustamam, D. Aldila ⓘD, and A. Yuwanda

Department of Mathematics, Universitas Indonesia, Depok 16424, Indonesia

Correspondence should be addressed to D. Aldila; aldiladipo@sci.ui.ac.id

Academic Editor: Lucas Jodar

A mathematical model of dengue diseases transmission will be discussed in this paper. Various interventions, such as vaccination of adults and newborns, the use of insecticides or fumigation, and also the enforcement of mechanical controls, will be considered when analyzing the best intervention for controlling the spread of dengue. From model analysis, we find three types of equilibrium points which will be built upon the dengue model. In this paper, these points are the mosquito-free equilibrium, disease-free equilibrium (with and without vaccinated compartment), and endemic equilibrium. Basic reproduction number as an endemic indicator has been found analytically. Based on analytical and numerical analysis, insecticide treatment, adult vaccine, and enforcement of mechanical control are the most significant interventions in reducing the spread of dengue disease infection caused by mosquitoes rather than larvicide treatment and vaccination of newborns. From short- and long-term simulation, we find that insecticide treatment is the best strategy to control dengue. We also find that, with periodic intervention, the result is not much significantly different with constant intervention based on reduced number of the infected human population. Therefore, with budget limitations, periodic intervention of insecticide strategy is a good alternative to reduce the spread of dengue.

1. Introduction

Dengue is the most rapidly growing disease in the world [1]. The disease is spread by *Aedes* mosquitoes and is therefore often referred to as a mosquito-borne viral disease. The disease has become endemic in more than 100 countries, including the Caribbean, Africa, the Americas, the Pacific, and Asia, including Indonesia [1]. As an endemic disease, dengue occurs regularly in subtropical and tropical regions of the world, and approximately 40% of people live in regions of the world where there is a risk of contracting it [2]. Dengue is a vector-borne disease transmitted from an infected human to a female *Aedes aegypti* mosquito by a bite. The mosquito, which needs regular meals of blood to mature its eggs, completes the cycle by biting a healthy human, transmitting the disease in one act [3].

Until now, the primary prevention for dengue has been control of mosquitoes, in both larval and adult forms. Larval control is carried out by larvicide treatment using long-lasting chemicals to kill larvae, which sure preferably have WHO clearance for use in drinking water [4]. Mechanical controls are also used to control larvae, with assistance from campaigns and educational programs carried out by governments. In Indonesia, such a program is known as 3M and consists of educating people about the importance of draining and shutting down and burying all tubs, buckets, or containers of water that which be used by female mosquitoes to breed and lay their eggs [5]. The larvae of *Aedes aegypti* can also grow in used goods that can hold water, and it is therefore recommended to make sure the environment around the house has no space that could allow mosquitoes to breed.

Adult mosquito control is achieved by the use of insecticide. Insecticide fumigation targets the vector *Aedes aegypti* mosquito as the main control of dengue epidemics. However, the long-term use of insecticides and larvicides poses several risks: one is resistance of the mosquito to the product, reducing its efficacy, while genetic mutation of the mosquito, making it less susceptible to the effects of the product, is another. Such products have also been linked to numerous adverse health effects including the worsening of asthma and respiratory problems [3, 6]. In Surabaya, Indonesia, larval mortality rates of under 80% indicate possible resistance of

Aedes aegypti to the insecticide temephos [7], and outside Indonesia, resistance to insecticide has been reported in multiple countries. In recent years, the frequency of kdr mutations associated with pyrethroid resistance has increased rapidly [8]. Pyrethroids have become the most frequently used public health insecticides globally due to their low cost and low toxicity to mammals [9], and they are of considerable concern when kdr is found in wild populations of vector mosquitoes [8]. With the many cases of *Aedes aegypti* mosquito's resistance to insecticide, it is therefore necessary to develop alternative strategies to slow its evolution.

Besides controlling dengue via control of mosquitoes population, one of the alternative strategies that is being used is dengue vaccine. In December 2015, the first vaccine against dengue by Sanofi Pasteur, Dengvaxia (CYD-TDV), was approved in three highly endemic countries: Mexico, Philippines, and Brazil [10]. This vaccine is the world's first dengue vaccine and is already licensed for individuals aged 9–45 years for the prevention of infectious disease caused by four dengue virus serotypes (DEN 1, DEN 2, DEN 3, and DEN 4) [11]. In Indonesia, so far, the government is still conducting clinical trials to determine its effectiveness. However, assessments of the public's acceptance of the dengue vaccine and its associated factors are widely lacking [12]. A lack of understanding about the importance of vaccination against dengue to the public will be able to reduce the success rate of vaccination interventions in various countries, especially in a country that is less intensive to educate people about the importance of dengue fever vaccination [13]. In 2017, *Dengvaxia* would have been applied if proven effective and suitable for the dengue serotypes which are pandemic in Indonesia [14].

The earliest mathematical models for dengue disease transmission are developed in [15, 16] which are closely related to the models for the transmission of malaria discussed in [17, 18]. The authors in [16] create the model for two types of viruses by allowing temporary cross immunity and increased susceptibility to the second infection due to the first infection. The intervention has not yet been used into the mathematical model [16]. In [19], the mathematical model with only insecticide campaign intervention is discussed. It has been shown that, with a steady insecticide campaign, it is possible to reduce the number of infected humans and mosquitoes and prevent an outbreak that could transform an epidemiological episode to an endemic disease [19]. A year after the research discussed in [19], it was updated [20], and the mathematical model for dengue was updated continuously with all controls included, that is, (1) proportion of larvicide, (2) proportion of adulticide, and (3) proportion of mechanical control. The results have shown that, even with a low, although continuous, index of control adulticide over time, the results are surprisingly positive [20]. However, it has been stated that to rely only on adulticide is a risky decision [20]. The research in [6, 7, 21] supports this claim, citing the problem of *Aedes aegypti* mosquito's resistance to insecticide. Under the new achievement in the field of vaccination technology with the discovery of the first vaccine against dengue by Sanofi Pasteur, the work in [22] devised two models, one assuming that unintentional vaccination

increases the infectious period and another assuming that unintentional vaccination leads to the development of symptoms. This argument is also supported by [3], in which the mathematical model is created with the vaccine as the new compartment, arguing that the vaccine must divide the human population into classes, that is, the perfect pediatric vaccine for newborns and perfect adult vaccine (conferring 100% protection throughout life), and also classes of human with imperfect vaccine effect [3]. Other mathematical models with different intervention were also introduced in [23] which discuss the use of mosquito repellent to reduce probability of success of infection in human population and in [4] which discuss the use of sterile mosquito strategy.

According to above explanation, it is important to find the best strategy for controlling dengue spreads for both short-term and long-term interventions. Therefore, a mathematical model of dengue disease transmission by using adult and newborn vaccines with waning immunity, the use of insecticides and larvicides, and mechanical control will be developed in the next section. Equilibrium points will be found, which ensure the existence of local stability. Basic reproduction numbers will be obtained as the main factor in whether the disease will become epidemic in a population or not. Numerical analysis for comparing the dynamic of infected humans and mosquitoes will be used to support the model interpretation.

2. Mathematical Model Construction

To construct our model, firstly we divide the human population into four compartments, that is,

> $S_h(t)$: susceptible (individuals who can be infected with dengue);

> $V_h(t)$: vaccinated (individuals who have had the vaccine injected into their bodies, making them resistant to infectious disease. However, the use of the vaccine does not provide perfect immunity. There will be a time when the vaccine does not work properly in the body or when the effect of the vaccine has begun to subside [3]);

> $In_h(t)$: infected (individuals who are infected with dengue. In this case, the infected human is incapable of transmitting the disease to other humans);

> $R_h(t)$: recovered (individuals who have recovered from dengue and have acquired temporal immunity to respective DEN virus).

On the other hand, we divide the mosquito population into three compartments, that is,

> $A_v(t)$: aquatic phase (the phase that includes the egg, larvae, and pupa stages, which live in water);

> $S_v(t)$: susceptible (mosquitoes that are able to infect with dengue);

> $In_v(t)$: infected (mosquitoes that have been infected with dengue by an infected human and are capable of transmitting dengue to humans).

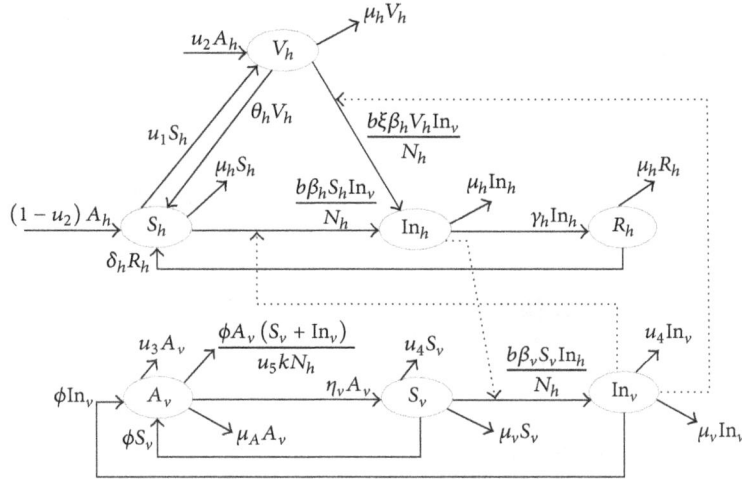

FIGURE 1: Mathematical model using vaccination of adults and newborns, fumigation and larvicide treatment, and enforcement of mechanical control.

Secondly, we have made some assumptions that we will use to describe a dynamic process in our model that we will construct: (1) There is no migration in either human or mosquito population. (2) Humans and mosquitoes are assumed to be born susceptible, there is no natural protection, and dengue is not passed onto the next generation (no vertical transmission) [19]. (3) The transmission process in susceptible and vaccinated humans is simply by the bite of an infected mosquito. Infected humans cannot transmit the virus to other susceptible or vaccinated humans [19]. (4) The death rate is considered to be a natural death rate in both populations. (5) Vaccinated human status is considered temporary because of the ability of the vaccine to subside over time [3]. (6) There is no recovered phase in mosquitoes due to their short lifespan [20]. (7) There is no resistant (immune) effect in mosquitoes to the use of synthetic fumigation, such as insecticides and larvicides [6, 7]; in this article, we assume that there is no resistant (immunity) effect to mosquitoes due to use of synthetic fumigation, such as insecticide and larvicide.

With the assumptions, variables, and transmission diagram given in Figure 1, the model is represented as a seven-dimensional system of differential equations which are given by

$$\frac{dS_h}{dt} = (1 - u_2) A_h - \frac{b\beta_h S_h \text{In}_v}{N_h} - u_1 S_h - \mu_h S_h + \delta_h R_h$$
$$\qquad + \theta_h V_h,$$

$$\frac{dV_h}{dt} = u_2 A_h + u_1 S_h - \frac{b\xi\beta_h V_h \text{In}_v}{N_h} - \mu_h V_h - \theta_h V_h,$$

$$\frac{d\text{In}_h}{dt} = \frac{b\beta_h \text{In}_v (S_h + \xi V_h)}{N_h} - \gamma_h \text{In}_h - \mu_h \text{In}_h,$$

$$\frac{dR_h}{dt} = \gamma_h \text{In}_h - \mu_h R_h - \delta_h R_h,$$

$$\frac{dA_v}{dt} = \phi \left(1 - \frac{A_v}{u_5 k N_h} \right) (S_v + \text{In}_v) - u_3 A_v - \eta_v A_v$$
$$\qquad - \mu_A A_v,$$

$$\frac{dS_v}{dt} = \eta_v A_v - \frac{b\beta_v S_v \text{In}_h}{N_h} - \mu_v S_v - u_4 S_v,$$

$$\frac{d\text{In}_v}{dt} = \frac{b\beta_v S_v \text{In}_h}{N_h} - \mu_v \text{In}_v - u_4 \text{In}_v,$$

$$(1)$$

with parameters description being the following:

N_h: total of human population

A_h: human per capita birth rate

b: average number of bites of humans by mosquitoes

β_h: average of successful transmission in human

β_v: average of successful transmission in mosquitoes

μ_h: human death rate

μ_v: mosquito death rate

μ_A: larval death rate

δ_h: the rate of change from R_h to S_h because of the disappearance of the temporal natural immunity

θ_h: the rate of change from V_h to S_h because of the disappearance of the temporal vaccine effect

ξ: reduction of β_h because of vaccine

γ_h: human recovery rate

ϕ: average number of eggs at each deposit

k: ratio for number of larvae per human

η_v: transition rate from A_v to S_v

u_1: adult vaccination rate

u_2: newborn vaccine

u_3: larvicide rate

u_4: fumigation rate

u_5: enforcement of mechanical control proportion to reduce k

In the next section, mathematical model analysis to find equilibrium points and their local stability criteria will be given.

3. Equilibrium Points and Local Stability

From the system of equation (1), we find three types of equilibrium points.

3.1. Mosquito-Free Equilibrium (MFE) Point. It is the equilibrium where the mosquito virus does not exist in the living environment, so any infectious disease never occurs at the MFE point (sterile conditions). This equilibrium is given by

$$S_h^1 = -\frac{A_h \left(\mu_h + \theta_h\right)\left(R_2 - 1\right)}{\mu_h \left(\mu_h + u_1 + \theta_h\right)},$$

$$V_h^1 = \frac{A_h \left(\mu_h u_2 + u_1\right)}{\mu_h \left(\mu_h + u_1 + \theta_h\right)},$$

$$\text{In}_h^1 = 0,$$

$$R_h^1 = 0,$$

$$A_v^1 = 0,$$

(2)

$$S_v^1 = 0,$$

$$\text{In}_v^1 = 0.$$

At the MFE point, it can be seen that S_h^1 can have either a positive or a negative value. However, from an epidemiological point of view, a population has biological meaning only if it has a nonnegative value. Therefore, to ensure that the MFE point exists, the positiveness for S_h^1 should be made; that is,

$$R_2 = \frac{\mu_h u_2}{\mu_h + \theta_h} < 1. \tag{3}$$

To guarantee local stability of equilibrium, it is necessary to ensure that all eigenvalues of system (1) in its Jacobian matrix evaluated in the MFE have negative values. The condition is

$$\frac{\eta_v \phi}{\left(\mu_v + u_4\right)\left(u_3 + \eta_v + \mu_A\right)} = R_1 < 1. \tag{4}$$

Please note that R_1 is known as the basic offspring number, which will guarantee the existence of mosquitoes in the population. The mosquito population will exist if, and only if, $R_1 \geq 1$ as will be discussed in the next equilibrium point.

3.2. Disease-Free Equilibrium (DFE) Point. DFE is an equilibrium where mosquito and human populations exist in the living environment, but the virus does not occur.

We have two types of DFE equilibrium point, which are DFE-1 and DFE-2. DFE-1 describes a condition where infected, recovered human groups and infected, recovered mosquito groups do not exist, while DFE-2 describes a condition where infected, recovered human groups, infected, recovered mosquito groups, and human vaccinated groups do not exist (special case without intervention of vaccines $(u_1 = u_2 = 0)$).

3.2.1. Disease-Free Equilibrium I (DFE-1) Point. This equilibrium is given by

$$S_h^2 = -\frac{A_h \left(\mu_h + \theta_h\right)\left(R_2 - 1\right)}{\mu_h \left(\mu_h + u_1 + \theta_h\right)},$$

$$V_h^2 = \frac{A_h \left(\mu_h u_2 + u_1\right)}{\mu_h \left(\mu_h + u_1 + \theta_h\right)},$$

$$\text{In}_h^2 = 0,$$

$$R_h^2 = 0,$$

(5)

$$A_v^2 = u_5 k N_h \left(R_1 - 1\right),$$

$$S_v^2 = \frac{u_5 k N_h \eta_v \left(R_1 - 1\right)}{\mu_v + u_4},$$

$$\text{In}_v^2 = 0,$$

which will exist if and only if

$$R_1 = \frac{\eta_v \phi}{\left(\mu_v + u_4\right)\left(u_3 + \eta_v + \mu_A\right)} > 1,$$

$$R_2 = \frac{\mu_h u_2}{\mu_h + \theta_h} < 1. \tag{6}$$

DFE-1 will be locally asymptotically stable if and only if

$$\frac{\eta_v \phi}{\left(\mu_v + u_4\right)\left(u_3 + \eta_v + \mu_A\right)} = R_1 > 1, \tag{7}$$

$$\frac{b^2 \beta_v u_5 k \beta_h A_h \left(u_1 \xi + \mu_h + \theta_h + \left(\mu_h + \theta_h\right)\left(\xi - 1\right) R_2\right)\left(R_1 - 1\right)}{\phi \left(\mu_v + u_4\right)^2 \left(\gamma_h + \mu_h\right) N_h \mu_h \left(\mu_h + u_1 + \theta_h\right)} \tag{8}$$

$$= \mathcal{R}_0 < 1.$$

Based on (8), it is known that

1. if $R_1 < 1$, then $\mathcal{R}_0 < 0$ will be obtained;
2. if $R_1 > 1$, then $\mathcal{R}_0 \in (0, 1)$ or $\mathcal{R}_0 \in (1, \infty)$ will be obtained.

3.2.2. Disease-Free Equilibrium II (DFE-2) Point. This equilibrium is given by

$$S_h^3 = \frac{A_h}{\mu_h},$$

$$V_h^3 = 0,$$

$$\text{In}_h^3 = 0,$$

$$R_h^3 = 0,$$

$$A_v^3 = u_5 k N_h (R_1 - 1),$$

$$S_v^3 = \frac{u_5 k N_h \eta_v (R_1 - 1)}{\mu_v + u_4},$$

$$\text{In}_v^3 = 0, \tag{9}$$

which only exist if and only if

$$R_1 = \frac{\eta_v \phi}{(\mu_v + u_4)(u_3 + \eta_v + \mu_A)} > 1. \tag{10}$$

To guarantee the local stability of equilibrium, it is necessary to ensure that all eigenvalues of system (1), evaluated in the Jacobi matrix on DFE-2 point, are negative. The condition is

$$\frac{A_h \beta_h \beta_v b^2 k u_5 \phi \eta_v}{(\mu_h N_h (\mu_v + u_4)(\mu_h + \gamma_h) \phi + b^2 u_5 \beta_h \beta_v k A_h (u_3 + \mu_A + \eta_v))(\mu_v + u_4)} = R_6 < 1,$$

$$\frac{\eta_v \phi}{(\mu_v + u_4)(u_3 + \eta_v + \mu_A)} = R_1 > 1. \tag{11}$$

3.3. Endemic Equilibrium (EE).

Endemic equilibrium describes a condition where all compartments, both human and mosquitoes, achieve coexistence. The endemic equilibrium point of system (1) is not in simple way to be written in explicit form. However, the existence of this equilibrium point might be written as equilibrium points that depend on values of In_v and In_h which are given by

$$S_h^* = \frac{(\text{In}_h^* \text{In}_v^* b \beta_h \xi + \mu_h \text{In}_h^* N_h (\theta_h + \mu_h))(\mu_h + \gamma_h) - N_h (b \beta_h A_h u_2 \xi \text{In}_v^*)}{(N_h \xi u_1 + b \beta_h \text{In}_v^* \xi + N_h \mu_h + N_h \theta_h) b \beta_h \text{In}_v^*},$$

$$V_h^* = \frac{N_h (u_2 A_h b \beta_h \text{In}_v^* + u_1 N_h \gamma_h \text{In}_h^* + u_1 N_h \mu_h \text{In}_h^*)}{(N_h \xi u_1 + b \beta_h \text{In}_v^* \xi + N_h \mu_h + N_h \theta_h) b \beta_h \text{In}_v^*},$$

$$R_h^* = \frac{\gamma_h \text{In}_h^*}{\mu_h + \delta_h}, \tag{12}$$

$$A_v^* = \frac{\text{In}_v^* (\mu_v + u_4)(N_h u_4 + N_h \mu_v + b \beta_v \text{In}_h^*)}{\eta_v b \beta_v \text{In}_h^*},$$

$$S_v^* = \frac{\text{In}_v^* N_h (\mu_h + u_4)}{b \beta_v \text{In}_h^*},$$

while In_v^* and In_h^* are taken from positive solution of

$$F_1 = (b \beta_v \text{In}_h^* + \mu_v N_h + u_4 N_h)(b \beta_v \text{In}_h^* \phi u_5 k N_h \eta_v$$

$$- (mu_v + u_4))(bk \text{In}_h^* N_h \beta_v \eta_v u_5 + bk \text{In}_h^* N_h \beta_v \mu_A u_5$$

$$+ bk \text{In}_h^* N_h \beta_v u_3 u_5 + b \phi \text{In}_h^* \text{In}_v^* \beta_v \tag{13}$$

$$+ \phi \text{In}_v^* N_h (\mu_v + u_4)) = 0,$$

$$F_2 = \left(A \text{In}_v^2 + B \text{In}_h + C \right) \text{In}_h + D \text{In}_v^2 + E \text{In}_v,$$

with

$$A = -b^2 \xi \beta_h^2 \mu_h (\delta_h + \gamma_h + \mu_h),$$

$$B = -b N_h \beta_h \mu_h (\xi \delta_h \gamma_h + \xi \delta_h \mu_h + \xi \delta_h u_1 + \xi \gamma_h \mu_h$$

$$+ \xi \gamma_h u_1 + \xi \mu_h^2 + \xi \mu_h u_1 + \delta_h \mu_h + \delta_h \theta_h + \gamma_h \mu_h$$

$$+ \gamma_h \theta_h + \mu_h^2 + \mu_h \theta_h),$$

$$C = -N_h^2 \mu_h (\mu_h + \theta_h + u_1)(\mu_h + \gamma_h)(\delta_h + \mu_h),$$

$$D = b^2 \xi A_h \beta_h^2 (\delta_h + \mu_h),$$

$$E = b A_h N_h \beta_h (\xi \mu_h u_2 + \xi u_1 - \mu_h u_2 + \mu_h + \theta_h)(\delta_h + \mu_h). \tag{14}$$

Substituting all parameters values from Table 1 into above couple of equations will give us existence of In_v and In_h numerically as shown in Figure 2. It can be seen that, as long as the intersection between F_1 and F_2 is in the first quadrant, we will have a positive endemic equilibrium.

For simple case when no intervention is given into system (1) ($u_1 = u_2 = u_3 = u_4 = 0$ and $u_5 = 1$), endemic equilibrium point is given by

$$(S_h^+, V_h^+, \text{In}_h^+, R_h^+, A_v^+, S_v^+, \text{In}_v^+), \tag{15}$$

TABLE 1: Parameters values.

Parameters	Value	Description
N_h	1000	Total of human population is assumed to be 1000 people.
μ_h	$1/(65 \times 365)$	Since human life expectation is approximately 65 years, we have $\mu_h = 1/(65 \times 365)$ [23].
A_h	$1000/(65 \times 365)$	Since in our model the total of the human population is constant, we have $A_h = N_h \mu_h = 1000/(65 \times 365)$.
ϕ	300	We assume that each female *Aedes aegypti* produces 300 eggs at each spawning.
β_h, β_v	0, 1	It is assumed that it needs 10 successful contacts to infect a human/mosquito with dengue [23].
θ_h	$1/60$	We assume that the effect of vaccination will have disappeared in 60 days.
γ_h	$1/14$	The natural recovery rate for the human population from dengue is 14 days [4].
ξ	0.1	With vaccination, the infection rate from mosquito to human population will be reduced by 90%.
μ_v	$1/30$	Life expectation of the mosquito population is 30 days [4].
μ_A	$0.75/21$	We assume that there is only a 25% chance that larvae might grow and become adult mosquitoes, with time to transition being 21 days.
η_v	$0.25/21$	Transition from aquatic phase to adult mosquito [4].
δ_h	$1/30$	Short-term immunity of humans to dengue after recovery is 30 days [23].
b	1	Mosquitoes only bite once a day [23].
k	2	We assume that the ratio between human and adult mosquitoes is 2; that is, each human related to 2 adult mosquitoes.

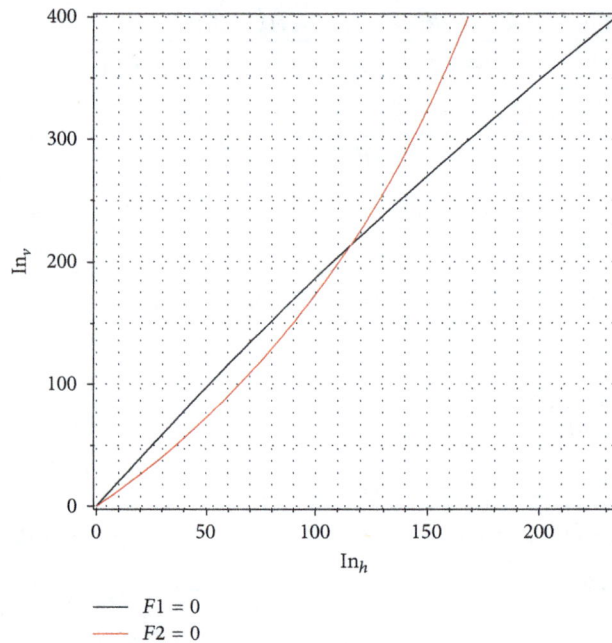

FIGURE 2: Existence of endemic equilibrium for In_v and In_h depending on couple of polynomial characteristics.

where

$$S_h^+ = -\frac{\mu_v (\mu_h + \gamma_h)\left(N_h \mu_h^2 \mu_v + (N_h \mu_v (\delta_h + \gamma_h) + b\beta_v A_h)\mu_h + b\beta_v A_h \delta_h\right)}{(R_4 - 1)\left(\mu_h^2 \mu_v + (\mu_v(\delta_h + \gamma_h) + bk\eta_v \beta_h)\mu_h + \mu_v \delta_h \gamma_h + bk\eta_v \beta_h(\delta_h + \gamma_h)\right)\mu_h \beta_v b},$$

$$V_h^+ = 0,$$

$$\text{In}_h^+ = \frac{bk A_h \eta_v \beta_h (R_3 - 1)(\mu_h + \delta_h)}{(R_4 - 1)\mu_h \left(\mu_h^2 \mu_v + (\mu_v(\delta_h + \gamma_h) + bk\eta_v \beta_h)\mu_h + \mu_v \delta_h \gamma_h + bk\eta_v \beta_h(\delta_h + \gamma_h)\right)},$$

$$R_h^+ = \frac{bkA_h\eta_v\beta_h\gamma_h\left(R_3 - 1\right)}{\left(R_4 - 1\right)\mu_h\left(\mu_h^2\mu_v + \left(\mu_v\left(\delta_h + \gamma_h\right) + bk\eta_v\beta_h\right)\mu_h + \mu_v\delta_h\gamma_h + bk\eta_v\beta_h\left(\delta_h + \gamma_h\right)\right)},$$

$$A_v^+ = -N_h k\left(R_5 - 1\right),$$

$$S_v^+ = -\frac{\left(R_4 - 1\right)\left(\mu_h^2\mu_v + \left(\mu_v\left(\delta_h + \gamma_h\right) + bk\eta_v\beta_h\right)\mu_h + \mu_v\delta_h\gamma_h + bk\eta_v\beta_h\left(\delta_h + \gamma_h\right)\right)\mu_h N_h^2}{\beta_h\left(N_h\mu_h^2\mu_v + \left(N_h\mu_v\left(\delta_h + \gamma_h\right) + b\beta_v A_h\right)\mu_h + b\beta_v A_h\delta_h\right)b},$$

$$In_v^+ = -\frac{A_h N_h b\beta_v\eta_v k\left(R_3 - 1\right)\left(\mu_h + \delta_h\right)}{\mu_v\left(N_h\mu_h\left(\delta_h + \gamma_h + \mu_h\right)\mu_v + b\beta_v A_h\left(\mu_h + \delta_h\right)\right)}.$$

$$(16)$$

It is seen that $S_h^+, V_h^+, In_h^+, R_h^+, A_v^+, S_v^+$, and In_v^+ points can have either a positive or a negative value. But as in the MFE and DFE case, an equilibrium point has biological meaning only if it has positive values. Therefore, to ensure that the EE point exists, the condition for $S_h^+, V_h^+, In_h^+, R_h^+, A_v^+, S_v^+$, and In_v^+ which has positive value needs to be made; that is,

R_3

$$= \frac{\left(\phi N_h\mu_h\left(\mu_h + \gamma_h\right)\mu_v + b^2\beta_h\beta_v kA_h\left(\mu_A + \eta_v\right)\right)\mu_v}{A_h b^2\beta_h\beta_v\eta_v k\phi}$$

$$< 1,$$

R_4

$$= \frac{b\beta_h k\mu_v\left(\mu_A + \eta_v\right)\left(\delta_h + \gamma_h + \mu_h\right)}{\left(b\beta_h k\left(\delta_h + \gamma_h + \mu_h\right)\eta_v + \mu_v\left(\mu_h + \delta_h\right)\left(\mu_h + \gamma_h\right)\right)\phi}$$

$$< 1,$$

$$R_5 = \frac{\mu_v\left(\mu_A + \eta_v\right)}{\phi\eta_v} < 1.$$

$$(17)$$

Numerical simulation, using data parameters in Table 1, is performed to show an example of the stability of endemic equilibrium points. The numerical simulation result of the equilibrium point stability can be seen in Table 2.

4. Basic Reproduction Number

4.1. Construction of Basic Reproduction Number. Basic reproduction number ($\mathscr{R}_0$) is defined as the expected number of secondary cases from one primary case in a virgin population during the infection period [24]. $\mathscr{R}_0$ can be taken from the spectral radius of the next-generation matrix. Please see [25] for further explanation about the construction of the next-generation matrix of the compartmental model in various disease models.

According to our model in system (1) and evaluating it at disease-free equilibrium (DFE) in (5), our next-generation matrix is given by

$$\mathscr{K} = \begin{pmatrix} 0 & -\frac{b\beta_h}{N_h\left(-\mu_v - u_4\right)}\left(\frac{A_h\left(\mu_h u_2 + u_1\right)\xi}{\mu_h\left(\mu_h + u_1 + \theta_h\right)} - \frac{A_h\left(\mu_h u_2 - \mu_h - \theta_h\right)}{\mu_h\left(\mu_h + u_1 + \theta_h\right)}\right) \\ \frac{b\beta_v u_5 k\left(\left(\mu_v + u_4\right)\left(u_3 + \mu_A + \eta_v\right) - \eta_v\phi\right)}{\phi\left(\mu_v + u_4\right)\left(-\gamma_h - \mu_h\right)} & 0 \end{pmatrix}. \quad (18)$$

The element of the next-generation matrix $\mathscr{K}$ can be interpreted as follows: the number of new infections in jth column is caused by one infection from ith row of $\mathscr{K}$. Please note that i and j for 1 and 2 represent In_h and In_v group. Therefore, for example, $\mathscr{K}_{2,1}$ represent the case that one infected mosquito will produce $b\beta_v u_5 k\left(\left(\mu_v + u_4\right)\left(u_3 + \mu_A + \eta_v\right) - \eta_v\phi\right)/\phi\left(\mu_v + u_4\right)\left(-\gamma_h - \mu_h\right)$ number of new infected people. On the other hand, $\mathscr{K}_{1,2}$ represent the case that one infected human will produce $-\left(b\beta_h/N_h\left(-\mu_v - u_4\right)\right)\left(A_h\left(\mu_h u_2 + u_1\right)\xi/\mu_h\left(\mu_h + u_1 + \theta_h\right) - A_h\left(\mu_h u_2 - \mu_h - \theta_h\right)/\mu_h\left(\mu_h + u_1 + \theta_h\right)\right)$ number of new infected mosquitos. Supported by dengue facts that the new infection in human and mosquito population cannot occur from contact between human and human or mosquito and mosquito, we have that $\mathscr{K}_{1,1}$ and $\mathscr{K}_{2,2}$ are equal to 0.

Finding the spectral radius of (18), our basic reproduction number associated with system (1) is given by

$\mathscr{R}_0$

$$= \frac{b^2\beta_v u_5 k\beta_h A_h\left(u_1\xi + \mu_h + \theta_h + \left(\mu_h + \theta_h\right)\left(\xi - 1\right)R_2\right)\left(R_1 - 1\right)}{\phi\left(\mu_v + u_4\right)^2\left(\gamma_h + \mu_h\right)N_h\mu_h\left(\mu_h + u_1 + \theta_h\right)}, \quad (19)$$

with R_1 and R_2 already defined in the previous section.

Please note that, according to the previous section, this $\mathscr{R}_0$ becomes a threshold number to guarantee the existence and local stability of the disease-free equilibrium point (see (5)) and endemic equilibrium point. We find that the disease-free equilibrium point will be locally asymptotically stable when $\mathscr{R}_0 < 1$. This situation will tend the system to possibility

TABLE 2: Numerical example to show existence and stability of equilibrium points for various values of parameters in Table 1 and $u_1 = 0$, $u_3 = 0$, $u_5 = 1$.

ϕ	u_4	u_2	$\mathscr{R}_0$	R_1	Stable equilibrium	Point
3	1	0,1	−0.0012	0.7258	MFE	($S_h = 999.7477$; $V_h = 0.25226$; $In_h = 0$; $R_h = 0$; $A_v = 0$; $S_v = 0$; $In_v = 0$)
300	1	0,1	0.0031	72.5807	DFE-1	($S_h = 999.7477$; $V_h = 0.2522598$; $In_h = 0$; $R_h = 0$; $A_v = 1972.4444$; $S_v = 22.7240$; $In_v = 0$)
300	1	0	0,0031	72.5807	DFE-2	($S_h = 1000$; $V_h = 0$; $In_h = 0$; $R_h = 0$; $A_v = 1972.444$; $S_v = 22.7240$; $In_v = 0$)
300	0	0,1	2,9962	2250,00	EE	($S_h = 494.6757$; $V_h = 0$; 2215; $In_h = 160$; 8530; $R_h = 344$; 2498; $A_v = 1999$; 1111; $S_v = 481$; 5783; $In_v = 232$; 3899)

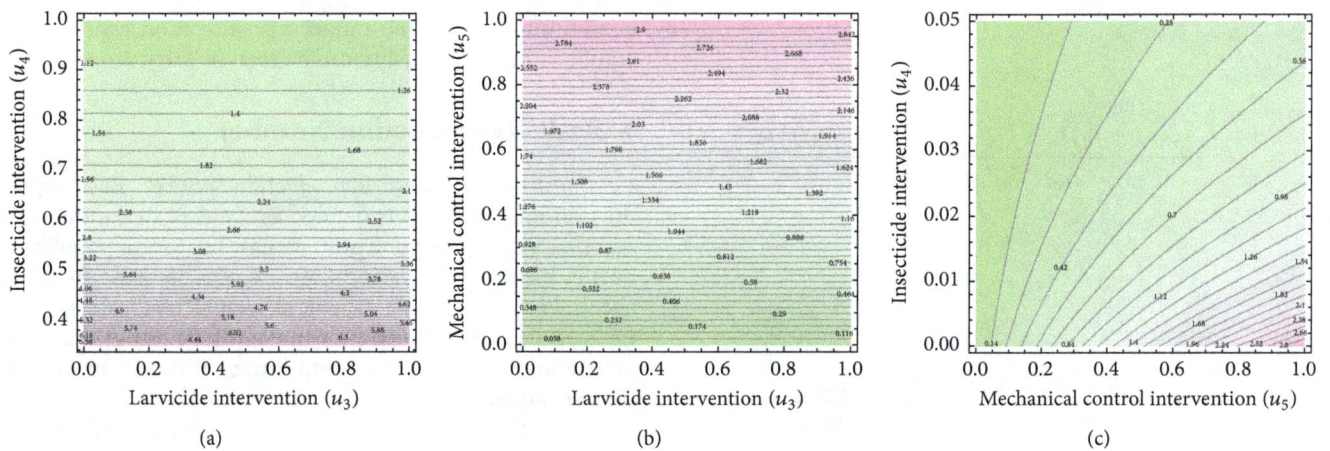

FIGURE 3: Sensitivity analysis of $\mathscr{R}_0$ with respect to dengue intervention in mosquito population.

equilibrium, that is, stable in MFE if $\mathscr{R}_1 < 1$ and stable in DFE if $\mathscr{R}_1 > 1$. On the other hand, if $\mathscr{R}_0 > 1$, the DFE or MFE will be unstable and the system will tend to endemic equilibrium point.

4.2. Sensitivity Analysis of Basic Reproduction Number.

In this subsection, a sensitivity analysis of the basic reproduction number will be performed to find and compare the most sensitive parameters (u_i) to determine the value of $\mathscr{R}_0$. In the first subsection, we will compare the sensitivity analysis of intervention in the mosquito population (u_i for $i = 3, 4, 5$) and in the next subsection we will compare the sensitivity of intervention in the human population (u_i for $i = 1, 2$).

4.2.1. Sensitivity Analysis in Mosquito Population.

As already stated in the previous section, we include larvicide, fumigation, and mechanical control in our model as u_3, u_4, and u_5, respectively. To find the sensitivity curve of basic reproduction number as shown in Figure 3 for u_i and u_j, we input all parameters into $\mathscr{R}_0$ except u_i and u_j and then plot its implicit equation.

In Figure 3(a), a comparison of the efficacy of larvicide and insecticide is performed and we find that insecticide is much more efficacious in reducing $\mathscr{R}_0$ than larvicide. In the next figure, Figure 3(b), we find that intervention using mechanical control is more efficacious in reducing $\mathscr{R}_0$. Finally, we compare the efficacy of insecticide and mechanical control, and we find that insecticide is much more efficacious in reducing $\mathscr{R}_0$. Therefore, from these three figures, we conclude that insecticide is the best way of controlling dengue spread, followed by mechanical control and larvicide, respectively.

4.2.2. Sensitivity Analysis in Human Population.

In this subsection, the same procedure is applied to find the sensitivity of $\mathscr{R}_0$ in Figure 4. It can be seen that the larger the intervention of vaccination we give, the smaller $\mathscr{R}_0$ will be, and reducing $\mathscr{R}_0$ with intervention of adult vaccination (u_1) is faster than with vaccination of newborns (u_2).

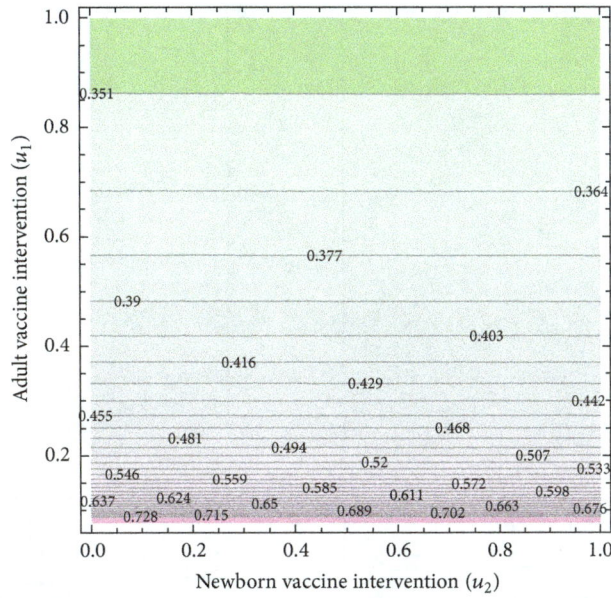

FIGURE 4: Sensitivity analysis of $\mathcal{R}_0$ with respect to dengue intervention in human population.

To back up the result given in the two previous figures in Figures 3 and 4, we will determine the value of each intervention in a single-intervention scenario. This means that if we only use vaccination intervention in the model, we will set the other parameters to 0 for u_2, u_3, u_4 and 1 for u_5. With this scenario, we need $u_1 = 0.04765, u_2 = 293.04, u_3 = 71.363, u_4 = 0.02436$, and $u_5 = 0.33364$ to reduce $\mathcal{R}_0$ to 0.99. It can be seen that u_4 is the smallest value of intervention to reduce $\mathcal{R}_0$ with respect to controlling the spread of dengue.

In the next section, we present some numerical experiments to show the long- and short-term behavior of our model in (1) with respect to the value of various interventions.

5. Numerical Experiment

To perform a numerical experiment in this section, we use the parameters values given in Table 1 and the initial condition given by

$$S_h(0) = 980;$$

$$V_h(0) = 0;$$

$$\text{In}_h(0) = 20;$$

$$R_h(0) = 0; \qquad (20)$$

$$A_v(0) = 2000;$$

$$S_v(0) = 980;$$

$$\text{In}_v(0) = 20.$$

As it can be seen from (20), total numbers of human ($S_h + V_v + \text{In}_h + R_h$) and adult mosquito ($S_v + \text{In}_v$) are 1000 in $t = 0$. The number of infected humans and mosquitoes is small to describe the situation when the infection of dengue

has just started. Using a value of u_i in the previous section to reduce $\mathcal{R}_0$ to 0.99, it can be seen that u_2 and u_3 will not satisfy the condition that u_i should be between 0 and 1. For the next simulation, therefore, we will only perform the dynamic behavior of infected groups (In_h, In_v) with respect to intervention of u_1, u_4, and u_5 as shown in Figures 5 and 6 for In_h and In_v, respectively.

Figures 5 and 6 show the dynamic of infected humans and mosquitoes in the short term ($t \in [0, 100]$) and long term ($t \in [300, 500]$). It can be seen that, without intervention, the number of infected humans and infected mosquitoes will tend to endemic equilibrium, since $\mathcal{R}_0 = 2.99 > 1$. After intervention is given until $\mathcal{R}_0 = 0.99 < 1$ (we take $u_1 = 0.04765$ or $u_4 = 0.02436$ and/or $u_5 = 0.33364$ to represent each simulation), the number of infected humans and mosquitoes will be decreased and pushed to the disease-free equilibrium point. It can also be seen in Figure 5 that intervention of adult vaccination in short-term simulations is the best way to reduce the number of infected humans to the lowest level, rather than other interventions, following this with fumigation and mechanical control interventions, respectively. Unfortunately, for long-term simulations, intervention by fumigation is the best way to reduce the number of infected humans, rather than an adult vaccination strategy. On the other hand, in both short- and long-term simulations, intervention by fumigation is the best way to reduce the number of infected mosquitoes, as shown in Figure 6.

The next simulation is performed to show the efficacy of u_4 as the best strategy for long-term intervention in both human and mosquito populations, as shown in Figure 7. It can be seen that an intervention of u_4 gradually from 0 to 0.1 will reduce $\mathcal{R}_0$ from 2.99 to 0.18. As a consequence, a smaller $\mathcal{R}_0$ will reduce the infected population and delay the

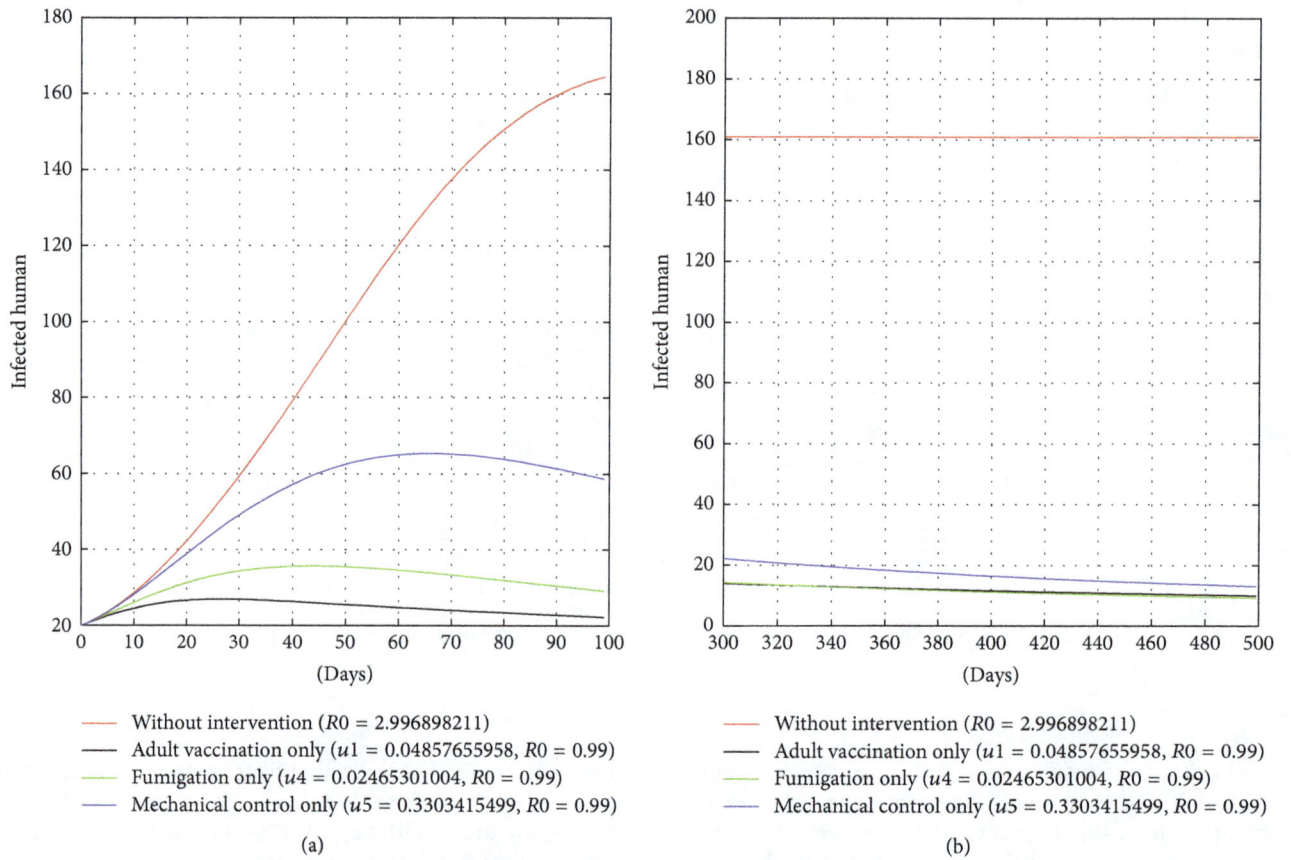

FIGURE 5: Dynamic of infected humans for short-term (a) and long-term (b) intervention.

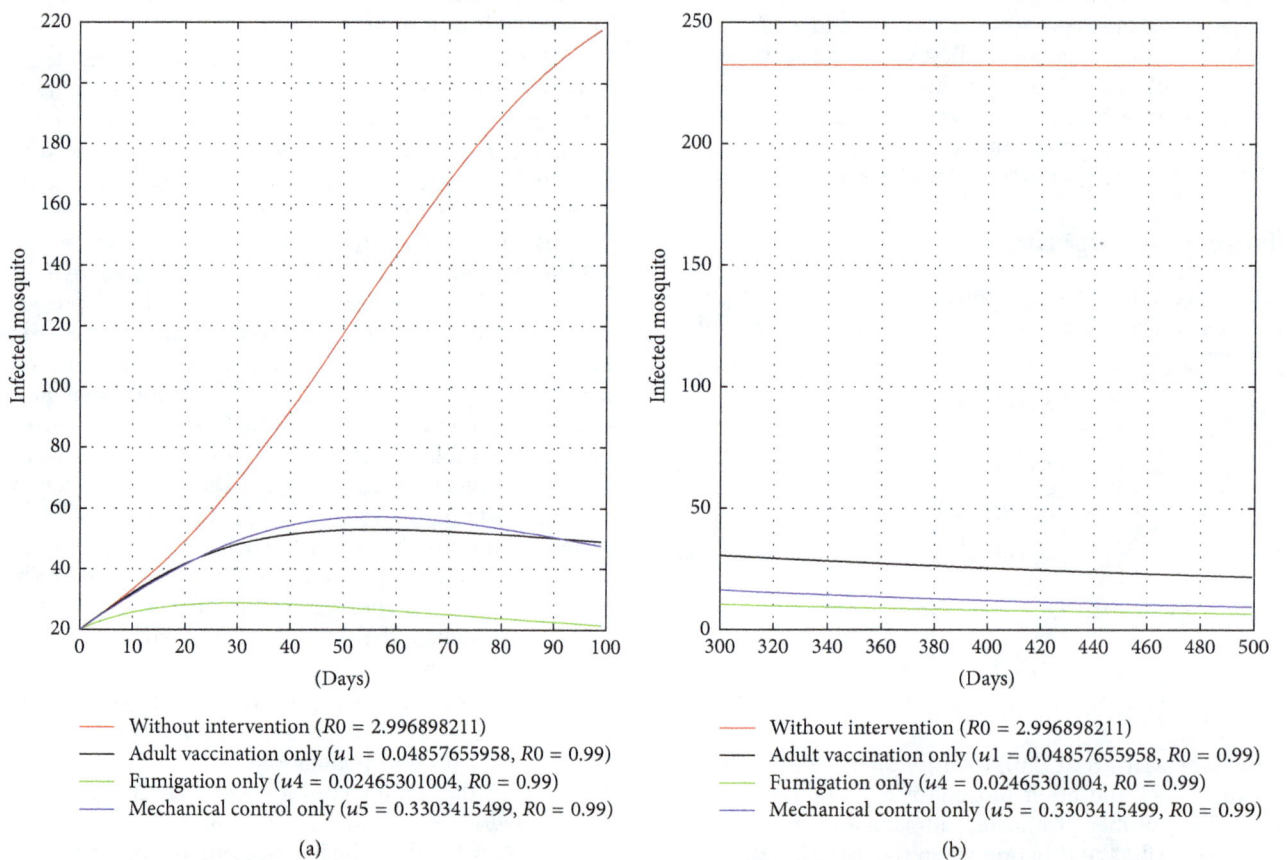

FIGURE 6: Dynamic of infected mosquitoes for short-term (a) and long-term (b) intervention.

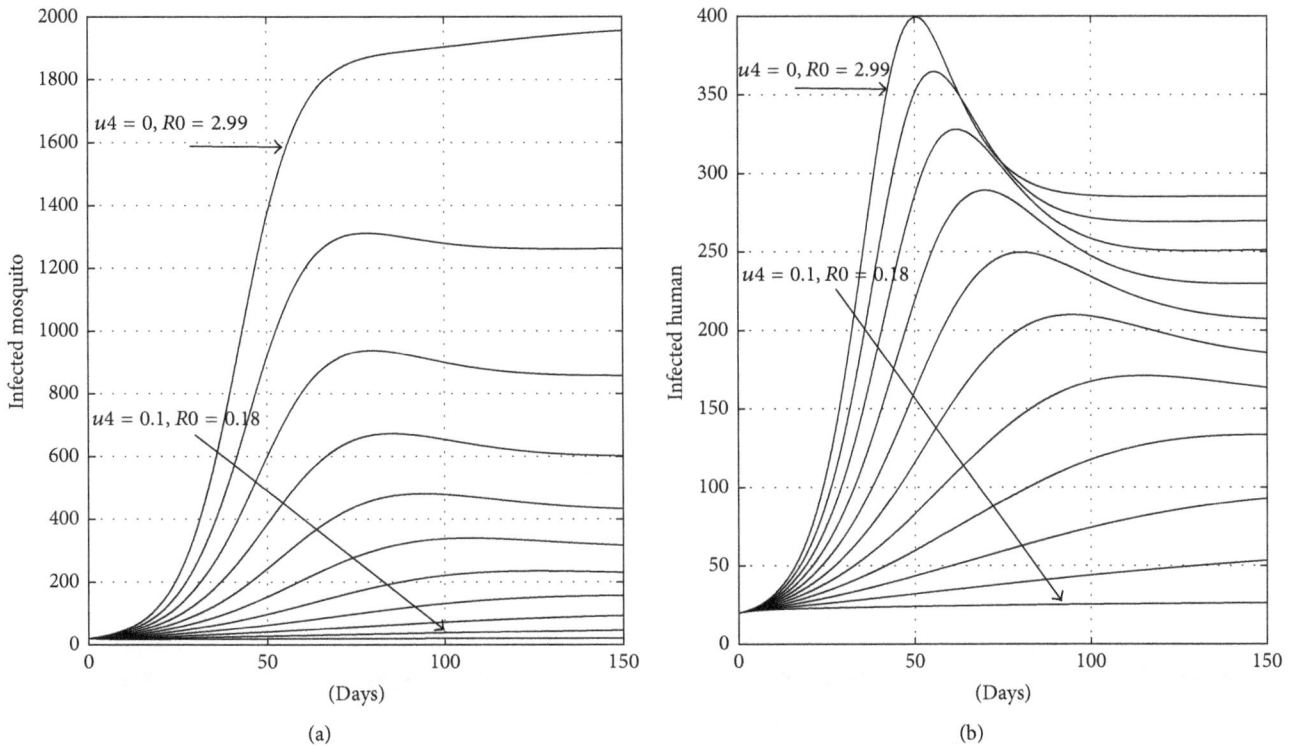

FIGURE 7: Sensitivity of u_4 with respect to the number of infected mosquitoes (a) and infected humans (b).

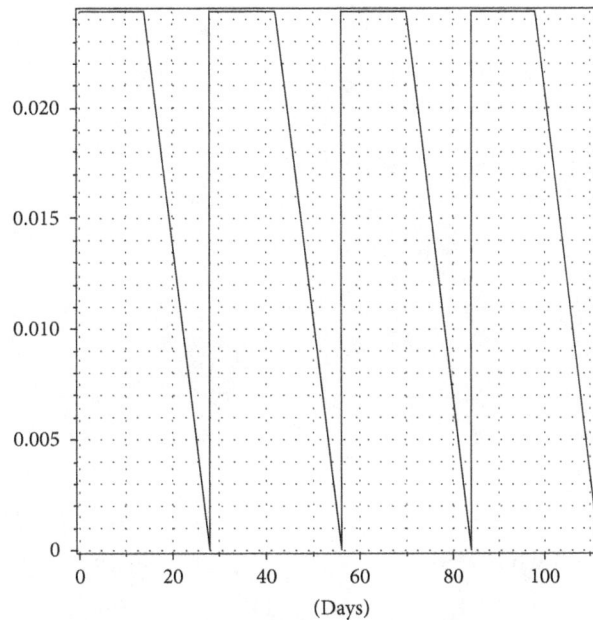

FIGURE 8: Biweekly fumigation strategy.

outbreak for some time. It can also be seen that it needs a proper value of u_4 in order that the dynamic of the infected population will never reach outbreak level.

The last simulation is performed to show the effect of periodic as opposed to constant fumigation intervention. For this purpose, the fumigation is implemented biweekly as a constant (0.02436), and the effect of fumigation will disappear linearly after two weeks, as illustrated in Figure 8. As a result, although constant intervention is much better at significantly reducing the number of infected and susceptible mosquitoes in the mosquito population, as shown in Figure 9, biweekly intervention is only slightly different from constant intervention in reducing the number of infected humans and increasing the number of susceptible humans, as shown in

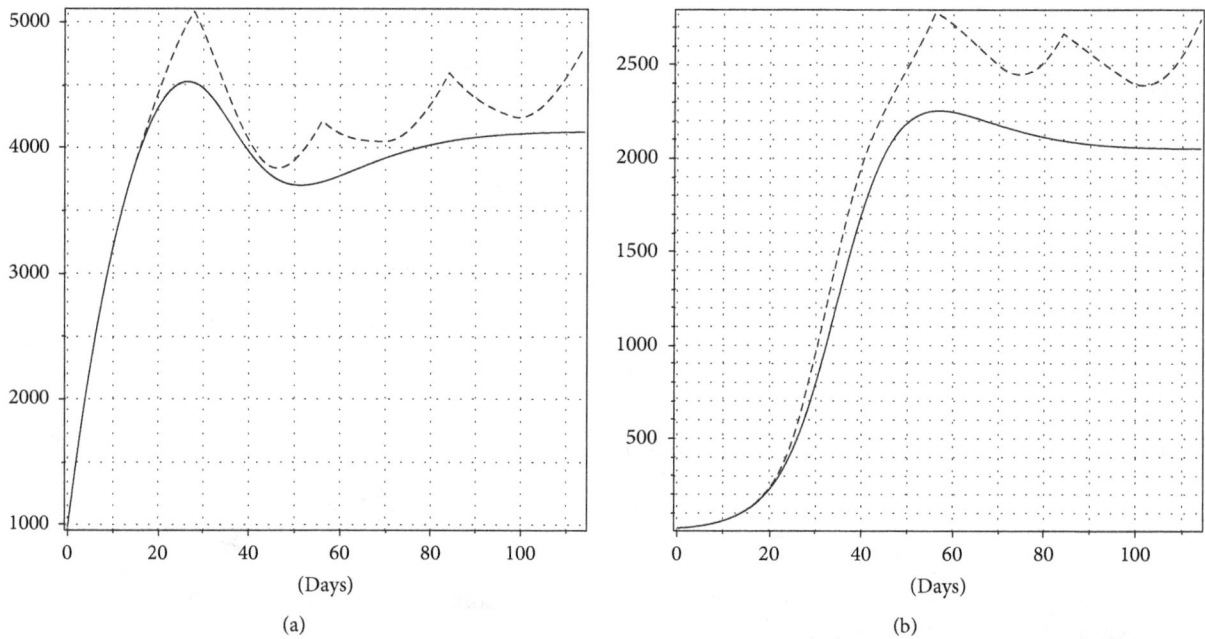

FIGURE 9: Susceptible (a) and infected (b) mosquito dynamics with constant (solid curve) and biweekly (dash curve) strategy.

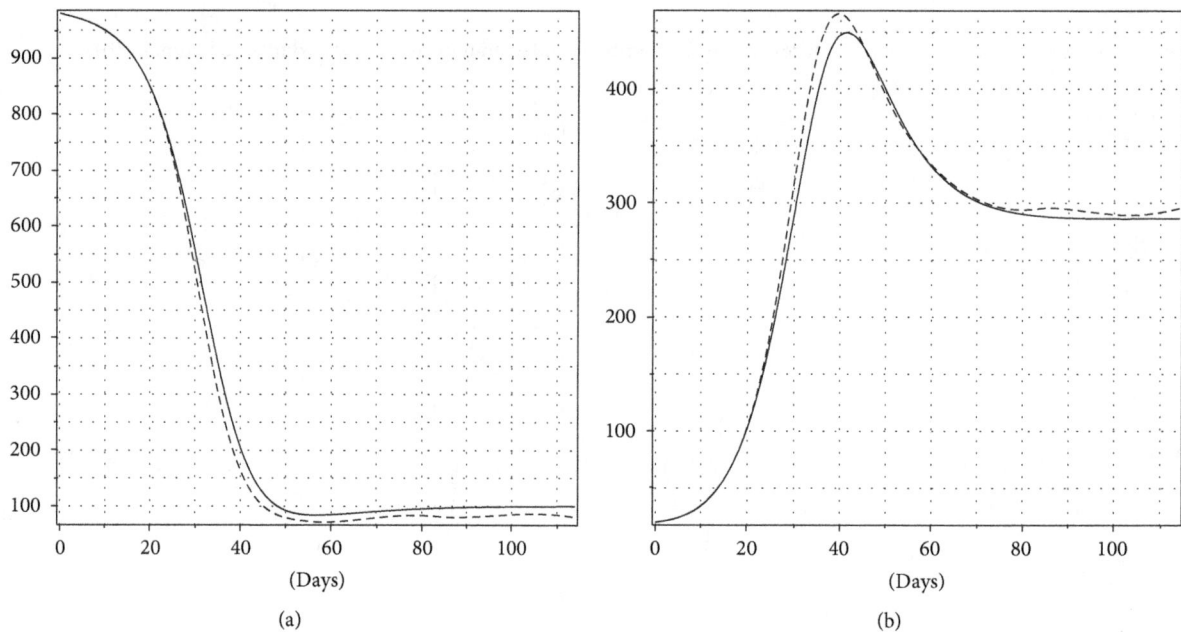

FIGURE 10: Susceptible (a) and infected (b) human dynamics with constant (solid curve) and biweekly (dash curve) strategy.

Figure 10. This result indicates that, rather than implementing fumigation constantly, which is more expensive, it would be better to implement a fumigation strategy periodically, since it involves a lower cost.

6. Conclusions

In this article, we have proposed a mathematical model of dengue spread, with various interventions, such as vaccination of adults and newborns in the human population, and larvicide, insecticide, and mechanical control of the mosquito population. Basic reproduction number and basic offspring as the endemic threshold for disease existence and mosquito existence, respectively, have been shown analytically. We find that disease-free equilibrium will be locally asymptotically stable if, and only if, basic reproduction number is smaller than one and will be unstable otherwise.

From sensitivity analysis and backed up with some numerical simulations, we find that fumigation is the best strategy for long-term intervention to reduce the infected populations of both mosquitoes and humans. But for short-term intervention, vaccination of the adult population is the best way to reduce the number of infected people. From numerical simulation, intervention using fumigation, apart from reducing the outbreak, can also delay an outbreak for some period of time.

We also find that although periodic intervention strategy of fumigation cannot reduce the number of infected humans and mosquitoes as efficiently as constant intervention, it is only slightly different. Therefore, if the government has a limited budget, then periodic intervention could be a good option to implement.

For future research, reconstructing the model in this article as an optimal control problem will be considered to show the effectiveness of interventions, not only based on a reduction of the number of infected humans and mosquitoes but also with the lowest cost for intervention purposes.

Conflicts of Interest

The authors declare that they have no conflicts of interest.

Acknowledgments

This research is funded by Ministry of Research and Higher Education (Kemenristek Dikti), with PUPT research grant, 2017.

References

[1] World Health Organization, *Dengue: Guidelines For Diagnosis, Treatment, Prevention and Control*, World Health Organization. ISBN 9789241547871, 2009.

[2] Education Nature, Dengue Transmission 2016. http://www.nature.com/scitable/topicpage/dengue-transmission-22399758.

[3] H. S. Rodrigues, M. T. T. Monteiro, and D. F. M. Torres, *Vaccination models and optimal control strategies to Dengue*, Universidade de Aveiro Departamento de Matematica, Portugal, 2013.

[4] D. Aldila, N. Nuraini, and E. Soewono, "Mathematical Model in Controlling Dengue Transmission with Sterile Mosquito Strategies," in *Proceedings of the AIP Conference*, Indonesia, 2015.

[5] R. I. Kementrian Kesehatan, "Demam Berdarah Dengue," *Buletin Jendela Epidemiologi*, vol. 2, pp. 11–14, August 2010.

[6] O. A. Aguirre-Obando, A. J. Pietrobon, A. C. D. Bona, and M. A. Navarro-Silva, "Contrasting patterns of insecticide resistance and knockdown resistance (kdr) in Aedes aegypti populations from Jacarezinho (Brazil) after a Dengue Outbreak," *Revista Brasileira de Entomologia*, vol. 60, no. 1, pp. 94–100, 2016.

[7] K. C. Mulyatno, A. Yamanaka, Ngadino, and E. Konishi, "Resistance of Aedes aegypti (L.) larvae to temephos in Surabaya, Indonesia," *Southeast Asian Journal of Tropical Medicine and Public Health*, vol. 43, no. 1, pp. 29–33, 2012.

[8] R. Deming, "Spatial Variation of Insecticide Resistance In The Dengue Vector *Aedes aegypti* Presents Unique Vector Control Challenges," *Parasites & Vectors*, vol. 9, no. 67, 2016.

[9] World Health Organization, *WHO Global Malaria Programme: World Malaria Report*, World Health Organization, Geneva, Switzerland, 2013.

[10] G. Olivera, N. Baurin, and L. Coudeville, "The Role Of Indirect Protection In The Assessment Of Dengue Vaccination Impact," in *Seventh Workshop Dynamical Systems Applied to Biology andNatural Sciences*, Evora, Portugal, 2016.

[11] S. Pasteur, "First Dengue Vaccine, Dengvaxia, Awarded the Vaccine Breakthrough Of 2016 at the 6th Annual Biopharma Industry Awards Event in Singapore," 2016, http://www.dengue.info/#overlay=content/first-dengue-vaccine-awarded-vaccine-breakthrough-2016-singapore-0.

[12] H. Harapan, S. Anwar, A. M. Setiawan, and R. T. Sasmono, "Dengue vaccine acceptance and associated factors in Indonesia: A community-based cross-sectional survey in Aceh," *Vaccine*, vol. 34, no. 32, pp. 3670–3675, 2016.

[13] J. Lee, V. Mogasale, J. K. Lim et al., "A Multi-country Study of the Household Willingness-to-Pay for Dengue Vaccines: Household Surveys in Vietnam, Thailand, and Colombia," *PLOS Neglected Tropical Diseases*, vol. 9, no. 6, p. e0003810, 2015.

[14] T. Wahyuni, "Berantas DBD dengan Jumatik dan Vaksin Dengvaxia," *CNN Indonesia*, 2016, http://www.cnnindonesia.com/gaya-hidup/20160112181200-255-103823/berantas-dbd-dengan-jumantik-dan-vaksin-dengvaxia/.

[15] L. Esteva and C. Vargas, "Analysis of a dengue disease transmission model," *Mathematical Biosciences*, vol. 150, no. 2, pp. 131–151, 1998.

[16] Z. Feng and J. X. Velasco-Hernandez, "Competitive exclusion in a vector-host model for the dengue fever," *Journal of Mathematical Biology*, vol. 35, no. 5, pp. 523–544, 1997.

[17] N. T. J. Bailey, *The Mathematical Theory of Infectious Diseases and its Application*, Griffin, London, UK, 1975.

[18] K. Dietz, "Transmission and control of arbovirus diseases," in *Proceedings of the Society for Industrial and Applied Mathematics*, D. Ludwig et al., Ed., p. 104, Philadelphia, PA, USA, 1974.

[19] H. S. Rodrigues, M. T. Monteiro, D. F. Torres, and A. Zinober, "Dengue disease, basic reproduction number and control," *International Journal of Computer Mathematics*, vol. 89, no. 3, pp. 334–346, 2012.

[20] H. S. Rodrigues, M. T. T. Monteiro, and D. F. M Torres, "Modeling and Optimal Control Applied to a Vector Borne Disease," in *Computational and Mathematical Methods in Science and Engineering (CMMSE) 2012*, J. Vigo-Aguiar, Ed., vol. 3, pp. 1063–1070, Murcia, Spain, July 2012.

[21] D. Aldila, N. Nuraini, E. Soewono, and A. K. Supriatna, "Mathematical Model of Temephos Resistance in Aedes aegypti Mosquito Population," in *Proceedings of theAIP Conference*, 2013.

[22] A. K. Supriatna, E. Soewono, and S. A. van Gils, "A two-age-classes dengue transmission model," *Mathematical Biosciences*, vol. 216, no. 1, pp. 114–121, 2008.

[23] D. Aldila, T. Götz, and E. Soewono, "An optimal control problem arising from a dengue disease transmission model," *Mathematical Biosciences*, vol. 242, no. 1, pp. 9–16, 2013.

[24] O. Diekmann and J. A. P. Heesterbeek, *Mathematical Epidemiology of Infectious Diseases, Model Building, Analysis and Interpretation*, John Wiley & Sons, Chichester, UK, 2000.

[25] O. Diekmann, J. A. P. Heesterbeek, and M. G. Roberts, "The construction of next-generation matrices for compartmental epidemic models," *Journal of the Royal Society Interface*, vol. 7, no. 47, pp. 873–885, 2010.

Permissions

The contributors of this book come from diverse backgrounds, making this book a truly international effort. This book will bring forth new frontiers with its revolutionizing research information and detailed analysis of the nascent developments around the world.

We would like to thank all the contributing authors for lending their expertise to make the book truly unique. They have played a crucial role in the development of this book. Without their invaluable contributions this book wouldn't have been possible. They have made vital efforts to compile up to date information on the varied aspects of this subject to make this book a valuable addition to the collection of many professionals and students.

This book was conceptualized with the vision of imparting up-to-date information and advanced data in this field. To ensure the same, a matchless editorial board was set up. Every individual on the board went through rigorous rounds of assessment to prove their worth. After which they invested a large part of their time researching and compiling the most relevant data for our readers.

The editorial board has been involved in producing this book since its inception. They have spent rigorous hours researching and exploring the diverse topics which have resulted in the successful publishing of this book. They have passed on their knowledge of decades through this book. To expedite this challenging task, the publisher supported the team at every step. A small team of assistant editors was also appointed to further simplify the editing procedure and attain best results for the readers.

Apart from the editorial board, the designing team has also invested a significant amount of their time in understanding the subject and creating the most relevant covers. They scrutinized every image to scout for the most suitable representation of the subject and create an appropriate cover for the book.

The publishing team has been an ardent support to the editorial, designing and production team. Their endless efforts to recruit the best for this project, has resulted in the accomplishment of this book. They are a veteran in the field of academics and their pool of knowledge is as vast as their experience in printing. Their expertise and guidance has proved useful at every step. Their uncompromising quality standards have made this book an exceptional effort. Their encouragement from time to time has been an inspiration for everyone.

The publisher and the editorial board hope that this book will prove to be a valuable piece of knowledge for researchers, students, practitioners and scholars across the globe.

List of Contributors

Driss Kiouach and Lahcen Boulaasair
MSTI (Modelling, Systems and Technologies of Information) Team, High School of Technology, Ibn Zohr University, Agadir, Morocco

Bakary Traoré, Boureima Sangaré and Sado Traoré
Department of Mathematics, Polytechnic University of Bobo Dioulasso, Bobo-Dioulasso 01, Burkina Faso

Sansao A. Pedro
Departamento de Matemática e Informática, Universidade Eduardo Mondlane, 257, Maputo, Mozambique

Sibaliwe Maku Vyambwera and Peter Witbooi
Department of Mathematics and Applied Mathematics, University of the Western Cape, Private Bag X17, Bellville 7535, South Africa

Steady Mushayabasa
University of Zimbabwe, Department of Mathematics, Harare, Zimbabwe

Anthony A. E. Losio
University of Zimbabwe, Department of Mathematics, Harare, Zimbabwe
University of Juba, Department of Mathematics, Juba, Central Equatoria, South Sudan

Rachel Waema Mbogo, Livingstone S. Luboobi and John W. Odhiambo
Institute of Mathematical Sciences (IMS), Strathmore University, Nairobi, Kenya

Jaharuddin
Department of Mathematics, Faculty of Mathematics and Natural Sciences, Bogor Agricultural University, Bogor, West Java 16680, Indonesia

Hadi Hermansyah
Department of Mechanical Engineering, Balikpapan State Polytechnic, Balikpapan, East Borneo 76126, Indonesia

Raid Kamel Naji and Reem Mudar Hussien
Department of Mathematics, College of Science, University of Baghdad, Baghdad, Iraq

Piotr Skrzypacz and Dongming Wei
School of Science and Technology, Nazarbayev University, 53 Kabanbay Batyr Ave., Astana 010000, Kazakhstan

Ahmed Sami Abdulghafour
Scientific Affairs Department, Aliraqia University, Baghdad, Iraq

Raid Kamel Naji
Department of Mathematics, College of Science, University of Baghdad, Baghdad, Iraq

Isaac Mwangi Wangari
Royal Melbourne Institute of Technology School of Mathematics and Geospatial Sciences, Melbourne, VIC, Australia

Lewi Stone
Royal Melbourne Institute of Technology School of Mathematics and Geospatial Sciences, Melbourne, VIC, Australia
Biomathematics Unit, Department of Zoology, Faculty of Life Sciences, Tel Aviv University, Tel Aviv, Israel

B. Mobisa
Department of Mathematics, Laikipia University, Nyahururu, Kenya

G. O. Lawi and J. K. Nthiiri
Department of Mathematics, Masinde Muliro University of Science and Technology, Kakamega, Kenya

Yudi Ari Adi
Department of Mathematics, Faculty of Mathematics and Natural Sciences, Ahmad Dahlan University, Yogyakarta 55166, Indonesia
Department of Mathematics, Faculty of Mathematics and Natural Sciences, Universitas Gadjah Mada, Yogyakarta 55281, Indonesia

Fajar Adi-Kusumo and Lina Aryati
Department of Mathematics, Faculty of Mathematics and Natural Sciences, Universitas Gadjah Mada, Yogyakarta 55281, Indonesia

Mardiah S. Hardianti
Department of Internal Medicine, Faculty of Medicine, Universitas Gadjah Mada, Yogyakarta 55281, Indonesia

Kevin Jan Duffy
Institute of Systems Science, Durban University of Technology, Durban 4000, South Africa

Obiora Cornelius Collins
Institute of Systems Science, Durban University of
Technology, Durban 4000, South Africa
School of Mathematics, Statistics and Computer
Science, University of Kwa Zulu-Natal, Durban 4000,
South Africa

**Jevgeṇijs Carkovs, Jolanta Goldšteine and Kārlis
Šadurskis**
Department of Probability Theory and Mathematical
Statistics, Riga Technical University, Kaļķu iela 1, Riga
LV-1658, Latvia

Mohammed Kizito and Julius Tumwiine
Department of Mathematics, Mbarara University of
Science and Technology, Mbarara, Uganda

A. Bustamam, D. Aldila and A. Yuwanda
Department of Mathematics, Universitas Indonesia,
Depok 16424, Indonesia

Index

www.ingramcontent.com/pod-product-compliance
Lightning Source LLC
Chambersburg PA
CBHW080623200326
41458CB00013B/4491

Radio Systems Engineering and Wireless Technology

Radio Systems Engineering and Wireless Technology

Edited by **Adrian Franel**

WILLFORD PRESS

New York

Published by Willford Press,
118-35 Queens Blvd., Suite 400,
Forest Hills, NY 11375, USA
www.willfordpress.com

Radio Systems Engineering and Wireless Technology
Edited by Adrian Franel

© 2016 Willford Press

International Standard Book Number: 978-1-68285-056-5 (Hardback)

Contents

Preface

Radio systems are widely used forms of wireless technologies across the world in today's time. They form a core area of electrical engineering with widespread applications. This book discusses the fundamentals as well as modern approaches of radio systems through detailed elaboration of topics such as electromagnetics, circuits, propagation, microwaves, etc. It is an essential guide for students who are looking for an elaborate reference on radio systems engineering. A number of latest researches have been included to keep the readers up-to-date with the global concepts in this area of study.

This book has been the outcome of endless efforts put in by authors and researchers on various issues and topics within the field. The book is a comprehensive collection of significant researches that are addressed in a variety of chapters. It will surely enhance the knowledge of the field among readers across the globe.

It gives us an immense pleasure to thank our researchers and authors for their efforts to submit their piece of writing before the deadlines. Finally in the end, I would like to thank my family and colleagues who have been a great source of inspiration and support.

Editor

Data Transparent and Polarization Insensitive All-Optical Switch Based on Fibers with Enhanced Nonlinearity

Matej KOMANEC, Pavel SKODA, Jan SISTEK, Tomas MARTAN

Department of Electromagnetic Field, Faculty of Electrical Engineering, Czech Technical University in Prague,
Technicka 2, Prague 6, 166 27, Czech Republic

komanmat@fel.cvut.cz, skodapav@fel.cvut.cz, sistekj@fel.cvut.cz, martant@fel.cvut.cz

Abstract. *We have developed a data transparent optical packet switch prototype employing wavelength conversion based on four-wave mixing. The switch is composed of an electro-optical control unit and an all-optical switching segment. To achieve higher switching efficiencies, Ge-doped silica suspended-core and chalcogenide arsenic-selenide single-mode fibers were experimentally evaluated and compared to conventional highly-nonlinear fiber. Improved connectorization technology has been developed for Ge-doped suspended-core fiber, where we achieved connection losses of 0.9 dB. For the arsenic-selenide fiber we present a novel solid joint technology, with connection losses of only 0.25 dB, which is the lowest value presented up-to-date. Conversion efficiency of -13.7 dB was obtained for the highly-nonlinear fiber, which is in perfect correlation with previously published results and thus verifies the functionality of the prototype. Conversion efficiency of -16.1 dB was obtained with arsenic-selenide fiber length reduced to five meters within simulations, based on measurement results with a 26 m long component. Employment of such a short arsenic-selenide fiber segment allows significant broadening of the wavelength conversion spectral range due to possible neglection of dispersion.*

Keywords

All-optical networks, optical switching, wavelength conversion, four-wave mixing, chalcogenide fibers.

1. Introduction

Increasing data traffic such as 3D multimedia data streams, full-HD videos and real-time data transfers imposes demands for all-optical network solutions, represented by optical burst or optical packet switching. With the rise of new modulation formats in optical communication such as dual-polarization quadrature phase-shift keying (DP-QPSK) [1] and m-ary quadrature amplitude modulation (m-QAM) [1], [2], optical networks will require modulation format transparent, polarization insensitive switching processes with switching speeds in orders of Tbit/s. Several solutions to optical packet switching have been proposed, e.g. based on optical gating [3], optical flip-flops [4] or micro-electro-mechanical systems (MEMS) [5]. All-optical processing of a 4-bit optical packet label was presented in [6], where several RF signals were imprinted on each label.

Optical packet switching based on wavelength conversion offers a viable solution for future optical packet switched networks. The major advantage of wavelength conversion based on four-wave mixing (FWM) stands in modulation format and data bitrate insensitivity. Highly-nonlinear fibers (HNLFs) have been exploited for wavelength conversion [7]. Specialty non-silica and microstructured fibers, e.g., chalcogenide fibers [8], bismuth fibers [9] and microstructured fibers thereof [10], [11], provide extremely high nonlinearities ($\gamma > 1000$ W^{-1}km^{-1}), promising enhanced conversion efficiencies while simultaneously decreasing component length, which can result in neglection of dispersive effects. Major drawback of these fiber stands in high coupling losses, when connected to a conventional silica fibers. Free-space optic approaches were proposed for coupling into arsenic-selenide fibers with only 37% coupling efficiency [12], but these are not suitable for real network application. Solid joints for arsenic-selenide fibers were presented [13] with 2.45 dB loss per joint achived by butt-coupling to silica fiber via 5 mm of high NA fiber and index matching oil to improve coupling. Furthermore in [13] arsenic-selenide single-mode fibers were measured and provided attenuation of ≈ 1 dB/m.

This paper presents results from the development of a prototype hybrid optical packet switch (OPS) based on optical fibers with enhanced nonlinearity (NLF) and electro-optical packet label processing. The aim was to provide transparent polarization insensitive data switching without exceeding the optical power limits in telecommunications (≤ 23 dBm). For maximal OPS transparency, wavelength conversion based on FWM was exploited. Novel solid joint technologies were developed for chalcogenide fibers to enable efficient broadband wavelenght conversion. Polarization sensitivity was significantly decreased by employing a specific polarization insensitive configuration. Wavelength conversion efficiencies were measured for all NLFs, considering limiting effects and NLFs insertion loss.

The paper first discusses theory of the FWM effect and demands placed on the OPS and its functionality. Afterwards a prototype configuration for experimental evaluation is described. Wavelength range is evaluated and frequency plan is proposed for spectrally efficient optical packet switching. 10 Gbit/s non-return zero (NRZ) was measured experimentally and bit-error rate (BER) and optical signal-to-noise ratio (OSNR) tests were carried out. Optimization of the arsenic-selenide fiber is presented. The paper concludes with a summary of achieved results and future enhancement possibilities.

2. Theoretical Background

The FWM effect considers two co-propagating waves, often denoted as pumps, at frequencies ω_{p1} and ω_{p2}. If the phase-matching condition is fulfilled, both waves are co-polarized and nonlinearity of the medium is sufficient, two new signals emerge at difference frequencies, which are denoted as idlers. In case of degenerate FWM (DFWM) only one pump is present, i.e., the frequencies of the two pumps are identical $\omega_{p1} = \omega_{p2}$. Then the pump propagates in the medium and if a phase-matched, co-polarized signal (data) is present at ω_{data}, only one idler is generated at frequency $\omega_p - \omega_{data}$. DFWM is exploited mostly for wavelength conversion or parametric amplification. For this work DFWM will be discussed and utilized.

Conversion efficiency η is one of the main factors for wavelength conversion via DFWM, i.e., for optical data switching. It is fundamentally dependent on optical fiber nonlinearity, optical fiber length and utilized optical power (pump peak power) as [14]:

$$\eta = \frac{P_{switched}(L)}{P_{input}(0)} = \left[\frac{\gamma P_{pump}}{g} \sinh(gL)\right]^2 \quad (1)$$

and

$$\gamma = \frac{2\pi n_2}{\lambda A_{eff}} \quad (2)$$

where $P_{switched}$ stands for the switched data peak power, P_{input} is the input data peak power, γ stands for fiber nonlinearity, g is the gain coefficient, P_{pump} is the pump peak power, A_{eff} stands for the effective mode area, n_2 is the nonlinear refractive index and L represents NLF effective length.

It is obvious that higher nonlinear refractive index n_2, i.e. nonlinear coefficient γ, results in more efficient nonlinear processes. On the other hand higher values of n_2 are in correlation with refractive index n, which implies higher refractive index contrast. When connecting the nonlinear fiber to conventional silica fiber (which is sooner or later unavoidable in any OPS setup) connection losses are then implied. Second factor which is interconnected with n_2 is material attenuation, which increases with higher n_2. As a representative of NLF with high n_2) a chalcogenide arsenic-selenide single-mode fiber (As$_2$Se$_3$ fiber) was selected for our measurements.

Nonlinear coefficient γ depends (apart from n_2) on A_{eff} and λ, which was considered fixed in vicinity of 1550 nm (C-band), therefore only A_{eff} can be minimized to increase nonlinearity. To achieve A_{eff} decrease, microstructured optical fibers [11] or fiber tapering [15], [16] can be employed. Both these methods again imply increased component attenuation. As a representative of microstructured NLF a highly Ge-doped microstructured suspended-core single-mode fiber (Ge-SCF) was employed in our measurements.

Modulation format transparency is closely related to bitrate and polarization insensitivity, e.g. when DP-QPSK is applied, the signal is propagated in both polarization axes and is also phase modulated. Proposed switching methodology of FWM is able to convert advanced modulation formats. Attention was paid to selected formats utilized in current networks and recognized by the ITU-T and IEEE respectively [17], [18] – 10 Gbit/s NRZ, 40 Gbit/s DPSK and 100 Gbit/s DP-QPSK.

Polarization insensitivity is a crucial parameter for efficient DFWM, where high switching efficiency has to be ensured to reduce switched signal amplification requirements. Amplification of the switched data decreases its OSNR performance. Various polarization insensitive configurations for DFWM wavelength conversion are discussed and proposed in [19], [20], [21].

3. Experimental Setup

An experimental 1x2 switch protype setup is depicted in Fig. 1. The developed OPS consisted of four functional segments. The first segment was composed of a data payload generator, label generator and a multiplexer (MUX) forming the optical packet, whose length was designed to be 100 ns, with a 80 ns 8-bit long label. The second segment was responsible for input data buffering. The third segment detected and processed the packet label and governed routing signal generation. This was realized by an opto-electronic controller (OEC) with an embedded FPGA (field-programmable gateway array) controlled remotely via Ethernet. Afterwards the buffered input data was multiplexed with the routing signal in athermal arrayed-waveguide grating (AWG) specifically developed for the purpose of this prototype, with stable function from -40 °C to +80 °C. In the fourth section the coupled routing signals and input data propagated through NLF and a switched data payload (in this paper denoted as switched data) appeared at a new wavelength defined by the DFWM process. Then routing signals and input data were attenuated in an optical bandpass filter (OBF) and furthermore demultiplexed in AWG. New label was attached to the switched data at the output of AWG thus forming a new optical packet. Polarization insensitive configuration with NLF will be discussed in detail in a separate section.

Vizualization of the OPS prototype is depicted in Fig. 2, with output for amplifier attachment (front panel) and

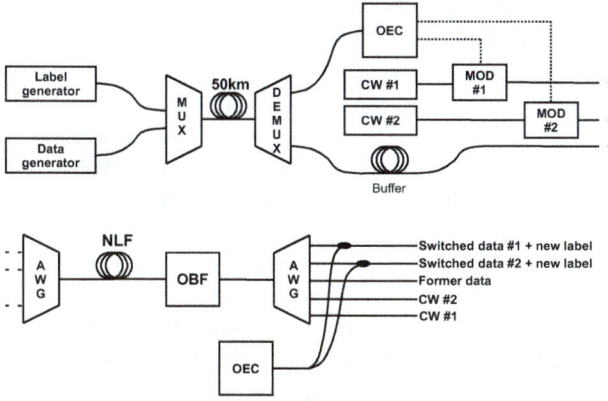

Fig. 1. Scheme of the proposed OPS setup, MOD – Mach-Zehnder modulator, MUX/DEMUX – multi/demultiplexer, PRBS – pseudo-random bit sequence, OEC – opto-electronic controller, NLF – fiber with enhanced nonlinearity, AWG – arrayed-waveguide grating.

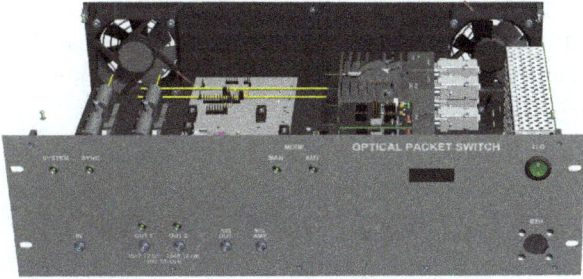

Fig. 2. Optical packet switch prototype - visualization.

various nonlinear fiber testing (rear panel). The prototype can be operated in either manual or automatic regime and can be controlled via Ethernet.

A standard commercialy available HNLF was chosen as a representative of a conventional NLF. Fiber dimensions were similar to standard SMF-28e fiber, γ was measured to be 11.35 $W^{-1}km^{-1}$, zero-dispersion wavelength in vicinity of 1559 nm and attenuation of 0.7 dB/km. For the measurements a 500 m long component was employed.

A suspended-core microstructured silica fiber with highly Ge-doped core provided by IPHT, Jena, Germany was exploited with a core diameter of 6.6 μm, γ of 21.0 $W^{-1}km^{-1}$, dispersion D of 32.6 ps/nm/km at 1550 nm and attenuation of 43.0 dB/km. For experimental tests a 100 m component was exploited with connection losses of 0.9 dB per junction.

Chalcogenide As_2Se_3 single-mode fiber with single-mode cut-off at approximately ~1300 nm, γ of approximately 1300 $W^{-1}km^{-1}$, dispersion D of -560 ps/nm/km at 1550 nm has been employed in our measurements. The fiber was prepared by the double-crucible technique developed at the Naval Research Laboratory [22]. It provided minimal core excentricity (below 1 μm), leading to simplified coupling to conventional fibers. A final component containing 26 m As_2Se_3 fiber spool with standard silica SMF-28e

fibers at output was developed at SQS Fiber optics, Czech Republic. Insertion loss (IL) of 16.0 dB for the whole component was obtained. A study of backreflected power was performed, which provided evidence of reflection at the input As_2Se_3 connection, therefore we assume ~0.25 dB loss per connection, which is the lowest loss according to authors knowledge for a As_2Se_3 single-mode fiber connection to conventional silica single-mode fiber reported world-wide up to date [13]. This leads to fiber attenuation of less than 0.6 dB/m (measured by the LUNA device), which is in contrast to manufacturer data of 0.72 dB/m at 1550 nm, furthermore this value is 0.4 dB lower than presented in [13]. It allows to develop As_2Se_3 components of ~1 m lengths with insertion loss <1.5 dB, which is a significant result for further exploitation of nonlinear processes based on arsenic-selenide fibers.

Table 1 summarizes material losses α_{mat} and component measured insertion losses α_{comp} for employed fibers (with connection to standard SMF-28e fiber outputs).

Fiber	α_{mat} [dB/m]	α_{comp} [dB]
500 m HNLF	$7 \cdot 10^{-4}$	0.55
26 m As_2Se_3	0.580	16.00
100 m Ge-SCF	0.043	6.10

Tab. 1. Nonlinearity, material attenuation and component insertion loss.

4. Measurement

For initial comparison, all three evaluated NLFs were tested at +1 channel detuning of the input data from the pump signal (considering DWDM 100 GHz grid). The highest conversion efficiency was expected from HNLF as HNLF provided the most suitable combination of low component IL (0.55 dB) and γL coefficient of 5.67. Second highest from As_2Se_3 fiber and the lowest conversion efficiency was expected from Ge-SCF. As_2Se_3 fiber provided significantly higher γL coefficient of 33.8, but on the other hand component IL of 16.0 dB. Ge-SCF γL coefficient of 2.1 combined with component IL of 6.1 dB promised low conversion efficiencies. For all measurements the pump was tuned from 6 dBm to 16 dBm peak power, whereas the input data was kept at the same optical power.

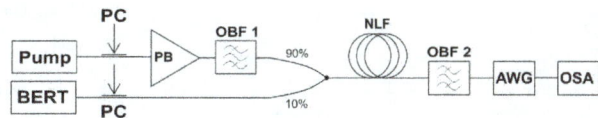

Fig. 3. Setup for conversion efficiency and wavelength range measurements.

Wavelength conversion efficiency η is independent of input data power (as defined in (1)), therefore only pump power played a critical role. In our case it was limited by the stimulated Brillouin scattering (SBS) effect (HNLF at 13.0 dBm, As_2Se_3 at 16.8 dBm and Ge-SCF at 26.6 dBm, considering pump-width of 1 MHz). Employed thin-film components (e.g. multiplexer - MUX or AWG) are power limited at 23 dBm, which was significant only for Ge-SCF, as all other evaluated fibers had their SBS thresholds below the 23 dBm value. The pump phase-modulation technique to suppress SBS can be utilized to increase pump peak power, but if advanced modulation formats should be employed, phase-modulation of the pump severely decreases switched signal parameters, as is presented in detail in [23].

Measurement setup for conversion efficiency and wavelength range evaluation is presented in Fig. 3. A BER tester with a SFP module at 1552.52 nm serves as the data source, whereas a tunable laser source (ID Photonics CoBrite) provides the pump signal. Both are polarization controller (PC) and the pump is separately amplified in a power booster (PB) and excess noise is filtered in an optical bandpass filter (OBF 1). A 90/10 coupler is exploited to preserve pump power. After propagation through NLF the idler signal is filtered in another optical bandpass filter (OBF 2) and sent into an AWG. Output is monitored by an optical spectral analyzer (OSA) - Yokogawa AQ6370.

For all measurements the obtained spectra after the NLF are presented. Conversion efficiency is then calculated as the difference between signal power (before conversion, i.e. NLF) and idler power (after conversion, i.e. NLF). HNLF measurements are presented in Fig. 4 with conversion efficiencies better than -15 dB for pump peak powers over 13 dBm. This result is in correlation with previously published results at similar pump peak powers [24].

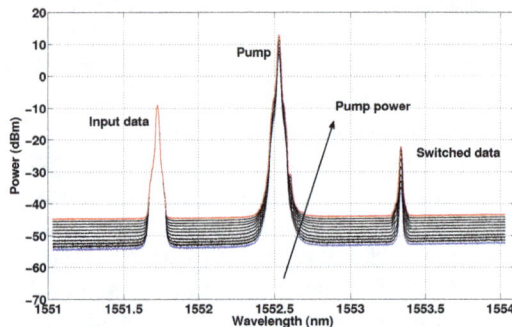

Fig. 4. Four-wave mixing in 500 m HNLF, pump power from 6 dBm to 16 dBm.

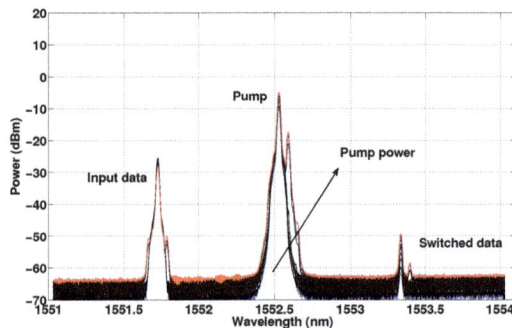

Fig. 5. Four-wave mixing in 26 m As_2Se_3 fiber, pump power from 6 dBm to 16 dBm.

Results for As_2Se_3 fiber are illustrated in Fig. 5. Chirped peaks appeared for the pump signal and the switched data payload. They were initiated by the reflection of the pump signal on the As_2Se_3/silica boundary and induced SBS on the reflected pump (in practice there was a conventional SBS backscattered signal and another SBS backscattered signal caused by the reflected pump, which then propagated in the same direction as the pump, data and idler). Conversion efficiencies were under -38 dB, where the major detrimental effect was implied by the component IL. As the effective length for sufficient nonlinear response of the As_2Se_3 fiber is 7.5 m, component IL reduction is possible to enhance nonlinear performance, therefore further simulations were carried out in Section 5.1.

For Ge-SCF, the conversion efficiency was below -31 dB and, as the effective length of Ge-SCF is more than 100 m, no component IL reduction was possible to enhance nonlinear performance. Further employment of Ge-SCF in the OPS was not considered. Evenmore when the pump peak power was at 16 dBm, Ge-SCF provided OSNR of switched data under 10 dB, which is below required values for typical optical communication. DFWM recorded spectra for Ge-PCF are illustrated in Fig. 6.

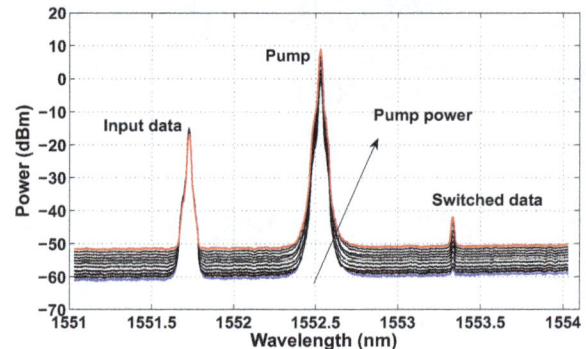

Fig. 6. Four-wave mixing in 100 m Ge-PCF, pump power from 6 dBm to 16 dBm.

Figure 7 then illustrates a comparison of all three NLFs and their conversion efficiencies. It can be observed that for pump peak powers over 12 dBm and 14 dBm for the HNLF and As_2Se_3 fiber respectively, SBS exhibits itself and starts to limit the conversion efficiency. In case of Ge-PCF the SBS treshold is still not trespassed. These results match the theoretical calculations.

4.1 Wavelength Range and Frequency Plan

With conversion efficiencies better than -15 dB at maximum pump peak power for HNLF with one channel detuning, additional measurements were carried out to evaluate conversion efficiency within wider wavelength range. In Fig. 8 conversion efficiency in dependence on channel detuning is presented with pump allocation at 1552.52 nm and with peak powers of 3 dBm, 8 dBm and 13 dBm. As can be observed flat conversion efficiency profiles were achieved

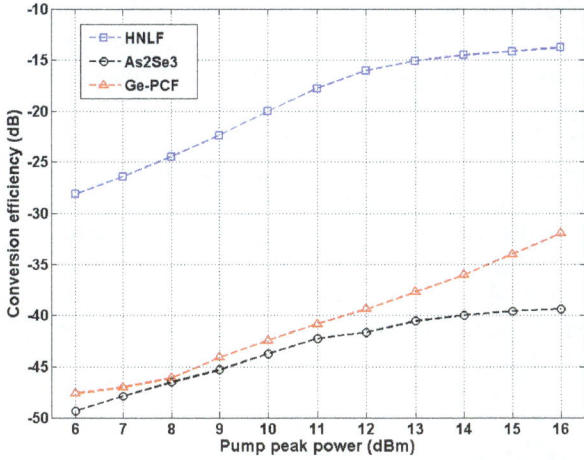

Fig. 7. Comparison of conversion efficiencies in dependence on pump peak power.

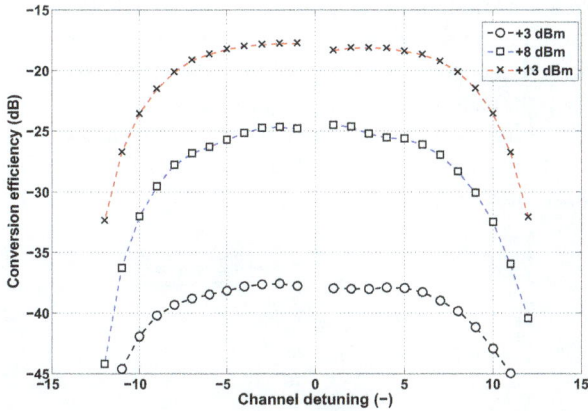

Fig. 8. HNLF - conversion efficiency in dependence on channel detuning and pump peak power, 100 GHz DWDM grid.

for ±8 channel detuning considering a 3 dB decrease limit, which is in perfect correlation with results published in [24]. The conversion efficiency increase between 8 dBm and 13 dBm pump was not as high as for 3 dBm to 8 dBm due to the onset of SBS.

For OPS based on DFWM it is significant to evaluate and decide for a spectrally efficient frequency plan. Attention was paid to facile filtering of the routing signals and input data, maximum conversion efficiency and frequency plan scalability. We considered a flat conversion profile of ±8 channels. The coarse frequency plan assumed routing signals (i.e. pumps) placed +3/+4 channels from the input data, switched data are then at +6/+8 channels from the former data signal for a 1x2 OPS. This setup enables utilization of commercially available filters (without requirement for filtering adjacent channels). If more outputs are required (1x4, 1x6 or 1x8 OPS), the setup can be expanded in a mirror-like fashion. Moreover in the dense option additional routing signals can be placed at +1/+5 channels, thus resulting in switched data at +2/+10 channels, but with significantly more complicated filtering or with higher requirements on AWG adjacent channel isolation than in the coarse variant.

4.2 BER and OSNR Tests

BER tests were performed to verify switching performance and to optimize the prototype subcomponents and cohesion between OEC and the optical part responsible for wavelength covnersion. A 10 Gbit/s NRZ (non-return zero) BER tester (BERT) was utilized as the input data source. As the pump source (Pump) a DFB diode was utilized. Polarization of both signals was controlled by polarization controllers (PCs) and coupled in a WDM multiplexer. Then both signals were amplified, where the total power booster (PB) output power varied from 10 dBm to 20 dBm, resulting as in previous measurements in 6 dBm to 16 dBm pump peak powers before HNLF. The polarization insensitive loop is described in detail in the following subsection. The switched signal was then filtered in an optical bandpass filter (OBF) to supress pump signal and former data signal and afterwards in AWG and amplified in a pre-amplifier (PRE-AMP). Finally the switched signal was detected at the BERT input. Complete setup is depicted in Fig. 9.

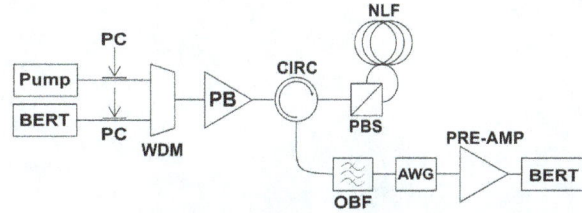

Fig. 9. Setup for OSNR, BER and polarization insensitivity measurements.

Obtained results are illustrated in Fig. 10, where the dependence of power booster total output power on BER is presented for HNLF. The best BER of 10^{-12} was achieved at 17 dBm, whereas for 16 dBm and 18 dBm BER values better than 10^{-10} were observed.

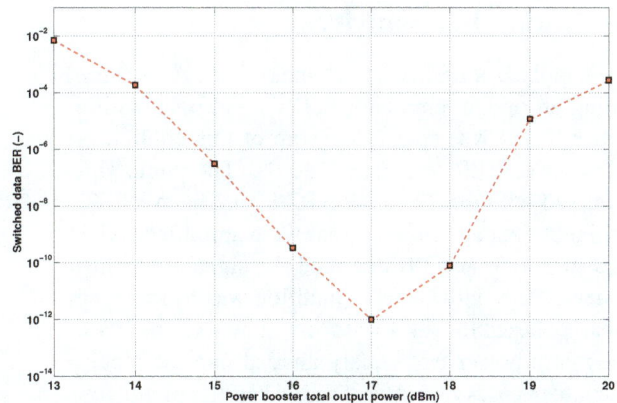

Fig. 10. Switched data BER dependence on the power booster total output power.

In Fig. 12 optical SNR (OSNR) and eye-diagram extinction ratio (ER) measuremend values of the switched signal are presented. Starting from 9 dBm pump peak powers, OSNR values exceeded 20 dB, but afterwards started to saturate around 25 dB with a decrease at 16 dBm pump peak power, i.e. 20 dBm power booster total output power, where ER was unmeasurable and also BER tests were lower than 10^{-4}.

This fact was accounted to high amplifier noise. The situation with best eye-diagram extinction ratio is presented in Fig. 11, with ER of 8.23 dB.

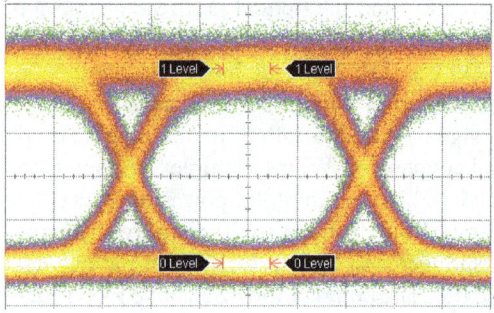

Fig. 11. Eye-diagram of the switched signal - extinction ratio of 8.23 dB.

From BER, OSNR and ER measurements, operation between 16 dBm to 18 dBm power booster total output power, i.e., pump peak power 12 dBm to 14 dBm, is desired. This is in good relation with previously calculated and measured SBS treshold.

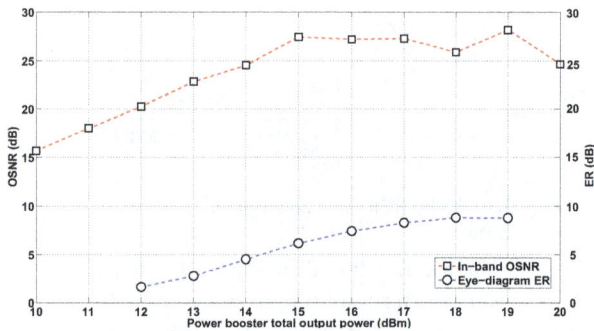

Fig. 12. Switched data in-band OSNR and eye-diagram ER for different power booster total output powers.

4.3 Polarization Insensitivity

For our polarization sensitivity measurements a setup incorporating an optical circulator (CIRC) and polarization beam-splitter (PBS) was selected. Setup of this configuration is identical to BER tests (see Fig. 9). The pump signal was coupled with the 10 Gbit/s NRZ signal from BERT. Then both input data and pump signal were amplified. NLF was placed in a loop and PBS divided the signals into fast and slow axes. The only critical condition was to keep the pump linear polarization in 45° to the slow axis of PBS, so that the pump power was equaly divided into the clockwise and counter-clockwise directions. Then the pump signal and input data were filtered (via OBF and AWG) and only switched data was detected at BERT.

In the experimental measurement power booster total output power was again tuned from 10 dBm to 20 dBm, resulting in 3 dBm to 13 dBm pump peak powers after PBS. By dividing the pump power, i.e. having two pumps with 3 dB lower peak powers, we operated closer to the SBS treshold in contrary to polarization sensitive configuration, where NLF was placed directly after power booster and in

front of optical bandpass filter. The inclusion of CIRC and PBS resulted in only 0.5 dB insertion loss. Conversion efficiency difference between polarization sensitive (the best situation with co-polarized waves was considered) and insensitive variant at pump peak powers over 10 dBm was around 3 dB. Polarization state of the input data was then varied in a 90° range with maximal difference in conversion efficiency of ±0.6 dB. Tuning the pump polarization state from the optimal 45° by less than 10° resulted only in insignificant variations of the output idler power. When the state of polarization was varied in the polarization sensitive setup, the switched data peak power values decreased rapidly.

5. Simulations

For additional comparison, simulations were carried out in Optiwave Optisystem software. Further polarisation features tested via simulation are presented in Fig. 13. For the polarization insensitive variant (blue) only small fluctuations of switched data peak power are observed (~0.01 dB). In case of the polarization sensitive setup (red), when the input data polarization state is tuned from +0° to +90° from the pump polarization state, almost 16 dB decrease in switched data peak power level is present.

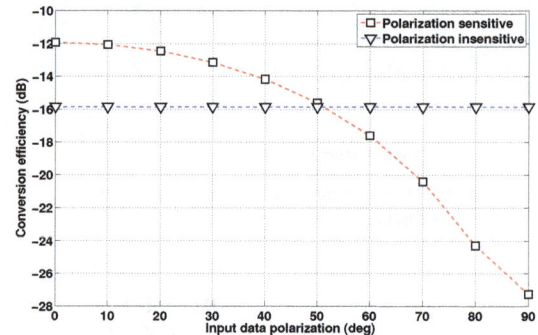

Fig. 13. Simulated switched data conversion efficiency in HNLF, when input data polarization is tuned by +0° to +90° with respect to the pump signal.

5.1 As$_2$Se$_3$ Fiber Optimization

Following As$_2$Se$_3$ fiber connectorization technology development, parameter characterization and primary conversion efficiency measurement further optimization was necessary. The component IL of 16.0 dB could be significantly reduced if only the effective length is utilized, which was calculated as 7.5 m for the employed As$_2$Se$_3$ fiber. Thus it was possible to decrease component IL to 4.9 dB, while maintaining the same nonlinear response. Conversion efficiency increased to -16.1 dB where the results were obtained within simulations considering a 5 m long As$_2$Se$_3$ fiber. Further analysis of As$_2$Se$_3$ fiber length in range from 1 m to 10 m suggests, that with employment of 3 m to 6 m long As$_2$Se$_3$ fiber segment, maximized switched data peak powers are obtained. Considering the aim of broadest wavelength range for the data switching, it is suitable to utilize

As_2Se_3 fiber segment of the shortest length, thus eliminating the effects of dispersion induced walk-off. Conversion efficiencies for different As_2Se_3 fiber lengths are presented in Fig. 14. In comparison with [7], we have demonstrated comparable conversion efficiencies to HNLF for As_2Se_3 fiber lengths of several meters, enabling employment of such components in the OPS prototype.

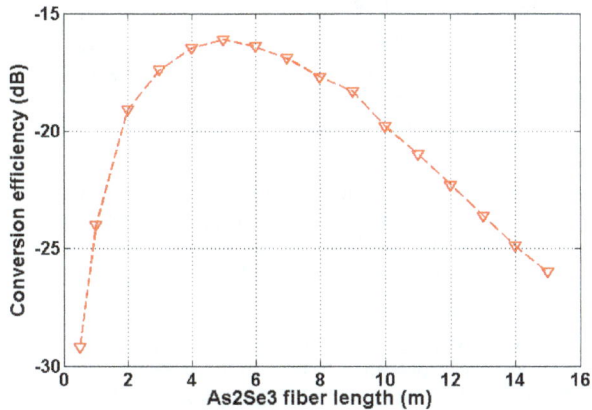

Fig. 14. Simulated switched data conversion efficiency in As_2Se_3 fiber with respect to fiber length.

6. Conclusion

Development of a data transparent and polarization insensitive optical packet switch prototype based on fibers with enhanced nonlinearity was presented. Technology of chalcogenide glass fiber processing and connectorization has been perfected. A single-mode arsenic-selenide fiber with cut-off at 1300 nm was connectorized and a component with 26 m length was developed. Attenuation of 0.6 dB/m was measured with connection losses of 0.25 dB to conventional silica fiber, thus presenting enhancement of current state-of-art parameters. Ge-doped suspended-core silica microstructured fiber by IPHT, Jena, Germany was also connectorized, providing connection loss of 0.9 dB.

Conversion efficiencies were measured for all utilized fibers. HNLF provided conversion efficiency of almost -13 dB. Conversion efficiency for the As_2Se_3 fiber was below -38 dB due to component insertion loss of 16 dB. Optimization within simulations with shorter lengths of As_2Se_3 fiber was carried out, where 5 m of As_2Se_3 fiber provided conversion efficiency of -16 dB. Ge-SCF showed almost no nonlinear response with conversion efficiency below -38 dB.

HNLF was employed for experimental optical packet switching, with modulation format transparent switching, sixteen 100 GHz DWDM channels flat conversion efficiency profile and polarization insensitive operation in the final proposed switching configuration. On the other hand SBS tresholds for HNLF were the lowest of all utilized fibers with enhanced nonlinearity. BER tests were performed for the polarization insensitive configuration with extinction ratio of more than 8 dB for the best eye-diagrams. BER better than 10^{-10} was observed. In-band OSNR measurements better than 25 dB were observed.

The introduced device follows amplification constrains given by long haul optical networks and combines good conversion efficiency with negligible polarization dependence. Functionality was tested and verified on HNLF, where simulations results imply significant enhancement in the device parameters, by employing the optimized As_2Se_3 fiber. Technical details have been protected under an utility model and in an ongoing patent process.

Acknowledgements

The authors thank SQS Fiber optics, Czech Republic for support and cooperation in the development of technological processes of chalcogenide fibers and IPHT Jena, Germany for providing the Ge-doped microstructured fiber under the framework COST Action TD1001. Furthermore we thank Petr Dvorak for his help with the development of the processing unit. This work was supported by the CTU grant No. SGS14/190/OHK3/3T/13.

References

[1] TIPSUWANNAKUL, E., LI, J., KARLSSON, M., ANDREKSON, P., Performance comparisons of DP-16QAM and duobinary-shaped DP-QPSK for optical systems with 4.1 bit/s/Hz spectral efficiency. *Journal of Lightwave Technology*, 2012, vol. 30, no. 14, p. 2307 - 2314.

[2] NAKAZAWA, M., OKAMOTO, S., OMIYA, T., KASAI, K., YOSHIDA, M., 256-QAM (64 Gb/s) coherent optical transmission over 160 km with an optical bandwidth of 5.4 GHz. *IEEE Photonics Technology Letters*, 2010, vol. 22, no. 3, p. 185 - 187.

[3] CALABRETTA, N., WANG, G., DITEWIG, T., RAZ, O., AGIS, F., DE WAARDT, S., DORREN, H. Scalable optical packet switches for multiple data formats and data rates packets. *IEEE Photonics Technology Letters*, 2010, vol. 22, no. 7, p. 483 - 485.

[4] HUYBRECHTS, K., MORTHIER, G., BAETS, R. Fast all-optical flip-flop based on a single distributed feedback laser diode. *Optics Express*, 2008, vol. 16, no. 15, p. 11405 - 11410.

[5] CHOU, H. F., HUANG, C.-H., BOWERS, J., TOUDEH-FALLAH, F., GUYREK, R. 3-D MEMS-based dynamically reconfigurable optical packet switch (DROPS). In *Proceedings of the International Conference on Photonics in Switching*. 2006, p. 1 - 3.

[6] LUO, J., DORREN, H. J. S., CALABRETTA, N. Optical RF tone in-band labeling for large-scale and low- latency optical packet switches. *Journal of Lightwave Technology*, 2012, vol. 30, no. 16, p. 2637 - 2645.

[7] NAKANISHI, T., HIRANO, M., OKUNO, T., ONISHI, M. Silica-based highly nonlinear fiber with g =30/W/km and its FWM-based conversion efficiency. In *Proceedings of the Optical Fiber Communication Conference and Exposition and The National Fiber Optic Engineers Conference*. Anaheim (CA, USA), 2006, p. OTuH7.

[8] SANGHERA, J. S., SHAW, L. B., PUREZA, P., NGUYEN, V. Q., GIBSON, D., BUSSE, L., AGGARWAL, I. D., FLOREA, C. M., KUNG, F. H. Nonlinear properties of chalcogenide glass fiber. *International Journal of Applied Glass Science*, 2010, vol. 1, no. 3, p. 296 - 308.

[9] CHANG, Y.-M., LEE, J., LEE, J. H. Bismuth nonlinear optical fiber for photonic ultrawideband radio- signal processing. *IEEE Journal of Selected Topics in Quantum Electronics*, 2012, vol. 18, no. 2, p. 891 - 898.

[10] FENG, X., POLETTI, F., CAMERLINGO, A., PARMIGIANI, F., PETROPOULOS, P., HORAK, P., PONZO, G. M., PETROVICH, M., SHI, J., LOH, W. H., RICHARDSON, D. J. Dispersion controlled highly nonlinear fibers for all-optical processing at telecoms wavelengths. *Optical Fiber Technology*, 2010, vol. 16, no. 6, p. 378 - 391.

[11] LE, S. D., NGUYEN, D. M., THUAL, M., BRAMERIE, L., SILVA, M. C., LENGLÉ, K., GAY, M., CHARTIER, T., BRILLAND, L., MÉCHIN, D., TOUPIN, P., TROLES, J. Efficient four-wave mixing in an ultra-highly nonlinear suspended-core chalcogenide As38Se62 fiber. *Optics Express*, 2011, vol. 19, no. 26, p. B653 - B660.

[12] SANGHERA, J., SHAW, L. B., PUREZA, P., NGUYEN, V. Q., GIBSON, D., AGGARWAL, I. D., FLOREA, C. M., KUNG, F. Progress of chalcogenide glass fibers. In *Proceedings of the Optical Fiber Communication Conference and Exposition and The National Fiber Optic Engineers Conference.* Anaheim (CA, USA), 2007, p. 1 - 3.

[13] TAEED, V. G., FU, L., PELUSI, M., ROCHETTE, M., LITTLER, I. C., MOSS, D. J., EGGLETON, B. J. Error free all optical wavelength conversion in highly nonlinear As-Se chalcogenide glass fiber. *Optics Express*, 2006, vol. 14, no. 22, p. 10371 - 10376.

[14] TOROUNIDIS, T. *Fiber Optic Parametric Amplifiers in Single and Multi Wavelength Applications.* PhD thesis. Gothenburg (Sweden): Chalmers University of Technology, 2006.

[15] DONG, L., THOMAS, B. K., FU, L. Highly nonlinear silica suspended core fibers. *Optics Express*, 2008, vol. 16, no. 21, p. 16423 - 16430.

[16] MAGI, E., YEOM, D.-I., NGUYEN, H., FU, L., EGGLETON, B. Enhanced Kerr nonlinearity in sub-wavelength diameter As2Se3 chalcogenide fibre tapers. In *Proceedings of the Optical Internet 2007 and the 2007 32nd Australian Conference on Optical Fibre Technology (COIN-ACOFT).* Melbourne (Australia), 2007, p. 1 - 3.

[17] *IEEE Standard for Information technology- Telecommunications and information exchange between systems-Local and metropolitan area networks- Specific requirements Part 3: Carrier Sense Multiple Access with Collision Detection (CSMA/CD) Access Method and Physical Layer Specifications Amendment 4: Media Access Control Parameters, Physical Layers and Management Parameters for 10 Gb/s Operation.* IEEE Std 802.3ae. IEEE, 2002.

[18] *IEEE Standard for Information technology- Telecommunications and information exchange between systems-Local and metropolitan area networks- Specific requirements Part 3: Carrier Sense Multiple Access with Collision Detection (CSMA/CD) Access Method and Physical Layer Specifications Amendment 4: Media Access Control Parameters, Physical Layersand Management Parameters for 40 Gb/s and 100 Gb/s Operation.* IEEE Std 802.3ba. IEEE, 2010.

[19] YU, C., PAN, Z., WANG, Y., SONG, Y., GURKAN, D., HAUER, M., WILLNER, A., STARODUBOV, D. Polarization insensitive four-wave mixing wavelength conversion using a fiber Bragg grating and a Faraday rotator mirror. In *Proceedings of the Optical Fiber Communications Conference (OFC).* 2003, p. 347 - 349.

[20] SAKAMOTO, T., SEO, K., TAIRA, K., MOON, N. S., KIKUCHI, K. Polarization-insensitive all-optical time division demultiplexing using a fiber four-wave mixer with a peak-holding optical phase-locked loop. *IEEE Photonics Technology Letters*, 2004, vol. 16, no. 2, p. 563 - 565.

[21] HU, H., MULVAD, H., GALILI, M., PALUSHANI, E., XU, J., CLAUSSEN, A., OXENLOWE, L., JEPPESEN, P. Polarization-insensitive 640 Gb/s demultiplexing based on four wave mixing in a polarization-maintaining fibre loop. *Journal of Lightwave Technology*, 2010, vol. 28, no. 12, p. 1789 - 1795.

[22] MOSSADEGH, R., SANGHERA, J. S., SCHAAFSMA, D., COLE, B. J., NGUYEN, V. Q., MIKLOS, R. E., AGGARWAL, I. D. Fabrication of single-mode chalcogenide optical fiber. *Journal of Lightwave Technology*, 1998, vol. 16, no. 2, p. 214.

[23] ELSCHNER, R., BUNGE, C. A., HUTTL, B., I COCA, A., SCHMIDT-LANGHORST, C., LUDWIG, R., SCHUBERT, C., PETERMANN, K. Impact of pump-phase modulation on FWM-based wavelength conversion of D(Q)PSK signals. *IEEE Selected Topics in Quantum Electronics*, 2008, vol. 14, no. 3, p. 666 - 673.

[24] ONISHI, M., OKUNO, T., KASHIWADA, T., ISHIKAWA, S., AKASAKA, N., NISHIMURA, M. Highly nonlinear dispersion-shifted fibers and their application to broadband wavelength converter. *Optical Fiber Technology*, 1998, vol. 4, no. 2, p. 204 - 214.

About Authors ...

Matej KOMANEC was born in 1984. He received MSc. and Ph.D. degrees from the Faculty of Electrical Engineering of the Czech Technical University in Prague where he is now a research employee. His professional interests cover fiber optics, nonlinear phenomena, specialty fibers and fiber sensing.

Pavel SKODA was born in 1982. He received MSc. degree from the Faculty of Electrical Engineering of the Czech Technical University in Prague where he is now a research employee. His professional interests cover optical fiber communication, all-optical networks, modulation formats and all-optical devices.

Jan SISTEK was born in 1974. He received MSc. and Ph.D. degrees from the Faculty of Electrical Engineering of the Czech Technical University in Prague where he is now an Assistant Professor. His professional interests cover optical fibers and associated electronics, microwave circuits and related measurement techniques.

Tomas MARTAN was born in 1976. He received MSc. and Ph.D. degrees at the Faculty of Electrical Engineering of the Czech Technical University in Prague where he is now an Assistant Professor. His professional interests cover specialty fibers, optical fiber sensing, optoelectronics and optical spectroscopy.

Optimization of Filter by using Support Vector Regression Machine with Cuckoo Search Algorithm

Mustafa İLARSLAN[1], Salih DEMIREL[2], Hamid TORPI[2], A. Kenan KESKIN[2], M. Fatih ÇAĞLAR[3]

[1] Turkish Air Force Academy, 34149, Yeşilyurt, Istanbul, Turkey
[2] Dept. of Electronics and Communication Engineering, Yıldız Technical University, Turkey
[3] Dept. of Electronics and Communication Engineering, Süleyman Demirel University, Turkey

m.ilarslan@hho.edu.tr, salihd@yildiz.edu.tr, torpi@yildiz.edu.tr, kkeskin@yildiz.edu.tr, mfcaglar@gmail.com

Abstract. *Herein, a new methodology using a 3D Electromagnetic (EM) simulator-based Support Vector Regression Machine (SVRM) models of base elements is presented for band-pass filter (BPF) design. SVRM models of elements, which are as fast as analytical equations and as accurate as a 3D EM simulator, are employed in a simple and efficient Cuckoo Search Algorithm (CSA) to optimize an ultra-wideband (UWB) microstrip BPF. CSA performance is verified by comparing it with other Meta-Heuristics such as Genetic Algorithm (GA) and Particle Swarm Optimization (PSO). As an example of the proposed design methodology, an UWB BPF that operates between the frequencies of 3.1 GHz and 10.6 GHz is designed, fabricated and measured. The simulation and measurement results indicate in conclusion the superior performance of this optimization methodology in terms of improved filter response characteristics like return loss, insertion loss, harmonic suppression and group delay.*

Keywords

Cuckoo Search Algorithm, optimization, ultra-wideband, band-pass filter, Support Vector Regression Machine (SVRM).

1. Introduction

In microwave and wireless systems, 3.1-10.6 GHz ultra-wideband (UWB) communication band, which has been dedicated by the Federal Communications Commission (FCC) since February 2002 [1], has a significant role because of its high data-rate, large channel capacity, low power consumption, immunity to multipath interference and coexistence with other wireless systems. With these advantages, UWB communication systems continue to attract attention as a popular research field in both academia and industry. It is incontestable that one of the essential components in UWB RF front-end modules is the band-pass filter (BPF) and hence, BPF design takes a critical role in UWB communication.

Band-pass filters are vital building blocks which allow a signal to pass at requested frequencies and repel the rest of frequencies in receiver and transmission systems. In RF and microwave communication systems, compactness, low insertion loss at the transmission band and high suppression at the rejection band are crucial parameters in order to design a well-suited BPF [2]. Different kinds of UWB BPF design methods and technologies have been investigated for many years in the literature [3-11]. As a filter realization technology, microstrips are thoroughly employed because of their low cost, easy fabrication and integration [12], [13].

Accurate and fast UWB BPF design is a difficult optimization problem as each part of the circuit highly affects the frequency response of the filter to a certain extent. Since there are no analytical models of elements, 3D EM simulators should be used in the design process which thereby consumes a lot of time and CPU resources.

In this paper, an accurate and faster methodology for designing a microstrip UWB BPF is demonstrated. This method uses the very popular Support Vector Regression Machine (SVRM) [14-16] which is trained by 3D EM simulator results of the desired microstrip shapes given as the basic design blocks and then combines the blocks to construct the BPF within an efficient and robust optimization process in accordance with the required design objectives. Thus, this method not only avoids the slowness of 3D EM simulators, but is also as accurate as these simulators. Moreover, the easy implementation of different kinds of shapes and technologies like strip lines is one of the prominent features of this novel design methodology. As an application example, an UWB BPF is designed by using three base fundamental shapes; Shunt Stub (SS), Etched Square Stub (ESS) and Defected Ground Structure (DGS). SWRM and Cuckoo Search Algorithm (CSA) are utilized together in the analysis and design of the BPF.

In recent years, Cuckoo Search Algorithm has become very popular as a new optimization method amongst the academic communities of various engineering disciplines [17], [18]. It was first proposed by Xin-She Yang and Suash Deb in 2009 and its performance was tested by using standard test functions. The results were superior when compared with other popular meta-heuristic optimization methods like Genetic Algorithm (GA) and Particle Swarm Optimization (PSO).

In this study, the training data sets for the desired basis shapes are created first by using the 3D Computer Simulation Technology Microwave Studio (CST MWS) and then the SVRM model of each shape is constructed with these data sets. Then, the output of the obtained SVRM models are fed into the CS optimization algorithm until the BPF design is optimized according to the design goals. The performance of the CSA is compared with the standard Meta-Heuristics; Genetic Algorithm (GA) and Particle Swarm Optimization (PSO) methods. Finally, the designed UWB BPF is fabricated and measured.

This paper is composed of 5 sections: After the introduction, the next section states the characteristics and design of the base elements and the SVRM model building. In the third section, the Cuckoo Search Algorithm (CSA) is presented for optimizing the defined cost function by making use of the SVRM models. The design procedure and fabrication of the UWB BPF is given in Section 4 and finally, Section 5 is for the conclusion.

2. Base Elements

In this study, 3 kinds of resonator types which are illustrated in Fig. 1-3 are used as the base elements for the UWB BPF design.

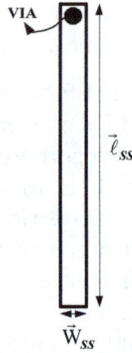

Fig. 1.   Shunt stub base element.

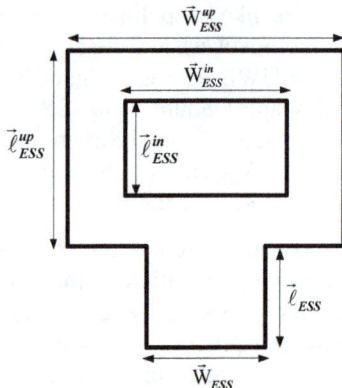

Fig. 2.   Etched Square stub base element.

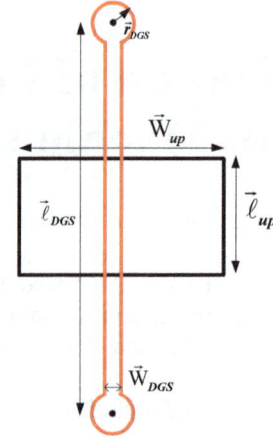

Fig. 3.   Defected Ground Structure with microstrip line base element.

Shunt Stub (SS) behaves like high pass, Etched Square Stub (ESS) works like band stop, while Defected Ground Structure (DGS) shows band stop and low pass characteristics. Operational parameters of the resonators are manipulated by their geometrical dimensions within the limitations of the filter. These limitations could arise from the design objectives such as compactness and the frequency response of the filter. In the design, RO-4350 material which has a dielectric permittivity (ε_r) of 3.48, substrate thickness (h) of 1.52 mm, copper thickness (t) of 35 μm and a tangent loss (tanδ) of 0.002 is utilized as the substrate.

3. Design Synthesis and Optimization Process

3.1 Mathematical Bases of Support Vector Regression

Given the training dataset $(\vec{x}_i, y_i)$, $i = 1,2,...,\lambda$ where $y_i \in R$ and λ is the size of the training data, Support Vector Regression Machine (SVRM) attempts to construct a continuous mapping function $f(\vec{x})$ from the independent p-dimensional input variable vector $\vec{x}$ to the dependent output variable y by linearly combining the results of a nonlinear transformation of the input samples:

$$f(\vec{x}) = \sum_{i=1}^{n_{sv}} (\alpha_i - \alpha^*_i) \, K(\vec{x}_i, \vec{x}) + b \qquad (1)$$

where n_{sv} is the number of the Support Vector (SV)s, $\alpha_i \geq 0$, $\alpha^*_i \geq 0$ are Lagrange multipliers and b is bias parameter, K is a kernel function which performs the nonlinear transformation and in practice, is directly defined. The measure of how well a sample is fitted by the function f is given by a so-called ϵ- insensitive loss function [14] described by

$$L(f(\vec{x}_i) - y_i) = |y_i - f(\vec{x}_i)|_\epsilon = \max|0, (|y_i - f(\vec{x}_i)| - \epsilon)| \,(2)$$

where ϵ is the radius of the regression tube and the distance among the predicted and target values for the training samples is defined as the empirical risk as follows:

$$R_{emp} = \frac{1}{\lambda} \sum_{i=1}^{\lambda} L(\vec{x}_i, y_i, f) \, . \tag{3}$$

Therefore, in SV regression, the goal is to minimize R_{emp}. In order to make support vector regression, minimization of the empirical risk formulation (3) is transformed into maximization of equation (4). Using the standard Lagrange multipliers technique, the aforementioned minimization problem can be transformed into Constrained Quadratic Programming (CQP) in which the following function must be maximized with respect to the Lagrange parameters (α, α^*) [14]:

$$Maximize\,W(\alpha, \alpha^*) =$$

$$-\frac{1}{2} \sum_{i,j=1}^{n_{sv}} (\alpha_i^* - \alpha_i)(\alpha_j - \alpha_j^*) K(\vec{x}_i, \vec{x}_j) - \sum_{i=1}^{n_{sv}} (\alpha_i + \alpha_i^*) + \sum_{i=1}^{n_{sv}} y_i(\alpha_i - \alpha_i^*)$$

$$\tag{4}$$

subject to:

$$0 \le \alpha_i \le C, 0 \le \alpha_i^* \le C, \quad \sum_{i=1}^{\lambda} \alpha_i = \sum_{i=1}^{\lambda} \alpha_i^* \tag{5}$$

where index i represents support vector elements of the training data and index j represents irrelevant elements remaining from the training data. The parameter $C > 0$ measures the trade-off between the capabilities of $f(\vec{x})$ to approximate the input samples and the error of the new samples. The CQP can be solved using standard optimization techniques subject to the conditions given by (5) and the result in λ Lagrange multiplier pairs (α_i, α_i^*). The parameter b can be computed by means of so-called Karush-Kuhn-Tucker conditions [14], [15], [16].

Since the insensitive loss function given by (2) applies the ϵ- tube selection process to the training dataset $(\vec{x}_i, y_i)$, $i = 1,2,..., \lambda$, thus only for the samples satisfying $|f(\vec{x}_i) - y_i| \ge \epsilon$, the Lagrangian multipliers (α_i, α_i^*) may be nonzero, and for the samples of $|f(\vec{x}_i) - y_i| < \epsilon$, the Lagrangian multipliers (α_i, α_i^*) vanish. The samples $(\vec{x}_i, y_i)$, $i = 1,2,..., n_{sv}$ that come with non-vanishing coefficients are called Support Vectors (SV). Therefore, we obtain a sparse expansion of the Lagrangian multipliers (α_i, α_i^*) in terms of the input variable vector $\vec{x}_i$. In other words, we perform generalization between the whole input $\vec{x}$ - and output y- domains using only a small subset of the training data that ensures enormous computational advantages [14].

3.2 3D EM Simulation-based SVRM Microstrip Modeling

In order to obtain accurate and fast design UWB BPF, the SVRM model of basic elements is employed. Black-box models of each element are created, including geometrical dimensions of elements as input parameters and S parameters of element as output parameters. Input variable vectors of SS, ESS and DGS models are defined as; $\vec{W}_{SS}, \vec{\ell}_{SS}, f$, $\vec{W}_{ESS}, \vec{\ell}_{ESS}, \vec{W}_{ESS}^{up}, \vec{\ell}_{ESS}^{up}, \vec{W}_{ESS}^{in}, \vec{\ell}_{ESS}^{in}, f$ and $\vec{W}_{DGS}$, $\vec{\ell}_{DGS}, \vec{W}_{up}, \vec{\ell}_{up}, \vec{r}, f$, respectively. Output parameters of the element models are the same; $|S_{11}|, \angle S_{11}, |S_{21}|, \angle S_{21}$, the magnitude and phase of S parameters. Since the SVRM model has one output, a parallel operation is run to compose the element models. Therefore, each element model contains four machines which have the same input because of the four output parameters. Radial kernel function is exploited for the SVM regression which is described by,

$$K(\vec{x}_i, \vec{x}) = e^{-\gamma \|\vec{x}_i - \vec{x}\|^2} \tag{6}$$

where γ is the variance of the kernel function and will be chosen in the training phase. The training dataset of base elements is obtained by CST Microwave Studio within the physical ranges given in Tab. 1.

Base Element	Input Variables	Min. Value	Max. Value	Interval	Data Number
SS	W_{SS} (mm)	0.2	1	0.2	5
	ℓ_{SS} (mm)	3	5	0.2	11
	f (GHz)	0.2	25.2	1	26
ESS	W_{ESS} (mm)	0.5	1	0.1	5
	ℓ_{ESS} (mm)	0.5	1	0.1	5
	W_{ESS}^{up} (mm)	1.7	2	0.1	4
	ℓ_{ESS}^{up} (mm)	1	1.3	0.1	4
	W_{ESS}^{in} (mm)	1	1.4	0.2	3
	ℓ_{ESS}^{in} (mm)	0.5	0.7	0.1	3
	f (GHz)	0.2	25.2	1	26
DGS	W_{DGS} (mm)	0.2	0.5	0.1	4
	ℓ_{DGS} (mm)	5	7	0.4	6
	W_{up} (mm)	1	2	0.2	6
	ℓ_{up} (mm)	2	4	0.2	11
	r (mm)	0.2	0.5	0.1	4
	f (GHz)	0.2	25.2	1	26

Tab. 1. Dimensional range of Base Elements for training data.

Total training data number of neural network for each frequency of SS, ESS and DGS are 55, 3.600 and 6.336, respectively. Furthermore, in Tab. 2, the accuracy of the models, the Support Vector numbers of $|S_{21}|$, are compared

Element	γ	C	ϵ	$\|S_{21}\|$ SVs Number	$\|S_{11}\|$ (%)	$\angle S_{11}$ (%)	$\|S_{21}\|$ (%)	$\angle S_{21}$ (%)
SS	0.001	10000	0.05	32	99.8	99.8	99.9	99.6
	0.001	10000	0.07	23	99.5	98.1	99.4	99.2
	0.001	10000	0.1	14	98.9	97.2	99.0	98.7
ESS	0.001	10000	0.05	1617	99.7	98.8	99.6	98.9
	0.001	10000	0.07	1272	99.3	98.6	99.1	98.8
	0.001	10000	0.1	820	98.1	97.9	98.3	98.5
DGS	0.001	10000	0.05	2915	99.2	98.1	99.0	97.5
	0.001	10000	0.07	2002	98.5	97.2	97.9	96.6
	0.001	10000	0.1	1223	97.2	96.1	96.8	96.0

Tab. 2. Accuracy with respect to SVRM parameters.

for different ϵ insensitive loss parameters. Tab. 2 gives the used SVRM parameters, selection tube radius ϵ, number of the SVs and the resulted accuracy for the SVRM model of the S parameters for 7 GHz. The numbers of SVs used to train the SVRM model for SS, ESS and DGS are 14, 820 and 1223 respectively, with the accuracy of at least 96.0%. The simulation results show that SVRM models of the elements are not only as accurate as the 3D EM simulation model, but also approximately 280 times faster than the CST model

3.3 Cuckoo Search Algorithm

Similar to other meta-heuristic optimization algorithms, it is a bio-inspired optimization algorithm based upon the obligate brood parasitism of some cuckoo species in nature which lay their eggs in the nests of other host birds. The Cuckoo Search which idealizes such breeding behavior was proposed by Xin-She Yang and Suash Deb in 2009 and since then, it has been applied extensively to various engineering optimization problems like antenna array optimization [19], data fusion in wireless sensor networks [20], and to multi-objective design optimization problems like reliable embedded system design [21]. It was also hybridized with quantum computing principles [22] and with power series [23] to obtain better performance.

In the CS, each egg in a nest represents a solution, and a cuckoo egg represents a new solution. The aim is to use the new and potentially better solutions (cuckoos) to replace a not-so-good solution (egg) in the nests. In the simplest form, each nest has one egg.

The CS is built upon the following three idealized rules:

- Each cuckoo lays one egg at a time and dumps its egg in a randomly chosen nest;

- The best nests with high-quality eggs will carry over to the next generation;

- The number of available host nests, n is fixed, and the egg laid by a cuckoo is discovered by the host bird with a probability $P_a \in (0,1)$. Discovering means that some set of worst nests (eggs) will be thrown away and their corresponding solutions will be discarded from further calculations.

Yang and Deb also discovered that the random-walk style search is better performed by Lévy flights rather than by simple random walk. Many studies have shown that the flight behavior of many animals and insects has demonstrated the typical characteristics of Lévy flights [17-24]. Lévy flight is defined as a random walk with the step-lengths based on a heavy-tailed probability distribution which enables CS to explore the whole solution space effectively. An important advantage of CS algorithm is its simplicity. In fact, compared with other population or agent-based meta-heuristic algorithms such as PSO and

GA, there is essentially only one parameter, P_a in CS as the population size (the number of available host nests, n) is fixed, making it very easy to implement and fast to converge.

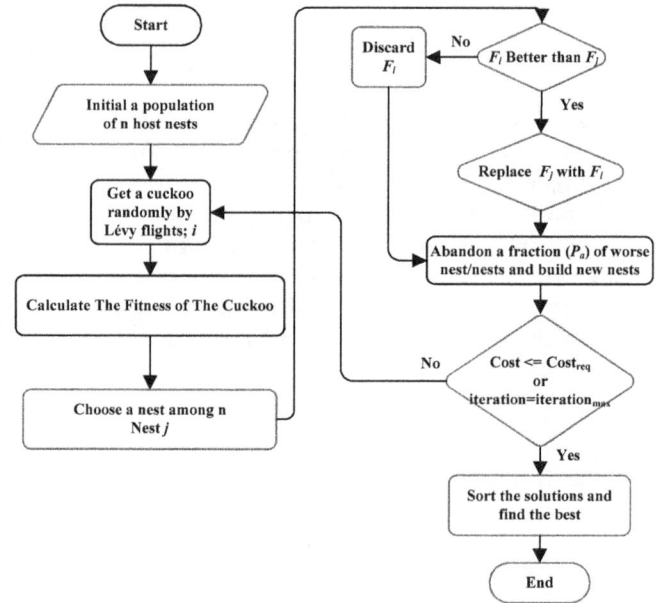

Fig. 4. Flow chart of the conventional Cuckoo Search Algorithm.

Fig. 4 is the flow chart indicating the main steps of the regular CS algorithm implementation [24].

3.4 Cost Function Evaluation and Updating Process

In microwave circuit design, two port structures could be demonstrated as cascaded connections of sub-structures. Thus, circuits can be solved by using ABCD parameters of each sub-structure. The total ABCD matrix of a circuit which is composed of cascaded n two-ports is described by;

$$\begin{bmatrix} A & B \\ C & D \end{bmatrix} = \begin{bmatrix} A_1 & B_1 \\ C_1 & D_1 \end{bmatrix} \begin{bmatrix} A_2 & B_2 \\ C_2 & D_2 \end{bmatrix} \cdots\cdots\cdots \begin{bmatrix} A_n & B_n \\ C_n & D_n \end{bmatrix}. \quad (7)$$

In our case, UWB BPF could be considered by connection of each basic element in cascade form. The frequency response of the filter is calculated using (7) and the ABCD parameters of base elements are transformed from S parameters [25] which are obtained by SVRM models per the elements' dimensions and frequency. There are 2 of SS, 3 of ESS and 4 of DGS in our BPF design making the $n = 9$. Therefore, the ABCD matrix of the filter is as follows (8-9)

$$[T]^n = \begin{bmatrix} A_n & B_n \\ C_n & D_n \end{bmatrix} \quad (8)$$

$$\begin{bmatrix} A & B \\ C & D \end{bmatrix}_{Filter} = [T]^1_{SS_1} [T]^2_{DGS_1} [T]^3_{ESS_1} [T]^4_{DGS_1} \cdots\cdots\cdots [T]^9_{SS_2} \quad (9)$$

S parameter of the filter is acquired using inverse transformation equations [25]. Meanwhile, there is no need to investigate full S parameters of base elements. The SVRM model results, which give us S_{11} and S_{21} of each elements, are enough to calculate ABCD parameters because of the reciprocity $S_{11} = S_{22}$ and $S_{12} = S_{21}$ [25].

In the design process, the optimal dimensions of the elements for the required UWB filter response are investigated by using the CS algorithm and SVRM models together under the analytical combination of ABCD parameters subject to cost function which is defined as follows (10)

$$\text{Cost Func.} = \omega_1 . \Lambda_1 + \omega_2 . \Lambda_2 + \omega_3 . \Lambda_3 , \qquad (10)$$

$$\Lambda_1 = \left(1 - |S_{11}|\right)\big|_{\Delta f_1} , \qquad (11)$$

$$\Lambda_2 = \left(1 - |S_{21}|\right)\big|_{\Delta f_2} , \qquad (12)$$

$$\Lambda_3 = \left(1 - |S_{11}|\right)\big|_{\Delta f_3} \qquad (13)$$

where Δf_1 is picked as 0.2-3.1 GHz to provide suppression at a lower band (11), Δf_2 is selected as 3.1-10.6 GHz to obtain pass band characteristics (12) and Δf_3 is taken between 10.6-20 GHz to suppress the second and third harmonic of the filter (13). Moreover, ω_1, ω_2, ω_3 (10) are chosen as 2/10, 1, 1/10, respectively. Optimization comes to the end when the iteration number is maximized or the cost value is minimized.

4. Design and Comparison of UWB Bandpass Filter

In this section, the UWB band pass filter design process is described and then the results of a specific design example are discussed. First, training data sets of bases elements are obtained with a 3D EM simulator in order to form fast and accurate SVRM models of SS, ESS and DGS within their physical limitations. Later, these models are employed in the CSA optimization process in order to obtain the required filter specifications which include rejection and pass band characteristics. The cost function of CSA is determined using analytical calculations of ABCD parameters for cascade-connected base element SVRM models. The CSA optimization process concludes when the iteration number or cost value reaches its limits. If the 3D model of basic elements is utilized instead of fast SVRM models, the duration of the optimization process would be extremely long. The design procedure of the filter is shown in Fig. 5.

RO-4350 material as mentioned in Section 2 is used for fabrication of the designed filter circuit. The aim is to design an UWB BPF that has an operational bandwidth between 3.1 GHz and 10.6 GHz. In order to achieve that, the dimensions of base elements of the filter are adjusted using the optimization process. After the optimization, the

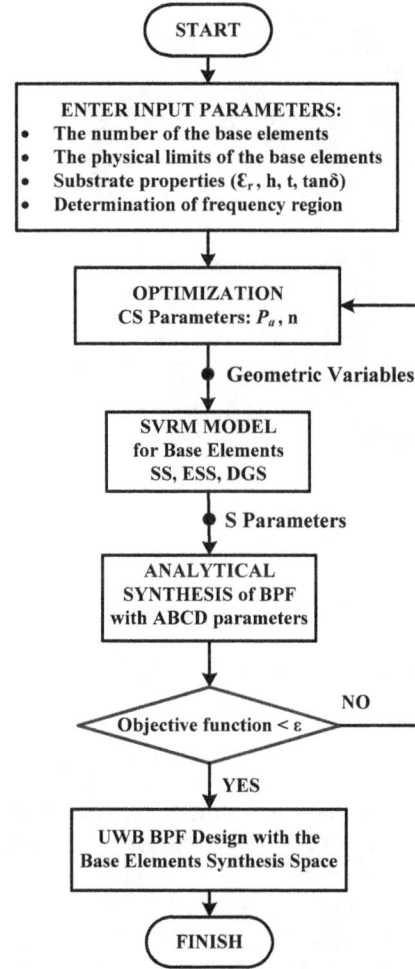

Fig. 5. General design procedure for the BPF.

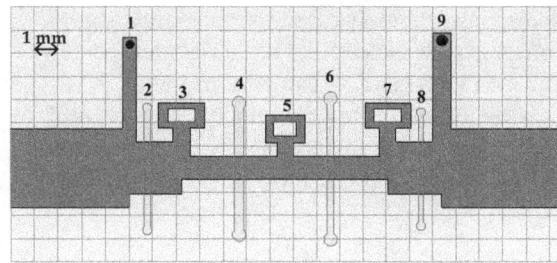

Fig. 6. Scaled filter drawing with grid background.

Base Element	Input Variables	1. Element	2. Element	3. Element	4. Element
SS	W_{SS}(mm)	0.6	0.8	-	-
	ℓ_{SS}(mm)	4	4.2	-	-
ESS	W_{ESS}(mm)	0.6	0.7	0.6	-
	ℓ_{ESS}(mm)	0.7	0.6	0.7	-
	W^{up}_{ESS}(mm)	2	1.7	2	-
	ℓ^{up}_{ESS}(mm)	1.1	1.2	1.1	-
	W^{in}_{ESS}(mm)	1.2	1	1.2	-
	ℓ^{in}_{ESS}(mm)	0.6	0.5	0.6	-
DGS	W_{DGS}(mm)	0.3	0.4	0.3	0.2
	ℓ_{DGS}(mm)	5.4	5.8	6.2	5
	W_{up}(mm)	1.6	1	1	1.6
	ℓ_{up}(mm)	2.4	3.8	3.8	2.4
	r (mm)	0.2	0.3	0.3	0.2

Tab. 3. Solution of the base elements.

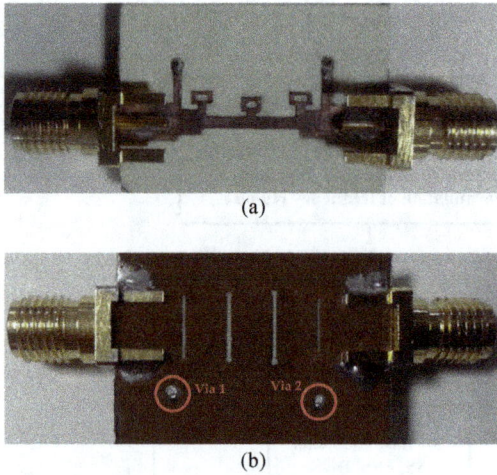

Fig. 7.   Photographs of the fabricated UWB BPF:
(a) top layer, (b) bottom layer.

designed circuit is manufactured and measured. The total size of the filter whose scaled drawing and the actual picture can be seen in Fig. 6 and Fig. 7 respectively is about 2.5 cm × 1.5 cm. The actual dimensions of each base element used in the design are given in Tab. 3.

The simulation and actual measurement results are in parallel with each other as given in Fig. 8. It is understood from the results that insertion loss, which increases with frequency, is better than -2 dB over the whole transition band and there is extra loss at the end of the transition band because of the SMA connectors utilized in the ports. Moreover, there is good suppression which is better than 10 dB until 25 GHz which contains the second and third harmonic at the rejection band. Return loss is under -10 dB at pass band. In addition, the low and high cut-off regions of frequency response show that the filter has good sharpness. Furthermore, the designed filter has a flat group delay over the whole operation band, as can be seen in Fig. 9.

In the CS optimization process, number of host nest (size of population), and fraction probability (P_a) is chosen as 50 and 0.25, respectively as they provided the best results.

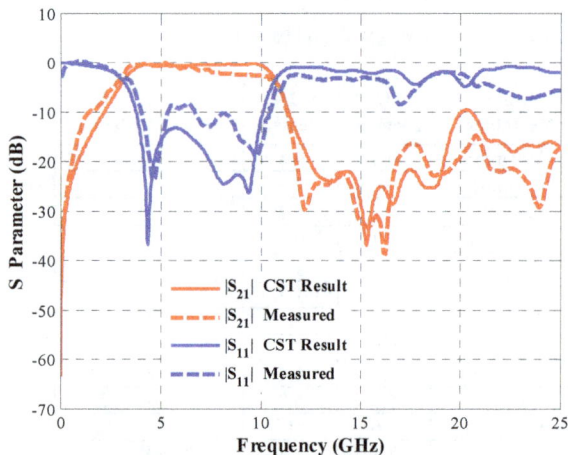

Fig. 8.   Simulated and measured S parameters of the designed UWB BPF.

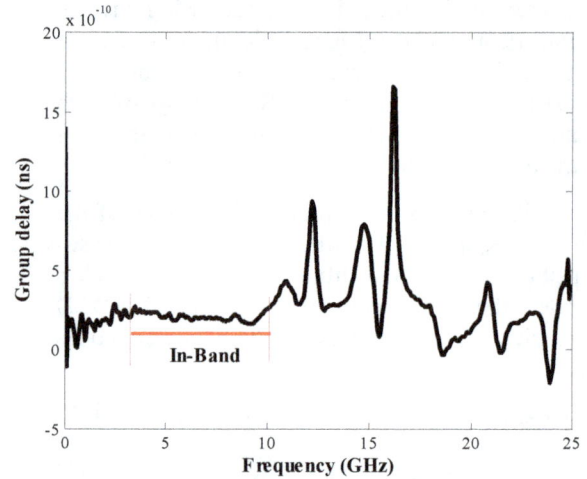

Fig. 9.   Measured group delay of the designed UWB BPF.

In order to compare it with other Meta-Heuristics, the PSO optimizer is constructed with the population (particle number) equal to 50, max/min velocity of ±0.1 and learning factors set to 2.0, respectively. Similarly, the GA optimizer has a population (chromosome) of 50, crossover probability of 0.8, and mutation probability of 0.1. The cost results of 30-time tries and 120 iteration of CSA and other standard Meta-Heuristic algorithms and benchmarking at 120th iteration for best try with the corresponding execution times for the same filter are shown in Tab. 4.

Algorithm	Worst (max)	Best (min)	Average (mean)	Execution Time (s)
CSA	2.687	0.266	0.813	131
GA	4.999	0.677	2.017	180
PSO	2.158	0.301	1.110	136

Tab. 4.   Comparison of the CSA performance with the Standard Meta-Heuristic Algorithms.

A desktop computer with 'Intel Core i7 CPU, 2.20 GHz Processor, 8 GB RAM' is used for the design and optimization process. At end of the optimization, the iteration number reaches 65 and the cost value of the operation is equal to 0.266. It is clear from Tab. 4 that CSA has a superior performance with respect to other popular Meta-Heuristics.

5. Conclusions

In this paper, a novel design methodology which uses fast and accurate SVRM models of base elements based on a 3D EM simulator is presented to design and analyze an UWB BPF. The outputs of the built SVRM models are used as input by a straightforward, simple and efficient CS algorithm under the rules of circuit theory to solve the filter response. The designed filter is manufactured and measured to show that the actual results are in coincidence with the simulation results. Furthermore, the performance of the CS algorithm is compared with other popular methods like GA and PSO to demonstrate the efficiency of the CS algorithm. The suggested methodology could be used for different kind of element shapes and filter types. Consequently, the proposed design methodology could be considered as an important contribution to the microwave design literature.

References

[1] Federal Communications Commission, Revision of Part 15 of the commission's rules regarding ultra wideband transmission system first report and order. *Tech. Rep.*, ET Docket 98-153, FCC02-48, FCC, Feb. 2002.

[2] MATTHAEI, G. L., YOUNG, L., JONES, E. M. T. *Microwave Filters, Impedance-matching Networks, and Coupling Structures.* Norwood: Artech House, 1980.

[3] OSKOUEI, H. D., FOROORAGHI, K., HAKKAK, M. Guided and leaky wave characteristics of periodic defected ground structures. *Progress In Electromagnetics Research*, PIER 73, 2007, p. 15–27.

[4] WU, B., LI, B., SU, T., LIANG, C. H. Equivalent-circuit analysis and lowpass filter design of split-ring resonator DGS. *Journal of Electromagnetic Waves and Applications*, 2006, vol. 20, no. 14, p. 1943–1953.

[5] AHN, D., PARK, J. S., KIM, C. S., KIM, J., QIAN, Y., ITOH, T. A design of the low-pass filter using the novel microstrip defected ground structure. *IEEE Trans. on Microw. Theory and Tech.*, Jan. 2001, vol. 49, no. 1, p. 86–93.

[6] LEE, J. K., KIM, Y. S. Ultra-wideband bandpass filter with improved upper stopband performance using defected ground structure. *IEEE Microw. Wireless Compon. Lett.*, Jun. 2010, vol. 20, no. 6, p. 316–318.

[7] DENG, H. W., ZHAO, Y. J., ZHANG, X. S., ZHANG, L., GAO, S. P. Compact quintuple-mode UWB bandpass filter with good out-of-band rejection. *Progress In Electromagnetics Research Letters*, 2010, vol. 14, p. 111–117.

[8] HUANG, J. Q., CHU, Q. X. Compact UWB band-pass filter utilizing modified composite right/left-handed structure with cross coupling. *Progress In Electromagnetics Research*, 2010, vol. 107, p. 179–186.

[9] CHOU, T. C., TSAI, M. H., CHEN, C. Y. A low insertion loss and high selectivity UWB bandpass filter using composite right/left-handed material. *Progress In Electromagnetics Research C*, 2010, vol. 17, p. 163–172.

[10] WANG, J. K., ZHAO, Y. J., QIANG, L., SUN, Q. A miniaturized UWB BPF based on novel scrlh transmission line structure. *Progress In Electromagnetics Research Letters*, 2010, vol. 19, p. 67–73.

[11] NAGHSHVARIAN JAHROMI, M., TAYARANI, M. Miniature planar UWB bandpass filters with circular slots in ground. *Progress In Electromagnetics Research Letters*, 2008, vol. 3, p. 87–93.

[12] PARK, J., KIM, J. P., NAM, S. Design of a novel harmonic-suppressed microstrip low-pass filter. *IEEE Microw. Wireless Compon. Lett.*, 2007, vol. 17, p. 424–426.

[13] ZHU, Y. Z., XIE, Y. J. Novel microstrip bandpass filters with transmission zeros. *Progress In Electromagnetics Research*, 2007, vol. 77, p. 29–41.

[14] VAPNIK, V. *The Nature of Statistical Learning Theory.* New York: Springer-Verlag, 1995.

[15] TOKAN, N. T., GÜNEŞ, F. Knowledge-based support vector synthesis of the microstrip lines. *Progress In Electromagnetics Research*, PIER 92, 2009, p. 65–77.

[16] GÜNEŞ, F., TOKAN, N. T., GÜRGEN, F. A knowledge-based support vector synthesis of the transmission lines for use in microwave integrated circuits. *Expert Systems with Applications*, 2010, vol. 37, p. 3302–3309.

[17] YANG, X. S., DEB, S. Cuckoo search via Lévy flights. In *Proc. of World Congress on Nature & Biologically Inspired Computing.* USA, 2009, p. 210–214.

[18] YANG, X. S., DEB, S. Engineering optimisation by cuckoo search. *Int. J. Mathematical Modelling and Numerical Optimisation*, 2010, vol. 1, no. 4, p. 330–343.

[19] KHODIER, M. Optimisation of antenna arrays using the cuckoo search algorithm. *IET Microwaves, Antennas & Propagation*, 2013, vol. 7, no. 6, p. 458–464.

[20] LAYEB, A. Hybrid quantum scatter search algorithm for combinatorial optimization problems. *Journal of Annals. Computer Science Series*, 2010, vol. 8, no. 2, p. 227–244.

[21] KUMAR, A., CHAKARVERTY, S. Design optimization for reliable embedded system using Cuckoo Search. In *3rd International Conference on Electronics Computer Technology.* 2011, p. 264–268.

[22] LAYEB, A. A novel quantum inspired cuckoo search for Knapsack. *International Journal of Bio-Inspired Computation*, 2011, vol. 3, no. 5.

[23] NOGHREHABADI, A., GHALAMBAZ, M., GHALAMBAZ, M., VOSOUGH, A. A hybrid Power Series – Cuckoo Search Optimization Algorithm to electrostatic deflection of micro fixed-fixed actuators. *International Journal of Multidisciplinary Sciences and Engineering*, July 2011, vol. 2, no. 4, p. 22–26.

[24] NAJMY, K., RANI, A., FAREQ, M., MALEK, A., SIEW-CHIN, N. Nature-inspired Cuckoo Search Algorithm for side lobe suppression in a symmetric linear antenna array. *Radioengineering*, September 2012, vol. 21, no. 3, p. 865–874.

[25] GONZALES, G. *Microwave Transistor Amplifiers: Analysis and Design.* 2nd ed. Prentice Hall, 1996.

About Authors ...

Mustafa ILARSLAN was graduated from the Middle East Technical University of Ankara in 1989, with a B.Sc. degree in Electrical & Electronics Engineering. He received M.Sc. degree in Electronics Engineering from the Osmangazi University of Eskisehir, Turkey. He has been the director of Aeronautics and Space Technologies Institute of TurAFA located in Istanbul, Turkey since March 2011. His research interests are aircraft and spacecraft avionics, systems engineering, radar and EW systems and technologies.

Salih DEMIREL has received M.Sc. and Ph.D. degrees in Electronics and Communication Engineering from Yıldız Technical University, Istanbul, Turkey in 2006 and 2009, respectively. He has been currently working as an Assistant Professor in the same department. His current research interests are among of microwave circuits especially optimization of microwave circuits, broadband matching circuits, device modeling, computer-aided circuit design, microwave amplifiers.

Hamit TORPI has received M.Sc. and Ph.D. degrees in Electronics and Communication Engineering from Yıldız Technical University, Istanbul, Turkey in 1990 and 1996, respectively. He has been currently working as an Assistant Professor in the same department. His current research interests are in the areas of multivariable network theory, device modeling, computer-aided microwave circuit design, monolithic microwave integrated circuits, and antennas.

A. Kenan KESKIN has received M.Sc. degree in Electronics and Communication Engineering from Yıldız Technical University, Istanbul, Turkey in 2012. He has been currently working as a Research Assistant and study-ing as a Ph.D. student in the same department. His current research interests are microwave circuits, computer-aided circuit design, UWB antennas, ground penetrating radars.

M. Fatih ÇAGLAR, received his B.Sc. degree in Electronics and Communication Engineering from the Istanbul Technical University in 1996 and M.Sc. degree in Electronics and Communication Engineering from the Suleyman Demirel University, in Isparta, in 1999. He had his Ph.D. degree from the Yıldız Technical University in Istanbul in Communication Engineering in 2007. His current research interests are among of RF/microwave circuits, especially modeling of microwave circuits, computer-aided circuit design and Artificial Neural Networks.

3

Pattern Synthesis of Dual-band Shared Aperture Interleaved Linear Antenna Arrays

Hua GUO, Chenjiang GUO, Jun DING

School of Electronics and Information, Northwestern Polytechnical University, Xi'an 710129, China

xdguohua@163.com, cjguo@nwpu.edu.cn, dingjun@nwpu.edu.cn

Abstract. *This paper presents an approach to improve the efficiency of an array aperture by interleaving two different arrays in the same aperture area. Two sub-arrays working at different frequencies are interleaved in the same linear aperture area. The available aperture area is efficiently used. The element positions of antenna array are optimized by using Invasive Weed Optimization (IWO) to reduce the peak side lobe level (PSLL) of the radiation pattern. To overcome the shortness of traditional methods which can only fulfill the design of shared aperture antenna array working at the same frequency, this method can achieve the design of dual-band antenna array with wide working frequency range. Simulation results show that the proposed method is feasible and efficient in the synthesis of dual-band shared aperture antenna array.*

Keywords

Pattern synthesis, linear arrays, dual-band, shared aperture, Invasive Weed Optimization.

1. Introduction

Design of the multifunctional antenna array is a key issue in the case of communication, remote sensing and electronic warfare, etc. Shared aperture antenna is one way to fulfill the multifunction of antenna array [1]. Two or more antenna subarrays that occupy the same area are known as shared aperture antennas or common aperture antennas [2]. If elements dedicated to different subarrays are interleaved in a shared aperture, the array is called interleaved or interlaced antenna array. Interleaving non-periodic subarrays provides a powerful and versatile tool to implement multifunctionality in antenna arrays [3].

The design of interleaved antenna array has been paid increasing attention in recent years. Many methods have been developed. In [1], a consistent strategy for the design of finite antenna arrays consisting of differently sized radiating elements is discussed. An effective and robust strategy for concurrently designing the transmitting and receiving antennas of a frequency modulated, continuous wave radar is discussed in [3]. A new method is described in [4] for adjusting the far-field polarization of an electronically steered phased-array antenna array. Three approaches to improving the efficiency of an array aperture by interleaving two arrays in the same aperture area is presented in [2]. The effect of allowing shared elements in interleaved thinned antenna arrays is investigated in [5]. An analytical technique based on almost difference sets (ADSs) for the design of interleaved linear arrays with well-behaved and predictable radiation features is proposed in [6].

However, the antenna arrays proposed above can only work at the same or similar frequencies. The working frequency range of the antenna arrays is narrow. Therefore, pattern synthesis of two interleaved linear arrays working at different frequencies is presented in this paper. The element positions are optimized by IWO and low PSLLs of the radiation patterns are obtained. IWO has been effectively used into the design of antennas [7], [8]. Usually, IWO outperforms the other optimization methods in the convergence rate as well as the final error level [9]. The rest of the paper is organized as follows. In Section 2, the mathematical model and the fitness function are given. How to determine the element positions of antenna arrays is depicted in Section 3. Section 4 describes the principles of IWO and the optimization procedure. Section 5 gives the simulation results and discussions. Finally, summary and conclusions are presented in Section 6.

2. Optimization Model

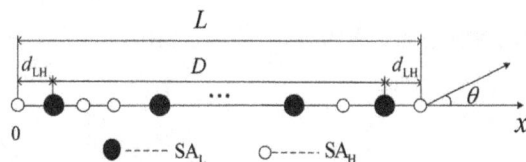

Fig. 1. Structure of a linear shared aperture antenna array.

The configuration of a linear shared aperture antenna array is as shown in Fig. 1. The length of the antenna array is L. The interleaved two subarrays are referred to as SA_L and SA_H, where the subscripts "L" and "H" indicate the lower and higher frequency subarrays. The normalized array factors of both subarrays can be given by

$$AF_L(\theta) = \frac{1}{N_L} \sum_{n=1}^{N_L} \exp(jk_L x_n^L \cos\theta) , \qquad (1)$$

$$AF_H(\theta) = \frac{1}{N_H} \sum_{n=1}^{N_H} \exp(jk_H x_n^H \cos\theta) \qquad (2)$$

where N_L and N_H are the element number of the lower and higher frequency subarrays, x_n^L and x_n^H are the nth element positions of both subarrays, $k_L = 2\pi/\lambda_L$ and $k_H = 2\pi/\lambda_H$ are wave numbers, λ_L and λ_H are wavelengths of the lower and higher working frequencies, θ is the angle measured from x-axis. The peak side lobe level of the radiation patterns can be calculated from

$$\text{PSLL} = \max\left\{\max|AF_H(\theta)|_{\theta\in S_H}, \max|AF_L(\theta)|_{\theta\in S_L}\right\} \qquad (3)$$

where S_L and S_H are the side lobe areas for the radiation patterns of both subarrays.

The objective is to find the best element positions of both subarrays that can minimize the PSLL of the radiation patterns. In order to eliminate the effect of mutual coupling, the adjacent array elements have a minimum spacing. As both subarrays work at different frequencies, the minimum spacing constrains of adjacent elements will be different. The minimum element spacings for both subarrays are given by d_L and d_H, respectively. The minimum spacing of the elements that belong to different subarrays is depicted by d_{LH}. So, the objective function can be depicted by

$$\begin{cases} \min\{\text{PSLL}\} \\ s.t.\ \min\left|x_i^L - x_j^L\right| \geq d_L \\ \min\left|x_m^H - x_n^H\right| \geq d_H, \min\left|x_i^H - x_m^L\right| \geq d_{LH} \\ i,j = 1,2\cdots N_L; m,n = 1,2\cdots N_H \\ i \neq j, m \neq n \end{cases} \qquad (4)$$

In the following procedure, the problem is changed into a maximization problem. So, the fitness function is defined by

$$f = \max\{|\text{PSLL}|\}. \qquad (5)$$

3. Interleaving of Array Elements

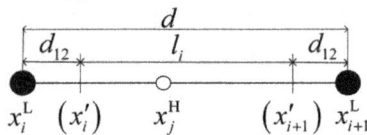

Fig. 2. Configuration of a segment of the array.

In this section, how to determine the array element positions of the two subarrays that satisfy those constrains given in (4) is proposed. Firstly, the element positions of the lower frequency subarray are determined. In order to efficiently use the whole array aperture, two elements are fixed at both sides of the antenna array. The positions of the first and last elements of the lower frequency subarray

are d_{LH} and $D+d_{LH}$, respectively. Only N_L-2 element positions need to be determined. As shown in Fig. 1, the aperture length of lower frequency subarray is D. As depicted in [10], the remaining region over the array aperture is given by

$$SP_L = D - (N_L - 1)d_L. \qquad (6)$$

Then, N_L-2 random real numbers among the range of $[0, SP_L]$ are calculated by

$$c_i^{L'} = SP_L \times r_i^L, i = 1,2,\cdots,N_L - 2 \qquad (7)$$

where r_i^L, $i = 1,2,\cdots,N_L-2$, are random numbers among the range of $[0,1]$. Then, $c_i^{L'}$, $i = 1,2,\cdots,N_L-2$, are sorted in ascending order and a new vector $\mathbf{C}^L = [c_1^L, c_2^L, \cdots, c_{N_L-2}^L]$ is obtained, where $c_1^L \leq c_2^L \leq,\cdots,\leq c_{N_L-2}^L$. Then, the element positions of lower frequency subarray can be obtained by

$$\begin{bmatrix} x_2^L \\ x_3^L \\ \cdots \\ x_{N_L-1}^L \end{bmatrix} = d_{LH} + \begin{bmatrix} c_1^L + d_L \\ c_2^L + 2\cdot d_L \\ \cdots \\ c_{N_L-2}^L + (N_L - 2)\cdot d_L \end{bmatrix}. \qquad (8)$$

It can be proved that the element spacing between x_i^L and x_{i+1}^L is $d_L+(c_{i+1}^L-c_i^L)$, $i = 2,3,\cdots,N_L-2$, which can satisfy the constrain of (4).

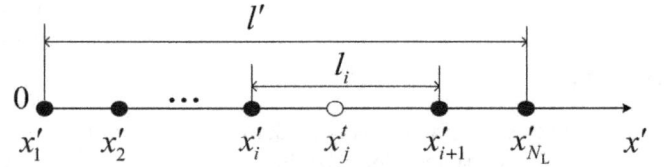

Fig. 3. Structure of the new coordinate system.

After the element positions of lower frequency subarray are calculated, the element positions of higher frequency subarray can be determined. As is shown in Fig. 2, if the adjacent element spacing of lower frequency subarray is less than $2d_{LH}$, no array elements of higher frequency subarray can be disposed between them. So, the length that can arrange the elements of higher frequency subarray among the range of x_i^L and x_{i+1}^L can be determined by

$$l_i = \begin{cases} 0, & x_{i+1}^L - x_i^L \leq 2d_{LH} \\ x_{i+1}^L - x_i^L - 2d_{LH}, & \text{else} \end{cases}, i = 1,2,\cdots,N_L - 1. \qquad (9)$$

The total length of the array aperture that can dispose higher frequency elements is given by

$$l' = \sum_{i=1}^{N_L-1} l_i. \qquad (10)$$

In order to determine the element positions of higher frequency subarray, a new coordinate system is determined (Fig. 3). The coordinate value of x_i', $i = 1,2,\cdots,N_L$, can be determined by

$$x_1' = 0, \quad x_{N_L}' = l', \quad x_i' = \sum_{n=1}^{i-1} l_n, \quad i = 2, 3, \cdots, N_L - 1. \quad (11)$$

There are N_H-2 elements of higher frequency subarray will be arranged in the length of l'. Similar to be shown above, a new parameter is given by

$$SP_H = l' - (N_H - 3)d_H. \quad (12)$$

Then, N_H-2 real random numbers among the range of $[0, SP_H]$ are calculated by

$$c_i^{H'} = SP_H \times r_i^H, \quad i = 1, 2, \cdots, N_H - 2 \quad (13)$$

where r_i^H, $i = 1, 2, \cdots, N_H-2$, are random numbers among the range of $[0,1]$. Then, $c_i^{H'}$, $i = 1, 2, \cdots, N_H-2$, are sorted in ascending order and a new vector $\mathbf{C}^H = [c_1^H, c_2^H, \cdots, c_{N_H-2}^H]$ is obtained, where $c_1^H \le c_2^H \le \cdots \le c_{N_H-2}^H$. Then, the element positions of higher frequency subarray in coordinate system x' can be calculated by

$$\begin{bmatrix} x_1' \\ x_2' \\ \cdots \\ x_{N_H-2}' \end{bmatrix} = \begin{bmatrix} c_1^H \\ c_2^H + d_H \\ \cdots \\ c_{N_H-2}^H + (N_H - 3) \cdot d_H \end{bmatrix}. \quad (14)$$

It can be proved that the element spacing between x_i' and x_{i+1}' is $d_H + (c_i^H - c_{i-1}^H)$, $i = 2, 3, \cdots, N_H-2$, which also satisfies the constrain proposed in (4). In order to fully utilize the whole array aperture, as is shown in Fig. 1, the first and last array element positions of higher frequency subarray are fixed to 0 and L, respectively. As is shown in Fig. 2, the rest element positions can be calculated by

$$\begin{cases} x_m^H = d_{LH} + x_i^L + (x_{m-1}' - x_i'), \quad x_i' \le x_{m-1}' < x_{i+1}' \\ i = 1, 2, \cdots, N_L - 1, \quad m = 2, 3, \cdots, N_H - 1 \end{cases}. \quad (15)$$

4. Optimization Strategy using IWO

4.1 Introduction to IWO

IWO is a numerical stochastic search algorithm that simulates the natural behavior of weed colonizing in the opportunity spaces for optimizing the function. This algorithm is simple. However, it has been shown to be effective in converging to an optimal result [11]. There are four steps of the algorithm which are described below:

1) Initialization

A certain number of weeds are randomly spread over the entire search space (K-dimension). The initial population of each generation is $\mathbf{X} = \{\mathbf{x}_1, \mathbf{x}_2, \cdots, \mathbf{x}_K\}$. Each search space has N elements.

2) Reproduction

Each number of the population $\mathbf{X}$ is allowed to produce weed seeds within a specified region centered at its own position. The number of seeds that are produced by $\mathbf{x}_k$, $k = 1, 2, \cdots, K$, depends on its relative fitness value in the population with respect to the best and worst fitness. The formula of weeds producing seeds is given by

$$weed_k = \left\lfloor \frac{f - f_{min}}{f_{max} - f_{min}}(s_{max} - s_{min}) + s_{min} \right\rfloor \quad (16)$$

where $\lfloor q \rfloor$ denotes the integer part of q, f is the current weed's fitness, f_{max} and f_{min} represent the best and worst fitness value of the current population, s_{max} and s_{min} are the maximum and minimum number of seeds that current population can produce, respectively.

3) Spatial distribution

The generated seeds are randomly distributed over the K-dimensional search space by normally distributed real random numbers which have zero mean and variance σ^2. The standard deviation σ is made to decrease over the generations in the following manner.

$$\sigma_{cur} = \sigma_{min} + \left(\frac{iter_{max} - iter}{iter_{max}} \right)^{nmi} (\sigma_{max} - \sigma_{min}) \quad (17)$$

where σ_{min} and σ_{max} are the minimum and maximum standard deviation, σ_{cur} is the standard deviation at the present time step, nmi represents the nonlinear modulation index. The maximum iteration number is $iter_{max}$.

4) Competitive exclusion

Some kind of competition between plants is needed for limiting the maximum number of plants in a colony. Initially, the plants in a colony will reproduce fast and all the produced plants will be included in the existing colony, until the number of plants in the colony reaches a maximum value p_max. However, it is expected that by this time the fitter plants have reproduced more seeds when compared to weaker plants. From then on, only the fittest plants up to p_max, among the existing ones and the reproduced ones, are taken in the colony and steps 2 to 4 are repeated until the maximum number of iterations have reached. So, the population size in each generation must be less than or equal to p_max. This method is known as competitive exclusion and is also a selection procedure of IWO.

4.2 Optimization Steps

In order to optimize the positions of the array elements by using IWO, the optimization procedure can be expressed as follows:

Step 1. The parameter values of the antenna arrays and IWO are given. A $N \times K$-dimensional matrix is chosen as the initial population to be optimized. Each dimension of the population can be depicted by r_i, $i = 1, 2, \cdots, N$, where $N = N_L + N_H - 4$ and $r_i \in [0,1]$. The first N_L-2 values are used to generate the element positions of lower frequency subarray while the rest N_H-2 values are used to generate the element positions of higher frequency subarray. Let $iter = 1$.

Step 2. The positions of the array elements are calculated by (8) and (15).

Step 3. The radiation patterns of the subarrays are calculated by (1) and (2). The peak side lobe level of the radiation patterns are determined by (3). The fitness value is defined by (5), which increases with the decrease of PSLL. The optimized parameters that can produce the best fitness are preserved as the ultimate result.

Step 4. The optimization parameters r_i, $i = 1, 2, \cdots, N$, are updated by IWO which has been introduced in Section 4.1.

Step 5. Let $iter = iter+1$, if $iter < iter_{max}$, go to step 2, otherwise, terminate iteration.

5. Optimization Results

In this section, several simulation results are given to show the feasibility and effectiveness of the proposed algorithm. The parameters used in above equations are given in Tab. 1. The minimum spacing constrains for the adjacent array elements of the two subarrays are chosen as follows: $d_L = \lambda_L/2$, $d_H = \lambda_H/2$, $d_{LH} = (\lambda_H+\lambda_L)/4$.

s_{min}	s_{max}	σ_{min}	σ_{max}	K	p_max	$iter_{max}$	nmi
0	10	10^{-3}	0.1	10	30	3000	3

Tab. 1. IWO parameter values.

Pattern Parameters	S-band	Ku-band	X-band	Ka-band
PSLL (dB)	−17.53	−17.56	−19.01	−19.03
MBW (°)	11.0	9.0	9.0	9.0
3dB BW (°)	3.62	0.72	3.08	0.88

Tab. 2. Parameters of the radiation patterns.

In order to demonstrate the effectiveness of IWO, the optimization results are compared with the results optimized by Particle Swarm Optimization (PSO) [9]. In case of PSO, as suggested in [9], both the cognitive rate (c_1) and the social rate (c_2) are set to 2.0 and the inertial weight is varied from 0.9 to 0.2. The number of sampling points for θ is 359. In order to obtain radiation patterns with low side lobe levels, the positions of the array elements are optimized. The algorithm is calculated 20 times and the best result is preserved as the ultimate result. A normal personal computer Intel Core i3 530 @2.93GHz CPU and 2GB of RAM is used and the algorithm is programmed by using MATLAB version 7.1.

5.1 S-band and Ku-band

In the first example, synthesis of S-band and Ku-band shared aperture antenna array is proposed. The central wavelengths of S-band and Ku-band are $\lambda_L = 10$ cm and $\lambda_H = 2$ cm, respectively. The element numbers of the two subarrays are chosen as $N_L = 20$ and $N_H = 45$. The total length of the whole array aperture is selected as $35\lambda_L/2$.

The central frequency ratio of the two working frequencies is $\mu = \lambda_L/\lambda_H = 5.0$ which is an integer.

For the best optimization result of the 20 calculations, the performance of IWO compared with Particle Swarm Optimization (PSO) is shown in Fig. 4. It can be seen that

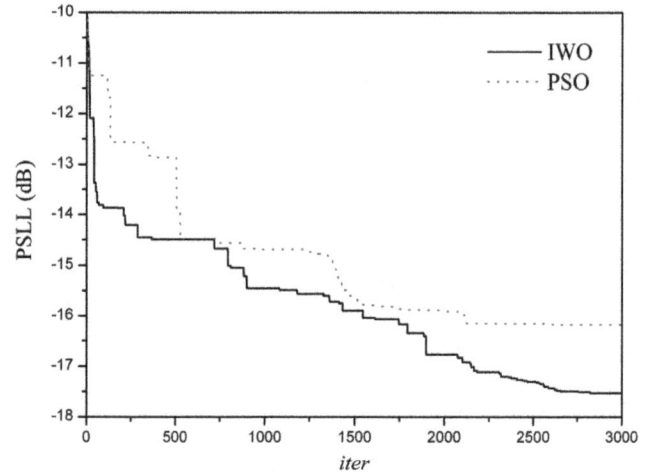

Fig. 4. PSLL versus iteration (S-band and Ku-band).

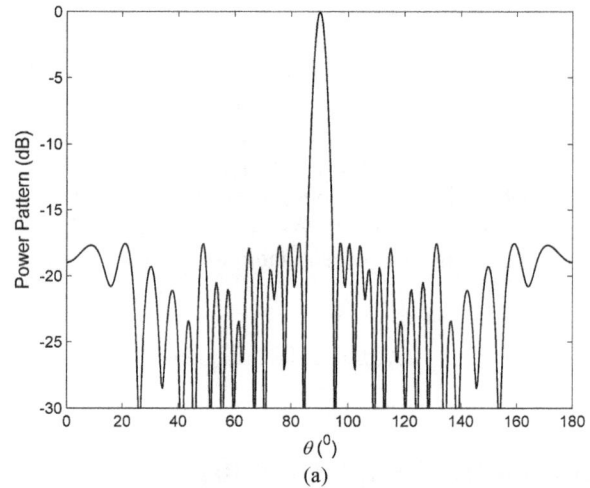

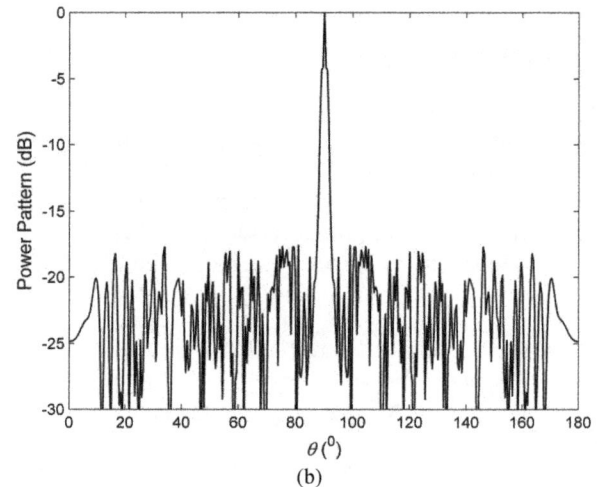

Fig. 5. Radiation patterns of the antenna array. (a) S-band; (b) Ku-band.

IWO can achieve a lower PSLL than PSO in this problem. The best results are PSLL = −17.53 dB optimized by IWO and PSLL = −16.16 dB synthesized by PSO. The radiation patterns of S-band subarray and Ku-band subarray are given in Fig. 5. The parameters of the radiation patterns, such as PSLL, main beam width (MBW) and 3 dB beam width (3 dB BW), are shown in Tab. 2. The element positions of both subarrays are given in Tab. 3 and Tab. 4. From Tab. 3 and Tab. 4, we can find that the minimum spacing of the adjacent elements for S-band subarray is 5.06 cm and which is 1.008 cm for Ku-band subarray. The minimum element spacing between S-band subarray and Ku-band subarray is 3.0 cm. They all fulfill those minimum spacing constrains given in (4). In order to show the advantage of the proposed method, the worst and average results of the 20 calculations are given. The worst and average results synthesized by IWO are PSLL = −16.60 dB and PSLL = −17.18 dB, respectively. The worst and average results synthesized by PSO are PSLL = −13.74 dB and PSLL = −15.44 dB. The computation time for a single optimization trail by IWO is about 3900 s while which is about 6100 s for PSO. From the results given above, it can be found that IWO is more effective and stable than PSO. Moreover, IWO is more timesaving than PSO.

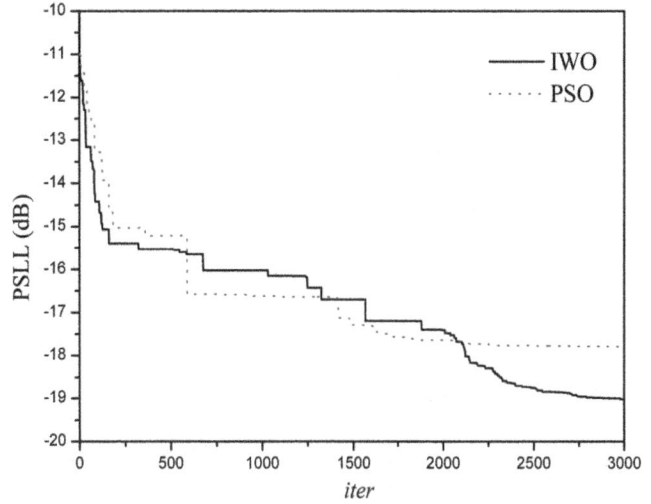

Fig. 6. PSLL versus iteration (X-band and Ka-band).

Element Number	Element Positions (cm)				
1-5	0.950	20.308	24.869	26.862	28.962
6-10	31.066	33.085	35.019	36.779	38.349
11-15	40.116	41.623	43.138	44.665	46.206
16-20	47.755	49.278	50.805	52.508	54.266
21-25	56.339	58.843	61.672	64.433	66.550

Tab. 5. Positions of X-band subarray elements.

Element Number	Element Positions (cm)				
1-5	3.000	50.637	58.511	66.030	73.243
6-10	80.359	88.063	93.759	98.819	105.040
11-15	111.466	117.259	123.243	128.447	133.794
16-20	138.865	144.926	154.303	164.937	172.000

Tab. 3. Positions of S-band subarray elements.

Element Number	Element Positions (cm)				
1-5	0.000	6.316	7.824	9.151	10.941
6-10	12.304	13.842	15.359	17.049	18.587
11-15	19.690	20.986	21.996	23.162	24.692
16-20	25.944	27.053	28.390	29.493	30.578
21-25	31.636	32.691	33.743	34.787	35.795
26-30	37.083	38.521	39.612	40.656	42.424
31-35	43.437	44.594	45.812	47.609	55.441
36-40	62.510	76.293	83.479	108.417	149.808
41-45	150.897	159.247	160.563	168.673	175.000

Tab. 4. Positions of Ku-band subarray elements.

Element Number	Element Positions (cm)				
1-5	0.000	1.908	2.650	3.380	3.993
6-10	4.523	5.154	5.787	6.236	6.711
11-15	7.130	7.623	8.080	8.587	9.027
16-20	9.628	10.030	10.441	11.016	11.516
21-25	12.049	12.452	12.873	13.344	13.751
26-30	14.166	14.571	14.976	15.384	15.888
31-35	16.447	16.950	17.350	17.759	18.256
36-40	18.688	19.173	21.484	21.912	22.465
41-45	22.952	23.486	23.892	30.106	57.506
46-50	59.818	60.415	62.819	63.411	67.500

Tab. 6. Positions of Ka-band subarray elements.

5.2 X-band and Ka-band

In this example, the synthesis result of X-band and Ka-band shared aperture antenna array is presented. The central working wavelengths of X-band and Ka-band are $\lambda_L = 3$ cm and $\lambda_H = 0.8$ cm. The element numbers of both subarrays are selected as $N_L = 25$ and $N_H = 50$. The length of the whole array aperture is selected as $45\lambda_L/2$. The central frequency ratio of the two working frequencies is $\mu = \lambda_L/\lambda_H = 3.75$.

Fig. 6 gives the convergence curves of the best result for IWO and PSO. From Fig. 6, we can observe that IWO has better performance than PSO. The best results for IWO and PSO are PSLL = −19.01 dB and PSLL = −17.78 dB, respectively. The parameters of the radiations are given in Tab. 2. Tab. 5 and Tab. 6 provide the element positions of X-band subarray and Ka-band subarray, respectively. From Tab. 5 and Tab. 6, we can find that the minimum element spacings of X-band subarray and Ka-band subarray are 1.01 cm and 0.4 cm, respectively. The minimum element spacing between X-band subarray and Ka-band subarray is 0.95 cm. The element spacings satisfy the distance constrains proposed in (4). The worst and average results of the 20 calculations for IWO are PSLL = −17.75 dB and PSLL = −18.18 dB. The worst and average results for PSO are PSLL = −16.73 dB and PSLL = −17.23 dB, respectively. In this example, it takes about 4600 s and 6800 s for a single trail by IWO and PSO, respectively. So, IWO can get a lower peak side lobe level and takes less algorithm time than PSO.

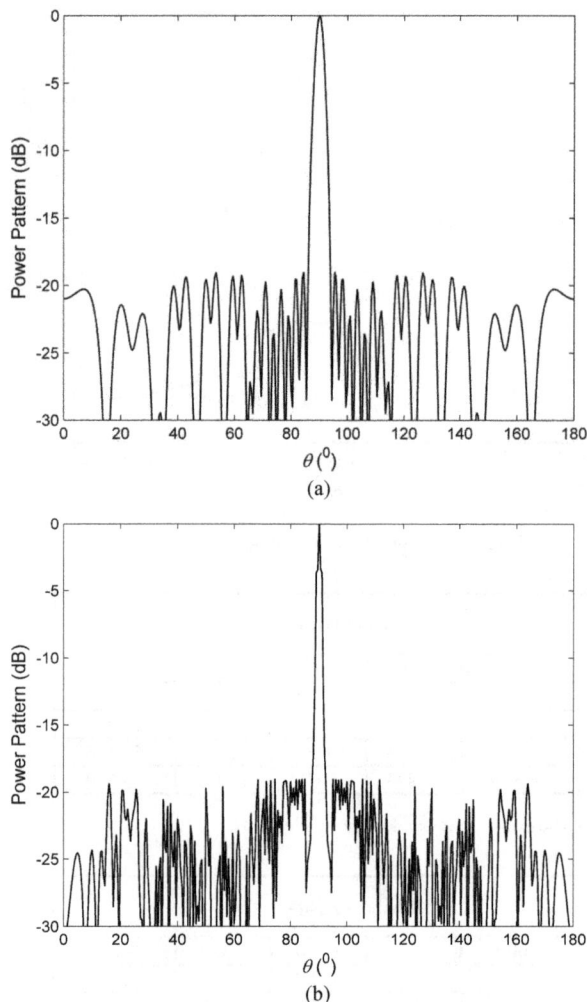

Fig. 7. Radiation patterns of the antenna array. (a) *X*-band; (b) *Ka*-band.

6. Conclusions

IWO is used in the synthesis of dual-band antenna arrays. The shared aperture interleaved linear antenna arrays working in a wide frequency range are considered here. In order to get radiation patterns with low side lobe levels, the element positions are optimized. The simulation results show that the PSLLs of the radiation patterns optimized by IWO are lower than those optimized by PSO. The PSLLs of the radiation patterns optimized by IWO are lower than −17 dB. Also, the minimum element spacing of the antenna arrays satisfies the designing constrains which will reduce the cross coupling effect of the adjacent elements.

Acknowledgements

This work was supported by the National Defense Pre-Research Foundation of China under Grant No. 9140A01010412HK03004.

References

[1] COMAN, C. I., LAGER, I. E., LIGHART, L. P. The design of shared aperture antennas consisting of differently sized elements. *IEEE Transactions on Antennas and Propagation*, 2006, vol. 54, no. 2, p. 376–383.

[2] HAUPT, R. L. Interleaved thinned linear arrays. *IEEE Transactions on Antennas and Propagation*, 2005, vol. 53, no. 9, p. 2858–2864.

[3] LAGER, I. E., TRAMPUZ, C., SIMEONI, M., LIGHART, L. P. Interleaved array antennas for FMCW radar applications. *IEEE Transactions on Antennas and Propagation*, 2009, vol. 57, no. 8, p. 2486–2490.

[4] SÁNCHEZ-BARBETTY, M., JACKSON, R. W., FRASIER, S. Interleaved sparse arrays for polarization control of electronically steered phased arrays for meteorological applications. *IEEE Transactions on Geoscience and Remote Sensing*, 2012, vol. 50, no. 4, p. 1283–1290.

[5] DU PLESSIS, W. P., KITCHING, C., BIN GHANNAM, A. Element sharing in interleaved antenna arrays. In *Proceedings of the 6th European Conference on Antennas and Propagation*. Prague (Czech Republic), 2012, p. 2230–2234.

[6] OLIVERI, G., MASSA, A. Fully interleaved linear arrays with predictable sidelobes based on almost difference sets. *IET Radar, Sonar and Navigation*, 2010, vol. 4, no. 5, p. 649–661.

[7] ROY, G. G., DAS, S., CHAKRABORTY, P., SUGANTHAN, P.N. Design of non-uniform circular antenna arrays using a modified invasive weed optimization algorithm. *IEEE Transactions on Antennas and Propagation*, 2011, vol. 59, no.1, p. 110–118.

[8] SEDIGHY, S. H., MALLAHZADEH, A. R., SOLEIMANI, M., RASHED-MOHASSEL, J. Optimization of printed Yagi antenna using invasive weed optimization (IWO). *IEEE Antennas and Wireless Propagation Letters*, 2010, vol. 9, p. 1275–1278.

[9] KARIMKASHI, S., KISHK, A. A. Invasive weed optimization and its features in electromagnetics. *IEEE Transactions on Antennas and Propagation*, 2010, vol. 58, no. 4, p. 1269–1278.

[10] CHEN, K. -S., HE, Z. -S., HAN, C. -L. A modified real GA for the sparse linear array synthesis with multiple constraints. *IEEE Transactions on Antennas and Propagation*, 2006, vol. 54, no. 7, p. 2169–2173.

[11] MEHRABIAN, A. R., LUCAS, C. A novel numerical optimization algorithm inspired from weed colonization. *Ecological Informatics*, 2006, vol. 1, p. 355–366.

About Authors ...

Hua GUO was born in 1981. He received his B.S. degree in Applied Physics from Xidian University (XDU) in 2004. In 2007, he obtained his M.S. degree in Radio Physics from XDU. He is now working towards the Ph.D. degree in Electromagnetic Fields and Microwave Techniques at Northwestern Polytechnical University (NWPU). His research interests include electromagnetic theory and calculation, array antenna design, phased arrays and radars.

Chenjiang GUO was born in 1963. He received his B.Eng. degree in Electronic Engineering in 1984 from Northwestern Polytechnical University (NWPU). He obtained his M.S. degree in Electromagnetic Fields and Microwave Techniques in 1987 from the NWPU. In 2007, he received his Ph.D. degree in Circuits and Systems from the NWPU. He is now a professor of Electromagnetic Fields and Microwave Techniques in the NWPU. His research interests include electromagnetic theory, antenna theory and design, microwave circuit design and electromagnetic compatibility (EMC).

Jun DING was born in 1964. She received her B.Eng. degree in Electronic Engineering in 1986 from Northwestern Polytechnical University (NWPU). She obtained her M.S. degree in Electromagnetic Fields and Microwave Techniques in 1989 from the NWPU. In 2005, she received her Ph.D. degree in Circuits and Systems from the NWPU. She is now a professor of Electromagnetic Fields and Microwave Techniques in the NWPU. Her research interests include electromagnetic calculation, antenna theory and design, microwave circuit design and electromagnetic compatibility (EMC).

4

Single-, Dual- and Triple-band Frequency Reconfigurable Antenna

Izni Husna IDRIS[1], Mohamad Rijal HAMID[1], Mohd Haizal JAMALUDDIN[2],
Mohamad Kamal A. RAHIM[1], James R. KELLY[3], Huda A. MAJID[1]

[1] UTM-MIMOS Center, Fakulti Kejuruteraan Elektrik, Universiti Teknologi Malaysia, Skudai, Johor, Malaysia
[2] Wireless Communication Center, Fakulti Kejuruteraan Elektrik, Universiti Teknologi Malaysia, Skudai, Johor, Malaysia
[3] Centre for Communication Systems Research, Faculty of Engineering and Physical Sciences, University of Surrey, Guildford, GU2 7XH, United Kingdom

ihusna2@live.utm.my, rijal@fke.utm.my, haizal@fke.utm.my, mkamal@fke.utm.my, j.r.kelly@surrey.ac.uk, huda_set@yahoo.com

Abstract. *The paper presents a frequency reconfigurable slot dipole antenna. The antenna is capable of being switched between single-band, dual-band or triple-band operation. The antenna incorporates three pairs of pin-diodes which are located within the dipole arms. The antenna was designed to operate at 2.4 GHz, 3.5 GHz and 5.2 GHz using the aid of CST Microwave Studio. The average measured gains are 1.54, 2.92 and 1.89 dBi for low, mid and high band respectively. A prototype was then constructed in order to verify the performance of the device. A good level of agreement was observed between simulation and measurement.*

Keywords

Frequency reconfigurable antenna, slot antenna.

1. Introduction

The demands on communications systems have increased dramatically in recent years. Modern communications systems often incorporate numerous radio transceivers each operating at a different frequency. It is no longer practical to provide a dedicated antenna element for each individual radio. Fortunately a frequency reconfigurable antenna, based on using a single radiating element is able to cover a variety of different frequencies and beam-width as reported in [1], [2]. In comparison with fixed multiband antennas, reconfigurable antennas increase the flexibility of a radio system [3]. Some fixed multi-band antennas are reported in [4]-[6]. In terms of gain, the performance of the fixed multi-band antennas is better than that of their reconfigurable counterparts. This can be attributed to losses within the switches of a reconfigurable antenna. However fixed multiband antennas can only operate at certain frequencies as compared to reconfigurable antenna that can access a range of operating frequencies. This is important for cognitive radio and spectrum aggregation both of which

will require antennas that can reconfigure their operating parameters. A fixed multiband antenna would not be adequate for use in these applications. Reconfigurable antenna is achieved through the use of microwave switches, such as MEMS or pin-diodes. The latter being the more popular choice. The reason for this is that MEMS switches are more expensive, afford lower reliability and require a higher bias voltage compared to pin-diodes [7]. Pin diodes, on the other hand, are based on more mature technology and are thus more readily available.

Broadly speaking the techniques for frequency reconfiguration can be divided into three main classes: 1) narrowband to another narrowband [8], 2) wideband to narrowband [9], and 3) multi-band to another multi-band [10]-[13]. An antenna that can be reconfigured from one narrowband to another narrowband is only capable of supporting one radio standard at a time. An antenna that can be reconfigured from wideband to narrowband, on the other hand, can support multiple radio standards at a time. However, the wideband operation inherently provides less interference rejection than narrowband operation. This limitation, is addressed by employing an antenna that can be reconfigured from one multi-band to another multi-band is preferred [10]-[13]. In [10], a varactor diode is used to tune the frequency location of the multi-band. In this way the antenna is able to cover a wide range of different frequencies. In [11], RF MEMS switches are used to achieve multiband reconfiguration in four different states. The antenna exhibits two dual-band states and two triple-band states. In this way the antenna is able to cover a wide frequency range from 0.8 GHz to 6 GHz. The antennas proposed in [12]-[13] can switch between single-band, dual-band, or multi-band operation.

The sickle-shaped slot antenna, reported in [12], incorporates six pin-diode switches. These switches are used to activate or deactivate certain slots. Consequently the antenna is able to provide seven different operating modes. The antenna proposed in [13] incorporates two switches. These switches enable it to provide three different operat-

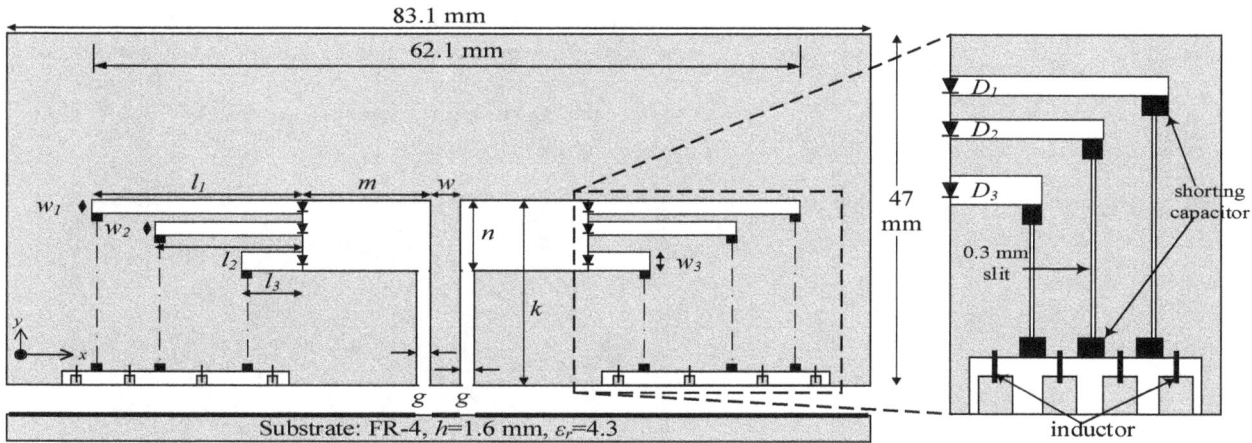

Fig. 1. The geometry of the proposed antenna.

ing states, namely: 2 dual-band and 1 triple-band mode. The antenna has applications in Bluetooth, WiMAX, and WLAN systems.

This paper proposes a new design for a frequency reconfigurable multi-band antenna based around a slot dipole. Three pairs of pin-diode switches are placed across, the dipole arms. It is possible to reconfigure the operating frequency of the antenna, by altering the states of the switches. In total, the proposed antenna is capable of operating in seven different states: three single-band, three dual-band, and one triple-band. The reminder of the paper is organized as follows: Section 2 describes the geometry of the antenna. Section 3 presents the simulation and measurement results. Finally Section 4 presents the conclusions.

2. Antenna Design

Fig. 1 describes the structure of the antenna along with the DC-bias circuitry used to control the microwave switches. The antenna is based on a design presented in [14] and consists of three pairs of slots, positioned in series. Then, the slots are merged into a rectangular box to have a stable and better impedance matching results. These slots are etched into the ground plane. The antenna is fed via a CPW transmission line.

It is fabricated on an FR-4 substrate having a relative permittivity (ε_r) of 4.3 and a thickness (h) of 1.6 mm. The size of the substrate is 83.1 mm × 47 mm. This size is chosen because it gives better S_{11} result after optimization was done. It is worth to note that the height of the substrate has a strong effect on the lowest resonance frequency. The geometrical parameters of the proposed antenna are given in Tab. 1. In comparison with a fixed multiband antenna, the proposed antenna can be reconfigured between a larger numbers of operating bands. The antenna also provides a degree of additional filtering, which would have the effect of reducing the level of interference at the receiver.

These factors represent significant advantages in comparison with [14].

Each dipole arms is approximately half-a-wavelength long at the corresponding resonant frequency. For example, the uppermost arm has a length of 62.1 mm which is approximately half-a-wavelength long at 2.45 GHz. The middle and lower arms are 46.6 mm and 36.7 mm long, respectively. To obtain the best results, the values of all of the geometrical parameters have been optimized using a computer simulation tool. This is necessary to account for the specific path taken by the surface currents as well as the effect of fringing fields. Three pairs of pin-diode switches are inserted into the antenna, as mentioned earlier. An Infineon BAR50-02V pin-diode was employed in the prototype antenna [15]. Fig. 2 shows the equivalent circuit for pin-diode in the on-state. Fig. 3 and Fig. 4 show the equivalent circuit for pin-diode in the off-state. At frequencies below 4 GHz the value of capacitor is 0.15 pF. At frequencies above 4 GHz the capacitor value is 0.12 pF. The pin-diode pairs are denoted D_1, D_2 and D_3. These pairs of diodes are located on the upper, middle, and lower pairs of dipole arms, respectively. The electromagnetic computer simulations results reported in this paper were obtained using CST Microwave Studio® 2012 [16].

Parameters	Dimensions (mm)	Parameters	Dimensions (mm)
g	0.50	n	5.00
k	19.00	w	3.00
l_1	16.55	w_1	0.50
l_2	8.80	w_2	0.50
l_3	3.85	w_3	1.00
m	13.00		

Tab. 1. Parameter value of the antenna design.

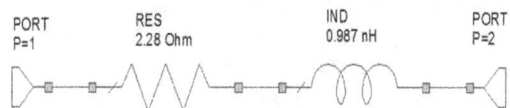

Fig. 2. Pin-diode on state equivalent circuit.

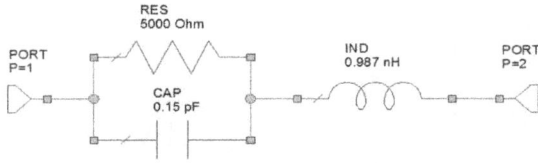

Fig. 3. Pin-diode off state equivalent circuit, frequency < 4 GHz.

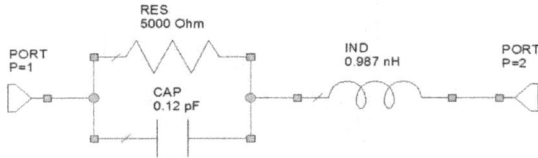

Fig. 4. Pin-diode off state equivalent circuit, frequency ≥ 4 GHz.

Tab. 2 describes the switch states required in order to obtain each of the seven different operating bands. States one, two, and three yield single-band operation. They are achieved by switching off only one pair of pin-diodes. The remaining pairs of diodes are switched on. To be specific, state one is achieved by switching off diode pair D_1. State two, on the other hand, is achieved by switching off diode pair D_2. In contrast, states four, five, and six are obtained by switching on only one pair of diodes. These states yield dual-band operation. To be specific, state five is achieved when only diode pair D_2 is switch on. Finally, if all of the pin-diodes are turned off, triple-band operation is achieved. This is referred to as state seven.

State	D_1	D_2	D_3	Bands
1	off	on	on	Single-band : (low)
2	on	off	on	Single-band : (mid)
3	on	on	off	Single-band : (high)
4	off	off	on	Dual-band : (low & mid)
5	off	on	off	Dual-band : (low & high)
6	on	off	off	Dual-band : (mid & high)
7	off	off	off	Triple-band : (low, mid & high)

Tab. 2. Switch configuration of the antenna design.

In the prototype, the biasing circuit comprises a number of 100 pF capacitor along with a 27 nH inductors. Each of the capacitors behaves as a DC block while each of the inductors serves as an RF choke. A series of 0.3 mm slits are introduced into the ground plane, as shown in the inset of Fig. 1. These slits serve as a means of isolating the bias voltages applied to the different switches from one another.

3. Results and Discussions

Fig. 5 shows a photograph of the fabricated antenna. Fig. 6 shows measured and simulated S_{11} curves corresponding to the seven operating bands. Inspection of these figures shows that there is generally a good level of agreement between measurement and simulation. The measured S_{11} for states one and two reveals resonances at 2.4 GHz and 3.5 GHz, respectively. For state one, a second passband also occurs around 5.8 GHz. This is also apparent on

inspection of the gain plot, shown in Fig. 9(a). When configured to operate in state seven, the antenna exhibits resonances at 2.4 GHz, 3.5 GHz and 5.2 GHz. This would enable the antenna to support Bluetooth, WiMAX and WLAN systems, respectively.

Fig. 5. Photograph of the proposed antenna.

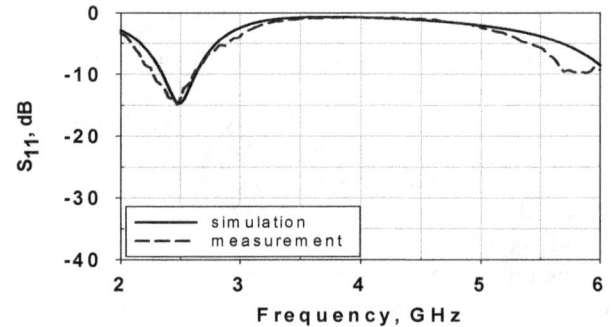

(a)

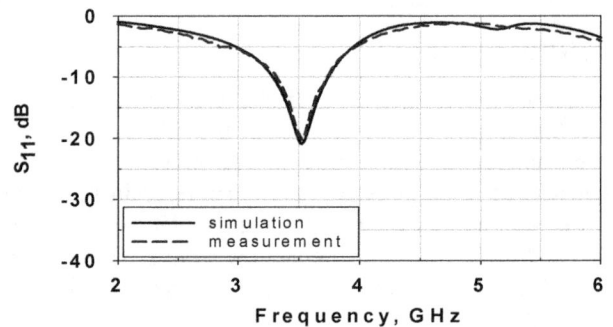

(b)

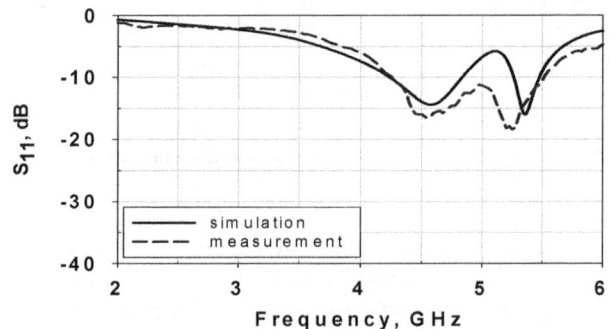

(c)

(d)

(e)

(f)

(g)

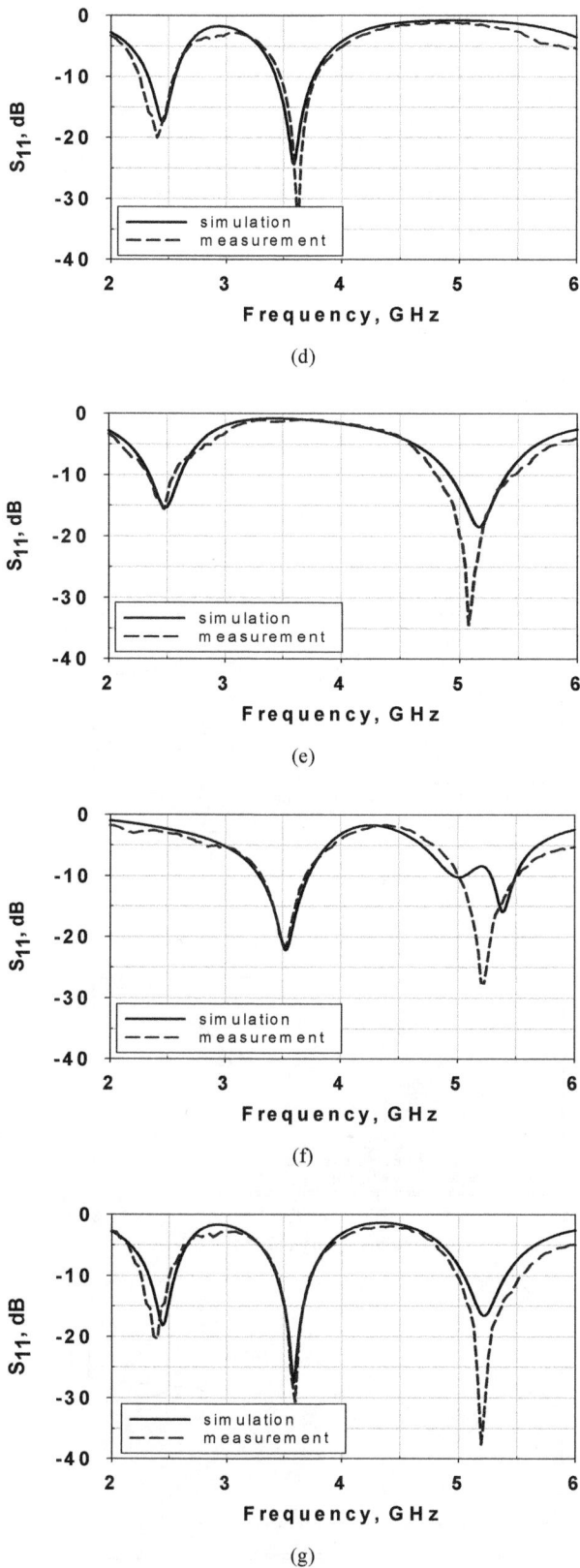

Fig. 6. Measured and simulated S_{11} of the antenna: (a) state one, (b) state two, (c) state three, (d) state four, (e) state five, (f) state six, (g) state seven.

Fig. 7 shows simulated current distributions associated with the proposed antenna. The figure illustrated the current distributions at three different frequencies. Fig. 7(b)

indicates that the current flows mainly along the longest arm at low frequency. At the middle range frequency; the current flows predominantly along the middle arm as shown in Fig. 7(c) whilst the high frequency current is concentrated along the shortest arm and shown in Fig. 7(d).

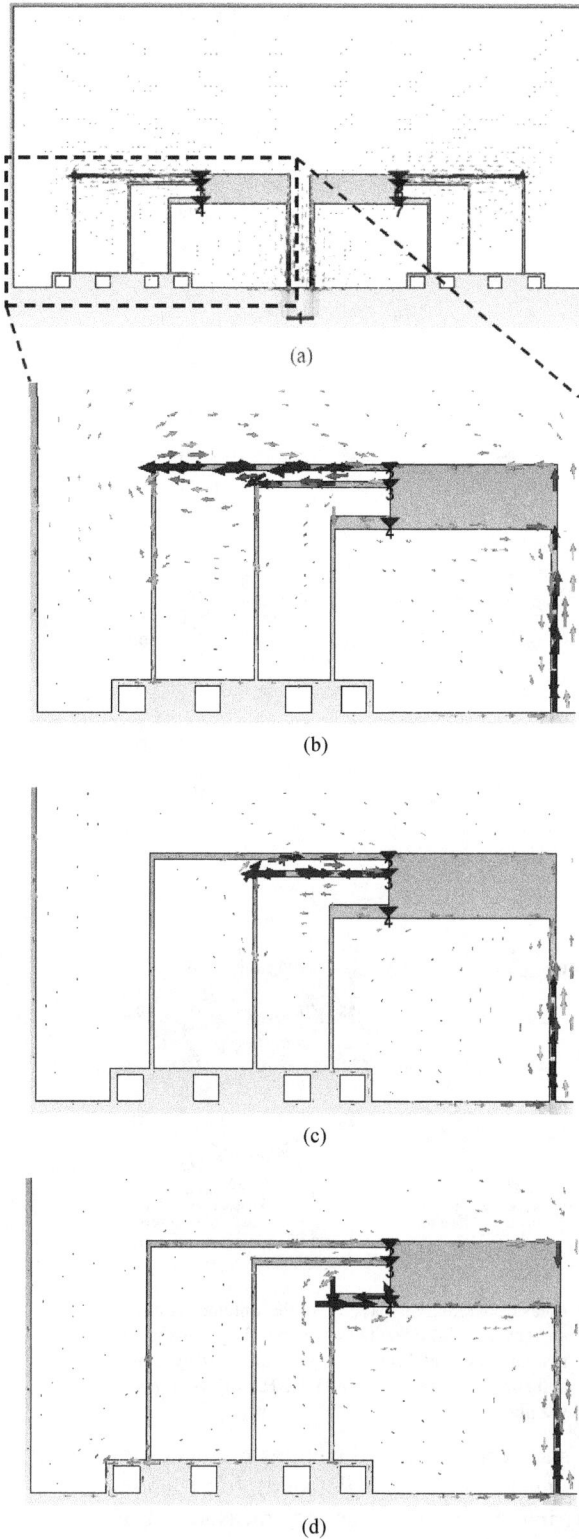

(a)

(b)

(c)

(d)

Fig. 7. Simulated current distribution for state seven (a) overall view at 2.4 GHz; zoom in view (b) at 2.4 GHz, (c) at 3.5 GHz and (d) at 5.2 GHz.

Fig. 8 shows the measured radiation patterns corresponding to the single-band and triple-band of operations. Fig. 8(a), Fig. 8(c) and Fig. 8(e) shows the x-z plane radiation patterns, while Fig. 8(b), Fig. 8(d) and Fig. 8(f) shows the y-z plane radiation patterns. The patterns were obtained at 2.4 GHz, 3.5 GHz, and 5.2 GHz. These results indicate that radiation patterns corresponding to different states but the same frequency are quite similar in terms of their shape. It is worth noting that the flexible wires, attached to the edge of the board, affected the radiation pattern especially at high frequency if they are scattered.

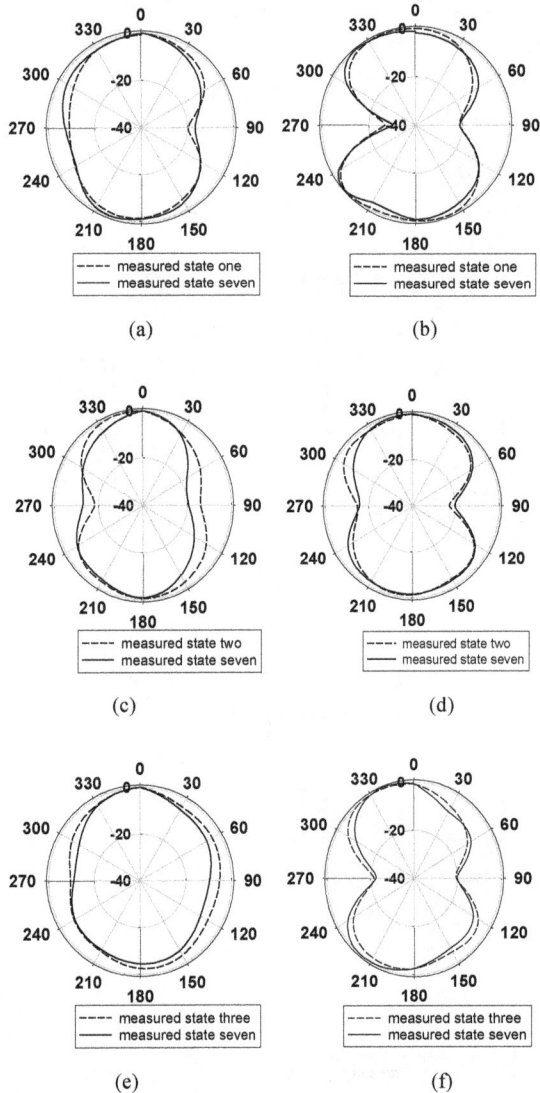

(a)   (b)

(c)   (d)

(e)   (f)

Fig. 8. Measured radiation pattern of the antenna: state one and seven at 2.4 GHz: (a) x-z plane, (b) y-z plane, state two and seven at 3.5 GHz: (c) x-z plane, (d) y-z plane, and state three and seven at 5.2 GHz: (e) x-z plane, (f) y-z plane.

Fig. 9 shows the measured gain, as a function of frequency, for states: one, two (single-band), five (dual-band), and seven (triple-band). The gain was measured in the 0° direction of x-z plane. From the graph, the passband gains for single and dual-band operation are seen to be comparable with those for triple-band operation. The measured

passband gain for state one is 1.58 dBi while that for state two is 3.00 dBi. The curve for state five is shown in Fig. 9(c). The measured gains are seen to be 1.52 dBi and 2.00 dBi at the centers of the low and high frequency passbands, respectively. For state seven, the measured gains are 1.53 dBi, 2.83 dBi and 1.78 dBi at the centers of the low, middle and high frequency passbands, respectively.

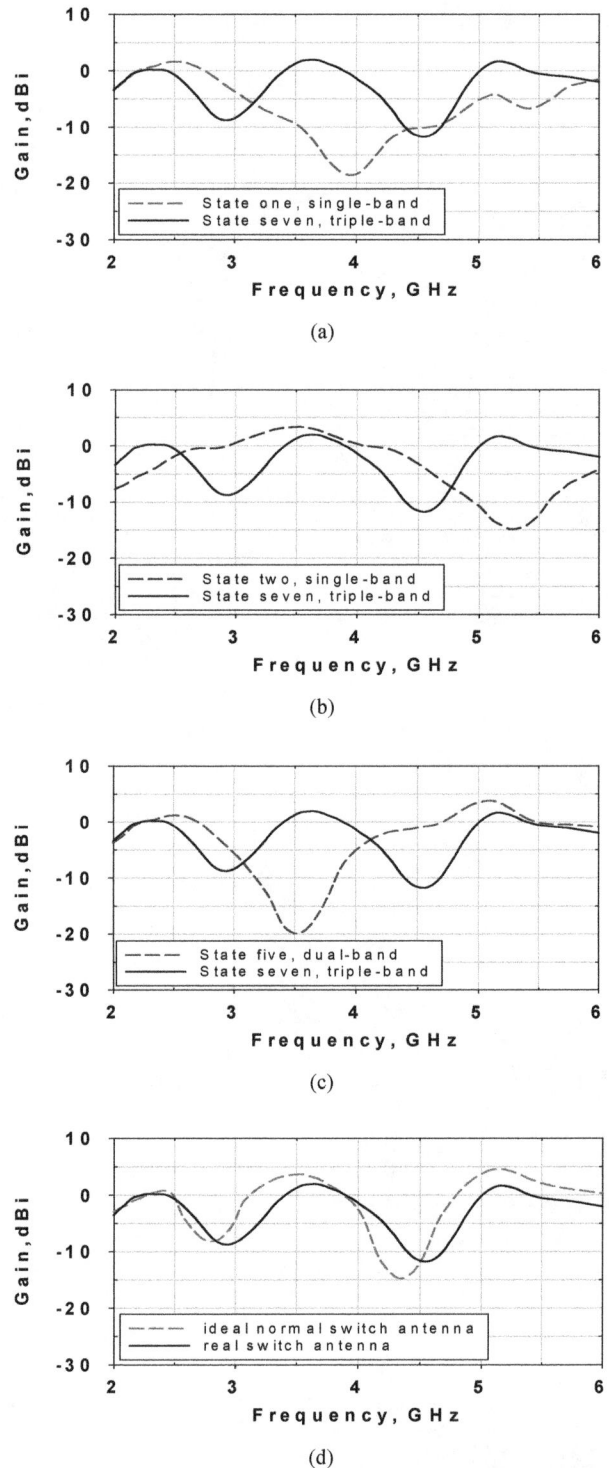

(a)

(b)

(c)

(d)

Fig. 9. Measured gain of the antenna: (a) state one and state seven, (b) state two and state seven, and (c) state five and state seven, (d) ideal and real switch.

Tab. 3 provides a summary of the gain measurement data. The effects of the pin-diode switches can be analyzed by comparing the gain performance of the proposed antenna with that of an antenna incorporating hardwired switches (i.e. idealized switches) as shown in Fig. 9(d). It is observed that the use of pin-diodes degrades the gain slightly, as one would expect.

State	Measured gain, dBi		
	Low band	Mid band	High band
1 (pin-diode)	1.58	-	-
2 (pin-diode)	-	3.00	-
5 (pin-diode)	1.52	-	2.00
7 (pin-diode)	1.53	2.83	1.78
Ideal switch	2.87	3.24	4.11

Tab. 3. Summary of gain measurement result.

4. Conclusion

This paper presents a design for novel frequency reconfigurable antenna. It is shown that frequency reconfigurability can be achieved by switching in and out selected pairs of dipole arms within the antenna. A prototype of the antenna was fabricated. This prototype incorporates pin-diode switches. The design was validated by comparing simulated and measured results. The flexibility to have different operating frequencies makes the proposed antenna suitable for use in electronic devices which must operate within multiple frequency bands.

Acknowledgements

The authors would like to thank the Ministry of Higher Education (MOHE) and MIMOS Center, Universiti Teknologi Malaysia (UTM) for supported this work, under grant reference number: Q.J130000.2523.04H83.

References

[1] MANTEUFFEL, D., ARNOLD, M., UHLIG, P. Considerations on configurable multi-standard antennas for mobile terminals realized in LTCC technology. *Radioengineering*, December 2009, vol. 18, no. 4, p. 395–401.

[2] CHAKER, H., MERIAH, S. M., BENDIMERAD, F. T. Optimization of micro strip array antennas using hybrid particle swarm optimizer with breeding and subpopulation for maximum side-lobe reduction. *Radioengineering*, December 2008, vol. 17, no. 4, p. 39–44.

[3] YANG, S., ZHANG, C., PAN, H. K. Frequency-reconfigurable antennas for multiradio wireless platforms. *IEEE Microwave Magazine*, 2009, vol. 10, p. 66–83.

[4] JOSEPH, S., PAUL, B., MRIDULA, S., MOHANAN, P. A novel planar fractal antenna with CPW-feed for multiband applications. *Radioengineering*, December 2013, vol. 22, no. 4, p. 1262–1266.

[5] LI, Y., ZHANG, Z., FENG, Z., ISKANDER, M. F. Design of penta-band omnidirectional slot antenna with slender columnar structure. *IEEE Transactions on Antennas and Propagation*, 2014, vol. 62, no. 2, p. 594–601.

[6] NASER-MOGHADASI, M., SADEGHZADEH, R. A., FAKHERI, M., ARIBI, T., VIRDEE, B. S. Miniature hook-shaped multiband antenna for mobile applications. *IEEE Antennas and Wireless Propagation Letters*, 2012, vol. 11, p. 1096–1099.

[7] YEOM, I., CHOI, J., KWOUN, S.-S., LEE, B., JUNG, C. Analysis of RF front-end performance of reconfigurable antennas with RF switches in the far field. *International Journal of Antennas and Propagation*, 2014, vol. 2014, article ID 385730, p. 1–14.

[8] MAJID, H. A., RAHIM, M. K. A., HAMID, M. R., ISMAIL, M. F. A compact frequency-reconfigurable narrowband microstrip slot antenna. *IEEE Antennas and Wireless Propagation Letters*, 2012, vol. 11, p. 616–619.

[9] HAMID, M. R., GARDNER, P., HALL, P. S., GHANEM, F. Switched-band Vivaldi antenna. *IEEE Transactions on Antennas and Propagation*, 2011, vol. 59, no. 5, p. 1472–1480.

[10] ABUTARBOUSH, H. F., NILAVALAN, R., NASR, K. M., AL-RAWESHIDY, H. S., BUDIMIR, D. Widely tunable multiband reconfigurable patch antenna for wireless applications. In *Proceedings of the Fourth European Conference on Antennas and Propagation (EuCAP)*. Barcelona (Spain), 2010, p. 1–3.

[11] BEMANI, M., NIKMEHR, S. A novel reconfigurable multiband slot antenna fed by a coplanar waveguide using radio frequency microelectro-mechanical system switches. *IEEE Microwave and Optical Technology Letters*, 2011, vol. 53, p. 751–757.

[12] SHAGATI, A. P., AZARMANESH, M., ZAKER, R. A novel switchable single- and multifrequency triple-slot antenna for 2.4-GHz Bluetooth, 3.5-GHz WiMAX, and 5.8-GHz WLAN. *IEEE Antennas and Wireless Propagation Letters*, 2010, vol. 9, p. 534–537.

[13] PAN, Y., LIU, K., HOU, Z. A novel printed microstrip antenna with frequency reconfigurable characteristics for Bluetooth/ WLAN/ WiMAX applications. *IEEE Microwave and Optical Technology Letters*, 2013, vol. 55, p. 1341–1345.

[14] CHEN, S.-Y., CHEN, Y.-C., HSU, P. CPW-fed aperture-coupled slot dipole antenna for tri-band operation. *IEEE Antennas and Wireless Propagation Letters*, 2008, vol. 7, p. 535–537.

[15] BAR50-02V, http://www.infineon.com/

[16] Computer Simulation Technology (CST) Studio Suite. ver. 2013, CST AG, *www.cst.com*

About Authors ...

Izni Husna IDRIS was born in Kuala Lumpur, Malaysia in 1989. She received her Bachelor degree in Electrical (Telecommunication) Engineering from Universiti Teknologi Malaysia (UTM) in 2012. She is now doing her Master's degree in Electrical (Telecommunication) Engineering in UTM. Her research interests include reconfigurable antenna and RF microwave communication systems.

Mohamad Rijal HAMID received the Ph.D. degree in Electrical Engineering from the University of Birmingham, UK, in 2011. He has been with the Faculty of Electrical Engineering (FKE), UTM, since 2001. His major research interest is reconfigurable antenna design for multimode wireless applications. Currently he is a Senior Lecturer at Communication Engineering Dept., FKE, UTM.

Mohd Haizal JAMALUDDIN received the Doctoral degree in Signal Processing and Telecommunications from the University de Rennes 1, France in 2009. His research interests include antenna design for millimeter wave applications, RF and microwave communication systems and specific antennas such as dielectric resonator, reflect array and dielectric dome antennas. He joined Universiti Teknologi Malaysia in 2003 as a Tutor at the Dept. of Electronic Engineering, Faculty of Electrical Engineering. Currently he is a Senior Lecturer at Wireless Communication Centre, Faculty of Electrical Engineering, Universiti Teknologi Malaysia.

Mohamad Kamal A. RAHIM received his Ph.D. degrees in Electrical Engineering from University of Birmingham UK in 2003. In 2005 he was appointed as a senior lecturer and in 2007 he was appointed as Assoc Professor at the faculty. Now he is the Professor in RF and Antenna at Faculty of Electrical Engineering Universiti Teknologi Malaysia. His research interest includes the areas of design of dielectric resonator antennas, microstrip antennas, small antennas, microwave sensors, RFID antennas for readers and tags, multi-function antennas, microwave circuits, EBG, artificial magnetic conductors, metamaterials, phased array antennas, computer aided design for antennas and design of millimeter frequency antennas. He has published over 200 articles in journals and conference papers.

James R. KELLY received the Ph.D. degree in Microwave Filters (2007) from Loughborough University, Leicestershire, England. Dr. James Kelly is currently a lecturer within the Centre for Communication Systems Research (CCSR) at the University of Surrey, Guildford, Surrey, England. His research interests include: body area networks, reconfigurable microwave circuits, microwave antennas, microwave filters, and metamaterials. He has published almost 60 academic papers in peer reviewed journals and conference proceedings. He frequently acts as a reviewer for various IEEE publications.

Huda A. MAJID received the Ph.D degree in Electrical Engineering from Universiti Teknologi Malaysia, in 2014. During the M.S. degree at Universiti Teknologi Malaysia, he conducted research focused on left-handed metamaterial and is currently developing frequency reconfigurable antenna for future wireless communications. His research interests concentrate mainly on reconfigurable antennas, metamaterials and textile antennas.

5

User Hand Influence on Properties
of a Dual-Band PIFA Antenna

Radek VEHOVSKÝ[1], Michal POKORNÝ[1], Kamil PÍTRA[2]

[1]Dept. of Radio Electronics, Brno University of Technology, Technická 12, 612 00 Brno, Czech Republic
[2]Military Research Institute, state enterprise, Veslarska 230, 637 00 Brno, Czech Republic

xvehov00@stud.feec.vutbr.cz, pokornym@feec.vutbr.cz, pitra@vvubrno.cz

Abstract. *This paper deals with the user hand influence on impedance matching and radiation pattern of a planar inverted-F antenna (PIFA). In the text, the PIFA structure is discussed in order to achieve broadband and multiband capability. Then, the dual-band PIFA antenna for operation at frequencies of GSM900 and GSM1800 systems is designed. In the next step we investigate the user influence in data mode (the user is typing a message or browsing with a phone). For this purpose the phantom hand was made of an agar based material. The first author's right hand was used as a template for the phantom. Numerical model was created by 3D scanning of the fabricated phantom. Finally, the comparison of the differences between simulations and measurement is presented.*

Keywords

Dual-band antenna, PIFA, user influence, human tissue, phantom hand.

1. Introduction

Nowadays, wireless communications devices are almost everywhere. Thanks to modern electronic components there is a continual trend of miniaturizing this equipment and similar demands are placed on their antennas. The radiation performance of small antennas is becoming more sensitive to the head and especially to the hand, since internal antennas were adopted into mobile devices.

In the data mode the index finger is mostly in close proximity of the antenna radiator. In order to evaluate the influence of the lossy dielectric object on the antenna performance, a hand model should be a necessary part of designing and testing of mobile devices. Due to hygienic limits, phantoms or numerical models are commonly used instead of real human hands. Several papers deal with this issue, but mainly covering only measurements or only simulations [1]–[4]. This paper deals with both, simulation and measurement, using a hand model and phantom which have precisely equal shape (corresponding with the first author's right hand). So the comparison of results is credi-

ble. The hand model is provided by scanning a phantom model using a 3D scanner. Agar was chosen as a tissue-equivalent material with additional water, whose dielectric properties are equivalent to biological muscle tissue [5].

The objectives of the paper are to design a dual-band PIFA antenna with the assistance of an electromagnetic field simulator, make a prototype and verify its properties by measurement. On this type of antenna, the user influence in data mode will be investigated. The phantom from the agar based material and its numerical equivalent will be used. Finally, the user hand influence on impedance matching, the radiation pattern and the radiation efficiency of PIFA, as well as SAR distribution in the hand will be presented.

2. Planar Inverted F-Antenna

PIFA is a commonly used antenna in mobile applications. It is called an inverted F-antenna because the side view of the antenna element is like the letter F with its face down.

The PIFA antenna is a modification of a conventional patch antenna. One of the main differences between them is their size. While the resonance length of an ordinary patch antenna is $\lambda/2$, PIFA reduces the resonance length to less than $\lambda/4$. It is accomplished by using a shorting pin located at the null-voltage point of a patch antenna.

It is known that planar antennas have a narrow bandwidth. This problem can be overcome by increasing substrate height, decreasing its ε_r or achieving multiband behavior of the antenna. In most cases, a thick air dielectric substrate is used. Multiband capabilities are often achieved by cutting slits on the patch.

2.1 Antenna Design

The aim of this chapter is to design a dual-band PIFA antenna for GSM900 and GSM1800 frequency ranges.

The first step of the antenna design is to determine approximate dimensions of the patch. The sum of width w and length l of the patch can be roughly expressed as follows [6]

$$w + l \approx \frac{\lambda}{4} \qquad (1)$$

where λ is the first resonant wavelength. The initial cutting of the patch slit, for achieving the second resonance, was adopted from [6].

The patch is made of IsoClad933 substrate and the ground plane is made of FR-4 substrate. Separation between substrates is 1 cm, supported by plastic distance posts. The shorting pin has diameter 1 mm. Feeding is provided by an SMA connector through the ground plane.

Simulations were performed in the electromagnetic field simulator CST Microwave Studio based on the FIT numerical method. The designed dimensions of the patch are shown in Fig. 1. The ground plane dimensions are $50 \times 120 \ \text{mm}^2$. The final prototype of the antenna was fabricated (see Fig. 2).

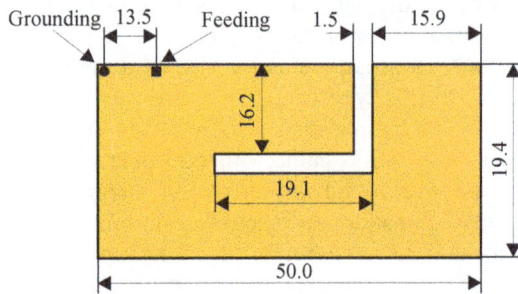

Fig. 1. Dual-band PIFA dimensions.

Fig. 2. Photo of the dual-band PIFA prototype.

2.2 Antenna Parameters

The impedance match measurement of the fabricated PIFA was carried out by the vector network analyzer Rohde&Schwarz ZVL3. For eliminating unwanted asymmetric currents flowing over the outside shield of the coaxial cable, ferrite chokes were used.

The comparison of the computed and measured frequency response of the reflection coefficient is shown in Fig. 3. Both characteristics show good impedance match in the required frequency bands, but some disagreement be-

tween them is obvious. Frequency shifts are caused mainly by inaccurate soldering of the connector and shorting pin.

In Fig. 4 the computed 3D radiation patterns for vertically and horizontally polarized waves are compared. It is obvious, considering antenna position in Fig. 4, that vertically polarized waves play a dominant role in antenna radiation. For vertically polarized waves at 925 MHz, antenna radiation is practically omnidirectional (like a dipole); at 1795 MHz, the antenna radiates primarily downwards. Horizontally polarized waves at 925 MHz are strongly suppressed; at 1795 MHz, the radiation pattern is approximately omnidirectional with gain around 0 dBi.

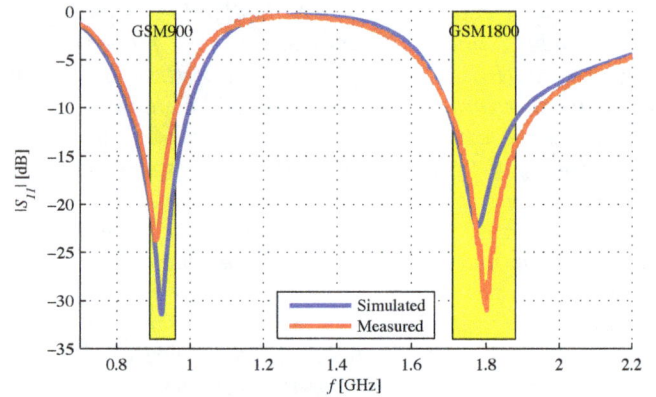

Fig. 3. Simulated and measured frequency response of the reflection coefficient of the PIFA.

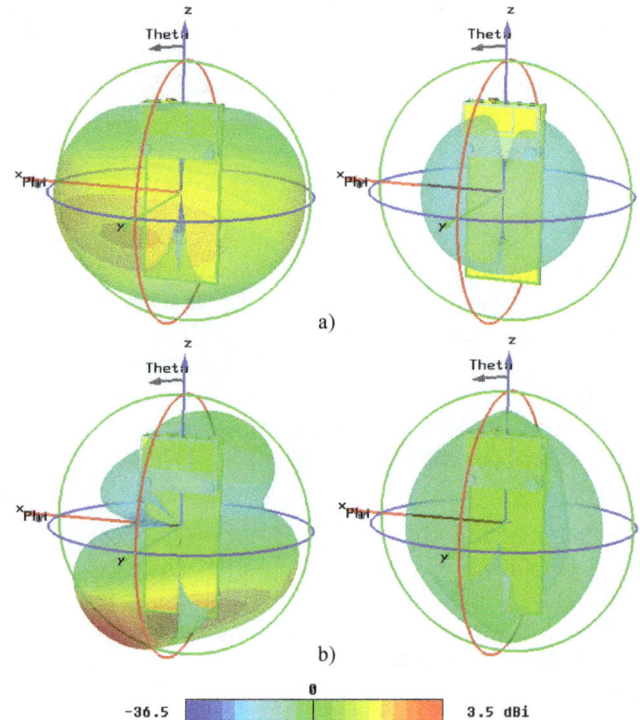

Fig. 4. Simulated radiation patterns at a) 925 MHz and b) 1795 MHz of the PIFA. Left pictures represent vertically polarized waves, the right pictures are computed for horizontally polarized waves.

3. Hand Models

This chapter deals with the description of hand models used for investigating user hand influence. The first author's right hand was used as a template for the models.

3.1 Phantom Fabrication

It is necessary to mold a cast from suitable material for creating a phantom hand with consistent shape of the original hand. This material should couple to the electromagnetic field produced by the device in the same way and to the same extent as a real human hand.

The first step was creating a plaster cast of the hand using a material named Efkocrea. Efkocrea is an elastic material, specially designated for making molds. After the cast dried out, the mold was destroyed and the plaster cast was varnished.

A reusable mold was divided into 7 horizontal layers. The separate layers were made by the product Lukopren N [7]. It is a silicone rubber which becomes vulcanized after being mixed with a catalyst.

The material for the phantom hand should have the required dielectric properties over a wide frequency range, be shapeable, and sturdiness is also desirable. Target values for the phantom hand material, i.e. the average permittivity and conductivity of dry and moistened palm skin, have been reported in [8] and are shown in Tab. 1.

f [MHz]	ε_r [-]	σ [S/m]
900	36.2	0.79
1800	32.6	1.26

Tab. 1. Average permittivity and conductivity of all tissues of the hand published in [8].

f [MHz]	ε_r [-]	σ [S/m]
900	50	0.4
1800	50	1.0

Tab. 2. Measured dielectric properties of agar based material.

Fig. 5. Fabrication process of reusable seven layer mold (left) and casting process of agar based material to this mold (right).

An available tissue-equivalent material suitable for phantom hand fabrication is agar, whose dielectric proper-

ties are equivalent to biological muscle tissue. Better agreement with this tissue can be performed by increasing the water content of agar [5]. The selected ratio of agar to distilled water is 3:4. Agar was melted down and mixed together with water in the chosen ratio. The measured parameters for the produced mixture are shown in Tab. 2. For the sake of high proportion of water in the mixture, a load-bearing structure of spruce wood was glued together to provide mechanical support. The process of fabricating the phantom based on agar is shown in Fig. 5.

3.2 Numerical Model Creation

The numerical model was created by 3D scanning of the plaster cast which was made beforehand. It is the best solution for precisely preserving the phantom dimensions and shape. The resolution of the model was reduced, in 3D editor, for maintaining lower computational demands. The final model consists of approximately 10 thousand triangles. After that, the file was imported into CST Microwave Studio, where it was located at the desired position relative to the antenna. A simplified wooden load-bearing structure was also modeled. The resulting structure is shown in Fig. 6.

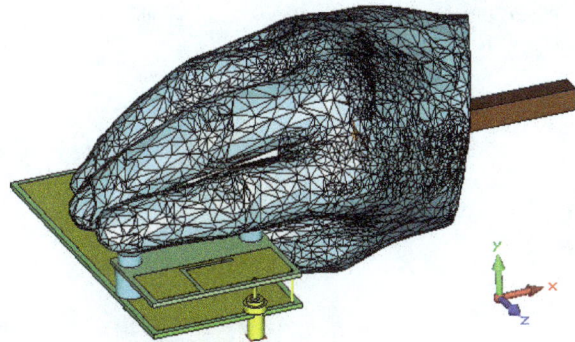

Fig. 6. Numerical model of the antenna with the phantom hand in the desired position.

4. The Effects of the User Hand

The effect of the user hand on impedance match was measured using the phantom hand described above. The phantom was placed in a typical position relative to the antenna in data mode (the same as in Fig. 6), which is shown in Fig. 7.

The comparison of the measured frequency response of the reflection coefficient between the antenna in free space and with phantom influence is shown in Fig. 8. It can be seen that the presence of the agar phantom brings slight impedance matching deterioration for both frequency bands. At the GSM900 band, it is also noticeable that the resonant frequency shifts towards the lower value, and that the higher resonant frequency is practically unchanged.

The simulation results with agar setting material prove sufficient agreement with measurements in the

GSM1800 band; in the GSM900 band, faint disagreement can be seen. The variance is probably due to big measurement error of the material parameters at low frequencies.

The comparison of the simulated frequency response of the reflection coefficient between materials (average from Tab. 1. and agar based) is also shown in Fig. 8. The correspondence between plots is obvious, which proves that the agar based material is properly selected.

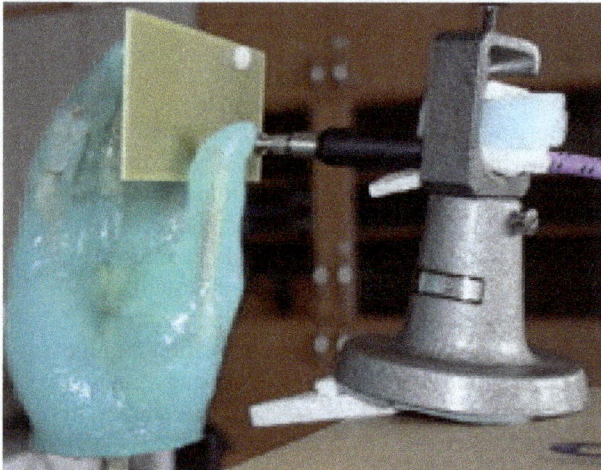

Fig. 7. Photo of the phantom prototype during antenna impedance match measurement.

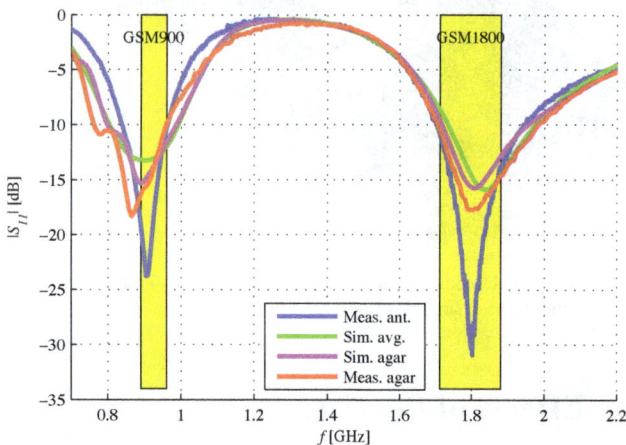

Fig. 8. User hand influence on impedance matching of the PIFA.

Figure 9 compares the computed radiation patterns between antenna in free space and with phantom influence for the xy plane. The xy plane is normal to the ground of the antenna as well as parallel with the fingers (see Fig. 6). Rotation angle corresponds to the angle Phi in Fig. 4, the antenna position, with respect to the Earth's surface, also agrees with Fig. 4.

For vertically polarized waves at the frequency of 925 MHz, attenuation due to the presence of the hand in the direction of the x axis is well observed. In this way, the antenna is surrounded by a hand which absorbs part of the radiation. On the contrary, at the frequency of 1795 MHz, the gain is larger in the x direction.

At the frequency of 925 MHz, horizontally polarized waves are not as strongly suppressed as in free space. At the frequency of 1795 MHz, also attenuation in the direction of the x axis is observed.

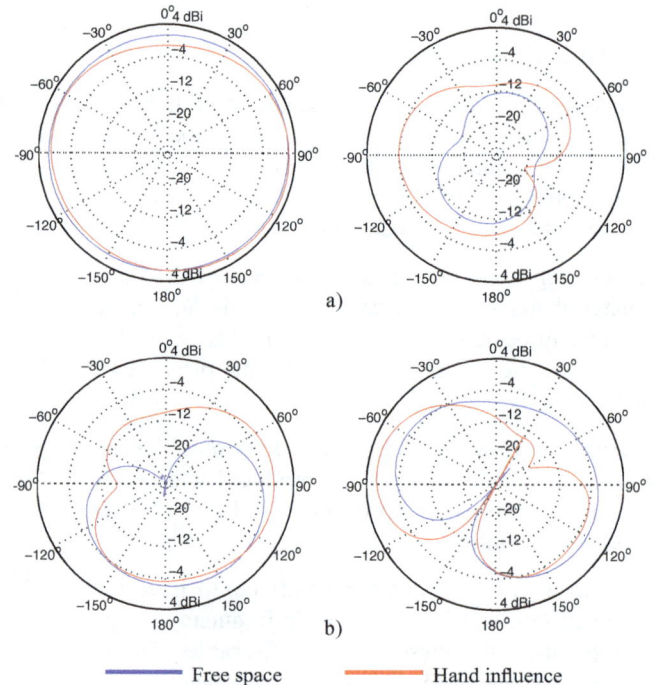

Fig. 9. Comparison of the calculated radiation characteristics of PIFA for the xy plane (see Fig. 6.) with the characteristics simulated in phantom hand presence at a) 925 MHz and b) 1795 MHz. Rotation angle corresponds to the angle Phi in Fig. 4. The left pictures represent vertically polarized waves, the right pictures are computed for horizontally polarized waves.

The comparison of the calculated radiation efficiency between the antenna in free space and with hand influence is shown in Tab. 3. It is obvious that the presence of human hand brings radiation efficiency decrease for both frequency bands. At the GSM1800 band, the decrease of radiation efficiency is more significant than that at the GSM900 band.

	f[MHz]	890	925	960	1710	1795	1880
PIFA	e_{cd} [%]	98.8	98.8	99.0	98.3	98.7	99.0
PIFA + hand	e_{cd} [%]	66.5	68.9	69.2	46.6	49.5	52.5

Tab. 3. Radiation efficiency (e_{cd}) calculation for only the PIFA and the PIFA with the hand model present. Lower, center and upper frequencies are selected for both GSM bands.

Figure 10 shows the simulated SAR distribution in the hand for the antenna in data mode. SAR is 10 g averaged, calculated for the same radiated power of 1 W, for both frequency bands.

Energy absorbed by the biological tissue mass is larger at the frequency of 1795 MHz, but all values meet the defined maximum limits very well. For mobile phones, the SAR limit (in Europe) is 2 W/kg averaged over the 10 g of tissue absorbing the most signal, [6].

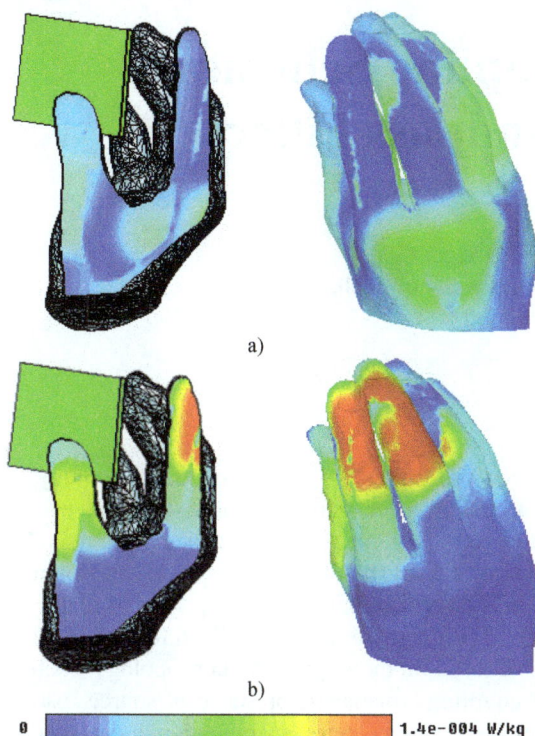

a)

b)

Fig. 10. Simulated SAR distribution in the hand at a) 925 MHz and b) 1795 MHz. Left pictures represent absorption inside the hand, the right pictures are 3D views to the surface of the hand.

5. Conclusion

The user hand influences on the properties of dual-band PIFA antennas are studied in this paper. The antenna covers the frequency bands of communication systems GSM900 and GSM1800 with impedance matching $|S_{11}| < -10$ dB at both bands.

The effect of the user hand on the antenna was measured using a hand model placed in a typical position of antenna in data mode. For this study, a phantom hand (corresponding to the first author's hand) and its equivalent numerical model were made. A suitable material of agar base has been developed and verified for phantom fabrication.

The presence of the phantom hand, in the measured position, brought slight impedance matching deterioration for both frequency bands. At the GSM900 band, it is evident that the resonant frequency shifts towards the lower value. The impact of the user hand on the radiation characteristics is quite different for both frequencies and polarizations. The radiation efficiency is influenced by the hand quite significantly, especially at the GSM1800 band. SAR distribution inside the hand meets the defined maximum limits very well.

This study examined the effect of a hand on one particular antenna. However, the presented methods, phantom made or numerical model, can serve well in the design and testing of any antenna.

Acknowledgements

The research was supported by the grant of the Czech Science Foundation no. P102/12/1274 and by the COST Action IC1102 (the grant no. LD12012). Measurements were performed in the laboratories supported by the SIX project; the registration number CZ.1.05/2.1.00/03.0072, the operational program Research and Development for Innovation.

References

[1] SAROLIC, A., SENIC, D., et al. Influence of human head and hand on PIFA antenna matching properties and SAR. In *19th International Conference on Software, Telecommunications and Computer Networks (SoftCOM)*, 2011.

[2] LI, C. H., OFLI, E., CHAVANNES, N., KUSTE, N. Effects of hand phantom on mobile phone antenna performance. *IEEE Transactions on Antennas and Propagation*, 2009, vol. 57, no. 9, p. 2763–2770.

[3] DAVOODI, D., SHARIFZAD, S. A Numerical study of the interaction between handset antennas and human head/hand in GSM 900, DCS, PCS and UMTS frequency bands. In *Progress in Electromagnetics Research Symp. Proc.* China, 2010, p. 733–737.

[4] BOYLE, K., YUN, Y., LIGHTHART, L. Analysis of mobile phone antenna impedance variations with user proximity. *IEEE Transactions on Antennas and Propagation*, 2007, vol. 55, p. 364–372.

[5] ZAJICEK, R., OPPL, L., VRBA, J. Broadband measurement of complex permittivity using reflection method and coaxial probes. *Radioengineering*, 2008, vol. 17, no. 1, p. 14–19.

[6] ZHANG, Zhijun. *Antenna Design for Mobile Devices.* 1st. ed. Singapore: John Wiley & Sons, 2011. ISBN 978-0-470-82446-7.

[7] Lukopren N products. [Online] Cited 2013-05-14. Available at: http://www.lucebni.cz/en/Produkty/Lukopren%20N.

[8] GABRIEL, C. Tissue equivalent material for hand phantoms. *Physics in Medicine and Biology*, 2007, vol. 52, p. 4205–4210.

About Authors ...

Radek VEHOVSKÝ was born in Opava, Czech Republic in 1989. He received his M.Sc. degree in Electrical Engineering from Brno University of Technology (BUT) in 2013. At present he is a Ph.D. student at the Dept. of Radio Electronics, BUT.

Michal POKORNÝ was born in Zlín, Czech Republic in 1983. He received his M.Sc. and Ph.D. degrees in Electrical Engineering from BUT in 2007 and 2011, respectively. At present he is a researcher at the Dept. of Radio Electronics, BUT. His research interests are antenna design and modeling of microwave semiconductor devices.

Kamil PÍTRA was born in Strakonice, Czech Republic in 1984. He received his M.Sc. and Ph.D. degrees in Electrical Engineering from BUT in 2010 and in 2014, respectively.

6

A Systematic Search for New Coupling Schemes of Cross-Coupled Resonator Bandpass Filters

Adam LAMECKI

Dept. of Microwave and Antenna Engineering, Gdansk University of Technology,
Narutowicza 11/12, 80-233 Gdansk, Poland

adlam@eti.pg.gda.pl

Abstract. *In this paper, a systematic approach to an extensive search for topologies of cross-coupled filters with generalized Chebyshev response is presented. The technique applies graph theory to find unique, nonisomorphic filter configurations, and tests whether a specific frequency response can be realized in a given set of topologies. The results of the search are then stored in a database of possible filter configurations.*

Keywords

Microwave filters, coupling matrix, generalized Chebyshev filters.

1. Introduction

Cross-coupled resonator filters with generalized Chebyshev response are currently one of the most commonly used filtering devices, and can be found in almost all high frequency communications systems. One advantage of this type of filter is the possibility of increasing the selectivity and/or equalizing the group delay response of the filter by introducing transmission zeros. The number and type (purely imaginary or complex pair) of transmission zeros that can be implemented in the design depends on the scheme of couplings between resonators—that is, on the filter topology. For certain types of topologies (e.g. those consisting of groups such as triplets or quadruplets, canonical filters [1], cul-de-sac [2] or extended-box configurations [3]), filtering properties such as type and maximum number of transmission zeros are well known. At the same time, new filter configurations continue to appear in the literature, some showing new resonator arrangements [4]-[14], [23], [24]. In any case, the catalog of filter topologies that can implement a certain type of frequency response is far from being complete, and adding new configurations to this catalog is of interest to filter designers.

In this paper, an approach to a systematic search for new filter topologies is presented. The technique begins with the generation of a set of unique candidate topologies for filters of a given order N. For each candidate topology, an extensive search for realizable responses is then performed. Finally, the topologies that can realize a particular type of generalized Chebyshev filtering response are stored in a database in the form of pairs (topology, type of realizable response). The database can be easily filtered by imposing various criteria on the requested filter topology, such as number of couplings, planarity, presence of source-load coupling. Then for a given electric specification, a database query can be executed to quickly find the topologies that can realize the desired response and a coupling matrix can be synthesized for each of these topologies. The results can also be filtered by additional criteria, like the number of couplings needed to realize the filter. For a filter designer, such a database provides a set of optional configurations, from which the designer can choose one to suit particular requirements in the best possible way, taking into account, for example, resonator arrangement or the number of couplings and the values of the computed coupling coefficients. This also enables one to compare different coupling schemes and select the optimal solution for given technology of filter implementation.

It should be noted that the functionality of the database is much different to (and to a certain extent complementary to) that offered by a package such as Dedale-HF [28]. For a given coupling topology and a given filtering characteristic, Dedale-HF can compute all possible coupling matrices that will realize the prescribed response. However, this can currently only be done for a relatively small set of known topologies. The present approach aims to find topologies with at least one solution, for a given number and position of transmission zeros, in the form of a coupling matrix. Once a new topology has been found, one can apply the techniques of Groebner bases [15] that are used by Dedale-HF to find all alternative solutions.

2. Generation of Candidate Topologies

The first step in constructing the catalog is to prepare the set of candidate topologies. The topology of the Nth-order cross-coupled filter can be presented in the form of a symmetric adjacency matrix P of size $(N+2 \times N+2)$ [26], whose nonzero elements $p_{i,j}$ $(i,j = 2\ldots N+2, i \neq j)$ correspond to nonzero couplings between resonators i and j, and $p_{1,j} = p_{j,1}$, $p_{i,N+2} = p_{N+2,i}$ $(i,j = 2\ldots N+1)$ correspond to one or more couplings between the source/load and resonators. Finally, the element $p_{(1,N+2)} = p_{(N+2,1)}$ corresponds to the direct source-load coupling. It should be noted that different adjacency matrices do not necessarily imply different topologies. For example, let us look at two matrices P_1 and P_2 that define the topology of a third-order filter (only the upper triangular elements are shown):

$$P_1 = \begin{bmatrix} 0 & 1 & 0 & 0 & 0 \\ & 0 & 1 & 1 & 0 \\ & & 0 & 1 & 0 \\ & & & 0 & 1 \\ & & & & 0 \end{bmatrix}, P_2 = \begin{bmatrix} 0 & 0 & 0 & 1 & 0 \\ & 0 & 1 & 1 & 0 \\ & & 0 & 1 & 1 \\ & & & 0 & 0 \\ & & & & 0 \end{bmatrix}. \tag{1}$$

Despite the different appearance, both matrices represent the same filter configuration, shown in Fig. 1. The difference comes from the different numbering of the nodes, but from the point of view of filter realization, the structures represent the same device. This observation implies that, in order to create a catalog of filter topologies, one must be able to generate a set of unique adjacency matrices which cannot be transformed into each another by a node renumbering. This problem can be readily solved using graph theory.

Matrix P can be regarded as an adjacency matrix of undirected, nonweighted, connected graph with $Q = N+2$ nodes (with two vertex labels, one for source/load and one for the resonators) and edges representing couplings. The problem of finding a set of unique topology candidates may be expressed in the language of graph theory as finding nonisomorphic graphs[1] with a given number of vertices and edges. The set must be composed of all possible nonisomorphic arrangements of resonators and couplings between them.

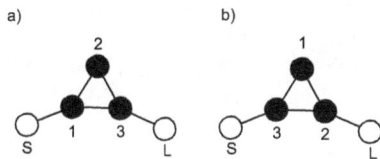

Fig. 1. Two isomorphic filter configurations with different adjacency matrix P.

Q	K
4	6
5	21
6	112
7	853
8	11 117
9	261 080
10	11 716 571

Tab. 1. Number of nonisomorphic connected graphs with $Q = N+2$ unlabeled nodes.

The procedure for selecting unique topology candidates is as follows: At first, the set of K nonisomorphic

graphs with $Q = N+2$ nodes is created. This is a problem of a combinatorial nature, and the size of this set grows rapidly as the number of vertices (nodes) increases, as shown in Tab. 2. In Fig. 2a), all six nonisomorphic graphs with 4 nodes ($N = 2$) are presented. Since two nodes must be used as source and load, these 6 configurations form an initial set of second-order filter candidates.

Looking at this set, it can be seen that in the first step the graphs with exactly $N+1$ edges can be removed, as they represent the inline and tree-like topologies which do not have cross-couplings.

At this stage, we can also define which nodes act as the source and the load, and which are regular resonators. This can be done by labeling graph vertices. Vertex labeling increases the number of possible combinations. Note that, for a graph with $N+2$ nodes, one obtains $\frac{(N+2)!}{2(N!)}$ possible combinations of source-load position, and each candidate topology has to be considered for different combinations of load/source location. This gives an augmented set, and the resulting graphs are further processed to remove isomorphic graphs from the set.

To illustrate this, all candidates that can be labeled by the described procedure for second-order filters (4-node graphs) are shown in Fig. 2b. It can be seen that labeling (adding source/load) increases the set of unique topologies from 4 to 10. Tab. 2 gives the total number of unique topologies with source/load for an increasing filter order N.

It can be seen that the count grows very rapidly, and so during the construction phase various additional requirements can be imposed on the set in order to reduce the total number of topology candidates:

- Limiting the number of couplings allowed by restricting the number of graph edges M that control the maximum number of interresonator couplings. Topologies with too many cross-couplings are not interesting from the practical viewpoint;

- Limiting the maximum number of couplings from source/load to resonators;

- Removing all candidates with source (load) connected only to load (source);

- Removing all candidates with direct source-load connection (optional for high-order filters);

- Eliminating nonplanar filter topologies.

These restrictions can be applied by postprocessing the generated topology candidates. In the case of the topology candidates of second order filters shown in Fig. 2b), one topology was removed from the set at this stage – namely, the one that has source (load) connected only to load (source). As a result, for a second-order cross-coupled filter, there are 9 possible resonator and coupling arrangements, as shown in Fig. 2c.

---

[1]Two graphs are nonisomorphic if they contain the same number of nodes (vertices) and cannot be transformed into each other by renumbering the nodes

a)

b)

c)

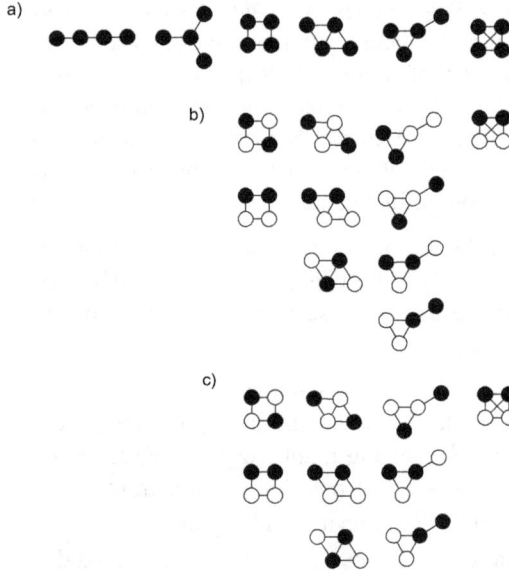

Fig. 2. Example graphs for second-order filter candidate dataset construction: a) initial set of nonisomorphic graphs with 4 nodes, b) basic set of 10 nonisomorphic candidates after adding labels and removing in-line and star topologies, c) final set of 9 nonisomorphic candidates for filter topology (the topology with source(load) connected only to load(source) was removed). White circle: source/load, black circle: resonators.

Q	N	K
4	2	10
5	3	83
6	4	836
7	5	11 144
8	6	230 176

Tab. 2. Number of nonisomorphic topology candidates for Nth order filters (graphs with $Q = N + 2$ nodes), labeled without restriction on filter topology

3. Definition of Response Types

The graph-theory methodology presented in the previous section yields a set of topology candidates. The next step is to determine whether the desired frequency response can be realized by any of these candidates. The low-pass prototype response of the generalized Chebyshev coupled-resonator filter is defined by the filter order, return loss level, and the set of transmission zeros (their number; their type, whether real, imaginary, or complex; their symmetry with respect to the center frequency; and their position with respect to passband, whether above or below). For a given filter order and topology, the number and type of realizable transmission zeros is one of the most important features. For instance, as far as the location of TZs with respect to center frequency is concerned, two variants of transmission zeros can be investigated:

- Asymmetric purely imaginary zeros at $s = j\omega_{0i}$ (counted as N_{asym}^{TZ});

- Symmetric pairs of purely imaginary zeros at $s = j\omega_{0i}$ and $s = -j\omega_{0i}$ (counted as N_{sym}^{TZ}).

The total number of transmission zeros is equal to $N_{Total}^{TZ} = N_{asym}^{TZ} + 2N_{sym}^{TZ}$, and for symmetric responses $N_{asym}^{TZ} = 0$. The admittance parameter $y_{21}(s)$ of the Nth-order coupled-resonator filter described by the coupling matrix M, of size $(N+2) \times (N+2)$, can be computed as

$$y_{21}(s) = j\left[-M - \omega W\right]_{(N+2,1)}^{-1} \qquad (2)$$

where W is a matrix similar to the identity matrix with w_{11} and $w_{N+2,N+2}$ equal to zero. The zeros of y_{21} can therefore be computed as the eigenvalues λ_i of the generalized eigenproblem in the following form [17], [18]

$$\hat{M}x - \lambda\hat{W}x = 0 \qquad (3)$$

where the matrices $\hat{M}$ and $\hat{W}$ are created from the matrices M and W by removing the last row and the first column. The problem is real-valued and asymmetric, and $\hat{W}$ is singular. For a given topology (nonzero pattern of M) the maximum number of finite eigenvalues is constant, and the position of the eigenvalues depends on values of nonzero m_{ij} elements. If a single solution of the coupling-matrix synthesis for a given set of zeros (eigenvalues), positions, and types is found, then it is clear that a continuum of other locations can also be realized due to the continuity of the eigenvalues with respect to the elements m_{ij}. The existence of continuum does not imply that it cover the whole ω axis. There might be up to N disjoint continuous sections. With this observation, it seems sufficient to test a filter configuration for various combinations of discrete transmission zeros.

Let us consider two types of transmission zeros:

- Purely imaginary zeros that can be located above or below the passband;

- Pairs of symmetrically located transmission zeros.

By limiting the number of purely imaginary zeros to 3, up to 5 different transmission characteristics can be obtained, as shown in Tab. 3. Only one characteristic is symmetric with respect to the center frequency. For asymmetric responses, the upper and lower band TZs can also be differentiated.

N_{Total}^{TZ}	N_{asym}^{TZ}	N_{sym}^{TZ}
1	1	0
2	2	0
2	0	1
3	3	0
3	1	1

Tab. 3. Example filter prototypes for filters with up to 3 transmission zeros with different numbers of purely imaginary transmission zeros N^{TZ}.

4. Topology Validation

To test if a candidate filter topology can realize a given type of response, a trial synthesis of the coupling matrix must be performed. If the synthesis succeeds, the topology is accepted. To perform the trial, a robust technique of coupling-matrix synthesis must be applied. Classical techniques based on rotations [1] are not suitable for this purpose, as it would be very difficult, if not impossible, to find the desired rotation sequence for all possible configurations generated by the procedure described in Sec. 2. Another technique for coupling-matrix synthesis based on multivariate polynomial systems [15] involves a polynomial equation that is very expensive to solve numerically. It is therefore not suitable in the case where tens of thousands of (topology, response) pairs are being processed. For this reason, techniques based on fast optimization algorithms are more suitable here. In particular, the author's experience suggests that techniques involving eigenvalue optimization [17], [18], [19] perform remarkably well in terms of speed and convergence, and for this reason they have been chosen for testing. The trial synthesis is first performed using analytical gradients [20], and if the procedure fails to converge within a prescribed number of iterations, an additional global optimization based on a particle swarm algorithm and using a zero-pole goal function [21] is performed.

At this stage, the procedure tries for each candidate topology to synthesize the coupling matrix with different types of response. In our work, responses with up to 6 transmission zeros were investigated. The search is composed of a series of trial coupling-matrix syntheses, starting from the most complex responses (with the highest number of transmission zeros) and progressing to the simplest (one asymmetric TZ). When the coupling-matrix synthesis succeeds for an assumed response (number of TZs), then the candidate is saved for further processing and the search process is restarted for the next candidate.

To speed up the database construction, the number of searches for realizable responses can be limited by incorporating a minimum path rule [2] or by the technique described in [25]. This allows one to estimate the maximum number of realizable transmission zeros for a given topology, thus limiting the number of response types that need to be checked. During synthesis, some of the couplings can be set to zero by the optimization procedure. When this occurs, the topology is discarded and the processing of the next candidate begins.

5. Final Caveats

The goal of the proposed methodology is to look for and create a catalog of the topologies of microwave cross-coupled bandpass filters. Two issues related to the proposed approach must be noted:

- During the trial synthesis, it was assumed that the transmission zeros are located close to the passband. This is a reasonable assumption, it is filters with high selectivity that we are interested in. As described in Section 3, using the continuity of eigenvalues, a certain continuum of other locations of TZs can also be real-

ized. However, the opposite is not true – if the candidate topology does not realize transmission zeros at a given set of test locations ω_{0i}, this does not imply that other zeros locations are not realizable. To the author's knowledge, the problem of whether an arbitrary topology can produce finite TZs of a particular type and at particular locations is still open.

- Despite the fact that a very robust technique was selected for the coupling-matrix synthesis, the convergence of the optimization method cannot be guaranteed. In consequence, there might appear cases in which the topology (which in theory is able to realize a given response) is missed during the search, due to the failure of the optimization technique. To minimize this risk, the optimization is performed a few times using different starting points.

With these remarks, it is noted that the proposed search does not ensure that *all* possible topologies are found, but it does allow the creation of a rather broad catalog of the filter topologies currently available.

6. Results

To date, a set of 27920 topology candidates for filters with orders from 2 to 6 has been processed using the technique outlined in the preceding sections. To create the candidates, the graph tools included in the *nauty* [27] software package were used. Statistical information about the numbers of graphs processed at each stage of this procedure is shown in Tab. 4. For practical reasons, it was assumed that source-load can be coupled to at most 2 resonators, and that direct source-load couplings are allowed. Note, that for $N > 3$, some of the candidate topologies correspond to non-planar graphs[2]. Such nonplanar graphs are shown in Fig. 3 for fourth-order filters.

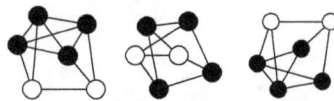

Fig. 3. All possible nonplanar topology candidates processed for fourth-order filters.

In the context of filter design, a nonplanar topology would lead to a device in which at least one coupling is crossed with another – such a crossing is usually difficult to realize when all resonators are in the same plane, but may be possible when the three-dimensional arrangement of resonators is allowed, like in the case of multi-layer LTCC and LCP packaging technology.

Currently, the database contains 7401 unique, nonisomorphic topologies. It took several weeks to generate such a dataset using parallel processing techniques, but once created it can be search in seconds. To give the reader some indication of the database's contents, all the topologies of second and third-order filters found are shown in Tab. 5 and

---

[2]Nonplanar graphs are those that cannot be drawn in the plane in such a way that no edges cross each other.

Tab. 6. Additionally in Tab. 7, the count of unique, nonisomorphic topologies found for different filter orders is shown. In Fig. 4, the response of third-order filters with 2 TZs below the passband is shown, along with one selected topology (out of six) found that realizes the response. In this case, the corresponding coupling matrix is:

$$M = \begin{bmatrix} 0 & 0 & 0 & 1.1062 & 0 \\ 0 & 0.7534 & 0.7295 & 0.8544 & 0 \\ 0 & 0.7295 & -0.0337 & -0.8057 & 1.0926 \\ 1.1062 & 0.8544 & -0.8057 & -0.2893 & 0.1733 \\ 0 & 0 & 1.0926 & 0.1733 & 0 \end{bmatrix}. \quad (4)$$

This topology allows two transmission zeros below the passband with a single negative cross-coupling to be realized, with the couplings from source/load to the resonators are all positive. The same response can be realized with the "extended-doublet" coupling scheme [12], which was also found using the technique (shown as the last topology in Tab. 6, in the category of two asymmetric zeros).

Filter order N	Count	Planar	Nonplanar
2	9	9	0
3	45	45	0
4	295	292	3
5	2314	2233	81
6	25257	22202	3055

Tab. 4. Number of topology candidates processed.

Zeros	Topology
One zero	
One pair	
Two zeros	

Tab. 5. Possible topologies of second-order filters found.

Zeros	Topology
One zero	
One zero pair	
Two asym. zeros	
Three asym. zeros	

Tab. 6. Possible topologies of third-order filters with up to 2 couplings from source/load to resonator found.

Filter order N	Count
2	7
3	18
4	106
5	671
6	6599

Tab. 7. Number of nonisomorphic topologies found.

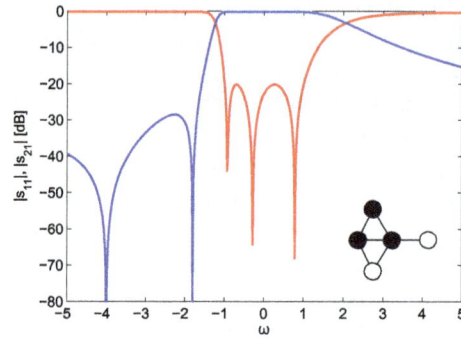

Fig. 4. Frequency response of the third-order filter with 2 purely imaginary TZs at $-j1.8$, $-j4$, and a possible coupling scheme.

Response type	Count
2 pairs	29
1 pair, 2 asym.	4
4 asym.	164
Total	197

Tab. 8. Number of topologies found for fifth-order filters with 4 purely imaginary TZs.

Higher-order filters offer a plethora of topologies. In order to present some early results, we selected a fifth-order filter with 4 asymmetric purely imaginary transmission zeros (see Fig. 5). Over 160 nonisomorphic topologies were found (see Tab. 8). A few possible configurations are shown in Fig. 6. A possible coupling matrix synthesized for the topology from Fig. 6e) and h) are shown in (5) and (6). It should be noted that many of the topologies found would be difficult to realize as physical filters, for example due to the high load of some resonators by a few strong couplings. However, some of them could provide a successful design.

Comparing the results presented here with coupling schemes available in Dedale-HF, it is worth noticing that the latter does not offer any topology that can realize a fifth-order asymmetric response with 4 TZs, while a few such topologies have been shown in this paper. However, the techniques implemented in Dedale-HF [15] could be used to find multiple solutions of the coupling-matrix synthesis problem for topologies found using the approach proposed here.

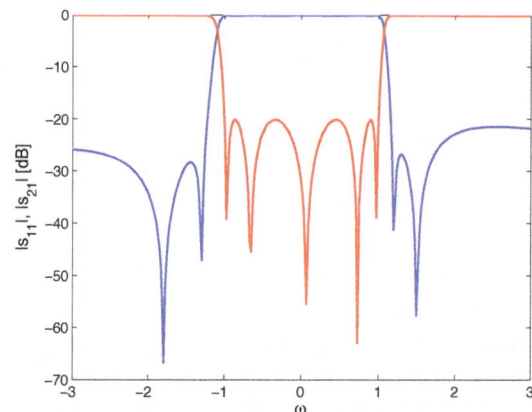

Fig. 5. Frequency response of the fifth-order filter with 4 purely imaginary TZs at $-j1.3$, $-j1.8$, $j1.2$ and $j1.5$.

$$M_1 = \begin{bmatrix} 0 & 0.9957 & 0 & 0 & 0 & 0 & 0 \\ 0.9957 & 0.0005 & -0.0809 & 0.8008 & 0 & 0 & 0.1458 \\ 0 & -0.0809 & 0.0245 & -0.4431 & 0.7416 & 0 & 0.9849 \\ 0 & 0.8008 & -0.4431 & -0.0282 & 0.1837 & 0.5257 & 0 \\ 0 & 0 & 0.7416 & 0.1837 & 0.1326 & 0.8039 & 0 \\ 0 & 0 & 0 & 0.5257 & 0.8039 & -0.2754 & 0 \\ 0 & 0.1458 & 0.9849 & 0 & 0 & 0 & 0 \end{bmatrix} \quad (5)$$

$$M_2 = \begin{bmatrix} 0 & 0 & 0 & 0.9199 & 0 & 0.3810 & 0 \\ 0 & -0.1483 & 0 & -0.7255 & 0.7255 & 0 & 0 \\ 0 & 0 & 0.8056 & 0 & 0 & 0.6607 & 0 \\ 0.9199 & -0.7255 & 0 & -0.2907 & -0.3850 & 0.3976 & 0 \\ 0 & 0.7255 & 0 & -0.3850 & -0.2907 & 0.3976 & 0.9199 \\ 0.3810 & 0 & 0.6607 & 0.3976 & 0.3976 & -0.2219 & 0.3810 \\ 0 & 0 & 0 & 0 & 0.9199 & 0.3810 & 0 \end{bmatrix} \quad (6)$$

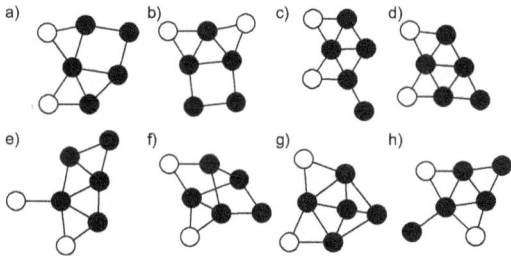

Fig. 6. 8 selected topologies found that can realize the response of Fig. 5.

7. Conclusion

In this paper, a novel systematic approach for exhaustively searching for new cross-coupled resonator filter topologies has been presented. So far, 7401 unique topologies for filters up to order 6 have been found and stored in a searchable database. Further work will be focused on the detailed analysis of the database records with the aim of selecting practically useful coupling schemes and their physical filter realizations.

Acknowledgements

This work was supported by the Polish State Committee for Scientific Research under Contract N N515 245637.

References

[1] CAMERON, R. J. General coupling matrix synthesis methods for Chebyshev filtering functions. *IEEE Transactions on Microwave Theory and Techniques*, 1999, vol. 47, no. 4, p. 433 - 442.

[2] CAMERON, R. J. Advanced coupling matrix synthesis techniques for microwave filters. *IEEE Transactions on Microwave Theory and Techniques*, 2003, vol. 51, no. 1, p. 1 - 10.

[3] CAMERON, R. J., HARISH, A. R., RADCLIFFE, C. J. Synthesis of advanced microwave filters without diagonal cross-couplings. *IEEE Transactions on Microwave Theory and Techniques*, 2002, vol. 50, no. 12, p. 2862 - 2872.

[4] XIAO, K., YE, L. F., ZHAO, F., CHAI, S.-L., LI, J. L.-W. Coupling matrix decomposition in designs and applications of microwave filters. *Progress in Electromagnetics Research*, 2011, vol. 117, p. 409 - 423.

[5] ROSENBERG, U., AMARI, S. Novel coupling schemes for microwave resonator filters. *IEEE Transactions on Microwave Theory and Techniques*, 2002, vol. 50, no. 12, p. 2896 - 2902.

[6] ROSENBERG, U., AMARI, S. Novel coupling schemes for microwave resonator filters. *IEEE MTT-S International Microwave Symposium Digest*. 2002, vol.3, p. 1605-160.

[7] MIRA, F., MATEU, J., COGOLLOS, S., BORIA, V. E. Design of ultra-wideband substrate integrated waveguide (SIW) filters in zigzag topology. *IEEE Microwave And Wireless Components Letters*, 2009, vol. 19, p. 281 - 283.

[8] LU, J.-C., LIAO, C.-K., CHANG, C.-Y. Microstrip parallel-coupled filters with cascade trisection and quadruplet responses. *IEEE Transactions on Microwave Theory and Techniques*, 2008, vol. 56, no. 9, p. 2101 - 2110.

[9] ZHOU, C. , GUO, Y.-X. , WANG, L., WU, W. Design of compact dual-band filter in multilayer LTCC with cross coupling. *Progress in Electromagnetics Research*, 2013, vol. 135, p. 515 - 525.

[10] LEE, K., LEE, T.-H., KIM, Y.-S., LEE, J. New negative coupling structure for substrate-integrated cavity resonators and its application to design of an elliptic response filter. *Progress in Electromagnetics Research*, 2013, vol. 137, p. 117 - 127.

[11] HO, M.-H., CHEN, P.-F. Suspended substrate stripline bandpass filters with source-load coupling structure using lumped and fullwave mixed approach. *Progress in Electromagnetics Research*, 2012, vol. 122, p. 519 - 535.

[12] LIAO, C.-K., CHI, P.-L., CHANG, C.-Y. Microstrip realization of generalized Chebyshev filters with box-like coupling schemes. *IEEE Transactions on Microwave Theory and Techniques*, 2007, vol. 55, no. 1, p. 147 - 153.

[13] MENG, W., WU, K.-L. A hybrid synthesis technique for N-tuplets microwave filters cascaded by resonator. In *Proceedings of the Asia Pacific Microwave Conference*. 2006, p. 1170 - 1173.

[14] TAMIAZZO, S., MACCHIARELLA, G. An analytical technique for the synthesis of cascaded N-tuplets cross-coupled resonators microwave filters using matrix rotations. *IEEE Transactions on Microwave Theory and Techniques*, 2005, vol. 53, no. 5, p. 1693 - 1698.

[15] CAMERON, R. J., FAUGERE, J. C., SEYFERT, F. Coupling matrix synthesis for a new class of microwave filter configuration. In *IEEE MTT-S International Microwave Symposium Digest*. 2005.

[16] LU, J.-C., LIN, Y.-W., CHANG, C.-Y. Design and implementation of the canonical-form-like coupling scheme with generalized Chebyshev responses. *Asia-Pacific Microwave Conference Proceedings (APMC)*. 2010, p. 115 - 118.

[17] LAMECKI, A., KOZAKOWSKI, P., MROZOWSKI, M. Fast synthesis of coupled-resonator filters. *IEEE Microwave and Wireless Components Letters*, 2004, vol. 14, no. 4, p. 174 - 176.

[18] KOZAKOWSKI, P., LAMECKI, A., MROZOWSKI, M. Eigenvalue approach to synthesis of prototype filters with source/load coupling. *IEEE Microwave And Wireless Components Letters*, 2005, vol. 15, no. 2, p. 98 - 100.

[19] LAMECKI, A., MROZOWSKI, M. Bandpass / bandstop coupling matrix synthesis based on rational representation of scattering parameters. In *Asia Pacific Microwave Conference Proceedings (APMC)*. 2010.

[20] KOZAKOWSKI, P., MROZOWSKI, M. Quadratic programming approach to coupled resonator filter CAD. *IEEE Transactions on Microwave Theory and Techniques*, 2006, vol. 54, no. 11, p. 3906 - 3913.

[21] KOZAKOWSKI, P., MROZOWSKI, M. Automated CAD of coupled resonator filters. *IEEE Microwave and Wireless Components Letters*, 2002, vol. 12, no. 12, p. 470 - 472.

[22] SZYDLOWSKI, L., LAMECKI, A., MROZOWSKI, M. Coupled-resonator filters with frequency-dependent couplings: Coupling matrix synthesis. *IEEE Microwave and Wireless Components Letters*, 2012, vol. 22, no. 6, p. 312 - 314.

[23] LESZCZYNSKA, N., SZYDLOWSKI, L., MROZOWSKI, M. A novel synthesis technique for microwave bandpass filters with frequency-dependent couplings. *Progress in Electromagnetics Research*, 2012, vol. 137, p. 35 - 50.

[24] JEDRZEJEWSKI, A., LESZCZYNSKA, N., SZYDLOWSKI, L., MROZOWSKI, M. Zero-pole approach to computer aided design of in-line SIW filters with transmission zeros. *Progress in Electromagnetics Research*, 2012, vol. 131, p. 517 - 533.

[25] AMARI, S. On the maximum number of finite transmission zeros of coupled resonator filters with a given topology. *IEEE Microwave and Guided Wave Letters*, 1999, vol. 9, no. 9, p. 354 - 356.

[26] WEST, D. B. *Introduction to Graph Theory*. Prentice Hall, 2001.

[27] McKAY, B., PIPERNO, A. *Nauty and Traces*. [Online] Available at: http://cs.anu.edu.au/people/bdm/nauty/

[28] *Dedale-HF*. [Online] Available at: http://www-sop.inria.fr/apics /Dedale

About Author ...

Adam LAMECKI received the M.Sc. degree and Ph.D. (with honors) in Microwave Engineering from the Gdańsk University of Technology (GUT), Gdańsk, Poland, in 2002 and 2007, respectively. He was a recipient of a Domestic Grant for Young Scientists awarded by the by Foundation for Polish Science in 2006. In 2008 he received Award of Prime Minister for the doctoral thesis and in 2011 a scholarship from Ministry of Science and Higher Education. His research interests include surrogate models and their application in the CAD of microwave devices, computational electromagnetics (mainly focused on finite element method) and filter design and optimization techniques.

7

High Input Impedance Voltage-Mode Biquad Filter Using VD-DIBAs

Winai JAIKLA [1], Dalibor BIOLEK [2,3], Surapong SIRIPONGDEE [1], Josef BAJER [2]

[1] Dept. of Engineering Education, Faculty of Industrial Education, King Mongkut's Institute of Technology Ladkrabang, Bangkok, 10520, Thailand
[2] Dept. of Electrical Engineering, Faculty of Military Technology, University of Defense Brno, Kounicova 65, 662 10 Brno, Czech Republic
[3] Dept. of Microelectronics, Faculty of Electrical Engineering and Communications, Brno University of Technology, Technická 10, 616 00 Brno, Czech Republic

kawinai@kmitl.ac.th, dalibor.biolek@unob.cz, kssurapo@kmitl.ac.th, josef.bajer@unob.cz

Abstract. *This paper deals with a single-input multiple-output biquadratic filter providing three functions (low-pass, high-pass and band-pass) based on voltage differencing differential input buffered amplifier (VD-DIBA). The quality factor and pole frequency can be electronically tuned via the bias current. The proposed circuit uses two VD-DIBAs and two grounded capacitors without any external resistors, which is suitable to further develop into an integrated circuit. Moreover, the circuit possesses high input impedance, providing easy voltage-mode cascading. It is shown that the filter structure can be easily extended to multi-input filter without any additional components, providing also all-pass and band-reject properties. The PSPICE simulation results are included, verifying the key characteristics of the proposed filter. The given results agree well with the theoretical presumptions.*

Keywords

Analog filter, VD-DIBA, voltage-mode, single input-multiple output.

1. Introduction

Analog active filter is one of the standard research topics in the circuit design. It is commonly utilized block for continuous-time analog signal processing. It is generally used in many fields, such as communications, measurement, instrumentation, and control systems [1]. Especially, the filters providing several functions within a single topology, namely the universal or multifunction filter, have been receiving considerable attention. One of the most popular analog filters is a single-input, multiple-output (SIMO) topology in which various transfer functions can be realized simultaneously. The SIMO topology can be found in many applications, for example in touch-tone telephone tone decoder, in phase-locked loop FM stereo demodulator, or in crossover network as a part of the three-way high-fidelity loudspeaker [2].

The design of analog circuits using active building blocks, taking into account several various criteria such as minimum number of active elements or others, has been receiving considerable attention. Biolek et al. [3] proposed several circuit ideas of building blocks for voltage-, current- and mixed mode applications. One of them is the voltage differencing differential input buffered amplifier (VD-DIBA). This device allows applications with interesting features, especially those providing the electronic controllability. It is obvious from the literature survey that a few circuits using VD-DIBA have been hitherto published, for instance the voltage-mode first-order allpass filter [4], inductance simulator [5], and multiple-input single-output (MISO) voltage-mode biquad filter [6].

This contribution presents a SIMO voltage-mode filter with high input impedance, employing VD-DIBAs. It is suitable for fabricating as a monolithic chip or also for off-the-shelf implementation, consisting of 2 active elements and 2 grounded capacitors. The proposed filter can provide three standard functions (low-pass, high-pass and band-pass). The quality factor and pole frequency can be electronically adjusted.

The paper is organized as follows: In Section 2, which follows this Introduction, the definition and features of the VD-DIBA are given, and the proposed filter is also presented. The non-ideal analysis is included in Section 3. The experimental results, namely SPICE simulations and measurements on a filter specimen, are illustrated in Section 4. Section 5 describes the filter extension to multi-input topology and transconductance type. The comparison with previous works is described in Section 6. Some concluding remarks are given in Section 7.

2. Theory and Principle

2.1 VD-DIBA Overview

The principle of VD-DIBA was introduced in [3]. The internal construction of VD-DIBA using commercially available ICs has been proposed in [4]. Its symbol and equivalent circuit are shown in Fig. 1(a) and (b), where V_+ and V_- are the voltage input terminals. The voltage is converted to the z-terminal current via a transconductance g_m, which can be tuned by the bias current. The difference of z- and v- terminal voltages is copied to the w terminal with the differential-input unity gain buffer. An ideal VD-DIBA has low-impedance w terminal and high-impedance v_+, v_-, z, and v terminals. The characteristics of VD-DIBA can be described as follows:

$$\begin{pmatrix} I_{v+} \\ I_{v-} \\ I_z \\ I_v \\ V_w \end{pmatrix} = \begin{pmatrix} 0 & 0 & 0 & 0 & 0 \\ 0 & 0 & 0 & 0 & 0 \\ g_m & -g_m & 0 & 0 & 0 \\ 0 & 0 & 0 & 0 & 0 \\ 0 & 0 & 1 & -1 & 0 \end{pmatrix} \begin{pmatrix} V_+ \\ V_- \\ V_z \\ V_v \\ I_w \end{pmatrix}. \tag{1}$$

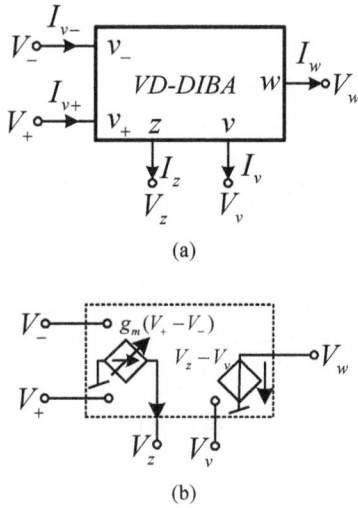

(a)

(b)

Fig. 1. VD-DIBA: (a) Symbol. (b) Equivalent circuit.

2.2 High Input Impedance Voltage-Mode Filter Using VD-DIBAs

The proposed second-order filter is illustrated in Fig. 2. It consists of two VD-DIBAs and two grounded capacitors. It is obvious that the proposed filter provides simultaneously three frequency responses (HP, LP and BP) with high input impedance property. Considering the ideal VD-DIBA, a routine analysis of the proposed filter provides the following voltage transfer functions:

$$\frac{V_{HP}}{V_{in}} = \frac{s^2}{s^2 + s\frac{g_{m2}}{C_1} + \frac{g_{m1}g_{m2}}{C_1C_2}}, \tag{2}$$

$$\frac{V_{LP}}{V_{in}} = \frac{\frac{g_{m1}g_{m2}}{C_1C_2}}{s^2 + s\frac{g_{m2}}{C_1} + \frac{g_{m1}g_{m2}}{C_1C_2}}, \tag{3}$$

and

$$\frac{V_{BP}}{V_{in}} = -\frac{s\frac{g_{m2}}{C_1}}{s^2 + s\frac{g_{m2}}{C_1} + \frac{g_{m1}g_{m2}}{C_1C_2}}. \tag{4}$$

The filter pole frequency (ω_0) and quality factor (Q) can be expressed as

$$\omega_0 = \sqrt{\frac{g_{m1}g_{m2}}{C_1C_2}}, \tag{5}$$

and

$$Q = \sqrt{\frac{C_1g_{m1}}{C_2g_{m2}}}. \tag{6}$$

It follows from (5) and (6) that the quality factor and pole frequency can be tuned electronically via transconductances.

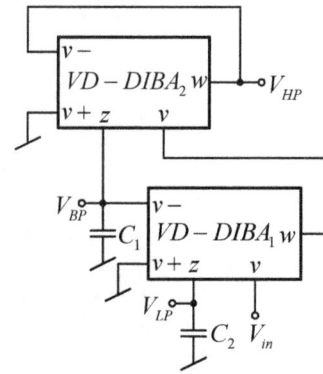

Fig. 2. Proposed voltage-mode filter.

The relative sensitivities of the proposed circuit can be found as plus or minus 0.5:

$$S_{g_{m1}}^{\omega_0} = S_{g_{m2}}^{\omega_0} = \frac{1}{2}; \ S_{C_1}^{\omega_0} = S_{C_2}^{\omega_0} = -\frac{1}{2}, \tag{7}$$

and

$$S_{C_1}^{Q} = S_{g_{m1}}^{Q} = \frac{1}{2}; \ S_{C_2}^{Q} = S_{g_{m2}}^{Q} = -\frac{1}{2}. \tag{8}$$

As a drawback, the LP and BP outputs are not of low impedance characters, thus they should be additionally buffered when applicable, or the filter topology should be modified as described in Section 5.

3. Non-ideal Case

In practice, the influences of voltage tracking errors from the unity-value gain of internal differential voltage buffer, and also the parasitic terminal impedances of VD-DIBA [4] will affect the filter performance. In this Section, these parameters will be taken into account. For non-ideal voltage buffer, its model is as follows:

$$V_w = \beta^+ V_z - \beta^- V_v. \tag{9}$$

Here β^+ and β^- are the voltage error gains from z and v terminals to w terminal. The influences of parasitic impedances of the $v+$, $v-$ and v terminals of VD-DIBA No. 2 and of $v+$ and v terminals of VD-DIBA No. 1 are negligible because of their connection to low-impedance outputs w, to the input voltage source, or to ground. The most important parasitic impedances are resistive and capacitive parts affecting the z terminals of VD-DIBAs, acting in parallel to C_1 and C_2. Let us denote them R_{z1}, C_{z1}, and R_{z2}, C_{z2}, respectively. Taking them into account together with (9), the transfer functions will be modified to the more general forms:

$$\frac{V_{HP}}{V_{in}} = \beta_1^- \beta_2^- \frac{\left(s + \dfrac{1}{R_{z1}C_1'}\right)\left(s + \dfrac{1}{R_{z2}C_2'}\right)}{D}, \quad (10)$$

$$\frac{V_{LP}}{V_{in}} = \beta_1^- \beta_2^- \frac{\dfrac{g_{m1}g_{m2}}{C_1'C_2'}}{D}, \quad (11)$$

and

$$\frac{V_{BP}}{V_{in}} = -\beta_1^- \beta_2^- \frac{\dfrac{g_{m2}}{C_1'}\left(s + \dfrac{1}{R_{z2}C_2'}\right)}{D} \quad (12)$$

where $C_1' = C_1 + C_{z1}$, $C_2' = C_2 + C_{z2}$, and

$$D = s^2 + s\frac{\omega_0^*}{Q^*} + \omega_0^{*2}, \quad (13)$$

$$\omega_0^{*2} = \frac{g_{m1}g_{m2}}{C_1'C_2'}\left(\beta_1^+ \beta_2^- + \frac{\beta_2^+}{g_{m1}R_{z2}} + \frac{1}{g_{m1}g_{m2}R_{z1}R_{z2}}\right), \quad (14)$$

$$Q^* = \sqrt{\frac{g_{m1}}{g_{m2}}\frac{C_1'}{C_2'}} \frac{\sqrt{\beta_1^+ \beta_2^- + \dfrac{\beta_2^+}{g_{m1}R_{z2}} + \dfrac{1}{g_{m1}g_{m2}R_{z1}R_{z2}}}}{\beta_2^+ + \dfrac{1}{g_{m2}R_{z1}} + \dfrac{C_1'}{C_2'}\dfrac{1}{g_{m1}g_{m2}R_{z1}R_{z2}}}. \quad (15)$$

It should be mentioned that the stray/parasitic z-terminal capacitances are absorbed by the external grounded capacitors as they appear in shunt with them. However, the parasitic resistances R_{z1} and R_{z2} not only affect the ω_0 and Q by they also add parasitic zeros to the HP and BP transfer functions. The product $\beta_1^- \beta_2^-$ of the voltage buffer gains affects the gain of all the filter sections. As a result, the effect of the finite low-frequency attenuation of BP and HP sections appears [7]. It can be described as follows:

$$\left.\frac{V_{in}}{V_{BP}}\right|_{f=0} = -\frac{\beta_2^+}{\beta_1^- \beta_2^-}\left(1 + \frac{1}{\beta_2^+ g_{m2}R_{z1}} + \frac{\beta_1^+ \beta_2^-}{\beta_2^+}g_{m1}R_{z2}\right), \quad (16)$$

$$\left.\frac{V_{in}}{V_{HP}}\right|_{f=0} = \frac{1}{\beta_1^- \beta_2^-}\left(1 + \beta_2^+ g_{m2}R_{z1} + \beta_1^+ \beta_2^- g_{m1}R_{z1}g_{m2}R_{z2}\right). \quad (17)$$

Note that these undesirable finite attenuations strongly depend on the $g_m R_z$ product. Consider unity gains of the

voltage buffers for the simplicity. Let us denote $g_m R_z = a$. Then (16) and (17) can be simplified to the forms

$$\left.\frac{V_{in}}{V_{BP}}\right|_{f=0} = -\left(1 + a + \frac{1}{a}\right), \quad (18)$$

$$\left.\frac{V_{in}}{V_{HP}}\right|_{f=0} = 1 + a + a^2. \quad (19)$$

Then one can see that $a = g_m R_z$ products of 10^1, 10^2, 10^3, and 10^4 result in the attenuations (18) of 20.9, 40.1, 60, 80 dB for BP output, and in the attenuations (19) of 40.9, 80.1, 120, 160 dB for HP output. Two rules of thumb can be applied: 1) The BP low-frequency attenuation can be increased by 20 dB via increasing the $g_m R_z$ product ten times. 2) The HP low-frequency attenuation is twice the size.

Similarly, Equations (14) and (15) for pole frequency and quality factor can be simplified as follows:

$$\frac{\omega_0^*}{\omega_0} = \sqrt{\frac{C_1 C_2}{C_1' C_2'}}\sqrt{1 + \frac{1}{a} + \frac{1}{a^2}}, \quad (20)$$

$$\frac{Q^*}{Q} = \sqrt{\frac{C_1' C_2}{C_1 C_2'}} \frac{\sqrt{1 + \dfrac{1}{a} + \dfrac{1}{a^2}}}{1 + \dfrac{1}{a} + \dfrac{C_1'}{C_2'}\dfrac{1}{a^2}}. \quad (21)$$

It follows from (20) and (21) that high values of a, which are necessary for suppressing parasitic low-frequency gains, ensure that the pole frequency and quality factor are not affected by the finite parasitic resistances R_z.

Note that the analysis of non-ideal case should include the current limits I_{max} of the internal g_m-stages of VD-DIBAs. Since these OTAs operate to grounding capacitors connected to z terminals of VD-DIBAs, the maximum voltages V_{max} at the filter outputs are limited by the values

$$V_{max} = \frac{I_{max}}{\omega C}. \quad (22)$$

In spite of the simplicity of (22), it describes well the potential limitations of the dynamic range of the filter due to nonlinear issues: The appropriate dynamic range can be ensured more problematically for high-frequency biquad, employing g_m stages with insufficient current-driving capability, especially with high working capacitances. However, the above parameters can be typical for non-on-chip filter prototyping from commercial ICs.

4. Experimental Results

The proposed universal filter in Fig. 2 was designed with the following parameters: $C_1 = C_2 = 10$ nF, $g_{m1} = g_{m2} = 10$ mS. The corresponding theoretical values of the pole frequency and quality factor are $f_0 = 159$ kHz and $Q = 1$. The relatively high capacitances were selected in order to provide reasonably high $g_m R_z$ product as explained below.

The VD-DIBA was implemented from commercial ICs as shown in Fig. 3. It consists of two basic blocks: the operational transconductance amplifier-OTA (MAX435) [8] as input stage and the differential-input buffer (AD8130) [9] as output stage. The transconductance gain (g_m) can be adjusted by external resistor R_g ($g_m = 4/R_g$) of the MAX435. The DC power supply voltages are ±5 V.

As obvious from Fig. 3, R_g was finally selected as 390 Ω. The corresponding values of g_m and f_0 are 10.26 mS and 163 kHz, respectively. The resistor R_{set} in Fig. 3 was designed according to the datasheet [8], providing the upper limit of 10 mA of the OTA current.

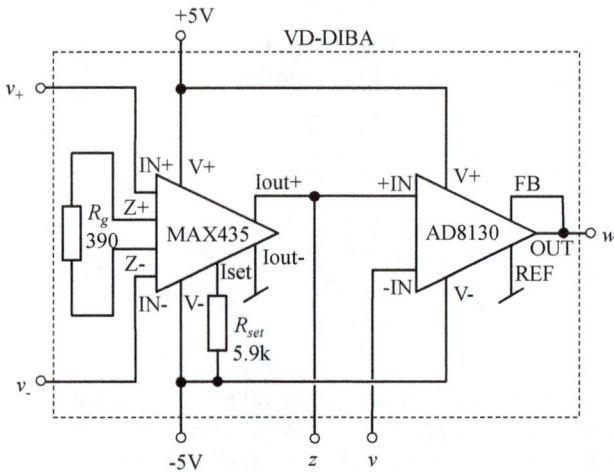

Fig. 3. VD-DIBA implementation by ICs MAX435 and AD8130.

In addition to excellent parameters of MAX435 (275 MHz bandwidth, 850 V/μs slew rate, 18 ns settling time, linear I/V characteristic), it provides rather low resistances R_z of the current outputs. The measured values are only 5 kΩ. It gives, together with the above value of g_m, the product $g_m R_z = 51.3$. According to (16) and (17), the estimated parasitic low-frequency attenuations at BP and HP outputs will be 34.4 dB and 68.6 dB, respectively. As shown below, these values correspond well with SPICE simulations and measurements. If higher values are requested, one should decrease R_z and increase capacitances accordingly, this way preserving the required pole frequency. As results from (20) and (21), $g_m R_z = 51.3$ causes negligible increase of pole frequency and decrease of the quality factor (below one per cent).

Before manufacturing the prototype, the designed filter was simulated in PSpice. The VD-DIBA was modeled via SPICE models of MAX435 [10] and AD8130 [11]. Since the model from [10] does not consider the output resistance R_z of the current outputs of MAX435, it was modeled by auxiliary resistor with $R_z = 5$ kΩ. The simulated frequency responses, compared with the characteristics for ideal case ($R_z \rightarrow \infty$) and with the characteristics measured via the network analyzer Agilent E4061B are

shown in Fig. 4. Note that the measured values were imported in PSpice by look-up-table-controlled sources. The extrapolated low-frequency attenuations are 33.8 dB at BP output and 67.1 dB at HP output. It is in a good agreement with the above mentioned values from the error analysis. The measured pole frequency deviates from the theoretical value by less than 7% due to tolerance of passive components.

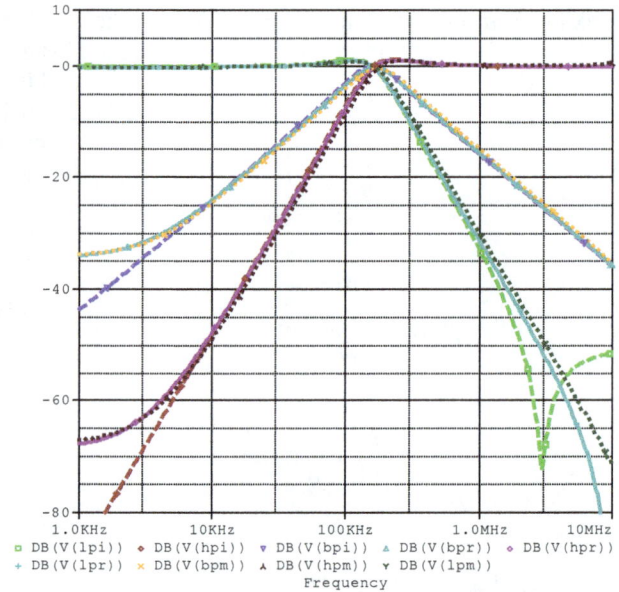

Fig. 4. Frequency responses of the designed filter; lpi, bpi, hpi: SPICE simulation, $R_z \rightarrow \infty$; lpr, bpr, hpr: SPICE simulation, $R_z = 5$ kΩ; lpm, bpm, hpm: measured.

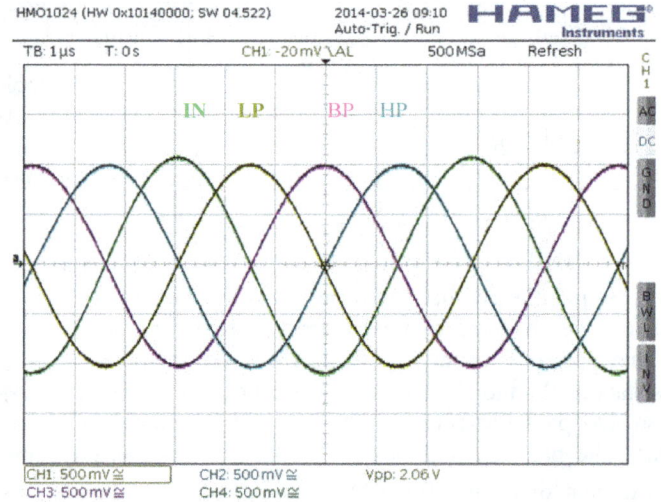

Fig. 5. Large-signal steady-state operation of the manufactured filter for sinusoidal 1.07 V/172 kHz excitation.

In addition to the above small-signal measurements, the manufactured filter was excited by sinusoidal signal in order to test the filter stability and signal limits due to nonlinear distortion. For large-signal operation, the pole frequency tends to be slightly increased due to nonlinear effects. Figure 5 shows the measured waveforms for sinusoidal 1.07 V/172 kHz signal. The THD measured are 0.76 %, 0.5 %, and 0.53 % for HP, BP, and LP outputs, respectively. The magnitudes above ca 1 V cause high increase of the nonlinear distortion due to current limitations of the g_m stages of VD-DIBAs. Such low dynamic range is caused by high capacitances used in the filter. For the upper current limit 10 mA for MAX435, $C = 10$ nF and frequency 172 kHz, the upper limit of the voltage according to (22) is 0.925 V. It is in very good agreement with the measurements.

5. Biquad Generalization

The filter from Fig. 2 can be generalized as shown in Fig. 6, considering five voltage inputs V_{i1} to V_{i5} and two current outputs I_{o1} and I_{o2}, the latter being added via a "z-copy" technique [3]. The types of filter sections, corresponding to input-output pairs, are summarized in Tab. 1. The highlighted three items on the first line denoted "V_{i1}" describe the features of the original filter from Fig. 2.

Fig. 6. Multi-input multi-output generalization of filter from Fig. 2.

	V_{o1}	V_{o2}	V_{o3}	V_{o4}	I_{o1}	I_{o2}
V_{i1}	HP	-BP	LP		$-HP^{(2)}$	$-BP^{(4)}$
V_{i2}	-HP	BP			$HP^{(2)}$	$-BP^{(4)}$
V_{i3}	$-BP^{(1)}$	LP			$BP^{(3)}$	
V_{i4}		HP	$-BP^{(1)}$	$-BP^{(1)}$		$-HP^{(4)}$
V_{i5}		BP	-LP	-LP	$HP^{(2)}$	$-BP^{(4)}$
$V_{i1}=V_{i3}=V_{i4}$		AP				
$V_{i3}=V_{i4}$		BR				

(1) BP gain is $g_{m2}C_1/(g_{m1}C_2)$, (2) HP gain is g_{m1}, (3) BP gain is $g_{m2}C_1/C_2$, (4) BP (HP) gain is g_{m2}; otherwise, the gains are equal to one.

Tab. 1. Low-pass (LP), band-pass (BP), high-pass (HP), all-pass (AP) and band-reject (BR) sections of multi-input multi-output filter in Fig. 6.

Note that particularly the column "V_{o2}" represents a promising extension of the biquad features: utilizing the output V_{o2}, all basic filter types can be implemented via a proper selection of the inputs, including allpass and band-reject sections (by interconnecting three or two inputs).

The additional current outputs of the internal OTAs of VD-DIBA No. 1 and 2 can easily serve as current outputs of BP and HP sections, enabling economical extension of the filter operation to the transconductance mode.

In the necessity of providing buffered, thus true volt-

age-mode HP, BP or LP outputs, one can use V_{o1} or V_{o4} terminals. For the operations when the input V_{i1} is not used, the feedback loop to the v terminal of VD-DIBA No. 2 should be led not from w but from z terminal of VD-DIBA No. 1. Since the output buffer is no longer included in the feedback loop, the filter stability and high-frequency behavior will be improved.

6. Comparison with Previous SIMO Voltage-Mode Filters

Literature survey shows that a lot of papers dealing with SIMO voltage-mode filter using various active building blocks have been published [12-30]. Considering the kinds of active elements, the filters in the above references employ: current feedback amplifier (CFA) [12], current conveyor (CCII) [13], [14], four terminal floating nullor (FTFN) and operational transconductance amplifier (OTA) [15], differential voltage current conveyor (DVCC) [16-19], differential different current conveyor (DDCC) [20-22], DDCC and OTA [23], OTA [24-26], differential difference current conveyor transconductance amplifier (DDCCTA) [27], [28], fully differential current conveyor (FDCCII) [29-31], DDCC and current controlled current conveyor (CCCII) [32].

The proposed circuit in Fig. 2 is compared with several SIMO voltage-mode filters from [12-32]. The results are shown in Tab. 2. It can be seen that it matches all the criteria in the best way among all other filters.

In addition to the above SIMO filters, the three-input single-output voltage-mode biquad utilizing one VDIBA (voltage differencing inverting buffered amplifier) [33], two capacitors, and one resistor has been published in [34] and also in [35]. Note that VDIBA differs from VD-DIBA by voltage inverter which replaces the differential-input buffer in VD-DIBA. The filter from [34], [35] pays a tax for employing only one active element: Electronic control is limited to one parameter, namely the transconductance of the VD-DIBA, thus the controlling range is smaller in comparison to (5), without any possibility not to disturb Q with tuning f_0. Since the single output is not of low-impedance nature, all the implemented filter types require additional voltage buffering.

7. Conclusions

The voltage-mode biquad filter has been presented in this contribution. The advantages of the proposed circuit are that: (i) it performs low-pass, high-pass, and band-pass, functions from the same simple circuit configuration; (ii) the quality factor and the pole frequency can be electronically controlled; (iii) the filter has high input impedance; (iv) the circuit uses only two VD-DIBAs, two grounded capacitors and no resistors, which is attractive for its IC implementation; (v) the functionality of the filter can be easily extended for providing all-pass and band-reject sec-

Ref	Active elements	No. of active elements	No. of R+C	Electronic control	Grounded capacitors only	High input impedance
[12]	CFA	5	5+2	No	Yes	Yes
[13]	CCII	3(cir.1), 4(cir.2)	6+2(cir.1), 7+2(cir.2)	No	Yes	No
[14]	CCII	4(cir.1&cir.2), 3(cir.3), 2(cir.4)	5+2	No	Yes	No
[15]	FTFN & OTA	2 & 2	4+2	Yes	Yes	Yes
[16]	DVCC	3	4+2	No	Yes	Yes
[17]	DVCC	3	3+2	No	Yes	Yes
[18]	DVCC	2	3+2	No	Yes	No
[19]	DVCC	1	2+3	No	No	Yes
[20]	DDCC	2	3+2	No	Yes	No
[21]	DDCC	2	3+2	No	Yes	No
[22]	DDCC	1	3+2	No	Yes	No
[23]	DDCC & OTA	1 & 2	0+2	Yes	Yes	Yes
[24]	OTA	5	0+2	Yes	Yes	Yes
[25]	OTA	8	0+2	Yes	Yes	Yes
[26]	OTA	4	0+2	Yes	Yes	Yes
[27]	DDCCTA	2	2+2	Yes	Yes	Yes
[28]	DDCCTA	3	0+2	Yes	Yes	Yes
[29]	FDCCII	2	2+2	No	Yes	Yes
[30]	FDCCII	1	3+2	No	Yes	No
[31]	FDCCII	1	2+2	No	Yes	Yes
[32]	DDCC & CCCII	1+1	1+2	Yes	Yes	Yes
Proposed filter	VD-DIBA	2	0+2	Yes	Yes	Yes

Tab. 2. Comparison of various SIMO voltage-mode filters.

tions via selecting various types of voltage inputs; (vi) the filter topology provides a possibility of its flexible modification and development, depending on the user's requirements, for example towards the transconductance mode of the operation.

The error analysis reveals that parasitic low-frequency gains of BP and HP sections can be suppressed via selecting the $g_m R_z$ product as high as possible. As results from the experiments described in Section 4, VD-DIBA should be implemented with R_z high enough. Otherwise, g_m must be designed too high, which results in large working capacitances. Then the corresponding low impedance level is a source of several troubles, especially low dynamic range of the voltage signals. For IC implementation, designing VD-DIBA with extra-high z-terminal impedance is thus a prerequisite for constructing high-performance biquads.

Acknowledgements

Research described in this paper was financially supported from the Faculty of Industrial Education, King Mongkut's Institute of Technology Ladkrabang (KMITL), Grant No. 2557-0203022; and from Project PRO-K217 of the University of Defense Brno.

References

[1] SEDRA, A. S., SMITH, K. C. *Microelectronic Circuits*. 3rd ed. Florida: Holt, Rinehart and Winston, 1991.

[2] IBRAHIM, M. A., MINAEI, S., KUNTMAN, H. A 22.5MHz current-mode KHN-biquad using differential voltage current conveyor and grounded passive elements. *International Journal of Electronics and Communications (AEÜ)*, 2005, vol. 59, no. 5, p. 311–318.

[3] BIOLEK, D., SENANI, R., BIOLKOVÁ, V., KOLKA, Z. Active elements for analog signal processing, classification, review and new proposals. *Radioengineering*, 2008, vol. 17, no. 4, p. 15–32.

[4] BIOLEK, D., BIOLKOVÁ, V. First-order voltage-mode all-pass filter employing one active element and one grounded capacitor. *Analog Integrated Circuits and Signal Processing*, 2009, vol. 65, no. 1, p. 123–129.

[5] PRASAD, D., BHASKAR, D. R., PUSHKAR, K. L. Realization of new electronically controllable grounded and floating simulated inductance circuits using voltage differencing differential input

buffered amplifiers. *Active and Passive Electronic Components*, 2011, vol. 2011, Article ID: 101432.

[6] PUSHKAR, K. L., BHASKAR, D. R., PRASAD, D. Voltage-mode universal biquad filter employing single voltage differencing differential input buffered amplifier. *Circuits and Systems*, 2013, vol. 4, p. 47–51.

[7] BIOLEK, D., BAJER, J., BIOLKOVÁ, V., KOLKA, Z., KUBÍČEK, M. Z Copy - Controlled Gain - Current Differencing Buffered Amplifier and its applications. *Int. Journal of Circuit Theory and Applications*. 2011, vol. 39, no. 3, p. 257–274.

[8] MAX435/MAX436 Wideband Transconductance Amplifiers, MAXIM, 19-0042; Rev. 1; 4/93.

[9] AD8129/AD8130 Low Cost 270 MHz Differential Receiver Amplifiers, Analog Devices, 2005. Available at: www.analog.com.

[10] MAX435 Family Macromodels, http://www.maximintegrated.com/design/tools/modeling-simulation/spice/operational-amplifiers/macro/MAX435.FAM

[11] AD8130 SPICE Macro-Model http://www.analog.com/en/specialty-amplifiers/differential-amplifiers/ad8130/products/mod-spice-models/resources.html

[12] ABUELMA'ATTI, M. T., AL-ZAHER, H. A. New universal filter with one input and five outputs using current-feedback amplifiers. *Analog Integrated Circuits and Signal Processing*, 1998, vol. 16, p. 239–244.

[13] HORNG, J. W., HOU, C., L., CHANG, C. M., CHUNG, W. Y., WEI, H. Y. Voltage-mode universal biquadratic filter with one input and five outputs using MOCCIIs. *Computers & Electrical Engineering*, 2005, vol. 31, p. 190–202.

[14] HORNG, J. W., HOU, C., L., CHANG, C. M., CHUNG, W. Y. Voltage-mode universal biquadratic filter with one input and five outputs. *Analog Integrated Circuits and Signal Processing*, 2006, vol. 47, p. 73–83.

[15] KUMAR, K., PAL, K. Voltage-mode universal biquadratic filter using FTFN and OTA. *Journal of Electrical and Electronics Engineering*, 2009, vol. 9, no. 2, p. 1083–1087.

[16] HORNG, J. W., HOU, C. L., CHANG, C. M., CHOU, H. P., LIN, C. T. High input impedance voltage-mode universal biquadratic filter with one input and five outputs using current conveyors. *Circuits System Signal and Processing*, 2006, vol. 25, no. 6, p. 767–777.

[17] MINAEI, S., YUCE, E. All-grounded passive elements voltage-mode DVCC-based universal filter. *Circuits System Signal and Processing*, 2010, vol. 29, p. 295–309.

[18] HORNG, J. W. Lossless inductance simulation and voltage-mode universal biquadratic filter with one input and five outputs using DVCCs. *Analog Integrated Circuits and Signal Processing*, 2010, vol. 62, p. 407–413.

[19] HORNG, J. W. Voltage-mode multifunction biquadratic filter employing single DVCC. *International Journal of Electronics*, 2012, vol. 99, no. 2, p. 153–162.

[20] CHEN, H. P. Universal voltage-mode filter using only plus-type DDCCs. *Analog Integrated Circuits and Signal Processing*, 2007, vol. 50, no. 2, p. 137–139.

[21] CHIU, W. Y., HORNG, J. W. Voltage-mode biquadratic filters with one input and five outputs using DDCCs. *Indian Journal of Engineering and Materials Sciences*, 2011, vol. 18, p. 97–101.

[22] CHIU, W. Y., HORNG, J. W. Voltage-mode highpass, bandpass, lowpass and notch biquadratic filters using single DDCC. *Radioengineering*, 2012, vol. 21, no. 1, p. 297–303.

[23] LEE, W. T., LIAO, Y. Z. New voltage-mode high-pass, band-pass and low-pass filter using DDCC and OTAs. *International Journal of Electronics and Communications (AEÜ)*, 2008, vol. 62, no. 9, p. 701–704.

[24] HORNG, J. W. Voltage-mode universal biquadratic filter with one input and five outputs using OTAs. *International Journal of Electronics*, 2002, vol. 89, p. 729–737.

[25] KUMNGERN, M., SUWANJAN, P., DEJHAN, K. Electronically tunable voltage-mode universal filter with single-input five-output using simple OTAs. *International Journal of Electronics*, 2013, vol. 100, no. 8, p. 1118–1133.

[26] KUMNGERN, M., DEJHAN, K. Voltage-mode low-pass, high-pass, band-pass biquad filter using CMOS OTAs. In *Proceedings of the IEEE International Instrumentation and Measurement Technology Conference*. 2009, p. 924–927.

[27] CHANNUMSIN, O., PUKKALANUN, T., TANGSRIRAT, W. Voltage-mode universal filter with one input and five outputs using DDCCTAs and all-grounded passive components. *Microelectronics Journal*, 2012, vol. 43, no. 8, p. 555–561.

[28] TANGSRIRAT, W., CHANNUMSIN, O., PUKKALANUN, T. Resistorless realization of electronically tunable voltage-mode SIFO-type universal filter. *Microelectronics Journal*, 2012, vol. 43, no. 8, p. 555–561.

[29] CHEN, H. P. Voltage-mode FDCCII-based universal filters. *International Journal of Electronics and Communications (AEÜ)*, 2008, vol. 62, no. 4, p. 320–323.

[30] LEE, C. N., CHANG, C. M. Single FDCCII-based mixed-mode biquad filter with eight outputs. *International Journal of Electronics and Communications (AEÜ)*, 2009, vol. 63, p. 736 to 742.

[31] MAHESHWARI, S., MOHAN, J., CHAUHAN, D. S. Novel cascadable all-pass/notch filters using a single FDCCII and grounded capacitors. *Circuits, Systems, and Signal Processing*, 2011, vol. 30, no. 3, p. 643–654.

[32] MANEEWAN, S., SREEWIROTE, B., JAIKLA, W. Electronically tunable voltage-mode universal filter using DDCC and CCCII. In *Proceedings of the International Conference on Circuits, System and Simulation*. 2011, p. 322–326.

[33] HERENCSÁR, N., KOTON, J., MINAEI, S., YUCE, E., VRBA, K. Novel resistorless dual-output VM all-pass filter employing VDIBA. In *Proceedings of the 7th International Conference on Electrical and Electronics Engineering–ELECO 2011*. Bursa (Turkey), 2011, p. 72–74.

[34] HERENCSÁR, N., CICEKOGLU, O., ŠOTNER, R., KOTON, J., VRBA, K. New resistorless tunable voltage-mode universal filter using single VDIBA. *Analog Integrated Circuits and Signal Processing*, 2013, vol. 76, p. 251–260.

[35] PUSHKAR, K. L., BHASKAR, D. R., PRASAD, D. Voltage-mode new universal biquad filter configuration using a single VDIBA. *Circuits Systems and Signal Processing*, 2014, vol. 33, p. 275–285.

About Authors ...

Winai JAIKLA was born in Buriram, Thailand. He received the B. S. I. Ed. degree in Telecommunication Engineering from King Mongkut's Institute of Technology Ladkrabang (KMITL), Thailand in 2002, M. Tech. Ed. in Electrical Technology and Ph.D. in Electrical Education from King Mongkut's University of Technology North Bangkok (KMUTNB) in 2004 and 2010, respectively. From 2004 to 2011 he was with Electric and Electronic Program, Faculty of Industrial Technology, Suan Sunandha Rajabhat University, Bangkok, Thailand. He has been with the Department of Engineering Education, Faculty of Industrial Education, King Mongkut's Institute of Technology Ladkrabang, Bangkok, Thailand since 2012. His research interests include electronic communications, analog signal processing and analog integrated circuits. He is a member of ECTI, Thailand.

Dalibor BIOLEK received the M.Sc. degree in Electrical Engineering from the Brno University of Technology, Czech Republic, in 1983, and the Ph.D. degree in Electronics from the Military Academy Brno, Czech Republic, in 1989, focusing in algorithms of the symbolic and numerical computer analysis of electronic circuits with a view to the linear continuous-time and switched filters. He is currently with the Department of EE, University of Defense Brno (UDB), and with the Department of Microelectronics, Brno University of Technology (BUT), Czech Republic. His scientific activity is directed to the areas of general circuit theory, frequency filters, and computer simulation of electronic systems. He has published over 300 papers and is author of three books on circuit analysis and simulation. At present, he is professor at the BUT and UDB in the field of Theoretical Electrical Engineering. Prof. Biolek is a member of the CAS/COM Czech National

Group of IEEE. He is also the president of Commission C of the URSI National Committee for the Czech Republic.

Surapong SIRIPONGDEE received the B. S. I. Ed. degree in Electronics and Computer and the M. Tech. Ed. electrical communications engineering from King Mongkut's Institute of Technology Ladkrabang (KMITL), Thailand in 1997 and 2002, respectively. He has been with the Department of Engineering Education, Faculty of Industrial Education, King Mongkut's Institute of Technology Ladkrabang, Bangkok, Thailand since 1997. His research interests include electronic communications, analog signal processing and analog integrated circuits.

Josef BAJER was born in 1982. In 2005, 2008, and 2011 he received the B.Sc. degree in Electrical and Special Aircraft Equipment, the M.Sc. degree in Avionic Systems, and Ph.D. degree in Electronic Systems at the Faculty of Military Technologies, University of Defense Brno (UDB), Czech Republic. His interests include analog and digital signal processing and applications of modern active elements working in current and hybrid modes.

8

Parallel Implementation of the Discrete Green's Function Formulation of the FDTD Method on a Multicore Central Processing Unit

Tomasz STEFAŃSKI[1], Sławomir ORŁOWSKI[1], Bartosz REICHEL[2]

[1]Faculty of Electronics, Telecommunications and Informatics, Gdansk University of Technology, Narutowicza 11/12, 80-233 Gdansk, Poland
[2]Faculty of Applied Physics and Mathematics, Gdansk University of Technology, Narutowicza 11/12, 80-233 Gdansk, Poland

tomasz.stefanski@pg.gda.pl, orlowski.slawomir@gmail.com, reichel@mif.pg.gda.pl

Abstract. *Parallel implementation of the discrete Green's function formulation of the finite-difference time-domain (DGF-FDTD) method was developed on a multicore central processing unit. DGF-FDTD avoids computations of the electromagnetic field in free-space cells and does not require domain termination by absorbing boundary conditions. Computed DGF-FDTD solutions are compatible with the FDTD grid enabling the perfect hybridization of FDTD with the use of time-domain integral equation methods. The developed implementation can be applied to simulations of antenna characteristics. For the sake of example, arrays of Yagi-Uda antennas were simulated with the use of parallel DGF-FDTD. The efficiency of parallel computations was investigated as a function of the number of current elements in the FDTD grid. Although the developed method does not apply the fast Fourier transform for convolution computations, advantages stemming from the application of DGF-FDTD instead of FDTD can be demonstrated for one-dimensional wire antennas when simulation results are post-processed by the near-to-far-field transformation.*

Keywords

Computational electromagnetics, discrete Green's function (DGF), finite-difference time-domain (FDTD) method, parallel processing.

1. Introduction

Recently, several computational techniques facilitating the finite-difference time-domain (FDTD) method [1] were developed based on the discrete Green's function (DGF) [2], [3], [4], [5]. DGF is the impulse response of a system of finite-difference equations defined on a discrete domain. The convolution of DGF with current sources exciting that domain allows to obtain the FDTD solution without executing the standard FDTD update procedure throughout the entire domain. Moreover, DGF-based computations do not need absorbing boundary conditions (ABCs) for simulations of the radiation and scattering problems in the FDTD grid. Therefore, DGF has been applied to FDTD simulations of antennas with savings in runtime and memory usage [6], [7], [8], [9].

In [6], the DGF-based scattering formulation of the FDTD method (DGF-FDTD) was developed for antenna simulations. It computes currents at conducting surfaces with the use of the march-on-in-time scheme (i.e., the evolution of antenna currents is computed one time step at a time based on currents computed for previous time steps). Similarly to the FDTD method, DGF-FDTD allows to obtain wideband frequency characteristics of an antenna in a single simulation run. Although DGF-FDTD solves the time-domain electric field integral equation, it is inherently discrete and the whole formulation is much more straightforward in comparison to other methods based on time-domain integral equations [9].

Recently, DGF-FDTD has been coupled with the FDTD method [10], [11], hence consistent with the discrete theory of electromagnetism hybridization of FDTD was developed. Because FDTD solutions have their own dispersion, anisotropy, and stability properties, the coupling of FDTD and integral-equation methods requires discrete equivalents to the integral operator and the Green's function (i.e., DGF) [4]. Using FDTD method hybridized with DGF, simulation scenarios involving interacting transmitters and scatterers can be tackled without computations of the field in free-space cells between these objects. Moreover, the transmitters and scatterers can be simulated separately and a system response can be obtained in terms of the diakoptics approach (i.e., as a response of interacting multi-port sub-systems) [12]. Currently, commercial FDTD solvers allow running two-stage simulations with the source of radiation simulated at the first stage and multiple simulations of the irradiation at the second stage [13]. Hence, many different objects weakly coupled with the source of radiation can be simulated with savings in runtime. Such FDTD simulations, as well as other applications of the diakoptics in the FDTD method, can take advantage of DGF-FDTD.

In spite of the mentioned above advantages of the DGF-FDTD method, its parallel implementations on modern computing architectures, such as multicore central processing units (CPUs) and graphics processing units (GPUs), have attracted little attention so far. Therefore, the parallel implementation of the DGF-FDTD method on CPU was developed. It has recently been reported that the application of the fast Fourier transform (FFT) for spatial convolution computations leads to favorable throughput of DGF-FDTD compared to the standard FDTD method [8]. The presented here parallel DGF-FDTD implementation does not employ FFT for acceleration of the convolution computations. In spite of that, advantages due to the application of DGF-FDTD instead of FDTD can be demonstrated for one-dimensional wire antennas (especially when simulation results are post-processed by the near-to-far-field (NTFF) transformation).

2. DGF-FDTD Method

The DGF-FDTD method represents FDTD update equations by means of the convolution of the current sources ($\mathbf{J}$, $\mathbf{M}$) and dyadic DGF ($\mathbf{G_{ee}}$, $\mathbf{G_{eh}}$, $\mathbf{G_{he}}$, $\mathbf{G_{hh}}$) [6]:

$$\begin{bmatrix} \mathbf{E} \mid_{ijk}^{n} \\ \eta \mathbf{H} \mid_{ijk}^{n} \end{bmatrix} =$$

$$\sum_{n'i'j'k'} \begin{bmatrix} \mathbf{G_{ee}} \mid_{i-i'j-j'k-k'}^{n-n'} & \mathbf{G_{eh}} \mid_{i-i'j-j'k-k'}^{n-n'} \\ \mathbf{G_{he}} \mid_{i-i'j-j'k-k'}^{n-n'} & \mathbf{G_{hh}} \mid_{i-i'j-j'k-k'}^{n-n'} \end{bmatrix} \begin{bmatrix} \eta \mathbf{J_{eq}} \mid_{i'j'k'}^{n'} \\ \mathbf{M_{eq}} \mid_{i'j'k'}^{n'} \end{bmatrix}$$

$$(1)$$

where:

$$\mathbf{J_{eq}} \mid_{ijk}^{n} = (s_x s_y s_z)^{-1} c \Delta t \mathbf{J} \mid_{ijk}^{n}, \qquad (2)$$

$$\mathbf{M_{eq}} \mid_{ijk}^{n} = (s_x s_y s_z)^{-1} c \Delta t \mathbf{M} \mid_{ijk}^{n}. \qquad (3)$$

In (1)–(3), $s_p = c\Delta t/\Delta p$ denotes the Courant number, c is the speed of light, Δt is the time-step size, Δp is the discretization-step size along the p-direction ($p = x, y, z$), η is the intrinsic impedance of free space, n is the time index, and i, j, k are the spatial indices in the grid. Equation (1) is referred to as the convolution formulation of the FDTD method [6]. If the length of DGF waveforms is equal to the number of time steps in the FDTD simulation, this formulation returns the same results as the direct FDTD method (assuming infinite numerical precision of computations).

Only the $\mathbf{G_{ee}}$ component of DGF is presented here for the sake of brevity. Its analytic closed-form expression in infinite free space takes the following form for the (i, j, k) cell [5]:

$$G_{ee,xz} \mid_{ijk}^{n} = \sum_{m=\alpha_x+\beta_x+\gamma_x}^{n-2} \binom{n+m}{2m+2} g_{xz} \mid_{ijk}^{m}, \qquad (4)$$

$$G_{ee,yz} \mid_{ijk}^{n} = \sum_{m=\alpha_y+\beta_y+\gamma_y}^{n-2} \binom{n+m}{2m+2} g_{yz} \mid_{ijk}^{m}, \qquad (5)$$

$$G_{ee,zz} \mid_{ijk}^{n} = -s_x s_y s_z U \mid^{n-1} \delta \mid_{ijk} +$$
$$\sum_{m=max(\alpha_{f,z}+\beta_{f,z}+\gamma_{f,z}-1,0)}^{n-2} \binom{n+m}{2m+2} f_{zz} \mid_{ijk}^{m+1} +$$
$$\sum_{m=\alpha_{h,z}+\beta_{h,z}+\gamma_{h,z}}^{n-2} \binom{n+m}{2m+2} h_{zz} \mid_{ijk}^{m} \qquad (6)$$

where:

$$g_{xz} \mid_{ijk}^{m} = -(-1)^{m+i+j+k} \sum_{\substack{\alpha+\beta+\gamma=m \\ \alpha \geq \alpha_x, \beta \geq \beta_x, \gamma \geq \gamma_x}} \binom{m}{\alpha, \beta, \gamma} \times$$
$$\binom{2\alpha+1}{\alpha+i+1}\binom{2\beta}{\beta+j}\binom{2\gamma+1}{\gamma+k} s_x^{2\alpha+2} s_y^{2\beta+1} s_z^{2\gamma+2}, \qquad (7)$$

$$g_{yz} \mid_{ijk}^{m} = -(-1)^{m+i+j+k} \sum_{\substack{\alpha+\beta+\gamma=m \\ \alpha \geq \alpha_y, \beta \geq \beta_y, \gamma \geq \gamma_y}} \binom{m}{\alpha, \beta, \gamma} \times$$
$$\binom{2\alpha}{\alpha+i}\binom{2\beta+1}{\beta+j+1}\binom{2\gamma+1}{\gamma+k} s_x^{2\alpha+1} s_y^{2\beta+2} s_z^{2\gamma+2}, \qquad (8)$$

$$f_{zz} \mid_{ijk}^{m} = -(-1)^{m+i+j+k} \sum_{\substack{\alpha+\beta+\gamma=m \\ \alpha \geq \alpha_{f,z}, \beta \geq \beta_{f,z}, \gamma \geq \gamma_{f,z}}} \binom{m}{\alpha, \beta, \gamma} \times$$
$$\binom{2\alpha}{\alpha+i}\binom{2\beta}{\beta+j}\binom{2\gamma}{\gamma+k} s_x^{2\alpha+1} s_y^{2\beta+1} s_z^{2\gamma+1}, \qquad (9)$$

$$h_{zz} \mid_{ijk}^{m} = -(-1)^{m+i+j+k} \sum_{\substack{\alpha+\beta+\gamma=m \\ \alpha \geq \alpha_{h,z}, \beta \geq \beta_{h,z}, \gamma \geq \gamma_{h,z}}} \binom{m}{\alpha, \beta, \gamma} \times$$
$$\binom{2\alpha}{\alpha+i}\binom{2\beta}{\beta+j}\binom{2\gamma+2}{\gamma+k+1} s_x^{2\alpha+1} s_y^{2\beta+1} s_z^{2\gamma+3}. \qquad (10)$$

Other terms denote: $\alpha_x = max(-i - 1, i)$, $\beta_x = |j|$, $\gamma_x = max(-k, k - 1)$, $\alpha_y = |i|$, $\beta_y = max(j, -j - 1)$, $\gamma_y = max(-k, k - 1)$, $\alpha_{f,z} = \alpha_{h,z} = |i|$, $\beta_{f,z} = \beta_{h,z} = |j|$, $\gamma_{f,z} = |k|$, $\gamma_{h,z} = max(|k| - 1, 0)$. $U \mid^n$ and $\delta \mid_{ijk}$ respectively denote the unit step and Kronecker delta functions. Expressions for other $\mathbf{G_{ee}}$ components can be obtained rotating the subscripts x, y, z and the corresponding summation indices.

Let us consider an antenna made of a perfect electric conductor (PEC) simulated inside the FDTD grid. For nodes in the grid belonging to PEC, the total electric field is equal to zero:

$$E_p^{total} \mid_{ijk}^{n} = E_p^{inc} \mid_{ijk}^{n} + E_p^{scat} \mid_{ijk}^{n} = 0 \quad (i, j, k, p) \in PEC \quad (11)$$

where E^{total}, E^{inc} and E^{scat} denote respectively total, incident and scattered electric field. (i, j, k, p) denotes the p-component of the field belonging to the (i, j, k) cell in the grid. With the use of (1), the scattered electric field can be obtained from currents induced on the antenna due to the incidence of the electric field:

$$E_p^{scat}\,|_{ijk}^n=$$
$$\sum_{(i',j',k',p')\in PEC}\sum_{n'=0}^{n-1}G_{ee,pp'}\,|_{i-i'j-j'k-k'}^{n-n'}\,\eta(J_{eq})_{p'}\,|_{i'j'k'}^{n'}. \tag{12}$$

Then, the equation relating the incident electric field and currents induced on the antenna can be obtained using (11)–(12):

$$E_p^{inc}\,|_{ijk}^n=$$
$$-\frac{\Delta t}{\varepsilon_0}\sum_{(i',j',k',p')\in PEC}\sum_{n'=0}^{n-1}G_{ee,pp'}\,|_{i-i'j-j'k-k'}^{n-n'}\,(s_xs_ys_z)^{-1}J_{p'}\,|_{i'j'k'}^{n'} \tag{13}$$

where $(i,j,k,p)\in PEC$. For $n=1$, the $\mathbf{G_{ee}}$ component of DGF (4)–(6) reduces to:

$$\mathbf{G_{ee}}\,|_{ijk}=-(s_xs_ys_z)\delta\,|_{ijk}\,\mathbf{I} \tag{14}$$

where $\mathbf{I}=\mathrm{diag}(1,1,1)$ denotes the unit dyad. Hence, one obtains from (13) the time-marching procedure for computations of the time evolution of the antenna currents based on the incident electric field:

$$J_p\,|_{ijk}^{n-1}=\frac{\varepsilon_0}{\Delta t}E_p^{inc}\,|_{ijk}^n+$$
$$\sum_{(i',j',k',p')\in PEC}\sum_{n'=0}^{n-2}G_{ee,pp'}\,|_{i-i'j-j'k-k'}^{n-n'}\,(s_xs_ys_z)^{-1}J_{p'}\,|_{i'j'k'}^{n'}. \tag{15}$$

These antenna currents can be employed for computations of radiation characteristics. In the developed code, the NTFF transformation is implemented based on formulation [1]. However, the far-field pattern is computed directly from the antenna currents. Therefore, for the NTFF computations, savings in runtime and memory usage are obtained in comparison to FDTD because the DGF-FDTD method does not need to employ the equivalence principle at a closed surface enclosing the antenna. Moreover, the computed antenna currents can be employed for excitation of the total-field scattered-field interface in FDTD simulations [11].

3. Parallel DGF-FDTD Solver

The method was implemented in double precision using the C programming language. A single iteration of the time-marching procedure of the developed parallel DGF-FDTD solver is implemented as presented in Fig. 1. In the parallel DGF-FDTD implementation, all tasks are executed by a set of parallel CPU threads. The OpenMP parallel programming standard was employed for implementation of the algorithm in software.

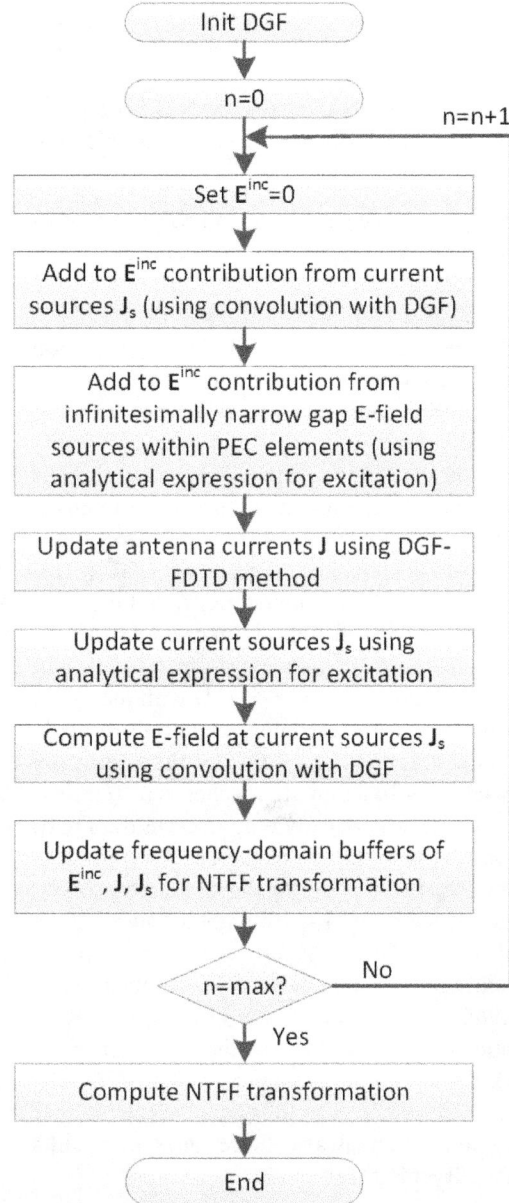

Fig. 1. Flowchart of the developed algorithm.

A single iteration of the time-marching procedure of the parallel DGF-FDTD method is implemented as follows (refer to Fig. 1):

- Buffer of the electric field incident at PEC elements ($\mathbf{E}^{inc}$) is set to zero with the use of parallel threads.

- Contributions to $\mathbf{E}^{inc}$ from current sources feeding the antenna are computed in parallel for each PEC element.

- Contributions to $\mathbf{E}^{inc}$ from infinitesimally narrow gap sources within PEC elements are computed in parallel for each PEC element.

- The DGF-FDTD update procedure (15) is executed in parallel for each PEC element.

- The current sources feeding the antenna are updated in parallel for each source.

- The electric field measured at the current sources feeding the antenna is computed in parallel for each source.

- The frequency-domain buffers for the NTFF transformation are updated in parallel for each PEC element and current source.

The method requires the generation of DGF waveforms (the init DGF step) corresponding to the currents at the antenna. The DGF generation is a part of the pre-processing stage or, alternatively, the DGF waveforms can be read from a file on a hard drive. Unfortunately, the DGF generation currently requires significant processor time. In the developed parallel DGF-FDTD solver, the hardware accelerated methods of the DGF generation [14], [15], [16] are available for antenna simulations. Although the DGF generation is an active topic of research in computational electromagnetics, these computations still remain a bottleneck for applications of the DGF-FDTD method. Therefore, DGF waveforms are truncated to speed up computations, and the windowing technique [17] is applicable for increasing the accuracy of results. However, such an approximation of DGF deteriorates the compatibility of DGF-FDTD with the direct FDTD method. Alternatively, the approximation of dyadic DGF can be obtained from scalar DGF by the truncation of scalar DGF when this function approaches zero (i.e., the steady state). This idea was already employed in the DGF-FDTD simulations [9] with the use of the DGF formulation derived based on scalar DGF [6]. In the developed solver, dyadic DGF is generated from (4)–(10) without intermediate computations of scalar DGF. Therefore, the latter approach to the DGF generation [9] cannot currently be applied in our DGF-FDTD solver. If the accuracy or stability of the DGF-FDTD computations is not satisfactory, then increasing the DGF window length is a solution to these problems. Finally, taking DGF waveforms whose length is equal to the number of time steps in a simulation always assures the same results as returned by the direct FDTD method.

The runtime scaling of the DGF-FDTD convolution computations executed over M PEC elements is of order $(M^2 n_s)$, where n_s denotes the DGF length (computational cost of the DGF generation is excluded from consideration). On the other hand, the direct FDTD computations require to update all cells in the three-dimensional domain containing the antenna. The runtime scaling of these computations is of order (N^3), where N^3 denotes the number of cells in a cubic domain. Therefore, the efficiency of the DGF-FDTD method is higher than the direct FDTD method if a small number of sparsely distributed PEC elements is simulated within a large domain.

For the sake of comparison, simulated antennas can be fed from electric current sources. The developed DGF-FDTD method can employ the one-cell gap model of source, similarly to the direct FDTD method, instead of the infinitesimally narrow gap model [18]. For this purpose, the computations of the incident electric field in the DGF-FDTD method [6] were modified to include contributions from such

current sources. As a result of simulation, the developed solver returns FDTD-compatible waveforms of the incident electric field and the antenna currents as well as far-field radiation patterns.

A graphical user interface (GUI) was developed for the DGF-FDTD solver, refer to Fig. 2, and integrated with the in-house written FDTD simulation tool [19]. It facilitates the edition of simulation parameters, drawing PEC elements, running simulations, and the presentation of results. The developed GUI also provides other functionalities that help in preparing and running DGF-FDTD simulations. In the developed GUI code, the OpenGL library was employed for visualization of computational domain and simulation results. It allows to place PEC elements into the domain, as well as rotate, shift and scale visualized objects. The architecture of the developed software package simplifies the development of its new features by using well defined design patterns along with own engine for data management [19].

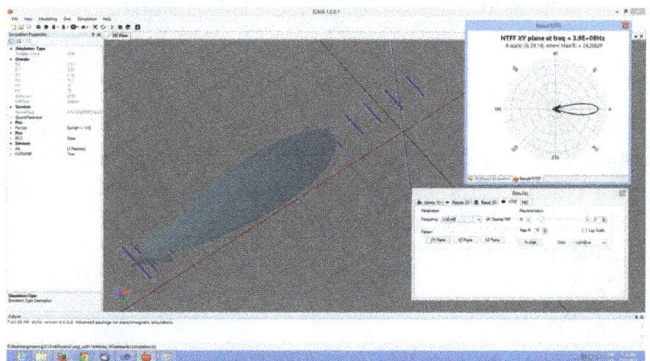

Fig. 2. GUI of the developed simulation tool.

4. Numerical Results

The method was tested on a machine with Intel i7-3770 3.4 GHz processor. The Courant numbers were taken as $s_x = s_y = s_z = 0.99/\sqrt{3}$ for the results presented here. In the presented investigations, the DGF waveforms were read from a file on a hard drive.

Fig. 3(a) presents comparison between waveforms computed with the use of DGF-FDTD and FDTD for a square loop antenna. It consists of 44 current elements (43 PEC elements and a current source at the feeding point). The spatial discretization in this simulation was taken as $\Delta x = \Delta y = \Delta z = 1$ mm. The number of time steps in the simulation was set to 600, which was equal to the DGF length. It allows to verify the correctness of the DGF-FDTD implementation for the DGF waveforms which are not distorted by the windowing. In this test, the size of FDTD domain in the reference simulation was sufficient to avoid reflections from imperfect ABC. The harmonic current source excited the antenna with the frequency set to 6.81 GHz. Fig. 3(b) presents the error between both methods. The correctness of the DGF-FDTD computations is validated by the error varying in the range -300 to -260 dB.

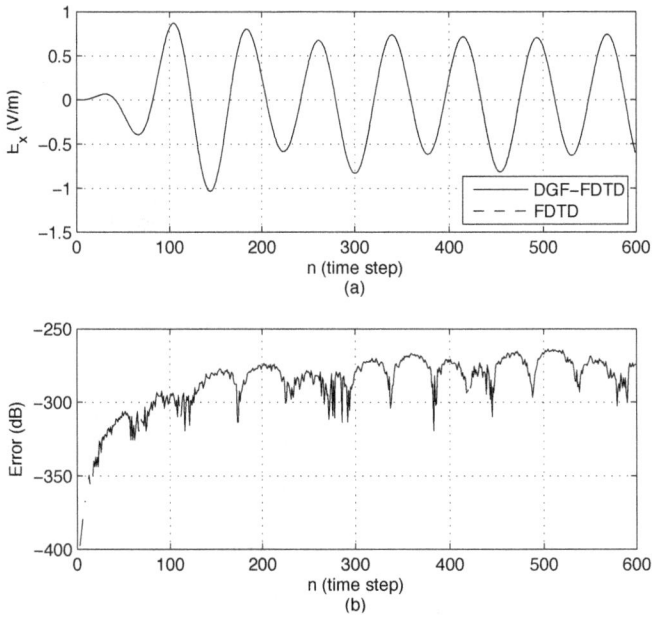

Fig. 3. (a) Electric field waveforms computed at the feeding point of the loop antenna simulated with the use of DGF-FDTD and FDTD. (b) Relative error between both methods. Discontinuity of the line means that both methods computed exactly the same result in double precision.

Arrays of Yagi-Uda antennas were considered for the sake of benchmarking of the parallel code. The spatial discretization in these simulations was taken as $\Delta x = \Delta y = \Delta z = 3$ cm. A single array element consisting of 143 current elements is presented in Fig. 4. The following configurations of antennas in the two-dimensional arrays were considered: 1×2, 1×3, 1×4, 2×3, 2×4. The distances between antennas in arrays were uniformly set to 24 cells. The modulated harmonic current source excited each antenna in arrays with the frequency band set to 350–450 MHz. The uniform amplitude and equal phase distributions were considered in the simulation scenario. The NTFF transformation was computed with the step size set to $1°$ for discrete polar and azimuthal angles. The number of time steps in simulations was set to 4000, whereas the DGF waveforms were truncated using the Hann's window with the length set to $n_s = 400$ samples. Such simulation parameters allowed to obtain stable and satisfactorily accurate results. However, parameters of the DGF truncation cannot be fixed and always depend on the simulated problem. Since the DGF waveforms were read from a file on a hard drive, the presented computing runtimes are independent of the methods of the DGF generation and truncation.

The correctness of the antenna-array simulations was verified by a comparison to the results of direct FDTD computations. For this purpose, characteristics of a single array element were computed. In Fig. 5, the comparison between electric field waveforms computed at the feeding point of a single Yagi-Uda antenna, simulated using DGF-FDTD and FDTD, is presented. As seen, waveforms simulated using the DGF-FDTD and FDTD methods overlap as long as the time step is less than n_s. It shows that the DGF length equal

to the number of time steps in a simulation allows to obtain results overlapping with results of the direct FDTD method. In Fig. 6, the comparison between far-field patterns computed using DGF-FDTD and FDTD is presented for a single Yagi-Uda antenna. As seen, the far-field patterns computed using the DGF-FDTD and FDTD methods overlap for simulation parameters taken as described above. The differences between DGF-FDTD and FDTD are insignificant, although FDTD employs the NTFF transformation implemented with the use of a single Huygens surface [1], [20]. The obtained results validate the correctness of the parallel implementation of DGF-FDTD in software.

In Fig. 7, execution runtimes are presented as a function of the number of current elements in the simulated antenna arrays. Runtimes were measured for the DGF-FDTD update loop, the NTFF transformation that is outside of the DGF-FDTD update loop, as well as the total execution runtimes of the code were measured. For the sake of comparison, results measured for the serial DGF-FDTD code are also presented. The developed parallel DGF-FDTD implementation is maximally 4.3 times faster than its serial implementation (2×4 array, 1144 current elements) when the total execution runtimes are compared. It is a satisfactory result for the code executed on CPU with 4 cores and 8 threads (hyper threading).

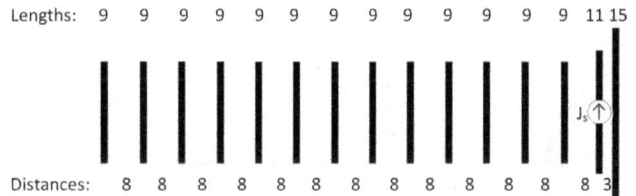

Fig. 4. Geometry of the Yagi-Uda antenna being an element of the considered antenna arrays (distances and lengths measured in grid cells).

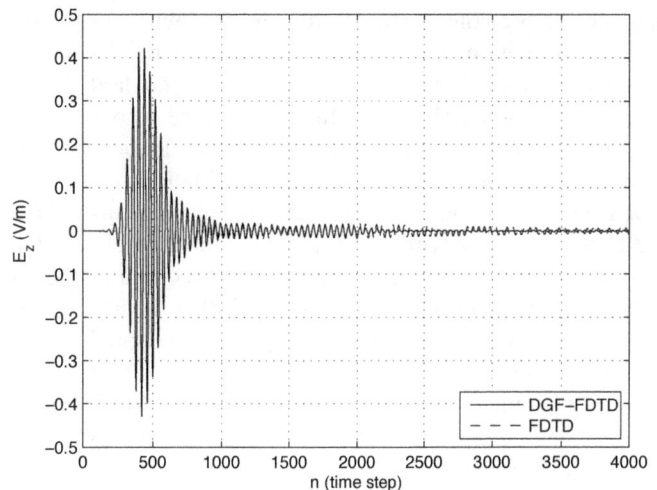

Fig. 5. Electric field waveforms computed at the feeding point of a single Yagi-Uda antenna.

(a)

DGF–FDTD
- - - FDTD

(b)

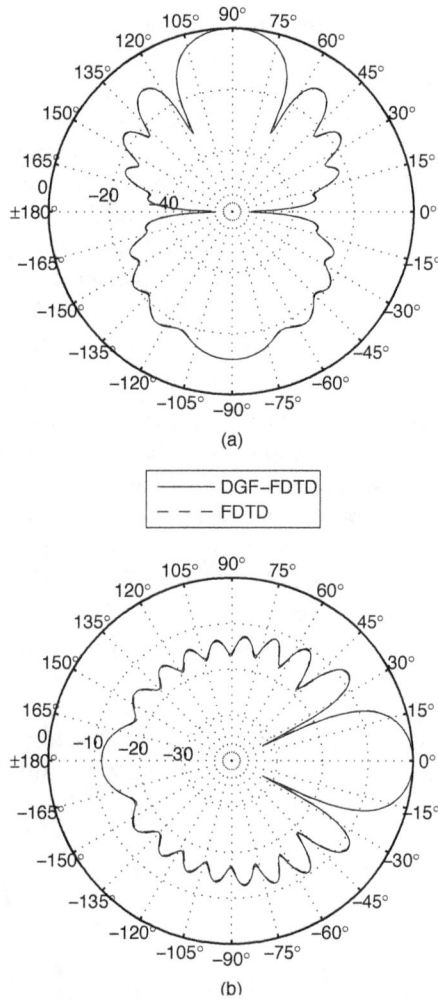

Fig. 6. Exemplary far-field patterns computed for a single Yagi-Uda antenna (values in dB, frequency of NTFF transformation was set to 390 MHz). (a) E-plane. (b) H-plane.

(a)

(b)

(c)

Fig. 7. Execution runtimes for serial and parallel implementations of the DGF-FDTD method. (a) Runtime of NTFF transformation. (b) Runtime of update loop. (c) Total execution runtime.

Execution runtimes were compared between DGF-FDTD and FDTD. Although the developed method does not apply FFT for convolution computations [8], advantages due to the application of DGF-FDTD instead of FDTD can be demonstrated for the considered one-dimensional wire antennas when simulation results are post-processed by the NTFF transformation. In the FDTD method, the standard NTFF transformation employs the Huygens surface whose area depends on the volume of the computational domain. On the other hand, DGF-FDTD computes the NTFF transformation from current sources feeding the antenna and currents induced on antenna wires. For a large number of observation points in the far-field zone, the NTFF computations may take significant processor time in FDTD. Therefore, DGF-FDTD was faster than FDTD in the presented tests for small antenna arrays. For instance, DGF-FDTD was maximally 5.9 times faster than FDTD for a single Yagi-Uda antenna (143 current elements). However, FDTD was faster than DGF-FDTD for 2 × 3 and 2 × 4 arrays (858 and 1144 current elements, respectively). Fortunately, the convolution computations can be accelerated using FFT, hence additional speedup of the DGF-FDTD method can still be obtained.

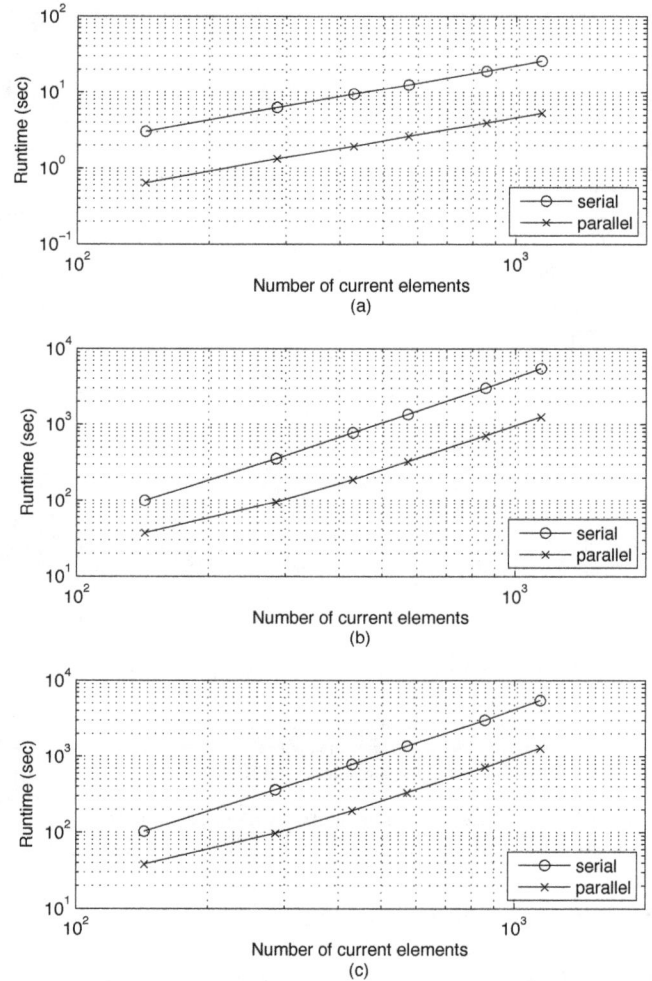

5. Conclusions

The parallel DGF-FDTD method was implemented on multicore CPU. The method is applicable to simulations of wire antennas made of PEC. Computed DGF-FDTD solutions are compatible with the FDTD grid enabling the perfect hybridization of FDTD with the use of time-domain integral equation methods. The computational efficiency of the developed parallel DGF-FDTD solver was investigated in antenna simulations. Although the developed method does not apply FFT for convolution computations, advantages due to the application of DGF-FDTD instead of FDTD can be demonstrated for one-dimensional antennas when simulation results are post-processed by the NTFF transformation. The presented implementation represents the intermediate step in the process of development of the accelerated DGF-FDTD solver executable on heterogeneous parallel processing systems. This topic is planned to be reported in the future.

Acknowledgments

This work was supported by the Polish National Science Center under Agreement DEC-2012/05/D/ST7/00141.

References

[1] TAFLOVE, A., HAGNESS, S. C. *Computational Electrodynamics: The Finite-Difference Time-Domain Method*, 3rd ed. Boston (USA): Artech House, 2005.

[2] VAZQUEZ, J., PARINI, C. G. Discrete Green's function formulation of FDTD method for electromagnetic modelling. *Electronics Letters*, 1999, vol. 35, no. 7, p. 554 - 555.

[3] HOLTZMAN, R., KASTNER, R. The time-domain discrete Green's function method (GFM) characterizing the FDTD grid boundary. *IEEE Transactions on Antennas and Propagation*, 2001, vol. 49, no. 7, p. 1079 - 1093.

[4] KASTNER, R. A multidimensional z-transform evaluation of the discrete finite difference time domain Green's function. *IEEE Transactions on Antennas and Propagation*, 2006, vol. 54, no. 4, p. 1215 - 1222.

[5] JENG, S.-K. An analytical expression for 3-D dyadic FDTD-compatible Green's function in infinite free space via z-transform and partial difference operators. *IEEE Transactions on Antennas and Propagation*, 2011, vol. 59, no. 4, p. 1347 - 1355.

[6] MA, W., RAYNER, M. R., PARINI, C. G. Discrete Green's function formulation of the FDTD method and its application in antenna modeling. *IEEE Transactions on Antennas and Propagation*, 2005, vol. 53, no. 1, p. 339 - 346.

[7] HOLTZMAN, R., KASTNER, R., HEYMAN, E., ZIOLKOWSKI, R. W. Ultra-wideband cylindrical antenna design using the Green's function method (GFM) as an absorbing boundary condition (ABC) and the radiated field propagator in a genetic optimization. *Microwave and Optical Technology Letters*, 2006, vol. 48, no. 2, p. 348 - 354.

[8] MIRHADI, S., SOLEIMANI, M., ABDOLALI, A. An FFT-based approach in acceleration of discrete Green's function method for antenna analysis. *Progress In Electromagnetics Research M*, 2013, vol. 29, p. 17 - 28.

[9] MIRHADI, S., SOLEIMANI, M., ABDOLALI, A. UWB antennas analysis using FDTD-based discrete Green's function approach. *IEEE Antennas and Wireless Propagation Letters*, 2013, vol. 12, p. 1089 - 1093.

[10] STEFAŃSKI, T. P. Hybrid technique combining the FDTD method and its convolution formulation based on the discrete Green's function. *IEEE Antennas and Wireless Propagation Letters*, 2013, vol. 12, p. 1448 - 1451.

[11] STEFAŃSKI, T. P. Application of the discrete Green's function-based antenna simulations for excitation of the total-field/scattered-field interface in the FDTD method. *Microwave and Optical Technology Letters*, 2014, vol. 56, no. 8, p. 1949 - 1953.

[12] DE HON, B. P., ARNOLD, J. M. Stable FDTD on disjoint domains - a discrete Green's function diakoptics approach. *Proceedings of the Second European Conference on Antennas and Propagation*. Edinburgh (UK), 2007, p. 1 - 6.

[13] BENKLER, S., CHAVANNES, N., KUSTER, N. Novel FDTD Huygens source enables highly complex simulation scenarios on ordinary PCs. *Proceedings of the IEEE Antennas and Propagation Society International Symposium*. Charleston (SC, USA), 2009, p. 1 - 4.

[14] STEFAŃSKI, T. P. Fast implementation of FDTD-compatible Green's function on multicore processor. *IEEE Antennas and Wireless Propagation Letters*, 2012, vol. 11, p. 81 - 84.

[15] STEFAŃSKI, T. P., KRZYŻANOWSKA, K. Implementation of FDTD-compatible Green's function on graphics processing unit. *IEEE Antennas and Wireless Propagation Letters*, 2012, vol. 11, p. 1422 - 1425.

[16] STEFAŃSKI, T. P. Implementation of FDTD-compatible Green's function on heterogeneous CPU-GPU parallel processing system. *Progress In Electromagnetics Research*, 2013, vol. 135, p. 297 - 316.

[17] STEFAŃSKI, T. P. Accuracy of the discrete Green's function formulation of the FDTD method. *IEEE Transactions on Antennas and Propagation*, 2013, vol. 61, no. 2, p. 829 - 835.

[18] WATANABE, S., TAKI, M. An improved FDTD model for the feeding gap of a thin-wire antenna. *IEEE Microwave and Guided Wave Letters*, 1998, vol. 8, no. 4, p. 152 - 154.

[19] ORŁOWSKI, S., STEFAŃSKI, T. P. Development of graphical user interface for modern FDTD simulation tool. In *Proceedings of the Progress in Electromagnetics Research Symposium*. Stockholm (Sweden), 2013, p. 1210 - 1214.

[20] MARTIN, T. An improved near- to far-zone transformation for the finite-difference time-domain method. *IEEE Transactions on Antennas and Propagation*, 1998, vol. 46, no. 9, p. 1263 - 1271.

About Authors...

Tomasz STEFAŃSKI received the M.Sc. degree in telecommunications and the Ph.D. degree in electronics engineering, from Gdansk University of Technology (GUT), Gdansk, Poland, in 2002 and 2007, respectively. He is currently leading projects founded by the Foundation for Polish Science and the National Science Center at the Faculty of Electronics, Telecommunications and Informatics at GUT. Before joining GUT in 2011, he was with the Swiss Federal Institute of Technology (ETH Zurich) conducting research on parallelization of electromagnetic solvers on modern computing architectures using OpenCL programming language. Between 2006 and 2009 he worked at the University of Glasgow developing parallel alternating direction implicit finite-difference time-domain (ADI-FDTD) full-wave solvers for general purpose high-performance computers and graphics processing units. His current research interests include computational electromagnetics, parallel processing and microwave engineering.

Sławomir ORŁOWSKI received the M.Sc. degree and the Ph.D. degree in physics from Nicolaus Copernicus University (NCU), Torun, Poland, in 2004 and 2009, respectively. He is currently working in research projects at the Faculty of Electronics, Telecommunications and Informatics at Gdansk University of Technology, Gdansk, Poland. Between 2008 and 2010 he worked at NCU developing software tools for optical coherence tomography. His current research interests include computational electromagnetics, software design, computer graphics, image and signal processing.

Bartosz REICHEL received the M.Sc degree in applied physics and the Ph.D. degree in theoretical physics from

Gdansk University of Technology (GUT), Gdansk, Poland, in 2004 and 2008, respectively. He is currently an Assistant Professor at the Faculty of Applied Physics and Mathematics at GUT. Since 2013, he is involved in research projects at the Faculty of Electronics, Telecommunications and Informatics at GUT focusing on computational electromagnetics and optimization of algorithms on graphics processing units. Between 2004 and 2008 he worked toward the Ph.D. degree investigating the propagation of short light pulses in waveguides. His current research interests include computational electromagnetics, parallel processing and electromagnetic scattering.

9

An Improved Split-Step Wavelet Transform Method for Anomalous Radio Wave Propagation Modeling

Asif IQBAL, Varun JEOTI

Dept. of Electrical and Electronics Engineering,Universiti Teknologi Petronas, Tronoh, Perak, 31750, Malaysia

asif.iqbal@msn.com, varun_jeoti@petronas.com.my

Abstract. *An Anomalous tropospheric propagation caused by ducting phenomenon is a major problem in wireless communication. Thus, it is important to study the behavior of radio wave propagation in tropospheric ducts. The Parabolic Wave Equation (PWE) method is considered most reliable to model anomalous radio wave propagation. In this work, an improved Split Step Wavelet transform Method (SSWM) is presented to solve PWE for the modeling of tropospheric propagation over finite and infinite conductive surfaces. A large number of numerical experiments are carried out to validate the performance of the proposed algorithm. Developed algorithm is compared with previously published techniques; Wavelet Galerkin Method (WGM) and Split-Step Fourier transform Method (SSFM). A very good agreement is found between SSWM and published techniques. It is also observed that the proposed algorithm is about six times faster than WGM and provide more details of propagation effects as compared to SSFM.*

Keywords

Radio wave propagation, parabolic equation method, Split Step Fourier transform, split step wavelet transform, wavelet Galerkin, tropospheric duct.

1. Introduction

Due to the varying nature of refractive index in troposphere, the radio waves do not follow the straight path as they do in free space. Since the radio refractive index varies spatially and diurnally, it is necessary to carefully take it into account before designing a wireless communication system. The gradient of refractive index causes the formation of layers or ducts. When radio waves are channeled through these ducts, they behave differently from normal environment. This non-standard propagation is also referred to as anomalous propagation. Due to ducting phenomenon and earth surface profile, radio wave is affected by reflection, refraction and diffraction mechanisms. Since the occurrence of these mechanisms is localized and geometry specific, it is very difficult to develop a uniform model to accurately predict the propagation in such varying environments. Various an-

alytical and numerical models have been developed to forecast the behavior of radio wave propagation in such environments. In numerical model, most reliable and widely used technique for tropospheric radio wave propagation modeling is the Parabolic Wave Equation (PWE) method [1]. In 1946, the PWE method was first used for tropospheric radio wave propagation modeling by Leonovich and Fock [2] and then, in 1984, it was extended [3] for the case of a two dimensional inhomogeneous atmosphere. With the passage of time, different versions of PWE are derived for many other areas [4], [5], [6], [7].

PWE is derived from the Helmholtz equation by taking into account only the forward propagation. As a full-wave model, PWE models have certain unique advantages. First, PWE can handle the refraction and diffraction effects simultaneously. Therefore, it is not only simple in calculations, the accuracy level is also high. Second, PWE can efficiently model the electromagnetic field distribution over irregular surfaces under non-uniform distribution of the refractive index of atmospheric structure. Third, PWE model results in iterative algorithms. Therefore, PWE model can be used for regional forecast of propagation path loss.

The vast application and extensive research shows the relative importance of the PWE model [4],[5], [6], [8]. It should be noted that the robustness of PWE solution depends on numerical technique used. For a given application, the selection of best numerical technique is warranted to provide the accurate results.

Due to complex boundary conditions and channel properties over earth surface, the rigorous solution of PWE is a challenging task and it has been investigated by many researchers for many years. PWE became an essential tool for modeling wave propagation after the development of SSFM by Tappert and Hardin [8]. SSFM takes the advantage of computationally efficient Fast Fourier Transform (FFT) algorithm [9], [10], [11]. In SSFM, diffraction and refraction effects are treated separately while it also allows larger range increments. The later property makes this algorithm attractive for large domain solutions. On the other hand, Finite difference (FD) method and Finite Element (FE) Method are also quite popular for solving PWE [4], [12], [13], [14], [15], [16]. Effective boundary handling makes FD and FE methods attractive to the numerical solution of PWE. However,

FD and FE methods are resource hungry in term of memory and processing for non-standard atmospheric conditions.

Besides the conventional numerical techniques mentioned above, another strong contender for the numerical solutions of Partial Differential Equations (PDEs) is the wavelet based methods. Briefly, a wavelet is a short duration oscillatory mathematical function. In 1980s, Morlet and Grossman used a French word ondeltte for these functions. Later, it was translated to wavelet by translating onde into wave [17]. Ingrid Daubechies constructed the class of compactly supported scaling and wavelet function in 1988 [18]. In early nineties, the attractive properties of wavelets had drawn researcher attention to the application of wavelets for numerical solution of PDEs. Afterwards, numerous linear and non-linear problems have been solved using wavelet based numerical technique. It was proved that the best localization properties of wavelets lead to efficient numerical method especially for the problem that involves the formation of shock, hurricane and turbulence etc. [19], [20], [21]. Normally, wavelet based techniques can be classified into three types: methods based on scaling function expansion, methods based on wavelet expansion and methods based on wavelet optimized finite differences [22].

Recently authors of this paper have developed a wavelet method based on scaling function expansion namely WGM for the solution of PWE [23]. But from numerical experiments, it is observed that WGM has similar characteristics as FE Method. It does not allow larger range increments for higher frequencies which leads to higher processing load. Due to the computational complexity problem encountered in WGM, the focus is now shifted towards SSWM. SSWM is also one of the popular techniques to model non-linear optical pulse propagation [24], [25]. SSWM is also used for the application of underwater acoustic propagation [26]. It is found that the complexity of SSWM is minimum as compared to SSFM and SSWM is considered to be more accurate and computationally more efficient for large domain solutions. Authors of this paper also showed the feasibility of SSWM for the solution of PWE for Perfectly Electric Conductive (PEC) ground surfaces [27], [28].

In this work, an improved formulation of SSWM is presented. The formulation of [27], [28] is generalized for impendence boundary conditions using Discrete Mixed Fourier Transform (DMFT). Unlike other sub-domain solutions, the proposed method not only allows larger range steps but also provides more accurate solution. It is seen to be as fast as SSFM.

This work mainly focuses on the modeling of tropospheric radio wave propagation under ducting phenomenon using wavelets based numerical technique. The model under consideration is a two-dimensional standard PWE. Furthermore, finitely conductive flat earth surface is assumed under range dependent/independent environment. The propagation path loss is computed for range independent and dependent environment cases. We also consider standard and ducting atmospheric conditions.

Since this work is still a fundamental research for the application of wavelet to model radio wave propagation, the topics like time dependent PWE, three dimensional (3D) PWE and/or wide angle PWE and experimental validation of proposed path loss propagation model are beyond the scope of this study. The rough surface modeling is also not studied in this work. Only the wavelet methods based on scaling function expansion are employed to solve PWE. Daubachies family of wavelet is used throughout this work. The application of other classes of wavelet and the comparison between different families of wavelet to model radio wave propagation is also beyond the scope of this study.

This paper is organized as follows: Problem formulation of tropospheric boundary value problem is given in section 2. A detailed SSWM formulation is given in section 3. Section 4 includes the numerical implementation of the SSWM algorithm. At the end, results are compared with those from WGM and SSFM for various environment conditions.

2. Problem Statement: Tropospheric Boundary Value Problem

To develop the outline of the method briefly, a fairly simple problem is selected to formulate. The model under consideration here is a two-dimensional standard PWE for the case of Tropospheric radio propagation over flat surfaces. The paraxial form of scalar wave equation in Cartesian coordinates system is given by [1]

$$\frac{\partial^2 u}{\partial z^2} - 2jk_0\frac{\partial u}{\partial x} + k_0^2(n^2-1)u = 0, \quad (1)$$

$$Z_{min} \leq z \leq Z_{max}, \; x \geq 0$$

where $k_0 = 2\pi/\lambda$ is the wave number in vacuum and m is the modified refractive index. Here, x-axis is the direction of propagation while z-axis is the height above ground level. The geometry of PWE problem for radio wave propagation modeling and range dependant refractivity profiles are illustrated in Fig. 1 (a) and (b) respectively. The limits of height and range axis are defined as; $Z_{min} \leq z \leq Z_{max}$ and $x \geq 0$. The whole domain is discretized into N_z no. of small grids. In Fig. 1, Δx and Δy are the discretizing size for range and height respectively. Bottom surface (Z_{min}) is finite or infinite conductive flat surface while an absorbing layer at the top of 'domain of interest' (Z_{req}). The bottom boundary is flat earth surface.

2.1 Boundary Conditions

Impendence Boundary Conditions:

Equation (1) should satisfy the boundary condition to handle the electromagnetic field at the earth surface. The boundary conditions at surface of a smooth, finitely conducting earth can be approximated by [16]

$$\alpha_1(x)\frac{\partial}{\partial z}u(x,z)|_{z=0} + \alpha_2(x)u(x,z)|_{z=0} = 0 \quad (2)$$

where $\alpha_1(x)$ and $\alpha_2(x)$ are constants. For a perfectly conducting surface, $\alpha_1(x) = 0$ (Dirichlet BC) and $\alpha_2(x) = 0$ (Neumann BC) for horizontal and vertical polarization respectively. For finitely conducting earth surface, $\alpha_1(x) = 1$, while $\alpha_2(x) = (j/\mu_0\omega)\eta$ for horizontal polarization and $\alpha_2(x) = -j\omega\varepsilon_s\eta$ for vertical polarization. Here, ε_s is the permittivity of the surface medium, μ_0 is the free-space permeability, and ω is the radial frequency.

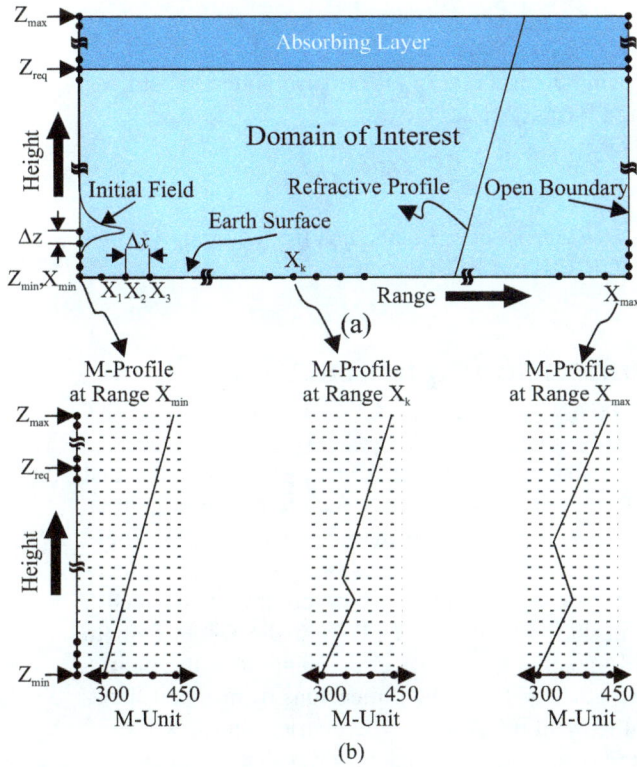

Fig. 1. (a) Geometry of the problem for radio wave propagation modeling (b) Refractivity profiles at specified ranges.

Absorbing Boundary Conditions

For absorbing boundary conditions, the maximum height of problem domain is infinite but only finite height can be handled with numerical methods. Truncation of the domain at finite height can cause strong artificial reflections. In order to stop these non-physical reflections, an absorbing layer, either perfectly matched layer (PML) termination, or non-local boundary condition (NLBC) is used [29], [1]. The absorbing layer in turn is implemented via adding a complex part to the refractive index or by using a window function.

A simple but effective filter is given by the Hanning window of the form of [1]

$$w_h(t) = \frac{1+\cos(\pi t)}{2}. \quad (3)$$

The Hanning window provides a smooth absorptive properties, since, $w_h(0) = 1$ and $w_h(1) = 0$, the derivative at the end points are zero. To make the absorption layers effective, the ratio of absorbing layer height and range increment need to be maximized. More details on domain truncation can be found in [1]. In this work, both complex part to the refractive index and window function is used to truncate the domain above 'domain of interest'.

3. Split-Step Wavelet Method

The SSWM methods have relatively same features as WGM [23] in discretizing altitude operator. However, in terms of the numerical implementation and boundary handling, both algorithms are different when applied to radio wave propagation modeling. The major difference between the SSWM and WGM is that SSWM uses image theory method with periodic wavelet functions, whereas WGM uses the fictitious domain approach with non-periodic wavelet functions. In the formulation of split-step wavelet method, the geometry of problem is modified by using image theory as shown in Fig. 2.

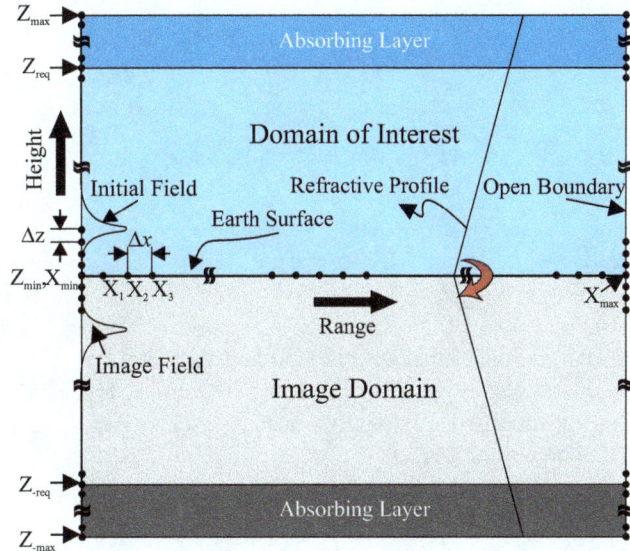

Fig. 2. Solution domain for SSWM.

Now the domain of integration with respect to z is changed from $[0 \quad Z_{max}]$ to $[Z_{-max} \quad Z_{max}]$. Due to symmetric extension of field and absorbing layer, we can write

$$u(x,Z_{-max}) = u(x,Z_{max}) \approx 0. \quad (4)$$

The equation mentioned above satisfies the definition of periodic boundary condition and it allows us to use integration over the whole z axis. However, due periodicity we can restrict to spatial domain to the interval $[Z_{-max} \quad Z_{max}]$.

Let us define the wavelet expansion for unknown function $u(x,z)$ in discrete form as

$$u(x,z) = \sum_{l=0}^{N_z-1} a_l(x)\varphi(z-l); \ z \in [Z_{-\max}, Z_{\max}] \quad (5)$$

where $a_l(x)$ are the unknown coefficient, $\varphi(z)$ the scaling function. In a fashion similar to WGM we obtain

$$\int_{Z_{-\max}}^{Z_{\max}} \varphi_k \left(\frac{\partial^2 u}{\partial z^2} + 2jk_0 \frac{\partial u}{\partial x} + k_0^2 (m^2-1) u \right) dz = 0. \quad (6)$$

After substituting the wavelet expansion of $u(x, z)$ into (6) and rearranging, we have

$$\sum_l \frac{\partial a_l(x)}{\partial x} \int_{Z_{-\max}}^{Z_{\max}} \varphi_k \varphi_l dz + \sum_l a_l(x) \frac{-j}{2k_0} \int_{Z_{-\max}}^{Z_{\max}} \varphi_k \frac{\partial^2 \varphi_l}{\partial z^2} dz$$

$$+ \sum_l a_l(x) \frac{-jk_0}{2} \int_{Z_{-\max}}^{Z_{\max}} (m^2-1) \varphi_k \varphi_l dz = 0. \quad (7)$$

In matrix notation, (7) can be written as

$$[I_{k,l}]\{\partial a_l(x)/\partial x\} + [L_{k,l} + S_{k,l}]\{a_l(x)\} = 0 \quad (8)$$

where

$$I_{k,l} = \delta_{k,l} = \int_{Z_{-\max}}^{Z_{\max}} \varphi_k \varphi_l dz \,,$$

$$L_{k,l} = \frac{-j}{2k_0} \int_{Z_{-\max}}^{Z_{\max}} \varphi_k \frac{\partial^2 \varphi_l}{\partial z^2} dz = \frac{-j}{2k_0} \left(\Omega_l^{0,2} \right) \,,$$

$$S_{k,l} = -\frac{jk_0}{2} \int_{Z_{-\max}}^{Z_{\max}} (m^2-1) \varphi_k \varphi_l dz,$$

$\delta_{k,l}$ is known as Kronecker delta function and $\Omega_l^{0,2}$ are the connection coefficient as described in previous sections. By considering the modified refractivity constant over an element, $S_{k,l}$ can be written as [30],

$$S_{k,l} = -\frac{jk_0}{2}(m^2-1) \int_{Z_{-\max}}^{Z_{\max}} \varphi_k \varphi_l dz. \quad (9)$$

The problem given in (7) is an initial value problem. The split-step method derives from the fact that the solution of problem (7) satisfies the identity

$$\{a_l(x+\Delta x)\} = \exp(L+S)\Delta x \{a_l(x)\}. \quad (10)$$

The exponential operator given in (10) can be split in two different ways, either

$$\{a_l(x+\Delta x)\} = \exp(L\Delta x)\exp(S\Delta x)\{a_l(x)\} \quad (11)$$

called asymmetric splitting or

$$\{a_l(x+\Delta x)\} = \exp(L\frac{\Delta x}{2})\exp(S\Delta x)\exp(L\frac{\Delta x}{2})\{a_l(x)\} \quad (12)$$

called symmetric splitting. It has been shown in [31] that the asymmetrical splitting is accurate to the order $O(\Delta x^2)$ while the symmetrical splitting is accurate to the order $O(\Delta x^3)$. More details on splitting the exponential operator. On the other hand, the accuracy of altitude operator is associated with the moments (M) of the chosen wavelet. In this formulation we are going to use asymmetric splitting and 3rd order moment Daubechies wavelets.

As $S_{k,l}$ is diagonal matrix, the exponential of $S_{k,l}$, required in (12), can be solved cheaply. However, since $L_{k,l}$ is not a diagonal matrix, instead it is a circulant matrix, there is a need to compute the exponential of operator $L_{k,l}$. Let,

$$P = e^Q; \ Q = L^{\Delta x}/2. \quad (13)$$

By using the fact that $L_{k,l}$ is circulant, one can compute (13) using FFT as [22]

$$P = \mathcal{F}^{-1} \exp(\Lambda_q) \mathcal{F} \quad (14)$$

where $\Lambda_q = diag(\widehat{q})$, $\widehat{q} = \mathcal{F}q$ and q is first column in Q matrix, and $\mathcal{F}$ is FFT.

4. Numerical Implementation of SSWM

In the formulation of SSWM, it was explained that the domain of computation is assumed to be periodic in an interval $[Z_{-\max} \ Z_{\max}]$. The matrices given in (8) should be scaled according to physical space. A detailed procedure for physical space mapping of differentiation matrix is given in the Section 7.4 of [22]. If the whole domain is divided in to N_z number of grids points then the indices $l = k = 0,\ldots,N_z-1$ and the dimensions of resultant linear system of (8) will be $N_z \times N_z$. The matrices in (8) will have a following structure for Daubechies wavelet of length 6

$$L_{k,l} = \frac{1}{(2Z_{\max})^2}
\begin{bmatrix}
\Omega_0^{0,2} & \Omega_{-1}^{0,2} & \cdots & \Omega_{-4}^{0,2} & \cdots & \Omega_4^{0,2} & \cdots & \Omega_1^{0,2} \\
\Omega_1^{0,2} & \Omega_0^{0,2} & \cdots & \Omega_{-3}^{0,2} & \cdots & 0 & \cdots & \Omega_2^{0,2} \\
\vdots & \vdots & \vdots & \vdots & \vdots & \vdots & \vdots & \vdots \\
\Omega_4^{0,2} & \Omega_3^{0,2} & \cdots & \Omega_0^{0,2} & \cdots & 0 & \cdots & 0 \\
0 & \Omega_4^{0,2} & \cdots & \Omega_1^{0,2} & \cdots & 0 & \cdots & 0 \\
\vdots & \vdots & \vdots & \vdots & \vdots & \vdots & \vdots & \vdots \\
0 & 0 & \cdots & 0 & \cdots & \Omega_1^{0,2} & \cdots & \Omega_{-4}^{0,2} \\
\Omega_{-4}^{0,2} & 0 & \cdots & 0 & \cdots & \Omega_0^{0,2} & \cdots & \Omega_{-3}^{0,2} \\
\vdots & \vdots & \vdots & \vdots & \vdots & \vdots & \vdots & \vdots \\
\Omega_{-1}^{0,2} & \Omega_{-2}^{0,2} & \cdots & 0 & \cdots & \Omega_3^{0,2} & \cdots & \Omega_0^{0,2}
\end{bmatrix}_{N_z \times N_z},$$

$$S_{k,l} = diag \begin{bmatrix} S_{0,0} \\ \vdots \\ S_{N_z-1,N_z-1} \end{bmatrix}_{N_z \times N_z}, \ I_{k,l} = [I]_{N_z \times N_z},$$

and

$$a_l(z) = [a_0 \quad \cdots \quad a_{N_z-1}]^T_{N_z \times 1}.$$

A symmetric extension of refractive profile is taken in extended region at bottom, as shown in Figure 3.3. We can write

$$S_{\hat{k},\hat{l}} = S_{k,l}; \ k,l = (\tfrac{N_z}{2}),...,N_z\text{-}1; \ \hat{k},\hat{l} = (\tfrac{N_z}{2}-1),...,0. \quad (15)$$

Similarly, the initial field is computed at $x = 0$. To satisfy the boundary condition over perfectly conducting surfaces in SSWM, unknown coefficients ($\{a_l(x)\}$) are computed for Dirichlet and Neumann BCs in accordance with image theory. Computational cost can be reduced by using discrete sine or cosine transforms (DST or DCT) for Dirichlet and Neumann BCs, respectively. Since, SSWM is inspired from SSFM, boundary handling cannot be done using conventional methods like fictitious domain approach and capacitance matrix method [21], [32], [33], [32], [23]. At this stage, no other method is available for handling the finite boundary conditions in SSWM. In order to account of finitely conductive surfaces, boundary handling is inspired from DMFT that was basically developed for SSFM to incorporate the impedance boundary conditions.

4.1 Discrete Mixed Fourier Transform (DMFT)

DMFT is basically developed to incorporate the boundary condition in the solution of SSFM [34]. An overview of DMFT formulation is given as follows:

Let us define an auxiliary function $w(x,z)$ for impedance boundary condition given in (2) by

$$w(x,z) = \frac{\partial u(x,z)}{\partial z} + \alpha u(x,m\Delta z) \quad 0 \le z \le \infty. \quad (16)$$

Backward difference formula is used to solve (16). The discrete form of (16) can be written as

$$w(x,m\Delta z) = \frac{u(x,m\Delta z) - u[x,(m-1)\Delta z]}{\Delta z} + \alpha u(x,m\Delta z) \quad (17)$$

where $m = 1...N-1$, with $w(x,0) = w(x,N\Delta z) = 0$ and Δz is the grid size for height operator. Let $r = (1+\alpha\Delta z)^{-1}$. Equation (17) can be written as

$$w(x,m\Delta z) = u(x,m\Delta z) - r[u(x,(m-1))\Delta z] \quad (18)$$

where, $m = 1...N-1$. DST of $w(x,m\Delta z)$ yields

$$W(x,g\Delta p) = \sum_{m=1}^{N-1} w(x,m\Delta z)sin\frac{gm\pi}{N}. \quad (19)$$

Inverse DST of $W(x,g\Delta p)$ is given by

$$w(x,m\Delta z) = \frac{2}{N}\sum_{g=1}^{N-1} W(x,g\Delta p)sin\frac{gm\pi}{N}. \quad (20)$$

Solution of (16) is given by

$$u(x,m\Delta z) = u_p(x,m\Delta z) + A(x)r^m. \quad (21)$$

Where the particular solution $u_p(x,m\Delta z)$ can be found by setting $u_p(0) = 0$ and using

$$u_p(x,m\Delta z) = w(x,m\Delta z) + ru_p[(x,(m-1))\Delta z]. \quad (22)$$

To completely specify the backward difference DMFT, the coefficient $A(x)$ can be computed as

$$A(x) = C(x) - G\sum_{m=0}^{N}{}'r^m u_p(x,m\Delta z), \quad (23)$$

$$C(x) = \sum_{m=0}^{N}{}'u(x,m\Delta z)r^m, \quad (24)$$

$$G = \frac{2(1-r^2)}{(1+r^2)(1-r^{2N})}. \quad (25)$$

$\sum'$ indicates that the $m = 0$ and $m = N$ are weighted with a factor of 1/2. The detailed formulation of DMFT can be found in [34], [35].

4.2 Use of DMFT Algorithm in SSWM

The step by step procedure to implement the DMFT is given as follows:

(1) Compute the coefficients $a_l(0,z)$ using initial field [23].
(2) Construct the discrete auxiliary function $w(x,m\Delta z)$ using (18).
(3) Perform DST of $w(x,m\Delta z)$ to compute $W(x,g\Delta p)$.
(4) Compute free space propagation using procedure given in (14) and extract one side of field.
(5) Multiply the free space propagator to obtain $W(x+\Delta x, g\Delta p)$.
(6) The coefficients $C(x)$ must also be propagated to the new range step

$$C(x+\Delta x) = C(x)\exp\left[i\Delta x\sqrt{k_0^2 + \left(\frac{\ln r}{\Delta z}\right)^2}\right].$$

(7) Perform the Inverse DST to obtain $w(x+\Delta x, m\Delta z)$.
(8) Solve for $u(x+\Delta x, m\Delta z)$ using (21)-(23).
(9) Finally, multiply the field distribution $u(x+\Delta x, m\Delta z)$ with environment propagator.
(10) Repeat step 2 to 9, until maximum required range achieved.
(11) After reaching the required range step, the required coefficients will be extracted from the extended domain solution.

It should be noted that, for perfectly conducting surfaces, DCT or DST can be used for vertical or horizontal polarized initial field respectively.

In summary, overall algorithm for SSWM implementation is given in Fig. 3.

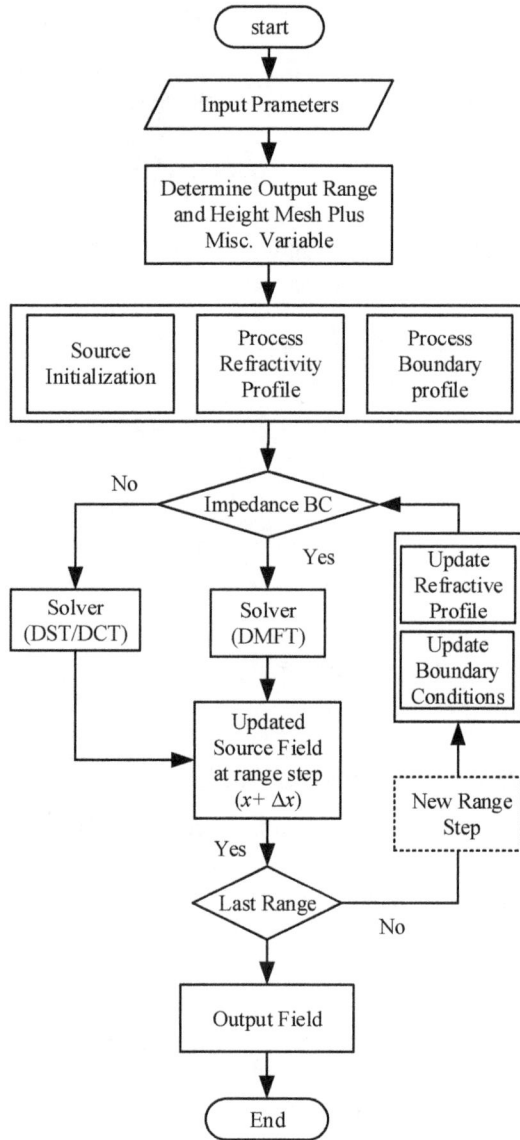

Fig. 3. SSWM Implementation Algorithm for PWE.

5. Results and Discussion

Simulation Experiments:

To illustrate the performance of proposed wavelet based techniques, three major types of numerical experiments, similar to [23], are performed to estimate the path loss in troposphere.

- In the first experiment, we take case of commonly available standard environment condition with Perfectly Electrical Conducting (PEC) BC. Though earth

is not PEC but for high conductive surfaces e.g. water it is justified.

- Second, we have chosen a case of propagation over water. Since, the evaporation duct exists almost all of the time, it is important to predict the behavior of radio wave propagation to design an effective maritime communication system. A detailed study is performed to examine the effect of evaporation duct height and antenna height.

SSWM results are validated by comparing with those of SSFM and WGM [23]. The testing performed here is just to show that the particular implementation of the SSWM, for modeling and simulation of tropospheric radio wave propagation, is valid and the models are suitable for their intended purpose within reasonable bound of accuracy. To show the relative difference between SSWM results and published techniques, Mean Relative Squared Difference (MRSD) is computed for each case with SSFM and WGM data. The formula used to calculate MRSD is given by [36],

$$MRSD =$$
$$\frac{1}{n}\sum_{i=0}^{k}\left(\frac{(SSFM \ or \ WGM \ Data_i) - (SSWM \ Data_i)}{SSFM \ or \ WGM \ Data_i}\right)^2. \quad (26)$$

5.1 Test Case 1: Propagation over Perfectly Electrical Conducting Earth Surface in Standard Environment

Simulation Setup:

In the first case, path loss results for 100 km transmission range over Perfectly Electrical Conducting (PEC) earth surface are obtained for standard environment at 5.8 Ghz. A refractivity profile used in this simulation is specified by a linear function as [37]

$$M(z) = 326.615 + 0.121433z, \qquad 0 \le z \le 100 \quad (27)$$

where z is measured in meters. Horizontally polarized transmitter antenna height is chosen at 25 m above the ground level. Beam-width is set to be 3 degrees. The vertical grid size is taken 0.054 m. The range increment is taken to be 125 m for SSFM and SSWM, and 1 m for WGM. The range increment for WGM is taken to be very small to avoid numerical oscillation problems [12]. The bottom boundary is assumed to be perfectly conducting flat earth surface. All simulations for SSWM and WGM are carried out with Daubechies wavelet of length 6. The parameters used in this simulation are summarized in Tab. 1.

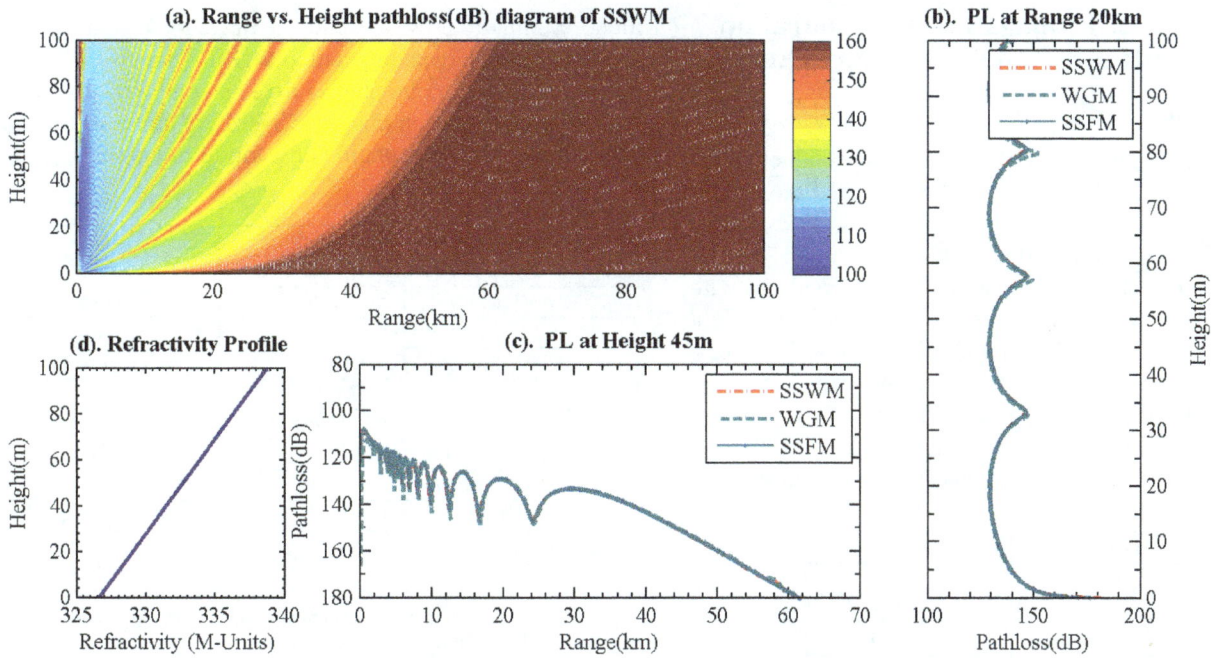

Fig. 4. (a) Range (km) vs. height (m) path loss (dB) diagram for standard environment. (b) Path loss vs. height for range of 20 km. (c) Path-loss vs. range for height of 45 m. (d) Standard environment refractive profile.

Parameter		Value
Frequency		5.8 GHz
Transmission range		100 km
Maximum height		100 m
Altitude grid size (Δz)		0.054 m
Range grid size Δx	SSFM, SSWM	125 m
	WGM	1 m
Antenna heigth		25 m
Beamwidth		3°
Bottom boundary		PEC
Wavelet Class	SSWM, WGM	Daubechies wavelet (D=6)

Tab. 1. Simulation parameters for propagation over PEC Earth surface in standard environment.

Discussion:

A 3D coverage map of height versus range propagation path loss (dB) over flat earth surface is shown in Fig. 4 (a). In Fig. 4 (a), the results from SSWM are presented and the results from other methods are used only for detailed comparison. The color code bar on the right side of Fig. 4 (a) shows the range of path loss (dB), i.e. [100 160]. For detailed comparison between developed technique, SSFM and WGM, the path loss data are extracted from the results obtained using different numerical schemes for the specific range and height and presented in Fig. 4 (b) and (c). Environment profile is shown in Fig. 4 (d).

From Fig. 4 (a), it can be seen that, for the case of standard refractivity gradient, energy moves away from the earths surface and the radio horizon is only few kilometers. Due to this, energy attenuates fast and beyond the horizon communication becomes a challenging task. A comparison

of propagation path loss results at range 20 km is shown in Fig. 4 (b). The results from both SSWM show a very good match with those from SSFM and WGM. Fig. 4 (c) illustrates the comparison of path loss versus range at height 45 m. A strong interference and deep nulls can be seen for the first few kilometers because of the reflections from the earth surface. This region is normally referred to as interference region. After 30 km, or beyond the horizon, the attenuation slope is very high. In this region, the major part of energy comes from the diffraction phenomenon. Due to this, it is referred to as diffraction region. In diffraction region, the attenuation reaches approximately 200 dB for standard environment. Therefore, a good communication after few kilometers is not possible for low altitude receiver antenna heights. Communication range can only be increased by increasing the altitude of our receiver or transmitter antenna.

For performance evaluation of the newly developed techniques SSWM, the MRSD is computed using (26) for the cases given in Fig. 4 (b) and (c). It is found that the MRSD is of the order of 10^{-5}, the detail of MRSD results is presented in Tab. 4.

It is also observed that, for large range increment sizes,

SSWM can produce good results while WGM suffers from oscillation problem, due to which the range increment size for WGM is taken quite small compared to SSWM. It is also observed that SSWM is about 18 times faster then WGM and the computation cost of SSWM is no different from that of SSFM and being a sub-domain solution, SSWM takes all advantages of WGM, as shown in Tab. 5.

5.2 Test Case 2: Propagation in Evaporation Duct over Finite Conductive Earth Surface

Simulation Setup:

This simulation is performed over finite conductive sea surface using 10.5 GHz. And we demonstrated the behavior of radio wave propagation in evaporation duct as shown in Fig. 5. The evaporation duct profile used in this simulation is given in Tab. 2. Simulations were carried out with horizontally polarized transmitter antenna with 15 m height above the ground level. Beam-width is set to be 2 degrees. The vertical grid size is taken 0.054 m. The range increment is 125 m for SSFM and SSWM and 1 m for WGM. The bottom boundary is assumed to be finitely conducting flat sea surface. The parameters used in this simulation are summarized in Tab. 3.

Height (m)	M-Units
0	357.021
0.135	334.332
0.223	332.730
0.368	331.169
0.607	329.673
1	328.273
1.649	327.007
2.718	325.920
4.482	325.061
7.389	324.488
11.76	324.293
12.182	324.294
20.086	324.623
33.115	325.720
54.598	328.010
100	332.186

Tab. 2. Refractive profile for evaporation duct.

Discussion:

Once more, a comparison between results is presented. In Fig. 5 (a), 3D converge map using SSWM shows the strong trapping of signals in evaporation duct. Fig. 5 (b) and (c) shows the detailed comparison of AREPS and from SSWM and WGM data at specified range and height. Fig. 5 (b) illustrates the path loss versus height at range 35 km while Fig. 5 (c) shows the path loss versus range at height 15 m. Environment profile used in this simulation is shown in Fig. 5 (d).

Again, for performance evaluation, MRSD is computed for SSWM using (26) for the cases given in Fig. 5 (b) and (c). Results given in Tab. 4 show that MRSE is of the order of 10^{-4}. Hence, a very good agreement between the results is found. Similar to previous case, the range increment size for SSWM is quite large compared to WGM. Therefore, SSWM saved much more computation time as compared to WGM while the computation cost of SSWM is similar to SSFM. The summary of computation cost for all test cases is given in Tab. 5.

Parameter		Value
Frequency		10.5 GHz
Transmission range		100 km
Maximum height		100 m
Altitude grid size (Δz)		0.054 m
Range grid size Δx	SSFM, SSWM	125 m
	WGM	1 m
Antenna heigth		15 m
Beamwidth		2°
Bottom boundary		Sea Surface
Wavelet Class	SSWM, WGM	Daubechies wavelet (D=6)

Tab. 3. Simulation parameters for propagation in evaporation duct over finite conductive Earth surface.

From Fig. 5 (a), it can also be seen that the radio waves are highly affected by this natural waveguide. In this case too, the same regions of interference, diffraction and troposcatter are present as in surface duct. In Fig. 5 (b), it can be observed that the propagation path loss is about 140 dB at observation height of 4 m and remains very stable for long distances. Fig. 5 (c) also shows another interesting feature of evaporation duct that, unlike previous cases, the interference region is shrunken to few kilometers and a very smooth attenuation can be observed with a rate much less than standard atmosphere. It also indicates that evaporation duct can be suitable for radio links using receiver antenna at few meters altitude. As the percentage occurrence of evaporation duct is more frequent as compared to surface-based and elevated duct.

Test Case	MRSD w.r.t.			
	SSFM		WGM	
	Range	Height	Range	Height
1	3.2×10^{-03}	1.2×10^{-05}	3.6×10^{-04}	1.1×10^{-04}
2	8.1×10^{-05}	1.0×10^{-04}	1.2×10^{-04}	5.1×10^{-05}

Tab. 4. Summary of MRSE results.

Test Case	Computation Time (Sec)		
	SSWM	SSFM	WGM
1	8.71	7.21	149.57
2	8.62	7.23	150.44

Tab. 5. Comparison of computation time for SSWM, SSFM and WGM.

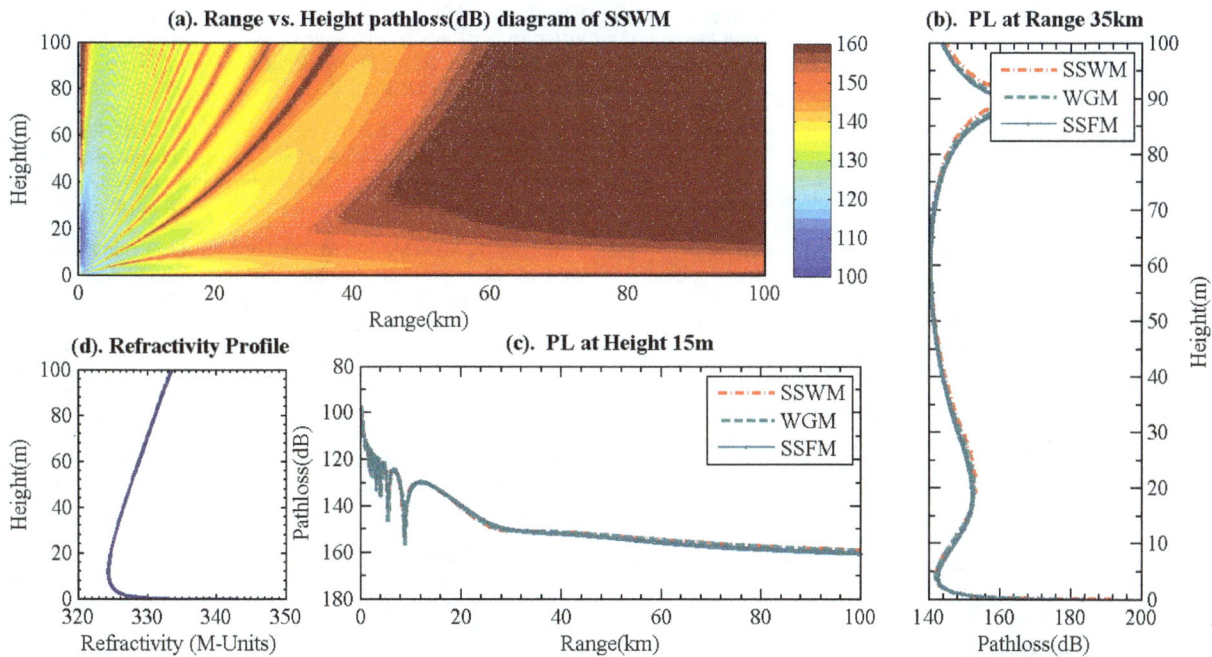

Fig. 5. (a) Range (km) vs. height (m) path loss (dB) diagram for evaporation duct. (b) Path loss vs. height for range of 35 km. (c) Path-loss vs. range for height of 15 m. (d) Evaporation duct refractive profile.

It can be observed from Tab. 5, the computation time for SSWM and SSFM is almost same. However, SSWM has better localization properties as compared to SSFM where the basis functions are entire-domain functions. Sub-domain basis functions gives us the opportunity to handle discontinuities more efficiently with same computational load while Fourier solutions may require very high order of harmonics to achieve the similar accuracy. It is also possible to extend sub-domain solution to adaptive gridding which may not possible in Fourier based solutions. The feasibility of SSWM for PWE also opens the door for hierarchical solutions for desired level of accuracy with less computations cost as compared to other finite element or difference methods.

6. Conclusion

In this paper, an improved SSWM is presented for the numerical solution of two dimensional PWE. It incorporate DCT/DST and DMFT for PEC and non-PEC boundary conditions respectively. The performance of SSWM is demonstrated for standard and ducting environment conditions and results were compared with those from AREPS. A strong agreement is found for all environment conditions. From results, it is also found that unlike other sub-domain solutions, the proposed method not only allows larger range steps but it also provides more accurate solution almost as fast as SSFM. In conclusion, this work can be a significant step towards the improvement of propagation model to predict the anomalous propagation in troposphere.

References

[1] LEVY, M. *Parabolic Equation Methods for Electromagnetic Wave Propagation.* Inst. of Engineering & Technology, 2000.

[2] LEONTOVICH, M., FOCK, V. Solution of the problem of propagation of electromagnetic waves along the Earth's surface by the method of parabolic equation. *Acad. Sci. USSR. J. Phys*, 1946, vol. 10, p. 13–24.

[3] KO, H., SARI, J., THOMAS, M., HERCHENROEDER, P., MARTONE, P. Anomalous propagation and radar coverage through inhomogeneous atmospheres. *AGARD Characteristics of the Lower Atmosphere Influencing Radio Wave Propagation 14 p(SEE N 84-24943 15-32)*, 1984.

[4] CLAERBOUT, J. *Fundamentals of Geophysical Data Processing.* New York: McGraw Hill, 1975.

[5] CLAYTON, R., ENGQUIST, B. Absorbing boundary conditions for acoustic and elastic wave equations. *Bulletin of the Seismological Society of America*, 1977, vol. 67, no. 6, p. 1529–1540.

[6] COLE, J. Modern developments in transonic flow. *SIAM Journal on Applied Mathematics*, 1975, vol. 29, no. 4, p. 763–787.

[7] HASEGAWA, A., TAPPERT, F. Transmission of stationary nonlinear optical pulses in dispersive dielectric fibers. i. anomalous dispersion. *Applied Physics Letters*, 1973, vol. 23, no. 3, p. 142–144.

[8] TAPPERT, F., HARDIN, R. Application of the split-step Fourier method to the numerical solution of nonlinear and variable-coefficient wave equations. *SIAM Review*, 1973, vol. 15, p. 423.

[9] COOLEY, J., TUKEY, J. An algorithm for the machine calculation of complex Fourier series. *Mathematics of Computations*, 1965, vol. 19, no. 90, p. 297–301.

[10] SINGLETON, R. An algorithm for computing the mixed radix fast Fourier transform. *IEEE Transactions on Audio and Electroacoustics*, 1969, vol. 17, no. 2, p. 93–103.

[11] OPPENHEIM, A., WEINSTEIN, C. Effects of finite register length in digital filtering and the fast Fourier transform. *Proceedings of the IEEE*, 1972, vol. 60, no. 8, p. 957–976.

[12] ISAAKIDIS, S., XENOS, T. Parabolic equation solution of tropospheric wave propagation using FEM. *Progress In Electromagnetics Research*, 2004, vol. 49, p. 257–271.

[13] DESHPANDE, V., DESHPANDE, M. Study of electromagnetic wave propagation through dielectric slab doped randomly with thin metallic wires using finite element method. *IEEE Microwave and Wireless Components Letters*, 2005, vol.15, no. 5, p. 306–308.

[14] ORAIZI, H., HOSSEINZADEH, S. A novel marching algorithm for radio wave propagation modeling over rough surfaces. *Progress In Electromagnetics Research*, 2006, vol. 57, p. 85-100.

[15] ARSHAD, K., KATSRIKU, F., LASEBAE, A. An investigation of wave propagation over irregular terrain and urban streets using finite elements, *Proceedings of the 6th WSEAS Int. Conference on Telecommunications and Informatics*, 2007, p. 105-110.

[16] APAYDIN, G., SEVGI, L. Numerical investigations of and path loss predictions for surface wave propagation over sea paths including hilly island transitions. *IEEE Transactions on Antennas and Propagation*, 2010, vol. 58, no. 4, p. 1302–1314.

[17] MEHRA, M., KEVLAHAN, N. An adaptive wavelet collocation method for the solution of partial differential equations on the sphere. *Journal of Computational Physics*, 2008, vol. 227, no. 11, p. 5610-5632.

[18] DAUBECHIES, I. Orthonormal bases of compactly supported wavelets. *Communications on Pure and Applied Mathematics*, 1988, vol. 41, no. 7, p. 909–996.

[19] GLOWINSKI, R., LAWTON, W., RAVACHOL, M., TENENBAUM, E. Wavelet solutions of linear and nonlinear elliptic, parabolic and hyperbolic problems in one space dimension. *Computing Methods in Applied Sciences and Engineering*, 1990, p. 55-120.

[20] LIANDRAT, J. *Resolution of the 1D regularized burgers equation using a spatial wavelet approximation*, ICASE Report No. 90-83, 1990.

[21] AMARATUNGA, K., WILLIAMS, J., QIAN, S., WEISS, J. Wavelet-Galerkin solutions for one dimensional partial differential equations. *International Journal for Numerical Methods in Engineering*, 1994, vol. 37, no. 16, p. 2703-2716.

[22] NIELSEN, O. *Wavelets in Scientific Computing*. Ph.D. thesis. Informatics and Mathematical Modelling, Technical University of Denmark, 1998.

[23] IQBAL, A., JEOTI, V. A novel wavelet-Galerkin method for modeling radio wave propagation in tropospheric ducts. *Progress In Electromagnetics Research B*, 2012, vol. 36, p. 35-52.

[24] PIERCE, I., WATKINS, L. Modelling optical pulse propagation in nonlinear media using wavelets. *Proceedings of the IEEE-SP International Symposium on Time-Frequency and Time-Scale Analysis*, 1996, p. 361-363.

[25] PASKYABI, M. B., RASHIDI, F. Split step wavelet Galerkin method based on parabolic equation model for solving underwater wave propagation. *In Proceedings of the 5th WSEAS International Conference on Wavelet Analysis and Multirate Systems*, USA, 2005, p. 1-7.

[26] LANDOLSI, T. Accuracy of the split-step wavelet method using various wavelet families in simulating optical pulse propagation. *Journal of the Franklin Institute*, 2006, vol. 343, no. 4-5, p. 458-467.

[27] IQBAL, A., JEOTI, V. A split step wavelet method for radiowave propagation modelling in tropospheric ducts. *2011 IEEE International RF and Microwave Conference*, 2011, p. 67-70.

[28] IQBAL, A., JEOTI, V. Numerical modeling of radio wave propagation in horizontally inhomogeneous environment using split-step wavelet method. *In 4th International Conference on Intelligent and Advanced Systems (ICIAS)*, 2012, p. 200-205.

[29] ANTOINE, X., ARNOLD, A., BESSE, C., EHRHARDT, M., SCHDLE, A. A review of transparent and artificial boundary conditions techniques for linear and nonlinear schrodinger equations. *Communications in Computational Physics*, 2008, vol. 4, no. 4, p. 729-796.

[30] SIRKOVA, I., MIKHALEV, M. Parabolic wave equation method applied to the tropospheric ducting propagation problem: a survey. *Electromagnetics*, 2006, vol. 26, no. 2, p. 155-173.

[31] KREMP, T. *Split-step wavelet collocation methods for linear and nonlinear optical wave propagation*. Ph.D. thesis, Institute of Photonics and Quantum Electronics, Karlsruhe Institute of Technology, 2002.

[32] QIAN, S., WEISS, J. Wavelets and the numerical solution of partial differential equations. *Journal of Computational Physics*, 1993, vol. 106, no. 1, p. 155-175.

[33] LU, D., OHYOSHI, T., ZHU, L. Treatment of boundary conditions in the application of wavelet-Galerkin method to an SH wave problem. *International Journal of the Society of Materials Engineering for Resources*, 1997, vol. 5, no. 1, p. 15-25.

[34] KUTTLER, J., DOCKERY, G. An improved-boundary algorithm for Fourier split-step solutions of the parabolic wave equation. *IEEE Transactions on Antennas and Propagation*, 1996, vol. 44, no. 12, p. 1592-1599.

[35] KUTTLER, J., JANASWAMY, R. Improved Fourier transform methods for solving the parabolic wave equation. *Radio Science*, 2002, vol. 37, no. 2, p. 1021.

[36] BALAGURUSAMY, E. *Numerical Methods*. New Delhi: Tata McGraw-Hill Pub.Co.Ltd, 1999.

[37] JANASWAMY, R. *A Rigorous Way of Incorporating Sea Surface Roughness Into the Parabolic Equation*, PN, 1995.

About Authors...

Asif IQBAL received his M.Sc. degree in Electrical & Electronics Engineering from Universiti Teknologi Petronas, Malaysia in 2012. Currently, he is full time Ph.D. student at Department of Electrical and Electronics Engineering, Universiti Teknologi Petronas, Malaysia. His research interest include fast and efficient numerical algorithms, radio propagation, wireless channel characterization and modelling. He is currently working on the channel sounder and emulators.

Varun JEOTI received his Ph.D. degree from Indian Institute of Technology Delhi India in 1992. He worked on several sponsored R&D projects in IIT Delhi and IIT Madras during 1980 to 1989. He was a Visiting Faculty in Electronics department in Madras Institute of Technology, Anna University for about 1 year from 1989 to 1990 and joined Delhi Institute of Technology for next 5 years till 1995. He moved to Electrical & Electronic Engineering school of Universiti Sains Malaysia in 1995 and joined Electrical & Electronic Engineering department of Universiti Teknologi PETRONAS in 2001. His research interests are in the area of signal processing, surface acoustic wave (SAW) devices, wireless SAW sensor network and wireless communication for maritime applications besides others.

10

Excessive Memory Usage of the ELLPACK Sparse Matrix Storage Scheme throughout the Finite Element Computations

Gökay AKINCI[1], *A. Egemen YILMAZ*[1], *Mustafa KUZUOĞLU*[2]

[1] Dept. of Electrical and Electronics Engineering, Ankara University, Ankara, Turkey
[2] Dept. of Electrical and Electronics Engineering, Middle East Technical University, Ankara, Turkey

gakinci@ankara.edu.tr, aeyilmaz@eng.ankara.edu.tr, kuzuoglu@metu.edu.tr

Abstract. *Sparse matrices are occasionally encountered during solution of various problems by means of numerical methods, particularly the finite element method. ELLPACK sparse matrix storage scheme, one of the most widely used methods due to its implementation ease, is investigated in this study. The scheme uses excessive memory due to its definition. For the conventional finite element method, where the node elements are used, the excessive memory caused by redundant entries in the ELLPACK sparse matrix storage scheme becomes negligible for large scale problems. On the other hand, our analyses show that the redundancy is still considerable for the occasions where facet or edge elements have to be used.*

Keywords

Finite element method, sparse matrix, edge elements, computational electromagnetics, ELLPACK.

1. Introduction

The Finite Element Method, which has been originally developed for static problems of structural mechanics and initially used by mechanical and civil engineers, is one of the most widely used numerical methods by the scientific community. It was first formulated in 1940s by Courant after a discussion regarding the versatility of piecewise approximations. In the 1950s, Argyris began putting together the many mathematical ideas (domain partitioning, assembly, boundary conditions, etc.) that form the basis of the Finite Element Method for aircraft structural analysis.

In its original form, the method depends on representation and approximate evaluation of continuous scalar functions at the corners of the subdomains (referred as "element"s) of the whole problem domain. Even though this formulation proves to be sufficient in handling most problems in the structural mechanics, it was compulsory to extend and generalize the method in order to compute vector functions. This yielded the definitions of the so-called edge and facet elements. Thanks to the definition of

edge and facet elements, it is possible to solve electromagnetic scattering and radiation problems as well as eddy current problems by means of the Finite Element Method. Tab. 1 enlists which element type is the most suitable one for evaluation of the major functions in electromagnetics.

A major advantage of the Finite Element Method is that it yields sparse matrices throughout the solution process. By means of special storage schemes, it is possible to store and solve very large scale matrices and matrix equations, respectively.

	Node Elements	Edge Elements	Facet Elements	Volume Elements
Types of Represented Functions	Scalar	Vector	Vector	Scalar
Representation Capability of Continuity	Total	Tangential Component	Normal Component	None
Physical Types of Represented Functions	Scalar Potential	Fields, Vector Potentials	Fluxes, Vector Densities	Scalar Densities
Examples from Electromagnetic Theory	Scalar Electric Potential (V or ϕ)	Vector Magnetic Potential **A**, Electric Field Intensity **E**, Magnetic Field Intensity **H**	Magnetic Flux Density **B**, Electric Field Density **D**, Current Density **J**	Charge Density (ρ)

Tab. 1. Element types and their representation capabilities.

ELLPACK sparse matrix storage scheme [1–3] is one of the most widely used and preferred schemes in practice due to its implementation ease. This scheme stores a matrix **A** with size $n \times n$ (1),

$$\mathbf{A} = \begin{bmatrix} a_{11} & 0 & a_{13} & 0 & 0 \\ a_{21} & a_{22} & 0 & 0 & 0 \\ 0 & a_{32} & 0 & a_{34} & a_{35} \\ 0 & 0 & a_{43} & a_{44} & 0 \\ 0 & a_{52} & 0 & a_{54} & a_{55} \end{bmatrix} \quad (1)$$

by means of the following two matrices (2),

$$data = \begin{bmatrix} a_{11} & a_{13} & * \\ a_{21} & a_{22} & * \\ a_{32} & a_{34} & a_{35} \\ a_{43} & a_{44} & * \\ a_{52} & a_{54} & a_{55} \end{bmatrix}, \quad indices = \begin{bmatrix} 1 & 3 & * \\ 1 & 2 & * \\ 2 & 4 & 5 \\ 3 & 4 & * \\ 2 & 4 & 5 \end{bmatrix} \quad (2)$$

By this definition, both data and indices matrices are of the sizes $n \times m$; where m is the maximum number of nonzero entries in a row of **A**.

The entries shown with symbols "*" are meaningless and redundant values held in the memory. These redundancies are nothing but the aspects referred as the "generosity" of the ELLPACK sparse matrix storage scheme.

We'll carry out an analysis of how generous the ELLPACK sparse matrix storage scheme behaves during the finite element computations [4-6]; in other words, for various element types of various shapes in 2-Dimension (2D) (Fig. 1 and Fig. 2), we'll try to compute the ratio of redundant (or meaningless entries shown with "*") entries in the whole matrix. We'll perform our analysis for node, edge, facet and volume elements for linear triangular [7] and quadrilateral [8] elements.

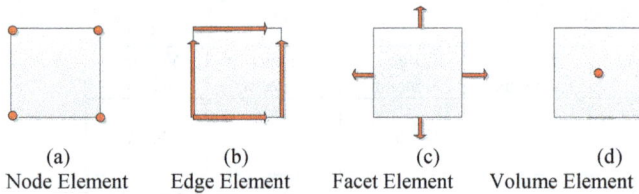

| (a) | (b) | (c) | (d) |
| Node Element | Edge Element | Facet Element | Volume Element |

Fig. 1. Quadrilateral elements.

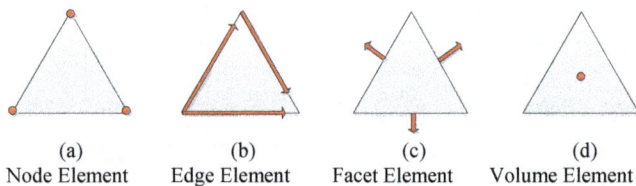

| (a) | (b) | (c) | (d) |
| Node Element | Edge Element | Facet Element | Volume Element |

Fig. 2. Triangular elements.

Assume that, we have a problem domain as shown as Fig. 3(a). In case the problem domain is homeomorphic to a rectangle; this domain in the xy-plane can be mapped to a rectangular domain in a uv-plane, as seen in Fig. 3(b). Hence, we'll carry out our further analysis in the uv-plane, without loss of generality.

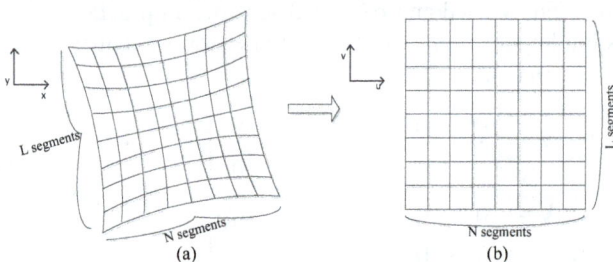

Fig. 3. Problem domain mapped to a regular rectangle in the uv-plane.

2. Quadrilateral Elements

First of all, we'll perform our analysis for quadrilateral node, edge, facet and volume elements, separately.

2.1 Quadrilateral Node Elements

If we consider that this mesh consists of node elements, then the total number of unknowns (i.e. number of nodes) will be $K_{quad\text{-}node} = (N + 1) \times (L + 1)$, which means that the size of the global system matrix obtained throughout the finite element solution will be $K_{quad\text{-}node} \times K_{quad\text{-}node}$.

Let us consider how much memory the ELLPACK sparse matrix storage scheme will allocate for storing this matrix. For this purpose, first, we have to consider the maximum number of nonzero entries in one row of this matrix.

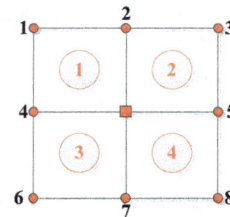

Fig. 4. A node shared by 4 elements.

As seen in Fig. 4, a node can maximally be shared by 4 elements. This means that this node will have 8 neighboring nodes; which means that the row corresponding to this node will have 8 off-diagonal nonzero entries in addition to 1 diagonal nonzero entry. Namely, in this case, the maximum number of nonzero entries in one row will be 9.

Hence data and indices matrices of ELLPACK will be of size $K_{quad\text{-}node} \times 9$. This means that each matrix will require $9 \times (N + 1) \times (L + 1)$ entries in the memory.

2.2 Quadrilateral Edge Elements

If we consider that this mesh consists of edge elements, then the total number of unknowns (i.e. number of edges) will be $K_{quad\text{-}edge} = 2NL + N + L$, which means that the size of the global system matrix obtained throughout the finite element solution will be $K_{quad\text{-}edge} \times K_{quad\text{-}edge}$.

As seen in Fig. 5, an edge can be shared at most by 2 elements. The row corresponding to this edge will have 6 off-diagonal nonzero entries in addition to 1 diagonal nonzero entry. Namely, in this case, the maximum number of nonzero entries in one row will be 7.

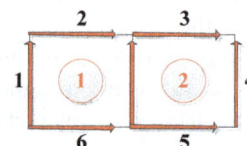

Fig. 5. An edge shared by 2 elements.

Hence data and indices matrices of ELLPACK will be of size $K_{quad\text{-}edge} \times 7$. This means that each matrix will require $7 \times K_{quad\text{-}edge} = 7 \times (2NL + N + L)$ entries in the memory.

2.3 Quadrilateral Facet Elements

All computations regarding the number of unknowns and number of nonzero entries will be equal to that of the edge elements.

Hence data and indices matrices of ELLPACK will be of size $K_{quad\text{-}facet} \times 7$. This means that each matrix will require $7 \times K_{quad\text{-}facet} = 7 \times (2NL + N + L)$ entries in the memory.

2.4 Quadrilateral Volume Elements

If we consider that this mesh consists of volume elements, then the total number of unknowns (i.e. number of centroids) will be $K_{quad\text{-}volume} = N \times L$, which means that the size of the global system matrix obtained throughout the finite element solution will be $K_{quad\text{-}volume} \times K_{quad\text{-}volume}$.

A centroid is owned by one element only; it is not shared. Hence,
- Only diagonal terms in the matrix,
- Only NL entries in the memory.

(ELLPACK or any other sparse matrix storage scheme is unnecessary. It is sufficient to compute and store the diagonal terms in an ordinary array)

3. Triangular Elements

We'll perform our analysis for triangular node, edge, facet and volume elements.

3.1 Triangular Node Elements

We can obtain a triangular element mesh from a quadrilateral mesh in a straightforward manner; which will yield $2NL$ elements. No new nodes are introduced, the total number of unknowns (number of nodes) will be identical to the quadrilateral case; that is $K_{tri\text{-}node} = K_{quad\text{-}node} = (N + 1) \times (L + 1)$.

A node can be shared at most by 6 elements as seen in Fig. 6. The matrix row corresponding to this node will have 6 off-diagonal nonzero entries in addition to 1 diagonal nonzero entry. Namely, in this case, the maximum number of nonzero entries in one row will be 7.

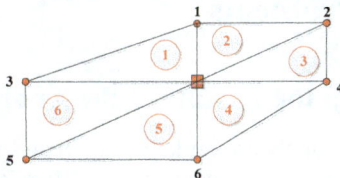

Fig. 6. A node shared by 6 elements.

Hence each matrix will require $7 \times K_{tri\text{-}node} = 7 \times (N + 1) \times (L + 1)$ entries in the memory.

3.2 Triangular Edge Elements

While conversion from the quadrilateral mesh to triangular mesh NL new edges are introduced; hence,

$$K_{tri\text{-}egde} = K_{quad\text{-}edge} + NL,$$
$$K_{tri\text{-}egde} = 2NL + N + L + NL = 3NL + N + L. \tag{3}$$

As seen in Fig. 7, an edge can be shared at most by 2 elements. The matrix row corresponding to this edge will have 4 off-diagonal nonzero entries in addition to 1 diagonal nonzero entry. Namely, in this case, the maximum number of nonzero entries in one row will be 5.

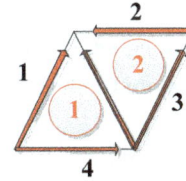

Fig. 7. An edge shared by 2 elements.

Hence each matrix will require $5 \times K_{tri\text{-}edge} = 5 \times (3NL + N + L)$ entries in the memory.

3.3 Triangular Facet Elements

All computations regarding the number of unknowns and number of nonzero entries will be equal to that of the edge elements and each matrix will require $5 \times K_{tri\text{-}facet} = 5 \times (3NL + N + L)$ entries in the memory, too.

3.4 Triangular Volume Elements

If we consider that this mesh consists of triangular volume elements, then the total number of unknowns (i.e. number of centroids) will be $K_{tri\text{-}volume} = 2NL$.

A centroid is owned by one element only; it is not shared. Hence,
- Only diagonal terms in the matrix,
- Only $2NL$ entries in the memory.

4. Exact Number of Nonzero Entries

We'll determine exact number of nonzero entries for each element type.

4.1 Quadrilateral Node Elements

There are 3 types of nodes inside the mesh.

Type 1 node: A node owned by only one element at the corner of the problem domain. There are 4 nodes of this sort. For such nodes, the corresponding row will have only

3 off-diagonal and 1 diagonal nonzero entries; i.e. 4 non-zero entries.

Type 2 node: A node owned by 2 elements at the border of the problem domain. There are $(2N + 2L - 4)$ nodes of this sort. For such nodes, the corresponding row will have 5 off-diagonal and 1 diagonal nonzero entries; i.e. 6 nonzero entries.

Type 3 node: A node owned by 3 elements. There are $(N - 1) \times (L - 1)$ nodes of this sort. For such nodes, each row will have 9 nonzero entries. Hence, the actual number of nonzero entries is equal to,

$$[4 \times 4] + [6 \times (2N + 2L - 4)] + [9 \times (N-1)(L-1)] \quad (4)$$
$$= 9NL + 3N + 3L + 1.$$

ELLPACK stores $9 \times (N + 1) \times (L + 1)$ entries and redundancy rate is

$$r = \frac{\text{number of redundant entries}}{\text{number of stored entries}}, \quad (5)$$
$$r = \frac{6N + 6L + 8}{9NL + 9N + 9L + 9}.$$

4.2 Quadrilateral Edge Elements

There are 2 types of edges.

Type 1 edge: An edge shared by only 1 element at the border of the problem domain. There are NL edges of this sort. Rows corresponding to these edges have 3 off-diagonal and 1 diagonal nonzero entries; i.e. 4 nonzero entries.

Type 2 edge: An edge shared by 2 elements. There are $NL + N + L$ edges of this sort. Rows corresponding to these edges have 6 off-diagonal and 1 diagonal nonzero entries; i.e. 7 nonzero entries. Hence, the exact number of nonzero entries will be $11NL + 7N + 7L$.

ELLPACK stores $14NL + 7N + 7L$ entries and redundancy rate is

$$r = \frac{3NL}{14NL + 7N + 7L}. \quad (6)$$

4.3 Quadrilateral Facet Elements

All computations regarding the number of unknowns, number of nonzero entries and redundancy rate will be equal to that of the edge elements.

4.4 Triangular Node Elements

There are 4 types of nodes inside the mesh.

Type 1 node: A node owned by only one element at the corner of the problem domain. There are 2 nodes of this sort. For such nodes, the corresponding row will have only 2 off-diagonal and 1 diagonal nonzero entries; i.e. 3 non-zero entries.

Type 2 node: A node shared by only 2 elements at the border of the problem domain. There are 2 nodes of this sort. For such nodes, the corresponding row will have only 3 off-diagonal and 1 diagonal nonzero entries; i.e. 4 non-zero entries.

Type 3 node: A node shared by 3 elements. There are $(2N + 2L - 4)$ nodes of this sort. For such nodes, the corresponding row will have only 4 off-diagonal and 1 diagonal nonzero entries; i.e. 5 nonzero entries.

Type 4 node: A node shared by 6 elements. There are $(NL - N - L + 1)$ nodes of this sort. For such nodes, the corresponding row will have only 6 off-diagonal and 1 diagonal nonzero entries; i.e. 7 nonzero entries. Hence, the exact number of nonzero entries will be $7NL + 3N + 3L + 1$.

ELLPACK stores $7NL + 7N + 7L + 7$ entries and redundancy rate is

$$r = \frac{4N + 4L + 6}{7NL + 7N + 7L + 7}. \quad (7)$$

4.5 Triangular Edge Elements

There are 2 types of edges inside the mesh.

Type 1 edge: An edge owned by only one element at the border of the problem domain. There are $2NL$ edges of this sort. For such nodes, the corresponding row will have only 2 off-diagonal and 1 diagonal nonzero entries; i.e. 3 nonzero entries.

Type 2 edge: An edge owned by 2 elements. There are $NL + N + L$ edges of this sort. For such nodes, the corresponding row will have only 4 off-diagonal and 1 diagonal nonzero entries; i.e. 5 nonzero entries. Hence, exact number of nonzero entries will be $11NL + 5N + 5L$.

ELLPACK stores $15NL + 5N + 5L$ entries and redundancy rate is

$$r = \frac{4NL}{15NL + 5N + 5L}. \quad (8)$$

4.6 Triangular Facet Elements

All computations regarding the number of unknowns, number of nonzero entries and redundancy rate will be equal to that of the edge elements.

5. Results and Their Implications for Real-Life Problems

5.1 Wrapping-Up the Results of the Analyses

Summarizing our analysis results in Fig. 8 and Fig. 9, we observe a very interesting aspect of the ELLPACK

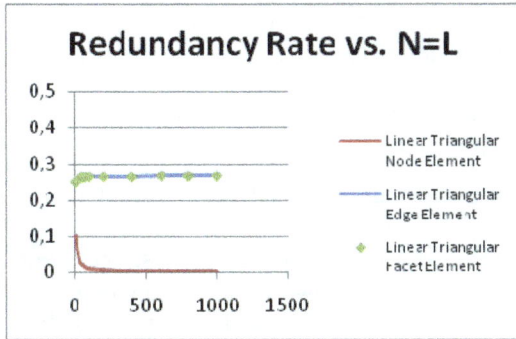

Fig. 8. Redundancy rate vs. different values of $N = L$ (for various types of triangular elements).

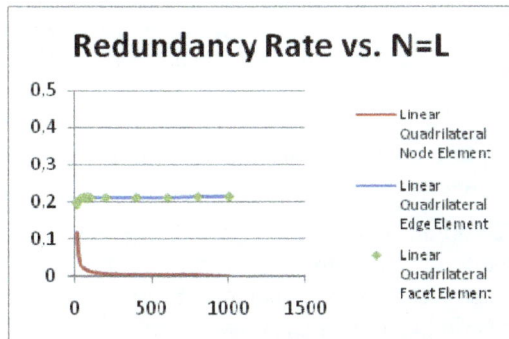

Fig. 9. Redundancy rate vs. different values of $N = L$ (for various types of quadrilateral elements).

sparse storage scheme. For node elements, as a matter fact, the redundancy of the scheme becomes quite negligible (at the order of 0.001%) for very large scale problems. This means that the scheme can be confidently implemented and used in case of node elements. For volume elements, the scheme can also be used safely since there is no redundancy.

On the other hand, for the edge and facet elements, which are quite frequently used in electromagnetic scattering problems, the redundancy rate converges to 26.6% for triangular edge/facet elements; and to 21.5% for quadrilateral edge/facet elements. This means that it should be noted that there will be a considerable amount of redundant resource usage of ELLPACK storage scheme for very large scale electromagnetic scattering problems.

5.2 Real Life Problems

5.2.1 Bistatic Radar Cross Section Computation for Isolated Scatterers

In order to give an idea about the situation in real life electromagnetic scattering problems, let us now consider some conventional problems modeled in two-dimensional space.

First, we will consider an isolated scatterer extending to infinity (or so long in that direction such that this assumption is valid) in one direction. For such problems, the main aim is to obtain the bistatic Radar Cross Section

(RCS) of the scatterer. For this purpose, an incident plane wave of an arbitrary direction at a specific frequency (or wavelength λ) is assumed onto the scatterer, and the scattered field caused by the scatterer at all directions (towards any elevation and azimuth angle) is computed by means of the Finite Element Method. As seen in Fig. 10, the scatterer can be thought as occupying a space modeled with $N_1 \times L_1$ elements. Here, it should be noted that it is common practice to choose the element size no more than 0.1λ in case of linear elements for an acceptable level of solution accuracy.

On the other hand, since the FEM is not suitable for direct use in open problem domains, the "Absorbing Boundary Conditions (ABCs)", particularly the so-called Perfectly Matched Layers (PMLs) [9] shall be applied to such problems in order to terminate the computational domain. The most common approach for PML implementation is the complex-coordinate stretching [10], and this method requires attaining a thickness of at least 3 elements to the PML for the convergence of the solution.

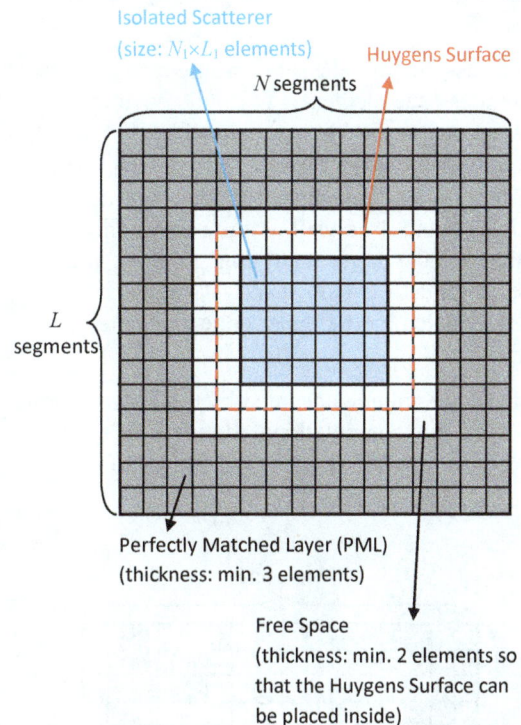

Fig. 10. Sample 2-D problem modeled via the Finite Element Method for computation of the RCS of an isolated scatterer.

Moreover, RCS computation requires far-field solutions. In order to preserve the computational resources, far-field is not computed directly. This is rather achieved by applying Huygens' Surface Equivalence Principle. For this purpose, a closed surface (over which the near field is known) totally inside the free space shall be constructed. By taking the necessary surface integral over this surface [11], [12], it is possible to compute the far-field from the near-field without extending the computational domain dramatically. Certainly, the cost of application of Huygens'

Surface Equivalence Principle is to surround the scatterer with at least 2-element free space in order to be able to choose a closed surface totally residing in free space, as seen in Fig. 10.

Summing up all the issues listed in the previous paragraphs, let us now consider an isolated scatterer with size $3\lambda \times 3\lambda$, i.e. a typical problem for which the Finite Element Method can be applied. First, let us assume that quadrilateral edge elements are used. For such a case, considering an element size of no more than 0.1λ, the scatterer shall be modeled with at least $N_1 \times L_1 = 30 \times 30$ elements. The space occupied by this scatterer shall be surrounded by (i) free space with a minimum thickness of two elements, and (ii) PML with a minimum thickness of two elements. This yields a computational domain of at least $N \times L = [30 + (2 \times 2) + (2 \times 3)] \times [30 + (2 \times 2) + (2 \times 3)] = 40 \times 40$ elements. The analysis held out Sections 3 and 4 as well as the results shown in Fig. 9 show that for this very typical scattering problem (which can be considered as a mid-scale problem), the redundancy rate of the ELLPACK sparse storage scheme is about 20.9 % if modeled by quadrilateral edge elements.

Now, let us assume that the same problem be modeled via linear triangular edge elements. With all considerations listed above, the computational domain shall be divided into at least $2 \times N \times L = 2 \times [30 + (2 \times 2) + (2 \times 3)] \times [30 + (2 \times 2) + (2 \times 3)] = 2 \times 40 \times 40$ elements. Again, the analysis held out Sections 3 and 4 as well as the results shown in Fig. 9 show that for the same scattering problem, the redundancy rate of the ELLPACK sparse storage scheme is about 26.2 % if modeled by triangular edge elements.

In case the scatterer of interest is a Perfect Electric Conductor (PEC), the volume occupied by the scatterer can be excluded from the computational domain as seen in Fig. 11. The PEC assumption imposes the fact that the total

electric and magnetic field inside the scatterer is exactly zero, and it is not required to spend any effort for that particular volume. Hence, exclusion of the relevant volume can be made conveniently in order to decrease the number of elements and the number of unknowns.

In this case, the computational domain contains at least $(40 \times 40) - (30 \times 30) = 700$ elements when quadrilateral edge elements are used. Again, the analysis held out Sections 3 and 4 as well as the results shown in Fig. 9 show that for this case, the redundancy rate of the ELLPACK sparse storage scheme is about 20.9 %.

Similarly, if triangular edge elements are used for the PEC scatterer, this will yield at least $2 \times [(40 \times 40) - (30 \times 30)] = 2 \times 700 = 1400$ elements. Eventually, this will yield a redundancy rate of 26.0 % for the ELLPACK sparse storage scheme.

5.2.2 Scattering Parameters' Computation for the Periodic Structures

Another common application of the Finite Element Method as regards the real life electromagnetic scattering problems is nothing but the computation of the scattering parameters for periodic structures. A singly-periodic structure, which can be modeled via 2-D finite element formulation, can be considered as an infinite series of a particular structure cascaded to each other. For such problems, the main aim is to obtain the reflection (the so-called S11 parameter) and the transmission (the so-called S21 parameter) characteristics of the periodic structure. For this purpose, usually a normal incident plane wave at a specific frequency (or wavelength λ) is assumed onto one side of the scatterer, and the scattered fields caused by the scatterer at forward and backward directions (towards any elevation and azimuth angle) are computed by means of the Finite Element Method. Analyses of such structures are quite crucial in real life especially at the design process of Frequency Selective Surfaces (FSSs), Radomes (Radar Domes) and Electromagnetic Band-Gap Structures (EBGs).

As seen in Fig. 12, the periodic structure can be thought as occupying a space modeled with $N_1 \times L_1$ elements. Again, the criterion for the element size to be no more than 0.1λ (in case of linear elements) is still valid for an acceptable level of solution accuracy. PMLs shall be applied to such problems in order to terminate the computational domain, as well. Similarly, attaining a thickness of at least 3 elements to the PML for the convergence of the solution is required. The far field is computed by means of Huygens' Surface Equivalence Principle. Eventually, the cost of application of Huygens' Surface Equivalence Principle is to place at least 2-element free space above and below the free periodic structure in order to be able to choose two surfaces (the former for computation of the scattered field and the latter for the transmitted field) totally residing in free space, as seen in Fig. 12.

The difference in modeling the periodic structures is at the step of imposing the so-called "Periodic Boundary

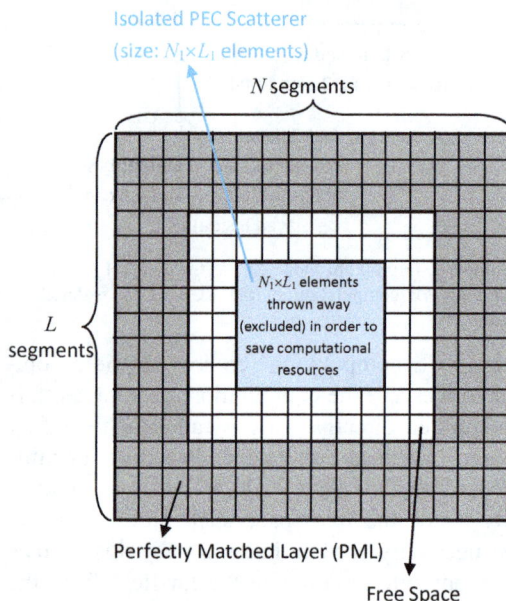

Fig. 11. The same 2-D problem for which the volume occupied by the isolated PEC scatterer is excluded.

Conditions" in order to apply Floquet's Theorem, by which the magnitude of the total fields at the Side Walls are set and forced to be identical, and the phase difference of the total fields are set to be constant [11], [12]. By this manner, it is possible and sufficient to carry out the computations only at a single period of the periodic structure (i.e. it is not necessary to carry out the computations over numerous periods, which would dramatically increase the computational cost).

Again, let us consider a periodic structure of which the cross section can be modeled via $N \times L_1$ quadrilateral elements as seen in Fig. 12. Assume that the thickness of the periodic structure is about 0.2λ (i.e. 2 elements). This is a valid assumption for practical cases since the FSSs, Radomes or EBGs are required to be as thin as possible. Also, it is common practice to adjust the period of such structures more than λ, particularly about 2λ (i.e. 20 elements).

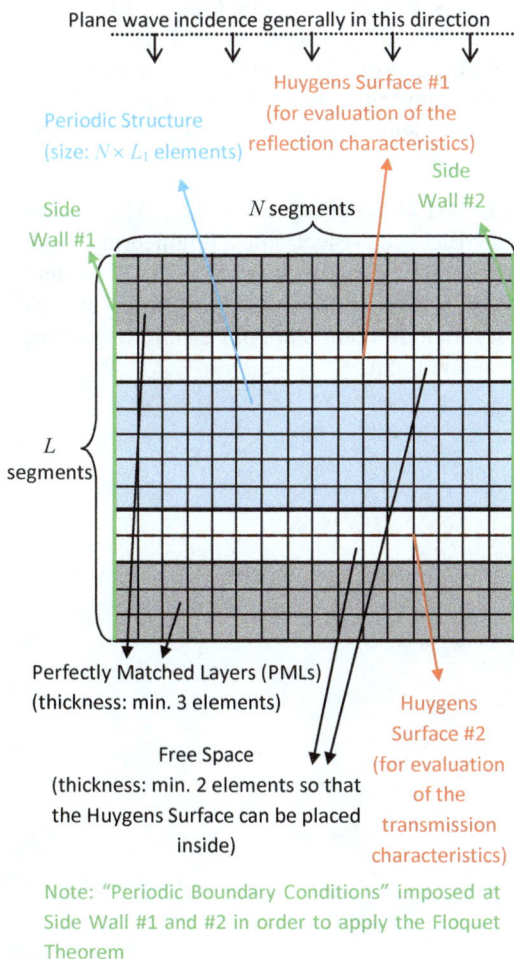

Fig. 12. Sample 2-D problem modeled via the Finite Element Method for computation of the scattering parameters (transmission and reflection characteristics) for a singly periodic structure.

Hence, the computational domain will contain at least $N \times L = 20 \times [2 + (2 \times 2) + (2 \times 3)] = 20 \times 12$ elements if quadrilateral elements are used. From the results of the previous sections and Fig. 9, it is apparent that the redun-

dancy rate of the ELLPACK storage scheme will be about 19.6 % for such a typical problem.

If the same problem is modeled by means of triangular edge elements, there will be at least $2 \times N \times L = 2 \times 20 \times [2 + (2 \times 2) + (2 \times 3)] = 2 \times 20 \times 12$ elements, which will yield a redundancy rate of 25.9 % for the ELLPACK storage scheme.

6. Conclusions

Due to its implementation ease, ELLPACK sparse matrix storage scheme is one of the most commonly used schemes in practice. On the other hand it has some amount of excessive memory usage. Even though this amount seems to be negligible at first glance, as demonstrated in the previous sections, it might be in the order of 1/5 in case of edge elements.

Particularly, throughout the finite element modeling and solution of the very large-scale electromagnetic scattering problems, ELLPACK sparse storage scheme have a redundancy rate about 26.6 % for triangular edge/facet elements; and about 21.5 % for quadrilateral edge/facet elements.

Moreover, the analyses held out in Section 5.2 demonstrate that for mid-scale real life problems (i.e. in which the electrically mid-size isolated scatterers or periodic structures are in concern) the redundancy rates are still quite significant: about 25.9 to 26.2 % for triangular edge/facet elements; and about 19.6 to 20.9 % for quadrilateral edge/facet elements.

Once again, it should be noted that all our analyses address any 2-D problem domain which is homeomorphic to a rectangle. Hence, the results can be generalized to any structured but non-uniform mesh consisting of triangular or quadrilateral mesh. On the other hand, what happens for the case of unstructured mesh (i.e. mesh with nodal discontinuities requiring the so-called "mesh refinement"), is still an open question. Even though it might not be possible to come up with generalized concrete formulation as in the structured mesh, our ongoing studies are focused on extraction of similar formulas especially for some popular benchmark electromagnetic scattering problems yielding unstructured mesh.

References

[1] VAZQUEZ, F., ORTEGA, G., FERNANDEZ, J. J., GARZON, E. M. Improving the performance of the sparse matrix vector product with GPUs. In *Proceedings of the IEEE 10th International Conference on Computer and Information Technology*. Bradford (United Kingdom), 2010, p. 1146–1151.

[2] BUATOIS, L., CAUMON, G., LEVY, B. Concurrent number cruncher: A GPU implementation of a general sparse linear solver. *International Journal of Parallel, Emergent and Distributed Systems*, 2009, vol. 24, no. 3, p. 205–223.

[3] MONAKOV, A., LOKHMOTOV, A., AVETISYAN, A. Automatically tuning sparse matrix-vector multiplication for GPU architectures. *Lecture Notes in Computer Science (LNCS)*, 2010, vol. 5952, p. 111–125.

[4] CENDES, Z. J. Vector finite elements for electromagnetics field computation. *IEEE Transactions on Magnetics*, 1991, vol. 27, no. 5, p. 3958–3966.

[5] COULOMB, J. L. Finite elements three dimensional magnetic field computation. *IEEE Transactions on Magnetics*, 1981, vol. 17, no. 6, p. 3241–3246.

[6] WANG, R., DEMERDASH, N. A. On the effects of grid ill-conditioning in three dimensional finite element vector potential magnetostatic field computations. *IEEE Transactions on Magnetics*, 1990, vol. 26, no. 5, p. 2190–2192.

[7] WU, J. Y., LEE, R. The advantages of triangular and tetrahedral edge elements for electromagnetic modeling with the finite-element method. *IEEE Transactions on Antennas and Propagation*, 1997, vol. 45, no. 9, p. 1431–1437.

[8] WARREN, G. S., SCOTT, W. R. An investigation of numerical dispersion in the vector finite-element method using quadrilateral elements. *IEEE Transactions on Antennas and Propagation*, 1994, vol. 42, no. 11, p. 1502–1508.

[9] BERENGER, J. P. A perfectly matched layer for the absorption of electromagnetic waves. *Journal of Computational Physics*, 1994, vol. 114, p. 185–200.

[10] CHEW, W. C, WEEDON, W. H. A 3D perfectly matched medium from modified Maxwell's equations with stretched coordinates. *Microwave and Optical Technology Letters*, 1994, vol. 7, no. 13, p. 599–604.

[11] VOLAKIS, J. L., CHATTERJEE, A., KEMPEL, L. C. *Finite Element Method for Electromagnetics – Antennas, Microwave Circuits, and Scattering Applications*. 1st ed. New York (NY, USA): IEEE Press, 1998.

[12] JIN, J. *The Finite Element Method in Electromagnetics*. 2nd ed. New York (NY, USA): Wiley Interscience, 2002.

About Authors ...

Gökay AKINCI was born in Kırıkkale, Turkey, in 1987. He received his M.Sc. degree in Electrical and Electronics Engineering from Kırıkkale University in 2011. He is currently working towards a Ph.D. degree at the Department of Electrical and Electronics Engineering in Ankara University. His current research interests are in the areas of parallel programming, sparse matrix and computational electromagnetics.

Asim Egemen YILMAZ received his B.Sc. degrees in Electrical-Electronics Engineering and Mathematics from the Middle East Technical University in 1997. He received his M.Sc. and Ph.D. degrees in Electrical-Electronics Engineering from the same university in 2000 and 2007, respectively. He is now with Ankara University, where he is an Associate Professor.

Mustafa KUZUOGLU received his B.Sc., M.Sc., and Ph.D. degrees in Electrical-Electronics Engineering from the Middle East Technical University in 1979, 1981, and 1986, respectively; where he is currently a Professor. His research interests include computational electromagnetics, inverse problems, and radar.

11

Fractal Metamaterial Absorber with Three-Order Oblique Cross Dipole Slot Structure and its Application for In-Band RCS Reduction of Array Antennas

Sijia LI, Xiangyu CAO, Jun GAO, Yuejun ZHENG, Di ZHANG, Hongxi LIU

Information and Navigation College, Air Force Engineering University, 710077, Xi'an, China

lsj051@126.com, xiangyucaokdy@163.com, gjgj9694@163.com, erikzhengyang@126.com, dee19910330@163.com, hongxi_liu517@163.com

Abstract. *To miniaturize the perfect metamaterial absorber, a fractal three-order oblique cross dipole slot structure is proposed and investigated in this paper. The fractal perfect metamaterial absorber (FPMA) consists of two metallic layers separated by a lossy dielectric substrate. The top layer etched a three-order oblique fractal-shaped cross dipole slot set in a square patch and the bottom one is a solid metal. The parametric study is performed for providing practical design guidelines. A prototype with a thickness of 0.0106λ (λ is the wavelength at 3.18 GHz) of the FPMA was designed, fabricated, measured, and is loaded on a 1×10 guidewave slot array antennas to reduce the in-band radar cross section (RCS) based on their surface current distribution. Experiments are carried out to verify the simulation results, and the experimental results show that the absorption at normal incidence is above 90% from 3.17 to 3.22 GHz, the size for the absorber is 0.1λ × 0.1λ, the three-order FPMA is miniaturized 60% compared with the zero-order ones, and the array antennas significantly obtain the RCS reduction without the radiation deterioration.*

Keywords

Guidewave slot array antennas, fractal perfect metamaterial absorber, surface current distribution, RCS reduction.

1. Introduction

Radar cross section reduction (RCSR) of the antennas has been a topic of immense strategic interest for the researchers. For out-of-band frequencies, it is well known that the RCS of antenna/array can be significantly reduced by placing the periodic resistive surface (PRS) [1], [2] and suitably shaped band-pass radome, such as frequency selective surfaces (FSS) [3], [4]. However, when the radome is transparent, no RCS reduction of the antenna will significantly take place for in-band frequencies.

Nowadays, the metamaterial absorber has been paid attention to for in-band RCS reduction. In [5], the application of electromagnetic band-gap (EBG) radar absorbing material loaded with lumped resistances to ridged waveguide slot antenna array to reduce its in-band RCS was investigated, in which the lumped resistive elements were used to better match the impedance of free space, and as the chief contributor of absorption. The design idea is different from this work in [5]. In [6], an artificial magnetic conductor (AMC) and perfect electric conductor (PEC) surface were combined together for in-band RCS reduction and radiation improvement of waveguide slot antenna based on the principle of passive cancellation. In [7], two different AMCs were analyzed for ultra-thin and broad-band RAM design. In [8], the reflection characteristic of a composite planar AMC surface has been investigated and fabricated for broadband RCS reduction. Similarly, a broadband RCS reduction for a waveguide slot antenna with orthogonal array of CSRR-AMC is presented in [9]. However, it is obvious that the RCS is reduced in boresight direction but increased in other directions in [6]-[10]. Perfect metamaterial absorber (PMA) with the ultrathin structure and the near-unity absorptivity was firstly proposed and demonstrated by Landy et. al. in [11], which has become an important aspect in the research of metamaterials. Later, to achieve polarization-insensitive absorption, wide incident angle absorption, broadband absorption, and multi-band absorption, many researchers make several efforts on the PMAs [12-20]. In 2013, a perfect metamaterial absorber with wide-angle and polarization-insensitive absorption was presented for RCS reduction of waveguide slot antenna in [16]. The maximum absorptivity of this absorber is 99.8% with a full width at half maximum (FWHM) of 220 MHz. Then, a perfect metamaterial absorber (PMA) is applied for RCS reduction of a circularly polarized tilted beam antenna in [17]. But these PMAs cannot be directly used for the reduction of the guidewave slot array antennas (GSAAs) due to the large structure and radiation deterioration, especially the gain reduction.

In this paper, an ultrathin and miniaturized FPMA with the three-order oblique cross dipole slot (TOOCDS) structure is presented and investigated for its application for RCS reduction of the 1×10 GSAAs without radiation property deterioration. Experimental results show that the absorption is above 90 % from 3.17 to 3.22 GHz at normal incidence and the size of the absorber is 0.1λ × 0.1λ. The three-order FPMA is miniaturized 60 % compared with the zero-order ones. The FPMA is loaded on the array antennas to reduce the RCS according to their surface current distribution. And the simulated results indicate that the array antennas significantly obtain the RCS reduction without the radiation deterioration.

2. Design and Analysis of FPMA

Fig. 1(a), 1(b) and 1(c) show the geometry of the ultra-thin and miniaturized FPMA, which is composed of two metallic layers separated by a lossy dielectric substrate. The top layer etched a fractal-shaped TOOCDS set in a square patch and the bottom one is a solid metal. The metal is copper with the conductivity of 5.8×10^7 S/m. FR4 is used as the substrate with the relative permittivity of 4.4 and loss tangent of 0.02, respectively. The absorber structure was simulated and optimized using HFSS 14.0 by considering a unit cell. Optimized parameters are shown in Fig. 1 (units: mm).

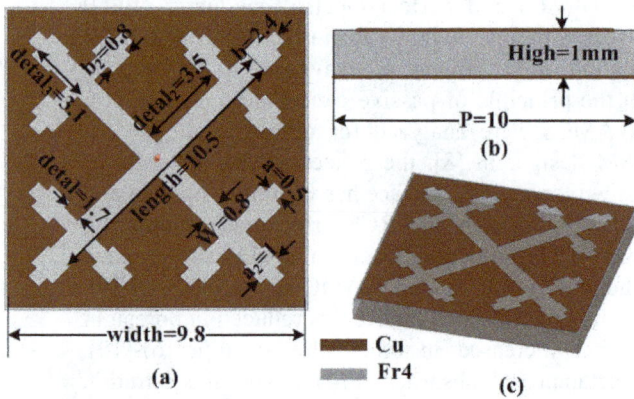

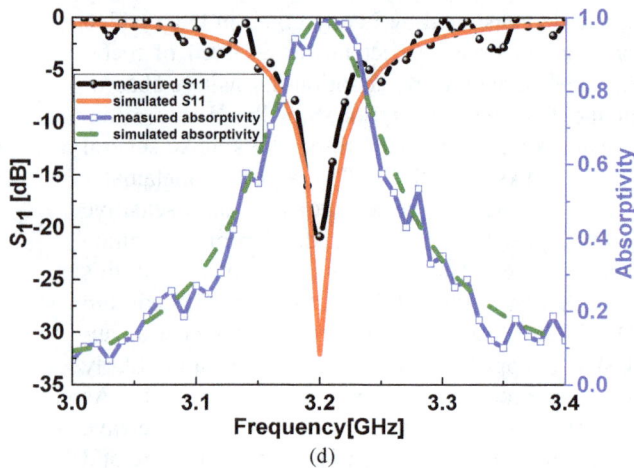

Fig. 1. Geometry and simulated absorptivity for the miniaturized FPMA unite cell. (a) Top view. (b) Side view. (c) Perspective view. (d) Simulated S_{11} and absorptivity for the FPMA with the optimized parameters.

The transmission is zero ($T = 0$) due to the copper ground plate without patterning in the bottom layer. The absorptivity (A) is defined as [11], [16], [18]

$$A = 1 - R = 1 - |S_{11}|^2 \qquad (1)$$

R represents the reflection of the FPMA.

The optimized absorptivity for the structure is reported in Fig. 1(d). The FPMA displays a great absorption capability. The designed idea of the FPMA is to adjust the effective $\varepsilon(\omega)$ and $\mu(\omega)$ independently by varying the dimensions of electric resonant component and magnetic resonant component in the unit cell so as to match the surface impedance to free space and achieve a large resonant dissipation at the meantime. And while, the tangent for the absorber is absorbing the microwave at the electric and magnetic resonant frequencies. Thus, the transmission and reflection are simultaneously minimized and absorption is maximized. From Fig. 1(d), we can see that the frequency range with experimental absorptivity larger than 90 % varies from 3.17 to 3.22 GHz with a FWHM of 130 MHz (3.14-3.27 GHz).

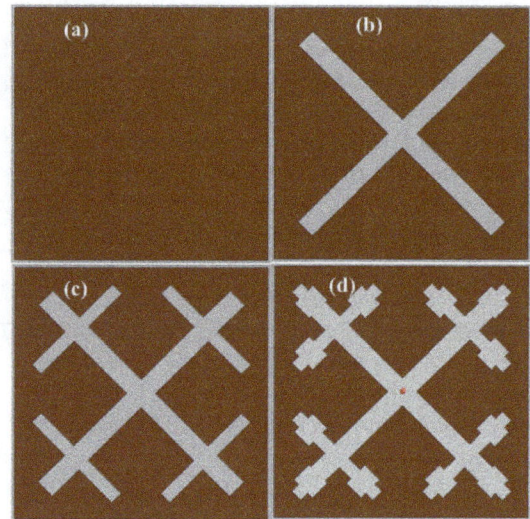

Fig. 2. Schematics of proposed fractal-shaped TOOCDS structure for FPMA under different cases. (a) Zero-order FPMA. (b) One-order FPMA. (c) Two-order FPMA. (d) Three-order FPMA.

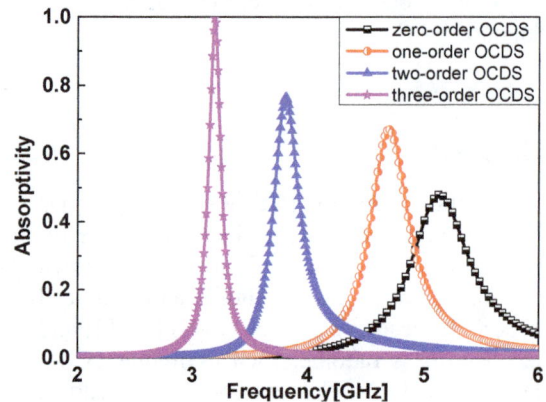

Fig. 3. Simulated results of the absorptivity for the proposed FPMA with different cases.

Fig. 2 gives the schematics of the proposed fractal TOOCDS structure under different cases and Fig. 3 shows the corresponding S_{11} results. The construction law of the fractal TOOCDS structure can be briefly described by the recursive principle as follows: in the first step, an oblique cross dipole slot (OCDS) originating from one central with a length of $detal_2$, followed by considering each end of these dipole as fresh starting points. Each of them produces a new OCDS with a length of $detal_3$. Taking this repeated procedure in a recursive fashion, the third and higher-order self-similar fractals are formed as a function factor F and iteration order n:

$$L_n = L_1/F^{(n-1)} . \qquad (2)$$

Note that L_3 is selected larger than L_2/F in this particular design to guarantee the small gap space which can be adjusted to tailor the magnetic resonant frequency and accordingly facilitates a balanced condition. Parameters for the structures are same with the ones in Fig. 1. From Fig. 2

and Fig. 3, we can see that the resonance frequency shifted from 5.12 to 3.2 GHz and the absorptivity increased from 0.48 to 0.99 as the order value increased from zero to three. It is obvious that the miniaturization for the fractal TOOCDS structure has been enhanced 60 % compared with the zero-order ones.

A parametric study was carried out and the simulated results for the proposed FPMA with different values of width, w, a, b, a_2, b_2, high and $\tan\delta$ is shown in Fig. 4. In the process of the study, the metallic layers were assumed to zero-thickness for reducing the computation time. It can be seen that the resonance frequency will be decreased as the width, a, b, a_2 and b_2 increase. The resonance frequency is increased as w changes from 0.7 to 1 mm. The optimized height is 1 mm. From Fig. 4, the b_2 is the minimum amount of elements that have to be illuminated to excite the absorber. The resonance frequency should be 3.2 GHz for its application on the RCS reduction for GSAAs and the S_{11} is minimized as the parameters optimization. The optimized parameters are shown in Fig. 1.

3. RCS Reduction of GSAAs

In this section, the ultrathin and miniaturized FPMA has been applied for RCS reduction on the GSAAs. The surface current distribution of the GSAAs is analyzed and the metal surface of the GSAAs is divided into the strong current area and the weak current area. The FPMA has been loaded on the weak current area for RCS reduction of the array antennas and the metal surface with the strong current is simultaneously retained to avoid the radiation deterioration. The geometry and the surface current distribution of the 1×10 GSAAs are shown in Fig. 5. The strong surface current area of the array antennas is near the two sides of the slots and the weak surface current area is around the array antennas from Fig. 5(d). So the FPMA is loaded on the weak current area to avoid the radiation deterioration in Fig. 6. The distance between the slots and the FPMA is optimized from 1 to 5 mm to minimize the gain reduction, and the optimized result is 3 mm.

To demonstrate the GSAAs loaded the proposed FPMA, an intensive simulation study is carried out and the results are discussed. Fig. 7. shows the simulated S_{11} and gain. The impedance bandwidth of 2.2 % ($S_{11} < -10$ dB) from 3.15 to 3.22 GHz is achieved for the common antennas (which is the GSAAs without FPMA) and the antennas loaded the FPMA obtained an impedance bandwidth of 2.25 % from 3.15 to 3.23 GHz, from which it can be found that the two antennas have the same resonant frequency. The gain, shown in Fig. 7, varies from 13.6 to 14.5 dBi for the two antennas at the working frequency (3.15 GHz to 3.23 GHz). The antennas loaded FPMA obtain a gain of 14.1 dBi, which are 0 to 0.68 dB lower than that of common antennas from 3.15 to 3.2 GHz and higher than that from 3.2 to 3.23 GHz. The radiation characters of the GSAAs loaded FPMA can be retained for the FPMA loaded on the weak surface current area.

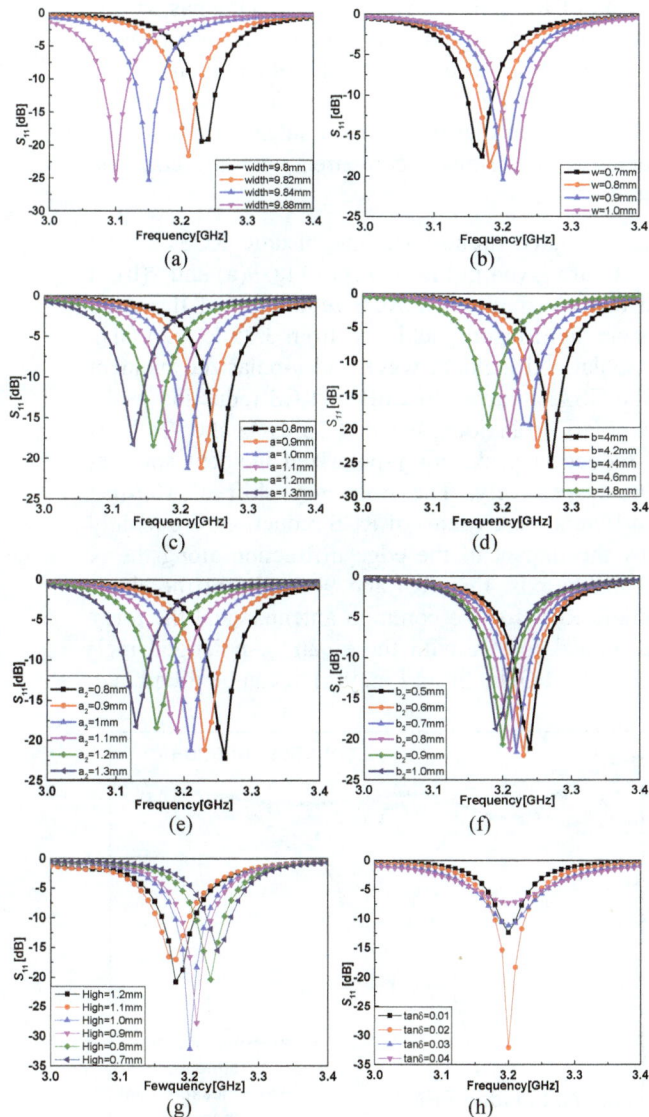

Fig. 4. The simulated results of S_{11} for the proposed FPMA with different parameters. (a) width, (b) w, (c) a, (d) b, (e) a_2, (f) b_2, (g) high, and (h) $\tan\delta$.

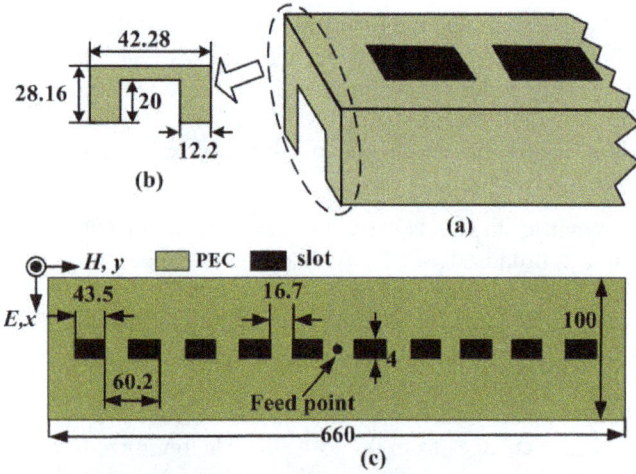

Fig. 5. The geometry and surface current distribution of the 1×10 guidewave slot array antennas (units: mm). (a) Geometry of array antennas. (b) Side view and (c) Top view for the array antennas. (d) Surface current distribution of the antennas at 0°, 90°, 180° and 270° at 3.18 GHz.

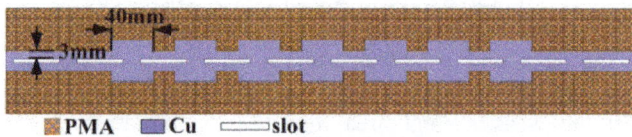

Fig. 6. The GSAAs loaded the FPMA based on the surface current distribution.

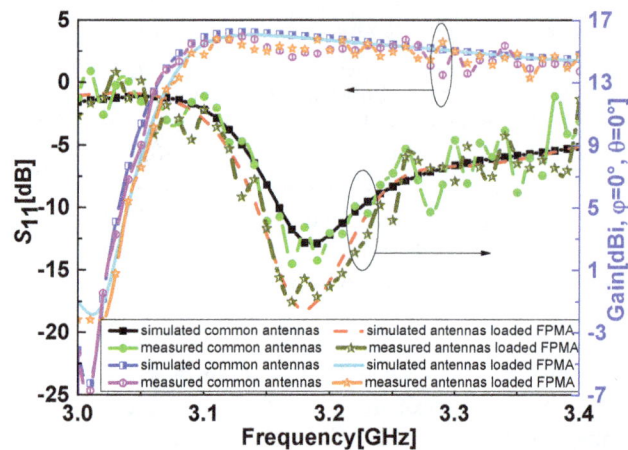

Fig. 7. Simulated S_{11} and gain for guidewave slot array antennas loaded FPMA and common antennas from 3.0 to 3.4 GHz.

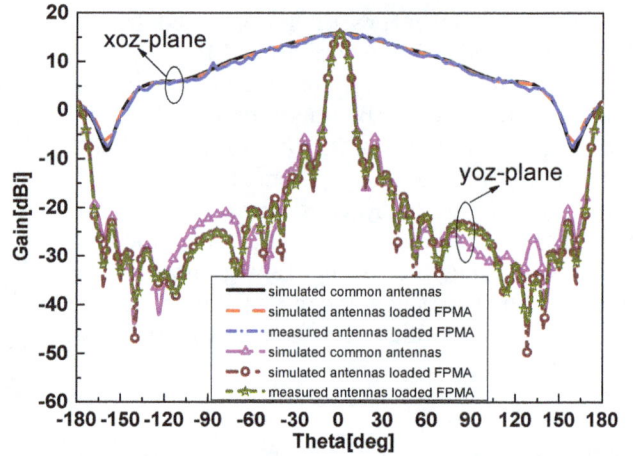

Fig. 8. Radiation patterns of the guidewave slot array antennas loaded FPMA and common antennas at 3.18 GHz.

Simulated radiation patterns of the array antennas at 3.18 GHz are given in Fig. 8. It can be seen that the radiation patterns of the guidewave slot array antennas loaded FPMA are the same as one of the common antennas at 3.18 GHz. The front-to-back ratio (FBR) remains to be better than 15.1 dB for the antennas loaded FPMA, which is 0.3 dB lower than that of common antennas. The radiation characters are retained compared with the common array antennas.

The simulated monostatic and bistatic RCSs of the two GSAAs are given in Fig. 9. From Fig. 9(a) and 9(b), it is found that the monostatic RCS reductions of the array antennas are significantly achieved from 3.15 to 3.25 GHz for the x-polarized incident wave and y-polarized incident wave. The GSAAs have -26.2 dB of RCS reduction peaks for the x-polarized incident wave at 3.2 GHz and -24.5 dB of RCS reduction peaks for y-polarized incident wave at 3.19 GHz, respectively. The frequency shift of 10 MHz and the different peak values of RCS reductions are mainly caused by the impact of the edge diffraction along the X axis and the Y axis. Fig. 9(c) and 9(d) present the simulated bistatic RCS for the common antennas and the array antennas loaded FPMA with the x- and y-polarized incident waves at 3.2 GHz. From Fig. 9(c), it can be found that

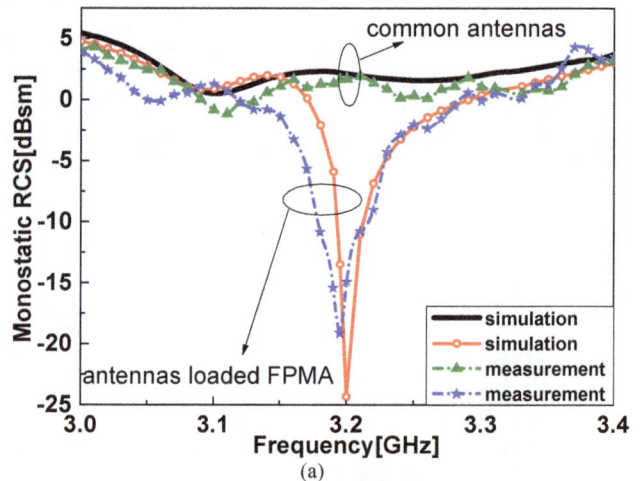

(a)

(b)

(c)

(d)

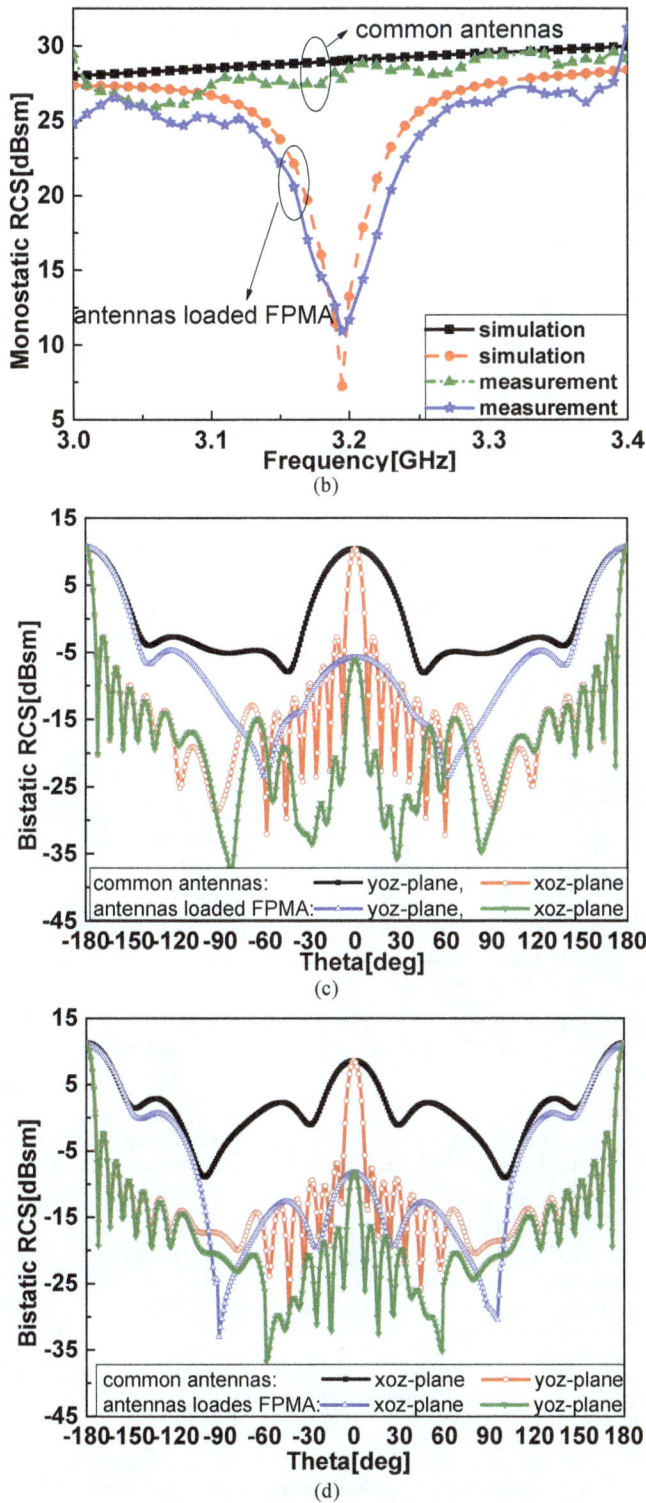

Fig. 9. Comparison of monostatic and bistatic RCS of the GSAAs loaded FPMA and common antennas. Simulated and measured monostatic RCS reduction with (a) *x*-polarized incident wave and (b) *y*-polarized incident wave impinging from normal direction. Simulated results of the bistatic RCS for antennas loaded FPMA with (c) *x*-polarized incident wave and (d) *y*-polarized incident wave at 3.2 GHz.

for the case of *x*-direction polarized incident wave, the bistatic RCS has been effectively reduced in angular ranges of [-110°, -80°] and [-50°, 50°] in xoz-plane, and

[-110°, 110°] in yoz-plane. From Fig. 9(d), we find the RCS has been effectively reduced in angular ranges of [-110°, 110°] in xoz-plane, and [-90°, 90°] in yoz-plane for *y*-direction polarized incident wave.

4. Fabrication and Measurement

To verify the simulation results, the prototypes of the FPMA, the proposed GSAAs loaded FPMA and the common array antennas are illustrated in Fig. 10 and have been experimentally studied. The proposed FPMA samples were respectively fabricated using an optical lithographic processes on a 1-mm-thick FR4 substrate with $\varepsilon_r = 4.4$ and $\tan\delta = 0.02$. The waveguide method [9], [11], [18] (closed system) has been used to verify the absorption. The measured results are shown in Fig. 1(d). The measured absorptivity larger than 90 % varies from 3.17 to 3.22 GHz with a FWHM of 110 MHz (3.14-3.25 GHz).

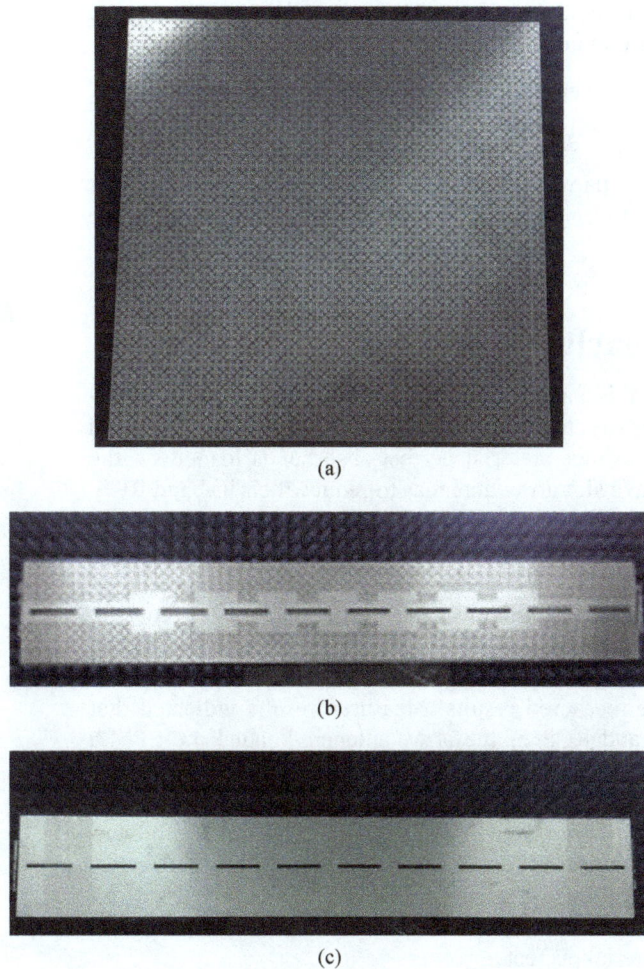

(a)

(b)

(c)

Fig. 10. The prototypes of array antennas. (a) The proposed FPMA. (b) The proposed guidewave slot array antennas loaded the FPMA. (c) The common array antennas.

The experimental S_{11} and gain are given in Fig. 7. As seen from the curve, the array antennas loaded the FPMA have a narrow impedance bandwidth of 2.5 % from 3.15 to

3.23 GHz. The proposed array antennas obtain an average gain value of 13 dBi across the operating bandwidth from 3.15 to 3.23 GHz and the gain reduction is less than 0.73 dB. The experimental results indicate that the reliability of the loaded FPMA for array RCS reduction is based on its surface current distribution. Fig. 8 shows the experimental radiation patterns results at 3.18 GHz for both the xoz-plane and yoz-plane. The patterns are stable cross the operating bandwidth. The FBR remains to be better than 10.5 dB for the antennas loaded the FPMA, which is 0.9 dB lower than that of common antennas.

The measured results of the monostatic RCS for two types of array antennas are shown in Fig. 9(a) and (b). The array antennas samples respectively have the monostatic RCS reduction peaks of -22.8 dB at 3.19 GHz for x-polarized incident wave and -20.1 dB at 3.2 GHz for y-polarized incident wave. The measured results indicate that the array antennas loaded FPMA significantly obtain the RCS reduction from 3.15 to 3.25 GHz for both the x-polarized and y-polarized incident wave. The measured and simulated results are in good agreement.

It is necessary to point out that the presented FPMA has the advantages of ultra-thin and miniaturized structure and perfect absorption. However, the limitation of the FPMA is narrow bandwidth of absorption. So the next work is to enhance the bandwidth of absorption for the FPMA.

5. Conclusion

In this paper, a fractal three-order oblique cross dipole slot structure is presented to miniaturize the PMA and the fractal metamaterial perfect absorber is loaded on the guidewave slot array antennas to reduce their in-band RCS. The surface current distribution of the array antennas divided into the strong and weak current areas is presented. The FPMA is only loaded on the weak current area. The RCS of the 1×10 guidewave slot array antennas is reduced. Simulated S-parameters, radiation patterns, gain, directivity, and RCS reduction of the array antennas are compared with the measured results. Measured results indicated that the in-band RCS of the array antennas loaded the FPMA has been dramatically reduced and the gain reduction is only about 0.73 dB. Experimentally and theoretically, it can be observed that the proposed FPMA has the advantages of the ultra-thin and miniaturized structure, the perfect absorption and the high RCS reduction. The next work is to enhance the bandwidth of absorption for the fractal metamaterial perfect absorber.

Acknowledgements

Authors thank the supports from the National Natural Science Foundation of China under Grant (No.61271100, No.61471389), Natural Science Foundational Research Fund of Shannxi Province (No.2010JZ6010 and No.2012JM8003), and the Doctoral Foundation Air Force Engineering University under Grant (No.KGD080914002). They also thank the reviewers for their valuable comments.

References

[1] GENOVESI, S., COSTA, F., MONORCHIO, A. Wide band radar cross section reduction of slot antennas arrays. *IEEE Trans. Antennas Propag.*, 2014, vol. 62, no. 1, p. 163–173.

[2] WENBO PAN, CHENG HUANG, PO CHEN, XIAOLIANG MA, CHENGGANG HU, XIANGANG LUO. A low-RCS and high-gain partially reflecting surface antenna. *IEEE Trans. Antennas Propag.*, 2014, vol. 62, no. 2, p. 945–949.

[3] ZHOU, H., QU, S.-B., LIN, B.-Q. Filter-antenna consisting of conical FSS radome and monopole antenna. *IEEE Trans. Antennas Propag.*, 2012, vol. 60, no. 6, p. 3040–3045.

[4] GENOVESI, S., COSTA, F., MONORCHIO, A. Low profile array with reduced radar cross section by using frequency selective surfaces. *IEEE Trans. on Antennas and Propagation*, 2012, vol. 60, no. 5, p. 2327–2335.

[5] LI, Y.-Q., ZHANG, H., FU, Y.-Q., YUAN, N.-C. RCS reduction of ridged waveguide slot antenna array using EBG radar absorbing material. *IEEE Antennas Wireless Propag. Lett.*, 2008, vol. 7, p. 473–476.

[6] TAN, Y., YAN, N., YANG, Y., FU, Y. Improved RCS and efficient waveguide slot antenna. *Electron. Lett.*, 2011, vol. 47, no. 10, p. 582–583.

[7] ZHANG, Y., MITTER, R., WANG, B. Z., HUANG, N. T. AMCs for ultra-thin and broadband RAM design. *Electron. Lett.*, 2009, vol. 45, no. 10, p. 484–485.

[8] HWANG, R. B., TSAI, Y. L. The reflection characteristics of a composite planar AMC surface. *Applied Phys. Lett. Advances*, 2012, no. 2, p. 012128.

[9] ZHAO, Y., CAO, X. Y., GAO, J., LI, W. Q. Broadband RCS reduction and high gain waveguide slot antenna with orthogonal array of CSRR-AMC. *Electron. Lett.*, 2013, vol. 49, no. 21, p. 1312-1313.

[10] IRIARTE GALARREGUI, J. C., TELLECHEA PEREDA, A., MARTINEZ DE FALCON, J. L., EDERRA, I., GONZALO, R., DE MAAGT, P. Broadband radar cross-section reduction using AMC technology. *IEEE Trans. Antennas Propag.*, 2013, vol. 61, no. 12, p. 6136-6143.

[11] LANDY, N. I., SAJUYIGBE, S., MOCK, J. J., SMITH, D. R., PADILLA, W. J. A perfect metamaterial absorber. *Phys. Rev. Lett*, 2008, vol. 100, p. 207402.

[12] LONG LI, YANG YANG, CHANGHONG LIANG. A wide-angle polarization-insensitive ultra-thin metamaterial absorber with three resonant modes. *J. Appl. Phys.*, 2011, vol. 110, p. 063702.

[13] LI SI-JIA, CAO XIANG-YU, GAO JUN, ZHENG QIU-RONG, ZHAO YI, YANG QIN. Design of ultrathin broadband perfect metamaterial absorber with low radar cross section. *Acta Phys. Sin.*, 2013, vol. 62, no. 19, p. 194101.

[14] WANGREN XU, SONKUSALE, S. Microwave diode switchable metamaterial reflector/absorber. *Applied Phys. Lett.*, 2013, vol. 103, p. 0301902.

[15] DEXIN YE, ZHENG WANG, ZHIYU WANG, KUIWEN XU, BIN ZHANG, JIANGTAO HUANGFU, CHANGZHI LI, LIXIN RAN. Towards experimental perfectly-matched layers with ultra-

thin metamaterial surfaces. *IEEE Trans. Antennas Propag.*, 2013, vol. 60, no. 11, p. 5162–5172.

[16] LIU, T., CAO, X. Y., GAO, J., ZHENG, Q. R., LI, W. Q., YANG. H. H. RCS reduction of waveguide slot antenna with metamaterial absorber. *IEEE Trans. Antennas Propag.*, 2013, vol. 61, no. 4, p. 2327–2335.

[17] SIJIA LI, XIANGYU CAO, TAO LIU, HUANHUAN YANG. Double-layer perfect metamaterial absorber and its application for RCS reduction of antenna. *Radioengineering*, 2014, vol. 23, no. 1, p. 222–228.

[18] SIJIA LI, JUN GAO, XIANGYU CAO, ZHAO ZHANG. Loaded metamaterial perfect absorber using substrate integrated cavity, *J. Appl. Phys.*, 2014, vol.115, p. 213703.

[19] HE-XIU XU, GUANG-MING WANG, MEI-QING QI, JIAN-GANG LIANG, JIAN-QIANG GONG, ZHI-MING XU. Triple-band polarization-insensitive wide-angle ultra-miniature metamaterial transmission line absorber. *Phys. Rev. B*, 2012, vol. 86, p. 205104.

[20] SIJIA LI, JUN GAO, XIANGYU CAO, WENQIANG LI, ZHAO ZHANG, DI ZHANG. Wideband, thin and polarization-insensitive perfect absorber based the double octagonal rings metamaterials and lumped resistances. *J. Appl. Phys.*, 2014, vol. 116, p. 043710.

About Authors ...

Sijia LI was born in Xi'an, Shaanxi province, P.R. China in 1987. He received the B. Eng. degree in Electronics and Information Engineering from Guangxi University, Nanning, China, in 2009 and the M. Eng. degree in Information and Telecommunication Engineering from Air Force Engineering University, Xi'an China, in 2011. He is currently working toward Ph.D degree in Electronic Science and Technology at the Information and Navigation college of Air Force Engineering University. He has been working with the Military Communication and Navigation Antenna and EMC Lab., since 2012. His research activity has been focused in the broadband and fractal perfect metamaterial absorber and its application for RCS reduction of antennas. He has authored and coauthored more than 40 scientific papers in major journals and international conferences. He received the excellent Master's dissertation in 2012. He is awarded the 2012 and 2013 Best Student Paper Prize (First Prize) of the National Graduate Mathematical Modeling Contest in Shanghai in December 2012 and in Changsha in December 2013. He was a recipient of the Best Paper Award at the Forth National Doctor Forum Symposium on Information Science of China in Guangzhou in October 2013. He is awarded the excellent student in the 2014 national summer school of the Microwave Material and Components. He is a reviewer of the Applied Physics Letter, Journal of Applied Physics, Transactions on Microwave Theory & Techniques, IEEE Transactions on Antennas & Propagation, and Microelectronics Journal.

Xiangyu CAO received the B. Sc and M.A. Sc degrees from Air Force, the Missile Institute in 1986 and 1989, respectively. She joined the Air Force Missile Institute in 1989 as an assistant teacher. She became an associate professor in 1996. She received Ph.D. degree in Missile Institute of Air Force Engineering University in 1999. From 1999 to 2002, she was engaged in postdoctoral research in Xidian University, China. She was a Senior Research Associate in the Dept. of Electronic Engineering, City University of Hong Kong from June 2002 to Dec 2003. She is currently a professor of Information and Navigation college of Air Force Engineering University of CPLA. She is the IEEE senior member from 2008. She has authored and coauthored more than 200 technical journal articles and conference papers, and holds one China soft patent. She is the coauthor of two books entitled, Electromagnetic Field and Electromagnetic Wave, and Microwave Technology and Antenna published in 2007 and 2008, respectively. Her research interests include smart antennas, electromagnetic metamaterial and their antenna applications, and electromagnetic compatibility. She is a reviewer of Applied Physics Letter, Journal of Applied Physics, IEEE Transactions on Antennas & Propagation, and IEEE Antennas Wireless Propagation Letter.

Jun GAO received the B.Sc and M.A.Sc degrees from Air Force the Missile Institute in 1984 and 1987, respectively. He joined the Air Force Missile Institute in 1987 as an assistant teacher. He became an associate professor in 2000. He is currently a professor of Information and Navigation College, Air Force Engineering University of CPLA. He has authored and coauthored more than 100 technical journal articles and conference papers, and holds one China soft patent. His research interests include smart antennas, electromagnetic metamaterial and their antenna applications.

Yuejun ZHENG was born in Yushan, Jiangxi province, in 1989. He graduated from Air Force Engineering University and received the B. S. degree in 2012. He is now a graduate student. His main research orientation is antenna radiation ameliorated and RCS reduction based on the metamaterial.

Di ZHANG was born in Baoding, Hebei province, in 1991. He received the B. S. degree from the Information and Navigation College, Air Force Engineering University in 2013. He is now a graduate student. His research interest is in electromagnetic metamaterial.

Hongxi LIU was born in Baotou, Inner Mongolia Autonomous Region, P. R. China in 1989. He received the B.S. degree in Electronic and Communication Engineering. Currently, He is now a graduate student. His recent research interests include the design of perfect metamaterial absorber design, and end-fire antennas.

12

Measurement and Calibration of a High-Sensitivity Microwave Power Sensor with an Attenuator

Yu Song MENG, Yueyan SHAN

National Metrology Centre, Agency for Science, Technology and Research (A*STAR)
1 Science Park Drive, Singapore 118221, Singapore

ysmeng@ieee.org, meng_yusong@nmc.a-star.edu.sg; shan_yueyan@nmc.a-star.edu.sg

Abstract. *In this paper, measurement and calibration of a high-sensitivity microwave power sensor through an attenuator is performed using direct comparison transfer technique. To provide reliable results, a mathematical model previously derived using signal flow graphs together with non-touching loop rule analysis for the measurement estimate (i.e., calibration factor) and its uncertainty evaluation is comparatively investigated. The investigation is carried out through the analysis of physical measurement processes, and consistent mathematical model is observed. Later, an example of Type-N (up to 18 GHz) application is used to demonstrate its calibration and measurement capability.*

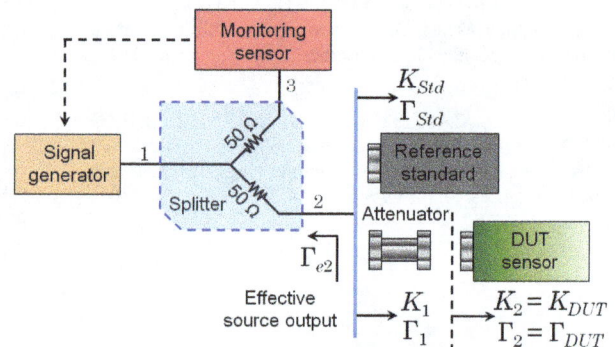

Fig. 1. Calibration of a high-sensitivity power sensor through an attenuator using direct comparison transfer technique.

Keywords

Direct comparison transfer, modeling, high-sensitivity, measurement uncertainty.

1. Introduction

Precise power measurements are essential for RF and microwave applications. Measurement instruments such as a microwave power sensor require accurate and traceable measurement capabilities, hence necessary calibrations. Direct comparison transfer technique has been widely accepted and implemented for calibrating a RF and microwave power sensor [1] - [5]. This method transfers the effective efficiency η_{Std} and the calibration factor K_{Std} of a reference standard to an uncalibrated power sensor which is the device under test (DUT), with the help of a power splitter [6] or a coupler which is used to minimize the source mismatch [7].

Sometimes, a DUT power sensor has an unmatched connector with the reference standards, and then an adaptor has to be used [3], [4], [5]. A generic model has been proposed in [5] to characterize the additional measurement error introduced by the adaptor, using signal flow graphs together with non-touching loop rule analysis [8]. Comparing to the general cases [1] - [4] where the DUT power sensor has a similar power range as the reference standard, there is another important application scenario where their power ranges are different (e.g., calibrating a high-sensitivity Agilent 8481D power sensor). An attenuator therefore has to be used, which acts as a 2-port adaptor in this case.

However in the literature, there is limited information reported for this calibration scenario with an attenuator. The mathematical model proposed for the calibrations using an adaptor in [5] could be a potential calibration model, although it still needs to be comparatively verified. Therefore in this paper, verification of the mathematical model for the calibration scenario with an adaptor and evaluation of its feasibility to calibrate a high-sensitivity power sensor with an attenuator as shown in Fig. 1 will be focused.

In the following, a brief description of the calibration system with an adaptor is given in Section 2 together with an introduction of the mathematical model derived using signal flow graphs together with non-touching loop rule analysis. In Section 3, through the analysis of physical measurement processes, the mathematical model for the calibration system with an adaptor in [5] is comparatively verified. The model is then implemented with a Type-N (up to 18 GHz) measurement system in Section 4 to evaluate its feasibility for calibrating a high-sensitivity power sensor with an attenuator as the 2-port adaptor. Finally, conclusions of this paper are given in Section 5.

2. Direct Comparison Transfer

2.1 A Brief Description

Direct comparison transfer technique for calibrating an RF and microwave power sensor through an adaptor/attenuator as shown in Fig. 1 consists of a signal generator and a 3-port power splitter which is used to minimize the source mismatch [7]. In this system, a monitoring sensor is connected to port 3 of the splitter. The effective efficiency η_{DUT} and the calibration factor K_{DUT} of a DUT are then estimated by alternately connecting a reference standard (with η_{Std} and K_{Std}) and the DUT (through an adaptor/attenuator) to port 2 of the splitter. For simplicity in the rest of this paper, we focus on the developments of the mathematical model for K_{DUT}. The same methodology can be applied to η_{DUT}.

2.2 Mathematical Model

For the calibration system with an adaptor between the DUT sensor and the splitter, a mathematical model has been proposed in [5] using signal flow-graphs together with non-touching loop rules as

$$K_{DUT} = K_{Std} \times \frac{P_{DUT}}{P_{3DUT}} \times \frac{P_{3Std}}{P_{Std}} \times \left| \frac{k_{2Std}}{k_{2DUT}} \right|^2$$

$$\times \left| \frac{1 - \Gamma_{DUT} S_{22A} - \Gamma_{e2} \Gamma_{A-DUT}}{S_{21A}(1 - \Gamma_{Std}\Gamma_{e2})} \right|^2 \quad (1)$$

where

- P_{DUT} and P_{3DUT} are the powers measured at port 2 using the DUT with an adaptor and that at port 3 using a monitoring sensor respectively,

- P_{Std} and P_{3Std} are the powers measured at port 2 using a reference standard and that at port 3 using the same monitoring sensor as for measuring P_{3DUT},

- k_{2Std} and k_{2DUT} are some unknown terms related to the leakage of cable and connector, linearity and frequency error etc. when the reference standard and the DUT are connected to port 2.

In this model (1),

$$\Gamma_{A-DUT} = S_{11A} + \Gamma_{DUT}S_{21A}S_{12A} - \Gamma_{DUT}S_{22A}S_{11A} \quad (2)$$

where Γ_{DUT} is the reflection coefficient of the DUT, and S_{lmA} is the scattering parameter (S-parameter) of the adaptor with $l, m = 1$ or 2. Γ_{Std} is the reflection coefficient of the reference standard. It is noted that Γ_{e2} is the equivalent source reflection coefficient at port 2 of the splitter and equal to [9]

$$\Gamma_{e2} = S_{22} - \frac{S_{21}S_{32}}{S_{31}}. \quad (3)$$

Here S_{pq} is the S-parameter of the 3-port power splitter with $p, q = 1, 2$ or 3.

However, this model (1) has not been comparatively validated due to limited information reported in the literature. Therefore in the following, a different interpreting way from [5] for deriving the mathematical model is focused, which is performed through the analysis of its corresponding physical measurement processes.

3. Physical Measurement Processes

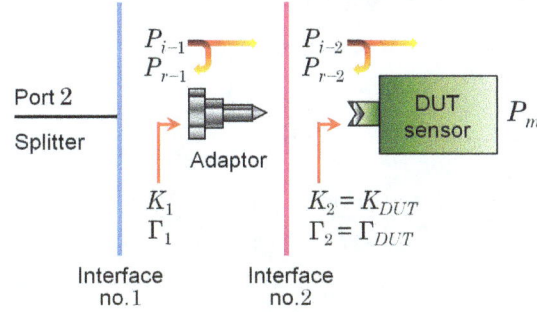

Fig. 2. Flow of the microwave power to a DUT power sensor with an adaptor.

Fig. 2 illustrates the physical processes of the incident microwave powers and their associated reflected powers due to the impedance mismatch at the connecting interfaces (i.e., no. 1 for splitter–adaptor interface and no. 2 for adaptor–DUT interface). Here P_{i-n} is the incident power at the nth interface ($n = 1$ or 2), and P_{r-n} is the associated reflected power. P_m is the measured power by the DUT power sensor and indicated on a power meter.

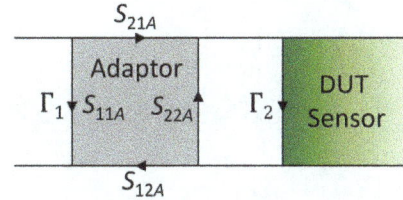

Fig. 3. A simplified flow-graph for an adaptor before a microwave power sensor corresponding to Fig. 2.

From the definition [9], the calibration factor K_2 of the DUT power sensor is

$$K_2 = \frac{P_m}{P_{i-2}}. \quad (4)$$

With Fig. 3, it is obtained that

$$P_{i-2} = P_{i-1} \left| \frac{S_{21A}}{1 - \Gamma_2 S_{22A}} \right|^2. \quad (5)$$

That is,

$$K_2 = \frac{P_m}{P_{i-1}} \left| \frac{1 - \Gamma_2 S_{22A}}{S_{21A}} \right|^2. \quad (6)$$

As the calibration factor K_1 for the DUT power sensor integrated with an adaptor is defined to be P_m/P_{i-1}, it is then derived,

$$K_2 = K_1 \left| \frac{1 - \Gamma_2 S_{22A}}{S_{21A}} \right|^2. \quad (7)$$

Moreover, the calibration factor K_1 at the interface with the microwave splitter (interface no. 1 as shown in Fig. 2) has been well-studied. Consistent conclusions have been achieved comparing the reported works in [5] with the method [1]. K_1 can be expressed as

$$K_1 = K_{Std} \times \frac{P_{DUT}}{P_{3DUT}} \times \frac{P_{3Std}}{P_{Std}} \times \left| \frac{k_{2Std}}{k_{2DUT}} \right|^2 \times \left| \frac{1 - \Gamma_1 \Gamma_{e2}}{1 - \Gamma_{Std}\Gamma_{e2}} \right|^2. \quad (8)$$

That is, for calibrating a DUT through an adaptor shown in Fig. 1, K_{DUT} of the stand-along DUT can be obtained from (7) and (8) as

$$K_{DUT} = K_2 = K_{Std} \times \frac{P_{DUT}}{P_{3DUT}} \times \frac{P_{3Std}}{P_{Std}} \times \left| \frac{k_{2Std}}{k_{2DUT}} \right|^2$$
$$\times \left| \frac{1 - \Gamma_1 \Gamma_{e2}}{1 - \Gamma_{Std} \Gamma_{e2}} \right|^2 \times \left| \frac{1 - \Gamma_2 S_{22A}}{S_{21A}} \right|^2. \quad (9)$$

For the item $|1 - \Gamma_1 \Gamma_{e2}|$ in (9), Γ_1 can be derived in terms of S-parameters of the adaptor and the reflection coefficient Γ_2 ($\Gamma_2 = \Gamma_{DUT}$) of the DUT power sensor as

$$\Gamma_1 = S_{11A} + \frac{S_{21A} \Gamma_2 S_{12A}}{1 - \Gamma_2 S_{22A}}. \quad (10)$$

Then,

$$|1 - \Gamma_1 \Gamma_{e2}|$$
$$= \left| 1 - \left\{ S_{11A} + \frac{S_{21A} \Gamma_2 S_{12A}}{1 - \Gamma_2 S_{22A}} \right\} \Gamma_{e2} \right|$$
$$= \left| \frac{1 - \Gamma_2 S_{22A} - \{S_{11A} - S_{11A} \Gamma_2 S_{22A} + S_{21A} \Gamma_2 S_{12A}\} \Gamma_{e2}}{1 - \Gamma_2 S_{22A}} \right|$$
$$= \left| \frac{1 - \Gamma_2 S_{22A} - \Gamma_{A-DUT} \Gamma_{e2}}{1 - \Gamma_2 S_{22A}} \right| \quad (11)$$

where $\Gamma_{A-DUT} = S_{11A} + \Gamma_{DUT} S_{21A} S_{12A} - \Gamma_{DUT} S_{22A} S_{11A}$ and $\Gamma_2 = \Gamma_{DUT}$ as indicated in Fig. 2.

Substituting $|1 - \Gamma_1 \Gamma_{e2}|$ in (9) with (11), it is found that the derived mathematical model (9) through the analysis of physical measurement processes as described above, is exactly the same as the one developed using the signal flow graphs together with non-touching loop rule analysis in [5] which is shown as (1) in this paper. That is, the mathematical model ((9) or (1)) has been comparatively validated and is suitable for the microwave power sensor calibration with an adaptor between the DUT sensor and the power splitter.

Moreover, our recent works [10] indicated that a high-sensitivity microwave power sensor (e.g., Agilent 8481D power sensor) could be a potential reference standard for calibrating a thermal voltage converter at a higher operating frequency. Accurate calibration of such a high-sensitivity microwave power sensor then becomes important. However, this type of sensor usually has a different power range comparing to the primary RF and microwave power standard. Therefore, an attenuator has to be used. In the following, the feasibility of the mathematical model ((9) or (1)) to calibrate such a high-sensitivity microwave power sensor with an attenuator as the 2-port adaptor will be evaluated and focused.

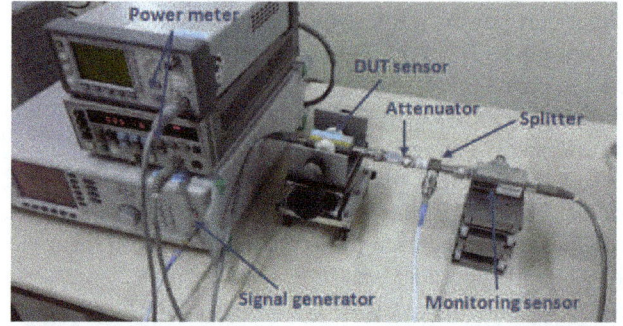

(a) The whole calibration system when the DUT sensor is connected to port 2 of the splitter through an attenuator

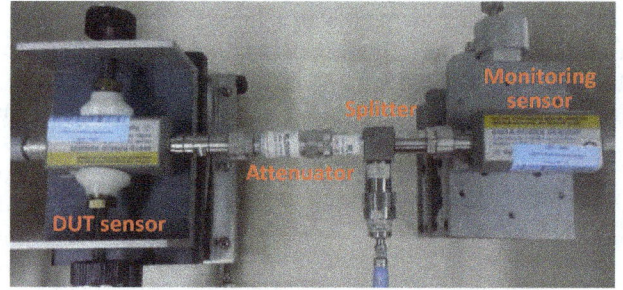

(b) Highlighted connections with the splitter

Fig. 4. Physical realization of direct comparison transfer for power sensor calibration with an attenuator as the 2-port adaptor.

4. Feasibility Study and Analysis

4.1 Practical Calibration System

Feasibility study of the mathematical model (9) (or (1)) for the calibration scenario shown in Fig. 1, is physically realized using a Type-N microwave power sensor calibration system where a 30 dB attenuator acts as the 2-port adaptor between a DUT power sensor and a splitter (similar to [11]). The practical calibration system is presented in Fig. 4.

As shown in Fig. 4, an Agilent 8481D power sensor (power range: 100 pW – 10 μW and frequency range: 10 MHz – 18 GHz) is the DUT sensor. The reference standard is a thermistor mount which is fitted with a Type-N connector and calibrated in term of the effective efficiency at 1 mW directly by means of a microwave micro-calorimeter. The microwave micro-calorimeter is a primary power standard with fixed output power which converts the absorbed microwave energy into the heat (i.e., thermalize the microwave energy).

Key referenced parameters (i.e., η_{Std} and K_{Std}) of the thermistor mount at 1 mW is then transferred to the Agilent 8481D sensor (with a smaller power range of [100 pW, 10μW]) through a 30 dB attenuator. It is noted that for the power leveling and monitoring at port 3 of the power splitter shown in Fig. 4, an Agilent 8481A sensor (power range: 1 μW – 100 mW and frequency range: 10 MHz – 18 GHz) is used.

4.2 Performance Evaluation and Analysis

The measurement estimate K_{DUT} is calculated accordingly using the mathematical model (9) (or (1)), while its associated measurement uncertainty is evaluated following an internationally recommended guideline, *Guide to the Expression of Uncertainty in Measurement* (GUM) [12] which has been widely accepted. For simplicity in the demonstration of uncertainty evaluation, the following relationship is used to represent the mathematical model,

$$y = f(x_1, x_2, x_3, ..., x_N). \tag{12}$$

Here, y is the estimate K_{DUT}, and x_1, x_2, x_3, ..., x_N represent the influencing quantities K_{Std}, Γ_{DUT}, Γ_{e2}, Γ_{Std}, S_{21A} etc.

According to the *Law of Propagation of Uncertainty* in the GUM [12], the combined standard uncertainty u_c associated with y (i.e., $u_c(y)$) can be obtained from the standard uncertainties of $x_1, x_2, x_3, ..., x_N$ through

$$u_c(y) = \sqrt{\sum_{i=1}^{N} \left[\frac{\partial f}{\partial x_i}\right]^2 u^2(x_i) + 2 \sum_{i=1}^{N-1} \sum_{j=i+1}^{N} \frac{\partial f}{\partial x_i} \frac{\partial f}{\partial x_j} u(x_i, x_j)} \tag{13}$$

where $u(x_i)$ is the associated standard uncertainty for i^{th} influencing quantity x_i. $u(x_i, x_j)$ is the covariance between x_i and x_j, and equal to

$$u(x_i, x_j) = r(x_i, x_j)u(x_i)u(x_j). \tag{14}$$

Here, $r(x_i, x_j)$ is the correlation coefficient between x_i and x_j. It is noted that the correlation coefficient r between the influencing quantities is relatively small in this calibration, and it is also inherently unreliable due to small sample size in practical calibrations as reported in [13]. Therefore in this study, $u_c(y)$ is evaluated with the assumption of zero correlation ($r = 0$) between the influencing quantities.

For the standard uncertainty $u(x_i)$ for x_i, it can be evaluated using either *Type A* or *Type B* method according to the GUM. For the *Type A* method, $u(x_i)$ is evaluated by the statistical analysis of series of observations; while for the *Type B* method, $u(x_i)$ is obtained from other information including previous measurement data, specifications from manufacturers, data provided in calibration and other certificates, and uncertainties assigned to reference data taken from handbooks, etc. Moreover for the complex-valued microwave quantities such as S-parameters and reflection coefficients, their standard uncertainties are evaluated with the assumption of zero correlation between their real and imaginary parts as we discussed in [14].

4.2.1 Evaluations Using the GUM and MCM Methods

In this paper, uncertainty evaluation using the GUM method is focused. This is because the uncertainty evaluation using the GUM method has been accepted and used in most of current routine calibration works. When evaluating

$u(x_i)$ following the *Type B* method, probability distribution for x_i needs to be prior-determined. Normally with the assumptions of Gaussian distributions for all the influencing quantities (ordinary cases), the measurement uncertainty of K_{DUT} can be evaluated with the mathematical model (9) (or (1)) accordingly.

To validate the assumed probability distributions, Monte Carlo method (MCM) as recommended in [15] is chosen for a comparison. In the Monte Carlo simulations, the characteristics of assumed Gaussian distributions for all the influencing quantities are directly from the measurement estimates.

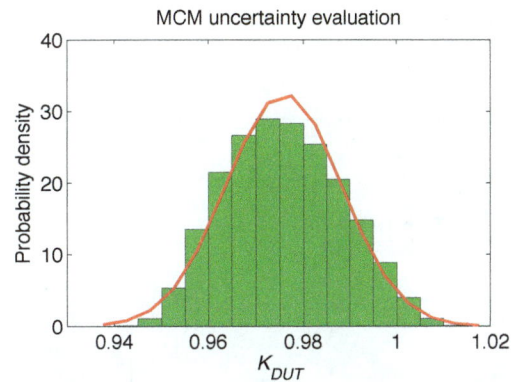

Fig. 5. Example of the simulated results using the Monte Carlo method with fitted distribution.

Fig. 5 shows an example of the simulated results using the MCM method. From Fig. 5, it is found that the representative distribution for K_{DUT} estimated using the MCM method also approximates to be Gaussian distributed. This is because the recommended guideline [15] is the *Law of Propagation of Distributions* essentially, which propagates the assigned probability distributions to the influencing quantities to the desired parameter (K_{DUT} in this study) as we discussed in [16]. For the MCM method, the measurement estimate and its associated uncertainty are determined from the experimental distribution as shown in Fig. 5.

The measurement estimate K_{DUT} and its associated combined standard uncertainty $u_c(K_{DUT})$ for the DUT power sensor (i.e., Agilent 8481D sensor) using the GUM and MCM methods are shown in Fig. 6 respectively. From Fig. 6, it is found that both the methods generate very close results especially at the lower frequencies. This indicates that the assumptions of Gaussian distribution for all the influencing quantities in (9) are suitable.

In the next subsection, feasibility of the mathematical model (9) (or (1)) to calibrate a high-sensitivity power sensor with an attenuator is further evaluated and compared to the calibration data from manufacturer. The uncertainty evaluated using the GUM method is used in performance comparison as it is implemented in most of current routine calibration works.

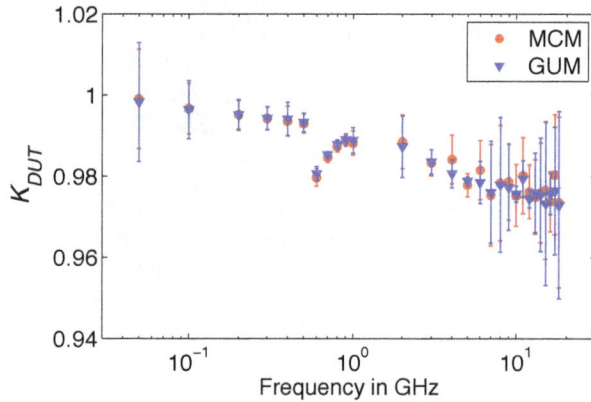

Fig. 6. The results estimated using the MCM and GUM methods (Uncertainty bars are shown for u_c).

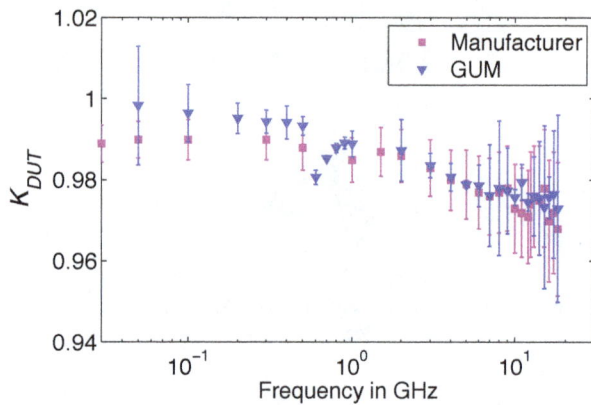

Fig. 7. The results provided by the manufacturer and estimated using the GUM (Uncertainty bars are shown for u_c).

4.2.2 Comparing the Results Following the GUM With the Data From Manufacturer

Fig. 7 presents the evaluated results using the GUM method against the calibration data from manufacturer. Comparing to the differences between the results from the GUM and MCM methods in Fig. 6, large discrepancies are observed between the results from the GUM method and from the manufacturer. To analyze their discrepancies quantitatively, error parameter E_n [17] as defined below is used,

$$E_n = \frac{\delta_A - \delta_B}{\sqrt{U_A^2 + U_B^2}} \qquad (15)$$

where δ_A and δ_B are the measurement estimates for K_{DUT} using the GUM method and the calibration data from manufacturer respectively, and U_A and U_B are their corresponding expanded uncertainties (equal to $2u_c$ at a confidence level of approximately 95% assuming a Gaussian distribution). According to [17], the discrepancies between the evaluated results are acceptable when $|E_n| \leq 1$.

The calculated $|E_n|$ for the discrepancies between the results from the GUM method and from the manufacturer is shown in Fig. 8. Only the data with same frequencies are

selected for analysis. It is observed from Fig. 8 that generally the calibration results from our method with uncertainty evaluation using the GUM method show very good agreements (all the $|E_n| \leq 0.5$) with the calibration data from manufacturer. These observations therefore demonstrate a good measurement and calibration capability of the mathematical model (9) (or (1)) to calibrate a high-sensitivity power sensor with an attenuator.

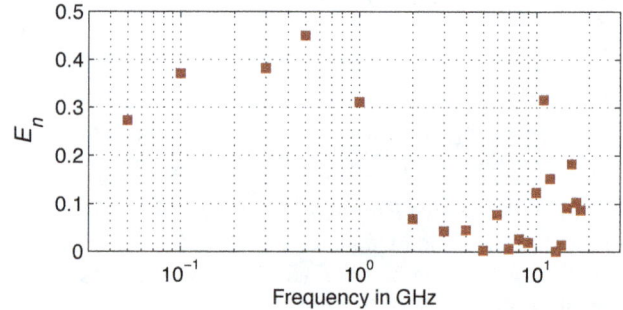

Fig. 8. The calculated $|E_n|$ with normalization for comparing the results from manufacturer and GUM as shown in Fig. 7.

Moreover it is found from Fig. 8 that extremely excellent agreements are achieved when the operating frequency $f \in [1, 10]$ GHz (i.e. $[10^0, 10^1]$ shown in Fig. 8) as all the $|E_n|$ in this region is ≤ 0.1. For other regions ($f \leq 1$ GHz and $f \geq 10$ GHz), the discrepancies between the results from the GUM method and from the manufacturer become larger. The main reasons for larger discrepancies at lower frequencies ($f \leq 1$ GHz) is due to the performance limitation of the microwave micro-calorimeter used in this study, while the larger discrepancies at higher frequencies ($f \geq 10$ GHz) might be because the reflection due to the impendence mismatch of additional attenuator with the splitter or the DUT sensor becomes significant at higher frequency and then affects our calibration results.

5. Conclusions

In this paper, we presented a feasibility study to calibrate a high-sensitivity microwave power sensor through an attenuator using direct comparison transfer technique. Our previously derived mathematical model using signal flow graphs together with non-touching loop rule analysis has been further investigated and validated with the analysis of physical measurement processes.

Performance of the mathematical model (9) (or (1)) was evaluated using the GUM and MCM Methods through a Type-N (up to 18 GHz) measurement system first, and then compared to the calibration data from manufacturer. Good agreements (all the $|E_n| \leq 0.5$) with the data from manufacturer have been achieved which demonstrates a good calibration and measurement capability of the mathematical model (9) (or (1)) to calibrate a high-sensitivity microwave power sensor with an attenuator.

References

[1] WEIDMAN, M. P. *Direct Comparison Transfer of Microwave Power Sensor Calibration.* NIST Technical Note 1379, 1996.

[2] GINLEY, R. A direct comparison system for measuring radio frequency power (100 kHz to 18 GHz). *Measure*, 2006, vol. 1, no. 4, p. 46 - 49.

[3] WALLIS, T. M., CROWLEY, T. P., LEGOLVAN, D. X., GINLEY, R. A. A direct comparison system for power calibration up to 67 GHz. In *Digest of the 2012 Conference on Precision Electromagnetic Measurements.* Washington DC (USA), 2012, p. 726 - 727.

[4] KANG, T. W., KIM, J. H., KWON, J. Y., et al. Direct comparison technique using a transfer power standard with an adapter and its uncertainty. In *Digest of the 2012 Conference on Precision Electromagnetic Measurements.* Washington DC (USA), 2012, p. 728 - 729.

[5] SHAN, Y., MENG, Y. S., LIN, Z. Generic model and case studies of microwave power sensor calibration using direct comparison transfer. *IEEE Transactions on Instrumentation and Measurement*, 2013, vol. 62, no. 6, p. 1834 - 1839.

[6] Agilent Application Note. *Differences in Application Between Power Dividers and Power Splitters.* [Online] 2007. Avaliable at: http://cp.literature.agilent.com/litweb/pdf/5989-6699EN.pdf

[7] ENGEN, G. F. Amplitude stabilization of a microwave signal source. *IRE Transactions on Microwave Theory and Techniques*, 1958, vol. 6, no. 2, p. 202 - 206.

[8] POZAR, D. M. *Microwave Engineering.* Addison-Wesley, 1993.

[9] FANTOM, A. *Radio Frequency and Microwave Power Measurement.* Peter Peregrinus Ltd., UK, 1990.

[10] SHAN, Y., MENG, Y. S., FILIPSKI, P. S. Evaluation of a calorimetric thermal voltage converter for RF-DC difference up to 1 GHz. *IEEE Transactions on Instrumentation and Measurement*, 2014, vol. 63, no. 2, p. 467 - 472.

[11] YHLAND, K., STENARSON, J., WINGQVIST, C. Power sensor linearity calibration with an unknown attenuator. In *Digest of the 2010 Conference on Precision Electromagnetic Measurements.* Daejeon (Korea), 2012, p. 769 - 770.

[12] BIPM, IEC, IFCC, ILAC, ISO, et al. *Evaluation of Measurement Data – Guide to the Expression of Uncertainty in Measurement.* JCGM 100:2008 (GUM 1995 with minor corrections). Joint Committee for Guides in Metrology, 2008.

[13] RIDLER, N. M., SALTER, M. J. An approach to the treatment of uncertainty in complex S-parameter measurements. *Metrologia*, 2002, vol. 39, no. 3, p. 295 - 302.

[14] MENG, Y. S., SHAN, Y. Measurement uncertainty of complex-valued microwave quantities. *Progress In Electromagnetics Research*, 2013, vol. 136, p. 421 - 433.

[15] BIPM, IEC, IFCC, ILAC, ISO, et al. *Evaluation of Measurement Data – Supplement 1 to the 'Guide to the Expression of Uncertainty in Measurement'– Propagation of Distributions Using a Monte Carlo Method.* JCGM 101:2008. Joint Committee for Guides in Metrology, 2008.

[16] ZHANG, Q., MENG, Y. S., SHAN, Y., LIN, Z. Direct comparison transfer of microwave power sensor calibration with an adaptor: modeling and evaluation. *Progress In Electromagnetics Research Letters*, 2013, vol. 38, p. 25 - 34.

[17] APLAC PT001. *Calibration Interlaboratory Comparisons.* Asia Pacific Laboratory Accreditation Cooperation – Proficiency Testing Committee, 2008.

About Authors...

Yu Song MENG received the B.Eng. (Hons.) and Ph.D. degrees in Electrical and Electronic Engineering from Nanyang Technological University, Singapore, in June 2005 and February 2010, respectively. From May 2008 to June 2009, he was a Research Engineer with the School of Electrical and Electronic Engineering, Nanyang Technological University, Singapore. Since July 2009, he has been with the Agency for Science, Technology and Research (A*STAR), Singapore. He was firstly with A*STAR's Institute for Infocomm Research as a Research Fellow, and then a Scientist I. In September 2011, he was transferred to A*STAR's National Metrology Centre where he is currently a Scientist II. His research interests include electromagnetic metrology, electromagnetic measurements and standards, and radio-wave propagations.

Yueyan SHAN received the B.Eng. degree from the Beijing Institute of Technology, China, the M.Eng. degree from Nanyang Technological University, Singapore, and the Ph.D. degree from The Hong Kong Polytechnic University, China. She participated in several R&D projects in radar signal design, wireless communication, automation and signal processing, test system design, as well as microwave measurements. In 1998, she joined the National Metrology Centre, Singapore. She contributed in the establishment of national time and frequency standard and the setup and development of the national microwave measurement standards and calibration systems. She is currently a Principal Metrologist and an Assistant Head of the Electrical Metrology Cluster.

13

Concentrated Ground Plane Booster Antenna Technology for Multiband Operation in Handset Devices

Cristina PICHER[1], Jaume ANGUERA[1,2], Aurora ANDÚJAR[1], Carles PUENTE[3], Adrián BUJALANCE[2]

[1] Technology Department, Fractus, Barcelona, Spain
[2] Electronics and Communications Department, Universitat Ramon Llull, Barcelona, Spain
[3] Signal Theory and Communications, Universitat Politècnica de Catalunya, Spain

jaume.anguera@fractus.com

Abstract. *The current demand in the handset antenna field requires multiband antennas due to the existence of multiple communication standards and the emergence of new ones. At the same time, antennas with reduced dimensions are strongly required in order to be easily integrated. In this sense, the paper proposes a compact radiating system that uses two non-resonant elements to properly excite the ground plane to solve the abovementioned shortcomings by minimizing the required Printed Circuit Board (PCB) area while ensuring a multiband performance. These non-resonant elements are called here ground plane boosters since they excite an efficient mode of the ground plane. The proposed radiating system comprises two ground plane boosters of small dimensions of 5 mm x 5 mm x 5 mm. One is in charge of the low frequency region (from 0.824 GHz to 0.960 GHz) and the other is in charge of the high frequency region (1.710 GHz–2.170 GHz). With the aim of achieving a compact configuration, the two boosters are placed close to each other in a corner of the ground plane of a handset device (concentrated architecture). Several experiments related to the coupling between boosters have been carried out in two different platforms (barphone and smartphone), and the best position and the required matching network are presented. The novel proposal achieves multiband performance at GSM850/900/1800/1900 and UMTS.*

Keywords

Handset antennas, multi-band, non-resonant antennas, ground plane modes.

1. Introduction

The current requirements in handset antenna design are related to multiband operation and miniaturization. A popular platform which is gaining popularity is the smartphone, which features bigger dimensions than conventional cellular barphones. However, the available space for the antenna is still limited due to the presence of large displays, batteries and related components (multiple cameras, hands-free speakers). The consequences of adding all these functionalities is a challenge for the antenna design because the antenna should have small dimensions in order to be easily integrated with such other elements of the device.

There are different ways of designing multiband and miniature antennas, which basically consist in shaping the geometry of different types of antennas like monopoles [1]–[5], PIFAs [6]–[9], and slots [10], or even adding parasitic elements to the radiating system [11], [12]. However, the antenna is not the only contributor to the radiation since the ground plane also plays a very important role in the overall radiating system [13]–[44].

As it was demonstrated in [14]–[20], the antenna bandwidth can be improved by achieving a ground plane length of approximately 0.4λ at the operating frequencies because the ground plane fundamental mode is efficiently excited. For example, in [17]–[25], slotted ground planes are used in order to achieve a longer electrical path in a shorter physical length with the aim of exciting the fundamental mode of the ground plane which is an efficient radiating mode.

Based on the same principle and knowing that the ground plane is the main contributor, there have been different studies which used the excitation of the ground plane as the principal means to obtain good radiation [29]–[45]. One way of achieving this excitation is through the use of coupling elements, which are strategically located in order to properly excite the ground plane and have a certain C-shape to obtain such coupling [35]–[42]. In [35], a monoband antenna system comprising a microstrip line and a coupling element of 2400 mm^3 was presented, and in [36], two coupling elements occupying a volume of 700 mm^3, a ground plane, and a matching circuitry were needed in order to obtain a quad-band radiating system. One element was required for the GSM850/900 bands and the other for the GSM 1800/1900 bands. Despite those couplers relied on a different principle than common handset antennas, they still featured a significant size similar to those (i.e. a size about the entire shortest edge of a mobile platform), therefore not providing a significant advantage compared to state-of-the-art antennas.

A different way of exciting the ground plane is by means of using very simple and small structures called ground plane boosters, as proposed in [29]–[32] without the need of using said aforementioned C-shaped structures. The ground plane boosters proposed here are solid metallic structures featuring a cube shape and connected with a feeding point to the ground plane. In [31], ground plane boosters have been used to properly excite the ground plane mode to obtain a multiband performance (GSM850/900/1800/1900, and UMTS) with a total volume of only 250 mm^3. In [32], planar booster elements (2D) were used to obtain LTE700, GSM850/900/1800/1900, UMTS, LTE2300/2500 and GPS with a total footprint of only 153 mm^2. These elements feature a non-resonant behavior with a very high quality factor (Q) compared to resonant antennas. Such a high Q explains the non-radiating nature of those elements. The name of ground plane booster was therefore proposed to describe such reactive, high-Q, non-radiating and non-resonating simple elements that are mainly used to excite the ground plane in order to obtain an efficient radiating structure [29]–[33], which can be obtained by exciting the first mode according to the characteristic modes theory [31], [44]. In particular, the modal significance has been computed for a ground plane having 100 mm × 40 mm (Fig. 1). When the modal significance is close to one, the mode is resonant and radiative mode whereas a mode having a modal significance close to zero means a reactive mode. For a ground plane having a 100 mm × 40 mm size, the main mode (J_1) dominates across the frequency region of 800 MHz to 2 GHz. Up to that point, J_1 and J_4 play an important role being still J_1 predominating up to 2.5 GHz. From current simulation shown in the following sections, it will be demonstrated that at the low frequency region the current distribution is mainly determined by J_1 and for the high frequency region is a linear combination of J_1 and J_4.

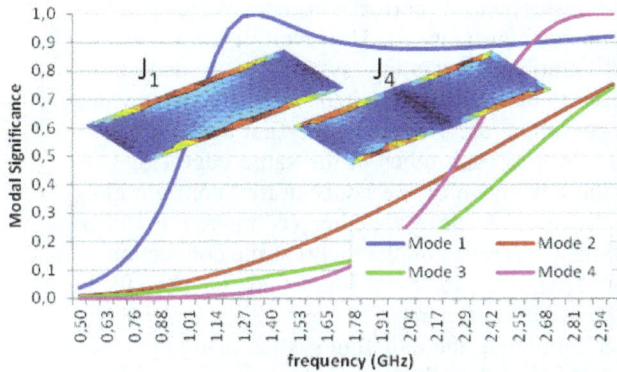

Fig. 1. Computed modal significance for the first eigen-modes for a ground plane of 100 mm × 40 mm. Current distribution associated to the first J_1 and fourth J_4 eigen vectors.

In the present study, such ground plane boosters have been used with the aim of achieving good radiation performance in several frequency bands in a novel configuration featuring a concentrated architecture [32]. In [31], a distributed configuration (Fig. 2a) was presented. Since the antenna engineer has a lot of constrains in terms of available area for the antenna, a further research has been carried out to obtain a different solution capable of reduc-

ing the required PCB area. In this case, a novel concentrated solution (Fig. 2-b) using two closely spaced ground plane boosters has been designed and analyzed. In principle, since both boosters are closely spaced, mutual coupling drastically degrades the performance compared to the distributed solution. Therefore, this paper analyzes different configurations and filter schemes capable of mitigating the mutual coupling while obtaining a multiband performance at the low and high frequency region in two different topologies of ground planes: a barphone and a smartphone. The latter has been included in order to establish the behavior of the proposed technology in one of the most popular devices of the current market demand.

The paper is divided as follows. Section 2 explains the behavior of the boosters that form the radiating system. Section 3 illustrates the measurements of the built prototype in order to validate the simulation results shown in the previous section. Finally, in Section 4, the conclusions are presented.

2. The Radiating System

The radiating structure of a first proposed design consists of a 100 mm × 40 mm ground plane, which is a typical size of a conventional barphone, and two metallic ground plane boosters of 5 mm × 5 mm × 5 mm located in one corner of the ground plane (Fig. 2-b). Owing to the reduced dimensions of the boosters and their close arrangement, the solution confines the multiband radiating solution of the mobile platform in an extremely concentrated configuration.

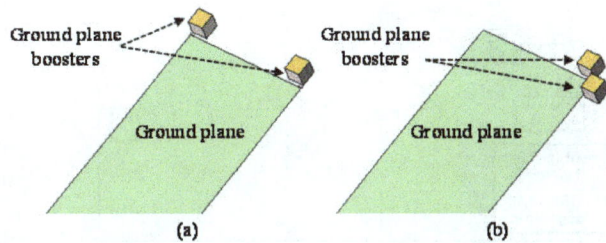

Fig. 2. (a) Two ground plane boosters distributed in each corner of the ground plane of a handset device; (b) concentrated configuration having both boosters close to each other (without being connected).

2.1 Functional Analysis

The position of the boosters and their separation is a critical factor that needs to be taken into account in the design process as it is explained next.

The radiating system needs to offer a multiband performance able to satisfy the current market demand. Thus, it should operate in the common cellular frequency bands, such as GSM850/900/1800/1900, and UMTS.

As discussed, the ground plane boosters are non-resonant and non-radiating elements, as they feature a high quality factor in each of the target frequency bands

($Q \approx 2250$ @0.9 GHz and $Q \approx 265$ @1.8 GHz) [31]. However, their position helps the radiation modes in the ground plane to be efficiently excited and provide good radiation performance. Therefore, the radiating structure comprises the ground plane boosters and the ground plane. A matching network is then added to provide the impedance matching at the desired frequency bands.

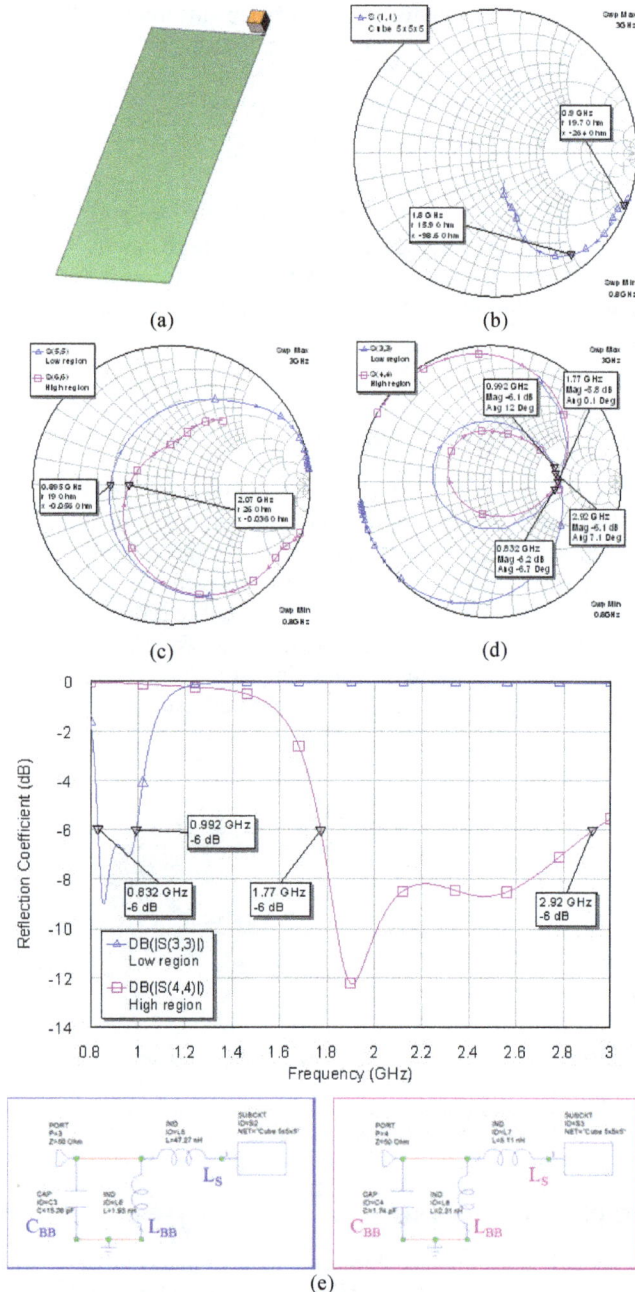

Fig. 3. Matching procedure applied at both frequency regions. (a) 3D view of the radiating structure. (b) Impedance of the radiating structure without any matching network. The ground plane booster has a strong capacitive behavior at both frequency regions. (c) The radiating structure resonates at 0.9 GHz or 2.0 GHz depending on the value of the series inductor. (d, e) The ground plane booster can be matched at both frequency regions offering good bandwidth results by means of different broadband matching networks.

First of all, it is necessary to understand the behavior of each booster in each frequency region independently, that is, having only one ground plane booster (Fig. 3-a). These ground plane boosters are highly reactive elements with a strong capacitive behavior across the desired frequencies (Fig. 3-b). In order to make them resonant, a series inductor (L_S) is added which results in an input impedance as an RLC series circuit (Fig. 3-c). After that, a broadband matching network (L_{BB} and C_{BB}) [45], [46] is added to enhance the bandwidth. The objective of the broadband matching network (a shunt LC circuit) is to create an input impedance loop that can be inscribed within the SWR = 3 circle of the Smith chart. With the proposed matching network, the bandwidth can be enhanced by a factor of 2.45 (SWR = 3), which is significant taking into account that only two components are used [47].

In the case of the booster for the low frequency region, good bandwidth results are obtained with a series inductor (L_S) of 47.3 nH, and a broadband matching network comprising a 1.9 nH inductor (L_{BB}) and a 15.3 pF capacitor (C_{BB}) (Fig. 3-d, e). For the booster in charge of the high frequency region, a similar matching network topology is applied, consisting of a series inductor of 5.1 nH (L_S) and a broadband matching network composed of a 2.3 nH inductor (L_{BB}) and a 1.7 pF capacitor (C_{BB}) (Fig. 3-d, e).

It has been found that the optimum position is in the shortest edge of the ground plane [31] because it is the place where the ground plane mode is efficiently excited.

With the objective to simultaneously cover the desired frequency bands, a modular radiating system is preferred. This means that two ground plane boosters are needed to match each one at one frequency region as in Fig. 2-b.

Each booster behaves correctly when it is alone and in the optimum position (Fig. 3). However, since a concentrated and multiband radiating system is needed, both boosters need to be working simultaneously and occupy the minimum possible area. Therefore, the first approach is to place them close to each other in the same edge (Fig. 4). Note that the advantage of the concentrated configuration having both boosters close together (compared to other configurations [27]) relies on the fact that the connection to a multiband front-end module is simplified since the feeding lines for each booster may be short. As it can be observed after simulating the radiating system using the electric scheme of Fig. 3-e, the required performance is not achieved (Fig. 4). In the low frequency region, the S_{11} is above -6 dB and the S_{21} presents unsatisfactory values (-0.28 dB), which means that this particular configuration as it is becomes completely useless for a modern multiband phone. In order to guarantee the same behavior as the one shown in Fig. 3, both boosters should have low S_{21} values ($S_{21} < $ -20 dB) in both frequency regions and therefore, be less coupled. In this sense, and as presented in the following section, the purpose of this paper relies on providing a proper isolation in order to guarantee the performance of this novel concentrated solution.

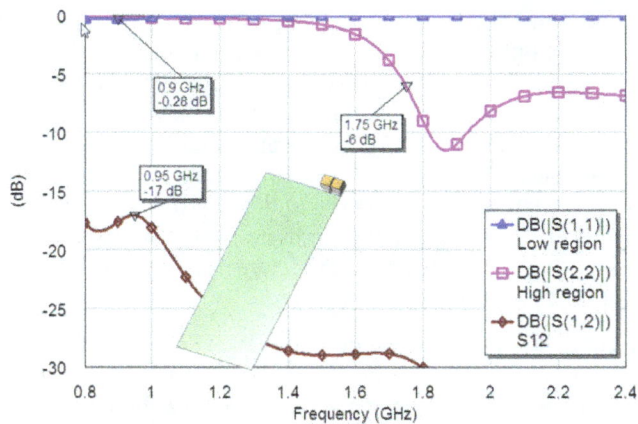

Fig. 4. Impedance response of the radiating system comprising two boosters close to each other. They use the same matching network as the one shown in Fig. 3. Multiband performance is not achieved in the low frequency region due to the strong mutual coupling.

2.2 Coupling Analysis

Commonly, the coupling of a determined circuit is directly measured by the S_{21} parameter. However, in this case, it would not be fair to rely on it because the boosters are non-resonant elements presenting very high impedance at the frequencies of interest and therefore they are totally mismatched when no matching network is used. Since they are not matched, they do not interfere to each other and therefore, the S_{21} is very low and therefore is not realistic [48]. For this reason, the maximum available gain (MAG) is calculated [49]. Maximum available gain takes into account conjugate matching at both ports.

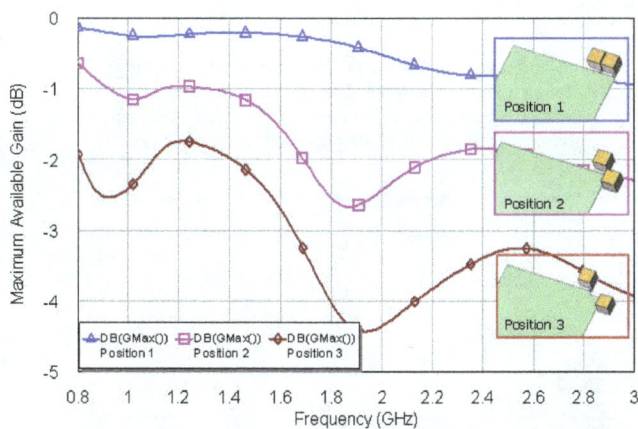

Fig. 5. Maximum available gain values obtained for the three experiments gathering different ground plane boosters' arrangements. The coupling decreases (the maximum available gain decreases) from position 1 (the worst) to position 3 (the best).

Without losing the objective of achieving a concentrated configuration, three different positions have been analyzed in terms of MAG. The first one (position 1) has both boosters next to each other in the same axis separated by 1.5 mm (Fig. 5). On the contrary, the second and third

positions (position 2 and 3) feature a different configuration because both boosters minimize the area of close contact as they are orthogonal to each other. In the second position, each booster is aligned with one of the edges of the ground plane (Fig. 5), and in the third position, they follow the same configuration but one of the boosters is located 5 mm away from the ground plane's corner (Fig. 5).

The best position is 'position 3' because it is the one where the radiating structure presents the lowest MAG between boosters, that is, less coupling. However, the configuration is not as concentrated as the other ones. On the contrary, 'position 1' presents the highest (worst) MAG values because the area of contact between boosters is larger. Hence, the chosen position is 'position 2' because it presents the best trade-off between compactness and coupling. This position shows lower MAG values compared to 'position 1' because the area of contact between boosters is lower.

2.3 Multiband Performance

Once the location for both boosters is chosen, it is important to establish the matching network needed for each frequency region (Fig. 6-a). As it has been explained in Section 2, each booster needs a series inductor in order to make it resonant at a given frequency within the band of operation, and a broadband matching network to achieve good bandwidth.

The booster located in the upper part of the shortest edge of the ground plane (Fig. 6-b) is responsible for the low frequency region, and the booster located next to the left side of the longer edge (Fig. 6-b) is in charge of the high frequency region. The radiating system has two input ports, one for each frequency region. This may be useful for front-end modules having separate entries for each frequency region/band as some used in some cellular phones. If needed, both ports can be merged into one by means of combining filters [31].

Based on the study of the MAG, 'position 2' reduces coupling compared to 'position 1'. However, they still do not behave as if they were alone and some kind of filtering to reduce coupling is needed. For the series inductor (L_{Slow}) corresponding to the low frequency region acts as high impedance at the high frequency region (its value of the order of 40 nH is a high impedance at the high frequency region), no filters are required to obtain better isolation at the high frequency region. However, the series inductor (L_{Shigh}) corresponding to the high frequency region is low (of the order of 12 nH) and it does not present high impedance in the low frequency region (12 nH is not a high impedance at the low frequency region). That is why a notch filter is needed in this case to achieve a good isolation at the low frequency region. In this sense and thanks to the lower coupling values between boosters obtained due to their relative location, only a simple notch filter (one stage) is added.

(a)

(b)

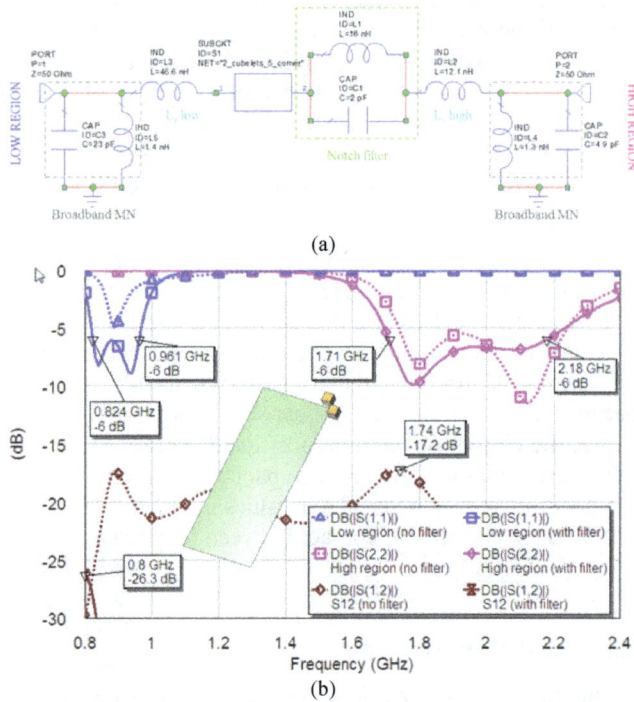

Fig. 6. (a) The matching network consists of 8 lumped elements. (b) Thanks to the notch filter (high impedance at frequencies around 0.9 GHz) and the broadband matching networks, multiband operation and good isolation between boosters is obtained. When no filter is used (dashed lines), coupling increases and operation in the low frequency region is lost.

After optimizing the components' values of the matching network, the multiband performance is achieved with occupying only 250 mm^3 (Fig. 6-a). Simulation results show that the radiating system is matched (SWR $\leq$ 3) from 0.824 GHz to 0.961 GHz (GSM850 and GSM900 as well as some LTE standard within this range such as LTE900) and from 1.71 GHz to 2.18 GHz (GSM1800, GSM1900, UMTS, as well as some LTE standards within this range such as LTE1800, 1900, 2100) (Fig. 6-b). S$_{21}$ is below -26 dB along the desired frequency bands (Fig. 6-b).

It is important to remark that without the filter it is not possible to keep the desired multiband behavior, especially in the low frequency region due to the mutual coupling (Fig. 6). Thanks to the notch filter, coupling decreases 10 dB with respect the solution without the notch filter and better performance from 0.824 GHz to 0.960 GHz is achieved.

Although the notch filter plays an important role in the performance of the radiating system, the position is also a key factor in terms of coupling. When both boosters are in 'position 1', despite having the notch filter, the multiband response gets worsen and out of the required specs, especially in the low frequency region, where GSM900 is lost (Fig. 7). Therefore, both the position and the filter play an important role in determining the multiband performance with highly isolated ports.

Since the smartphone platform is gaining interest, a new experiment considering a ground plane of

120 mm × 50 mm, which is a typical size of said platform, has been carried out. Using the same ground plane booster architecture and the same matching network, the performance of the new radiating system has been also simulated (Fig. 8). As it is observed in Fig. 8, the impedance performance of the concentrated ground booster antenna technology in the new platform is not altered in the low frequency region and even enhanced in the high frequency region, which means that the proposed architecture is a standard solution for both types of platforms.

Current distributions on the ground plane (Fig. 9) also show that the predominant mode (first mode) of the radiating structure at 0.9 GHz presents a current distribution similar to that produced by a half-wavelength dipole, that is, null at the short edges of the ground plane and maximum at the center on the long edge. At 1.8 GHz, although the current distribution slightly differs from the one obtained in the low frequency region, the first mode J$_1$ is still predominant (main current follows the y-direction) and also the J$_4$ mode appears.

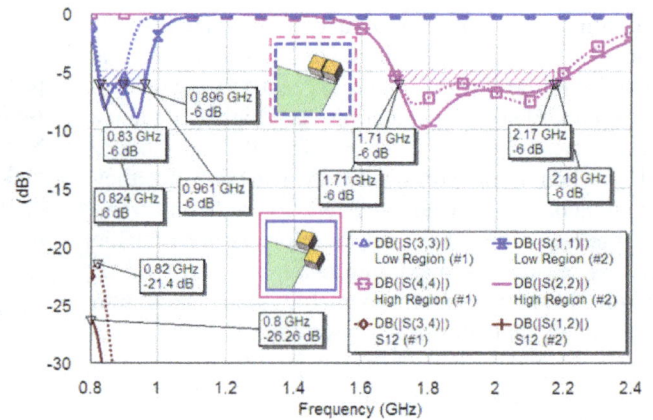

Fig. 7. The radiating system with the optimized matching network that uses a notch filter has less bandwidth in the low frequency region when boosters are in 'position 1' compared to the proposed 'position 2'.

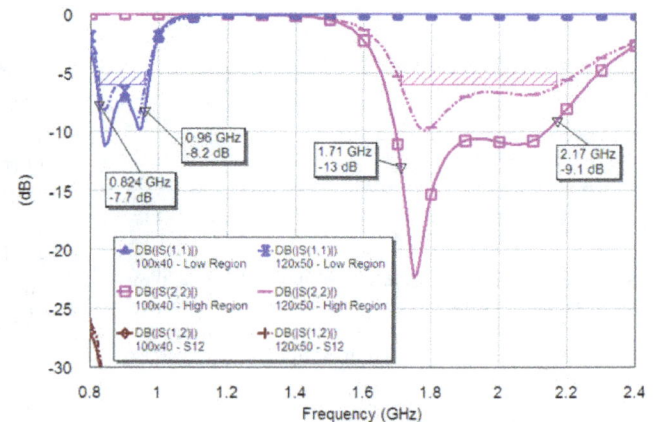

Fig. 8. The multiband behavior is perfectly achieved using a smartphone platform of 120 mm × 50 mm. Impedance matching is enhanced in the case of the largest ground plane, especially in the high frequency region. Both platforms offer good isolation at both frequency regions (> 26 dB).

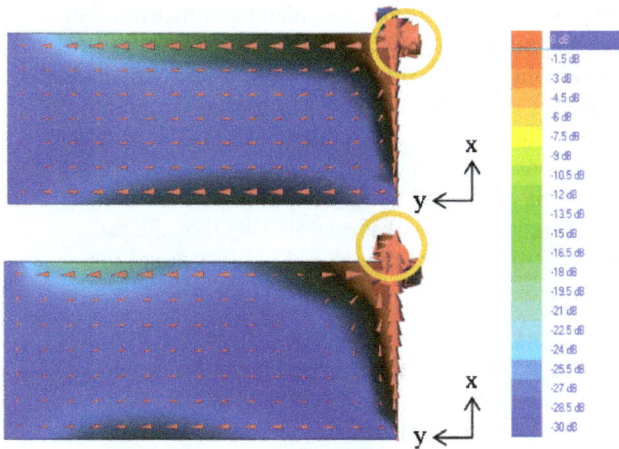

Fig. 9. Current distribution of the radiating system (also taking into account the matching network) at 0.9 GHz (top) and 1.8 GHz (bottom) (current max: 4 A/m). The circle indicates the booster with stronger excitation.

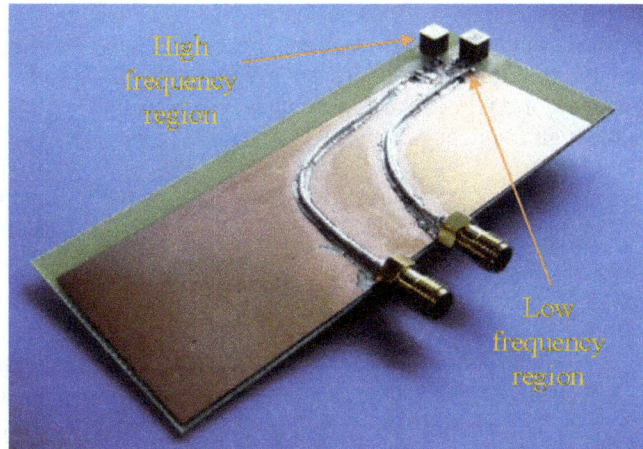

Fig. 10. Built prototype. Ground plane is 100 mm × 40 mm printed over a 1 mm thick FR4 substrate. Ground plane boosters are 5 mm × 5 mm × 5 mm. The same scheme is used for the ground plane of 120 mm × 50 mm representative of the smartphone.

It is also useful to see the current distribution in both frequency regions to show that both boosters work closely to each other without being interfered (Fig. 9). At 0.9 GHz, the main player of the radiation is the booster aligned with the longest edge of the ground plane (Fig. 9, top) because it is more excited. Since the other booster has density current values below -20 dB, good isolation is demonstrated. At 1.8 GHz, the same phenomenon is observed (Fig. 9, bottom) although now, the main player is the booster aligned with the shortest edge of the ground plane.

3. Measurements

With the purpose of validating the simulation results, two prototypes made of FR4 with solid metallic boosters has been built and measured. Each ground plane booster of 5 mm × 5 mm × 5 mm is located (orthogonal disposal) in one side of a corner of the 100 mm × 40 mm and 120 mm × 50 mm ground plane respectively featuring the concentrated solution (Fig. 10).

The radiating system has been matched following the teachings of the simulation results. The final matching network of the built prototype consists of 9 lumped elements from Murata (high Q tight tolerance). For the low frequency region, a series inductor ($L_{Slow} = 36$ nH) and a broadband matching network ($L_{BB} = 2.2$ nH and $C_{BB} = 15$ pF) are needed. For the high frequency region, a notch filter ($L_N = 18$ nH and $C_N = 1.8$ pF), a series inductor ($L_{Shigh} = 8.7$ nH), and a broadband matching network ($L_{BB} = 1.3$ nH, $C_{BB} = 5.1$ pF) with some fine-tuning components ($C_S = 7$ pF, $C_P = 1.5$ pF) are used.

The measured reflection coefficient (Fig. 11) in both prototypes is below -6 dB from 0.819 GHz to 0.969 GHz, and from 1.71 GHz to 2.2 GHz, which means that the radiating system may be operative in many frequency bands such as GSM850, GSM900, GSM1800, GSM1900, and UMTS. S_{21} is below -21 dB among the low and high frequency regions in both prototypes, ensuring a good isolation between ports, which agrees with the simulation.

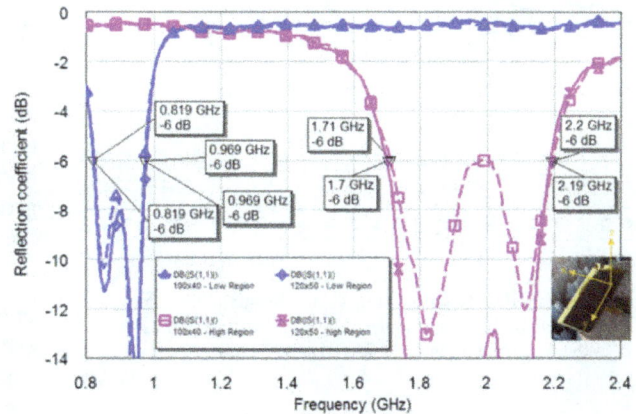

Fig. 11. The radiating system is matched from 0.824 GHz to 0.960 GHz and from 1.71 GHz to 2.17 GHz in both platforms.

The efficiency has been measured using 3D pattern integration using the anechoic chamber Satimo Stargate 32. Regarding the radiation efficiency values (η_r) of the 100 mm × 40 mm prototype (Fig. 12 and Fig. 13), acceptable results have been obtained among the low and high frequency region, offering average values of 49 % and 64 %, respectively. The total antenna efficiency (η_a) (Fig. 12 and Fig. 13 which also takes into account the mismatch losses ($\eta_a = \eta_r \cdot (1 - |S_{11}|^2)$), reaches a peak of 55 % in the low frequency region and 63 % in the high frequency region, which is a high value considering the small volume of the booster element.

Regarding the smartphone prototype, the efficiency results (Fig. 12 and Fig. 13) are considerably better than in the previous prototype of 100 mm × 40 mm because the ground plane fundamental mode is better excited due to its larger size (more close to 0.4λ at the operating frequencies [14], [46]). Based on the characteristic mode analysis in [46], the smartphone platform presents better modal significance values (closer to 1) in the low frequency region than the barphone platform. This results in a better efficiency of the smartphone case. In this case, the impedance matching is similar but the radiation efficiency is much higher because the whole radiating system radiates more efficiently. Therefore, the smartphone provides an average total antenna efficiency of 51 % in the low frequency region and a 70 % in the high frequency region.

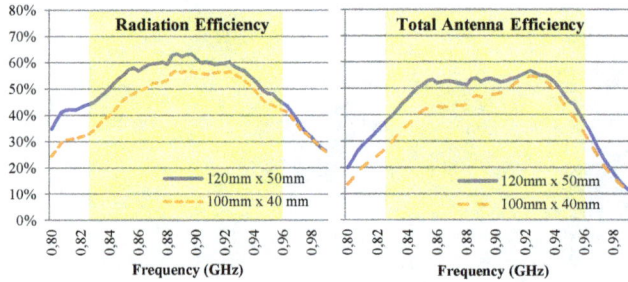

Fig. 12. Measured efficiency values in the low frequency region of both prototypes. Highlighted area indicates the frequencies range of 0.824 GHz–0.960 GHz.

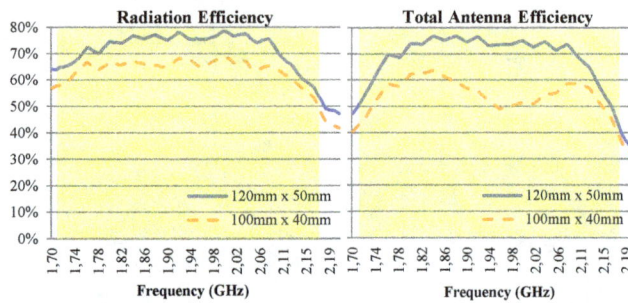

Fig. 13. Measured efficiency values in the high frequency region of both prototypes. Highlighted area indicates the frequency range of 1.71 GHz–2.17 GHz.

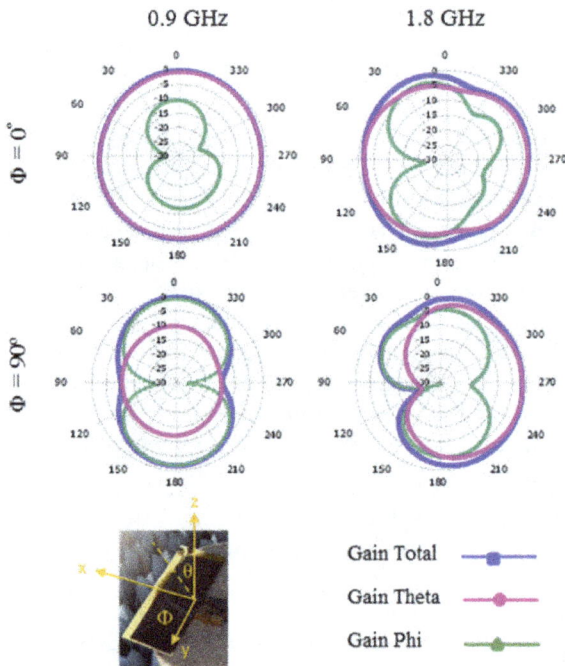

Fig. 14. Measured radiation patterns at 0.9 GHz and 1.8 GHz. The radiation patterns of the 120 mm × 50 mm ground plane feature a similar shape but they have not been included for the sake of brevity.

Finally, radiation patterns have also been measured in the same anechoic chamber. The main cuts normalized to the maximum gain ($\varphi = 0°$ and $\varphi = 90°$) are shown at 0.9 GHz and 1.8 GHz. Both prototypes present an omnidirectional behavior in $\varphi = 0°$ and a deep in the y-axis at both frequency regions (Fig. 14) which is the typical dipole type pattern for this kind of handset platforms. Although a slightly different shape is observed in 1.8 GHz, the radiation pattern is still omnidirectional and no directional lobes are observed.

It is remarkable the small size of only 250 mm^3, which is much smaller than other existing antennas providing also operability at 5 frequency bands, such as typical PIFA antennas with an average volume of 6600 mm^3 and monopole antennas with an average volume of 2200 mm^3 [47]. This makes this solution interesting for the new generation of wireless handheld devices including smartphones.

4. Conclusions

A new radiating system based on ground plane booster technology has been presented. Basically, this technology uses simple elements called ground plane boosters for exciting the efficient radiation mode of the ground plane, which presents high radiation efficiency at the frequencies of interest.

Although each ground plane booster can satisfactory operate in one frequency region, when they work close to each other with the purpose of obtaining a concentrated and a multiband radiating system, coupling between them increases and its operability becomes severely degraded. In order to overcome such a limitation, a new architecture where boosters are orthogonally disposed in a corner of the PCB has been proposed. This new solution satisfies the two main requirements: low coupling between boosters and compactness.

While the orthogonal arrangement of the boosters with respect to the ground plane improves the coupling, it is still not sufficient to guarantee operation since the input impedance and isolation levels are still out of bounds. In this sense, the proposed radiating system for the 100 mm × 40 mm platform further includes a matching network with a notch filter. Such a notch filter, when combined with the specific arrangement of boosters and the ground plane, results in a significant improvement of the overall performance, particularly in the low frequency region.

The present research has also been focused on analyzing how this technology behaves in a smartphone platform and it has been concluded that there is no need to change the radiofrequency system because the new platform does not affect the impedance response, while at the same time the efficiency becomes significantly improved. That is, the proposed architecture performs robustly across different platforms, which suggest the potential advantage of a standardized solution. This would in turn become a very significant advantage as mobile devices currently require a customized platform and a specific antenna system. The measured results demonstrate that good matching and satisfactory efficiency values can be obtained at least from 0.824 GHz to 0.96 GHz and from 1.71 GHz to 2.17 GHz, which means that the proposed radiating system can operate in many frequency bands such as GSM850, GSM900, GSM1800, GSM1900, and UMTS, as well as in many LTE standards within the above frequency range. Furthermore, the volume is only 250 mm^3 compared to the average of 6600 mm^3 of PIFA and 2200 mm^3 of monopoles used in many commercial phones.

Based on the obtained results, it is concluded that the proposed technique is a useful and promising antenna solution offering a concentrated configuration, operating in multiple frequency bands, and only requiring a small volume (250 mm^3) of the handset device.

Acknowledgements

This work was financially supported by Fractus under Grant 2009 TEM00028 (Agència de Gestió d'Ajuts Universtaris i Recerca, Generalitat de Catalunya).

References

[1] WONG, K. L. Planar antennas for wireless communications. *Wiley Series in Microwave and Optical Engineering*. New Jersey: John Wiley & Sons, 2003.

[2] TUNG, H.C., FANG, C.Y., WONG, K.L. Dual band inverted-L monopole antenna for GSM/DCS mobile phone. In *IEEE Antennas and Propagation Society International Symposium*, vol. 3, 2002.

[3] WONG, K.L., LEE, G.Y., CHIOU, T.W. A low-profile planar monopole antenna for multiband operation of mobile handsets. *IEEE Transactions on Antennas and Propagation*, January 2003, vol. 51, no. 1, p. 121–125.

[4] ANGUERA, J., PUENTE, C., MARTÍNEZ, E., ROZAN, E. The fractal Hilbert monopole: A two-dimensional wire. *Microwave and Optical Technology Letters*, 2003, vol. 36, no. 2, p. 102–104.

[5] LIN, S.Y. Multiband folded planar monopole antenna for mobile handset. *IEEE Transactions on Antennas and Propagation*, 2004, vol. 52, no. 7, p. 1790–1794.

[6] LI, Z. RAHMAT-SAMII, Y. Optimization of PIFA-IFA combination in handset antenna designs. *IEEE Transactions on Antennas and Propagation*, 2005, vol. 53, no. 5, p. 1770–1778.

[7] MARTÍNEZ, M., LETSCHKE, O., GEISSLER, M., HEBERLING, D., MARTÍNEZ, A. M, SÁNCHEZ, D. Integrated planar multiband antennas for personal communication handsets. *IEEE Transactions on Antennas and Propagation*, 2006, vol. 54, no. 2, p. 384–391.

[8] KIM, B., PARK, S., YOON, Y., OH, J., LEE, K., KOO, G. Hexaband planar inverted-F antenna with novel feed structure for wireless terminals. *IEEE Antennas Wireless Propagation. Letters*, 2007, vol. 6, p. 66–69.

[9] HSIEH, H., LEE, Y., TIONG, K., SUN, J. Design of a multiband antenna for mobile handset operations. *IEEE Antennas Wireless Propagation Letters*, 2009, vol. 8, p. 200–203.

[10] LIN, C. WONG, K. L. Printed monopole slot antenna for internal multiband mobile phone antenna. *IEEE Transactions on Antennas and Propagation*, 2007, vol. 55, no. 12, p. 3690–3697.

[11] RISCO, S., ANGUERA, J., ANDÚJAR, A., PÉREZ, A., PUENTE C. Coupled monopole antenna design for multiband handset devices. *Microwave and Optical Technology Letters*, 2010, vol. 52, no. 2, p. 359–364.

[12] SUGIMOTO, S., IWASAKI, H. Wide band planar monopole antenna with asymmetric parasitic elements. In *Proceedings of the Fourth European Conference on Antennas and Propagation (EuCAP)*. Barcelona (Spain), April 2010, p. 1–4

[13] VAINIKAINEN, P., OLLIKAINEN, J., KIVEKÄS, O., KELANDER, I. Resonator-based analysis of the combination of mobile handset antenna and chassis. *IEEE Transactions on Antennas and Propagation*, 2002, vol. 50, no. 10, p. 1433–1444.

[14] SALONEN, P. Effect of groundplane size on radiation efficiency and bandwidth of dual-band U-PIFA. In *IEEE Antennas and Propagation Society International Symposium*. Columbus (OH, USA), June 2003, vol. 3, p. 70–73.

[15] HOSSA, R., BYNDAS, A., BIALKOWSKI, M. E. Improvement of concentrated terminal antenna performance by incorporating open-end slots in ground plane. *IEEE Microwave and Wireless Components Letters*, 2004, vol. 14, no. 6, p. 283–285.

[16] ABEDIN, M. F., ALI, M. Modifying the ground plane and its effect on planar inverted-F antennas (PIFAs) for mobile phone handsets. *IEEE Antennas and Wireless Propagation Letters*, 2003, vol. 2, no. 1, p. 226–229.

[17] ANGUERA, J., SANZ, I., SANZ, A., GALA, D., CONDES, A., PUENTE, C., SOLER, J. Enhancing the performance of the handset antennas by means of groundplane design. In *IEEE International Workshop on Antenna Technology Small Antennas and Novel Metamaterials*. New York (USA), March 2006, p. 29–32.

[18] CABEDO, M., ANTONINO, E., RODRIGO, V., SUÁREZ, C., Análisis Modal de un Plano de Masa Radiante Doblado y con una Ranura para Terminales Móviles. In *XXI Simposium Nacional de la Unión Científica Internacional de Radio, URSI '06*. Oviedo (Spain), September 2006 (in Spanish).

[19] TANG, I.T., LIN, D.B., CHEN, W.L., HORNG, J.H., LI, C.M. Compact five-band meandered PIFA by using meandered slots structure. In *IEEE Antennas and Propagation Society International Symposium*. June 2007, p. 653–656.

[20] CABEDO, A., ANGUERA, J., PICHER, C., RIBÓ, M., PUENTE, C. Multi-band handset antenna combining a PIFA, slots, and ground plane modes. *IEEE Transactions on Antennas and Propagation*, 2009, vol. AP-57, no. 9, p. 2526–2533.

[21] PICHER, C., ANGUERA, J., CABEDO, A., PUENTE, C., KAHNG, S. Multiband handset antenna using slots on the ground plane: Considerations to facilitate the integration of the feeding transmission line. *Progress In Electromagnetics Research C*, 2009, vol. 7, p. 95–109.

[22] ANGUERA, J., SANZ, I., MUMBRÚ, J., PUENTE, C. Multi-band handset antenna with a parallel excitation of PIFA and slot radiators. *IEEE Transactions on Antennas and Propagation*, 2010, vol. AP-58, no. 2, p. 348–356.

[23] RAZALI, A. R., BIALKOWSKI, M. E. Coplanar inverted-F with open-end ground slots for multiband operation. *IEEE Antennas and Wireless Propagation Letters*, 2009, vol. 8, p. 1029–1032.

[24] ANGUERA, J., PUENTE, C. Shaped ground plane for radio apparatus. *Patent Appl. WO2006/070017*, Dec. 29, 2005.

[25] ANGUERA, J., ANDÚJAR, A., PUENTE, C., MUMBRÚ, J. Antennaless wireless device. *Patent Appl. WO2010/015365*, July 2009.

[26] ANGUERA, J., ANDÚJAR, A., PUENTE, C., MUMBRÚ, J. Antennaless wireless device capable of operation in multiple frequency regions. *Patent Appl. WO2010/015364*, July 2009.

[27] ANDÚJAR, A., ANGUERA, J., PUENTE, C. Ground plane boosters as a compact antenna technology for wireless handheld devices. *IEEE Transactions on Antennas and Propagation*, 2011, vol. 59, no. 5, p. 1668–1677.

[28] ANGUERA, J., PICHER, C., ANDÚJAR, A., PUENTE C. Concentrated antennaless wireless device providing operability in multiple frequency regions. *Patent Appl. US61/671906*, July 2012.

[29] ANGUERA, J., ANDÚJAR, A., GARCÍA, C. Multiband and small coplanar antenna system for wireless handheld devices. *IEEE Transactions on Antennas and Propagation*, 2013, vol. 61, no. 7, p. 3782–3789.

[30] ANGUERA, J., PICHER, C., ANDÚJAR, A., KAHNG, S., PUENTE, C. Compact multiband antenna system for smartphone platforms. In *Proc. of the European Conference on Antennas and Propagation EuCAP*. Gothenburg (Sweden), April 2013.

[31] SCHROEDER, W. L., FAMDIE, C. T., SOLBACH, K. Utilization and tuning of the chassis modes of a handheld terminal for the design of multiband radiation characteristics. *IEEE Wideband Multi-Band Antennas Arrays*, Sept. 2005, p. 117–121.

[32] VILLANEN, J., OLLIKAINEN, J., KIVEKÄS, O., VAINIKAINEN, P. Compact antenna structures for mobile handsets. In *IEEE VTC2003 Fall Conference*. Orlando (Florida, USA), 2003.

[33] VILLANEN, J. HOLOPAINEN, J., KIVEKÄS, O., VAINIKAINEN, P. Mobile broadband antennas. In *URSIGA 2005 Conference*. New Delhi (India), October 2005.

[34] VILLANEN, J., OLLIKAINEN, J., KIVEKÄS, O., VAINIKAINEN, P. Coupling element based mobile terminal antenna structures. *IEEE Transactions on Antennas and Propagation*, 2006, vol. 54, no. 7, p. 2142–2153.

[35] HOLOPAINEN, J., VILLANEN, J., ICHELN, C., VAINIKAINEN, P. Mobile terminal antennas implemented by using direct coupling. In *Proc. of the European Conference on Antennas and Propagation EuCAP*. Nice (France), November 2006, p. 1–6.

[36] RAHOLA, J., OLLIKAINEN, J. Optimal antenna placement for mobile terminals using characteristic mode analysis. In *Proc. of the European Conference on Antennas and Propagation EuCAP*. Nice (France), November 2006.

[37] VILLANEN, J., ICHELN, C., VAINIKAINEN, P. Coupling element-based quadband antenna structure for mobile terminals. *Microwave and Optical Technology Letters*, 2007, vol. 49, no. 6.

[38] OZDEN, S., NIELSEN, B. K., JORGENSEN, C. H., VILLANEN, J., ICHELN, C., VAINIKAINEN, P. Quad-band coupling element antenna structure. *U.S. Patent 7 274 340*, Sep. 25, 2007.

[39] HOLOPAINEN, J., VALKONEN, R., KIVEKÄS, O., ILVONEN, J., VAINIKAINEN, P. Broadband equivalent circuit model for capacitive coupling element-based mobile terminal antenna. *IEEE Antennas and Wireless Propagation Letters*, 2010, vol. 9, p. 716 to 719.

[40] CABEDO-FABRÉS, M., ANTONINO-DAVIU, E., VALERO-NOGUEIRA, A., BATALLER, M. F. The theory of characteristic modes revisited: A contribution to the design of antennas for modern applications. *IEEE Antennas and Propagation Magazine*, Oct. 2007, vol. 49, no. 5, p. 52–68.

[41] HAI ZHAO, GUI LIN, BECKMAN, C. Design of a coupling element based penta-band mobile phone antenna. In *Antennas and Propagation Conference LAPC 2009*. Loughborough (UK), November 2009, p. 209-212.

[42] ANGUERA, J., PUENTE, C., BORJA, C., FONT, G., SOLER, J. A systematic method to design single-patch broadband microstrip patch antennas. *Microwave and Optical Technology Letters*, Nov. 2001, vol. 31, no. 3, p. 185–188.

[43] ANDÚJAR, A., ANGUERA, J., PUENTE, C. A systematic method to design broadband matching networks. In *Proceedings of*

the 4th European Conference on Antennas and Propagation EuCAP 2010. Barcelona (Spain), 2010.

[44] RAHOLA, J., OLLIKAINEN, J. Analysis of isolation of two-port antenna systems using simultaneous matching. In *Proceedings of the European Conference on Antennas and Propagation EuCAP 2007*. Edinburgh (UK), November 11-16, 2007.

[45] POZAR, D. M. *Microwave Engineering*. 2nd edition. John Wiley and Sons, 1998.

[46] ANGUERA, J., ANDÚJAR, A. Ground plane contribution in wireless handheld devices using radar cross section analysis. *Progress In Electromagnetics Research M*, 2012, vol. 26, p. 101–114.

[47] ROWEL, C., LAM, E. Y. Mobile-phone antenna design. *IEEE Antennas and Propagation Magazine*, Aug. 2012, vol. 54, no. 4, p. 14–34.

About Authors ...

Cristina PICHER was born in Sabadell, Barcelona, in 1984. She received her Technical Engineer Degree in Telecommunication Systems in 2006, her Engineer degree in Telecommunication in 2009, and her Master of Science in Telecommunication Networks in 2009, all of them from Ramon Llull University (URL, Barcelona). She began her investigation in miniature and multiband antennas in 2005 at Fractus (Barcelona) when she started her final degree projects in the Electronic and Telecommunication Dept. of URL in collaboration with the Dept. of Technology and IPR. In 2008 and 2009 she was awarded by the Ministry of Science and Innovation Spain: "Arquímedes" Introduction to Scientific Research program for her research in "Miniature and multiband techniques combining handset antenna and ground plane design" and "Multiband handset antennas with slotted ground planes: human head interaction". From 2009-2012, she worked as a R&D Engineer in Fractus, where she was doing her Ph.D. in the field of small and multiband antennas for handset and wireless devices. In September 2010, she started leading projects in the antenna field for handset and wireless applications in a context of Industry-University collaboration between Fractus and the Dept. of Electronics and Communications of Ramon Llull University-Barcelona, Spain. In 2012, she joined Vodafone Spain, where she is working as a Design and Engineering Specialist in projects related to the deployment and optimization of the 3G/4G network. She has published more than 25 papers in scientific journals, international, and national conferences. She is author of 3 patents in the antenna field.

Jaume ANGUERA was born in Vinaròs, Spain, in 1972. He received the Technical Engineering degree in Electronic Systems and Engineering degree in Electronic Engineering, both from Ramon Llull University (URL), Barcelona, Spain, in 1994 and 1998, respectively, and the Engineering and Ph.D degrees in Telecommunication Engineering, both from the Polytechnic University of Catalonia (UPC), Barcelona, Spain, in 1998 and 2003, respectively. In 1997-1999 he joined the Electromagnetic and Photonic Engineering Group of the Signal Theory and Communications Dept. of the UPC as a researcher in microstrip fractal-shaped antennas. In 1999, he was a researcher at Sistemas

Radiantes, Madrid, Spain, where he was involved in the design of a dual-frequency dual-polarized fractal-inspired microstrip patch array for mobile communications. In the same year, he became an Assistant Professor at the Dept. of Electronics and Telecommunications, URL, where he is currently teaching antenna theory and electromagnetics. Since 1999, he is with Fractus, Barcelona, Spain, where he holds the position of R&D manager. At Fractus he lead projects on antennas for base station systems, antennas for automotive and currently managing handset and wireless antenna projects. His current research interests are multiband and small antennas, broadband matching networks, diversity and MIMO antenna systems, electromagnetic dosimetry, genetic optimized antennas, and antennas for wireless devices. From 2003 to 2004, he was with Fractus-Korea (Republic South of Korea) as the leading manager for developing projects in the area of miniature and multiband antennas for handset and wireless applications with Korean companies (Samsung and LG). Since 2001, he has been leading research projects in the antenna field for handset and wireless applications in a frame of Industry-University collaboration: Fractus and Dept. of Electronics and Telecommunications, Universitat Ramon Llull-Barcelona, Spain resulting in the direction of more than 70 bachelor and master theses, some of them awarded by Spanish institutions and companies. He holds more than 95 granted invention patents (most of them licensed) and 30 more pending patents in the antenna field. He is author of more than 170 journals, international and national and conference papers. He is a Senior Member IEEE. Dr.Anguera was member of the fractal team that in 1998 received the European Information Technology Grand Prize for the fractal-shaped antenna application to cellular telephony. 2003 Finalist to the Best Doctoral Thesis (Fractal and Broadband Techniques on Miniature, Multifrequency, and High-Directivity Microstrip Patch Antennas) on UMTS (prize awarded by Telefónica Móviles España). New faces of Engineering 2004 (awarded by IEEE and IEEE foundation). In the same year 2004, he received the Best Doctoral Thesis in "Network and BroadBand Services" (XXIV Prize Edition "Ingenieros de Telecomunicación") awarded by Colegio Oficial de Ingenieros de Telecomunicación and the Company ONO. He is a reviewer for IEEE Transaction and Antennas and Propagation, IEEE Antennas and Wireless Propagation Letters, IEEE Antennas and Propagation Magazine, Progress in Electromagnetic Research, IEE Electronics Letters, and ETRI journal (Electronics and Telecommunications Research Institute). He is editor of International Journal on Antennas and Propagation (IJAP). His biography is listed in Who'sWho in the World, Who'sWho in Science and Engineering, Who'sWho in Emerging Leaders and in IBC (International Biographical Center, Cambridge-England).

Aurora ANDÚJAR was born in Barcelona, Spain, 1984. She received the Bachelor's degree in Telecommunication Engineering specializing in Telecommunication Systems in 2005, the Master degree in Telecommunications Engineering in 2007 and the Master of Science in Telecommunication Engineering and Management in 2007, and the Ph.D in Telecommunication Engineering in 2013 from the Polytechnic University of Catalonia (UPC), Barcelona, Spain. In 2004-2005 she received a research fellowship in the field of Electromagnetic Compatibility from the Signal Theory and Communications Dept. at UPC. In 2005 she worked as a Software Test Engineer for applications intended for handset wireless devices. In 2006 she worked as a Software Engineer designing a load simulation tool for testing Digital Campus in academic environments and developing improvements in the performance of web servers referred to the management of static and dynamics contents. From 2007 up to date she is working as R&D Engineer in Fractus, Barcelona, where she actively contributes to the prosecution and growth of the patent portfolio of the company. She is also involved in several projects in the field of small and multiband handset antenna design. Since 2009 she is leading research projects in the antenna field for handheld wireless devices in the collaborative university-industry framework. She has published more than 60 journals, international and national and conference papers. She is also author of 10 invention patents in the antenna field. She has directed 16 bachelor and master theses. Dr. Andújar is member of the COIT (Colegio Oficial de Ingenieros de Telecomunicación) and AEIT (Asociación Española de Ingenieros de Telecomunicación). She is editor of International Journal on Antennas and Propagation (IJAP).

Carles PUENTE was born in Badalona, Spain. He is a co-founder of Fractus and leads its antenna technology research team, with responsibility for the company's intellectual property portfolio development and antenna development. Carles is a university professor at the Universitat Politècnica de Catalunya (UPC) where he started researching fractal-shaped antennas while a student in the late 1980s. He gained his MSc from the University of Illinois at Urbana-Champaign, USA in 1994 and his PhD from UPC in 1997. From 1994 to 1999 he worked with the Faculty of Electromagnetic and Photonic Engineering at UPC on pioneering developments of fractal technology applied to antennas and microwave devices. Dr. Puente was awarded with the Best Doctoral Thesis in Mobile Communications 1997 by the COIT, the European Information Society Technology Grand Prize from the European Commission in 1998 and the Premi Ciutat de Barcelona in 1999. He and his team at Fractus were awarded with the Technology Pioneer distinction by the World Economic Forum in 2005. He has authored more than 50 invention patents and over 90 scientific publications in fractal and related antenna technologies.

Adrián BUJALANCE was born in Esplugues de Llobregat, Spain, in 1984. He received the BSc in Telecommunication Engineering from Universitat Ramon Llull - La Salle, Barcelona, Spain, in 2013, where he is currently working toward the MSc in Telecommunication Engineering. Since 2010, he has worked as a research collaborator at the Technology & IPR Dept. of Fractus. He has published 5 journal papers. His research interests include design and implementation techniques of compact radiating systems for mobile handset devices.

14

Dielectric Properties Determination
of a Stratified Medium

Paiboon YOIYOD, Monai KRAIRIKSH

Faculty of Engineering, King Mongkut's Institute of Technology Ladkrabang, Bangkok 10520, Thailand

S3610116@kmitl.ac.th, kkmonai@kmitl.ac.th

Abstract. *The method of detection of variation in dielectric properties of a material covered with another material, which requires nondestructive measurement, has numerous applications and the accurate measurement system is desirable. This paper presents a dielectric properties determination technique whereby the dielectric constant and loss factor are extracted from the measured reflection coefficient. The high frequency reflection coefficient shows the effect of the upper layer, while the dielectric properties of the lower layer can be determined at the lower frequency. The proposed technique is illustrated in 1-11 GHz band using 5 mm-thick water and 5% saline solution. The fluctuation of the dielectric properties between the high frequency and the low frequency, results from the edge diffraction in the material and the multiple reflections at the boundary of the two media, are invalid results. With the proposed technique, the dielectric properties of the lower layer can be accurately determined. The system is validated by measurement and good agreement is obtained at the frequency below 3.5 GHz. It can be applied for justifying variation of the material in the lower layer which is important in industrial process.*

Keywords

Dielectric properties, stratified medium, reflection coefficient, diffraction coefficient

1. Introduction

The dielectric properties determination of a covered material or stratified medium by using a microwave technique is essential and has numerous applications. One of the determination techniques of dielectric properties is the free space technique whose main advantages include distinguishing capability of inhomogeneous materials, nondestructiveness, non-contact, and no machinery that fits the sample required despite sophisticated procedure to obtain accurate results [1], [2]. The techniques need an insertion of a perfectly conducting plate behind the plate of unknown material. The comprehensive reviews of dielectric properties measurement techniques and nondestructive testing using both millimeter and microwave are respectively shown in [3]. Such measurement techniques together with nondestructive testing are employed in a number of applications, such as moisture content detection [4–6], determination of liquids [7], skin cancer detection [8], surface crack detection [9], corrosion detection [10], [11], in which some sensors require contact [7–11] while the other are contactless [5], [6]. Apart from the contactless measurement, certain techniques require embedding of a scatterer in the medium under test [12], [13] which inevitably results in limitation in some applications as in agricultural applications [14], [15].

The work in [16] utilized a free space technique to characterize ripeness of mango but sample of mango must be prepared in a planar sample holder. Therefore, it is unsuitable in practice. A number of the free space techniques widely employed in the past, the sole magnitude measurement technique is less complex and thus inexpensive; nevertheless, it requires transmission measurement [17]. In addition, the use of millimeter wave reflectometer [18] is one of the attractive solutions. Fruit testing nevertheless must be carried out by measuring through the peel nondestructively. The methods in [19], [20] estimate the dielectric properties and thickness of multilayer object but they are contact measurement. The contactless measurement can however be accomplished by the technique of through-wall measurement [21]. From the aforementioned statement, it is desirable to estimate dielectric properties of a stratified object, consisting of the upper layer, lower layer, and the thickness of the upper layer. The issue is that the low variation of dielectric properties of the lower layer has little effect on the total variation of the measured results. Therefore, it is necessary to propose an accurate measurement technique to estimate the dielectric properties and thickness of the upper layer, which leads to the accurate estimation of the dielectric properties of the lower layer.

To develop a measurement system, a measurement technique for determination of not merely the dielectric properties of upper and lower layers but also the thickness of upper layer must be well established. To this end, in this paper the reflection measurement in wideband to determine the dielectric properties of the upper layer with high frequencies in which depth of penetration is shorter than the depth of the upper layer is introduced. The measured material is supposed to be in a data base for comparing dielectric properties of the upper layer when measured at high frequency, and then the values at the lower frequency can be obtained. The application of interest is measurement of material in a container in industrial process. The relevant work is the reflection measurement for estimating thickness of snow on the road [22]. Using the full band dielectric properties, it is possible to calculate depth of penetration of wave through the upper layer and thus velocity of wave in the upper layer. As a result, the thickness of the upper layer is determined. The suitable frequency at the lower frequency band is selected from the frequency response of the dielectric properties. Finally, the dielectric properties of the lower layer can be determined. To obtain a tangible insight into the aforesaid technique, the planar structure is investigated in this paper.

The organization of this paper begins with the introduction in Sec. 1. Section 2 discusses the principles of the proposed technique. The calculation results of the measurement technique are illustrated in Sec. 3. Validation of the calculation results, together with the discussions, is carried out by measurements shown in Sec. 4. Section 5 discusses the limitation of the system and addresses the factors affecting the required bandwidth. This paper is then ended with a conclusion in Sec. 6.

2. Dielectric Properties Determination of a Stratified Medium

Let us consider a stratified medium of planar structure with upper layer thickness d as shown in Fig. 1. The upper medium has dielectric properties of μ_1, ε_1 and σ_1. The lower layer is infinite extent in thickness and possesses dielectric properties of μ_2, ε_2 and σ_2. The width and length of the structure are finite size with the width of W. This structure is illuminated by a uniform plane wave in free space in which dielectric properties are μ_0, ε_0 and σ_0, where σ_0 is zero. Fig. 1 depicts the various components of waves for derivation of the reflection coefficient. The transmitting and receiving antennas are at $O(x',y',z')$ and $P(x,y,z)$, res-

pectively, where position of y' is same as y at $W/2$, $O(x',y',z')$ and $P(x,y,z)$ are almost identical position to act as a monostatic radar. The transmitter transmits incident electric wave E_i on the upper surface at $z = -d$. With perfect isolation between the two antennas, the receiving antenna receives reflected and diffracted waves from four edges (D_1, D_2, D_3 and D_4). The reflected wave is the sum of the reflection at the upper surface and the multiple reflections between the upper surface and the interface between dielectric 1 and dielectric 2. The reflection coefficient at $z = -d$ ($\Gamma_{in}(z = -d)$) can be expressed as (1).

$$\Gamma_{in}(z=-d)=\Gamma_{01}+T_{01}T_{10}\sum_{i=1}^{n}\Gamma_{12}^{i}\Gamma_{10}^{i-1}e^{-2i\gamma_1 d}+D \quad (1)$$

where Γ_{01} and Γ_{10} are the reflection coefficients between the free space and the upper layer for propagation in upward direction and downward direction, respectively, T_{01} and T_{10} are the transmission coefficients between the free space and the upper layer for propagation in upward direction and downward direction, respectively, Γ_{12} is the reflection coefficient at the boundary of the two media, γ_1 is propagation constant in dielectric 1 ($\gamma_1 = \alpha_1 + j\beta_1$), i represents the ith reflection, n is the number of multiple reflections in the material, and D is diffraction coefficient as shown in (2).

$$D=\frac{E_{D_1}+E_{D_2}+E_{D_3}+E_{D_4}}{E_i}. \quad (2)$$

Diffracted electric field intensity on each side from [23] is expressed in rectangular coordinate as follows

$$E_{D_k}(x,y,z)=\pm E_i\left\{T\left(\sqrt{\varepsilon_{r1}}\sin\left(\tan^{-1}\left(\frac{z}{x}\right)\right)-\sqrt{\varepsilon_{r1}-\cos^2\left(\tan^{-1}\left(\frac{z'}{x'}\right)\right)}\right)\right.$$

$$\times\frac{e^{-j\frac{\pi}{4}}}{2\sqrt{2\pi k_d}}\frac{F_i\left[2k_d\left(\sqrt{(x^2+z^2)}\right)\cos^2\left(\frac{\tan^{-1}\left(\frac{z}{x}\right)-\cos^{-1}\left(\frac{\cos\left(\tan^{-1}\left(\frac{z'}{x'}\right)\right)}{\sqrt{\varepsilon_{r1}}}\right)}{2}\right)\right]}{\sqrt{\varepsilon_{r1}}\cos\left(\tan^{-1}\left(\frac{z}{x}\right)\right)+\cos\left(\tan^{-1}\left(\frac{z'}{x'}\right)\right)}$$

$$+T\left((1-R)\cos\left(\tan^{-1}\left(\frac{z'}{x'}\right)\right)-(1+R)\sqrt{\varepsilon_{r1}}\cos\left(\tan^{-1}\left(\frac{z}{x}\right)\right)\right)\frac{e^{-j\frac{\pi}{4}}}{2\sqrt{2\pi k_d}}$$

$$\times\frac{F_i\left[2k_d\left(\sqrt{(x^2+z^2)}\right)\cos^2\left(\frac{\cos^{-1}\left(\sqrt{1-\frac{\cos^2(\tan^{-1}(z'/x'))}{\varepsilon_{r1}}}\right)-\left(\tan^{-1}\left(\frac{z}{x}\right)-\frac{\pi}{2}\right)}{2}\right)\right]}{\sqrt{\varepsilon_{r1}}\cos\left(\tan^{-1}\left(\frac{z}{x}\right)+\frac{\pi}{2}\right)-\sqrt{\varepsilon_{r1}-\cos^2\left(\tan^{-1}\left(\frac{z'}{x'}\right)\right)}}\right\}$$

$$(3)$$

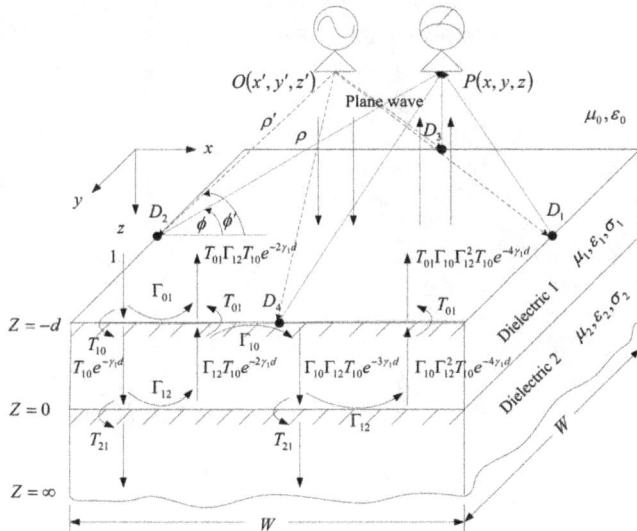

Fig. 1. Geometry of the problem.

where E_i is the incident electric field intensity, T and R are Fresnel transmission coefficient and reflection coefficient, respectively, $k = 1, 2, 3$ and 4 for electric field intensity on each side, $(+)$ is for perpendicular polarized and $(-)$ is for parallel polarized diffraction coefficients. ρ is a distance from the diffraction point to the observation point $\rho = (x^2 + z^2)^{1/2}$, ϕ is an angle between the line from the diffraction point to the observation point with respect to x-axis $\phi = \tan^{-1}(z/x)$, ρ' is a distance from source point to diffraction point $\rho' = (x'^2 + z'^2)^{1/2}$, ϕ' is an angle between the line from the source point to the diffraction point with respect to x-axis $\phi' = \tan^{-1}(z'/x')$, and ε_{r1} is relative permittivity of dielectric 1.

Since every reflection and transmission term in the summation has a very low magnitude, therefore the summation shrinks very rapidly. Hence, the summation in (1) can be truncated as shown in (4a).

$$\Gamma_{in} = \Gamma_{01} + \frac{T_{01}T_{10}\Gamma_{12}e^{-2\gamma_1 d}}{1 - \Gamma_{10}\Gamma_{12}e^{-2\gamma_1 d}} + D \qquad (4a)$$

where $\Gamma_{10} = -\Gamma_{01}$, $T_{10} = 1 + \Gamma_{01}$ and $T_{01} = 1 + \Gamma_{10} = 1 - \Gamma_{01}$. Hence

$$\Gamma_{in} = \frac{\Gamma_{01} + \Gamma_{12}e^{-2\gamma_1 d}}{1 + \Gamma_{01}\Gamma_{12}e^{-2\gamma_1 d}} + D. \qquad (4b)$$

Rewriting (4b), one can solve for Γ_{12} as shown in (5). In order to get d, we use the reflection coefficient in time domain. The detail is explained in Sec. 3.1.

$$\Gamma_{12} = \frac{\Gamma_{in} - \Gamma_{01} - D}{(1 - \Gamma_{01}(\Gamma_{in} - D))e^{-2\alpha_1 d}e^{-j2\beta_1 d}}. \qquad (5)$$

α_1 and β_1 are respectively the attenuation constant and phase constant in the upper medium. For very high frequencies, attenuation through the upper layer is great enough such that $\Gamma_{in} \approx \Gamma_{01}$. The intrinsic impedance of the upper layer, η_1, can be directly calculated and then curve fitted to the database of the material to estimate η_1 values for lower frequencies. Using η_1 to characterize other parameters including η_2, the intrinsic impedance of the lower layer derived from (5), can be solved.

$$\eta_2 = \frac{\eta_1[(\Gamma_{in} - \Gamma_{01}) + (1 - \Gamma_{in}\Gamma_{01})e^{-2\alpha_1 d}e^{-j2\beta_1 d}]}{(1 - \Gamma_{in}\Gamma_{01})e^{-2\alpha_1 d}e^{-j2\beta_1 d} - (\Gamma_{in} - \Gamma_{01})} \qquad (6)$$

or

$$\frac{\sigma_2 + j\omega\varepsilon_2}{j\omega\mu_2} = 1 \Big/ \left[\frac{\eta_1[(\Gamma_{in} - \Gamma_{01}) + (1 - \Gamma_{in}\Gamma_{01})e^{-2\alpha_1 d}e^{-j2\beta_1 d}]}{(1 - \Gamma_{in}\Gamma_{01})e^{-2\alpha_1 d}e^{-j2\beta_1 d} - (\Gamma_{in} - \Gamma_{01})}\right]^2 \qquad (7)$$

where $\eta_2 = x + jy$. Hence

$$\frac{\varepsilon_0\varepsilon_{r2}''}{j\mu_0\mu_{r2}} + \frac{\varepsilon_0(\varepsilon_{r2}' - j\varepsilon_{r2}'')}{\mu_0\mu_{r2}} = \frac{1}{(x + jy)^2} \qquad (8)$$

and

$$\frac{\varepsilon_0\varepsilon_{r2}'}{\mu_0\mu_{r2}} - j\frac{2\varepsilon_0\varepsilon_{r2}''}{\mu_0\mu_{r2}} = \frac{x^2 - y^2 - j2xy}{(x^2 - y^2)^2 + (2xy)^2}. \qquad (9)$$

By equating the real part of the left hand side to that of the right hand side of (9), and the imaginary part is also equated in the same manner, the dielectric properties can be derived from the real part x and imaginary part y of η_2

$$\frac{\varepsilon_0\varepsilon_{r2}'}{\mu_0\mu_{r2}} = \frac{x^2 - y^2}{(x^2 - y^2)^2 + (2xy)^2}, \qquad (10)$$

$$\frac{2\varepsilon_0\varepsilon_{r2}''}{\mu_0\mu_{r2}} = \frac{2xy}{(x^2 - y^2)^2 + (2xy)^2}. \qquad (11)$$

The dielectric constant and loss factor of the lower layer are expressed in a closed form as shown in (12) and (13), respectively.

$$\varepsilon_{r2}' = \frac{\mu_0\mu_{r2}(x^2 - y^2)}{\varepsilon_0((x^2 - y^2)^2 + (2xy)^2)}, \qquad (12)$$

$$\varepsilon_{r2}'' = \frac{\mu_0\mu_{r2}(2xy)}{2\varepsilon_0((x^2 - y^2)^2 + (2xy)^2)}. \qquad (13)$$

From the above expressions, it can be seen that by measuring the reflection coefficient (Γ_{in}) at the upper surface in a wideband, the reflection coefficient between the free space and the upper layer (Γ_{01}) can be found from the high frequency which is unable to penetrate to the lower layer. Then, the dielectric properties of the upper layer at low frequency can be determined from the database. The above procedure is shown in a block diagram in Fig. 2.

Using the full band dielectric properties, it is possible to calculate the depth of penetration through the upper layer and consequently velocity of wave in the upper layer. From the result of reflection coefficient, the thickness d of the upper layer is determined by inverse Fourier transform

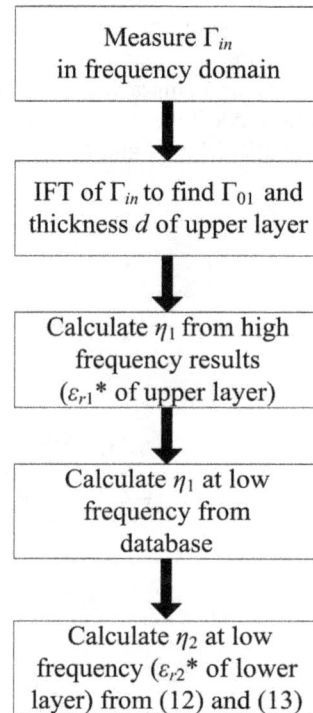

Fig. 2. Block diagram of the proposed technique.

(IFT). Using the high frequency for determining $\varepsilon_{r1}{}^* = \varepsilon_{r1}' - j\varepsilon_{r1}''$), then η_1 at low frequency is determined from the database of the material. Finally, the dielectric properties $\varepsilon_{r2}{}^* = \varepsilon_{r2}' - j\varepsilon_{r2}''$ of the lower layer η_2 are determined. This principle can be illustrated in Sec. 3.

3. Calculation Results

3.1 Determination of Thickness of the Upper Layer

At high frequency, the intrinsic impedance of the upper layer is calculated from (14)

$$\eta_1 = \eta_0 \left(\frac{1 + \Gamma_{in}}{1 - \Gamma_{in}} \right). \tag{14}$$

From the obtained dielectric properties at the high frequency band, the full band one can be obtained from the database of the material. For illustration, the material size of 18λ for the frequency of 6 GHz, the upper and lower layers are water and 5% saline solution [24], respectively. The resultant dielectric properties and depth of penetration of water and 5% saline solution are listed in Tab. 1.

Fig. 3(a) depicts the magnitude of reflection coefficient in frequency domain which was calculated using (4). The magnitude of reflection coefficient varies in a similar manner to damped sinusoidal below 5.5 GHz due to the effect of reflection at the upper surface and at the interface of the materials. Then, the steady response can be observed between 5.5 and 8.5 GHz since reflection takes place only at the upper surface. Note that the fluctuation at frequency higher than 8.5 GHz is caused by diffraction at the four edges of the material.

An inverse Fourier transforms of Γ_{in} provides the time-domain reflectometry as in Fig. 3(b). The first peak at 0 ns represents the reflection at the upper surface whereas the other peaks of the response represent the reflection between the upper and lower layers. Observing the time t, these peaks occur and accounting for the velocity of wave in the upper layer, a thickness of the upper layer can be determined from

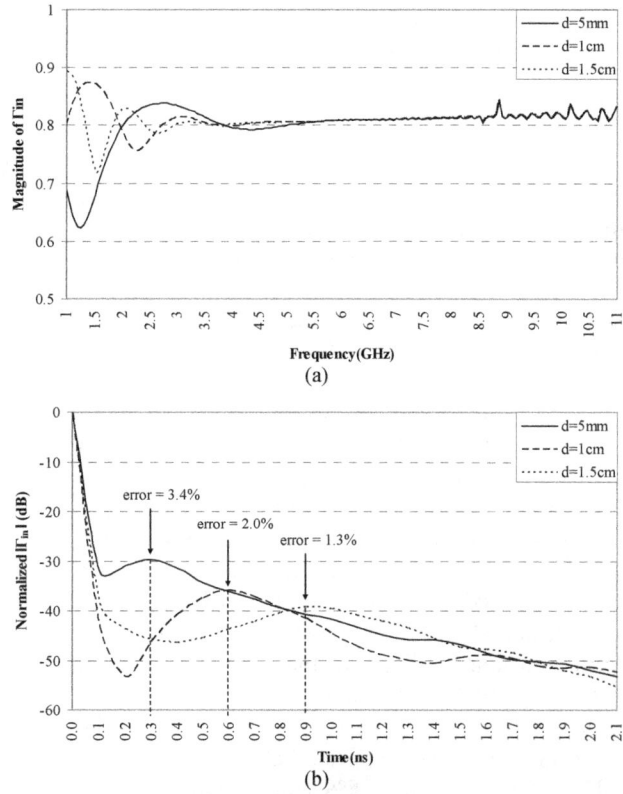

Fig. 3. Magnitude of reflection coefficient: (a) Frequency domain, (b) Time domain.

$$d = v_1 \cdot \frac{t}{2} \tag{15}$$

where v_1 is velocity of wave in the upper layer calculated from $v_1 = 1/(\mu\varepsilon_1)^{1/2}$ and t is the total time wave traveling forth and back. This velocity is obtained from the dielectric properties of water at the frequency corresponding to the time the peaks occur (at $f = 1/t$; dielectric properties of water are $75.754 - j12.358$, $77.488 - j6.625$ and $77.895 - j4.417$). v_1 at these frequencies are 0.344×10^8 m/s, 0.341×10^8 m/s and 0.340×10^8 m/s.

The estimated thickness for 5 mm, 1 cm and 1.5 cm layers are 5.15 mm (error 3.4%), 1.02 cm (error 2.0%) and 1.52 cm (error 1.3%), respectively. From Tab. 1, for the thickness of the upper layer of 5 mm, the frequencies lower than 5 GHz can penetrate to the lower layer. One can hence determine the dielectric properties of the lower layer. From the above results, it can be concluded that for the unknown upper and lower media, one can not only determine the dielectric properties and thickness of the upper layer by wideband measurement with the resulting thickness taken from inversed Fourier transform to time domain response, but also determine dielectric properties of the lower layer from the lower frequency.

With the above procedure, the different materials were investigated. It was found that when the upper and lower layers were respectively 5% saline solution and water, the frequency range for wideband measurement decreased since saline solution is lossier than water.

Freq. (GHz)	Water	dp (mm)	5% saline	dp (mm)
1	77.960-j3.969	106	63.164-j141.956	3.51
2	77.230-j7.738	27	62.611-j75.443	2.83
3	76.144-j11.323	12	61.789-j55.085	2.46
4	74.741-j14.963	7.05	60.727-j46.141	2.14
5	73.066-j17.820	4.61	59.455-j41.464	1.86
6	71.164-j20.682	3.28	58.008-j39.268	1.62
7	69.082-j23.267	2.47	56.419-j38.006	1.42
8	66.866-j25.569	1.94	54.722-j37.358	1.24
9	64.588-j27.592	1.58	52.949-j37.051	1.1
10	62.196-j29.344	1.32	51.128-j36.923	0.98
11	59.814-j30.838	1.12	49.285-j36.879	0.88

Tab. 1. Dielectric properties and depth of penetration of water and 5% saline solution.

3.2 Dielectric Properties of Material with Different Dimensions

In practice, the size of the material is finite. Therefore, the effect of size contributes to the reflection coefficient and dielectric properties. In this regard, the size of the material was varied; the upper and lower layers are water ($d = 5$ mm) and 5% saline solution, respectively. Fig. 4 shows the determined dielectric properties of the material of interest with the size of 18λ and 42λ for the frequency of 6 GHz. Note that the fluctuations of the determined dielectric properties between 4.5-10.5 GHz are due to the edge diffraction in the material and the multiple reflection at the boundary of the two media that affect calculation of dielectric properties. The dielectric properties in this frequency band are invalid results. Those above 10.5 GHz are the dielectric properties of the upper layer whereas those below 4.5 GHz are for the lower layer. For the size of the material of 18λ and 42λ, the similar determined dielectric properties are obtained. The dielectric properties of material are not significantly affected by the size of the material. For dielectric constant in Fig. 4(a), the dielectric constant below 4.5 GHz approaches the dielectric constant of 5% saline solution and the one above 10.5 GHz approaches the dielectric constant of water. For loss factor, the frequency below 5 GHz approaches that of 5% saline solution and the one above 10.5 GHz approaches that of water as seen in Fig. 4(b).

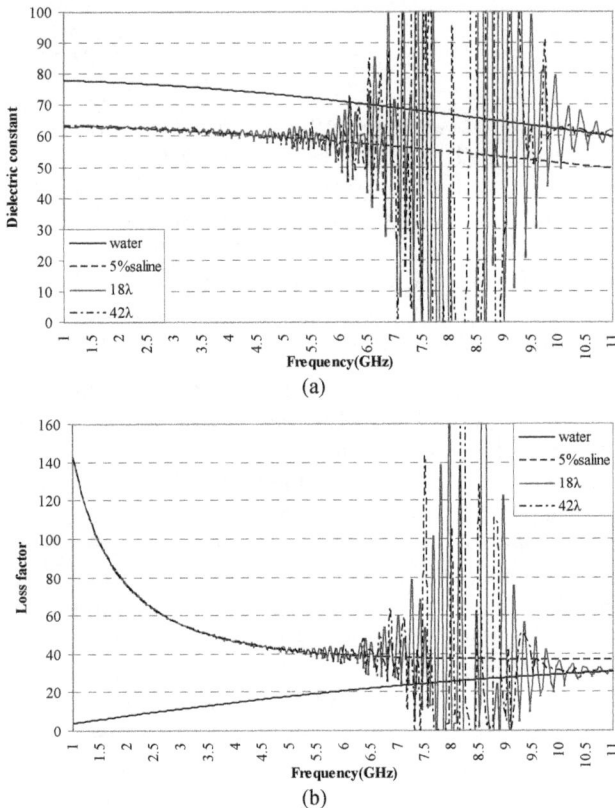

Fig. 4. Dielectric properties for different material dimensions ($d = 5$ mm, $W = 18λ$, 42λ at 6 GHz): (a) Dielectric constant, (b) Loss factor.

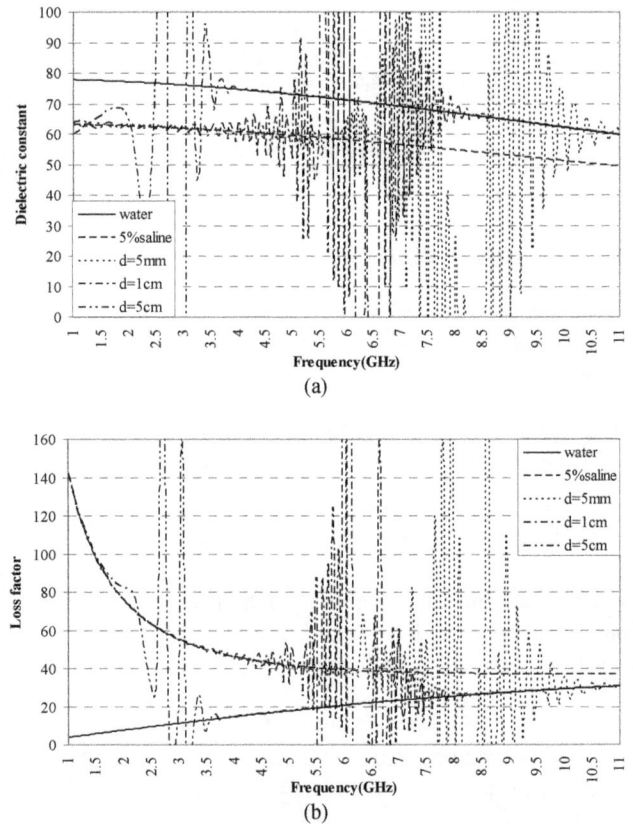

Fig. 5. Dielectric properties for different thickness of the upper layer ($W = 18λ$ at 6 GHz): (a) Dielectric constant, (b) Loss factor.

Let us consider the variation of dielectric properties for various thicknesses of the upper layer of water, lower layer of 5% saline solution and material size of 18λ at 6 GHz, respectively. Fig. 5(a) shows variation of dielectric constant whereas Fig. 5 (b) shows variation of loss factor; for d equal to 5 mm, 1 cm and 5 cm. Obviously, the obtained dielectric properties of material depends on the thickness of the upper layer. The thin layer has a wide range of low frequency of the lower layer. On the other hand, the thick layer has a wide range of high frequency of the upper layer. The transition between the low and the high frequency responses has fluctuation which is related to thickness of the upper layer. Consider Fig. 5(a) the values of dielectric constant of 5% saline solution are obtained below 5 GHz ($d = 5$ mm), 3.5 GHz ($d = 1$ cm) and 1.25 GHz ($d = 5$ cm). On the other hand, the values of water are obtained above 10.5 GHz ($d = 5$ mm), 8 GHz ($d = 1$ cm) and 4 GHz ($d = 5$ cm), respectively. In Fig. 5(b), the values of loss factor of 5% saline solution are obtained below 5.5 GHz ($d = 5$ mm), 4.25 GHz ($d = 1$ cm) and 1.75 GHz ($d = 5$ cm). The values of water are obtained above 10.5 GHz ($d = 5$ mm), 8 GHz ($d = 1$ cm) and 3.75 GHz ($d = 5$ cm), respectively. It should be pointed out that the thickness of the upper layer has significant effect on frequency range of determined dielectric properties.

3.3 Variation of Dielectric Properties of the Lower Layer

In this illustration, the upper layer is water ($d = 5$ mm) and the material size is 18λ for the frequency of 6 GHz. When the material of the lower layer is changed from 5% to 15% saline solution, the frequency response is in the same fashion. From Fig. 6, the saline solution with higher concentration possesses lower dielectric constant and higher loss factor. Clearly, we can determine the variation of dielectric properties of the lower layer from the proposed technique.

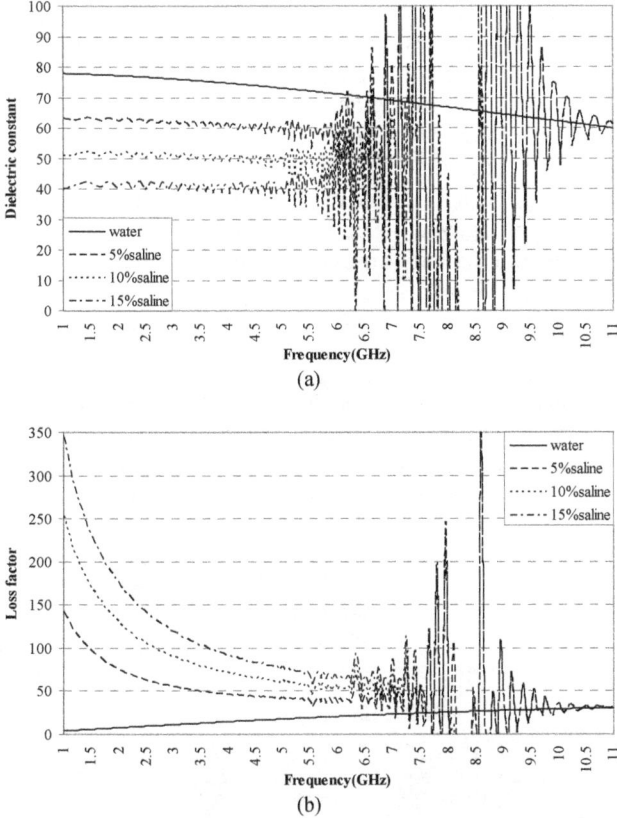

(a)

(b)

Fig. 6. Variation of dielectric properties of lower layer ($d = 5$ mm, $W = 18\lambda$ at 6 GHz): (a) Dielectric constant, (b) Loss factor.

4. Experimental Results

4.1 Experimental Setup

A plastic container of 60 cm × 90 cm × 52 cm in size, which the aperture area is 0.54 m², was first filled with 5% saline solution until reaching 40 cm in depth from the bottom. A large plastic sheet (thickness of 0.1 mm) was laid over the surface of saline solution. Its edges are wrapped on the edges of the plastic container. A plastic sheet was pressed to delete air bubbles. This plastic sheet can be used as a flat separator between saline solution and water before water of thickness of 5 mm was filled on top of the plastic sheet. The temperatures of water and the 5% saline solution

were 25 °C. This container was surrounded by wave absorbers, and two conical log antennas were used to transmit and receive microwave signal. The photograph of the measurement setup is depicted in Fig. 7. The frequency was varied from 1 GHz to 11 GHz, with the transmitting power of 10 mW using a vector network analyzer. The polarization of the transmitting and receiving antennas was respectively right-hand and left-hand circular polarization as these are the available wideband antennas in our laboratory. The antennas were separated by a wave absorber to decouple the antennas (measured S_{21} of −50 dB). The distance from the antennas to the surface of water was 1 m. It was calculated from the largest dimension of the antenna at the center frequency [25]. Note that this distance is shorter than the distance calculated from the largest dimension of the sample. Hence, the planar wavefront is not ensured. This may affect the accuracy of the upper layer thickness determination. A vector network analyzer was open, short and load calibrated and used for reflection measurement. The method to obtain a linear polarized wave from the circular polarized wave can be explained as follows:

$\vec{E}_{receive}$ is the received wave at the receiving antenna.

$$\vec{E}_{receive} = \Gamma E_i \hat{a}_w \cdot \hat{a}_{ant}$$
$$= \Gamma[E_x \hat{a}_x + jE_y \hat{a}_y] \cdot [\hat{a}_x \mp j\hat{a}_y] \quad (16)$$

where Γ is the complex reflection coefficient of the object under test, E_i is incident electric wave, $\hat{a}_w$ is polarization vector of the incident wave, and $\hat{a}_{ant}$ is polarization vector of receiving antenna. When the transmitted right-hand circular polarized wave (RHCP) is received by a right-hand circular polarized antenna (RHCP), the received wave is

$$\Gamma[(E_x + E_y)]. \quad (17)$$

The wave received by the left-hand circular polarized antenna (LHCP) can be expressed by (16).

$$\Gamma[(E_x - E_y)]. \quad (18)$$

Therefore, the linear polarized wave can be found from (17) ± (18) and then divided by 2.

Fig. 7. Experimental setup.

Note that receiving the transmitted RHCP wave by a RHCP antenna corresponds to measuring S_{11} whereas receiving by LHCP antenna corresponds to measuring S_{21}.

This assumption is realizable since the RHCP and LHCP antennas are connected to ports 1 and 2 of the network analyzer, respectively. In addition, the reflected wave from the material under test is in the main beam direction of the antennas which polarization is almost purely circular polarization.

4.2 System Calibration

The system was calibrated by placing a conducting plate (made of copper with a thickness of 1.44 mm that does not contribute significantly to phase error) on the surface of water (see Fig. 8(a)). The total reflection coefficient (Γ_{total_PEC}) is the summation of mutual coupling (S_{21}) and reflection from the conducting plate which is assumed to be perfect conductor ($\Gamma_{in\_copper}=-1$). S_{21} can be determined from

$$S_{21} = \Gamma_{total\_PEC} - \Gamma_{in\_copper}.$$ (19)

After removing the conducting plate, the reflection coefficient of the material under test was measured as shown in Fig. 8(b). The total reflection coefficient is the summation of the mutual coupling of the antennas (S_{21}) and the reflection coefficient of water and saline solution interface. Substituting S_{21} in (19) one can find the reflection coefficient between saline solution and water as shown in (20).

$$\Gamma_{in\_water,5\%saline} =$$
$$= \Gamma_{total\_water,5\%saline} - \left(\Gamma_{total\_PEC} - \Gamma_{in\_copper}\right)$$ (20)

The calibrated-measured reflection coefficient can be found from the measured reflection coefficients from the material under test and the reflection of the conducting plate.

4.3 Experimental Results

The calibrated-measured results were substituted on the left side of (4b) and the dielectric constant and loss factor were determined from (12) and (13).

Fig. 9 shows of the measured dielectric properties. Figs. 9(a) and (b) are respectively for dielectric constant and loss factor when the water is 5 mm thick. Clearly, the dielectric properties approach those of water at frequencies higher than 10.7 GHz as the depth of penetration of the wave is shorter than the thickness of water.

The comparisons of the measured results show the error of thickness is in the order of 5.6%. The discrepancy can be attributed from residue transmission at the high frequency and the effect of a plastic sheet separating the two media. The error from the determined dielectric properties of the upper layer results in error in velocity of wave in the upper layer and hence the depth of the upper layer. Nevertheless, a good agreement of the determined dielectric constant can be accomplished with slight error. For the loss factor, the fluctuations around the actual values are attributed from the limited signal-to-noise ratio in the experiment. It is recommended to utilize sufficiently high signal-to-noise ratio in practical use.

Comparing the measured results in Fig. 9 to the calculation results in Fig. 4, the good agreement is obtained. The dielectric properties of the lower layer approach those of 5% saline. While those in Fig. 4 are obtained at fre-

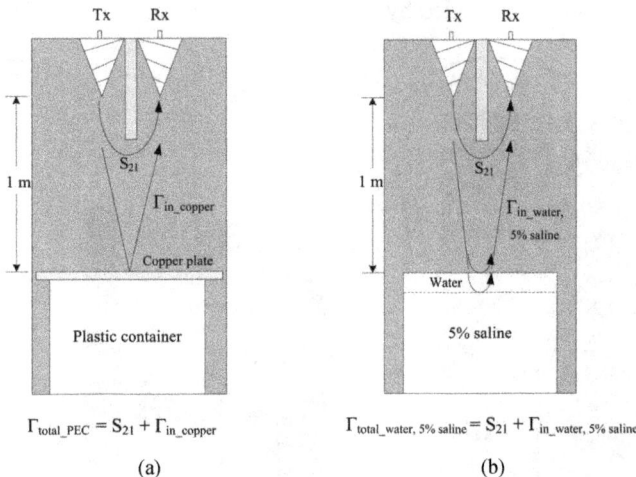

Fig. 8. Calibration process: (a) Conducting plate measurement. (b) Water and 5% saline solution measurement.

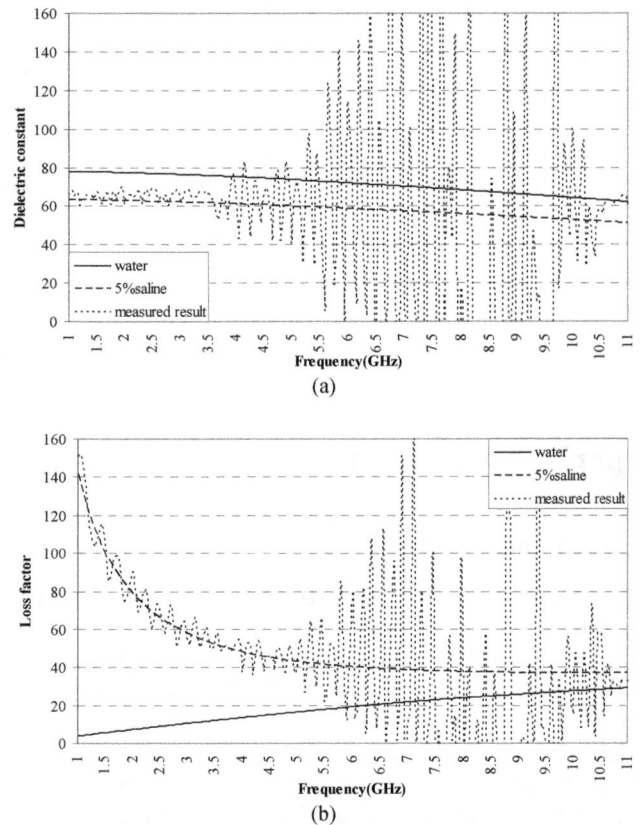

Fig. 9. Determined dielectric properties for $d = 5$ mm: (a) Dielectric constant. (b) Loss factor.

quency lower than 4.5 GHz, the results in Fig. 9 are obtained at frequency lower than 3.5 GHz. The difference results from the calculation results in Fig. 4 neglects the multiple reflections in the upper layer. In addition, the size of the measured material is not exactly the same as the one in calculation. Hence, the effect of edge diffraction is different.

5. Discussion

From the results in the previous sections, it is worth mentioning about the limitation of the system and the required bandwidth for the specific measurement.

The main objective of this work is to determine the unknown dielectric properties of the lower material at the lower frequency band. The upper material is the known material which is determined from the measured reflection coefficient at the higher frequency band. Then, from these dielectric properties at higher frequency band, the dielectric properties at lower frequency band can be found from the wideband database, which is already measured and cataloged as the priori data. With the dielectric properties at the lower frequency band, they lead to the determination of the dielectric properties of the lower material.

It is also essential to discuss about the required bandwidth of the measurement system. The bandwidth is directly related to the dielectric properties and thickness of the upper material. It is obtained from the inverse Fourier transform of the frequency domain measurement to time domain. Then, the thickness is obtained from (15). The thickness d contributes to the accuracy of the determined dielectric properties of the lower material, as seen in (7).

For the high loss material in the upper layer, the depth of penetration is shallow. For instance, when the upper layer is water the minimum d is 5 mm with error of 3.4%. For d less than 5 mm error is in excess of 10%. On the other hand, for the upper layer of 5% saline, the minimum d is 3 mm. The d thinner than 3 mm, error is excessively high. Hence, dielectric properties and thickness of the upper layer play a key role in the required bandwidth.

Furthermore, for the thick upper layer the required high frequency that wave does not penetrate to the lower layer decreases. For instance, for the same condition that the upper layer is water and the lower layer is 5% saline, the thicker upper layer requires lower frequency band. The system design for the measurement bandwidth can be accomplished by calculation using the proposed technique.

6. Conclusion

This paper has presented a dielectric properties determination technique for a stratified medium. The issue of interest in this work is a two-layer planar dielectric structure in which both the dielectric properties for the lower

layer and the thickness of the upper layer are unknown. The reflection coefficients of both layers were derived and the dielectric constant and loss factor of the lower medium were extracted. The technique started with wideband measurement to obtain the frequency domain response. The steady response on high frequency exhibited the effect of only the upper layer since wave could not penetrate to the lower layer. Therefore, a single layer was considered at this frequency band and dielectric properties could be accurately determined. Then, the full band dielectric properties were utilized for determining the depth of penetration. The full band dielectric properties and the time domain response can be obtained from the inverse Fourier transform. Hence, the thickness of the upper layer was determined. Using the derived dielectric properties extraction expressions, the dielectric properties of the lower layer can be determined. The measured results in 1-11 GHz band using 5 mm-thick water and 5% saline solution validated the proposed technique at the frequency below 3.5 GHz. The good agreement is accomplished and the proposed technique can be applied for justifying material in industrial process.

Acknowledgements

This work was supported by Thailand Research Fund under the Royal Golden Jubilee Ph.D. Program, Grant no. PHD/0323/2551 and King Mongkut's Institute of Technology Ladkrabang Research Fund (Grant no. KREF115061). The authors faithfully appreciate Mr. Vichit Lohprapan for kindly proofreading the manuscript.

References

[1] GHODGAONKAR, D. K., VARADAN, V. V., VARADAN, V. K. A free-space method for measurement of dielectric constants and loss tangents at microwave frequencies. *IEEE Transactions on Instrumentation and Measurement*, 1989, vol. 37, no. 3, p. 789–793. DOI: 10.1109/19.32194

[2] ZOUGHI, R., BAKHTIARI, S. Microwave nondestructive detection and evaluation of disbounding and delamination in layered-dielectric-slabs. *IEEE Transaction on Instrumentation and Measurement*, 1990, vol. 39, no. 6, p. 1059–1063. DOI: 10.1109/19.65826

[3] VENKATESH, M. S., RAGHAVAN, G. S. V. An overview of dielectric properties measuring techniques. *Canadian Biosystems Engineering*, 2005, vol. 47, p. 7.15–7.30.

[4] SINGH, D., YAMAGUCHI, Y., YAMADA, H., SINGH, K. P. Response of microwave on bare soil moisture and surface roughness by X-band scatterometer. *IEICE Transactions on Communications*, 2000, vol. E83-B, no. 9, p. 2038–2043.

[5] KHARKOVSKY, S., AKAY, M. F., HASAR, U. C., ATIS, C. D. Measurement and monitoring of microwave reflection and transmission properties of cement-based specimens. *IEEE Transactions on Instrumentation and Measurement*, 2002, vol. 51, no. 6, p. 1210–1218. DOI: 10.1109/TIM.2002.808081

[6] THAKUR, K. P., HOLMES, W. S. Noncontact measurement of moisture in layered dielectrics from microwave reflection spectroscopy using an inverse technique. *IEEE Transactions on Microwave Theory and Techniques*, 2004, vol. 52, no. 1, p. 76–82. DOI: 10.1109/TMTT.2003.821243

[7] HASAR, U. C., WESTGAT, C. R., ERTUGRUL, M. Permittivity determination of liquid materials using waveguide measurements for industrial applications. *IET Microwave, Antennas and Propagation*, 2010, vol. 4, no. 1, p. 141–152. DOI:10.1049/iet-map.2008.0197

[8] MEHTA, P., CHAND, K., NARAYANSWAMY, D., BEETNER, D. G., ZOUGHI, R., STOECKER, W. V. Microwave reflectrometry as a novel diagnostic tool for detection of skin cancers. *IEEE Transactions on Instrumentation and Measurement*, 2006, vol. 55, no. 4, p. 1309–1316. DOI: 10.1109/TIM.2006.876566

[9] SEKIGUCHI, H., SHIRAI, H. Electromagnetic scattering analysis for crack depth estimation. *IEICE Transactions on Electronics*, 2003, vol. E86-C, no. 9, p. 2224–2229.

[10] GHASR, M. T., KHARKOVSKY S., ZOUGHI, R., AUSTIN, R. Comparison of near-field millimeter-wave probes for detecting corrosion precursor pitting under paint. *IEEE Transactions on Instrumentation and Measurement*, 2005, vol. 54, no. 4, p. 1497–1504. DOI: 10.1109/TIM.2005.851086

[11] GHASR, M. T., CARROL, B., KHARKOVSKY, S., AUSTIN, R., ZOUGHI, R. Millimeter-wave differential probe for nondestructive detection of corrosion precursor pitting. *IEEE Transactions on Instrumentation and Measurement*, 2006, vol. 55, no. 5, p. 1620–1627. DOI: 10.1109/TIM.2006.880273

[12] HUGHES, D., ZOUGHI, R. A novel method for determination of dielectric properties of materials using a combined embedded modulated scattering and near-field microwave techniques. Part I-Forward model. *IEEE Transactions on Instrumentation and Measurement*, 2005, vol. 54, no. 6, p. 2389–2397. DOI: 10.1109/TIM.2005.858132

[13] HUGHES, D., ZOUGHI, R. A novel method for determination of dielectric properties of materials using a combined embedded modulated scattering and near-field microwave techniques. Part II-Dielectric property recalculation. *IEEE Transactions on Instrumentation and Measurement*, 2005, vol. 54, no. 6, p. 2398–2401. DOI: 10.1109/TIM.2005.858133

[14] VENKATESH, M. S., RAGHAVAN, G. S. V. An overview of microwave processing and dielectric properties of agri-food materials. *Journal of Biosystems Engineering*, 2004, vol. 88, no. 1, p. 1–18. DOI: 10.1016/j.biosystemseng.2004.01.007

[15] NELSON, S. O. Agricultural applications of dielectric measurements. *IEEE Transactions on Dielectrics and Electrical Insulation*, 2006, vol. 13, no. 4, p. 688–702. DOI: 10.1109/TDEI.2006.1667726

[16] ABUDUL KHALID, M. F., RAMLI, A. S., BABA, N. H., SAAD, H. A novel preliminary study on microwave characterization of siamese mangoes ripeness at K-band. In *Proceedings of International RF and Microwave Conference*. Kuala Lumpur (Malaysia), 2008, p. 143–147.

[17] HASAR, U. C. A fast and accurate amplitude-only transmission-reflection method for complex permittivity determination of lossy materials. *IEEE Transactions on Microwave Theory and Techniques*, 2008, vol. 56, no. 9, p. 2129–2135. DOI: 10.1109/TMTT.2008.2002229

[18] ODA, M., MASE, A., UCHINO, K. Non-destructive measurement of sugar content in apples using millimeter wave reflectometry and artificial neural networks for calibration. In *Proceedings of the 25th Asia-Pacific Microwave Conference*. Melbourne (Australia), 2011, p. 1386–1389.

[19] GHASR, M. T., SIMMS, D., ZOUGHI, R. Multimodal solution for a waveguide radiating into multilayered structures-dielectric prop-erty and thickness evaluation. *IEEE Transactions on Instrumentation and Measurement*, 2009, vol. 58, no. 5, p. 1505–1513. DOI: 10.1109/TIM.2008.2009133

[20] SEAL, M. D., HYDE, M. W., HAVRILLA, M. J. Nondestructive complex permittivity and permeability extraction using a two-layer dual-waveguide probe measurement geometry. *Progress In Electromagnetics Research*, 2012, vol. 123, p. 123–142. DOI:10.2528/PIER11111108

[21] CHARVAT, G. L., KEMPEL, L. C., ROTHWELL, E. J., COLEMAN, C. M., MOKOLE, E. L. A through-dielectric radar imaging system. *IEEE Transactions on Antennas and Propagation*, 2010, vol. 58, no. 8, p. 2594–2603. DOI: 10.1109/TAP.2010.2050424

[22] OSA, K., SUMANTYO, J. T. S., NISHIO, F. An application of microwave measurement for complex dielectric constants to detecting snow and ice on road surface. *IEICE Transactions on Communications*, 2011, vol. E94-B, no. 11, p. 2987–2990. DOI: 10.1587/transcom.E94.B.2987

[23] GENNARELLI, G., RICCIO, G. A uniform asymptotic solution for the diffraction by a right-angle dielectric wedge. *IEEE Transactions on Antennas and Propagation*, 2011, vol. 59, no. 3, p. 898–903. DOI: 10.1109/TAP.2010.2103031

[24] STOGRYN, A. Equations for calculating the dielectric constant of saline water. *IEEE Transactions on Microwave Theory and Techniques*, 1971, vol. MTT-19, no. 8, p. 733–736. DOI: 10.1109/TMTT.1971.1127617

[25] KRAUS, J. D., MARHEFKA, R. J. *Antennas for All Applications*. 3rd ed. New York: McGraw-Hill, 2002.

About the Authors ...

Paiboon YOIYOD was born in Phetchaburi in 1981, Thailand. He received the Bachelors degree in Electrical Engineering with first class honor from Rangsit University and Masters degree in Telecommunications Engineering from King Mongkut's Institute of Technology Ladkrabang in 2005 and 2010, respectively. He is currently pursuing the Ph.D. degrees in Electrical Engineering. He joined in production engineer, Pioneer Manufacturing (Thailand) Co.,Ltd. in 2005 as an engineer. In 2006, he joined the Rangsit University as an engineer. His current research interests include dielectric measurements, sensors, and measurement systems.

Monai KRAIRIKSH was born in Bangkok, Thailand. He received the B.Eng., M.Eng. and D.Eng. degrees in Electrical Engineering from King Mongkut's Institute of Technology Ladkrabang (KMITL), Thailand in 1981, 1984, and 1994, respectively. He was a visiting research scholar at Tokai University in 1988 and at Yokosuka Radio Communications Research Center, Communications Research Laboratory (CRL) in 2004. He joined the KMITL and is currently a Professor at the Department of Telecommunication Engineering. He has served as the Director of the Research Center for Communications and Information Technology during 1997-2002. His main research interests are in antennas for mobile communications and microwave in agricultural applications. Dr. Krairiksh was the chairman of the IEEE MTT/AP/Ed joint chapter in 2005 and 2006. He served as the General Chairman of the 2007 Asia-Pacific Microwave Conference, and the advisory committee of the 2009 International Symposium on Antennas and Propagation. He was the President of the Electrical Engineering/Electronics, Computer, Telecommunications and Information Technology Association (ECTI) in 2010 and 2011 and was an editor-in-chief of the ECTI Transactions on Electrical Engineering, Electronics, and Communications. He was recognized as a Senior Research Scholar of the Thailand Research Fund in 2005 and 2008 and a Distinguished Research Scholar of the National Research Council of Thailand. He has been a distinguished lecturer of IEEE Antennas and Propagation Society during 2012-2014.

15

A Coplanar Waveguide Fed Hexagonal Shape Ultra Wide Band Antenna with WiMAX and WLAN Band Rejection

Tapan MANDAL[1], Santanu DAS[2]

[1] Dept. of Information Technology, Govt. College of Engineering and Textile Technology, Serampore, Hooghly, India, PIN-712201
[2] Dept. of Electronics & Tele-Comm. Engineering, Bengal Engineering and Science University, Shibpur, Howrah, India, PIN-711103

tapanmandal20@rediffmail.com, santanumdas@yahoo.com

Abstract. *In this paper, a coplanar waveguide (CPW) fed hexagonal shape planar antenna has been considered for ultra-wide band (UWB). This antenna is then modified to obtain dual band rejection. The Wireless Local Area Network (WLAN) and Wireless Microwave Access (WiMAX) band rejections are realized by symmetrically incorporating a pair of L-shape slots within the ground plane as well as a couple of I-shape stubs inserted on the bottom side of radiating patch. The proposed antenna has stop bands of 5.05–5.92 GHz and 3.19–3.7 GHz while maintaining the wideband performance from 2.88– 13.71 GHz with reflection coefficient of ≤ -10 dB. The antenna exhibits satisfactory omni-directional radiation characteristics throughout its operating band. The peak gain varies from 2 dB to 6 dB in the entire UWB frequency regions except at the notch bands. Surface current distributions are used to analyze the effects of the L-slot and I-shape stub. The measured group delay has small variation within the operating band except notch bands and hence the proposed antenna may be suitable for UWB applications.*

Keywords

Hexagonal planar antenna, CPW fed, UWB, WLAN band, WiMAX band.

1. Introduction

Ultra–Wide Band (UWB) technology becomes more and more important owing to many wireless applications such as multimedia communications, sensor networks, ground penetrating radar, medical imaging and precision localization systems. Due to the characteristics like high data transmission rates, high precision ranging, low complexity, easy connection and high security, UWB technology has been used in large consumer devices as laptops, digital cameras, high definition TVs and bio-medical sensors. As an essential part of the UWB system, the UWB antenna should be designed with low profile, low volume, low cost, large bandwidth and good omni directional radiation patterns and constant group delay.

In 2002, the US-Federal Communications Commission (US-FCC) approved the UWB frequency band from 3.1 to 10.6 GHz for commercial communication applications [1]. Since then, considerable research efforts have been paid into UWB communication technology. This type of UWB antenna has been realized by using either microstrip line [2–5] or coplanar waveguide (CPW) [6–10] feeding structure. In particular, CPW fed antennas have many salient features like less radiation loss, less dispersion and easy integration with monolithic microwave integrated circuits (MMIC). Therefore, CPW fed UWB antennas are currently under consideration for numerous applications.

Many of UWB antennas have been offered for various applications in the last decade [6–15]. Planar UWB monopole antennas with rectangular [6], disk [7], [8], elliptical [9], hexagonal [10] and triangular [11] shapes have been reported. The CPW fed hexagonal monopole antennas are found to have -10 dB return loss bandwidth for UWB application [10] with large size structure. A compact hexagonal wide slot antenna with microstrip fed monopole for UWB application has been described [2] but its structure is complex.

However, the existing Wireless Local Area Network (WLAN: IEEE 802.11a) and Wireless Microwave Access (WiMAX: IEEE 802.16) service bands of 5.15–5.825 GHz and 3.3–3.7 GHz are responsible for the performance degradation of UWB system because of the interference. These existing narrow bands may cause interferences with the UWB systems. To prevent this problem, UWB antennas with band rejection characteristic is desirable. In the conventional design, band stop filters are added at the end of the antenna or the devices. Thereby the size of the antenna is increased. Several UWB antennas with band rejection characteristic have been proposed [4–6], [8], [11–15]. For printed monopole antennas, the familiar methods to achieve band-notch function are etching slots on the metallic patch, feeder or the ground plane in different shapes such as C-shape [6], [8], [14], U-shape [4], L-shape [5], [11], [12], I-shape [13] slot etc. However, most of the notch band UWB antennas are formed by etching half wavelength or quarter wavelength slot on radiating patch.

UWB antennas with dual notch bands have been surfaced in [15–21]. The configuration is either in complex [14], [17–19], [21] or large in size [15]. An effective technology is to insert open circuited stubs into the UWB antenna. The band notch performance is achieved by placing parasitic strips in close proximity to the antenna [18–20]. However, most of UWB antennas [2–11], [14] do not have more than single notch band which reveals that potential interference from other narrow band may still exist.

In this paper, a CPW-fed simple hexagonal monopole antenna with WLAN and WiMAX band notch characteristics is proposed. The design initially begins with a regular hexagonal monopole antenna (RHMA). The band rejection characteristics are obtained by incorporating a pair of L-slots on the ground plane away from the radiating patch and I-shape stubs on the other side of the patch. All the simulations in terms of impedance bandwidth, input impedance, gain, efficiency are carried out using Method of Moment based IE3D simulation software [22]. In this paper, all reflection coefficient measurements are taken out with the help of Agilent Technologies Vector Network Analyzer (N5230A).

2. Antenna Configuration

2.1 Prototype Antenna Geometry

Fig. 1 shows the geometry of the prototype antenna. It is printed on a substrate with dielectric constant ε_r of 4.4 and height h of 1.59 mm. As shown in Fig. 1, there are three sections which comprise the structure of the antenna: CPW transmission-line, rectangular shape ground plane and hexagonal shape radiating patch. A 50Ω CPW transmission line is designed with a strip width W_f of 4.1 mm. The lower band edge frequency f_l has been determined for UWB using formulas [8], [10] given as

$$f_l = \frac{c}{\lambda} = \frac{7.2}{(H + r + T)} \tag{1}$$

where H is the height of the hexagon and r is the radius of an equivalent cylindrical monopole antenna in cm and T is the gap between ground plane and patch in cm. With reference to configurations in Fig. 1, the dimensions of height H and radius r of the equivalent cylindrical monopole antenna are obtained by equating their areas as follows:

$$r = \frac{3L_p}{4\pi}, \tag{2}$$

$$H = \sqrt{3}L_p. \tag{3}$$

2.2 Band – Notch Antenna Design

In another form (Fig. 2) the RHMA is made on the same substrate with the dimensions same as before. A notch band in the frequency range of 5.15–5.825 GHz is

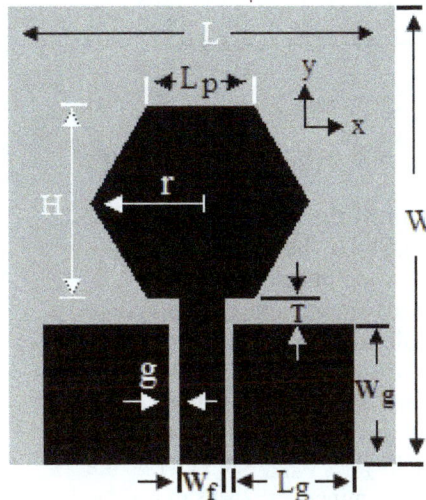

Fig. 1. Geometry of prototype antenna.

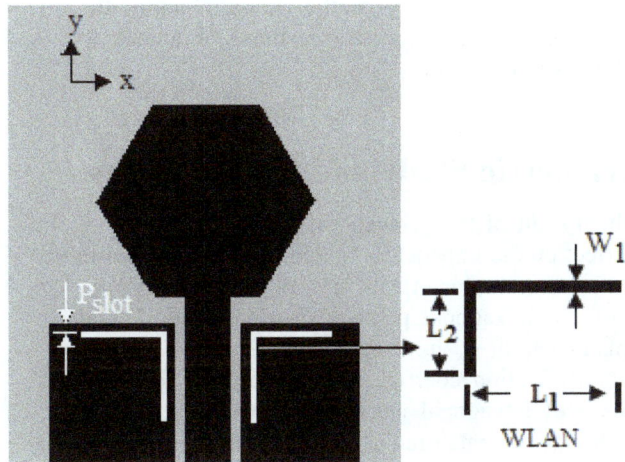

Fig. 2. Geometry of prototype with L-slot antenna.

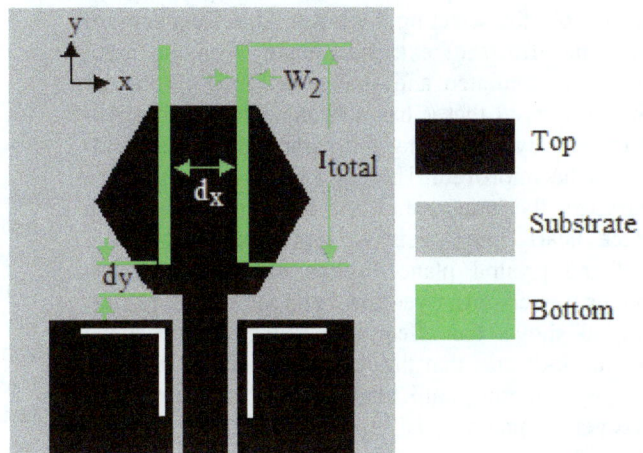

Fig. 3. Geometry of proposed antenna.

obtained by symmetrically inserting a pair of thin L-slots near the feed line within the ground plane. Usually, the length of the each slot is made approximately equal to half the guided wavelength λ_g at the desired notch frequency of the band. This is given by

$$L_{total\_WLAN} = \frac{\lambda_g}{2} = \frac{c}{2 f_{notch\_WLAN}\sqrt{\dfrac{(\varepsilon_r + 1)}{2}}}, \qquad (4)$$

$$L_{total\_WLAN} = L_1 + L_2 \qquad (5)$$

where L_{total_WLAN} is the length of the slot. For the center notch frequency $f_{notch\_WLAN} = 5.5$ GHz the length of the slot is calculated as 16.76 mm. The optimum slot width W_1 is found to be 0.3 mm by way of simulation. In addition to this WiMAX stop band is realized by symmetrically incorporating a couple of narrow I-shape parasitic strips (stubs) on bottom side of the radiating patch (Fig. 3). Each parasitic strip has length of half wavelength at desired notch frequency. For center notch $f_{notch\_WiMAX} = 3.5$ GHz, the length of the strip can be determined as 26.13 mm. The optimum strip width W_2 is found to be 1 mm. From (4), (5), the total length of the L-shape slots and I-shape stubs may be obtained at the beginning of the design. Finally the position and lengths are adjusted by using simulator to achieve the desired results.

3. Parametric Study and Observations

In theory, all of the geometrical parameters of Fig. 1 have an effect on the impedance matching. However, some of them may have major effects than others. Specifically, the size of the hexagonal patch, extrusion depth T and ground plane length L_g have considerable effects on the bandwidth. Reflection coefficient characteristics for various radius r of hexagonal patch are shown in Fig. 4. It indicates that as the patch radius increases, the lower edge frequency moves towards the left side of the plot. The main reason is the increase of radius which in turn increases the height H of the monopole antenna. Thus patch plays a significant role for selecting the lower edge frequency of the band. The effects of extrusion depth T on the input impedance are simulated and shown in Fig. 5. From the result, it is observed that T has a strong effect for impedance matching. By optimizing the parameter T, the bandwidth could be improved. The gap creates a capacitance that neutralizes the inductive effects of the radiating patch to produce nearly pure resistive input impedance. The length of the ground plane L_g has important role on matching characteristics over the band of prototype antenna. Fig. 6 shows the effect of varying ground plane length which indicates that the reflection coefficient characteristics is changing significantly while the higher and lower frequency matching is improved. However, at middle of the frequency band, the performance degrades. In this case an optimum value of L_g is taken to be 11.6 mm as a compromise of all frequency matches perfectly. The simulation responses of prototype antenna cover the entire UWB frequency range for $r = 13.6$ mm, $T = 1.375$ mm, $L_g = 11.6$ mm, $g = 0.4$ mm and $W_f = 4.1$ mm.

Next parametric study and discussion are carried out for the antenna with WLAN band rejection (Fig. 2). The

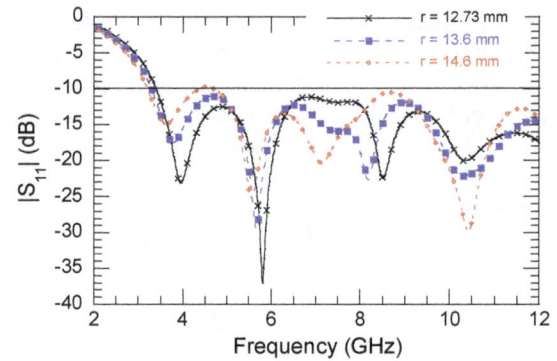

Fig. 4. Simulated reflection coefficient for different radius r.

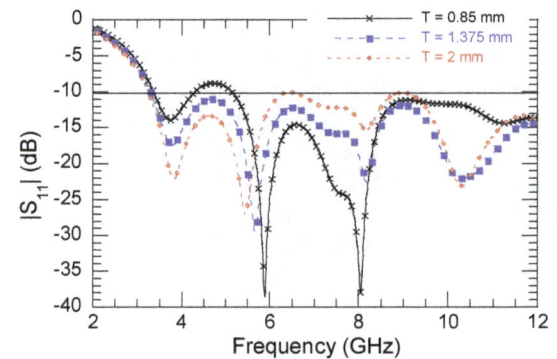

Fig. 5. Simulated reflection coefficient for different T.

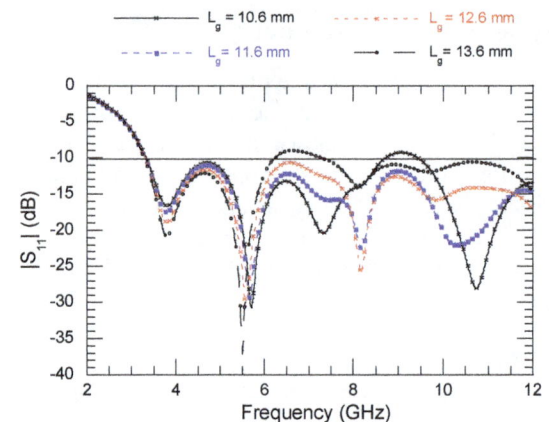

Fig. 6. Simulated reflection coefficient for different L_g.

total length of the inverted L-shape slots etched from the ground plane is deduced as in (4). Two L-shape slots are introduced symmetrically with respect to the fed line of the planer monopole as shown in Fig. 2. A pair of notch frequencies of two slots are coupled together and a band stop characteristics with improved notching and rejection bandwidth is obtained. Here, the reflection coefficient characteristics for various values of slot length L_{total_WLAN} and width W_1 are shown in Fig. 7 to Fig. 8 respectively. Fig. 7 indicates that the notch band moves downwards with higher peak as the length of the L-slots is increased. It is observed from the parametric study that the resonant frequency of the notch-band depends on the length of the slot and notch bandwidth depends upon width of the slot. This property provides a great freedom to the designers to select the notch band for the antennas as is evident from Fig. 7 and Fig. 8.

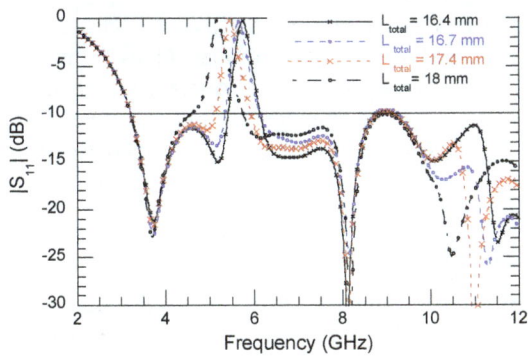

Fig. 7. Reflection coefficient for various lengths of L-slot.

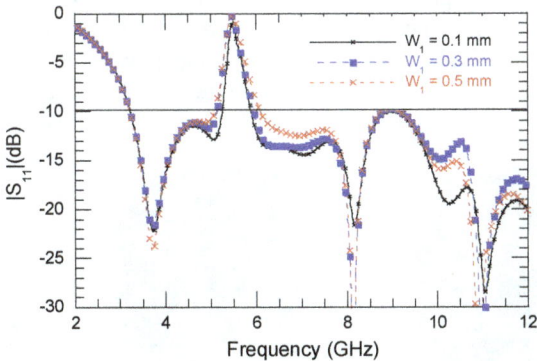

Fig. 8. Reflection coefficient for various width of L-slot.

Fig. 9 shows the reflection coefficient characteristics with changing the position of the slots with other parameters remaining constant. It can be seen that the peak value at notch frequency (5.5 GHz) decreases while the value of P_{slot} rises. An optimum P_{slot} is chosen to be 1 mm by way of parametric study. Due to the presence of L-slots inside the ground plane, maximum current flows back to the feeding part. Therefore, negligible amount of currents radiates from the antenna and degenerates radiation around 5.15 GHz to 5.93 GHz.

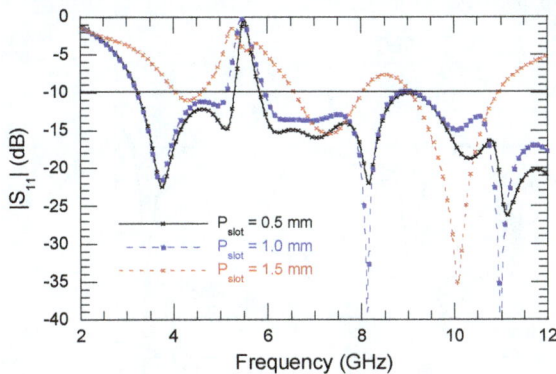

Fig. 9. Effect of position of the L-slots on reflection coefficient.

Furthermore to realize the WiMAX stop band in UWB region, a pair of I-shape strips (stubs) have been incorporated on the other side of substrate. The reflection coefficient plots for various length and width are illustrated in Fig. 10 and Fig. 11 respectively. It is observed from Fig. 10 that the notch band moves downwards with higher

peak as the length of strips is increased. Similarly, it is observed that an increase in width W_2 of resonator results in moving downwards of the center frequency of the notch. Strips length has greater impact than strips width on shifting the frequency as is evident from Fig. 10 and Fig. 11.

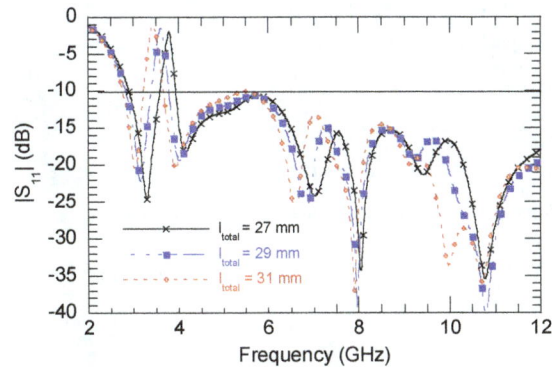

Fig. 10. Reflection coefficient for various length of I-shape stub.

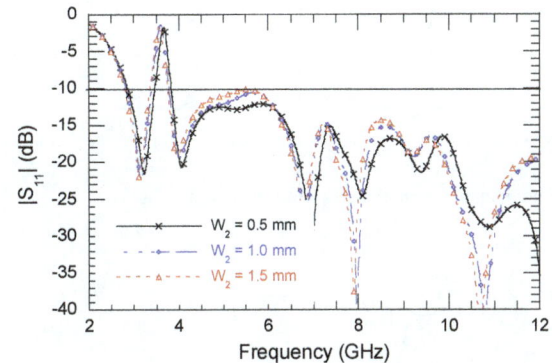

Fig. 11. Reflection coefficient for various width of I-shape stub.

To show the dependency of reflection coefficient on the position of the stub, parametric studies have been done for the same. Fig. 12 and Fig. 13 give the variation in peak value of reflection coefficient with changing the position of the I-strips while other parameters remaining the same. As d_x and d_y value (Fig. 3) rises, the notch BW increases. Therefore, these d_x and d_y have significant freedom for controlling the notch BW. The optimum values of d_x and d_y are chosen to be 11 mm and 3.15 mm respectively. The reflection coefficient characteristics for all the configurations are illustrated in Fig. 14 for a comparison. The detailed design dimensions are given in Tab. 1.

Parameters	L	W	r	H	L_p	L_g
Dimensions	36	48	13.6	23.55	13.6	11.6
Parameters	g	W_f	T	L_1	L_2	W_1
Dimensions	0.4	4.1	1.375	8.4	8.2	0.3
Parameters	l_{total}	W_2	d_x	d_y	p_{slot}	W_g
Dimensions	29	1	11	3.125	1	13.5

Tab. 1. Parameters value of the proposed antenna (dimensions are in mm).

The simulated reflection coefficient characteristics of the proposed antenna reveal stop bands of 0.38 GHz (3.32 to 3.7 GHz) and 0.78 GHz (5.15–5.93 GHz) within the UWB frequency span.

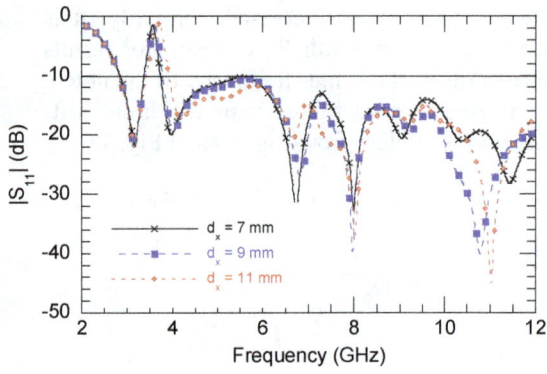

Fig. 12. Simulated reflection coefficient for various value of d_x.

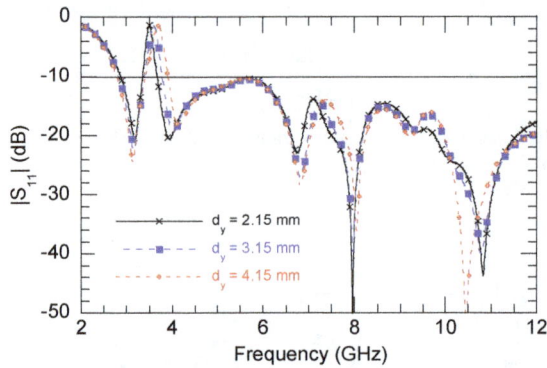

Fig. 13. Simulated reflection coefficient for various value of d_y.

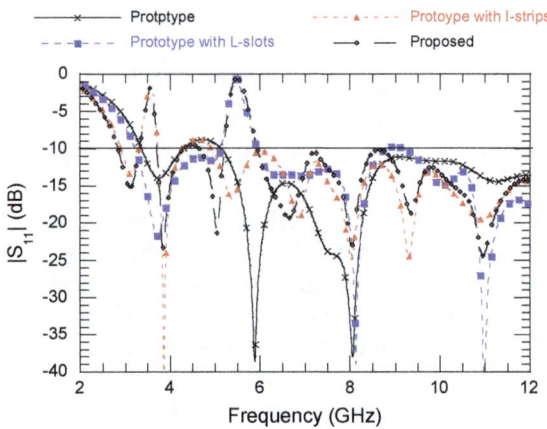

Fig. 14. Reflection coefficient characteristics of all configurations for comparison.

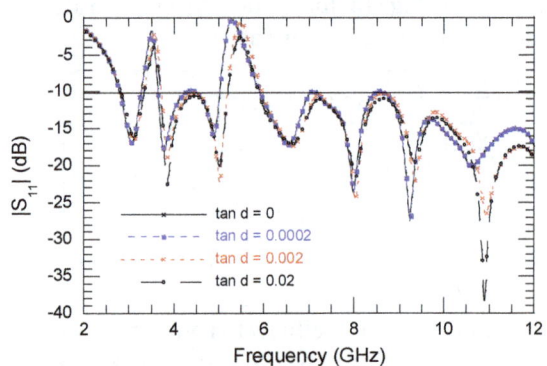

Fig. 15. Reflection coefficient characteristics for various value of loss tangent (tan δ).

Fig. 15 indicates the reflection coefficient characteristics of the proposed antenna for various values of loss tangent. From the characteristics, it is observed that the peak value of reflection coefficient at notch frequency decreases while the loss tangent of the substrate increases.

4. Experimental Results

Photograph of the prototype and proposed antennas are shown in Fig. 16. The simulated and measured performances of prototype antenna (Fig. 16(a)) are plotted in Fig. 17 for comparison. It is observed from the plot that the simulated response provides BW of 10.37 GHz (3.15 to 13.52 GHz) where the experimental response yields 10.54 GHz (3.06 GHz – 13.60 GHz) impedance band. Therefore, a reasonable good agreement between simulation and measurement is achieved.

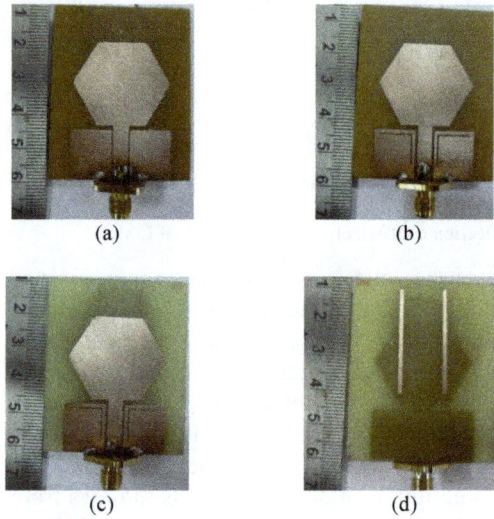

Fig. 16. Photograph of fabricated structure: (a) prototype (top plane), (b) prototype with L-slots (top plane), (c) proposed (top plane), (d) proposed (bottom plane).

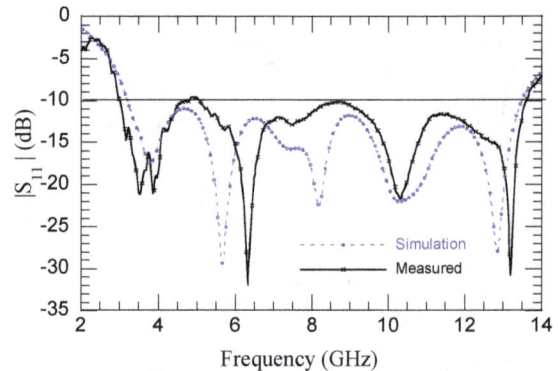

Fig. 17. Comparison between simulation and measured results of the prototype antenna.

Comparison between simulation and measured results of the prototype with L-slots antenna (Fig. 16 (b)) is illustrated in Fig. 18. The simulated reflection coefficient characteristic (Fig. 18) reveals stop band of 0.78 GHz (5.15

to 5.93 GHz) for $S_{11} \le$ -10 dB within the frequency span from 3.15 GHz to 13.08 GHz. The measured characteristic shows 0.86 GHz (5.15–6.01 GHz) stop band which covers the entire WLAN band. The measured result reasonably agrees with the simulated result which has sharp frequency stop band of WLAN band after the L-slots are inserted on the ground plane. The discrepancies between measured and simulated results may be due to the fabrication tolerance of the antenna. The simulated and measured responses of the proposed antenna (Fig. 16(c, d)) are shown in Fig. 19. The simulated reflection coefficient characteristic reveals stop bands of 0.38 GHz (3.32–3.7 GHz) and 0.78 GHz (5.15 to 5.93 GHz) for $S_{11} \le$ -10 dB within the frequency span from 2.84 GHz to 13.36 GHz. The measured characteristic shows 0.52 GHz (3.19–3.7 GHz) and 0.87 GHz (5.05 to 5.92 GHz) stop bands which cover the entire WiMAX and WLAN band in frequency span of 2.88–13.71 GHz. The loss tangent of the substrate and dimensional mismatch between simulated and physical structures may cause the difference between the simulated and measured peak value at notch frequency.

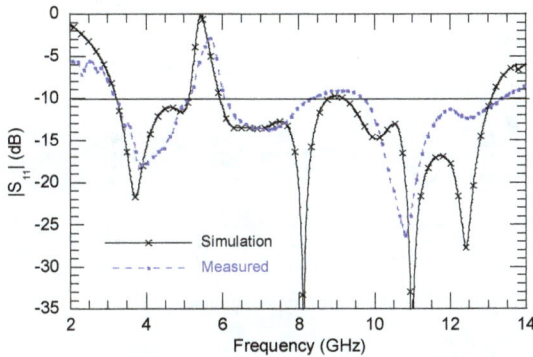

(i) at 3.1 GHz (a) E-plane pattern   (b) H-plane pattern

(ii) at 9 GHz (a) E-plane pattern   (b) H-plane pattern

Fig. 20. Measured radiation pattern characteristics for the prototype (Fig. 16(a)) and the proposed antenna (Fig. 16(c, d)).

The H-plane radiation patterns are omnidirectional, whereas in E-plane, it is figure of eight because of small ground plane on the same side of the patch. It is to be noted that adding slots in the ground plane as well as strips on the other side substrate does not significantly alter the radiation patterns of the antenna. The results of Fig. 20 show that the radiation patterns are reasonably stable throughout ultra-wide band. Thus this can be considered as an UWB antenna with WiMAX, WLAN notch band characteristics for UWB applications.

Fig. 18. Comparison between simulation and measured results of the prototype antenna with L-slots.

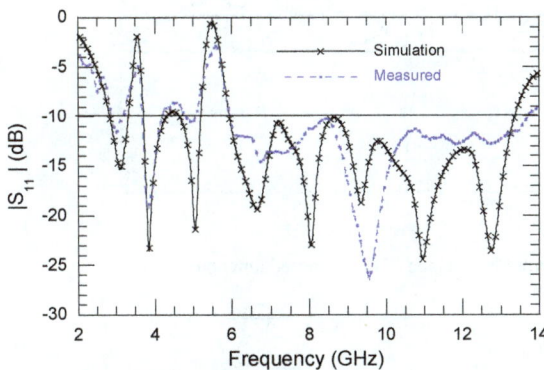

Fig. 19. Comparison between simulation and measured results of the proposed antenna.

5. Radiation Pattern

The radiation patterns are shown in Fig. 20. The antenna is printed in the X–Y plane and it is Y-polarized because the monopole is in the Y-direction. Therefore, the E-plane for this antenna is the YZ-plane and the H-plane is the XZ-plane.

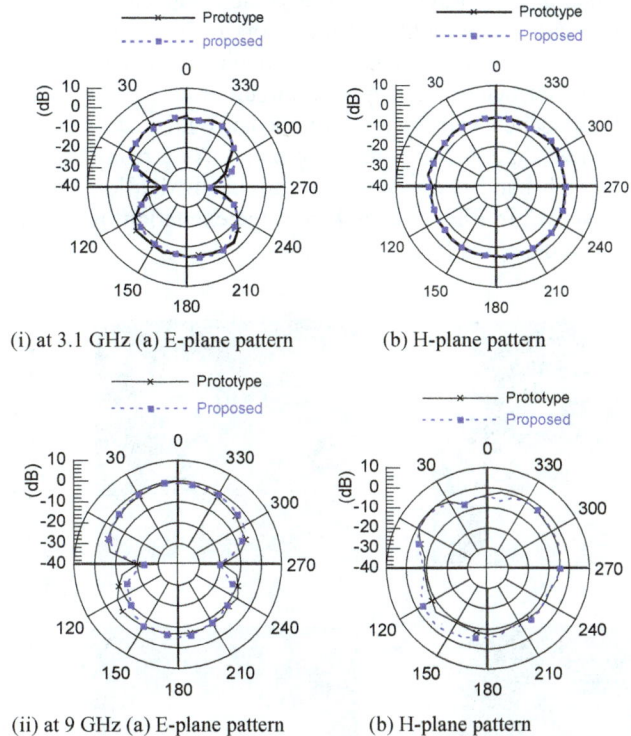

6. Surface Currents and Input Impedance, Gain and Efficiency

Fig. 21 displays the current distribution of the proposed antenna. The current distribution at 3.1 GHz as shown in Fig. 21(a) indicates the excitation of lower order mode at this frequency.

The current distribution in Fig. 21(b) clearly indicates the formation of standing waves at notch frequency as the current is confined at the region of I-shape stubs. It is confirmed that the I-shape stubs effectively reflects the signal power back to the input port and thus mismatching occurs. As shown in Fig. 21(c), at another notch frequency, the current distribution is concentrated and oppositely directed on the interior and exterior sides of the slot. The current has high density at the edges of slot and low density at the corner point of the L-slot. The top of the slot is modeled as a transmission line short circuit mode. The resultant equivalent load at the corner of the slot is high impedance (open circuit) due to the length from the top to the corner along the slot is approximately λ/4 at the notch frequency. Therefore, the slot behaves as an open-circuited load with large input impedance, causing a total impedance mismatch between the feed line and the radiating patch. As a result the desired notch band is created. The current distribution at 8 GHz as shown in Fig. 21(d) indicates the excitation of higher order mode at this frequency.

pedance varies around 50 Ω while its imaginary part has small values and fluctuates around zero. This is mainly because of a continuous coupling obtained between the hexagonal patch and the ground plane and hence the matching is achieved over the entire UWB region. Fig. 23 shows the input impedance plot of the proposed antenna with notch frequencies. The impedance of the structure changes acutely at 3.5 GHz and 5.5 GHz as is evident from the very large value of its imaginary part. This causes large reflection at these frequencies resulting in steep rise of magnitude of S_{11}. In the notch bands, most of the power fed into the antenna is reflected back which leads to a decrease of the radiation efficiency and hence the antenna gain. The simulated gain is plotted in Fig. 24 in which two sharp drops occur at around 3.5 GHz and 5.5 GHz as expected. The similar variation can be seen in the antenna efficiency curve plotted in Fig. 25. This characteristic can make sure the ability of the proposed antenna to reject the

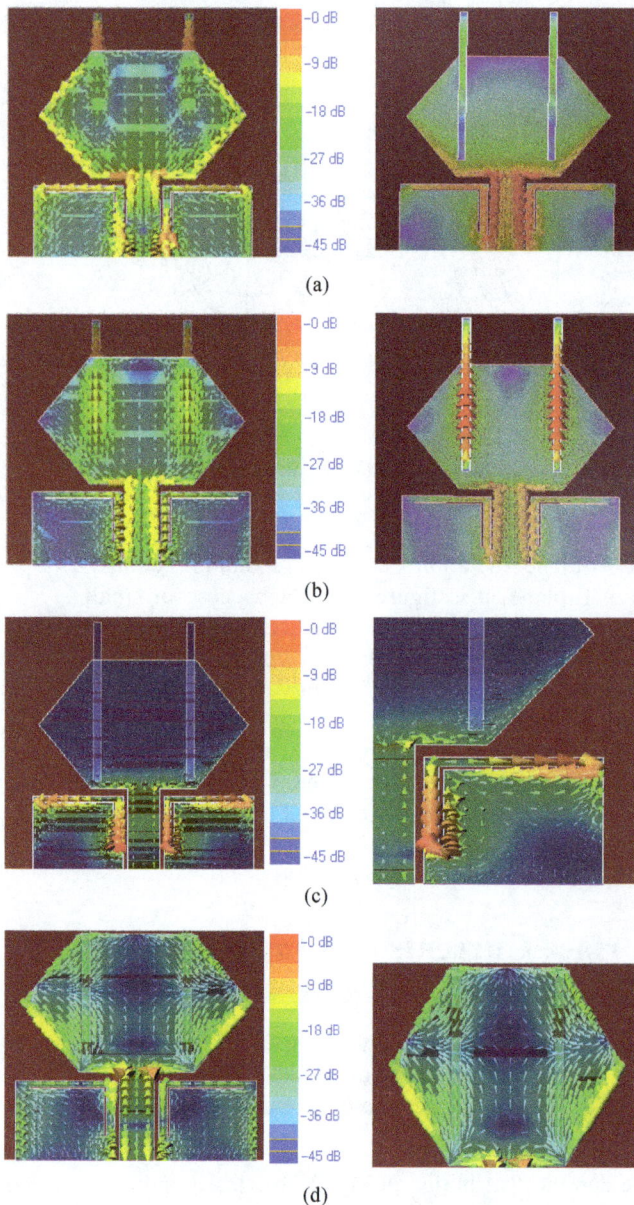

Fig. 21. Simulated current distribution at:(a) 3.1 GHz (left side - top plane; right side - bottom plane), (b) 3.5 GHz (left side - top plane; right side - bottom plane), (c) 5.5 GHz (left side –top plane; right side – zoomed of a slot) and (d) 8 GHz (left side –top plane; right side – zoomed of only patch).

The real and imaginary parts of the input impedance verses frequency of the prototype antenna is shown in Fig. 22. It is observed that the real part of the antenna im-

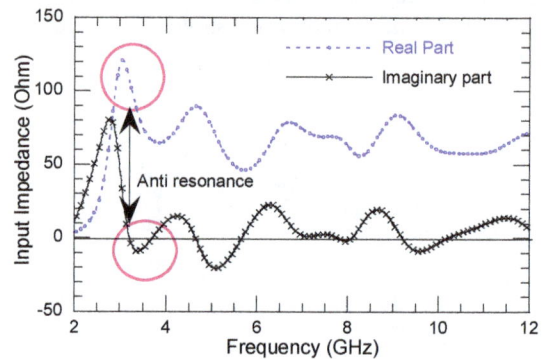

Fig. 22. Input impedance of the prototype antenna.

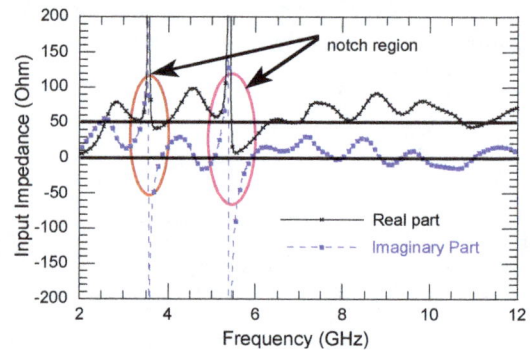

Fig. 23. Input impedance of the proposed antenna.

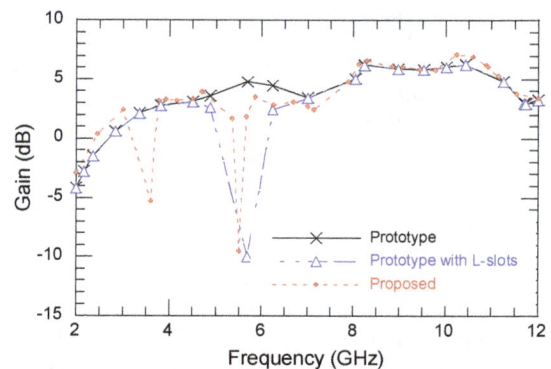

Fig. 24. Simulated gain versus frequency plot.

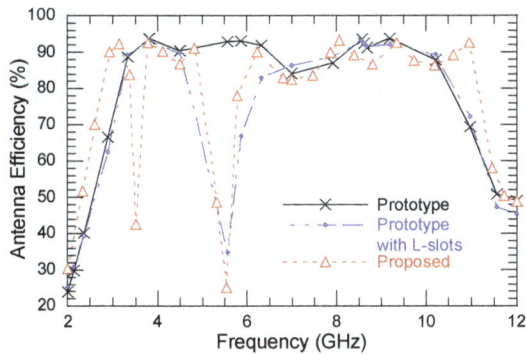

Fig. 25. Simulated antenna efficiency versus frequency plot.

interference effectively. The proposed antenna gain varies from 2.0 – 6.0 dB over the whole operating range except in notch bands. It provides antenna efficiency more than 80 % except at notch bands.

7. Transfer Function and Time Domain Study

A pair of proposed antennas is used as transmitting and receiving antenna to measure the magnitude of S_{21} and group delay. The transmitting and receiving antenna are placed at a distance of 170 mm. The measured magnitude of S_{21} and group delay of the proposed antenna are depicted in Fig. 26. The transfer function is relatively flat

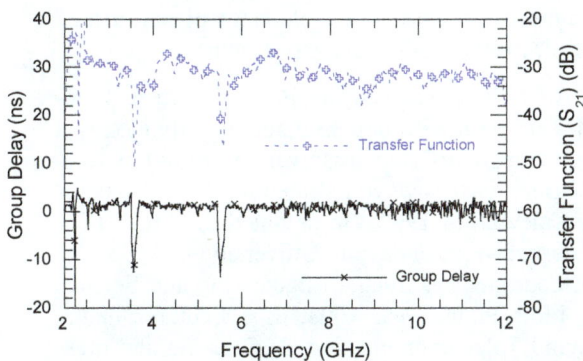

Fig. 26. The measured transfer function and group delay of the antenna systems (left side Y axis – Group delay; right side Y axis - S_{21}).

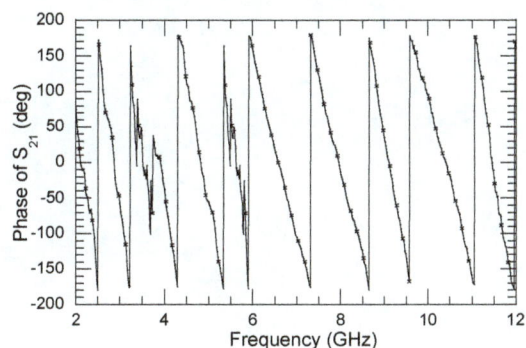

Fig. 27. The measured phase of S_{21} for the proposed antenna system.

(variation less than 10 dB) over the operating band except the notch bands. It is observed that the variation of group delay of the proposed antenna is virtually constant across the whole UWB band except at the two notch bands region. The measured phase of the transfer function is shown in Fig. 27. It is observed that the phase of S_{21} is relatively linear from 3 GHz to 12 GHz excluding the notch bands in face to face mode. The measured group delay corresponds well to the phase of S_{21}, so it proves that the antenna has a good time-domain characteristic and a small pulse distortion as well.

8. Conclusion

In this paper, a simple ultra-wideband planar hexagonal printed monopole antenna with dual narrow band-notch characteristics is presented. To realize the WLAN and WiMAX rejection bands in UWB, a couple of half wave length L-slots are added in the ground plane as well as a pair of I-shape strips placed on the other side of the patch. Input impedance, surface current distributions are built at the same time for analysis and explanation of dual narrow band-notch characteristics. The antenna has been successfully optimized, fabricated and measured. This antenna has ultra wide-band performance in the frequency band from 2.88 to 13.71 GHz for magnitude of $S_{11} \leq$ -10 dB with dual excellent rejection bands. The radiation patterns are observed to be nearly omni-directional over the entire UWB frequency band. The proposed antenna peak gain varies between 2.0 dB and 6 dB at all frequencies other than the notch bands. Satisfactory time domain results have been found in respect of group delay and phase. Therefore the proposed antenna is expected to be a good candidate in various wireless applications.

References

[1] First Report and Order, "Revision of part 15 of the commission's rule regarding ultra-wideband transmission system FCC 02- 48", *Federal Communications Commission*, 2002.

[2] REZA GHADERI, M., MOHAJERI, F. A compact hexagonal wide slot antennas with microstrip fed monopole for UWB applications. *IEEE Antenna and Wireless Propagation Letters*, 2011, vol. 10, p. 682–685.

[3] ZHANG, K., WANG, T., CHENG, L. L. Analysis of band notched UWB printed monopole antennas using a novel segmented structure. *Progress In Electromagnetics Research C*, 2013, vol. 34, p. 13–27.

[4] MANDAL, T., DAS, S. Ultra wide band printed hexagonal monopole antennas with WLAN band rejection. *Microwave and Optical Technology Letters*, 2012, vol. 54, p. 1520–1525.

[5] ABDO ABDEL MONEM SHAALAN, RAMADAN, M. I. Design of a compact hexagonal monopole antenna for ultra-wideband applications. *Journal of Infrared, Millimeter, and Terahertz Waves*, 2010, vol. 31, p. 958–96.

[6] YI-CHENG LIN, KUAN-JUNG HUNG. Compact ultra-wide band rectangular aperture antenna and band-notched designs. *IEEE*

Transactions on Antennas and Propagation, 2006, vol. 54, no. 11, p. 3075-3081.

[7] LIANG, J., GUO, L., CHIAU, C. C., CHEN, X., PARINI, C.G. Study of CPW-fed circular disc monopole antenna. *IEE Proceedings on Microwaves, Antennas and Propagation*, 2005, vol. 152, no. 6, p. 520–526. doi:10.1049/ip-map:20045179.

[8] HABIB, M. A., BOSTANI DJAIZ, A., NEDIL, M., YAGOUB, M. C. E., DENIDNI, T. A. Ultra wideband CPW-fed aperture antenna with WLAN band rejection. *Progress In Electromagnetics Research*, 2010, vol. 106, p. 17–31.

[9] PENGCHENG LI, JIANXIN LIANG, XIAODONG CHEN. Study of printed elliptical/circular slot antennas for ultra-wide band applications. *IEEE Transactions on Antennas and Propagation*, 2006, vol. 54, no. 6, p. 1670–1675.

[10] RAY, K. P., TIWARI, S. Ultra wide band printed hexagonal monopole antennas. *IET Microwaves, Antennas and Propagation*, 2010, vol. 4, no. 4, p. 437–445. DOI: 10.1049/iet-map.2008.0201.

[11] GHATAK, R., BISWAS, B., KARMAKAR, A., PODDAR, D. R. A circular fractal UWB antenna based on Descartes circle theorem with band rejection capability. *Progress In Electromagnetics Research C*, 2013, vol. 37, p. 235–248.

[12] SUN, A., YIN, Y. Z., JING, S. H., YANG, Y., LIU, B. W., LI, Z. Broadband CPW-fed antenna with band-rejected characteristics for WLAN/WiMAX operation. *Progress In Electromagnetics Research C*, 2011, vol. 22, p. 47–54.

[13] ZHOU, D., GAO, S., ZHU, F., ABD – ALHAMEED, R. A., XU, J. D. A simple and compact planar ultra wide-band antenna with single or dual band notched characteristics. *Progress In Electromagnetic Research C*, 2012, vol. 123, p. 47–65.

[14] FEI YU, CHUNHUA WANG. A CPW-fed novel planar ultra-wideband antenna with a band-notch characteristic. *Radioengineering*, 2009, vol. 18, no. 4, p. 551–555.

[15] MANDAL, T., DAS, S. Design and analysis of a coplanar waveguide fed ultra wideband hexagonal open slot antenna with WLAN and WiMAX band rejection. *Microwave and Optical Technology Letter*, 2014, vol. 56, no. 2, p. 434–443.

[16] CHATTOPADHYAY, K., DAS, S., DAS, S., BHADRA CHAUDHURI, S. R. Ultra-wideband performance of printed hexagonal wide slot antenna with dual band-notched characteristics. *Progress In Electromagnetics Research C*, 2013, vol. 44, p. 83–93.

[17] MA, X. L., SHAO, W., HE, G. Q. A novel dual narrow band-notch band CPW fed UWB slot antenna with parasitic strips. *Applied Computational Electromagnetic Society*, 2012, vol. 27, no. 7, p. 581–588.

[18] KIM, K. H., PARK, S. O. Analysis of the small band rejected antenna with the parasitic strip for UWB. *IEEE Transactions on Antennas and Propagation*, 2006, vol. 54, no. 6, p. 1688–1692.

[19] KIM, D. O., JO, N. I., JANG, H. A., KIM, C. Y. Design of the ultra wide band antenna with a quadruple-band rejection characteristics using a combination of the complementary split ring resonators. *Progress In Electromagnetic Research*, 2011, vol. 112, p. 93–107.

[20] ISLAM, M.T., AZIM, R., MOBASHSHER, A. T. Triple band notched planar UWB antenna using parasitic strips. *Progress In Electromagnetic Research*, 2012, vol. 129, p. 161–179.

[21] YINGSONG LI, WENXING LI, TAO JIANG. Implementation and investigation of a compact circular wide slot UWB antenna with dual notched band characteristics using stepped impedance resonators. *Radioengineering*, 2012 vol. 21, no.1, p. 517–527.

[22] Zeland IE3D™ software.

About Authors...

Tapan MANDAL (1977) received the B. Tech. degree in Electronics and Communication Engineering from Kalyani Govt. Engineering College of Kalyani University in 2001. He has done M.E. Degree from Bengal Engineering College (D.U.), Shibpur India, in 2003. From 2004 to 2007, he worked as a Lecturer at the University Institute of Technology, Burdwan University. Since 2007, he is associated with the Department of Information Technology of Govt. College of Engineering and Textile Technology, Serampore, India and presently holds the post of an Asst. Professor. His current research interests include the planar printed antenna and electromagnetic band gap structures.

Santanu DAS (1968) received the B. E. degree in Electronics and Telecommunication Engineering from Bengal Engineering College of Calcutta University (India) in 1989 and M.E. degree in Microwave Engineering from Jadavpur University, Calcutta, in 1992. He obtained the Ph.D. (Engineering) degree from Jadavpur University in 1998. He joined the Department of Bengal Engineering and Science University, India in the year 1998, as a Lecturer in the Electronics and Telecommunication Engineering and presently holds the post of Professor. His current research interests include microstrip circuits, FSS, antenna elements and arrays. He is a life member of the Institution of Engineers, India.

16

A CPW-fed Triple-band Antenna for WLAN and WiMAX Applications

Xiaolin YANG, Fangling KONG, Xuelin LIU, Chunyan SONG

Dept. of Physics and Electronics, University of Electronic Science and Technology of China,
No.4, Section 2, North Jianshe Road, 610054 ChengDu, SiChuan

yxlin@uestc.edu.cn, 770311315@qq.com, 595476221@qq.com, 331020368@qq.com

Abstract. *In this letter, a compact printed antenna fed by coplanar waveguide for triple-band is presented. The proposed antenna consists of two rectangular metallic loops in front and a slit square ring on the backside. Tri-band has been achieved, which can be easily tuned by adjusting the sizes of the rectangles. An analysis of equal lumped circuit mechanism as well the triple band operation is provided. Key parameters to tune the resonant frequencies have been identified. The overall dimension of the proposed antenna is 30×26 mm^2. Simulated results show that the presented antenna can cover three separated impedance bandwidths of ~13% at 300 MHz (2.2–2.5 GHz), ~14% at 500 MHz (3.3–3.8 GHz), and ~15% at 800 MHz (5.1–5.9 GHz), which are well applied for both 2.4/5.2/5.8-GHz WLAN bands and 3.5/5.5-GHz WiMAX bands.*

Keywords

Triple-band, coplanar waveguide, equivalent circuit.

1. Introduction

With the development of wireless communication technology, demand for designing an antenna with multi-band operation performance has increased rapidly since such an antenna is vital for integrating more than one communication standards in a single compact system which can effectively promote the portability of a modern personal communication system. Furthermore, mitigate the size of antenna is particularly important for the antenna designer. Owing to its properties, such as low cost, light weight, simple fabrication and compact size, monopole planar antennas are considered to be the best choice for multiband applications.

Many kinds of multi-band antennas have been reported in papers. A tri-band antenna with complicated ground and multiple metallic strips is reported in [1]. Some multi-band antennas are realized by cutting slots such as H-shaped [2], U-shaped [3]–[5], and many other shapes [6]–[8] in the radiating patch or ground, which can introduce different electrical paths. Strips with different length are also used in some letters, and different resonant frequency can be obtained by adjusting the length of them

[9]–[11]. However, these methods have drawbacks, namely increasing the physical size of antenna and creating undesired radiation from parasitic elements.

Recently, several multiple bands antennas have been developed. A quadruple-band elliptical antenna for WiMAX applications is reported in [13]. It was found good performance at four bands. However, it may be difficult to adjust and analysis each band dependently for they are introduced by one resonant element (i.e. the sector slot). Moreover, a triple-band antenna is studied in [14]. The reflection coefficient at three bands reaches -25 dB. But for the frequency outside the multi-bands the band-notched performance is bad (S11 < -5 dB) and the size is a little big for its stereo structure.

In this paper, a monopole antenna with coplanar waveguide (CPW) feed line is presented, which exhibits multi-frequency functionality by using two square electrical paths of different lengths and a parasitic split rectangular loop in backside. These three specific shaped structures and their coupling property introduce two distinct resonant modes, and the slit square loop on backside not only plays a role of band-notched function, but also improves the impedance matching at 5.5 GHz. The measured results show good agreement with the simulated ones, which demonstrates that the antenna shows a good multiband characteristic to satisfy the requirement of WLAN in the 2.4/5.2/5.8-GHz bands and WiMAX in the 3.5/5.5-GHz bands. Details of the designed antenna are described in the paper, and both simulated and measured results are presented. Good multi-bands and notched-bands characteristics are obtained when compared to the design in [14]. Furthermore, the size is reduced significantly.

2. Antenna Design and Implementation

Fig. 1(a) shows the geometry of the presented triple-band antenna. The photograph of the fabricated antenna is displayed in Fig. 1(b). A 50 Ω CPW transmission line (with width of 3.8 mm and a gap of 0.2 mm between the feed line and the ground plane) and two square loops are printed on top of the substrate whose dielectric constant is 2.65 and the thickness is 0.8 mm, and a slit rectangular ring is printed on the back side of the substrate. The overall size of the proposed antenna reaches only 30×26 mm^2.

Their reflection coefficient is shown in Fig. 2(b) respectively. Because the length of the first square loop is $L_1 + W_3/2$ which is about one-fourth wavelength at 3 GHz but half wavelength at 5.8 GHz. Similarly to the half wavelength dipole antenna, dual-band is introduced. Compared to the Step1, a notched band at 2.7 GHz and one-band at 3.55 GHz are introduced after adding the rectangular ring to the Step1. The length of the slot between the two rectangular rings is $L_1 - 2W_1 + W_3/2$ which is about quarter wavelength at 2.7 GHz. It can be equaled to an open-ended quarter wavelength transmission line which operates as a series resonance circuit. To the frequency at 3.55 GHz, the length

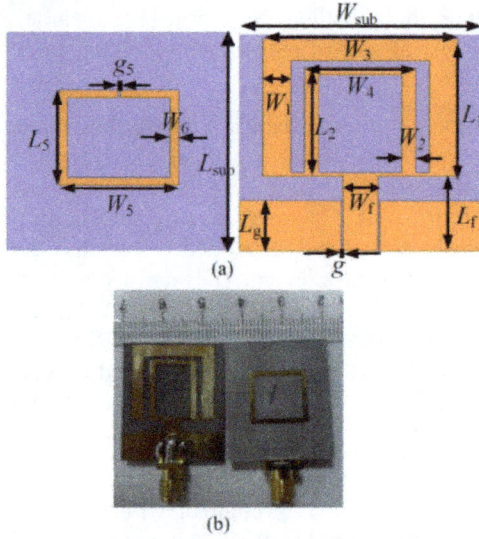

of the inner rectangular loop is larger than quarter wavelength but a little less than half wavelength. So it can be treated as a parallel circuit with finite resistance. The reflection coefficient becomes larger than -10 dB at 5.6 GHz, which is about -40 dB as for Step1. Due to the similar reason, the length of the slot equals to half wavelength at 5.6 GHz, a parallel resonance circuit with finite resistance is introduced.

To further widen the bandwidth at 3.6 GHz, a series resonance circuit is recommended. The corresponding conceptual equivalent circuit model is shown in Fig. 3. Z_{A1} represents the complex input impedance of Step1. R_{eqi}, L_{eqi} and C_{eqi} (i = 1, 2, 3, 4) are the resistor, inductor and capacitor values of the series and parallel resonant circuits respectively.

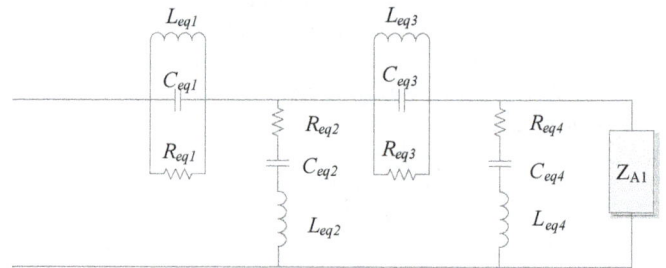

Fig. 1. (a) Geometry of the proposed antenna.
(b) Photograph of tri-band antenna.

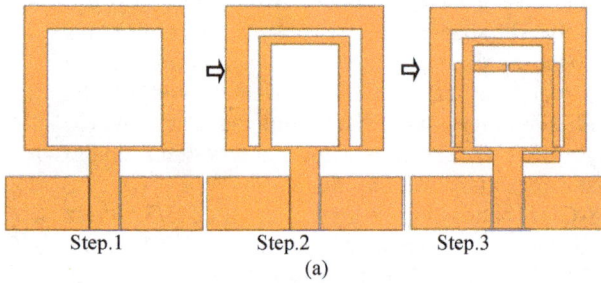

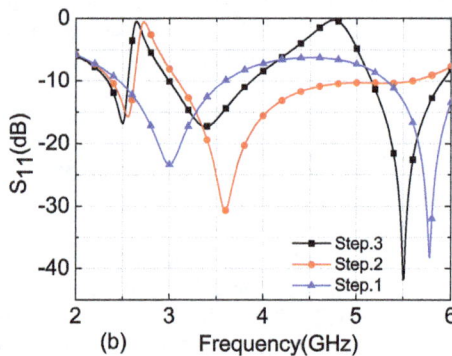

Fig. 2. (a) Process of designing ultimate antenna.
(b) Simulated reflection coefficient of Step 1 to 3.

Fig. 3. Equivalent circuit model of Step1 to Step 2.
R_{eq1} = 43.3 Ohm, C_{eq1} = 5.5 pF, L_{eq1} = 0.38 nH,
R_{eq2} = 0.0001 Ohm, C_{eq2} = 0.108 pF, L_{eq2} = 29.2 nH,
R_{eq3} = 48 Ohm, C_{eq3} = 3.71 pF, L_{eq3} = 0.21 nH,
R_{eq4} = 104.96 Ohm, C_{eq4} = 0.38 pF, L_{eq4} = 4.96 nH.

All the values of the circuit are calculated by the following formulas [12]:

The value of the series resonant circuit is given by

$$BW = \frac{1}{Q} = \frac{R}{\omega_0 L} = 2\frac{\Delta\omega}{\omega_0}, \qquad (1)$$

$$\omega_0 = \frac{1}{\sqrt{LC}}. \qquad (2)$$

Here BW equals to the -3dB bandwidth.

As for the parallel resonant circuit, the resonant frequency and bandwidth can be predicted by using the formulas given in:

$$BW = \frac{1}{Q} = \frac{1}{\omega_0 RC} = 2\frac{\Delta\omega}{\omega_0}, \qquad (3)$$

$$\omega_0 = \frac{1}{\sqrt{LC}}. \qquad (4)$$

After the initial values of R_{eqi}, L_{eqi} and C_{eqi} (i = 1, 2, 3, 4) are calculated by using the equations (1)-(4), the equivalent circuit is built, tuned and optimized in ADS 2009 software package. The calculated and simulated input impedance and reflection coefficient of the equivalent circuit is compared in Fig. 4(a)–(c), in which a reasonable agreement can be observed.

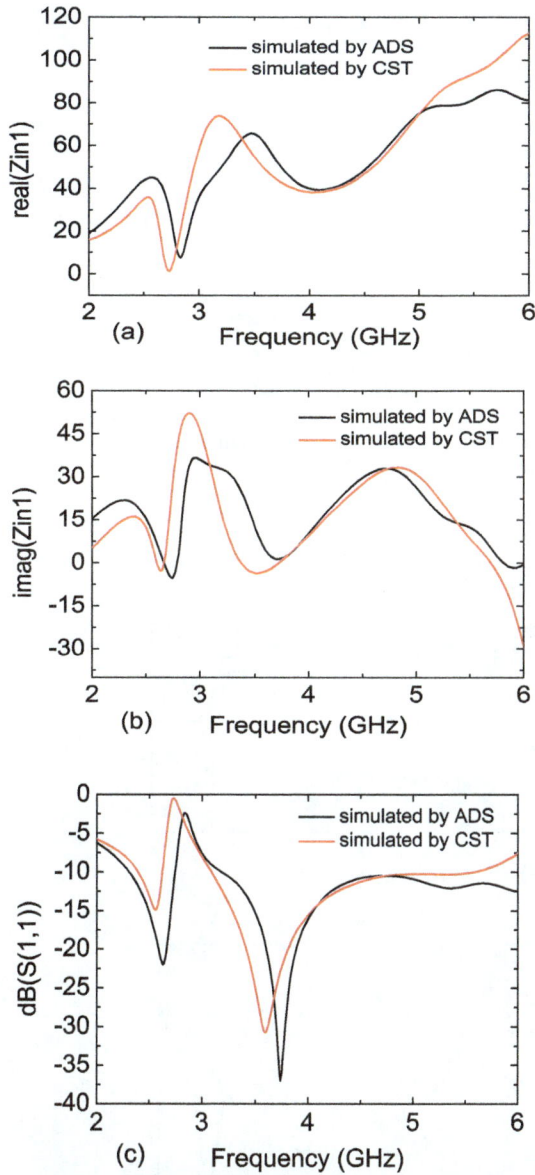

(a)

(b)

(c)

Fig. 4. Impedance and reflection coefficient of Step 2: (a) real part, (b) imaginary part, (c) reflection coefficient.

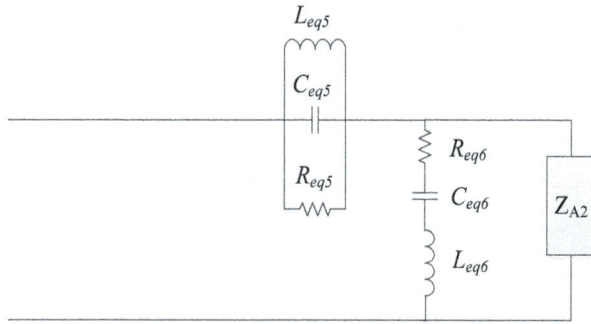

Fig. 5. Equivalent circuit model of Step 2 to Step 3. R_{eq1} = 1376.9 Ohm, C_{eq1} = 5.9 pF, L_{eq1} = 0.19 nH, R_{eq2} = 73.26 Ohm, C_{eq2} = 0.216 pF, L_{eq2} = 5.32 nH.

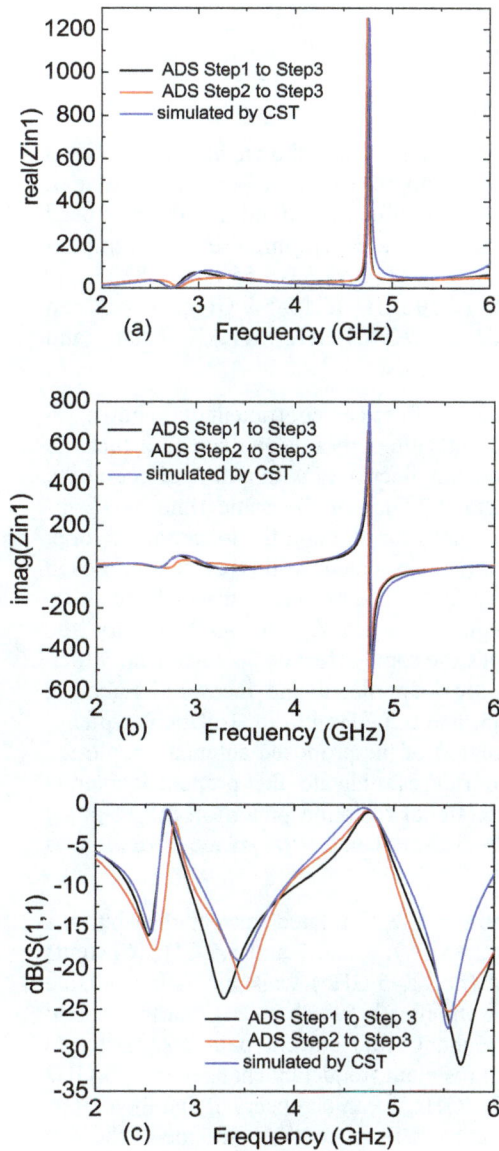

(a)

(b)

(c)

Fig. 6. Impedance and reflection coefficient of Step 3: (a) real part, (b) imaginary part, (c) reflection coefficient.

By adding an open-ended rectangle ring on the back of Step 2, a stop band at 4.7 GHz and another band at 5.5 GHz are formed. As analyzed above, the length of the rectangle on the back is $L_5 + W_5/2$, which equals to half wavelength at 4.7 GHz. Treating it as an open-ended half wavelength transmission line, a parallel resonance circuit can be achieved. A series resonance circuit is added to match the impedance at 5.5 GHz. The equivalent circuit model is shown in Fig. 5. Z_{A2} represents the complex input impedance of Step 2. R_{eqi}, L_{eqi} and C_{eqi} are the resistor, inductor and capacitor values of the series and parallel resonant circuits respectively. The calculated and simulated input impedance and reflection coefficient of the equivalent circuit is depicted in Fig. 6. Good agreement is obtained.

Putting all the circuits together, as shows in Fig. 7, the results simulated by ADS and CST coincide with each other as described in Fig. 6. All the values of the circuit remain the same except L_{eq1} = 0.38 nH, C_{eq3} = 3.71 pF, L_{eq4} = 4.96 nH owing to the effect between rectangles on the back and front.

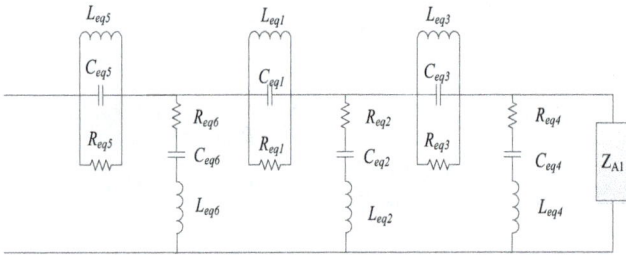

Fig. 7. Equivalent circuit model of Step 1 to Step 3.

Parameters	Value (mm)	Parameters	Value (mm)
L_{sub}	30	W_{sub}	26
L_1	19	W_1	3
L_2	14.5	W_2	1.5
L_g	7	W_3	21
L_5	13	W_4	12
L_f	10.4	W_5	14
g_5	0.3	W_6	1
W_f	3.8		

Tab. 1. Optimal dimensions of the designed antenna.

The optimal parameters are shown in Tab. 1. Fig. 8 shows the simulated and measured reflection coefficient. According to the results, it can be found that the proposed compact antenna effectively covers three separated impedance bandwidths of 220 MHz (2.33–2.55 GHz), 880 MHz (3.0–3.88 GHz) and 750 MHz (5.15–5.9 GHz), which can well satisfy both 2.4/5.2/5.5-GHz WLAN bands and 3.5/5.5-GHzWiMAX bands.

The simulated reflection coefficient is shown in Fig. 9(a) for different values of L_2. One can find that the band-notched frequency decreases with a little effect on the higher band around 3.7 GHz at the same time as L_2 increases, which elongates the slot and the inner square loop. Fig. 9(b) presents the simulated results of the proposed antenna with different L_5. The second notched-band shifts toward lower frequency when L_5 increased. As for the parameter W_5, it has the same effect as parameter L_5, which is shown in Fig. 9(c). Simulated and measured radiation patterns at 2.5, 3.5, and 5.5 GHz in the H-plane (xz-plane) and E-plane (yz-plane) of the proposed antenna are plotted in Fig. 10. These figures indicate the proposed antenna provides omni-directional radiation pattern in the xz-plane at the desired frequency bands. yz -plane radiation is bidirectional.

Fig. 11 illustrates the simulated current distributions on the antenna at 2.45, 2.7, 3.7, 4.7 and 5.5 GHz. Currents for the first band (i.e. 2.45 GHz) walk along the outside loop and the inner one mostly which means that the second rectangular loop enhance the current path comparing to Step1 and then the resonant frequency changes from 3 GHz to 2.45 GHz. At 2.7 GHz, it can be observed that the majority of the electric current flows on the both side of the slot equally and oppositely, which results in a notched band at this frequency. The electric current distributes on the inner square loop mostly at 3.7 GHz.

While at 4.7 GHz, the direction of the current on the back is contrary to that on the front patch. Then the notched band at 4.7 GHz is introduced. For 5.5 GHz it concentrates

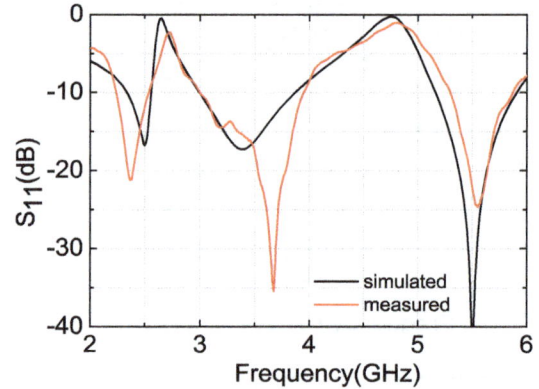

Fig. 8. Simulated and measured reflection coefficient of the antenna.

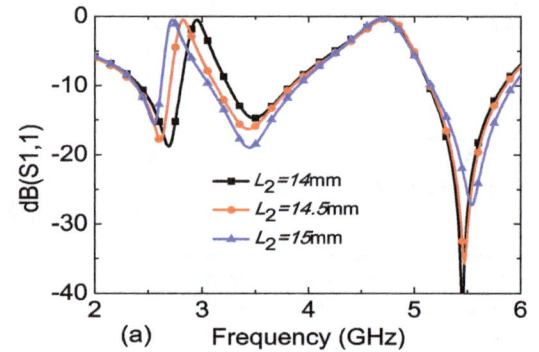

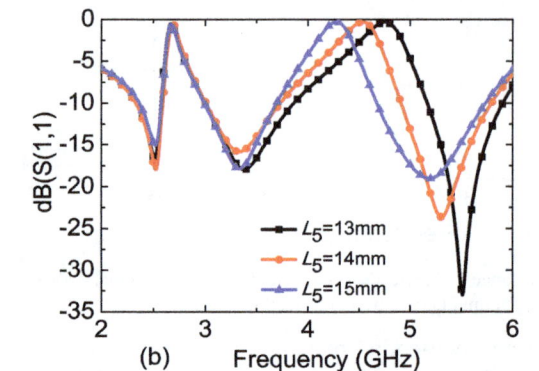

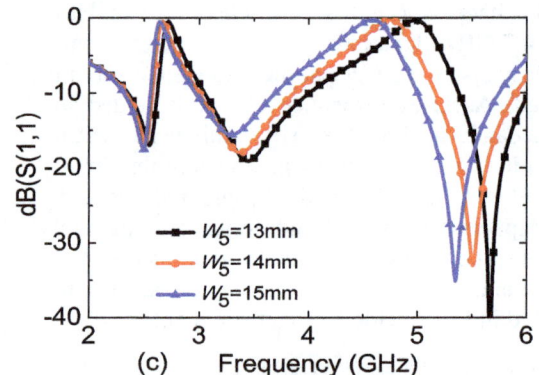

Fig. 9. Simulated reflection coefficient of the antenna with parameters: (a) L_2, (b) L_5, (c) W_5.

(a) E-plane (b) H-plane

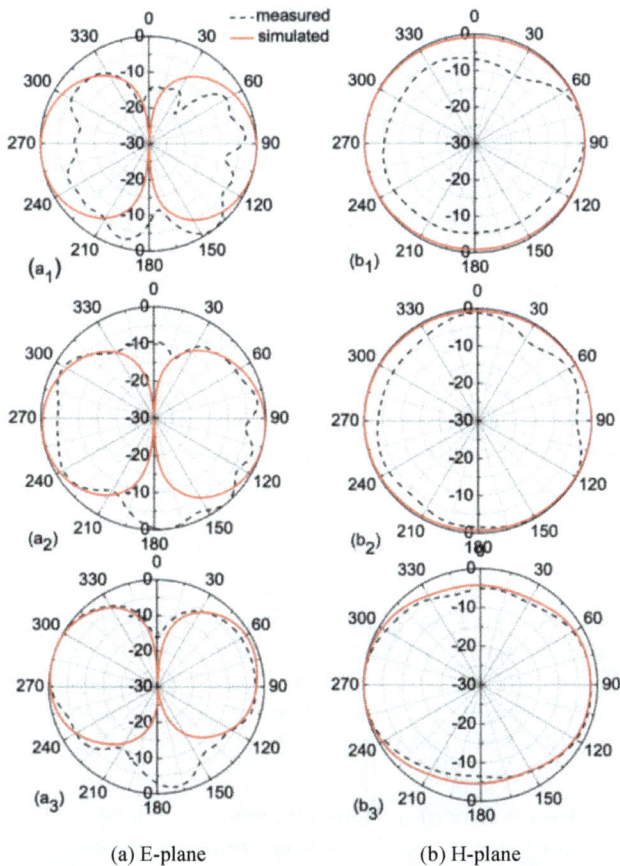

Fig. 10. (a) Simulated and measured radiation patterns of the antenna for E-plane at (a_1) 2.5 GHz, (a_2) 3.5 GHz, (a_3) 5.5 GHz. (b) Simulated and measured radiation patterns of the antenna for H-plane at (b_1) 2.5 GHz, (b_2) 3.5 GHz, (b_3) 5.5 GHz.

around the square loop on the backside of the substrate mainly for coupling. Fig. 12 depicts the overall efficiency and the measured antenna gain versus frequency. The peak gain varies in a range of 1.08–1.39 dB of the lower band and 2.35–3.49 dB of the middle band and 2.46–3.28 dB of the upper band. The gain increases as frequency goes up, which is caused by the distortion of radio pattern. As a consequence, the omni-directional antenna develops into a quasi-directional one plotted in Fig. 10. The total efficiency of the three bands is 62% at 2.45 GHz, 70% at 3.5 GHz and 79% at 5.5 GHz respectively. The loss in connectors, the twisted cables and measurement tolerance may become bigger than the one in the lower band. So the efficiency varies slightly although antenna gain increases.

3 Conclusion

A triple-band monopole planar antenna with two square loops and a slit square loop for WLAN and WiMAX systems has been proposed and studied in this paper. Parameters sweep and the corresponding equivalent circuits' analysis are completed. There is a good agreement between simulated and measured results which indicate the multi-band planar antenna is suitable for applications as an internal antenna for portable devices which work during the frequency band of WLAN and WiMAX.

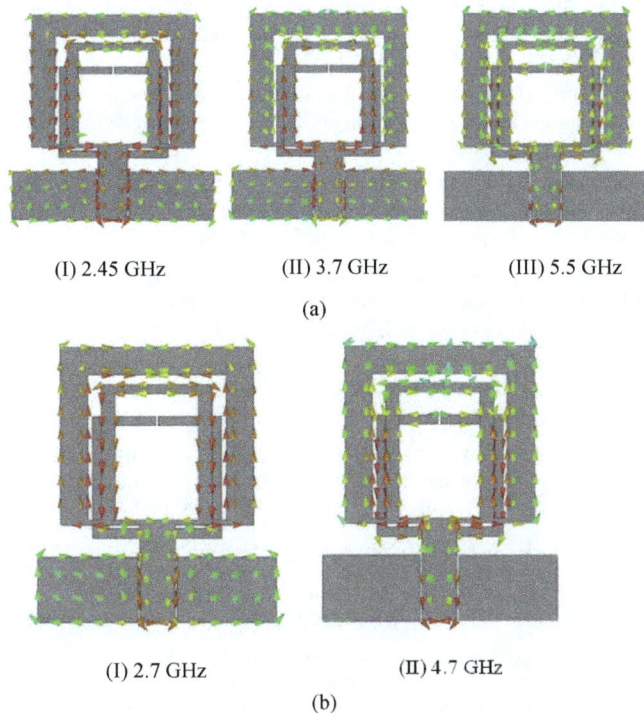

(I) 2.45 GHz (II) 3.7 GHz (III) 5.5 GHz

(a)

(I) 2.7 GHz (II) 4.7 GHz

(b)

Fig. 11. (a) Surface current distribution for three pass bands: (I) 2.45 GHz, (II) 3.7 GHz, (III) 5.5 GHz. (b) Surface current distribution for two notched-bands: (I) 2.7 GHz and (II) 4.7 GHz.

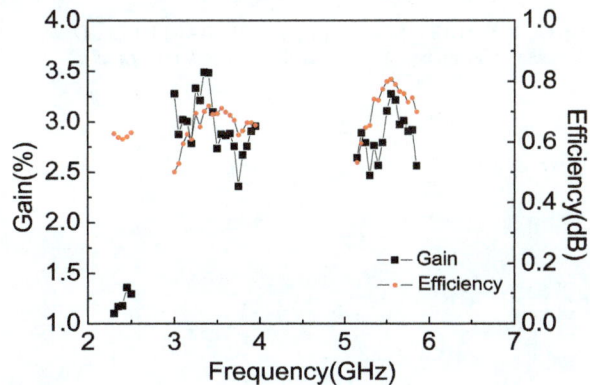

Fig. 12. Measured gain and efficiency of the antenna.

References

[1] CHIANG, M.J., WANG, S., HSU, C.C. Compact multi-frequency slot antenna design incorporating embedded arc-strip. *IEEE Antennas Wireless Propag. Lett.*, 2012, vol. 11, p. 834–837.

[2] CHANG, T.H., KIANG, J.F. Compact multi-band h-shaped slot antenna. *IEEE Trans. Antennas Propag.*, 2013, vol. 61, no. 8, p. 4345–4399.

[3] LIU, H.W., KU, C.H., YANG, C.F. Novel CPW-fed planar monopole antenna for WiMAX/WLAN applications. *IEEE Antennas Wireless Propag. Lett.*, 2010, vol. 9, p. 240–243.

[4] MOK, W.C., WONG, S.H., LUK, K.M., LEE, K.F. Single-layer single-patch dual-band and triple-band patch antennas. *IEEE Trans. Antennas Propag.*, 2013, vol. 61, no. 8, p. 4341–4344.

[5] BAE, H.R., SO, S.O., CHO, C.S., LEE, J.W., KIM, J. A crooked U-slot dual-band antenna with radial stub feeding. *IEEE Antennas Wireless Propag. Lett.*, 2009, vol. 8, p. 1345–1348.

[6] LIU, W.C., WU, C.M., TSENE, Y.J. Parasitically loaded CPW-fed monopole antenna for broadband operation. *IEEE Trans. Antennas Propag.*, 2011, vol. 59, no. 6, p. 2415–2419.

[7] ANTONIADES, M. A., ELEFTHERIADES, G. V. A compact multiband monopole antenna with a defected ground plane. *IEEE Antennas Wireless Propag. Lett.*, 2008, vol. 7, p. 652–655.

[8] ZHAI, H.Q., MA, Z.H., HAN, Y., LIANG, C.H. A compact printed antenna for triple-band WLAN/WiMAX applications. *IEEE Antennas Wireless Propag. Lett.*, 2013, vol. 12, p. 65–68.

[9] WONG, K.L., LEE, G.Y., CHIOU, T.W. A low-profile planar monopole antenna for multiband operation of mobile handsets. *IEEE Trans. Antennas Propag.*, 2003, vol. 51, no. 1, p. 121–125.

[10] BAEK, J.G., HWANG, K.C. Triple-band unidirectional circularly polarized hexagonal slot antenna with multiple L-shaped slits. *IEEE Trans. Antennas Propag.*, 2013, vol. 61, no. 9, p. 4831–4835.

[11] DENG, C.P., LIU, X.Y., ZHANG, Z. K., TENTZERIS, M. M. A miniascape-like triple-band monopole antenna for WLAN applications. *IEEE Antennas Wireless Propag. Lett.*, 2012, vol. 11, p. 1330–1333. DOI: 10.1109/LAWP.2012.2227292.

[12] FUGUO ZHU, GAO, S., HO, A.T.S., ABD-ALHAMEED, R.A., SEE, C.H., BROWN, T.W.C., JIANZHOU LI, GAO WEI, JIA-DONG XU. Multiple band-notched UWB antenna with band-rejected elements integrated in the feed line. *IEEE Trans. Antennas Propag.*, 2013, vol. 61, no. 8, p. 3952–3960. DOI: 10.1109/TAP.2013.2260119.

[13] SHARMA, V., SAXENA, V. K., SHARMA, K. B., BHATNA-GAR, D. Multi-band elliptical patch antennas with narrow sector slot for WiMAX applications. *International Journal of Microwave and Optical Technology*, 2012, vol. 7, no. 2. p. 89–96.

[14] LOSITO, O., BOZZETTI, M., DIMICCOLI, V., BARLETTA, D. Multiple sector ring monopole antenna. In *Proceedings of the 6th European Conference on Antenna and Propagation*. Prague (Czech Rep.), 2012, p. 804–807.

About Authors ...

Xiaolin YANG was born in Gansu, China, in 1974. He received the Ph.D. at Applied Physics from the Lanzhou University in 2005. He is currently an Associate Professor in School of Physical Electronics, University of Electronic Science and Technology of China (UESTC). His current research interests are microwave circuits and systems and ultra-wideband antenna.

Fangling KONG was born in Jiangxi Province, China. She received the Bachelor degree in Physics from GanNan Normal University, in 2012, and is currently working towards the M.S degree in Electronic and Communication Engineering at UESTC. Her research interests are in UWB antenna.

Xuelin LIU was born in Hunan Province, China. She received the Bachelor degree in Physics from HuNan Normal University, in 2012, and is currently working towards the M.S degree in Electronic and Communication Engineering at UESTC. Her research interests are in UWB antenna.

Chunyan SONG was born in Heibei Province, China. She received the Bachelor degree in Jiaotong from Southwest Jiaotong University, in 2009, and is currently working towards the M.S degree in Electronic and Communication Engineering at UESTC. Her research interests are in UWB antenna.

17

First-Order Statistics Prediction for a Propagation Channel of Arbitrary Non-Geostationary Satellite Orbits

Milan KVICERA, Pavel PECHAC

Faculty of Electrical Engineering, Czech Technical University in Prague, Technicka 2, 166 27 Prague, Czech Republic

kvicemil@fel.cvut.cz, pechac@fel.cvut.cz

Abstract. *A method enabling the prediction of the first-order statistics of received signal levels for arbitrary non-geostationary satellite orbits based on a reference dataset for a wide range of elevation angles is introduced for azimuth-independent scenarios and high elevation angles. The method is further validated by experimental data obtained during measurements with a remote-controlled airship utilized as a pseudo-satellite. These experimental trials were performed at a frequency of 2.0 GHz at two scenarios at Stromovka Park in Prague, Czech Republic, in August 2013 and March 2014. An excellent match between the predicted and actual cumulative distribution functions of received signal levels was identified for both scenarios.*

Keywords

Satellite-to-Earth propagation, channel measurements, modeling, vegetation

1. Introduction

Central to the design of a satellite system are the constellation and selected orbit which determine the elevation and azimuth angles at which a user shall receive a direct signal. The elevation and azimuth angle toward a particular satellite can remain the same for a fixed receiver, as in case of a geostationary Earth orbit, or change dramatically when considering for example a highly elliptical orbit. This has a strong influence on the corresponding satellite-to-Earth propagation channel as, based on user surroundings, the signal may be shadowed differently determined by the direction of the incoming signal. To predict such behavior, a number of various propagation channel models for different scenarios can be found in the literature, see for example [1–8]. Generally, such models need to be based on available experimental data; however, obtaining suitable experimental data is demanding. It is common to utilize a so-called pseudo-satellite which may be in a form of a transmitter (Tx) placed on a helicopter [9–11] or a remote-controlled airship [7], [12], [13], at a crane or the upper-most point in the surroundings for fixed-elevation measurements [14–16], or even collect data from an existing satellite [17–19].

However, it is not feasible to obtain experimental data for all the combinations of azimuth and elevation angles observed by a user on Earth when considering a particular non-geostationary satellite system. Instead, respecting high costs of experimental campaigns when using a pseudo-satellite, pre-defined flight paths, such as a star-pattern [7], circle [11], or hemisphere [10], are chosen for selected scenarios.

Considering azimuth-independent scenarios identified in the text as regular, such as a densely vegetated area, only one set of reference experimental data in a vast range of elevation angles at an arbitrary azimuth should be sufficient to predict received signal characteristics for arbitrary non-geostationary satellite orbits leading to less demanding experimental campaigns. Such a novel approach would follow [20], where a probability density function (PDF) of elevation angles between a user and a low Earth orbit satellite is utilized to obtain resulting rain attenuation time series. However, such an experiment has not been documented in the literature and needs to be validated. Thus, a series of measurements at 2.0 GHz at Stromovka Park in Prague, Czech Republic, were carried out in 2013 and 2014. Throughout these trials, a remote-controlled airship was utilized as a pseudo-satellite following pre-defined flight paths according to sub-satellite points of selected Galileo and Iridium satellites. Both the left- (LHCP) and right-handed (RHCP) circularly polarized signals were transmitted from the airship towards a receiver located at two different scenarios. Unlike [21], where four different distributions were fitted to the first-order statistics of experimental data obtained previously at Stromovka Park with a low sampling rate of 100 Hz, this paper presents a method how to obtain a cumulative distribution function (CDF) of received signal levels for various non-geostationary satellite orbits based on a reference dataset for the case of azimuth-independent scenarios. Details of the experimental campaign are provided in Sec. 2, the data processing method is described in Sec. 3, while the results and discussion are given in Sec. 4.

2. Measurement Setup and Trials

The measurement setup used during the trials was as follows. A remote-controlled airship carried a Tx, the same type as in [7], connected to an LHCP and an RHCP planar wideband antenna attached to a positioner enabling an instant pointing towards the receiver (Rx) location based on the airship GPS coordinates. Unmodulated continuous wave signals with a fixed output power of 27 dBm were transmitted at frequencies of 2.00106 GHz and 2.00086 GHz by the LHCP and RHCP antenna, respectively. Unlike [7] and [12], to obtain the received signal levels of both the co-polarized and cross-polarized components of the transmitted signals, a dual-polarized rectangular patch antenna was connected by an H-hybrid and two power splitters to a sensitive, custom-made, four-channel receiver with a low noise floor of -126 dBm for a measurement bandwidth of 12.5 kHz. The receiver provided a 10-kHz sampling rate and its first two channels were tuned to 2.00106 GHz and the remaining two were tuned to 2.00086 GHz. The height of the upwards-pointing receiving antenna was 1.5 meters and the altitude of the airship was kept approximately 200 meters above ground level at a near-constant speed of 8 m/s. Similar to [7], recorded signal levels were recalculated to a uniform distance of 20 km to eliminate the influence of free space loss for different distances between the Tx and Rx. Further, data obtained during periods of airship pitch and roll of more than 15 degrees were removed as they represent gusty conditions during which the Tx antenna positioner did not perfectly keep the direction towards Rx.

The airship followed pre-defined flight paths over the vegetated area of Stromovka Park according to the selected typical sub-satellite points of the Galileo PFM and Iridium 98 satellite for the location of Prague (50.08° N, 14.43° E), see Fig. 1. Two scenarios, marked as A and B in Fig. 1 and shown in more detail in Fig. 2, were selected to represent regular scenarios independent of azimuth: a receiver located inside coniferous trees and within a group of tall deciduous trees, respectively. It should be noted that all the flight paths in Fig. 1 refer to scenario A and were thus slightly shifted for the case of scenario B. By performing the measurements in July 2013 and March 2014, represen-

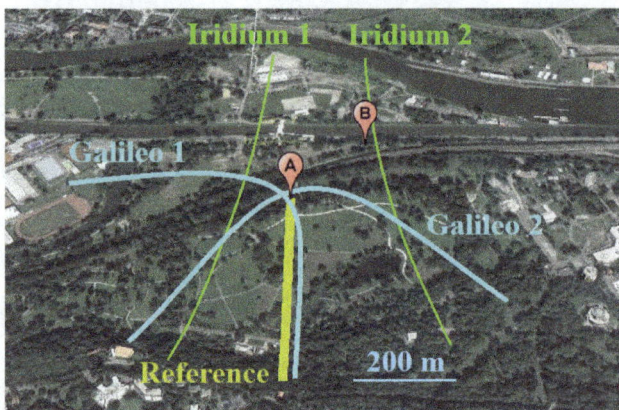

Fig. 1. Regular scenarios A and B at Stromovka Park together with the pre-defined airship flyovers simulating the Iridium 98 satellite (two almost north-south thin lines), Galileo PFM satellite (two wider curved lines) and the reference north-south flyover (widest line). (Image from Google Earth).

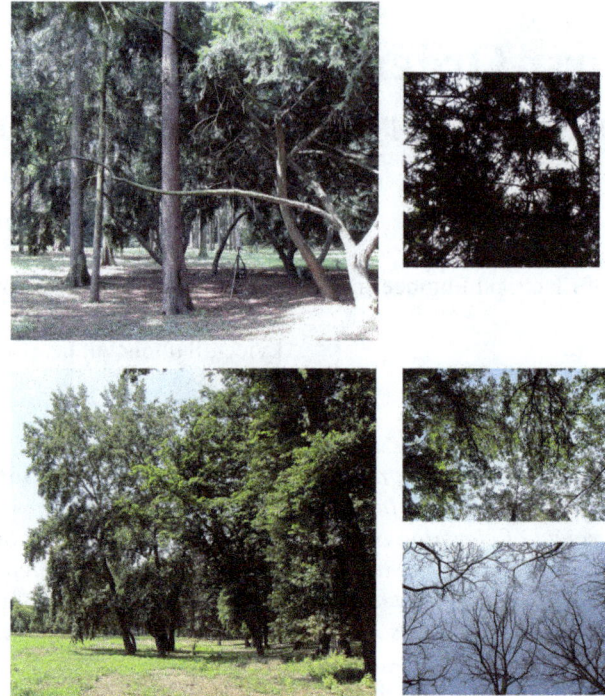

Fig. 2. Scenarios A (top left) and B (bottom left) in detail during the summer season together with the upwards views from the Rx during the summer and winter season (top right and bottom right). As scenario A is evergreen, only one upwards view is shown for both seasons.

tative experimental data were obtained for the actual satellite passes during both the summer and winter season. As scenario A is evergreen, it demonstrates the repeatability of the measurements while defoliation of scenario B represents different propagation conditions. Thus only these two scenarios can be utilized to obtain a statistically significant amount of experimental data and validate the method presented below in Sec. 3.

It should be noted that although the utilized frequencies are at the upper part of L-band, higher than the actual Galileo or Iridium frequencies, the method introduced in the next section is limited by the azimuthal symmetry of a particular scenario rather than by a selected frequency. In addition, even though the airship simulated actual non-geostationary satellite systems, arbitrary flyovers could have been selected.

3. Data Processing

The experimental data were processed in the following way. The measurements were aimed at high elevation angles and, thus, the lowest elevation considered was 30 degrees from the Rx point of view. On the other hand, to avoid an insufficient amount of data for high elevation angles, the maximum considered elevation was 80 degrees. For the analysis described below, only the co-polarized components of the received LHCP and RHCP signals were considered.

As a subsequent step, the entire range of elevation angles achieved during the north-south reference flyover (shown by the thick line in Fig. 2) was divided into five-degree intervals and a PDF $pdf_{reference,i}(r)$ of received signal levels r in dB was calculated by means of histogram for each elevation interval i. In contrast to [20], where segments of constant time are utilized, our approach requires constant elevation intervals as the simulated flyovers do not follow the actual time duration of the particular satellite pass. The choice of the interval size respected a statistically significant number of data samples (about 80000 and 40000 for the lowest and highest elevation interval, respectively) and, in addition, propagation conditions and antenna patterns are considered as not significantly varying within such an interval, similar to [7].

Consequently, for the Galileo or Iridium flyovers shown in Fig. 1, a PDF of the elevation intervals $pdf_\Theta(\theta)$ achieved during a particular flyover was calculated and thus the probability $p_{\Theta,i}(\theta)$ of each elevation interval i was obtained. Then, following (1), a PDF of signal levels received during the reference flyover within an elevation interval i, $pdf_{reference,i}(r)$, was multiplied by $p_{\Theta,i}(\theta)$ so that a PDF of received signal level $pdf_{predicted,i}(r)$ was predicted for this elevation interval. After that, a PDF of received signal levels $pdf_{predicted}(r)$ for the whole range of elevation angles from 30 degrees up to 80 degrees was obtained according to (2), simply as a sum of the PDFs for every elevation interval.

$$pdf_{predicted,i}(r) = p_{\Theta,i}(\theta) \cdot pdf_{reference,i}(r) \text{ for } i = 1...n, \quad (1)$$

$$pdf_{predicted}(r) = \sum_{i=1}^{n} pdf_{predicted,i}(r) \quad (2)$$

where n is the total number of elevation intervals. Obtaining a corresponding CDF is then straightforward by using (3)

$$\Pr\{r \le r_{th}\} = \int_{-\infty}^{r_{th}} pdf_{predicted}(r)dr. \quad (3)$$

4. Results and Discussion

To analyze and quantify the quality of the overall match of the predicted and actual received signal level CDF for a particular satellite flyover, we calculated the mean and standard deviation of their difference for the whole range of percentages with a step of 0.01. Corresponding results are then given in Tab. 1 with the exception of Iridium flyover number 2 for scenario B during the summer season as the experimental data are not available.

Based on this table it is evident that overall similar results were obtained for scenarios A and B during the summer and winter season. Absolute values of maximum and minimum mean of differences are below 5 dB and about 0 dB, respectively. Together, with maximum standard deviations below 2.5 dB, these values indicate an excellent match of the predicted and actual CDFs. Such results also reflect the fact that even scenario B, consisting of only five

Scenario		Mean ± standard deviation (dB)			
		Galileo 1	Galileo 2	Iridium 1	Iridium 2
A, sum.	LHCP	-1.4 ± 1.6	0.8 ± 1.5	1.1 ± 2.4	-2.5 ± 0.8
	RHCP	-1.8 ± 1.3	1.2 ± 0.8	-0.2 ± 2.1	-2.1 ± 0.7
A, winter	LHCP	3.6 ± 1.2	1.9 ± 1.4	3.0 ± 1.7	0.0 ± 0.8
	RCHP	2.4 ± 0.9	1.2 ± 1.4	0.9 ± 2.2	1.2 ± 1.1
B, sum.	LHCP	-2.9 ± 0.8	-0.6 ± 1.8	0.8 ± 2.1	Not avail.
	RHCP	-4.9 ± 0.9	-3.1 ± 2.0	-3.8 ± 1.3	Not avail.
B, winter	LHCP	-2.3 ± 1.0	-2.9 ± 2.7	-1.4 ± 1.5	-2.3 ± 2.2
	RHCP	-3.6 ± 1.0	-1.7 ± 2.1	-2.7 ± 1.3	0.8 ± 1.7

Tab. 1. Mean and standard deviation of difference between predicted and measured CDF.

deciduous high-rose trees, with non-uniformly distributed sparse trunks and branches without leaves in the winter season, can still be considered as regular, although it could be assumed to be more dependent on azimuth than scenario A.

It should be noted that a certain level of inaccuracy is introduced into the predicted results due to the Rx antenna pattern not being perfectly azimuth independent. However, respecting the Rx antenna design, the maximum difference of gain for both the LHCP and RHCP polarization can be identified for the lowest elevations as only about 2 dB for an elevation of 30 degrees.

To illustrate results from Tab. 1, Figs. 3 and 4 represent selected actual, predicted and reference signal level CDFs for the Galileo satellite flyover number 2, RHCP signal at scenario A and B during the summer season, respectively. For these cases, the mean value of the differences is 1.2 dB and -3.1 dB, whereas the standard deviation of the differences is 0.8 dB and 2.0 dB, respectively. Based on Figs. 3 and 4, such values lead to a very good visual match of the predicted and actual signal level CDF, even for the worse case.

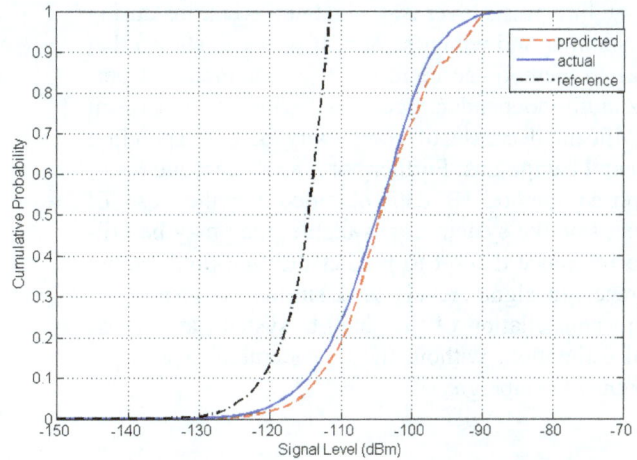

Fig. 3. An example of a match between the predicted and actual signal level CDF for the case of the RHCP signal during the Galileo satellite flyover number 2 at scenario A within the summer season. It corresponds to values of (1.2±0.8) dB taken from Tab. 1.

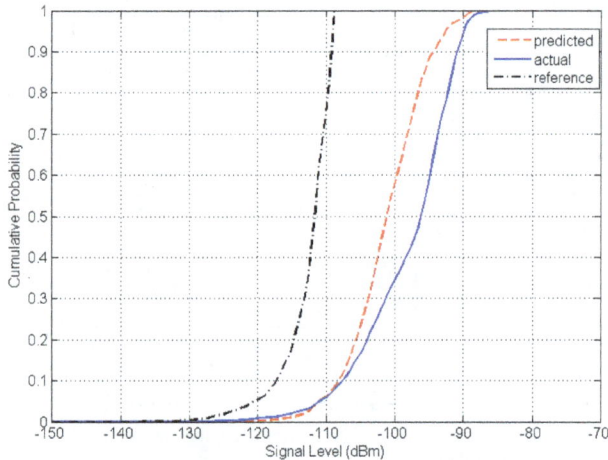

Fig. 4. An example of a match between the predicted and actual signal level CDF for the case of the RHCP signal during the Galileo satellite flyover number 2 at scenario B within the summer season. It corresponds to values of (-3.1±2.0) dB taken from Tab. 1.

To summarize, based on Tab. 1, it can be stated that an overall very good match of the prediction was achieved for both scenarios A and B during summer and winter. Based on the large experimental dataset which has been processed, the presented prediction method can be considered as successfully validated for the case of high elevation angles and regular scenarios.

5. Conclusion

We have presented a method enabling the prediction of the first-order statistics of signal levels received at high elevation angles considering arbitrary satellite orbits and azimuth-independent scenarios. This method is based on utilizing a reference dataset for a wide range of elevation angles at one azimuth only. The validity of this method was demonstrated at a frequency of 2.0 GHz on a significant number of experimental datasets obtained by using a remote-controlled airship simulating selected Galileo and Iridium satellite passes over two vegetated scenarios during both the summer and winter season. It was also shown that the proposed method can be utilized for common, not perfectly azimuth-independent scenarios, which would result in a significant decrease of complexity of corresponding experimental campaigns. Furthermore, as a consequence of the proposed method, if data obtained for the case of an existing satellite system are available, they may be utilized as a reference dataset to predict the first-order statistics of received signal levels at a similar frequency for a different constellation of the satellite system, at least as an initial estimation, without the considerable expense of a measurement campaign.

Acknowledgements

This publication was supported by the European social fund within the framework of realizing the project „Support of inter-sectoral mobility and quality enhancement of research teams at the Czech Technical University in Prague", CZ.1.07/2.3.00/30.0034.

References

[1] JOST, T., CARRIE, G., PEREZ-FONTAN, F., WANG, W. FIEBIG, U.-C. A Deterministic satellite-to-indoor entry loss model. *IEEE Transactions on Antennas and Propagation*, 2013, vol. 61, no. 4, p. 2223–2230. DOI: 10.1109/TAP.2012.2232898

[2] CHEFFENA, M., PEREZ-FONTAN, F. Land mobile satellite channel simulator along roadside trees. *IEEE Antennas and Wireless Propagation Letters*, 2010, vol. 9, p. 748–751. DOI: 10.1109/LAWP.2010.2060465

[3] AIT-IGHIL, M., LEMORTON, J., PEREZ-FONTAN, F., LACOSTE, F., THEVENON, P., BOURGA, C., BOUSQUET, M. SCHUN - a hybrid land mobile satellite channel simulator enhanced for multipath modelling applied to satellite navigation systems. In *Proceedings of the 7th European Conference on Antennas and Propagation*. Gothenburg (Sweden), 2013, p. 692–696.

[4] CARRIE, G., PEREZ-FONTAN, F., LACOSTE, F., LEMORTON, J. A generative MIMO channel model: encompassing single satellite and satellite diversity cases. In *Proceedings of the 6th European Conference on Antennas and Propagation*. Prague (Czech Rep.), 2012, p. 2454–2458. DOI: 10.1109/EuCAP.2012.6206627

[5] JEANNIN, N., PEREZ-FONTAN, F., MAMETSA, H.-J., CASTANET, L. Physical-statistical model for the LMS channel at Ku/Ka Band. In *Proceedings of the 5th European Conference on Antennas and Propagation*. Rome (Italy), 2011, p. 3571–3575.

[6] ABELE, A., PEREZ-FONTAN, F., BOUSQUET, M., VALTR, P., LEMORTON, J., LACOSTE, F., CORBEL, E. A new physical-statistical model of the land mobile satellite propagation channel. In *Proceedings of the 4th European Conference on Antennas and Propagation*. Barcelona (Spain), 2010, p. 1-5.

[7] KVICERA, M., PECHAC, P. Building penetration loss for satellite services at L-, S- and C-band: measurement and modeling. *IEEE Transactions on Antennas and Propagation*, 2011, vol. 59, no. 84, p. 3013–3021. DOI: 10.1109/TAP.2011.2158963

[8] ARNDT, D., HEYN, T., HEUBERGER, A., PRIETO-CERDEIRA, R., EBERLEIN, E. State modeling of the land mobile satellite channel with angle diversity. In *Proceedings of the 6th European Conference on Antennas and Propagation*. Prague (Czech Rep.) 2012, p. 3130–3144. DOI: 10.1109/EuCAP.2012.6206141

[9] MOLNAR, B., FRIGYES, I., BODNAR, Z., HERCZKU, Z., KORMANYOS, Z., BERCES, J., PAPP, I., JUHASZ, L. Characterisation of the satellite-to-indoor channel based on narrow-band scalar measurements. In *Proceedings of the 8th IEEE International Symposium on Personal, Indoor and Mobile Radio Communications*. Helsinki (Finland), 1997, p. 1014–1018.

[10] PEREZ-FONTAN, F., HOVINEN, V., SCHONHUBER, M., PRIETO-CERDEIRA, R., DELGADO-PENIN, J. A., TESCHL, F., KYROLAINEN, J., VALTR, P. Building entry loss and delay spread measurements on a simulated HAP-to-indoor link at S-band. *EURASIP Journal on Wireless Communications and Networking*, 2008, vol. 2008, no. 5, 6 p. DOI:10.1155/2008/427352

[11] TESCHL, F., PEREZ-FONTAN, F., SCHONHUBER, M., PRIETO-CERDEIRA, R., TESCHL, R. Attenuation of spruce, pine, and deciduous woodland at C-band. *IEEE Antennas and Wireless Propagation Letters*, 2012, vol. 11, p. 109–112. DOI: 10.1109/LAWP.2012.2184253

[12] HORAK, P., PECHAC, P. Excess loss for high elevation angle links shadowed by a single tree: measurements and modeling.

IEEE Transactions on Antennas and Propagation, 2012, vol. 60, no.7, p. 3541–3545. DOI: 10.1109/TAP.2012.2196944

[13] STEINGASS, A., LEHNER, A. Measuring the navigation multipath channel – a statistical analysis. In *Proceedings of the 17th International Technical Meeting of the Satellite Division of he Institute of Navigation (ION GNSS 2004)*. Long Beach (CA, USA), 2004, p. 1157–1164.

[14] LACOSTE, F., LEMORTON, J., CASADEBAIG, L., ROUSSEAU, F. Measurements of the land mobile and nomadic satellite channels at 2.2 GHz and 3.8 GHz. In *Proceedings of the 6th European Conference on Antennas and Propagation*. Prague (Czech Rep.), 2012, p. 2422–2426. DOI: 10.1109/EuCAP.2012.6206356

[15] JOST, T., WANG, W., FIEBIG, U.-C., PEREZ-FONTAN, F. Comparison of L- and C-Band satellite-to-indoor broadband wave propagation for navigation applications. *IEEE Transactions on Antennas and Propagation*, 2011, vol. 59, no. 10, p. 3899–3909. DOI: 10.1109/TAP.2011.2163753

[16] KING, P. R., STAVROU, S. Low elevation wideband land mobile satellite MIMO channel characteristics. *IEEE Transactions on Wireless Communications*, 2007, vol. 6, no. 70, p. 2712–2720. DOI: 10.1109/TWC.2007.051018

[17] TESCHL, F., HOVINEN, V., PEREZ-FONTAN, F., SCHONHUBER, M., PRIETO-CERDEIRA, R. Narrow- and wideband land mobile satellite channel statistics for various environments at Ku-band. In *Proceedings of the 6th European Conference on Antennas and Propagation*. Prague (Czech Rep.), 2012, p. 2464–2468. DOI: 10.1109/EuCAP.2012.6206670

[18] JOST, T., WANG, W., SCHUBERT, F., ANTREICH F., FIEBIG, U.-C. Channel sounding using GNSS signals. In *Proceedings of the 5th European Conference on Antennas and Propagation*. Rome (Italy), 2011, p. 3724–3728.

[19] VELTSISTAS, P., KALABOUKAS, G., KONITOPOULOS, G., DRES, D., KATIMERTZOGLOU, E., CONSTANTINOU, P. Satellite-to-indoor building penetration loss for office environment at 11 GHz. *IEEE Antennas and Wireless Propagation Letters*, 2007, vol. 6, p. 96–99. DOI: 10.1109/LAWP.2007.893070

[20] ARAPOGLOU, P.-D. M., PANAGOPOULOS, A. D. A tool for synthesizing rain attenuation time series in LEO Earth observation satellite downlinks at Ka band. In *Proceedings of the 5th European Conference on Antennas and Propagation*. Rome (Italy), 2011, p. 1467–1470.

[21] KOUROGIORGAS, C. I., KVICERA, M., SKRAPARLIS, D., KORINEK, T., SAKARELLOS, V. K., PANAGOPOULOS, A. D., PECHAC, P. Modeling of first-order statistics of the MIMO dual polarized channel at 2 GHz for land mobile satellite systems under tree shadowing. *IEEE Transactions on Antennas and Propagation*, 2014, vol. 62, no. 10, p. 5410–5415. DOI: 10.1109/TAP.2014.2346186

About the Authors ...

Milan KVICERA was born in 1983. He received the M.Sc. degree and the Ph.D. degree in Radio Electronics from the Czech Technical University in Prague, Czech Republic, in 2008 and 2012, respectively. He is currently working as a postdoc researcher in the Department of Electromagnetic Field, Czech Technical University in Prague. His research interests are focused on radiowave propagation and satellite communication.

Pavel PECHAC received the M.Sc. degree and the Ph.D. degree in Radio Electronics from the Czech Technical University in Prague, Czech Republic, in 1993 and 1999 respectively. He is currently a Professor in the Department of Electromagnetic Field, Czech Technical University in Prague. His research interests are in the field of radiowave propagation and wireless systems.

18

High-Isolation Dual-Polarized Microstrip Antenna via Substrate Integrated Waveguide Technology

Sijia LI [1], Xiangyu CAO [1], Jun GAO [1], Pengcheng GAO [2]

[1] Information and Navigation College, Air Force Engineering University, No.1 Fenghao Street, 710077 Xi'an, China
[2] National Key Lab. of Antenna and Microwave Technology, Xidian University, No.2 Taibai Road 710071, Xi'an, China

lsj051@126.com, xiangyucaokdy@163.com, gjgj9694@163.com, 634672961@qq.com

Abstract. *A dual-polarized microstrip antenna with high-isolation is proposed based on the substrate-integrated waveguide (SIW) technology in this paper. According to the SIW technology, the metalized holes (MHs) are inserted into a substrate of the proposed antenna and the electric fields of the feeding parts are enclosed, so the isolation of the antenna is enhanced. The bandwidth is improved due to the MHs in the four sides of the antenna. A prototype of the proposed antenna has been fabricated and measured. Experimental results indicate that the antenna obtains the isolation more than 40 dB and achieves the impedance bandwidths of 21.9 % and 23.8 % (11.8–14.6 GHz and 11.65–14.8 GHz for two ports) of the reflection coefficients less than -20 dB. The cross polarization with the main lobe remains less than -30 dB and the half-power beam width is about 70° for the proposed antenna. Meanwhile, the front-to-back ratio remains to be better than 20 dB. A good agreement between the measured and simulated results validates the proposed design.*

Keywords

Substrate-integrated waveguide, isolation, dual-polarized microstrip antenna, aperture coupled antenna.

1. Introduction

Nowadays, dual-polarized microstrip antennas have popular application in wireless mobile and satellite communication systems, synthetic aperture radars and radio frequency identification systems due to their salient features, such as lightweight, compact, and low profile configurations. The well-known aperture-coupled antenna first proposed and demonstrated by Pozar [1], has been developed into a variety of dual-polarization slot-coupled patch antennas reported in [2–5]. Many efforts have been made on the dual-polarization slot-coupled patch antennas to achieve sufficient bandwidth, low back radiation, low levels of cross-polarization, high efficiency, and high isolation [6–13]. However, the isolation of dual-polarized antennas needs to be more enhanced according to their applications.

In general, two popular techniques for achieving dual polarization using aperture-coupled patch antennas are: (i) crossed slots located at the center of the patch and (ii) two off-center orthogonal coupling slots. The first technique requires a relatively complicated feed arrangement or a multilayer construction to reduce the coupling between two feed lines. Further, several detailed studies on reducing the isolation are proposed in literatures [14–18]. A "T" slot and edge-slots loaded lower patch are presented to improve isolation in [14]. The antenna obtains a 7.8 % bandwidth at -10 dB reflection coefficient and the 30 dB isolation. Obviously, the structure is not advisable to design the broadband antenna. Then the application of air bridge in the dual-polarized antenna has reported in [15–17]. The isolations of these antennas are about than 34 dB, and bandwidth achieves 20 %. A dual-polarized microstrip patch antenna fed by quasi-cross-shaped slot with the ports isolation higher than 35 dB has been discussed in [18]. The profile increases due to the structure of U-shaped folded two feed lines and the antenna covers the band of 3.3–3.9 GHz with the reflection coefficient of -10 dB. It is obvious that the bandwidth of the dual-polarized antenna is not wide enough.

In this paper, a novel design of a high isolation, wideband dual-polarized microstrip antenna is presented, designed and fabricated based on the substrate-integrated waveguide (SIW) technology. The isolation for the antenna is enhanced because the electric fields of the feeding parts are enclosed by the metalized holes (MHs). The bandwidth is improved due to the MHs. Simulated and experimental results of the constructed prototype show that the high isolation of 40 dB can be obtained, and wide bandwidth and dual polarization are achieved for the proposed antenna theoretically and experimentally. This paper is organized as follows: Section 2 presents the design of the proposed antenna structure. The simulated and experimental results are provided and discussed in Section 3. Finally, brief conclusions are given in Section 4.

2. Design and Analysis

The proposed antenna is based on the conventional aperture coupled antenna. The first step is to design an aperture coupled antenna according to the microstrip theory [1–3]. Second step is to insert the MHs into the substrates based on the SIW technology. The last step is the parameters optimization to enhance the isolation. According to the guideline, a high-isolation dual-polarized antenna is proposed. As shown in Fig. 1, two "H" shaped slots are placed under the square radiating patch to excite two orthogonal modes.

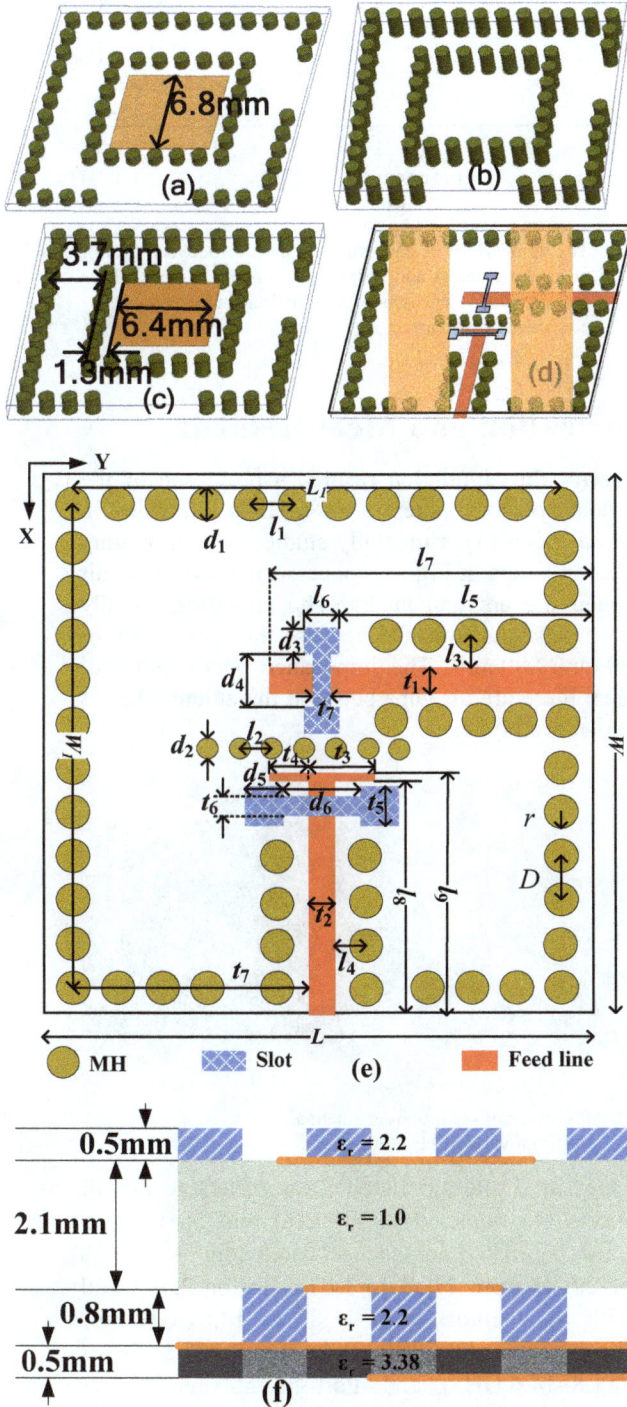

Fig. 1. Geometry of the proposed antenna and the detailed design parameters. (a) Top layer. (b) Second layer. (c) Third layer. (d) Bottom layer. (e) Perspective view for bottom layer. (f) Side view of the antenna without MHs.

One of the feed lines is left open and the other is terminated by a "T" junction. The proposed antenna radiates 0° polarized waves when Port I is excited and Port II is connected to a match load. Similarly when Port II excites, Port I is connected to a match load, it is 90° polarized. The performance is simulated by Ansoft HFSS 12.0, and its detailed optimum dimensions are depicted in Fig. 1. The metal portion of the antenna is modeled as lossy copper with a conductivity of 5.8×10^7 S/m.

Fig. 1 and Tab. 1 give the configuration of the proposed dual-linear polarized coupled antenna with optimized parameters. On one side of an RT/duroid 4003C substrate (the thickness is 0.5 mm and permittivity is 3.38), the feed lines are printed. On the other side it is a ground plane embedded with a coupling aperture. Lots of MHs with the diameter of d_2 between the two feed slots and of d_1 in the four substrates are inserted separately into the substrate to avoid the strong coupling and the leak of electromagnetic wave. The square radiating patch is printed on a 0.8-mm-thick Taconic TLY substrate with permittivity of 2.2. Above the Taconic TLY substrate, it is the substrate with thickness of 2.1 mm and permittivity of 1.0. In fabrication, it is the foam material and $\varepsilon_r = 1.04$. The holes are fabricated by printed circuit board techniques and the metallic columns with a radius of 0.5 mm are used to instead of the MHs. The top square patch is printed on the bottom side of a top Taconic TLY substrate with thickness of 0.5 mm. Outside MHs are inserted into the four substrates for reducing the leakage of the electromagnetic wave for the proposed antenna. The MH with radius (r, $r = 0.5\ d_1$) and distances (l_1) shown in Fig. 1(e) is embedding into the bottom substrate, and it can be described as follows.

$$0.05\frac{c_0}{f_H} < D < 0.25\frac{c_0}{f_H}\ , \qquad (1)$$

$$D/4 < r < D/2 \qquad (2)$$

where c_0 is the speed of light and f_H is the low frequency of bandwidth for the proposed antenna. From (1) and (2), we can design 0.5 mm for the radius ($d_1/2 = 0.5$ mm) and 1.5 mm for the distance between the two MHs ($l_1 = 1.5$ mm). The radius of MHs between two slots in the RT/duroid 4003C substrate is 0.3 mm ($d_2/2 = 0.3$ mm) and the distance is 0.9 mm ($l_2 = 0.9$ mm).

To realize the effect of the MHs, Fig. 2 shows the electric field distributions of the feeding ports for the proposed antenna at 12 GHz, 13.5 GHz, and 14.5 GHz when the two ports are both feeding. It is obvious that the electric field flows along the edges of the feed lines and the slots. The different electric fields for two feeding parts are separated by the MHs, and the electric field is enclosed. So the isolation is enhanced due to the MHs. The equivalent impedances for the proposed antenna and the conventional antenna (it is same with the proposed antenna without MHs) are given in Fig. 3. The capacitive character for the proposed antenna is improved and the imaginary part of the equivalent impedance is decreased through embedding the

MHs into the four sides of the proposed antenna. So the impedance bandwidth for this antenna is improved due to the MHs compared with the conventional antenna.

l_1	1.5	t_1	1.1	W_1	16.5
l_2	0.9	t_2	0.9	L_1	16.5
l_3	0.95	t_3	1.85	d_1	1
l_4	0.95	t_4	0.85	d_2	0.6
l_5	8.8	t_5	0.9	d_3	0.6
l_6	0.9	t_6	0.2	d_4	2.7
l_7	11.1	t_7	0.2	d_5	0.6
l_8	8.6	W	18.5	d_6	2.8
l_9	8.8	L	18.5	d_7	7.95

Tab. 1. Parameters of the proposed antenna (unit: mm).

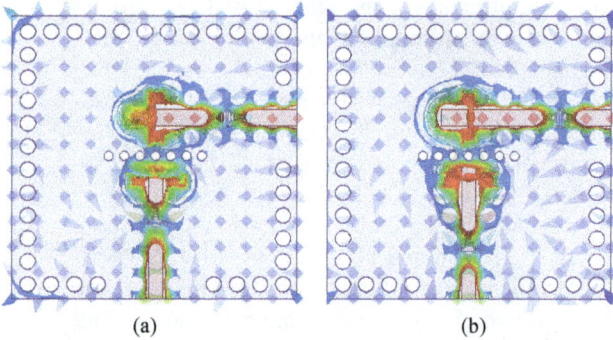

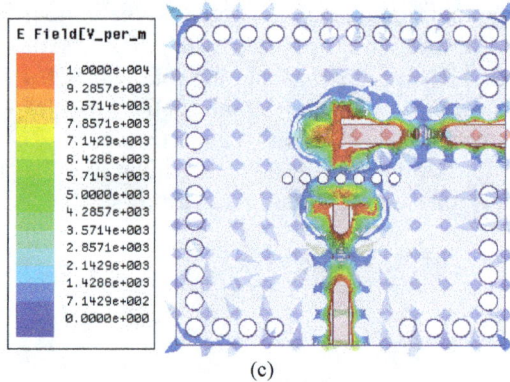

Fig. 2. The electric field distributions of the proposed antenna at (a) 12 GHz, (b) 13.5 GHz, and (c) 14.5 GHz when the two input ports are all feeding.

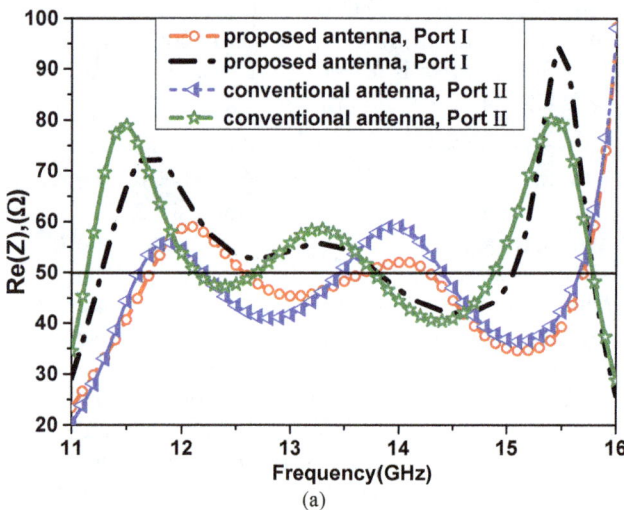

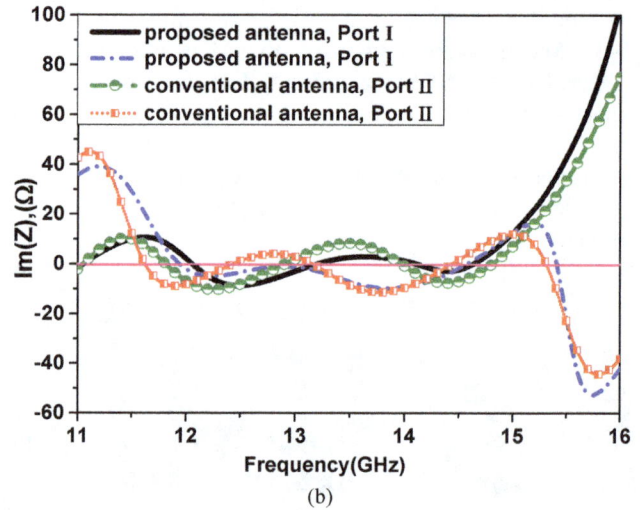

(a)

(b)

Fig. 3. The comparison of equivalent impedance between the proposed antenna and conventional antenna. (a) Real parts of equivalent impedance. (b) Imaginary parts of equivalent impedance.

3. Fabrication and Measurement

To verify the simulation results, a prototype of the proposed dual-polarized antenna via MHs is illustrated in Fig. 4 and has been experimentally studied. The mounting for the antenna layers in Fig. 4 is necessary to avoid the air gaps between the antenna layers. The mountings hardly affect the antenna performance because the size of prototype is 80 mm × 80 mm. The impedance bandwidth will reduce when there are air gaps between the antenna layers in the antenna prototype.

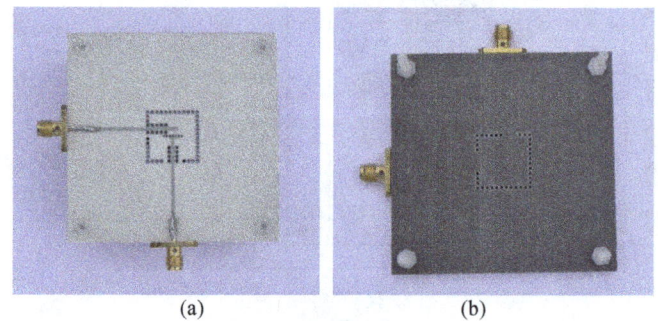

Fig. 4. Prototype of the proposed antenna.
(a) Front view. (b) Bottom view.

The measured and simulated S-parameters results of the two ports are depicted in Fig. 5(a) and 5(b). From Fig. 5(a), the simulated impedance bandwidth is 22.1 % with S_{11} < -20 dB from 11.75 to 14.7 GHz for Port I and 22.5 % with S_{22} < -20 dB from 11.82 to 14.81 GHz for Port II. The experimental results show that Port I achieves 21.9 % (11.8–14.6 GHz) and Port II obtains 23.8 % (11.65–14.8 GHz) impedance bandwidth of reflection coefficients less than -20 dB. The S_{12} of the conventional antenna is above -23 dB on the whole bandwidth from Fig. 5(b). When the MHs are loaded, the S_{12} can be improved under -38 dB in simulation. Measured S$_{12}$ is lower than -40 dB over the frequency range. It is observed from the plots that the measurements have good agreement with the simulations.

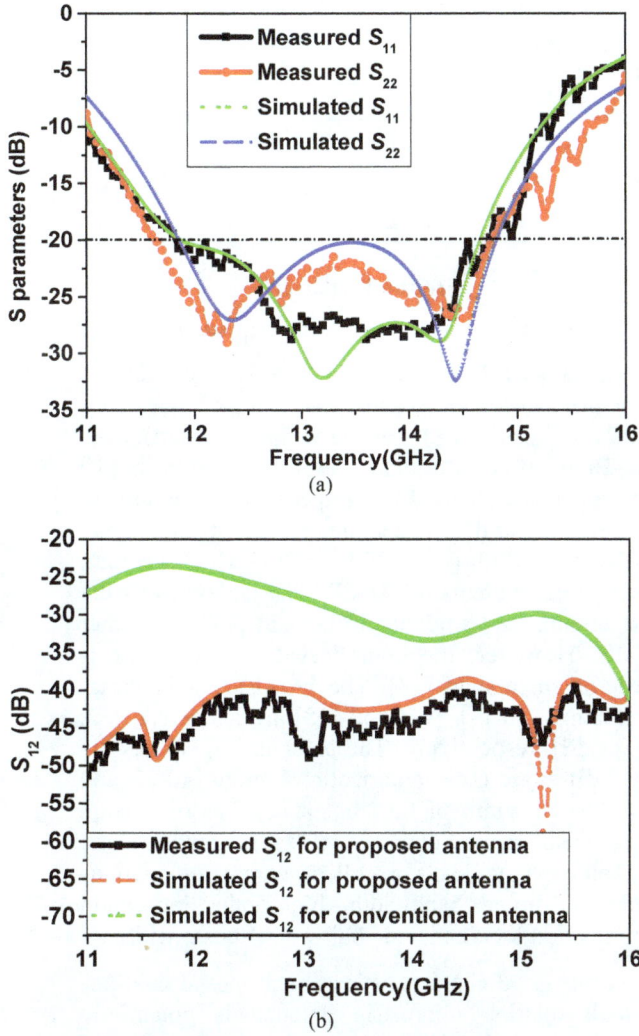

(a)

(b)

Fig. 5. Measured and simulated S parameters for the proposed antenna. (a) S$_{11}$ and S$_{22}$ results. (b) Measured and simulated S$_{12}$ for the proposed antenna and the simulated S$_{12}$ for the conventional antenna.

Fig. 6 presents the simulated and experimental radiation patterns for xoz-plane and yoz-plane at center frequency (13.5 GHz) of Port I and Port II for the proposed antenna. From Fig. 6, we observe that the cross-polarization with the main lobe remains less than -30 dB while the front-to-back ratio (FBR) remains to be better than 20 dB, and the half-power beam width is 72° both in xoz-plane and yoz-plane for Port I. Correspondingly, it is shown that the cross-polarization components are all -36 dB down from the copolarization components on boresight within 3-dB beamwidth of 70° and FBR of 22 dB in xoz-plane and yoz-plane for Port II at the same frequency. In conclusion, experimental results indicate that the isolation is more than 40 dB and the impedance bandwidth is above 21 % of reflection coefficients less than -20 dB in frequency range.

The measured results are in good agreement with the corresponding simulations. The proposed antenna achieves the low cross-polarization, low back radiation characteristics, wide bandwidth and high isolation.

The gain of the proposed antenna is measured and simulated. The experimental and simulated results are shown in Fig. 7. The simulated gain varies between 8.2 and 8.8 dBi with 0.6 dB variation for Port I and Port II. Within the operating frequency range, the experimental results show that the gain of the proposed antenna is above 7.9 dBi and with 0.7 dB variation for two ports. The gain differences of the dual polarizations are due to the different reflection coefficient for two ports and the difference between measured and simulated results mainly comes from the measurement tolerance and fabrication. In addition, the foams ($\varepsilon_r = 1.04$) were not considered in simulation and may also influence the gain for the proposed dual-polarized antenna.

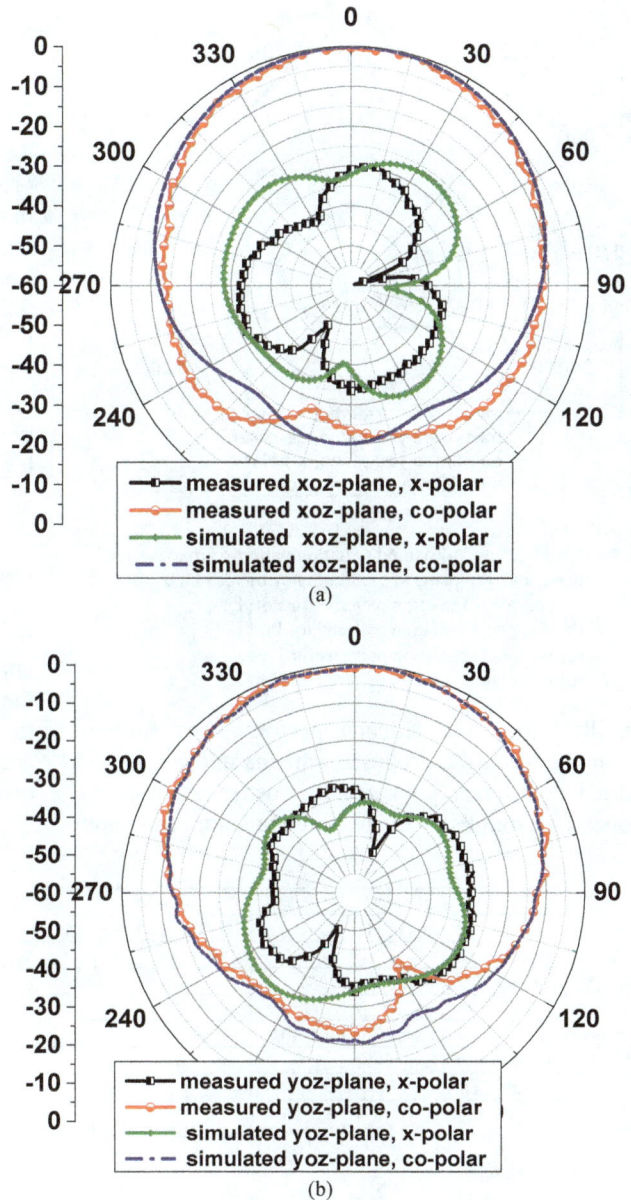

(a)

(b)

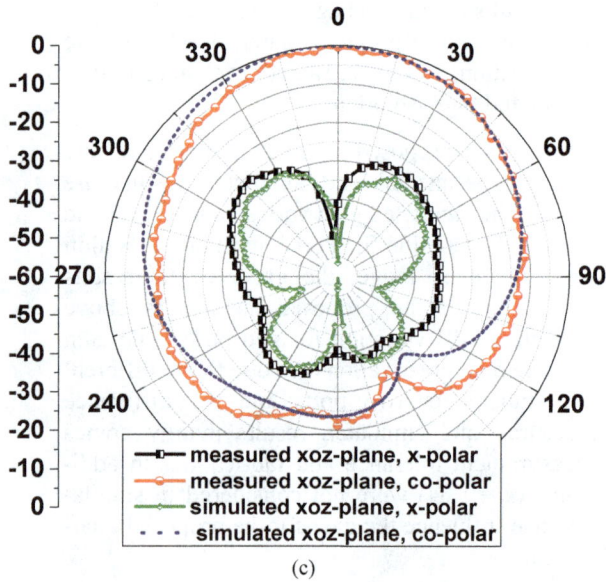

(c)

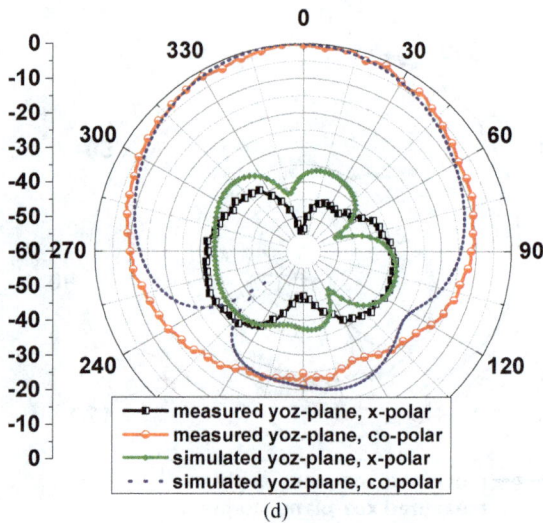

(d)

Fig. 6. Simulated and measured radiation patterns for xoz-plane and yoz-plane at the frequency of 13.5 GHz: (a) xoz-plane radiation patterns for Port I. (b) yoz-plane radiation patterns for Port I. (c) xoz-plane radiation patterns for Port II. (d) yoz-plane radiation patterns for Port II.

For illustrating the antenna performances further, comparisons between the proposed antenna and the previous reported dual-linear polarized microstrip antennas in term of operating frequency range, antenna gain, projection

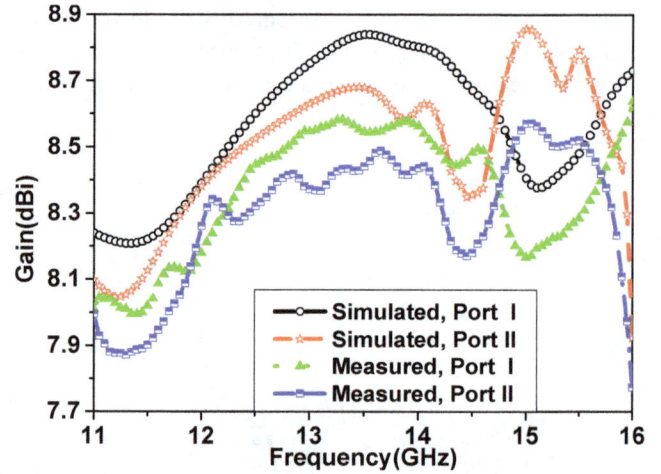

Fig. 7. Simulated and measured gain of the proposed antenna.

area, height and FBR are given in Tab. 2. According to Tab. 2, the proposed antenna achieves the impedance bandwidth of 21.9 % and 23.8 % at the reflection coefficients less than -20 dB. It is better than that of the antennas in [3], [18]. The antenna obtains 43 % impedance bandwidth in [17], but the standard is reflection coefficients less than -10 dB. The bandwidth is less 10 % in [17] when it is with the reflection coefficients of -20 dB. The isolations of the proposed antenna and antennas in [3] and [18] are higher than 40 dB. However, the complicated feed network is employed for antenna in [18]. The broadside gain variations of antennas in [3], [17] and [18] are about 7.4, 8.2, 9.3 and 9.5 dBi, respectively. The antenna in [17] has the FBR of 19 dB, main cross-polarization lobe of -40 dB and half-power beam width of 60°, but it consists of a large rectangular metal ground (1.18λ × 1.18λ). In summary, the proposed antenna achieves excellent parameter such as high isolation, broad bandwidth, high gain, low main cross-polarization lobe and wide half-power beam width.

Due to its good characteristics, the proposed dual-polarized high-isolation microstrip antenna is potentially suitable to be used with differential microwave and radio frequency circuits for various wireless communication applications. It is needed to point out that the complexity of the proposed antenna is obviously increased due to the metalized holes compared to the traditional dual-polarized antennas. Especially the radius of the metalized hole and its distance are mainly affected the isolation and impedance bandwidth of the dual-polarized antenna.

Antennas	Proposed	Ref [3]	Ref [17]	Ref [18]
Operating Range (GHz, -20dB)	21.9%; 23.8%	14.2%; 14.5%	45%; 43%(-10dB)	16.2%; 16.9%(-10dB)
Isolation (dB)	40	40	38	50
Broadside Gain (dBi)	7.9-8.6	7.4	8.5-10	9-10
FBR	20	13	19	10
Main cross-polarization lobe (dB)	-30	-25	-40	-17
Half-power beam width	70°	45°	60°	45°
Height (λ)	0.18	0.13	0.21	0.11

Tab. 2. Performance comparisons between the proposed and referenced antennas.

4. Conclusion

In this paper, a Ku band dual-linear polarized and aperture coupled antenna with high isolation is proposed based on the substrate-integrated waveguide technology. The metalized holes have been utilized in the substrate of the antenna to enhance the isolation. The prototype of the proposed antenna has been fabricated and measured. Good agreement is observed between the simulation and the measurement. Experimental results indicate that the isolation is more than 40 dB and the bandwidth is obtained 21.9 % for Port I and 23.8 % for Port II of reflection coefficients less than -20 dB. The gain of the antenna is above than 7.8 dBi. The antenna achieves a half-power beam width of 70° and the main cross-polarization lobe of -30 dB. Due to the merits of high isolation and broad bandwidth, the proposed dual-polarized antenna shows a potential use for the satellite communication.

Acknowledgements

Authors thank the supports from the National Natural Science Foundation of China under Grant (No. 61271100, No. 61471389), Natural Science Foundational Research Fund of Shannxi Province (No. 2010JZ6010 and No. 2012JM8003), and the Doctoral Foundation Air Force Engineering University under Grant (No. KGD080914002). They also thank the reviewers for their valuable comments.

References

[1] POZAR, D. M. A reciprocity method of analysis for printed slot and slot-coupled microstrip antennas. *IEEE Transactions on Antennas Propagation*, 1986, vol. 34, no. 12, p. 1439–1446.

[2] CHIOU, T.-W., WONG, K.-L. Broad-band dual-polarized single microstrip patch antenna with high isolation and low cross polarization. *IEEE Transactions on Antennas Propagation*, 2002, vol. 50, no. 3, p. 399–401.

[3] SIM, C. Y. D., CHANG, C. C., ROW, J. S. Dual-feed dual-polarized patch antenna with low cross polarization and high isolation. *IEEE Transactions on Antennas Propagation*, 2009, vol. 57, no. 10, p. 3405–3409.

[4] GAO, S.-C., LI, L.-W., LEONG, M.-S., YEO, T.-S. Dual-polarized slot coupled planar antenna with wide bandwidth. *IEEE Transactions on Antennas Propagation*, 2003, vol. 51, no. 3, p. 441–448.

[5] YONG-XIN GUO, KWAI-MAN LUK. Dual-polarized dielectric resonator antennas. *IEEE Transactions on Antennas Propagation*, 2003, vol. 51, no. 5, p. 1120–1123.

[6] CHANGLIANG DENG, PINGHUI LI, WENQUAN CAO. A high-isolation dual-polarization patch antenna with omnidirectional radiation patterns. *IEEE Antennas and Wireless Propagation Letters*, 2012, vol. 11, p. 1273–1276.

[7] VALLOZZI, L., VAN TORRE, P., HERTLEER, C., ROGIER, H., MOENECLAEY, M., VERHAEVERT, J. Wireless communication for firefighters using dual-polarized textile antennas integrated in their garment. *IEEE Transactions on Antennas and Propagation*, 2010, vol. 58, no. 4, p. 1357–1368.

[8] GAO, S., SAMBELL, A. Dual-polarized broad-band microstrip antennas fed by proximity coupling. *IEEE Transactions on Antennas and Propagation*, 2005, vol. 53, no. 1, p. 526–530.

[9] QUAN XUE, SHAO WEI LIAO, JIAN HUA XU. A differentially-driven dual-polarized magneto-electric dipole antenna. *IEEE Transactions on Antennas and Propagation*, 2013, vol. 61, no. 1, p. 425–430.

[10] XU LIN QUAN, RONG LIN LI. A broadband dual-polarized omnidirectional antenna for base stations. *IEEE Transactions on Antennas and Propagation*, 2013, vol. 61, no. 2, p. 942–947.

[11] SERRA, A. A., NEPA, P., MANARA, G. A wide-band dual-polarized stacked patch antenna. *IEEE Antennas and Wireless Propagation Letters*, 2007, no. 6, p. 141–143.

[12] QI WU, SCARBOROUGH, C. P., MARTIN, B. C., SHAW, R. K. et al. A Ku-band dual polarization hybrid-mode horn antenna enabled by printed-circuit-board metasurfaces. *IEEE Transactions on Antennas and Propagation*, 2013, vol. 61, no. 3, p. 1089–1098.

[13] MICHAIL, G. CH., UZUNOGLU, N. K. Dual-frequency and dual-polarization multilayer microstrip antenna element. *Microwave and Optical Technology Letters*. 2004, vol. 42, no. 4, p. 311–315.

[14] UZ ZAMAN, A., MANHOLM, L., DERNERYD, A. Dual polarized microstrip patch antenna with high port isolation. *Electronics Letters*, 2007, vol. 42, no. 10, p. 612–613.

[15] BARBA, M. A high-isolation, wideband and dual-linear polarization patch antenna. *IEEE Transactions on Antennas Propagation*, 2008, vol. 56, no. 5, p. 1472–1476.

[16] KABOLI, M., MIRTAHERI, S. A., ABRISHAMIAN, M. S. High isolation X-polar antenna. *IEEE Antennas Wireless Propagation Letters*, 2010, vol. 9, p. 401–404.

[17] BIAO LI, YING-ZENG YIN, WEI HU, YANG DING, YANG ZHAO. Wideband dual-polarized patch antenna with low cross polarization and high isolation. *IEEE Antennas Wireless Propagation Letters*, 2012, vol. 11, p. 427–430.

[18] JIE LU, ZHENQI KUAI, XIAOWEI ZHU, NIANZU ZHANG. A high-isolation dual-polarization microstrip patch antenna with quasi-cross-shaped coupling slot. *IEEE Transactions on Antennas Propagation*, 2011, vol. 59, no. 7, p. 2713–2717.

About Authors ...

Sijia LI was born in Xi'an, Shaanxi province, P.R. China in 1987. He received B. Eng. degree from Guangxi University, Nanning, China, in June, 2009 and M. Eng. degree from the Telecommunication Engineering Institute, Air Force Engineering University of CPLA, Xi'an China, in 2011. Currently, he is working toward Ph.D degree at the Information and Navigation College, Air Force Engineering University. His research activity has been focused in the dual-polarization antennas, tilted beam antennas, perfect metamaterial absorber design and its application for RCS reduction of antennas. He has authored and co-authored more than 40 technical journal articles and international conference papers. He received the excellent Master's dissertation in 2012. He is awarded the 2012 and 2013 Best Student Paper Prize (First Prize) of the National Graduate Mathematical Modeling Contest in Shanghai in December 2012 and in Changsha in December 2013. He was a recipient of the Best Paper Award at the Forth National Doctor Forum Symposium on Information Science of China in Guangzhou in October 2013. He is awarded the excellent student in the 2014 national summer school of the Microwave Material and Components. Now, he is a reviewer of the Applied Physics Letter, Journal of Applied Physics, IEEE Transactions on Microwave Theory and Techniques, IEEE Transactions on Antennas & Propagation, Microelectronics Journal, and Journal of Electromagnetic Waves and Applications.

Xiangyu CAO received the B.Sc and M.A.Sc degrees from the Air Force Missile Institute in 1986 and 1989, respectively. She joined the Air Force Missile Inst. in 1989 as an assistant teacher. She became an associate professor in 1996. She received Ph.D. degree in the Missile Inst. of Air Force Engineering University in 1999. From 1999 to 2002, she was engaged in postdoctoral research in Xidian University, China. She was a Senior Research Associate in the Department of Electronic Engineering, City University of Hong Kong from June 2002 to Dec 2003. She is currently a professor of School of Information and Navigation, Air Force Engineering University of CPLA. She has authored and coauthored more than 200 technical journal articles and conference papers, and holds one China soft patent. Her research interests include computational electromagnetic, smart antennas, electromagnetic metamaterial and their antenna applications, and electromagnetic compatibility.

Jun GAO received the B.Sc and M.A.Sc degrees from the Air Force Missile Inst. in 1984 and 1987, respectively. He joined the Air Force Missile Institute in 1987 as an assistant teacher. He became an associate professor in 2000. He is currently a professor of School of Information and Navigation, Air Force Engineering University of CPLA. He has authored and coauthored more than 60 technical journal articles and conference papers, and holds one China soft patent. His research interests include smart antennas, electromagnetic metamaterial and their antenna applications.

Pengcheng GAO received the B.S. degree from the National Key Lab. of Antenna and Microwave Technology, Xidian University, China, in 2011. Currently, he is working toward M. Eng. degree at the National Key Lab. of Antenna and Microwave Technology, Xidian University. His research interest is in the substrate integrated waveguide technology, electromagnetic metamaterial and their antenna applications.

19

Inverse Problem Solution in Landmines Detection Based on Active Thermography

Barbara SZYMANIK

West Pomeranian University of Technology in Szczecin, Faculty of Electrical Engineering,
Sikorskiego 37, 70-313 Szczecin, Poland

szymanik@zut.edu.pl

Abstract. *Landmines still affect numerous territories in the whole world and pose a serious threat, mostly to civilians. Widely used non-metallic landmines are undetectable using metal detector. Therefore, there is an urging need to improve methods of detecting such objects. In the present study we introduce relatively new method of landmines' detection: active infrared thermography with microwave excitation. In this paper we present the optimization based method of solving inverse problem for microwave heating. This technique will be used in the reconstruction of detected landmines geometric and material properties.*

Keywords

Microwave heating, landmines detection, active thermography, inverse problems.

1. Introduction

Nowadays, metal detector is still the most popular device used in demining. It is able to detect landmines containing metallic parts, nevertheless it is almost completely useless in case of the most common, nonmetallic (with bakelite, PVC, and polyethylene casings) devices. Therefore, it is extremely important to work on improving the methods of landmines' detection and removal. Currently the intensive investigations of several new methods of landmine detection are conducted [1]. In the present study, we introduce the relatively new method of landmines' detection: active infrared thermography with microwave excitation [2], [3], which can be considered complementary to the metal detector. Microwave enhanced infrared thermography combines two phenomena: microwave heating and thermal imaging. The volumetric microwave heating induces the thermal contrast between the landmine and soil. Thermal patterns obtained at the ground's surface are observed using sensitive thermovision camera. This method is able to detect an object, determine its size and approximate its location.

Active infrared thermography with microwave excitation can be used to detect objects buried below the ground, regardless of what material they are composed of.

It is undoubtedly the basic advantage of this method. However buried stones, branches, and various types of waste can produce similar thermal signatures to antipersonnel landmines while heated by microwaves. It is therefore important to not only detect an object that can be possibly a landmine, but also to determine some of its parameters to classify the object to the group of potentially dangerous [4]. In this paper the method of solving inverse problem for microwave heating will be presented. The proposed technique will be used in estimation of chosen geometric and material properties of landmines. Both numerical and experimental results will be shown.

2. Numerical Modeling – Forward Problem

The phenomenon of microwave heating can be simulated using finite element method (FEM). The propagation of microwave through a dielectric material is governed by the electromagnetic wave equation [5]:

$$\nabla \times \left(\frac{1}{\mu_r} \nabla \times \vec{E} \right) - \left(\varepsilon_r - j \frac{\sigma}{\omega \varepsilon_0} \right) k_0^2 \vec{E} = 0 \quad (1)$$

where μ_r is the relative permeability, ε_0, ε_r are permittivity of vacuum and material, respectively, σ is the material conductivity, k_0 is the wave number and $\vec{E}$ is electric field vector. The heat transfer equation can be written as follows:

$$\rho C_p \frac{\partial T}{\partial t} - \nabla(k \cdot \nabla T) = p \quad (2)$$

where ρ, C_p, k indicate material's density, specific heat capacity and thermal conductivity, respectively and $p = 2\pi f \varepsilon_0 \varepsilon'' E^2$ is the volumetrically dissipated power, dependent on wave frequency f and dielectric properties of material.

2D simulations were conducted using commercial software COMSOL Multiphysics to obtain the data which may be used in inverse problem using optimization method (described in the subsequent section). The proposed geometry is presented in Fig. 1.

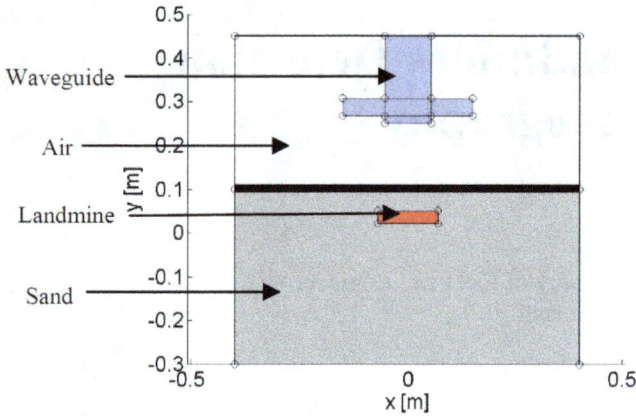

Fig. 1. 2D numerical model geometry.

3. Inverse Problem in Infrared Landmine Detection

The chosen parameters of the landmine can be reconstructed by solving the inverse problem [6]. In our case, the inverse problem is to determine the parameters of the landmines buried in the sand on the basis of the temperature distribution along the sand surface. This distribution can be obtained from the sequence of thermal images. In Fig. 1 one may see black line indicating the boundary from which the temperature distributions were read. The main assumption of our method is a strong relationship between the geometric parameters and material properties of buried landmines and the shape of lines presenting temperature distributions along marked boundary.

The total time of observation was set to 1200 seconds, with 600 seconds of microwave heating. The temperature distribution was measured during the whole time every 40 seconds. As the result we received 30 linear temperature distributions to work with. Reconstructed parameters are: object's size (width and height), the depth at which the object is buried and the position of the object in the relation to the centre of waveguide aperture (geometric parameters are shown in Fig. 2). Additionally, one material parameter – the value of imaginary part of complex dielectric permittivity, was also taken into account. This parameter was chosen after the study of the system's sensitivity to modification of all of the material parameters (i.e. density, specific heat, thermal diffusivity, complex dielectric permittivity).

The optimization algorithm, composed of combined Genetic Algorithms (GA) [7] and Pattern Search (PS), is presented in Fig. 3. The minimized function is defined as summed minimal square errors between the optimal temperature distribution and the distributions obtained in every iteration of the algorithm:

$$MSE_T = \sum_{l=1}^{30} \frac{1}{N}\left(\sum_{k=1}^{N}\left(f_{desT}(x_k, l\cdot\Delta t) - f_T(x_k, l\cdot\Delta t)\right)^2\right) (3)$$

where $f_{des\,T}$ indicates the desired temperature distribution, f_T

is the distribution obtained in every iteration, Δt indicates the time step, $x-$ coordinate of the measurement point, N is the number of measurement points.

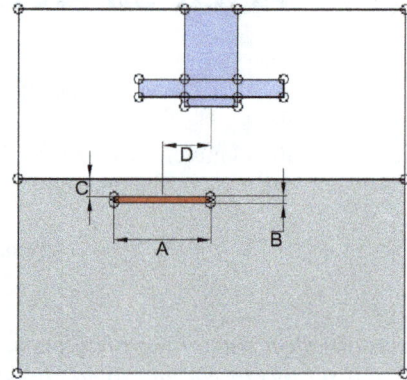

Fig. 2. Estimated geometrical properties: A - width, B - height, C - depth of burial, D - position against the waveguide.

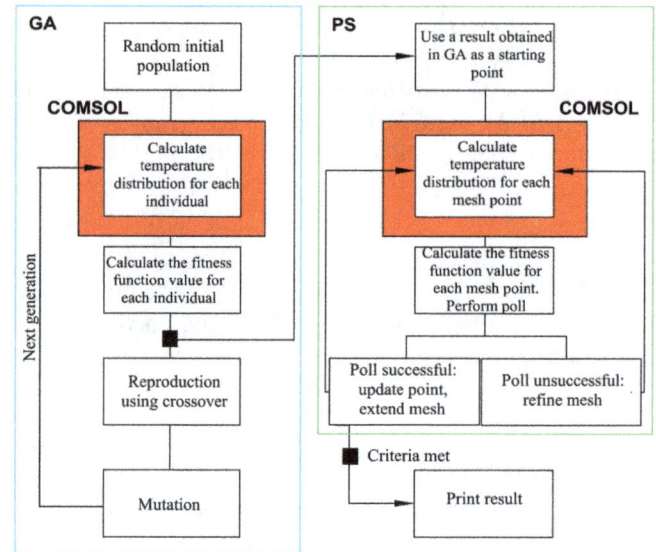

Fig. 3. Flowchart of optimization algorithm.

4. Results – Numerical Data

In order to verify the proposed algorithm's efficiency, the analysis including four sets of numerical data was conducted. The optimization goal in the first case was a landmine of width 0.2 m, height 0.01 m, buried at 0.2 m below the ground and located 0.1 m from the waveguide aperture centre, and in the second case a landmine of width 0.14 m, height 0.03 m, buried at 0.05 m below the ground and located centrally under the waveguide. The imaginary part of complex dielectric permittivity was set to 0.23 in the second case. The first set of numerical data was used to prove the efficiency of the proposed algorithm of estimating the landmine properties in case when there is an offset between the feed and the landmine. In the second case besides the geometric properties complex dielectric permittivity was also estimated. In the third and fourth case the Gaussian noise of 5 % was added to both previous data sets. In Fig. 4 the optimal temperature distributions obtained as a result of forward problem solution for all of the cases are presented.

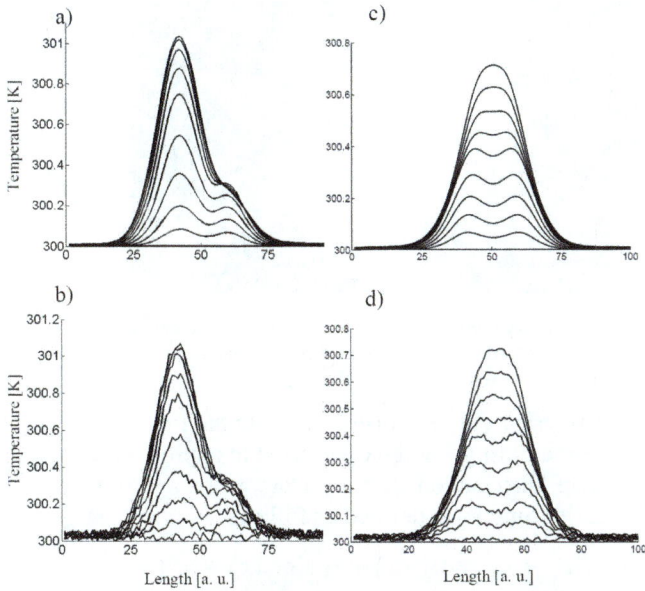

Fig. 4. The optimal temperature distributions. a) and b) for the mine of width 0.2 m, height 0.01 m, buried at 0.2 m below the ground and located 0.1 m from the waveguide aperture centre for the case without noise and with noise, respectively c) and d) for the mine of width 0.14 m, height 0.03 m, buried at 0.05 m below the ground and located centrally under the waveguide for the case without noise and with noise, respectively.

In all cases, the genetic algorithm using the population of 10 individuals was used. The generations number for GA was set to 10. The solution from GA was used as the starting point in the pattern search algorithm, which ran in 50 iterations. As the result of the optimization process is the reconstructed geometrical object parameters. Additionally, for the second case (the mine of width 0.14 m, height 0.03 m, buried at 0.05 m below the ground and located centrally under the waveguide) the imaginary part of complex dielectric permittivity was found. Figure 5 presents the comparison between optimal temperature distribution and those obtained in the optimization process for all sets of data. The results for the four test cases are gathered in Tab. 1–4. It may be noticed that in the case of data with added noise the results are significantly worse: especially the object width was estimated with maximum 90% error. However, in the case of relatively large objects like landmines, the correct estimation of location parameters (depth, position against the aperture center) seems to be more important. Moreover the imaginary part of dielectric permittivity was estimated with error of absolute value equal to 4 %. Therefore, it can be expected that the method of parameter estimation may be used not only for the exact location of the object, but also to verify the type of material from which the object is constructed.

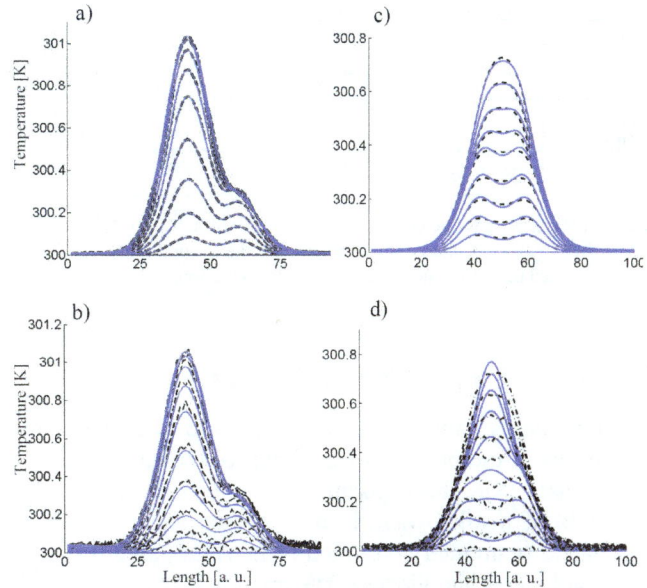

Fig. 5. Comparison between goal temperature (blue line) and reconstructed temperature (black line) distributions along ground surface. a) and b) for the mine of width 0.2 m, height 0.01 m, buried at 0.2 m below the ground and located 0.1 m from the waveguide aperture centre for the case without noise and with noise, respectively. c) and d) for the mine of width 0.14 m, height 0.03 m, buried at 0.05 m below the ground and located centrally under the waveguide for the case without noise and with noise, respectively.

Parameter	Desired value	Estimated value	Error
Width [m]	0.2 m	0.19 m	-5%
Height [m]	0.01 m	0.01 m	0%
Depth [m]	0.025 m	0.025 m	0%
Position against aperture [m]	0.1 m	0.1 m	0%

Tab. 1. Estimated parameters for the mine of width 0.2 m, height 0.01 m, buried at 0.2 m below the ground and located 0.1 m from the waveguide aperture centre. Without noise.

Parameter	Desired value	Estimated value	Error
Width [m]	0.2 m	0.38 m	90%
Height [m]	0.01 m	0.01 m	0%
Depth [m]	0.025 m	0.026 m	4%
Position against aperture [m]	0.1 m	0.108 m	8%

Tab. 2. Estimated parameters for the mine of width 0.2 m, height 0.01 m, buried at 0.2 m below the ground and located 0.1 m from the waveguide aperture centre. With noise.

Parameter	Desired value	Estimated value	Error
Width [m]	0.14	0.14	0%
Height [m]	0.03	0.06	100%
Depth [m]	0.05	0.06	20%
Position against aperture [m]	0	0	
ε''	0.23	0.24	4%

Tab. 3. Estimated parameters for the mine of width 0.14 m, height 0.03 m, buried at 0.05 m below the ground and located centrally under the waveguide. Without noise.

Parameter	Desired value	Estimated value	Error
Width [m]	0.14	0.07	-50%
Height [m]	0.03	0.06	100%
Depth [m]	0.05	0.04	20%
Position against aperture [m]	0	0	
ε''	0.23	0.19	-4%

Tab. 4. Estimated parameters for the mine of width 0.14 m, height 0.03 m, buried at 0.05 m below the ground and located centrally under the waveguide. With noise.

5. Results − Experimental Data

The next stage of the study was devoted to evaluate the effectiveness of the proposed algorithm for the case of experimental data [8]. The experimental setup, shown in Fig. 6, consists of a microwave heating device and thermovision camera (computer controlled FLIR A325, able to record thermal images and videos). The microwave heating device system is based on a magnetron, generating microwaves (frequency 2.45 GHz) of maximum power of 1000 W. The magnetron is connected to the rectangular waveguide with a proper flange. The waveguide is placed above the container with sand, in which the inert landmines are buried. Sand surface is observed using a thermovision camera FLIR A325.

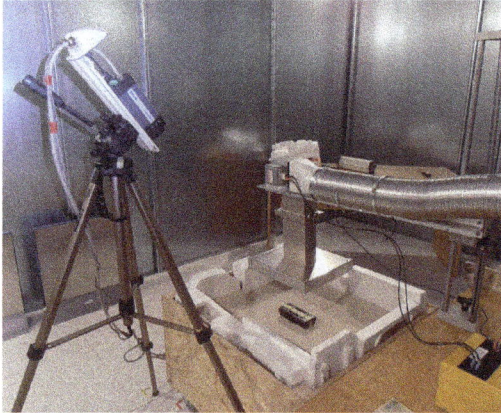

Fig. 6. Experimental setup.

In the experiment, inert landmine PMA-1 (width 0.14 m, height 0.03 m), with bakelite casing was used. The landmine was buried to a depth of five centimeters. The sand with buried landmine was heated for 10 minutes and then the system was observed with themovision camera for another ten minutes of natural cooling. During this time, the sequence of 600 thermograms, presenting the temperature distribution on the surface of the sand, was recorded. In each thermogram, the temperature values were collected along a single line passing through the central part of the heated space, as shown in Fig. 7. As a result, the 600 temperature distributions (shown in Fig. 8) were obtained.

It can be noticed, that obtained linear distributions are noisy. As it was observed in the previous section, the proposed algorithm of parameters estimation gives distinctly worse results for data with added noise. Therefore, the

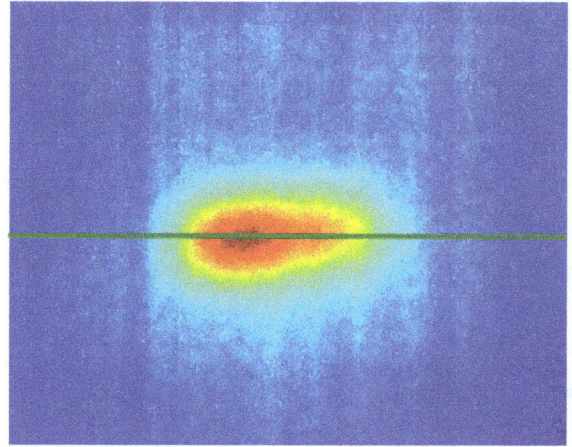

Fig. 7. Exemplary thermogram obtained for PMA-1 landmine. Temperature values were collected from marked green line.

procedure of filtering out the noise is used. In the first step, the Matlab Curve Fitting Toolbox was used to approximate the temperature distribution curves. Data were approximated by Fourier functions, defined as follows:

$$f(x) = a_0 + a_1 \cdot \cos(wx) + b_1 \sin(wx) + ... + \\ + a_5 \cos(5wx) + b_5 \sin(5wx) \tag{4}$$

where the parameters $a_0, ..., a_5, b_1, ..., b_5, w$ were selected using non-linear least squares method.

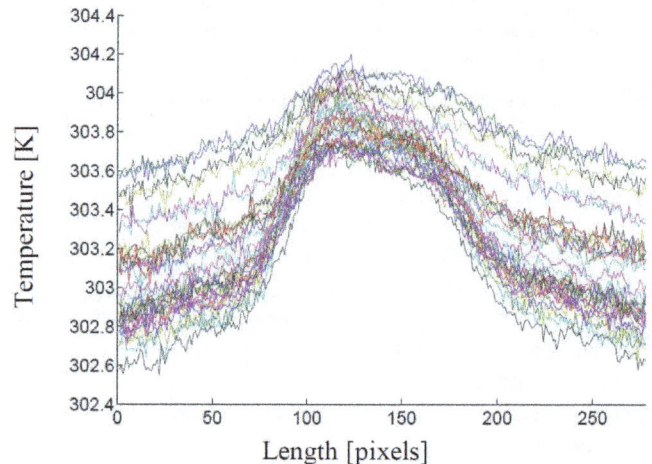

Fig. 8. Temperature distributions collected from thermograms.

As a result of approximation, 600 curves presented in Fig. 9 were obtained. Presented data was used as input to the optimization procedure.

The parameter estimation procedure was carried out just like in the case of numerical data, described in the previous section. Again the genetic algorithm using the population of 10 individuals was used. The generations number for GA was set to 10. The solution from GA was used as the starting point in the pattern search algorithm, which ran in 50 iterations.

Figure 10 shows the comparison between the exemplary temperature distributions obtained in the optimization

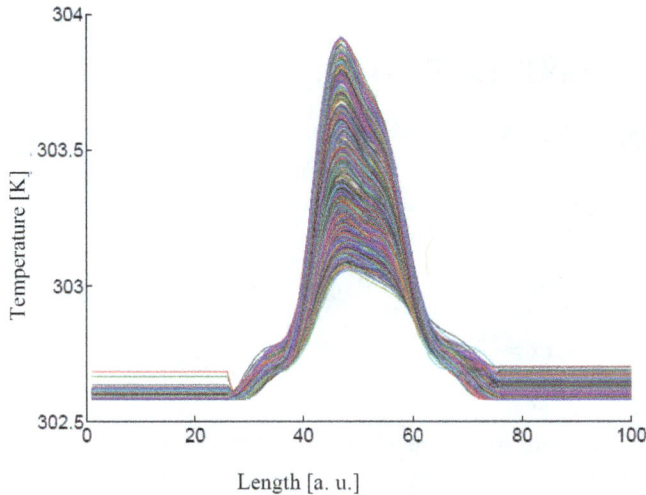

Fig. 9. Approximated temperature distributions.

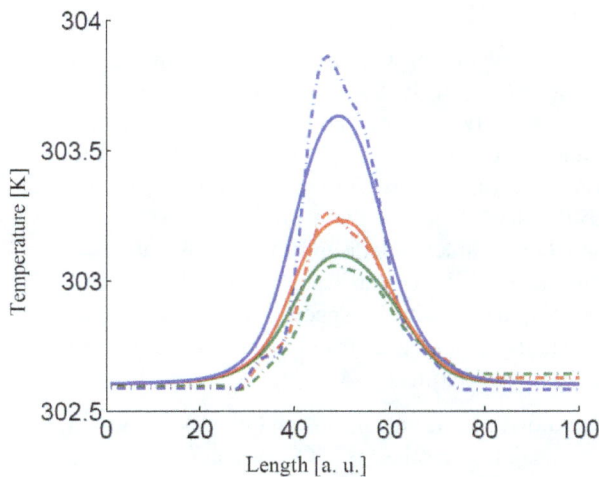

Fig. 10. The comparison between the exemplary temperature distributions obtained in the optimization process (solid lines) and those obtained experimentally (dashed lines).

process and those obtained experimentally. In Tab. 5 the estimated parameters are gathered. The good agreement between the results of estimation and the actual parameters of the detected object may be noticed. In the case of the geometrical parameters, the largest error occurs in the estimation of the mine height, while the width, the depth of burial and location of mine are estimated with a small error. Imaginary part of the complex dielectric permittivity was estimated to 0.28 (compared to $\varepsilon''= 0.23$, for bakelite). Estimation of material properties is essential in this study, since it can be used to determine the approximate type of material.

Parameter	Actual value	Estimated value	Error
Width [m]	0.14	0.13	-7%
Height [m]	0.03	0.08	167%
Depth [m]	0.05	0.06	20%
Position against aperture [m]	0	0.02	5%
ε''	0.23	0.28	22%

Tab. 5. Estimated parameters for the case of experimental data.

6. Conclusions

It was shown, that the proposed algorithm may be used to estimate chosen parameters of the landmines. Solution of the inverse problem allowed the estimation of geometric parameters and one material parameter of the object. The imaginary value of the complex dielectric permittivity may be used to approximate determination of the type of material. It may be especially useful to distinguish between variety types of plastic (used to landmines' casings construction) and other materials, which often contaminate the minefields (like rocks, metal parts and wood chunks). The estimation of the burial depth seems to be very important, since this information may be crucial for the safety of deminers.

References

[1] MACDONALD, J., LOCKWOOD, J. R. *Alternatives for Landmine Detection.* RAND, 2003.

[2] MENDE, H., DEJ, B., KHANNA, S., APPS, R., BOYLE, M., ADDISON, F. Microwave enhanced IR detection of landmines using 915 MHz and 2450 MHz. *Defence Research Reports*, no. DRDC-OTTAWA-TM-2004-266. Ottawa (Canada): Defence R&D, 2004.

[3] MALADEGUE, X. *Theory and Practice of Infrared Technology for Nondestructive Testing.* New York (USA): John Wiley and Sons, 2001.

[4] THANH, N. T., SAHLI, H., HAO, D. N. Infrared thermography for buried landmine detection: Inverse problem setting. *IEEE Transactions on Geoscience and Remote Sensing*, 2008, vol. 46, p. 3987–4004.

[5] SZYMANIK, B., GRATKOWSKI, S. Numerical modelling of microwave heating in landmines detection. *International Journal of Applied Electromagnetics and Mechanics*, 2011, vol. 37, no. 2-3, p. 215–229.

[6] SZYMANIK, B. Objects' parameters reconstruction in landmines' detection based on active thermography. In *International Interdisciplinary PhD Workshop IIPhDW 2013.* Brno (Czech Republic), 8-11 September, 2013.

[7] GOLDBERG, D. E. *Genetic Algorithms in Search, Optimization and Machine Learning.* Addison-Wesley Publishing Company Inc. 1989.

[8] SZYMANIK, B. Zastosowanie aktywnej termografii podczerwonej ze wzbudzeniem mikrofalowym do wykrywania niemetalicznych min lądowych. (The use of active infrared thermography with microwave excitation for detection of non-metallic landmines.) *PhD Thesis.* West Pomeranian University of Technology, Poland, 2013 (in Polish).

About Author ...

Barbara SZYMANIK was born in 1982. She received her master's degree in Mathematics in 2006 and in Physics in 2009. In 2013 she defended her doctoral thesis and received her PhD degree in Technical Sciences. Her scientific interests are mainly NDT of materials using active infrared thermography.

20

Invariance of the Null Distribution of the Multiple Coherence

Radoslav BORTEL, Pavel SOVKA

Faculty of Electrical Engineering, Czech Technical University in Prague, Technická 2, 166 27 Prague, Czech Republic

bortelr@feld.cvut.cz, sovka@feld.cvut.cz

Abstract. *In this paper we investigate the invariance of the null distribution of the multiple coherence (MC) to the statistics of the examined signals. We show that when the MC is computed between a group of signals $x_i[n]$, $i = 1, \ldots, K$ and a signal $y[n]$, the null distribution of the MC is independent of the distribution of $x_i[n]$ and $y[n]$ if at a given frequency the joint distribution of the spectra of the segments of $x_i[n]$ and $y[n]$ is rotationally symmetric with respect to the rotation of the spectra of the segments of $x_i[n]$ or $y[n]$.*

The significance of this result lies in the improvement of the multiple coherence analysis. Hitherto, the null distribution of the MC was known only for signals with the multivariate Gaussian distribution; therefore, an MC estimate could be evaluated for its statistical significance only in this limited case. With the results presented in this paper, it will be possible to evaluate the statistical significance of MC estimates for much wider class of signals.

Keywords

Multiple coherence, null distribution, invariance.

1. Introduction

The multiple coherence (MC) is a measure that indicates a relationship between a single signal $y[n]$ and a group of signals $x_i[n]$ ($i = 1 \ldots K$). It has found numerous applications for example in signal detection (e.g. [19], [20]), blood flow analysis (e.g. [13], [14]), brain signal processing (e.g. [3], [8], [9], [11], [17], [18]), data acquisition (e.g. [15], [16]), geology (e.g. [10]) and many other fields.

A crucial part of the multiple coherence analysis is the evaluation of the MC estimate for its statistical significance. A statistical test is employed to detect if the value of the MC estimate is high enough to indicate that the coupling between the examined signals is statistically significant and not just a random occurrence.

To perform such statistical testing the null distribution of the MC estimate needs to be known. In the past, the null distribution was derived under the condition that the ex-

amined signals are multivariate Gaussian [2], [5], [6], [7], [19], [20]. However, the real world signals are often non-Gaussian; therefore, it would be desirable to know the null distribution of the multiple coherence for a more general class of signals.

A partial achievement was already attained in the special case, where $K = 1$. In this case, the MC reflects the connection between two signals, and becomes equivalent to the magnitude squared coherence (MSC). The null distribution of an MSC estimate was shown to have a distribution invariant to the distribution of the second signal if the signals are statistically independent and the first signal is stationary Gaussian [12]. Later, the assumption of Gaussianity was relaxed, requiring only the spectra of the segments of one of the signals to have a distribution that is spherically symmetric at a given frequency [4].

To our knowledge it was never analytically examined, if such generalization applies to MC estimates for $K > 1$. However, such generalization would widen the applicability of the known statistical tests, and be beneficial in cases, where the examined signals are non-Gaussian, or their statistical distribution is unknown.

In this paper we therefore analyze the null distribution of the MC. We show that this distribution is in fact invariant to the distribution of the signals $x_i[n]$ and $y[n]$ if at a given frequency the joint distribution of the spectra of the segments of $x_i[n]$ and $y[n]$ is rotationally symmetric with respect to the rotation of the spectra of segments of $x_i[n]$ or $y[n]$.

2. Methods

2.1 MC Definition

If $x_i[n]$ and $y[n]$ are zero mean, wide sense stationary, we can define their power and cross-spectral densities as

$$\mathbf{S}_{XX}(\Omega) = \mathcal{F}\{\mathbf{R}_{xx}[k]\}, \tag{1}$$
$$\mathbf{S}_{XY}(\Omega) = \mathcal{F}\{\mathbf{R}_{xy}[k]\}, \tag{2}$$
$$S_{YY}(\Omega) = \mathcal{F}\{R_{yy}[k]\} \tag{3}$$

where $\mathcal{F}$ denotes the discrete time Fourier transform, and $\mathbf{R}_{xx}[k]$, $\mathbf{R}_{xy}[k]$, $R_{yy}[k]$ are the covariance matrices

$$\mathbf{R}_{xx}[k] = E[\mathbf{x}^*[n]\mathbf{x}^T[n+k]], \tag{4}$$

$$\mathbf{R}_{xy}[k] = E[\mathbf{x}^*[n]y[n+k]], \tag{5}$$

$$R_{yy}[k] = E[y^*[n]y[n+k]] \tag{6}$$

where $\mathbf{x}[n] = [x_1[n], \ldots, x_K[n]]^T$, and $*$ denotes the complex conjugate.

The multiple coherence between $x_i[n]$ and $y[n]$ is given as [2]

$$|\gamma_{xy}(\Omega)|^2 = \frac{\mathbf{S}_{XY}^H(\Omega)\mathbf{S}_{XX}^{-1}(\Omega)\mathbf{S}_{XY}(\Omega)}{S_{YY}(\Omega)} \tag{7}$$

where ^{H} denotes the conjugate transpose. In the following text we shall assume that $|\gamma_{xy}(\Omega)|^2$ exists for at least some Ω.

2.2 MC Estimation

If $x_i[n]$ and $y[n]$ are finite length records of a wide sense stationary process, the MC estimate can be computed in the following way [2].

First, the signals are segmented

$$x_{li}[n] = x_i[(l-1)M+n], \tag{8}$$

$$y_l[n] = y[(l-1)M+n] \tag{9}$$

where M denotes the segment length, and $l = 1, \ldots, L$, where L is the number of segments. Next, let $X_{li}(\Omega)$ and $Y_l(\Omega)$ be the discrete Fourier transforms of $x_{li}[n]$ and $y_l[n]$, respectively (each Fourier spectrum has M frequency points). Also, let us define

$$\mathbf{X}(\Omega) = [X_{li}(\Omega)]_{\substack{l=1\ldots L \\ i=1\ldots K}}, \tag{10}$$

$$\mathbf{Y}(\Omega) = [Y_1(\Omega), \ldots, Y_L(\Omega)]^T. \tag{11}$$

Now, we can compute the estimates of the cross and power spectral densities

$$\widehat{\mathbf{S}}_{XX}(\Omega) = \frac{1}{L}\mathbf{X}^H(\Omega)\mathbf{X}(\Omega), \tag{12}$$

$$\widehat{\mathbf{S}}_{XY}(\Omega) = \frac{1}{L}\mathbf{X}^H(\Omega)\mathbf{Y}(\Omega), \tag{13}$$

$$\widehat{S}_{YY}(\Omega) = \frac{1}{L}\mathbf{Y}^H(\Omega)\mathbf{Y}(\Omega), \tag{14}$$

with which the MC estimate $|\widehat{\gamma}_{xy}(\Omega)|^2$ will be given as

$$|\widehat{\gamma}_{xy}(\Omega)|^2 = \frac{\widehat{\mathbf{S}}_{XY}^H(\Omega)\widehat{\mathbf{S}}_{XX}^{-1}(\Omega)\widehat{\mathbf{S}}_{XY}(\Omega)}{\widehat{S}_{YY}(\Omega)} \tag{15}$$

$$= \frac{\mathbf{Y}^H(\Omega)\mathbf{X}(\Omega)(\mathbf{X}(\Omega)^H\mathbf{X}(\Omega))^{-1}\mathbf{X}(\Omega)^H\mathbf{Y}(\Omega)}{\mathbf{Y}^H(\Omega)\mathbf{Y}(\Omega)}. \tag{16}$$

2.3 Null Distribution of MC Estimate

The null distribution of the MC estimate is defined as

$$G_0(g) = P\left[|\widehat{\gamma}_{xy}(\Omega)|^2 < g \,\Big|\, |\gamma_{xy}(\Omega)|^2 = 0\right] \tag{17}$$

where $G_0(g)$ is the cumulative distribution function (CDF) of the null distribution of the MC estimate, and $P[.|.]$ denotes the conditional probability operator.

In this section we will analyze the invariance of (17). For this purpose we will assume $x_i[n]$ and $y[n]$ to be finite length records of a strict sense stationary random process. Further, to simplify the derivation we will drop the argument (Ω), and assume it implicitly.

Also, we will use the following notation. Let $f_{\mathbf{XY}}(\mathbf{X}, \mathbf{Y})$ be the joint probability density function of $\mathbf{X}$ and $\mathbf{Y}$, let $f_{\mathbf{X}}(\mathbf{X})$ be the probability density function of $\mathbf{X}$, and let $f_{\mathbf{Y}}(\mathbf{Y})$ be the probability density function of $\mathbf{Y}$. In the following subsections we will analyze (17) assuming rotational symmetry of $f_{\mathbf{XY}}(\mathbf{X}, \mathbf{Y})$ with respect to the rotation of $\mathbf{X}$ or $\mathbf{Y}$.

2.3.1 Rotational Symmetry of $f_{\mathbf{XY}}(\mathbf{X}, \mathbf{Y})$ with Respect to Rotation of Y

In this section we will assume that $f_{\mathbf{XY}}(\mathbf{X}, \mathbf{Y})$ is invariant to the rotation of the argument $\mathbf{Y}$ – that is we will assume that $f_{\mathbf{XY}}(\mathbf{X}, \mathbf{Y}) = f_{\mathbf{XY}}(\mathbf{X}, \mathbf{BY})$ if $\mathbf{B}$ is a rotation matrix (i.e. $\det(\mathbf{B}) = 1$ and $\mathbf{B}^H = \mathbf{B}^{-1}$). This means that $f_{\mathbf{XY}}(\mathbf{X}, \mathbf{Y})$ and $f_{\mathbf{Y}}(\mathbf{Y})$ can also be expressed as $f_{\mathbf{XY}}(\mathbf{X}, \mathbf{Y}) = f'_{\mathbf{XY}}(\mathbf{X}, |\mathbf{Y}|)$ and $f_{\mathbf{Y}}(\mathbf{Y}) = f'_{\mathbf{Y}}(|\mathbf{Y}|)$, where $f'_{\mathbf{XY}}$ and $f'_{\mathbf{Y}}$ are new functions, and $|.|$ denotes the L_2 norm.

With these assumptions the MC estimate CDF (17) can be expressed as

$$\int_C f'_{\mathbf{XY}}(\mathbf{X}, |\mathbf{Y}|)d\mathbf{X}d\mathbf{Y} \tag{18}$$

where C is a set of $\mathbf{X}$ and $\mathbf{Y}$ for which the following holds

$$\mathbf{X} \in \mathbb{C}^{LK}, \tag{19}$$

$$\mathbf{Y} \in \mathbb{C}^{L}, \tag{20}$$

$$|\widehat{\gamma}_{xy}|^2 = \frac{\mathbf{Y}^H\mathbf{X}(\mathbf{X}^H\mathbf{X})^{-1}\mathbf{X}^H\mathbf{Y}}{\mathbf{Y}^H\mathbf{Y}} < g \tag{21}$$

where $\mathbb{C}$ denotes the set of the complex numbers. Integral (18) can be rearranged

$$\int_{\mathbb{C}^{LK}}\left(\int_{C_1} f'_{\mathbf{XY}}(\mathbf{X}, |\mathbf{Y}|)d\mathbf{Y}\right)d\mathbf{X} \tag{22}$$

where C_1 denotes a set of $\mathbf{Y}$ for which conditions (20) and (21) hold.

Now, let $\mathbf{X} = \mathbf{R_X}\mathbf{S_X}\mathbf{T}_X^H$ be the singular value decomposition of $\mathbf{X}$, where $\mathbf{R_X}$ is a LxL orthonormal matrix, $\mathbf{S_X}$ is a LxK diagonal matrix, and $\mathbf{T_X}$ is a KxK orthonormal matrix. We can make a substitution $\mathbf{Y} = \mathbf{R_X}\mathbf{a}$ (where $\mathbf{a} =$

$[a_1, \ldots, a_L]^T$). The numerator in (21) will then become

$$\mathbf{Y}^H \mathbf{X} (\mathbf{X}^H \mathbf{X})^{-1} \mathbf{X}^H \mathbf{Y} \tag{23}$$

$$= \mathbf{a}^H \mathbf{R}_{\mathbf{X}}^H \mathbf{R}_{\mathbf{X}} \mathbf{S}_{\mathbf{X}} \mathbf{T}_{\mathbf{X}}^H (\mathbf{T}_{\mathbf{X}} \mathbf{S}_{\mathbf{X}}^H \mathbf{R}_{\mathbf{X}}^H \mathbf{R}_{\mathbf{X}} \mathbf{S}_{\mathbf{X}} \mathbf{T}_{\mathbf{X}}^H)^{-1} \mathbf{T}_{\mathbf{X}} \mathbf{S}_{\mathbf{X}}^H \mathbf{R}_{\mathbf{X}}^H \mathbf{R}_{\mathbf{X}} \mathbf{a}$$

$$= \mathbf{a}^H \mathbf{S}_{\mathbf{X}} \mathbf{T}_{\mathbf{X}}^H \mathbf{T}_{\mathbf{X}}^{-H} (\mathbf{S}_{\mathbf{X}}^H \mathbf{S}_{\mathbf{X}})^{-1} \mathbf{T}_{\mathbf{X}}^{-1} \mathbf{T}_{\mathbf{X}} \mathbf{S}_{\mathbf{X}}^H \mathbf{a} \tag{24}$$

$$= \mathbf{a}^H \mathbf{S}_{\mathbf{X}} (\mathbf{S}_{\mathbf{X}}^H \mathbf{S}_{\mathbf{X}})^{-1} \mathbf{S}_{\mathbf{X}}^H \mathbf{a} = \mathbf{a}^H \mathbf{\Gamma} \mathbf{a} = \sum_{l=1}^{K} |a_l|^2 \tag{25}$$

where the expression $\mathbf{S}_{\mathbf{X}} (\mathbf{S}_{\mathbf{X}}^H \mathbf{S}_{\mathbf{X}})^{-1} \mathbf{S}_{\mathbf{X}}^H$ was denoted as $\mathbf{\Gamma}$, and for $\mathbf{X}$ full rank it is a diagonal matrix

$$\mathbf{\Gamma} = \mathrm{diag}\{[\underbrace{1, \ldots, 1}_{K}, 0, \ldots, 0]\}. \tag{26}$$

If $\mathbf{X}$ is rank deficient, $\mathbf{\Gamma}$ is different (some of the trailing ones get replaced by zeros), but the set of the rank deficient matrices $\mathbf{X}$ has zero measure in $\mathbb{C}^{KL}$, does not affect the integral (22), and so needs not be considered.

The denominator in (21) will be

$$\mathbf{Y}^H \mathbf{Y} = \mathbf{a}^H \mathbf{R}_{\mathbf{X}}^H \mathbf{R}_{\mathbf{X}} \mathbf{a} = \mathbf{a}^H \mathbf{a} = \sum_{l=1}^{L} |a_l|^2. \tag{27}$$

With this notation the entire condition (21) will be transformed into

$$\frac{\sum_{l=1}^{K} |a_l|^2}{\sum_{l=1}^{L} |a_l|^2} < g. \tag{28}$$

Consequently, (22) will become

$$\int_{\mathbb{C}^{LK}} \left(\int_{\mathcal{D}_1} f'_{\mathbf{XY}}(\mathbf{X}, |\mathbf{a}|) d\mathbf{a} \right) d\mathbf{X} \tag{29}$$

where $\mathcal{D}_1$ is a set of $\mathbf{a}$ for which (28) holds. Because $\mathcal{D}_1$ is independent of $\mathbf{X}$, we can now change the order of the integration, and we will get

$$\int_{\mathcal{D}_1} \left(\int_{\mathbb{C}^{LK}} f'_{\mathbf{XY}}(\mathbf{X}, |\mathbf{a}|) d\mathbf{X} \right) d\mathbf{a} = \int_{\mathcal{D}_1} f'_{\mathbf{Y}}(|\mathbf{a}|) d\mathbf{a}. \tag{30}$$

So at this point we can see that the MC estimate CDF does not depend on the distribution of $\mathbf{X}$, which means that it does not depend on the distribution of $x_i[n]$.

Now, we will show that (30) does not depend even on $f'_{\mathbf{Y}}(.)$. For this purpose we will denote

$$a_l = z_{2l-1} + j z_{2l}, \quad z_l \in \mathbb{R}, \tag{31}$$

$$\mathbf{a} = [z_1 + j z_2, z_3 + j z_4, \ldots, z_{2L-1} + j z_{2L}]^T, \tag{32}$$

$$\mathbf{z} = [z_1, \ldots, z_{2L}]^T \tag{33}$$

where $\mathbb{R}$ denotes the set of real numbers, and j is the imaginary unit. Next, we will substitute $\mathbf{z}$ with the spherical coordinates [1]

$$z_l = r \left(\prod_{k=1}^{l-1} \sin \alpha_k \right) \cos \alpha_l, \quad l = 1, \ldots, 2L-1, \tag{34}$$

$$z_{2L} = r \prod_{k=1}^{2L-1} \sin \alpha_k. \tag{35}$$

This substitution can be expressed in a simpler form as

$$z_l = r c_l(\boldsymbol{\alpha}), \text{ where } \boldsymbol{\alpha} = [\alpha_1, \ldots, \alpha_{2L-1}]^T \tag{36}$$

where $c_l(.)$ is just a shorthand notation for the products of the trigonometric functions in (34) and (35). Also, note that we have $|\mathbf{a}| = |\mathbf{z}| = r$. Assuming that r is greater than zero, the inequality in (28) will become

$$\frac{\sum_{l=1}^{K} |a_l|^2}{\sum_{l=1}^{L} |a_l|^2} = \frac{\sum_{l=1}^{2K} z_l^2}{|\mathbf{a}|^2} = \frac{\sum_{l=1}^{2K} r^2 c_l^2(\boldsymbol{\alpha})}{r^2} = \sum_{l=1}^{2K} c_l^2(\boldsymbol{\alpha}) < g, \tag{37}$$

which does not depend on r, and therefore in our integration r will not be bound by any condition (it will range from 0 to ∞).

The Jacobian of substitution (34), (35) is [1]

$$J = r^{2L-1} \prod_{k=1}^{2L-2} \sin^{2K-1-k} \alpha_k, \tag{38}$$

which we abbreviate by

$$J = h(\boldsymbol{\alpha}) r^{2L-1}. \tag{39}$$

Integral (30) thus becomes

$$\int_{\mathcal{G}} \left(\int_0^{\infty} f'_{\mathbf{Y}}(r) h(\boldsymbol{\alpha}) r^{2L-1} dr \right) d\boldsymbol{\alpha} \tag{40}$$

where $\mathcal{G}$ is a properly constructed set of values of $\boldsymbol{\alpha}$, which we do not need to express here. This expression can be rearranged into

$$\int_{\mathcal{G}} h(\boldsymbol{\alpha}) d\boldsymbol{\alpha} \int_0^{\infty} f'_{\mathbf{Y}}(r) r^{2L-1} dr. \tag{41}$$

This formula expresses $G_0(g)$ in a form that allows to show its invariance on $f'_{\mathbf{Y}}(.)$. Note that the second integral does not depend on g, and will evaluate into a scaling constant (which for any $f'_{\mathbf{Y}}(.)$ must be such that $G_0(g) = 1$ for $g > 1$, because MC estimate lies within interval $\langle 0, 1 \rangle$, and $G_0(g)$ is a CDF). The shape of $G_0(g)$ is therefore given solely by the first integral, which does not depend on $f'_{\mathbf{Y}}(.)$. Consequently, (41) is invariant to $f'_{\mathbf{Y}}(.)$, which means that $G_0(g)$ does not depend on the distribution of $y[n]$.

2.3.2 Rotational Symmetry of $f_{\mathbf{XY}}(\mathbf{X}, \mathbf{Y})$ with Respect to Rotation of $\mathbf{X}$

In this subsection we will assume that $f_{\mathbf{XY}}(\mathbf{X}, \mathbf{Y}) = f_{\mathbf{XY}}(\mathbf{BX}, \mathbf{Y})$ if $\mathbf{B}$ is a rotation. This means $f_{\mathbf{XY}}(\mathbf{X}, \mathbf{Y})$ and $f_{\mathbf{X}}(\mathbf{X})$ can also be expressed as $f_{\mathbf{XY}}(\mathbf{X}, \mathbf{Y}) = f''_{\mathbf{XY}}(|\mathbf{X}_1|, \ldots, |\mathbf{X}_K|, \mathbf{Y})$ and $f_{\mathbf{X}}(\mathbf{X}) = f'_{\mathbf{X}}(|\mathbf{X}_1|, \ldots, |\mathbf{X}_K|)$, where $f''_{\mathbf{XY}}$ and $f'_{\mathbf{X}}$ are new functions, and $\mathbf{X}_i$ denotes the i-th column of $\mathbf{X}$.

With these assumptions MC estimate CDF (17) can be expressed as

$$G_0(g) = \int_{\mathcal{C}} f''_{\mathbf{XY}}(|\mathbf{X}_1|, \ldots, |\mathbf{X}_K|, \mathbf{Y}) d\mathbf{X} d\mathbf{Y}. \quad (42)$$

This integral can be rearranged into

$$\int_{\mathbb{C}^L} \left(\int_{\mathcal{C}_2} f''_{\mathbf{XY}}(|\mathbf{X}_1|, \ldots, |\mathbf{X}_K|, \mathbf{Y}) d\mathbf{X} \right) d\mathbf{Y} \quad (43)$$

where $\mathcal{C}_2$ denotes the set of $\mathbf{X}$, for which conditions (19) and (21) hold.

Now, we will use steps somewhat similar to those used in the previous section.

Let $\mathbf{Y} = \mathbf{Q_Y} \mathbf{R_Y}$ be the QR decomposition of $\mathbf{Y}$, where $\mathbf{Q_Y}$ is a $L \times L$ orthonormal matrix, and $\mathbf{R_Y} = [\sigma, 0, \ldots, 0]^T$. We will make a substitution $\mathbf{X} = \mathbf{Q_Y} \mathbf{A}$ (where $\mathbf{A} = [a_{li}]_{li}$ is a $L \times K$ matrix). The numerator in (21) will thus become

$$\mathbf{Y}^H \mathbf{X} (\mathbf{X}^H \mathbf{X})^{-1} \mathbf{X}^H \mathbf{Y} \quad (44)$$
$$= \mathbf{R_Y^H} \mathbf{Q_Y^H} \mathbf{Q_Y} \mathbf{A} (\mathbf{A}^H \mathbf{Q_Y^H} \mathbf{Q_Y} \mathbf{A})^{-1} \mathbf{A}^H \mathbf{Q_Y^H} \mathbf{Q_Y} \mathbf{R_Y} \quad (45)$$
$$= \mathbf{R_Y^H} \mathbf{A} (\mathbf{A}^H \mathbf{A})^{-1} \mathbf{A}^H \mathbf{R_Y} \quad (46)$$
$$= {}_1\mathbf{A} (\mathbf{A}^H \mathbf{A})^{-1} {}_1\mathbf{A}^H |\sigma|^2 \quad (47)$$

where ${}_1\mathbf{A}$ denotes the first row of $\mathbf{A}$. The denominator in (21) will be

$$\mathbf{Y}^H \mathbf{Y} = \mathbf{R_Y^H} \mathbf{Q_Y^H} \mathbf{Q_Y} \mathbf{R_Y} = \mathbf{R_Y^H} \mathbf{R_Y} = |\sigma|^2. \quad (48)$$

Thus, the entire condition (21) will transform into

$${}_1\mathbf{A} (\mathbf{A}^H \mathbf{A})^{-1} {}_1\mathbf{A}^H < g. \quad (49)$$

Consequently, (43) will become

$$\int_{\mathbb{C}^L} \left(\int_{\mathcal{D}_2} f''_{\mathbf{XY}}(|\mathbf{A}_1|, \ldots, |\mathbf{A}_K|, \mathbf{Y}) d\mathbf{A} \right) d\mathbf{Y} \quad (50)$$

where $\mathbf{A}_i$ denotes the i-th column of $\mathbf{A}$ (note that $|\mathbf{X}_i| = |\mathbf{A}_i|$ due to the orthonormality of $\mathbf{Q_Y}$), and $\mathcal{D}_2$ is a set of $\mathbf{A}$ for which (49) holds.

Because $\mathcal{D}_2$ is independent of $\mathbf{Y}$, we can now change the order of the integration, and we will get

$$\int_{\mathcal{D}_2} \left(\int_{\mathbb{C}^L} f''_{\mathbf{XY}}(|\mathbf{A}_1|, \ldots, |\mathbf{A}_K|, \mathbf{Y}) d\mathbf{Y} \right) d\mathbf{A}$$
$$= \int_{\mathcal{D}_2} f'_{\mathbf{X}}(|\mathbf{A}_1|, \ldots, |\mathbf{A}_K|) d\mathbf{A}. \quad (51)$$

So at this point we can see that the MC estimate CDF does not depend on the distribution of $\mathbf{Y}$, which means it does not depend on the distribution of $y[n]$.

Now, we will show that (51) does not depend even on $f_{\mathbf{X}}(.)$. For this purpose we will denote

$$a_{li} = z_{2l-1,i} + jz_{2l,i}, \quad z_{li} \in \mathbb{R}, \quad (52)$$
$$\mathbf{A} = [z_{2l-1,i} + jz_{2l,i}]_{\substack{l=1\ldots L \\ i=1\ldots K}}, \quad (53)$$
$$\mathbf{Z} = [z_{li}]_{\substack{l=1\ldots 2L \\ i=1\ldots K}}. \quad (54)$$

Next, we will substitute $\mathbf{Z}$ with the spherical coordinates

$$z_{li} = r_i \left(\prod_{k=1}^{l-1} \sin \alpha_{ki} \right) \cos \alpha_{li}, \quad \begin{array}{l} l = 1, \ldots, 2L-1 \\ i = 1, \ldots, K \end{array}, \quad (55)$$
$$z_{2L,i} = r_i \prod_{k=1}^{2L-1} \sin \alpha_{ki}, \quad i = 1, \ldots, K. \quad (56)$$

This substitution can be expressed in a simpler form as

$$z_{li} = r_i c'_l(\boldsymbol{\alpha}_i), \text{ where } \boldsymbol{\alpha}_i = [\alpha_{1,i}, \ldots, \alpha_{2L-1,i}]^T \quad (57)$$

where $c'_l(.)$ is just a shorthand notation for the product of the trigonometric functions in (55) and (56). (57) can also be written as

$$\mathbf{Z} = \mathbf{CR}, \text{ where} \quad (58)$$
$$\mathbf{C} = [c'_l(\boldsymbol{\alpha}_i)]_{\substack{l=1\ldots 2L \\ i=1\ldots K}}, \quad \mathbf{R} = \text{diag}\{[r_1, \ldots, r_K]\}. \quad (59)$$

Further, note that using $L \times L$ identity matrices $\mathbf{I}_L$, the matrix $\mathbf{A}$ can be expressed as

$$\mathbf{A} = [\mathbf{I}_L, j\mathbf{I}_L] \mathbf{Z} = [\mathbf{I}_L, j\mathbf{I}_L] \mathbf{CR} = \mathbf{C}_c \mathbf{R}, \quad (60)$$

where we denoted $\mathbf{C}_c = [\mathbf{I}_L, j\mathbf{I}_L] \mathbf{C}$. Thus, the condition in (49) will become

$${}_1\mathbf{A} (\mathbf{A}^H \mathbf{A})^{-1} {}_1\mathbf{A}^H = {}_1\mathbf{C}_c \mathbf{R} (\mathbf{R}^H \mathbf{C}_c^H \mathbf{C}_c \mathbf{R})^{-1} \mathbf{R}^H {}_1\mathbf{C}_c^H \quad (61)$$
$$= {}_1\mathbf{C}_c \mathbf{R} \mathbf{R}^{-1} (\mathbf{C}_c^H \mathbf{C}_c)^{-1} \mathbf{R}^{-H} \mathbf{R}^H {}_1\mathbf{C}_c^H \quad (62)$$
$$= {}_1\mathbf{C}_c (\mathbf{C}_c^H \mathbf{C}_c)^{-1} {}_1\mathbf{C}_c^H < g \quad (63)$$

where ${}_1\mathbf{C}_c$ denotes the first row of $\mathbf{C}_c$, and r_i were assumed to be greater than zero. Note that (63) is independent of r_i, and therefore in our integration r_i will not be bound by any condition (it will range from 0 to ∞). Also, note that we have $|\mathbf{A}_i| = |\mathbf{Z}_i| = r_i$, for $i = 1, \ldots, K$ (where $\mathbf{Z}_i$ denotes the i-th column of $\mathbf{Z}$).

The Jacobian of substitution (55), (56) is [1]

$$J = \prod_{i=1}^K \left(r_i^{2L-1} \prod_{k=1}^{2L-2} \sin^{2L-1-k} \alpha_{ki} \right) \quad (64)$$
$$= \left(\prod_{i=1}^K r_i^{2L-1} \right) \left(\prod_{i=1}^K \prod_{k=1}^{2L-2} \sin^{2L-1-k} \alpha_{ki} \right), \quad (65)$$

which we abbreviate by

$$J = h'(\boldsymbol{\alpha}) \prod_{i=1}^K r_i^{2L-1} \quad (66)$$

where $h'(\boldsymbol{\alpha})$ is just a shorthand notation for the expression in the second bracket in (65). Integral (51) thus becomes

$$\int_{\mathcal{E}} \left(\int_0^\infty \cdots \int_0^\infty f_{\mathbf{X}}'(r_1,\ldots,r_K) h'(\boldsymbol{\alpha}) \prod_{i=1}^K r_i^{2L-1} dr_1 \ldots dr_K \right) d\boldsymbol{\alpha} \tag{67}$$

where $\mathcal{E}$ is a properly constructed set of values of $\boldsymbol{\alpha}$, which we do not need to express here. This integral can be rearranged into

$$\int_{\mathcal{E}} h'(\boldsymbol{\alpha}) d\boldsymbol{\alpha} \int_0^\infty \cdots \int_0^\infty f_{\mathbf{X}}'(r_1,\ldots,r_K) \prod_{i=1}^K r_i^{2L-1} dr_1 \ldots dr_K . \tag{68}$$

This formula expresses $G_0(g)$ in a form that allows to show its invariance on $f_{\mathbf{X}}'(.)$. Note that all the integrals except the first one do not depend on g, and will evaluate into a scaling constant, which for any $f_{\mathbf{X}}'(.)$ must be such that $G_0(g) = 1$ for $g > 1$ (because $G_0(g)$ is a CDF). The shape of $G_0(g)$ is therefore given solely by the first integral in (68), which does not depend on $f_{\mathbf{X}}'(.)$. Consequently, (68) is invariant to $f_{\mathbf{X}}'(.)$, which means that $G_0(g)$ does not depend on the distribution of $x_i[n]$.

2.3.3 Alternative Rotational Symmetry in Real and Imaginary Parts of X or Y

If we denote

$$X_{l,i} = U_{2l-1,i} + jU_{2l,i}, \quad Y_l = V_{2l-1} + jV_{2l} \tag{69}$$

$$\mathbf{U} = [U_{li}]_{\substack{l=1\ldots2L \\ i=1\ldots K}}, \quad \mathbf{V} = [V_1,\ldots,V_{2L}]^T \tag{70}$$

we have

$$f_{\mathbf{XY}}(\mathbf{X},\mathbf{Y}) = f_{\mathbf{UV}}(\mathbf{U},\mathbf{V}) \tag{71}$$

where $f_{\mathbf{UV}}(\mathbf{U},\mathbf{V})$ is the joint PDF of $\mathbf{U}$ and $\mathbf{V}$. We also have $|\mathbf{Y}| = |\mathbf{V}|$ and $|\mathbf{X}_l| = |\mathbf{U}_l|$ (where $\mathbf{U}_l$ denotes the l-th column of $\mathbf{U}$).

Now, if $f_{\mathbf{UV}}(\mathbf{U},\mathbf{V}) = f_{\mathbf{UV}}(\mathbf{U},\mathbf{B}_r\mathbf{V})$, where $\mathbf{B}_r$ is a real rotation matrix, we can write

$$f_{\mathbf{XY}}(\mathbf{X},\mathbf{Y}) = f_{\mathbf{UV}}(\mathbf{U},\mathbf{V}) = f_{\mathbf{UV}}'(\mathbf{U},|\mathbf{V}|) = f_{\mathbf{UV}}'(\mathbf{U},|\mathbf{Y}|) . \tag{72}$$

Consequently, (keeping in mind that the L_2 norm is invariant to a rotation) if $f_{\mathbf{UV}}(\mathbf{U},\mathbf{V})$ is rotationally symmetric with respect to the rotation of $\mathbf{V}$, then $f_{\mathbf{XY}}(\mathbf{X},\mathbf{Y})$ is rotationally symmetric with respect to the rotation of $\mathbf{Y}$.

A similar argument applies to the rotational symmetry in $\mathbf{U}$. If $f_{\mathbf{UV}}(\mathbf{U},\mathbf{V}) = f_{\mathbf{UV}}(\mathbf{B}_r\mathbf{U},\mathbf{V})$, we can write

$$f_{\mathbf{XY}}(\mathbf{X},\mathbf{Y}) = f_{\mathbf{UV}}(\mathbf{U},\mathbf{V}) = f_{\mathbf{UV}}'(|\mathbf{U}_1|,\ldots,|\mathbf{U}_K|,\mathbf{V})$$
$$= f_{\mathbf{UV}}'(|\mathbf{X}_1|,\ldots,|\mathbf{X}_K|,\mathbf{Y}), \tag{73}$$

which shows that $f_{\mathbf{XY}}(\mathbf{X},\mathbf{Y})$ is rotationally symmetric with respect to the rotation of $\mathbf{X}$.

Consequently, instead of the rotational symmetry of $f_{\mathbf{XY}}(\mathbf{X},\mathbf{Y})$ with respect to the rotation of $\mathbf{X}$ or $\mathbf{Y}$, we could require the rotational symmetry of $f_{\mathbf{UV}}(\mathbf{U},\mathbf{V})$ with respect to the rotation of $\mathbf{U}$ or $\mathbf{V}$.

2.3.4 Explicit Expression for the Null Distribution

The abovementioned rotational symmetries will be achieved at a given Ω if the real and imaginary parts of $\mathbf{X}(\Omega)$ and $\mathbf{Y}(\Omega)$ are zero mean independent Gaussian with equal variance.

Consequently, the explicit expression for the null distribution is equal to the one derived for this kind of Gaussian random variables (e.g. in [7])

$$G_0(g) = Beta(g,K,L) \tag{74}$$

where $Beta(.,K,L)$ denotes the CDF of the Beta distribution with parameters K and L.

2.3.5 Necessity of Null Assumption

The assumption that $|\gamma(\Omega)|^2 = 0$ did not actually enter the derivation in Sec. 2.3.2 and Sec. 2.3.1. However, our results are still limited to the null distribution of the MC estimate.

Note that if $x_i[n]$ and $y[n]$ are $M \cdot L$ samples long records of wide sense stationary signals, the number of signals K is fixed and the number of segments L tends to infinity, then CDF (74) limits to the Heaviside step function. This means that the MC estimate limits to zero. This can happen only if the cross spectral density estimates in (13) limits to zero. Because (13) is an unbiased estimator of the cross spectral density, this can happen only if the true cross spectral densities of $x_i[n]$ and $y[n]$ are zero, in which case the true MC is also zero. Therefore, the MC estimate can have distribution (74) only if the true MC of $x_i[n]$ and $y[n]$ is zero. Since we have showed that our assumption of the rotational symmetries provides the MC estimate with distribution (74), our assumption also implies that the true MC is zero. Therefore, our results are applicable only for the null distribution of the MC estimate.

3. Conclusions

3.1 Results and Corollaries

When the multiple coherence estimate is computed between a group of signals $x_i[n]$, $i = 1,\ldots,K$ and a signal $y[n]$, these signals are segmented into L segments $x_{li}[n]$ and $y_l[n]$, and $X_{li}(\Omega)$, $Y_l(\Omega)$ are the discrete Fourier transforms of the individual segments, then the null distribution of the multiple coherence estimate at a given frequency Ω is invariant to the distribution of $x_i[n]$ and $y[n]$ if the joint distribution the spectra of the segments has either one of the following rotational symmetries:

i) $f_{\mathbf{XY}}(\mathbf{X},\mathbf{Y}) = f_{\mathbf{XY}}(\mathbf{BX},\mathbf{Y})$

ii) $f_{\mathbf{XY}}(\mathbf{X},\mathbf{Y}) = f_{\mathbf{XY}}(\mathbf{X},\mathbf{BY})$

where $\mathbf{B}$ is a rotation matrix ($\det(\mathbf{B}) = 1$ and $\mathbf{B}^H = \mathbf{B}^{-1}$), $\mathbf{X} = [X_{li}(\Omega)]_{\substack{l=1\ldots L \\ i=1\ldots K}}$, $\mathbf{Y} = [Y_1(\Omega),\ldots,Y_L(\Omega)]^T$, and

$f_{\mathbf{XY}}(\mathbf{X}, \mathbf{Y})$ is the joint PDF of $\mathbf{X}$ and $\mathbf{Y}$.

A less general corollary of this finding is that the null distribution of the multiple coherence estimate will be independent of the distribution of $x_i[n]$ and $y[n]$ if $x_i[n]$ and $y[n]$ are independent for each $i = 1,\ldots,K$,[1] and if at least one of the following conditions hold

i) the distribution of $[X_{1i}(\Omega),\ldots,X_{Li}(\Omega)]^T$ is rotationally symmetric for each i and a given Ω,

ii) the distribution of $[Y_1(\Omega),\ldots,Y_L(\Omega)]^T$ is rotationally symmetric for a given Ω.

A simpler (but even less general) corollary is that the null distribution of the multiple coherence estimate will be independent of the distribution of $x_i[n]$ if $y[n]$ is Gaussian and independent of $x_i[n]$, *or* the null distribution of MC will be independent of the distribution of $y[n]$ if $x_i[n]$ are multivariate Gaussian and independent of $y[n]$.

3.2 Hypothesis Testing

Note that even though we have provided our derivation for a fairly general assumption of the rotational symmetries of $f_{\mathbf{XY}}(\mathbf{X}, \mathbf{Y})$, we do not imply that these rotational symmetries should be used directly as a null hypothesis in the hypothesis testing. Assuming these rotational symmetries automatically implies the fact that the true MC is equal to zero (Sec. 2.3.5). For the hypothesis testing, we recommend to use one of the corollaries stated above. For example, the hypothesis testing can be performed in the following manner:

Prior knowledge: the distribution of $[X_{1i}(\Omega),\ldots,X_{Li}(\Omega)]^T$ is rotationally symmetric for each i, *or* the distribution of $[Y_1(\Omega),\ldots,Y_L(\Omega)]^T$ is rotationally symmetric.

Null hypothesis: $x_i[n]$ and $y[n]$ are independent for each $i = 1,\ldots,K$,

If the null hypothesis holds, than the MC estimate should have distribution (74). Now, if the MC estimated from the data exceeds a chosen quantile of this distribution the null hypothesis appears unlikely, and we can choose to reject it, and conclude that the signals are most likely dependent.

3.3 Final Remarks

The significance of this result lies in the improvement of the multiple coherence analysis. Hitherto, the applicability of the formula for the null distribution (74) was guaranteed only if all the examined signals were multivariate Gaussian. With the findings presented in this paper, it will be possible to apply these formulas to a much wider class of signals. This, we believe, will be fairly beneficial in numerous fields that use the multiple coherence for signal detection or examination of relationship between signals.

Acknowledgements

This work was supported by the research grant Spatial Filtering Techniques for High Density Electroencephalographic Recordings GAČR P102/11/P109.

References

[1] BLUMENSON, L. E. A derivation of *n*-dimensional spherical coordinates. *The Americal Mathematical Montly*, 1960, vol. 67, no. 1, p. 63 - 66.

[2] BRILLINGER, D. R. *Time Series: Data Analysis and Theory*. Society for Industrial and Applied Mathematics, 2001.

[3] FELIX, L. B., MIRANDA DE SÁ, A. M. F. L., INFANTOSI, A. F. C., YEHIA, H. C. Multivariate objective response detectors (MORD): statistical tools for multichanel EEG analysis during rhytmic stimulation. *Annals of Biomedical Engineering*, 2007, vol. 35, no. 3, p. 443 - 452.

[4] GISH, H., COCHRAN, D. Invariance of the magnitude-squared cohrence estimate with respect to second-channel statistics. *IEEE Transactions on Acoustics, Speech, and Signal Processing*, 1987, vol. 35, no. 12, p. 1774 - 1776.

[5] GOODMAN, N. R. Statistical analysis based on a certain multivariate complex Gaussian distribution (an introduction). *Annals of Mathematical Statistics*, 1963, vol. 34, no. 1, p. 152 - 177.

[6] JAMES, A. T. Distributions of matrix variates and latent roots derived from normal samples. *Annals of Mathematical Statistics*, 1964, vol. 35, no. 2, p. 475 - 501.

[7] KHARTI, C. G. Classical statistical analysis based on a certain multivariate complex Gaussian distribution. *Annals of Mathematical Statistics*, 1965, vol. 36, no. 1, p. 98 - 114.

[8] MIRANDA DE SÁ, A. M. F. L., FELIX, L. B., INFANTOSI, A. F. C. A matrix-based algorithm for estimating multiple coherence of a periodic signal and its application to the multichannel EEG during sensory stimulation. *IEEE Transactions on Biomedical Engineering*, 2004, vol. 51, no. 7, p. 1140 - 1146.

[9] MIRANDA DE SÁ, A. M. F. L., INFANTOSI, A. F. C., MELGES, D. B. A multiple coherence-based detector for evoked responses in the EEG during sensory stimulation. In *30th Annual International IEEE EMBS Conference*. Vancouver (British Columbia, Canada), 2008, p. 3516 - 3519.

[10] MOSKALSKI, S., TORRES, R. Influences of tides, weather, and discharge on suspended sediment concentration. *Continental Shelf Research*, 2012, vol. 37, p. 36 - 45.

[11] NEDUNGADI, A. G., DING, M., RANGARAJAN, G. Block coherence: a method for measuring the interdependence between two block of neurobiological time series. *Biological Cybernetics*, 2011, vol. 104, no. 3, p. 197 - 207.

[12] NUTTALL, A. H. Invariance of distribution of coherence estimate to second-channel statistics. *IEEE Transactions on Acoustics, Speech, and Signal Processing*, 1981, vol. 29, no. 1, p. 120 - 122.

---

[1] In this case the MC estimate has the null distribution at all frequencies.

[13] PANERAI, R. B., EAMES, P. J., POTTER, J. F. Multiple coherence of cerebral blood flow velocity in humans. *American Journal of Physiology - Heart and Circulatory Physiology*, 2006, vol. 291, no. 1, p. H251 - H259.

[14] PENG, T., ROWLEY, A. B., AINSLIE, P. N., POULIN, M. J., PAYNE, S. J. Multivariate system identification for cerebral autoregulation. *Annals of Biomedical Engineering*, 2008, vol. 36, no. 2, p. 308 - 320.

[15] RICHARDS, T. C. Dynamic testing of A/D converters using the multiple coherence function. *IEEE Transactions on Instrumentation and Measurement*, 2008, vol. 57, no. 11, p. 2596 - 2607.

[16] RICHARDS, T. C. Dynamic testing of data acquisition channels using the multiple coherence function. KARAKEHAYOV, Z. (Ed.) *Data Acquisition Applications*. Intech, 2012, p. 51 - 78.

[17] SALVADOR, R. A simple view of the brain through a frequency-specific functional connectivity measure. *NeuroImage*, 2008, vol. 39, no. 1, p. 279 - 289.

[18] SALVADOR, R., ANGUERA, M., GOMAR, J. J., BULLMORE, E. T., POMAROL-CLOTET, E. Conditional mutual information maps as descriptors of net connectivity levels in the brain. *Frontiers in Neuroinformatics*, 2010, vol. 4, no. 115.

[19] TRUEBLOOD, R. D., ALSPACH, D. L. Multiple coherence. In *11th Asilomar Conference on Circuits, Systems and Computers*. Pacific Grove (CA, USA), 1977, p. 327 - 333.

[20] TRUEBLOOD, R. D., ALSPACH, D. L. Multiple coherence as a detection statistic. *NOSC Technical Report 265*. San Diego: Naval Ocean Systems Center, 1978.

21

A Novel Multi-permittivity Cylindrical Dielectric Resonator Antenna for Wideband Applications

Ubaid ULLAH[1], Mohd Fadzil AIN[1], Mohamadariff OTHMAN[1], Ihsan ZUBIR[1], Nor Muzlifah MAHYUDDIN[1], Zainal Ariffin AHMAD[2], Mohd Zaid ABDULLAH[1]

[1]School of Electrical and Electronic Engineering, University Sains Malaysia, 14300 Nibong Tebal Pulau Pinang
[2] School of Material and Mineral Resource Engineering, University Sains Malaysia, 14300 Nibong Tebal Pulau Pinang

xs2ubaid@gmail.com, ee.mfadzil@usm.my, andikalusia83@yahoo.com, ihsan_zubir@yahoo.com.my, ee.muzlifah@usm.my, zainal@usm.my, mza@usm.my

Abstract. *In this paper, a novel multi-permittivity cylindrical dielectric resonator antenna for wideband application is presented. The multi-permittivity cylinder is formed by combining two different permittivity material sectors in such a way that each sector (with constant permittivity) is 90 degree apart. A direct microstrip line coupling terminated with T-stub at the open end is used to excite the multi-permittivity cylindrical dielectric resonator. The angular position of the multi sector dielectric resonator with respect to the longitudinal axis of the microstrip line and length of the additional strip at the open end of the feeding circuit is key parameters for wideband operation of the antenna. By optimizing all parameters of the proposed antenna, wideband impedance bandwidth of 56 % (12.1 GHz to 21.65 GHz) is achieved. The average gain of the antenna throughout the bandwidth is 5.9 dB with good radiation properties in both E-plane and H-plane. A well matched simulation and experimental results show that the antenna is suitable for wideband applications.*

Keywords

Dielectric resonator antenna (DRA), multi-permittivity dielectric resonator, wideband antenna.

1. Introduction

There is fast growing demand for modern day communication systems for wireless and radar applications to be employed in a wider range of frequency, which signifies the importance of wideband antenna. In the recent past abundant amount of research is carried out in the field of wideband antenna and tremendous development is made [1-5]. Since the emergence of dielectric resonator antennas (DRAs) back in 1983, it has been studied extensively by different researchers and vast numbers of articles are published covering different aspects of DRA i.e. low profile structures, excitation of DRA using different feeding schemes, compactness, polarization, theoretical and mathematical analysis, wideband operation and array DRAs. It has been proven that DRA offers high radiation efficiency, small size, wide band operation, flexible excitation mechanism and ease of fabrication. A very well written review on DRA can be noted in [6] which comprehensively address almost all features of DRA. In this paper, the attention of the readers is drawn towards wideband DRA. Bandwidth of DRA is mainly controlled by the dielectric constant of the material. The relation between DRA and dielectric constant is inverse, which means DRA with low permittivity will have wider bandwidth and hence lower radiation efficiency. Several wideband DRAs are reported in the literature, in which wideband performance of the DRA is achieved by manipulating different shapes of the dielectric resonators, using various feeding schemes, multiple dielectric resonators, modified DRA structures, hybrid designs, and by exciting DRA in multiple modes [7-10]. A comparison is done between previously published works in terms of antenna maximum dimension, type of feeder employed and impedance bandwidth, summary is given in Tab. 1.

In this paper, a multi-permittivity cylindrical dielectric resonator antenna (MCDRA) consists of four sectors, excited by a modified 50 Ω microstrip line on top of the small conducting ground plane is presented. The cylindrical dielectric resonator is formed by placing four 90° pie shape sectors in such a way that, two similar permittivity sectors are positioned in non-adjacent quadrant. With this setup, more than 50% impedance bandwidth is achieved with a single element DRA. The proposed design is simulated using computer simulation technology (CST 2014) and verified using high frequency structure simulator (HFSS). To validate the design in a real world the antenna prototype is fabricated and characterized. A close agreement between simulation and experimental results is obtained. In the following section configuration of the proposed antenna is described, followed by a parametric study and subsequently simulated and measured results are presented.

DRA Type	Dimensions (mm)	Feed type	Resonant frequency (GHz)	Impedance bandwidth (%)	Refrences
Rectangular	50 x 50	Aperture	5.8	20	[1]
Rectangular	50 x 50	Slot	2.4	28.6	[11]
Cylindrical	115 x 115	Microstrip fed	2.35	14.65	[12]
Rectangular	150 x 150	Differently feed	2.4	10.4	[13]
Bowtie	60 x 60	Coaxial probe	5.5	49.4	[14]
Rectangular	30 x 30	Coaxial probe	3.4	25	[15]
Half Cylinder	140 x 110	Microstrip Line	2.4	7.45	[16]

Tab. 1. Summary of some of the selected dielectric resonator antennas.

2. Antenna Geometry and Design Configuration

Configuration and prototype of the proposed MCDRA are shown in Fig. 1. The corresponding dimensions of the geometry shown in Fig. 1 are defined as: a = radius of the multi-permittivity cylinder, d = depth of cylinder, L_m = length of microstrip line, W_m = width of microstrip line, s = length of microstrip line stub, Ls = length of substrate, Ws = width of substrate, ε_s = permittivity of substrate, ε_{r1} = high permittivity sectors, ε_{r2} = low permittivity sectors and θ is the angle of the position of MCDRA with respect to the longitudinal axis of the microstrip feed line. As shown in Fig. 1. A multi-permittivity cylindrical dielectric resonator is loaded over a modified microstrip line, placed on top of a small conducting ground plane. The antenna is designed and analyzed in CST® which utilizes finite integration technique in the time domain. Subsequently the design is verified in HFSS® which employs finite element method in the frequency domain. It is observed that the impedance matching of the proposed antenna is vastly dependent on the angular position 'θ' of the multi-permittivity cylinder with respect to the longitudinal axis (V-axis) of the microstrip feed line as shown in Fig. 1(b).

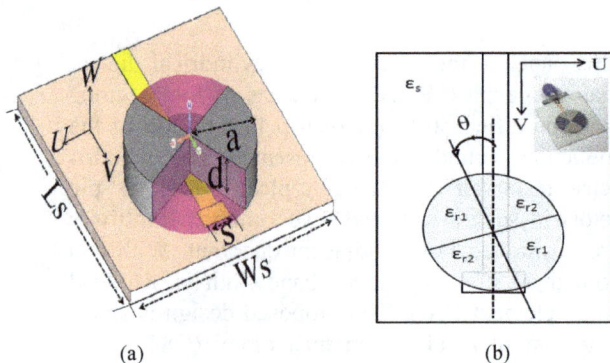

(a)   (b)

Fig. 1. Illustration of the proposed wideband ECDRA: (a) perspective view, (b) top view.

Further improvement in the impedance bandwidth is attributed to the addition of stub at the open end of the microstrip line, adjusting length 's' of the stub and the position of the stub along the U-axis of the feed line. Assortment of an appropriate combination of different permittivity sectors also plays a vital role in achieving wideband operation of the MCDRA. In the next section, a parametric study is done to evaluate the effect of varying the aforementioned parameters on the impedance bandwidth of the antenna.

3. Parametric Study of MCDRA

Parametric study of the proposed MCDRA is carried out in CST to achieve optimum values of all the parameters for practical design of the antenna. As mentioned in the preceding section the angular position 'θ' illustrated in Fig. 1(b) of the multi-permittivity DRA with respect to the longitudinal axis of the microstrip feed line plays an important role in accomplishing wideband MCDRA. Initially the multi-permittivity cylindrical resonator is rotated in an anticlockwise direction with 10° angle step size and the resulted S_{11} is evaluated. Afterwards the step size is further reduced to fine tune the impedance bandwidth of the antenna. Figure 2 shows variations in impedance bandwidth of the antenna relative to the change in angle theta 'θ' and length 's' of the additional stub of the microstrip line. Strength of electric field and magnetic field inside MCDRA is evaluated with each transformation of angle and length of the stub. As high permittivity material has a high quality factor and therefore can store more of the electrometric energy inside, this helps in achieving a strong coupling to the feeding structure. While low permittivity resonator has the tendency to operate in wider bandwidth, hence it is important to position the resonator in such a way that maximum field strength is achieved in all sectors of the MCDRA. By placing the multi-permittivity cylinder at the position where θ = 27.5° the resonator is excited in an asymmetric manner and maximum bandwidth is achieved. The optimum values of all the parameters for which the optimum performance of the antenna in terms of impedance bandwidth is achieved are listed in Tab. 2.

Parameter	Value	Parameter	Value
a	7 mm	W_m	1.898 mm
d	2.5 mm	ε_s	3.35
s	3 mm	ε_{r1}	15
Ls, Ws	25, 24 mm	ε_{r2}	10.2
L_m	20 mm	θ	27.5

Tab. 2. Optimized value for multi-permittivity cylindrical DRA.

(a)

(b)

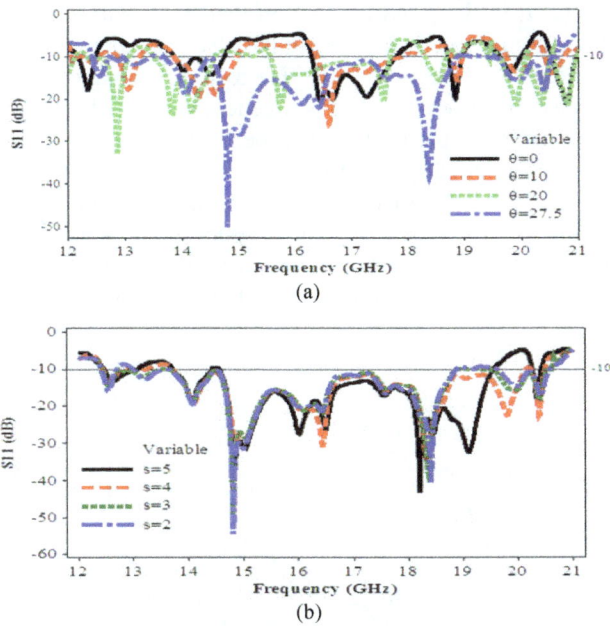

Fig. 2. Illustration of impedance bandwidth and VSWR of the antenna: (a) S_{11} for different values of angle (theta), (b) S_{11} for various values of length of strip 's'.

4. Results and Discussion

The proposed antenna is simulated using two commercially used electromagnetic simulation software's to confirm the wide band operation of the antenna in both time domain and frequency domain. To further validate the simulation results a prototype of the antenna shown in Fig. 1(b) is fabricated and measured. As mentioned earlier, two different permittivity pie shape sectors are placed adjacent to form a cylinder. The high permittivity sector is of magnesium titanium oxide doped with cobalt ($MgTiO_3$ + Co) to achieve epsilon of 15. For lower permittivity sector, Rogers RT6010 substrate is used with epsilon of 10.2. Comparison of simulated and measured impedance bandwidth of the antenna is illustrated in Fig. 3. The simulated impedance bandwidth ranges from 12.3 GHz to 20.7 GHz which is almost 51 %. The measured impedance bandwidth increases considerably to 56 % (12.1 GHz - 21.65 GHz), which is possibly because of the small air gaps that remain in the fabricated multi-permittivity cylindrical dielectric resonator.

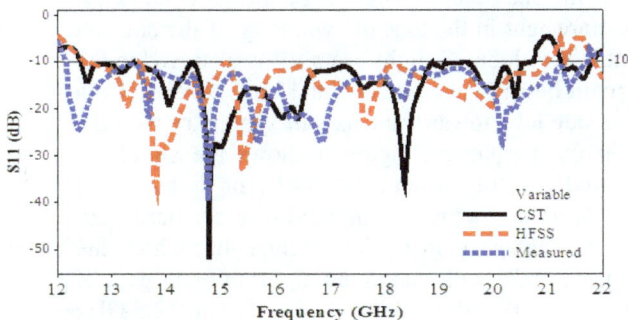

Fig. 3. Optimized impedance bandwidth of the MCDRA.

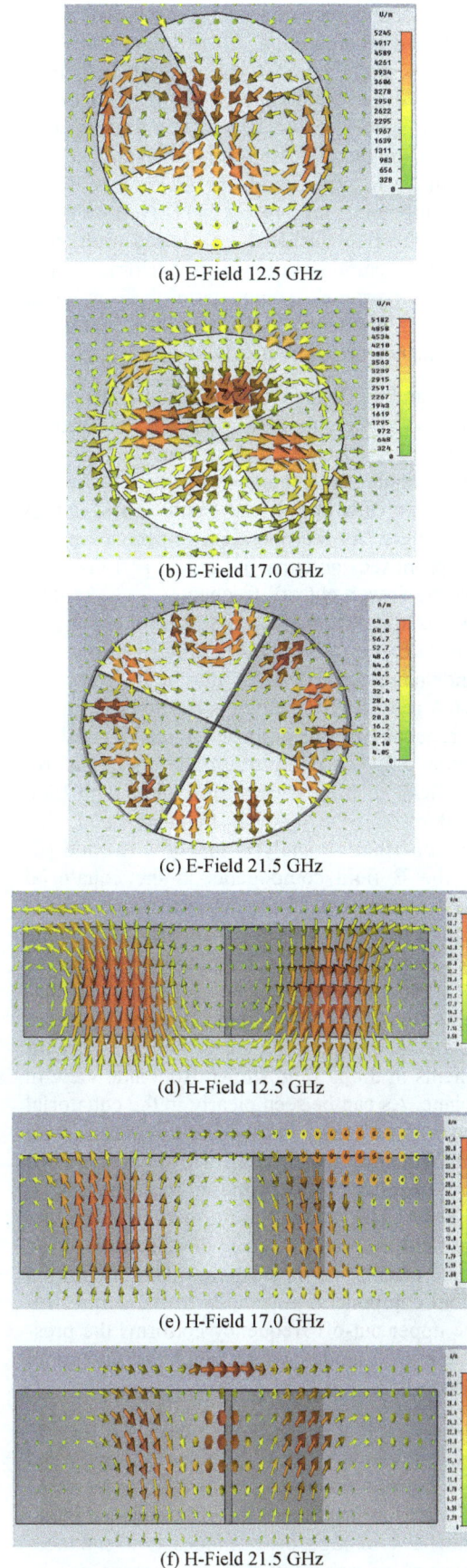

(a) E-Field 12.5 GHz

(b) E-Field 17.0 GHz

(c) E-Field 21.5 GHz

(d) H-Field 12.5 GHz

(e) H-Field 17.0 GHz

(f) H-Field 21.5 GHz

Fig. 4. Illustration of electric and magnetic field distribution in the proposed multi-permittivity cylindrical DRA at different frequencies.

To study wideband response of the MCDRA mode analysis is performed in simulation. It was found that more than one mode was excited in MCDRA which helped in the increase in impedance bandwidth. When the cylindrical dielectric resonator is energized using direct microstrip line coupling it excites mode of a transverse magnetic (TM) type or hybrid mode in which both the electric field and magnetic field components are present [17]. To provide an approximation of the field inside the proposed multi permittivity cylinder eigenmode analysis is performed. The field patterns for our proposed multi permittivity cylinder at the lower cut-off frequency, centre frequency and the upper cut-off frequency are presented here.

The E field and H field distribution in the proposed DRA are illustrated in the equatorial plane and the meridian plane respectively. E field is shown from the top view of the resonator in equatorial plane cuts while the H field is shown from the side view and the resonator cross section appears as a rectangle. As the operating impedance bandwidth of the proposed antenna is from 12.1 GHz to 21.65 GHz, so the lower cut-off frequency is 12.1 GHz, centre frequency is approximately 17.0 GHz and upper cut-off frequency is 21.65 GHz. Figure 4 (a) and (d) shows the electric (E) and magnetic (H) fields distribution, respectively in the DRA at 12.5 GHz, which is on the brink of the lower cut-off frequency of our proposed antenna. As there is a direct relation between permittivity and operating frequency of the DRA, so the dielectric resonator with high permittivity will resonate at a lower frequency and vice versa. The field distribution analysis in Fig. 4 (a) and (d) exhibited that the E field components in the equatorial plane and H field components in the meridian plane existed mostly in the high permittivity sectors of the DRA at the lower cut-off of frequency. The field components near the centre frequency shown in Fig. 4(b) and (e) and upper cut-off frequency shown in Fig. 4(c) and (f) illustrate more variation in E fields in equatorial plane and presence of hybrid components in the H field shown from side view in the meridian plane. As can be seen clearly in the equatorial plane the number of variations in the E field increases with the increase in frequency which confirms the presence of high order modes excited in the proposed multi-permittivity cylindrical DRA. The color of the arrows indicates the intensity of the field in the DRA while the direction of the arrows shows the orientation of the fields. The orientation of magnetic field components especially at the centre frequency and the upper cut-off frequency confirms the presence of intricate hybrid modes, which helps the antenna to operate in wideband.

For efficient coupling of the cylindrical DRA the microstrip line is terminated with a T-stub at the open end which helped in exciting the antenna with high order mode and improves its performance in terms of impedance bandwidth. The drawback of high order modes is it enhances the level of cross polarization but still this antenna can be used for many applications where cross polarization level is not a concern and low profile structures are required.

Further, as stated in the previous section that, the rotation angle 'θ' have a vital role in improving the electromagnetic energy coupling from the microstrip line to the resonator. As the dielectric resonator is of inhomogeneous nature due to sectors of different permittivity, proper positioning of the DR on top of microstrip line for efficient energy transfer is important. The parametric study exhibited that the angular position of the resonator with respect to the feed line can improve the reflection coefficient of the antenna and optimum value for maximum power transformation is achieved.

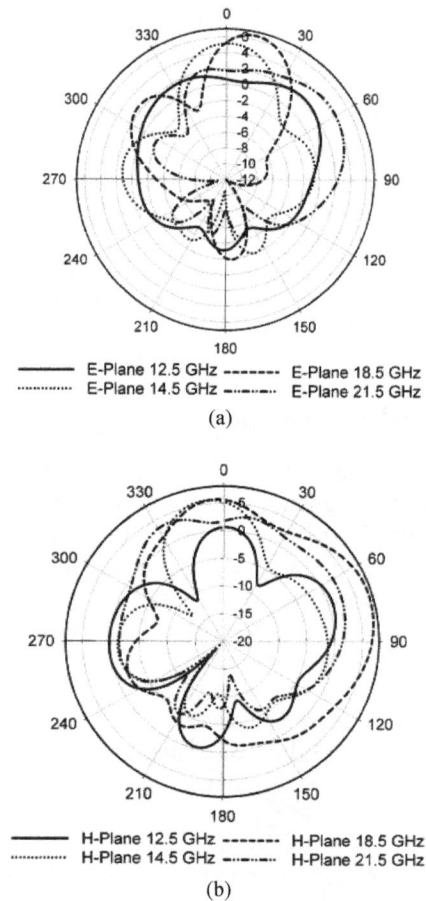

(a)

(b)

Fig. 5. Radiation pattern of the MCDRA at different frequencies: (a) E- plane, (b) H-plane.

The E-plane and H-plane radiation pattern of the antenna is shown in Fig. 5 at different frequencies. It is observed that the maximum radiation of the antenna is towards the boresight in the axis of symmetry of the antenna with a small back lobe. A slight beam tilt is observed in the H-plane main beam of the antenna at 18.5 GHz and also a shift in the side lobe towards the axis of symmetry with the increase in the frequency. Figure 6 shows the simulated and measured gain of the antenna. The gain is evaluated using gain absolute methods compared to a standard horn antenna. The average gain of the antenna throughout the bandwidth is 5.9 dB with maximum and minimum gain of approximately 9 dB and 3.5 dB at 13.1 GHz and 12.5 GHz respectively.

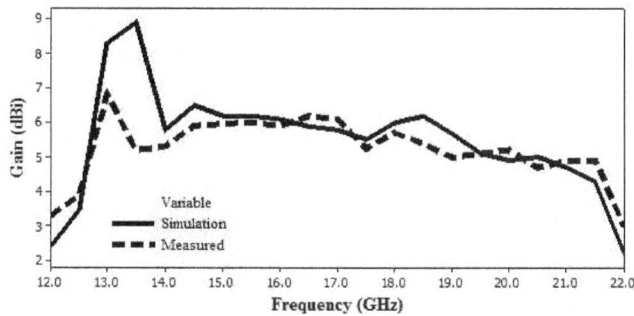

Fig. 6. Simulation and measure gain of the proposed antenna.

5. Conclusion

A novel multi-permittivity cylindrical dielectric resonator antenna for wideband applications has been designed and studied theoretically and experimentally. A parametric study of the antenna has been performed, so as to achieve optimum dimension and wideband operation of the antenna. It has been found that more than 50% impedance bandwidth can be realized with a relatively simple and low profile structure of the antenna.

Acknowledgements

This work was financially supported by Global Fellowship Scheme of Universiti Sains Malaysia and USM Research University (RU) Grant under project no. 1001/PELECT/814117/854004.

References

[1] TZE-HSUAN, C., YU-CHING, H., WEI-FANG, S., JEAN-FU, K. Wideband dielectric resonator antenna with a tunnel. *IEEE Antennas and Wireless Propagation Letters*, 2008, vol. 7, p. 275–278.

[2] CHU, L. C. Y., GUHA, D., ANTAR, Y. M. M. Conformal strip-fed shaped cylindrical dielectric resonator: Improved design of a wideband wireless antenna. *IEEE Antennas and Wireless Propagation Letters*, 2009, vol. 8, p. 482-485.

[3] YONGMEI, P., KWOK-WA, L. Wideband circularly polarized trapezoidal dielectric resonator antenna. *IEEE Antennas and Wireless Propagation Letters*, 2010, vol. 9, p. 588–591.

[4] CHAUDHARY, R. K., KUMAR, R., SRIVASTAVA, K. V. Wideband ring dielectric resonator antenna with annular-shaped microstrip feed. *IEEE Antennas and Wireless Propagation Letters*, 2013, vol. 12, p. 595–598.

[5] KHALILY, M., KAMARUDIN, M. R., JAMALUDDIN, M. H. A novel square dielectric resonator antenna with two unequal inclined slits for wideband circular polarization. *IEEE Antennas and Wireless Propagation Letters*, 2013, vol. 12, p. 1256–1259.

[6] KHALILY, M. R., MOKAYEF, M., DANESH, SH., GHAHFE-ROKHI, S. E. A. A new wideband circularly polarized dielectric resonator antenna. *Radioengineering*, 2014, vol. 23, p. 175–180.

[7] KHALILY, M., RAHIM, M. K. A., KISHK, A. A. Bandwidth enhancement and radiation characteristics improvement of rectangu-

lar dielectric resonator antenna. *IEEE Antennas and Wireless Propagation Letters*, 2011, vol. 10, p. 393–395.

[8] MESSAOUDENE, I., DENIDNI, T. A., BENGHALIA, A. Ultrawideband DRA integrated with narrow-band slot antenna. *Electronics Letters*, 2014, vol. 50, p. 139–141.

[9] FENG, W. Y., DENIDNI, T. A., Z. SHENG, Q., GAO, W. Band-notched UWB rectangular dielectric resonator antenna. *Electronics Letters*, 2014, vol. 50, p. 483–484.

[10] DHAR, S., GHATAK, R., GUPTA, B., PODDAR, D. R. A wideband Minkowski fractal dielectric resonator antenna. *IEEE Transactions on Antennas and Propagation*, 2013, vol. 61, p. 2895 to 2903.

[11] MENG, Z., JIN, P. Wideband hybrid circularly polarised rectangular dielectric resonator antenna excited by modified cross-slot. *Electronics Letters*, 2014, vol. 50, p. 1123–1125.

[12] PRAVEEN KUMAR, A. V., HAMSAKUTTY, V., YOHANNAN, J., MATHEW, K. T. A wideband conical beam cylindrical dielectric resonator antenna. *IEEE Antennas and Wireless Propagation Letters*, 2007, vol. 6, p. 15–17.

[13] BIN, L., KWOK-WA, L. On the differentially fed rectangular dielectric resonator antenna. *IEEE Transactions on Antennas and Propagation*, 2008, vol. 56, p. 353–359.

[14] THAMAE, L. Z., ZHIPENG, W. Broadband bowtie dielectric resonator antenna. *IEEE Transactions on Antennas and Propagation*, 2010, vol. 58, p. 3707–3710.

[15] PAN, Y. M., LEUNG, K. W. Wideband omnidirectional circularly polarized dielectric resonator antenna with parasitic strips. *IEEE Transactions on Antennas and Propagation*, 2012, vol. 60, p. 2992–2997.

[16] PRAVEEN KUMAR, A. V., HAMSAKUTTY, V., YOHANNAN, J., MATHEW, K. T. Microstrip line-fed half-cylindrical dielectric resonator antenna for 2.4-GHz WLAN application. *Microwave and Optical Technology Letters*, 2006, vol. 48, p. 724–726.

[17] PETOSA, A. *Dielectric Resonator Antenna Handbook*. 1st ed. London: Artech House, 2007.

About Authors ...

Ubaid ULLAH was born in Malakand, Pakistan in 1983. He received his BS in Electrical Engineering from Pakistan in 2010 and MS degree in Electronic Engineering from Universiti Sains Malaysia in 2012. He is currently working towards his PhD in Universiti Sains Malaysia. His research interests include Dielectric Resonator Antenna, Wideband DRA, microwave circuit, LTCC based antenna in package, applied electromagnetic and small antenna. He has published several articles in ISI indexed journal's and some well reputed international conferences.

Mohd Fadzil AIN received his BS in Electronic Engineering from Universiti Teknologi Malaysia, Malaysia in 1997, MS in RF and Microwave from Universiti Sains Malaysia (USM), Malaysia in 1999, and PhD in RF and Microwave from University of Birmingham, United Kingdom in 2003. He joined USM in 2003. He is actively involved in technical consultancy with several companies in repairing microwave equipment. His current research interests include wireless circuit design, LTCC based antenna in package, rain propagation, microwave link and dielectric resonator antenna.

Nor Muzlifah MAHYUDDIN received her B.Eng degree in Electric-Telecommunication from Universiti Teknologi Malaysia, in 2005 and M.Sc. degree in Electronics System Design from Universiti Sains Malaysia, in 2006. In addition, she also received a PhD degree in Microelectronics System Design from Newcastle University, Newcastle upon Tyne, United Kingdom, in 2011. She has worked in Agilent Technologies, Penang, Malaysia as an intern. She is currently working as a lecturer in Universiti Sains Malaysia, starting from March 2012. She has produced several papers in the topic of low-swing signaling scheme. Her current research interests are in the field of RF and microwave engineering, reliability and signal integrity. The topic of interests includes the modeling design of split-ring resonator in high performance application, the impact of variability on the design of microstrip-based circuits, and the power integrity in the high performance circuits. Dr. Nor Muzlifah Mahyuddin is currently a member of IEEE and involved in the Communications Society (ComSoc). Subsequently, she is also a member of IET and professional member of Association for Computing Machinery (ACM). She is also registered under Board of Engineers Malaysia.

Mohamadariff OTHMAN received his BS in Electronic Engineering from Universiti Multimedia, Malaysia in 2006 and MS in RF and Microwave from Universiti Sains Malaysia, Malaysia in 2009. Currently, he is pursuing PhD in dielectric resonator antenna (DRA) in Universiti Sains Malaysia, Malaysia. His current research interests include solid dielectric and thick film fabrication, dielectric characterization, wideband DRA and patch antenna.

Ihsan Ahmad ZUBIR was born in Perak, Malaysia, in 1985. He received the B.Eng. degree in Electrical and Electronics Engineering from Universiti Malaysia Pahang, Malaysia, in 2008, M.S. degree in Electrical and Electronics Engineering from University Science of Malaysia, Malaysia in 2012 and currently he is pursuing Ph.D. degree in Communication Engineering in University Science of Malaysia, Malaysia. His research interests include dielectric resonator antenna, wireless transceiver and RF circuits design.

Zainal Arifn AHMAD received his BS in Material Engineering from Universiti Sains Malaysia, Malaysia and MS from University of Manchester, Institute of Science and Technology (UMIST). He received his PhD from University of Sheffield. His current research interests include ZTA ceramic for cutting insert, LTCC based circuits, metal-ceramic joining, crystal glaze ceramic, TCP bioceramic and dielectric ceramic for antenna. He is currently serving as a senior Professor in School of Material in mineral resource engineering, University Sains Malaysia.

Mohd Zaid ABDULLAH graduated from University Sains Malaysia (USM) with a B. App. Sc. degree in Electronic in 1986 before joining Hitachi Semiconductor as a test engineer. In 1989 he commenced an MSc in Instrument Design and Application at UMIST. He remained in Manchester conducting research in Electrical Impedance Tomography at the same university, and received his PhD degree in 1993. He joined USM in the same year. His research interests include microwave tomography, digital image processing, computer vision and ultra wide band sensing. He has published numerous research articles in international journals and conference proceedings. One of his papers was awarded The Senior Moulton medal for the best article published by the Institute of Chemical Engineering in 2002. From 2001 to 2006, he was an associate professor and deputy dean of the USM's School of Electrical and Electronic Engineering. He was promoted to full professor later in 2006 and at present a dean of the school.

22

Optimization of Excitation in FDTD Method and Corresponding Source Modeling

Bojan DIMITRIJEVIC, Bojana NIKOLIC, Slavoljub ALEKSIC, Nebojsa RAICEVIC

Faculty of Electronic Engineering, University of Nis, A. Medvedeva 14, 18 000 Nis, Serbia

{bojan.dimitrijevic, bojana.nikolic, slavoljub.aleksic, nebojsa.raicevic}@elfak.ni.ac.rs

Abstract. *Source and excitation modeling in FDTD formulation has a significant impact on the method performance and the required simulation time. Since the abrupt source introduction yields intensive numerical variations in whole computational domain, a generally accepted solution is to slowly introduce the source, using appropriate shaping functions in time. The main goal of the optimization presented in this paper is to find balance between two opposite demands: minimal required computation time and acceptable degradation of simulation performance. Reducing the time necessary for source activation and deactivation is an important issue, especially in design of microwave structures, when the simulation is intensively repeated in the process of device parameter optimization. Here proposed optimized source models are realized and tested within an own developed FDTD simulation environment.*

Keywords

Finite difference time domain, source modeling, excitation, optimization

1. Introduction

The finite difference time domain (FDTD) method currently draws significant scientific attention as one of the most efficient methods for analysis and characterization of wide range of electromagnetic problems [1]. The proper excitation and source modeling in the FDTD computational domain is especially important issue in every application of FDTD simulation.

Introduction of discrete internal sources is usually done by applying either hard source or soft source excitation. The hard source excitation is consisted in assigning specific value of certain electric (or magnetic) field component at a single or several grid points in every time step through an appropriate time function [2]. The soft source excitation is introduced by adding the appropriate time function to the field value obtained in regular update equation [1].

Although it is not a physical reality, the plane wave excitation has enormously large significance in many theoretical and analytical considerations. For this reason it is very important to introduce the same excitation in simulation environments and to enable the comparative analysis of the results. The necessity of plane-wave source arose originally with the first FDTD modeling in the field of defense and bioelectromagnetics [1]. Considering scattering problems, where the particular structure of interest is far away from the radiation source and the incident wave can be considered as a plane wave, Yee was the first to introduce the initial-condition approach [3]. However, today mainly accepted approach for plane-wave excitation is total-field/ scattered-field (TF/SF) formulation [4], [5]. The TF/SF technique showed very good performance in FDTD modeling of long-duration pulsed or continuous wave excitation and it is widely used in guided-wave simulations [1]. The TF/SF technique has been extensively studied in the literature and many modifications and improvements of this basic method can be found [6–9].

However, another way of plane wave excitation modeling includes adding or assigning of an electric (or magnetic) field value at specific positions in one plane, unlike the commonly used TF/SF technique, where corrections are made in both electric and magnetic field components (displaced in time and space for a half time step) on the boundary surface. The advantage of this direct approach is its simplicity. Its main difficulty is, however, the existence of wave propagation in undesirable direction. But since very effective boundary conditions like the convolutional perfectly matched layer (CPML) [10] are available, this is no longer an obstacle to its application. The considerations regarding this approach can be found in [11].

The excitation modeling in FDTD formulation significantly affects the simulation performance. A sudden excitation of the domain causes undesirable numerical variations in whole computational domain. This problem is usually resolved by slow introduction of the source excitation, using the appropriate shaping functions in time. A number of time functions for slow introduction of source excitation are available in the literature [12]. However, a gradual raise of the excitation signal is time consuming and can be a significant difficulty in applications where the intensive repetition of simulations is required. For this reason, a certain compromise between the required time and satisfactory simulation performance must be achieved.

The optimization presented in this paper is conducted in order to minimize the propagation of the undesirable energy through the computational domain. As a result of the optimization process, very simple and closed-form optimal function is obtained. Numerically obtained optimal shaping function is compared with the solutions that can be found in the literature and some general remarks are derived.

2. Source Modeling

In hard source modeling, instead of calculating field value using FDTD update equations, the field value in the specific grid points is assigned using a time function. If the source is at position (i, j, k), it is then

$$E^n_{v(i,j,k)} = E^n_{source}, \; v = (x, y, z). \qquad (1)$$

In the same manner as in the case of the point source, the plane wave source can be also modeled as a hard source, applying (1) on a group of points belonging to a specific surface or even volume. Regardless of the case, hard source model is equivalent to the ideal voltage source and for this reason acts like an electric wall, causing reflections of any wave arrived to the source location [2], [13]. This means that such a source, more specifically in the case of plane wave excitation separates the computational domain in two independent regions without mutual interaction. Hard source modeling is commonly applied in excitation of guided-wave structures. In [14] the analytical solution for the FDTD hard source has been derived, so its application is broaden to the validation of FDTD codes or FDTD schemes.

One possible solution to the reflective behavior of the hard sources is to combine it with the regular FDTD update equations [1]. Namely, the hard source can be turned on only while there is a signal excitation. The simulation setup in that case should provide that the signal excitation ends before any scattered or reflected wave returns to the source position. After this period, the usual FDTD update of the field components can be applied. The main difficulty of this approach is that it can't be used for continuous excitation.

In soft source modeling, source excitation is added to the value obtained in applied FDTD update equations. If the source is at position (i, j, k), it is then

$$E^n_{v(i,j,k)} = E^n_{v(i,j,k)FDTDupdate} + E^n_{v \, source}, \; v = (x, y, z). \; (2)$$

The main advantage of the soft source is the fact that it is transparent to the incoming waves and it allows the different incident fields to interact [1], [15]. For soft source modeling of the plane wave it can be either TF/SF method or method of direct adding/assigning of electric (magnetic) field value [11] applied. When point soft source modeling is concerned, the additional source component is usually introduced through the current density $J^{t+1/2}_{v(i,j,k)}$, which is

defined at the same spatial position as the resulting electric field component but the time step is the same as the one of the magnetic field. If the source is at position (i, j, k), it is then

$$\frac{\Delta E^{t+1}_{v(i,j,k)}}{\Delta t} = \frac{1}{\varepsilon_0} \frac{\Delta H^{t+1/2}_{w(i,j,k)}}{\Delta u} - \frac{1}{\varepsilon_0} \frac{\Delta H^{t+1/2}_{u(i,j,k)}}{\Delta w} - \frac{1}{\varepsilon_0} J^{t+1/2}_{v(i,j,k)} \; (3)$$

$$v = (x, y, z), \; u = (z, x, y), \; w = (y, z, x),$$

$$i = \{0, ..., N_x - 1\}, \; j = \{0, ..., N_y - 1\}, \; k = \{0, ..., N_z - 1\},$$

$$n = \{0, ..., N_t - 1\}$$

where Δt is the time step, Δx, Δy and Δz are spatial steps along x, y and z axis, respectively. Total numbers of elementary cells along x, y and z axis are denoted by N_x, N_y and N_z, respectively. Total number of elementary time steps is denoted by N_t. The electric magnetic permittivity in vacuum is denoted by ε_0.

Since the FDTD cell is usually much shorter than one-tenth of the main wavelengths of interest, physically the soft source current acts as a Hertzian dipole antenna [14]. The point soft source physically corresponds to real current (or voltage) source.

One of the main difficulties regarding the soft source modeling is constant deposit of charges and generation of the charge-associated fields [16], [1]. This can be circumvented by using matched voltage or current sources. By applying the source resistance R_S, voltage source at position (i, j, k) can be introduced in update equation of the corresponding field component as

$$E^{t+1}_{v(i,j,k)} = \frac{1-b}{1+b} E^t_{v(i,j,k)} +$$

$$+ \frac{\Delta t / \varepsilon_0}{1+b} \left(\frac{\Delta H^{t+1/2}_{w(i,j,k)}}{\Delta u} - \frac{\Delta H^{t+1/2}_{u(i,j,k)}}{\Delta w} - \frac{V^{t+1/2}_{S \, v(i,j,k)}}{R_S \Delta u \Delta w} \right) \quad (4)$$

$$b = \frac{\sigma \Delta t}{2\varepsilon_0} + \frac{\Delta t \Delta v}{2 R_S \varepsilon_0 \Delta u \Delta w}$$

where σ is the electric conductivity.

It is shown in [14] that the fields radiated by hard and soft source models are identical and the relation that connects excitation $E^n_{v \, source}$ from (1) and $J^{t+1/2}_{v(i,j,k)}$ from (2) can be expressed as

$$E^n_{v \, source} = -\frac{1}{3\varepsilon_0} \int_{-\infty}^{t_n} J_S(t) dt, \qquad (5)$$

$$J_S(t_{n+1/2}) = J^{n+1/2}_{v(i,j,k)}.$$

3. Excitation Modeling

A proper excitation of the FDTD computational domain is important in every FDTD application. The abrupt source introduction yields intensive numerical variations at high frequencies which propagate through the whole computational domain. A generally accepted solution is to slowly introduce the source. In order to fulfill this, it is necessary for the excitation function to satisfy following conditions: at zero time step the excitation function value must be zero (if fields are initiated at zero) and the excitation function must be smooth. The most commonly used time functions for slow introducing of pulsed source excitation available in the literature are given in Tab. 1.

Broad band Gaussian with DC component	$f(t) = \exp\left(-(t-t_0)^2/t_w^2\right),\ 0 < t < 2t_0$
Broad band Gaussian without DC component – Gaussian derivative	$f(t) = \dfrac{-2}{t_w}(t-t_0)\exp\left(-(t-t_0)^2/t_w^2\right)$
Blackman – Harris window	$b(t) = \sum_{n=0}^{3} a_n \cos\left(\dfrac{2\pi n t}{T}\right),\quad 0 < t < T$ $a = \begin{bmatrix} 0.353222222 \\ -0.488 \\ 0.145 \\ -0.010222222 \end{bmatrix},\ T = \dfrac{1.55}{f_{bw}}$ f_{bw} - half bandwidth of the pulse
Differentiated Blackman – Harris window	$d_b(t) = -\sum_{n=0}^{3} a_n n \sin\left(\dfrac{2\pi n t}{T}\right),\ 0 < t < T$
Raised cosine [17]	$z(t) = 0.5\left(1 - \cos\left(\dfrac{2\pi t}{T}\right)\right),\ 0 < t < T_r/2$ T_r - period of ramped cosine ($T_r \approx 3T$)

Tab. 1. Time functions used for pulsed excitation [12].

The raised cosine [17] is considered to be the most suitable for excitation of FDTD domain and it is the preferred choice, especially compared to linear and exponential ramps [15]. However, the presented excitation functions (Tab. 1) are not designed to meet the specific requirements of FDTD formulation. In order to get a better insight in FDTD nature, we shell start from the case of z-polarized plane wave, propagating along y axis. Ampere's and Faraday's law in that case have the form

$$\frac{\partial H_x}{\partial t} = \frac{1}{\mu_0}\frac{\partial E_z}{\partial y}, \quad (6)$$

$$\frac{\partial E_z}{\partial t} = \frac{1}{\varepsilon_0}\frac{\partial H_x}{\partial y} - \frac{J_S}{\varepsilon_0}. \quad (7)$$

If one differentiates (1) over space variable y and (2) over time t, it yields

$$\frac{\partial^2 H_x}{\partial y \partial t} = \frac{1}{\mu_0}\frac{\partial^2 E_z}{\partial y^2}, \quad (8)$$

$$\frac{\partial^2 E_z}{\partial t^2} = \frac{1}{\varepsilon_0}\frac{\partial^2 H_x}{\partial t \partial y} - \frac{1}{\varepsilon_0}\frac{\partial J_S}{\partial t}. \quad (9)$$

Substituting (8) in (9), it is obtained

$$\frac{\partial^2 E_z}{\partial t^2} = \frac{1}{\varepsilon_0}\frac{1}{\mu_0}\frac{\partial^2 E_z}{\partial y^2} - \frac{1}{\varepsilon_0}\frac{\partial J_S}{\partial t}. \quad (10)$$

Equation (10) indicates that the second time derivative of the excitation function should be the one to investigate, since it causes the propagation through the computational domain.

4. Numerical Optimization of FDTD Domain Excitation Function

In order to minimize the propagation of the undesirable energy through the FDTD computational domain that consequently appears during its excitation, the minimization of the second time derivative of the excitation function should be performed.

Without loss of generality, the excitation function will be analyzed in its normalized form. Considering (10) and [17], in order to have the desirable properties, the excitation function f(x) should satisfy the following criteria

1. $f(0) = 0$ and $f(1) = 1$;

2. $f(x)$ is an odd function with respect to the point (1/2, 1/2) in Cartesian coordinate system;

3. the first derivative of $f(x)$ is continuous function and $f'(0) = 0$ (in order to avoid large values in the second derivative).

Two possible solutions are considered.

4.1 Polynomial Optimization

For the purpose of optimization, the excitation function will be presented as a linear combination of basic functions which satisfy conditions 1, 2 and 3. Considered basic functions are in polynomial form and given as

$$f_n(x) = \begin{cases} 0, & x < 0 \\ 2^{n-1}x^n, & 0 \le x \le \frac{1}{2}, n = 2,3,\dots \\ 1 - 2^{n-1}(1-x)^n, & \frac{1}{2} \le x \le 1 \\ 1, & x > 1 \end{cases} \quad (11)$$

If functions $f_n(x)$ fulfill conditions 1, 2 and 3, then their linear combination

$$f(x) = C_2 f_2(x) + C_3 f_3(x) + \dots \quad (12)$$

also fulfills the same condition 3. However, there is an additional requirement for $f(x)$, in order to satisfy the conditions 1 and 2

$$\sum_{i=2}^{+\infty} C_i = 1. \quad (13)$$

Applying numerical iterative minimization of the mean square value of the second derivative $\partial^2 f(x)/\partial x^2$ ($\min_{C_i, i=2,3,4,\dots} \overline{f''^2(x)}$), it is obtained

$$C_2 = \frac{3}{2}, \ C_3 = \frac{1}{2}, \ C_4 = C_5 = C_6 = \ldots = 0. \quad (14)$$

Thus, the optimal excitation function has the form

$$f_{opt}(x) = \begin{cases} 0, & x < 0 \\ 3x^2 - 2x^3, & 0 \le x \le 1 \\ 1, & x > 1 \end{cases} \quad (15)$$

4.2 Trigonometric Optimization

If the basic functions in series expansion of the excitation function are

$$f_m^{(M)}(x) = \begin{cases} 1, & x < 0 \\ \cos(2m+1)\pi x, & 0 \le x \le 1 \\ -1, & x > 1 \end{cases} \quad (16)$$

the excitation function is then

$$f^{(M)}(x) = \frac{1}{2} - \sum_{m=0}^{M} a_{2m+1} f_{2m+1}^{(M)}(x) \quad (17)$$

where a_{2m+1} are series coefficients, which should fulfill the criterion

$$\sum_{m=0}^{M} a_{2m+1} = \frac{1}{2}. \quad (18)$$

Applying numerical iterative minimization of the mean square value of the second derivative $\partial^2 f^{(M)}(x)/\partial x^2$ in this case, a function with expansion coefficients presented in Tab. 2 for different M values is obtained. The mean square values of the second derivative of function $f^{(M)}(x)$ are also given.

5. Optimization Results

In Fig. 1 excitation function obtained in the optimization process using polynomial basic functions (denoted as Opt), as well as the ones using trigonometric basic functions for different values of M are presented. The curve that corresponds to the value $M = 0$ is actually the excitation function that is widely used in the literature and known as raised cosine.

In Fig. 2 one can observe the second derivatives of the functions from Fig. 1. It can be seen from Fig. 2 that the second derivative of the excitation function $M = 0$ (raised cosine) significantly deviates from the second derivative of the optimal function obtained using polynomial expansion. It can be also observed that with the increase of M the second derivative of the function with trigonometric expansion converges to the one of the optimal function with polynomial expansion. This confirms that the same optimal result is obtained regardless of the applied type of basic functions in optimization process. Since the solution

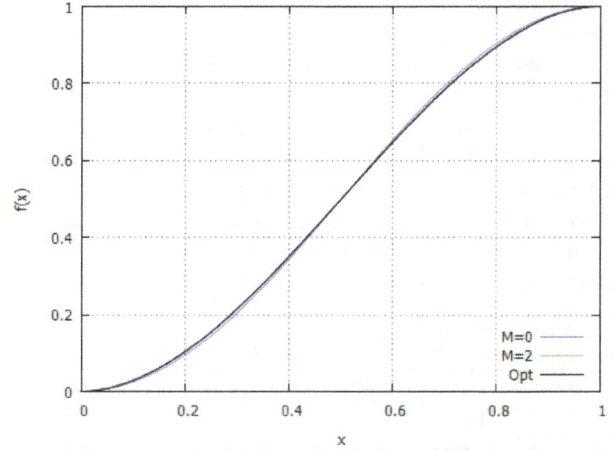

Fig. 1. The proposed optimized excitation functions.

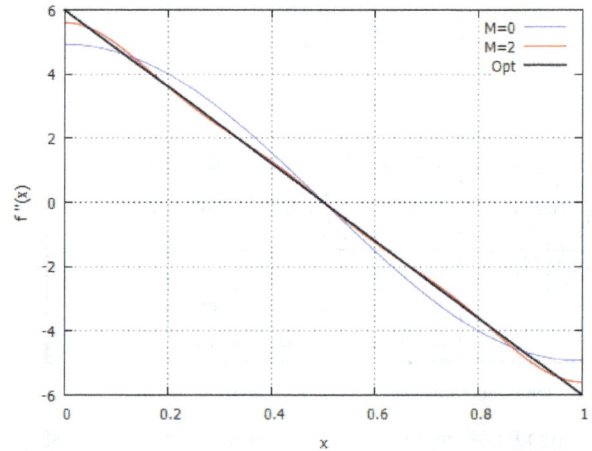

Fig. 2. Second derivative of the proposed optimized excitation functions.

obtained in polynomial form is simple and in a closed form, we propose function (15) as the excitation function for FDTD domain.

Optimality of the obtained function shape is verified in an own developed FDTD simulation environment [18].

6. Optimal Excitation Signals

Using the obtained optimal excitation function (Fig. 3), two pulsed signals are proposed for efficient excitation of FDTD domain. The total pulsed signal retains the optimal properties only if it is formed using the proposed optimal function as segments that are appropriately symmetrically extended or scaled in time and amplitude. Thus, the proposed pulsed function has the form

$$f_p(x) = \begin{cases} 0, & x < 0 \\ 3x^2 - 2x^3, & 0 \le x \le 1 \\ 3(2-x)^2 - 2(2-x)^3, & 1 \le x \le 2 \\ 0, & x > 2 \end{cases} \quad (19)$$

M	$\overline{f''^2(x)}$	a_1	a_3	a_5	a_7	a_9	$\cdots$
0	12.1761	0.5	-	-	-	-	-
1	12.0276	0.493902	0.00609756	-	-	-	-
2	12.0086	0.493123	0.00608793	7.88996 x 10^{-4}	-	-	-
3	12.0037	0.492920	0.00608543	7.88672 x 10^{-4}	2.05298 x 10^{-4}	-	-
4	12.0019	0.492846	0.00608452	7.88554 x 10^{-4}	2.05267 x 10^{-4}	7.51175 x 10^{-5}	-
$\to \infty$	12.0000	0.492767	0.00608354	7.88427 x 10^{-4}	2.05234 x 10^{-4}	7.51054 x 10^{-5}	$\cdots$

Tab. 2. Expansion coefficients in $f^{(M)}(x)$.

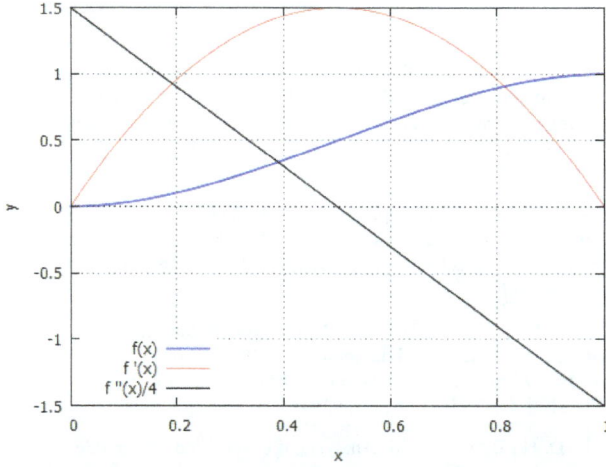

Fig. 3. Proposed optimal excitation function and its first and second derivative.

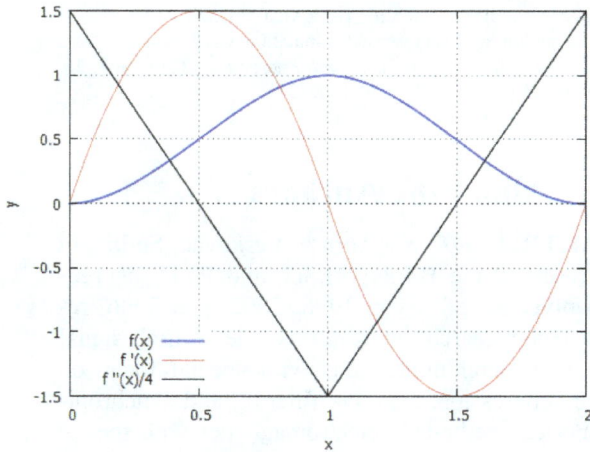

Fig. 4. Proposed optimal pulsed excitation function with DC component and its first and second derivative.

Function (19) along with its first and second derivative is presented in Fig. 4.

In case the pulsed signal with no DC component is required, it should be also obtained as symmetrically extended or adequately scaled optimal function (15). The pulsed signal with no DC component shouldn't be formed as the first derivative of the optimal pulsed function (Fig. 4), because in that case the resulting function would change its nature and wouldn't have optimal properties any more. Thus, we propose the pulsed function with no DC component in the form

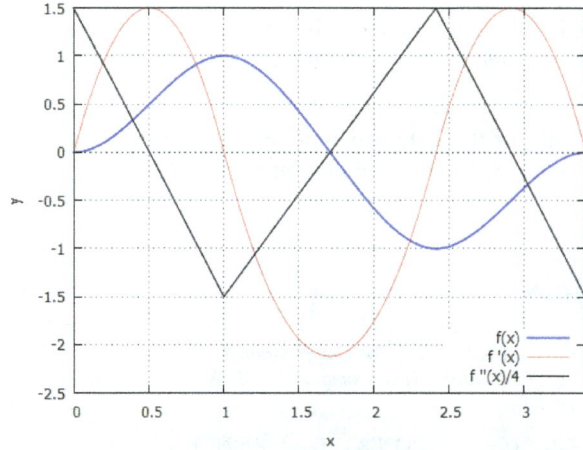

Fig. 5. Proposed optimal pulsed excitation function without DC component and its first and second derivative.

$$f_d(x) =$$
$$= \begin{cases} 0, & x < 0 \\ 3x^2 - 2x^3, & 0 \le x \le 1 \\ 3(1+\sqrt{2}-x)^2 - \sqrt{2}(1+\sqrt{2}-x)^3 - 1, & 1 \le x \le 1+\sqrt{2} \\ -3(2+\sqrt{2}-x)^2 + \sqrt{2}(2+\sqrt{2}-x)^3, & 1+\sqrt{2} \le x \le 2+\sqrt{2} \\ 0, & x > 2+\sqrt{2} \end{cases}$$

(20)

It is interesting to mention that in (20) in the function segment $1 \le x \le 1+\sqrt{2}$ the required transition is from 1 to -1, thus the optimal function (15) should be scaled by $\sqrt{2}$ over time and by 2 over amplitude in order to keep its optimal characteristics. Function (20), along with its first and second derivative, is presented in Fig. 5.

7. Conclusion

The first part of this paper contains the overview of the relevant principles in source modeling in FDTD, with special focus on differences between hard and soft sources and on different source geometry (plane wave sources and point sources). The second part of the paper is dedicated to the source optimization, more specifically to the optimization of the excitation time function, which has a significant influence on the behavior of the generator, regardless of its type. Generally accepted and the most frequently used excitation functions are listed. However, neither of them is designed primarily for FDTD application, taking into account specificities of FDTD method.

The optimization problem in this work is defined in order to minimize the propagation of the undesirable energy through the computational domain. This is accomplished by minimizing the mean square value of the second time derivative of the excitation function. As a result of the optimization process, very simple and closed-form optimal function is obtained.

In addition, two functions for pulsed excitation of FDTD domain, with and without DC component, are proposed.

Optimality of the obtained function shape is verified in the own developed FDTD simulation environment.

References

[1] TAFLOVE, A., HAGNESS, S. B. *Computational Electrodynamics: The Finite-Difference Time-Domain*. 3rd ed. Boston: Artech House, 2005.

[2] BUECHLER, D. N., ROPER, D., DURNEY, C. H., CHRISTENSEN, D. A. Modeling sources in the FDTD formulation and their use in quantifying source and boundary condition errors. *IEEE Transactions on Microwave Theory and Techniques*, 1995, vol. 43, p. 810–814. DOI: 10.1109/22.375228

[3] YEE, K. S. Numerical solution of initial boundary value problems involving Maxwell's equations in isotropic media. *IEEE Transactions on Antennas and Propagation*, 1966, vol. 14, no. 3, p. 302–307. DOI: 10.1109/TAP.1966.1138693

[4] MEREWETHER, D. E., FISHER, R., SMITH, F. W. On implementing a numeric Huygen's source scheme in a finite difference program to illuminate scattering bodies. *IEEE Transactions Nuclear Science*, 1980, vol. 27, no. 6, p. 1829–1833. DOI: 10.1109/TNS.1980.4331114

[5] UMASHANKAR, K. R., TAFLOVE, A. A novel method to analyze electromagnetic scattering of complex objects. *IEEE Transactions on Electromagnetic Compatibility*, 1982, vol. 24, no. 4, p. 397–405. DOI: 10.1109/TEMC.1982.304054

[6] WATTS, M. E., DIAZ, R. E. Perfect plane-wave injection into a finite FDTD domain through teleportation of fields. *Electromagnetics*, 2003, vol. 23, p. 187–201. DOI: 10.1080/02726340390159504

[7] OGUZ, U., GUREL, L., ARIKAN, O. An efficient and accurate technique for the incident-wave excitations in the FDTD method. *IEEE Microwave and Guided Wave Letters*, 1998, vol. 46, no. 6, p. 869–882. DOI: 10.1109/22.681215

[8] TAN, T., POTTER, M. Optimized analytic field propagator (O-AFP) for plane wave injection in FDTD simulations. *IEEE Transactions on Antennas and Propagation*, 2010, vol. 58, no. 3, p. 824–831. DOI: 10.1109/TAP.2009.2039310

[9] GUIFFAUT, C., MAHDJOUBI, K. A perfect wideband plane wave injector for FDTD method. In *Proceedings of the IEEE International Symposium on Antennas and Propagation*. Salt Lake City (UT, USA), 2000, vol. 1, p. 236–239. DOI: 10.1109/APS.2000.873752

[10] RODEN, J. A., GEDNEY, S. D. Convolution PML (CPML): An efficient FDTD implementation of the CFS–PML for arbitrary media. *Microwave and Optical Technology Letters*, 2000, vol. 27, no. 5, p. 334–339.

[11] MANSOURABADI, M., POURKAZEMI, A. FDTD hard source and soft source reviews and modifications. *Progress In Electromagnetics Research C*, 2008, vol. 3, p. 143–160. DOI: 10.2528/PIERC08032302

[12] GEDNEY, S. D. *Introduction to the Finite-Difference Time-Domain (FDTD) Method for Electromagnetics*. Course text, Morgan and Claypool Publishing, 2011. [Online] Cited 2015-03-03. Available at: http://www.engr.uky.edu/~gedney/courses/ee624.

[13] INAN, I. M., MARSHALL, R. A. *Numerical Electromagnetics – The FDTD Method*. Cambridge: Cambridge University Press, 2011.

[14] COSTEN, F., BERENGER, J.-P., BROWN, A. Comparison of FDTD hard source with FDTD soft source and accuracy assessment in Debye media. *IEEE Transactions on Antennas and Propagation*, 2009, vol. 57, no. 7, p. 2014–2022. DOI: 10.1109/TAP.2009.2021882

[15] KALIALAKIS, C. *Finite Difference Time Domain Analysis of Microstrip Antenna-Circuit Modules*. PhD Thesis. School of Electronic and Electrical Engineering, The University of Birmingham, July 1999.

[16] WAGNER, C. L., SCHNEIDER, J. B. Divergent fields, charge and capacitance in FDTD simulations. *IEEE Transactions on Microwave Theory and Techniques*, 1998, vol. 46, no. 12, p. 2131–2136. DOI: 10.1109/22.739294

[17] ROPER, D. H., BAIRD, J. M. Analysis of overmoded waveguides using the finite difference time domain method. In *Digest Proceedings of the IEEE Microwave Theory and Techniques Society Symposium*. Albuquerque (USA), 1992, p. 401–404. DOI: 10.1109/MWSYM.1992.187997

[18] NIKOLIC, B., DIMITRIJEVIC, B., RAICEVIC, N, ALEKSIC, S. Implementation of FDTD based simulation environment. *Facta Universitatis. Ser.: Electronics and Energetics*. 2013, vol. 26, p. 121–132. DOI: 10.2298/FUEE1302121N

About the Authors ...

Bojan DIMITRIJEVIĆ was born in Leskovac, Serbia, in 1972. He received the B.E.E., M.Sc., and Ph.D. degrees from the University of Niš in 1998, 2002, and 2006, respectively. His research interests include digital signal processing in telecommunications with special focus on interference suppression, adaptive filtering and synchronization, numerical methods in electromagnetics with special focus on FDTD method and signal, material and component modeling.

Bojana NIKOLIĆ was born in Niš, Serbia in 1982. She received the Dipl. – Ing. and Ph.D. degrees in Telecommunications from the Faculty of Electronic Engineering in Niš in 2007 and 2012, respectively. Her research interests include FDTD numerical modeling in electromagnetics and wireless communications.

Slavoljub ALEKSIĆ was born in Berčinac, Serbia in 1951. He received Dipl. – Ing., M. Sc. and Ph.D. degrees in Theoretical Electrical Engineering from the Faculty of Electronic Engineering, University of Niš, Serbia in 1975, 1979 and 1997, respectively. His researching areas are: electromagnetic field theory, numerical methods in electromagnetics, lightning protection systems, low-frequency EM fields, microstrip transmission lines with isotropic, anisotropic and bianisotropic media, cable joints and cable terminations, permanent magnets analysis, power lines.

Nebojša RAIČEVIĆ was born in Niš, Serbia in 1965. He received his the Dipl. – Ing., M.Sc. and Ph.D. degrees from the Faculty of Electronic Engineering of Niš, Serbia, in 1989, 1998 and 2010, respectively. His research interests include: cable terminations and joints, numerical methods for electromagnetic problems solving, microstrip transmission lines with isotropic, anisotropic and bianisotropic media, analysis of metamaterial structures, electromagnetic compatibility, nonlinear electrostatic problems, magnetic field calculation of coils and permanent magnets.

23

RF MEMS Based Tunable Bowtie Shaped Substrate Integrated Waveguide Filter

Muhammad Zaka Ur REHMAN [1, 2], Zuhairi BAHARUDIN [1], Mohd Azman ZAKARIYA [1],
Mohd Haris Md. KHIR [1], Muhammad Taha JILANI [1]

[1] Dept. of Electrical and Electronics Engineering, Universiti Teknologi Petronas, Tronoh, Perak, Malaysia
[2] Dept. of Physics, COMSATS Institute of Information Technology, Park Road, Islamabad, Pakistan

zaka_g01951@utp.edu.my, zuhairb@petronas.com.my

Abstract. *A tunable bandpass filter based on a technique that utilizes substrate integrated waveguide (SIW) and double coupling is presented. The SIW based bandpass filter is implemented using a bowtie shaped resonator structure. The bowtie shaped filter exhibits similar performance as found in rectangular and circular shaped SIW based bandpass filters. This concept reduces the circuit footprint of SIW; along with miniaturization high quality factor is maintained by the structure. The design methodology for single-pole triangular resonator structure is presented. Two different inter-resonator couplings of the resonators are incorporated in the design of the two-pole bowtie shaped SIW bandpass filter, and switching between the two couplings using a packaged RF MEMS switch delivers the tunable filter. A tuning of 1 GHz is achieved for two frequency states of 6.3 and 7.3 GHz. The total size of the circuit is 70 mm x 36 mm x 0.787 mm (L x W x H).*

Keywords

Substrate integrated waveguide (SIW), tunable filter, bowtie filter, RF MEMS, double coupling

1. Introduction

Filters have received a particular attention with the advent of various wireless systems, this interest has dramatically increased with the introduction and development of new millimeter waves applications over the past decade. Various applications have been recently proposed including wireless local area networks [1], radars [2], intelligent transportation systems [3] and imaging sensors [4]. Efficient filters demand has also increased with the development of chipsets operating at 60 GHz or higher frequencies by a number of semiconductor industries [5].

Filters based on Substrate Integrated Waveguide (SIW) structures are achieved through incorporating the rectangular waveguide structure into the microstrip substrate [6]. SIWs are dielectric filled and are formed from the substrate material utilizing two rows of conducting vias

connecting bottom and top metal plates, these vias are embedded in dielectric filled substrate; hence providing easy combination with other planar circuits and a reduction in size. The size reduction along with involving dielectric filled substrate instead of air-filled reduces the quality factor (Q), but the entire circuitry including waveguide and microstrip transitions can be realized by using printed circuit board (PCB) technology or other techniques, like LCP [7] and LTCC [8].

The design of an SIW bandpass filter can either utilize a design methodology based on coupling matrix method [9], or it can also follow a methodology used for designing air filled waveguide filters. The design of an SIW filter based on the methodology adopted in a rectangular waveguide, a shunt inductive coupling realization is adopted. Vias of irregular diameters placed in the center of the cavity may possibly occur in an inductive post filter; which is based on a requirement of control couplings. Large couplings might occur in the use of a small diameter. The utilization of shunt inductive vias at the couplings of the filter realizes a shunt inductive coupling filter as depicted in Fig. 1 (a) or an iris (aperture) coupling post as shown in Fig. 1 (b). A detailed literature on the development of SIW filters has been reported in [10].

A three pole structure of a SIW bandpass filter based on shunt inductive vias is shown in Fig. 1 (a). It utilizes four coupling vias centered in the guide and two microstrip to SIW transitions at the input and output. The two vias at the transitions facilitate in input and output couplings, while the centered vias are for coupling between the resonators.

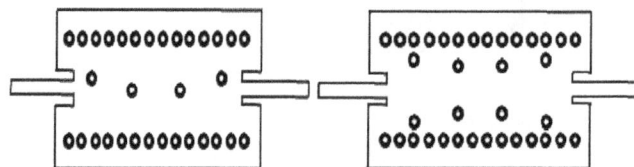

Fig. 1. (a) Shunt inductive coupling post filter. (b) Iris coupling based filter.

An SIW bandpass filter based on iris coupling posts is shown Fig. 1 (b); the apertures form three resonators. The filter's structure is such that the three cavities of half wavelength are formed in the center while SIW to microstrip transition are on the two edges of the filter; such a filter operational at 60 GHz has been presented in [11].

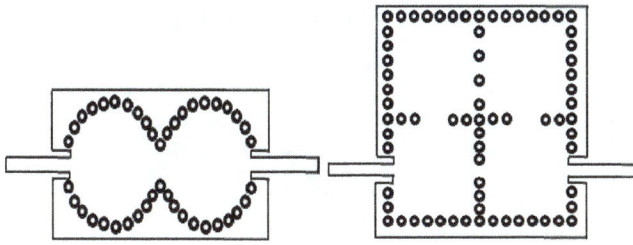

Fig. 2. Cavity filters with (a) circular cavities; (b) rectangular cavities.

Cavity filters with circular cavities[12] as presented in Fig. 2(a) and rectangular cavities [13] as depicted in Fig. 2(b) has been observed in literature. These variants of SIW allow more design variations and transmission zeros are also introduced due to cross coupling, better selectivity is also presented by these designs. Various SIW filters structures have been proposed in the literature; however there still exists a need to further miniaturize the structure. Furthermore the cavities are only either in circular or rectangular shape; therefore a triangular shaped cavity would reduce the circuit footprint.

In this paper, a tunable bandpass filter based on triangular cavity is presented. The tunable filter utilizes packaged RF MEMS switches which can be directly soldered on the filter circuit to switch the filter at two distinct frequencies. The tunable filter is designed through additional incorporation of a switchable extended coupling mechanism. As a validation of the proposed tunable filter, the design, fabrication and measured response of the two pole bowtie shaped bandpass filter are presented. The proposed bandpass filter configuration is suitable for integration with planar devices and its small footprint area allows other devices to be easily integrated on a single board.

2. Triangular SIW Cavity Filter

Design and implementation of SIW filters are being performed through defined practical methods so far. The most common technique is to form the SIW cavity through metallic sidewalls [14], a resemblance of which is displayed in Fig. 3. A dielectric substrate having thickness of H forms the cavity and it is of length L and width W. The bottom and top of the cavity are constructed through placing metallic plates and conducting posts/vias going through the substrate connecting the top and bottom plates; hence forming the sidewalls of the cavity. The vias are of diameter d and the separation between two neighboring vias is given as s. The choice of diameter and separation between the two vias forms the basis of the SIW filters; therefore these should be selected in a manner that minimum radiation loss is exhibited.

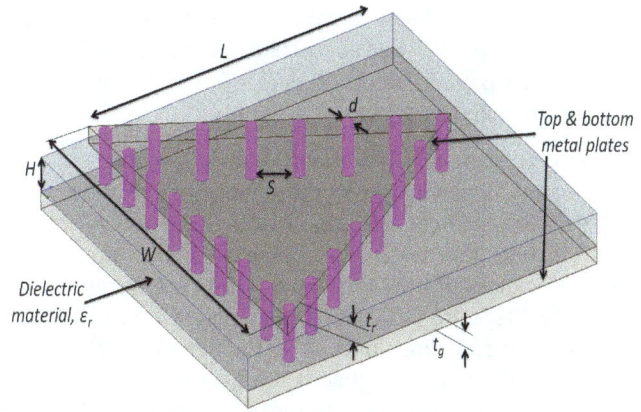

Fig. 3. Substrate Integrated Waveguide structure geometry.

The Deslandes and Wu [14] study reveals two primary design rules for SIW structures as given in (1) and (2); these rules are followed in order to ensure the same design and modeling methodology adopted for rectangular waveguides. These rules pertain to the diameter d of the via posts and the via post spacing s:

$$d < \frac{\lambda_g}{5},$$ (1)

$$s \leq 4d$$ (2)

where λ_g is the guided wavelength. In the design d and s are chosen to be 0.8 mm and 2 mm respectively, these values ensure less radiation losses and the SIW cavity acts closely to a triangular waveguide. For the TE_{101} mode, the dimensions of the SIW resonator structure are calculated by using the relation in (3) [14]

$$f_{TE_{101}} = \frac{c}{2\sqrt{\mu_r \varepsilon_r}} \sqrt{\left(\frac{1}{W_{eff}}\right)^2 + \left(\frac{1}{L_{eff}}\right)^2}.$$ (3)

W_{eff} and L_{eff} denote the effective width and length of the SIW resonator, respectively, and are given as:

$$W_{eff} = W - \frac{d^2}{0.95s}, \quad L_{eff} = L - \frac{d^2}{0.95s}$$ (4)

where W and L are the real width and length of the SIW resonator, μ_r and ε_r are the relative permeability and permittivity of the substrate respectively, and c is the velocity of light in free space. In this design the width and length of the triangular resonator structure is computed using (3) and (4) as displayed in Fig. 3. Utilizing this method the cavity is designed for the specifications laid out in Tab. 1.

Parameter	Value
Tunable center frequency	7.3 GHz
Chebyshev response filter order	2
Passband ripples (dB)	0.01
Passband bandwidth at -3 dB	> 300 MHz

Tab. 1. Design specifications of the tunable bandpass filter.

The printed circuit of triangular resonator and its subsequent bowtie shaped two pole filter is etched over using Rogers RT/Duriod 5880 material substrate having dielectric constant $\varepsilon_r = 2.2$ and substrate height H of 787 μm, and a dissipation factor of $\tan\delta = 9 \times 10^{-4}$. The copper thickness denoted as t_r and t_g is 17.5 μm.

The proposed triangular SIW cavity filter's resonance frequency is mainly dependent upon the dimensions of the cavity and the arrangement of vias forming the cavity walls. Theoretically, the resonance frequency does not depend on the thickness of the substrate H. However, it has been observed in literature [15] that it does play a role on the loss (mainly on radiation loss). The thicker the substrate the lower is the loss or higher is the Quality factor. Therefore, keeping in view the fabrication limitations a relatively thicker substrate is utilized for the SIW cavity.

2.1 Single Inter-resonator and I/O Coupling

In order to accomplish two pole bandpass SIW filter, once the triangular resonator is created for a specific resonant mode, the design methodology closely resembles conventional simulation-based microstrip filter design [14]. Two single resonators are coupled through inter-resonator coupling dimensions. In this coupling, a resonator length and coupling openings in both the resonators are used to couple the two resonators.

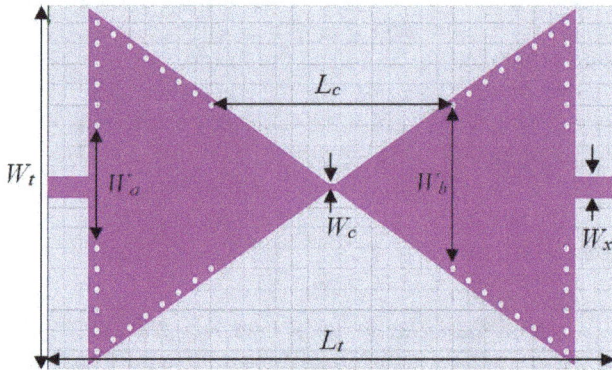

Fig. 4. Bowtie shaped two pole bandpass filter dimensions.

In Fig. 4 the dimensions are labeled and the values are given in Tab. 2, while the fabricated two pole bandpass filter with single coupling is depicted in Fig. 7.

External quality factor and coupling coefficients are calculated from derived expressions based on lowpass prototype parameters [14].

$$K_{1,2} = \frac{FBW}{\sqrt{g_1 g_2}}, \quad Q_{e1} = \frac{g_0 g_1}{FBW}, \quad Q_{e2} = \frac{g_2 g_3}{FBW} \quad (5)$$

where FBW is the fractional bandwidth, and g_0, g_1, g_2, g_3 are the Chebyshev lowpass prototype values which are used to calculate the external quality factors Q_{e1} and Q_{e2}. The coupling coefficient $K_{1,2}$ computed based on second order Chebyshev response prototype parameters is compared with the simulated coupling coefficient obtained through weak coupling. The coupling coefficient's value is dependent of two parameters; the coupling opening width

denoted as W_b and inter-resonator combining width denoted as W_c. For 0.01 dB passband ripple, (5) results in values of $K_{1,2} = 0.1402$ and $Q_{e1} = Q_{e2} = 7.48$, then widths of W_c and W_b are selected from a closely matched coupling coefficients values. The values obtained through various combinations of the dimensions are depicted in Fig. 5.

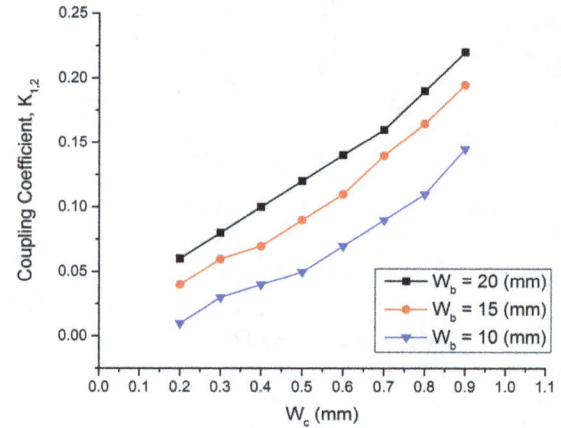

Fig. 5. Variation of inter-resonator dimensions against coupling coefficient.

The microstrip to SIW transitions at the input and output ports are of width W_x, the dimensions of the feed lines are computed through transmission line calculator. However the input and output coupling openings denoted as W_a are selected as a result of comparison of the simulated extracted external quality factors computed through (6) and external quality factors calculated on basis of theoretical LPF prototype parameters as given in (5)

$$Q_{ext} = \frac{f_0}{\Delta f_{-3dB}}. \quad (6)$$

A selection of dimension resulting in a close match of theoretical and simulated external quality factors dictates the I/O coupling as depicted in Fig. 6.

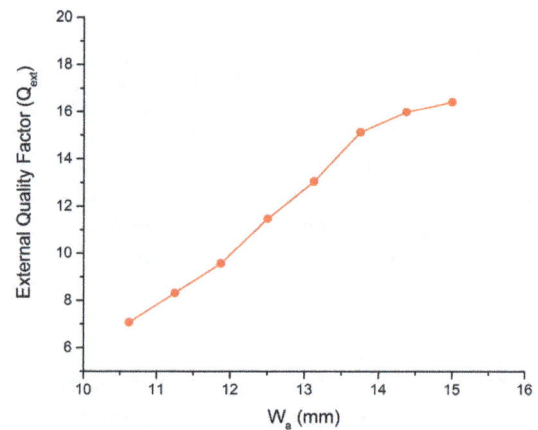

Fig. 6. External Q-factor for input/output coupling dimension.

The coupling is a result of iterations and adjustments to the dimensions of the coupling areas of the filter through full-wave simulations until desired filter and response is achieved.

Fig. 7. Fabricated two pole fixed bandpass filter.

2.2 Double Coupling and Tunable Structure

An extended coupling overlaying the single coupling builds the basis of frequency shifting. The switching between the extended coupling and the single coupling forms tunable bandpass filter mechanism. The extended coupling is placed on top and bottom of the single coupling along with vias placed in the center of each length. The overlaying double coupling is shown in Fig. 8(a), while the notation values are displayed in Tab. 2.

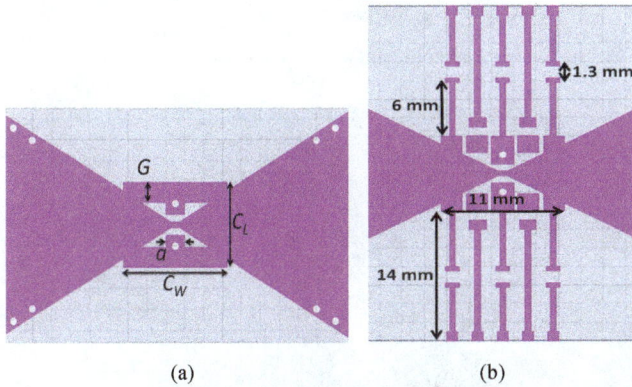

(a)   (b)

Fig. 8. (a) Overlaying double coupling, (b) Dimensions of bias circuitry.

Notation	Value (mm)	Notation	Value (mm)
W_t	36.0	W_c	0.5
L_t	70.0	W_x	2.0
L	35.0	L_c	30.0
W	30.0	C_w	11
W_a	12.0	C_L	8
W_b	15.9	G, a	2

Tab. 2. Tunable filter parameters dimensions.

The tunable filter is constructed based on packaged RF MEMS switches, a total of four packaged RMSW201 RADANT MEMS [16]. The filters are actuated with a voltage of 90 V, and an onboard bias circuitry is built to place the switches. The bias lines of 0.5 mm thickness with spacing of 1.5 mm are placed on the board to actuate the RF MEMS switches, the bias circuitry and its corresponding dimensions are shown in Fig. 8(b).

Fig. 9 shows the operational diagram of the RF MEMS switches along with its bias circuitry, a voltage of

90 V is applied at the gate of the switch, while GND is connected through 100 kΩ resistors to the drain and source of the switch. The switches are placed on pads created on top of the board in between the coupling area, an on state of the switch refers to forming the double coupling whereas the off state refers to single coupling. The pads are wire-bonded to the gate, drain and source of the RF MEMS switches.

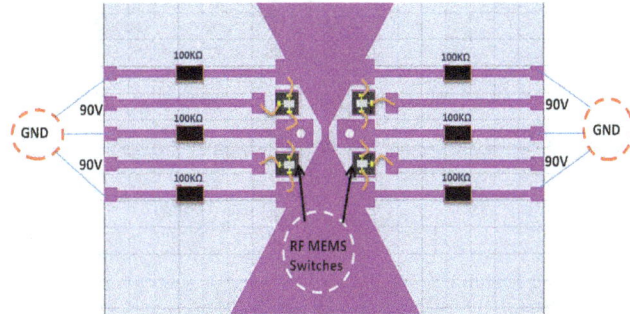

Fig. 9. The functionality and placement of RF MEMS and bias resistors (the figure is rotated).

3. Results and Discussion

The simulations to obtain the filter responses from the designed structures are conducted using ANSYS High Frequency Structure Simulator (HFSS). In addition Agilent Vector Network Analyzer (VNA) is utilized for the measurements of the fabricated filters.

The fixed filter described in the specifications in Tab. 1 and its corresponding design structure shown in Fig. 4 is realized with the responses shown in Fig. 10. The simulated S_{21} and S_{11} response of the bowtie shaped bandpass filter reveals that the S_{11} value at the center frequency of 7.3 GHz is less than -30 dB, whereas the S_{21} response is greater than -0.5 dB and the passband bandwidth at -3 dB is greater than 400 MHz. The measured S_{21} response at the center frequency of 7.3 GHz is better than -1.7 dB and its corresponding S_{11} response at the center frequency is less than -20 dB. The passband bandwidth at -3 dB is greater than 350 MHz.

The desired tunable filter with the double coupling shown in Fig. 8 and Fig. 9 is realized with the responses shown in Fig. 11 and Fig. 12. The details of the passband are depicted in Fig. 11(a) and Fig. 12(a) for the simulated and measured S_{21} responses, respectively.

Fig. 11 shows the response of the simulated two pole tunable filter designed for two resonant frequencies of 6.3 and 7.3 GHz. Response of the filter is obtained through realizing the resonator structure designed using the equations presented in Sec. 2.

The simulated S_{21} and S_{11} response of the bowtie shaped tunable bandpass filter reveals that when the switch is disconnected a two pole bandpass response is achieved at 6.3 GHz, whereas when the RF MEMS switch connects

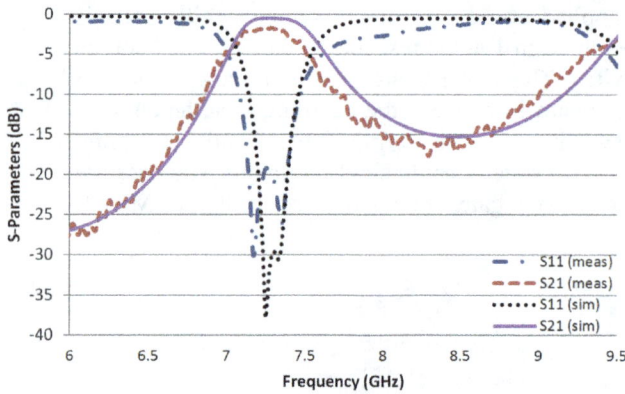

Fig. 10. Response of the fixed bowtie bandpass filter.

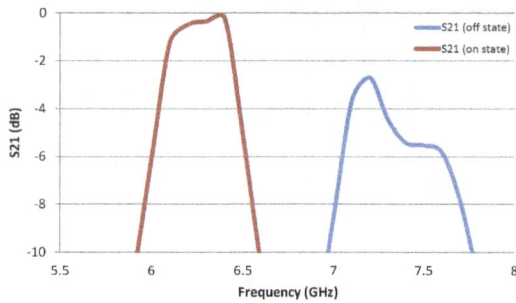

Fig. 11. (a) Passband details for the simulated S_{21} response.

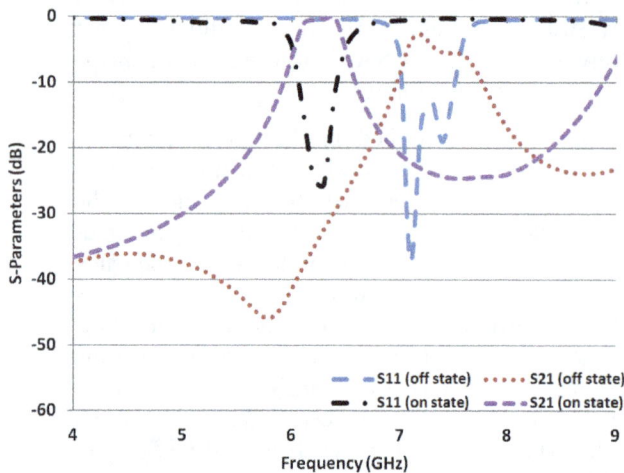

Fig. 11. (b) Simulated response of the tunable bandpass filter.

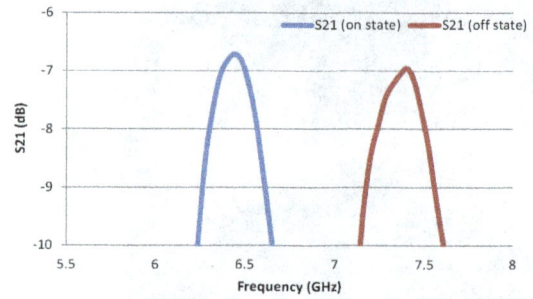

Fig. 12. (a) Passband details for the measured S_{21} response.

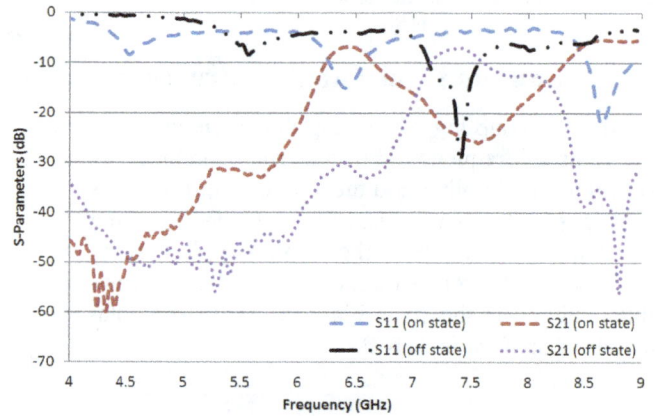

Fig. 12. (b) Measured response of the two pole tunable filter.

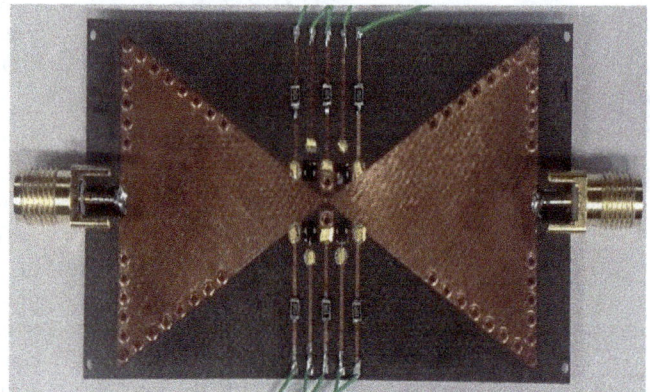

Fig. 13. Fabricated bowtie shaped tunable two pole filter.

Key Parameters	Simulated	Measured
Passband center frequency (GHz)	6.3, 7.3	6.44, 7.4
Passband S_{11} (dB)	< 25,20	< 14, 18
Passband S_{21} (dB)	> -0.5,-3.5	> -7,-7.5
Stopband rejection (dB)	> 25	> 25
Passband bandwidth at -3 dB level	0.4 GHz	0.435 GHz

Tab. 3. Summary of the tunable bandpass filter performance.

the double layered coupling a resonant frequency of 7.3 GHz is achieved. The passband bandwidth at -3 dB for both the resonant frequencies is observed to be greater than 400 MHz.

Measured S_{21} and S_{11} response of the two pole bandpass filter is presented in Fig. 12. The measured responses show similar trends in terms of frequency tuning between the two resonant modes as observed in the simulated responses. The S_{21} responses at the center frequencies of 6.44 and 7.4 GHz are better than -7 dB and their corresponding S_{11} responses are less than -14 dB and -18 dB respectively. The passband bandwidth at -3 dB is greater than 400 MHz, whereas the lower and upper stopband rejections are better than -25 dB. These responses are summarized in Tab. 3.

The measured insertion loss response includes the losses of the SMA connectors at the input and output of the tunable bandpass filter. The upper passband produced due to cancellation of the TE_{101} and TE_{201} modes is distanced enough from the filter passband, hence a pure Chebyshev response is observed as the design was based on Chebyshev LPF prototype parameters. However high order resonant modes presents spurious at 10 GHz. Consequently suppression of spurious is achievable by employing low-pass filters at the input and output of the resonators. Therefore the cut-off for the lowpass filters has to be matched with the upper passband of the bandpass filter.

4. Conclusion

A tunable bowtie shaped bandpass filter based on RF MEMS switches utilizing triangular structure SIW is proposed in this paper, the filter exhibits good performance and a miniaturized version of the SIW structure is exploited in the design process. The tunable filter is unique in terms of the tuning mechanism exploiting the coupling area. The RF MEMS two states switchable bandpass filter is designed, simulated, fabricated and the measured filter response shows two distinct frequency states at 6.4 and 7.4 GHz with a nearly constant bandwidth of 400 MHz. The simple structure of the filter allows readily integration with planar circuits and devices.

Acknowledgements

This work was supported by the Fundamental Research Grant Scheme (FRGS) of the Department of Higher Education, Malaysia. The authors would like to thank Dr. John Ojur Dennis and his student Mr. Farooq Ahmed of Nano fabrication facility in Universiti Teknologi PETRONAS, Malaysia for providing their guidance and wirebonding facility.

References

[1] JAMES, J., PENGBO SHEN, NKANSAH, A., XING LIANG, GOMES, N. J. Millimeter-wave wireless local area network over multimode fiber system demonstration. *IEEE Photonics Technology Letters*, 2010, vol. 22, no. 9, p. 601–603. DOI: 10.1109/LPT.2010.2043249

[2] LI WANG, GLISIC, S., BORNGRAEBER, J., WINKLER, W., SCHEYTT, J. C. A single-ended fully integrated SiGe 77/79 GHz receiver for automotive radar. *IEEE Journal of Solid-State Circuits*, 2008, vol. 43, no. 9, p. 1897–1908. DOI: 10.1109/JSSC.2008.2003994

[3] YAN, X., ZHANG, H., WU, C. Research and development of intelligent transportation systems. In *Proceedings of the 11th International Symposium on Distributed Computing and Applications to Business Engineering & Science*. Guilin (China), 2012, p. 321–327.

[4] WILSON, J. P., SCHUETZ, C. A., MARTIN, R., DILLON, T. E., YAO, P., PRATHER, D. W. Polarization sensitive millimeter-wave imaging sensor based on optical up-conversion scaled to a distributed aperture. In *Proceedings of the 37th International conference on Infrared, Millimeter, and Terahertz Waves*. Wollongong (Australia), 2012, vol. 1, p. 23–28.

[5] NIKNEJAD, A. M., HASHEMI, H. *Millimetre-wave Silicon Technology: 60 GHz and beyond*. Springer, 2008.

[6] HIROKAWA, J., ANDO, M. Single-layer feed waveguide consisting of posts for plane TEM wave excitation in parallel plates. *IEEE Transactions on Antennas and Propagation*, 1998, vol. 46, no. 5, p. 625–630. DOI: 10.1109/8.668903

[7] KI SEOK YANG, PINEL, S., IL KWON KIM, LASKAR, J. Lowloss integrated-waveguide passive circuits using liquid-crystal polymer System-on-Package (SOP) technology for millimeter-wave applications. *IEEE Transactions on Microwave Theory and Techniques*, 2006, vol. 54, no. 12, p. 4572–4579. DOI: 10.1109/TMTT.2006.886004

[8] JUNFENG XU, ZHI NING CHEN, XIANMING QING, WEI HONG. 140-GHz planar broadband LTCC SIW slot antenna array. *IEEE Transactions on Antennas and Propagation*, 2012, vol. 60, no. 6, p. 3025–3028. DOI: 10.1109/TAP.2012.2194673

[9] MOKHTAARI, M., BORNEMANN, J., RAMBABU, K., AMARI, S. Coupling-matrix design of dual and triple passband filters. *IEEE Transactions on Microwave Theory and Techniques*, 2006, vol. 54, no. 11, p. 3940–3946. DOI: 10.1109/TMTT.2006.884687

[10] REHMAN, M. Z. U., BAHARUDIN, Z., ZAKARIYA, M., KHIR, M., KHAN, M., WENG, P. W. Recent advances in miniaturization of substrate integrated waveguide bandpass filters and its applications in tunable filters. In *Proceedings of Business Engineering and Industrial Applications Colloquium*. (Malaysia), 2013, p. 109–114. DOI: 10.1109/BEIAC.2013.6560093

[11] SUNG TAE CHOI, KI SEOK YANG, TOKUDA, K., YONG HOON KIM. A V-band planar narrow bandpass filter using a new type integrated waveguide transition. *IEEE Microwave and Wireless Components Letters*, 2004, vol. 14, no. 12, p. 545–547. DOI: 10.1109/LMWC.2004.837386

[12] TANG, H. J., HONG, W., HAO, Z. C., CHEN, J. X., WU, K. Optimal design of compact millimetre-wave SIW circular cavity filters. *Electronics Letters*, 2006, vol. 41, no. 19, p. 1068–1069. DOI: 10.1049/el:20052251

[13] XIAO-PING CHEN, KE WU. Substrate integrated waveguide cross-coupled filter with negative coupling structure. *IEEE Transactions on Microwave Theory and Techniques*, 2008, vol. 56, no. 1, p. 142–149. DOI: 10.1109/TMTT.2007.912222

[14] DESLANDES, D., WU, K. Accurate modeling, wave mechanisms, and design considerations of a substrate integrated waveguide. *IEEE Transactions on Microwave Theory and Techniques*, 2006, vol. 54, no. 6, p. 2516–2526. DOI: 10.1109/TMTT.2006.875807

[15] ALI KHAN, A., MANDAL, M. K., SANYAL, S. Unloaded quality factor of a substrate integrated waveguide resonator and its variation with the substrate parameters. In *Proceedings of International Conference on Microwave and Photonics*. Dhanbad (India), 2013, p. 1–4. DOI: 10.1109/ICMAP.2013.6733496

[16] RADANT MEMS, MA, USA. *RMSW201 SPST RF MEMS switch (datasheet)*. 2 pages. [Online] Cited 2014-05-10. Available at: http://www.radantmems.com/radantmems.data/Library/RMSW201.pdf

About the Authors ...

Muhammad Zaka Ur REHMAN was born in Lahore, Pakistan in 1986. He is serving as a Lecturer in COMSATS Inst. of Information Technology (CIIT), Islamabad, Pakistan. He received his B.S. degree in Electronics from CIIT in 2007, the MSc Degree in Digital Signal Processing in Communication Systems from Lancaster University, UK, in 2010, and is currently working towards the Ph.D. degree in Electrical Engineering (with an emphasis on RF and Microwave Circuits) at the Universiti Teknologi PETRONAS, Perak, Malaysia. His research interests include RF MEMS for microwave applications, substrate integrated waveguide structures and reconfigurable filters design.

Zuhairi BAHARUDIN is a faculty member at Universiti Teknologi PETRONAS, Malaysia. He obtained his Diploma and B.Eng. Hons. Electrical from the University Technology MARA, Shah Alam, Malaysia and subsequently his M.Eng in Electrical Power Engineering from the University of South Australia, Australia. He received his Ph.D. degree in Electrical Engineering from the Universiti Teknologi PETRONAS in 2010. His research interests are in effects of harmonics on power systems, applications of artificial intelligence of load forecasting, and power economics operation and control.

Mohd Azman Bin ZAKARIYA received bachelors in Electrical Engineering from Universiti Teknologi Malaysia, and Master of Science in Communications and Signal Processing from University of Newcastle upon Tyne, UK. He is a lecturer in Universiti Teknologi PETRONAS, Malaysia. He is also working towards his Ph.D. from Universiti Sains Malaysia. His research interests include dielectric resonator antennas, defected ground structure.

Mohd Haris Md KHIR received the B. Eng. degree in Electrical and Electronic Engineering from Universiti Tek-

nologi MARA, Selangor, Malaysia, in 1999, the Masters of Science degree in Computer & System Engineering from Rensselaer Polytechnic Institute, New York, USA, in 2001, and the Ph.D. degree in Systems Engineering from Oakland University, Michigan, USA, in 2010. Since 2002, he has been with Universiti Teknologi PETRONAS, Perak, Malaysia, where he is currently a Senior Lecturer in Electrical & Electronic Engineering Department. His research interests include Micro-Electro-Mechanical Systems (MEMS) sensors/actuators design and fabrication based on CMOS and MUMPS technologies. He has successfully fabricated a number of MEMS devices such as accelerometers, micro-mirror, micro switches, energy harvester, electromagnetic sensors, gas sensors, and thermal electric generator (TEG) system.

Muhammad Taha JILANI received bachelor's degree in Electrical Technology followed by masters in Telecommunication, in 2007 & 2009 respectively. He is currently working toward Ph.D. in the area of dielectric material characterization using RF and microwave frequencies at Universiti Teknologi PETRONAS, Malaysia.

24

Design of a Wideband Inductively Coupled Loop Feed Patch Antenna for UHF RFID Tag

Mohd Saiful Riza BASHRI, Muhammad Ibn IBRAHIMY, S. M. A MOTAKABBER

Dept. of Electrical and Computer Engineering, International Islamic University Malaysia, 53100 Kuala Lumpur, Malaysia

mohdsaifulriza@yahoo.com, ibrahimy@iium.edu.my, amotakabber@iium.edu.my

Abstract. *A planar wideband patch antenna for ultra-high frequency (UHF) radio frequency identification (RFID) tag for metallic applications is presented in this research work. Three different shape patches are inductively coupled to a triangle loop to form wide impedance bandwidth for universal application UHF (860–960 MHz) RFID. The structure of the proposed antenna exhibits planar profile to provide ease of fabrication for cost reduction well suited for mass production. The simulation of the antenna was carried out using Finite Element Method (FEM) based software, Ansoft HFSS v13. The simulated and measured impedance bandwidth of 113 MHz and 117 MHz (Return Loss ≥ 6 dB) were achieved to cover the entire UHF RFID operating frequency band worldwide. The simulated and measured radiation patterns at the operating frequency of 915 MHz are in good agreement. Moreover the simulated maximum antenna gain at the bore sight direction in free space and when mounted on 200 × 200 mm² metal plate are -5.5 dBi and -9 dBi respectively which is enough to provide reasonable read range over the entire UHF RFID system operating band.*

Keywords

Complex impedance matching, patch antenna, radio frequency identification (RFID), metallic object, ultra high frequency (UHF)

1. Introduction

Recently, Radio Frequency Identification (RFID) technology is gaining traction in various sectors due to its numerous advantages such as it does not require line of sight, high read distance, fast date rate and large storage capacity as compared to conventional barcode technology [1]. Some of the sectors utilizing RFID are supply chain management, logistics, access control and real time location service (RTLS) etc. RFID in its basic form consists of tag, attached to an object to be tracked and reader whose function is to read the information contained inside the tag memory. Generally, RFID can be categorized into several types based on their operating frequencies, power source and protocols that govern its communication. Low frequency (LF) and high frequency (HF) systems are operated based on near-field communication thus having limited read range up to only 1 meter. As for ultra-high frequency (UHF) and microwave systems, the interaction between tag and reader is accomplished via propagating electromagnetic wave hence able to provide longer read range. As such, UHF based RFID technology is rapidly becoming the preferred solution.

Tag is made up of a microchip and an antenna connected together. To operate the microchip, ample power is needed. Due to cost factors, most systems employ passive tag where there is no on board power source such as battery to provide the power to the microchip. To circumvent this matter, tag antenna extracts the energy from the incident radio wave emitted by the reader to be delivered to the microchip. In addition, in the absence of transmitter on the tag, a special modulation technique is utilized in RFID called backscattering modulation [1]. In this method, the electromagnetic wave from the reader is modulated and reflected back to the reader. The modulation is performed by the tag microchip by changing its input impedance between two states which are matched and mismatched to the antenna input impedance to represent the binary code '1' and '0' of the information to be transmitted to the reader. The corresponding high (mismatched) and low (matched) power of the reflected wave received is then demodulated by the reader.

Antenna design is of great importance in passive UHF RFID system to ensure tag is able to operate properly [2], [3]. Although numerous works have been done in designing tag antenna, there are still many open issues that require further studies and research in order to truly exploit its potential. One of the issues is performance degradation of commonly used label typed dipole tag antenna [4–6] when placed on metal surface due to cancellation of tangential electric current at the boundary between the antenna and the metal surface [7], [8]. One of the many attempts to mitigate the problem is to separate the antenna and the metal surface by using a foam spacer to create constructive interference between the incoming and reflected signal. However, it results in thicker antenna structure which is unsuitable for RFID applications.

Another method that has been widely adopted is the use of microstrip patch antenna due to its grounded structure. When mounted on metal objects, the metal plane will act as an extension of its ground plane hence giving little effect to the antenna performance. Several microstrip antennas for UHF RFID tag have been proposed by [9–13]. However, they exhibit narrow bandwidth. To operate worldwide, the required impedance bandwidth should be able to cover the whole frequency range of UHF RFID band (860–960 MHz) [14]. List of operating frequency of several countries is shown in Tab. 1. Several solutions to improve the impedance bandwidth of patch antenna for RFID were presented by [15–19]. However, the structures of the presented antennas require multi or cross-layered configuration which will add significant manufacturing cost to the antenna fabrication due to additional process required. Moreover, the impedance bandwidth performances of the antenna were evaluated based on the half-power bandwidth (Return Loss ≥ 3 dB) that accounts for only half of the power accepted by the tag antenna to be actually delivered to the tag's microchip. Several complete planar patch antenna have been proposed by [20–23] although with limited bandwidth.

Region/country	Operating frequency, f (MHz)
North America	902–928
Europe	865–868
China	917–922
Japan	916–921 & 952–956
Australia	918–926
Hong Kong	865–868 & 920–925
Taiwan	922–928

Tab. 1. List of operating frequency for several major countries.

This letter proposes a planar wideband microstrip patch RFID tag antenna design for metallic applications. The wide impedance bandwidth is achieved by utilizing three radiating elements to excite three resonances close to each other. The complex impedance matching with the referenced microchip, Alien Higgs-3, with impedance value of $Z_c = 31 - j212\ \Omega$ and sensitivity, P_{th} of -18 dBm is realized by using inductively coupled triangle loop structure. The structure of the proposed antenna does not incorporate any via hole or shorting wall/plate which further simplify its fabrication process. The proposed antenna design concept and configuration will be explained in Sec. 2. Section 3 demonstrates the simulation and measurement results while conclusions are drawn in Sec. 4.

2. Antenna Design and Configuration

Several important factors for designing antenna for UHF RFID tag have been comprehensively presented by [2], [3]. The aim of this research is to design metal mountable tag antenna for use in metallic applications where typical label-type dipole antenna suffers performance deg-

radiation like shift in operating frequency that leads to impedance mismatch and distorted radiation pattern. Moreover, to realize a universal tag that is able to operate across the world, a wideband characteristic is required which is quite challenging for patch antenna due to its inherent narrow bandwidth. To begin with the antenna design, Alien Higgs-3 was selected as a referenced microchip [24]. The impedance of the microchip, Z_c, is $31 - j212\ \Omega$ at 915 MHz. Typically, most antenna is designed to match with 50 Ω characteristic impedance of feeding line such as coaxial cable. However, for tag antenna, its impedance must be conjugate matched with the impedance of the microchip which is connected to [25]. This is very crucial particularly for passive UHF RFID system where the tag itself does not possess its own power source to operate the microchip [2] and all the power needed is extracted from the electromagnetic signal emitted by the reader. To ensure sufficient power is delivered to the microchip, impedance matching is crucial. The evaluation of the matching efficiency can be evaluated based on return loss, RL as expressed in (1) below [26]

$$RL(\text{dB}) = -20\log_{10}|\Gamma| \tag{1}$$

where Γ is the reflection coefficient at the antenna input terminal. Γ can be calculated as shown in (2) below [25]

$$\Gamma = \frac{Z_c - Z_{in}^*}{Z_c + Z_{in}^*} \tag{2}$$

where Z_{in} is the antenna input impedance.

There are several impedance matching techniques that have been proposed such as T-matching network, inductive coupled feed loop, nested loop, open end microstrip line shorted to ground, proximity-coupled feed and open stub feed [2], [12], [23], [27]. In this work, an inductively coupled loop structure in the form of triangle was used for complex impedance matching with the referenced microchip. The resulting input impedance seen at the input terminal of the antenna due to the triangle feed loop is given by (3) [27]

$$Z_{in} = Z_{loop} + \frac{(2\pi f M)^2}{Z_A} \tag{3}$$

where $Z_{loop} = j2\pi f L_{loop}$ is the input impedance of the feed loop and Z_A is the antenna impedance without the matching element. Based on (3), resistance and reactance at the input terminal of the antenna can be calculated based on (4) and (5)

$$R_{in}(f_0) = \frac{(2\pi f_0 M)^2}{R_A(f_0)}, \tag{4}$$

$$X_{in}(f_0) = 2\pi f_0 L_{loop}. \tag{5}$$

It can be seen from (4) and (5) that the input resistance depends on the mutual inductance between the feeding loop and the patches while the reactance value is solely contributed by the loop's inductance. The mutual coupling, M is then determined by the size of the loop and its distance from the patches. As for the loop inductance, it is mainly affected by its aspect ratio. The inductance of the triangle loop feed network, L_{loop} can be approximated using (6) [28]

$$L_{loop} \approx N^2 \frac{\mu_0 \mu_r}{2\pi} \begin{bmatrix} 2c\ln\left(\dfrac{2c}{0.5s}\right) + b\ln\left(\dfrac{2c}{0.5s}\right) \\ -2(b+c)\sinh^{-1}\left(\dfrac{b^2}{\sqrt{4b^2c^2-b^4}}\right) \\ -2c\sinh^{-1}\left(\dfrac{2c^2-b^2}{\sqrt{4b^2c^2-b^4}}\right) - (2c+b) \end{bmatrix} \tag{6}$$

where b, c, and s are the dimension of the triangle loop. N, μ_0, and μ_r are number of turns of the loop, permittivity of free space and effective permittivity of the substrate. The approximate geometrical dimension of the matching loop calculated based (5) and (6) is shown in Tab. 2.

The impedance bandwidth of the proposed antenna is enhanced by utilizing coplanar multi-resonator configuration. Three patches of different shapes were constructed to be the radiating elements. The patches resonate at three different frequencies closed to each other to form a wide impedance bandwidth to cover the entire frequency range of UHF RFID. The radiating elements of the antenna are composed of one narrow rectangular patch and two meandered patches. The physical length, L of the patches can be initially approximated using the closed form expression as shown in (7)[29]

$$L = \frac{1}{2f_r\sqrt{\varepsilon_{reff}}\sqrt{\mu_0\varepsilon_0}} - 2\Delta L \tag{7}$$

where f_r, ε_{reff} and ΔL are the resonant frequency, effective dielectric constant and extension of patch length due to fringing field effect. The effective dielectric constant is calculated as in (8) [30]

$$\varepsilon_{reff} = \frac{(\varepsilon_r+1)}{2} + \frac{(\varepsilon_r-1)}{2}\left[1+\frac{12h}{W}\right]^{-1/2} \tag{8}$$

where W is the width of the patch. In this design, the width of the patches is chosen to be less than the effective width to reduce the overall size of the antenna. Nevertheless, a good balance between gain performance and size is observed. Then, all three patches are fed by the triangle feed loop at each side as illustrated in Fig. 1. The antenna design was simulated using commercial electromagnetic simulator Ansoft HFSS v13 based on the approximated calculation of the antenna parameters. Afterwards, parametric refinement on the antenna parameter for the matching loop and the radiating patches geometry were carried out. The input resistance was matched by varying the distances, $d1$, $d2$ and $d3$ of the patches from the loop. As for the reactance part, the geometry of the triangle loop, b, c and s were varied to get $X_{in} = 212j$ Ω required to cancel the capacitive

value of the microchip. FR-4 epoxy glass substrate with a dielectric constant, ε_r of 4.4 and thickness of 1.6 mm was used due to its cheap cost [30]. It also lowers the Q-value of the antenna thus contributes to increased bandwidth. The optimal design parameter is tabulated in Tab. 2. The antenna was then fabricated using photolithography and etching technique. The final prototype of the antenna is shown in Fig. 2. A test fixture shown in Fig. 3 was used to probe the antenna. The input impedance of the proposed antenna was then measured using the two port differential probe technique proposed by [31], [32] due the balanced feed structure of the antenna. The impedance measurement setup is shown in Fig. 4. The input impedance of the antenna was then extracted from the measured S-parameters.

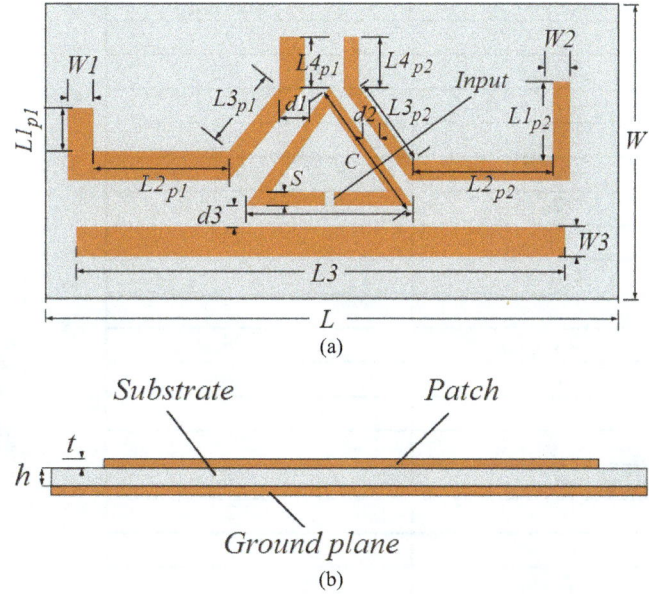

Fig. 1. Structure of the antenna. (a) Top view and (b) side view.

Fig. 2. Prototype of the antenna.

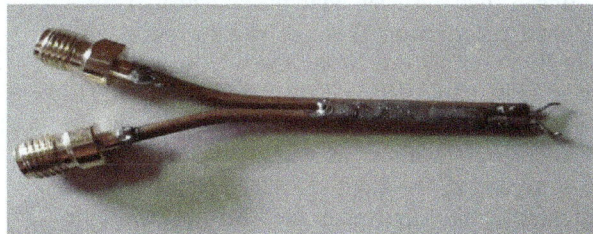

Fig. 3. Prototype of test fixture.

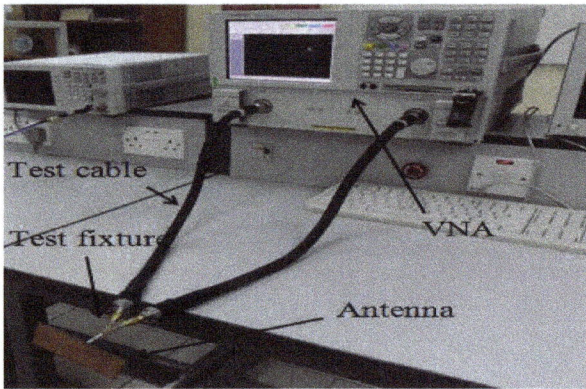

Fig. 4. Impedance measurement setup.

Parameter	Value (mm)
$W1$	7
$L1_{p1}$	10
$L2_{p1}$	37
$L3_{p1}$	10
$L4_{p1}$	12
$W2$	5
$L1_{p2}$	15
$L2_{p2}$	37
$L3_{p2}$	14
$L4_{p2}$	14
$W3$	5
$L3$	86
c	29.4
b	34
s	2
$d1$	2
$d2$	1
$d3$	2
t	0.0358
h	1.6
Ground plane and substrate	130 x 63

Tab. 2. Optimized design parameter of the proposed antenna.

3. Results and Discussion

The simulation and measurement results of the design are presented in this section to verify the initial deduction. The surface current density of the antenna illustrated in Fig. 5 to show the proposed antenna resonates at three different resonant frequencies 882 MHz, 908 MHz and 949 MHz. It can also be seen that the antenna exhibits linear horizontal polarization although slight cross polarization might be observed due to the meandering structure of the patches. The simulated and measured input impedance of the antenna against the conjugate impedance of the referenced microchip is shown in Fig. 6. By varying the parameter, d_1, d_2 and d_3 the input resistance of the antenna can be matched to the input resistance of the microchip over the UHF RFID frequency band while the size of the triangle loop feed, b, c and s can be adjusted to present the

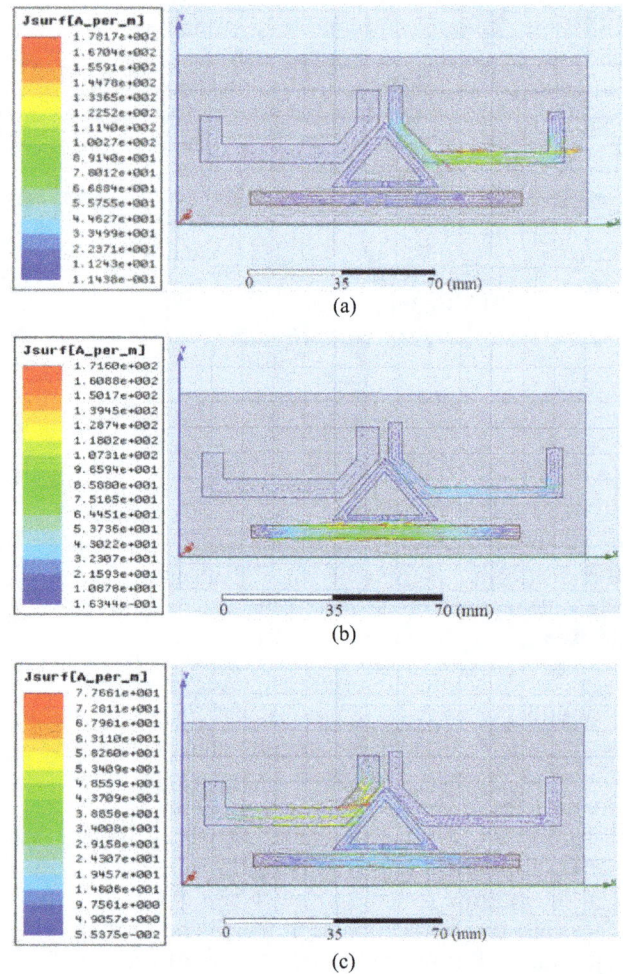

Fig. 5. Surface current distribution at three resonant frequencies. (a) 882.4 MHz, (b) 908.6 MHz and (c) 949.3 MHz.

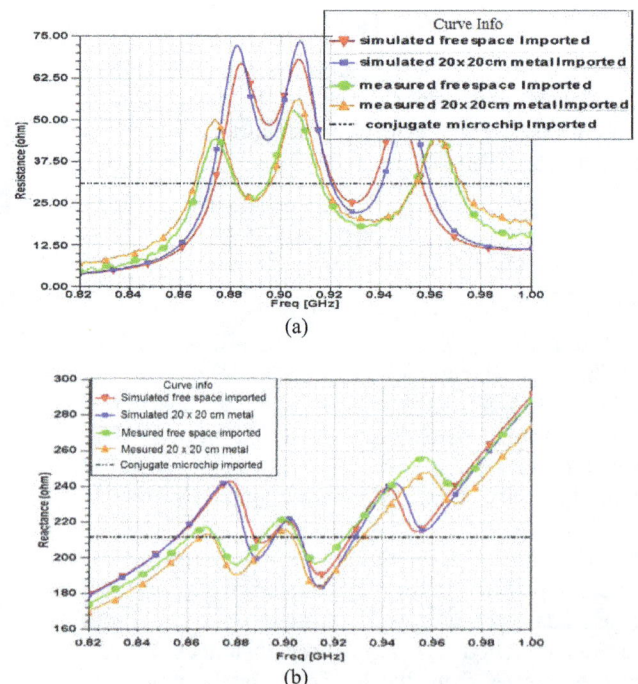

Fig. 6. Simulated and measured (a) resistance and (b) reactance value of the input impedance of the antenna against conjugate impedance of referenced microchip.

required reactance to cancel the capacitive value of the microchip. To observe the effect of metal surface to its performance, the antenna was simulated on top of metal plane with dimension of 200×200 mm^2 in addition to the free space scenario. For actual measurement, the antenna was then attached to a metal plate of the same size and the impedance of the antenna was measured experimentally.

The slight difference between the simulated and measured input impedance of the proposed antenna shown in Fig. 6 was due to the fabrication inaccuracy as well as the surrounding effects when the measurement was taken.

To evaluate the impedance bandwidth of the antenna, equation (1) is used to calculate the return loss of the antenna [12], [15], [16], [22], [33]. The simulated and measured return loss for both free space and mounted on metal plane is shown in Fig. 7. Despite the difference of input impedance between the simulation and measurement, both return loss performance for both cases are 113 MHz and 117 MHz, well over the required 100 MHz for the entire UHF RFID operating frequency band. As for comparison of the antenna performance on free space and when it is mounted on a metal plate, a slight shift on the resonant frequency is observed on the simulation result. However, based on the measurement, the antenna input impedances on free space and when it was attached on the metal surface is almost identical. Hence, it is evident that the proposed antenna works well when being mounted on the metallic surface. In order to further investigate the performance of the antenna, the peak gain of the antenna over the UHF RFID operating frequency band was simulated and the result is shown in Fig. 8 while the simulated antenna efficiency is shown in Fig. 9. It is seen that the efficiency of the antenna is quite low around 30% thus result in low gain. This is primarily due to the thin and lossy nature of the substrate. On the other hand, the use of the low permittivity substrate would result in increase of antenna size and cost. The peak gain of the antenna is more uniform across the operating frequency of 860–960 MHz when mounted on the metal plate as compared to free space condition. This is likely because of the reduced backside radiation due to reflection by the larger metallic surface as opposed to smaller size ground plane without the metal plate. Nevertheless, the low gain for tag antenna was not uncommon as reported in previous works since a read range of about 1 to 3 meters is enough for some of the RFID applications [17], [22], [34].

In addition, the far field radiation patterns of the antenna were measured in the anechoic chamber and compared to simulation results. The simulated and measured normalized E-field and H-field pattern at the operating

frequency of 915 MHz are depicted in Fig. 10 respectively. It can be safely concluded the radiation pattern for both simulation and measurement results are in good agreement.

To further evaluate the performance of the antenna, the theoretical maximum read range of the antenna was calculated using Friis transmission equation (8) at three different operating frequencies as listed in Tab. 2 [3]. The

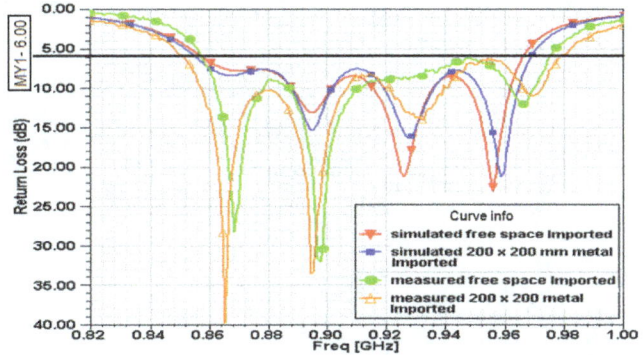

Fig. 7. Simulated and measured return loss of the antenna.

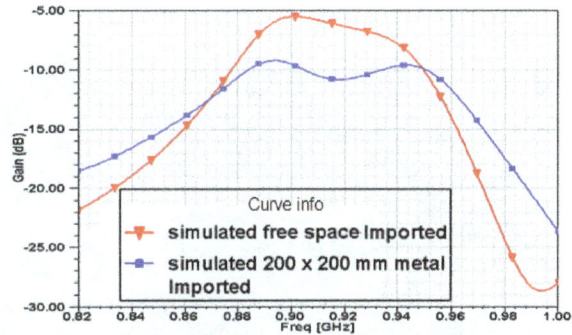

Fig. 8. Simulated peak gain of the antenna.

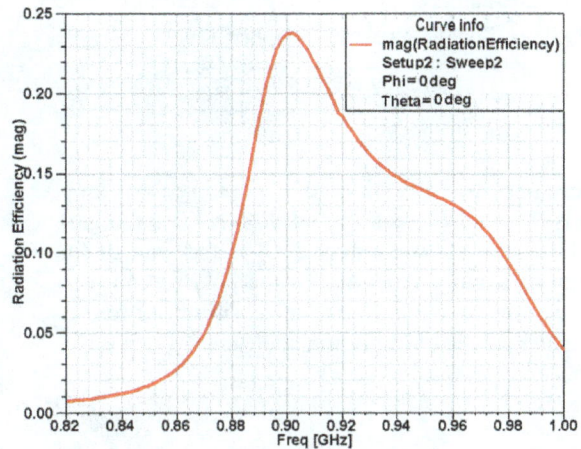

Fig. 9. Simulated antenna efficiency.

Country/ Region	Center freq., f_c (MHz)	EIRP (W)	Simulated peak gain (dBi)		Calculated read range (m)	
			Free space	200×200 mm^2 metal sheet	Free space	200×200 mm^2 metal sheet
Europe	886	3.3	-13.28	-11.50	2.68	3.33
North America	915	4	-6.05	-10.56	6.18	3.61
Japan	954	4	-11.54	-12.06	2.99	2.80

Tab. 3. Theoretical calculated read range of the antenna.

(a)

(b)

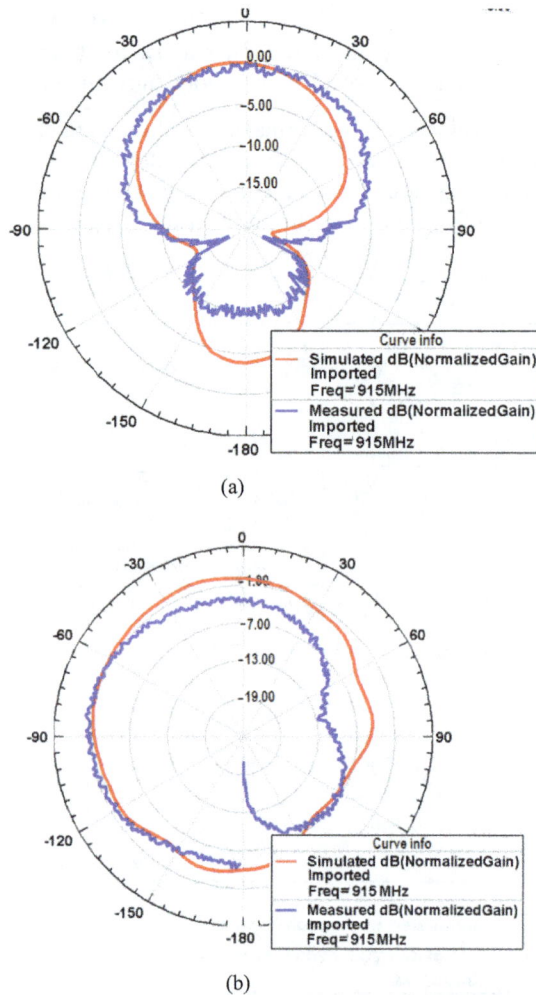

Fig. 10. Simulated and measured normalized radiation pattern at 915 MHz. (a) E-plane and (b) H-plane field pattern.

results show the antenna is able to give minimum read range of at least 2 meter worldwide as shown in Tab. 3.

$$r = \frac{\lambda}{4\pi} \sqrt{\frac{P_t G_t (\theta, \phi) G_r (\theta, \phi) p\tau}{P_{th}}} . \qquad (9)$$

λ is the free-space wavelength of the operating frequency, P_t represents the reader's transmitted power, G_t is the gain of the reader's transmitting antenna, G_r is the gain of the receiving tag antenna, p accounts for the polarization mismatch between the antenna, P_{th} is the sensitivity of microchip and τ is the power transmission coefficient given by

$$\tau = 1 - |\Gamma|^2 . \qquad (10)$$

4. Conclusion

A new wideband microstrip antenna for tagging metallic objects is presented in this paper. The designed tag antenna demonstrates simulated and measured impedance bandwidth of 113 MHz and 117 MHz based on 6 dB return loss. Moreover, the antenna exhibits planar configuration without any multi or cross-layered configuration which

significantly reduces the fabrication cost especially for mass production. For future work, the antenna will be integrated with the referenced microchip to experimentally measure the read range of the antenna and comparison will be made with the theoretical results.

Acknowledgements

This research has been supported by the Ministry of Higher Education of Malaysia through the Fundamental Research Grant Scheme FRGS13-027-0268.

References

[1] DOBKIN, D. M. *The RF in RFID : Passive UHF RFID in Practice*. Massachusetts: Elsevier Inc., 2008.

[2] MARROCCO, G. The art of UHF RFID antenna design: impedance-matching and size-reduction techniques. *IEEE Antennas and Propagation Magazine,* 2008, vol. 50, no. 1, p. 66 to 79. DOI: 10.1109/MAP.2008.4494504

[3] RAO, K. V. S., et al. Antenna design for UHF RFID tags: A review and a practical application. *IEEE Transactions on Antennas and Propagation*, 2005, vol. 53, no. 12, p. 3870–3876. DOI: 10.1109/TAP.2005.859919

[4] YANG, B., FENG, Q. A folded dipole antenna for RFID tag. In *Proceedings of International Conference on Microwave and Millimeter Wave Technology*. Nanjing (China), 2008, p. 1047–1049. DOI: 10.1109/ICMMT.2008.4540601

[5] CHOI, Y., et al. Design of modified folded dipole antenna for UHF RFID tag. *Electronics Letters*, 2009, vol. 45, p. 387–389. DOI: 10.1049/el.2009.0198

[6] MONTI, G., et al. Broad-band dipole for RFID applications. *Progress In Electromagnetics Research C*, 2010, vol. 12, p. 163 to 172. doi:10.2528/PIERC10012606

[7] PROTHRO, J. T., et al. The effects of a metal ground plane on RFID tag antennas. In *Proceedings of the IEEE Antennas and Propagation Society International Symposium*, 2006. DOI: 10.1109/APS.2006.1711302

[8] GHANNAY, N., et al. Effects of metal plate to RFID tag antenna parameters. In *Proceedings of the Mediterrannean Microwave Symposium 2009*. Tangiers (Morrocco), 2009, p. 1–3. DOI: 10.1109/MMS.2009.5409801

[9] TASHI, T., et al. A complete planner design of microstrip patch antenna for a passive UHF RFID tag. In *Proceedings of the 17th International Conference on Automation and ComputingICAC 2011*. Huddersfield (UK), 2011, p. 12–17.

[10] SON, H.-W., et al. Design of wideband RFID tag antenna for metallic surfaces. *Electronics Letters*, 2006, vol. 42, no. 5, p. 263 to 265.DOI: 10.1049/el:20064323

[11] CHOI, W., et al. An RFID tag using a planar inverted-F antenna capable of being stuck to metallic objects. *ETRI Journal*, 2006, vol. 20, p. 216–218. DOI: DOI: 10.4218/etrij.06.0205.0082

[12] SON, H.-W., JEONG, S.-H. Wideband RFID tag antenna for metallic surfaces using proximity-coupled feed. *IEEE Antennas and Wireless Propagation Letters*, 2011, vol. 10, p. 377–380. DOI: 10.1109/LAWP.2011.2148151

[13] MO, L., QIN, C. Tunable compact UHFRFID metal tag based on CPW open stub feed PIFA antenna. *International Journal of Antennas and Propagation*, 2012, 8 p. DOI: 10.1155/2012/167658

[14] Regulation, U. *Regulatory status for using RFID in the UHF spectrum*, Cited 28-09-2012. Available: http://www.gs1.org/docs /epcglobal/UHF_Regulations.pdf

[15] MO, L., et al. Broadband UHF RFID tag antenna with a pair of U slots mountable on metallic objects. *Electronics Letters*, 2008, vol. 44, p. 1173–1174. DOI: 10.1049/el:20089813

[16] HUANG, J. Z., et al. A compact broadband patch antenna for UHF RFID tags. In *Proceedings of the Asia Pacific Microwave Conference APMC2009*. Singapore, 2009, p. 1044–1047. DOI: 10.1109/APMC.2009.5384364

[17] LAI, M., et al. Low-profile broadband RFID tag antennas mountable on metallic objects. In *Proceedings of the IEEE Antennas and Propagation Society International Symposium*. Toronto (Canada), 2010, 4 p. DOI: 10.1109/APS.2010.5561167

[18] TAN, L. R., WU, R. X. Miniaturized broadband tag antenna for multi-standard UHF RFID applications. In *Proceedings of the IEEE International Conference on Microwave Technology and Computational Electromagnetics*. Beijing (China), 2011, p. 274 to 276. DOI: 10.1109/ICMTCE.2011.5915510

[19] LU, J.-H., ZHENG, G.-T. Planar broadband tag antenna mounted on the metallic material for UHF RFID system. *IEEE Antennas and Propagation Magazine*, 2011, vol. 10, p. 1405–1408. DOI: 10.1109/LAWP.2011.217899

[20] EUNNI, M. S. M., DEAVOURS, D. D. A novel planar microstrip antenna design for UHF RFID. *Journal of Systemics, Cybernetics and Informatics*, 2007, vol. 5, no. 1, p. 6–10.

[21] TASHI, et al. Design and simulation of UHF RFID tag antennas and performance evaluation in presence of a metallic surface. In *Proceedings of the 5th International Conference on Software, Knowledge Information, Industrial Management and Applications*. Benevento (Italy), 2011, 5 p. DOI: 10.1109/SKIMA.2011.6089974

[22] CHO, H.-G., et al. Design of an embedded-feed type microstrip patch antenna for UHF radio frequency identification tag on metallic objects. *IET Microwaves, Antennas & Propagation*, 2010, vol. 4, p. 1232–1239. DOI:10.1049/iet-map.2009.040

[23] MO, L., QIN, C. Planar UHF RFID tag antenna with open stub feed for metallic objects. *IEEE Transactions on Antennas and Propagation*, 2010, vol. 58, no. 9, p. 3037–3043. DOI: 10.1109/TAP.2010.2052570

[24] Alien. *Alien Higgs 3 EPC Class 1 Gen 2 RFID Tag IC*. Cited 20-10-2012. Available: http://www.alientechnology.com/docs/ products/Alien-Technology-Higgs-3-ALC-360.pdf

[25] LOO, C.-H., et al. Chip impedance matching for UHF RFID tag antenna design. *Progress In Electromagnetics Research*, 2008, vol. 81, p. 359–370. DOI:10.2528/PIER08011804

[26] BIRD, T. S. Definition and misuse of return loss. *IEEE Antennas and Propagation Magazine*, 2009, vol. 51, p. 166–167.

[27] SON, H.-W., PYO, C.-S. Design of RFID tag antennas using an inductively coupled feed. *Electronics Letters*, 2005, vol. 41, no. 18, p. 994–996. DOI: 10.1049/el:20051536

[28] GROVER, F. W. *Inductance Calculations: Working Formulas and Tables*. New York: D. Van Nostrand, 1946.

[29] Balanis, C. A. *Antenna Theory Analysis and Design*. 3rd ed. New Jersey: John Wiley & Sons, 2005.

[30] KUMAR, G., RAY, K. P. *Broadband Microstrip Antenna*. Noorwood, MA: Artech House, 2003.

[31] KUO, S.-K., et al. An accurate method for impedance measurement of RFID tag antenna. *Progress In Electromagnetics Research*, 2008, vol. 83, p. 93–106. DOI:10.2528/PIER08042104

[32] QING, X. et al. Impedance characterization of RFID tag antennas and application in tag co-design. *IEEE Transactions on Microwave Theory and Techniques*, 2009, vol. 57, p. 1268–1274. DOI: 10.1109/TMTT.2009.2017288

[33] XU, L., et al. UHF RFID tag antenna with broadband characteristic. *Electronics Letters*, 2008, vol. 44, no. 2, p. 79–80. DOI: 10.1049/el:20083009

[34] LU, J.-H., HUNG, K.-T. Planar inverted-E antenna for UHF RFID tag onmetallic objects with bandwidth enhancement. *Electronics Letters*, 2010, vol. 46, no. 17, p. 1182–1183. DOI: 10.1049/el.2010.0817

About the Authors ...

Mohd Saiful Riza BASHRI was born in Selangor, Malaysia, in May 1985. He received the B.Sc. degrees in Communication Engineering from International Islamic University Malaysia (IIUM) in 2009. He worked for Telekom Malaysia (TM) from 2009 to 2011 as an assistant manager which oversaw the planning of submarine cable network in the South East Asia region. Currently he is pursuing his master degree at IIUM. His research interests include antenna design, RFID and microwave devices.

Muhammad Ibn IBRAHIMY was born in Sirajgonj, Bangladesh, in August 1962. He received the B.Sc. and M.Sc. degrees in Applied Physics and Electronics from the University of Rajshahi, Bangladesh, in 1985 and 1986, respectively, and the Ph.D. degree in Biomedical Signal Processing from the National University of Malaysia, in 2001. From 2001 to 2003, he was a Postdoctoral Fellow in the Dept. of Electrical and Electronic Engineering at the Mie University of Japan. He is now Associate Professor in the Dept. of Electrical and Computer Engineering, IIUM. Dr. Ibrahimy is a senior member of the Inst. of Electrical and Electronics Engineers (IEEE), a member of the Bangladesh Computer Society (BCS) and the Bangladesh Electronics Society (BEC). His research interests are Analog and Digital Electronic System Design, Medical and Industrial Instrumentation, Biomedical Signal Processing, VLSI Design, RFID and Computer Networks in Telemedicine. He has published about 50 research articles in peer reviewed journals, conferences and 2 books.

S. M. A. MOTAKABBER was born in Naogaon, Bangladesh, in May 1966. He received the B.Sc. and M.Sc. degrees in Applied Physics and Electronics from the University of Rajshahi, Bangladesh, in 1986 and 1987, respectively, and the Ph.D. degree in Electrical, Electronic and Systems Engineering from the National University of Malaysia, in 2011. From 1993 to 2011, he was an Associate Professor in the Dept. of Applied Physics and Electronic Engineering, University of Rajshahi, Bangladesh. He is now an Assistant Professor in the Dept. of Electrical and Computer Engineering, IIUM. Dr. Motakabber is a member of the Inst. of Electrical and Electronics Engineers (IEEE), a life member of the Bangladesh Association for Advancement of Science (BAAS) and the Bangladesh Electronics Society (BEC). His research interests are Analog and Digital Electronic System Design, Medical and Industrial Instrumentation, VLSI Design, RFID, Robotics, Automation and Computer Control Systems. He has published about 20 research articles in peer reviewed journals, conferences and 1 book.

25

An Electronically Reconfigurable Patch Antenna Design for Polarization Diversity with Fixed Resonant Frequency

Mohamed Nasrun OSMAN[1], Mohamad Kamal A. RAHIM[1], Peter GARDNER[2], Mohamad Rijal HAMID[1], Mohd Fairus MOHD YUSOFF[1], Huda A. MAJID[1]

[1]Communication Engineering Dept., Faculty of Electrical Engineering, Universiti Teknologi Malaysia, 81310 Skudai, Johor, Malaysia
[2]School of Electronic, Electrical and Computer Engineering, University of Birmingham, Edgbaston, Birmingham, B15 2TT, United Kingdom.

mnasrun2@live.utm.my, mkamal@fke.utm.my, p.gardner@bham.ac.uk, rijal@fke.utm.my, fairus@fke.utm.my, huda2@live.utm.my

Abstract. *In this paper, an electronically polarization reconfigurable circular patch antenna, with fixed resonant frequency and operating at Wireless Local Area Network (WLAN) frequency band (2.4–2.48 GHz,) is presented. The structure of the proposed design consists of a circular patch as a radiating element fed by a coaxial probe, cooperated with four equal-length slits etched on the edge along the x-axis and y-axis. A total of four switches were used and embedded across the slits at specific locations, thus controlling the length of the slits. By activating and deactivating the switches (ON and OFF) across the slits, the current on the patch is changed, thus modifying the electric field and polarization of the antenna. Consequently, the polarization excited by the proposed antenna can be switched into three types; linear polarization, left-hand circular polarization or right-hand circular polarization. This paper proposes a simple approach that enables switching the polarizations and excites at the same operating frequency. Simulated and measured results of the ideal case (using copper strip switches) and the real case (using PIN diode switches) are compared and presented to demonstrate the performance of the antenna.*

Keywords

Polarization reconfigurable antenna, circular patch, slit perturbations

1. Introduction

Antenna is a key and critical component in wireless telecommunication systems. The evolution of the antenna has rapidly grown and been extensively investigated. It leads to a change of system requirements, hence making it more complex. Adjustments of the operating system scenario are needed in order to meet current trends and demands from end users - such as light weight and compacted devices with enhanced performance. Conventional antenna may face restrictions and limitations to adapt to the new adjustments, due to their characteristics of being fixed and inflexible. One solution to overcome this restriction is the use of reconfigurable antenna.

Reconfigurable antennas, or multi-functional antennas, have recently become quite an active and important topic of research all over the globe. This is due to the fact that the characteristics and properties of the reconfigurable antenna - such as operating frequency/bandwidth, far-field radiation pattern and polarization [1] - can be altered dynamically using external control, hence providing additional functionality and versatility to the systems. Flexibility and reconfigurability features could result in the adaptability of the antenna to address complex system requirements and new specifications. Furthermore, through this concept, the number of components, sizes and hardware complexities could be reduced [2] - hence the cost would be more economically efficient. These benefits would render reconfigurable antennas to become a highly desired feature in modern radio-frequency (RF) systems for wireless and satellite communications.

Reconfigurability can be achieved in different ways, such as through deliberate modification of the antenna physical structure, radiating edge and feeding network. These changes will alter the current flows and distributions on the antenna structure, hence modifying the performance characteristics. To achieve this reconfigurability feature, mechanical [3], optical (photoconducting switches) [4] or electronic (RF switching devices) [5] switches can be used. Among these types of switches, RF PIN diodes and RF MEMS are typical components used in designing the reconfigurable antenna. The PIN diode needs less complicated biasing components, and the lower cost results in it becoming a more popular choice amongst researchers. Even though PIN diodes also have a few drawbacks, such as high insertion loss and lower efficiency, the simplicity in biasing is a key criterion for the selection. The equivalent circuit of the PIN diode for forward bias and reverse bias is shown in Fig. 1.

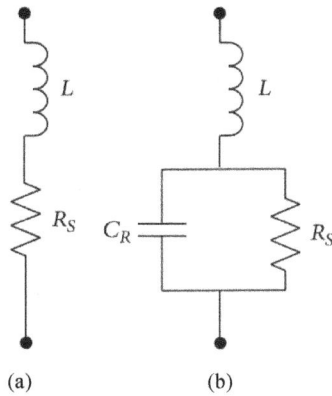

Fig. 1. PIN diode equivalent circuit. (a) forward bias; (b) reverse bias.

Recently, polarization reconfigurable antennas, which can incorporate and offer various polarizations from a single antenna, have attracted considerable attention. This is because the antenna has the ability to reduce the multipath fading [7], and the realization of frequency re-use [8]. In addition, this type of antenna is experimentally proven to improve the capacity offered in the multiple-input-multiple-output (MIMO) systems [9], [10]. For instance, the polarization reconfigurable antenna might take place between linear polarization (LP) and circular polarization (CP), between orthogonal LPs [11], [12], or between left-hand (LHCP) and right-hand circular polarization (RHCP) [13], [14]. Ideally, the main goals in designing reconfigurable antennas are to have the characteristics to be separable and independent. This is because the major challenge to achieve this kind of reconfigurability is to accomplish, without significant changes, on the frequency response and impedance matching. Therefore, careful design approaches are required in order to achieve this feature.

A few papers on the polarization reconfigurable antenna between LP and CP are reported in the literature. Reconfigurable polarization is achieved in [15], through the control of two crossing diagonal slots on the square patch. A similar technique is also used in [16]. In [17], the authors incorporated PIN diodes on the inset of the U-shaped slotline. However, the design required switchable matching stubs in order to obtain good reflection coefficient, which adds more size and complexity to the structure. Furthermore, reference [18] proposed an interesting design approach to achieve polarization reconfigurability features by creating loop slots in the ground plane. The switches were embedded on every slot of the ground plane, which altered the desired polarization sense. Even though the proposed solutions [16-18] managed to obtain a polarization reconfigurable antenna, however the changes of the polarization sense were excited at a different resonant frequency.

Hence, this paper proposes polarization reconfigurable antenna with fixed operating frequency. Details of the design, development and analysis of the polarization reconfigurable circular patch antenna with switchable slit length are presented. Two pairs of slits were located at the edge of the circular radiating patch, positioned on the x-axis and y-axis. The polarization reconfigurability feature of the an-

tenna was executed by placing four switches at specific positions across the slits. The change of the state of the switches consequently alters the length of the slit, hence creating a length difference between the slit in the x-axis and the y-axis. This, accordingly, will determine the types of polarization excited by the antenna - either LHCP, RHCP or LP. Interestingly, all types of polarization senses produced are operated at the same resonant frequency.

Work done in [19] proposed quite a similar approach. However, the polarization was only able to be reconfigured between LHCP and RHCP, without consideration made for the LP. Despite that, the biasing technique applied in this paper is much simpler and requires smaller dimensions.

This manuscript is divided into two main sections based on the type of switch adopted: "Design A" (ideal diode) and "Design B" (PIN diode). The simulated and measured results for both designs are presented, compared and analyzed.

2. Stage 1: Proof of Concept (Copper Strips)

In this section, the initial stage of the design started with the employment of ideal switch. The ideal switch is represented by copper strips or metal pins, with dimensions of 1 mm × 1 mm. The presence of the copper strips indicates the switch in the 'ON' state, meanwhile the absence denotes the 'OFF' state condition.

2.1 Antenna Geometry and Operation

The schematic geometry of the proposed antenna is illustrated in Fig. 2. This consists of a circular patch with radius of r on the finite fully grounded Taconic RF-35 substrate (dimensions of $L \times W$ in mm) with a dielectric constant, ε_r of 3.52, thickness, h of 1.52 mm and tangent loss, $\tan\delta$ of 0.0018. Four slits, which have a length of L_s mm and width of W_s mm are etched at the corner of the circular patch, located along the x-axis and y-axis of the structure with 90° apart from each other. Each slit will have a switch located across it, thus altering the length of the slits. As shown in Fig. 2, the switches are placed at the distance of L_p from the center of the patch. The location of the switch is crucial as it determines the optimum result for the axial ratio. The feed location is positioned diagonally and is fed from the back of the structure through subminiature (SMA) probe. Parameter d is optimized to achieve good impedance matching.

The design procedure generally started with determining the value of unloaded Q_o of the patch, which depends on the radius r, thickness of the substrate h, and the dielectric constant ε_r. By using (1), the value of slit perturbation is determined [20].

$$\left|\frac{\Delta S}{S}\right| = \frac{1}{1.841Q_0} \qquad (1)$$

where S is area of patch and ΔS is the area of slit perturbation. Then, the dimension of the slit is optimized using parameter sweep in order to obtain good circular polarization.

The approach and mechanism of the proposed antenna can be explained using the cavity model method. Due to the diagonal feeding on the structure and perturbation segments, the two near degenerated orthogonal resonant modes TM_{01} and TM_{10} are excited simultaneously. The existence of the perturbation slits will drive the surface current at the edge of the structure to move along it. The lengthened path is only affected to the current travel in the perpendicular direction to the L_s. However, from the parallel direction, the small width, W_s of the perturbation slits will slightly affect the current that is coming from that route. For example, when the pair of slits is cut on the x-axis of the patch, only TM_{01} mode will be affected without giving much effect on the TM_{10} mode, and vice versa.

Due to the presence of the copper strips, the slit length will be shortened, thereby causing an effective surface current to flow across it (shortest distance) instead of travelling around the slits. Consequently, the length difference between the slit on the x-axis and on the y-axis will provide the phase delay between both orthogonal resonant modes. At a specific length difference, the two orthogonal degenerated resonant modes will have the same amplitude and in-phase quadrature, which the CP is excited. Meanwhile, LP is excited when two switches on the x-axis and y-axis are ON, which means no phase difference between both orthogonal resonant modes.

Overall, this proposed antenna works in three polarization modes; namely LHCP, RHCP and LP. The type of excited polarization depends on the configurations of four switches. The switching conditions of all switches, with the respected modes, are tabulated in Tab. 1. The other configurations are not presented, as they provide the same response and redundancy results due to the geometry symmetry. Fig. 3 shows the physical structure used in the simulation, and the photograph of the prototype is shown in Fig. 4.

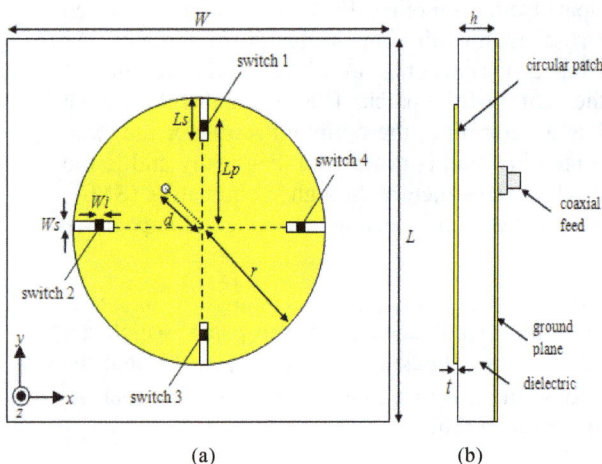

Fig. 3. The physical structure of the proposed antenna used in the simulation: (a) C1-LHCP; (b) C2-RHCP; (c) C3-LP.

Configuration	State of switch				Polarization mode
	SW1	SW2	SW3	SW4	
C1-LHCP	ON	OFF	ON	OFF	LHCP
C2-RHCP	OFF	ON	OFF	ON	RHCP
C3-LP	ON	ON	OFF	OFF	LP

Tab. 1. Switching configurations of the proposed antenna.

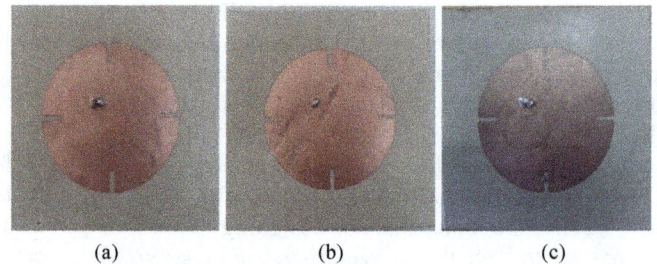

Fig. 4. Front view photograph of fabricated antenna prototypes. (a) C1-LHCP. (b) C2-RHCP. (c) C3-LP.

2.2 Result and Analysis of the Proposed Antenna

The prototype of the antenna is tested and analyzed. S-parameters and far-field test is measured using a Rohde & Schwartz ZVL network analyzer and in an anechoic chamber, respectively. The simulation is carried out using Computer Simulation Technology (CST) Microwave Studio full wave simulator.

Figure 5 shows the comparison between simulated and measured reflection coefficient of the proposed antenna for all switch configurations. It can clearly be seen that for C1-LHCP and C2-RHCP, the two degenerated resonant modes (TM_{01} and TM_{10}) are excited with close frequency at 2.45 GHz/2.49 GHz and 2.45 GHz/2.50 GHz, respectively. Meanwhile, the resonant frequency for C3-LP is 2.48 GHz. The measured -10 dB bandwidths (BW) of S_{11} are 71 MHz (2.438 ÷ 2.509 GHz), 84 MHz (2.437 GHz to 2.521 GHz) and 43 MHz (2.462 ÷ 2.505 GHz) for C1-LHCP, C2-RHCP and C3-LP, respectively. A good impedance matching is achieved for all configurations. It is observed that the measured center frequency is slightly higher than the simulated one. This frequency shift may be due to fabrication uncertainty.

The simulated and measured result of the axial ratio of two CP operations is shown in Fig. 6. For C1-LHCP, the

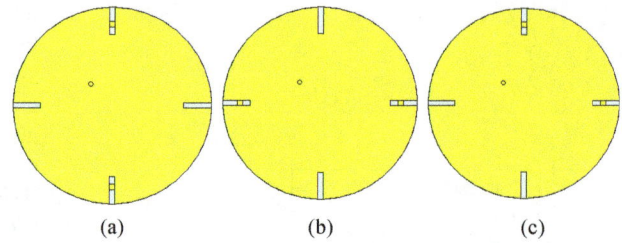

Fig. 2. Geometry of the proposed antenna: (a) front view; (b) side view.

(a)

(b)

(c)

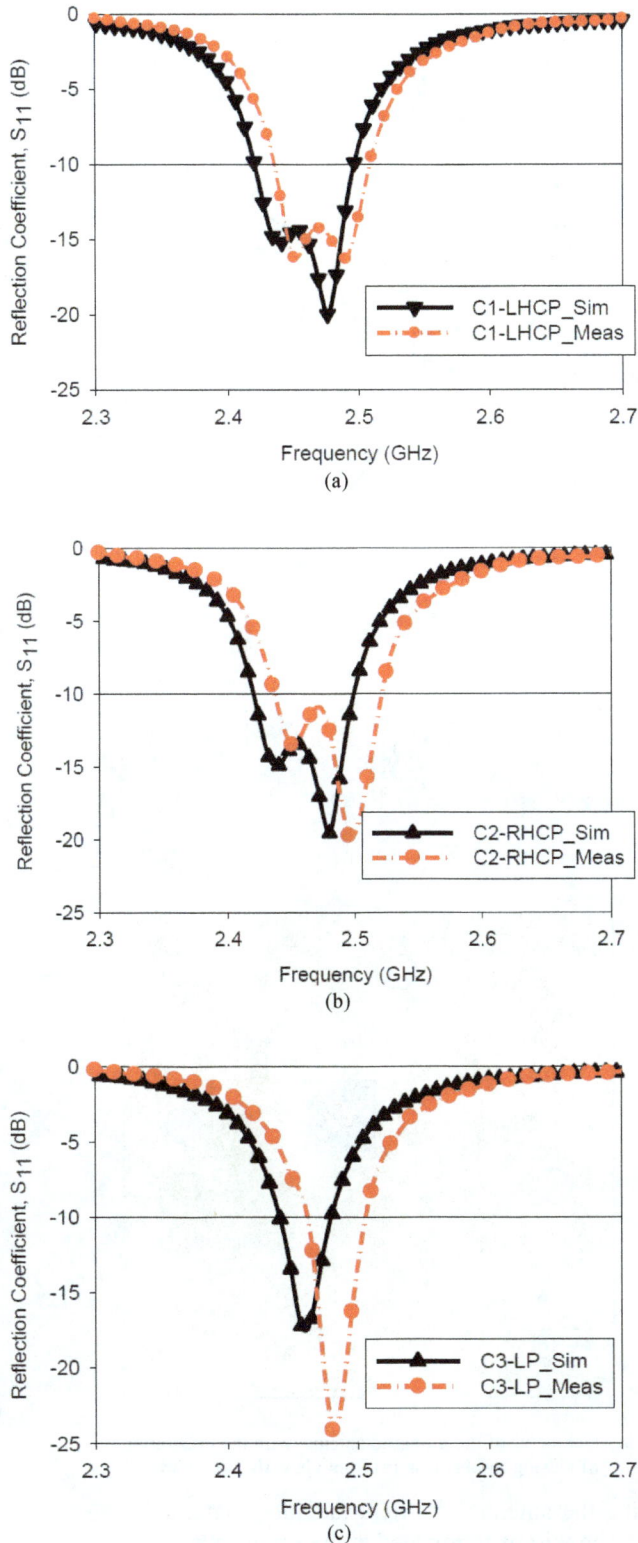

Fig. 5. Comparison of simulated and measured reflection coefficient of the proposed antenna: (a) C1-LHCP; (b) C2-RHCP; (c) C3-LP.

measured 3dB axial ratio bandwidth is 19 MHz (2.456 ÷ 2.475 GHz), or 0.77% with respect to the center frequency of 2.468 GHz. In addition, the BWCP for C2-RHCP is 18 MHz (2.463 ÷ 2.481 GHz), or 0.73% with respect to frequency of 2.472 GHz. The phenomenon variance of the reflection coefficient result leads to the axial

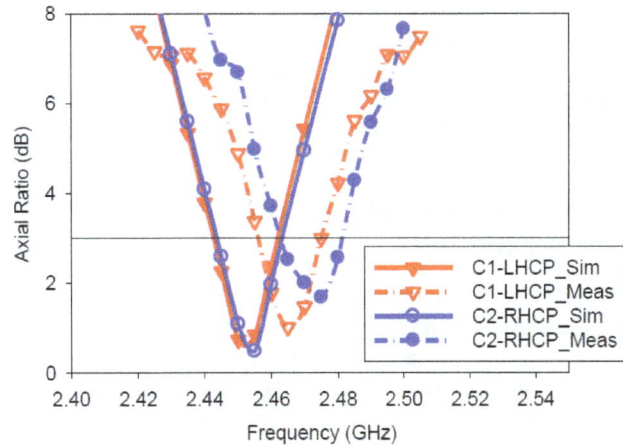

Fig. 6. Simulated and measured axial ratio.

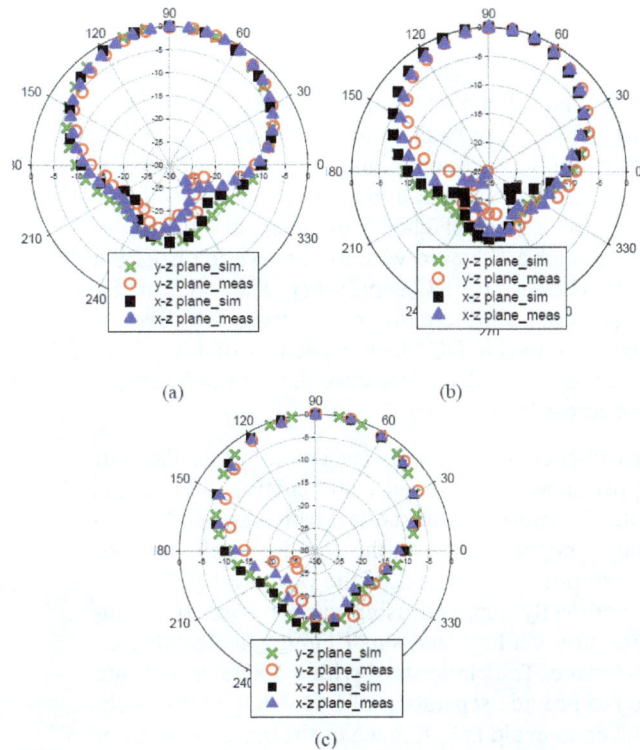

(a)   (b)

(c)

Fig. 7. Simulated and measured normalized far-field radiation pattern at frequency of 2.47 GHz: (a) C1-LHCP; (b) C2-RHCP; (c) C3-LP.

ratio to be shifted as well. The far-field radiation of the proposed antenna is investigated. Figure 7 presents the simulated and measured normalized radiation pattern (*y-z* plane and *x-z* plane) at 2.47 GHz. Good radiation pattern performance, with broadside form, is achieved for all types of polarization modes at the respective center frequency.

3. Stage 2: Antenna Design and Operation; Realization of Active Switches (PIN Diodes)

In this section, the copper strips are replaced by RF PIN diode. In this design, the RF PIN diode being used is

Infineon BAR 50-02L. For activating and deactivating the diodes, the antenna structure is modified by integrating a biasing circuit into the structure. To activate the diode, forward bias current is supplied to the ON-state, and the diode is left unbiased for the OFF state condition.

3.1 Modified Structure of the Antenna with Integrated Biasing Circuit

Figure 8 illustrates the antenna structure with modifications, including biased line and lump elements for the purpose of DC biasing. To accommodate the biasing network, minor adjustment was made on the design and re-simulation was done using the CST software in order to obtain the optimum results. The comparison of the dimension dissimilarity, between without biasing network (design A) and with consideration of biasing network (design B), is tabulated in Tab. 2.

In this structure, L-shape narrow slots with width of $W_b = 0.3$ mm and a length of 9 mm ($L_{b1} + L_{b2}$) are created to separate between the positive and negative areas of the DC voltage supply. Additional lump elements, RF capacitors and RF inductors are introduced into the design for a specific purpose. Five RF chokes inductors are placed between the radiator and biased wire to improve the isolation between RF current and DC supply [7]. The value of all inductors is 27 nH. Furthermore, considering the effect of the narrow slots, twelve DC block capacitors of 100 pF are mounted across it in order to preserve the continuity of the RF current across the radiator.

The cathodes of the diodes are connected to the four small copper areas and supplied with negative DC voltage. Meanwhile, the anodes of the diodes, which are soldered to the remaining copper region on the circular patch, are connected to the positive terminal of the positive DC battery. The DC is directly supplied using copper wire to the radiator through a via-hole and small circle slots created on the ground plane. The circle slots, with a diameter of L_c are introduced to provide separation of DC voltage. This technique is taken to avoid the presence of the biased wire from

(a)

Enlargement

* V-: negative voltage, V+: positive voltage

(b)

Fig. 8. Geometry of the proposed antenna with the integration of biasing mechanism: (a) front view, (b) back view.

interrupting the antennas far-field radiation pattern. In the meantime, the wire is terminated at the small square copper located outside the circular patch. For the activation of the diodes, the external switching control board is constructed. The switching board is composed of copper wire, 9V DC battery, a dip switch to control the biasing and the resistor for current limitation. The resistors with a value of 100 Ω are chosen so that the forward biased current 90 mA is obtained (ON-state) and left unbiased in the OFF-state condition. The photograph of the fabricated antenna is shown in Fig. 9.

Parameter	Value (mm)		Parameter	Value (mm)	
	A	B		A	B
L	55	55	W_b	-	0.3
W	55	55	W_i	1	-
h	1.524	1.524	L_s	5	5
t	0.035	0.035	L_p	15.3	17
d	5.5	5.5	L_{b1}	-	5
r	17.9	17.9	L_{b2}	-	4
W_s	1	1	L_c	-	4

Tab. 2. Dimension of the designed antennas (Design A and Design B).

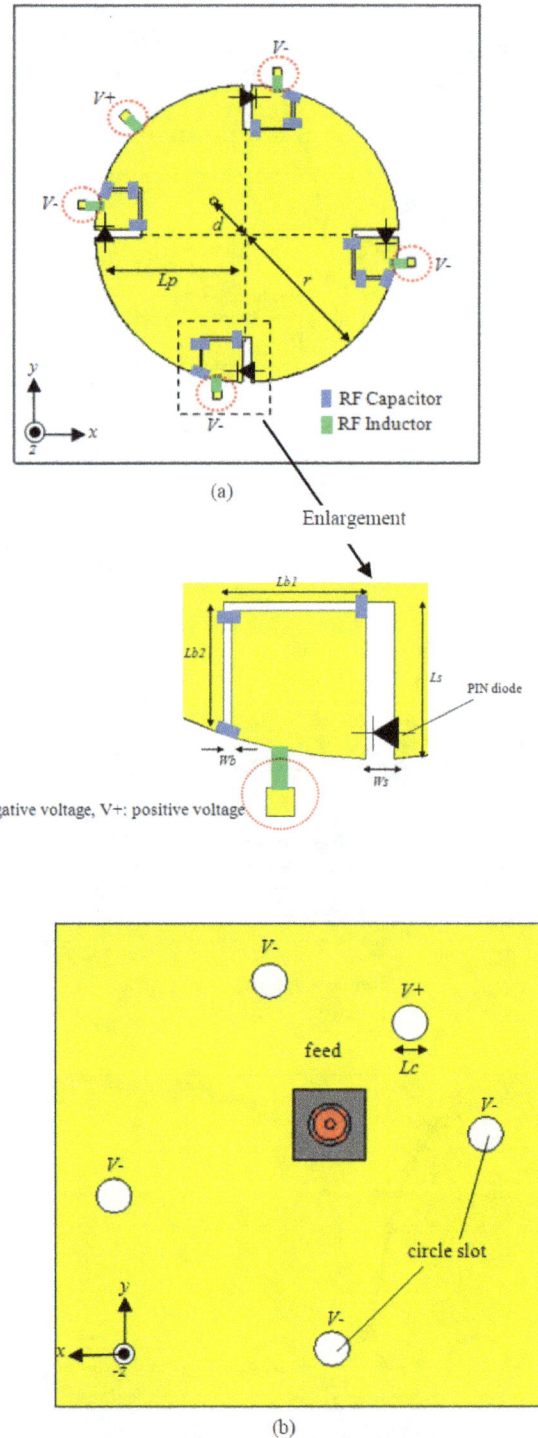

(a)

(b)

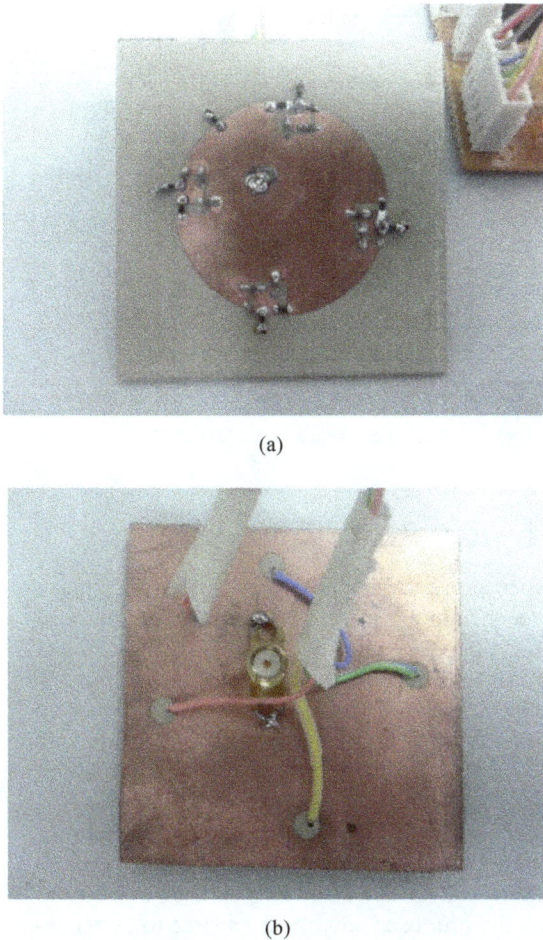

Fig. 9. Photograph of fabricated antenna prototypes: (a) front view, (b) back view.

(a)

(b)

Fig. 10. Comparison between the simulated and measured reflection coefficient result of the modified structure. (a) Circular polarization. (b) C3-LP.

3.2 Results and Analysis of the Modified Antenna

For accurate simulation results, the PIN diode (Infineon BAR 50-02L) was included in the design and simulated using CST. This step was done by the utilization of touchstone block that contains s2p file which can be obtained from the manufacturer. The touchstone file consists of the information s-parameter of the diode for the ON and OFF state condition.

The optimized antenna was measured using the similar equipment and set up used for the measurement of the ideal diode. The simulated and measured reflection coefficients of the modified structure for all switch configurations are presented in Fig. 10. Good measured impedance matching was achieved for C3-LP, with -10 dB reflection coefficient bandwidth is in the frequency range of 2.469 to 2.515 GHz, or 46 MHz.

However, the reflection coefficient for CP operations is slightly above -10 dB. The impedance bandwidths, based on -10 dB reflection coefficient is 40 MHz (2.494 GHz to 2.534 GHz) and 48 MHz (2.5÷2.548 GHz) for C1-LHCP and C2-RHCP, respectively.

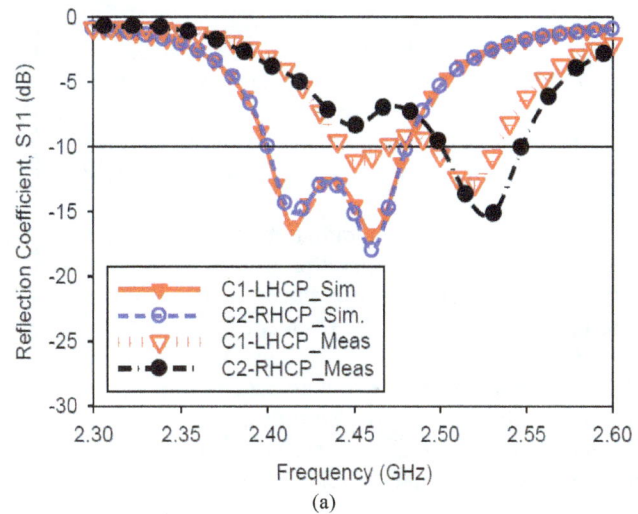

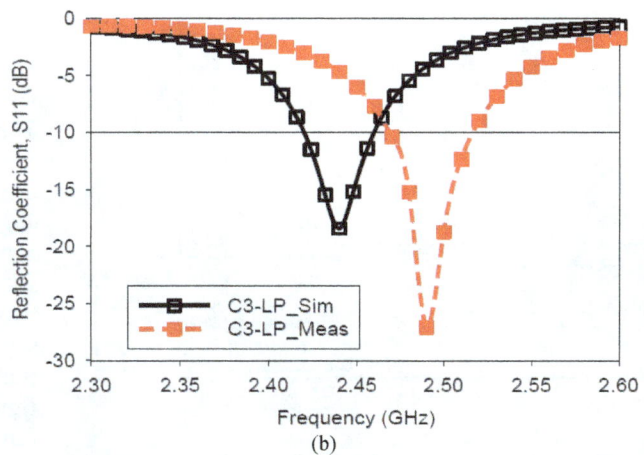

It can be seen that measurement is shifted to the higher frequency. It is also observed that the good impedance matching was obtained except for C2-RHCP configuration. The lowest value of the magnitude of S_{11} for C2-RHCP is -6.7 dB obtained at a frequency of 2.473 GHz. The slight discrepancies between simulation and measurement results could be due to fabrication tolerance, SMA port or parasitic effects from the PIN diode. As mentioned before, the position of the diode in the slits is highly crucial as it determines the axial ratio result and could affect the impedance of the antenna.

Figure 11 shows the simulated and measured results for axial ratio. The measured BW_{CP}, as referred to 3 dB axial ratio bandwidth, is 31 MHz (2.483 ÷ 2.452 GHz) and 22 MHz (2.474 ÷ 2.452 GHz) for C1-LHCP and C2-RHCP, respectively. It has to be noted that the BW_{CP} is totally covered by the BW for linear polarization. Since the proposed antenna performs polarization reconfiguration at a single operating frequency, thus the available working bandwidth is mainly determined by BW_{CP}. Due to symmetry structure, the performances for the both CP operations are almost identical for reflection coefficient and axial ratio result, like the case of ideal diode. The frequency shifted of the reflection coefficient cause the variance between simulated and measured result for the axial ratio. This is due to the shift of the two degenerated orthogonal modes which will consequently shift the frequency of the excited CP mode.

The measured far-field radiation patterns in the y-z plane and x-z plane for all operation modes are plotted in Fig. 12. Measured results show that the antenna has a broadside radiation pattern at the resonant frequency and 3-dB beamwidth of more than 90° is obtained. In addition, the graph depicts that the proposed antenna has almost the same pattern in both planes. It is also noticed that the gain for C3-LP is higher than the CP modes. The comparison of results between design A and design B is summarized in Tab. 3.

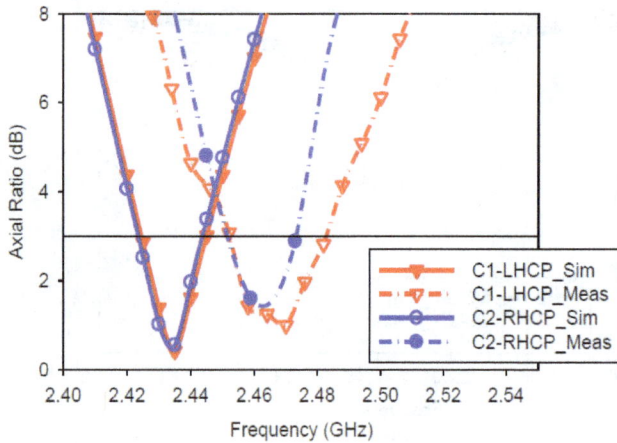

Fig. 11. Simulated and measured axial ratio of the modified structure.

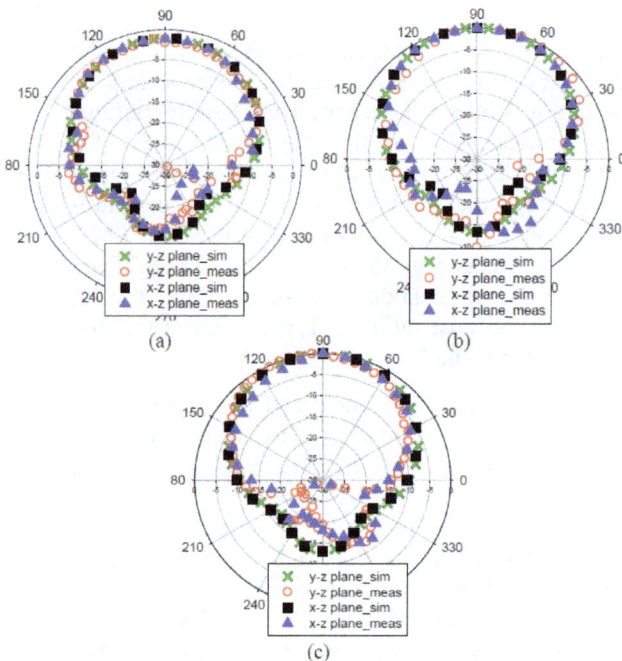

Fig. 12. Measured normalized radiation pattern of the modified structure: (a) C1-LHCP at 2.47 GHz; (b) C2-RHCP at 2.47 GHz; (c) C3-LP at 2.49 GHz.

Conf.	Polar.	Meas. S₁₁ (dB)		Meas. Bandwidth (MHz)			
		Design A	Design B	-10dB BW		3-dB BW (Axial Ratio)	
				A	B	A	B
C1-LHCP	LHCP	-16.48	-12.60	71	40	19	31
C2-RHCP	RHCP	-19.83	-15.25	84	48	18	22
C3-LP	LP	-24.06	-26.62	43	46	NA	

*A: Design A (Ideal diode) *B: Design B (PIN diode) *NA: Not applicable

Tab. 3. Comparison between Design A and Design B.

4. Conclusions

This paper presents a simple technique to achieve polarization reconfigurable antenna, with a fixed resonant frequency, that operates within WLAN frequency band. The study begins with the use of copper strips as a switch for proof of the concept. Then, minor modifications were made on the structure in order to accommodate the employment of real switches (RF PIN diode) and switching network. Using four switches embedded across the slits, the slit length is altered, hence producing a phase difference between two degenerated orthogonal resonant modes. Consequently, depending on the switching state, the polarization excited by the proposed antenna is capable to be reconfigured between three types - LHCP, RHCP or LP - that functioned at the same operating frequency. The results of reflection coefficient, axial ratio and radiation pattern are presented and comparisons are made between all different bias conditions. The results demonstrate good impedance matching, and broadside radiation pattern. The discrepancies of the results might be due to parasitic effects of the PIN diode, or fabrication tolerance. Small in dimension, and with unique features, the proposed antenna is potentially suitable in wireless applications for limited space and multipath rich environments; such as to serve as an access point in an indoor scenario for WIFI application.

Acknowledgements

The authors thank the Ministry of Higher Education (MOHE) for supporting the research work, Research Management Centre (RMC), School of Postgraduate (SPS) and Communication Engineering Dept., Universiti Teknologi Malaysia (UTM) for the support of the research under grant no QJ13000.7123.04H38, 4L811, 05H35 and 4F360.

References

[1] BYUN, S.-B., LEE, J.-A., LIM, J.-H., YUN, T.-Y. Reconfigurable ground-slotted patch antenna using PIN diode switching. *ETRI Journal*, Dec. 2007, vol. 29, no. 6, p. 832–834.

[2] KUMAR, S., GOLI, H., BASKARAN, P., ANGELA, P. P. Novel reconfigurable microstrip antenna. In *IEEE 3rd International Conference on Industrial and Information Systems*. Kharagpur (India), Dec. 2008, p. 1–5, DOI: 10.1109/ICIINFS.2008.4798448

[3] RUVIO, G., AMMANN, M. J., CHEN, Z. N. Wideband reconfigurable rolled planar monopole antenna. *IEEE Transactions on Antennas and Propagation*, 2007, vol. 55, no. 6, p. 1760–1767. DOI: 10.1109/TAP.2007.898575

[4] PANAGAMUWA, C. J., CHAURAYA, A. C., VARDAXOG-LOU, J. C. Frequency and beam reconfigurable antenna using photoconducting switches. *IEEE Transactions on Antennas and Propagation*, Feb. 2006, vol. 54, no. 2, p. 449–454. DOI: 10.1109/TAP.2005.863393

[5] NIKOLAOU, S., KIM, B., VRYONIDES, P. Reconfiguring antenna characteristics using PIN diodes. In *The 3rd European Conference on Antennas and Propagation EuCap2009*. Berlin (Germany), 2009, p. 3748–3752.

[6] ISMAIL, M. F., A. RAHIM, M. K., MAJID, H. A. The investigation of PIN diode switch on reconfigurable antenna. In *IEEE International RF and Microwave Conference RFM 2011*. Seremban (Negeri Sembilan, Malaysia), December 2011, p. 234–237. DOI: 10.1109/RFM.2011.6168737

[7] FERRERO, F., LUXEY, C., JACQUEMOD, G., STARAJ, R., FUSCO, V. Polarisation-reconfigurable patch antenna. In *International Workshop on Antenna Technology: Small and Smart Antennas Metamaterials and Applications IWAT 2007*. Cambridge (UK), 2007, no. 1, p. 73–76. DOI: 10.1109/IWAT.2007.370083

[8] LIM, J.-H., BACK, G.-T., YUN, T.-Y. Polarization-diversity cross-shaped patch antenna for satellite-DMB systems. *ETRI Journal*, Apr. 2010, vol. 32, no. 2, p. 312–318. DOI: 10.4218/etrij.10.0109.0280

[9] PIAZZA, D., MOOKIAH, P., D'AMICO, M., DANDEKAR, K. R. Pattern and polarization reconfigurable circular patch for MIMO systems. In *European Conference on Antennas and Propagation EuCap 2009*. Berlin (Germany), 2009, p. 1047–1051.

[10] PIAZZA, D., MOOKIAH, P., D'AMICO, M., DANDEKAR, K. R. Experimental analysis of pattern and polarization reconfigurable circular patch antennas for MIMO systems. *IEEE Transactions on Vehicular Technology*, 2010, vol. 59, no. 5, p. 2352–2362. DOI: 10.1109/TVT.2010.2043275

[11] LI, Y., ZHANG, Z., CHEN, W., FENG, Z. Polarization reconfigurable slot antenna with a novel compact CPW-to-slotline transition for WLAN application. *IEEE Antennas and Wireless Propagation Letters*, 2010, vol. 9, p. 252–255. DOI: 10.1109/LAWP.2010.2046006

[12] AMINI, M. H., HASSANI, H. R., NEZHAD, S. M. A. A single feed reconfigurable polarization printed monopole antenna. In *European Conference on Antennas and Propagation EuCap 2012*. Prague (Czech Rep.), 2012, p. 1–4. DOI: 10.1109/EuCAP.2012.6205882

[13] LAI, C.-H., HAN, T.-Y., CHEN, T.-R. Circularly-polarized reconfigurable microstrip antenna. *Journal of Electromagnetic Waves and Applications*, 2009, vol. 23, no. 2-3, p. 195–201. DOI: 10.1163/156939309787604373

[14] WANG, C.-C., CHEN, L.-T., ROW, J.-S. Reconfigurable slot antennas with circular polarization. *Progress in Electromagnetics Research Letters*, 2012, vol. 34, p. 101–110. DOI:10.2528/PIERL12072410

[15] MONTI, G., CORCHIA, L., TARRICONE, L. Patch antenna with reconfigurable polarization. *Progress in Electromagnetics Research C*, 2009, vol. 9, p. 13–23.

[16] HYUN, D.-H., BAIK, J.-W., LEE, S. H., KIM, Y.-S. Reconfigurable microstrip antenna with polarisation diversity. *Electronics Letters*, 2008, vol. 44, no. 8, p. 509–511. DOI: 10.1049/el:20080125

[17] KIM, B., PAN, B., NIKOLAOU, S., KIM, Y.-S., PAPAPOLYMEROU, J., TENTZERIS, M. M. A novel single-feed circular microstrip antenna with reconfigurable polarization capability. *IEEE Transactions on Antennas and Propagation*, 2008, vol. 56, no. 3, p. 630–638. DOI: 10.1109/TAP.2008.916894

[18] YANG, X.-X., SHAO, B.-C., YANG, F., ELSHERBENI, A. Z., GONG, B. A polarization reconfigurable patch antenna with loop slots on the ground plane. *IEEE Antennas and Wireless Propagation Letters*, 2012, vol. 11, p. 69–72. DOI: 10.1109/LAWP.2011.2182595

[19] CHEN, Y. B., CHEN, T. B., JIAO, Y. C., ZHANG, F. S. A reconfigurable microstrip antenna for switchable polarization. *Journal of Electromagnetic Waves and Applications*, 2006, vol. 20, no. 10, p. 1391–1398. DOI: 10.1163/156939306779276820

[20] GARG, R., BHARTIA, P., BAHL, I., ITTIPIBOON, A. *Microstrip Antenna Design Handbook*, Artech House Publishers, January 2001.

About the Authors ...

Mohamed Nasrun OSMAN was born in November 1987. He received his Electrical Engineering degree major telecommunication from Universiti Teknologi Malaysia in 2010. He is currently working towards his PhD degree in Electrical Engineering at the same university. His research interests include reconfigurable antenna design, RF design and wireless MIMO system.

Mohamad Kamal A RAHIM was born in Alor Star Kedah Malaysia on 3rd November, 1964. He received the B Eng degree in Electrical and Electronic Engineering from University of Strathclyde, UK in 1987. He obtained his Master Engineering from University of New South Wales, Australia in 1992. He graduated his PhD in 2003 from University of Birmingham, U.K., in the field of Wideband Active Antenna. From 1992 to 1999, he was a lecturer at the Faculty of Electrical Engineering, Universiti Teknologi Malaysia. From 2005 to 2007, he was a senior lecturer at the Department of Communication Engineering, Faculty of Electrical Engineering, Universiti Teknologi Malaysia. He is now a Professor at Universiti Teknologi, Malaysia. His research interest includes the design of active and passive antennas, dielectric resonator antennas, microstrip antennas, reflect-array antennas, electromagnetic band gap (EBG), artificial magnetic conductors (AMC), left-handed metamaterials and computer aided design for antennas.

Peter GARDNER graduated in Physics from the University of Oxford in 1980 with first class honors. He gained eight years of industrial experience in research and development of active microwave components for Ferranti International, Poynton, Cheshire and later as an independent consultant. He obtained his MSc in 1990 and PhD in 1992. He joined the School of Electronic and Electrical Engineering at Birmingham in 1994, as a lecturer, and currently is a Head of School for School of Electronic, Electrical and Computer Engineering. His research interests include active and adaptive antenna, microwave amplifier, transmitter linearization and micromachine antennas.

Mohamad Rijal HAMID received the M.Sc. degrees in Communication Engineering from the Universiti Teknologi

Malaysia, Johor Bahru, Malaysia, in 2001 and the PhD Degree at the University of Birmingham, Birmingham, United Kingdom. He has been with Universiti Teknologi Malaysia at the Faculty of Electrical Engineering, UTM since 1999. Currently his position is a Senior Lecturer. His major research interest is reconfigurable antenna design for multimode wireless applications.

Mohd Fairus MOHD YUSOFF is a graduate faculty member of the Faculty of Electrical Engineering, UTM. He joined UTM in 2002 as a Tutor. He received his Bachelor in Engineering (Electrical-Telecommunication) in 2002 and Master of Electrical Engineering (Electrical-Electronics and Telecommunications) in 2005 from Universiti

Teknologi Malaysia. He obtained his PhD in 2012 from University of Rennes 1, France in area of Signal Processing and Telecommunication. His main research interest and areas is antenna design, millimeter waves and microwave devices.

Huda A. MAJID obtained his first degree in Electrical Engineering majoring Telecommunication in 2007 and his M.Sc in 2010 from Universiti Teknologi Malaysia, Johor Bahru. He received his PhD degree from the same institution in 2013. His research interest includes design of metamaterials, reconfigurable antennas, RF and microwave. He is currently working as a Postdoctoral Research Fellow since January 2014.

26

On the Design and Performance Analysis of Low-Correlation Compact Space-Multimode Diversity Stacked Microstrip Antenna Arrays for MIMO-OFDM WLANs over Statistically-Clustered Indoor Radio Channels

Asuman SAVASCIHABES[1], Ozgur ERTUG[1], Erdem YAZGAN[2]

[1] Telecommunications and Signal Processing Laboratory, Dept. of Electrical and Electronics Engineering,
University of Gazi, Ankara, Turkey
[2]Dept. of Electrical and Electronics Engineering, University of Hacettepe, Ankara, Turkey

ahabes@nny.edu.tr, ertug@gazi.edu.tr, yazgan@hacettepe.edu.tr

Abstract. *The support of high spectral efficiency MIMO spatial-multiplexing communication in OFDM-based WLAN systems conforming to IEEE 802.11n standard requires the design and use of compact antennas and arrays with low correlation ports. For this purpose, compact space-multimode diversity provisioning stacked circular multimode microstrip patch antenna arrays (SCP-ULA) are proposed in this paper and their performance in terms of spatial and modal correlations, ergodic spectral efficiencies as well as compactness with respect to antenna arrays formed of vertically-oriented center-fed dipole elements (DP-ULA) and dominant-mode operating circular microstrip patch antennas (CP-ULA) are presented. The lower spatial and modal correlations and the consequent higher spectral efficiency of SCP-ULA with ML detection over statistically-clustered Kronecker-based spatially-correlated NLOS Ricean fading channels with respect to DP-ULA and CP-ULA at significantly lower antenna and array sizes represents SCP-ULA as a promising solution for deployment in terminals, modems and access points of next-generation high-speed 802.11n MIMO-OFDM WLAN systems.*

Keywords

IEEE 802.11n MIMO-OFDM WLAN, spectral efficiency, spatial correlation, multimode antenna, spatial-multiplexing, Kronecker channel model, NLOS Ricean fading

1. Introduction

In the concurrent and next-generation communication systems, the spectral efficiency and transmission quality can be vastly enhanced by multiple-input multiple-output (MIMO) communication techniques [1]. In communication systems employing MIMO spatial-multiplexing, higher data rates can be achieved when there are a large number of scatterers between the transmit and receive antennas i.e. rich-scattering environment. However, the spatial correlation between transmit and receive antenna ports that is dependent on antenna-specific parameters such as the radiation patterns, the distance between the antenna elements as well as the channel characteristics such as unfavorable spatial distribution of scatterers and angular spread severely degrades the capacity and quality achievable by MIMO spatial-multiplexing systems.

The space consumption of MIMO antennas is especially vital in applications such as access points, modems and end-user terminal equipments (laptops, PDAs etc.) of WLAN and WIMAX systems. When regularly spaced antenna elements are used in MIMO systems, the correlation between the antenna elements in a space diversity system and hence the channel capacity and transmission quality are dependent on the distance between antenna array elements, the number of antenna elements and the array geometry. However, due to the physical constraints and the concerns on ergonomics and aesthetics, the distance between antenna elements in practice cannot be extended beyond a certain level which limits the use of space-only diversity MIMO spatial-multiplexing systems to achieve the desired spectral efficiencies and transmission qualities. As an alternative solution to achieve compactness in MIMO systems, the use of pattern diversity [2], [3], multimode diversity [4], [5], and polarization diversity [6], [7] techniques in conjunction with space diversity are proposed in the literature.

Besides polarization diversity that is well-known, multimode and pattern diversity techniques that are less addressed in antenna engineering community are achieved by using higher-order mode generation in antenna structures and in general microstrip, biconical, helical, spiral, sinuous and log-periodic antenna structures are amenable to higher-order mode generation. In this manner, the higher-order modes generated in a single antenna structure with directional radiation patterns resulting in low spatial correlation in angle space are used as diversity ports in a MIMO system within a compact space. In pattern diversity on the other hand that is slightly different than multimode diversity, orthogonal radiation patterns generated on distinct antennas that are co-located at the phase-centers are generated and used as diversity ports.

In this work, a multimode stacked circular microstrip patch antenna used in a uniform linear array structure (SCP-ULA) for MIMO-OFDM WLAN systems conforming to IEEE 802.11n standard is designed and the associated spatial power correlation, ergodic spectral efficiency and compactness with respect to omnidirectional dipole (DP-ULA) and circular microstrip uniform linear arrays (CP-ULA) operating in the dominant isotropic TM01 mode are analyzed. Section 2 represents the 802.11n MIMO-OFDM WLAN details as well as the associated system and statistically-clustered Kronecker-based correlated channel models for MIMO spatial-multiplexing. Section 3 is dedicated to the design procedure of stacked circular microstrip antenna for IEEE 802.11n MIMO-OFDM WLAN communication in HFSSv.11@TM CAD program and the analysis of the marginal and superimposition radiation patterns as well as S-parameters and VSWR variations versus frequency respectively. In Section 4, the gains of SCP-ULA with respect to CP-ULA and DP-ULA in terms of spatial/modal correlations, ergodic spectral efficiency as well as compactness are presented. Finally, Section 5 concludes the paper.

2. IEEE 802.11N WLANS and Associated System/Channel Models

The wireless local area network (WLAN) technology for medium-range indoor/outdoor wireless communications standardized by IEEE P802.11 working group has emerged from pre-802.11 standards towards spectrally-efficient and multipath-robust OFDM modulation based 802.11b and 802.11a/g with data rates increasing up to a maximum of 54 Mbps for 802.11a/g. These standards are limited to the use of single transmit and receive antenna at the access points and modems as well as laptops/PDAs in WLANs for end-users forming a SISO-OFDM (single-input single-output OFDM) channel. In the last decade, with the proliferation of MIMO spatial-multiplexing technology using multiple transmit and receive antennas achieving much higher data rates without sacrificing either bandwidth or transmit power with respect to SISO systems, IEEE 802.11 standard is later extended with the version 802.11n incorporating MIMO capability with the first amendment published in 2009 [8] proposing operation at lower and upper ISM bands of 2.4 GHz and 5.8 GHz with the corresponding bandwidths of 20 MHz and 40 MHz respectively. By the use of MIMO spatial-multiplexing technology, higher-order 64/128/256-QAM modulations, a 40 MHz channel bandwidth at 5.8 GHz that is double that of legacy IEEE

802.11 a/b/g systems, a cyclic prefix (CP) of 400 ns that is half of legacy systems which reduces symbol time and hence increases data rates, more efficient OFDM structure with 52 subcarriers with respect to 48 subcarriers in legacy systems, and frame-aggregation/block-acknowledgement protocol with higher packet sizes at the MAC layer [9], [10], IEEE 802.11n standard sets forth the basis for multi-media-enabling high-throughput next-generation WI-FI networks.

The transmitter and receiver block diagrams of an IEEE 802.11n based MIMO-OFDM WLAN system is illustrated in Fig. 1. The binary data is first encoded by channel encoder after which the encoded bits are multiplexed into sub-streams, modulated, and transmitted from each antenna. At the receiver side, after a digital representation of N received signals is obtained by ADCs, the CP is removed and N-point DFT is performed per receiver branch. Since the MIMO-OFDM system turns into a narrow band flat-fading MIMO channel per sub-carrier over multipath fading channels, the received signal vector per sub-carrier is given by:

$$\mathbf{y}(i,k) = \mathbf{H}(i)\mathbf{s}(i,k) + \mathbf{n}(i,k) \qquad (1)$$

where i and k are sub-carrier and symbol indices respectively. Here, $\mathbf{H}(i)$ represents $N \times M$ dimensional channel matrix for the ith sub-carrier, $\mathbf{s}(i, k)$ represents modulated transmit symbol vectors for the ith sub-carrier and $\mathbf{n}(i, k)$ represents additive white Gaussian noise at the ith subcarrier and kth symbol index with N independent and identically distributed (i.i.d) zero-mean complex elements with variance of σ_n^2.

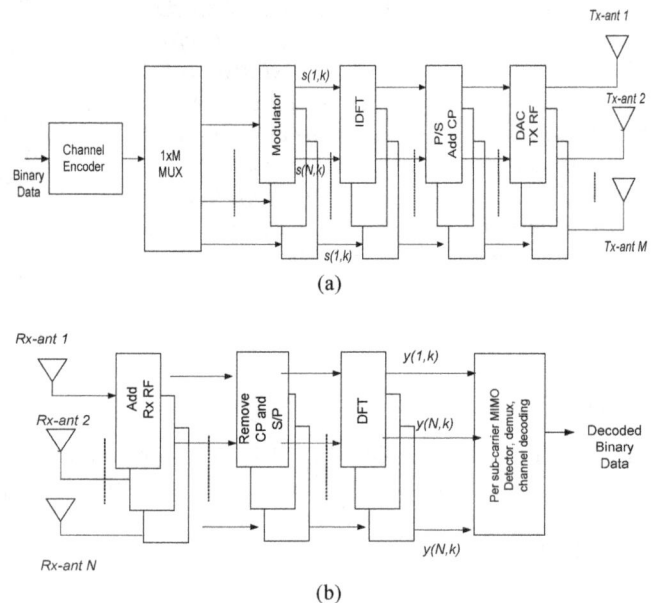

(a)

(b)

Fig. 1. Transmitter and receiver diagrams of a MIMO-OFDM WLAN system: (a) Transmitter. (b) Receiver.

Amongst the MIMO channel models proposed in the literature such as deterministic models (ray-tracing, recorded impulse response etc.) or stochastic models (geometric ring, parametric, correlation-based etc.), we deploy as the channel model the statistically-clustered Kronecker channel model for the characterization of the spatially-correlated MIMO channel in this work which has also been standardized for MIMO-OFDM IEEE 802.11n WLAN systems in [9]. Kronecker model assumes seperability between transmit/receive spatial correlations and the same model is also used for the performance analysis of indoor MIMO systems with stacked circular microstrip patch antennas with pattern diversity in [3].

The general geometry of the statistically-clustered Kronecker model we employ throughout the sequel representing the clusters and the transmission paths is represented in Fig. 2. For simplicity, only reflection from a single cluster is assumed in this work for the evaluation of the correlation and consequently spectral efficiencies, and the extension of the model to include a higher number of clusters is straightforward due to the additivity of spatial correlations.

The spatially-correlated channel matrix $\mathbf{H}_c$ in Kronecker model is formulated as:

$$\mathbf{H}_c = \sqrt{\frac{K}{1+K}}\mathbf{H}_{LOS} + \sqrt{\frac{1}{1+K}}\mathbf{H}_{NLOS} \qquad (2)$$

and includes Rayleigh fading for $K = 0$, Ricean fading for higher K values and also AWGN channel for $K = \infty$ where $\mathbf{H}_{LOS}$ and $\mathbf{H}_{NLOS}$ are LOS (line-of-sight) and NLOS (nonline-of-sight) components respectively, K is the Ricean K-factor given by the ratio of the LOS component's power over NLOS component's power [1] and the NLOS channel matrix is defined as [9]:

$$\mathbf{H}_{NLOS} = \mathbf{R}_r^{1/2}\mathbf{H}_w\mathbf{R}_t^{1/2} \qquad (3)$$

where $\mathbf{H}_w$ denotes Wishart-type random matrix, $\mathbf{R}_r$ and $\mathbf{R}_t$ denote the transmit and receive spatial power correlation matrices respectively. The LOS component of the channel is also assumed to be rank one, and defined as [11]:

$$\mathbf{H}_{LOS} = \mathbf{a}\left(\Omega_{LOS,r}\right)\cdot\mathbf{a}\left(\Omega_{LOS,t}\right)^H \qquad (4)$$

where $\mathbf{a}(\Omega)$ is the array response as a function of the solid angle $\Omega = (\phi, \theta)$, while $\Omega_{LOS,t}$ and $\Omega_{LOS,r}$ are AoA/AoD corresponding to the LOS component at the transmitter and receiver respectively.

3. Design of Multimode Stacked Circular Patch (SCP) Microstrip Antenna

The multimode stacked circular microstrip patch antenna (SCP) proposed for IEEE 802.11n MIMO-OFDM WLAN application in this work has the upper antenna in the stack excited at the TM11 mode and the bottom antenna excited at the TM21 mode to meet the compactness requirements since the radius of a circular microstrip patch

antenna scales up with the mode number m excited given by the formula [12]:

$$a = \frac{\chi_m\lambda}{2\pi\sqrt{\varepsilon_r}} \qquad (5)$$

where χ_m indicates the first zero of the derivative of the first kind Bessel function of order m $J_m(x)$ as presented in Tab. 1.

	TM_{01}	TM_{11}	TM_{21}	TM_{31}	TM_{41}	TM_{51}	TM_{61}
χ_m	3.82	1.84	3.04	4.18	5.29	6.38	7.46

Tab. 1. χ_m for different modal orders.

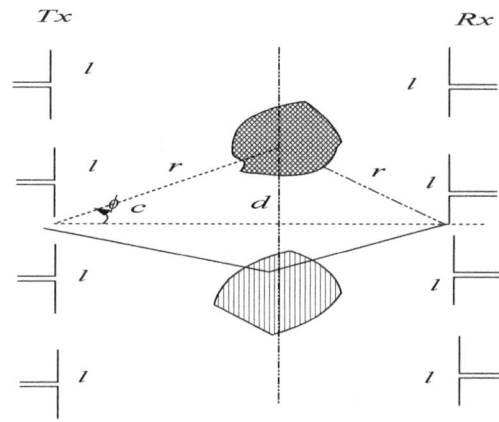

Fig. 2. Geometry of the clustered channel model representing clusters and transmit/receive antenna arrays (4×4 array) where ϕ_c is the mean A.O.D. of the cluster.

The far-field radiation patterns for a circular microstrip patch antenna excited at the m-th mode is further given by [13]:

$$\vec{E}_m = \frac{e^{-jk_f r}}{r}\left(E_{m,\theta}\,\vec{\theta} + E_{m,\phi}\,\vec{\phi}\right), \qquad (6)$$

$$E_{m,\theta} = \frac{j^m V_m^0 k_f a}{2}\left[J_{m+1}(z) - J_{m-1}(z)\right]\cos(m(\phi - \phi_0)), \qquad (7)$$

$$E_{m,\phi} = -\frac{j^m V_m^0 k_f a}{2}\left[J_{m+1}(z) + J_{m-1}(z)\right]\cos(\theta)\sin(m(\phi - \phi_0)) \qquad (8)$$

where $k_f = 2\pi/\lambda$ is the wavenumber, $J_m(x)$ is the Bessel function of first kind of order m, a is the patch radius, V_m^0 is the peak input voltage of the mth mode, ϕ_0 is the reference azimuth angle for the feed of the circular patch, and $z = k_f a \sin(\theta)$. Via these far-field radiation patterns, neglecting elevation spread that is acceptable for indoor propagation environments and assuming the look-direction coincident with broadside is $\theta = \pi/2$, only θ components of the far-field radiation pattern dependent on the azimuth angle ϕ in (6) remains. Azimuth plane radiation patterns for mode orders $m = 1, 2, 3$ are presented in Fig. 3.

(a)

(b)

(c)

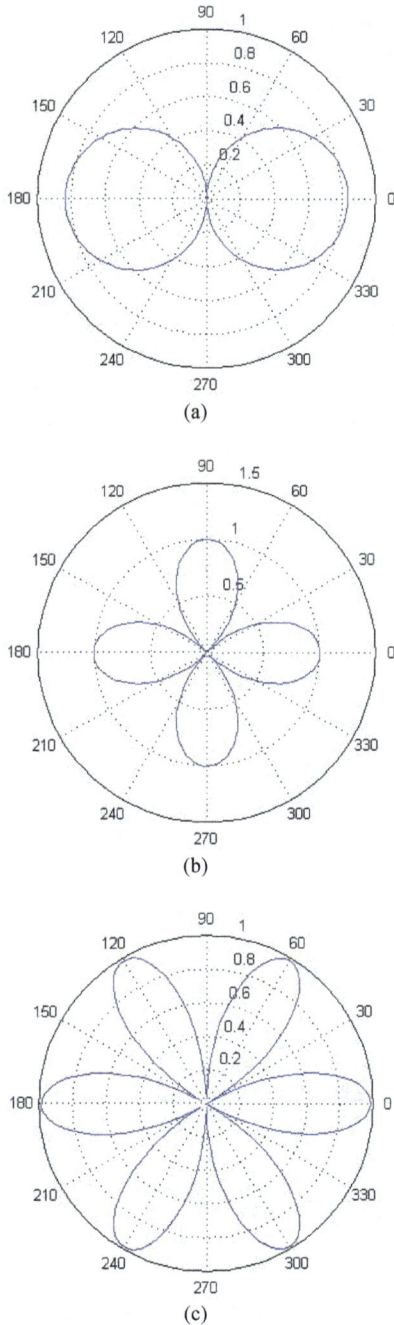

Fig. 3. Radiation patterns of multimode circular patch antennas: (a) TM11 mode, (b)TM21 mode, (c)TM31 mode.

The SCP antenna fed by microstrip feed lines designed in HFSSv.11 3D EM design and analysis software is presented in Fig. 4. To stay at the proper point of the trade-off between antenna bandwidth improvement and the antenna efficiency decrease/pattern distortion as well as to keep mutual coupling low, the distances between ground and bottom antenna as well as between circular antennas are kept as 0.5 mm. To ensure high radiation efficiency, the dielectric constant of the substrate is further chosen low as $\varepsilon_r = 2.2$.

The theoretical radius of the top and bottom antennas in SCP antenna excited by TM11 and TM21 modes via (5)

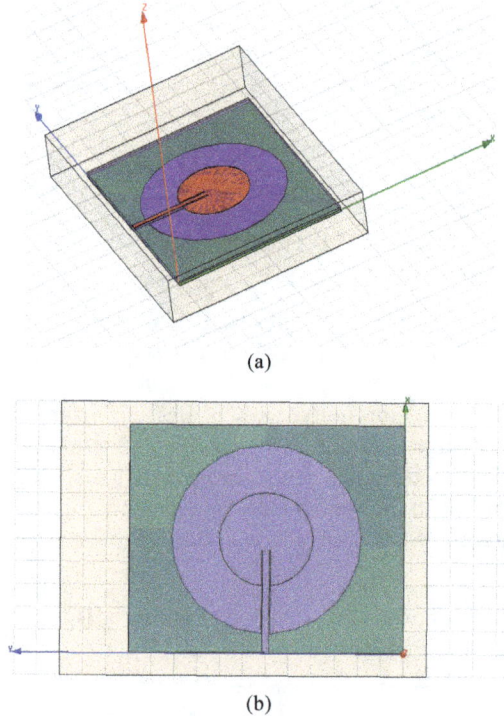

(a)

(b)

Fig. 4. The side (a) and the top (b) view of the SCP antenna with microstrip feed lines in HFSS.

are respectively given by 10.2 mm and 17 mm respectively. However, including the fringing effect [7], the actual effective radius of the circular patch antenna that is larger than estimated by (7) is obtained by the formula:

$$a_e = a\left(1 + \frac{2h}{\pi a \varepsilon_r}\left(\ln(\frac{a}{2h}) + (1{,}41\varepsilon_r + 1{,}77) + \frac{h}{a}(0{,}26\varepsilon_r + 1{,}65)\right)\right)^{1/2}$$

(9)

through which the effective radiuses of TM11 and TM21 modes are found to be 10.6 mm and 17.4 mm respectively. To operate both stack elements at the same resonant frequency, the radius of the antenna of TM21 mode is further modified and set as 20.67 mm in HFSSv.11@TM while the radius of the antenna operating on TM11 mode is used according to the theoretical value of 10.26 mm.

For the impedance matching of the SCP antenna at 802.11n upper ISM band WLAN operating frequency of 5.8 GHz for a 50 Ω load, the radiation resistances for both stack antennas have to be evaluated first for TM11 mode and TM21 modes. The real Poynting vector is given by [13]:

$$P + jQ = \frac{1}{2}\iint_{size}\left(\vec{E}_{ax}\vec{H}_a^*\right)\hat{z}\,dxdy$$

(10)

and integrating the real part of (10):

$$P_r = \frac{1}{2\eta_0}\int_0^{2\pi}\int_0^{\pi/2}\left(|E_\theta|^2 + |E_\varphi|^2\right)r^2\sin\theta d\theta d\varphi$$

. (11)

The relation between (11) and the radiation resistance R_r is given by:

$$P_r = \frac{1}{2}G_r(E_0 h)^2 = \frac{1}{2}G_r V_0^2, \qquad (12)$$

$$R_r = \frac{1}{G_r} \qquad (13)$$

where G_r is the radiation conductance. In this manner, the radiation resistance of elements operating on TM11 mode and TM21 mode are obtained as 342.57 Ω and 156.13 Ω respectively. In the stack antenna the radiation resistance of TM21 mode is further obtained as 131.21 Ω and the radiation resistance of TM11 mode is found as 315.04 Ω in HFSSv.11@TM.

Antenna input impedances are described by the equations:

$$Z_{in} = R_{in} + jX_{in}, \qquad (14)$$

$$R_{in} = \frac{1}{G_{in}}, \qquad (15)$$

$$G_{in} = G_r + G_d + G_c \qquad (16)$$

where Z_{in}, is input impedance, R_{in} is input resistance, X_{in} is input reactance, G_{in} is input conductance, G_d is dielectric conductance and G_c is the conductor conductance respectively. Since the antenna patch used in the design is selected as a zero-loss conductor, there exists no conductor resistance. On the other hand, the dielectric conductance value can be calculated via [14] as:

$$G_d = \frac{\varepsilon_{m0} \tan \delta}{4\mu_0 h f_r}\left[(ka_r)^2 - m^2\right] \qquad (17)$$

where $\varepsilon_{m0} = 2$ for $m = 0$ and $\varepsilon_{m0} = 1$ for $m \neq 0$, and h is height of dielectric substrate of microstrip line. The effective loss tangent of the dielectric material with $\varepsilon_r = 2.2$ is determined as $\delta = 0.0009$. Using these values, the dielectric conductances for TM11 and TM21 modes are obtained as $G_d = 1.4925 \times 10^{-4}$ and $G_d = 6.0936 \times 10^{-4}$ respectively. These values are quite small so that it is rational not to take dielectric conductances into account for G_{in} calculations and to state that the input resistances of the antenna operating at TM11 mode is 342.57 Ω, and the input resistance of the antenna operating at TM21 mode is 156.13 Ω. To adapt these resistance values to 50 Ω, microstrip feed line widths for both elements in the stacked antenna are determined as 1.5 mm via the guideline equations in [15] such that both stack elements are matched to 50Ω input resistance.

In the design of SCP antenna, the antenna reactances are further calculated via

$$X_{in} = -jw\mu_0 h\left[\frac{1}{\pi a_e^2 k^2} + \right.$$

$$\left. \sum_{m=2}^{\infty}\frac{j_0^2 k_{0m}\rho_0}{\pi a^2 j_0^2 k_{0m}a_e(k^2 - k_{0m}^2)} + \frac{2}{\pi}\sum_{m=1}^{\infty}\left(\frac{\sin(n\Delta)}{n\Delta}\right)^2\right] \qquad (18)$$

and found as -j162.7 Ω and -j107.13 Ω for TM11 and TM21 modes respectively. In the stack antenna, the evaluation of the reactances of antenna with TM21 mode is obtained as j77.316 Ω while that of TM11 mode is found as j123.25 Ω that are quite high values for matching and can be minimized by optimizing the length of the microstrip feed lines.

The characteristic impedance of the feed line can be found depending on the size of the microstrip feed line by using:

$$Z_0 = \begin{cases} \dfrac{\eta}{2\pi\sqrt{\varepsilon_{re}}}\ln\left(\dfrac{8h}{L} + 0.25\dfrac{L}{h}\right), & (L/h \leq 1) \\[3mm] \dfrac{\eta}{\sqrt{\varepsilon_{re}}}\left\{\dfrac{L}{h} + 1.393 + 0.667\ln\left(\dfrac{L}{h} + 1.444\right)\right\}^{-1} & (L/h \geq 1) \end{cases}$$

$$(19)$$

where L is microstrip feed line width. Effective dielectric constant is also calculated via:

$$\varepsilon_{re} = \frac{\varepsilon_{re} + 1}{2} + \frac{\varepsilon_{re} - 1}{2}F(L/h). \qquad (20)$$

According to the desired characteristic impedance value, the ratio of the microstrip feedline width L to dielectric substrate height h is calculated via:

$$\frac{L}{h} = \begin{cases} \dfrac{8\exp(A)}{\exp(2A)-2}, & (L/h < 2) \\[3mm] \left\{B-1-\ln(2B-1)+\dfrac{\varepsilon_r-1}{2\varepsilon_r}\left[\ln(B-1)+0.39-\dfrac{0.61}{\varepsilon_r}\right]\right\}, & (L/h \geq 1) \end{cases}$$

$$(21)$$

where A and B are:

$$A = \frac{Z_0}{60}\sqrt{\frac{\varepsilon_r+1}{2}} + \frac{\varepsilon_r-1}{\varepsilon_r+1}\left(0.23 + \frac{0.11}{\varepsilon_r}\right),$$

$$B = \frac{377\pi}{2Z_0\sqrt{\varepsilon_r}}. \qquad (22)$$

With the height of the dielectric substrate $h = 0.5$ mm, the line width L is calculated as 1.55 mm via (22) for the microstrip feed lines and in this manner, the antenna reactances of TM11 mode and TM21 modes are minimized to j11.513 Ω and j4.5072 Ω respectively with the corresponding antenna input impedances finally obtained and used in S-parameter and VSWR analysis as $Z_{in1} = 45.507 + j11.513$ Ω and $Z_{in2} = 50.678 + j4.5072$ Ω for TM11 and TM21 modes respectively.

The S-parameters versus frequency of the designed SCP antenna are analyzed in Fig. 5. The resonance of both modes are achieved around 5.8 GHz that is suitable for IEEE 802.11n MIMO-OFDM WLANs operating in upper-ISM band and the operating bandwidth of SCP antenna is measured to be around 55 MHz via S11 and S22 return-loss S-parameters that is sufficiently adequate for IEEE 802.11n WLAN communication which requires 40 MHz

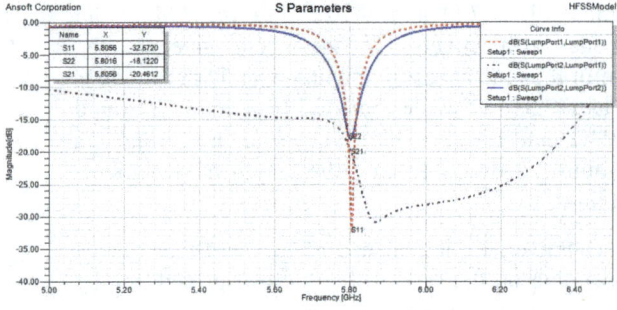

Fig. 5. S-parameters versus frequency of SCP antenna.

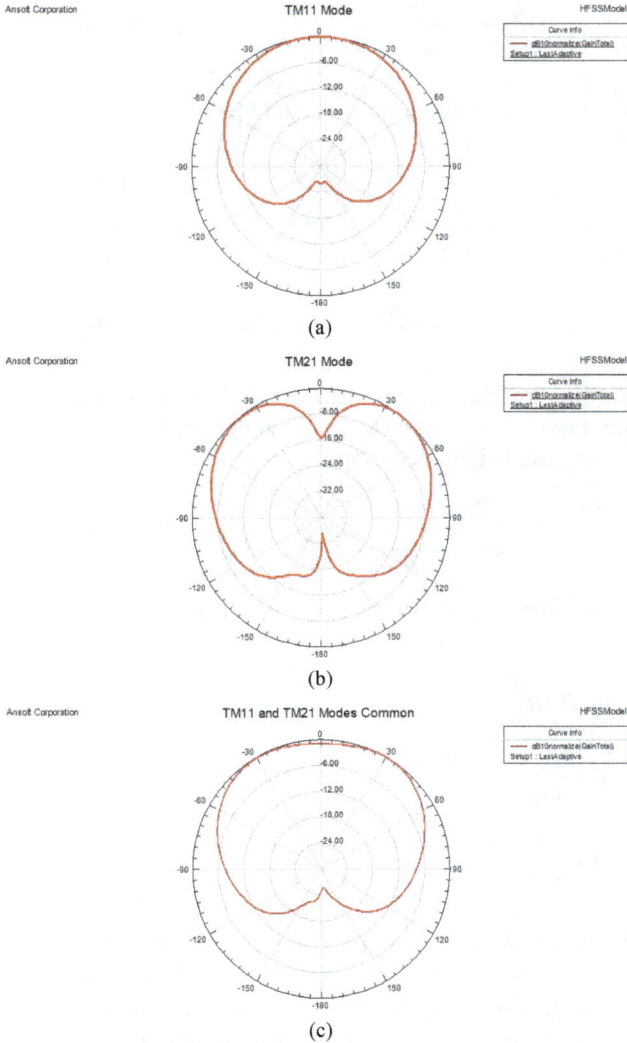

(a)

(b)

(c)

Fig. 6. Radiation patterns of SCP antenna: (a) TM11 alone, (b) TM21 alone, (c) TM11 and TM21 superimposed.

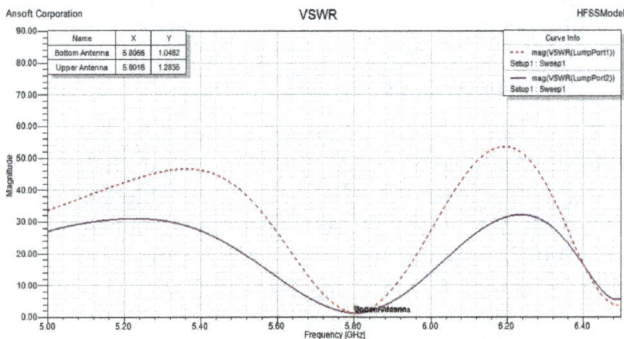

Fig. 7. VSWR plots versus frequency for SCP antenna.

bandwidth. Furthermore, the S21 mutual-coupling S-parameter in Fig. 4 is well below approximately -20 dB within the operating bandwidth around 5.8 GHz letting the omittance of pattern distortion due to the coupling effects between the collocated antennas. This fact is also well apparent in Fig. 6 where the radiation patterns of both modes are presented marginally and in superimposed form. Due to the low mutual coupling achieved, the radiation pattern distortion imposed by TM21 mode on TM11 mode is negligible in the superimposed radiation pattern in Fig. 6(c).

The VSWR plots of the SCP antenna for both TM11 and TM21 modes versus frequency are further presented in Fig. 7. The VSWR values for the TM11 and TM21 modes at the resonance frequency of 5.8 GHz is measured in HFSSv.11@TM as 1.0482 and 1.2835 respectively that are very close to the theoretical lower bound of unity validating the reliability of the impedance matching of SCP antenna.

4. Spatial/Modal Correlations and Spectral Efficiency Analysis of SCP-ULA

To evaluate the achievable gains with multimode SCP antenna used as a uniform linear array (SCP-ULA) for IEEE 802.11n MIMO-OFDM WLAN communications when employed at access points, modems or end-user terminal equipments, the spatial/modal power correlations of the SCP-ULA is compared with the spatial/modal power correlations of uniform linear arrays of center-fed dipole antennas of length $\lambda/2$ (DP-ULA) and the spatial/modal power correlations of uniform linear arrays of isotropic TM01 dominant-mode operating circular patch antennas in Fig. 8 for a statistically-clustered indoor propagation environment with a single-cluster of angular spread σ_c^2 and uniformly-distributed mean angle of arrival (AoA) and angle of departure (AoD) with respect to array broadsides. The spatial/modal normalized complex correlation coefficient of all three types of antennas is given by:

$$\rho = \frac{\int_{-\pi}^{\pi} E_1(\phi)E_2^*(\phi)PAS(\phi;\phi_c,\sigma_c^2)\exp(-jk_f d(n_2-n_1)\sin(\phi))d\phi}{\sqrt{\int_{-\pi}^{\pi}|E_1(\phi)|^2 PAS(\phi;\phi_c,\sigma_c^2)d\phi}\cdot\sqrt{\int_{-\pi}^{\pi}|E_2(\phi)|^2 PAS(\phi;\phi_c,\sigma_c^2)d\phi}}$$

(23)

and the power azimuth spectrum PAS $(\phi;\phi_c,\sigma_c^2)$ that defines the distribution of power over the sub-multipath components within a cluster is used as truncated-Laplacian density with mean azimuth angle ϕ_c and angular spread σ_c^2 the form of which is given by [16]:

$$PAS(\phi;\phi_c,\sigma_c^2) = \frac{e^{-\left|\frac{\sqrt{2}(\phi-\phi_c)}{\sigma_c}\right|}}{\sqrt{2}\sigma_c\left(1-e^{-\frac{\sqrt{2}\pi}{\sigma_c}}\right)} .$$

(24)

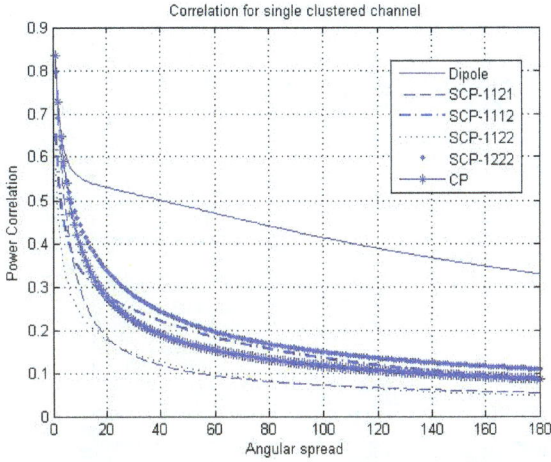

Fig. 8. Power correlation values of DP-ULA, CP-ULA and SCP-ULA averaged over azimuth angle ϕ for a 2×2 array.

The power correlation value that is strictly related to SNR scaling for MIMO channels is then given by $\rho_p = |\rho|^2$ and the power correlation values are averaged over uniformly-distributed mean azimuth angles over azimuth angle ϕ.

The power correlation values of the SCP antenna for all angular spread values and spatial/modal combinations in the densest configuration where stack antennas are nearby in the most compact scenario allowed by the physical radius of the bottom antenna for 2×2 configuration as presented in Fig. 8 are much lower than that of the two nearest antennas spaced at the patch radius of SCP bottom antenna in $\lambda/2$- length DP-ULA and CP-ULA which dictates spectral efficiency gains for MIMO spatial-multiplexing.

In Fig. 9, the power correlations of SCP-ULA versus mean azimuth angle for two different values of low (5°) and high (20°) angular spread are presented. For all angular spread values, DP-ULA has the maximum correlation values in comparison with CP and SCP at broadside, where the cluster is at the center of the antenna arrays and the oscillations of the patch antennas SCP-1122 (upper antenna radiated at TM11 and lower antenna radiated at TM22 mode), and SCP-1222 (upper antenna radiated at TM12 and lower antenna radiated at TM22 mode) decrease directly with the increase in angular spread. Besides, the maximum value of the correlations for all antennas is inversely proportional with angular spread.

The ergodic spectral efficiencies in bps/Hz of MIMO-OFDM WLAN system with DP-ULA, CP-ULA and SCP-ULA versus angular spread for 2×2, 4×4, 6×6 and 8×8 configurations are further presented in Fig. 10 over single-cluster Kronecker-model NLOS Rayleigh fading channel via [17]:

$$\eta_e = E\left\{ \frac{1}{N_c} \sum_{l=1}^{N_c} \log_2\left(\det\left(\mathbf{I}_N + \frac{SNR}{M} \mathbf{H}_c \mathbf{H}_c^{H} \right) \right) \right\} \quad (25)$$

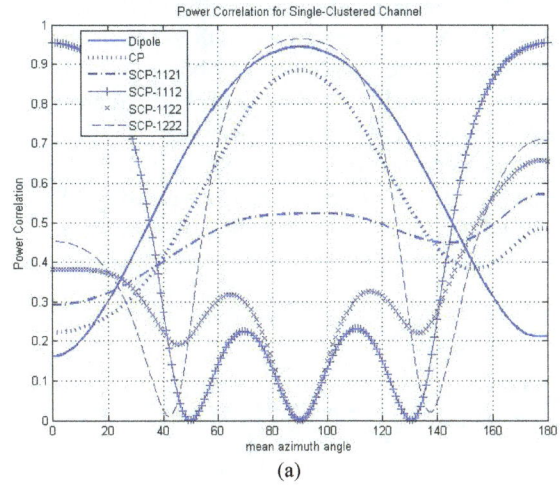

(a)

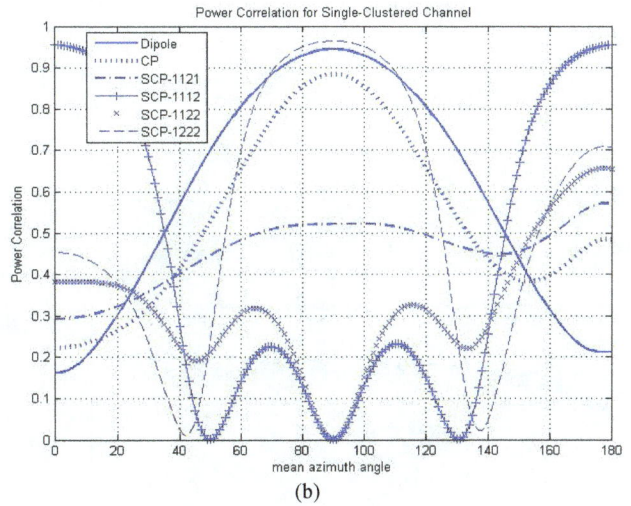

(b)

Fig. 9. Power correlation values of DP-ULA, CP-ULA and SCP-ULA versus mean azimuth angle (a): $\sigma_c^2 = 5°$, (b) $\sigma_c^2 = 20°$.

where the correlated channel matrix $\mathbf{H}_c$ has the Kronecker form [18-20]; $\mathbf{H}_c = \mathbf{R}_{RX}^{1/2} \mathbf{H}_w \mathbf{R}_{TX}^{1/2}$, in terms of the receive and transmit normalized power correlation matrices $\mathbf{R}_{RX}$ and $\mathbf{R}_{TX}$ and the elementwise-independent $N \times M$ Wishart-type random matrix $\mathbf{H}_w$ with elements distributed as $CN(0.1)$.

For 2×2 and 4×4 configurations, the SCP-ULA has much higher spectral efficiency than DP-ULA and slightly higher spectral efficiency than CP-ULA. On the other hand, for 6×6 and 8×8 configurations, CP-ULA and SCP-ULA have nearly the same spectral efficiencies which is also higher than DP-ULA. Furthermore, based on the spectral efficiency results presented in Fig. 10, data rates achievable with SCP-ULA conforming to IEEE 802.11n standards with 40 MHz bandwidth at 5.8 GHz for 2×2, 4×4, 6×6 and 8×8 array configurations are 164 Mbps, 324 Mbps, 480 Mbps, 644 Mbps respectively for high angular spreads that are much higher than standard 54 Mbps data rate achievable with SISO-OFDM 802.11 a/g WLAN systems.

The ergodic spectral efficiency of SCP-ULA, CP-ULA and DP-ULA versus mean azimuth angle of the cluster with respect to the antenna array boresights are further presented in Fig. 11 for low and high angular spreads of $\sigma_c^2 = 50$ and $\sigma_c^2 = 200$ respectively. In both cases, the ergodic spectral efficiency of SCP-ULA is the highest when the cluster is at the boresight; i.e. $\phi_c = 90°$ followed by CP-ULA and DP-ULA. On the other hand, as the cluster moves towards endfires with respect to the

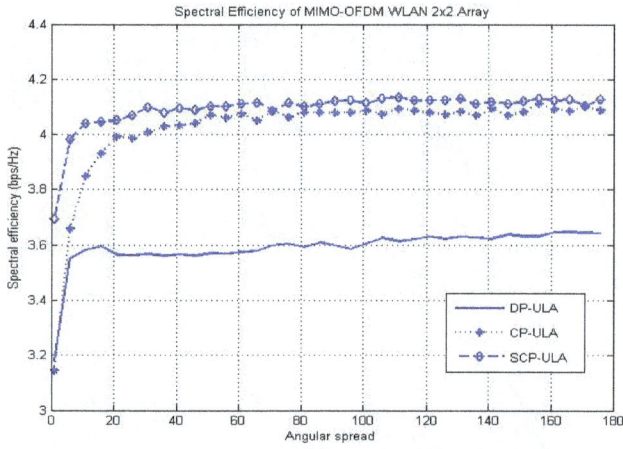

(a)

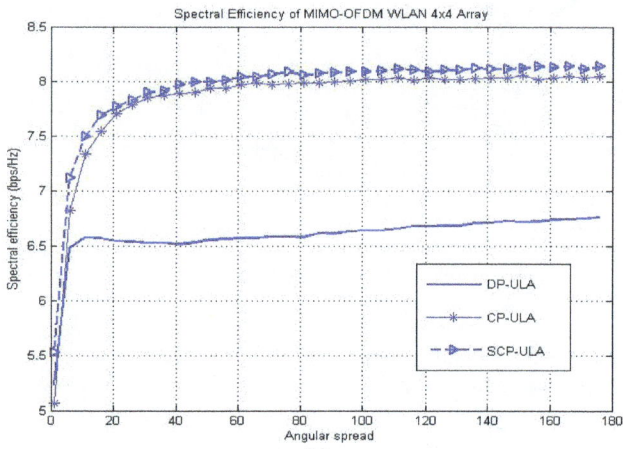

(d)

Fig. 10. Ergodic spectral efficiency of MIMO-OFDM WLAN over Kronecker-based statistically-clustered single-cluster NLOS Rayleigh fading channel with DP-ULA, CP-ULA and SCP-ULA versus angular spread at SNR = 10 dB: (a) 2×2, (b) 4×4, (c) 6×6, (d) 8×8 array.

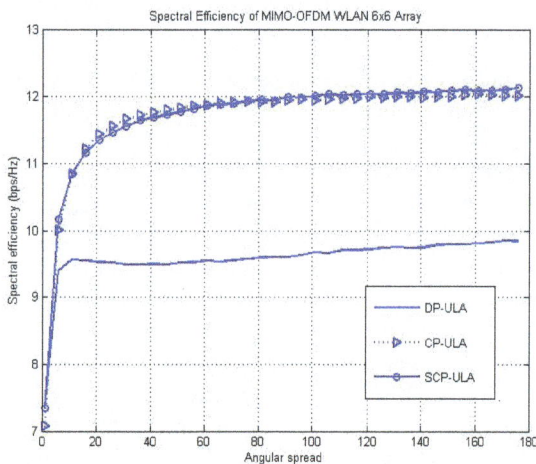

(b)

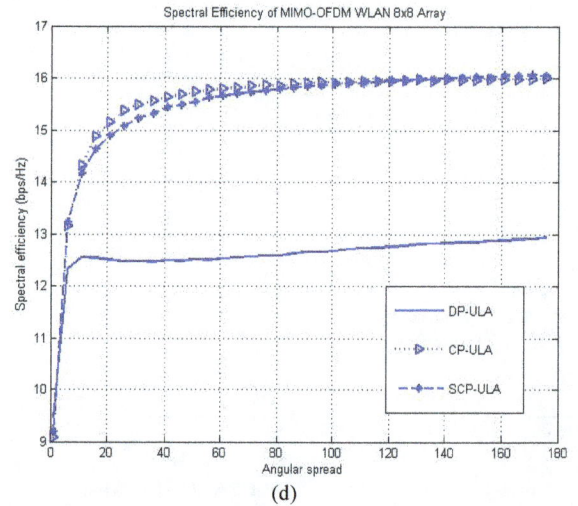

(a)

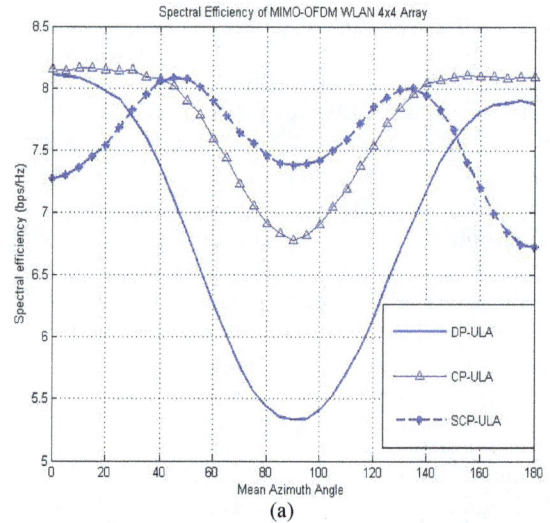

(c)

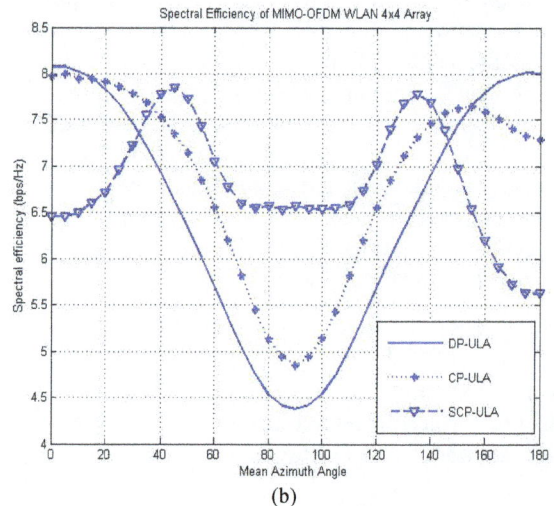

(b)

Fig. 11. Ergodic spectral efficiency of MIMO-OFDM WLAN over NLOS fading channel versus mean azimuth angle for 4×4 configuration of DP-ULA, CP-ULA and SCP-ULA at (a) $\sigma_c^2 = 5°$, (b) $\sigma_c^2 = 20°$.

antenna array boresights; i.e. $\phi_c = 0°$ and $\phi_c = 180°$, CP-ULA and DP-ULA respectively presents higher spectral efficiencies with respect to SCP-ULA. Despite this lower spectral efficiency of SCP-ULA with respect to CP-ULA and DP-ULA at endfires, the clusters in practical scenarios are mainly concentrated around boresight of antenna arrays in indoor propagation mediums and the range of mean azimuth angles around boresight over which SCP-ULA dominates CP-ULA and DP-ULA in terms of spectral efficiency is where the clusters reside realistically.

Most importantly, these lower correlation and higher spectral efficiency gains are achieved at 33.3% and 49.9%, 40% and 50%, and 42,9% and 50% denser space in compactness with SCP-ULA with respect to DP-ULA and CP-ULA for 4×4, 6×6 and 8×8 configurations respectively as tabulated in Tab. 2 and Tab. 3.

(Type)/(N×M)	2×2 (%)	4×4 (%)	6×6(%)	8×8(%)
DP-ULA	0	33.3	40.0	42.9
CP-ULA	49.7	49.9	50.0	50.0

Tab. 2. Compactness gain of SCP-ULA with respect to CP-ULA and DP-ULA.

(TYPE)/(N×M)	2×2 (mm)	4×4 (mm)	6×6 (mm)	8×8 (mm)
DP-ULA	21.2	63.6	106	148.4
CP-ULA	42.4	84.8	127.2	169.6
SCP-ULA	16.9	33.8	50.7	67.6

Tab. 3. Physical array length of SCP-ULA, CP-ULA and DP-ULA for minimum distance between elements.

5. Conclusions

In this paper, we proposed space-multimode stacked circular patch uniform linear antenna arrays (SCP-ULA) for deployment in size-constrained access points, modems and end-user terminal equipments of IEEE 802.11n MIMO-OFDM WLAN systems at 5.8 GHz. The performance of SCP-ULA in terms of spatio-modal power correlations, ergodic spectral efficiencies and compactness are compared with these of dominant-mode operating circular patch antenna arrays (CP-ULA) and center-fed dipole arrays (DP-ULA). The higher or comparable spectral efficiencies and lower spatio-modal correlations achieved by SCP-ULA over Kronecker-based statistically-clustered single-cluster NLOS Rayleigh fading channels in much more compact space requirements present space-multimode SCP-ULA antennas as a favorable solution for deployment in spatially-compact equipments of next-generation high-speed indoor IEEE 802.11n MIMO-OFDM WLAN systems.

References

[1] TELATAR, I. E. Capacity of multi-antenna Gaussian channels. *European Transactions on Telecommunications*, 1999, vol. 10, no. 6, p. 585–596. DOI: 10.1002/ett.4460100604

[2] FOSCHINI, G. J., GANS, M. On limits of wireless communications in a fading environment when using multiple antennas. *Wireless Personal Communications*, 1998, vol. 6, p. 311–355. DOI: 10.1023/A:1008889222784

[3] FORENZA, A., HEATH, R. W. Benefit of pattern diversity via two-element array of circular patch antennas in indoor clustered MIMO channels. *IEEE Transactions on Communications*, 2006, vol. 54, no. 5, p. 943–954. DOI: 10.1109/TCOMM.2006.873978

[4] SANCHEZ-FERNANDEZ, M., RAJO-IGLESIAS, E., QUEVEDO-TERUEL, O., PABLO-GONZALEZ, M. L. Spectral efficiency in MIMO systems using space and pattern diversities under compactness constraints. *IEEE Transactions on Vehicular Technology*, 2008, vol. 57, no. 3, p. 1637–1645. DOI: 10.1109/TVT.2007.909279

[5] SVANTESSON, T. Correlation and channel capacity of MIMO systems employing multimode antennas. *IEEE Transactions on Vehicular Technology*, 2002, vol. 51, no. 6, p. 1304–1312. DOI: 10.1109/TVT.2002.804856

[6] MUKHERJEE, A., KWON, H. M. Compact multi-user wideband MIMO system using multiple-mode microstrip antennas. In *Proceedings of 65th Vehicular Technology Conference VTC 2007 Spring*. Dublin (Ireland), April 2007, p. 584–588. DOI: 10.1109/VETECS.2007.131

[7] WALDSCHMIDT, C., KUHNERT, C., SCHULTEIS, S. WIESBECK, W. Compact MIMO-arrays based on polarization-diversity. In *Proceedings of IEEE Antennas and Propagation Symposium 2003*. Columbus (USA), 2003, vol. 2, p. 499–502. DOI: 10.1109/APS.2003.1219284

[8] *IEEE 802.11n-2009–Amendment 5: Enhancements for Higher Throughput*. IEEE-SA. 29 October 2009. DOI: 10.1109/IEEESTD.2009.5307322.

[9] *IEEE P802.11n Wireless LANs: TGn Channel Models*. IEEE-SA. May 10 2004. doc.: IEEE 802.11-03/940r4.

[10] Wi-Fi CERTIFIED n: Longer-Range, Faster-Throughput, Multimedia-Grade Wi-Fi® Networks, Wi-Fi Alliance. Sept. 2009.

[11] PAULRAJ, A., NABAR, R., GORE, D. *Introduction to Space-Time Wireless Communications*. New York: Cambridge University Press, 2003.

[12] SVANTESSON, T. On the capacity and correlation of multi-antenna systems employing multiple polarizations. In *Proc. of IEEE Antennas and Propagation Symposium 2002*. San Antonio (USA), 2002, vol. 3, p. 202–205. DOI: 10.1109/APS. 2002.1018190

[13] VAUGHAN, R. G. Two-port higher mode circular microstrip antennas. *IEEE Transactions on Antennas and Propagation*, 1988, vol. 36, no. 3, p. 309–321. DOI: 10.1109/8.192112

[14] GARG, R., BHARTIA, P., BAHL, I. *Microstrip Antenna Design Handbook*. Artech House, 2001.

[15] BALANIS, C.A. *Antenna Theory: Analysis and Design*. 2nd ed. United States of America: John Wiley & Sons Inc., 1997.

[16] BAHL, J. *Lumped Elements for RF and Microwave Circuits*. Boston: Artech House, 2003.

[17] VAN ZELST, A., VAN NEE, R., AWATER, G. A. Space-division multiplexing for OFDM systems. In *IEEE Vehicular Technology Conference 2000*. Tokyo (Japan), 2000, vol. 2, p. 1070–1074. DOI: 10.1109/VETECS.2000.851289

[18] CHE-NEE CHUAH, KAHN, J. M., TSE, D. N. C. Capacity of multi-antenna array systems in indoor wireless environments. In *Proceedings of GLOBECOM'98*. Sidney (Australia), 1998, vol. 4, p. 1894–1899. DOI: 10.1109/GLOCOM.1998.775873

[19] GOLDSMITH, A. *MIMO Wireless Communications*. Cambridge University Press, 2007.

[20] VAN ZELST, A. Space-division multiplexing algorithms. In *10th Mediterranean Electrotechnical Conference* 2000, MELECON 2000. Cyprus, 2000, vol. 3, p. 1218–1221. DOI: 10.1109/MELCON.2000.879755

About the Authors ...

Asuman SAVAŞCIHABEŞ has received her Ph.D. degree in Electrical Engineering in 2013 specializing on Telecommunications and Signal Processing at Gazi University, Ankara, Turkey. She is currently working as an assistant professor in Electrical and Electronics Engineering Department of Nuh Naci Yazgan University, Kayseri, Turkey. Her research interests include wireless communication systems, MIMO communication and transmit/receive diversity of MIMO-OFDM.

Özgür ERTUG was born in Ankara, Turkey in 1975. He received his B.Sc. degree in 1997 from University of Southern California, USA, M.Sc. degree from Rice University in 1999 and Ph.D. degree from Middle East Technical University in 2005. He is currently working as an assistant professor in Electrical and Electronics Engineering Department of Gazi University. His main research interests lie in algorithm and architecture design as well as theoretical and simulation-based performance analysis of wireless communication systems especially in the physical layer.

Erdem YAZGAN received the B.S. and M.S. degrees from the Middle East Technical University, Ankara, Turkey, in 1971 and 1973, respectively, and the Ph.D. degree from Hacettepe University, Ankara, Turkey, in 1980, all in Electrical Engineering. Since 1990, she has been a Professor with the Department of Electrical Engineering, Hacettepe University. In 1989, she was a Visiting Professor with Essex University, Essex, U.K. In 1994, she was with the Electroscience Laboratory, Ohio State University, Columbus. Her research interests include HF propagation, low altitude radar systems, mobile communications, MICs, reflector and microstrip antennas, Gaussian beam solutions, conformal mapping, and medical electronics.

27

Low-Profile Fully-Printed Multifrequency Monopoles Loaded with Complementary Metamaterial Transmission Line

Xue LI [1,2], Jinwen TIAN [1]

[1] Inst. for Pattern Recognition and Artificial Intelligence, Huazhong University of Science and Technology,
Wuhan 430074, China
[2] Defense Forces Academy, Zhengzhou 450052, China

ghoul_gargoyle@163.com, jwtian@mail.hust.edu.cn

Abstract. *The design of a new class of multifrequency monopoles by loading a set of resonant-type complementary metamaterial transmission lines (CMTL) is firstly presented. Two types of CMTL elements are comprehensively explored: the former is the epsilon negative (ENG) one by loading complementary split ring resonators (CSRRs) with different configurations on the signal strip, whereas the latter is the double negative (DNG) one by incorporating the CSRRs and capacitive gaps. In both cases, the CMTLs are considered with different number of unit cells. By cautiously controlling the geometrical parameters of element structure, five antenna prototypes coving different communication standards (GSM, UMTS, DMB and WiMAX) are designed, fabricated and measured. Numerical and experimental results illustrate that the zeroth-order resonance frequencies of the ENG and DNG monopoles are in desirable consistency. Moreover, of all operating frequencies the antennas exhibit fairly good impedance matching performances better than -10 dB and quasi-omnidirectional radiation patterns.*

Keywords

Metamaterial transmission line, planar monopole, multi-frequency antenna, complementary, zeroth-order resonator (ZOR)

1. Introduction

Over recent years, there has been a renewed interest in using artificial metamaterial (MTM) transmission line (TL) in the design of microwave devices and components [1]-[20] due to its abnormal electromagnetic (EM) properties that are hardly realized in nature. As to the planar microstrip monopoles [9]-[20], these antennas in general can be classified into three categories according to the loading manner and working mechanism of the MTM TL elements. The monopoles of the first category are based on the double negative (DNG) MTM TLs [9]-[16] which are also termed as composite right/left handed (CRLH) TLs. These antennas can be engineered with broadband or multi-frequency operation because the operating modes can be arbitrarily controlled by the well-conducted dispersion curve of CRLH elements. The second category is the monopole or dipole antennas built on the meta-surface made of periodically arranged left handed (LH) particles [17]. Through this type of loading, the antenna performances are improved in terms of both enhanced radiation behavior and broadened impedance matching bandwidth. The third category is the monopole by introducing LH MTMs [18] or split ring resonators (SRRs) [19] along it or even by embedding complementary SRRs (CSRRs) in the monopoles [20]. In this regard, the miniaturization and band-notch characteristic in an ultrawide operation band were achieved in virtue of the subwavelength resonance of the loaded elements.

Although compactness and multifunction are realized in aforementioned monopoles made of DNG MTM TLs, the mostly reported structures are confined to the nonresonant-type TL elements by using chip components which are restricted to low frequency operation and are not efficient radiators [9]-[12]. As to the rarely reported distributed TL monopoles [13], [14], the shunt inductors are commonly realized by grounded vias which would degrade the antenna gain due to the metallic losses. Moreover, further miniaturization is still a pressing task since the compactness is of great importance to portable and handheld antennas. These issues make an improved and alternative strategy that can be easily characterized and experimentally implemented a pressing task. The goal of this paper is thus to explore a novel avenue in the implementation of fully printed monopoles with simultaneous compact and multifunctional feature. The fundamentals and working mechanism of the monopoles using resonant-type complementary MTM TL (CMTL) will be firstly introduced. Then the multifrequency monopoles made of epsilon negative (ENG) and DNG CMTLs by etching the CSRRs on the signal strip is proposed, characterized and eventually fabricated.

2. Resonant-type Microstrip-fed ENG and DNG Monopoles

2.1 Fundamentals and Theoretical Background

The crucial obstacle of pushing the resonant-type CMTL element such as CSRRs in the monopole application is the structure incompatibility because the CSRRs commonly require a ground plane while the host monopole is unable to afford. In this paper, the resonant-type structures utilized for the monopoles are inspired from [16], [21], where the CSRRs are etched on the signal strip for filter and divider applications. Fig. 1 plots the sketch of the conventional and proposed conceptual microstrip-fed monopoles, respectively and the corresponding equivalent circuit model. As a first step towards the design of a DNG monopole, four essential parameters are defined: the phase φ_{CMTL} induced by the CMTL element, the phase φ_{Mono} caused by the host monopole, resonant modes (indices) n and the number of utilized CMTL elements N. Similar to the nonresonant-type CRLH monopole, the overall possible eigenfrequencies of the DNG monopoles should satisfy the following resonant condition [13], [16]

$$\varphi_{\text{total}} = N \times \varphi_{\text{CMTL}} + \varphi_{\text{Mono}} = n\pi/2, \quad n = 0, \pm 1, \pm 3, \ldots \quad (1)$$

where φ_{total} is the phase shift across the entire antenna. Observation from (1) indicates that the relation of the φ_{Mono} and φ_{CMTL} plays an important role in determination of the resonant modes. Note that the stop band of the ENG TL element in the guided wave design can also be employed for a radiated wave design in the case of an open circuit (infinitely large series impedance) or a resonance (often referred to as zeroth-order resonance, ZOR) of the shunt branch in the equivalent circuit model. In both ENG and DNG cases, the current has no place to go except through the shunt branch of the circuit.

2.2 Monopoles by Loading Single CMTL Element

Figure 2 shows the flow chart of the formation of the proposed DNG CMTL element evolved from [21]. As can

be seen, the element is composed of CSRRs etched on the square patch in the upper metallic strip and a square ring with or without two splits in the top and bottom region. The square ring connected to the feedline is formed by introducing an additional gap (closed slot) outside the CSRRs. The CSRRs still response to the axial electric-field component and thus afford the resonant negative permittivity. The resonant effect of the CSRRs is accounted by the parallel tank formed by L_p and C_p in the circuit model. The capacitive gaps at both sides are modeled by the capacitance C_g, whereas the inductive effect of the host TL is modeled by the inductance L_s. Both C_g and L_s contribute to the negative permeability. The DNG CMTL element can be ENG CMTL element by closing the splits of the square ring (defined as a hybrid ENG resonator) or by removing the gap (CSRRs left only). Since the length of the gap has been extended significantly relative to [21], a larger C_g and in turn enhanced LH characteristics will be engineered. In this particular design, since the ground plane of the CMTL element is not placed beneath the CSRRs but is coplanar with the host monopole, the fringing capacitances of the gaps C_f are weakened and thus are neglected for convenience. Therefore, the proposed CMTL element can be easily integrated with the planar monopoles of which the ground plane remains unaltered. The C_f has nothing to do with the resonant modes but results in slightly reduced resonant frequencies, which will be corroborated by extensive calculations in the upcoming section.

Fig. 3 portrays the layout of the resulting monopoles based on CSRRs-loaded CMTL elements with different configurations. The CMTL element is loaded at the end of

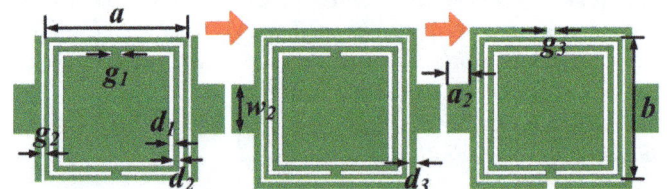

Fig. 2.   Flow chart of the formation of the proposed ENG (the middle one) and DNG (the last one) CMTL element evolved from previous CRLH element (the first one) [21].

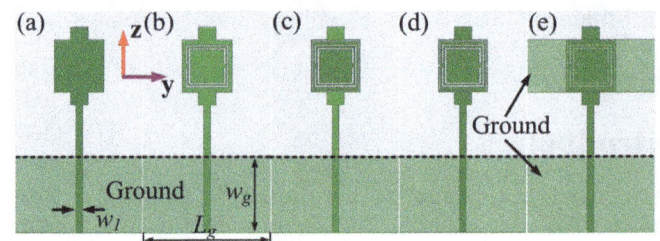

Fig. 3.   Layouts of the proposed single-cell microstrip-fed monopoles. (a) Conventional antenna (Case 1); (b) ENG antenna (Case 2); (c) hybrid ENG antenna (Case 3); (d) DNG antenna (Case 4); (e) hybrid ENG antenna with ground underneath the CMTL element (Case 5). The geometrical parameters of these antennas (in millimeter: mm) are $a = 11.5$, $b = 10.8$ $g_1 = 0.72$, $g_2 = 0.2$ $g_3 = 0.4$, $d_1 = d_2 = 0.36$, $d_3 = 0.2$, $w_1 = 1.6$, $w_2 = 4$, $L_g = 30$, $w_g = 18$, and $a_2 = 1.5$.

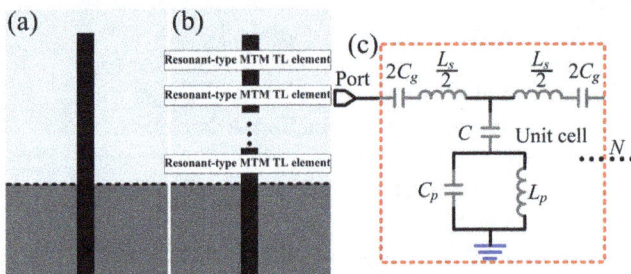

Fig. 1.   Sketch of microstrip-fed monopoles: (a) Conventional monopole. (b) Conceptual monopole loaded with resonant-type CMTL elements and (c) equivalent circuit model.

the host monopole. A total of six cases are considered for comprehensive analysis. They are orderly the conventional monopole without CMTL loading, the ENG monopoles loaded by the CSRRs only or by aforementioned hybrid resonator, the DNG monopole, the hybrid ENG monopole with a ground plane, and the monopole with the gap only (not shown here for brevity of contents). In the fifth case, the ground beneath the hybrid CMTL element is not connected to that of the host monopole.

The design flow starts from the conventional microstrip-fed monopole where the ground plane $L_g \times w_g$ is finely designed to obtain a desirable impedance matching over a wide bandwidth and a normal monopolar radiation with high efficiency. The length of the host monopole is designed to operate at GSM band (centered at 1.8 GHz). All designs are conducted in the commercial full-wave finite-element-method (FEM) EM field simulator Ansoft HFSS (Version of 13.2) and are all layouts are built on the commonly available 1mm-thick F4B substrate with dielectric constant of $\varepsilon_r = 2.2$ and loss tangent of $\tan\delta = 0.001$. Fig. 4 depicts the simulated reflection coefficients S_{11} of the proposed single-cell MTM-inspired monopoles. To illustrate the effects of the gap, the results of the monopole with the gap only are also provided.

Following Fig. 4, we conclude that the fundamental reflection dip (around GSM 1.8 GHz) corresponds to the operating frequency f_0 of the host monopole, whereas the second reflection dip (covering the Satellite Digital Mobile Broadcasting (DMB) band, 2605÷2655 MHz) corresponds to the ZOR frequency f_{M1} of the CSRRs while the third reflection dip (around 4.65 GHz, f_{S2}) corresponds to the resonance of the gap in hybrid DNG antenna case. As previously discussed, C_f is negligible when the ground is removed, which finds strong support from the almost constant trend of the reflection response except for the slightly reduced operating frequencies in case 5 due to the enhanced C_f. A further comparison between case 3 and the case with the gap only also indicates that the CSRRs interact with the gap, leading to the slightly reduced f_{S2} in case 3. Moreover, the uniform E-field distributions (not shown for brevity of contents) are clearly observed in CSRRs at f_{M1} while are not shown at residual frequencies, indicating a ZOR mode and strong radiation of CSRRs. This feature distinguishes the monopoles from any previous ones by loading chip components which are always not an efficient radiator. The relatively wider bandwidth of the DNG antenna around f_{M1} relative to both ENG antennas is due to that the excited $n = +1$ mode is located in close proximity to the $n = 0$ mode (ZOR mode). Note that the $n = +1$ mode never occurs in the ENG case. Moreover, the $f_0 = 2.14$ GHz (UMTS band) in the DNG antenna case has been shifted upwards. This is because the CMTL element provides less phase shift than conventional monopole. This means that the DNG monopole affords slightly shorter actual length. Most importantly, the ZOR frequency is observed almost the same in all ENG and DNG cases. This is because the ZOR frequency is only dependent on the L_p and C_p relative to the dimensions of the CSRRs.

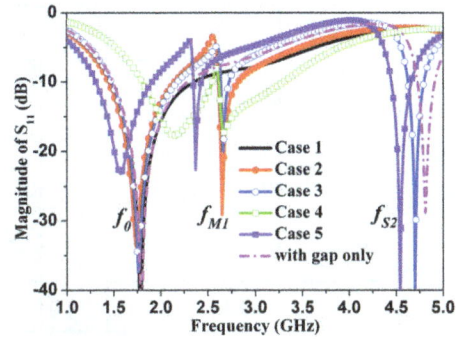

Fig. 4. Simulated reflection coefficients of the proposed single-cell MTM-inspired antennas.

To examine the far-field radiation characteristics, Fig. 5 depicts the 3-D patterns of the MTM-inspired antennas at different frequencies for an intuitionistic view. Following the figure, the monopolar or quasi-monopolar patterns are clearly observed in all cases. The little broken uniformity of the patterns in xoy plane at f_{S2} is due to the radiation of the gap which also radiates and thus facilitates the spatial power to be re-synthesized. The detailed performances of these antennas can be referred to Tab. 1. The relatively low efficiency of the ENG antennas at f_{M1} with respect to that at f_0 and f_{S2} is due to that f_{M1} is located very close to the low efficiency dip (not shown for brevity) at which no radiation occurs. However, it is improved in the DNG case due to double negative permittivity and permeability around f_{M1}, which enables strong radiation. The low antenna gain in all cases is due to the omnidirectional patterns and small ground dimensions of the host monopole. Further simulation results indicate that the gain is enhanced by an average of 0.7 dB when L_g increases per 10 mm in some specific range.

For verification, we have fabricated three antenna prototypes (case 2, case 3 and case 4, see orderly in Fig. 6) whose footprints occupy the same area of $1\times30\times52.6$ mm^3. Fig. 6 compares the simulated and measured (through a N5230C vector network analyzer) reflection coefficients while Fig. 7 gives the measured radiation patterns in two

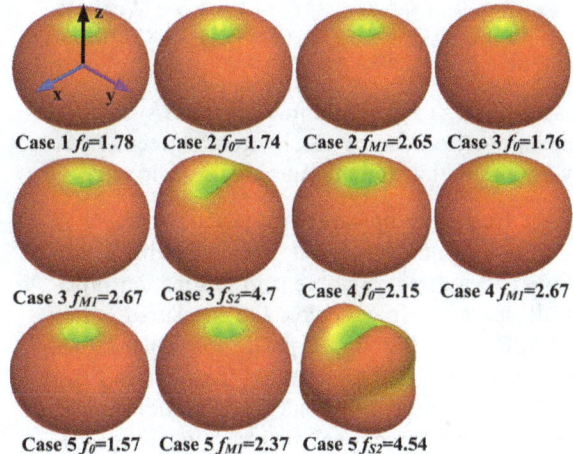

Case 1 f_0=1.78 Case 2 f_0=1.74 Case 2 f_{M1}=2.65 Case 3 f_0=1.76

Case 3 f_{M1}=2.67 Case 3 f_{S2}=4.7 Case 4 f_0=2.15 Case 4 f_{M1}=2.67

Case 5 f_0=1.57 Case 5 f_{M1}=2.37 Case 5 f_{S2}=4.54

Fig. 5. Simulated 3-D radiation patterns of the single-cell MTM-inspired monopoles at different operating frequencies.

Antennas	Antenna gain (dB)			BW (GHz)			Radiation efficiency (%)		
	f_0	f_{M1}	f_{S2}	f_0	f_{M1}	f_{S2}	f_0	f_{M1}	f_{S2}
Case 1	-1.44	-	-	0.9	-	-	97.5	-	-
Case 2	-1.64	0.59	-	0.7	0.21	-	98.3	82.6	-
Case 3	-1.44	0.38	0.8	0.7	0.09	0.17	97.3	75.8	89.1
Case 4	0.23	0.64	-	0.75	0.67	-	95.8	88.1	-

Tab. 1. Comparison of performances of the single-cell MTM-inspired monopoles. Note: the bandwidth (BW) is defined by the -10 dB reflection coefficients. The efficiency is defined as the radiated power divided by the accepted power at the center frequencies shown in Fig. 5.

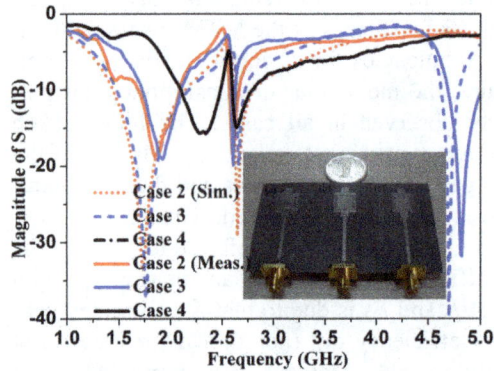

Fig. 6. Simulated and measured reflection coefficients of the fabricated antennas.

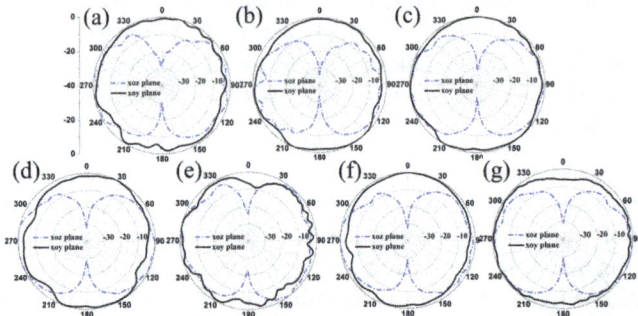

Fig. 7. Measured co-polarized patterns of the (a)-(b) ENG, (c)-(e) hybrid ENG and (f)-(g) DNG antenna at different frequencies. (a) f_0 = 1.85 GHz, (b) f_{M1} = 2.6 GHz, (c) f_0 = 1.92 GHz, (d) f_{M1} = 2.61 GHz, (e) f_{S2} = 4.81 GHz, (f) f_0 = 2.28 GHz, (g) f_{M1} = 2.64 GHz.

principle planes through a far-field measurement system in an anechoic chamber. As can be seen, a good agreement of results between simulation and measurement is achieved.

The slight frequency upwards in the measurement is attributable to the nonideal substrate that is utilized and also to the tolerances that are inherent in fabrication process. Across all operating frequencies, the measured S_{11} is better than -10 dB. From Fig. 7, the quasi-monopolar radiations are clearly illustrated at all selected frequencies from the nearly null radiation at the broadside direction in *xoz* plane while quasi-omnidirectional patterns in *xoy* plane. Thus far, the multifrequency operation of the proposed antennas by loading single CMTL element has been unambiguously demonstrated.

2.3 Monopoles by Loading Dual CMTL Elements

In this section, we will explore the effects of the number of the CMTL elements on the antenna performances. For simplicity, dual CMTL elements are considered without loss of generality. Fig. 8 plots the proposed monopoles by loading dual-cell CMTL elements with the same geometrical parameters and footprints as those shown in Fig. 3. The reflection coefficients are compared in Fig. 9. As can be seen, the impedance matching and bandwidth are almost without deterioration when additional element is loaded for ENG antennas. By contrary, the matching of the DNG antenna deteriorates significantly especially for the not well excited resonance of the host monopole. Nevertheless, the ZOR frequency still exhibits in all ENG and DNG

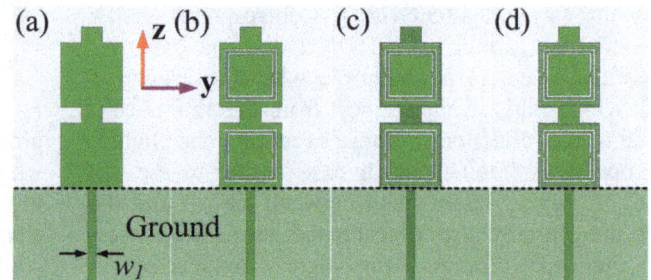

Fig. 8. Layouts of the proposed monopoles by loading dual-cell CMTL elements. (a) Conventional antenna; (b) ENG antenna; (c) Hybrid ENG antenna; (d) DNG antenna.

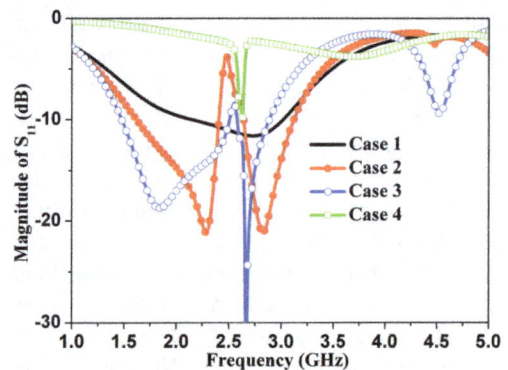

Fig. 9. Simulated reflection coefficients of the proposed dual-cell MTM-inspired monopoles.

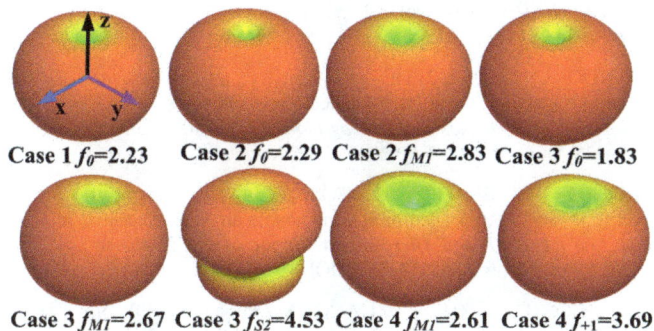

Fig. 10. Simulated 3-D radiation patterns of the dual-cell MTM-inspired monopoles at different operating frequencies.

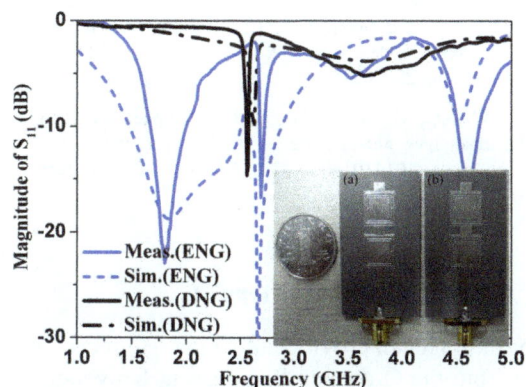

Fig. 11. Simulated and measured reflection coefficients of the dual-cell monopoles.

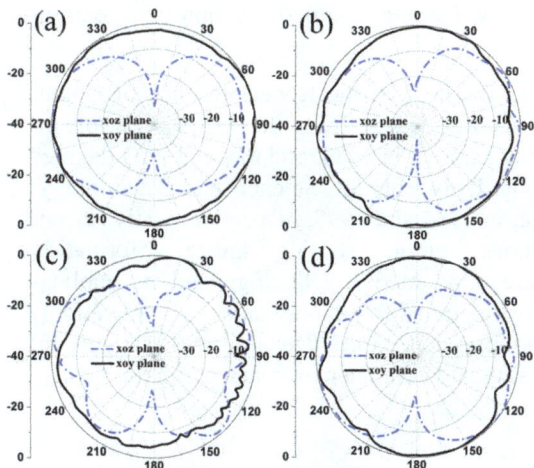

Fig. 12. Measured co-polarized patterns of the (a)-(b) hybrid ENG and (c)-(d) DNG antennas at different frequencies. (a) $f_0 = 1.8$ GHz, (b) $f_{M1} = 2.69$ GHz, (c) $f_{S2} = 4.61$ GHz, (d) $f_0 = 2.56$ GHz.

antennas around 2.6 GHz, further demonstrating that ZOR mode is independent while the input impedance is seriously dependent on the number of MTM elements [22]. In DNG case, the $n = +1$ mode around 3.5 GHz (WiMAX band) is weakly excited. Fig. 10 gives the 3-D far-field radiation patterns. Almost the same phenomena are expected as those shown in Fig. 5 except for the antenna gain which has been considerably improved when additional CMTL element is loaded. The reason is because that the CSRRs

contribute to the major radiation of the monopole as previously discussed.

For verification, the hybrid ENG and DNG antennas are fabricated. Fig. 11 plots the comparison of reflection coefficients between simulations and measurements while Fig. 12 depicts the measured radiation patterns in two principle planes. A desirable agreement of results is also observed. The reason for slight deviations especially for the narrower bandwidth in the measurement case is due to that the CMTL dimensions of prototypes deviate from those in the simulation model. Nevertheless, the measured S_{11} is better than -10 dB over all operating frequencies. From Fig. 12, the measured typical monopolar patterns further confirm the effectiveness of dual-cell elements in the monopole design.

3. Conclusion

The use of resonant-type ENG and DNG CMTLs in the design of low-profile multifrequency monopoles is presented. Results reveal that at all operating frequencies the monopoles exhibit quasi-monopolar radiation patterns and fairy good impedance matching performances with return loss better than 10 dB. Moreover, the ZOR frequency is independent on the number of unit cells and is observed as the same for both ENG and DNG antennas. These antennas feature compact, low profile, completely uniplanar, and are without any metallic vias and lumped loadings, rendering easy avenue toward monopoles with multifunction and high integration by using simple photolithography. Other variations of CSRRs and geometrical parameters can be explored for arbitrary manipulation of frequency ratio and further improvement of performances.

References

[1] ELEFTHERIADES, G. V., BALMAIN, K. G. Negative Refraction Metamaterials: Fundamental Principles and Applications. Hoboken, NJ: Wiley, 2005.

[2] XU, H.-X., WANG, G.-M., LIANG, J.-G., PENG, Q. Novel CRLH TL based on fractal geometry and series power divider application. Acta Physica Sinica, 2012, vol. 61, no. 7, p. 074101.

[3] XU, H.-X., WANG, G.-M., LIANG, J.-G. Novel designed CSRRs and its application in tunable tri-band bandpass filter based on fractal geometry. Radioengineering, 2011, vol. 20, no. 1, p. 312 to 316.

[4] CALOZ, C., ITOH, T. Electromagnetic Metamaterials: Transmission Line Theory and Microwave Applications: The Engineering Approach. Hoboken, NJ: Wiley, 2006.

[5] XU, H.-X., WANG, G.-M., ZHANG, C.-X., WANG, X. Characterization of composite right/left handed transmission line. Electronics Letters, 2011, vol. 47, no. 18, p. 1030–1032. DOI: 10.1049/el.2010.3707

[6] MARQUES, R., MARTIN, F., SOROLLA, M. Metamaterials with Negative Parameters: Theory, Design, and Microwave Applications. Hoboken, NJ: Wiley, 2008.

[7] XU, H.-X., WANG, G.-M., ZHANG, C.-X., LIANG, J.-G. Novel design of compact microstrip diplexer based on fractal-shaped composite right/left handed transmission line. *Journal of Infrared and Millimeter Waves*, 2011, vol. 30, no. 5, p. 390–396. DOI: 10.3724/sp.j.1010.2011.00390

[8] XU, H.-X., WANG, G.-M., CHEN, X., LI, T.-P. Broadband balun using fully artificial fractal-shaped composite right/left handed transmission line. *IEEE Microwave and Wireless Components Letters*, 2012, vol. 22, no. 1, p. 16–18. DOI: 10.1109/LMWC.2011.2173929

[9] KIM, D., KIM, M. Narrow-beamwidth T-shaped monopole antenna fabricated from metamaterial wires. *Electronics Letters*, 2008, vol. 44, no. 3, p. 180–182. DOI: 10.1049/el:20082854

[10] ANTONIADES, M. A., ELEFTHERIADES, G. V. A folded-monopole model for electrically small NRI-TL metamaterial antennas. *IEEE Antennas and Wireless Propagation Letters*, 2008, vol. 7, p. 425–428. DOI: 10.1109/LAWP.2008.2008773

[11] JI, J. K., KIM, G. H., SEONG, W. M. Bandwidth enhancement of metamaterial antennas based on composite right/left-handed transmission line. *IEEE Antennas and Wireless Propagation Letters*, 2010, vol. 9, p. 36–39. DOI: 10.1109/LAWP.2010.2041628

[12] ZHU, J., ANTONIADES, M. A., ELEFTHERIADES, G. V. A compact tri-band monopole antenna with single-cell metamaterial loading. *IEEE Transactions on Antennas and Propagation*, 2010, vol. 58, no. 4, p. 1031–1038. DOI: 10.1109/TAP.2010.2041317

[13] IBRAHIM, A., SAFWAT, A. M. E., EL-HENNAWY, H. Triple-band microstrip-fed monopole antenna loaded with CRLH unit cell. *IEEE Antennas and Wireless Propagation Letters*, 2011, vol. 10, p. 1547–1550. DOI: 10.1109/LAWP.2011.2181813

[14] ANTONIADES, M. A., ELEFTHERIADES, G. V., ROGERS, E. S. A broadband dual-mode monopole antenna using NRI-TL metamaterial loading, *IEEE Antennas and Wireless Propagation Letters*, 2009, vol. 8, p. 258–261. DOI:10.1109/LAWP.2009.2014402

[15] KOKKINOS, T., FERESIDIS, A. P. Low-profile folded monopoles with embedded planar metamaterial phase-shifting lines. *IEEE Transactions on Antennas and Propagation*, 2009, vol. 57, p. 2997–3008. DOI: 10.1109/TAP.2009.2028605

[16] XU, H.-X., WANG, G.-M., LV, Y.-Y., QI, M.-Q., GAO, X., GE, S. Multifrequency monopole antennas by loading metamaterial transmission lines with dual-shunt branch circuit. *Progress in Electromagnetics Research*, 2013, vol. 137, p. 703–725.

[17] ELSHEAK, D. N., ISKANDER, M. F., ELSADE, H. A., ABDAL-LAH, E. A., ELHENAWY, H. Enhancement of ultra-wideband microstrip monopole antenna by using unequal arms V-shaped slot printed on metamaterial surface. *Microwave and Optical Technology Letters*, 2010, vol. 52, no. 10, p. 2203–2209. DOI: 10.1002/mop.25447

[18] PALANDOKEN, M., GREDE, A., HENKE, H. Broadband microstrip antenna with left-handed metamaterials. *IEEE Transactions on Antennas and Propagation*, 2009, vol. 57, no. 2, p. 331–338. DOI: 10.1109/TAP.2008.2011230

[19] BARBUTO, M., BILOTTI, F., TOSCANO, A. Design of a multifunctional SRR-loaded printed monopole antenna. *International Journal of RF and Microwave Computer-Aided Engineering*, 2012, vol. 22, no. 4, p. 552–557. DOI: 10.1002/mmce.20645

[20] LIU, J., GONG, S., XU, Y., ZHANG, X., FENG, C., QI, N. Compact printed ultra-wideband monopole antenna with dual band-notched characteristics. *Electronics Letters*, 2008, vol. 44, no. 12, p. 710–711. DOI: 10.1049/el:20080931

[21] GIL, M., BONACHE, J., MARTIN, F. Synthesis and applications of new left handed microstrip lines with complementary split-ring resonators etched on the signal strip. *IET Microwaves Antennas and Propagation*, 2008, vol. 2, no. 4, p. 324–330. DOI:10.1049/iet-map:20070225

[22] XU, H.-X., WANG, G.-M., GONG, J.-Q. Compact dual-band zeroth-order resonance antenna. *Chinese Physics Letters*, 2012, vol. 29, no.1, p. 014101-1–014101-4. DOI: 10.1088/0256-307X/29/1/014101

About the Authors ...

Xue LI was born in Henan province of China. He received his MS degree from the Ordnance Engineering University, Shijiazhuang, China, in 2003, and MS degree from the PLA Information Engineering University, Zhengzhou, China, in 2006. He is now pursing his PhD degree. His research interests include computer vision and pattern recognition.

Jinwen TIAN was born in the Hebei province of China. He received his BS degree from the Daqing Petroleum Institute, Daqing, China, in 1983, and MS and PhD degrees from Huazhong University of Science and Technology, Wuhan, China, in 1994 and 1998, respectively. His current interest includes remote sensing image information processing, streaming media technology and its applications, the wavelet transform theory and its applications, image data compression, target detection and recognition, augmented reality and computer software simulation, etc.

Permissions

All chapters in this book were first published in Radioengineering, by Spolecnost pro radioelektronicke inzenyrstvi; hereby published with permission under the Creative Commons Attribution License or equivalent. Every chapter published in this book has been scrutinized by our experts. Their significance has been extensively debated. The topics covered herein carry significant findings which will fuel the growth of the discipline. They may even be implemented as practical applications or may be referred to as a beginning point for another development.

The contributors of this book come from diverse backgrounds, making this book a truly international effort. This book will bring forth new frontiers with its revolutionizing research information and detailed analysis of the nascent developments around the world.

We would like to thank all the contributing authors for lending their expertise to make the book truly unique. They have played a crucial role in the development of this book. Without their invaluable contributions this book wouldn't have been possible. They have made vital efforts to compile up to date information on the varied aspects of this subject to make this book a valuable addition to the collection of many professionals and students.

This book was conceptualized with the vision of imparting up-to-date information and advanced data in this field. To ensure the same, a matchless editorial board was set up. Every individual on the board went through rigorous rounds of assessment to prove their worth. After which they invested a large part of their time researching and compiling the most relevant data for our readers.

The editorial board has been involved in producing this book since its inception. They have spent rigorous hours researching and exploring the diverse topics which have resulted in the successful publishing of this book. They have passed on their knowledge of decades through this book. To expedite this challenging task, the publisher supported the team at every step. A small team of assistant editors was also appointed to further simplify the editing procedure and attain best results for the readers.

Apart from the editorial board, the designing team has also invested a significant amount of their time in understanding the subject and creating the most relevant covers. They scrutinized every image to scout for the most suitable representation of the subject and create an appropriate cover for the book.

The publishing team has been an ardent support to the editorial, designing and production team. Their endless efforts to recruit the best for this project, has resulted in the accomplishment of this book. They are a veteran in the field of academics and their pool of knowledge is as vast as their experience in printing. Their expertise and guidance has proved useful at every step. Their uncompromising quality standards have made this book an exceptional effort. Their encouragement from time to time has been an inspiration for everyone.

The publisher and the editorial board hope that this book will prove to be a valuable piece of knowledge for researchers, students, practitioners and scholars across the globe.

List of Contributors

Matej KOMANEC
Department of Electromagnetic Field, Faculty of Electrical Engineering, Czech Technical University in Prague, Technicka 2, Prague 6, 166 27, Czech Republic

Pavel SKODA
Department of Electromagnetic Field, Faculty of Electrical Engineering, Czech Technical University in Prague, Technicka 2, Prague 6, 166 27, Czech Republic

Jan SISTEK
Department of Electromagnetic Field, Faculty of Electrical Engineering, Czech Technical University in Prague, Technicka 2, Prague 6, 166 27, Czech Republic

Tomas MARTAN
Department of Electromagnetic Field, Faculty of Electrical Engineering, Czech Technical University in Prague, Technicka 2, Prague 6, 166 27, Czech Republic

Mustafa İLARSLAN
Turkish Air Force Academy, 34149, Yeşilyurt, Istanbul, Turkey

Salih DEMIREL
Dept. of Electronics and Communication Engineering, Yıldız Technical University, Turkey

Hamid TORPI
Dept. of Electronics and Communication Engineering, Yıldız Technical University, Turkey

A. Kenan KESKIN
Dept. of Electronics and Communication Engineering, Yıldız Technical University, Turkey

M. Fatih ÇAĞLAR
Dept. of Electronics and Communication Engineering, Süleyman Demirel University, Turkey

Hua GUO
School of Electronics and Information, Northwestern Polytechnical University, Xi'an 710129, China

Chenjiang GUO
School of Electronics and Information, Northwestern Polytechnical University, Xi'an 710129, China

Jun DING
School of Electronics and Information, Northwestern Polytechnical University, Xi'an 710129, China

Izni Husna IDRIS
UTM-MIMOS Center, Fakulti Kejuruteraan Elektrik, Universiti Teknologi Malaysia, Skudai, Johor, Malaysia

Mohamad Rijal HAMID
UTM-MIMOS Center, Fakulti Kejuruteraan Elektrik, Universiti Teknologi Malaysia, Skudai, Johor, Malaysia

Mohd Haizal JAMALUDDIN
Wireless Communication Center, Fakulti Kejuruteraan Elektrik, Universiti Teknologi Malaysia, Skudai, Johor, Malaysia

Mohamad Kamal A. RAHIM
UTM-MIMOS Center, Fakulti Kejuruteraan Elektrik, Universiti Teknologi Malaysia, Skudai, Johor, Malaysia

James R. KELLY
Centre for Communication Systems Research, Faculty of Engineering and Physical Sciences, University of Surrey, Guildford, GU2 7XH, United Kingdom

Huda A. MAJID
UTM-MIMOS Center, Fakulti Kejuruteraan Elektrik, Universiti Teknologi Malaysia, Skudai, Johor, Malaysia

Radek VEHOVSKÝ
Dept. of Radio Electronics, Brno University of Technology, Technická 12, 612 00 Brno, Czech Republic

Michal POKORNÝ
Dept. of Radio Electronics, Brno University of Technology, Technická 12, 612 00 Brno, Czech Republic

Kamil PÍTRA
Military Research Institute, state enterprise, Veslarska 230, 637 00 Brno, Czech Republic

Adam LAMECKI
Dept. of Microwave and Antenna Engineering, Gdansk University of Technology, Narutowicza 11/12, 80-233 Gdansk, Poland

Winai JAIKLA
Dept. of Engineering Education, Faculty of Industrial Education, King Mongkut's Institute of Technology Ladkrabang, Bangkok, 10520, Thailand

Dalibor BIOLEK
Dept. of Electrical Engineering, Faculty of Military Technology, University of Defense Brno, Kounicova 65, 662 10 Brno, Czech Republic
Dept. of Microelectronics, Faculty of Electrical Engineering and Communications, Brno University of Technology, Technická 10, 616 00 Brno, Czech Republic

Surapong SIRIPONGDEE
Dept. of Engineering Education, Faculty of Industrial Education, King Mongkut's Institute of Technology Ladkrabang, Bangkok, 10520, Thailand

Josef BAJER
Dept. of Electrical Engineering, Faculty of Military Technology, University of Defense Brno, Kounicova 65, 662 10 Brno, Czech Republic

Tomasz STEFAN'SKI
Faculty of Electronics, Telecommunications and Informatics, Gdansk University of Technology, Narutowicza 11/12, 80-233 Gdansk, Poland

Sławomir ORŁOWSKI
Faculty of Electronics, Telecommunications and Informatics, Gdansk University of Technology, Narutowicza 11/12, 80-233 Gdansk, Poland

Bartosz REICHEL
Faculty of Applied Physics and Mathematics, Gdansk University of Technology, Narutowicza 11/12, 80-233 Gdansk, Poland

Asif IQBAL
Dept. of Electrical and Electronics Engineering,Universiti Teknologi Petronas, Tronoh, Perak, 31750, Malaysia

Varun JEOTI
Dept. of Electrical and Electronics Engineering,Universiti Teknologi Petronas, Tronoh, Perak, 31750, Malaysia

Gökay AKINCI
Dept. of Electrical and Electronics Engineering, Ankara University, Ankara, Turkey

A. Egemen YILMAZ
Dept. of Electrical and Electronics Engineering, Ankara University, Ankara, Turkey

Mustafa KUZUOĞLU
Dept. of Electrical and Electronics Engineering, Middle East Technical University, Ankara, Turkey

Sijia LI
Information and Navigation College, Air Force Engineering University, 710077, Xi'an, China

Xiangyu CAO
Information and Navigation College, Air Force Engineering University, 710077, Xi'an, China

Jun GAO
Information and Navigation College, Air Force Engineering University, 710077, Xi'an, China

Yuejun ZHENG
Information and Navigation College, Air Force Engineering University, 710077, Xi'an, China

Di ZHANG
Information and Navigation College, Air Force Engineering University, 710077, Xi'an, China

Hongxi LIU
Information and Navigation College, Air Force Engineering University, 710077, Xi'an, China

Yu Song MENG
National Metrology Centre, Agency for Science, Technology and Research (A*STAR) 1 Science Park Drive, Singapore 118221, Singapore

Yueyan SHAN
National Metrology Centre, Agency for Science, Technology and Research (A*STAR) 1 Science Park Drive, Singapore 118221, Singapore

Cristina PICHER
Technology Department, Fractus, Barcelona, Spain

Jaume ANGUERA
Technology Department, Fractus, Barcelona, Spain
Electronics and Communications Department, Universitat Ramon Llull, Barcelona, Spain

Aurora ANDÚJAR
Technology Department, Fractus, Barcelona, Spain

Carles PUENTE
Signal Theory and Communications, Universitat Politècnica de Catalunya, Spain

Adrián BUJALANCE
Electronics and Communications Department, Universitat Ramon Llull, Barcelona, Spain

Paiboon YOIYOD
Faculty of Engineering, King Mongkut's Institute of Technology Ladkrabang, Bangkok 10520, Thailand

Monai KRAIRIKSH
Faculty of Engineering, King Mongkut's Institute of Technology Ladkrabang, Bangkok 10520, Thailand

Tapan MANDAL
Dept. of Information Technology, Govt. College of Engineering and Textile Technology, Serampore, Hooghly, India, PIN-712201

Santanu DAS
Dept. of Electronics & Tele-Comm. Engineering, Bengal Engineering and Science University, Shibpur, Howrah, India, PIN-711103

Xiaolin YANG
Dept. of Physics and Electronics, University of Electronic Science and Technology of China, No.4, Section 2, North Jianshe Road, 610054 ChengDu, SiChuan

Fangling KONG
Dept. of Physics and Electronics, University of Electronic Science and Technology of China, No.4, Section 2, North Jianshe Road, 610054 ChengDu, SiChuan

Xuelin LIU
Dept. of Physics and Electronics, University of Electronic Science and Technology of China, No.4, Section 2, North Jianshe Road, 610054 ChengDu, SiChuan

Chunyan SONG
Dept. of Physics and Electronics, University of Electronic Science and Technology of China, No.4, Section 2, North Jianshe Road, 610054 ChengDu, SiChuan

Milan KVICERA
Faculty of Electrical Engineering, Czech Technical University in Prague, Technicka 2, 166 27 Prague, Czech Republic

Pavel PECHAC
Faculty of Electrical Engineering, Czech Technical University in Prague, Technicka 2, 166 27 Prague, Czech Republic

Sijia LI
Information and Navigation College, Air Force Engineering University, No.1 Fenghao Street, 710077 Xi'an, China

Xiangyu CAO
Information and Navigation College, Air Force Engineering University, No.1 Fenghao Street, 710077 Xi'an, China

Jun GAO
Information and Navigation College, Air Force Engineering University, No.1 Fenghao Street, 710077 Xi'an, China

Pengcheng GAO
National Key Lab. of Antenna and Microwave Technology, Xidian University, No.2 Taibai Road 710071, Xi'an, China

Barbara SZYMANIK
West Pomeranian University of Technology in Szczecin, Faculty of Electrical Engineering, Sikorskiego 37, 70-313 Szczecin, Poland

Radoslav BORTEL
Faculty of Electrical Engineering, Czech Technical University in Prague, Technická 2, 166 27 Prague, Czech Republic

Pavel SOVKA
Faculty of Electrical Engineering, Czech Technical University in Prague, Technická 2, 166 27 Prague, Czech Republic

Ubaid ULLAH
School of Electrical and Electronic Engineering, University Sains Malaysia, 14300 Nibong Tebal Pulau Pinang

Mohd Fadzil AIN
School of Electrical and Electronic Engineering, University Sains Malaysia, 14300 Nibong Tebal Pulau Pinang

Mohamadariff OTHMAN
School of Electrical and Electronic Engineering, University Sains Malaysia, 14300 Nibong Tebal Pulau Pinang

Ihsan ZUBIR
School of Electrical and Electronic Engineering, University Sains Malaysia, 14300 Nibong Tebal Pulau Pinang

Nor Muzlifah MAHYUDDIN
School of Electrical and Electronic Engineering, University Sains Malaysia, 14300 Nibong Tebal Pulau Pinang

Zainal Ariffin AHMAD
School of Material and Mineral Resource Engineering, University Sains Malaysia, 14300 Nibong Tebal Pulau Pinang

Mohd Zaid ABDULLAH
School of Electrical and Electronic Engineering, University Sains Malaysia, 14300 Nibong Tebal Pulau Pinang

Bojan DIMITRIJEVIC
Faculty of Electronic Engineering, University of Nis, A. Medvedeva 14, 18 000 Nis, Serbia

Bojana NIKOLIC
Faculty of Electronic Engineering, University of Nis, A. Medvedeva 14, 18 000 Nis, Serbia

Slavoljub ALEKSIC
Faculty of Electronic Engineering, University of Nis, A. Medvedeva 14, 18 000 Nis, Serbia

Nebojsa RAICEVIC
Faculty of Electronic Engineering, University of Nis, A. Medvedeva 14, 18 000 Nis, Serbia

Muhammad Zaka Ur REHMAN
Dept. of Electrical and Electronics Engineering, Universiti Teknologi Petronas, Tronoh, Perak, Malaysia
Dept. of Physics, COMSATS Institute of Information Technology, Park Road, Islamabad, Pakistan

Zuhairi BAHARUDIN
Dept. of Electrical and Electronics Engineering, Universiti Teknologi Petronas, Tronoh, Perak, Malaysia

Mohd Azman ZAKARIYA
Dept. of Electrical and Electronics Engineering, Universiti Teknologi Petronas, Tronoh, Perak, Malaysia

Mohd Haris Md. KHIR
Dept. of Electrical and Electronics Engineering, Universiti Teknologi Petronas, Tronoh, Perak, Malaysia

Muhammad Taha JILANI
Dept. of Electrical and Electronics Engineering, Universiti Teknologi Petronas, Tronoh, Perak, Malaysia

Mohd Saiful Riza BASHRI
Dept. of Electrical and Computer Engineering, International Islamic University Malaysia, 53100 Kuala Lumpur, Malaysia

Muhammad Ibn IBRAHIMY
Dept. of Electrical and Computer Engineering, International Islamic University Malaysia, 53100 Kuala Lumpur, Malaysia

S. M. A MOTAKABBER
Dept. of Electrical and Computer Engineering, International Islamic University Malaysia, 53100 Kuala Lumpur, Malaysia

Mohamed Nasrun OSMAN
Communication Engineering Dept., Faculty of Electrical Engineering, Universiti Teknologi Malaysia, 81310 Skudai, Johor, Malaysia

Mohamad Kamal A. RAHIM
Communication Engineering Dept., Faculty of Electrical Engineering, Universiti Teknologi Malaysia, 81310 Skudai, Johor, Malaysia

Peter GARDNER
School of Electronic, Electrical and Computer Engineering, University of Birmingham, Edgbaston, Birmingham, B15 2TT, United Kingdom

Mohamad Rijal HAMID
Communication Engineering Dept., Faculty of Electrical Engineering, Universiti Teknologi Malaysia, 81310 Skudai, Johor, Malaysia

Mohd Fairus MOHD YUSOFF
Communication Engineering Dept., Faculty of Electrical Engineering, Universiti Teknologi Malaysia, 81310 Skudai, Johor, Malaysia

Huda A. MAJID
Communication Engineering Dept., Faculty of Electrical Engineering, Universiti Teknologi Malaysia, 81310 Skudai, Johor, Malaysia

Asuman SAVASCIHABES
Telecommunications and Signal Processing Laboratory, Dept. of Electrical and Electronics Engineering, University of Gazi, Ankara, Turkey

Ozgur ERTUG
Telecommunications and Signal Processing Laboratory, Dept. of Electrical and Electronics Engineering, University of Gazi, Ankara, Turkey

Erdem YAZGAN
Dept. of Electrical and Electronics Engineering, University of Hacettepe, Ankara, Turkey

Xue LI
Inst. for Pattern Recognition and Artificial Intelligence, Huazhong University of Science and Technology, Wuhan 430074, China
Defense Forces Academy, Zhengzhou 450052, China

Jinwen TIAN
Inst. for Pattern Recognition and Artificial Intelligence, Huazhong University of Science and Technology, Wuhan 430074, China

www.ingramcontent.com/pod-product-compliance
Lightning Source LLC
Chambersburg PA
CBHW080704200326
41458CB00013B/4953